ENCYCLOPEDIA

OF

FOOD
MICROBIOLOGY

ENCYCLOPEDIA
OF
FOOD
MICROBIOLOGY

Editor-in-Chief

RICHARD K. ROBINSON

Editors

CARL A. BATT

PRADIP D. PATEL

ACADEMIC PRESS

A Harcourt Science and Technology Company

**San Diego   San Francisco   New York   Boston
London   Sydney   Tokyo**

This book is printed on acid-free paper

Academic Press
A Harcourt Science and Technology Company
24–28 Oval Road, London NW1 7DX, UK
http://www.hbuk.co.uk/ap/

Academic Press
525 B Street, Suite 1900, San Diego, California 92101-4495, USA
http://www.apnet.com

ISBN 0-12-227070-3

A catalogue for this encyclopedia is available from the British Library

Library of Congress Catalog Card Number: 98-87954

Access for a limited period to an on-line version of the Encyclopedia of Food Microbiology is included in the purchase price of the print edition.
This on-line version has been uniquely and persistently identified by the Digital Object Identifier (DOI)

10.1006/rwfm.2000

By following the link

http://dx.doi.org/10.1006/rwfm.2000

from any Web Browser, buyers of the Encyclopedia of Food Microbiology will find instructions on how to register for access.

Typeset by Selwood Systems, Midsomer Norton, Bath
Printed in Great Britain by The Bath Press, Bath

00 01 02 03 04 05 BP 9 8 7 6 5 4 3 2 1

EDITORIAL ADVISORY BOARD

FOREWORD

Public concern about food safety has never been greater. In part this is due to the ever increasing demand from consumers for higher and higher standards. But new food-borne pathogens like *E. coli* O157 have emerged in recent years to become important public health problems, and changes in production and manufacturing sometimes reopen doors of opportunity for old ones. A powerful reminder that food scientists have much unfinished business to attend to is provided by the succession of food scares that generate strong stories for the media.

Experience tells us that science must underpin all approaches to food safety, whether through the application and implementation of well-tried approaches or the development of new or improved methods. Microbiologists have had a central role in this since the high quality work of pioneers like van Ermengem on botulism and Gaffky on typhoid more than a century ago. The large amount of important data that has accumulated since then joins with the current rapid rates of technological and scientific advance to make the need for a structured and authoritative source of information a very pressing one. It is provided by this encyclopedia.

These are exciting times for food microbiologists. Expectations are high that as scientists we will soon provide answers to the many problems still posed by microbes – from spoilage to food poisoning. Approaches like HACCP are making everyone think hard about how best to apply the data we have to develop better ways for reducing and eliminating food-borne pathogens. The pace of scientific developments continues to accelerate and more and better methods are available for the detection and enumeration of microbes than ever before. The microbes themselves continue to evolve and so present moving targets. The solid foundation presented by the mass of information in this encyclopedia provides the launching pad and guide for meeting these challenges.

It could be said that a penalty of working in food microbiology is that because the subject is broad-ranging, mature and dynamic, its practitioners, teachers and students have to know about many things in breadth and depth. For most of us, of course, this is not a penalty but an attractive bonus because of its intellectual challenge. I am particularly pleased to be associated with the encyclopedia because it will help us all to meet this test with confidence. I wish it every success.

Professor H Pennington
Department of Medical Microbiology
University of Aberdeen

PREFACE

Although food microbiology and food safety have, in recent times, become major concerns for governments around the world, equally important is the fact that, without yeasts and bacteria, popular meals like bread and cheese would not exist. Consequently, a knowledge of the relationship between foodstuffs and the activities of bacteria, yeasts and mycelial fungi has become a top priority for everyone associated with food and its production. Farmers have concerns related to produce harvesting and storage, food processors have to generate wholesome retail products that are both free from pathogenic organisms and have a satisfactory shelf life and, last but not least, food handlers and consumers need to be aware of the procedures necessary to ensure that food is safely prepared and stored.

In order for these disparate groups to operate successfully, accurate and objective information about the microbiology of foods is essential, and this encyclopedia seeks to provide a source of such information. In some areas, introductory articles are provided to guide readers who may be less familiar with the subject but, in general, superficiality has been avoided. Thus, the coverage has been developed to include details of all the important groups of bacteria, fungi, viruses and parasites, the various methods that can be employed for their detection in foods, the factors that govern the behaviour of the same organisms, together with an analysis of likely outcomes of microbial growth/metabolism in terms of disease and/or spoilage. A further series of articles describes the contribution of microorganisms to industrial fermentations, to traditional food fermentations from the Middle or Far East, as well as during the production of the fermented foods like bread, cheese or yoghurt that are so familiar in industrialized societies. The division of these topics into 358 articles of approximately 4000 words, has meant that the contributing authors have been able to handle their specialist subject(s) in real depth.

Obviously, another group of editors might have approached the project in a different manner, but we feel confident that this encyclopedia will provide readers at all levels of expertise with the data being sought. A point enhanced, perhaps, by the inclusion at the end of each article of a list for further reading, comprising a selection of review articles and key research papers that should encourage further exploration of any selected topic. If this confidence is borne out in practice, then the efforts of the contributors, the members of the Editorial Board and the editorial team from Academic Press will be well rewarded, for raising the scientific profile of food microbiology is long overdue.

R.K. Robinson, C.A. Batt, P.D. Patel
Editors

INTRODUCTION

The advent of antibiotics gave the general public, and many professional microbiologists as well, the feeling that bacterial diseases were under control, and the elimination of smallpox and the control of polio suggested that even viruses posed few problems. However, this complacency has received a nasty jolt over the last decade, and the emergence of HIV and multiple-drug-resistant strains of bacteria has become a major concern for the medical profession. The food industry has been similarly shaken by the appearance of new, and potentially fatal, strains of *Escherichia coli*, a species that for over 100 years was regarded as little more than a nuisance. Equally unexpected was the devastating impact of BSE, and fresh reports of the activities of so-called 'emerging food-borne pathogens' are appearing with alarming regularity.

In some cases, it has been possible to understand, with the advantages of hindsight, why a particular species of bacterium, fungus or protozoan has become a major risk to human health while, on other occasions, the vagaries of nature have left the 'experts' totally bemused. However, even in these latter situations, control over the threat posed to food supplies has to be instituted, but the ability of the food industry, in conjunction with Public Health and other bodies, to develop effective responses can only be as good as the scientific knowledge available. In the case of food microbiology, this background has to be derived from a wide range of sources. Thus, agricultural practices may alter the biochemistry of a crop and, perhaps, its microflora as well; the microflora of any given foodstuff and/or processing facility will have specific characteristics that need to be understood before control is possible; techniques must be available to monitor a retail food for microorganisms that would pose a risk to the consumer. As the procedures necessary to monitor these various facets become ever more sophisticated, so fewer microbiologists can claim total competence, and the need for a specialist source of outside knowledge increases.

It is this latter need that the *Encyclopedia of Food Microbiology* seeks to satisfy for, within this work, a busy microbiologist can find details of all the important genera of food-borne bacteria and fungi, how the same genera may react in different foods and under different environmental conditions, and how to detect the growth and/or metabolism of the same organisms in foods using classical or modern techniques. In order to place this information into a broader context, the reader can explore the latest advice concerning food standards/specifications, or the role of monitoring systems like HACCP in achieving product targets for specific microorganisms; potential concerns over viruses and protozoa are also evaluated in the light of current knowledge. Readers interested in fermented foods will find the pertinent information in a similarly accessible form; indeed, purchasers of the print version of the encyclopedia will be entitled to register for access to the on-line version as well. This form allows the user the benefit of extensive hypertext linking and advanced search tools, adding value to the encyclopedia as a reference source, teaching aid and text for general interest.

It is inevitable, of course, that short articles written to a tight deadline may have omissions, but it is to be hoped that such faults are minimal and, in any event, more than compensated for through the careful selections of further reading. If this optimism is justified, then the major credit rests with the authors of each article. They are all recognized as experts in their fields, and their willing participation has been much appreciated by the editors. The role of the Editorial Advisory Board merits a special mention as well, for their constructive criticisms of the list of articles, their suggestions for authors and their expert refereeing of the manuscripts has provided a solid foundation for the entire enterprise.

However, the finest manuscripts are of little value to the scientific community until they have been published, and the editorial team at Academic Press – Carey Chapman (Editor-in-Chief), Tina Holland (Associate Editor), Nick Fallon (Commissioning Editor), Laura O'Neill (Editorial Assistant), Tamsin Cousins (Production Project Manager), Richard Willis (Freelance Project Manager), Emma Parkinson (Electronic Publishing Developer), Peter Lord (Publishing Services Manager), Emma Krikler (Picture Researcher) – have been outstanding in their support of the project. Obviously, each member of the team has made an important contribution, but it must be recorded that the role of Tina Holland has been absolutely invaluable. Thus, not only has Tina coordinated the numerous inputs from the editors, referees and authors, but even found time to help the editors with the location of authors; the editors acknowledge this unstinting assistance with much gratitude.

R.K. Robinson, C.A. Batt, P.D. Patel
Editors

GUIDE TO USE OF THE ENCYCLOPEDIA

Structure of the Encyclopedia

The material in the Encyclopedia is arranged as a series of entries in alphabetical order. Some entries comprise a single article, whilst entries on more diverse subjects consist of several articles that deal with various aspects of the topic. In the latter case the articles are arranged in a logical sequence within an entry.

To help you realize the full potential of the material in the Encyclopedia we have provided three features to help you find the topic of your choice.

1. Contents Lists

Your first point of reference will probably be the contents list. The complete contents list appearing in each volume will provide you with both the volume number and the page number of the entry. On the opening page of an entry a contents list is provided so that the full details of the articles within the entry are immediately available.

Alternatively you may choose to browse through a volume using the alphabetical order of the entries as your guide. To assist you in identifying your location within the Encyclopedia a running headline indicates the current entry and the current article within that entry.

You will find 'dummy entries' where obvious synonyms exist for entries or where we have grouped together related topics. Dummy entries appear in both the contents list and the body of the text. For example, a dummy entry appears for Butter which directs you to Milk and Milk Products: Microbiology of Cream and Butter, where the material is located.

Example

If you were attempting to locate material on Dairy Products via the contents list.

DAIRY PRODUCTS *see BRUCELLA*: Problems with Dairy Products; CHEESE: In the Market Place; Microbiology of Cheese-making and Maturation; Mould-ripened Varieties; Role of Specific Groups of Bacteria; Microflora of White-brined Cheeses; FERMENTED MILKS: Yoghurt; Products from Northern Europe; Products of Eastern Europe and Asia; PROBIOTIC BACTERIA: Detection and Estimation in Fermented and Non-fermented Dairy Products

At the appropriate location in the contents list, the page numbers for articles under *Brucella*, etc. are given.

If you were trying to locate the material by browsing through the text and you looked up Dairy Products then the following information would be provided.

DAIRY PRODUCTS *see BRUCELLA*: Problems with Dairy Products; **CHEESE**: In the Market Place; Micro-biology of Cheese-making and Maturation; Mould-ripened Varieties; Role of Specific Groups of Bacteria; Microflora of White-brined Cheeses; **FERMENTED MILKS**: Yoghurt; Products from Northern Europe; Products of Eastern Europe and Asia; **PROBIOTIC BACTERIA**: Detection and Estimation in Fermented and Non-fermented Dairy Products.

Alternatively, if you were looking up *Brucella* the following information would be provided.

BRUCELLA

Contents
Characteristics
Problems with Dairy Products

2. Cross References

All of the articles in the Encyclopedia have an extensive list of cross references which appear at the end of each article, e.g.:

ATP BIOLUMINESCENCE/Application in Dairy Industry.
See also: **Acetobacter. ATP Bioluminescence**: Application in Meat Industry; Application in Hygiene Monitoring; Application in Beverage Microbiology. **Bacteriophage-based Techniques for Detection of Food-borne Pathogens. Biophysical Techniques for Enhancing Microbiological Analysis**: Future Developments. **Electrical Techniques**: Food Spoilage Flora and Total Viable Count (TVC). **Immuno-magnetic Particle-based Techniques**: Overview. **Rapid Methods for Food Hygiene Inspection. Total Viable Counts**: Pour Plate Technique; Spread Plate Technique; Specific Techniques; MPN; Metabolic Activity Tests; Microscopy. **Ultrasonic Imaging**: Non-destructive Methods to Detect Sterility of Aseptic Packages. **Ultrasonic Standing Waves**.

3. Index

The index will provide you with the volume number and page number of where the material is to be located, and the index entries differentiate between material that is a whole article, is part of an article or is data presented in a table. On the opening page of the index, detailed notes are provided.

4. Colour Plates

The colour figures for each volume have been grouped together in a plate section. The location of this section is cited both in the contents list and before the *See also* list of the pertinent articles.

5. Contributors

A full list of contributors appears at the beginning of each volume.

CONTRIBUTORS

Lahsen Ababouch
Department of Food Microbiology and Quality Control
Institut Agronomique et Vétérinaire Hassan II
Rabat
Morocco

D Abramson
Agriculture & Agri-Food Canada
Cereal Research Centre
195 Dafoe Road
Winnipeg
Manitoba
R3T 2M9
Canada

Ann M Adams
Seafood Products Research Center
US Food and Drug Administration
PO Box 3012
22201 23rd Drive SE
Bothell
WA 98041–3012
USA

Martin R Adams
School of Biological Sciences
University of Surrey
Guildford
GU2 5XH
UK

G E Age
PO Box 553
Wageningen
The Netherlands

M Ahmed
Food Control Laboratory
PO Box 7463
Dubai
United Arab Emirates

Imad Ali Ahmed
Central Food Control Laboratory, Ajman Municipality
PO Box 3717
Ajman
UAE

Peter Ahnert
Department of Biochemistry
Ohio State University
Columbus
OH 43210
USA

William R Aimutis
Land O' Lakes Inc.
PO Box 674101
St Paul
Minnesota
55164–0101
USA

J H Al-Jedah
Central Laboratories
Ministry of Public Health
Qatar

Cameron Alexander
Macromolecular Science Department
Institute of Food Research,
Reading Laboratory
Earley Gate
Whiteknights Road
Reading
RG6 6BZ
UK

Marcos Alguacil
Departmento de Genética
Facultad de Ciencias, Universidad de Málaga
Spain

M Z Ali
Central Laboratories
Ministry of Public Health
Qatar

M D Alur
Food Technology Division
Bhabha Atomic Research Centre
Mumbai 400085
India

R Miguel Amaguaña
US Food and Drug Administration
Washington, DC
USA

Vilma Moratade de Ambrosini
Centro de Referencia para Lactobacilos and Universidad
Nacional de Tucumán
Casilla de Correo 211
(4000)-Tucuman
Argentina

Wallace H Andrews
US Food and Drug Administration
Washington, DC 20204
USA

Dilip K Arora
Department of Botany
Banaras Hindu University
Varanasi 221 005
India

B Austin
Department of Biological Sciences
Heriot-Watt University
Riccarton
Edinburgh EH14 4AS
Scotland, UK

Aslan Azizi
Iranian Agricultural Engineering Research Institute
Agricultural Research Organization
Evin Tehran
Iran

S De Baets
Laboratory of Industrial Microbiology and Biocatalysis
Department of Biochemical and Microbial Technology
Faculty of Agricultural and Applied Biological Sciences
University of Gent
Coupure links 653
B-9000
Gent
Belgium

Les Baillie
Biomedical Sciences
DERA
CBD Porton Down
Salisbury
Wiltshire
UK

Gustavo V Barbosa-Cánovas
Biological Systems Engineering
Washington State University
Pullman
Washington 99164–6120
USA

J Baranyi
Institute for Food Research
Reading
UK

Eduardo Bárzana
Departamento de Alimentos y Biotecnología
Facultad de Química
Universidad Nacional Autónoma de México
Mexico City 04510
Mexico

Carl A Batt
Department of Food Science
Cornell University
Ithaca
NY 14853
USA

Derrick A Bautista
Saskatchewan Food Product Innovation Program
Department of Applied Microbiology and Food Science
University of Saskatchewan
Canada

S H Beattie
Hannah Research Institute
Ayr KA6 5HL
UK

H Beck
Department for Health Service
South Bavaria
Veterinärstrasse 2
85764 Oberschleissheim
Germany

Reginald Bennett
FDA
Center for Food Safety and Applied Nutrition
Washington, DC
USA

Marjon H J Bennik
Agrotechnological Research Institution (ATO-DLO)
Bornsesteeg 59
6709 PD
Wageningen
The Netherlands

Merlin S Bergdoll (dec)
Food Research Institute
University of Wisconsin-Madison
Madison, WI
USA

R G Berger
Food Chemistry
University of Hannover
Germany

K Berghof
BioteCon Gesellschaft für Biotechnologische
Entwicklung und Consulting
Hermannswerder Haus 17
14473 Potsdam
Germany

P A Bertram-Drogatz
Mediport VC Management GmbH
Wiesenweg 10
12247 Berlin
Germany

Gail D Betts
Campden and Chorleywood Food Research Association
Chipping Campden
Gloucestershire
GL55 6LD
UK

R R Beumer
Wageningen Agricultural University
Laboratory of Food Microbiology
Bomenweg 2
NL 6703 HD Wageningen
The Netherlands

Rijkelt R Beumer
Wageningen University and Research Centre
Department of Food Technology and Nutritional Sciences
Bomenweg 2
NL 6703 HD Wageningen
The Netherlands

Saumya Bhaduri
Microbial Food Safety Research Unit
Eastern Regional Research Center
Agricultural Research Service
US Department of Agriculture
600 East Mermaid Lane
Wyndmoor
PA 19038
USA

Deepak Bhatnagar
Southern Regional Research Center
Agricultural Research Service
US Department of Agriculture
LA
USA

J R Bickert
Halosource Corporation
First Avenue South
Seattle
WA 98104
USA

Hanno Biebl
GBF – National Research Centre for Biotechnology
Braunschweig
Germany

Clive de W Blackburn
Microbiology Unit
Unilever Research Colworth
Colworth House
Sharnbrook
Bedford
UK

I S Blair
Food Studies Research Unit
University of Ulster at Jordanstown
Shore Road
Newtownabbey
Co. Antrim
Northern Ireland
BT37 9QB

G Blank
Department of Food Science
University of Manitoba
Winnipeg
MB
Canada

Hans P Blaschek
Department of Food Science and Human Nutrition
University of Illinois
488 Animal Science Lab
1207 West Gregory Drive
Urbana
IL 61801
USA

D Blivet
AFSSA
Ploufragan
France

R G Board
South Lodge
Northleigh
Bradford-on-Avon
Wiltshire
UK

Enne de Boer
Inspectorate for Health Protection
PO Box 9012
7200 GN Zutphen
The Netherlands

Christine Bonaparte
Department of Dairy Research and Bacteriology
Agricultural University
Gregor Mendel-Str. 33
A-1180 Vienna
Austria

Kathryn J Boor
Department of Food Science
Cornell University
Ithaca
NY 14853
USA

A Botha
Department of Microbiology
University of Stellenbosch
Stellenbosch 7600
South Africa

W Richard Bowen
Biochemical Engineering Group
Centre for Complex Fluids Processing
University of Wales Swansea
Singleton Park
Swansea
SA2 8PP
UK

Catherine Bowles
Leatherhead Food Research Association
Leatherhead
Surrey
UK

Patrick Boyaval
INRA
Laboratoire de Recherches de Technologie Laitière
65 rue de Saint-Brieuc
35042
Rennes Cedex
France

F Bozoğlu
Department of Food Engineering
Middle East Technical University
Ankara
Turkey

Astrid Brandis-Heep
Philipps Universität
Fachbereich Biologie
Laboratorium für Mikrobiologie
D-35032 Marburg
Germany

Susan Brewer
Department of Food Science and Human Nutrition
University of Illinois
Urbana
Illinois
USA

Aaron L Brody
Rubbright Brody Inc.
PO Box 956187
Duluth
Georgia
30095–9504
USA

Bruce E Brown
B. E. Brown Associates
328 Stone Quarry Priv.
Ottawa
Ontario
K1K 3Y2
Canada

G Bruggeman
Laboratory of Industrial Microbiology and Biocatalysis
Department of Biochemical and Microbial Technology
Faculty of Agricultural and Applied Biological Sciences
University of Gent
Coupure links 653
B-9000
Gent
Belgium

Andreas Bubert
Department for Microbiology
Theodor-Boveri Institute for Biosciences
University of Würzburg
Am Hubland
97074 Würzburg
Germany

Ken Buckle
Department of Food Science and Technology
The University of New South Wales
Sydney
Australia

Lloyd B Bullerman
Department of Food Science and Technology
University of Nebraska
PO Box 830919
Lincoln
NE 68583–0919
USA

Justino Burgos
Food Technology Section
Department of Animal Production and Food Science
University of Zaragoza
Spain

Frank F Busta
Department of Food Science and Nutrition
University of Minnesota
St Paul
Minnesota 55108
USA

Daniel Cabral
Departmento de Ciencias Biológicas
Facultad de Ciencias Exactas y Naturales
Pabellon II 4to piso – Ciudad Universitaria
1428 Buenos Aires
Argentina

María Luisa Calderón-Miranda
Biological Systems Engineering
Washington State University
Pullman
Washington 99164–6120
USA

Geoffrey Campbell-Platt
Gyosei Liaison Office
Gyosei College
London Road
Reading
Berks RG1 5AQ
UK

Iain Campbell
International Centre for Brewing and Distilling
Heriot-Watt University
Edinburgh
EH14 4AS
Scotland

Frédéric Carlin
Institut National de la Recherche Agronomique
Unité de Technologie des Produits Végétaux
Site Agroparc
84914
Avignon
Cedex 9
France

Brigitte Carpentier
National Veterinary and Food Research Centre
22 rue Pierre Curie
F-94709
Maisons-Alfort Cedex
France

Maria da Glória S Carvalho
Departamento de Microbiologia Médica
Instituto de Microbiologia
Universidade Federal do Rio de Janeiro
Rio de Janeiro 21941
Brazil

O Cerf
Alfort Veterinary School
7 Avenue du Général de Gaulle
F-94704
Maisons-Alfort Cedex
France

Lourdes Pérez Chabela
Universidad Autónoma Metropolitana-Iztapalapa
Mexico
Apartado Postal 55–535
CP 09340 Mexico DF
Mexico

Perng-Kuang Chang
Southern Regional Research Center
Agricultural Research Service
US Department of Agriculture
LA
USA

E A Charter
Canadian Inovatech Inc.
31212 Peardonville Road
Abbotsford
BC V2T 6K8
Canada

Parimal Chattopadhyay
Department of Food Technology and Biochemical
Engineering
Jadavpur University
Calcutta-700 032
India

Yusuf Chisti
Department of Chemical Engineering
University of Almería
E-04071 Almería
Spain

Thomas E Cleveland
Southern Regional Research Center
Agricultural Research Service
US Department of Agriculture
LA
USA

Dean O Cliver
University of California, Davis, School of Veterinary
Medicine
Department of Population Health and Reproduction
One Shields Avenue
Davis
California 95616–8743
USA

T E Cloete
Department of Microbiology and Plant Pathology
Faculty of Biological and Agricultural Sciences
University of Pretoria
Pretoria 0002
South Africa

Roland Cocker
Cocker Consulting
Bergeendlaan 16
1343 AR Almere
The Netherlands

Timothy M Cogan
Dairy Products Research Centre
Teagasc
Fermoy
Ireland

David Collins-Thompson
Nestlé Research and Development Center
210 Housatonic Avenue
New Milford
Connecticut
USA

Janet E L Corry
Division of Food Animal Science
Department of Clinical Veterinary Science
University of Bristol
Langford
Bristol
BS40 5DT
UK

Aldo Corsetti
Institute of Dairy Microbiology
Faculty of Agriculture of Perugia
06126 S. Costanzo
Perugia
Italy

Polly D Courtney
Department of Food Science and Technology
Ohio State University
2121 Fyffe Road
Columbus
OH 43210
USA

M A Cousin
Department of Food Science
Purdue University
West Lafayette
Indiana
47907–1160
USA

N D Cowell
Elstead
Godalming
Surrey
GU8 6HT
UK

Julian Cox
Department of Food Science and Technology
The University of New South Wales
Sydney
Australia

C Gerald Crawford
US Department of Agriculture
Agricultural Research Service
Eastern Regional Research Center
600 E. Mermaid Lane
Wyndmoor
PA 19038
USA

Theresa L Cromeans
Department of Environmental Sciences and Engineering
School of Public Health
University of North Carolina
North Carolina
USA

Kofitsyo S Cudjoe
Department of Pharmacology
Microbiology and Food Hygiene
Norwegian College of Veterinary Medicine
PO Box 8146 Dep
0033 Oslo
Norway

David Cunliffe
Macromolecular Science Department
Institute of Food Research
Reading Laboratory
Earley Gate, Whiteknights Road
Reading
RG6 6BZ
UK

Ladislav Čurda
Department of Dairy and Fat Technology
Prague Institute of Chemical Technology
Czech Republic

G J Curiel
Unilever Research Vlaardingen
PO Box 114
3130 AC Vlaardingen
The Netherlands

G D W Curtis
Bacteriology Department
John Radcliffe Hospital
Oxford
UK

Michael K Dahl
Department of Microbiology
University of Erlangen
Staudtstrasse 5
91058 Erlangen
Germany

Crispin R Dass
The Heart Research Institute Ltd
145 Missenden Road
Camperdown
Sydney
NSW 2050
Australia

E Alison Davies
Technical Services & Research Department
Aplin & Barrett Ltd (Cultor Food Science)
15 North Street
Beaminster
Dorset
DT8 3DZ
UK

Brian P F Day
Campden and Chorleywood Food Research Association
Chipping Campden
Gloucestershire
GL55 6LD
UK

J M Debevere
Laboratory of Food Microbiology and Food Preservation
Faculty of Agricultural and Applied Biological Sciences
University of Ghent
Coupure Links 654
9000 Ghent
Belgium

Joss Delves-Broughton
Technical Services and Research Department
Aplin & Barrett Ltd (Cultor Food Science)
15 North Street
Beaminster
Dorset
DT8 3DZ
UK

Stephen P Denyer
Department of Pharmacy
The University of Brighton
Cockcroft Building
Moulescoomb
Brighton
BN2 4GJ
UK

P M Desmarchelier
Food Safety and Quality
Food Science Australia
PO Box 3312
Tingalpo DC
Queensland 4173
Australia

Janice Dewar
CSIR Food Science and Technology
PO Box 395
Pretoria 001
South Africa

Vinod K Dhir
Biotec Laboratories Ltd
32 Anson Road
Martlesham Heath
Ipswich
Suffolk
IP5 3RD
UK

M W Dick
Department of Botany
University of Reading
Reading
RG6 6AU
UK

Vivian M Dillon
Department of Biology and Biochemistry
University of Bath
Bath
UK

Eleftherios H Drosinos
Department of Food Science and Technology
Laboratory of Microbiology and Biotechnology of Foods
Agricultural University of Athens
Iera Odos 75
Athens
Greece

F M Dugan
USDA–ARS Western Regional Plant Introduction Station
Washington State University
Washington
USA

B Egan
Marine Biological and Chemical Consultants Ltd
Bangor
UK

H M J van Elijk
Unilever Research Vlaardingen
PO Box 114
3130 AC Vlaardingen
The Netherlands

Hartmut Eisgruber
Institute for Hygiene and Technology of Foods of Animal
Origin, Veterinary Faculty
Ludwig-Maximilians University
80539 Munich
Germany

Phyllis Entis
QA Life Sciences, Inc.
6645 Nancy Ridge Drive
San Diego
CA 92121
USA

John P Erickson
Microbiology – Research and Development
Bestfoods Technical Center
Somerset
New Jersey
USA

Douglas E Eveleigh
Department of Microbiology
Rutgers University
Cook College
76 Lipman Drive
New Brunswick
NJ 08901-8525
USA

Richard R Facklam
Streptococcus Laboratory
Respiratory Diseases Branch
Division of Bacterial and Mycotic Diseases
Centres for Disease Control and Prevention
Mail Stop CO-2
Atlanta
GA 30333
USA

M Fandke
BioteCon Gesellschaft für Biotechnologische
Entwicklung und Consulting
Hermannswerder Haus 17
14473 Potsdam
Germany

Nana Y Farkye
Dairy Products Technology Center
Dairy Science Department
California Polytechnic State University
San Luis Obispo
CA 93407
USA

Manuel Fidalgo
Departmento de Genética
Facultad de Ciencias, Universidad de Málaga
Spain

Christopher W Fisher
Department of Food Science and Human Nutrition
University of Illinois
Urbana
IL 61801
USA

G H Fleet
CRC for Food Industry Innovation
Department of Food Science and Technology
The University of New South Wales
Sydney
New South Wales 2052
Australia

Harry J Flint
Rowett Research Institute
Greenburn Road
Bucksburn
Aberdeen
UK

Samuel Formal
Department of Microbiology and Immunology
Uniformed Services University of the Health Sciences
F Edward Hébert School of Medicine
4301 Jones Bridge Road
Bethesda
MD 20814
USA

Pina M Fratamico
US Department of Agriculture
Agricultural Research Service
Eastern Regional Research Center
600 E. Mermaid Lane
Wyndmoor
PA 19038
USA

Colin Fricker
Thames Water Utilities
Manor Farm Road
Reading
RG2 0JN
UK

Daniel Y C Fung
Department of Animal Sciences and Industry
Kansas State University
Manhattan
Kansas 66506
USA

H Ray Gamble
United States Department of Agriculture
Agricultural Research Service
Parasite Biology and Epidemiology Laboratory
Building 1040, Room 103, BARC-East
Beltsville
MD 20705
USA

Indrawati Gandjar
Department of Biology
Faculty of Science and Mathematics
University of Indonesia
Jakarta
Indonesia

Mariano García-Garibay
Departamento de Biotechnología
Universidad Autónoma Metropolitana
Iztapalapa, Apartado Postal 55–535
Mexico City 09340
Mexico

María-Luisa García-López
Department of Food Hygiene and Food Technology
University of León
24071-León
Spain

S K Garg
Department of Microbiology
Dr Ram Manohar Lohia Avadh University
Faizabad 224 001
India

A Gasch
BioteCon Gesellschaft für Biotechnologische
Entwicklung und Consulting
Hermannswerder Haus 17
14473 Potsdam
Germany

Michel Gautier
Ecole Nationale Supérieure d'Agronomie
Institut National de la Recherche Agronomique
65 rue de SrBrieuc
35042
Rennes cédex
France

Gerd Gellissen
Rhein Biotech GmbH
EichsFelder Str. 11
40595 Düsseldorf
Germany

N P Ghildyal
Fermentation Technology and Bioengineering
Department
Central Food Technological Research Institute
Mysore 570013
India

M Gibert
Institut Pasteur
Unité Interactions Bactéries Cellules
28 rue du Dr Roux
75724 Paris
Cedex 15
France

Glenn R Gibson
Microbiology Department
Institute of Food Research
Reading
UK

M C te Giffel
Wageningen Agricultural University
Laboratory of Food Microbiology
Bomenweg 2
NL 6703 HD Wageningen
The Netherlands

A Gilmour
Food Science Division (Food Microbiology)
Department of Agriculture for Northern Ireland
Agriculture and Food Science Centre
Newforge Lane
Belfast
BT9 5PX
Northern Ireland, UK

Giorgio Giraffa
Istituto Sperimentale Lattiero Caseario
Via A. Lombardo
11 – 26900 Lodi
Italy

R W A Girdwood
Scottish Parasite Diagnostic Laboratory
Stobhill Hospital
Glasgow
GL21 3UW
UK

Andrew D Goater
Institute of Molecular and Biomolecular Electronics
University of Wales
Dean St
Bangor
Gwynedd
LL57 1UT
UK

Marco Gobbetti
Instituto di Produzioni e Preparazioni Alimentari
Facoltà di Agraria di Foggia
Via Napoli 25
71100 Foggia
Italy

Millicent C Goldschmidt
Department of Basic Sciences
Dental Branch
The University of Texas Health Center at Houston
6516 John Freeman Avenue
Houston
Texas 77030
USA

Lorena Gómez-Ruiz
Departamento de Biotechnología
Universidad Autónoma Metropolitana
Iztapalapa, Apartado Postal 55–535
Mexico City 09340
Mexico

Katsuya Gomi
Division of Life Science
Graduate School of Agricultural Science
Tohoku University
Japan

M Marcela Góngora-Nieto
Biological Systems Engineering
Washington State University
Pullman
Washington 99164–6120
USA

S Gonzalez
Universidad Nacional de Tucumán, Argentina
Cerela–Conicet
San Miguel de Tucumán
Argentina

Silvia N Gonzalez
Centro de Referencia para Lactobacilos (Cerela) and
Universidad Nacional de Tucumán
Chacabuco 145 (4000)
Tucumán
Argentina

Leon G M Gorris
Unit Microbiology and Preservation
Unilever Research Vlaardingen
PO Box 114
3130 AC Vlaardingen
The Netherlands

Grahame W Gould
17 Dove Road
Bedford
MK41 7AA
UK

M K Gowthaman
Fermentation Technology and Bioengineering
Department
Mysore 570013
India

Lone Gram
Danish Institute for Fisheries Research
Department of Seafood Research
Technical University of Denmark Bldg 221
DK-2800 Lyngby
Denmark

AGE Griffioen
Stichting EFFI
PO Box 553 Wageningen
The Netherlands

Mansel W Griffiths
Department of Food Science
University of Guelph
Guelph
Ontario
N1G 2W1
Canada

C Grönewald
BioteCon Gesellschaft für Biotechnologische
Entwicklung und Consulting
Hermannswerder Haus 17
14473 Potsdam
Germany

Isabel Guerrero
Universidad Autónoma Metropolitana-Iztapalapa
Mexico
Apartado Postal 55–535
CP 09340 Mexico DF
Mexico

G C Gürakan
Middle East Technical University
Ankara
Turkey

Carlos Horacio Gusils
Centro de Referencia para Lactobacilos and Universidad
Nacional de Tucumán
Casilla de Correo 211
(4000)-Tucuman
Argentina

Thomas S Hammack
US Food and Drug Administration
Washington, DC 20204
USA

S A S Hanna,
48 Kensington Street
Newton
MA 02460
USA

Karen M J Hansen
Saskatchewan Food Product Innovation Program
University of Saskatchewan
Saskatoon
SK
S7N 5A8
Canada

J Harvey
Food Science Division (Food Microbiology)
Department of Agriculture for Northern Ireland
Agriculture and Food Science Centre
Newforge Lane
Belfast
BT9 5PX
Northern Ireland, UK

Wilma C Hazeleger
Wageningen University and Research Centre
Department of Food Technology and Nutritional Sciences
Bomenweg 2
NL 6703 HD Wageningen
The Netherlands

G M Heard
CRC for Food Industry Innovation
Department of Food Science and Technology
University of New South Wales
Sydney
New South Wales 2052
Australia

Nidal Hilal
Biochemical Engineering Group
Centre for Complex Fluids Processing
Department of Chemical and Biological Process Engineering
University of Wales Swansea
Singleton Park
Swansea SA2 8PP
UK

G Hildebrandt
Institute for Food Hygiene
Free University of Berlin
Germany

Colin Hill
Department of Microbiology and National Food Biotechnology Centre
University College
Cork
Ireland

A D Hitchins
Center for Food Safety and Applied Nutrition
Food and Drug Administration
Washington, DC
USA

Jill E Hobbs
George Morris Centre
345, 2116 27th Avenue NE
Calgary
Alberta
T2E 7A6
Canada

Ailsa D Hocking,
CSIRO Food Science Australia
Riverside Corporate Park
North Ryde
New South Wales 2113
Australia

Cornelis P Hollenberg
Institut für Microbiology
Heinrich-Heine-Universität Düsseldorf
40225 Düsseldorf
Germany

Richard A Holley
Department of Food Science
University of Manitoba
Winnipeg
Manitoba
R3T 2N2
Canada

Wilhelm H Holzapfel
Institute of Hygiene and Toxicology
Federal Research Centre for Nutrition
Bundesforschungsanstalt
Haid-und-Neu-Str. 9
D-7613 Karlsruhe
Germany

Rolf K Hommel
Cell Technologie Leipzig
Fontanestr. 21
Leipzig
D-04289
Germany

Dallas G Hoover
Department of Animal and Food Sciences
University of Delaware
Newark
DE 19717–1303
USA

Thomas W Huber
Medical Microbiology and Immunology
Texas A&M College of Medicine
Temple
Texas
USA

Robert Hutkins
Department of Food Science and Technology
University of Nebraska
338 FIC
Lincoln
NE 68583–0919
USA

Cheng-An Hwang
Nestlé Research and Development Center
210 Housatonic Avenue
New Milford
Connecticut
USA

John J Iandolo
Department of Microbiology and Immunology
University of Oklahoma Health Sciences Center
Oklahoma City
OK 73190
USA

Y Iimura
Department of Applied Chemistry and Biotechnology
Yamanashi University
Kofu
Japan

Charlotte Nexmann Jacobsen
Department of Dairy and Food Research
Royal Veterinary and Agricultural University
Rolighedsvej 3,0
1958 Frederiksberg C
Denmark

Mogens Jakobsen
Department of Dairy and Food Research
Royal Veterinary and Agricultural University
Rolighedsvej 3,0
1958 Frederiksberg C
Denmark

Dieter Jahn
Institute for Organic Chemistry and Biochemistry
Albert Ludwigs University Freiburg
Albertstr. 21
79104 Freiburg
Germany

B Jarvis
Ross Biosciences Ltd
Daubies Farm
Upton Bishop
Ross-on-Wye
Herefordshire
HR9 7UR
UK

Ian Jenson
Gist-brocades Australia Pty, Ltd
Moorebank
NSW
Australia

Juan Jimenez
Departmento de Genética
Facultad de Ciencias, Universidad de Málaga
Spain

Karen C Jinneman
Department of Veterinary Science and Microbiology
University of Arizona
Tucson
AZ 85721
USA

Juan Jofre
Department of Microbiology
University of Barcelona
Spain

Eric Johansen
Department of Genetics and Microbiology
Chr. Hansen A/S
10–12 Bøge Allé
DK-2970
Hørsholm
Denmark

Nick Johns
Independent Research Consultant
15 Collingwood Close
Steepletower
Hethersett
Norwich NR9 3QE
UK

Eric A Johnson
Department of Food Microbiology
Food Research Institute, University of Wisconsin
Madison
WI
USA

Clifford H Johnson
US Environmental Protection Agency
Cincinatti
Ohio
USA

Rafael Jordano
Department of Food Science and Technology
Campus Rabanales, University of Córdoba
E-14071 Córdoba
Spain

Richard Joseph
Department of Food Microbiology
Central Food Technological Research Institute
Mysore
570 013
India

Vinod K Joshi
Department of Post-harvest Technology
Dr YSP University of Horticulture and Foresty
Nauni
Solan-173 230
India

Vijay K Juneja
United States Department of Agriculture
Eastern Regional Research Center
600 East Mermaid Lane
Wyndmoor
Pennsylvania
USA

G Kalantzopoulos
Department of Food Science and Technology
Agricultural University of Athens
Greece

Chitkala Kalidas
Field of Microbiology
Department of Food Science
Cornell University
Ithaca NY 14853
USA

A Kambamanoli-Dimou
Department of Animal Production
Technological Education Institute
Larissa
Greece

Peter Kämpfer
Institut für Angewandte Mikrobiologie
Justus-Liebig-Universität Giessen
Senckenbergstr. 3
D-35390 Giessen
Germany

N G Karanth
Fermentation Technology and Bioengineering
Department
Mysore 570013
India

Embit Kartadarma
Department of Food Science and Technology
The University of New South Wales
Sydney
Australia

K L Kauppi
University of Minnesota
Department of Food Science and Nutrition
St Paul
USA

C A Kaysner
US Food and Drug Administration
22201 23rd Drive SE
Bothell
Washington 98021
USA

William A Kerr
Department of Economics
University of Calgary
2500 University Drive NW
Calgary
Alberta
T2N 1N4
Canada

Tajalli Keshavarz
Department of Biotechnology
University of Westminster
115 New Cavendish Street
London
W1M 8JS
UK

George G Khachatourians
Department of Applied Microbiology and Food Science
University of Saskatchewan
Saskatoon
Canada

W Kim
Department of Microbiology
University of Georgia
Athens
Georgia
USA

P M Kirk
CABI Bioscience UK Centre (Egham)
Bakeham Lane
Egham
Surrey
TW20 9TY

Todd R Klaenhammer
Departments of Food Science and Microbiology
Southeast Dairy Foods Research Center
Box 7624
North Carolina State University
Raleigh
NC 27695–7624
USA

Hans-Peter Kleber
Institut für Biochemie
Fakultöt für Biowissenschaften
Pharmazie und Psychologie
Universität Leipzig
Talstr. 33
Leipzig
D-04103
Germany

Thomas J Klem
Department of Food Science
Cornell University
USA

Wolfgang Kneifel
Department of Dairy Research and Bacteriology
Agricultural University
Gregor Mendel-Str. 33
A-1180 Vienna
Austria

Barb Kohn
VICAM LP
313 Pleasant Street
Watertown
MA 02172
USA

C Koob
BioteCon Gesellschaft für Biotechnologische
Entwicklung und Consulting
Hermannswerder Haus 17
14473 Potsdam
Germany

P Kotzekidou
Department of Food Science and Technology
Faculty of Agriculture
Aristotle University of Thessaloniki
PB 250
GR 540 06
Thessaloniki
Greece

K Krist
Meat and Livestock Australia
Sydney
Australia

Pushpa R Kulkarni
University Department of Chemical Technology
University of Mumbai
Matunga
Mumbai 400 019
India

Madhu Kulshreshtha
Division of Plant Pathology
Indian Agricultural Research Institute
New Delhi 11012
India

Susumu Kumagai
Department of Biomedical Food Research
National Institute of Infectious Diseases
Toyama 1–23–1
Shinjuku-ku
Tokyo 162–8640
Japan

G Lagarde
Inovatech Europe B.V.
Landbouwweg
The Netherlands

Keith A Lampel
US Food and Drug Administration
Center for Food Safety and Applied Nutrition HFS-327
200 C St SW
Washington
DC 20204
USA

S Leaper
Campden and Chorleywood Food Research Association
Chipping Campden
Gloucestershire
GL55 6LD
UK

J David Legan
Microbiology Department
Nabisco Research
PO Box 1944
DeForest Avenue
East Hanover
NJ 017871
USA

J J Leisner
Department of Veterinary Microbiology
Royal Veterinary and Agricultural University
Stigbøjlen 4
DK-1870 Frederiksberg C
Denmark

H L M Lelieveld
Unilever Research Vlaardingen
PO Box 114
3130 AC Vlaardingen
The Netherlands

D F Lewis
Food Systems Division
SAC
Auchincruive
Ayr KA6 5HW
Scotland
UK

M J Lewis
Department of Food Science and Technology
University of Reading
UK

E Litopoulou-Tzanetaki
Department of Food Science, Faculty of Agriculture
Aristotle University of Thessaloniki
54006
Thessaloniki
Greece

Aline Lonvaud-Funel
Faculty of Œnology
University Victor Segalen Bordeaux 2
351, Cours de la Libération
33405 Talence Cedex
France

S E Lopez
Departamento de Ciencias Biológicas
Facultad de Ciencias Exactas y Naturales
Pabellon II 4to piso – Ciudad Universitaria
1428 Buenos Aires
Argentina

G Love
Centre for Electron Optical Studies
University of Bath
Claverton Down
Bath
BA2 7AY
UK

Robert W Lovitt
Biochemical Engineering Group
Centre for Complex Fluids Processing
Department of Chemical and Biological Process
Engineering
University of Wales Swansea
Singleton Park
Swansea
SA2 8PP
UK

Majella Maher
National Diagnostics Centre
National University of Ireland
Galway
Ireland

R H Madden
Food Microbiology
Food Science Department
Department of Agriculture for Northern Ireland and
Queen's University of Belfast
Newforge Lane
Belfast
BT9 5PX
Northern Ireland

T Mahmutoğlu
TATKO TAS
Gayrettepe
Istanbul
Turkey

K A Malik
Chairman
Pakistan Agricultural Research Council
Islamabad
Pakistan

Miguel Prieto Maradona
Department of Food Hygiene and Food Technology
University of León
24071-León
Spain

Scott E Martin
Department of Food Science and Human Nutrition
University of Illinois
486 Animal Sciences Laboratory
1207 West Gregory Drive
Urbana
IL 61801
USA

L Martínková
Laboratory of Biotransformation
Institute of Microbiology
Academy of Sciences of the Czech Republic
Prague
Czech Republic

Tina Mattila-Sandholm
VTT Biotechnology and Food Research
Tietotie 2
Espoo
PO Box 1501
FIN-02044 VTT
Finland

D A McDowell
Food Studies Research Unit
University of Ulster at Jordanstown
Shore Road
Newtownabbey
Co. Antrim
Northern Ireland
BT37 9QB

Denise N McKenna
Microbiology – Research and Development
Bestfoods Technical Center
Somerset
New Jersey
USA

M A S McMahon
Food Studies Research Unit
University of Ulster at Jordanstown
Shore Road
Newtownabbey
Co. Antrim
Northern Ireland
BT37 9QB

T A McMeekin
School of Agricultural Science
University of Tasmania
Hobart
Australia

Luis M Medina
Department of Food Science and Technology
Campus Rabanales
University of Córdoba
E-14071 Córdoba
Spain

Aubrey F Mendonca
Iowa State University
Department of Food Science and Human Nutrition
Ames
Iowa
USA

James W Messer
US Environmental Protection Agency
Cincinnati
Ohio
USA

M C Misra
Fermentation Technology and Bioengineering
Department
Central Food Technological Research Institute
Mysore 570013
India

Vikram V Mistry
Dairy Science Department
South Dakota State University
Brookings
South Dakota 57007
USA

D R Modi
Department of Microbiology
Dr Ram Manohar Lohia Avadh University
Faizabad 224 001
India

Richard J Mole
Biotec Laboratories Ltd.
32 Anson Road
Martlesham Heath
Ipswich
Suffolk
IP5 3RD
UK

M C Montel
Station de Recherches sur la Viande
INRA
63122 Saint Genès Champanelle
France

M Moresi
Istituto di Tecnologie Agroalimentari
Università della Tuscia
Via S C de Lellis
01100 Viterbo
Italy

André Morin
Imperial Tobacco Limited
3810 rue St-Antoine
Montreal
Quebec H4C 1B5
Canada

Maurice O Moss
School of Biological Sciences, University of Surrey
Guildford
GU2 5XH
UK

M A Mostert
Unilever Research Vlaardingen
PO Box 114
3130 AC Vlaardingen
The Netherlands

Donald Muir
Hannah Research Institute
Ayr
KA6 5HL
Scotland, UK

Maite Muniesa
Department of Microbiology
University of Barcelona
Spain

E A Murano
Center for Food Safety and Department of Animal
Science
Texas A&M University
Texas
USA

M J Murphy
CBD Porton Down
Salisbury
SP4 0JQ
UK

K Darwin Murrell
Agricultural Research Service
US Department of Agriculture
Beltsville
Maryland 20705
USA

C K K Nair
Radiation Biology Division
Bhabha Atomic Research Centre
Mumbai 400 085
India

Motoi Nakao
Horiba Ltd
Miyanohigashimachi
Kisshoin
Minami-ku
Kyoto
Japan
601–8510

A W Nichol
Charles Sturt University
NSW
Australia

D S Nichols
School of Agricultural Science
University of Tasmania
Hobart
Australia

Poonam Nigam
Biotechnology Research Group
School of Applied Biological and Chemical Sciences
University of Ulster
Coleraine BT52 1SA
UK

M de Nijs
TNO Nutrition and Food Research Institute
Division of Microbiology and Quality Management
PO Box 360
3700 AJ Zeist
The Netherlands

S H W Notermans
TNO Nutrition and Food Research Institute
PO Box 360
3700 AJ Zeist
The Netherlands

Martha Nuñez
Centro de Referencia par Lactobacilos (Cerela)
Chacabuco 145 (4000)
Tucumán
Argentina

George-John E Nychas
Department of Food Science and Technology
Laboratory of Microbiology and Biotechnology of Foods
Agricultural University of Athens
Iera Odos 75
Athens
11855
Greece

R E O'Connor-Shaw
Food Microbiology Consultant
Birkdale
Queensland
Australia

Louise O'Connor
National Diagnostics Centre
National University of Ireland
Galway
Ireland

Triona O'Keeffe
Department of Microbiology and National Food
Biotechnology Centre
University College
Cork
Ireland

Rachel M Oakley
United Biscuits (UK Ltd)
High Wycombe
Buckinghamshire
HP12 4JX
UK

Yuji Oda
Department of Applied Biological Science
Fukuyama University
Fukuyama
Hiroshima 729–0292
Japan

Lucy J Ogbadu
Department of Biological Sciences
Benue State University
Makurdi
Nigeria

Guillermo Oliver
Centro Referencia para Lactobacilos and Universidad
Nacional de Tucumán
Casilla de Correo 211
(4000)-Tucuman
Argentina

Ynes R Ortega
Seafood Products Research Center
US Food and Drug Administration
PO Box 3012
22201 23rd Drive SE
Bothell
WA 98041–3012
USA

Andrés Otero
Department of Food Hygiene and Food Technology
University of León
24071-León
Spain

Kozo Ouchi
Kyowa Hakko Kogyo Co. Ltd
1–6–1 Ohtemachi
Chiyoda-ku
Tokyo 100–8185
Japan

Barbaros H Özer
Department of Food Science and Technology
Faculty of Agriculture
University of Harran
63040
Şanlıurfa
Turkey

Dilek Özer
GAP Regional Development Administration
Şanlıurfa
Turkey

J Palacios
Universidad Nacional de Tucumán, Argentina
Cerela-Conicet
San Miguel de Tucumán
Argentina

Ashok Pandey
Laboratorio de Processos Biotecnologicos
Universidade Federal do Parana
Departmento de Engenharia Quimica
CEP 81531-970 Curitiba-PR
Brazil

Photis Papademas
Department of Food Science and Technology
University of Reading
Whiteknights
Reading
Berkshire
RG6 6AP
UK

A Pardigol
BioteCon Gesellschaft für Biotechnologische
Entwicklung und Consulting
Hermannswerder Haus 17
14473 Potsdam
Germany

E Parente
Dipartimento di Biologia, Difesa e Biotecnologie Agro-
Forestali
Università della Basilicata
Via N Sauro 85
85100 Potenza
Italy

Zahida Parveen
University of Huddersfield
Department of Chemical and Biological Sciences
Queensgate
Huddersfield
HD1 3DH
UK

P Patáková
Faculty of Food and Biochemical Technology
Institute of Chemical Technology
Prague
Czech Republic

Pradip Patel
Science and Technology Group
Leatherhead Food Research Association
Randalls Road
Leatherhead
Surrey
KT22 7RY
UK

Margaret Patterson
Food Science Division
Department of Agriculture for Northern Ireland and The
Queen's University of Belfast
Agriculture and Food Science Centre
Newforge Lane
Belfast
BT9 5PX
UK

P A Pawar
Fermentation Technology and Bioengineering
Department
Central Food Technological Research Institute
Mysore 570013
India

Janet B Payeur
National Veterinary Services Laboratories
Veterinary Services
Animal and Plant Health Inspection Service
Department of Agriculture
1800 Dayton Road
Ames
IA 50010
USA

Gary A Payne
Department of Plant Pathology
North Carolina State University
Raleigh
North Carolina
USA

Ron Pethig
Institute of Molecular and Biomolecular Electronics
University of Wales
Dean St
Bangor
Gwynedd
LL57 1UT
UK

L Petit
Unité Interactions Bactéries Cellules
Institut Pasteur
28 rue du Dr Roux
75724 Paris
Cedex 15
France

William A Petri Jr
Department of Medicine, Division of Infectious Diseases
University of Virginia Health Sciences Center
MR4, Room 2115, 300 Park Place
Charlottesville
VA 22908
USA

M R A Pillai
Isotope Division
Bhabha Atomic Rsearch Centre
Mumbai 400 085
India

D W Pimbley
Leatherhead Food Research Association
Randalls Road
Leatherhead
Surrey
KT22 7RY
UK

J I Pitt
CSIRO Food Science Australia
Riverside Corporate Park
North Ryde
New South Wales 2113
Australia

M R Popoff
Institut Pasteur
Unité Interactions Bactéries Cellules
28 rue du Dr Roux
75724 Paris
Cedex 15
France

U J Potter
Centre for Electron Optical Studies
University of Bath
Claverton Down
Bath
BA2 7AY
UK

B Pourkomailian
Department of Food Safety and Preservation
Leatherhead Food RA
Randalls Road
Surrey
UK

K Prabhakar
Department of Meat Science and Technology
College of Veterinary Science
Tirupati 517 502
India

W Praphailong
National Center for Genetic Engineering and
Biotechnology
Rajdhevee
Bangkok
Thailand

M S Prasad
Fermentation Technology and Bioengineering
Department
Mysore 570013
India

J. C du Preez
Department of Microbiology and Biochemistry
University of the Orange Free State
PO Box 339
Bloemfontein 9300
South Africa

Barry H Pyle
Montana State University
Bozeman
Montana
USA

Laura Raaska
VTT Biotechnology and Food Research
PO Box 1501
FIN-02044 VTT
Finland

Moshe Raccach
Food Science Program
School of Agribusiness and Resource Management
Arizona State University East
Mesa
Arizona 85206–0180
USA

Fatemeh Rafii,
Division of Microbiology
National Center for Toxicological Research, US FDA
Jefferson
AR
USA

M I Rajoka,
National Institute for Biotechnology and Genetic
Engineering (NIBGE)
PO Box 577
Faisalabad
Pakistan

Javier Raso
Biological Systems Engineering
Washington State University
Pullman
Washington 99164-6120
USA

K S Reddy
Department of Meat Science and Technology
College of Veterinary Science
Tirupati 517 502
India

S M Reddy
Department of Botany
Kakatiya University
Warangal
506 009
India

Wim Reybroeck
Department for Animal Product Quality and
Transformation Technology
Agricultural Research Centre CLO-Ghent
Melle
Belgium

V G Reyes
Food Science Australia
Private Bag 16
Sneydes Road
Werribee
Victoria
VIC 3030
Australia

E W Rice
US Environmental Protection Agency
Cincinnati
Ohio 45268
USA

Jouko Ridell
Department of Food and Environmental Hygiene, Faculty
of Veterinary Medicine
University of Helsinki
Finland

R K Robinson
Department of Food Science
University of Reading
Whiteknights
Reading
Berkshire RG6 6AP
UK

Hubert Roginski
Gilbert Chandler College
The University of Melbourne
Sneydes Road
Werribee
Victoria
3030
Australia

Alexandra Rompf
Institute for Organic Chemistry and Biochemistry
Albert Ludwigs University Freiburg
Albertstr. 21
79104 Freiburg
Germany

T Ross
School of Agricultural Science
University of Tasmania
Hobart
Australia

T Roukas
Department of Food Science and Technology
Aristotle University of Thessaloniki
Greece

M T Rowe
Food Microbiology
Food Science Department
Department of Agriculture for Northern Ireland and
Queen's University of Belfast
Newforge Lane
Belfast
BT9 5PX
Northern Ireland

W Michael Russell
Departments of Food Science and Microbiology
Southeast Dairy Foods Research Center
Box 7624
North Carolina State University
Raleigh
NC 27695–7624
USA

G Salvat
AFSSA
Ploufragan
France

R Sandhir
Department of Biochemistry
Dr Ram Manohar Lohia Avadh University
Faizabad 224 001
India

Robi C Sandlin
Department of Microbiology and Immunology
Uniformed Services University of the Health Sciences
F Edward Hébert School of Medicine
4301 Jones Bridge Road
Bethesda
MD 20814
USA

Jesús-Angel Santos
Department of Food Hygiene and Food Technology
University of León
24071-León
Spain

A K Sarbhoy
Division of Plant Pathology
Indian Agricultural Research Institute
New Delhi 110012
India

David Sartory
Severn Trent Water
Shrewsbury
UK

Joanna M Schaenman
Department of Medicine
Division of Infectious Diseases
University of Virginia Health Sciences Center
MR4, Room 2115, 300 Park Place
Charlottesville
VA 22908
USA

Barbara Schalch
Institute of Hygiene and Technology of Food of Animal Origin
Ludwig-Maximilians-University Munich
Veterinary Faculty
Veterinärstr. 13
81369 Munich
Germany

P Scheu
BioteCon Gesellschaft für Biotechnologische Entwicklung und Consulting
Hermannswerder Haus 17
14473 Potsdam
Germany

Bernard W Senior
Department of Medical Microbiology
University of Dundee Medical School
Ninewells Hospital
Dundee
DD1 9SY
UK

Gilbert Shama
Department of Chemical Engineering
Loughborough University
UK

Arun Sharma
Food Technology Division
Bhabha Atomic Research Centre
Mumbai 400 085
India

M Shin
Faculty of Pharmaceutical Sciences
Kobe Gakuin University
Kobe
Japan

J Silva
Universidad Nacional de Tucumán, Argentina
Cerela–Conicet
San Miguel de Tucumán
Argentina

Dalel Singh
Microbiology Department
CCS Haryana Agricultural University
Hisar
125 004
India

Rekha S Singhal
University Department of Chemical Technology
University of Mumbai
Matunga
Mumbai 400 019
India

xxxvi **Contributors**

Emanuele Smacchi
Institute of Industrie Agranie (Microbiologia)
Faculty of Agriculture of Perugia 06126 S. Constanzo
Perguia
Italy

Christopher A Smart
Macromolecular Science Department
Institute of Food Research
Reading Laboratory
Earley Gate
Whiteknights Road
Reading R66 6BZ
UK

H V Smith
Scottish Parasite Diagnostic Laboratory
Stobhill Hospital
Glasgow
G21 3UW
Scotland, UK

O Peter Snyder
Hospitality Institute of Technology and Management
670 Transfer Road
Suite 21A
St Paul
MN 55114
USA

Mark D Sobsey
Department of Environmental Sciences and Engineering
School of Public Health
University of North Carolina
North Carolina
USA

Carlos R Soccol
Laboratorio de Processos Biotecnologicos
Departamento de Engenharia Quimica
Universidade Federal do Parana
CEP 81531–970
Curitiba-PR
Brazil

M El Soda
Department of Dairy Science and Technology
Faculty of Agriculture
Alexandria University
Alexandria
Egypt

R A Somerville
Neuropathogenesis Unit
Institute for Animal Health
West Mains Road
Edinburgh
EH9 3JF
UK

N H C Sparks
Department of Biochemistry and Nutrition
Scottish Agricultural College
Auchincruive
Ayr
Scotland

M Van Speybroeck
Laboratory of Industrial Microbiology and Biocatalysis
Department of Biochemical and Microbial Technology
Faculty of Agricultural and Applied Biological Sciences
University of Gent
Coupure links 653
B-9000
Gent
Belgium

D J Squirrell
CBD Porton Down
Salisbury
SP4 0JQ
UK

E Stackebrandt
DSMZ – German Collection of Microorganisms and Cell Cultures
Brunswick
Germany
Deutsche Sammlung von Mikroorganisem und Mascheroder
Weg 1 B
38124, Braunschweig
Germany

Jacques Stark
Gist-brocades Food Specialties
R&D
Delft
The Netherlands

Colin S Stewart
Rowett Research Institute
Greenburn Road
Bucksburn
Aberdeen
UK

G G Stewart
International Centre for Brewing and Distilling
Heriot-Watt University
Riccarton
Edinburgh
Scotland
EH14 4AS
UK

Gordon S A B Stewart (dec)
Department of Pharmaceutical Sciences
The University of Nottingham
University Park
Nottingham
NG7 2RD
UK

Duncan E S Stewart-Tull
University of Glasgow
Glasgow
G12 8QQ
UK

A Stolle
Institute of Hygiene and Technology of Food of Animal Origin
Ludwig-Maximilians-University Munich
Veterinary Faculty
Veterinärstr. 13
81369
Munich
Germany

Liz Straszynski
Alcontrol Laboratories
Bradford
UK

M Stratford
Microbiology Section
Unilever Research
Colworth House
Sharnbrook
Bedfordshire
MK44 1LQ
UK

M Surekha
Department of Botany
Kakatiya University
Warangal
506 009
India

B C Sutton
Apple Tree Cottage
Blackheath
Wenhaston
Suffolk
IP19 9HD
UK

Barry G Swanson
Food Science and Human Nutrition
Washington State University
Pullman
Washington 99164–6376
USA

Jyoti Prakash Tamang
Microbiology Research Laboratory
Department of Botany
Sikkim Government College
Gangtok
Sikkim 737 102
India

A Y Tamime
Scottish Agricultural College
Auchincruive
Ayr
UK

J S Tang
American Type Culture Collection
10801 University Blvd
Manassas
VA 20110-2209
USA

Chrysoula C Tassou
National Agricultural Research Foundation
Institute of Technology of Agricultural Products
S. Venizelou 1
Lycovrisi 14123
Athens
Greece

S R Tatini
University of Minnesota
Department of Food Science and Nutrition
1334 Eckles Ave
St Paul
MN 55108
USA

D M Taylor
Neuropathogenesis Unit
Institute for Animal Health
West Mains Road
Edinburgh
EH9 3JF
UK

John R N Taylor
Cereal Foods Research Unit
Department of Food Science
University of Pretoria
Pretoria 0002
South Africa

Lúcia Martins Teixeira
Departamento de Microbiologia Médica
Instituto de Microbiologia
Universidade Federal do Rio de Janeiro
Rio de Janeiro 21941
Brazil

Paula C M Teixeira
Escola Superior de Biotecnologia
Rua Dr António Benardino de Almeida
4200 Porto
Portugal

J Theron
Department of Microbiology and Plant Pathology
Faculty of Biological and Agricultural Sciences
University of Pretoria
Pretoria 0002
South Africa

Linda V Thomas
Aplin & Barrett Ltd
15 North Street
Beaminster
Dorset
DT8 3DZ
UK

Angus Thompson
Technical Centre
Scottish Courage Brewing Ltd
Sugarhouse Close
160 Canongate
Edinburgh
EH8 8DD
UK

Ulf Thrane
c/o Eastern Cereal and Oilseed Research Centre
K.W. Neatby Building, FM 1006,
Agriculture and Agri-Food Canada
Ottowa
Ontario K1A 0C6
Canada

Mary Lou Tortorello
National Center for Food Safety and Technology
US Food and Drug Administration
6502 South Archer Road
Summit-Argo
Illinois 60501
USA

Hau-Yang Tsen
Department of Food Science
National Chung Hsing University
Taichung
Taiwan
Republic of China

Nezihe Tunail
Department of Food Engineering
Faculty of Agriculture
University of Ankara
Diṣkapì
Ankara
Turkey

D R Twiddy
Consultant Microbiologist
27 Guildford Road
Horsham
West Sussex
RH12 1LU
UK

N Tzanetakis
Department of Food Science
Faculty of Agriculture
Aristotle University of Thessaloniki
54006
Thessaloniki
Greece

C Umezawa
Faculty of Pharmaceutical Sciences
Kobe Gakuin University
Kobe
Japan

F Untermann
Institute for Food Safety and Hygiene
University of Zurich
Switzerland

Matthias Upmann,
Institute of Meat Hygiene
Meat Technology and Food Science
Veterinary University of Vienna
Veterinärplatz 1
A-1210 Vienna
Austria

Tümer Uraz
Ankara University
Faculty of Agriculture
Department of Dairy Technology
Ankara
Turkey

M R Uyttendaele
Laboratory of Food Microbiology and Food Preservation
Faculty of Agricultural and Applied Biological Sciences
University of Ghent
Coupure Links 654
9000 Ghent
Belgium

E J Vandamme
Laboratory of Industrial Microbiology and Biocatalysis
Department of Biochemical and Microbial Technology
Faculty of Agricultural and Applied Biological Sciences
University of Gent
Coupure links 653
B-9000
Gent
Belgium

P T Vanhooren
Laboratory of Industrial Microbiology and Biocatalysis
Department of Biochemical and Microbial Technology
Faculty of Agricultural and Applied Biological Sciences
University of Gent
Coupure links 653
B-9000
Gent
Belgium

L Le Vay
School of Ocean Sciences
University of Wales
Bangor
UK

P H In't Veld
National Institute of Public Health and the Environment
Microbiological Laboratory for Health Protection
PO Box 1
3720 BA Bilthoven
The Netherlands

Kasthuri Venkateswaran
Jet Propulsion Laboratory
National Aeronautics and Space Administration
Planetary Protection and Exobiology, M/S 89–2, 4800
Oakgrove Dr.
Pasadena
CA 91109
USA

V Venugopal
Food Technology Division
Bhabha Atomic Research Centre
Mumbai 400 085
India

Christine Vernozy-Rozand
Food Research Unit National Veterinary School
Lyon
France Ecole Nationale Véténaire de Lyon
France

B C Viljoen
Department of Microbiology and Biochemistry
University of the Orange Free State
Bloemfontein
South Africa

Birte Fonnesbech Vogel
Danish Institute for Fisheries Research
Department of Seafood Research
Technical University of Denmark Bldg 221
DK-2800 Lyngby
Denmark

Philip A Voysey
Microbiology Department
Campden and Chorleywood Food Research Association
Chipping Campden
Gloucestershire
GL55 6LD
UK

Martin Wagner
Institute for Milk Hygiene
Milk Technology and Food Science
University for Veterinary Medicine
Veterinärplatz 1
1210 Vienna
Austria

Graeme M Walker
Reader of Biotechnology
Division of Biological Sciences
School of Science and Engineering
University of Abertay Dundee
Dundee
DD1 1HG
Scotland

P Wareing
Natural Resources Institute
Chatham Maritime
Kent
ME4 4TB
UK

John Watkins
CREH Analytical
Leeds
UK

Ian A Watson
University of Glasgow
Glasgow
G12 8QQ
UK

Bart Weimer
Center for Microbe Detection and Physiology
Utah State University
Nutrition and Food Sciences
Logan
UT 84322–8700
USA

Irene V Wesley
Enteric Diseases and Food Safety Research
USDA, ARS, National Animal Disease Center
Ames IA 50010
USA

W B Whitman
Department of Microbiology
University of Georgia
Athens
Georgia
USA

Martin Wiedmann
Department of Food Science
Cornell University
Ithaca
NY 14853
USA

R C Wigley
Boghall House
Linlithgow
West Lothian
EH49 7LR
Scotland

R Andrew Wilbey
Department of Food Science
University of Reading
Whiteknights
Reading
UK

F Wilborn
BioteCon Gesellschaft für Biotechnologische
Entwicklung und Consulting
Hermannswerder Haus 17
14473 Potsdam
Germany

A G Williams
Hannah Research Institute
Ayr
KA6 5HL
UK

Alan Williams
Campden and Chorleywood Food Research Association
Chipping Campden
Gloucestershire GL55 6LD
UK

J F Williams
Department of Microbiology
Michigan State University
East Lansing
MI 48824
USA

Michael G Williams
3M Center
260–6B-01
St Paul
MN55144–1000
USA

Caroline L Willis
Public Health Laboratory Service
Southampton,
UK

F Y K Wong
Food Science Australia
Cannon Hill
Queensland
Australia

Brian J B Wood
Reader in Applied Microbiology
Dept. of Bioscience and Biotechnology
University of Strathclyde
Royal College Building
George Street
Glasgow
G1 1XW
Scotland

S D Worley
Department of Chemistry
Auburn University
Auburn
AL 36849
US

Atte von Wright
Department of Biochemistry and Biotechnology
University of Kuopio
PO Box 1627
FIN-70211 Kuopio
Finland

Chris J Wright
Biochemical Engineering Group
Centre for Complex Fluids Processing
Department of Chemical and Biological Process
Engineering
University of Wales Swansea
Singleton Park
Swansea
SA2 8PP
UK

Peter Wyn-Jones
Sunderland University
UK

Hideshi Yanase
Department of Biotechnology
Faculty of Engineering
Tottori University
4–101 Koyama-cho-minami
Tottori
Tottori 680–0945
Japan

Yeehn Yeeh
Institute of Basic Science
Inje University
Obang-dong
Kimhae 621–749
South Korea

Seyhun Yurdugül
Middle East Technical University
Department of Biochemistry
Ankara
Turkey

Klaus-Jürgen Zaadhof
Institute for Hygiene and Technology of Foods of Animal
Origin
Veterinary Faculty
Ludwig-Maximilians University
80539 Munich
Germany

Gerald Zirnstein
Centers for Disease Control
GA
USA

Cynthia Zook
Department of Food Science and
University of Minnesota
St Paul
MN 55108
USA

CONTENTS

VOLUME 1

VOLUME 2

VOLUME 3

N

Accreditation Schemes *see* **Laboratory Management**: Accreditation Schemes.

ACETOBACTER

Rolf K Hommel, Cell Technologie Leipzig, Germany
Peter Ahnert, Department of Biochemistry, Ohio State University, Columbus, USA

Acetic acid bacteria have been used for making vinegar, their best-known product, since Babylonian times. For most of this time, vinegar was obtained by fermentation from alcoholic solutions without understanding of the natural process. A number of researchers established the microbial basis of this process in the beginning of the nineteenth century, such as Kützing, Lafare and Boerhaave. In 1822 Persoon performed the first biological study of surface films of wine and beer and proposed the name *Mycoderma*. Later Kützing (1837) isolated bacteria from naturally fermented vinegar for the first time. Considering them to be a kind of algae, he named them *Ulvina aceti*. Pasteur established the causal connection between the presence of *Mycoderma aceti* and vinegar formation in the first systematic studies on acetic acid fermentation. These discoveries and following studies resulted in better understanding and new methods (Pasteur method) of vinegar formation.

Characteristics of the Genus *Acetobacter*

The classification of protobacteria by DNA–rRNA hybridization studies places acetic acid bacteria in the rRNA superfamily IV (synonymous: alpha group). Acetic acid bacteria, formerly classified into the family Pseudomonadaceae, constitute the family Acetobacteraceae consisting of only two closely related genera, *Acetobacter* and *Gluconobacter*, each of which is a separate rRNA branch. The family Acetobacteraceae represents strictly aerobic chemoorganotrophic bacteria, able to carry out a great variety of incomplete oxidations and living in or on plant materials, such as fruits and flowers. Some members of this family are plant pathogens. None display any pathogenic effect toward mammals, including humans.

Based on physiological criteria the present nomenclature of the genus *Acetobacter* subdivides it into four species: *A. aceti*, *A. liquefaciens*, *A. pasteurianus* and *A. hansenii*. DNA–rRNA hybridization studies indicate the presence of three additional species: *A. diazotrophicus*, *A. methanolicus* and *A. xylinum*. Based on DNA–DNA hybridization a new species, *A. europaeus*, has been proposed; strains of this species show very low similarity to other species of the genus.

Acetobacter are Gram-negative rods. Old cells may become Gram-variable. Cells appear singly, in pairs, or in chains and they are motile by peritrichous flagella or non-motile. There is no endospore formation.

Acetobacter spp. are obligate aerobes except for *A. diaztrophicus* and *A. nitrocaptans* which belong to the diverse group of free-living aerobic or microaerophilic diazotrophs.

The metabolism is respiratory and never fermentative. Single amino acids do not serve as sole source of nitrogen and carbon. Essential amino acids are not known. Depending on growth substrates, some strains may require *p*-aminobenzoic acid, niacin, thiamin, or pantothenic acid as growth factors. The temperature range is 5–42°C with optima between 25 and 30°C.

Acetobacter strains show a moderate to high acid tolerance. The pH range is between pH 4 and pH 7, with optima between pH 5.4 and pH 6.3. Strains used in making vinegar are more resistant toward acidic pH. The minimum accepted by *A. acidophilus* is pH 2.6. The maximum is pH 4.2. The internal pH closely follows the external (*A. aceti*). At or below pH 5.0 the membrane potential of a cell is normally uncoupled, resulting in free proton exchange across the cytoplasmic membrane, thus depriving ATP synthesis of its driving force. However, the formation of acetic acid (or other acids) proceeds via membrane-bound dehydrogenases. These processes are closely connected to irreversible ATP-yielding reactions, sufficient to keep the energy metabolism alive. In *A. aceti*, a gene encoding citrate synthase is involved in acetic acid tolerance. This enzyme is assumed to play a central role in supplying sufficient ATP to protect the cell against accumulation of acetic acid.

Ethanol concentrations higher than 8% and 10% inhibit strains *A. aceti* and *A. xylinum*, respectively. Some strains, for example spoilers of saké, tolerate a higher ethanol content. In general, the ethanol tolerance in *Acetobacter* is higher than in *Gluconobacter*.

The high direct oxidative capacity for sugars, alcohols and steroids is a special feature of *Acetobacter*. This ability is used in vinegar fermentation, food processing, chemical synthesis, and even in enantioselective oxidations, for example with *A. pasteurianus*. Examples of other reactions are the formation of 2,5-dioxogluconic acid by *A. melanogenum* and *A. carinus*, the oxidations of ethanediol to glycolic acid, of lactate to acetoin, of glycerol to dihydroxyacetone, for example polyols in which two secondary *cis*-arranged hydroxyl groups in D-configuration may be oxidized to ketoses. Two strains, *A. rancens* and *A. peroxidans*, are reported to oxidize n-alkanes, mainly by monoterminal attack yielding the corresponding fatty alcohols and fatty acids.

Acetobacter are equipped with two sets of enzymes, catalysing the same oxidation reactions. Enzymes in the first set are bound in the cytoplasmic membrane, the active site facing the periplasm. Enzymes in the second set are located in the cytoplasm and are NADP-dependent. The latter enzymes display neutral or alkaline pH optima. Membrane-bound enzymes show acidic optima. The high oxidative capacity of *Acetobacter* is attributed to membrane-bound proteins such as alcohol dehydrogenase, aldehyde dehydrogenase, glucose dehydrogenase and sorbitol dehydrogenase. The specific activities of these enzymes are up to three orders of magnitude higher than those of their cytoplasmic counterparts. Most membrane-bound enzymes share the prosthetic group pyrroloquinoline quinone (PQQ; **Fig. 1**). The substrates do not need to be transported into the cell for oxidation. Electrons generated are transferred by the reduced form of PQQ either directly to a ubiquinone (Q_9) of the respiratory chain or via a cytochrome

c (subunit of some alcohol dehydrogenases) to the terminal ubiquinol oxidase which is either cytochrome a_1 or cytochrome o. Energy is gained by these oxidations but they are not contributing to carbon metabolism. For instance, the oxidation of one mole ethanol to one mole acetic acid yields six moles of ATP. It is assumed that these systems function as ancillary energy-generating pathways when the energy demand of the cell is high. N_2-fixing cells of *A. diazotrophicus* contain three-times higher enzyme levels of quinoprotein glucose dehydrogenase than under non-N_2-fixing conditions. Flavin (FAD) is an additional covalently bound prosthetic group present in the membrane-bound gluconate dehydrogenase. It is also linked directly to the respiratory chain.

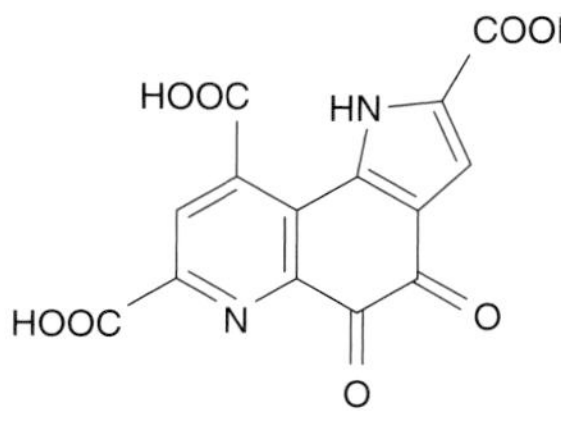

For intracellular sugar metabolism, *Acetobacter* possesses the hexose monophosphate pathway and a complete tricarboxylic acid cycle. Glycolysis is absent or rudimentary. In *A. xylinum* sugar metabolism proceeds (as in *Gluconobacter*) via the Entner–Doudoroff pathway.

In addition to the typical growth substrates, such as sugars or ethanol, *A. methanolicus* also accepts methanol. The major assimilatory pathway of this facultative methylotrophic bacterium proceeds via the ribulose monophosphate cycle. In contrast, most of the Gram-negative methanol-utilizing bacteria that contain the ribulose monophosphate cycle are obligate methylotrophs.

Some *Acetobacter* species, e.g. *A. xylinum*, synthesize bacterial cellulose. The fibres may be regarded as part of the glycocalyx and serve to maintain these highly aerobic organisms at the liquid–air interface. This exopolysaccharide, (β-glu1$\rightarrow$4β-glu)n, is excreted into the medium and then rapidly aggregates as microfibrils yielding a surface pellicle. Bacterial cellulose is produced either in static cultures, or in submerged, fed-batch cultures with low share conditions. Yields up to $28\,\mathrm{g\,l^{-1}}$ of dry polysaccharide may be obtained after selection of high yielding strains. The product, cellulose I form, is very pure. It does not contain hemicelluloses, lignins or pectic substances, and is therefore used mainly in medicine as wound dressings for patients with burns, extended loss of tissue, etc.

The majority of *Acetobacter* species have 1–8 plasmids varying in size from 1.5 to 95 kb. Isolates from some German submerse vinegar processes have 3–11 plasmids, isolates from surface fermentation processes 3–7 plasmids (2–70 kb). Plasmid profile analysis has become a powerful tool for controlling homogeneity, stability, identity etc. of the microbial populations in production processes. However, many strains used industrially are highly variable, changing their phenotypic and other properties in just a few generations.

Figure 1 Structure of pyrroloquinoline quinone (PQQ).

This phenomenon could not be correlated with plasmid profiles. *Acetobacter* contains four ribosomal RNA operons on the chromosome.

Recombinant DNA techniques have been adapted to *Acetobacter*. Host–vector systems and transformation methods have been developed. Transformation systems are available for *A. aceti* and *A. xylinum*. For instance, bacteriophages specific for *Acetobacter* lead to a complete stop of submerged fermentation. Morphologically different phage types are described which were isolated from vinegar fermentations in Europe. They belong to the Bradley's group A and to the Myoviridae. The high number of phages in disturbed acetic acid fermentations suggests their responsibility for production problems.

The classical, well known and well-studied, niches of *Acetobacter* are in making vinegar and spoilage of beer and wine (see below). *Acetobacter* spp. were originally associated with plants and soils. Preferred habitats, such as fruits and flowers, are rich in sugars, alcohols and/or acids. Fermenting fruits, in particular, are excellent sources of sugar and ethanol. Various *Acetobacter* spp. have been isolated from apricots, almonds, beets, bananas, figs, guavas, grapes, mandarins, mangoes, oranges, pomegranates, pears, peaches, persimmons, pineapples, plums, strawberries and tomatoes. *A. aceti*, *A. xylinum* and *A. pasteurianus* were predominantly associated with ripe grapes. Local priorities have been found: *A. pasteurianus* accounted for 75% of the strains in isolates from Southern France. Especially high numbers were found on damaged grapes. *Acetobacter* spp. have been isolated from the immature spadix of the palm tree. *A. xylinum* was present on the leaflets of the palm tree and in the surrounding air. Acidic acid bacteria, such as *A. aceti* and *A. pasteurianus*, hibernate in dried and injured apples and in spring they are able to spread to flowers. The nitrogen fixing *A. diazotrophicus* has settled the stem and roots of sugarcane in Brazil. Different *Acetobacter* spp. have been isolated from cocoa bean flora. *Acetobacter* is the causal agent of bacterial rot in pears and apples, resulting in different shades of browning and tissue degradation. Pears are more susceptible to bacterial brown rot. *Acetobacter* spp. have been isolated from decaying apple tissue and from the larvae and adults of apple maggots. Artificial inoculation of 100 cells may induce apple brown rot. Although the optimum temperature is about 25°C, rotting also proceeds at 4°C. The pink disease of pineapple fruit is caused by *A. liquefaciens*. Through the open flowers the bacterium enters the internal nectary and placental regions, invades the ovaries and starts to grow in the ripening tissue. The dilution of nectar by rain during flowering is a prerequisite, because undiluted nectar is not a growth substrate.

Methods of Detection

Strains of both *Acetobacter* and *Gluconobacter* are present in the same habitat. Members of the latter genus are normally co-isolated. For routine isolation of *Acetobacter* from natural or artificial habitats, culture media of low pH, containing 2–4% ethanol as energy source, are recommended. Aerobic growth is optimal between 25°C and 30°C. As low cell counts are expected, enrichment cultures become necessary. For such purposes beer has been recommended. However, preservatives added to the beer may limit success. Many specific enrichment cultures adapted to individual sources are described in older literature.

Yeast water–glucose medium is recommended for isolation and purification. It contains yeast water (supernatant of autoclaved bakers' or brewers' yeast, 200 g l⁻¹), and glucose (20 g l⁻¹), pH 5.5–6.0. This composition is also very useful for the enrichment of solid media (agar concentration: 15–30 g l⁻¹).

Wort medium comprises malt powder diluted with tap water to 8% soluble solids; for solid medium the pH should be 5.5–6.

Peptone glucose agar comprises bacto-peptone or bacto-tryptone (5 g l⁻¹), glucose (20 g l⁻¹), primary potassium phosphate (1 g l⁻¹) in tap water; agar concentration: 15–20 g l⁻¹.

To enhance the growth of some strains, the addition of yeast extract (3–5 g l⁻¹) may be useful. Further enhancement may be achieved by the addition of 100 ml of filtered and freshly prepared tomato juice to 1 litre of culture medium.

An isolation procedure to differentiate between *A. pasteurianus*, *A. aceti* and *Gluconobacter oxydans* has been developed by Frateur, which uses three to five different culture media for each species. It includes enrichment in liquid media (30°C) and subsequent agar plating.

Isolation of production strains from vinegar tends to be difficult due to the strains being highly adapted to the production conditions (see below). A mixture (4 ml) of vinegar to be tested and pasteurized vinegar is added to tubes with 15 ml of solid medium (yeast water, 100 g l⁻¹ glucose, 30 g l⁻¹ CaCO₃, 20 g l⁻¹ agar). Bacterial growth proceeds in the interphase. Alternatively, yeast extract–calcium lactate agar or wort agar containing 1.5% ethanol or 5 g l⁻¹ yeast extract, and 25 g l⁻¹ agar may be used.

Acetobacter settling on flowers or fruits may be efficiently enriched in a broth containing 50 g l⁻¹ glucose, 10 g l⁻¹ yeast extract, and 0.1 mg l⁻¹ cycloheximide at 30°C. The ring or pellicle formed after 2–8 days is plated out on a solid medium which may also serve for further purification of the acid-forming colonies: $50 \, g \, l^{-1}$ glucose, $10 \, g \, l^{-1}$ yeast extract, $30 \, g \, l^{-1}$ $CaCO_3$ and $25 \, g \, l^{-1}$ agar.

In cider making, various culture media are recommended for successful isolation of acetic acid bacteria from orchard soil, apples, pomace, juice, fermenting juice, cider or from the factory equipment. The media are based on apple juice–yeast extract made from apple juice with low tannin content, pH 4.8 and $30 \, g \, l^{-1}$ agar. The addition of $0.1 \, mg \, l^{-1}$ actidione is recommended to suppress yeasts and moulds. Incubation is carried out at 28°C for 3–5 days. Alternatively, a broth composed of 500 ml of sweet cider, 500 ml deionized water, $12 \, g \, l^{-1}$ yeast extract, and $2 \, g \, (NH_4)_3PO_4$ (pH 5), yields good results.

Strains of *A. diazotrophicus* can be isolated by stepwise enrichment in different media, including semisolid ones. *Acetobacter* strains may be held in or on a variety of media, such as beer (without preservation agents) or wort. Recommended conservation media are summarized in **Table 1**. Optimum growth is obtained at 25–30°C. Agar cultures should be kept at 4°C and transferred monthly. Most strains stay alive lyophilized for several years and some for longer than 10 years.

Phenotypic identification of Acetobacteraceae is based on general properties which are partially shared with *Gluconobacter* and some members of the genus *Frateuria* (superfamily II syn. gamma subclass). One of the properties used to further identify *Acetobacter* is the oxidation of acetate and lactate to CO_2 and H_2O (overoxidation) at neutral and acidic pH. This is detected on medium composed of ethanol (3%), $CaCO_3$ ($20 \, g \, l^{-1}$), and agar ($20 \, g \, l^{-1}$). The appearance of acetic acid reveals clear zones around the colonies and overoxidation results in (re-)precipitation of $CaCO_3$. Alternatively, bromcresol green ($0.022 \, g \, l^{-1}$) may be added to a medium composed of yeast extract ($30 \, g \, l^{-1}$), ethanol (2%) and agar ($20 \, g \, l^{-1}$). The colour shifts from green to yellow as acid is formed. Overoxidation results in a change of the indicator to blue.

Bacteria belonging to Acetobacteraceae may be Gram-negative or Gram-variable (namely older cells), are strictly aerobic and oxidize ethanol to acetic acid in neutral or acidic media. Cells are ellipsoidal to rod-shaped (0.6–0.8×1–$4 \, \mu m$), have a respiratory type of metabolism, are oxidase negative and acidify glucose below pH 4.5. They do not form endospores, liquefy gelatin, reduce nitrate or form indole. The rapid phenotypic identification of *Acetobacter* is based on the following features:

- peritrichous flagella

Table 1 Common media for maintenance and cultivation of *Acetobacter*

Medium	Component	Amount
Glucose–yeast extract agar ('Acetobacter/Gluconobacter agar') pH 7.5 ± 0.2, 25°C	Glucose	100 g
	Yeast extract	10 g
	$CaCO_3$	20 g
	Agar	25 g
	Distilled/deionized water	1 l
'Acetobacter agar'	Glucose	50 g
	Yeast extract	10 g
	$CaCO_3$	30 g
	Agar	25 g
	Tap water	1 l
'Acetobacter medium' pH 7.0 ± 0.2, 25°C	Glucose	3 g
	Autolysed yeast	10 g
	$CaCO_3$	10 g
	Agar	15 g
	Distilled/deionized water	1 l
Mannitol agar (YPM agar: $12 \, g \, l^{-1}$ agar)	Mannitol	25 g
	Yeast extract	5 g
	Peptone	3 g
	Agar	15 g
	Distilled/deionized water	1 l
Yeast–glucose agar pH 7.0 ± 0.2, 25°C	Glucose	20 g
	Yeast extract	10 g
	Agar	15 g
	Distilled/deionized water	1 l
Potato glycerol agar 25°C	Glycerol	20 g
	Glucose	5 g
	Yeast extract	10 g
	Agar	15 g
	Supernatant of freshly sliced and boiled potatoes ($200 \, g \, l^{-1}$)	1 l

- oxidation of acetic acid and lactic acid to CO_2 and H_2O
- no H_2S formation
- growth factors may or may not be required
- specific ubiquinone types.

The ubiquinones are Q_9 and Q_9 with minor component of Q_8 (*A. aceti, A. pasteurianus*); some strains have Q_{10} or Q_{10} with minor component Q_9 (*A. methanolicus, A. diazotrophicus, A. xylinum, A. liquefaciens*). *Acetobacter* strains grow at pH 5 and prefer ethanol or lactate over glucose for growth. Further differentiation among *Acetobacter* species can be achieved as shown in **Table 2**.

The phenotypic identification may be affected by spontaneously occurring mutations even in taxonomically important properties. Mutants of *A. aceti* exist that are unable to oxidize ethanol. In these cases

Table 2 Phenotypical differences among *Acetobacter* species (Reproduced with permission from Swings 1992)

Feature	A. aceti	A. lique-faciens	A. pasteur-ianus	A. hansenii	A. xylinum	A. methano-licus	A. diazo-trophicus
Formation of							
Water-soluble brown pigments on GYC[a] medium	–	+	–	–	–	–	+
γ-Pyrones from D-fructose	–	+	–	–	–	–	+
5-Oxogluconic acid from D-glucose	+	d	–	d	+	–	–
2, 5-Dioxogluconic acid from D-glucose	–	+	–	–	–	–	+
Ketogenesis from glycerol	+	+	–	+	+	(+)	d
Growth factor required	d	d	+	d	+	+	–
Growth on carbon sources							
Ethanol	+	+	d	–	–	(+)	+
Methanol	–	–	–	–	–	+	–
Sodium acetate	+	d	d	–	–	(+)	+
Growth on L-amino acids in the presence of D-mannitol as carbon source							
L-Asparagine	d	+	–	+	–	–	+
L-Glutamine	d	+	–	+	–	–	–
Formation of H_2S	–	–	–	–	–	–	–
Growth in presence of 30% D-glucose	–	–	–	–	–	–	+
Ubiqinone type	Q_9	Q_{10}	Q_9	ND	Q_{10}	Q_{10}	Q_{10}
G+C content (mol%)	56–60	62–65	53–63	58–63	55–63	62	61–63

Symbols: +, 90% or more of the strains positive; (+), weakly positive reaction; d, 11–89% of the strains positive; –, 90% or more of the strains negative; ND, not determined.
[a] Glucose–yeast extract cycloheximide.

DNA–rRNA hybridization studies and polymerase chain reaction (PCR) studies will be more useful for exact taxonomic classification.

Importance to the Food Industry

Food Processes

Acetobacter spp. are used in different processes of making foods and food additives. Well established are the fermentations to produce one special product, such as acetic acid or gluconic acid. These oxidation reactions are backed by the high oxidative capacity of enzymes bound in the cytoplasmic membrane with the active centre directed into the periplasm. Other processes also use such enzymes but are more complex with regard to the microbial population and the substrates used.

Vinegar is the most popular product of *Acetobacter*. This process is discussed in detail in a separate article. In acetic acid fermentation, mixtures of highly adapted predominantly *Acetobacter* spp. are used, which are not derived from pure cultures. These strains display an extremely strong tolerance to acetic acid. The most important detectable species are *A. pasteurianus*, *A. lovaniensis*, *A. ascendens*, *A. paradox*, *A. aceti*, *A. xylinum* and *A. orleanensis*.

Difficulties may arise during isolation, subsequent cultivation under artificial conditions, and preservation due to the high adaptation as demonstrated with *A. polyoxogenes* isolated in Japan and with *A. acidophilus*. Both could not be maintained in strain collections and were lost. *A. europaeus* isolated from production facilities in Switzerland requires acetic acid for growth on agar plates and tolerates 4–8%. Specialized industrial strains tolerate pH values down to 2.6. DNA–DNA hybridization studies shows nearly no difference between isolates from different commercial processes (90–100% hybridization). However, a comparison of highly productive strains from German plants and those from strain collections showed large differences. Values below 45% were obtained with definite strains (*A. pasteurianus*, *A. aceti*, *G. oxydans*). The DNA–DNA similarities of *A. europaeus* strains isolated from industrial processes versus strains from collections are below 22%.

Membrane-bound quinoproteins, i.e. alcohol and aldehyde dehydrogenases, are the enzymatic basis of acetic acid formation (**Fig. 2**). These enzymes are more active and stable under acidic conditions than those of *Gluconobacter*. Prevention of overoxidation of acetic acid to CO_2 and H_2O requires a constant high concentration of ethanol. Lack of ethanol and oxygen damage acetic acid bacteria populations. Even

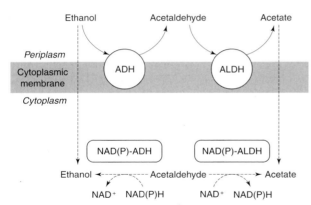

Figure 2 Scheme of ethanol oxidation by acetic acid bacteria. The formation of acetic acid via the quinoprotein alcohol dehydrogenase (ADH) and aldehyde dehydrogenase (ALDH) yields 6 mol of ATP per mol of ethanol. The cytoplasmic pyridine nucleotide-dependent counterparts are alcohol dehydrogenase (NAD(P)-ADH) and aldehyde dehydrogenase (NAD(P)-ALDH). The preferred direction of the reversible reaction of the NAD(P)-ADH is indicated by the arrow. (Reproduced with permission from Matsushita et al (1994).)

the application of pure oxygen or oxygen-enriched air harms the process.

Aspergillus niger and *Gluconobacter oxydans* (syn. *Acetobacter suboxydans*) are the preferred microorganisms in the production of gluconic acid, which has useful properties, such as an extremely low toxicity, low corrosiveness and formation of water-soluble complexes with divalent and trivalent metal ions. On this basis gluconic acid has found many applications in the food, textile and tanning industries and in medicine. According to German law gluconic acid is considered a food. The quinoprotein-dependent glucose oxidase also occurs in *Acetobacter*. *Acetobacter methanolicus* possesses a similar enzyme. It grows on methanol and oxidizes glucose that is not assimilated to gluconic acid. Glucose ($150-250 \, g \, l^{-1}$) is nearly completely (98%) converted with a productivity $> 30 \, g \, l^{-1} \, h^{-1}$.

Acetobacter spp. are involved in a number of natural fermentations. A typical tropical beverage, palm wine, is made from the fermentation of palm sap. It is a result of a mixed alcoholic, lactic acid and acetic acid fermentation. The microbial consortium is complex. At first, yeasts and *Zymomonas* ferment the available sugar to ethanol, which is partly converted into acetic acid by *Acetobacter* spp., which appear after two to three days of fermentation and can be isolated from the final product.

Acetobacter strains have been isolated from cocoa wine, made by fermentation of cocoa seeds. Its alcohol content of 9–12% is higher than that of palm wine.

Species of *Acetobacter*, *A. rancens*, *A. xylinum*, *A. ascendens*, *A. lovaniensis* and others, are involved in cocoa seed fermentation. Seeds in the fruit mass are

naturally fermented by a high supply of air in up to 13 days. Yeasts, moulds and bacteria are involved. Ethanol produced by yeasts is oxidized by *Acetobacter* spp. Seeds are killed in the presence of acetic acid and by a temperature of 50°C. The fruit mass is then degraded and the typical flavour and brownish colour of the bean are developed.

Nata is a dessert delicacy in Southeast Asia. This gelatin-like, firm, creamy-yellow to pinkish substance is composed of a form of cellulose formed by bacteria from sugared fruit juices. *Acetobacter hansenii* is responsible for this process. Nata is usually grown on fruit juice and the floating mat is candied, while still chewable, to produce gumdrops-like treats.

A symbiosis culture of a yeast with *Acetobacter* (*A. xylinum*) is the so-called 'tea fungus'. In this process a slightly sweet, alcoholic, aromatic and acidic beverage (kombucha, Ma-Gu etc.) is made by fermenting sweetened (sucrose, $5-150 \, g \, l^{-1}$) black tea. Health effects attributed to it include in vitro antimicrobial activity, improved athletic performance, and enhanced sleep and pain thresholds. The yeasts yield ethanol; the alcohol content never exceeds $10 \, g \, l^{-1}$. *A. xylinum* initially oxidizes ethanol to acetic acid and glucose to gluconic acid. Acetic acid concentrations may rise to levels as high as $30 \, g \, l^{-1}$ if the tea is allowed to ferment for up to 30 days. The usual concentration of acetic acid is $10 \, g \, l^{-1}$ (after 3–5 days). Gluconic acid is also present in substantial quantities, about $20 \, g \, l^{-1}$. Different yeasts are involved, like *Brettanomyces*, *Candida*, *Pichia*, *Saccharomyces*, *Schizosaccharomyces*, *Torulaspora*, *Zygosaccharomyces*, depending on origin and culture conditions. The production of cellulose by *A. xylinum* forms a compact, granular and gelatinous surface film in which yeast and bacterial cells are housed. Both benefit from the floating mat which eases aeration for these aerobic microorganisms.

Food Spoiling

Acetobacter may cause both considerable economic profits and losses. The latter aspect results from the spoiling activity in many products that provide sufficient conditions for growth. The best example is the aerobic acidification of wine, the origin of vinegar making. *Acetobacter* spp. are present in the wine must, on the surface of grapes, and on injured grapes. Acetic acid formation may only proceed when adequate oxygen is available, i.e. on fruits, in juices and in mashes. Under the anaerobic conditions of wine making, vinegary spoilage is rare. Its occurrence is indicated by increased concentrations of acetic acid, ethyl acetate and D-lactate. There is a positive correlation between acetic acid and ethyl acetate. Sweet white wines spoiled by dextran-producing *Aceto-*

bacter spp. often turn viscous and slimy. Usually, in all alcoholic beverages containing less than 15% ethanol, formation of acetic acid is possible. The higher the alcohol content, the more resistant the beverage will be toward bacterial infection. Alcohol concentrations of 10% inhibit strains of *A. aceti* and *A. xylinum*. Infections by *Acetobacter* are indicated by an increase of volatile and non-volatile free organic acids and a decrease of glucose and ethanol.

Spoilage of saké, containing up to 24% ethanol, by *Acetobacter pasteurianus* has been reported. The saké smelled like acetic acid. Cloudiness of tequila during the summer is due to *Acetobacter* spp. Even cider-making may be affected by acetic acid bacteria which enter the facilities with injured apples etc. The pH of apple juice of 3.2–4.2 only allows growth of acid-tolerant microorganisms, such as *Acetobacter* (*A. aceti*, *A. pasteurianus*) and *Gluconobacter*. Spoilage of cider by these bacteria may cause acidification and the so-called 'cider sickness' of sweet ciders, characterized by an unpleasant odour and taste (acetaldehyde) and the formation of adducts of acetaldehyde with polyphenols giving a milky, colloidal precipitate. Acidification of beer may be caused by *A. xylinum* and *A. pasteurianus*. Their infection of draft beer may result in formation of slime, accompanied by turbidity and loss of alcohol content due to formation of acetic acid. This makes the beer ropy and causes strong alterations in flavour (vinegar) and colour.

Acetic acid bacteria are also found in unusual habitats. Moist flour (> 13% humidity) can be settled by a microbial community and subsequently acetic acid formation may start. Meat conserved by lactic acid was alkalized by *A. pasteurianus* and *A. aceti* by overoxidation of lactonic acid, which allowed settlement of pathogenic and toxigenic bacteria that were excluded before by the low pH and by the acid.

See also: **Biochemical and Modern Identification Techniques**: Introduction. **Cider (Hard Cider)**. **Ecology of Bacteria and Fungi in Foods**: Influence of Redox Potential and pH. **Fermentation (Industrial)**: Production of Xanthan Gum. **Fermented Foods**: Origins and Applications; Fermented Vegetable Products; Fermentations of the Far East. *Gluconobacter*. **History of Food Microbiology**. **Metabolic Pathways**: Release of Energy (Aerobic). **Spoilage of Plant Products**: Cereals and Cereal Flours. **Spoilage Problems**: Problems caused by Bacteria. **Vinegar**. **Wines**: Microbiology of Wine-making.

Further Reading

Adachi T (1968) *Acetic Acid Bacteria. Classification and Biochemical Activities*. Tokyo: University of Tokyo Press.

Beppu T (1993) Genetic organization of *Acetobacter* for acetic acid fermentation. *Antonie van Leeuwenhoek* 64: 121–135.

Cannon RE and Anderson SM (1991) Biogenesis of bacterial cellulose. *Critical Reviews in Microbiology* 17: 435–447.

Griffin AM, Edwards KJ, Gasson MJ and Morris VJ (1996) Identification of structural genes involved in bacterial exopolysaccharide production. *Biotechnology and Genetic Engineering Reviews* 13: 1–18.

Matsushita K, Toyama H and Adachi O (1994) Respiratory chains and bioenergetics of acetic acid bacteria. *Advances in Microbial Physiology* 36: 247–301.

Swings J (1992) The Genera *Acetobacter* and *Gluconobacter*. In: Balows A, Trüper HG, Dworkin M, Harder W and Schleifer KH (eds) *The Prokaryotes*, 2nd edn. P. 2269. New York: Springer.

ACINETOBACTER

Peter Kämpfer, Institut für Angewandte Mikrobiologie, Justus-Liebig-Universität Giessen, Germany

Introduction

Bacteria of the genus *Acinetobacter* are ubiquitous organisms which can be isolated from soil, water, sewage and a wide variety of foodstuffs. They belong to the typical psychrophilic spoilage flora and can grow even under refrigerated conditions. *A. johnsonii* and *A. lwoffii* are the species predominantly isolated from various foods. This article summarizes the characteristics of the genus and relevant species, the recommended methods of detection and enumeration in foods, the difficulties in identification of genomic species and the importance of the genus and individual species in the food industry including biotechnical applications.

Characteristics of the Genus and Relevant Species

Gram-negative non-fermentative bacteria belonging to the genus *Acinetobacter* have been classified previously under a variety of names. At least 15 different 'generic' names have been used to describe these organisms, e.g. *Bacterium anitratum, Herellea vaginicola* and *Mima polymorpha, Achromobacter, Alcaligenes, Micrococcus calcoaceticus,* and *Moraxella glucidolytica* and *Moraxella lwoffii,* just to list the most well known. The name *Acinetobacter* was proposed in 1954 for a genus encompassing a heterogeneous collection of Gram-negative, non-motile, oxidase-positive (*Moraxella*) and oxidase-negative saprophytic organisms that could be distinguished from other bacteria by their lack of pigmentation. As a consequence of extensive nutritional studies, the oxidase-negative strains were shown to be different from the oxidase-positive strains of the genus *Moraxella,* and in 1971 the Subcommittee on the Taxonomy of *Moraxella* and Allied Bacteria recommended that the genus *Acinetobacter* should include only the oxidase-negative organisms. In *Bergey's Manual of Systematic Bacteriology* (1984) the genus *Acinetobacter* is classified in the family Neisseriaceae, with only one species, *A. calcoaceticus,* comprising two varieties, var. *anitratus* (formerly *Herellea vaginicola*) and var. *lwoffii* (formerly *Mima polymorpha*). Further phylogenetic studies have resulted in the proposal that members of the genus should be classified in the new family Moraxellaceae, which at present includes *Moraxella, Acinetobacter, Psychrobacter* and related organisms, and which constitutes a discrete phylogenetic branch within the gamma subdivision of the class Proteobacteria on the basis of rRNA–DNA hybridization studies and 16S rRNA sequencing (overall 16S rDNA sequence similarities of more than 94%).

Genus

The genus *Acinetobacter* is now defined as including strictly aerobic, non-motile, Gram-negative (but sometimes difficult to destain), oxidase-negative and catalase-positive diplococcoid rods, having a DNA G+C content of 29–47 mol%.

Members of the genus *Acinetobacter* are ubiquitous, free-living saprophytes which can be isolated from soil, water and various foods (especially meat).

Bacteria of the genus *Acinetobacter* form short, plump rods, typically $1–1.5 \times 1.5–2.5$ μm in the logarithmic growth phase of growth, often becoming more coccoid in the stationary phase. They are non-fastidious, non-fermentative organisms which are easy to cultivate and able to utilize a large variety of organic substrates as sole carbon sources. Furthermore, they can grow at a wide range of temperatures, forming smooth, sometimes mucoid colonies on solid media, although clinical isolates prefer 37°C, and some environmental isolates grow best at incubation temperatures of 20–30°C. The genus is distributed widely but as revealed by epidemiological analyses there are significant population differences between the acinetobacters found in clinical and other environments. In clinical environments they can be isolated as commensals from the skin of hospital staff and patients, and they have also been recognized as nosocomial pathogens involved in outbreaks of hospital infection, especially within high-dependency or intensive care units.

Species

Delineation of species within the genus *Acinetobacter* has resulted in a complex situation. Traditionally, a microbial species has been considered to be a group of strains showing a high degree of phenotypic similarities. But it is now accepted that genomic relationships obtained by DNA–DNA hybridization provide the best information for the designation of species (genomic species). A species should include strains with approximately 70% or higher DNA–DNA relatedness and ΔT_m values of 5°C or less (ΔT_m is the difference between the melting temperature of a homologous hybrid and the melting temperature of a heterologous hybrid). A formal species name can be given to genomic species which can be differentiated by phenotypic properties.

Extensive investigations on strains belonging to the genus *Acinetobacter* using the DNA–DNA pairing have resulted in the recognition of (to date) 18 DNA–DNA homology groups (genomic species) (**Table 1**).

The independently described genomic species groups 13–17 of Bouvet and Jeanjean and 13–15 of Tjernberg and Ursing have, in part, common numbering. Only one genomic species (14 *sensu* Tjernberg and Ursing and 13 *sensu* Bouvet and Jeanjean) has been shown to correspond. In this article, the suffixes BJ and TU (which are widely used in the literature) are used to avoid confusion.

Seven of the genomic species have been given formal species names: *A. calcoaceticus, A. lwoffii* (these two 'old' species were amended), *A. baumannii, A. junii, A. johnsonii, A. haemolyticus,* and *A. radioresistens* (Table 1). Genomic species 1, 2 and 3 of Bouvet and

Table 1 Numbers of *Acinetobacter* genomic species as defined by different authors (adapted from Bergogne-Bérézin et al 1996)

Species	Genomic species numbers after different authors		
	Bouvet and Grimont (1986) Bouvet and Jeanjean (1989)	Nishimura et al (1987, 1988)	Tjernberg and Ursing (1989)
A. calcoaceticus	1	1	1
A. baumannii	2	1	2
	3	–	3
	–	–	13
A. haemolyticus	4	4	4
A. junii	5	–	5
	6	4	6
A. johnsonii	7	3	7
A. lwoffii	8	2	8
	9	–	8
	10	–	10
	11	–	11
A. radioresistens	12	5	12
	13	–	14
	14	–	–
	15	–	–
	16	–	–
	17	–	–
	–	–	15

Grimont and group 13 of Tjernberg and Ursing have been shown to have an extremely close relationship and are referred to by some research groups as the 'A. calcoaceticus–A. baumannii complex'. Groups 5 (A. junii), 7 (A. johnsonii) and 8/9 (A. lwoffii) have often been found in specimens from nonclinical environments including food.

Ecology

Members of the genus *Acinetobacter* are ubiquitous organisms easily isolated from soil, water, sewage and spoiled food using appropriate enrichment techniques. They are often isolated from a wide variety of foodstuffs, including eviscerated chicken carcasses, various poultry and other meats, and milk products. Apart from organisms of the authentic genus *Pseudomonas* and several other Gram-negative non-fermentative organisms, they are responsible for the economically important spoilage of foods such as bacon, chickens, eggs and fish, even when stored under refrigerated conditions or after irradiation treatment. Furthermore, *Acinetobacter* is considered to be a normal inhabitant of human skin and has been implicated as a presumed causal or contributory agent in some infectious disease processes. It can be isolated particularly from moist skin areas such as toe webs, the groin, and the axilla. Although their pathogenicity is generally low, they may cause occasional serious opportunistic infections, including septicaemia, pneumonia, and meningitis, particularly in patients with reduced natural resistance. Hospital reservoirs of the organism may include waterbaths, disinfectants, room humidifiers, peritoneal dialysis fluid, wet mattresses, respirometers, and the hands of hospital staff. Unfortunately there have been few studies in which *Acinetobacter* isolates from environmental sources have been identified at the genomospecies level. From the few data in the literature, it is obvious, as far as food isolates are concerned, that there is a clear difference in the distribution of genomic species between clinical and food isolates. In food, genomic species 7 (A. johnsonii) and 8 (A. lwoffii) are found predominantly. These genomic species can also often be isolated from the environment, especially water and waste water.

Genomic species 2 (A. baumannii) and the unnamed genomic species 3 and 13TU seem to be the species found most commonly in specimens from hospitalized patients. As a consequence, it appears that there is a significant population difference between the *Acinetobacter* species found in clinical and other environments. Each population appears to be characterized by predominant groups of genomic species.

Recommended Methods of Detection and Enumeration in Foods, Including Formulations of Standard/Popular Media and Cultural Techniques/Conditions

Detection

For the examination of foods for the presence of certain types and numbers of microorganisms, in most cases isolation and cultivation procedures are indispensable. None of the methods in common use (i.e. standard plate counts and most probable number methods for viable cells) allows the exact quantification of certain microorganisms (including acinetobacters) in foods or food products. Each of these methods has its specific limitations. Recently, a direct approach has been developed for the detection of microorganisms which is independent of cultivation. The application of genus- or species-specific rRNA-targeted oligonucleotide probes for *in situ* identification of microorganisms without cultivation bears a great potential for the study of microorganisms on foods unbiased by the limitations of cultivation (**Fig. 1**). The detection of members of the genus *Acinetobacter* in aquatic habitats using this approach has been successful. It can be assumed, that this approach (in addition to other molecular techniques, see below) will also become important in food microbiology.

Isolation

Isolation of members of the genus *Acinetobacter* can be accomplished using a wide variety of standard and commercially available laboratory media including nutrient agar, trypticase soy (TS) agar, brain–heart infusion agar and MacConkey agar. Nutrient agar with the addition of sheep or human blood may be useful for the detection of haemolysin-producing strains. Several defined media consisting of a mineral base with one or several carbon sources (acetate, pyruvate, or lactate) have been used for specific purposes. Because *Acinetobacter* belongs to the saprophytic, psychrophilic spoilage flora of foods, like *Pseudomonas* and other Gram-negative non-fermentative organisms, the methods for isolation specified in the articles on total viable counts should be consulted.

A more specific enrichment culture procedure for isolating members of the genus from soil and water may be as follows. Liquid enrichment cultures containing 20 ml of Baumann's enrichment medium are inoculated with a 5 ml sample of water or a filtered 10% soil suspension and vigorously aerated at either 30°C or room temperature. Cultures are examined microscopically after 24 or 48 h and streaked onto suitable isolation (e.g. nutrient- or TS agar) media. *Acinetobacter* strains have a slightly acid pH optimum for growth, and can be favoured by vigorous aeration at pH 5.5–6.0.

Baumann's enrichment medium comprises (per litre): sodium acetate (trihydrate), 2 g; KNO_3, 2 g; $MgSO_4 \cdot 7H_2O$, 0.2 g; dissolved in 0.04 M KH_2PO_4–Na_2HPO_4 buffer (pH 6.0) containing 20 ml per litre of an appropriate mineral base.

Selective growth of *Acinetobacter* isolates can be accomplished by suppressing growth of other microorganisms in the specimens. A selective medium for

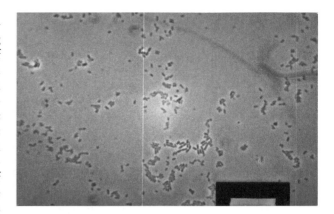

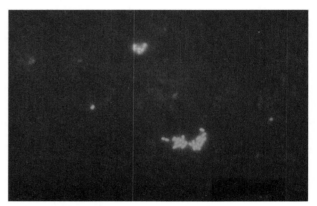

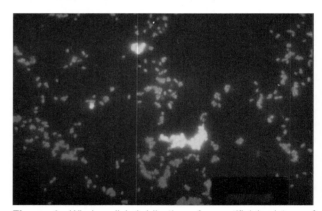

Figure 1 Whole-cell hybridization of an artificial mixture of *Acinetobacter* sp. HG 7 strain 68 and *Pseudomonas putida OUS* 82 with a Fluoresceine-labelled probe (green) specific for members of the genus *Acinetobacter* and a Rhodamine-labelled probe (red) specific for all gamma-subclass Proteobacteria. Phase contrast (top), Fluoresceine epifluorescence visualized with Zeiss filter set 09 (middle), Fluoresceine plus Rhodamine epifluorescence visualized with Zeiss filter sets 09 and 15 (bottom). Acinetobacters show higher signal intensities because they were stained with both probes whereas pseudomonads were stained with the gamma-subclass probe only. Individual micrographs represent the identical microscopical field. The scale bar equals 20 μm. (Photomicrographs by Alexander Neef, Institute for Applied Microbiology, Justus-Liebig-University Giessen, Germany).

Acinetobacter species, comprising sugars, bile salts and bromcresol purple, is available commercially as Herellea Agar (Difco). The improved selectivity of this medium, a modification containing vancomycin, ampicillin, cefsulodin, sugars and phenylalanine has been suggested. Selection of *Acinetobacter* can also be achieved by enrichment cultivation.

Holton's selective medium comprises (per litre): Agar, 10 g; casein pancreatic digest, 15 g; peptone, 5 g; NaCl, 5 g; desiccated ox-bile, 1.5 g; fructose, 5 g; sucrose, 5 g; mannitol, 5 g; phenylalanine, 10 g; phenol red, 0.02 g, adjusted to pH 7.0. After autoclaving, the following filter-sterilized ingredients are added (final concentration in g l^{-1}): vancomycin, 0.01 g; ampicillin, 0.061 g; cefsulodin, 0.03 g.

After overnight incubation at 37°C, red colonies are tested for negative oxidase reaction and negative phenylalanine deamination (10% ferric chloride method). These colonies can be regarded as presumptive *Acinetobacter* isolates.

The optimum growth temperature for most strains is 30–35°C. Although most strains will grow reasonably well at 37°C, some environmental strains have considerably lower optimum growth temperatures and may be unable to grow at 37°C. It has been shown, that food isolates belonging to *Acinetobacter* genomic species 3, 5, 7, 8/9 and 10, grew at 5°C, whereas not all strains grew at 37°C. In order to meet the requirements of acinetobacters and other non-fermenting Gram-negative organisms, a general cultivation temperature of 30°C is recommended. In some cases, however, the selection of a lower temperature or a combination of temperatures may be advisable.

Identification at the Genus Level and Metabolic Characters

Identification at the genus level can be made with the following criteria: Gram-negative coccoid rods (sometimes difficult to destain), negative oxidase reaction (using Kovac's reagent; acinetobacters lack cytochrome *c*, but they do contain cytochromes *a* and *b*), non-fermenting D-glucose in O-F test, non-motile in hanging drop preparation test, aerobic growth, positive in catalase test, nitrate reduction in most cases negative, and hydrolysis of Tween in most cases positive.

Identification to the genus level can also be done by a transformation assay. With this test, DNA samples from any *Acinetobacter* strain can transform auxotroph *Acinetobacter* test strains to prototrophy, indicating a genetic relationship between the tested and competent strain.

Many *Acinetobacter* isolates resemble saprophytic pseudomonads and other Gram-negative non-fermentative organisms in their common ability to utilize a large number of organic compounds as sole sources of carbon and energy. For this reason they also play a significant role in degrading a variety of organic pollutants. However, most isolates cannot utilize glucose as a carbon source, although occasional rare strains are able to do so via the Entner–Doudoroff pathway. Many *Acinetobacter* isolates, however, show acidification of media containing sugars, including glucose, via an aldose dehydrogenase. Normally, all the enzymes of the tricarboxylic acid cycle and the glyoxylate cycle are present. The majority of strains are not able to reduce nitrate to nitrite (tested in the conventional nitrate reduction assay). However, both nitrate and nitrite can be used as nitrogen sources via an assimilatory nitrate reductase.

The colonies of *Acinetobacter* spp. on solid media are often smooth, sometimes mucoid, and comparable in size to those of pseudomonads or enterobacteria. The colour of colonies may vary from greyish to pale yellow.

Identification at the Genomic Species Level

Different DNA–DNA hybridization methods have been used to clarify the taxonomy of the genus and to allocate *Acinetobacter* isolates to genomic species. These methods include a nitrocellulose filter method, the S1 endonuclease method, the hydroxyapatite method and a quantitative bacterial dot filter method. In general, all these methods are time-consuming and laborious and for identification purposes can only be applied in special situations. However, as outlined below, the majority of phenotypic methods do not allow a clear identification of all *Acinetobacter* genomic species.

Identification by Physiological Tests Several schemes of physiological/biochemical tests for identification of acinetobacters to the genomic species have been devised, including tests for sugar acidification and detection of haemolysis and other specific enzymic activities, growth temperature and carbon source utilization tests. These schemes are often combined with computer-assisted identification methods based on probability calculations. A set of tests useful for phenotypic identification of most (but not all) genomic species is shown in **Table 2**. No single test or even a few tests can be used for the unambiguous identification of *Acinetobacter* genomic species and, therefore, additional molecular methods (see below) or DNA–DNA hybridization studies are highly recommended.

Several commercially available identification systems, such as API 20NE, API LAB Plus, Biolog, have included *Acinetobacter* in their databases. In the

Table 2 Selected phenotypic tests for differentiation of *Acinetobacter* genomic species (results of genomic species 14BJ–17BJ are based on only one strain) (adapted from Bergogne-Bérézin et al 1996)

	Genomic species												
	1,2,3	4,6	5	7	8/9[b]	10	11	12	14TU	13BJ	15BJ	16BJ	17BJ1
	13TU[a]				15TU[b]					14BJ[a]			
Most discriminatory tests													
Utilization of:													
Azelate	(+)[c]	–	–	(–)	V	V	+	(+)	–	+	–	(+)[b]	–
β-Alanine	(+)	(–)	–	(–)	–	+	+	–	V	+	+	–	+
L-Leucine	(+)	(+)	V	(–)	–	–	–	+	(–)	+	(+)	(+)	(+)
L-Phenylalanine	(+)	–	–	–	V	–	–	+	+	+	+	+	+
L-Tryptophan	(+)	–	–	–	–	–	–	(–)	+	+	+	–	+
4-Hydroxybenzoate	(+)	(+)	–	(–)	–	V	(+)	(–)	+	+	+	–	+
Phenylacetate	(+)	–	–	–	(+)	–	V	(+)	+	–	+	+	+
Growth at 37°C	(+)	(+)	+	–	V	+	–	+	(–)	+	+	+	+
Acid from glucose	(+)	V	–	–	V	+	–	(–)	+	+	–	–	–
Gelatinase	–	(+)	–	–	–	–	–	–	+	+	+	+	+
Haemolysis of:													
Sheep blood	–	(+)	V	–	–	–	–	–	+	+	+	+	+
Human blood	–	(+)	V	–	–	–	–	–	+	+	+	+	+
Additional tests													
Utilization of:													
Citrate	(+)	V	V	V	(–)	+	V	–	+	+	–	+	+
Glutarate	(+)	(–)	(–)	V	V	+	+	+	V	+	–	(+)	(+)
L-Arginine	+	(+)	(+)	V	(–)	–	–	(+)	(+)	+	+	+	+
L-Aspartate	(+)	V	V	V	–	+	+	V	V	–	–	(+)	–
L-Histidine	+	+	(+)	–	–	+	+	–	+	+	+	+	+
Phenylpyruvate	(+)	–	–	–	V	–	–	(+)	+	+	+	+	+
Protocatechuate	+	+	(–)	V	–	+	+	–	+	+	+	+	+
Vanillate	(+)	(+)	–	(–)	–	V	V	–	–	+	–	+	–
Hydrolysis of:													
pNP-β-D xylopyranoside[d]	(+)	V	–	–	–	V	–	–	–	–	–	+	+
γ-L-Glutamate-pNA[d]	(+)	V	–	–	–	–	–	–	V	–	–	+	+

[a]Phenotypically very similar genomic species were grouped.
[b]Genomic species 8 (*A. lwoffii*) and 9 are considered to be one species.
[c]+, all strains positive; (+), 80% of the tested strains or more positive; V, 20–79% of the tested strains positive; (–), 20% of the tested strains or less positive; –, all strains negative.
[d]pNP, *p*-nitrophenyl; pNA, *p*-nitroanilide.

majority of cases, isolates are identified correctly to the genus level with these systems. A correct identification at the genomic species level, however, is difficult and possible in most cases for only few selected species.

Identification by Chemical Analyses of Cellular Components

The chemical analysis of different cell components is also increasingly used for identification of bacteria. Analytical methods such as gas chromatography (GC), high-pressure gas-liquid chromatography (HPLC), infrared spectroscopy or pyrolysis mass spectrometry are applied. From studies on polyamine and fatty acid patterns of representative strains of *Acinetobacter* genomic it is obvious that differentiation of genomic species was not possible. With a study comparing cell envelope protein electrophoretic profiles a better differentiation could be achieved, but this method requires rigorous stand-

ardization of electrophoretic conditions and cannot be recommended for routine use.

Identification by Molecular Methods

Molecular identification methods are increasingly developed and being validated against DNA–DNA hybridization data. These methods include restriction analysis of 16S rRNA genes amplified by the polymerase chain reaction (PCR) fingerprinting, tRNA fingerprinting, amplified ribosomal DNA restriction analysis, analysis of amplified 16S–23S rDNA intergenic spacer regions, analysis of DNA sequences of genes encoding small subunit rRNA or gyrase subunit B, or direct sequencing of PCR-amplified 16S rDNA.

By using PCR fingerprinting, a combination of patterns generated by five restriction enzymes allows the identification of most *Acinetobacter* genomic species. A further method is based on the restriction of purified chromosomal DNA with two enzymes, ligation of adaptors, and selective amplification with specific primers, one of which is radioactively labelled. The amplification products are separated on sequencing gels and analysis of patterns with approximately 50 bands is preferably performed by a computer-assisted reading system. Also with this method a separation of all genomic species of *Acinetobacter* is possible. Characterization of microorganisms by ribotyping comprises isolation and purification of DNA, digestion by restriction enzymes, separation of resulting fragments by agarose gel electrophoresis, and transfer to nylon membrane by vacuo-blotting. The resulting selective banding patterns are visualized by hybridization with a nonradioactive labelled probe of cDNA from rRNA of *Escherichia coli*. Genomic species specific bandings could be obtained with this method. The direct sequencing of PCR-amplified 16S rDNAs from strains of all genomic species have shown the presence of unique sequence motifs, that can be used for genomic species differentiation.

Conclusions Extensive studies on the physiological/biochemical properties of large numbers of strains validated by DNA homology studies have shown that some genomic groups of *Acinetobacter* can only be identified unambiguously by DNA–DNA hybridization. This is most obvious for the genomic species of the *A. calcoaceticus–A. baumannii* complex. It is recommended that identification should be considered and stated to be presumptive unless an extensive set of assimilation tests or DNA hybridization tests are used. The additional application of high resolution molecular methods (16S rRNA sequencing, AFLP, ribotyping and others) is advisable.

Epidemiological Typing

Epidemiological typing is important for the study of the clinical significance but also geographical spread of specific strains. A large number of methods is now available to discriminate *Acinetobacter* strains, with or without any reference to genomic species. Again, the application of several methods, preferably including at least one molecular method, is highly recommended before designating apparently indistinguishable isolates as the same strain. Although the prevalence of acinetobactors in hospitals is generally low, and these bacteria are associated more commonly with colonization than with infection, several typing methods have been developed and applied in these areas, including biotyping, serotyping, phage typing, and bacteriocin typing (as phenotypic and traditional methods) in addition to comparative typing ('fingerprinting') methods, which include gel electrophoresis of proteins or DNA fragments (mainly genotypic methods).

Details of Procedures Specified in National/International Regulations or Guidelines

To date, no procedure has been specified in National/International regulations or guidelines for the specific detection and isolation of *Acinetobacter*. Because the genus *Acinetobacter* belongs to the typical psychrophilic spoilage flora of food, it is recommended that the procedures and media specified for pseudomonads and similar organisms are used.

For species identification it is essential to use standardized procedures and any identification at the species level should be interpreted with care (see above).

Importance of the Genus and Individual Species in the Food Industry

The genus *Acinetobacter* belongs to the bacterial genera which are considered to be normally found on many foods and food products, especially refrigerated fresh products. Because the genus is considered to be ubiquitous, the primary food-source environments are soil and water. However, plants and plant products, animal hides, human skin and air and dust can also be considered as sources to foods. Acinetobacters have approximate minimum a_w values for growth of about 0.96. They do not grow in conditions with low moisture content but some are able to grow at low temperatures and can even multiply at temperatures < 5°C. Because *Acinetobacter* can be consistently isolated from a wide variety of foods and food products it is necessary to keep the initial numbers low to improve the shelf life of the product. As far as food isolates are concerned, only a few studies have identified organisms belonging to the genus *Acinetobacter* at the genomic species level. It seems, however, that genomic species 7 (*A. johnsonii*) and 8 (*A. lwoffii*) are predominantly found.

Meats and Poultry

The genus *Acinetobacter* belongs to those genera most frequently reported on fresh meat and poultry. Because the internal tissues of healthy slaughter animals are normally free of bacteria at the time of slaughter, the stick knife, animal hide, hand of handlers (skin), storage containers and environments (water) must be considered as primary sources and routes. Ground meats have a greater surface area and therefore the growth of aerobic organisms, including

Acinetobacter, is favoured (**Fig. 2**). The generally low temperature favours a typical spoilage flora. On organs, like livers, kidneys, hearts, tongues of bovine, porcine and ovine origins the surface numbers of microorganisms range from $\log_{10}$ 2–5 cm^{-2}. *Acinetobacter* has been reported to belong to the initial biota, although micrococci, streptococci and coryneforms were found to be dominant.

Most investigations dealing with the spoilage of fresh red meats have been reported for beef, although there is general agreement that pork, lamb, veal and other meats spoil in a similar way.

Ground beef or hamburger beef is spoiled almost exclusively by bacteria, with *Pseudomonas* spp. and *Acinetobacter* spp. playing a major role. The temperature of incubation is the main reason that only selected genera of bacteria are found on spoiled meat. It has been reported that several of *Acinetobacter* strains isolated from meat can grow successfully at low (refrigerator) temperatures. On the surface of refrigerated meat with a pH of around 5.6, enough carbohydrates are present to allow growth up to 10^8 microorganisms per square centimetre. Psychrophilic microorganisms, including *Acinetobacter*, readily utilize organic acids and amino acids as sources of carbon, nitrogen and energy. In poultry, pseudomonads seem to be dominant, but *Acinetobacter* have been frequently reported in these spoilage processes (around 20% of all organisms may be *Acinetobacter* spp.) together with other Gram-negative bacteria of the genera *Alcaligenes*, *Moraxella*, *Serratia*, *Pantoea* and others. *Acinetobacter* spp. can also be found on processed meat and poultry. Vacuum packaging leads to a significant change of the gaseous environment with an almost total removal of O_2 and an increased level of CO_2 which is inhibitory for several microorganisms. It has been reported that *Acinetobacter* is able to survive under these conditions.

Cured hams are also spoiled by organisms fermenting the sugars contained in the curing solutions. This results in 'sours' of various types, depending on their location within the ham. Among other genera, *Acinetobacter* has been implicated as the cause of ham sours.

Seafood

Organisms belonging to the genus *Acinetobacter* have also been found on spoiled fish and seafood products, either fresh or frozen. On frozen catfish fillets, salmon steaks, oysters, crustaceans, molluscs and various other seafoods the surface numbers of microorganisms were in the range $\log_{10}$ 2–6 cm^{-2}. A study of the microorganisms of raw Pacific shrimps showed that *Acinetobacter* comprised 4–24% of the microflora. This proportion increased after blanching and

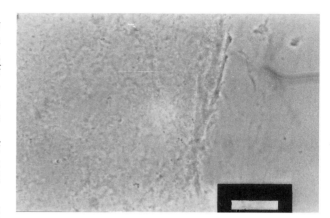

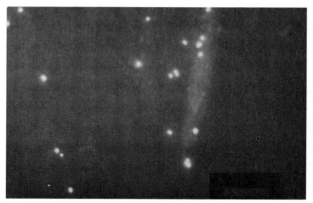

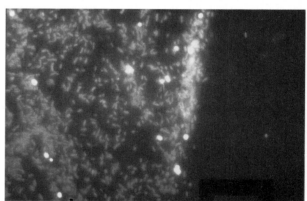

Figure 2 Detection of acinetobacters within a sample of spoiled tartar after seven weeks of storage at 4°C. *In situ* hybridization of an ethanol-fixed sample with a Fluoresceine-labelled probe specific for members of the genus *Acinetobacter* and a Rhodamine-labelled probe binding to all bacteria. Exposures show phase contrast (top), Fluoresceine epifluorescence (middle) and Fluoresceine plus Rhodamine epifluorescence (bottom). Acinetobacters bound both probes and show therefore brighter signals whereas other bacteria are stained with the red-labelled probe only. Individual micrographs represent the identical microscopical field. Green epifluorescence has been analysed with Zeiss filter set 09, red epifluorescence with Zeiss filter set 15. The scale bar of 20 μm applies to all panels (Photomicrographs by Alexander Neef, Institute for Applied Microbiology, Justus-Liebig-University Giessen, Germany).

machine peeling of 16–35%. Fish spoilage is primarily caused by Gram-negative organisms belonging to asporogeneous Gram-negative rods, including *Pseudomonas* and *Acinetobacter*. Many fish-spoilage bacteria are capable of growing at temperatures 0–1°C. The number of *Acinetobacter* seems to decrease with storage-duration.

Dairy and Miscellaneous Food Products

Acinetobacter spp. were also found to be involved in the spoilage of cottage cheese (slimy curd), although *Alcaligenes* seems to be the most frequently found causative genus. Furthermore, it has been reported that organisms of the genus are involved in the spoilage of eggs (rotting). Colourless rots are often caused by Gram-negative non-fermentative rods such as *Pseudomonas*, *Acinetobacter* and other genera.

Importance of the Genus in Other Environments

Water and Soil

As already stated, *Acinetobacter* species are ubiquitous organisms which can be isolated from soil, water, and sewage. It has been assumed that *Acinetobacter* may constitute as much as 0.001% of the total heterotrophic aerobic population of soil and water (often *A. lwoffii*, *A. junii* and *A. johnsonii* are isolated from environmental samples). Even from heavily polluted waters and soils they can be frequently isolated and they play an important role in the mineralization of organic compounds. In several studies the phenomenon of enhanced biological phosphorus removal from waste water treatment plants is attributed to the genus *Acinetobacter*. It has been shown, however, by detailed analyses of the bacterial population using 16S rRNA-targeted oligonucleotide probes, that the members of the genus constitute only 5–10% of the bacterial community in such systems.

Clinical Environment

In the clinical environment, *Acinetobacter* strains play a significant role in colonization and infection of patients admitted to hospital intensive care units. *Acinetobacter* species (*A. baumannii*, genomic species 3 and 13TU are predominantly found in clinical specimens) can be considered as opportunistic pathogens, which are responsible for nosocomial infections, including septicaemia, pneumonia, endocarditis, meningitis, skin and wound sepsis, and urinary tract infection. Although the organism is associated predominantly with nosocomial infection, in some cases, community-acquired infections have been reported, which is indicative of the primary pathogenicity of

these organisms. The main sites of infection are the lower respiratory tract and the urinary tract, with rates of infection ranging from 15–30% of total infections caused by acinetobacters. Acinetobacters are considered originally to be relatively low-grade pathogens. Some characteristics of these bacteria seem to play a role in enhancing the virulence of strains involved in infections: (1) the presence of a polysaccharide capsule formed of L-rhamnose and D-glucose; (2) the property to adhere to human epithelial cells in the presence of fimbriae, which correlates with twitching motility, and/or capsular polysaccharide; (3) the production of certain enzymes (butyrate esterase, caprylate esterase and leucine arylamidase), which seem to be involved in damaging tissue lipids; and (4) the potentially toxic role of the lipopolysaccharide component of the cell wall and the presence of lipid A. *Acinetobacter* species play a predominant role as agents of nosocomial pneumonia, particularly ventilator-associated pneumonia. They have, furthermore, been described as being responsible for difficult-to-treat infections which can be explained by their frequent multiple resistances to a number of antibiotics. Several different resistance mechanisms have been found in *Acinetobacter* and they have been found to develop antibiotic resistance extremely rapidly in response to challenge with new antibiotics. *Acinetobacter* spp. have already developed resistance to broad-spectrum cephalosporins, 4-quinolones and, to some extent, the carbapenems. Mechanisms of gene transfer for mobilization of resistance genes between different members of the genus (and possibly to unrelated bacteria of greater pathogenicity) have been reported.

Biotechnological Applications

Selected strains of *Acinetobacter* are capable of utilizing a wide variety of hydrophobic growth substrates including crude oil, gas oil, several triglycerides, and middle-chain-length alkanes. This is possible because of the ability of emulsan production by some strains. The biopolymer emulsan is the extracellular form of a polyanionic, cell-associated heteropolysaccharide with the property of stabilizing emulsions of hydrocarbons in water. Purified emulsan has a number of potential applications in the petroleum industry, including viscosity reduction during pipeline transport following formation of heavy oil/water emulsions, and production of fuel oil/water emulsions for direct combustion with dewatering. Furthermore, the application of emulsan in the food industry is important and it has been found that some bioemulsifiers (among them one emulsan of a selected *Acinetobacter* strain) produced emulsification activities better than the known food emulsifiers gum arabicum and carboxymethylcellulose.

Further strains *Acinetobacter* were isolated which were able to produce extracellular polymers (termed biodispersans) capable of dispersing limestone in water. The active component was identified as an anionic polysaccharide. Limestone is used in a wide variety of commercial industries, and purified biodispersan may have potential applications in several manufacturing processes producing common products such as paints, ceramics, and paper.

The biodegradation of organic industrial pollutants like different aromatic and aliphatic compounds in biological remediation processes are further biotechnological applications. The biodegradation (biotransformation) abilities of *Acinetobacter* can be extended to toxins, like ochrotoxin. It has been found that one strain of *Acinetobacter* was able to transform the initial ochrotoxin A into a less toxic product.

Because members of the genus *Acinetobacter* are easy to isolate, cultivate, and manipulate genetically in the laboratory, it can be expected that potential field of application will be extended and possibly become very important.

See also: **Fish**: Spoilage of Fish. **Meat and Poultry**: Spoilage of Meat; Spoilage of Cooked Meats and Meat Products. **National Legislation, Guidelines & Standards Governing Microbiology**: Canada; European Union; Japan. ***Pseudomonas***: Introduction. **Total Viable Counts**: Pour Plate Technique; Spread Plate Technique; Specific Techniques; MPN; Metabolic Activity Tests; Microscopy.

Further Reading

Baumann P (1968) Isolation of *Acinetobacter* from soil and water. *J. Bacteriol.* 96: 39–42.

Bergogne-Bérézin E, Joly-Guillou ML and Towner KJ (eds) (1996) *Acinetobacter – Microbiology, Epidemiology, Infections, Management*. Boca Ratan, New York, London, Tokyo: CRC Press.

Bouvet PJM and Jeanjean S (1989) Delineation of new proteolytic genomic species in the genus *Acinetobacter*. *Res. Microbiol.* 140: 291–299.

Bouvet PJM and Grimont PAD (1986) Taxonomy of the genus *Acinetobacter* with the recognition of *Acinetobacter baumannii* sp. nov., *Acinetobacter haemolyticus* sp. nov., *Acinetobacter johnsonii*, sp. nov., and *Acinetobacter junii* sp. nov. and emended descriptions of *Acinetobacter calcoaceticus* and *Acinetobacter lwoffii*. *Int. J. Syst. Bacteriol.* 36: 228–240.

Jay JT (1996) *Modern Food Microbiology*. New York: Chapman & Hall, International Thomson.

Juni E (1978) Genetics and physiology of *Acinetobacter*. *Annu. Rev. Microbiol.* 32: 349–371.

Nishimura Y, Kano M, Ino T et al (1987) Deoxyribonucleic acid relationship among the radiation-resistant *Acinetobacter* and other *Acinetobacter*. *J. Gen. Appl. Microbiol.* 33: 371–376.

Nishimura Y, Ino T and Iizuka H (1988) *Acinetobacter radioresistens* sp. nov. isolated from cotton and soil. *Int. J. Syst. Bacteriol.* 38: 209–211.

Tjernberg I and Ursing J (1989) Clinical strains of *Acinetobacter* classified by DNA–DNA hybridization. *APMIS* 79: 595–605.

Towner KJ (1992) The genus *Acinetobacter*. In: Balows A, Trüper HG, Dworkin M, Harder W and Schliefer K-H (eds) *The Prokaryotes*. P. 3137. New York: Springer-Verlag.

Wagner M, Erhart R, Manz W et al (1994) Development of an rRNA-targeted oligonucleotide probe specific for the genus *Acinetobacter* and its application for in situ monitoring in activated sludge. *Appl. Env. Microbiol.* 60: 792–800.

ADENYLATE KINASE

M J Murphy and **D J Squirrell**, CBD Porton Down, Salisbury, UK

Introduction

Adenosine triphosphate bioluminescence utilizes the firefly luciferase reaction which emits light in the presence of ATP. Since ATP is present in all living material, and is rapidly degraded following cell death, it can be used as a cell marker to monitor biomass. In typical assays, the limit of detection is about 10^4 bacteria or 100 yeast cells. The assay is simple, taking about 1 min to perform, and produces results that are easy to interpret.

Since the 1980s ATP bioluminescence has become an accepted method for rapid microbiological testing and has found application in a variety of fields. It has been used for the detection and enumeration of microorganisms in urine and blood, textiles, process water, soil, the air, plastic packaging material, food, milk and drinking water. It is also a useful tool for microbial adhesion studies as well as the investigation of microbial susceptibility to biocides, antibiotics and bacteriophages.

The most common application is for hygiene monitoring. The availability of single shot disposable reagents which interface neatly with portable luminometers makes the assay very simple to perform. The aim of most hygiene monitoring tests is to ensure that cleaning of equipment and surfaces has been adequately carried out before food processing operations are started. Tests using ATP bioluminescence allow non-specialist cleaning staff to carry out their own quality assurance on the factory floor. The detection of ATP from non-microbial sources is seen as an advantage, since food residues can act as a growth medium for very low numbers of microorganisms that might have gone undetected or that might be present in the environment. The quality of raw materials can also be ascertained prior to use. Rapid testing enables remedial action to be taken promptly, before contamination of food products occurs.

However, despite its considerable advantages over conventional microbiological techniques, ATP bioluminescence has had a relatively limited impact, in 1997 accounting for less than 2% of the overall microbiological testing market. This is largely because many applications require greater sensitivity, and the correlation between ATP readings and colony forming units from traditional plate counts is relatively poor.

These limitations may be overcome by using adenylate kinase (AK) rather than ATP as the cell marker. Like ATP, the presence of AK can be detected by a simple bioluminescent assay. However, it has the potential to offer much greater sensitivity than the measurement of ATP alone and is also far more reliable as an indicator of cell numbers.

Adenylate Kinase

Adenylate kinase (AK) is a constitutively expressed enzyme which comprises about 0.1% of cell protein. It catalyses the equilibrium reaction:

$$Mg^{2+} \cdot ATP + AMP \longleftrightarrow ADP + Mg^{2+} \cdot ADP$$

(Equation 1)

where ATP, ADP and AMP are the adenosine tri-, di- and monophosphates respectively.

In vivo the reaction serves to maintain the balance of adenylates in the cell, usually proceeding to the right to rephosphorylate the AMP which is produced by various phosphatase reactions. The ADP formed can then be further phosphorylated to ATP. However, by adding an excess of ADP the reaction in vitro can be driven in the opposite direction. This generates ATP which can then be measured using the luciferin/luciferase reaction:

$$2ADP \xrightarrow{\quad Mg^{2+} \quad} \textbf{ATP} + AMP$$

(Equation 2)

$$ATP + LH_2 + E \xrightarrow{\quad Mg^{2+} \quad} LH_2\text{-}AMP \cdot E + PP_i$$

(Equation 3)

$$LH_2\text{-}AMP \cdot E + O_2 \longrightarrow L + CO_2 + AMP + \textbf{Light}$$

(Equation 4)

where LH_2 is luciferin, E is luciferase, PP_i is inorganic pyrophosphate and L is oxyluciferin.

Adenylate kinase (ATP : AMP phosphotransferase, EC 2.7.4.3) is found in virtually all eukaryotic and prokaryotic cells, the exception being some sulphur bacteria in which AMP-sulphate is used instead of ATP for energy transfer. In eukaryotes it is predominantly found in the space between the inner and outer mitochondrial membranes and in Gram-negative bacteria in the periplasmic space. It is the only enzyme produced by cells for the purpose of rephosphorylating AMP to ADP and, as such, is essential for life. (Exceptionally, other kinases may have limited nonspecific AK activity which can allow very limited growth in cells in which the gene for AK has been deleted.) It is a stable protein with a relatively long intracellular lifetime. The molecular mass of AK is 20–25 kDa. In bacteria it is usually made up of 214 or 217 amino acids. Typical AK equilibrium ($[ADP]^2/[ATP] \cdot [AMP]$) and Michaelis ($K_m$ for ADP) constants are 2.26 and 1.58 mmol l^{-1} respectively.

A medium-sized bacterium contains about 10^{-21} moles of AK in comparison with about 10^{-18} moles of ATP, so there is approximately 1 molecule of AK for every 1000 molecules of ATP. The enzyme has a turnover number of around 40 000, which means that with just 1 min incubation it can generate 40 times more ATP for bioluminescent signal production than would be possible from the ATP naturally present on its own. The amplification reaction requires only a single substrate and provides a linear increase in the amount of ATP over time. In theory, if a count of 10^3 cells is taken as the limit of detection for a conventional ATP assay, an AK assay with a 25 min incubation should allow single bacterial cells to be detected. Raised backgrounds from contaminating ATP and AK prevent this from being easily achieved, but it has been demonstrated.

From an operator's viewpoint there is very little difference between the two assays (**Fig. 1**). By premixing the ADP with a detergent-based extractant, the extraction and amplification steps needed in the AK assay can be performed simultaneously. The only

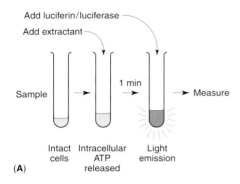

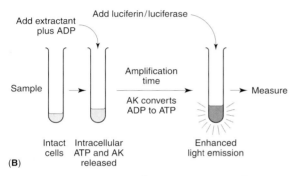

Figure 1 Comparison of (**A**) ATP-based and (**B**) AK-based bioluminescent assays.

variation between the two assays may be an extended incubation time (of about 5 min) for the AK assay.

Kinases other than AK, such as pyruvate kinase, could potentially be used as ATP-generating cell markers. Several have been tried, but none has been found to give results comparable with AK. One reason for this is that all of the other kinases require two substrates: a phosphate donor *and* ADP. Obtaining a high degree of purity in the reagents is consequently made more difficult. Adenylate kinase uses two ADP molecules with one as the phosphate acceptor and the other as the phosphate donor. The AK approach thus has a unique advantage in terms of simplicity. Other reasons for the greater usefulness of AK over other kinases include its enzymatic properties (it has a high turnover number), its characteristics as a protein (it is very small for an enzyme and is particularly robust) and the critical nature of its metabolic function.

Materials and Methods

Certain requirements must be addressed in the development of reagents for AK assays. There are both upper and lower limits to the concentrations of ADP substrate that can be used. A very high ADP concentration is inhibitory to the firefly luciferase reaction, reducing the light output for a given amount of AK, while a very low concentration provides too little substrate for conversion to ATP. Concentrations of

between $10 \, \mu\text{mol} \, l^{-1}$ and $1 \, \text{mmol} \, l^{-1}$ work out to be appropriate for most purposes.

The ADP substrate must be highly purified to minimize the background signals which any extraneous ATP will cause. Commercially available 'high purity' ADP contains around 0.1% ATP. With the ADP concentrations above, contamination by ATP at the 0.1% level would give levels of $100 \, \text{nmol} \, l^{-1}$ to $1 \, \mu\text{mol} \, l^{-1}$ in the reagents. This would be equivalent to the amount of AK-generated ATP from 10^4 or 10^5 bacterial cells. The degree of extra purification needed to optimize the reagents is therefore of the order of 10 000-fold to 100 000-fold. Fortunately this is relatively easy to achieve using anion exchange chromatography where the ADP elutes before the contaminating ATP. The inclusion of EDTA in the ADP preparation prevents premature conversion of ADP by extraneous AK. As a result, magnesium, which is needed as a cofactor for both the AK and luciferase reactions, has to be added in excess immediately prior to the assay. This is most conveniently done by including it in the sample buffer.

The luciferase preparations used for the light-generating reaction should contain minimal levels of contaminating AK to reduce background signals. Attention is therefore required in the production of the luciferase enzyme itself and of reagent components such as bovine serum albumin (BSA) which are commonly added as bulking or stabilizing agents.

As in ATP bioluminescence, an extraction step is needed. This enables the ADP substrate to contact the target AK, and the subsequently generated ATP to react with the luciferase. Extraction is usually carried out using a detergent. This needs to be carefully selected to avoid inactivating either the AK or the luciferase. The concentration chosen for the detergent has to be a balance between maximizing disruption of cell membranes, and thus extraction of the AK, and minimizing inactivation of its enzymatic activity. **Table 1** compares the effects of different detergents on the light emission from AK assays. Freeze-thaw lysis, a technique that efficiently releases AK with minimal inactivation, may be used as a control method when evaluating detergent performance.

Table 1 Net luminescence readings from AK assays of 2×10^4 *Escherichia coli* cells carried out with a range of detergents at 0.05% and 0.25% concentration

Detergent	0.05%	0.25%
Chlorhexidine diacetate	184 000	61 000
Tween 20	140 000	127 000
Triton X-100	217 000	178 000
IGEPAL CA 630	232 000	190 000
N, N', N'-Polyoxyethylene (10)-N'-tallow-1,3-diaminopropane	323 000	278 000

However, since it is a time-consuming method and requires specialist equipment, it is unsuitable for general use.

Reagents kits for AK assays, containing highly purified ADP, fully optimized extractants and luciferase reagents low in contaminating AK, are being developed for commercial use.

Assay Capability

Adenylate kinase offers at least 10 times, and usually 100 times, greater sensitivity than conventional ATP assays (**Fig. 2**). Because the amplification kinetics are linear, and therefore controlled, it is possible to increase the sensitivity of AK assays by extending the incubation time (**Fig. 3**). Increasing the incubation time in ATP assays has no effect, since the majority

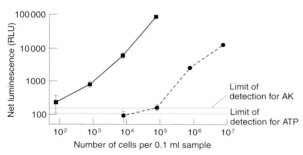

Figure 2   Comparison of ATP (circles) and AK (squares) assays for *Escherichia coli*. Cells were freshly harvested from culture and washed to remove extracellular AK or ATP. One minute incubation time was allowed for extraction and amplification. Cell numbers were determined by traditional plate counts. Data points are mean values ± 1 SD ($n = 3$).

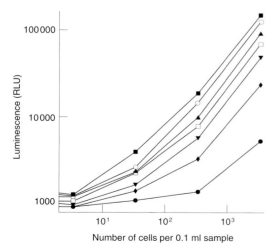

Figure 3   Adenylate kinase assay for *Escherichia coli* showing the effect of increasing incubation time. Cells were freshly harvested from an overnight broth culture and diluted in magnesium acetate buffer. Incubation times were 1 min (●); 5 min (◆); 10 min (▼); 15 min (□); 20 min (▲); 30 min (○) and 40 min (■).

of the ATP is extracted within the first 10–20 s of the usual 1 min incubation. The most convenient combination of speed and sensitivity for AK is achieved with a 5 min incubation. Using this, fewer than 200 cells of bacteria such as *Escherichia coli* or *Pseudomonas aeruginosa* can be readily detected. The assay is reproducible and is relatively unaffected by changes in growth medium or conditions.

By using small reaction volumes in combination with incubation times of up to 1 h, limits of detection approaching the single-cell level become possible. At this point, sampling statistics rather than assay sentivity become limiting and the technique compares favourably with polymerase chain reaction (PCR) and growth assays. Such small-volume assays can, in effect, be carried out on filter membranes and the full capability of the AK approach may be realized in this format.

Because ATP is the principal energy medium in cells, its concentration fluctuates widely within bacteria according to their metabolic status. Levels of ATP may be reduced by around 50% in freshly harvested cells merely by centrifugal washing in buffered saline solution (**Fig. 4A**). Cell ATP reserves may be further depleted (by over 95%) when the cells are stressed by nutrient limitation or exposure to biocidal or biostatic treatments. The ATP bioluminescence assay signals will therefore depend upon, in anything other than the most precisely prepared samples, variable intracellular ATP concentrations. This leads to inaccuracies when attempting to quantify the bacterial loading of a sample. So when the standard hygiene monitoring approach using ATP as the cell marker is adopted, caution must be exercised in setting pass and fail levels to take these variations into account.

In contrast, AK as a constitutively expressed enzyme is present at relatively constant levels regardless of the energy status of the cell. It therefore provides a much more reliable measure of the number of bacteria present than ATP (Fig. 4B). Nevertheless, several factors still affect the correlation between cell numbers, determined as colony forming units (cfu), and the net bioluminescent signal from an AK assay. Firstly, the plate counts obtained will tend to be lower than the number of cells present in the sample because a proportion of the cells may be in a 'viable nonculturable' state, (that is, they are not dead, but are unable to grow on the growth medium provided). Secondly, a cfu may represent one cell only, or it may represent several cells if they are stuck together or have simply come to rest on the agar plate closer to one another than the diameter of an individual colony. Both of these factors mean that traditional agar plate counts underestimate contamination levels in terms of the real cell population.

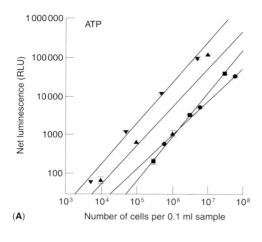

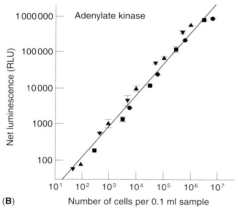

Figure 4 Comparison of assays for *Escherichia coli*. (**A**) ATP; (**B**) adenylate kinase. Freshly harvested cells, washed (▲) and unwashed (▼); starved cells, washed (●) and unwashed (■). Data points are mean values ± 1 SD (*n* = 3).

The third factor affecting the correlation between AK levels and viable cell numbers is a dependence upon the causes that result in a cell becoming non-viable. Chemical treatments (i.e. disinfection) generally destroy AK at the same time as killing cells. More subtle causes of death, such as antibiotic treatments, may allow enzymatically active AK to be present in non-viable cells. If this happens, an AK assay may provide an overestimation of the number of live cells. Taking this together with the underestimation obtained from plate counting means that an AK assay may indicate the presence of more cells than a traditional cfu count would suggest. However, which indicator provides the better measure of 'contamination' will depend upon the reason for carrying out the assay in the first place; AK may be particularly useful where safety critical monitoring is required.

Although the intracellular concentration of AK is relatively constant, cell size will inevitably vary in a growing population of dividing cells. This factor will give rise to a limited degree of imprecision in the correlation between AK assay results and cell numbers and therefore with cfu measurements. This factor will equally affect ATP-based assays.

Matrix Effects

As a new technique, AK bioluminescence has mostly been tested on relatively 'clean' laboratory samples. However, in industrial applications, the source of the sample on which the test is being carried out could have a large effect on the end result. Adenylate kinase is present in virtually all living matter, not just bacteria. This means that extraneous organic material and the sample itself could contribute towards the signal. This applies to plant sources, e.g. fruit or vegetable juices, as well as meat and dairy produce.

The detection of non-microbial AK may be considered to be advantageous when checking the effectiveness of cleaning regimes used for food preparation surfaces. However, it may cause problems, for example, when trying to pick up spoilage organisms in food samples. When total microbial loading is to be determined, the non-microbial AK may either be removed by suitable sample pre-treatment methods or accounted for by subtracting the results from control assays carried out on uncontaminated samples, the latter approach being limited by the endogenous AK levels. Some comparative measurements of these in various foods and beverages are given in **Table 2**. The levels can be seen to vary greatly, with those in meat being especially high. As might be expected, cooking and ultra-high temperature (UHT) treatment cause a reduction in AK levels.

Table 2 Pairwise comparisons of adenylate kinase (AK) levels in various samples, and between extracellular and total AK levels in the same samples. Extracellular AK activity measurements were determined using ADP substrate without extractant and total levels were obtained using ADP plus extractant; signal as rate of increase of light output per minute is given in brackets

Pairs of measurements for comparison	Ratio
Extracellular AK versus total AK	
Guinness: free AK (223); total AK (1225)	1 : 5.5
Raw chicken: free AK (1.9×10^8); total AK (9.4×10^8)	1 : 10.6
Pasteurized milk: free AK (44 782); total AK (65 617)	1 : 1.5
Different samples compared for free AK levels	
Beverages: Guinness, free AK (223); Heineken, free AK (6398)	1 : 29
Milk: UHT, free AK (28 598); pasteurized, free AK (44 782)	1 : 1.6
Chicken: raw, free AK (1.9×10^8); cooked, free AK (3.3×10^7)	5.8 : 1

Potential Applications

The main application areas for AK-based cell detection methods in the food industry are rapid contamination screening, sterility checking and quality testing for:

- materials that should contain little endogenous AK or ATP (e.g. process water, CIP (clean-in-place) rinse water)
- samples where endogenous free AK and ATP can be removed by filtration or other methods (e.g. beverages such as beer, cider and soft drinks)
- hygiene monitoring assays to allow the monitoring of low-level contamination by microorganisms and other organic matter with high sensitivity and reproducibility (e.g. raw materials, air quality and food preparation surfaces)
- assays for specific organisms (using e.g. antibody capture, phage lysis or short enrichment steps).

Such applications exploit the advantages which the high sensitivity of the AK approach can provide and will expand the areas of microbiological testing where rapid testing is feasible. Assays with AK as a marker should come to supplement rather than replace direct ATP measurements, which will remain appropriate where background contamination is inherently high, extreme sensitivity is not needed, or false positives are more of a problem than false negative results. Some examples follow.

Hygiene Monitoring

The method for hygiene monitoring using AK should be essentially the same as the widely used ATP method, lending itself equally well to the use of single-shot disposables. A typical AK-based test method would be:

1. Swab area of interest with swab moistened with magnesium acetate buffer.
2. Place swab in a cuvette with ADP and extractant.
3. Incubate at room temperature for 5 min.
4. Add bioluminescence reagent.

Table 3 Environmental samples tested from various locations around a laboratory, using AK as a hygiene monitor

Sample location	RLU at 1 min	RLU at 5 min	Change in RLU (min^{-1})
Door handle	8 947	13 477	906
Hand	246 863	473 241	45 276
Laboratory floor	75 065	190 850	23 157
Laboratory bench	306 318	526 861	44 109
'Clean' bench	4 877	8 450	715
Cleaned bench	3 302	3 583	56
Computer screen	5 954	10 069	823
Inside of autoclave	416	512	19
Control (no swabbing)	316	414	20

RLU, relative light units.

5. Measure light output using a portable luminometer.

Table 3 shows the AK levels found in a microbiology laboratory. Typical sensitivities achievable using AK for hygiene monitoring are demonstrated in **Table 4**, which shows the signals obtained using this method for three different microorganisms.

Specific Assays

In cases where identification of the presence of particular organisms is required rather than measurement of the total microbial loading, the AK assay may be used as the end point in immunoassays. One approach is to use magnetic beads. These are commercially available, pre-coated with antibodies specific for certain bacteria such as *Listeria monocytogenes*, *E. coli* O157 and *Salmonella* sp. In standard tests they are used to isolate cells from complex suspensions, such as food, prior to enrichment. Culture on agar plates or biochemical identification procedures are then employed to determine the level of contamination in the original sample. The procedures are time-consuming, taking 24–72 h to complete.

Low numbers of organisms captured on magnetic beads may be directly detected using AK. Contamination from other organisms or the matrix are

Table 4 The sensitivity of detection for *Candida albicans*, *Staphylococcus aureus* and *Escherichia coli* using the same hygiene monitoring procedure as Table 3

C. albicans		S. aureus		E. coli	
Cfu per swab	RLU change (5 min)	Cfu per swab	RLU change (5 min)	Cfu per swab	RLU change (5 min)
18 000	165 939	450 000	104 959	750 000	441 949
1 800	36 980	45 000	10 453	75 000	155 356
180	3 467	4 500	1 373	7 500	11 700
18	479	450	268	750	1 508
1.8	254	45	137	75	421
0.18	231	4.5	152	7.5	179
0	170	0	178	0	114

Cfu, colony forming units; RLU, relative light units.

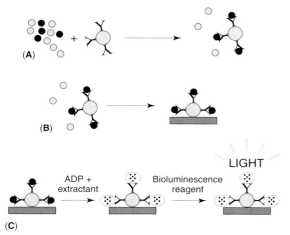

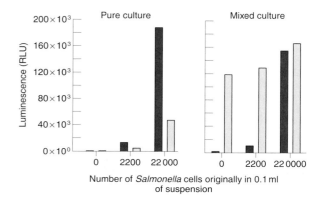

Figure 5 Magnetic bead immunoassay for selective immobilization of microorganisms from food and beverage samples, prior to generic lysis and detection with AK bioluminescence. (**A**) Target cells are removed from suspension by mixing with magnetic beads pre-coated with specific antibody. (**B**) Beads are immobilized by applying a magnetic field. Unwanted material is removed by washing. (**C**) The captured cells are lysed and ADP is added. After incubation, bioluminescence reagent is added and the light output is measured.

Figure 6 Magnetic bead immunoassay for *Salmonella typhimurium*. Mixed culture assay carried out in the presence of a constant 3.5×10^5 vegetative *Bacillus subtilis* var. *niger* cells. Solid bars, signal associated with bound material (specific signal); open bars, signal associated with the supernatant (non-target cells and unbound target cells).

eliminated during the separation. Once immobilized the bacteria can be lysed in the same way as a non-specific assay (**Fig. 5**). The rate at which the bacteria will bind to the beads depends upon the complexity of the matrix. For liquid samples, binding is rapid, taking less than 5 min. In a more complex medium the incubation will need to be increased to allow the organisms time to 'find' the antibodies. Nonetheless results will be obtained in a maximum of 1 h – a considerable time saving on conventional methods, and without the need to start with a pure culture. A typical assay procedure is:

1. Immobilize washed beads and remove supernatant.
2. Add sample to beads, resuspend and incubate at room temperature.
3. Immobilize beads and remove supernatant (containing unbound material).
4. Wash beads to remove any loosely bound material.
5. Resuspend beads in magnesium acetate buffer.
6. Add ADP plus extractant and incubate for 5 min.
7. Immobilize beads and transfer the supernatant (containing extracted AK and generated ATP) to a cuvette containing bioluminescence reagent.
8. Measure light output using a portable luminometer.

Figure 6 gives the results of a magnetic bead immunoassay for *Salmonella* in aqueous suspension. They demonstrate the assay's capability of detecting fewer

than 10^4 *Salmonella* cells, in the presence of a 10-fold excess of *Bacillus subtilis* var. *niger*, in a total assay time of 15 min.

Another approach for detecting the presence of specific microorganisms using AK is to produce specific lysis of the target cells. This may be achieved using lytic bacteriophages or commercially available lysins such as colicin for *E. coli* and lysostaphin for *Staphylococcus aureus*.

Bacteriophage-mediated lysis of target cells may be achieved by adding to a culture a phage specific for the organism of interest. If the target organisms are present the phage will infect them and cause the cells to be lysed. This releases all the intracellular components, including AK, into suspension. By adding ADP (with no extractant) the activity of this released AK can be measured (**Fig. 7**). A typical procedure is:

1. Mix sample with culture medium and incubate at 37°C for 1–2 h.
2. Split sample into two: infect one with bacteriophage and leave the other as an uninfected control.

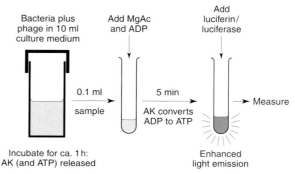

Figure 7 Bacteriophage/AK assay for *Escherichia coli*. MgAc, magnesium acetate buffer.

3. Incubate at 37°C.
4. At timed intervals remove a sample from each culture into a cuvette containing magnesium acetate buffer.
5. Add ADP (but no extractant) and incubate for 5 min.
6. Add bioluminescence reagent and measure light output.

Using bacteriophages, fewer than 10^3 log phase *E. coli* cells can be detected in around 30 min under laboratory conditions (**Fig. 8**). The bacteria must be in log phase in order to be receptive to phage infection, so a short culture step of about 2 h would be required for 'real' samples which will probably contain stationary phase or otherwise stressed cells. The time course of the phage lysis is specific to the phage/host combination used. The coliphage used for the experiment in Figure 8 starts to induce lysis after about 20 min. Newport phage which is specific for *Salmonella newport* takes about 40 min to reach this point. This means that assay time will vary depending on the target organisms and bacteriophage chosen.

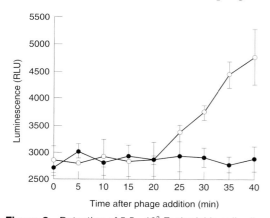

Figure 8 Detection of 5.5×10^3 *Escherichia coli* cells per ml by bacteriophage-mediated release of adenylate kinase: ○, infected culture; ●, uninfected control. Data points are mean values ± 1 SEM ($n = 3$).

Bacteriophages vary in their specificity. Some are capable of infecting bacteria from the same genus whereas others are strain-specific within a particular species, e.g. MS-2 and HIRH strains of *E. coli*. This allows flexibility in designing assays to be as specific or as broad in range as required by the particular application.

Specific capture and specific lysis steps can be combined to give extra confidence in assays for a particular organism. This is exemplified in an assay for *Staphylococcus aureus* in which antibodies are coated onto sticks, which are dipped into liquid samples, to provide specific immobilization of the target cells. Lysostaphin is then used as an *S. aureus* specific lysin,

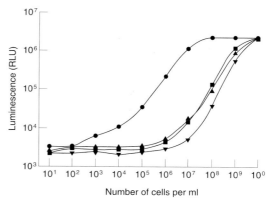

Figure 9 Specificity of AK release achievable using lysostaphin ●, *Staphylococcus aureus* + lysostaphin; ■, *S. aureus*, free AK; ▲, *E. coli* + lysostaphin; ▼, *E. coli*, free AK.

which provides efficient and rapid lysis of the target cells, leaving any nonspecifically bound cells intact (**Fig. 9**).

This prototype dipstick *S. aureus* assay uses polystyrene 'paddle' type immunosticks coated with monoclonal anti-*S. aureus* antibody. The proposed procedure is:

1. Add a 1 ml sample to the tube containing the dipstick.
2. Incubate at room temperature for 40 min.
3. Wash with phosphate-buffered saline.
4. Add assay reagents (ADP, magnesium acetate and lysostaphin).
5. Incubate for 20 min.
6. Transfer sample to a cuvette.
7. Add bioluminescence reagent.
8. Place in a luminometer and measure the light emission.

Table 5 shows the results obtained from the prototype assay for a range of *E. coli* and *S. aureus* concentrations.

Table 5 Prototype *Staphylococcus aureus* dipstick assay. Signals are obtained by incubating immunosticks with either *Escherichia coli* or *Staphylococcus aureus*. Lysostaphin is used to release adenylate kinase prior to assay.

	Cfu (ml⁻¹)	RLU
Culture medium blank	0	35 198
E. coli (control)	1.2×10^7	57 003
	1.2×10^5	60 896
	1.2×10^3	40 757
S. aureus	4.4×10^7	1 977 733
	4.4×10^5	613 449
	4.4×10^3	39 825

Cfu, colony forming units; RLU, relative light units.

Table 6 Comparison of ATP and adenylate kinase bioluminescent assays for the detection of microorganisms

	Adenylate kinase	ATP
Nature of cell marker	Enzyme	Metabolite
Essential for viability	Yes	Yes
Amount per average bacterial cell	10^{-21} mol	10^{-18} mol
Assay time	5 min	1 min
Incubation time-dependent	Yes	No
Relative light output per bacterium	0.8	0.01
Intracellular levels	Constant	Variable
Correlation with cell number	Good	Approximate
Assay detection limit (0.1 ml sample volume)		
bacteria	100	1000–10 000[a]
yeasts	10	100

[a] Detection limit for bacteria using ATP bioluminescence dependent upon energy status of the cell.

Conclusion

Bioluminescence assay based on AK is a fast-developing technique in rapid microbiological testing which, through increased sensitivity and better correlation with the traditional plate count method, overcomes some of the limitations of solely ATP-based bioluminescence. A summary comparison of the two bioluminescent approaches is given in **Table 6**. Inevitably, food samples will contain their own AK which will tend to swamp that from contaminating microorganisms. The applicability of the AK approach in rapid testing is thus restricted to cleanliness monitoring (i.e. testing for the absence of AK); testing samples that have an inherently low level of AK; testing samples that can be easily treated to remove endogenous AK; and specific testing. In the short term AK bioluminescence will probably be most used in the first three application areas. In the longer term it is likely that the specific assays will predominate, because in comparison with techniques such as ELISA, methods based on AK can be quick, simple and sensitive. Particularly promising are the possibilities for test methods with double specificity achieved by combining a specific capture step with a specific lysis step. These should allow significant opportunities for replacing traditional microbiological testing with rapid methods.

See also: **Biophysical Techniques for Enhancing Microbiological Analysis**: Future Developments. **Electrical Techniques**: Food Spoilage Flora and TVC. **Sampling Regimes & Statistical Evaluation of Microbiological Results**. *Staphylococcus*: Detection by Cultural and Modern Techniques. **Total Viable Counts**: Pour Plate Technique; Spread Plate Technique; Specific Techniques; MPN; Metabolic Activity Tests; Microscopy.

Further Reading

Atkinson DE (1970) Enzymes as control elements in metabolic regulation. In: Boyer PD (eds) *The Enzymes, Structure and Control*. P. 472. London: Academic Press.

Barman TE (1985) *The Enzyme Handbook*. Vol. 1, P. 488. New York: Springer.

Blasco R, Murphy MJ, Sanders MF and Squirrell DJ (1998) Specific assays for bacteria using bacteriophage mediated release of adenylate kinase. *Journal of Applied Microbiology* 84: 661–666.

Brolin SE, Borglund E and Agren (1979) Photokinetic microassay of adenylate kinase using the firefly luciferase reaction. *Journal of Biochemical and Biophysical Methods* 1: 163–169.

Murphy MJ, Squirrell DJ, Sanders MF and Blasco R (1996) The use of adenylate kinase for the detection and identification of low numbers of micro-organisms. In: Hastings JW, Kricka LJ and Stanley PE (eds) *Bioluminescence and Chemiluminescence: Molecular Reporting with Photons*. P. 319. Chichester: John Wiley.

Squirrell DJ and Murphy MJ (1995) Adenylate kinase as a cell marker in bioluminescent assays. In: Campbell AK, Kricka LJ and Stanley PE (eds) *Bioluminescence and Chemiluminescence: Fundamentals and Applied Aspects*. P. 486. Chichester: John Wiley.

Squirrell DJ and Murphy MJ (1997) Rapid detection of very low numbers of micro-organisms using adenylate kinase as a cell marker. In: Stanley PE, Simpson WJ and Smither R (eds) *A Practical Guide to Industrial Uses of ATP-luminescence in Rapid Microbiology*. Lingfield: Cara Technology.

Aerobic Metabolism *see* **Metabolic Pathways**: Release of Energy (Aerobic); Release of Energy (Anaerobic).

AEROMONAS

Contents

Introduction

Detection by Cultural and Modern Techniques

Introduction

I S Blair, **M A S McMahon** and **D A McDowell**, Food Studies Research Unit, University of Ulster at Jordanstown, Co. Antrim, Northern Ireland

Genus Characteristics

Aeromonas is a member of the family Vibrionaceae, which includes four other genera, namely *Vibrio*, *Photobacterium*, *Plesiomonas* and *Enhydrobacter*. Differentiation among the members of Vibrionaceae is based on a number of biochemical and biophysical characteristics (**Table 1**). These characteristics adequately differentiate between most of the constituent genera of this family, however differentiation between *Aeromonas* and *Plesiomonas* genera requires the additional biochemical analyses presented in **Table 2**.

The members of the genus *Aeromonas* are chemo-organotrophic facultative anaerobes demonstrating both respiratory and fermentative metabolism. These bacteria are Gram-negative bacilli of 0.3–1 μm diameter and 1–3.5 μm length which are usually present singly, in pairs or in short chains. Members of this genus can grow over a wide range of environmental conditions, for example, pH values from 4.0 to 10.0 (with an optimum around neutrality) and salt concentrations up to 6.5%. Most members of the genus are mesophiles with an optimal growth temperature

Table 2 Differentiation between *Aeromonas* and *Plesiomonas*

Characteristic	Aeromonas[a]	Plesiomonas
Monotrichous flagella (liquid media)	+[b]	−
Lopotrichous flagella (liquid media)	−[c]	+
KCN, growth	d[d]	−
L-Arabinose, utilization	d	−
Salacin, acid	d	−
Sucrose, acid	+	−
D-Mannitol, acid	+	−
myo-Inositol, breakdown	−	+
Voges–Proskauer	d	−
D-Glucose, gas	d	−

[a]Except *A. salmonicida*.
[b]+, typically positive.
[c]−, typically negative.
[d]d, differs among species/strains.

of 28°C. Some *Aeromonas* can grow at temperatures ranging from 42°C to 4°C. The capacity to grow at such extreme temperatures varies among strains and seems to be closely related to the source of isolation, or to environmental adaptation. Most members of the genus are motile, and flagella, if present, are singular and polar. Some biogroups of *A. salmonicida* are non-motile.

The species which make up the genus *Aeromonas* have very diverse characteristics. This diversity, combined with the absence of widely accepted standardized identification schemes, has led to ambiguity in the identification of presumptive aeromonad isolates. Identification of presumptive *Aeromonas* isolates to

Table 1 Differentiation of the genera within the family Vibrionaceae[a]

Characteristics	Vibrio	Photobacterium	Aeromonas	Plesiomonas	Enhydrobacter
Sheathed polar flagella	+	−	−	−	−
Motility	+	+	[+][e]	+	−
Na+ required for/stimulates growth	+	+	−	−	−
Lipase	[+][b]	D	[+]	−	+
D-Mannitol utilization	[+][c]	−	[+]	−	−
Sensitivity to O/129[d]	+	+	−	+	−
Mol% G+C of DNA	38–51	40–44	57–63	51	66
Accumulation of poly-β-hydroxybutyrate/inability to utilize poly-β-hydroxybutyrate	−	+			

[a]Symbols: +, all species positive; [+], most species positive; −, all species negative; D, some species positive some species negative.
[b]*V. nereis*, *V. anguillarum* biovar II, *V. costicola* are negative.
[c]*V. nereis*, *V. anguillarum* biovar II, *V. marinus* are negative.
[d]There are some exceptions.
[e]Except *A. media* and *A. salmonicida*.

the genus level is based on a series of simple biochemical tests, although the tests and/or combinations of tests used are not universally accepted. Typically, identification to genus level involves a negative Gram stain, positive oxidase and positive catalase test results and resistance to the vibriostatic agent O/129. Additional criteria may include the presence of DNAase and the absence of growth in the presence of 6.5% salt.

Taxonomic identification within the genus *Aeromonas* has been the subject of considerable debate, and differentiation/classification at species level is somewhat confused. There is disagreement about the allocation of members of the *Aeromonas* genus to individual species, with some workers suggesting there are eight separate species including the non-motile psychrophilic *A. salmonicida*. This distribution is not universally accepted and other workers have suggested that the genus *Aeromonas* should be subdivided into seven species (**Fig. 1**). These inconsistencies may be in part due to uncertainties caused by variations in biochemical and physiological growth characteristics. As a result of these difficulties most clinical laboratories report *Aeromonas* isolates as members of:

- *A. caviae* group, which includes *A. caviae*, *A. eucrenophila* and *A. media*
- *A. hydrophila* group, which includes *A. hydrophila* and a motile biogroup of *A. salmonicida*
- *A. sobria* group, which includes *A. sobria* and *A. veronii*

A variety of methodologies have been used to differentiate aeromonad species. The majority of these are based on biochemical tests, although the number and range of biochemical tests used varies considerably. Over 300 biochemical tests have been evaluated for their ability to speciate accurately presumptive *Aeromonas* isolates, although only 30 tests are necessary for the accurate identification to species level. Further work has suggested that a much

smaller number of tests (as few as five) will allow identification of 98% of *Aeromonas* species. Perhaps the greatest problem associated with the study of *Aeromonas* is the lack of a universally accepted taxonomy at species level. The biochemical tests listed in **Table 3** would seem to offer the best compromise between accuracy and complexity in the speciation of aeromonads.

The responses of the various *Aeromonas* species during these tests are well documented. Decarboxylase reactions are particularly important in the speciation of aeromonads, but these tests need to be carefully applied as the results obtained are significantly affected by the incubation temperatures and the media used. Such methodological difficulties have raised concerns in relation to the classification of species, based solely on biochemical characteristics. Nevertheless, the recent establishment of several new species – *A. eucrenophila*, *A. jandaei*, *A. media*, *A. schubertii*, *A. trota* and *A. veronii* – have been based on single biochemical characteristics (**Table 4**).

To avoid the above difficulties associated with biochemical speciation, alternative methodologies have

Table 3 Biochemical tests for determining species of *Aeromonas*

Aesculin hydrolysis	Growth on KCN broth
Histidine utilization	Arginine utilization
Mannitol fermentation	Salacin fermentation
Sucrose fermentation	Voges–Proskauer
Gas from glucose	H_2S from cysteine
Lysine decarboxylase	Ornithine decarboxylase
Arginine dihydrolase	Ampicillin resistance

Table 4 Biochemical speciation of recently identified *Aeromonas* strains

Species	Key biochemical characteristic
A. veronii	Ornithine decarboxylase positive
A. schubertii	Mannitol utilization negative
A. jandaei	Sucrose utilization negative
A. trota	Ampicillin sensitive

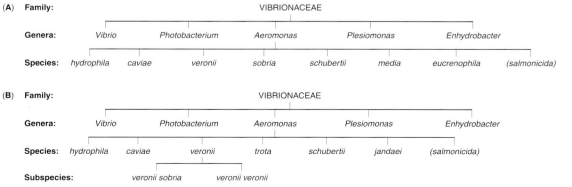

Figure 1 Classification of the genus *Aeromonas* according to (**A**) Holt et al (1994) and (**B**) Merino et al (1995).

been used to speciate *Aeromonas*. Cluster analysis has differentiated the genus into three major phenons, equivalent to *A. hydrophila*, *A. sobria* and *A. caviae*. DNA–DNA hybridization studies of these phenons have shown that each phenon contains a number of hybridization groups (HGs) or genospecies; *A. hydrophila* belongs to HG 1, 2 and 3, *A. caviae* to groups 4, 5 and 6, and *A. sobria* to groups 7, 8/10, 9 and 13. However, due to phenotypic differences within the hybridization groups, they can each contain more than one biotype (or subspecies).

Other methods have been used in the characterization and speciation of aeromonads including combinations of biochemical tests, haemagglutination, cytotoxicity and mouse lethality doses. Characterization has also been carried out using electrophoretic comparison of soluble protein profiles, and outer membrane proteins. Characterization by the use of DNA restriction endonuclease analysis has provided genotypic classification. The development and application of molecular methods such as 16S rRNA sequencing and ribotyping can assist in taxonomic and epidemiological investigations.

Importance to the Food Industry

The Prevalence of *Aeromonas* in Foods

Aeromonas spp. are ubiquitous in the environment, and have been isolated from a range of environmental sources including fresh water, salt water and chlorinated drinking water. There is some evidence that animal faeces can also be a source, although this is not universally accepted. *Aeromonas* species may be readily recovered from most raw foods, including meats, fish, poultry, fresh vegetables and ready-to-eat products (**Table 5**). This diversity of occurrence may be an indication of environmental contamination, most probably from water, animal faeces or food handlers. Although the organism occurs worldwide, the incidence of different species of *Aeromonas* found in foods varies between and within countries. The most prevalent *Aeromonas* isolate from food in Europe is *A. hydrophila*, whereas in Japan, *A. caviae* is the most prevalent food isolate. These observations suggest either that there are distinct differences in the geographical incidence of particular species of *Aeromonas* in foods, or that differences among the isolation methods used in different countries result in different isolation rates.

Aeromonas species such as *A. hydrophila*, are important to the food industry because of (1) their psychrotrophic nature and (2) their ability to express a range of virulence factors under refrigerated storage conditions. *Aeromonas* species such as *A. hydrophila* can be regarded as both spoilage and potentially pathogenic organisms.

Psychrotrophic Nature Consumer demand for less processed, more 'natural' and convenient foods, has placed greater emphasis on refrigeration as a primary means of restricting the growth of spoilage and pathogenic organisms. Although the majority of the *Aeromonas* strains present in foods have optimum temperatures for growth which are characteristic of mesophiles, most *Aeromonas* species are capable of growth at refrigeration temperatures, i.e. 4–5°C, and are, therefore, more accurately classified as psychrotrophic mesophiles. Most environmental/food *Aeromonas* isolates can grow at these temperatures, however, only 50% of clinical isolates are capable of growth at refrigeration temperatures. Populations of *Aeromonas*, present in foods as part of the normal flora, have been shown to increase by 10–1000-fold during 7–10 days of storage at 5°C. *A. hydrophila* grows slowly at 0°C and some strains have been reported to grow at temperatures as low as –3.5°C. This organism is not significantly inactivated by

Table 5 Prevalence of *Aeromonas* species in retail foods

Food product	% positive	A. hydrophila	A. sobria	A. caviae	Reference
Fish (freshwater)	100	50	33.3	16.7	Sierra et al (1995)
Raw milk	59.7	39.5	9.3	34.9	Kirov et al (1993)
Pasteurized milk	3.8	28.6	42.9	28.6	Kirov et al (1993)
Cold smoked fish	10.9	66.6	0	33.3	Gobat and Jemmi (1993)
Hot smoked fish	14.3	100	0	0	Gobat and Jemmi (1993)
Cooked/frozen shellfish	8.8	100	0	0	Gobat and Jemmi (1993)
Minced meat	94.1	75.7	24.3	27	Gobat and Jemmi (1993)
Ham	38.2	47	11.7	41.3	Gobat and Jemmi (1993)
Raw chicken	84.4	38.7	40.8	20.5	Gobat and Jemmi (1993)
Whipped cream	50	N.D.[a]	N.D.	N.D.	Knochel and Jeppesen (1990)
Cream cakes/deserts	6	N.D.	N.D.	N.D.	Knochel and Jeppesen (1990)
Mayonnaise/salads	10	N.D.	N.D.	N.D.	Knochel and Jeppesen (1990)
Vegetables	51	37	0	63	Nishikawa and Kishi (1988)

[a]N.D., Not determined.

subzero storage, and has, for example, remained viable in oysters stored at –72°C for periods of up to 18 months.

Concerns stimulated by the combination of the psychrotrophic nature, ubiquity in foods and potential pathogenicity of *Aeromonas* spp. have led to extensive research into the growth characteristics of these organisms at refrigeration temperatures. The rates and limits of growth of strains of *A. hydrophila* at refrigeration temperatures are dependent on their sources and previous histories (equivalent to the pre-incubation temperature in vitro). Some reports indicate that strains pre-incubated at a higher temperature (e.g. 37°C) and subsequently incubated at a refrigeration temperature (5°C), have considerably lower growth rate than strains pre-incubated at 10°C.

Virulence Factors   The main virulence factors of *Aeromonas* species that can be associated with disease are the expression of extracellular proteins such as exotoxins and exoenzymes, endotoxin (lipopolysaccharide (LPS) layer), the presence of S-layers, the presence of fimbriae or adhesins and the production of capsular layers.

Expression of Extracellular Proteins   A number of studies have examined the ability of *Aeromonas* food isolates to express extracellular proteins associated with both virulence and spoilage. Most environmental/food isolates express exoenzymes such as proteinases, which cause spoilage in foods and contribute to pathogenicity by causing direct tissue damage. *Aeromonas* spp. that do not express proteinases show reduced virulence, indicating that proteinases contribute to the pathogenic nature of *A. hydrophila*. It has also been suggested that aeromonad proteinases are also required for the activation of other virulence factors such as aerolysin.

Two types of exotoxins have been reported in *Aeromonas*: (1) cytotoxins, including aerolysins, phospholipases and haemolysins; and (2) enterotoxins. Many *Aeromonas* species also express other extracellular proteins such as amylases, nucleases and lipases, although the role of these proteins in pathogenicity remains to be determined.

Haemolytic and proteolytic activities have been readily detected in *A. hydrophila* cultures grown for 3–10 days, at 5°C in laboratory media and food slurries. The vegetative cells of this species and many of their exoenzymes and exotoxins are relatively easily destroyed by heat challenge (56°C for 10 min). Therefore, unheated or recontaminated ready-to-eat foods, may pose the greatest health risk due to infection and/or intoxication. *Aeromonas* species are also reported to express a number of heat-stable exotoxins and exoenzymes which may survive heat treatment.

Endotoxin (lipopolysaccharide)   LPS from *Aeromonas* species, like endotoxins from other Gram-negative bacteria, has a range of pathogenic effects on animals. The secretion of some exotoxins has been reported to be dependent on the presence of the O-antigen LPS, where strains lacking the O-antigens LPS (rough strains), secrete less toxin than those strains rich in O-antigen LPS (smooth strains). The possession of O-antigen LPS has been reported to be temperature dependent (smooth at 20°C, rough at 37°C), with some strains possessing LPS being more virulent when grown at low temperatures.

Presence of S-layer   In *Aeromonas* spp., as with a wide range of other bacteria, the possession of an S-layer is associated with pathogenicity. The S-layer increases the capacity of organisms to adhere to the gut mucosa, which may explain why serotype O:11 *Aeromonas* species (*A. hydrophila* and *A. veronii* biotype *sobria*), which possess an S-layer, are most frequently associated with gastroenteritis.

Fimbriae/Adhesins   *Aeromonas* species possess filamentous (fimbriae) and non-filamentous (outer membrane, S-layer synonymous) adhesins, which are intimately involved in the adhesion of the bacterium to host cells. The production of such adhesins is increased during liquid culture at low temperatures (5°C).

Capsular Layer   *Aeromonas hydrophila* and *A. veronii* biotype *sobria* have been reported to produce a capsular polysaccharide during growth in glucose-rich media. Preliminary studies indicated that possession of this capsule enhanced the virulence of these strains.

Factors Affecting Growth and Survival of Aeromonads in Foods at Low Temperature

Many factors influence the growth and survival of aeromonads in food products at refrigeration temperatures, including salt/$NaNO_2$ concentrations, pH, gaseous environment and background microflora. These factors interact such that individual limits of growth are less important (in, for example, the design of hazard analysis critical control point (HACCP) and other control systems) than their overall cumulative and/or synergistic effects on the survival and or growth of *Aeromonas* spp.

Salt Concentration/pH   The growth of *Aeromonas* spp. is optimum in 1–2% NaCl, although some strains of *Aeromonas* may tolerate concentrations of NaCl in excess of 5% under otherwise ideal conditions. *Aeromonas* spp. can grow in the pH range 4.0–10.0

(optimum 7.2) at optimal growth temperatures (28°C) although growth is significantly reduced below pH 5–5.5. The tolerance of this organism to salt and pH is greatly reduced at refrigeration temperatures.

Atmosphere   A number of studies have investigated the growth of *Aeromonas* species, especially *A. hydrophila*, under modified atmospheres at refrigeration temperatures and there has been considerable debate on the ability of *A. hydrophila* to grow at low temperatures in atmospheres with increased CO_2. It is generally agreed that the combination of increased CO_2 with refrigeration temperatures reduces the growth rate of *A. hydrophila*. Some authors report that strains of *A. hydrophila* are capable of growth in atmospheres containing greater than 20% CO_2, at temperatures as low as 0°C. However, under such conditions, growth rates are significantly lower than under aerobic conditions. Other authors report that although *Aeromonas* remained viable at 0–5°C, growth does not occur under modified atmospheres of greater than 20% CO_2. There is some evidence that elevated CO_2 concentrations affect the physiological functions of *A. hydrophila* cells as manifested in the production of greatly elongated cell structure (filamentation). Cells may also form into long chains when cultured under 100% CO_2. These structural and organizational changes are rapidly reversed if cultures are removed from high CO_2 atmospheres. The mechanism and significance of these conformational changes remain to be elucidated.

The effect of CO_2 on other species of *Aeromonas* commonly found in food products has not been addressed to date. Several authors have cautioned the use of modified atmospheres to extend the shelf life of foods, as the ability of *Aeromonas* spp. to grow at significant rates under such conditions may lead to the consumption of foods containing high levels of aeromonads. Multiple-hurdle strategies including pH, salt levels and elevated CO_2 may need to be applied to decrease the survival/growth of *Aeromonas* spp. present in foods packaged under modified atmosphere and to ensure food safety.

Background Microflora   There is conflicting evidence as to the ability of *Aeromonas* strains to effectively compete with the normal background flora present on foods, during storage under refrigeration. Most *A. hydrophila* strains can grow and actively compete with background flora in foods at refrigeration temperatures, implying that refrigeration does not in itself, provide adequate means of restricting *Aeromonas* growth. However, other aeromonad strains can grow at 5°C, but do not compete effectively with other psychrotrophs such as *Pseudomonas*

fragi. Thus, refrigeration should be sufficient to control this potentially pathogenic group.

Association of *Aeromonas* Species with Pathogenicity

Aeromonas spp. are opportunistic pathogens causing a wide range of conditions including septicaemia, meningitis and localized wound infections in susceptible hosts, e.g. the immunocompromised, infants and the elderly. Some strains are clearly enteropathogenic and linked to gastroenteritis, with symptoms ranging from mild diarrhoea to a life-threatening cholera-like disease. Enteropathogenic strains of *Aeromonas* such as *A. hydrophila* and *A. veronii* biotype *sobria* have been isolated from approximately 2% of acute gastrointestinal cases in Britain and in India. Serological evidence of the involvement of *Aeromonas* spp. in diarrhoeal illness includes the presence of antibodies to either the infecting *Aeromonas* or one of its virulence factors. Serotype O:11 (usually *A. veronii* biotype *sobria* and *A. hydrophila*) is one of the most frequently isolated serotypes associated with diarrhoeal illness attributed to *Aeromonas* in the UK. Serotype O:34 (*A. hydrophila*, *A. veronii* biotypes *veronii* and *sobria*; and *A. schubertii*) is also frequently isolated from patients with gastroenteritis. The enteropathogenic nature of *A. caviae* is not universally accepted, and most authors regard *A. cavaie*, which is usually exotoxin negative, as a non-pathogenic species. However, reports that some strains of this species can indeed express cytotoxins and enterotoxins, indicate that it should be regarded as a potential enteropathogen.

Despite strong evidence for the association of *Aeromonas* spp. with diarrhoeal illness, a carrier rate in healthy people of up to 3.2% has been reported and no significant difference can be demonstrated between the isolation rate in symptomatic and non-symptomatic patients. Human challenge studies, have found no evidence of colonization with *Aeromonas* strains or diarrhoeal episodes associated with the challenge. To date there are few published cases in which *Aeromonas* species have been directly associated with food-borne gastroenteritis (**Table 6**). This low rate of association of *Aeromonas* with gastroenteritis may be due to under-reporting of illness to, or by, clinicians. Further uncertainty is introduced by the fact that faecal samples from patients with gastroenteritis are not routinely cultured for the presence of *Aeromonas* spp. Thus the true epidemiological distribution and occurrence of gastroenteritis caused by members of this genus remains unknown.

In conclusion, although the genus *Aeromonas* and its constituent species have been widely recognized for some considerable time, their increasing importance

Table 6 Incidence of gastroenteritis categorically associated with *Aeromonas* spp.

Location	Food isolate	Number ill	Reference
Dinner party (Nigeria)	Land snails	1	Todd et al (1989)
Public house (Scotland)	Prawns	>20	Todd et al (1989)
Restaurant (England)	Oysters	3	Todd et al (1989)
Not specified (USA)	Oysters	472	Abeyta and Wekell (1986)
Not specified (USA)	Oysters	7	Abeyta and Wekell (1986)
Not specified (Italy)	Cockles	1	Bernardeschi et al (1988)

in response to changes in food processing and storage practices means that they should perhaps be redefined as emergent pathogens. Considerable research is necessary to clarify the numerous anomalies which impede the classification and speciation of this genus. Until the criteria associated with differentiation within this genus are simplified and agreed, meaningful investigation of the epidemiology and pathogenicity of the group will continue to be hampered by misclassification.

See also: **Aeromonas**: Detection by Cultural and Modern Techniques. **Bacteria**: Classification of the Bacteria – Traditional. **Biochemical and Modern Identification Techniques**: Food Poisoning Organisms. **Chilled Storage of Foods**: Use of Modified Atmosphere Packaging. **Shellfish (Molluscs and Crustacea)**: Contamination and Spoilage. **Water Quality Assessment**: Routine Techniques for Monitoring Bacterial and Viral Contaminants; Modern Microbiological Techniques.

Further Reading

Abeyta CS and Wekell MM (1986) Potential source of *Aeromonas hydrophila*. *Journal of Food Safety* 9(1): 11–23.

Abeyta CS Jr, Palumbo A and Stelma GN Jr (1994) *Aeromonas hydrophila* Group. In: Hui YH, Gorham JR, Murrell KD and Cliver DO (eds) *Foodborne Disease Handbook. Diseases Caused by Bacteria.* Vol. 1, p. 1. New York: Marcel Dekker.

Altwegg M and Geiss HK (1989) *Aeromonas* as a human pathogen. *Critical Reviews in Microbiology* 16(4): 253–286.

Bernardeschi P, Bonechi I and Cavallini G (1988) *Aeromonas hydrophila* infection after cockles ingestion. *Haematologica* 73(6): 548–549.

Gobat P and Jemmi T (1993) Distribution of mesophilic *Aeromonas* species in raw and ready to eat fish and meat products in Switzerland. *International Journal of Food Microbiology* 20: 117–120.

Holt JG, Krieg NR, Sneath PHA, Staley JT and Williams SY (1994) *Bergey's Manual of Determinative Bacteriology.* Pp. 190 and 253. Baltimore: Williams & Wilkins.

Janda JM (1991) Recent advances in the study of the taxonomy, pathogenicity, and infectious syndromes associated with the genus *Aeromonas. Clinical Microbiology Reviews* 4: 397–410.

Kirov SM (1993) The public health significance of *Aeromonas* spp. in foods: a review. *International Journal of Food Microbiology* 20: 179–198.

Kirov SM, Ardestani EK and Hayward LJ (1993) The growth and expression of virulence factors at refrigeration temperature by *Aeromonas* strains isolated from foods. *International Journal of Food Microbiology* 20: 159–168.

Knochel S and Jeppesen NC (1990) Distribution and characteristics of *Aeromonas* in food and drinking water in Denmark. *International Journal of Food Microbiology* 10: 318–321.

Merino S, Rubires X, Knochel S and Tomas JM (1995) Emerging pathogens: *Aeromonas* spp. *International Journal of Food Microbiology* 28(2): 157–168.

Nishikawa YK and Kishi T (1988) Isolation and characterisation of motile *Aeromonas* from human food and environmental specimens. *Epidemiology and Infection* 101: 213–223.

Schofield GM (1992) A review: emerging food borne pathogens and their significance in chilled foods. *Journal of Applied Bacteriology* 72: 267–273.

Sierra ME, Gonzalez-Fandos E, Garcia-Lopez ML, Fernandez MCG and Prieto M (1995) Prevalence of *Salmonella*, *Yersinia*, *Aeromonas*, *Campylobacter* and coldgrowing *Escherichia coli* on freshly dressed lamb carcasses. *Journal of Food Protection* 58(11): 1183–1185.

Todd LS, Hardy JC, Stringer MF and Bartholomew BA (1989) Toxin production by strains of *Aeromonas hydrophila* grown in laboratory media and prawn puree. *International Journal of Food Microbiology* 9(3): 146–156.

Detection by Cultural and Modern Techniques

B Austin, Department of Biological Sciences, Heriot-Watt University, Edinburgh, Scotland, UK

Aeromonads are common inhabitants of eutrophic freshwater and estuarine waters, usually with a salinity of $\leqslant 15\permil$, with maximal numbers occurring in temporate climates during the warmer summer months. Also, representatives occur in chlorinated and unchlorinated potable (and bottled) water. Initially, two major groups of aeromonads were recognized, namely the motile, mesophilic aquatic group encompassing the *Aeromonas hydrophila* complex which may be cultured at 37°C, and the non-motile psychrophilic group (growth occurs at ≤30°C), notably the fish pathogen *A. salmonicida*. Currently, 16 hybridization groups (HG) accommodating 14 phenospecies of *Aeromonas* are recognized in the family Aeromonadaceae. For over a century *Aeromonas* spp. have been implicated with animal diseases, for example red leg disease of frog and furunculosis of salmonid fish, caused by *A. hydrophila* and *A. salmonicida*, respectively. Since the 1950s, aeromonads have been increasingly associated with human disease, especially in association with intestinal and extra-intestinal infections leading to diarrhoea. In addition, there is recent sporadic involvement of *Aeromonas* with septicaemias, and bone/joint, eye, respiratory tract and wound infections in humans. Consequently, there has been an upsurge of interest in culturing and detection methods, especially for species associated with human diseases. It is not surprising that many of the methods used for the examination of food have been adapted from clinical microbiology. Yet, some of these methods are not necessarily appropriate for use with foods. Nevertheless, it is fairly straightforward to recover *Aeromonas*, although the recovery of organisms does not necessarily infer problems insofar as aeromonads are common in the natural (especially aquatic) environment.

This article describes the approaches to culturing, and serological and molecular methods for the detection of *Aeromonas* of relevance to food (including fish and shellfish and potable waters) (**Fig. 1**). The organisms of most interest to food microbiologists are loosely referred to as the *A. hydrophila* complex, and include *A. caviae*, *A. sobria* and *A. veronii* biotype

sobria, and to some extent *A. jandaei* and *A. schubertii*. It is apparent that isolates involved with disease processes may produce any of a wide range of pathogenicity/toxigenicity factors, including proteases, haemolysins and enterotoxins. Yet, the recovery of an aeromonad from food does not imply the presence of a pathogenic strain.

At the outset of a bacteriological investigation, the type of information required needs to be clarified:

- Is a global bacterial count needed? In this case the most probable numbers technique (MPN) or spread/pour plating will be appropriate.
- Is identification of the various aeromonad species requested? In this case culturing and conventional or modern identification processes (phenotypic, serologic or molecular) should be used.
- Is a mere indication of the presence of certain specified organisms required? In this case serology or molecular biology techniques would be used.

Detection by Culturing

Commonly Used Media

- *Aeromonas* (Ryans) agar: 0.2% (w/v) L-arginine hydrochloride, 0.3% (w/v) bile salts no. 3, 0.08% (w/v) ferric ammonium citrate, 0.25% (w/v) inositol, 0.15% (w/v) lactose, 0.35% (w/v) L-lysine hydrochloride, 0.5% (w/v) proteose peptone, 0.5% (w/v) sodium chloride, 1.067% (w/v) sodium thiosulphate, 0.3% (w/v) sorbose, 0.375% (w/v) xylose, 0.3% (w/v) yeast extract, 1.25% (w/v) agar, 0.004% (w/v) bromthymol blue, 0.004% (w/v) thymol blue, 5 mg l^{-1} ampicillin; pH 8.0; dissolve by boiling; autoclaving is not required. *Aeromonas* forms dark-green colonies of 0.5–1.5 mm in diameter with dark centres.
- Alkaline peptone water (APW): 1% (w/v) peptone, 1% (w/v) sodium chloride; pH 8.5–9 (typically at pH 8.5).
- Ampicillin–dextrin agar (ADA): 1% (w/v) dextrin, 0.01% (w/v) ferric chloride hexahydrate, 0.02% (w/v) magnesium sulphate heptahydrate, 0.2% (w/v) potassium chloride, 0.3% (w/v) sodium chloride, 0.5% (w/v) tryptose, 0.2% (w/v) yeast extract, 1.5% (w/v) agar, 0.004% (w/v) bromthymol blue, 10 mg l^{-1} ampicillin, 100 mg l^{-1} sodium deoxycholate; pH 8.0. *Aeromonas* spp. develop as yellow, circular, convex colonies.
- Bile salts–brilliant green agar (BBG): 1% (w/v) proteose peptone, 0.5% (w/v) Lab Lemco beef extract, 0.5% (w/v) sodium chloride, 0.85% (w/v) bile salts no. 3, 1.5% (w/v) agar, 0.000033% (w/v) brilliant green, 0.0025% (w/v) neutral red; pH 7.2; dissolve

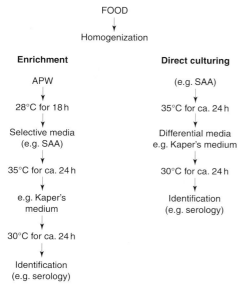

Figure 1 Stages in the recovery of *Aeromonas* from food.

by heating; autoclaving is not required. *Aeromonas* produces whitish colonies on this medium.

- Bile salts–brilliant green–starch agar (BBGS): 1% (w/v) proteose peptone, 0.5% (w/v) Lab Lemco beef extract, 0.5% (w/v) sodium chloride, 0.5% (w/v) bile salts, 1% (w/v) soluble starch, 1.5% (w/v) agar, 0.005% brilliant green; pH 7.2; dissolve by heating; autoclaving is not required. After flooding with Lugol's iodine, putative *Aeromonas* may be visualized by the presence of clearing (indicative of starch degradation) around the colonies.

- *Meso*-inositol–xylose agar (MIX): 0.01% (w/v) ammonium ferric citrate, 0.2% (w/v) potassium chloride, 0.3% (w/v) sodium chloride, 0.02% (w/v) magnesium sulphate heptahydrate, 1% (w/v) *meso*-inositol, 0.3% (w/v) yeast extract, 0.15% (w/v) bile salts no. 3, 0.5% (w/v) xylose, 1.5% (w/v) agar, 0.0005% (w/v) bromthymol blue, 20 mg l^{-1} ampicillin; pH 7.2. *Aeromonas* produces convex, circular blue–green colonies.

- Peptone–beef extract–glycogen agar (PBG): 1% (w/v) beef extract, 0.5% (w/v) glucose, 1% (w/v) peptone, 0.5% (w/v) sodium chloride, 0.004% (w/v) bromthymol blue, 1.5% (w/v) agar, and 2% (w/v) agar for overlay. Presumptive *Aeromonas* appear as yellow colonies with yellow haloes in the otherwise green medium. Ellipsoidal colonies may be seen if they are buried in the medium.

- Rimler Shotts medium (RS): 0.05% (w/v) L-lysine hydrochloride, 0.65% (w/v) L-ornithine hydrochloride, 0.35% (w/v) maltose, 0.68% (w/v) sodium thiosulphate, 0.03% (w/v) L-cysteine hydrochloride, 0.003% (w/v) bromthymol blue, 0.08% (w/v) ferric ammonium citrate, 0.1% (w/v) sodium deoxycholate, 0.0005% (w/v) novobiocin, 0.3% (w/v) yeast extract, 0.5% (w/v) sodium chloride, 1.35% (w/v) agar; pH 7.0: after boiling to dissolve the ingredients, autoclaving is not required. *Aeromonas* develop as yellow colonies after incubation of spread plates of RS at 30°C for 24 h.

- Rippey–Cabelli agar (mA): 0.1% (w/v) ferric chloride hexahydrate, 0.02% (w/v) magnesium sulphate heptahydrate, 0.2% (w/v) potassium chloride, 0.3% (w/v) sodium chloride, 0.5% (w/v) trehalose, 0.5% (w/v) tryptose, 0.2% (w/v) yeast extract, 1.5% (w/v) agar, 0.004% (w/v) bromthymol blue, 1% (v/v) ethanol, 20 mg l^{-1} ampicillin, 100 mg l^{-1} sodium deoxycholate, pH 8.0. *Aeromonas* spp. develop as yellow, circular, convex colonies.

- Starch–ampicillin agar (SAA): 0.1% (w/v) beef extract, 1% (w/v) proteose peptone no. 3, 0.5% (w/v) sodium chloride, 0.1% (w/v) starch, 1.5% (w/v) agar, 25 mg l^{-1} of phenol red, 10 mg l^{-1} of ampicillin. Putative *Aeromonas* colonies are 3–

5 mm in diameter, and are yellow to honey pigmented. After flooding the plates with full or half strength Lugol's iodine, *Aeromonas* colonies will be surrounded by a clear zone, indicating such hydrolysis.

- Tryptone–soya–ampicillin broth (TSAB): tryptone soya broth containing 30 mg l^{-1} ampicillin.

- Xylose–deoxycholate–citrate agar (XDCA): 1.25% nutrient broth no. 2, 0.5% (w/v) sodium citrate, 0.5% (w/v) sodium thiosulphate, 0.1% (w/v) ferric ammonium citrate (brown), 0.25% (w/v) sodium deoxycholate, 1.2% (w/v) agar, 1% (w/v) xylose, 0.0025% (w/v) neutral red; pH 7.0; dissolve by heating; autoclaving is not required. *Aeromonas* develop as colourless colonies.

Nonselective Approach to Culturing

Aeromonads have been routinely isolated from a wide range of habitats, and cultured, without pre-enrichment on nonselective media, especially brain–heart infusion agar (BHIA) or tryptone–soya agar (TSA) with incubation for 48 h at 37°C and 15–25°C for mesophiles and psychrophiles, respectively. With this approach, non-distinctive 2–3 mm diameter round, raised/convex shiny cream/off-white colonies develop. Such colonies could represent many taxa, and it is important to carry out further tests, as outlined below, before suggesting the presence of aeromonads. It should be emphasized that *A. media*, *A. salmonicida* and some isolates of *A. hydrophila* may form a brown water-soluble pigment on BHIA and TSA. When mixed populations are likely to occur, such as in water and food, enrichment and/or selective methods are inevitably necessary (**Table 1**). However, there is no consensus about the most suitable media to use. As a compromise, those media that are most commonly used are highlighted.

Enrichment Techniques

Enrichment techniques are used when low populations of aeromonads and the presence of injured/dormant cells may occur. Such damaged cells do not grow readily following direct transfer to solid media. Consequently an initial recovery phase in liquid medium is desirable to enhance the likelihood of their recovery. APW has been used most commonly, especially with ice cream, raw meat, poultry, seafood, vegetables and water. Other enrichment broths include RS broth medium, TSAB and 0.2% (w/v) teepol broth.

Methods

Food If available, 25 g of the food sample is added to 225 ml of APW or another enrichment broth with

Table 1   Media used for the recovery of *Aeromonas* spp.

Category of sample	Medium	Application
Diseased fish	Congo red agar	Spread plating
	Coomassie brilliant blue agar	Spread plating
Food	*Aeromonas* (Ryans) medium	Streak plate
	Alkaline peptone water	Enrichment culture, most probable numbers technique
	Ampicillin–dextrin agar	Spread plating, membrane filtration
	Bile salts–brilliant green agar	Streak plate
	Bile salts–brilliant green–starch agar	Spread plating
	Bile salts–irgasan–brilliant green agar	Spread plating, streak plate
	Blood–ampicillin agar	Streak plate
	MacConkey agar	Spread plating, streak plate
	Modified starch–ampicillin agar	Spread plating
	Peptone–beef extract–glycogen agar	Pour plating
	Rimler Shotts medium	Spread plating, membrane filtration
	Rippey–Cabelli agar	Membrane filtration
	Rippey–Cabelli–mannitol agar	Streak plate
	Rippey–Cabelli–trehalose agar	Streak plate
	Starch–ampicillin agar	Spread plating
	Starch–DNA–ampicillin agar	Spread plating
	Thiosulphate–citrate–bile salt–sucrose agar	Streak plate
	Tryptone–soya–ampicillin broth	Enrichment culture, most probable numbers technique
	Xylose–deoxycholate–citrate agar	Streak plate
Water	*Aeromonas* (Ryans) medium	Streak plating
	Alkaline peptone water	Enrichment culture, most probable numbers technique
	Bile salts–brilliant green agar	Streak plate
	Bile salts–brilliant green–starch agar	Spread plating
	Ampicillin–dextrin agar	Spread plating, membrane filtration
	Dextrin–fuchsin sulphite agar	Spread plating, membrane filtration
	Glutamate–starch–penicillin agar	Spread plating
	Glutamate–starch–phenol red agar	Spread plating
	MacConkey trehalose agar	Spread plating
	Meso-inositol–xylose agar	Membrane filtration
	Peptone–beef extract–glycogen agar	Pour plating
	Rimler Shotts medium	Spread plating, membrane filtration
	Rippey–Cabelli agar	Membrane filtration
	Starch–glutamate–ampicillin–penicillin agar	Spread plating
	Tryptone–xylose–ampicillin agar	Membrane filtration
	Xylose–ampicillin agar	Membrane filtration
	Xylose–deoxycholate–citrate agar	Streak plate

incubation at 28°C overnight. Then, 0.1 ml quantities are transferred to selective media, such as blood ampicillin agar (sheep blood agar supplemented with 30 ml l⁻¹ of ampicillin), usually with incubation at 35°C for 24 h. A separate approach has been adopted for chicken and ground meat whereby 10 g quantities of the meat are washed in 90 ml volumes of 0.1% (w/v) peptone water, and 10 ml of the washings transferred to 90 ml amounts of TSAB with incubation at 30°C for 24 h. Then, a loopful of the broth culture is streaked for single colony isolation on a suitable selective medium, such as ADA (this is a modified version of mA) or SAA, with further incubation at 30°C for 24 h.

Drinking Water   Only low populations of *Aeromonas*, i.e. < 10 colony forming units (cfu) ml⁻¹, are normally associated with drinking water. To determine the presence of viable cells, 50–100 ml of water is passed through membrane filters (pore size = ≤ 0.45 µm), and then transferred to 10–25 ml volumes of APW or 0.2% (w/v) teepol broth, with overnight incubation at 25°C. Thereafter, inocula are transferred to selective media, such as BBG, MIX or XDCA with anaerobic incubation. This allows the growth of the facultatively anaerobic (= fermentative) aeromonads, which will need to be identified further (**Table 2**).

Aquatic Animals (especially Oysters)   TSAB enrichment in combination with PBG after incubation for 24 h at 35°C, has been reported to give the highest recovery of *Aeromonas* from oysters.

Most Probable Numbers Technique (MPN)

APW and TSAB are most commonly used for estimating bacterial numbers in three or five-tube series MPN. Essentially, the methods mirror that of the

Table 2 Identification of *Aeromonas* spp. by conventional phenotypic traits

Species	Motility	Growth at 37°C	Arginine dihydrolase	Lysine decarboxylase	Gas production from glucose	Voges–Proskauer reaction	Acid from cellobiose	Acid from sucrose	Aesculin degradation
A. caviae	+	+	+	−	−	−	.	+	+
A. eucrenophila	+	+	+	−	+	−	.	v	+
A. hydrophila	+	+	+	+	+	+	.	+	+
A. jandaei	+	+	+	+	+	+	.	−	−
A. media	−	+	+	−	−	−	.	+	+
A. salmonicida	−	−	+	v	v	v	.	+	+
A. schubertii	+	+	−	+	v	−	.	+	−
A. sobria	+	+	−	+	v	−	.	+	−
A. trota	+	+	+	+	v	−	+	v	−
A. veronii									
bv.sobria	+	+	+	+	+	+	.	+	−
bv. veronii	+	+	−	+	+	+	.	+	+

+, −, v and . correspond to ⩾80, ⩽20, 21–79% and unstated positive responses, respectively.

MPN used to assess the presence of coliforms in water. Thus, known volumes of the material are inoculated into replicates of the liquid medium, and a positive result is indicated by the presence of turbidity after incubation for 24–48 h at 35°C. It is then necessary to subculture loopfuls of the turbid broths onto suitable solid media, such as BBG or SAA, followed by identification of the organisms present.

Selective Isolation Techniques

In developing selective media, use has been made of the general inability of aeromonads to ferment inositol or xylose, or the ability to attack trehalose, in contrast to the commonly occurring Enterobacteriaceae representatives. Also, ampicillin, bile salts, brilliant green, cefsulodin, deoxycholate, ethanol, irgasan, novobiocin and Pril have often been adopted as selective agents. As a note of caution, not all aeromonads are capable of growth on selective media. For example, some strains are susceptible to ampicillin, and would therefore be incapable of growth on media containing ampicillin, e.g. SAA.

Food Interest has focused predominantly on the recovery of aeromonads from meat (including fish and shellfish), milk, cheese and vegetables (including salads) because of the perceived involvement of the bacteria with food spoilage and human disease (foodborne pathogens). Techniques have centred on membrane filtration, pour or spread plating or MPN. The most commonly used selective media include RS agar for oysters, mA for poultry, SAA (and a modified derivative) and BBGS (this medium overcomes some of the problems associated with swarming by *Proteus* spp.) for a wide range of foods (of animal and plant origin). In addition, thiosulphate–citrate–bile salts–sucrose agar (a medium designed originally for the

recovery of vibrios) has been used with seafoods, and MacConkey agar and XDCA (both used for the selective isolation of Enterobacteriaceae representatives) have been used for some foods, especially seafood and ice cream.

Generally, 10–25 g quantities of the samples are homogenized in diluent (APW, 0.1% (w/v) peptone water, phosphate-buffered saline, peptone–saline, peptone–Tween 80 or TSAB), 10-fold dilutions prepared in fresh diluent, and known volumes, i.e. 0.1–1.0 ml, inoculated onto SAA, *Aeromonas* (Ryans) medium, bile salts–irgasan–brilliant green agar (BIBG) or blood–ampicillin agar and incubated aerobically at 28°C or 35°C for 24 h. This approach enables a detection limit of 100 *Aeromonas* cfu g^{-1} of food. Some workers have argued that BIBG provides the best selectivity, while supporting the growth of *Aeromonas* (but not necessarily of *A. caviae*), which appear as colonies of 1–2 mm in diameter. As starch-containing media, e.g. SAA and BBGS, will be flooded with Lugol's iodine to determine starch degradation, it is essential to quickly subculture prospective aeromonads onto fresh media for further characterization. Certainly, the use of Lugol's iodine is a disadvantage, insofar as contamination may occur during transfer of the colonies to fresh media.

Another approach has been to filter 1.0 ml volumes of appropriate dilutions of the foodstuff through hydrophobic grid membrane filters, before transfer to mA supplemented with (0.5%, w/v) trehalose agar followed by incubation at 35°C for 20 h. Yellow colonies, which may be regarded as trehalose-positive, are highlighted, and the membrane transferred to mA supplemented with (0.5%, w/v) mannitol agar for re-incubation at 37°C for 2–3 h. Colonies, which remain yellow, are again highlighted, before the filter is transferred to a filter pad saturated with the oxidase test reagent, i.e. 1% (w/v) tetra-*p*-phenylenediamine dihydrochloride solution. The development of a purple colour within 15 s is indicative of the production of oxidase. The trehalose, mannitol and oxidase-positive colonies are considered to represent the *A. hydrophila* complex.

Water PBG is a medium differential and selective for *Aeromonas*. In use, 0.1–1.0 ml of appropriate dilutions of the water are incorporated into molten cooled (to 50°C) PBG, mixed thoroughly and allowed to set for 1 h at room temperature, after which an overlay of 2% water agar is added uniformly to the surface, and allowed to set. Incubation may be at 15–37°C for 5–7 days. A positive reaction for the oxidase test differentiates aeromonads from enterics.

Dextrin–fuchsin sulphite agar (DFS) in conjunction with membrane filtration has been used to quantify aeromonads from aqueous samples. The samples are filtered through membrane filters, and the filters are inverted onto plates of DFS, which are then overlayered with molten cooled DFS, thereby trapping the filter between layers of the medium. Red colonies, which develop after incubation at 30°C for 24 h are indicative of *Aeromonas*.

RS has been used successfully for the recovery of aeromonads from water. In this medium, maltose comprises the carbon source, lysine and ornithine function for the detection of carboxylase activity, and cysteine detects hydrogen sulphide production (by enterics which may also grow on the medium). Selection is achieved by the presence of novobiocin and sodium deoxycholate. pH changes are measured by the response to bromthymol blue.

With the increasing awareness of the possible presence of damaged bacterial cells that may need special recovery methods, a modified glutamate–starch–phenol red (GSP) agar with trace glucose (to enhance recovery of the stressed cells) and 20 mg ml^{-1} of ampicillin has been advocated to reduce the concomitant growth of contaminants, such as pseudomonads. Aeromonad colonies are yellow, whereas other organisms produce pinkish colonies.

Fish – *Aeromonas salmonicida* Coomassie brilliant blue agar (CBB; tryptone–soya agar supplemented with 100 mg l^{-1} of Coomassie brilliant blue) and Congo red (CRA; tryptone–soya agar supplemented with 30 ml l^{-1} of Congo red agar) have been used with success to readily differentiate A-layer positive (virulent) isolates from fish. Thus, after incubation of swabbed material (from kidney, spleen or surface lesions – furuncles or ulcers) for 48 h at 25°C, A-layer-positive colonies appear as dark blue or deep red on CBB and CRA respectively.

Differential Media

The recovery of putative *Aeromonas* by using the above-mentioned techniques raises the question of identification. One approach, which has been commonly adopted for food and drink microbiology, has been to use differential media, such as formulated for *A. hydrophila*. For this organism, a single tube medium was developed (Kaper's medium; 0.5% (w/v) proteose peptone, 0.3% (w/v) yeast extract, 1% (w/v) tryptone, 0.5% (w/v) L-ornithine hydrochloride, 0.1% (w/v) mannitol, 1% (w/v) inositol, 0.04% (w/v) sodium thiosulphate, 0.05% (w/v) ferric ammonium citrate, 0.002% (w/v) bromcresol purple, 0.3% (w/v) agar: pH 6.7)), and is suitable for determining motility, inositol and mannitol fermentation, ornithine decarboxylase and deamination, and the production of H$_2$S and indole. Thus, *A. hydrophila* gives an alkaline reaction on the top of the medium, acid production in the butt, motility, and indole but not H$_2$S production (H$_2$S production may occur on the top).

Identification of *Aeromonas*

Typically *Aeromonas* comprise Gram-negative rod-shaped cells that are catalase and oxidase positive, degrade DNA, and are resistant to the vibriostatic agent, O/129. Speciation may be achieved by means of commercial kits, such as the API 20E rapid identification system. However, some shortcomings of rapid systems have been identified, insofar as environmental isolates, including those from food and potable water, may be mis-identified or not listed in the published profile index. Nevertheless, considerable success has resulted with other approaches, such as automated biochemical identification systems, e.g. the Abbott Quantum II system, with a spectrophotometer at 492.6 nm, and a sample cartridge containing 20 inoculated biochemical chambers. Also, species of relevance to food may be differentiated on the basis of key phenotypic traits (see Table 2).

Modern Methods

Serology

A wide range of serological methods have been used to detect and differentiate *Aeromonas*. The comparatively simple slide agglutination technique using monospecific polyclonal antisera usually produced in rabbits is still used for confirming the identity of pure and mixed cultures. However, for auto-agglutinating cultures, a mini-passive agglutination test has been recommended. This technique involves the use of sheep erythrocytes sensitized with O-antigen from the named *Aeromonas* species (extracted with hot physiological saline). Then, this reacts with diluted antiserum. A disadvantage is that false negative results may sometimes be obtained, especially with cultures that have been maintained in laboratory conditions for prolonged periods. Hence, old cultures may not be suitable for use in serology.

The first of the rapid diagnostic procedures developed for *Aeromonas* namely *A. salmonicida*, was the latex agglutination test, which is particularly useful insofar as it permits the detection of bacterial cells in animal tissues, and enables diagnosis usually within 2 min. Essentially, latex particles (usually of 0.81 μm in diameter) are coated with the antiserum (or immunoglobulins). Coated latex particles are stable, and may be stored at 4°C for many months. In use, a drop of the latex suspension is placed on black plasticized paper, and an equal volume of the unknown antigen (in suspension or solution) added, with gentle mixing for 2 min, whereupon a positive result is indicated by clumping of the latex. As with all serological procedures the use of negative and positive controls is desirable. The system has been commercialized for *A. salmonicida*. The method has the advantage in that positive diagnoses may be possible from tissues unsuitable for culturing, e.g. frozen or chemically preserved products. A variation of this test is the India ink immunostaining reaction, named after its originator as the Geck test, which allows diagnosis within 15 min. This is a microscopic technique, in which the precise mode of action is unknown, although Geck suggested that it could be regarded as an immunoadsorption method. Essentially, India ink is mixed with antiserum, which is mixed with the antigen on a microscope slide, and visualized at a magnification of ×400 and/or ×1000. A positive reaction may be clearly seen as bacterial cells outlined with the India ink. Unfortunately, negatives are difficult to visualize. A co-agglutination test, using specific antibody-sensitized staphylococci suspensions – akin to the latex agglutination reaction – has met with some success, and permitted results to be recorded within 60 min.

Indirect and direct fluorescent antibody techniques (FAT) are useful for the detection of cells in pure and mixed culture and in aqueous samples and sections of solid material. Indeed, FAT, particularly indirect-FAT, is regarded by some workers as superior to culturing for the diagnosis of individual *Aeromonas* species, although low numbers of cells are difficult to visualize and may be confused with autofluorescing debris.

The development of monoclonal antibodies has increased the specificity and sensitivity of serological tests, especially the enzyme-linked immunosorbent assay (ELISA). Both tube and dipstick ELISA systems, based on horseradish peroxidase or alkaline phosphatase conjugates have been used for detecting the presence of various *Aeromonas* spp. An advantage of the dipstick ELISA is that it can be self-contained, enabling use outside the laboratory environment. Effective systems have been developed for the detection and diagnosis of *A. salmonicida* infections in fish. In particular, high titre ($\geqslant 1 : 64\,000$) monoclonal antibodies are mixed with coating buffer (0.015 M Na_2CO_3, 0.035 M $NaHCO_3$, 0.003 M NaN_3; pH 9.6) in the ratio of 1 : 100. Then plastic or glass sticks are immersed into the antibody solution for 24 h at 37°C, after which the antibody-coated dipsticks (=probes) are air-dried, and stored at 4°C or room temperature, until required. Then, the probes are placed in suspensions/homogenates containing *Aeromonas* for 10 min at room temperature, followed by thorough washing for 2–3 min in tap water or buffer, depending on the system, before transfer for 10 min at room temperature to enzyme conjugate, i.e. a 10^{-2} dilution of purified (using protein A-Sepharose) monoclonal antibody (4 mg) which has been conjugated at room temperature with alkaline phosphatase (5000 units of bovine Type VII-T from bovine intestinal mucosa) before dialysing for 18 h at 4°C with two changes of phosphate-buffered saline (PBS), followed by the addition of glutaraldehyde to 0.2% (v/v) with further incubation at room temperature for 2 h. Thereafter, the conjugate is dialysed against PBS and then 0.05 M Tris buffer supplemented with 1 mM $MgCl_2$ at pH 8.0, before the addition of bovine serum albumin and NaN_3 to 1% (w/v) and 0.02% (w/v), respectively. After thorough washing for 2–3 min to remove unbound conjugate, the probes are transferred to enzyme substrate, i.e. 20 mg of *p*-nitrophenyl phosphate, disodium hexahydrate dissolved in 50 ml of diethanolamine buffer (96 mg l^{-1} diethanolamine, 48 mg l^{-1} $MgCl_2$; pH 9.6), for 10 min, whereupon a positive reaction is indicated by a yellow coloration. Reliable diagnoses are achieved within an hour.

Immunomagnetic separation, especially of *A. salmonicida*, has been evaluated. Thus, monoclonal antibodies developed to antigens, such as lipopolysaccharide, have been adsorbed to magnetic particles, which are introduced to aqueous suspensions containing the organism. In the presence of magnets, the beads and thus the bacteria are then recovered. Subsequently, the bacteria may be cultured or detected in ELISA-based systems. Less success is achieved with viscous liquids or solid samples, the latter of which will require an initial separation phase to prepare suspensions of the bacteria.

Dot blot immunoassays have been developed, and used successfully to detect antigens, especially of the fish pathogens.

The different serological techniques have proponents as well as opponents. Generally, direct and indirect (latex) agglutination requires 10^7 cfu ml^{-1} for positive results to be recorded, and are less sensitive than FAT or ELISA, which are capable of detecting 10^3–10^4 cfu ml^{-1}. However, there is a view that latex and co-agglutination techniques permit more positive samples to be detected than indirect-FAT.

Molecular Methods

DNA probe technology has been used with some taxa, e.g. *A. salmonicida*. Certainly, some species-specific probes have been developed. Specific DNA fragments specific to *A. salmonicida* have been incorporated into a polymerase chain reaction (PCR) technique enabling a sensitivity of detection of < 10 bacterial cells. Again, the evidence suggests that PCR detected *Aeromonas* more often than culturing.

Other Methods

Some other approaches have been tried, including multilocus, enzyme electrophoresis and pyrolysis mass spectrometry. However, these methods have not been adopted widely with *Aeromonas*.

Future Perspectives

The significance of *Aeromonas* in food and potable water has been overshadowed by other supposedly more important organisms, such as *Escherichia* and *Salmonella*. Unless there is a dramatic change in the perceived importance of *Aeromonas* in food, it seems unlikely that there will be much emphasis on the development of new methodologies. Instead, attention will remain focused on the fish and human pathogenic taxa. With these, there is likely to be increasing interest in the development and commercialization of sensitive molecular methods, certainly involving PCR.

See also: **Aeromonas**: Introduction. **Fish**: Spoilage of Fish. **Sampling Regimes & Statistical Evaluation of Microbiological Results**. **Shellfish (Molluscs and Crustacea)**: Contamination and Spoilage. **Waterborne Parasites**: Detection by Conventional and Developing Techniques.

Further Reading

Austin B and Austin DA (1989) *Methods for the Microbiological Examination of Fish and Shellfish*. Chichester: Ellis Horwood.

Austin B, Bishop I, Gray C, Watt B and Dawes J (1986) Monoclonal antibody-based enzyme-linked immunosorbent assays for the rapid diagnosis of clinical cases of enteric redmouth and furunculosis in fish farms. *Journal of Fish Diseases* 9: 469–474.

Carnahan AM and Altwegg M (1996) 1. Taxonomy. In: Austin B, Altwegg M, Gosling PJ and Joseph S (eds) *The Genus* Aeromonas. P. 1. Chichester: Wiley.

Farmer JJ III, Arduino MJ and Hickman-Brenner FW (1991) The genera *Aeromonas* and *Plesiomonas*. In: Balows A, Trüper HG, Dworkin M, Harder W and Schleifer K-H (eds) *The Prokaryotes*, 2nd edn. P. 3012. New York: Springer-Verlag.

Gehat PF and Jemmi T (1995) Comparison of seven selective media for the isolation of mesophilic *Aeromonas* species in fish and meat. *International Journal of Food Microbiology* 24: 375–384.

Holmes P and Sartory DP (1993) An evaluation of media for the membrane filtration enumeration of *Aeromonas* from drinking water. *Letters in Applied Microbiology* 17: 58–60.

Kaper J, Seidler RJ, Lockman H and Colwell RR (1979) Medium for the presumptive identification of *Aeromonas hydrophila* and Enterobacteriaceae. *Applied and Environmental Microbiology* 38: 1023–1026.

Meyer NP (1996) 2. Isolation and enumeration of aeromonads. In: Austin B, Altwegg M, Gosling PJ and Joseph S (eds) *The Genus* Aeromonas. P. 3984. Chichester: Wiley.

Palumbo SA (1996) 11. The *Aeromonas hydrophila* group in food. In: Austin B, Altwegg M, Gosling PJ and Joseph S (eds) *The Genus* Aeromonas. P. 287. Chichester: Wiley.

Palumbo S, Abeyta C and Stelma G (1992) *Aeromonas hydrophila* group. In: Vanderzant C and Splittstoesser DF (eds) *Compendium of Methods for the Microbiological Examination of Foods*, 3rd edn. P. 497. Washington, DC: American Public Health Association.

Aflatoxin *see* **Mycotoxins**: Classification.

ALCALIGENES

Thomas J Klem, Department of Food Science, Cornell University, USA

Introduction

The genus *Alcaligenes* consists of motile Gram-negative rod or coccal bacteria. *Bergey's Manual of Determinative Bacteriology*, 9th edition, lists seven species in the genus: *A. eutrophus*, *A. latus*, *A. faecalis*, *A. paradoxus*, *A. piechaudii*, *A. xylosoxidans* subsp. *xylosoxidans* and *A. xylosoxidans* subsp. *denitrificans*. Metabolism is strictly respiratory, although most strains (*A. eutrophus*, *A. faecalis*, and both subspecies of *A. xylosoxidans*) are capable of anaerobic respiration using nitrate or nitrite as terminal electron acceptors. *Alcaligenes* spp. are chemoorganotrophic organisms, able to use a wide variety of carbon sources for growth. They are common in soil and water environments, but are also found as normal inhabitants of vertebrate intestinal tracts and in clinical samples as a result of opportunistic infection.

Improvements in bacterial identification have resulted in changes to the classification of many genera, and *Alcaligenes* is no exception. Based on 16S rRNA sequencing and hybridization studies, *A. eutrophus* is now placed in a proposed new genus, *Ralstonia*, with the new name *Ralstonia eutropha*. However, in this review the familiar designation *A. eutrophus* will be used. *A. piechaudii* is now placed in the related genus *Achromobacter* as *Achromobacter piechaudii*. The status of *A. xylosoxidans* subsp. *denitrificans* as a member of the genus *Alcaligenes* or *Achromobacter* is currently under question. In addition to these changes, a proposed new member of the genus has been described, *Alcaligenes defragrans*. This new species metabolizes monoterpene substrates, and can utilize nitrate or nitrite as electron acceptor.

Species in the genus *Alcaligenes* are of most interest in the area of biotechnology. They produce a plastic-like storage material which serves as a model system for the industrial production of biodegradable polymers. As soil-dwelling microbes, they are often found in sites contaminated with organic and inorganic compounds that present threats to human welfare. Some isolates have adapted to metabolize or neutralize these health hazards, and thus show potential in the development of biodegradation processes or as biosensors.

The role of *Alcaligenes* spp. in the food and health industries is more complex. Enzymes and a polysaccharide isolated from *Alcaligenes* have been used in the commercial production of amino acids or as a food additive, respectively. The polysaccharide, curdlan, even exhibits potential as a treatment against certain immune diseases. On the other hand, these ubiquitous bacteria have the potential to contaminate food supplies or become opportunistic pathogens. Due to this latter possibility, diagnostic tests need to carefully distinguish *Alcaligenes* from its pathogenic relative *Bordetella*.

Biotechnological Applications of *Alcaligenes*

Biodegradable Plastics

A. eutrophus and *A. latus* produce a high-molecular-weight, biodegradable polymer known as polyhydroxybutyrate (PHB). Purified PHB has the physical properties of brittle plastic, and in recent decades this find has opened a new field in the study of naturally produced plastics as alternatives to petroleum-based materials. The primary focus is on reducing space requirements for solid waste disposal, but PHB could have potential use in the medical field, where its degradability may make it useful for slow release drug delivery.

In both *A. eutrophus* and *A. latus*, a single operon contains the three genes required for PHB synthesis. *phbA*, a β-ketothiolase, joins two acetyl-CoA molecules to create acetoacetyl-CoA, which is reduced to (*R*)-β-hydroxybutyryl-CoA by *phbB*, an NADPH-requiring acetoacetyl-CoA reductase. Molecules of (*R*)-β-hydroxybutyryl-CoA form the monomeric assembly units of PHB, which are polymerized through ester linkage by *phbC*, PHB synthase (**Fig. 1**).

A. eutrophus produces PHB under conditions of nitrogen, oxygen, or phosphorus limitation, whereas *A. latus* produces the polymer continuously throughout growth. The polymer is stored in the form of cytoplasmic granules and is believed to act as a carbon reserve and source of reducing equivalents. PHB synthesis in *A. latus* appears to play a basic role in cellular metabolism as a way to regenerate NAD$^+$ from NADH, either by using an NADH-dependent

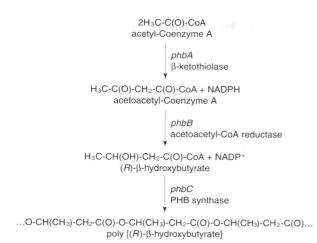

2H₃C-C(O)-CoA
acetyl-Coenzyme A

phbA
β-ketothiolase

H₃C-C(O)-CH₂-C(O)-CoA + NADPH
acetoacetyl-Coenzyme A

phbB
acetoacetyl-CoA reductase

H₃C-CH(OH)-CH₂-C(O)-CoA + NADP⁺
(*R*)-β-hydroxybutyrate

phbC
PHB synthase

...O-CH(CH₃)-CH₂-C(O)-O-CH(CH₃)-CH₂-C(O)-O-CH(CH₃)-CH₂-C(O)...
poly [(*R*)-β-hydroxybutyrate]

Figure 1 The biosynthetic pathway for polyhydroxybutyrate synthesis in *Alcaligenes eutrophus* and *A. latus*.

acetoacetyl-CoA reductase, or by transferring protons from NADH to NADP⁺.

In a nutrient-rich environment, PHB is enzymatically degraded to acetyl-CoA, which enters the primary metabolic pathways and is ultimately mineralized to carbon dioxide. Degradation is initiated by a depolymerase encoded by the *phbZ* gene. Intracellular and extracellular depolymerases exist, but each enzyme class is only capable of degrading a specific type of polymer: either non-structured, amorphous PHB granules in the cell, or ordered, crystalline extracellular PHB. Therefore, the function of the extracellular enzymes will be a critical factor in determining the success of the commercial use of biodegradable plastics. One of the model systems for this work is the extracellular depolymerase secreted by *A. faecalis*.

PHB is similar to only one type of the numerous plastics used commercially. Much effort has been devoted to the biosynthesis of polymers with different physical properties by incorporating monomeric units other than hydroxybutyrate. A glucose-utilizing mutant of *A. eutrophus* grown with glucose and propionic acid has been shown to make a copolymer consisting of hydroxybutyryl (HB) and hydroxyvaleryl (HV) monomers. The HB–HV copolymer is less brittle than PHB, and thus has wider applicability. The compound has already been marketed in Europe and the US for certain specialized products. *A. latus* is also exploited for production of HB–HV, as polymer yield per cell is greater than with *A. eutrophus*.

Pioneering work with *Alcaligenes* has created a burgeoning field devoted to the development of microbial-based biodegradable plastics. Numerous prokaryotes from many different genera are known to produce variations on the PHB polymer with respect to chain length of the monomeric unit. Ultimately,

the higher cost of producing degradable plastics by bacterial fermentation will impose limitations on the use of this approach. Research in the field is now oriented toward making PHB in plants with the goal of producing a cost-effective alternative to traditional plastics.

Bioremediation

Alcaligenes is phylogenetically located among the beta subgroup of proteobacteria, which also contains the genus *Pseudomonas*. Therefore it is not surprising that species in this genus are quite metabolically diverse like the pseudomonads. *A. eutrophus*, *A. faecalis*, *A. xylosoxidans* (both subspecies), and *A. paradoxus* strains are known to utilize an array of aromatic compounds, as well as many heavy metals, for carbon and energy. Many of these compounds are xenobiotic, and represent lingering environmental health hazards. Research on xenobiotic-metabolizing isolates is directed towards exploiting their biodegradative potential to redeem sites that would otherwise remain contaminated, or be prohibitively expensive to reclaim by other technology.

Alcaligenes spp. contain chromosomally encoded genes for the catabolism of phenol and catechol via the meta-cleavage pathway. Subsequently, some species acquired additional genes that extended their metabolic capabilities to include polychlorinated biphenyls (PCBs), and other halogenated aromatic compounds. For example, strains of *A. eutrophus* have been isolated that contain a transposon for the degradation of PCBs, and others with a plasmid for the degradation of the herbicide 2,4-dichlorophenoxyacetic acid. Degradation of these molecules typically proceeds by a series of dehalogenations and aromatic ring oxidations which converts the molecule into catechol, followed by entry into the meta-cleavage pathway and further metabolism to acetate and pyruvate.

Heavy metals are another class of common contaminants at toxic waste sites, often found in conjunction with aromatic compounds. *A. eutrophus* CH34 was isolated from metal-contaminated soil and found to contain two megaplasmids that encode for resistance to Co⁺⁺, Ni⁺⁺, CrO₄⁻⁻, Hg⁺⁺ and Tl⁺ (plasmid pMOL28), and resistance to Cd⁺⁺, Co⁺⁺, Cu⁺⁺, Zn⁺⁺, Hg⁺⁺, and Tl⁺ (plasmid pMOL30).

Resistance to heavy metal toxicity proceeds via a metal ion/proton antiporter efflux system. The best studied operons are *czcCBA* (Co⁺⁺, Zn⁺⁺, Cd⁺⁺) in pMOL30, and *cnrCBA* (Co⁺⁺, Ni⁺⁺) in pMOL28. These operons contain structural genes for the transmembrane cation/proton antiporters which are essential for the efflux of metal ions from the cytoplasm. Metal ion expulsion and proton influx results in a localized increase in pH near the cell. Respired carbon dioxide is converted into bicarbonate by the alkaline conditions, which forms a precipitate with the metal, preventing reentry to the cell.

Bioreactors featuring *Alcaligenes* have shown promise in field studies for the removal of metals from solution by bioprecipitation. Lab-scale experiments have been performed using engineered strains of *Alcaligenes* able to catabolize xenobiotic compounds, as well as remove metals. In this case, the bacteria are able to consume the contaminating xenobiotic as sole carbon source, thus reducing the cost of providing an outside substrate. Work has also been done to develop *Alcaligenes* as biosensors for heavy metals, by fusing reporter genes to those normally expressed in the presence of metals.

Relevance to the Food Industry

Enzymatic Production of Amino Acids

L-Glutamate is used to enhance the flavour of foods, particularly as the monosodium salt found in fermented sauces, such as soy. It is produced commercially by fermentation of several species of bacteria. During the fermentation process, a significant amount of flavourless by-product, pyroglutamate (5-oxoproline), is produced. 5-Oxoprolinase is an enzyme that converts 5-oxoproline back into L-glutamate in an ATP-dependent reaction. The enzyme is ubiquitous in nature, but recently a non-ATP utilizing derivative was found in a strain of *A. faecalis*. The enzyme may find practical use in the food industry for increasing yields of L-glutamate.

L-Lysine is an essential amino acid and a common dietary supplement. In one production method, a combination of chemical and enzymatic reactions are used to synthesize the amino acid. A racemic mixture of the cyclic lysine derivative D,L α-amino-β-caprolactam is chemically synthesized. The racemate is treated with a hydrolase from *Candida lumicole*, which produces L-lysine from the L-isomer. The remaining D-isomer is racemized by α-amino-ε-caprolactam racemase from *A. faecalis*, and the process is repeated until all the material is converted into L-lysine.

Curdlan

In stationary phase growth, *A. faecalis* secretes an exopolysaccharide composed of linear, unbranched D-glucose molecules in a β-1,3 glycosidic linkage. This form of polysaccharide is synthesized by several bacterial species and is known by the common name curdlan. Synthesis of curdlan is believed to occur through the polymerization of UDP-glucose units, and two loci involved in curdlan synthesis have been

cloned from *A. faecalis*. Curdlan has potential as a food additive, and may even have medical applications.

The property of curdlan that has the most promise in regard to the food industry is the ability of the polysaccharide to form a stable gel. In aqueous solution curdlan is insoluble, but becomes soluble upon heating. Increasing the pH or additional heating of a curdlan solution causes a change of phase to a solid gel. This change is the result of the previously disordered glucan chains assuming an ordered triple helical structure. The gel exhibits stability across a wide pH range, and retains its physical properties on freezing and thawing. Curdlan is currently used in Japan as a food stabilizer and thickener.

Curdlan also exhibits properties that may prove useful in a clinical setting. A sulphated form of the polysaccharide has been shown to prevent human immunodeficiency virus (HIV) from binding to the CD4 receptor of T cells in vitro. Clinical trials are underway to demonstrate in vivo effectiveness. Certain viral and bacterial infections are known to cause a rise in levels of the hormone tumour necrosis factor (TNF). TNF is normally involved in the host immune response, but in some disease states a rise in the hormone is observed, and this causes problems such as inflammation and endotoxic shock. Curdlan sulphate has been found to prevent exaggerated levels of TNF expression, retaining the beneficial action of the hormone without the side effects.

Food Microbiology

In the field of food science, *Alcaligenes* is recognized as a potential contaminant of dairy products, meats and seafood. This is a particular concern in the dairy industry, because food items (milk especially) can be kept under refrigeration for extended lengths of time. *Alcaligenes* spp., among others, originate in milk samples during processing. Once introduced, they can grow on prolonged storage under refrigeration, even though optimum growth occurs at 20–30°C. These psychrotrophic organisms can give off-flavours to milk and reduce its keeping quality, if the storage time before sterilization was long enough to allow for bacterial growth.

Methods for the enumeration of psychrotrophic organisms have been developed, but there is no accepted standard methodology due to the diverse nature of dairy products. The primary emphasis is on keeping bacterial levels low, and not on identification.

Detection Methods

The genera *Alcaligenes* and *Bordetella* have been shown to be closely related on the basis of 16S rRNA sequence, fatty acid composition, and biochemical properties. *B. pertussis* and *B. parapertussis*, the causative agents of whooping cough, are isolated from human samples, whereas *B. bronchiseptica* and *B. avium* are animal and bird pathogens, respectively. Recently, two new species, *B. hinzii* and *B. holmesii*, have been isolated from human blood, and are the only members of the genus not associated with respiratory infections. *A. faecalis* and both subspecies of *A. xylosoxidans* can be found as saprophytic inhabitants of human and animal intestinal tracts and are sources of nosocomial infection. Although not usually pathogenic, they may be opportunistic invaders in a compromised host. Therefore, it is critical to be able to distinguish confidently between the two genera when dealing with veterinary or clinical isolates.

Several types of solid media, primarily Bordet-Gengou (BG) agar, have been developed for the isolation of bordetellae. *Alcaligenes* spp. will grow on some of these, but may be distinguished by their colony morphology. Rapid and sensitive polymerase chain reaction (PCR) tests have been developed to distinguish between *B. pertussis*, *B. parapertussis*, and *B. bronchiseptica* once initial characterization based on biochemical tests has been performed. Some of the properties distinguishing *Bordetella* spp. from *Alcaligenes* spp. are listed in **Table 1**.

Multidrug resistance is common in many bacterial genera, and these organisms can rapidly spread throughout the world, primarily due to the ease of travel. Therefore, it is urgent that diagnostic tools be developed for epidemiological typing. Advances in PCR, gel electrophoresis, and automated ribotyping

of organisms have resulted in the ability to classify species and variants within species to a degree not previously possible. *A. xylosoxidans* subsp. *xylosoxidans* has become more prominent as an opportunistic pathogen, particularly in the hospital environment. Many isolates are resistant to the common classes of β-lactam, aminoglycoside and quinolone antibiotics. A combination of antibiotic selectivity and pulsed-field gel electrophoresis (PFGE) of digested chromosomal DNA has been developed as an effective method for distinguishing variants of this strain.

Conclusion

Bacteria in the genus *Alcaligenes* are most commonly found in the environment. As environmental conditions change, the bacteria must be able to respond rapidly to survive. This flexibility has endowed species of the genus with capabilities that may prove useful in numerous industrial applications. In nutrient poor environments, cells make a storage compound, PHB, that has served as a model system for the development of biodegradable polymers. Other strains have adapted to neutralize or utilize potentially toxic chemicals as sources of carbon and energy, which may be exploited for the development of bioremediation processes. An exopolysaccharide, curdlan, secreted during stationary phase perhaps to act as an antidesiccant, may have several uses in the food and health care industries.

Beneficial applications aside, there are some health risks associated with *Alcaligenes* spp. They can become opportunistic pathogens under certain circumstances, as they are already present in the body as inhabitants of the intestinal tract. *Alcaligenes* are genetically related to species in the genus *Bordetella*,

Table 1 Differential characteristics of *Bordetella* and *Alcaligenes* spp.[a]

	Growth on MacConkey agar	Motility	Citrate utilization	Nitrate reduction	Oxidase	Urease
B. pertussis	−	−	−	−	+	−
B. parapertussis	+	−	v[b]	−	−	+
B. avium	+	+	v	−	+	−
B. bronchiseptica	+	+	+	+	+	+
B. hinzii	+	+	+	−	+	v
B. holmesii	+	−	−	−	−	−
A. faecalis	+	+	+	−	+	−
A. xylosoxidans subpp. *xylosoxidans*	+	+	+	+	+	−
A. xylosoxidans subpp. *denitrificans*	+	+	+	+	+	v

[a]Information from Collier L, Balows A and Sussman M (1998) *Topley and Wilson's Microbiology and Microbial Infections*, 9th edn. London: Oxford University Press; Lennette EH, Balows A, Hausler Jr WJ and Shadomy HJ (1985) *Manual of Clinical Microbiology*, 4th edn. Washington DC: American Society for Microbiology Press; Roop RM (1990) *Bordetella* and *Alcaligenes*. In: Carter GR and Cole JR (eds) *Diagnostic Procedures in Veterinary Bacteriology and Mycology*, 5th edn. New York: Academic Press.
[b]v, Some strains positive, some negative.

therefore it is important to clearly distinguish between the two. Bordetellae can be reliably identified through a combination of classical biochemical tests and newer molecular biology techniques. Molecular biology methods are also used for the epidemiological typing of multidrug resistant strains of *A. xylosoxidans* subsp. *xylosoxidans*.

See also: **Biofilms**. **Fermentation (Industrial)**: Production of Xanthan Gum; Colours/Flavours Derived by Fermentation. **Fermented Foods**: Fermentations of the Far East.

Further Reading

Brauneg G, Lefebvre G and Genser KF (1998) Polyhydroxyalkanoates, biopolyesters from renewable resources: physiological and engineering aspects. *Journal of Biotechnology* 65: 127–161.

Knippschild M and Ansorg R (1998) Epidemiological typing of *Alcaligenes xylosoxidans* subsp. *xylosoxidans* by antibacterial susceptibility testing, fatty acid analysis, PAGE of whole-cell protein and pulsed-field gel electrophoresis. *Zentralblatt für Bakteriologie* 288: 145–157.

Sutherland IW (1990) *Biotechnology of Microbial Exopolysaccharides*. Cambridge: Cambridge University Press.

Taghavi S, Mergeay M, Nies D and van der Lelie D (1997) *Alcaligenes eutrophus* as a model system for bacterial interactions with heavy metals in the environment. *Research in Microbiology* 148: 536–551.

Takeda-Hirokawa N, Neoh LP, Akimoto H et al (1997) Role of curdlan sulfate in the binding of HIV-1 gp120 to CD4 molecules and the production of gp120-mediated TNF-alpha. *Microbiology and Immunology* 41: 741–745.

Ale *see* **Lager**.

Algae *see* **Single-cell Protein**: The Algae.

ALTERNARIA

S E Lopez and **D Cabral**, Departamento de Ciencias Biológicas, Facultad de Ciencias Exactas y Naturales, Ciudad Universitaria, Buenos Aires, Argentina

Introduction

Alternaria is one of the most widespread genera in the world. Almost any kind of substrate can support these saprophytic species: from flour to leather, from bottled water to textiles. It can grow on food, clothes, materials, goods and paper. As a facultative pathogen it can be found in a variety of cultures, crops and manufactured products.

Many species have been described based on morphological and/or host differences. Isolation and characterization of field and laboratory cultures show their variation and the influence of environmental factors. As a consequence, the taxonomy of *Alternaria* is confusing, with synonymy in different systems.

Different *Alternaria* species produce toxic compounds as secondary metabolites when grown on different substrates. These toxins cause disease in living plants, and may also induce health disorders in animals and humans. Isolation and characterization of these compounds, together with toxicity in vertebrates, have been the main focus of recent research. Most information relates to *Alternaria alternata* (**Fig. 1**).

Characteristics of the Genus and Relevant Species

Classification

Alternaria are Fungi Imperfecti belonging to the order Hyphomycetes, the family Dematiaceae, genus *Alternaria* Nees. The type species is *A. alternata* (syn. *A. tenuis*).

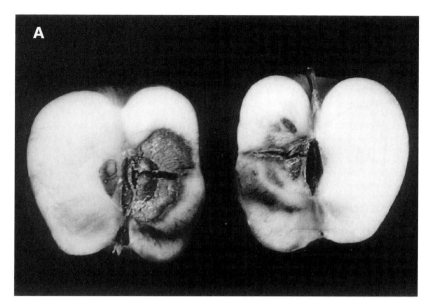

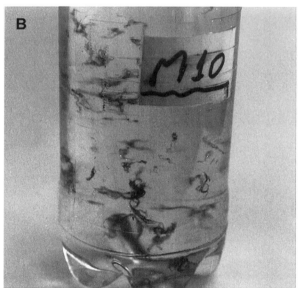

Figure 1 *Alternaria alternata* infection (**A**) Mouldy heart in Red Delicious apple produced by *A. alternata*. (**B**) bottled water with *A. alternata* contamination.

Description of Genus

The genus *Alternaria* is characterized by dark colonies, colour grey to blackish-brown or black. Conidia are typically dictyosporic dry, sometimes forming simple or branching chains, smooth or verrucose walls, arising on conidiophores which usually become geniculate and showing scars after they are detached. Conidia are formed by blastic ontogeny as outgrowths of protoplasm through a defined apical pore in the conidiogenous cell, ovoid to subclavate, narrowing to the distal portion. Some species form a defined beak. This type of spores (porospores) are also found in two other genera, *Stemphylium* and *Ulocladium*, with which *Alternaria* may be confused. The former is distinguished by its percurrent proliferation, and the latter by the narrow base of conidia which differs from the broad base in *Alternaria*.

Approximately 73 species have been described, growing saprophytically on all kind of substrates and pathogenically on vegetable host and stored grains, seeds and fruits.

Taxonomy in the genus has been developed by many mycologists. One of the characters used to segregate species is the presence or absence of chains and the number of conidia in them: they can be Longicatenatae (ten spores or more); Brevicatenatae (three to five) or Noncatenatae (single spores). Another character used is the formation of an apical beak in the conidia and the nature of the transition from body to beak and septation. A pseudorostrum can be present when conidia proliferate as secondary conidiophores to originate new conidial chains. There are many other variations in morphology, depending on the strain, the cultural conditions and the host range.

Type Species

The type species is:

Alternaria alternata (Fr.) Keissler, 1912, *Beih. Bot. Zbl.*, 29: 434. *Torula alternata* Fr., 1832, *Syst. mycol.* 3:500. *A. tenuis* C. G. Nees, 1816/17, *Syst. Pilze Schwamme*: 72.

The epithet *alternata* should be used instead of *tenuis*, because the last name is invalid.

Colonies filamentous, grey, dark brown or black, growing fast in potato dextrose agar (PDA) or malt extract agar (MEA). Conidiophores single or in small groups, straight or curved, sometime geniculate, 3–6 nm × 20–50 nm, with scars.

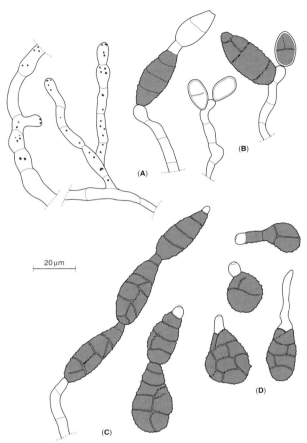

Figure 2 *Alternaria alternata*. (**A**) Mycelium. (**B**) Conidiogenous cells with young conidia. (**C**) Chains of matured conidia. (**D**) Conidial germination.

Conidia ellipsoidal, ovoid, obclavate, obpiriform, 9–18 nm × 20–63 nm, often with a short conical or cylindrical beak 2–5 nm in diameter tapering to the apex or blunt, about one third or one quarter of the conidial length. Body with 1–8 transverse septa and 1–2 longitudinal septa, sometimes oblique, in each cell, golden-brown to brown, beak paler (**Fig. 2**).

Cosmopolitan, plurivorous on soil, fibres, food, around 90 species of vegetal hosts among trees, shrubs and crops.

Species Identification

In the search for relevant characters to reach a confident species identification, it is useful to consider the following:

- colony and vegetative hyphae: not variable enough among species or strains to be considered in keys
- conidium: morphology and spatial patterns of sporulation
- conidiophores: only useful in extreme variations
- host or substrate: mixed species in nature are usual. This association can be helpful, but has to be

used with discretion and together with other considerations.

Other Important Species

A selection of species is listed in **Table 1**. Important species include:

- *A. brassicae* (Berk.) Sacc.: on *Abelmoschus, Arnoracia, Brassica, Lepidium, Lunaria, Raphanus, Ricinus, Syringa, Triticum, Zea*
- *A. brassicicola* (Schwein.) Wiltshire: on *Armoracia, Brassica, Crambe, Lunaria, Raphanus, Rosa*
- *A. citri*: on *Citrofortunella, Fortunella, Murraya, Phoenix, Prunus, Psidium*
- *A. tenuissima*: on *Amaranthus, Cajanus, Cichorium, Fragaria, Glycine, Lycopersicon, Nicotiana, Passiflora, Phaseolus, Pittosporum, Prunus, Santolina, Tragopogon, Vaccinium, Viola*.

Patterns of Sporulation

Simmons proposed a three-dimensional point of view for species characterization from examination of sporulating mycelia under transmitted light at × 50 magnification. He defined six patterns, based on chain length, branching, conidial size, elevation over substrate, secondary conidiophores and overlapping characters (**Table 2**).

Group 1 Chains thin, formed by 5–10 conidia, without branching (or exceptionally branched).

Group 2 Chains formed by about 10 conidia, unbranched, conidia broader than group 1.

Group 3 Tree-like branching, with primary conidiophore long, dark and erect on substrate, originating short chains toward its apex.

Group 4 Chains well branched on short primary conidiophores, looking like a shrub.

Group 5 Chains of 10–20 conidia, without or minor branching, conidia thin as in group 1.

Group 6 Short chains of broad conidia, arising in secondary conidiophores, conidia grouped in loose tufts.

Table 1 Selected *Alternaria* species from plants used as food or feed

Species	Host/substrate	Conidia[a]	Chains[b]	Beak	Common diseases
A. alternata[c]	Many plants (ca. 115 spp., over 43 families, and substrates	Small to medium	Long	Present or absent	Many kinds of diseases
A. brassicae	Brassica, Raphanus, Lepidium, Lunaria, Ricinus, Triticum, Zea	Large	Absent	Present	Blight of Cruciferae
A. brassicicola	Armoracia, Brassica, Crambe, Lunaria, Raphanus	Large	Long	Absent	Leaf spot of cabbage and cauliflower
A. cheiranthi	Cheiranthus, Spinacia	Large	Short	Absent	Leaf spot
A. citri[d]	Citrofortunella, Citrus, Fortunella, Phoenix, Prunus, Psidium	Small	Present	Absent	Brown spot of citrus fruit
A. cucumerina	Citrullus, Cucumis, Cucurbita, Cyamopsis	Large	Absent	Present	Leaf blight of cucurbits
A. dauci	Daucus, Petroselinium	Large	Absent	Present	Blight of carrots
A. gaisen	Pyrus	Large	Absent	Present	Black spot of pear
A. helianthi	Helianthus	Large	Absent	Absent	Leaf spot, head blight
A. helianthinficiens	Helianthus	Large	Absent	Present	Seed spot
A. infectoria	Cereals	Small	Present	Absent	Leaf spot of wheat
A. longipes	Nicotiana	Large	Present	Present	Tobacco brown spot
A. longissima	Sorghum	Large	Present or absent	Absent	Spot of grains
A. peponicola	Cucurbitaceae	Small	Present	Absent	Decay of cucurbits
A. petroselini	Petroselinium	Large	Absent	Absent	Leaf spot
A. porri	Allium	Large	Absent	Present	Purple blotch
A. radicina	Cucurbita, Daucus, Petroselinium	Small	Absent	Absent	Black rot of carrot
A. raphani	Brassica, Matthiola, Raphanus	Large	Present	Absent	Black pod blotch
A. solani	Amaranthus, Capsicum, Datura, Lycopersicon, Petunia, Solanum	Large	Absent	Present	Early blight of potato and tomato
A. sonchi	Cichorium, Emilia, Lactuca, Sonchus	Median to large	Absent	Short	Round spot of leaves
A. tenuissima	Amaranthus, Fragaria, Glycine, Lycopersicon, Nicoriana, Viola, Phaseolus, Prunus, Vaccinium	Small	Present	Absent	Spots on leaves and fruits
A. triticina	Triticum	Median	Present	Absent	Yellow leaf spot
A. zinniae	Eupathorium, Helianthus, Zinnia	Medium	Absent	Present	Leaf spot

[a]Conidial size: small 30–40 μm; median 50–100 μm; large > 100 μm.
[b]Short chains, 5–10 conidia; long chains; > 10 conidia.
[c]Syn. *A. tenuis*.
[d]Syn. *A. alternata* f. sp. *citri*.

Table 2 *Alternaria* grouping by their macroscopic appearance at × 50 magnification. From Simmons and Roberts (1993)

Group	Chains	Conidial number	Size	Branching	Conidiophore primary	Conidiophore secondary
1	Short to moderate	5–10	Thin	None	Short	None
2	Moderate	Around 10	Broad	None	Short	None
3	Short to moderate	Around 5–10	Thin or broad	Arborescent	Long-erect	Short
4	Short to moderate	Around 5–10	Broad?	Bush-like	Short	Short
5	Moderate to long	10–20	Thin	None or minor	Short	Short
6	Short to moderate	5–10	Broad	Sparse	Short	Coarse, long
Other	Long	20	Thin or broad	Abundant	Diverse	Diverse

Strains cultured on potato-carrot agar for 7–10 days and observed by transmitted light at × 50.

Table 3 Optimal temperatures and water activity for germination, growth and sporulation of selected *Alternaria* species

Species	Germination		Growth		Sporulation	
	Optimum temperature (°C)	Water activity	Optimum temperature (°C)	Water activity	Optimum temperature (°C)	Water activity
A. alternata	15–30	0.84	25–30	0.85	25–27	0.90
A. brassica	21–28	0.84–0.90	22–24		22–23	
A. brassicicola	28–35		20–27		22	
A. citri			25	< 0.91		
A. cucumerina	25–28		27–30		27	
A. dauci	22–24				25–30	0.96
A. helianthi	25–28		25		15–20	
A. longipes	22–25		25–29			
A. porri	21–30		23–30		25	0.75–0.80
A. ricini	25–30		28		20	
A. solani	20–28	0.92	19–28		20–26	
A. triticina	15–27				25	

Physiological Aspects

Environmental Factors Affecting Growth and Dispersal

Alternaria may persist on fresh foods after harvest and may cause deterioration of fruit and vegetables. In stored cereal grains, it persists when low humidity levels prevent the growth of *Penicillium* and *Aspergillus* species. If the humidity level increases above a water activity level of 0.15, *Alternaria* generally dies. The concern about *Alternaria* development on foodstuffs is related not only to damage to the products but also to the production of mycotoxins and their probable toxicity for humans and animals.

Factors determining growth, sporulation and metabolite production are variable depending on substrates and strain investigated (**Table 3**).

Enzymes

Other metabolites related to pathogenic activity, the enzymes, have been studied, although in only a few pathosystems and with incomplete results. Their actions are particularly during the important penetration and colonization of the host by the saprotrophic fungus.

Some tests have confirmed enzymatic processes, using in vitro assays. Although in vitro results do not always repeat in vivo behaviour, the presence of the following systems has been demonstrated: cellulase in filtrates of *A. solani* and *A. alternata* against tomato; pectin methylgalacturonase by *A. solani* and pectin methyl-lyase by *A. alternata* in tomato; pectin methylesterase by *A. porri*; and cutinases during germination of *A. alternata* and during wall degradation by *A. brassicicola*. *Alternaria citri* produced pectin methylesterase and polygalacturonase during tissue maceration of citrus fruits.

Ecological Aspects

Alternaria alternata is a common and cosmopolitan component of the phyllosphere, rhizosphere and carposphere of most plants, with a gradation in virulence from phylloplane or rhizoplane saprophyte, endophyte or latent pathogen to strongly necrotrophic pathogen. Often this species coexists with other *Alternaria* spp. such as *A. tenuissima* or *A. infectoria*.

As Saprophyte

Alternaria alternata have been frequently reported colonizing the surfaces of leaves, twigs, fruit, grains and seeds in the early stages of these structures' development, following colonization by bacteria and yeasts. Although it is an epiphyte, eventually asymptomatic subepidermal host tissue penetration or substomatal chamber colonization occurs. Confined to the substomatal area, where it develops as a single 5–7-celled fusiform hypha, *Alternaria* may benefit from nutrient leakage from the host, or derive protection from desiccation and mycophagous invertebrates.

The mycelium of *Alternaria* is also common under the outer pericarp of nearly all kernels of wheat and barley. Despite various degrees of discoloration, the barley seeds may retain high germinability and good seedling vigour, and thus colonization in itself may not be harmful.

These colonization patterns are typical of opportunistic saprophytes. The frequent recovery of *Alternaria* from surface-sterilized tissue and tolerance of fungicide applications is probably due to this particular strategy. In this habitat *Alternaria* competes spatially with other fungi, as *Cochliobolus* and *Acremonium* and antagonistically with *Cladosporium*, *Epicoccum* and *Fusarium*, e.g. its incidence is decreased when *F. culmorum* is abundant. It also interacts with insects and mites, providing substrate for its growth and reproduction.

As Pathogen

The genus has been recorded in many kinds of plants all over the world. A short life cycle, easily detachable spores dispersed by wind, dark, resistant mycelium and conidia are some of the properties of pathogenic species. They can affect leaves, stems, flowers, fruits, tubercles and roots of plants in the field or in storage. The symptomatologies produced are as diverse as the plants and organs affected. *Alternaria* species produce spots or blights in leaves; spots and cankers in stems and twigs; damping off in seedlings; and fruit, tuber, root and seed rots.

In most cases, sporulation occurs 7–10 days after host colonization and conidia can be produced for several weeks, resulting in a high infection rate. The range of temperatures of these diseases is wide; for example, *A. solani* can grow at temperatures between 1°C and 45°C, and germinate at 4–35°C. The presence of melanin in spores and mycelial walls protects these structures against radiation effects and adverse environmental conditions, determining resistance. All these characteristics are advantageous for disease establishment and dispersal, attributes of an effective pathogen.

After Harvesting

Owing to the presence of *Alternaria* species in all kinds of substrates and their ubiquity in a wide range of conditions, they are common environmental contaminants. Spores are therefore likely to be present in harvested fruits and grains, and in storage may germinate in propitious temperature and humidity conditions.

The spores deposited on tomatoes, apples, olives, pepper, citrus and potatoes surfaces in the field are able to germinate in the cold storage conditions usually applied to this kind of fleshy fruits and tubers. Wounds on the pericarp are the common route of infection and colonization. Cereals, grains and nuts are generally stored in dry conditions which limit development of moulds such as *Penicillium* and *Aspergillus*, decreasing competition for *Alternaria* species.

Besides product spoilage, which diminishes the value of the commodities, mycotoxin production can occur.

Alternaria Metabolites

Alternaria species are known to produce more than 70 secondary metabolites with antibiotic, mycotoxic and phytotoxic properties, belonging to several classes of chemicals. These compounds were mainly isolated from *A. alternata* and related species (*A. alternata* complex or pathotypes). Most are phytotoxins, some of which play an important role in the pathogenesis of plants. A few, known as mycotoxins, have a potential toxicological risk as food and feed contaminants.

Phytotoxins

Based on their host range of action, two groups of phytotoxins may be distinguished: host-specific toxins (HSTs) and general phytotoxins.

Host-specific Toxins The concept of pathogenesis is based on HSTs. Studies of toxins and their effects on plants are one of the keys to understanding host–pathogen interactions. One pathogen may synthesize a compound that damages one plant but has little or no effect on others. Isolates that produce the toxin are pathogenic to susceptible host plants, but those that fail to produce it are not. The HSTs are released from germinating spores at the infection site, and lead to successful colonization of the pathogen into host tissue. The virulence and host specificity of a given pathogen is based on production of a distinctive HST. The production of a specific HST appears to be essential for most *Alternaria* pathogens.

Since 1933 when the host-specific toxicity of culture filtrates of *A. alternata* isolates (*A. gaisen* syn. *A. kikuchiana*) pathogenic to Japanese pear cultivar 'Nijisseiki' was found, several pathogens that produce HSTs have been reported. Most are from the genera *Alternaria* and *Helminthosporium*, which are often called 'saprophytic pathogens'. The *Alternaria* HSTs known to date are listed in **Table 4**.

General Phytotoxins General phytotoxins are non-specific toxins (NSTs): zinniol, alternaric acid, alternariol, alternariol methyl ether, tenuazonic acid, radicinil, radicinol, tentoxin, altenuene, isoaltenuene, alternaric acid, curvularin, altertoxin, alteichin, alterlosin, altersolanol and bostricin are some of them.

Mycotoxins

Some of the *Alternaria* NSTs, as well as being phytotoxic, may exhibit antibacterial, antifungal, antiviral, insecticidal, cytotoxic and fetotoxic properties, in addition to animal toxicity. However, relatively little is known about *Alternaria* metabolites and toxicology, because past research has focused on mycotoxins produced by *Aspergillus*, *Penicillium* and *Fusarium*.

The first report of *Alternaria*-produced toxins in food and feed was in 1960. *Alternaria* strains isolated from mouldy grains were incriminated in human toxicosis in the Soviet Union in the 1940s, confirmed by their toxicity in experimental animals which were fed with grains infected with those strains.

The genus has been implicated as the aetiological

Table 4 Host-specific toxins of selected *Alternaria* species

HST	Pathogen	Disease	Host	Effect
AM toxin I, II, III	*A. alternata* f. sp *mali*	*Alternaria* blotch of apple	Apple	Chloroplast and plasma membrane changes
ACR toxin, ACRL toxin	*A. citri*	Brown spot of citrus	Lemon	
ACT toxin, ACTG toxins A to F	*A. citri*	Brown spot of citrus	Tangerine, mandarin, Japanese pear	Plasma membrane alteration
AK toxin I, II	*A. gaisen*	Black spot of Japanese pear	Japanese pear	K loss, membrane depolarization
AF toxin I, II, III	*A. alternata* f. sp strawberry pathotype	Black spot of strawberry	strawberry (II) Japanese pear (I, III)	vein necrosis, K loss
AL toxin (AAL toxin)	*A. alternata* f. sp. *lycopersici*	*Alternaria* steam canker	Tomato	Necrosis
AT toxin	*A. longipes*	Brown spot of tobacco	Tobacco	

agent in haemorrhagic syndrome in poultry. *Alternaria longipes*, a common pathogen of tobacco, could be a factor in lung disease.

In Namibia, the presence of *A. alternata* and *Phoma sorghina* in millet and sorghum has been associated with onyalai, a human haematological disease. Although animals fed with both species or treated with tenuazoic acid did reproduce some symptoms, the evidence was not conclusive.

Several mycotoxins have been isolated and identified from various *Alternaria* species, although few data are available about the toxicity of purified metabolites (**Table 5**).

Methods of Detection in Food

The presence of *Alternaria* mycelium, and sometimes sporophores and spores, can be detected in plant material, food, feed and beverages by direct examination with the unaided eye or under stereoscopic or light microscopy, or by indirect techniques. Analyses may be done early in the field, to detect the living plant as an epiphyte or colonizing the tissues as weak or latent pathogen, using phylloplane and endophytic techniques.

Phylloplane Detection

Alternaria species, mainly *A. alternata*, are widely distributed components of the phylloplane surfaces. Identification and enumeration of filamentous fungi in this habitat are difficult, and require direct and culture techniques.

Direct Methods Collodion films or clear adhesive tape pressed over the surface of leaves produce replicas which can be observed under different magnifications. Another procedure is the scanning electron microscopy of pieces of dehydrated leaves. The inconvenience of this methodology is the impossibility of distinguishing between species if they are not sporulate or if the mycelium is not readily identifiable.

Culture Techniques Washing of plant structures to recover propagules and culturing of the resulting water, and culture of pieces of plants after superficial sterilization, can produce colonies of fungi. Neither technique discriminates between active or inactive organisms or between species, but combined with direct methods they allow enumeration and further knowledge of populations in the phylloplane.

Endophytic Detection

Opportunistic invasion of host parenchyma or growth within substomatal chambers is observed by tissue clearing under light microscopy. This can be performed on leaves and stems, either on the whole structure or on individual tissues. Samples are boiled for 5–10 min in a lactophenol-ethanol mixture (1:2 v/v) and stored overnight in the mixture at room temperature, then stained with 0.05% trypan blue in lactophenol for 30–45 min at 60°C. Dilution and plating techniques can be used for grain and other commodities. For harvested and manufactured products, direct examination, plating and dilution techniques are commonly used.

Techniques for Studying Seeds

Using ×10 to ×100 magnifications or more in the inspection of damaged seeds, mycelium and propagules can be detected over the surface or inside. The inner side of the outer pericarp layers of barley, wheat and other grains is examined by removing small pieces of the tissue and putting them on drops of agar on van Tieghem cells where they can be directly observed under high magnification, or isolated and identified when the mycelia grow out.

General counting media of high water activity and minimal levels of inhibitory compounds are suitable for enumeration and isolation of *Alternaria* spp.

Table 5 Potentially hazardous *Alternaria* mycotoxins, involved in food contamination

Common name	Fungal source	Natural occurrence	Toxic to	Chemical compound
Altenuene (ALT)	*A. alternata*	Olives, sunflower seeds, peppers, tomatoes, apples, sorghum, ragi	Bacteria mammalian cells, mice (cytotoxic)	Dibenzo-α-pyrone
Tenuazonic acid (AT)	*A. alternata, A. citri, A. gaisen*[a], *A. mali*[b], *A. longipes, A. tenuissima*	Olives, sunflower seeds, mandarins, peppers, sorghum, tobacco, rice, ragi, melons, processed tomatoes and apple	Mice, monkeys, viruses, bacteria, fungi, dogs, chickens, guinea pigs, rabbits, rats (potentially carcinogenic)	Tetramic acid
Alternariol (AOH)	*A. alternata, A. dauci, A. cucumerina, A. gaisen*[a] *A. solani A. tenuissima*	Olives, sunflower seeds, peppers, tomatoes, mandarins, apples, sorghum, wheat, rye, triticale, oats, barley, pecans	Bacteria, mammalian cells, mice (fetotoxic, teratogenic, cytotoxic)	Dibenzo-α-pyrone
Alternariol monomethyl ether (AME)	*A alternata, A. dauci, A. cucumerina*	Olives, sunflower seed, peppers, tomatoes, mandarins, apples, sorghum wheat, rye, triticale, oars, barley, ragi, pecans, melons	Bacteria, mammalian cells, mice (mutagenic, cytotoxic)	Dibenzo-α-pyrone
Altertoxin I (ATX-I)	*A. alternata, A. mali*[b]	Sorghum	HeLa cells, bacteria, mice (cytotoxic, antifungic, tumour-promoting activity)	Perylene derivatives
Altertoxin II (ATX-II)	*A. alternata, A. mali*[b]		Bacteria, HeLa cells, mice (mutagenic, antifungal activity)	
Altertoxin III (ATX-III)	*A. alternata*		Tumour-promoting, mutagenic activity	

[a]Syn. *A. kikuchiana.*
[b]*A. alternata* f. sp. *mali.*

Plating and Dilution Methods

Plating and dilution techniques are the same for all filamentous fungi. Both are effective when some contamination value is needed, and are useful in examinations of particulate foods.

See also: **Fungi**: Classification of the Deuteromycetes. **Metabolic Pathways**: Production of Secondary Metabolites – Fungi. **Mycotoxins**: Classification; Occurrence; Detection and Analysis by Classical Techniques; Toxicology. **Spoilage of Plant Products**: Cereals and Cereal Flours. **Total Counts**: Microscopy. **Total Viable Counts**: Microscopy.

Further Reading

Beuchet LR (1987) *Food and Beverage Mycology*, 2nd edn. AVI. New York: Van Nostrand Reinhold.
Chelkowski J and Visconti A (1992) *Alternaria. Biology, Plant Diseases and Metabolites*. Amsterdam: Elsevier.
Ellis MB (1971) *Dematiaceous Hyphomycetes*. Kew: Commonwealth Mycological Institute.
Ellis MB (1976) *More Dematiaceous Hyphomycetes*. Kew: Commonwealth Mycological Institute.

Joly P (1964) *Le Genre Alternaria. Recherches Physiologiques. Biologiques et Systématiques*. Paris: Lechevalier.
Miller JD and Trenholm HL (1994) *Mycotoxins in Grain. Compounds Other Than Aflatoxin*. St Paul: Eagan Press.
Neergaard P (1945) *Danish species of Alternaria and Stemphylium*. Copenhagen: Munksgaard.
Nishimura S and Kohmoto K (1983) Host specific toxins and chemical structures from *Alternaria* species. *Annual Review of Phytopathology* 21: 87–116.
Pitt JY and Hocking AD *Fungi and Food Spoilage*, 2nd edn. London: Blackie.
Raghubir PS and Salunkhe DK (1991) *Mycotoxins and Phytoalexins*. Boca Raton: CRC Press.
Rotem J (1994) *The Genus Alternaria. Biology, Epidemiology and Pathogenicity*. Minnesota: APS Press.
Samson RA, Hocking AD, Pitt JI and King AD (1990) *Modern Methods in Food Mycology* Amsterdam: Elsevier.
Samson RA, Hoekstra ES, Frisvad JC and Filtenborg O (1996) *Introduction to Food-borne Fungi*. Baern: CBS.
Simmons EG and Roberts RG (1993) *Alternaria* themes and variations (73). *Mycotaxon* 48: 109–140.

> **Antimicrobial Packaging** *see* **Chilled Storage of Foods**: Packaging with Antimicrobial Properties.

> **Antimicrobial Systems** *see* **Natural Antimicrobial Systems**: Preservative Effects During Storage; Antimicrobial Compounds in Plants; Lysozyme and Other Proteins in Eggs; Lactoperoxidase and Lactoferrin.

ARCOBACTER

Irene V Wesley, Enteric Diseases and Food Safety Research, USDA, National Animal Disease Center, Ames, USA

Introduction

The rRNA superfamily VI of the Proteobacteria was proposed in 1991 to encompass the genera *Campylobacter*, *Helicobacter* and *Arcobacter*. The microbes are microaerophilic, fastidious, Gram-negative and spiral-shaped. Motility is by means of polar flagella, which may be single or multiple, sheathed or naked (**Table 1**).

Arcobacter was originally designated *Campylobacter cryaerophila* (Latin: loving cold and air). The microbe was first recovered from aborted bovine and porcine fetuses, although attempts to induce experimental abortions in animals have met with limited success. *Arcobacter* spp. have been isolated from water, cattle, swine, birds, including ostriches, reptiles and meats. *Arcobacter* spp. have been associated with cases of human enteritis in Canada, Thailand, Taiwan, South Africa, Italy, France, Germany and the US. This article summarizes the characteristics of this newly described genus, its distribution in food animals and meats, current methods of isolation and the potential importance of *Arcobacter*, especially *A. butzleri*, to the food industry.

Characteristics of the Genus

Arcobacter (Latin: arc-shaped bacterium) differs from other *Campylobacter* species. *Campylobacter* grows in low-oxygen (microaerobic) environments at 37°C. In contrast, *Arcobacter* grows in atmospheric oxygen (aerotolerant) and at 15°C (cryophilic), which is lower than the temperatures used for cultivation of *Campylobacter*.

Taxonomy

Four species are recognized: *A. butzleri*, *A. cryaerophilus*, *A. skirrowii* and *A. nitrofigilis* (**Table 2**). *A. butzleri* is regarded as the primary human pathogen. *A. nitrofigilis* is the type strain. It is a nitrogen-fixing bacterium that is restricted to the roots of *Spartina alterniflora*, a salt marsh plant.

Methods of Detection

The ability to grow in air (aerotolerant) and at 15–30°C distinguishes *Arcobacter* spp. from other *Campylobacter* spp. Whereas *C. jejuni* grows optimally at 42°C, few *Arcobacter* display this thermotolerance.

Table 1 Characteristics of members of rRNA superfamily VI

Genus	Growth at 15°C	Oxygen tolerance	Flagella
Arcobacter	+	Aerotolerant	Single Unsheathed
Campylobacter	–	Microaerophilic	Single Unsheathed
Helicobacter	–	Microaerophilic	Multiple Sheathed

Table 2 Summary of *Arcobacter* spp. and host distribution

Bacterium	Source	Target organ
A. butzleri	Humans	Intestine
	Livestock	Intestine, placenta, fetus
A. cryaerophilus	Humans	Intestine
	Livestock	Intestine, placenta, fetus
A. skirrowii	Livestock	Intestine Reproductive tract
A. nitrofigilis	*Spartina alterniflora*	Roots

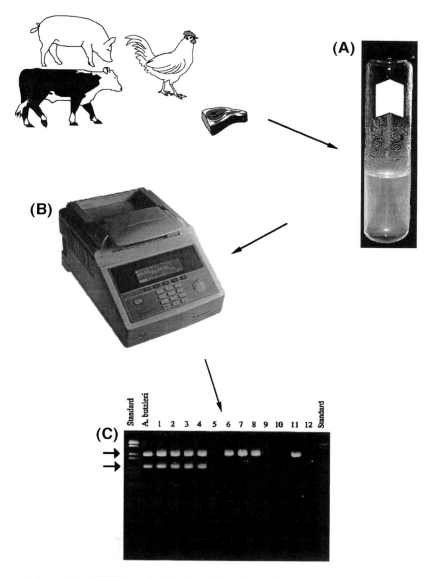

Figure 1 Polymerase chain reaction (PCR)-based detection of *Arcobacter* in meats and livestock. (**A**) Samples are enriched in EMJH-P80 media (1 week, 30°C). (**B**) After incubation, template DNA and primers are placed in a thermal cycler for multiplex PCR amplification. (**C**) The PCR products or amplicons are detected by gel electrophoresis. The presence of a 1223 bp PCR amplicon (top horizontal arrow) is specific for *Arcobacter. A. butzleri* exhibits two PCR products: the 1223 bp amplicon characteristic of the genus and a smaller (686 bp, bottom horizontal arrow) product (lanes 1, 2, 3, 4). *Arcobacter* species other than *butzleri* are identified by a single 1223 amplicon (lanes 6, 7, 8 and 11). No PCR product indicates the absence of *Arcobacter* in the enrichment (lanes 5, 9, 10 and 12).

There is no standard method for the isolation of *Arcobacter*, which hampers comparison of field survey data. In general, *Arcobacter* have been recovered in media designed for *Campylobacter* but at lower incubation temperatures. *Arcobacter* spp. were originally isolated from aborted livestock fetuses in EMJH-P80, a media formulated for *Leptospira*. Following EMJH-P80 enrichment (25°C, 5–7 days), an aliquot is removed and examined by dark-field microscopy for typical *Campylobacter*-like motility. For isolation, an aliquot from the enrichment is plated directly or filtered (0.45 µm) on to the surface of a blood agar plate, the filtrate streaked for colony isolation, and the plate incubated (25–30°C, in air, 2–3 days). The appearance of typical *Campylobacter*-like colonies following incubation in air at 25–30°C is characteristic of *Arcobacter*.

Polymerase chain reaction (PCR) can be performed directly from the EMJH-P80 enrichment, as depicted in **Figure 1**. Livestock faecal or meat samples are enriched in EMJH-P80 media (3–5 days, 30°C, in air). After incubation, DNA is extracted, added to primers targeting specific DNA sequences, and placed in a thermal cycler for PCR amplification. The PCR products or amplicons are detected by gel electrophoresis. In a multiplex PCR, two different genes are amplified which may yield two bands upon electrophoresis. The presence of a single band (1223 bp), resulting from amplification of sequences encoding the 16S rRNA of the genus, identifies the microbe as *Arcobacter*. All members of the genus exhibit this product. The presence of an additional smaller (686 bp) amplicon, due to amplification of sequences encoding the 23S rRNA genes, is specific of *A. butzleri*. Thus, *A. butzleri* exhibits both the genus- (1223 bp) and species-specific (686 bp) amplicons. The assay provides genus and species identification and bypasses potential misidentification based on dark-field microscopy.

Species Identification

Biochemical tests to phenotype *Arcobacter* species are limited. All isolates hydrolyse indoxyl acetate, as do a number of *Campylobacter* spp., including *C. jejuni* and *C. coli*. Nitrate reduction, growth in glycine and susceptibility to nalidixic acid and cephalothin have been used for species identification. In practicality, few tests are as useful in species identification and as simple to perform as the catalase test. The test distinguishes the two species of clinical interest: *A. cryaerophilus* exhibits a strong catalase reaction while that of *A. butzleri* is weak.

A serotyping scheme consisting of 65 groups has been developed for *Arcobacter* and assigns the majority of human isolates to serogroup 1. In one study, up to 22 different serotypes, including serotype 1, were identified in 162 poultry isolates of *A. butzleri*.

PCR assays to detect all members of the genus *Arcobacter* and for each of the species of *Arcobacter* have been reported. A multiplex PCR assay to identify simultaneously *Arcobacter* and *A. butzleri* in livestock and foods directly from enrichment has been detailed. The PCR assays are highly specific, rapid, simple to perform and reproducible. The assay can be performed directly from EMJH-P80 enrichment and can identify *A. butzleri*, thus eliminating the need to employ the limited number of phenotypic tests reported for the species.

DNA-based fingerprinting of field strains may be useful in tracking the spread of *Arcobacter* contamination. PCR-mediated DNA fingerprinting confirmed the identity of *A. butzleri* isolates recovered from a nursery school outbreak. The fingerprints of *A. butzleri* indicated a single source of contamination and person-to-person transmission. PCR fingerprinting may also characterize *A. butzleri* isolates

from meat. Nearly 86 unique PCR fingerprints were obtained from over 100 *A. butzleri* field strains recovered from mechanically separated turkey. The multiple DNA fingerprints, like the diverse serotypes recovered from poultry, suggested numerous sources of contamination.

Importance of *Arcobacter* in Livestock and Foods

Infections in Humans

Identity of serotypes from poultry and from clinical faecal specimens has suggested that consumption of contaminated poultry may be a source for human infections. In addition, human infections have been epidemiologically linked to consumption of contaminated water in developing countries; person-to-person transmission has also been documented in a nursery school outbreak.

The distribution of *Arcobacter* in food animals and meat products is summarized in **Figure 2**. *Arcobacter* species have been cultured from clinically healthy cattle, poultry and pigs. The single reported isolation from sheep was from faeces of a lamb with diarrhoea. It was identified as *A. skirrowii* and is the type strain for the species.

The distribution of *Arcobacter* spp. in lamb, seafoods, shellfish and raw milk is unknown (indicated by a question mark in Fig. 2).

Cattle and Beef

Since first described from aborted bovine fetuses, *Arcobacter* spp. have been cultured from faeces of calves with diarrhoea, cows with mastitis, as well as from clinically healthy dairy cows. PCR assay has expedited field surveys to screen rapidly for meats and faecal samples of healthy livestock (Fig. 1). Using this strategy, *Arcobacter* was detected in approximately

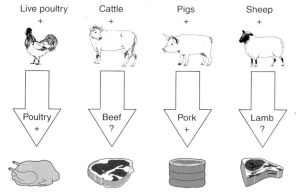

Figure 2 Distribution of *Arcobacter* in food animals and meat. The presence of *Arcobacter* is indicated by +. A question mark (?) indicates insufficient data.

11% of dairy cattle faecal samples by PCR. By comparison, using a similar approach *C. jejuni* was identified in 43% of clinically healthy cattle.

Few surveys are available for beef, and this is reflected by a question mark (?) in Figure 2. In the single study reported to date, *Arcobacter* was found in less than 2% of minced beef samples (*n* = 68) examined in the Netherlands.

Swine and Pork

Arcobacter is present in both healthy and clinically ill pigs. It has been found in aborted porcine fetuses, normal fetuses obtained from a slaughterhouse, in rectal, preputial or vaginal swabs taken from pigs from a herd with reproductive problems and from specific pathogen-free (SPF) animals. Using the PCR-based strategy outlined in Figure 1, *Arcobacter* was detected in 46% of faecal samples obtained from clinically healthy swine. In that same survey, *C. coli* was detected in approximately 70% of pigs.

Arcobacter has been recovered from naturally infected hogs as well as from pork. *Arcobacter* was detected in 56% of nearly 300 ground pork samples obtained from five Iowa slaughter facilities. Recoveries from the five plants ranged from 0% to 90%. However, using an *Arcobacter*-selective broth and medium, the organism was isolated in only 0.5% of nearly 200 retail purchased pork samples in the Netherlands.

Chicken and Poultry Products

For poultry *Arcobacter* was detected in fewer than 2% of caecal samples before processing. In a limited survey of 200 chickens, *A. butzleri* was recovered in 27% of birds. Whereas 53% of older birds (56 weeks old) were positive, *Arcobacter* was found in only 1% of birds less than 2 months old.

Retail purchased poultry may be a significant reservoir of *A. butzleri*. *Arcobacter* was detected in 24% of approximately 200 retail purchased poultry in the Netherlands. In France, *A. butzleri* was recovered from 81% of nearly 200 poultry carcasses examined. nearly half of the poultry isolates in that study were of serogroup 1, which is the serogroup associated with human infection. *A. butzleri*, predominantly of serotype 1, was recovered from 97% of poultry carcasses (*n* = 125) obtained from five Canadian processing plants. In the US, *A. butzleri* was detected in 57% of mechanically separated ground turkey samples. Differences in recovery rates could be attributed to multiple factors, such as hygienic conditions in each country, bias in plant selection, or to differences in the sensitivity of isolation methods.

Methods of Inactivation

Because of their phylogenetic relatedness, inactivation strategies developed for *C. jejuni* in foods may be adequate for *Arcobacter*. By comparing D_{10} values (the irradiation dose which reduces by 10-fold the number of viable bacteria), *A. butzleri* (0.27 kGy) was found to be slightly more resistant to irradiation than *C. jejuni* (0.18 kGy). Thus, irradiation doses (1.5–4.5kGy) approved by the US Food and Drug Administration for meats would kill *Campylobacter* as well as *Arcobacter*.

The thermotolerance of *Arcobacter* may be similar to that of *C. jejuni*. However, no studies have been reported on the survival of *Arcobacter* following heating, refrigeration or freezing. *Arcobacter*, like *Campylobacter* and *Helicobacter*, is inactivated by standard chlorination procedures used for water treatment plants.

See also: **Campylobacter** : Introduction. **Helicobacter**.

Further Reading

Atabay HI and Corry JEL (1997) The prevalence of campylobacters and arcobacters in broiler chickens. *J. Appl. Microbiol.* 83: 619–626.

Bastyns K, Cartuyvels D, Chapelle S, Vandamme P, Goossens H and De Wachter R (1995) A variable 23S rDNA region is a useful discriminating target for genus-specific and species-specific PCR amplification in *Arcobacter* species. *Syst. Appl. Microbiol.* 18: 353–356.

Collins CI, Wesley IV and Murano EA (1996) Detection of *Arcobacter* spp. in ground pork by modified plating methods. *J. Food Prot.* 59: 448–452.

de Boer E, Tilburg JJHC, Woodward DL, Lior H and Johnson WM (1996) A selective medium for the isolation of *Arcobacter* from meats. *Lett. Appl. Microbiol.* 23: 65–66.

Harmon KM and Wesley IV (1997) Multiplex PCR for the identification of *Arcobacter* and differentiation of *Arcobacter butzleri* from other arcobacters. *Vet. Microbiol.* 58: 215–227.

Kiehlbauch JA, Brenner DJ, Nicholson MA et al (1991) *Campylobacter butzleri* sp. nov. isolated from humans and animals with diarrheal illness. *J. Clin. Microbiol.* 29: 376–385.

Nachamkin I (1995) *Campylobacter* and *Arcobacter*. In: Murray PR, Baron EHJ, Pfaller MA, Tenover FC and Yolken RH (eds) *Manual of Clinical Microbiology*. P. 483. Washington, DC: American Society for Microbiology Press.

Neill SD, Campbell JN, O'Brien JJ, Weatherup STC and Ellis WA (1985) Taxonomic position of *Campylobacter cryaerophila* sp. nov. *Int. J. Syst. Bacteriol.* 35: 342–356.

Vandamme P, Falsen E, Rossau R et al (1991) Revision of *Campylobacter*, *Helicobacter*, and *Wolinella* taxonomy: emendation of generic descriptions and proposal of

Arcobacter gen. nov. *Int. J. Syst. Bacteriol.* 41: 88–103.

Vandamme P, Vancanneyt M, Pot B et al (1992) Polyphasic taxonomic study of the emended genus *Arcobacter* with *Arcobacter butzleri* comb. nov. and *Arcobacter skirrowii* sp. nov., an aerotolerant bacterium isolated from veterinary specimens. *Int. J. Syst. Bacteriol.* 42: 344–356.

Vandamme P, Giesendorf BAJ, Van Belkum et al (1993)

Discrimination of epidemic and sporadic isolates of *Arcobacter butzleri* by polymerase chain reaction-mediated DNA fingerprinting. *J. Clin. Microbiol.* 31: 3317–3319.

Wesley IV (1997) *Helicobacter* and *Arcobacter*: potential human foodborne pathogens? *Trends Food Sci. Technol.* 8: 293–299.

ARTHROBACTER

Marco Gobbetti, Istituto di Produzioni e Preparazioni Alimentari, Facoltà di Agraria di Foggia, Italy
Emanuele Smacchi, Institute of Industrie Agranie Microbiologia, Faculty of Agriculture of Perugia, Italy

Arthrobacter spp. often populate several foods. In some cases food quality is not affected, in others the effect is not completely known or may be underestimated. They certainly contribute to the biodegradation of some pesticides and to the ripening of smear surface-ripened cheeses.

This article describes the taxonomy, metabolism and genetics as well as the role in foods of *Arthrobacter* spp.

General Characteristics

Arthrobacter is a genus of mainly soil bacteria whose major distinguishing feature is a rod–coccus growth cycle. Irregular rods in young cultures are replaced by stationary-phase coccoid cells, which when transferred to fresh medium produce outgrowths to give irregular rods again. Coccoid cells may assume large and morphologically aberrant forms when in conditions of severe nutritional stress. Both rod and coccoid forms are Gram-positive but may be easily decolorized. Gram-negativity may appear in mid-exponential- to stationary-phase cells. Cells do not form endospores; they are non-motile or motile by one subpolar or a few lateral flagella, obligate aerobes and catalase-positive. Their metabolism is respiratory, never fermentative (little or no acid is formed from sugars) and the nutrition is non-exacting. The G+C content of the DNA is in the range of 59–66 mol% (actinomycete branch) and the cell-wall peptidoglycan contains lysine as diamino acid.

A total of 15 species of *Arthrobacter (sensu stricto)* are recognized in *Bergey's Manual of Systematic Bacteriology* (**Table 1**). The *incertae sedis* species do not have the necessary chemotaxonomic data for inclusion in the genus. Two groups of species, *A. globiformis/A. citreus* and *A. nicotianae/A. sulfureus*, are accepted on the basis of DNA–DNA homology,

Table 1 *Arthrobacter* species as reported in *Bergey's Manual of Systematic Bacteriology*

Sensu stricto	A. globiformis
	A. crystallopoietes
	A. pascens
	A. ramosus
	A. aurescens
	A. histidinolovorans
	A. ilicis
	A. ureafaciens
	A. atrocyaneus
	A. oxydans
	A. citreus
	A. nicotianae
	A. protophormiae
	A. uratoxydans
	A. sulfureus
Incertae sedis	A. mysorens[a]
	A. picolinophilus[a]
	A. radiotolerans[a]
	A. siderocapsulatus[a]
	A. duodecadis[b]
	A. variabilis[b]
	A. viscosus[b]
	A. simplex[c]
	A. tumescens[c]

[a] The chemotaxonomic data necessary to allow the inclusion of these species in the genus *Arthrobacter* are lacking.
[b] Species named *Arthrobacter* which have been shown to contain *m*-diaminopimelic acid in the cell wall peptidoglycan.
[c] Species named *Arthrobacter* which have been shown to contain LL-diaminopimelic acid in the cell wall peptidoglycan.

16S rRNA cataloguing studies, peptidoglycan structure, teichoic acid content and lipid composition (**Table 2**). However, based on 16S rDNA studies, the separation of *Arthrobacter* into two groups of species has no phylogenetic validation. Members of the second group (e.g. *A. nicotianae*) are significantly more closely related to certain members of the first group (e.g. *A. globiformis*) than members of the first

Table 2 Some characteristics of the *Arthrobacter* species *sensu stricto*[a]

Characteristics	A. globiformis	A. crystallopoietes	A. pascens	A. ramosus	A. aurescens	A. histidinolovorans	A. ilicis	A. ureafaciens	A. atrocyaneus	A. oxydans	A. citreus	A. nicotianae	A. protophormiae	A. uratoxydans	A. sulfureus
Peptidoglycan type	Lys-Ala	Lys-Ala	Lys-Ala	Lys-Ala	Lys-Ala-Thr-Ala	Lys-Ala-Thr-Ala	Lys-Ala-Thr-Ala	Lys-Ala-Thr-Ala	Lys-Ser-Ala	Lys-Ser-Thr-Ala	Lys-Thr-Ala	Lys-Ala-Glu	Lys-Ala-Glu	Lys-Ala-Glu	Lys-Glu
A3α variation[b]	+	+	+	+	+	+	+	+	+	+	+	-	-	-	-
A4α variation[b]	-	-	-	-	-	-	-	-	-	-	-	+	+	+	+
MK-9(H$_2$)[c]	+	+	+	+	+	+	+	+	+	+	+	-	-	-	-[d]
MK-8[c]	-	-	-	-	-	-	-	-	-	-	-	+	+	+	-[d]
Teichoic acid	-	-	-	-	-	-	-	-	-	-	-	+	+	+	+
Cell wall sugars	Gal, Glu	Gal, Glu	Gal, Glu	Gal, Rha, Man	Gal (Man)	Gal, Glu	Gal, Rha, Man	Gal, (Man)	Gal, Glu (Man)	Gal, Glu	Gal	Gal, Glc (one strain)	ND	ND	Gal, Glc (one strain)
DNA homology[e]	100%[f]	16%[f]	35%[f]	25%[f]	22%[f]	36%[f]	ND	31%[f]	24%[f]	50%[f]	18%	100%[g]	39%[g]	ND	22%[g]
Starch hydrolysis	+	-	+	-	+	-	-	-	+	-	-	+	-	-	-
Motility	-	-	-	+	-	-	+	-	+	-	+	-	-	+[h]	+[h]

a Symbols: +, 90% or more of the strains are positive; −, 90% or more of the strains are negative; ( ), conflicting reports on occurrence; ND, no data.

b Within the type A peptidoglycan (cross-linkage between positions 3 and 4 of the peptide subunits), two groups occur: A3α variations (the interpeptide bridge of peptidoglycan contains only monocarboxylic acids and/or glycine) and A4α variations (the interpeptide bridge always contains a dicarboxylic acid and in most strains also alanine).

c MK-9(H$_2$), dihydrogenated menaquinones with nine isoprene units as major components. MK-8, unsaturated menaquinones with eight isoprene units as major components.

d A. sulfureus either contains MK-9 as the major menaquinone or comparable amounts of MK-9 and MK-10.

e Homology index is expressed as % of ^{3}H-DNA bound to a certain disc DNA relative to the homologous reaction.

f DNA homologies of named strains versus A. globiformis DSM 20125.

g DNA homologies of named strains versus A. nicotianae DSM 20123.

h A. uratoxydans, rods, motile by peritrichous flagella or non-motile. A. sulfureus, rods motile by one or few lateral flagella or non-motile.

group to each other (e.g. *A. globiformis* vs. *A. citreus*). Nutritional versatility is characteristic: carbohydrates, organic acids, amino acids, aromatic compounds and nucleic acids are used as carbon and energy sources. A comparison between some metabolic properties of *A. globiformis* and *A. nicotianae* is reported in **Table 3**. With the exception of biotin, vitamins or other organic growth factors are not required. Arthrobacters mainly use inorganic nitrogen. *A. citreus* is a notable exception as it uses a more limited range of compounds as energy and carbon sources and requires complex growth factors in addition to a siderophore such as ferrichrome or mycobactin for growth. The optimum temperature for growth is 25–30°C, and most arthrobacters grow in the range of about 10–35°C. Many strains also grow at 5°C and a few at 37°C. Growth at 37°C is influenced by the culture medium.

On a phylogenetic basis (homology within the 16S ribosomal gene) the *Arthrobacter* species could not be separated from members of the genus *Micrococcus* and will be included in Section XXIII – Class: Actinobacteria of the new edition of *Bergey's Manual of Systematic Bacteriology*.

Table 3 Comparison of some metabolic properties of *Arthrobacter globiformis* DSM 20124 and *Arthrobacter nicotianae* DSM 20123[a]

Metabolic properties	A. globiformis DSM 20124	A. nicotianae DSM 20123
Utilization of		
4-Aminobutyrate	+	+
5-Aminovalerate	–	+
Malonate	+	+
4-Hydroxybenzoate	+	+
Glyoxylate	+	–
2-Ketogluconate	+	–
L-Leucine	–	+
L-Asparagine	+	+
L-Arginine	+	
L-Histidine	+	+
L-Xylose	+	+
D-Ribose	+	+
L-Arabinose	+	+
D-Galactose	+	+
L-Rhamnose	+	–
D-Xylitol	+	–
m-Inositol	+	–
2,3-Butylenglycol	–	+
Glycerol	+	+
Nicotine	–	+
Hydrolysis of		
Xanthine	+	–
Caseine	+	+

[a] Symbols: +, 90% or more of the strains are positive; –, 90% or more of the strains are negative.

Ecology

Arthrobacters are numerically important among the indigenous bacterial flora of soils and rhizospheres. Nutritional versatility, extreme resistance to drying and starvation ensure their predominance in soils of different geographical locations. Soil acidity decreases cell viability. Psychrotrophic strains are abundant in terrestrial subsurface environments and occur in glacier silts. Isolates have been found in oil brines raised from soil layers at ca. 200–700 m depth.

Arthrobacter spp. are relatively common on the aerial surface of plants including flowers. Marine and freshwater fish and other seafoods contain arthrobacters. They occur in shark spoilage, eviscerated freshwater fish, fish-pen slime and shrimp. Sewage, brewery waste, deep poultry litter, dairy waste activated sludge and the surface of smear surface-ripened cheeses are other populated habitats.

Arthrobacter spp. *sensu stricto* have not been isolated from clinical sources.

Culture Media

Arthrobacter spp. are normally isolated from soil by plating on non-selective media because they are an appreciable proportion of the aerobic, cultivable population. Soil extract agar is largely used because it is sufficiently poor in carbon and energy sources. Possible modifications could be the addition of low concentrations of yeast extract and glucose to give higher counts, the incorporation of nystatin and cycloheximide to suppress fungi growth, and salt to reduce growth of Gram-negative bacteria.

The isolation of arthrobacters in selective medium (**Table 4**) gives counts several times higher than those on nutritionally poor medium. The combination of 0.01% cycloheximide and 2.0% NaCl is effective in inhibiting fungi and most *Streptomyces* spp., *Nocardia* spp. and Gram-negative bacteria. Methyl red

Table 4 Isolation of *Arthrobacter* spp. by selective medium of Hagedorn and Holt[a]

Compound	Quantity
Trypticase soy agar	0.4%
Yeast extract	0.2%
NaCl	2.0%
Cycloheximide	0.01%
Methyl red[b]	150 μg ml^{-1}
Agar	1.5%

[a] Plate counts are made by spreading 0.1 ml amounts of suitable dilutions over the surface of sterile medium in Petri dishes. Peptone solution (0.5%, wt/vol) is used as the diluent.
[b] The methyl red is filter-sterilized and added aseptically to the autoclaved, cooled medium. The medium is adjusted to the pH of the particular soil being examined.

$(150\,\mu g\,ml^{-1})$ inhibits other Gram-positive bacteria but does not affect arthrobacters.

Metabolism and Enzymes

Carbohydrate dissimilation by *Arthrobacter* spp. falls into two groups. *A. globiformis*, *A. ureafaciens* and *A. crystallopoietes* primarily use the Embden–Meyerhof–Parnas and, to a lesser extent, the hexose monophosphate (HMP) pathways. *A. pascens* and *A. atrocyaneus* use the Entner–Doudoroff and HMP pathways. The pyruvate formed is oxidized by the tricarboxylic acid cycle and the cytochrome system mediates the terminal electron transport. When acetate is used as the carbon and energy source, arthrobacters must produce tricarboxylic acid cycle intermediates for biosynthetic purposes and should have mechanisms to produce acceptor molecules for C_2 units. The glyoxalate cycle serves this purpose: key enzymes of this cycle have been found in *Arthrobacter* spp. when grown on acetate plus glycine. *A. globiformis* grows on glycine as the sole carbon and energy source and converts this amino acid through serine into pyruvate. Pyruvate is converted into C_4 dicarboxylic acids for the tricarboxylic acid cycle and also into phosphoenolpyruvate, as a precursor of carbohydrates. Also nucleic acids (both DNA and RNA) are decomposed to produce uric acid, allantoin and urea. Arthrobacters from soil, but not from cheese and fish, use both uric acid and allantoin as the sole sources of carbon, energy and nitrogen.

Arthrobacters carry out heterotrophic nitrification. Cells must be provided with a carbon compound to produce energy and to synthesize the carbon-containing products of nitrification. Ammonium is converted into an amide, which is then oxidized to acetohydroxamic acid. The latter is rapidly converted by a reversible reaction into free hydroxylamine, but it is also slowly oxidized to nitrosoethanol. Nitrite and nitrate are late products in this sequence. Nitrite and nitrate are also formed from aliphatic nitro-compounds. Arthrobacters isolated from soil respire nitrate in the presence of oxygen, but, in contrast to other soil bacteria, do not synthesize periplasmic-type nitrate reductase.

Soil arthrobacters grown with excess glucose and a limiting amount of NH_4^+, HPO_4^- or SO_4^- are particularly rich in storage polysaccharides such as glycogen. Glycogen enables survival for prolonged periods of nitrogen depletion and at the same time provides energy and intermediates for protein synthesis when inorganic nitrogen is available. Glycogen has an exceptionally high degree of branching.

Extracellular polysaccharides are commonly produced by arthrobacters. Polysaccharides may consist of glucose, galactose, and uronic acid or mannuronic acid. Strains synthesize β-fructofuranosidase which transfers the fructosyl residues of sucrose to aldoses or ketoses, to produce hetero-oligosaccharides. Such compounds serve to protect against predation by protozoa in natural environments, and are never used as a carbon and energy source by producers.

The majority of *Arthrobacter* spp. isolated from soil, milk, cheese and activated sludge are highly proteolytic. When actively growing in the soil arthrobacters produce extracellular proteinases. Synthesis is repressed by high amino acid concentration. Enzymes are very stable. The proteinase of *A. ureafaciens* consists of a single peptide chain of 221 amino acid residues cross-linked by two disulphide bridges which, in part, explain its stability. Proteinases may have very high temperature optima, ca. 70°C, and milk-clotting properties. Two extracellular serine proteinases with molecular masses of about 53–55 and 70–72 kDa have been purified from *A. nicotianae* isolated from smear surface cheese. The enzymes differ with respect to temperature optimum (55–60 and 37°C), tolerance to low values of pH and temperature, heat stability, sensitivity to EDTA and sulphydryl blocking agents and hydrophobicity. Peptidases have been less studied than proteinases. An aminopeptidase of broad specificity, a proline iminopeptidase with activity against long peptides with a free N-terminal proline, and an imidodipeptidase (prolidase), which hydrolyses only dipeptides, were found in cell extracts of soil *Arthrobacter* spp.

Arthrobacters, especially those isolated from soil, have enzymes which enable them to degrade unusual and polymeric compounds. Strains which use levoglucosan (1,6-anhydro-β-D-glucopyranose) possess a levoglucosan dehydrogenase. Glucose is produced from levoglucosan by three steps: dehydrogenation, intramolecular hydrolysis and NADH-dependent reduction. Levoglucosan dehydrogenase catalyses the initial step. This pathway is distinct from those reported for soil yeasts and fungi. A levanase, which rapidly hydrolyses levan (β-D-fructose polymer) in an endo-type manner to produce a series of levan-oligosaccharides, was found. Pectolytic activity as well as the capacity to degrade another polyuronide such as alginic acid seems rather rare in arthrobacters. Methylamine oxidase, used to assimilate carbon in the form of methylamine or ethylamine, is synthesized by methylotroph *Arthrobacter* strains. It oxidizes primary amines methyl-, ethyl-, propyl-, butyl-, ethanol- and benzylamine, but not tyramine, spermine, putracine, trimethylamine and dimethylamine. Only O_2 acts as a reoxidizing substrate for this enzyme. A maltooligosyl-trehalose synthase, which converts maltooligosaccharide into maltooligosyl trehalose by intramolecular transglycosylation, and a malto-oligosyl-trehalose trehalohydrolase, which hydrolyses the α-1,4-glucosidic bond between maltooligosyl and trehalose moieties, have been found in *Arthrobacter* spp. which accumulate trehalose. *A. globiformis* produces an inulinase which degrades inulin through an exo-type reaction. Choline oxidase has been found in *A. pascens* and *A. globiformis*. This is a cytosolic flavoprotein, hydrogen-peroxide-forming oxidase which oxidizes choline to produce glycine betaine by a two-step reaction with betaine aldehyde as the intermediate. Betaine acts as a non-toxic osmolyte, highly compatible with metabolic functions at high cytoplasmic concentrations and contributes to turgor adjustment in cells subjected to osmotic stress.

Genetics and Bacteriophages

Several genes of *Arthrobacter* spp. have been cloned and sequenced. The focus here is on genes for food enzymes.

A 5.1 kbp genomic DNA fragment was cloned from trehalose-producing *Arthrobacter* sp. strain Q36. Sequence analysis revealed two open reading frames (ORFs) of 2325 and 1794 bp, encoding maltooligosyl-trehalose synthase and maltooligosyl-trehalose trehalohydrolase. Enzymes have several regions common to the α-amylase family. Some arthrobacters infrequently produce β-galactosidase when grown in lactose minimal media. The gene has similarities with the *Escherichia coli lacZ* gene. When DNA was transformed into an *E. coli* host, three fragments each encoding a different β-galactosidase isoenzyme were obtained. The nucleotide sequence of the smallest fragment has no total similarity with the *lacZ* family but has regions similar to β-galactosidase isoenzymes from *Bacillus stearothermophilus* and *B. circulans*. The gene encoding inulin fructotransferase was sequenced in *A. globiformis* S14-3 and *Arthrobacter* sp. H65-7. The two genes share only 49.8% homology and the sequence analysis of the *ift* gene from strain H65-7 consists of a single ORF of 1314 bp that encodes a signal peptide of 32 amino acids and a mature protein of 405 amino acids. The gene encoding an extracellular isomalto-dextranase (*imd*) was isolated from the chromosome of *A. globiformis* T6 and expressed in *E. coli*. A single ORF consisting of 1926 bp that encodes a polypeptide composed of a signal peptide of 39 amino acids and a mature protein of 602 amino acids was found. The primary structure has no significant homology with the structures of any other reported carbohydrases and the enzyme differs in that it is capable of hydrolysing dextran by releasing only isomaltose units from dextran chains. Isomaltose inhibits the biosynthesis of mutan which is the major component of dental plaque and may be of significant importance in the prevention of dental caries. The *pcd* plasmid gene for phenylcarbamate hydrolase was sequenced in *A. oxydans* P52. It has significant homology with esterases of eukaryotic origin.

A host–vector system based on pULRS8 containing the kanamycin resistant gene, *kan* (Tn5), was used for transforming *Arthrobacter* sp. strain MIS38 by electroporation. Electrotransformation was optimized; a square wave pulse of $1\,kV\,cm^{-1}$ electric field strength for 0.5 ms duration yielded 3×10^5 transformants per microgram plasmid DNA. The host–vector system expressed a lipase gene of *Arthrobacter* sp. MIS38 in other strains.

Oligonucleotide probes for cheese surface bacteria, including *Arthrobacter/Micrococcus*, were developed. This is an important contribution to identify the smear microflora. Sequences were chosen from sites of the 16S rRNA. Because of the intermixing of some *Arthrobacter* and *Micrococcus* species and the significant heterogeneity of this cluster, it was not possible to design an *Arthrobacter/Micrococcus* specific oligonucleotide for colony-hybridization that fits all the species and at the same time excludes related species (e.g. *Dermatophilus congolensis*). However, the few species from other genera also targeted do not live in the same habitat, namely the cheese surface.

A total of 17 bacteriophages, active against 19 soil arthrobacters, have been detected in concentrated samples of river water and sewage. Bacteriophages have not been found in either concentrated or unconcentrated soil extracts due to the greater viral retention capacity of the soil and to the fluctuations in the phage sensitivity of soil bacteria. Electron microscopic studies showed morphologies characteristic of Bradley's groups B and C. The G+C content of bacteriophages was in the range 60.2–65.3% which agrees with the G+C range of arthrobacters. Isolation of bacteriophages for *A. globiformis* depends on the nutritional features of the soil. Indigenous host cells in non-amended soil are present in a non-sensitive sphaeroid state, with the cells becoming sensitive to the phage in a rate-limiting fashion as outgrowth occurs.

Role in Foods

Arthrobacters are frequently encountered in foods. They may occur as uneffective inhabitants but when at high cell concentrations they may indicate inadequate hygiene. They play an important role in biodegrading pesticides and in ripening of smear surface-ripened cheeses.

Vegetables and Biodegradation of Pesticides

Arthrobacters are largely distributed among the indigenous bacterial flora of soils. They are not limited to any particular soil but are found in sandy, clay, peaty, grassland and tropical soils. They occur in the rhizosphere and in the epiphytic part of the plants. In the rhizosphere they release growth factors and auxin, but are also sensitive to soil bacteriostasis, especially to wheat root secretions. Arthrobacters may abundantly populate vegetables during and after food processing. *A. globiformis* is largely found in healthy sugarbeet roots stored at 5°C. Since *Arthrobacter* spp. may be associated with the seed before the fruit opens, they may spread to the aerial parts of many higher plants (e.g. soybean). Arthrobacters are found in ready-to-use vegetables. Most of the isolates in frozen peas, beans and corn correspond to *Arthrobacter* spp. Blanched vegetables may still contain arthrobacters. Since cells do not survive blanching, airborne contamination of the surfaces of processing equipment could be another source of infection.

Arthrobacters play a role in controlling some soil-borne pathogens and are capable of participating in the degradation of various pesticides, in polluted temperate and cold environments. *Arthrobacter* spp. have been recovered during culture of the causal organism of pitch canker of Southern pines, *Fusarium moniliforme* var. *subglutinans*. Electron microscopic observations revealed that hyphae of the pathogen fungus growing near *Arthrobacter* spp. were enlarged, producing many vesicular-like structures. The surface of these hyphae was warped and wrinkled in comparison with normal hyphae. Arthrobacters are chitinolytic bacteria. Enzymes, capable of hydrolysing polymers, lyse fungal hyphae and hence inhibit growth. Inhibition of *Aspergillus* spp. and *Penicillium* spp. was shown in stored cereal grains.

Polychlorinated phenols such as 4-chlorophenol may be accidentally released into the environment. *A. ureafaciens* degrades 4-chlorophenol through the elimination of the chloro-substituent and the production of the hydroquinone as transient intermediates. Other para-substituted phenols are metabolized through the hydroquinone pathway. Picolinic acid (2-carboxylpyridine), structurally similar to the herbicide picloram (4-amino-3,5,6-trichloropicolinic acid) and the photolytic product of another herbicide, diquat (1,1′-dimethyl-4,4′-bipyridylium ion), is used as carbon and energy sources by *A. picolinophilus* (species *incertae sedis*). Diazinon [phosphorothioic acid O, O-diethyl O-(2-isopropyl-6-methyl-6-pyrimidinyl)] added to paddy water for controlling stem borer, leafhopper and planthopper pests of rice is degraded by arthrobacters. One isolate from treated paddy water metabolized it in the presence of ethyl alcohol or glucose. *A. oxydans*, isolated from soil, degrades the phenylcarbamate herbicides phenmedipham methyl (3-methylcarbaniloyloxy) carbanilate and desmedipham (3-ethoxycarbonylaminophenyl-phenylcarbamate), by hydrolysing their central carbamate linkages (carbamate hydrolase). Phenmedipham and desmedipham are hydrolysed at comparable rates, whereas phenisopham, a compound with an additional alkyl substitution at the carbamate nitrogen, is not hydrolysed. In some cases, synthetic aromatic compounds, such as *m*-chlorobenzoate, may only account for incomplete degradation. The benzoate-oxidizing enzyme of *Arthrobacter* spp. produces 4-chlorocatechol from *m*-chlorobenzoate which is not further degraded by catechol-metabolizing enzymes. Co-metabolism could result from an accumulation of some toxic product, or an inability of the organism to carry the metabolism to a stage where the carbon could be assimilated.

Arthrobacters may contribute significantly to waste water treatment and ground-water remediation problems because of the accidental release of gasoline in the environment. There are difficulties with the biological treatment of gasoline oxygenates such as methyl *t*-butyl ether. Ethers tend to be quite resistant to biodegradation by microorganisms. Pure cultures of *Arthrobacter* spp. are able to degrade methyl *t*-butyl ether by using it as a carbon source.

Meat, Eggs and Fish

Catalase-positive bacteria with a rod–coccus growth cycle such as *Corynebacterium*, *Microbacterium* and *Arthrobacter* are often isolated in fresh beef. They are also recovered from turkey giblets and traditional bacon stored aerobically. Microbial and chemical changes in aerobically stored bacon fall into two phases, the first of microbial growth and reduction of nitrate to nitrite, and the second in which most of the accumulated nitrite is broken down to unknown products. *Arthrobacter–Corynebacterium* are mainly associated with the last phase of bacon storage. Poultry litter contains yellow strains and strains growing on citrate plus ammonia, classified as *A. citreus* and *A. aurescens*.

Arthrobacter spp., together with *Pseudomonas* spp., are the most prevalent bacteria found in liquid egg. A study conducted on microbial contamination of egg-shells and egg packing materials showed that arthrobacters comprised approximately 13% of the total number of isolates. Dust, soil and faecal material are the most common sources of contamination. Arthrobacters do not cause spoilage of shell eggs, and in liquid egg may not affect keeping quality but may indicate the possibility of contamination by spoilage organisms present in soil or faecal material.

Arthrobacter spp. together with *Moraxella, Pseudomonas, Acinetobacter* and *Flavobacterium–Cytophaga* spp. are the microorganisms predominantly associated with raw Pacific and Gulf coast shrimps. In peeled shrimp, the number and composition of the microflora vary but arthrobacters may remain constant. They are isolated in greater proportion from plants that used minimal washing of raw shrimp. Pond-reared shrimps also contain arthrobacters, and the pond water frequently yields over 90% of such bacteria. Arthrobacters are commonly isolated from Dungeness crab (*Cancer magister*) meat, both from retails and intestine. They increase in proportion during processing of crab meat, since they populate the brine and are less sensitive to cooking, but they do not multiply in refrigerated crab meat.

Milk and Cheese

Arthrobacter spp. are part of the microflora of raw milk and in some cases constitute the most predominant of the non-spore forming Gram-positive rod-shaped bacteria. Psychrotrophic strains increase during long-term storage of refrigerated raw milk. Some psychrotrophic isolates of *Arthrobacter* spp. synthesize a β-galactosidase with similarities to that of *Escherichia coli* but which differed in the optimal temperature, ca. 20°C lower. Removal of lactose from refrigerated milk or whey was proposed as a use of this β-galactosidase to produce low-lactose products during shipping and storage.

Bacteria such as *Brevibacterium, Arthrobacter, Micrococcus* and *Corynebacterium* spp. are dominant at the end of ripening of smear surface-ripened cheeses such as Limburger, Brick, Münster, Saint-Paulin, Appenzeller, Trappist, Taleggio and Quartirolo. During the initial stages of ripening, the surface microflora is dominated by yeasts and moulds which cause an increase in pH due to a combination of lactate utilization and ammonia production, enabling the growth of acid-sensitive bacteria such as *Arthrobacter* spp. Low-molecular-weight compounds (peptides, amino acids, fatty acids, etc.) are produced on the surface through the coupled action of various extracellular hydrolases produced by the smear microflora. The diffusion of these compounds to the interior of the cheese is required for the development of the characteristic qualities of these cheeses. Arthrobacters, together with yeasts and *Brevibacterium linens*, were the main microorganisms found in Limburger cheese during ripening; yeasts dominate up to 9 days of ripening, *B. linens* reaches its highest level in 35-day-old cheese but *Arthrobacter* spp. account for ca. 78% of the total count. Grey and greenish-yellow *Arthrobacter* spp. have the highest proteolytic activity among the surface microflora. Coryneform bacteria from 21 brick cheeses (including Limburger, Romadur, Weinkäse and Harzer) from six German dairies were identified as *A. nicotianae, B. linens, B. ammoniagenes* and *Corynebacterium variabilis*. After an initial variability of the surface microflora of the Tilsiter cheeses from 14 Austrian cheese plants, the decrease of the yeast cell numbers is followed by the growth of a mixed population composed of *A. citreus, A. globiformis, A. nicotianae, A. variabilis, B. linens* and *B. ammoniagenes*. The yellow–green coloration of the Taleggio cheese surface is mainly caused by *A. globiformis* and *A. citreus*. Even though the role during ripening is only partially known, other Italian cheeses such as Quartirolo and Robiola contain arthrobacters on the cheese surface. Also mould surface-ripened cheeses such as Brie and Camembert from 20 days until the end of ripening are largely populated by *Brevibacterium* and *Arthrobacter* spp. together with fungal hyphae and yeast cells.

Mixed cultures suitable for surface ripening have been developed. Cultures (single or mixed species of *Arthrobacter* and yeasts) are added as starters during the manufacture of Tilsit cheese. *A. citreus* has a significant effect in cheese proteolysis. Again in Tilsit cheese, a starter composed of *Lactobacillus helveticus, L. delbrueckii, B. linens, Arthrobacter* spp. and *Geotrichum candidum* or *Debaryomyces hansenii* was used. Mixed cultures of arthrobacters with yeasts and micrococci are used for red smear cheeses.

Despite their high cell numbers in the smear, the role of extracellular enzymes of *Arthrobacter* spp. is probably underestimated with respect to *B. linens*. Two extracellular serine proteinases of *A. nicotianae* ATCC 9458 show characteristics which indicate a significant contribution to proteolysis on the surface of smear-ripened cheeses: high activity at the pH and temperature of cheese ripening, tolerance to NaCl and extensive activity on α_{s1}- and, especially, β-caseins. Since only plasmin and, probably, cathepsin D have a certain role in β-casein hydrolysis in cheeses, the activity of *Arthrobacter* enzymes should be fundamental.

Four strains of *A. nicotianae* isolated from red smear cheese showed inhibition to *Listeria* spp. Inhibition is more effective against *L. innocua* and *L. ivanovii* than against *L. monocytogenes*. The inhibitory compound loses activity upon heating, and has a molecular mass greater than 12–14 kDa.

Miscellaneous Biotechnological Potentialities

Arthrobacters are a commercially important host for the production of valuable bioproducts. Some species are used to produce phytohormones, riboflavin, and α-ketoglutaric acid. Coryneform bacteria, including

Arthrobacter spp., are the most important microbial group for the commercial production of amino acids (e.g. glutamic acid). A mutant of *Arthrobacter*, strain DSM 3747, was used for the production of L-amino acids from D,L-5 monosubstituted hydantoins. *Arthrobacter* sp. MIS38 isolated from oil spills produces no glycolipids and only a lipopeptide. The lipopeptide (arthrofactin) is an effective biosurfactant. Arthrofactin is at least five times more effective than surfactin (the best-known lipopeptide biosurfactant). Moreover, arthrofactin is a better oil remover than synthetic surfactants, such as Triton X-100 and sodium dodecyl sulphate. The potential of arthrobacters has been evaluated for the production of flavour metabolites, precursors and enhancers, and has found useful application in the synthesis of terpenes and sweeteners such as D-xylose.

Involvement in Clinical Specimens

Strains of *Arthrobacter* spp. (*sensu stricto*) have never been described as causing disease in humans. However, in recent years, clinical microbiologists have begun to fully recognize the enormous diversities of coryneform bacteria in clinical specimens. Over a 6-year period, Swiss and Swedish clinical bacteriology laboratories isolated strains, which, through 16S rDNA gene sequence and peptidoglycan analyses, were unambiguously assigned to the *Arthrobacter* spp. Based on their unique cellular fatty acid patterns, two new species, *A. cumminsii* sp. nov. and *A. woluwensis* sp. nov., were proposed. *A. cumminsii* might be the most frequently encountered *Arthrobacter* in clinical specimens since it was isolated from patients with urinary tract and deep tissue infections, and external otitis. *A. cumminsii* seems to be a microorganism with no or rather low pathogenicity since cases of severe, life-threatening infections were not observed in patients. It might be the only bacterial agent for selected cases of urinary tract infections. It is likely that *A. cumminsii* is part of the normal human skin and mucosa membrane flora, in particular, in the genitourinary tract.

See also: **Brevibacterium**. **Cheese**: Mould-ripened Varieties; Microflora of White-brined Cheeses. **Fish**: Spoilage of Fish. **Meat and Poultry**: Spoilage of Meat. **Milk and Milk Products**: Microbiology of Liquid Milk.

Further Reading

Bae HS, Lee JM and Lee ST (1996) Biodegradation of 4-chlorophenol via hydroquinone pathway by *Arthrobacter ureafaciens* CPR706. *FEMS Microbiology Letters* 145: 125–129.

Frändberg E and Schnürer J (1998) Antifungal activity of chitinolytic bacteria isolated from airtight stored cereal grain. *Canadian Journal of Microbiology* 44: 121–127.

Funke G, Pagano-Niederer M, Sjödén B and Falsen E (1998) Characteristics of *Arthrobacter cumminsii*, the most frequently encountered *Arthrobacter* species in human clinical specimens. *Applied and Environmental Microbiology* 36: 1539–1543.

Jones D and Keddie RM (1992) The Genus *Arthrobacter*. In: Balows A, Trüper HG, Dworkin M, Harder W and Schleifer KH (eds) *The Prokariots*. P. 1283. New York: Springer-Verlag.

Keddie RM, Collins MD and Jones D (1986) Genus *Arthrobacter*. In: Sneath PH, Mair NS, Sharpe ME and Holt JG (eds) *Bergey's Manual of Systematic Bacteriology*. Vol. 2, p. 1288. Baltimore: Williams & Wilkins.

Koch C, Rainey FH and Stackebrandt E (1994) 16S rDNA studies on members of *Arthrobacter* and *Micrococcus*: and aid for their future taxonomic restructuring. *FEMS Microbiology Letters* 123: 167–172.

Kolloffel B, Burri S, Meile L and Teuber M (1997) Development of 16S rRNA oligonucleotide probes for *Brevibacterium*, *Micrococcus/Arthrobacter* and *Microbacterium/Aureobacterium* used in dairy starter cultures. *Systematic and Applied Microbiology* 20: 409–417.

Smacchi E, Fox PF and Gobbetti M (1999) Purification and characterization of two extracellular proteinase from *Arthrobacter nicotianae* 9458. *FEMS Microbiology Letters* 170: 327–333.

ASPERGILLUS

Contents
Introduction
Aspergillus oryzae
Aspergillus flavus

Introduction

Perng-Kuang Chang, **Deepak Bhatnagar** and **Thomas E Cleveland**, Southern Regional Research Center, Agricultural Research Service, US Department of Agriculture, New Orleans, USA

Morphological Characteristics of *Aspergillus*

The genus *Aspergillus* contains more than 100 recognized species and belongs to the Deuteromycetes (Fungi imperfecti). **Figure 1** illustrates the basic morphological structures of *Aspergillus*. *Aspergillus* species produce asexual spores (conidiospores or conidia) on a specialized structure called an aspergillum. The aspergillum consists of a swollen vesicle bearing either one or two layers of synchronously formed specialized cells. The specialized cells bearing the conidia are called phialides. An aspergillum with only phialides is referred to as uniseriate. When a second layer of specialized cells (metulae) are present between the vesicle and phialides, the aspergillum is referred to as biseriate. The aspergillum is borne on a long stipe, the basal part of which forms the 'foot cell'

characteristic of *Aspergillus*. The aspergillum, stipe and foot cell together are called the conidiophore.

In addition to the typical conidial state (anamorphic) characteristic of the genus, some species also reproduce sexually and have ascosporic states (teleomorphic). The morphological event associated with sexual reproduction is the appearance of cleistothecia. Cleistothecia are closed ascocarps (fruiting bodies) containing thousands of asci, all of which originate from a single dikaryotic cell. The asci contain the mecospores called ascospores. Other structures found in *Aspergillus* include Hülle cells which are thick-walled cells frequently associated with cleistothecia, and sclerotia which are asexually formed compact aggregates of mycelia resistant to unfavourable conditions and capable of remaining dormant for long periods. The colour of the sclerotia varies from yellow to brown or black depending on the species.

Isolation Methods and Identification Media

In nature, *Aspergillus* species are abundant and grow saprophytically in numerous natural substrates over a wide range of climate conditions. *Aspergillus* species have long been known to be common contaminants of human foods and animal feeds.

For the detection of *Aspergillus* as well as other fungi in foods and feeds, dilution plating and direct plating methods are routinely used. The dilution plating method includes preparation of a food homogenate, followed by serial dilution and plating out in triplicate; either the pour-plate method or the spread-plate method can be used. The direct plating method is more efficient than the dilution method for detecting individual mould species. The food sample may be surface-disinfected before the fungi are enumerated. If a surface-disinfected procedure is used, samples are disinfected with 5% sodium hypochlorite or with 70% ethanol/water solution. If a non-surface-disinfected procedure is used, samples are held at −20°C to kill mites and insects that might interfere with analysis.

Potato dextrose agar or malt extract agar are common media for isolation and enumeration of *Aspergillus* species. To control bacterial growth,

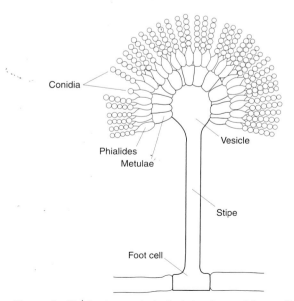

Figure 1 The basic morphological structures of *Aspergillus*.

molten agar medium is acidified with 10% tartaric acid solution to pH 3.5 ± 0.1. Alternatively, antibiotics such as chlortetracycline, chloramphenicol, gentamycin or streptomycin are added to inhibit bacterial growth. To control 'spreader moulds' (*Rhizopus*, *Mucor*, etc.), rose bengal, dichloran or NaCl (7.5%) are used. Most recently, the petrifilm dry rehydratable film method developed by the 3M company has been adopted by AOAC International (AOAC, Association of Official Analytical Chemists).

The genus may be recognized on the above media with low power magnification ($\times10-30$). Further distinction between *Aspergillus* species is based on morphological characteristics. Most *Aspergillus* species are readily identified when grown on special media, such as Czapek agar (CZ), Czapek yeast extract agar (CYA) and malt extract agar (MEA). For teleomorphic species, cornmeal agar, oatmeal agar, and malt agar are commonly used. For xerophilic species, CZ, CYA and MEA supplemented with 20% sucrose, or dichloran 18% glycerol (DG18) agar are recommended. Culture agar plates with three inoculation points are generally incubated at 25°C for 5–7 days in the dark for anamorphic species and 10–14 days for teleomorphic species.

Identification of *Aspergillus* Species

Morphology is the mainstay of *Aspergillus* identification and classification. The important identification keys to *Aspergillus* species include shape and colour of conidial heads, characteristics of conidiophores, roughness of conidial surface, presence of metulae and the presence of Hülle cells. Sclerotial production is also an important diagnostic character for some *Aspergillus* species. In addition to conventional taxonomic techniques, a number of different schemes have also been developed to distinguish between closely related *Aspergillus* species that are of economic and industrial significance. These techniques include: (1) colony size/colour on special media; (2) secondary metabolite profile; (3) isozyme pattern; (4) restriction fragment length polymorphism (RFLP); (5) polymerase chain reaction (PCR); (5) electrophoretic karyotypes; (7) gene sequence analysis; and (8) enzyme-linked immunosorbent assay (ELISA).

Aspergillus in Foods and Feeds

The occurrence of *Aspergillus* in foods and feeds is dependent on the substrate and environmental factors such as water activity, temperature, pH, redox potential, presence of preservatives and microbial com-

petition. **Table 1** summarizes *Aspergillus* species and teleomorphs of *Aspergillus* (*Eurotium* and *Neosartorya*) associated with various foods.

Because of their soil-borne nature and ability to grow at low water activity, *Aspergillus* species are common contaminants of various cereal grains. The presence of *Aspergillus* and *Eurotium* in processed, dehydrated foods is mostly due to their ability to grow at lower water activities (a_w 0.7–0.8). In meat products, contamination is often due to the added spices that are commonly contaminated with these xerophilic species. In pasteurized milk and cheeses, the source of contamination is usually factory related, for example, from bottling facilities or unclean air.

The teleomorphic genera that are of importance in food are *Eurotium* and *Neosartorya*. *Eurotium* species produce bright-yellow cleistothecia and have heads without metulae. *Neosartorya* species produce white cleistothecia and also have heads without metulae. All *Eurotium* species are xerophilic; they are spoilage moulds found in stored grains, nuts, feeds and spices. The most common species are *E. chevalieri*, *E. repens*, *E. rubrum* and *E. amstelodami*. *Neosartorya fischeri* produces heat-resistant ascospores and has been associated with spoilage in thermally processed fruit products.

Mycotoxin Production and Aspergillosis by *Aspergillus*

Aspergillus species produce a variety of mycotoxins that pose a potential threat to human and animal health. The most important toxigenic species and their toxins are listed in **Table 2**. The most significant mycotoxins are aflatoxins produced by *A. flavus*, *A. parasiticus* and *A. nomius* (they are also reported to be produced by *A. tamarii*), sterigmatocystin produced by *A. versicolor* and *Emericella nidulans*, cyclopiazonic acid produced by *A. flavus*, *A. tamarii* and *A. versicolor*, and ochratoxin A produced by *A. ochraceus* and related aspergilli.

Aflatoxin B_1 is one of the most potent hepatocarcinogens. Contamination of agricultural commodities with aflatoxins has long been implicated in human hepatocellular carcinoma (HHC). Recently, an association of HHC and dietary exposure to aflatoxins has been established from independent studies on patients living in high risk areas, such as Qidong, China and sub-Saharan Africa. The United States and many other countries have set legal limits of aflatoxin B_1 in various food items. Sterigmatocystin is also carcinogenic, causing hepatocellular carcinoma in test animals. Cyclopiazonic acid causes severe gastrointestinal upset and neurological disorders, and degeneration and necrosis of organs, especially

Table 1 Species of *Aspergillus* or teleomorphs of *Aspergillus* found in foods and feeds

Food	Aspergillus/Eurotium/Neosartorya *species*
Corn	*A. flavus, A. niger, A. restrictus, E. rubrum, E. amstelodami, E. chevalieri*
Wheat	*A. restrictus, A. candidus, A. versicolor, A. ochraceus, A. niger, A. sydowii, Eurotium* spp.
Barley	*A. fumigatus, A. niger, A. flavus, A. sydowii*
Rice	*A. restrictus, A. candidus, A. fumigatus, A. versicolor, A. ochraceus, A. terreus, E. repens, E. rubrum, E. amstelodami, E. chevalieri*
Nuts	*A. flavus, A. niger, A. parasiticus, A. ficuum, A. restrictus, A. penicillioides*
Spices	*A. flavus, A. fumigatus, A. niger, A. ochraceus, E. halophilicum*
Coffee	*A. ochraceus*
Dried fruits	*A. niger, E. herbariorum*
Jams/preserves	*Eurotium* spp.
Cured ham	*A. glaucus, A. fumigatus, A. niger, A. flavus*
Cured meats	*A. versicolor, A. flavus, A. niger, A. tamarii, A. wentii, A. restrictus, A. fumigatus, E. repens, E. rubrum, E. amstelodami*
Dried meats	*A. flavus, A. parasiticus, A. versicolor, A. sydowii, E. repens, E. rubrum, E. amstelodami, E. chevalieri*
Dried seafood products	*A. restrictus, A. niger, A. versicolor, A. flavus, A. wentii, E. repens, E. rubrum, E. amstelodami, E. chevalieri*
Dried-salted fish	*A. clavatus, A. niger, A. flavus, A. fumigatus, A. glaucus, A. restrictus, A. sydowii, A. wentii, E. repens, E. rubrum, E. amstelodami*
Smoked dried fish	*A. flavus, A. niger, A. ochraceus, A. tamarii*
Onions	*A. niger*
Tomatoes	*A. niger*
Strawberries	*N. fischeri*
Cheeses	*A. versicolor*

Table 2 Mycotoxin production by the genus *Aspergillus* and by related teleomorphs

Species	Mycotoxins
A. aculeatus	Aculeasins, neoxaline, secalonic acid D
A. caespitosus	Fumitremorgins, verruculogen
A. candidus	Candidulin, terphenyllin, xanthoascin
A. carneus	Citrinin
A. clavatus	Ascladiol, clavatol, cytochalasin E, kojic acid, patulin, tryptoquivalins, xanthocillins
A. flavus	Aflatoxin B_1 and B_2, aflatrem, aflavinins, aspergillic acids, cyclopiazonic acid, kojic acid, neoaspergillic acids, β-nitropropionic acid, paspalinins
A. fumigatus	Fumigaclavins, fumagillin, fumigatin, fumitoxins, fumitremorgins, gliotoxin, kojic acid, spinulosin, tryptoquivalins, verruculogen
A. niger	Malformins, naphthoquinones, nigragillin
A. nomius	Aflatoxin B_1, B_2, G_1 and G_2, aspergillic acids
A. ochraceus	Emodin, kojic acid, neoaspergillic acids, ochratoxins, penicillic acid, secalonic acid A, viomellein, xanthomegnin
A. oryzae	Aspergillomarasmin, kojic acid, β-nitropropionic acid, oryzacidin
A. parasiticus	Aflatoxin B_1, B_2, G_1 and G_2, aflavinine, aspergillic acids, kojic acid, neoaspergillic acids
A. sydowii	Nidulotoxin, sterigmatocystin, griseofulvin
A. tamarii	Cyclopiazonic acid, fumigaclavin A
A. terreus	Citreoviridin, citrinin, gliotoxin, mevinolin, patulin, quadrone, terrein, terreic acid, terretonin, territrems, terredionol, terramide A
A. ustus	Austamid, austdiol, austins, asutocystine, kojic acid, sterigmatocystin, xanthocillin X
A. versicolor	Nidulotoxin, sterigmatocystin
A. wentii	Emodin, kojic acid, β-nitropropionic acid, wentilacton, physicon
Em. nidulans	Nidulotoxin, sterigmatocystin
Eu. amstelodami	Physicon, sterigmatocystin
Eu. chevalieri	Emodin, physicon, gliotoxin, xanthocillin X
Eu. herbariorum	Sterigmatocystin
Eu. pseudoglaucum	Sterigmatocystin, citrinin
N. fischeri	Canescin, mevinolins, terrein, trypacidin, tryptoquicalins
N. quadricincta	Cyclopaldic acid

A., *Aspergillus*; *Em.*, *Emericella*; *Eu.*, *Eurotium*; *N.*, *Neosartorya*.

digestive tract, liver, kidney and heart of test animals. Ochratoxin A causes hepatic and renal tumours; it also has fetotoxic, teratogenic and immuno-suppressive effects in test animals.

In addition to causing mycotoxicoses, several *Aspergillus* species are important agents of mycoses (aspergillosis) in animals and humans. *A. fumigatus* is the most frequent *Aspergillus* species isolated from animal aspergillosis. It regularly affects significant numbers of poultry and livestock. Other opportunistic pathogens include *A. flavus* and *Emericella nidulans* (*A. nidulans*). In humans, *A. fumigatus* is the primary agent of invasive pulmonary aspergillosis especially in immunocompromised hosts.

Enzymes and Organic Acids from *Aspergillus*

Aspergillus species are good producers of many food-grade and industrial enzymes. **Table 3** lists the major industrial enzymes, their applications, and the corresponding *Aspergillus* species that produce these enzymes. Direct use of *Aspergillus* in food fermentations and industrial processes is another way to harness their catalytic capacities. *A. oryzae* and *A. sojae* are koji moulds widely used in soy sauce, miso and sake production in Asia. *A. oryzae* has GRAS (generally regarded as safe) status. They provide the enzymes for transforming raw materials of wheat, soybean or rice; they also contribute to the colour, flavour, aroma and texture of the fermented products.

Aspergillus species accumulate considerable amounts of organic acids. The biochemical pathways of carbon utilization in *Aspergillus* have been extensively studied for many years. The organic acids produced by *Aspergillus* can be categorized into two groups. One group of acids is derived from sugars by oxidation (gluconic acid, kojic acid) and another group is associated with the tricarboxylic acid cycle (citric acid, itaconic acid, malic acid).

Gluconic acid (produced by *A. niger*) is the oxidative product of D-glucose by glucose oxidase. Gluconic acid is a general chelating agent; it is primarily used in cleaning and metal-finishing processes. Itaconic acid (produced by *A. itaconicus* and *A. terreus*) is used as a copolymer in the plastics industry. Of the organic acids, citric acid produced by *A. niger* is the most important in economic terms. *A. niger* grows on a variety of carbon sources, including mono-, di- and polysaccharides and organic acids. The two primary substrates used for commercial citric acid production are sucrose (molasses) and dextrose. Citric acid has GRAS status from the US Food and Drug Administration. Its use by the food industry ranges from pH adjustment for gelation control in jams and jellies to the control of acid flavour in beverages (fruit juice, soft drinks, wine, etc.) and confectionery. It chelates heavy metals, such as iron and copper, to prevent oxidative deterioration in flavour, grey or brown discoloration of fresh or canned products, and to retain vitamin C and reduce enzymatic browning in frozen foods. Citric acid is also widely used in pharmaceuticals (effervescent salts, blood anticoagulant), cosmetics (creams, plaque inhibitor) and industrial applications (detergents, cleaners).

Table 3 Industrial enzymes from *Aspergillus* species

Enzyme	Sources	Uses
α-Amylase, glucoamylase	*A. oryzae, A. niger, A. awamori*	Breadmaking, enhancement of characteristics of baked goods, starch saccharification, production of high-fructose corn syrup, production of light beers, production of alcoholic beverages
Pectinase	*A. niger, A. alliaceus*	Fruit juice clarification, increase of grape juice extraction in winemaking
Lipase	*A. niger, A. awamori*	Cheese ripening, flavour development of cheese products, flavour modification
Protease	*A. niger, A. oryzae, A. melleus, A. saitoi*	Dough conditioning, beer chillproofing, meat tenderizing, soybean protein modification
Catalase	*A. niger*	Removal of hydrogen peroxide formed in irradiated foods and commercial cake baking, removal of hydrogen peroxide added in cheese and milk
Glucose oxidase	*A. niger*	Removal of oxygen in high-fat products, canned foods, and bottled or carbonated drinks, enzymatic production of gluconic acid, diagnostic kits for glucose measurement
Cellulase	*A. niger*	Beer brewing, fruit processing, improvement of food texture, modification of textiles
Phytase	*A. ficuum, A. niger*	Increase of availability of organic phosphorus in animal feeds
Tannase	*A. niger, A. tamarii*	Hydrolysis of flavonols in instant and fermented tea

Concluding Remarks

Aspergillus species possess a metabolic versatility that has both positive and negative impacts on our daily life. They are used in the fermentation industry for the production of enzymes (α-amylase, lipase, protease: *A. niger*, *A. oryzae*), organic acids (citric acid: *A. niger*), therapeutics (lovastatin: *A. terreus*; asperlicin: *A. allicaeus*) and fermented beverages and foods (sake: *A. oryzae*; soy sauce, miso: *A. oryzae*, *A. sojae*). More recently, the use of *Aspergillus* species (*A. niger*, *A. oryzae* and *E. nidulans*) as hosts for the production of extracellular fungal and mammalian proteins has been exploited. On the other hand, they are aetiological agents of aspergillosis in animals and humans (*A. fumigatus*, *A. flavus*, *A. niger*), contaminants of foods and feeds, and mycotoxin producers (*A. flavus* and *A. parasiticus*). Recent discoveries at the genetic and molecular biological level of *Aspergillus* are enhancing our ability to harness and explore the beneficial attributes as well as to prevent *Aspergillus* contamination and mycotoxin production for a safer and more economically viable food supply.

See also: **Biochemical and Modern Identification Techniques**: Food Spoilage Flora (i.e. Yeasts and Moulds). **Fungi**: Overview of the Classification of the Fungi; Food-borne Fungi – Estimation by Classical Culture Techniques; Classification of the Deuteromycetes. **Mycotoxins**: Classification; Occurrence; Detection and Analysis by Classical Techniques; Immunological Techniques for Detection and Analysis; Toxicology. **Spoilage Problems**: Problems Caused by Fungi.

Further Reading

Arora DK, Mukerji KG and Marth EH (eds) (1991) *Handbook of Applied Microbiology – Foods and Feeds*. New York: Marcel Dekker.

Bennett JW and Klich MA (eds) (1992) *Aspergillus – Biology and Industrial Applications*. Stoneham, MA: Butterworths-Heinemann.

Doyle MP, Beuchat LR and Montville TJ (eds) (1997) *Food Microbiology – Fundamentals and Frontiers*. Washington, DC: ASM Press.

Eaton DL and Groopman JD (eds) (1994) *The Toxicology of Aflatoxins: Human Health, Veterinary and Agricultural Significance*. New York: Academic Press.

Klich MA (1992) *Laboratory Guides to Common* Aspergillus *Species* Port Jervis, New York: Lubrecht and Cramer Ltd.

Pitt JI and Hocking AD (1997) *Fungi and Food Spoilage*. Sydney: Chapman & Hall.

Powell KE, Renwick A and Peberdy JF (eds) (1994) *The Genus Aspergillus – From Taxonomy and Genetics to Industrial Application*. New York: Plenum.

Samson RA and Pitt JI (eds) (1990) *Modern Concepts in Penicillium and Aspergillus Classification*. New York: Plenum.

Smith JE and Henderson RS (eds) (1991) *Mycotoxins and Animal Foods*. Boca Raton, FL: CRC Press.

US Food and Drug Administration (1992) *Bacteriological Analytical Manual*, 7th edn. Arlington, VA: AOAC International.

Vanderzant C and Splittstoesser DF (eds) (1992) *Compendium of Methods for the Microbiological Examination of Foods*. Washington, DC: American Public Health Association.

Aspergillus oryzae

Katsuya Gomi, Division of Life Science, Graduate School of Agricultural Science, Tohoku University, Japan

Characteristics of the Species

Aspergillus oryzae plays a pivotal role in oriental food manufacturing, such as saké, shoyu (soy sauce) and miso (soybean paste). For thousands of years, it has been used for making fermented food and beverages. In addition, *A. oryzae* has been used in the production of industrial enzymes for food processing. *A. oryzae* is, therefore, accepted as a microorganism having GRAS (generally regarded as safe) status.

A. oryzae is an aerobic filamentous fungus, and belongs to the *Aspergillus* subgenus *Circumdati* section *Flavi*, previously known as the *A. flavus* group. *Aspergillus* section *Flavi* contains industrially important species such as *A. oryzae* as well as agronomically and medically significant fungi such as *A. flavus* and *A. parasiticus*, which produce a potent carcinogenic substance, aflatoxin. Taxonomically, *A. oryzae* is very closely related to *A. flavus*, *A. parasiticus* and *A. sojae* which has also been used for shoyu fermentation for a long time. Despite such close relatedness, *A. oryzae* and *A. sojae* never produce aflatoxins and are used in fermented food manufacturing. Thus, it is of great importance to differentiate these four species accurately, although recent taxonomical studies on *Aspergillus* section *Flavi* have some controversial aspects.

A. oryzae is isolated from soils and plants, particularly rice. *A. oryzae* is named after its occurrence in nature and cultivation industrially on rice, *Oryza sativa*. *A. oryzae* has an optimal growth temperature of 32–36°C ($\pm$ 1°C) and is unable to grow above 44°C. It has an optimal growth pH of 5–6 and can germinate at pH 2–8. It has been reported that *A. oryzae* could grow in corn flour with a water content of about

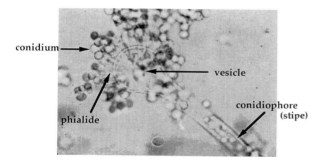

Figure 1 Conidiophore structures of *A. oryzae*

16%. It can generally grow on media with a water activity (a_w) above 0.8, but rarely grows below 0.8.

Like most other fungi, *A. oryzae* grows vegetatively as haploid multinucleate filaments, designated hyphae or mycelia. Hyphae of *A. oryzae* extend at the apical tips and multiply by branching, so that the colony covers the surface of the solidified agar medium after several days of incubation. Hyphal growth keeps on going in liquid medium as long as the hyphae are not exposed to air atmosphere. However, conidiophore structures, which bear asexual reproductive spores called conidia **Figure 1**, are produced when hyphae are transferred on to solidified agar medium. When grown on the surface of an agar medium the colony is initially white because of the vegetative hyphal growth, then turns to yellowish green as a large amount of conidia form. In most strains of *A. oryzae*, the colour of fresh culture or conidia is yellowish green, but that of old culture is brown, sometimes green with brown shades. Conidial heads are usually globose to radiate, 100–200 μm in diameter. In *A. oryzae*, a sexual life cycle has not been found as in other industrially important filamentous fungi such as *A. niger* and *Penicillium chrysogenum*. Conidia of *A. oryzae* are haploid, but multinucleate (conidia have mostly 2–4 or more nuclei) in contrast to uninucleate conidia of *A. nidulans* or *A. niger*. This makes genetic manipulation of *A. oryzae* more difficult, compared with *A. niger*. Conidia are large, 5–8 μm in diameter, and spherical to slightly oval. Conidial walls are mostly smooth to finely roughened. Most strains of *A. oryzae* have only phialides on the vesicles (uniseriate sterigmata), but some contain metulae and phialides (biseriate sterigmata). Stipes of conidiophores are colourless and mostly roughened, occasionally smooth to less roughened. They are long, in the range 1–5 mm.

Due to the difficulties of classical genetic analyses, little was known of the genetics of *A. oryzae*, or its fine genetic map. Recently, however, by means of pulsed-field gel electrophoresis (PFGE), electrophoretic karyotyping of an *A. oryzae* strain, RIB40, has been accomplished. *A. oryzae* chromosomes pre-

pared from protoplasts were separated by PFGE and gave seven ethidium bromide (EtdBr)-stained bands. The sizes of these were estimated roughly as 7.0, 5.2, 5.0, 4.5, 4.0, 3.7 and 2.8 Mbp, in comparison with the chromosome size of *Schizosaccharomyces pombe*. Of these seven chromosomal bands, the smallest was assumed to be a doublet suggested by the fluorescence intensity of EtdBr stain and the results of Southern blot analyses with 100 random clones isolated from *A. oryzae*. Consequently, it is likely that *A. oryzae* has eight chromosomes and the genome size is approximately 35 Mbp. The number of chromosomes is the same as that of *A. nidulans* and *A. niger*, and the genome size also is consistent with that of *A. nidulans* (31 Mbp) and *A. niger* (36–39 Mbp). In addition, 13 genes including rDNA of *A. oryzae* were hybridized to the chromosomal bands, and at least one gene was assigned to an individual chromosome. The related fungus *A. flavus* has also been electrophoretically karyotyped and was assumed to have eight chromosomes and an estimated 36 Mbp genome size. When chromosomes derived from genealogically different strains of *A. oryzae* were separated on PFGE the results revealed slightly different patterns. This indicated that changes in genome organization such as a chromosomal translocation have often occurred in *A. oryzae* intraspecifically. Also, this fact means that the electrophoretic karyotype cannot be used for distinguishing *A. oryzae* from *A. flavus*, although it would be expected to be a promising taxonomic criterion.

Importance in the Food Industry

For a thousand years, *Aspergillus oryzae* and *A. sojae* have been widely used in the Orient, particularly in Japan, as the starter for preparation of koji, which is a solid-state culture of the mould grown on cereal grains such as rice, wheat or barley. Thus these fungi are called koji mould. Koji is a complex enzyme preparation (which is comparable to malt in brewing) including amylolytic and proteolytic enzymes for making fermented foods and beverages, i.e. saké, shoyu, miso etc. The koji also contributes to the colour, flavour and aroma which are important for the overall character of the fermented products. The koji mould is the most important factor in determining the distinctive quality of these fermented foods.

In ancient times for food fermentations in the production of saké or shoyu, koji preparation was dependent on passive inoculation by *A. oryzae* associated with the rooms and facilities in the factory, which originated from the atmosphere or more likely from rice husks and kernels. Once koji with good qualities had been made it was conserved successively as a seed culture of *A. oryzae* for koji making. Very old publications in Japan described the use of rice smut, in which a plant pathogenic fungus *Ustilaginoidea virens* is mainly found but from which *A. oryzae* is also isolated, as a seed culture of the mould. Whereas *U. virens* cannot grow on steamed rice grains because of its low productivity of proteolytic enzymes, *A. oryzae* is able to grow rapidly and thus predominates in koji preparation. Nowadays, seed cultures of *A. oryzae* with favourable characteristics for the fermented products are available as conidiospores of the mould (*tane-koji*) from *tane-koji* manufacturers in Japan who provide them for saké, shoyu and miso production.

The most important role of *A. oryzae* in the food industry is as a source of a variety of enzymes for hydrolysing the raw materials used for fermentation. Of the hydrolytic enzymes produced by *A. oryzae*, the most important enzymes in saké fermentation are α-amylase and glucoamylase, which play crucial roles in starch solubilization and saccharification. α-Amylase of *A. oryzae* is known as Taka-amylase A, which has been extensively studied. The enzyme, a glycoprotein consisting of a single polypeptide chain of 478 amino acid residues, has been characterized by X-ray crystallography. The genes encoding α-amylase have been cloned and sequenced from genealogically unrelated strains of *A. oryzae*. Interestingly, all strains have two or three copies of α-amylase-encoding genes (*amyA*, *amyB* and *amyC*). These multiple genes have nearly identical nucleotide sequences in the coding and 5′-flanking regions and significant divergences only in the 3′-flanking region. All of these *amy* genes are functional in the mould. The observation that there are multiple functional *amy* genes in *A. oryzae* could explain the reason why this mould is a high producer of α-amylase. As mentioned above, *A. oryzae* is very closely related to *A. sojae*, and the aflatoxigenic fungi, *A. flavus* and *A. parasiticus*. Among them, *A. oryzae* is known as an α-amylase hyper-producer and has thus been used for the industrial production of the enzyme. Southern analysis of a number of strains of the four species showed that all strains of *A. oryzae* examined have multiple (two or three) genes, whereas all the strains tested of the other three species have a single gene. It is, therefore, suggested that the ability of *A. oryzae* to make rapid use of available starch due to the high productivity of α-amylase could have been selected preferentially during long-term cultivation on rice grains in saké fermentation. Furthermore, since the finding that multiple *amy* genes are a distinctive feature of *A. oryzae*, this may be used as a taxonomical criterion to dif-

ferentiate *A. oryzae* from the other closely related *Aspergillus* spp. As the isolated phage or cosmid clones contained only individual *amy* genes they could not be clustered on a single chromosome. The three *amy* genes were found by PFGE Southern blot analysis to be dispersed on three different chromosomes.

Another amylase, glucoamylase, produced by *A. oryzae* is also important in saké fermentation, because the rate of alcohol fermentation is dependent on the concentration of glucose in the *moromi* mash. The multiple forms of glucoamylase purified from rice-koji, did not adsorb to or digest raw starch, in contrast to that of the *A. niger* group. The glucoamylase of *A. niger* exists in two major forms; one larger form is able to adsorb onto and digest raw starch, whereas the smaller form has no activity to raw starch but digests soluble starch or maltodextrins. The smaller glucoamylase is likely generated by limited proteolysis in the C-terminal region of the larger one. Very recently, it has been demonstrated that *A. oryzae* has at least two glucoamylase-encoding genes, designated *glaA* and *glaB*. The *glaA* gene encodes glucoamylase with a starch-binding domain and an ability to digest raw starch. On the other hand, the *glaB* gene product has no starch-binding domain nor digestibility of raw starch. Interestingly, the *glaB* gene has been shown to be specifically expressed in solid-state culture, for instance, in rice-koji preparation, but had undetectable expression in submerged culture. In particular, the expression level of *glaB* is much higher than that of *glaA* in rice-koji making, with the result that the *glaB*-encoded glucoamylase predominates in rice-koji. This indicates the possibility of the existence of several kinds of genes other than the *glaB* gene being specifically expressed under solid-state culture conditions, whereas useful enzymes and metabolites are produced for fermented food processing.

In addition to amylolytic enzymes, proteolytic, cellulolytic, and xylanolytic enzymes are important in shoyu making. The genes encoding these enzymes have been isolated and sequenced. Furthermore, a number of genes encoding other extracellular and intracellular proteins including industrially important enzymes have been cloned and their structures and functions have been investigated (**Table 1**).

Although *A. oryzae* produces a copious amount of useful enzymes, its safety in food fermentation is most important. Since there have been no reports of invasive growth or infections by *A. oryzae* in healthy humans, it is generally recognized as non-pathogenic. All the *A. oryzae* isolates so far examined for aflatoxin production have been proven non-aflatoxigenic. On the other hand, some strains of *A. oryzae* have been found to produce mycotoxins, cyclopiazonic acid and kojic acid. However, these compounds are generally

Table 1   Genes isolated from *Aspergillus oryzae*

Genes for extracellular enzymes	Genes for intracellular proteins
α-Amylase (*amyA, amyB, amyC*)	Acetamidase (*amdS*)
Glucoamylase (*glaA, glaB* (solid-culture-specific))	3-Phosphoglycerate kinase (*pgkA*)
α-Glucosidase (*agdA*)	Glyceraldehyde 3-phosphate dehydrogenase
Alkaline protease (*alpA*)	Calmodulin (*cmdA*)
Aspartic (acid) protease (*pepA* (*pepO*))	Nitrate reductase (*niaD*)
Neutral protease II (*mep20*)	Alcohol dehydrogenase
Carboxypeptidase O	Enolase (*enoA*)
Cellulase (β-1,4-glucanase) (*celA, celB, celC*)	Protein disulphide isomerase (*pdiA*)
β-Galactosidase (*lacA*)	RNA polymerase II
Ribonuclease T1, T2 (*rntA, rntB*)	Nuclease O (*nucO*)
Nuclease S1 (*nucS*)	Nuclease O inhibitor
Polygalacturonase (*pgaA, pgaB, pecA, pecB*)	Pyruvate decarboxylase (*pdcA*)
Pectin lyase (*pelA, pelB*)	β-Tubulin (*benA*)
Xylanase (*xynF1, xynG1*)	Actin-related protein (*arpA*)
Lipolytic enzyme (*cutL*)	Translation elongation factor (*tef1*)
Monodiacyl lipase (*mdlB*)	Orotidine-5′-phosphate decarboxylase (*pyrG*)
Tyrosinase (*melO*)	Ornithine carbamoyltransferase (*argB*)
Tannase	Activator of *amdS* (*amdR*)
	Catabolite repressor (*creA*)
	Activator for conidiation (*brlA*)
	Activator of amylolytic enzymes (*amyR*)
	CAAT binding protein complex (*hapB, hapC, hapE*)

not much produced in the koji preparation and are easily decomposed by yeasts in the fermentation process.

Method for Detection and Identification of *A. oryzae*

A. oryzae is primarily found in the Orient, especially in Japan and China, where it is widely used for the fermentation of food and alcoholic beverages. It is also isolated from soils or plants in subtropical regions. *A. oryzae* and *A. sojae* are koji moulds used for food fermentation, and never produce aflatoxin, but are taxonomically conceived as domesticated fungi of *A. flavus* and *A. parasiticus* which are aflatoxin-producers. Because of their close relatedness, accidental use of an *A. flavus* or *A. parasiticus* strain in food fermentation processes may result in aflatoxin contamination in food products. Therefore, it is important to differentiate *A. oryzae* and *A. sojae* from aflatoxigenic *A. flavus* and *A. parasiticus* and to certify the non-aflatoxigenic properties of the strain used.

As in other filamentous fungi, *A. oryzae* isolates are obtained by dilution plating and direct plating methods from foods, soil and plant materials. Identification of an *A. oryzae* isolate is principally based on morphological properties such as the diameter, colour, and texture of the colony, the size and surface texture of conidia, the structure of conidial heads (biseriate or uniseriate), and the length and surface texture of stipes. Morphological characters of the isolates are examined after growth on Czapek agar

(CZ) or Czapek yeast extract agar (CYA) at 25°C and 37°C, and malt extract agar (MEA) at 25°C for usually 7 days. Macroscopical and microscopical observations on a number of isolates belonging to four taxa, namely *A. oryzae*, *A. sojae*, *A. flavus* and *A. parasiticus*, have not identified one character alone which will allow differentiation of the four species, since each species has a high degree of intraspecific variation and there is interspecific overlap of each character. However, the most reliable characteristic for differentiating *A. flavus/oryzae* from *A. parasiticus/sojae* is the texture of conidial walls. Conidial walls of *A. flavus/oryzae* are smooth to finely roughened, whereas those of *A. parasiticus/sojae* are definitely rough. To further distinguish *A. oryzae* and *A. flavus*, a combination of several characters is required. For example, the conidial diameters of *A. oryzae* are slightly larger than those of *A. flavus*, and colonies on CZ or CYA become brown with age in most *A. oryzae* strains but remain green in *A. flavus*. In addition, conidiophores of *A. oryzae* are mostly longer than those of *A. flavus*. The taxonomic key for differentiation of the four species is described in **Table 2**.

Although taxonomy using the morphological features shown in Table 2 is established to differentiate the four species, it is time-consuming and requires experience for accurate identifications. It is sometimes impossible to determine the species because of the intra- and interspecific variety of morphological characters. In particular, there are some reports that 'wild' isolates of *A. flavus* may change morphologically to

Table 2 Taxonomic key for identification of *Aspergillus oryzae* and other closely related species

A.	Conidial walls are smooth to finely roughened
1.	Conidial diameters usually 4–8.5 μm; conidia usually greyish yellow to olive brown in age; conidiophores predominantly > 0.8 mm long
	A. oryzae
2.	Conidial diameters usually 3–6 μm; conidia remaining green in age; conidiophores predominantly < 0.8 mm long
	A. flavus
B.	Conidial walls are consistently coarsely roughened
1.	Conidia diameters usually 5–8 μm; conidia usually pale brown in age; conidia not ornamented with dark-coloured tubercles
	A. sojae
2.	Conidial diameters usually 3–6 μm; conidia remaining green in age
	A. parasiticus

resemble 'domesticated' *A. oryzae* with successive transfers.

Besides morphological methods, biochemical and molecular biological techniques have been used to identify *Aspergillus* section *Flavi*. These include iso-enzyme pattern, DNA complementarity, restriction fragment length polymorphism (RFLP), and random amplified polymorphic DNA (RAPD). DNA complementarity is used to compare the relatedness of the species by measuring the rate and extent of re-association of DNA from two species. The degree of nuclear DNA complementarity was 100% similarity between *A. oryzae* and *A. flavus*, 91% similarity between *A. sojae* and *A. parasiticus*. DNA complementarity between *A. flavus* and *A. parasiticus* was also high at 70%. This indicated that the four taxa may be divided into two groups, namely, *A. flavus/oryzae* group and *A. parasiticus/sojae* group, consistent with the morphological difference in conidial wall texture. However, this method cannot differentiate *A. oryzae* from *A. flavus*. Isoenzyme patterns obtained by polyacrylamide gel electrophoresis have also been used for taxonomic study. When electrophoretic mobility patterns of several kinds of enzymes including extracellular and intracellular enzymes from isolates of *Aspergillus* section *Flavi* were compared, it was possible to distinguish *A. flavus/oryzae* from *A. parasiticus/sojae* but impossible to differentiate *A. oryzae* from *A. flavus*.

In the RFLP method, DNA from different species is digested with a restriction endonuclease following electrophoresis and the resulting DNA fragment patterns are then compared. Total DNA of the representative strains of the four related species in *Aspergillus* section *Flavi* was digested with various restriction enzymes and separated by agarose gel electrophoresis. One enzyme, *Sma*I, produced interspecifically distinctive cleavage patterns of the four species as shown in **Figure 2**. All species had a 1.8 kb band which stained strongly with EtdBr. *A. oryzae* also showed major bands at 3.0 kb and 1.0 kb, whereas *A. flavus* showed a major band at 4.0 kb. Similarly, *A. sojae* had 3.4 kb and 1.0 kb bands whereas *A. parasiticus* had a 4.4 kb band. It was

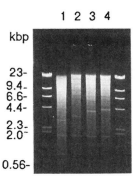

Figure 2 Distinctive cleavage patterns produced by the RFLP method

confirmed by using several strains of each species that these restriction patterns were intraspecific. Thus the species-specific differences in the length of the band at 3–5 kb could be used for differentiation of the four species. This method is very simple and rapid, no experience is required and the identification can be done within 3 days. Furthermore, since the species-specific bands can be readily distinguished from each other, the isolates can be identified even if it is difficult to differentiate them by conventional morphological methods. This RFLP method, therefore, provides an excellent adjunct to other taxonomic keys to distinguish the four industrially important species.

Recently, *A. sojae* and *A. parasiticus* have also been differentiated from each other by a RAPD method.

As described above, the koji-moulds, *A. oryzae* and *A. sojae*, can be clearly distinguished from aflatoxigenic fungi, *A. flavus* and *A. parasiticus*, morphologically and molecularly. However, because of a high degree of DNA similarity among the four species, it is important to confirm that the strain to be used in food fermentation does not produce aflatoxin by chemical analysis, particularly when strain improvement has been done by mutagenesis or a strain has been newly isolated from natural habitat. Production of aflatoxins can be assessed by several methods. The standard methods for quantitation of aflatoxins are based on aflatoxin production on solid or liquid cultures followed by extraction with solvents, separation and detection by thin-layer chromatography (TLC). As a qualitative method, small agar plugs from plate cultures are spotted directly onto a TLC plate and analysed by TLC. Alternatively, simple methods for the detection of the aflatoxins produced on agar plates under long-wave UV light have been developed as described below.

Detection of the Fluorescence on APA Medium

Conidia of the isolate are inoculated on to the agar-solidified APA medium (3% sucrose, 1% $(NH_4)H_2PO_4$, 0.1% K_2HPO_4, 0.05% KCl, 0.05% $MgSO_4$. $7H_2O$, 0.001% $FeSO_4$. $7H_2O$, 5×10^{-4} M $HgCl_2$, 0.05% corn steep liquor, 2% agar, pH 5.5) and then grown at 28°C for 7–10 days. Aflatoxin-producing strains are detected by blue or green fluorescence of the aflatoxin diffusing around the colony at 365 nm.

UV Adsorption Detected by UV Photography

The isolate is inoculated on GY agar medium (2% glucose, 0.5% yeast extract, 2% agar) and grown at 28°C in the dark for 3 days. Plastic Petri dishes are placed upside down on a black background and photographed under long-wave UV light (365 nm) with a camera equipped with a UV lens and UV interference filter. In the UV photographs, non-aflatoxigenic strains appear as white colonies, whereas aflatoxin-producing strains are observed as grey or black colonies because of UV absorption by the aflatoxin produced.

Molecular Characterization of Non-aflatoxigenicity of *A. oryzae*

Although *A. oryzae* and *A. flavus* can be distinguished by morphological and RFLP methods, it must be accepted that these two species are very closely related on the molecular level. Therefore, why is aflatoxin produced by *A. flavus* and not by *A. oryzae*? Aflatoxins are synthesized initially by condensation of acetate units to form norsolorinic acid, which is then converted into aflatoxins through a biosynthetic pathway involving at least 16 enzymes. So far, several genes encoding enzymes involved in aflatoxin biosynthesis have been cloned and well characterized from *A. flavus* and *A. parasiticus*. These are *pksA* coding for polyketide synthase, *nor-1* for a reductase converting norsolorinic acid into averantin, *verA* for an enzyme converting versicolorin A into sterigomatocystin, and *omtA* for an *O*-methyltransferase involved in the conversion of sterigmatocystin into *O*-methyl-sterigmatocystin. In addition, a transcriptional activator gene, *aflR*, responsible for the expression of the pathway genes has also been isolated. More recently,

these five genes have been shown to be clustered on about 60 kb region of the chromosome DNA in both aflatoxigenic fungi. Furthermore, *A. nidulans*, which produces sterigmatocystin but not aflatoxin has been shown to have 25 genes which are possibly required for the biosynthesis of sterigmatocystin within a 60 kb DNA fragment. Since there are so many genes involved in the synthesis of aflatoxin, it is most likely that *A. oryzae* has lost one or more of the genes required for aflatoxin biosynthesis.

To determine the existence of the genes involved in the aflatoxin synthetic pathway in koji-moulds, Southern blot analyses have been done using the *aflR* and *omtA* genes as hybridization probes. All strains of *A. flavus*, *A. parasiticus* and *A. sojae* tested showed hybridized hands with both genes, whereas none of the *A. oryzae* tested hybridized to the *aflR* gene and one of the three strains did not hybridize to the *omtA* gene. Another experiment in which the *verA* gene was used as a hybridization probe showed that the *verA* homologue was detected in 38 of 46 strains of *A. oryzae* examined, but not in eight strains. The transcripts of the *verA* homologue could not be detected in the strains of *A. oryzae* with the *verA* examined by reverse transcription-polymerase chain reaction (RT-PCR) under the conditions of aflatoxin production. These results indicate that *A. oryzae* does not produce aflatoxin because at least one of the genes required for aflatoxin biosynthesis is absent and/or is transcriptionally blocked. In addition, homologous genes involved in aflatoxin biosynthesis exist in all isolates of another koji mould, *A. sojae*, examined so far, but some of the genes are not transcribed even under aflatoxin-producing conditions. Molecular biological techniques have thus revealed the non-aflatoxigenicity of koji moulds. Nevertheless, the strain of *A. oryzae* that lacks one or more of the aflatoxin biosynthetic genes, preferably the regulatory gene, *aflR*, should be used in the fermentation process, to ensure the non-aflatoxigenicity of the strain used on the molecular level.

See also: **Aspergillus**: Introduction; *Aspergillus flavus*. **Fermented Foods**: Fermentations of the Far East. **Fungi**: The Fungal Hypha; Overview of the Classification of the Fungi; Food-borne Fungi – Estimation by Classical Cultural Techniques; Classification of the Deutero-mycetes. **Genetic Engineering**: Modification of Yeast and Moulds. **Genetics of Microorganisms**: Fungi.

Further Reading

Bennett JW and Klick MA (1992) *Aspergillus: Biology and Industrial Applications*. Boston: Butterworth-Heinemann.

Gomi K, Tanaka A, Iimura Y and Takahashi K (1989) Rapid differentiation of four related species of *koji* molds by agarose gel electrophoresis of genomic DNA digested with *Sma*I restriction enzyme. *Journal of General and Applied Microbiology* 35: 225–232.

Hara S, Fennell DI and Hesseltine CW (1974) Aflatoxin-producing strains of *Aspergillus flavus* detected by fluorescence of agar medium under ultraviolet light. *Applied Microbiology* 27: 1118–1123.

Hata Y, Ishida H, Ichikawa E et al (1998) Nucleotide sequence of an alternative glucoamylase-encoding gene (*glaB*) expressed in solid-state culture of *Aspergillus oryzae*. *Gene* 207: 127–134.

Kinghorn JR and Turner G (1992) *Applied Molecular Genetics in Filamentous Fungi*. Edinburgh: Blackie.

Kitamoto K, Kimura K, Gomi K and Kumagai C (1994) Electrophoretic karyotype and gene assignment to chromosomes of *Aspergillus oryzae*. *Bioscience, Biotechnology and Biochemistry* 58: 1467–1470.

Klick MA and Mullaney EJ (1987) DNA restriction enzyme fragment polymorphism as a tool for rapid differentiation of *Aspergillus flavus* from *Aspergillus oryzae*. *Experimental Mycology* 11: 170–175.

Klick MA and Pitt JI (1988) Differentiation of *Aspergillus flavus* from *A. parasiticus* and other closely related species. *Transactions of British Mycological Society* 91: 99–108.

Klick MA and Pitt JI (1988) *A Laboratory Guide to the Common* Aspergillus *Species and Their Teleomorphs*. North Ryde, Australia: CSIRO Division of Food Processing.

Klick MA, Yu J, Chang P-K et al (1995) Hybridization of genes involved in aflatoxin biosynthesis to DNA of aflatoxigenic and non-aflatoxigenic aspergilli. *Applied Microbiology and Biotechnology* 44: 439–443.

Kusumoto K, Yabe K, Nogata Y and Ohta H (1998) *Aspergillus oryzae* with and without a homolog of aflatoxin biosynthetic gene *ver-1*. *Applied Microbiology and Biotechnology* 50: 98–104.

Powell KE, Renwick A and Peberdy JF (1994) *The Genus* Aspergillus: *from Taxonomy and Genetics to Industrial Application*. New York: Plenum Press.

Samson RA and Pitt JI (1990) *Modern concepts in* Penicillium *and* Aspergillus *classification*. New York: Plenum Press.

Yabe K, Ando Y, Ito M and Terakado N (1987) Simple method for screening aflatoxin-producing molds by UV photography. *Applied and Environmental Microbiology* 53: 230–234.

Aspergillus flavus

Deepak Bhatnagar and **Thomas E Cleveland**, Southern Regional Research Center, Agricultural Research Service, USDA, New Orleans, USA

Gary A Payne, Department of Plant Pathology, North Carolina State University, Raleigh, USA

Filamentous fungal species in *Aspergillus* section *Flavi*, more commonly known as the *Aspergillus flavus* group, are of great economic importance. Several species, including *Aspergillus flavus*, not only cause food spoilage, but also produce potent toxins and are pathogenic to humans and animals. Other species such as *A. oryzae* and *A. sojae* are used in food fermentations. Undoubtedly, the most important fungi in this group are the aflatoxigenic moulds, *A. flavus* Link ex Fries, and *A. parasiticus* Speare, and the recently described but much less frequently reported species, *A. nomius*, because these species occur worldwide and cause preharvest aflatoxin contamination in corn, peanuts, cottonseed and treenuts (**Fig. 1**). These seed-inhabiting fungi also contaminate a wide variety of crops after harvest, during handling and during processing. Fungi of this group lack a sexual stage of development, and propagate vegetatively through asexual spores (conidiospores or conidia) or the vegetative cells (filamentous mycelia (**Fig. 2**)).

The three aflatoxigenic species are closely related and have many similarities but, among other characteristics, they can be distinguished by the toxins produced by them (**Figs 3 and 4**). *A. flavus* can produce aflatoxins B_1 and B_2 and cyclopiazonic acid, but only about half the natural isolates are toxigenic. *A. parasiticus* produces aflatoxin G_1 and G_2, in addition to B_1 and B_2, but not cyclopiazonic acid, and

Figure 1 *Aspergillus flavus* growth on a corn ear. (Photo courtesy: Patrick Dowd.) (See also color **Plate 1**.)

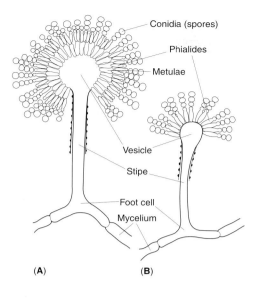

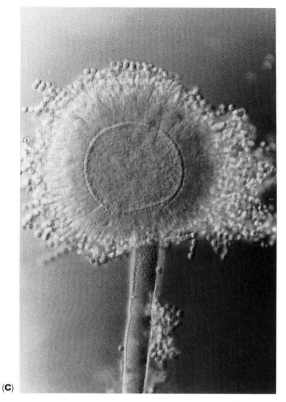

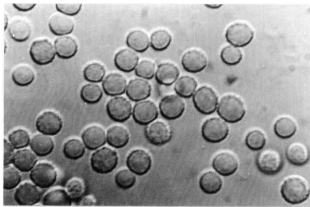

Figure 2 The conidiophore is attached to the mycelium by a characteristically foot-shaped structure. Conidial heads, (**A**) biseriate and (**B**) uniseriate, are characteristic of *Aspergillus flavus*. Electron micrographs of *A. flavus*: (**C**) conidiophore (magnification ×330), (**D**) conidia or spores (magnification ×1875). (Photos courtesy: John Pitt.)

almost all isolates are toxigenic. *A. nomius* is morphologically similar to *A. flavus* but, like *A. parasiticus*, produces B and G aflatoxins without producing cyclopiazonic acid.

Biology and Habitat of *A. flavus*

Aspergillus flavus is found in temperate regions of the world as well as in subtropical regions, whereas *A. parasiticus* is adapted to warmer environments such as the tropical and subtropical regions. Both fungi are present in soil, but airborne conidia of *A. flavus* are more commonly found than those of *A. parasiticus*. Thus, *A. parasiticus* is more commonly associated with peanuts than crop species with above ground seeds. However, *A. flavus* is the most common cause of aflatoxin contamination in peanuts as well as field crops.

From an agronomic perspective, *A. flavus* and *A. parasiticus* are plant pathogens, but living tissue is only a minor substrate for these soil-borne filamentous fungi. From an ecological perspective, *A. flavus* and *A. parasiticus* are considered to be saprophytic. They grow on a wide variety of substrates

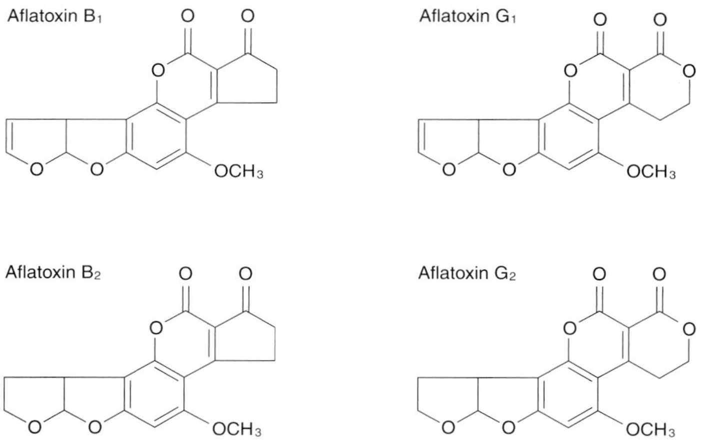

Figure 3 Major toxins produced by *Aspergillus flavus* (aflatoxins B_1 and B_2) and *A. parasiticus* (aflatoxins B_1, B_2, G_1 and G_2).

including decaying plant and animal debris found in the soil, and compete with the other soil microflora. The two major factors that influence soil populations of these two fungi are soil temperature and soil moisture. *A. flavus* and *A. parasiticus* can grow at temperatures of 12–48°C and at water activity (a_w) as low as 0.80. The optimum temperature for growth is 25–42°C. Fungal growth and conidial germination are ideal at water activity greater than 0.90, and completely inhibited at $a_w < 0.75$. Thus, these organisms are semithermophilic and semixerophytic.

In interactions of *A. flavus* and *A. parasiticus* with their hosts, the primary source of inoculum (predominantly conidia) appears to be the soil. The survival structure of the fungi has not been determined, but existing data suggest that fungal mycelium in debris is likely the primary soil propagule. The presence of sclerotia (highly melanized, compacted mycelial bodies) in infected tissue and in the soil in the southern US argues that these structures may also play an important role in the survival of *A. flavus*. In the case of peanuts, populations of the fungi in the soil are important in the epidemiology of the disease, as the pods appear to be infected once they enter the soil. In other hosts discussed, airborne populations of inoculum appear to be important.

Pre-harvest Contamination

Temperature and moisture have a significant effect on the host–pathogen interaction because they directly affect both the host plant and the fungus. Under conditions optimum for these fungi, i.e. high temperature and low moisture, these fungi thrive and outcompete other soil and plant microflora. Under such conditions, the fungi are able to produce abundant inoculum, and the inoculum is easily dispersed in the air as conidia or windblown debris. These conditions allow *A. flavus* to outcompete other micro-

flora on the kernel surfaces, placing them in an ideal position to colonize seeds that are injured or have a compromised defence system. Under drought conditions, many of the physiological defence systems of the host plant are compromised due to high temperatures and water stress. Further, these conditions often lead to cracks in the seed, which allow the fungi to breach the seed's structural barriers.

Injury, especially that caused by insects, is very important in the epidemiology of the disease. Sporulation of *A. flavus* and *A. parasiticus* is often observed on developing seeds damaged by insects. Injury not only provides an easy entry site for the fungus, but it also causes dehydration of the kernels thus creating a more favourable environment for these fungi. In many cases the damage caused by insects can be severe. In peanuts, however, it has been shown that only minor injury is needed to increase aflatoxin contamination.

Post-harvest Contamination

Species of *Aspergillus* can also rot improperly stored grain and contaminate the grain with mycotoxins. The two major environmental conditions that dictate the success of this infection are temperature and moisture similar to pre-harvest contamination. Properly dried grain does not support growth of aflatoxigenic fungal species. Insect activity in stored products also creates favourable microclimates for fungal growth. Once fungal growth starts, the water of metabolism from the fungus provides sufficient water for further growth and mycotoxin development.

Diversity in *A. flavus* Populations

Aspergillus flavus is composed of several vegetative compatibility groups (VCGs). Hyphal fusion (under very stringent conditions) is limited to strains within the same VCG, and thus genetic exchange across VCGs is restricted. For this reason, physiological and morphological traits are typically more uniform within a VCG than within the species as a whole. Strains of *A. flavus* from different VCGs differ in several characters, including enzyme production, virulence, sclerotial morphology and other physiological traits.

A. flavus also has been divided into two strains, S and L. Isolates belonging to the S group produce numerous small sclerotia and fewer conidia than those of the L strain. The L strain is composed of the so called 'typical' isolates of *A. flavus* which produce larger and fewer sclerotia. The S strains are not restricted to one VCG, but rather belong to a number of VCGs.

Detection and Differentiation of *A. flavus* in Foods and Feeds

Use of Growth Media

The toxigenic *A. flavus* and *A. parasiticus* are very closely related to *A. oryzae* and *A. sojae*, species that are used in the manufacture of fermented foods and do not produce toxins. Obviously, accurate differentiation of related species within section *Flavi* is important in order to determine the potential for toxin production in food and feed.

Generally, detection of *A. flavus* in foods and feeds is carried out by using traditional microbiological plating methods, either by surface spread or direct plating of kernels and seeds. Media used for detection include potato dextrose agar (PDA), acidified PDA and PDA with antibiotics such as chlortetracycline, chloramphenicol, oxytetracycline, gentamycin and streptomycin. Because these fungi are semixerophytic, a selective medium containing up to 7% sodium chloride has been used to isolate *A. flavus*.

A differential medium, called *Aspergillus* differential medium (ADM), contains ferric citrate (0.05%) as the differential ingredient. This compound reacts with *A. flavus* metabolites such as kojic acid and aspergillic acid (Fig. 4) to produce a bright orange–yellow pigment on the reverse side of the colony. Dichloran and chloramphenicol have been added to ADM to make a new medium called *Aspergillus flavus* and *parasiticus* agar (AFPA). This medium contains peptone, 10 g; yeast extract, 20 g; ferric ammonium citrate, 0.5 g; chloramphenicol, 100 mg; agar, 15 g; distilled water, 1 l; and dichloran, 2 mg (the final pH of 6.2). Cultures on AFPA are routinely incubated at 30°C for 42–48 h. Dichloran inhibits spreading of fungi, and chloramphenicol inhibits bacterial contamination. *A. flavus* and *A.*

parasiticus are identified on this medium by production of typical yellow to olive green spores and a bright orange reverse. This medium permits rapid identification of *A. flavus* and *A. parasiticus* (within 3 days) because these fungi grow rapidly at 30°C. Another advantage of the use of this medium is the isolation and identification of potentially aflatoxigenic fungi from other aspergilli. For example, *Aspergillus niger* produces a yellow but not orange reverse colour, and after 48 h of incubation *A. niger* starts to develop its dark brown to black conidia, which easily distinguish it from *A. flavus*. *A. ochraceus* grows relatively slowly at 30°C and the yellow colour appears after 48 h.

The use of a bleomycin-containing medium has been utilized by the fermentation industry to separate aflatoxigenic *A. parasiticus* from the kojic mould *A. sojae*. Growth of both species is reduced by the presence of bleomycin, but *A. sojae* isolates barely germinate or produce microcolonies.

ELISA Methods

Another potential detection method for *A. flavus* is the use of immunoassays. Enzyme-linked immunosorbent assays (ELISA) have been developed to detect different mould species in foods based on the production of extracellular polysaccharides that are cell bound and immunologically active. However, more work needs to be done to develop specific kits for aflatoxigenic species such as *A. flavus*.

Toxin Production

Screening isolates for aflatoxin production can also be used to differentiate *A. flavus*. Cultures can be grown on coconut-cream agar and observed under UV light, or simple agar plug techniques coupled with thin layer chromatography (TLC) can be used to screen cultures for aflatoxin production as an aid to identification.

Morphological Characteristics

Because media can influence the morphology and colour of *Aspergillus*, identification of *Aspergillus* species requires growth on media developed for this purpose. Czapek agar, a defined medium based on mineral salts, or a derivative such as Czapek yeast extract agar, and malt extract agar or Czapek yeast extract – 2% sucrose agar can be a useful aid in identifying species of *Aspergillus*.

The morphology of the asexual reproductive structures is the predominant characteristic that distinguishes the species of *Aspergillus*. Species of *Aspergillus* are identified based on the arrangement of the conidial head, the shape and size of the vesicle, the texture and length of the stipe, the shape, texture,

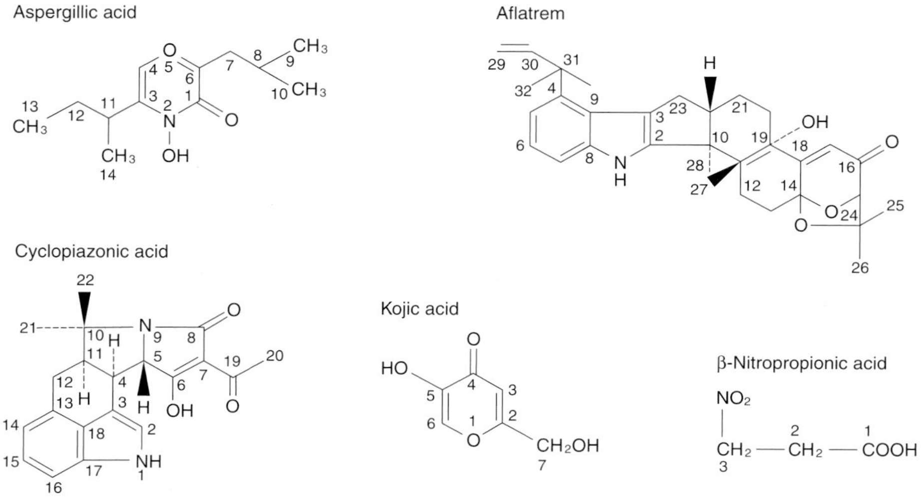

Aspergillic acid

Aflatrem

Cyclopiazonic acid

Kojic acid

β-Nitropropionic acid

Figure 4 Other minor toxic secondary metabolites produced by *Aspergillus flavus*.

and colour of the conidia and the presence or absence of metulae. The stipe, vesicle, phialides and conidia form a structure called the conidiophore.

Stipe The stipe, which is also known as the stalk, is a thick-walled hyphal branch which arises perpendicularly from the foot cell (Fig. 2). The foot cell is a specialized cell characteristic of aspergilli; however, its absence does not prove that the isolate is not from the *Aspergillus* group. The stipes in the *A. flavus* group are rough and hyaline (non-pigmented).

Vesicle The aerial tip of the developing stipe swells to form a structure known as a vesicle. Vesicles of fungi in the *A. flavus* group are elongated to globose. The shape varies somewhat with the composition of the substrate. The diameter is in the range 10–65 μm.

Metulae and Phialides Conidiogenous cells, referred to as phialides (formerly termed sterigmata), develop on the vesicle surface. In some species of *Aspergillus* the phialides are the first layer of cells on the surface of the vesicle. In other species a layer of supporting cells, metulae, form on the surface of the vesicle and give rise to the phialides. Conidia always form by budding of the cytoplasm from the phialide cells. Thus, conidia form in chains with the youngest conidium adjacent to the phialide. Species lacking metulae (e.g. *A. parasiticus*) are termed uniseriate; species with metulae and phialides (e.g. *A. flavus*) are termed biseriate (Fig. 2).

Conidia Conidial colour and microscopic morphology are important in species identification. Conidia (singular: conidium), also called spores, are asexual reproductive structures. Conidia in *Aspergillus* species are single-celled structures that may be uni- or multinucleate. Ornamentation of conidia is the most effective criterion for distinguishing *A. flavus* from *A. parasiticus* (**Table 1**). Conidia from *A. flavus* are smooth to slightly roughened, whereas conidia from *A. parasiticus* are rough or echinulate. The colour of the conidia determines the colour of the conidial head, which in *A. flavus* is green or olive green.

Sclerotia In some cases the characteristics of sclerotia are used in taxonomy. Several groups of aspergilli such as the *A. flavus*, *A. ochraceus*, *A. niger* and *A. candidus* groups produce resistant survival structures called sclerotia (singular: sclerotium). A sclerotium is a hard compact mass of hyphae with a darkened (melanized) rind capable of surviving unfavourable environmental conditions. Sclerotia vary in size and shape, and their colour ranges from yellow to brown or black. *A. flavus* produces large, globose sclerotia whereas *A. nomius* produces vertically elongated sclerotia.

DNA Methods

The morphological methods are time consuming, often requiring 2–3 weeks for accurate identification. Various methods using biochemical differences between the fungi of this group have been developed. Even the degree of nuclear DNA complementarity to determine the relatedness between different members of the section *Flavi* has been used. These methods are used for separating some toxigenic fungi (such as *A. flavus*) from food fermentation fungi (such as *A. oryzae*). More recently, DNA restriction fragment length polymorphism (RFLP) has been used as a tool for rapid differentiation of *A. flavus* from other fungi of the section *Flavi*. These methods have been successfully utilized for limited purposes. However, as more deciphering of the *Aspergillus* genome occurs, differentiation of fungi of the *A. flavus* group may be readily available based on specific DNA profiles.

Economic Significance of *Aspergillus flavus*

Toxin Production

The toxigenic species of the *Aspergillus flavus* group produce a number of toxins (**Table 2**). The best known of these are the aflatoxins. Other toxic compounds produced by *A. flavus* are cyclopiazonic acid, kojic acid, β-nitropropionic acid, aspertoxin, aflatrem and flavutoxin. These secondary metabolites are described in detail.

Aflatoxins Aflatoxin production is the consequence of a combination of fungal species, substrate and environment. The factors affecting aflatoxin production can be divided into three categories: physical, nutritional, and biological factors.

Aflatoxins (Fig. 3) are extremely potent naturally occurring carcinogens that occur in feed for livestock as well as in food for human consumption. Aflatoxin B_1 is the most carcinogenic of the aflatoxins as well as the most abundant, and thus receives the most attention in mammalian toxicology. In fact, aflatoxin B_1 is second in carcinogenicity only to the most carcinogenic family of chemicals known, the synthetically derived polychlorinated biphenyls (PCBs). Aflatoxin B_1 is a hepatocarcinogen in rats and trout, and can induce carcinomas when ingested at rates below $1\ \mu g\ kg^{-1}$ body weight. Significant emphasis has focused on pre-harvest control of aflatoxin con-

tamination, because that is when the fungi first colonize host-plant tissues.

Although *A. flavus* and *A. parasiticus* are considered 'weak' parasites, under favourable environmental conditions they can colonize and infect living plant tissue, and contaminate seeds with aflatoxin. Even in cases of serious aflatoxin contamination the percentage of seeds infected is often low. Because high levels of aflatoxin can be produced in individual seeds, and the tolerance level for aflatoxin contamination is low, even a small number of infected seeds can be economically important because it can result in the rejection of the entire lot of a commodity.

Aflatoxin contamination has received significant publicity since the incidence of these compounds in food and feed is ubiquitous and has occurred in many parts of the world: this has resulted in serious food safety and economic implications for the entire agriculture industry. Aflatoxin content in foods and feeds is, therefore, regulated in many countries. Of the countries that attach a numerical value to their tolerance, the difference between the limits varies significantly. A guideline of 20 parts aflatoxin per billion parts of food or feed substrate (ppb) is the maximum allowable limit imposed by the US Food and Drug Administration for interstate shipment of foods and feeds. European countries are expected to introduce more stringent guidelines that may restrict aflatoxin levels in imported foods (3–5 ppb).

Aflatoxin synthesis has no obvious physiological role in primary growth and metabolism of the organism and is, therefore, considered to be a 'secondary' process. It is known that protein synthesis and con-

Table 1 Key characteristics of aflatoxin-producing fungi

Characteristic	A. flavus	A. parasiticus	A. nomius
Conidiophore arrangement (Metulae)	Mostly biseriate	Mostly uniseriate	Mostly biseriate
Conidia	Almost smooth to moderately roughened, variable in size (3–8 μm)	Conspicuously roughened, less variation in size (4–7 μm)	Similar to *A. flavus*
Sclerotia	Large, globose	Large, globose	Small, elongated
Colony colour	Green	Dark yellow–green	Green
Aflatoxins produced	B_1, B_2	B_1, B_2, G_1, G_2	B_1, B_2, G_1, G_2
Cyclopiazonic acid produced	Yes (most isolates)	No	No

Table 2 Significant mycotoxins produced by the *Aspergillus flavus* group and their toxic effects

Mycotoxin(s)	Toxicity	Species producing
Aflatoxins B_1 and B_2	Acute liver damage, cirrhosis, carcinogenic (liver), teratogenic, immunosuppressive	*A. flavus*, *A. parasiticus*, *A. nomius*
Aflatoxins G_1 and G_2	Effects similar to those of B aflatoxins: G_1 toxicity less than that of B_1 but greater than that of B_2	*A. parasiticus*, *A. nomius*
Cyclopiazonic acid	Degeneration and necrosis of various organs, tremorgenic, low oral toxicity	*A. flavus*

sequently growth decline during the aflatoxin-producing phase (idiophase). As yet, there is no confirmed biological role of aflatoxins in the ecological survival of the fungal organism. However, since aflatoxins are toxic to certain potential competitor microbes in the ecosystem, a survival benefit to the producing fungi is implied. It should be noted, however, that aflatoxin per se is a poor antibiotic. Theories have also been proposed about a possible biological role of aflatoxins or related compounds as deterrents to insect feeding activity on fungal conidia and overwintering structures.

The mode of action, metabolism and biosynthesis of aflatoxins has been extensively studied, particularly in the last decade. The chemical binding of the liver enzyme-activated aflatoxin molecule to animal DNA, causing mutations and possible carcinogenesis, has been elucidated. The chemistry, biochemistry and molecular biology of synthesis of aflatoxins B_1 and B_2 has been understood in significant detail.

Several agronomic practices can reduce preharvest aflatoxin contamination of certain crops. These include the use of pesticides, altered cultural practices (such as irrigation), and the use of resistant varieties. However, such procedures have demonstrated only a limited potential for reducing aflatoxin levels in the field, especially in years of drought when environmental conditions favour the contamination process. Broad areas are being studied for control of aflatoxin contamination. These include: (a) fundamental molecular and biological mechanisms that regulate the biosynthesis of aflatoxin by the fungi, and the ecological and biological factors that influence toxin production in the field; and (b) biochemistry of host-plant resistance to aflatoxin and/or aflatoxigenic fungi. Knowledge in these areas has already aided significantly in the development of novel methods to manipulate the chain of events in aflatoxin contamination.

Cyclopiazonic Acid Cyclopiazonic acid (CPA) is an indole-tetramic acid mycotoxin produced predominantly by several *Penicillium* spp. However, most *A. flavus* (but not *A. parasiticus*) strains produce this compound. CPA is widely distributed in the environment and has been detected in naturally contaminated agricultural raw materials and mixed animal feeds.

The toxicity of CPA has been demonstrated in many animal species, including chicken, rabbit, dog, pig and rat. Treated animals show severe gastrointestinal distress and neurological disorders after ingestion of food contaminated with cyclopiazonic acid. Affected organs, in particular the digestive tract, liver, kidney and heart, show degenerative changes and necrosis.

The co-occurrence of cyclopiazonic acid along with aflatoxins in naturally contaminated agricultural products has not yet been adequately studied.

Miscellaneous **A. flavus** Metabolites

Aspergillic Acid This mycotoxin, on ingestion by mice, produces severe convulsions followed by death. Aspergillic acid is produced by *A. flavus* and *A. sojae*, hydroxyaspergillic acid by *A. flavus*, *A. sojae* and *A. oryzae*, and neoaspergillic acid by *A. flavus*, *A. ochraceus* and *A. sclerotiorum*.

Aflatrem The tremorgen, aflatrem, has been isolated from various strains of *A. flavus* and adds another facet of toxicity to the several toxic metabolites of this fungus. Aflatrem has the ability to produce a hypertensive state in dosed animals with initial inactivity followed by response to auditory and tactile stimuli. Attempts at movement produce marked tremors over the entire body.

Aspertoxin Aspertoxin, a molecule closely related to sterigmatocystin (a precursor of aflatoxins), has been isolated from *A. flavus* and shown to be embryotoxic in the chicken. It is not considered of relevance in animal feedstuffs.

Kojic Acid Kojic acid is produced by various strains of *Aspergillus* and *Penicillum*. It is found in very low concentration in traditional Japanese foods such as miso, soy sauce and saké. Kojic acid is also used as an additive for preventing enzymatic browning and for cosmetics. There is only very limited information about kojic acid toxicity. Kojic acid has been subjected to mutagenicity testing and found to be a weak mutagen in bacteria but not in eukaryotic systems. Reports also suggest that it can induce sister chromatid exchange and chromosomal aberrations.

β-Nitropropionic Acid *A. flavus* is considered by some to be one of the most active heterotrophic nitrifying microorganisms. β-Nitropropionic acid (BNPA) is probably involved in the nitrification pathway of *A. flavus*, and is suggested as a key intermediate in formation of nitrates. The toxicity of BNPA is not established, but inorganic nitrates per se are relatively nontoxic to humans and animals except when reduced to nitrites prior to ingestion or reduced within the gastrointestinal tract prior to absorption.

Aspergillus flavus as Allergens and Animal Pathogens

Several allergic and infective conditions of humans and certain other vertebrates are caused by *Aspergillus* species. These include allergic bronchiopulmonary aspergillosis and invasive pulmonary aspergillosis. The most common cause of most of these conditions is *Aspergillus fumigatus*. However, other aspergilli, including members of the *A. flavus* group are sometimes implicated.

Ecological Benefits

Although *A. flavus* group fungi are not commonly recognized as beneficial, these ubiquitous organisms become dominant members of the microflora under certain circumstances and exert multiple influences on both biota and environment. These fungi are important degraders of crop debris and may play roles in solubilizing and recycling crop and soil nutrients. *A. flavus* can even degrade lignin. As insect pathogens, these fungi may serve to limit pest populations and have been considered potential agents to replace chemical pesticides.

Many insects typically carry *A. flavus* isolates internally. Excretion of large quantities of diverse enzymes is a characteristic of the *A. flavus* group. Insect use of enzymes excreted by the fungus that degrade or detoxify plant products can result in a symbiotic *A. flavus*–insect relationship.

Conclusion

The *Aspergillus* section *Flavi* comprises a metabolically diverse group of fungi. Species within this group are either prized for their many industrial applications or feared for the toxins they produce. Of the latter group, *A. flavus* is the best known species. It occurs in warm temperate and subtropical climates all over the world. Although the fungus is not a very aggressive pathogen, under weather conditions conducive for its growth. *A. flavus* can colonize seeds in the field and contaminate them with aflatoxin. Because of its ability to grow at low water activity, *A. flavus* is also well adapted to colonize seeds of grain and oil crops in storage where exposure of seed to moisture is purposely limited. Control methods have been developed for post-harvest control of aflatoxin contamination, but there are no effective control strategies to prevent aflatoxin accumulation in the field when conditions are favourable for the fungus. Several tools are now available for the detection of *A. flavus* and for distinguishing *A. flavus* from related species used in the fermentation industry. Newer molecular tools being developed promise to make this distinction even easier. *A. flavus* itself is composed of a diverse population, and isolates of the fungus may differ in several morphological and physiological traits. As few as half of the strains from some locations produce aflatoxin. A better understanding of the population genetics of *A. flavus* and the genetics of secondary metabolism in this fungus is helping in the development of new control strategies to eliminate pre-harvest aflatoxin contamination resulting in a safer, economically viable food and feed supply.

See also: **Aspergillus**: Introduction. **Enzyme Immunoassays**: Overview. **Mycotoxins**: Classification. **Nucleic Acid-based Assays**: Overview. **Spoilage of Plant Products**: Cereals and Cereal Flours. **Spoilage Problems**: Problems Caused by Fungi.

Further Reading

Bennett JW and Klich MA (eds) (1992) *Aspergillus – Biology and Industrial Applications*. Stoneham, Massachusetts: Butterworth-Heinemann.

Bhatnagar D, Lillehoj EB and Arora DK (eds) (1992) *Handbook of Applied Mycology*, vol. 5. *Mycotoxins in Ecological Systems*. New York: Marcel Dekker.

Cotty PJ, Bayman P, Egel DS and Elias KS (1994) Agriculture, Aflatoxins and Aspergillus. In: Powell KA, Renwick JF and Peberdy JF (eds) *The Genus Aspergillus*. P. 1. New York: Plenum Press.

Eaton DL and Groopman JD (eds) (1994) *The Toxicology of Aflatoxins: Human Health, Veterinary and Agricultural Significance*. New York: Academic Press.

Gourama H and Bullerman LB (1995) *Aspergillus flavus* and *Aspergillus parasiticus*: aflatoxigenic fungi of concern in foods and feeds: a review. *Journal of Food Protection* 58: 1395–1404.

Hocking AD, Doyle MP, Beuchat, MR and Montville TJ (eds) (1997) *Toxigenic Aspergillus Species in Food Microbiology – Fundamentals and Frontiers* Washington, DC: ASM Press.

Klich MA and Pitt J (1992) *Laboratory Guides to Common Aspergillus Species and Their Teleomorphs*. North Ryde, NSW, Australia: CSIRO Division of Food Processing.

Payne GA and Brown MP (1998) Genetics and physiology of aflatoxin biosynthesis. *Annual Review of Phytopathology* 36: 329–362.

Powell KE, Renwick JF and Peberdy JF (eds) (1994) *The Genus Aspergillus – from Taxonomy and Genetics to Industrial Application*. New York: Plenum Press.

Samson RA and Pitt JI (eds) (1990) *Modern Concepts in Penicillium and Aspergillus Classification*. New York: Plenum Press.

Sinha KK and Bhatnagar D (eds) (1998) *Mycotoxins in Agriculture and Food Safety*. New York: Marcel Dekker.

Smith JE and Henderson RS (eds) (1991) *Mycotoxins and Animal Foods*. Boca Raton: CRC Press.

Atomic Force Microscopy *see* **Microscopy**: Atomic Force Microscopy.

ATP BIOLUMINESCENCE

Contents

Application in Meat Industry

Derrick A Bautista, Department of Applied Microbiology and Food Science, University of Saskatchewan, Canada

Introduction

Meat production is a major contributor to the world's food supply. Unfortunately, meat products have been documented as one of the predominant sources of food-borne illness. Outbreaks of food-borne illnesses involving infectious *Escherichia coli*, *Salmonella* spp. and *Campylobacter jejuni* are an indication that food industries may require a better means of assessing microbial levels in food products. One of the problems associated with conventional microbiological techniques is the incubation time required to obtain results (e.g. more than 24 h). Present methodologies are inadequate for the needs of the food industry in determining quality and safety of their products. Although there are 'systems' approaches to drive the momentum of safe food production, there is a need for protocols that can evaluate the hygienic condition of the processing system effectively, accurately and in 'real time'. Unfortunately, conventional microbiological techniques are inadequate methods for real-time analysis of food production systems.

There have been several developments in microbiology designed to speed up the determination of microbial populations in food samples. One approach that may be of practical use to the meat industry is 5′-adenosine triphosphate (ATP) bioluminescence. This has been used to determine microbial levels in a variety of food products and has been shown to be as reliable as plate count procedures. The major advantage of the ATP bioluminescence assay is the speed of the test – results are produced within 15–60 min – which would allow a manufacturer to assess the hygiene of food products and equipment during a production run.

The ATP Bioluminescence Reaction

Adenosine triphosphate is a universal energy transfer molecule that is found in all living cells. It is a nucleotide identical to the molecule found in RNA. The phosphate bonds of the molecule are the major source of energy release (**Fig. 1**). It is typically used for synthesis of amino acids, protein synthesis, active transport systems, etc. Research has suggested that the level of ATP in a sample could be used to measure biomass. However, several assumptions are made if this were applied to bacteria:

- all living organisms contain ATP
- ATP is neither associated with dead cells nor absorbed onto surfaces
- the level of ATP among taxa is fairly consistent given a set environmental conditions and metabolic activity.

Quantifying the level of intracellular ATP in a sample gives an indirect measurement of the number of cells in that sample. An easy method to quantify ATP levels is to rely on the production of light from the bioluminescence assay.

Bioluminescence is the biological production of light from various animals and fungi. A common occurrence in nature is the intermittent glow from the American domestic firefly (*Photinus pyralis*). Research has shown that ATP is a major constituent of the bioluminescence reaction from the firefly and that the evolution of light is directly proportional to the amount of ATP present.

In the firefly, light production is a catalytic reaction between luciferin and luciferase which is fuelled by

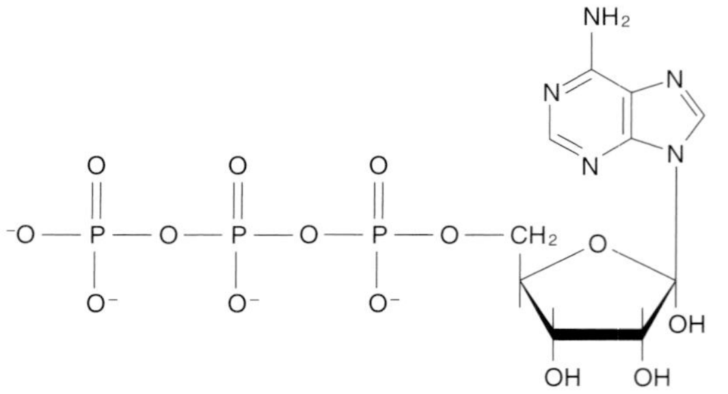

Figure 1 Chemical structure of adenosine 5′-triphosphate.

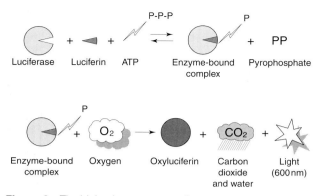

Figure 2 The bioluminescence reaction.

ATP. In more detail, luciferase combines with luciferin to form an unstable enzyme-bound luciferyl adenylate molecule. The molecule will react with oxygen to form oxyluciferin, CO_2, water and light at 600 nm (**Fig. 2**). The reaction is stoichiometric (i.e. 1 ATP molecule yields 1 photon of light), enabling evolution of light to be used as an index of ATP level.

Use of ATP Bioluminescence in Food Assays

A means of quantifying microorganisms in a sample can be developed based on the assumptions concerning the ATP content in microorganisms and the stoichiometry of bioluminescence reaction.

The ATP content of microorganisms can be used as a fuel source for the luciferase enzyme in place of the ATP found in firefly tails. Therefore, the light output generated from this bioluminescence reaction should be proportional to the total ATP found in the microbial population. The reaction would be instantaneous and can be easily monitored by a light-measuring device such as a luminometer (**Fig. 3**). However,

extraction of microbial ATP from food samples may prove more of a challenge since most food products will contain a certain amount of non-microbial ATP. Earlier research showed that determination of microbial content in food was difficult because of interference by background ATP from food products. This was especially true for meat products. Background ATP concentrations can be equivalent to ATP levels found in bacterial populations of 1×10^5 colony forming units (cfu) per millilitre or more. Therefore, it is imperative that the background ATP must be minimized in food samples to increase the sensitivity of the ATP bioluminescence assay.

Most ATP bioluminescence assay kits for food purposes employ some system of minimizing non-microbial ATP. This may include a non-ionic detergent (e.g. Trition X-100) to break open somatic cells, sonication, low-speed centrifugation or chromatographic techniques (e.g. ion exchange resins). In some protocols, filtration or an apyrase (an ATP hydrolysing enzyme) may be used to ensure reduction of background ATP from the sample.

After concentration of cells, bacteria can be combined with acids (e.g. trichloroacetic acid), organic solvents, strong cationic detergents or boiling buffers to release intracellular ATP from microorganisms. An illustration of the ATP assay can be found in **Figure 4**.

Raw Meat Materials Of the several ATP bioluminescence assay kits developed for food applications, very few have been developed specifically for meat products. In most cases, kits originally developed for other food products or other applications have been used (e.g. raw milk quality, fruit juice, hygiene

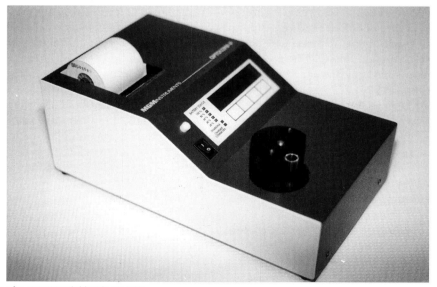

Figure 3 Example of a commercial luminometer.

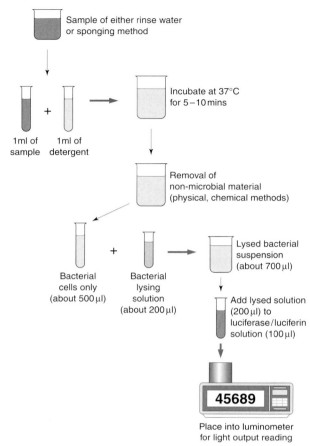

Figure 4 A typical ATP bioluminescence assay for food products.

monitoring). The main problem found by all researchers who applied the technique to raw meat products was the level of background and somatic ATP from sample preparation.

Conventional microbiological sampling protocol for meat products requires homogenization prior to microbial analysis by plate count. Unfortunately, meat tissues can contain a large number of ATP molecules and homogenization can release a tremendous amount of background ATP. The large flux of background ATP cannot be accommodated by most ATP bioluminescence assays. The interference associated with this problem can be so great that the reliability of the test within a meat product or between samples cannot be assured. To overcome the problem with homogenization, several sampling protocols have been developed that attempt to minimize the amount of free ATP.

The 'rinse-bag' method is one approach that is widely used by many researchers. Samples of meat (i.e. excised sample) or poultry carcasses are placed into a stomacher bag with an aliquot of diluent (sterile distilled water, 0.1% peptone, etc.). The stomacher

bag is either mechanically or manually shaken for 2 min to rinse the bacteria off the sample. The liquid is then used for microbiological analysis.

Another method employs the use of sterile sponge (Nasco, Fort Atkinson, US) pre-wetted with a diluent (usually 0.1% peptone). The sponge is swabbed aseptically over a target area on the animal carcass, placed in a bag containing diluent and homogenized to liberate any microorganisms. The liquid expressed from the sponge is used for microbiological analysis.

One company (Celis-Lumac, Cambridge, UK) has developed a swabbing procedure to determine total viable counts on meat carcasses. A cotton-tipped baton is moistened with a wetting solution and used to swab $25\,cm^2$ of a carcass surface. The baton is placed into a buffer and mixed to liberate the cellular material. The suspension is used for microbiological analysis.

In all cases, each strategy attempts to remove surface bacteria only and minimizes the evolution of background ATP from the meat tissues. However, some background ATP will still be present and additional steps are required to remove this non-microbial ATP; these may include a prefiltration step and/or detergents. However, strong detergents, such as Tween 80, should be avoided in any diluent with the rinse-bag method since they may adversely affect the conformation of the luciferase enzyme and thereby inhibit bioluminescence output. If they must be used, the sample must be thoroughly rinsed free of the detergent to avoid any adverse effect on the luciferase enzyme.

Several researchers have successfully increased the sensitivity of the test down to $100–1000\,cfu\,ml^{-1}$. In each protocol, the key to the success of the assay was the clarification step that eliminated or degraded non-microbial ATP and/or other interfering components (e.g. lipids or organic acids). One system used lipase in addition to somatic detergents and a coarse prefiltration step prior to analysis of microbial ATP. Other protocols may require a patented detergent/hydrolysing agent which effectively degrades somatic cells and any free ATP.

Correlations between the modified ATP bioluminescence assays and plate counts have been very good (r 0.80–0.95, $P < 0.05$) using the rinse-bag, sponge or swab methods (**Fig. 5**, **Table 1**).

Finished Meat Products   The application of ATP bioluminescence to finished meat products (e.g. cooked ham) may be easier than assaying comminuted raw meat materials. This is due partly to the low levels of ATP and the destruction of live cells after cooking procedures. The added advantage of cooking procedures makes the removal of non-microbial ATP

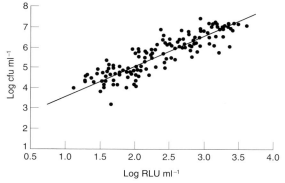

Figure 5 Relationship between ATP bioluminescence readings and plate count for determining the microbial load in poultry 'carcass rinse' samples (*n* = 149). RLU, relative light units. Courtesy of IAMFES, Inc.

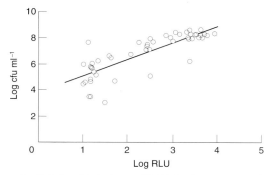

Figure 6 Relationship between ATP bioluminescence and plate count for determining the microbial load in a ham product (*n* = 50). RLU, relative light units.

less of an ordeal than with raw materials. In fact, homogenized samples may be used instead of the rinse-bag or sponging methods. Correlation between the ATP bioluminescence assay and plate count for finished meat products can be good (*r* 0.82, S_{yx} 0.59, *P* < 0.001) (**Fig. 6**).

BactoFoss: Automation of the ATP Bioluminescence System An automated system, BactoFoss, has been developed by FOSS (MN, USA). It uses several protocols for extracting non-microbial ATP and quantifying microbial ATP in a self-contained, automated system.

A meat sample (about 10 g) is combined with a diluent (0.85% NaCl, 0.1% peptone) and homo-

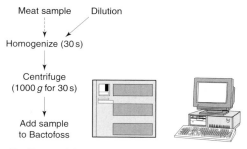

Figure 7 Protocol for preparing meat samples for the BactoFoss automated luminometer.

genized in a stomacher for 30 s. To clarify the suspension, an aliquot of the homogenized mixture is centrifuged (350 *g* for 30 s) to precipitate meat tissues. The supernatant is used with the BactoFoss, which automatically takes out the necessary sample volume and performs the measurement (**Fig. 7**). The procedure for microbial ATP analysis by the machine consists of:

● Intake of sample.
● Lysing of somatic cells.
● Washing of debris and non-microbial cells.
● Extraction of microbial ATP (with detergent).
● Measuring of light.
● Output of results.

Microbiological Analysis of Pork and Beef Products using the BactoFoss One study correlated bioluminescence results for pork and beef samples analysed using the BactoFoss machine with standard aerobic plate procedures. The pork samples (*n* = 70) had microbial levels of contamination between 3×10^3 cfu g^{-1} and 5×10^7 cfu g^{-1}; the beef samples (*n* = 65) had microbial levels between 7×10^2 cfu g^{-1} and 7×10^9 cfu g^{-1}. All samples were analysed simultaneously by both the BactoFoss and standard aerobic plate count procedures (**Fig. 8**, **Fig. 9**). The BactoFoss has a calibration feature which increases accuracy of the ATP bioluminescence assay. Under calibration mode, a correlation coefficient of 0.93 and residual standard deviation (S_{yx}) of 0.23 log cfu g^{-1} were obtained for pork samples between 1×10^5 cfu g^{-1} and 5×10^7 cfu g^{-1}. For beef samples a correlation coefficient of 0.94 was achieved under calibration mode.

Table 1 Correlation between plate counts and ATP bioluminescence assays using various methods to remove somatic ATP

Meat product	Method used	No. of samples	Correlation coefficient
Beef	Rinse-bag	159	0.83
Beef	Sponge	400	0.92
Beef	Swabbing	111	0.95
Poultry	Rinse-bag	149	0.85
Pork	Sponge	320	0.93
Pork	Swabbing	71	0.93

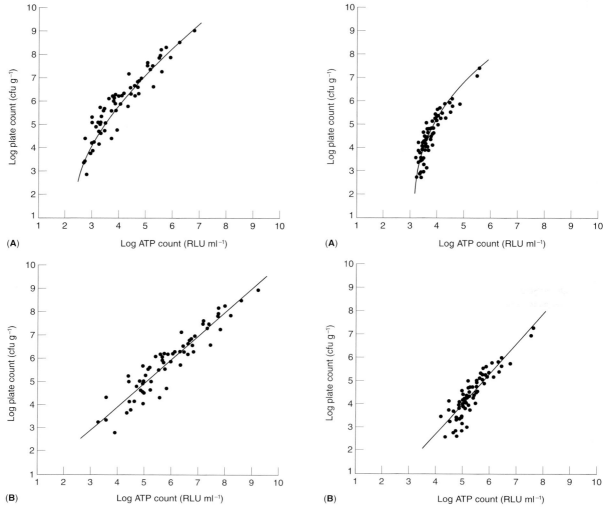

Figure 8 Relationship between ATP bioluminescence and plate count for determining the microbial load in beef samples (*n* = 65). (**A**) without calibration feature. (**B**) with calibration feature. Reproduced from Bautista et al (1998), courtesy of CRC Press, Inc.

Figure 9 Relationship between ATP bioluminescence and plate count for determining the microbial load in pork samples (*n* = 70). (**A**) without calibration feature. (**B**) with calibration feature. Courtesy of CRC Press, Inc.

Detection of *Escherichia coli* O157:H7

Several attempts have been made to use ATP bioluminescence assays for detecting specific types of bacteria, mainly pathogenic. One approach included a selective pre-enrichment procedure prior to the assay to propagate target bacteria. By using the ATP bioluminescence with selective pre-enriched medium, a large generation of light would indicate the presence of target bacteria. This method has been somewhat successful in vitro. However, target bacteria on meat samples were not as readily detectable.

Another approach is to use serological techniques to capture and concentrate target bacteria or their byproducts and then use the ATP bioluminescence assay to detect them. GEM Biomedical (Hamden, Conn.; USA) have developed an immunocapture method that involves a test tube coated with specific antibodies for O antigen of *Escherichia coli* O157. Initially, meat or other food sample is added to a broth supplemented with a selective agent (e.g. EC broth with novobiocin) and incubated at 42°C for 4 h. A sample is withdrawn and added to a tube coated with the antibody for the somatic O antigens found on *E. coli* O157. If *E. coli* O157 or other O157 types are present in the food sample, the target bacteria (i.e. bacterial antigens) will become attached to the coated tube. The tube is aspirated and washed several times to remove debris and non-target antigens. A conjugate of the primary antibody is added that has been covalently coupled to the luciferase enzyme. After a second washing step to remove unbound conjugate antibodies, bioluminescence reagents (luciferin and ATP) are added. The mixture is placed into a luminometer

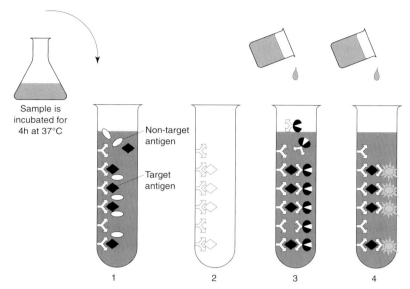

Figure 10 Bioluminescence immunoassay kit for detecting *E. coli* O157:H7. 1, load tube; 2, wash tube – antigens remain attached; 3, add conjugate antibody and wash; 4, engage bioluminescence reaction.

and any light signal production will be indirectly correlated with the bound antigens. From start to finish, the entire assay takes about 7 h (**Fig. 10**).

Using pure cultures of *E. coli* O157, results show that the sensitivity is approximately 8×10^3 cfu ml^{-1} for several types of nutrient broth. Using meat samples inoculated with various levels of *E. coli* O157, the sensitivity of this assay was 10–100 cfu g^{-1}. The system has also been developed for generic *E. coli* and *Salmonella* spp.

The Role of ATP Bioluminescence in Meat Processing

Currently, ATP bioluminescence assays are used for validation of hygiene for sanitation programmes in hazard analysis critical control point (HACCP) systems where it is necessary to assess the overall contamination of both food residues and microflora.

Some researchers have investigated the application of ATP bioluminescence to real-time monitoring of process waters during chill immersion. The technique involved concentrating bacteria from prechill or chill water by a filtration method and analysing the microbial ATP by lysis and bioluminescence. When compared with plate count methods, the ATP bioluminescence assay produced similar results (*r* 0.85), but in a fraction of the time (< 15 min). It was suggested that the modified ATP bioluminescence assay would be useful for immediate action where problems of contaminated process water exist, and that recycling of water in the chill immersion areas

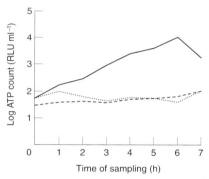

Figure 11 Determination of microbial contamination of poultry process waters using the ATP bioluminescence assay. Solid line, scald water; dotted line, prechill; dashed line, chill. Courtesy of Poultry Science Association, Inc.

could be regulated according to the results of this assay. This application may improve the economy of water usage during chill immersion (**Fig. 11**).

The ATP assay could also be useful for the validation of hygiene of carcass surfaces during poultry processing allowing determination of the level of cleanliness of the bird in real time. In this application, a swabbing procedure was used to determine the level of cleanliness of birds at several points along a processing line. The areas were analysed by both plate count and ATP bioluminescence assay, then compared for interpretation (**Table 2, Fig. 12**). With the plate count method, contamination levels on poultry carcasses were consistent (*P* > 0.20) between stations 1 and 10, but there was a significant (*P* < 0.001) drop in contamination at stations 11 and 12. It was proposed that the cleanliness of the carcasses was

Table 2 Stations analysed for carcass cleanliness at a poultry processing plant

Station	Location
1	Shower area 1
2	Shower area 2
3	Evisceration
4	Inspection station 1
5	Crop removal
6	Neck cutter
7	Decapitation
8	Carcass vacuuming
9	Inside/outside carcass washing
10	Inspection station 2
11	Prechilling of carcasses
12	Chilling of carcasses

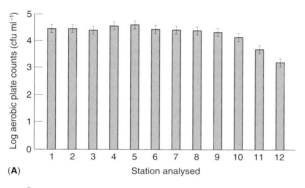

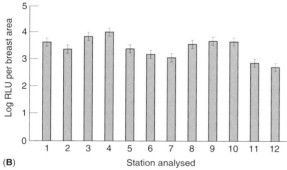

Figure 12 The microbiological quality of key areas in a poultry processing facility by (**A**) plate count and (**B**) the ATP bioluminescence assay. Reproduced from Olsen (1991), courtesy of Elsevier Science, Inc.

improved at stations 10 and 11 owing to a dilution effect during the prechilling and chilling areas. With the ATP bioluminescence assay, similar significant ($P < 0.001$) reductions were observed for station 11 and 12, but some variation between stations 1 and 10 was also observed. Levels of ATP were higher on carcasses sampled at stations 3 and 4 and stations 8–10. The higher levels of ATP at stations 3 and 4 were associated with the evisceration process. At stations 8–10 the higher ATP levels were attributed to removal of the head and crop. The results suggest that the ATP bioluminescence assay could have an immediate application as an effective feedback mechanism to allow

for correction when contamination levels are inappropriate.

Advantages and Disadvantages of ATP Bioluminescence

Conventional microbiological techniques using cultural media require 24–72 h of incubation time before results can be interpreted. In that span of time, meat products can be further processed, packaged and delivered to the retailer before a problem can be detected. This can contribute to poor product quality due to large numbers of spoilage microflora, or foodborne illness due to large numbers of pathogenic bacteria. Therefore, conventional microbiological techniques should be re-evaluated as the industry's standard for ensuring food quality and safety.

The short turnover time is the major advantage of the ATP bioluminescence assay. This advantage could come into play when raw material requires microbiological analysis, allowing poor-quality materials to be identified quickly and removed from the production process. Meat processors could therefore be assured that the microbiological quality of raw materials for further processing will always be high, and that finished products will be of the best quality and have good shelf life. This reason alone has interested many meat processors in incorporating the ATP bioluminescence assay into their quality assurance programmes.

The protocol of the test can be easily understood and requires very little training. In fact, proper 'aseptic technique' is all that is required to perform the test. There is no need for special facilities or any additional equipment. The manufacturers of the ATP bioluminescence assays provide the necessary equipment (luminometer, pipette aids, tube holders, etc.) to allow the user to begin testing immediately. In addition, most manufacturers have an excellent level of support for their systems to accommodate questions or problems.

However, the ATP bioluminescence assay at present is being used as an equivalent to total viable microbial count procedure. This may be an inconvenience when specific bacteria such as pathogens need to be identified and, especially, quantified. Another disadvantage of the ATP bioluminescence assay is the lower detectable limit. Research has shown that the lowest level of microorganisms that can be accurately determined by this technique is approximately 1×10^3 cfu ml^{-1}; this is because of the inability of these assays to remove non-microbial ATP completely from food samples. Further research must be directed towards the total and consistent removal of non-microbial ATP from food samples to improve the sensitivity and variability of the assay.

Table 3 Measure of agreement between the ATP bioluminescence assays and plate counts for determining surface contamination on beef and poultry carcasses

Cutoff level by plate count (log cfu ml⁻¹)	Corresponding ATP count [a] (log cfu ml⁻¹)	Observed agreement	κ value[b]
Beef carcasses			
4.0	3.43	74.1	0.50
5.0	3.98	90.0	0.82
6.0	4.52	82.3	0.71
Poultry carcasses			
4.0	1.31	93.9	0.15
5.0	2.03	88.6	0.74
6.0	2.61	88.6	0.76

[a] Based on linear relationship; $y = mx + b$.
[b] Values between 0.50 and 0.60 indicate good agreement; values greater than 0.70 indicate very good agreement.

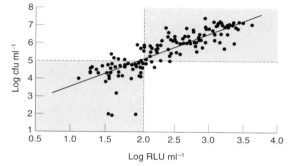

Figure 13 An example of predictive quartiles used to set up a platform rejection test for rinse water at 1×10^5 cfu ml⁻¹. The shaded area represents observed agreement between the ATP bioluminescence assay and the plate count.

In their present form, ATP bioluminescence assays may be inadequate if governing regulations require exact numbers of bacteria. However, most microbiological analysis by food companies does not totally rely on the exact enumeration of populations of microorganisms. Instead, quality assurance has interpreted results based on the acceptable level of microorganisms in particular food products. Therefore, the ATP bioluminescence assay should be use in the same capacity as conventional microbial analysis, that is based on the criteria that correspond to the appropriate cutoff limit (**Table 3**, **Fig. 13**). In one study of this approach, using a simple linear relationship, predictive quartiles were set at 1×10^4 cfu cm⁻², 1×10^5 cfu cm⁻² and 1×10^6 cfu cm⁻², and with corresponding ATP count levels. Using the equivalent cutoff of 1×10^5 cfu cm⁻² of the ATP bioluminescence assay for beef samples, 90% of the samples were accurately assigned to the predictive quartile. Using the equivalent cutoff of 1×10^6 cfu cm⁻², beef samples were accurately predicted 82.3% of the time by the ATP bioluminescence assay. Another method used a statistical calculation as an indication of agreement between the two types of tests. A κ test value of 50–60% indicates good agreement and values above 70%

indicate very good agreement. Based on this test, the agreement between the ATP bioluminescence assay and plate count was shown to be satisfactory.

See also: **Acetobacter**. **ATP Bioluminescence**: Application in Dairy Industry; Application in Hygiene Monitoring; Application in Beverage Microbiology. **Bacteriophage-based Techniques for Detection of Food-borne Pathogens**. **Biophysical Techniques for Enhancing Microbiological Analysis**: Future Developments. **Electrical Techniques**: Food Spoilage Flora and Total Viable Count (TVC). **Rapid Methods for Food Hygiene Inspection**. **Immunomagnetic Particle-based Techniques**: Overview. **Total Viable Counts**: Pour Plate Technique; Spread Plate Technique; Specific Techniques; MPN; Metabolic Activity Tests; Microscopy. **Ultrasonic Imaging**. **Ultrasonic Standing Waves**.

Further Reading

Bautista DA, Sprung W, Barbut S and Griffiths MW (1998) A sampling regime based on an ATP bioluminescence assay to assess the quality of poultry carcasses at critical control point during processing. *Food Research International* 30: 803–809.

Bautista DA, Vaillancourt JP, Clarke RA, Renwick S and Griffiths MW (1994) Adenosine triphosphate bioluminescence as a method to determine microbial levels in scald and chill tanks at a poultry abattoir. *Poultry Science* 73: 1673–1678.

Bautista DA, Vaillancourt JP, Clarke RA, Renwick S and Griffiths MW (1995) Rapid assessment of the microbiological quality of poultry carcasses using ATP bioluminescence. *Journal of Food Protection* 58: 551–554.

Olsen O (1991) Rapid food microbiology: application of bioluminescence in the dairy and food industry – a review. In: Nelson WH (ed.) *Physical Methods for Microorganisms Detection*. P. 64. Boca Raton: CRC Press.

Siragusa GR, Cutter CN, Dorsa WJ and Koohmaraie M (1995) Use of a rapid microbial ATP bioluminescence assay to detect contamination on beef and pork carcasses. *Journal of Food Protection* 58: 770–775.

Stanley PE (1989) A concise beginner's guide to rapid microbiology using adenosine triphosphate (ATP) and luminescence. In: Stanley PE, McCarthy BJ and Smither R (eds) *ATP Luminescence: Rapid Methods in Microbiology.* P. 1. Oxford: Blackwell.

Stannard CJ and Gibbs PA (1986) Rapid microbiology: application of bioluminescence in the food industry – a review. *Journal of Bioluminescence and Chemiluminescence* 1: 3–10.

Application in Dairy Industry

Wim Reybroeck, Department for Animal Product Quality and Transformation Technology, Agricultural Research Centre CLO-Ghent, Melle, Belgium

Due to its complex biochemical composition and high water content, milk is an excellent nutrient medium for spoilage organisms, the so-called saprophytic bacteria. Their multiplication depends mainly on temperature and competing microorganisms and on their metabolic products. The possible presence of pathogenic microorganisms in raw milk and dairy products can lead to a potential food-borne health disease hazard.

Bioluminescent measurement of adenosine 5′-triphosphate (ATP), mostly very simple and rapid, is suitable for monitoring most critical control points in routine use.

Why measure ATP?

When a cellular constituent is used to reflect the presence of living microorganisms, it must fulfil two fundamental criteria: being present in all living cells and disappearing after the death of the cell. ATP is present in all cells, prokaryote and eukaryote alike (**Fig. 1**). It constitutes an indispensable intermediate in energy transfer. Most ATP is found inside living cells, a smaller amount is found in extracellular fluids. When cells are dying, production of ATP is stopped and existing ATP is quickly degraded. Measurement of ATP is therefore fundamental to the study of living processes.

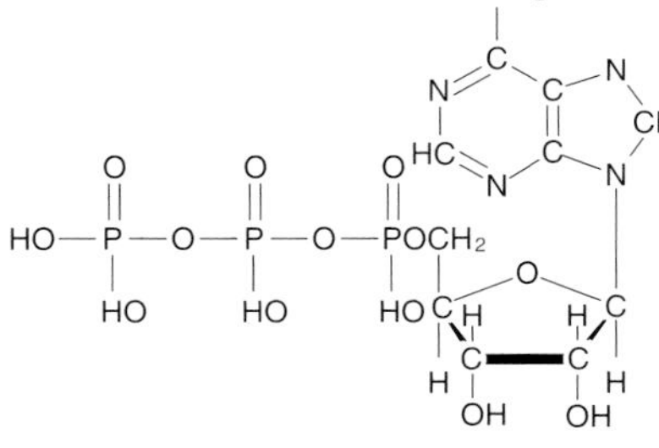

When using ATP to estimate the number of microorganisms, it is assumed that the intracellular content of ATP is fairly constant in all bacterial cells involved. Generally, bacteria contain about 1 fg (femtogram (fg) = 10^{-15} g) ATP per cell or about 1.6×10^{-18} mol ATP.

ATP Bioluminescence

The firefly luciferin-luciferase assay is the most sensitive method available for measuring ATP, based on the following high-quantum-yield reactions:

$$\text{Luciferase}$$

D-luciferin + ATP $\dashrightarrow$ Luciferyl-AMP + PPi

Luciferyl-AMP + O_2 $\dashrightarrow$ Oxyluciferin + AMP + CO_2 + light

Luciferin is converted with ATP into a form capable of being catalytically oxidized by luciferase. The oxyluciferin in an excited state returns to a stable situation with the emission of light (yellow-green, (λ_{max} = 565 nm) and the formation of CO_2 and AMP. With constant concentrations of luciferase/luciferin and oxygen, the intensity of light emitted is directly proportional to the ATP concentration.

The amount of ATP can thus be indirectly quantified by measurement of the light intensity with a luminometer. The result is expressed in relative light units (RLUs).

The quantity of ATP reacting per unit of time and the intensity of light produced is influenced by different factors. Therefore it is necessary to work with purified luciferase under controlled conditions (pH, buffers, temperature).

Besides chemical quenching, the result can also be influenced by optical quenching, caused by absorption of a part of the emitted light by the colour or the turbidity of the sample. Quenching effects can be corrected using internal standardization.

There are many possible assay schemes for ATP analysis using firefly luciferase bioluminescence. Sample preparation is highly important, and depends on what ATP-containing component is to be measured and the physical properties of the sample. The sample preparation procedure should minimize inhibition of luciferase/luciferin light output by sample components or assay conditions.

Figure 1 Adenosine 5′-triphosphate (ATP).

It is important that the ATP measurement procedure is designed so that the proper ATP pool is measured. Often, biological samples contain many different types of cells, as well as free ATP in solution.

ATP in Milk and Dairy Products

ATP present in milk has been shown to be of both cellular and extracellular origin, and can be classified into three different groups:

- *Free ATP.* Milk contains free ATP, in solution or partly associated with the calcium phosphate citrate caseinate micelles. The quantity of free ATP is mostly limited due to hydrolysis by ATPases present in raw milk.
- *Somatic ATP.* Raw milk also contains somatic cells. The cell types found are epithelial cells, neutrophil granulocytes, lymphocytes and monocytes. The average somatic cell count in farm milk is about 200 000–300 000 ml^{-1}, with an ATP content of about 300 fg per cell, resulting in an important amount of somatic ATP in raw milk. There is also a direct relation between the number of somatic cells in raw milk and the level of ATPases found. In heat-treated (UHT) milk the somatic ATP is degraded.
- *Microbial or bacterial ATP.* To estimate the bacteriological quality of raw milk, control on spoilage bacteria or monitoring of sterility, the microbial ATP is the target ATP pool. The sample must be pretreated in order to remove or eliminate all non-bacterial ATP. Often, selective lysing of somatic cells with non-ionic detergents is used, followed by an enzymatic hydrolysis of the unwanted ATP with apyrase or a physical step like filtration.

ATP Bioluminescence Applications in Dairy Industry

Screening of Hygiene of Dairy Farms and Milk Tankers

Control of milking equipment and storage tanks at dairy farms is useful, and gives a good indication of cleanliness. Swabs are taken from clusters, rubber ware and milk lines in order to detect not only the presence of viable microorganisms, but also milk residues, giving accurate information about the total hygiene. The test method has potential value for troubleshooting and optimization of cleaning systems.

The test procedure is simple now, thanks to the availability of single-dose, ready-to-use, self-contained swabs and small fully portable luminometers. The assay can therefore also be performed at the farm, since no aseptic laboratory conditions are required.

ATP bioluminescence is a practical, useful and rapid technique for assessing the hygienic status of milk tankers. Swabs are taken from different sampling points (internal surfaces of manhole lid, vessel roof, vessel side wall, flexible hose and air elimination vessel) and there is an assessment of the total ATP load of the rinse water. Cut-off RLU values used to represent clean/dirty criteria for judging the hygiene of the area tested are different depending on the kit system used. As cut-off standards, 50 RLU (Celsis.Lumac, Cambridge, UK) and 500 RLU (Biotrace, Bridgend, UK) are used in practice. Using ATP methods to monitor tanker cleaning efficiency facilitates a rapid response to hygiene problem-solving and potentially contributes positively to the improvement of the quality of transported raw milk.

Assessment of the Bacteriological Quality of Raw Milk

The quality of raw materials used in processing largely influences the final product quality. The presence and multiplication of saprophytic bacteria in raw milk might change the milk composition and affect the manufacturing process (e.g. yield of cheese production). Moreover, the flavour of the raw milk may be adversely influenced and heat-stable bacterial enzymes may continue to act in the product, particularly during long storage periods, even at low temperature. The stability and/or flavour of cream and UHT milk may be changed. The extent of these changes is mainly dependent on the numbers, species and activity of the respective bacteria of the microflora.

Low numbers of saprophytic microorganisms and very low numbers or absence of pathogenic microorganisms including mastitis pathogens are the main criteria for milk of high hygienic value. In the EU the legal requirement for the bacteriological quality of raw milk is described in the Milk Hygiene Council Directive 92/46 EEC of 16 June 1992, laying down the health rules for the production and placing on the market of raw milk, heat-treated milk and milk-based products. In Chapter 4 of Annex A standards to be met for collection of raw milk from the production holding or for acceptance of treatment or processing establishments are defined. Raw cow's milk, intended for the production of milk-based products, must meet the following bacteriological standard: plate count 30°C (per ml) $\leq 100\,000$ (geometric average over a period of 2 months, with at least two samples a month).

As a consequence there is an increased demand from the dairy industry for a simple, sensitive and rapid method for assessing the bacteriological quality of raw milk upon arrival at the dairy plant in order

Table 1   Assay protocol of the Celsis.Lumac milk bacteria kit

50 µl raw milk in luminescence cuvette
+ 100 µl NRS/Somase (somatic extractant/apyrase)
   5 min incubation at room temperature (enzymatic hydrolysis
   of non-microbial ATP)
+ 100 µl L-NRB (bacterial extractant)
   30 s extraction of bacterial ATP
+ 100 µl Lumit-PM (luciferin/luciferase)
   10 s integration of light signal in luminometer (Biocounter)

The result is recorded in relative light units. The corresponding cfu ml^{-1} can be read from the correlation curve.

to meet European regulations and internal quality standards. Methods based on ATP bioluminescence or direct epifluorescent filter technique (DEFT) are suitable for this aim.

Different tests based on ATP bioluminescence have been developed for this purpose. All tests using a protocol without a bacterial concentration step lack sensitivity because of the presence of large amounts of non-bacterial ATP in raw milk (**Table 1**).

Measurement was also hindered by the quenching effect of milk on luminescence triggered by firefly luciferase (**Table 2**). A better selectivity and sensitivity was obtained with the introduction of a filtration or centrifugation step in the ATP protocol. In this way a detection level of 1–4 × 10^4 cfu ml^{-1} could be obtained. However during trials most of these protocols were found to be cumbersome and labour-intensive.

The Bactofoss (Foss Electric, Hillerød, Denmark) was the first automatic instrument for estimating the total bacterial count based on ATP bioluminescence. Before analysis the milk is homogenized with a Bactofoss Homogenizer 60 (Foss Electric, Hillerød, Denmark) in order to separate bacterial clumps and fat globules. The sample is then placed in the sample holder on the Bactofoss and measurement (90 s) is automatically performed, followed by a rinsing cycle (90 s). According to the internal calibration used, the light signal measured is mathematically transformed into an estimate of bacteria (cfu ml^{-1}).

Results of trials indicated serious underestimation

Table 3   Protocol of the Celsis.Lumac Raw Milk Microbial Kit

Filtration funnel
Mounting of 8.3 mm nylon membrane filter (0.8 µm NHZ)
Filling of temperature-controlled filtration funnel (40°C)
   300 µl prewarmed non-ionic detergent L-NRS®
   300 µl raw milk
   4 min incubation = extraction of somatic ATP
Vacuum filtration: 93.3 kPa (700 mmHg)
Rinsing filter: 600 µl Ringer ($\frac{1}{4}$ strength) 40°C

Transfer of filter to cuvette

Biocounter
Cuvette in Biocounter™ M2500
Automatic injection 200 µl NRM™
   30 s extraction of bacterial ATP
Automatic injection 100 µl Lumit®-QM
   10 s integration
Result in relative light units (RLU)

Rinsing filtration funnel
   2 × 600 µl ATP-free sterile distilled water

of the total viable counts in many milk samples. This could possibly be caused by an incomplete extraction of the ATP out of the bacteria (protective fat layer). Consequently, this instrument is no longer used for this application, despite its perfect mechanical functioning.

The Promega kit (Enliten™ Milk Total Viable Organisms Assay) for the assessment of the total bacterial count of raw milk uses a special clarifying agent to concentrate the bacterial cells in a pellet by centrifugation (5 min; 12 000 g). After washing and redissolving the pellet, the bacterial ATP is extracted and assayed in a bioluminescent way. However, the method did not prove useful for instant-checking of the microbiological quality of the load of incoming milk tankers because working in batches is not practical.

The Raw Milk Microbial Kit (Celsis.Lumac, Cambridge, UK), is a simplified and user-friendly semi-automatic ATP method replacing the more cumbersome ATP-F test (**Table 3**). The first part of this assay is performed in a temperature-controlled fil-

Table 2   Commercial ATP methods for judging the bacteriological quality of raw milk

Name	Reference	Test time (min)	Sensitivity (cfu ml^{-1})
Protocols without concentration of the milk bacteria			
Milk Bacteria Kit (ATP platform test)[a]	Bossuyt 1982	5	3 ×5^5
Protocols with concentration of the bacteria by filtration			
ATP-F test	Van Crombrugge et al 1989	7	2 × 10^4
Bactofoss	Eriksen and Olsen 1989	3	1 × 10^4
Biotrace Milk Microbial ATP Kit	Griffiths et al 1991	7	2 × 10^4
Raw Milk Microbial Kit[a]	Reybroeck and Schram 1995	6	1 × 10^4
Protocols with concentration of the bacteria by centrifugation			
Enliten milk total viable organisms assay	Pahuski et al 1991	7	1 × 10^4

[a]Still commercially available as a kit.

tration system (Biofiltration System, Celsis.Lumac, Cambridge, UK). After lysing the somatic cells and concentrating the bacteria from the milk on a nylon membrane filter using vacuum filtration (under pressure 93.3 kPa), the minifilter is transferred to a flat-bottomed luminescence cuvette and introduced into a luminometer (Biocounter M-2500, Celsis.Lumac, Cambridge, UK). After automatic injection of the cationic detergent, bacterial ATP is extracted and assayed by automatic addition of luciferase and luciferine. The intensity of emitted light is integrated over 10 s and expressed in RLU.

A single sample is tested in less than 6 min. A filtration system with two separate filtration funnels allows analysis of at least 15 samples in an hour by one operator. Interpretation of the result is easy with a correlation curve cfu/RLU or using cut-off RLU values.

A bacterial concentration of 10^4 cfu ml^{-1} can be considered as the detection limit. The standard deviation of repeatability (s_r) of the ATP filtration method is 0.048 log cfu ml^{-1}, irrespective of the number of bacteria in the milk. There is a good relationship between the bacterial concentration and the relative light intensity obtained, as indicated by a low standard error of estimate (s_{yx}) and a good bacteriological grading of the milk samples. No influence was found of fat, protein or somatic cells on the ATP test. Obstruction of the membrane filter could occasionally occur with a high concentration of somatic cells.

An ATP reading of 4700–5700 RLU can be used as qualitative cut-off value to differentiate raw milk with total bacterial counts greater than or less than 10^5 cfu ml^{-1}. In one study it was reported that approximately only 1 in every 75 samples might be falsely accepted or rejected. The conversion line and the cut-off value depend on the dominant microflora present in the raw milk and other factors such as frequency of milk collection. In milk stored for a longer time at low temperature, the activity of the microorganisms decreases, resulting in a lower ATP content per cell.

It is therefore advisable for each user to determine a conversion line by analysing several milk samples with both the reference and the ATP method.

Different investigators have demonstrated that the Raw Milk Microbial kit is a practical and reliable cost-effective technique for assessing the hygienic quality of raw milk.

Shelf-life Prediction of Dairy Products

The keeping quality of correctly pasteurized milk and cream may be severely reduced by microbial growth, mainly of Gram-negative psychrotrophic bacteria, able to grow at low temperatures. These heat-sensitive organisms are supposed to be absent after pas-

teurization, but can be present by post heat treatment contamination before and during the filling stage. Gram-positive bacteria and spores are more thermoduric and hence can survive the pasteurization process but normally heat-shocked and injured cells do not start to grow if the dairy product is stored refrigerated (<7°C) from production down the chain to the consumer. All protocols (**Table 4**) for rapid detection (within 28 h of production) of post-pasteurization contamination are based on an ATP measurement after a pre-incubation step in order to allow the present Gram-negative bacteria to grow to a measurable number. During pre-incubation, the growth of Gram-positives can be suppressed by addition of selective inhibitors. A few examples of ATP assays are described in **Table 5** and **Table 6**.

ATP Assay for Psychrotroph Proteases

Raw milk stored at low temperature for prolonged periods can contain a high number of psychrotrophic bacteria. Psychrotrophs, mainly *Pseudomonas* strains (e.g. *P. fluorescens*), which themselves are killed in the heating process may produce extracellular heat stable enzymes capable of breaking down proteins and fats. Heat resistant proteases affect the stability of the milk during the UHT heat treatment and during subsequent ambient storage (off flavours and gelation). Lipases can result in cheese yield losses and cheese bitterness.

The amount of psychrotrophic protease can be measured using a bioluminescent assay based on the principle that protease degrades luciferase and hence reduces the amount of light emitted. The decrease in light emission is directly proportional to the protease concentration in the sample.

Sterility Testing of UHT Dairy Products

To some extent, UHT treatment is replacing sterilization because heat denaturation is minimized due to the short high-temperature period. Aseptic filling is an essential part of the UHT process. To ensure that sterility has been achieved, a proportional number (0.1–0.3% of the total production run) of original packages are incubated for 48 h at 28–30°C and checked afterwards by traditional microbiological methods, pH measurement or ATP bioluminescence (**Table 7**).

Methods based on ATP measurements are not only fast but also have the advantage of detecting spoilage microorganisms, difficult to culture on traditional media or bacteria not changing the product pH during growth. The limit of detection is less than 10 organisms per package (initial). In practice a result exceeding 2 × RLU background measurement is mostly used as cut-off level. To obtain such a level an increase of

Table 4 Shelf-life prediction tests for pasteurized milk: pre-incubation procedures and test time

Authors	Inhibitors	Time/temperature	Assay time (min)
Waes and Bossuyt (1982)	0.06% benzalkon A 50% 0.002% crystal violet	24 h/30°C	46
Shelf Life Prediction Test (Charm Sciences)[a]	–	18–27 h/21°C	4
Pasteurized Milk Screen Kit (Celsis.Lumac)[a]	0.06% quaternary ammonium salt 0.002% inhibitive dye	24 h/25°C	16

[a]Commercially available as a kit.

Table 5 Assay protocol of the Celsis.Lumac Pasteurized Milk Screen Kit

50 μl pre-incubated milk in luminescence cuvette
+ 100 μl NRS/Somase (somatic extractant/apyrase)
 15 min incubation at room temperature (enzymatic hydrolysis of non-microbial ATP)
+ 100 μl L-NRB (bacterial extractant)
 30 s extraction of bacterial ATP
+ 100 μl Lumit-PM (luciferin/luciferase)
 10 s integration of light signal in luminometer (Biocounter)
Result < 100 RLU: absence of Gram-negative bacteria
100 RLU < result < 150 RLU: reconfirmation after 6 h extra sample incubation
Result > 150 RLU: presence of Gram-negative bacteria

Table 6 Assay protocol of the Charm Shelf Life Prediction Test for Pasteurized Milk

Labelled test tube
+ Reagent E tablet (apyrase)
+ 100 μl Diluent E (buffer)
+ 100 μl pre-incubated milk
 5 s mixing
 3 min incubation at 35°C (enzymatic hydrolysis of non-microbial ATP)
+ content of SLP liquid vial (bacterial extractant)
 5 s mixing
 let particles settle
Transfer of sample liquid into SLP tablet vial (luciferin/luciferase), leaving the solids on bottom of test tube
 (10 s) shaking until tablet is completely broken apart
5 s measurement of light signal in luminometer (LUMinator T)
RLU ⩾ control point: positive microbial activity, keeping quality problems may be expected
Control point: 2 × (average 6 negative control samples)

Table 7 Assay protocol of the Celsis.Lumac Milk Sterility Test Kit

50 μl pre-incubated milk in luminescence cuvette
+ 10 μl Somase (0.0125 U apyrase)
 15 min incubation at room temperature (enzymatic hydrolysis of non-microbial ATP)
+ 100 μl L-NRB (bacterial extractant)
 30 s extraction of bacterial ATP
+ 100 μl Lumit-PM (luciferin/luciferase)
 10 s integration of light signal in luminometer (Biocounter)
Result < 2 × RLU$_{control}$: sterility; result > 2 × RLU$_{control}$: non-sterility

about 5×10^4 cfu ml^{-1} of microorganisms is needed. Sometimes two or three sample aliquots are combined in the same luminescence cuvette, reducing the reagent costs. Complete automated tests are now available with the use of an autosampler and an auto-luminometer. The Biotrace Dairy Kit assay (Biotrace, Bridgend, UK) for example is automatically performed in the wells of a microplate (**Table 8**).

Sterility control of UHT cream, infant formula milk, chocolate milk, soya milk, dairy dessert products and UHT yoghurt drinks is also possible with the application of similar ATP test procedures. Dairy products which contain fruit with an acidic pH can also be assayed by buffering the sample before the apyrase is added. A fast check of sterility allows a fast release for sale, saving storage costs and reducing the need for large warehouse facilities.

Monitoring of (Starter) Culture Activity

ATP measurement is a good parameter for assessing metabolic activities of living cells. A practical application for the dairy industry is the determination of the activity of starter cultures. Starter cultures tested are bacterial strains used for yoghurt and cheese production.

The measuring protocol is based on limiting the quenching effects (optical effects by turbidity of the coagulated milk and a pH effect) by diluting the disturbing factors and correcting the pH. By choosing an appropriate dilution (1/100 or 1/1000) of the culture

Table 8 Assay protocol of the Biotrace Dairy Kit

Pipette 50 μl pre-incubated UHT milk (e.g. 30°C for 48 h) into a microwell
Place micotitre plate in the Microbial Luminescence System (MLS) instrument
Press the 'UHT Assay' button
The following steps are carried out automatically:
 ATPase is injected into each sample
 Non-microbial ATP is destroyed by apyrase (15 min incubation)
 Extractant is added and microbial ATP is released (30 s incubation)
 Enzyme is added and microbial ATP levels are measured
 Results are analysed by system software
Pass/fail results are presented on screen

sample it is possible to obtain a good linearity between 10^6–10^9 cfu ml^{-1}. The Dairy Products Sterility Test Kit (Celsis.Lumac, Cambridge, UK) or the Yogurt Culture Activity Test (Charm Sciences, Malden, MA, US) can be used.

Detection of Inhibitory Substances

The occurrence of residues of antimicrobials and the presence of inhibitors in milk, both resulting from mastitis therapy of cows, is still a significant problem for the dairy industry. To ensure both a technological and toxicological safety, milk has to be tested. For screening microbiological inhibitor tests and fast substances/group-specific immunotests are commonly applied. The test principle of microbial inhibition tests is usually based on the detection of growth inhibition noticed visually by interpreting the colour change of a pH indicator.

The effect of inhibitors on the growth activity of test strains can also be monitored rapidly with ATP measurements with a good sensitivity (0.007 IU ml^{-1} of penicillin G). No commercial kits are available at present.

Recently the Valio T102-test based on a detection by bacterial luminescence was developed by Valio (Helsinki, Finland), using a genetic modified *Strepto-coccus salivarius* ssp. *thermophilus* strain. Instead of acid production, the light production is monitored, allowing the test time to be reduced from 4.5 h to about 2 h, with nearly identical sensitivity.

Determination of Somatic Cell Count and Mastitis Control

Mastitis, an inflammation of the cow's udder, is the most important health problem in dairy cows. Once an infection is established, an increase in the concentration of somatic cells in milk occurs.

In the EU, following the standards defined in Chapter 4 of Annex A of the Milk Hygiene Council Directive 92/46 EEC of 16 June 1992, raw cow's milk intended for the production of milk-based products must have a somatic cell count (per ml) $\leqslant 400\,000$ (geometric average over a period of 3 months, with at least one sample a month).

Taking into consideration that the somatic cells present in the milk are living cells and thus contain ATP, it is possible to evaluate the somatic cell count in milk by measuring the (somatic) ATP level. It has been demonstrated that the ATP content in the somatic cells in raw milk samples correlates well with somatic cell count (fluoro-opto-electronic cell count). ATP was proven to be a good indicator of mastitis, important in mastitis control.

Besides the possibility of using common ATP reagents (BioOrbit, Turku, Finland; Celsis.Lumac,

Table 9 Charm SomaLite procedure

Remove the plunger from a 3 ml syringe

Add 2 ml Buffer S to a syringe with mounted filter

Add 1.0 ml milk directly to the Buffer S in the syringe

Insert plunger and slowly filter dropwise; the somatic cells are captured on the filter surface, while microbial cells and free ATP pass through

Rinse the filter with 2 ml Buffer S

Dry the filter by fully raising and depressing the plunger three times

Position the syringe/filter over the microtube containing the barrier and tablet (enzyme)

Pour liquid (extractant) of microtube into the syringe and filter dropwise into tube containing the barrier and tablet

Cap the tube and tap vigorously until barrier dissolves and tablet disperses in the liquid

Shake tube for extra 10 s

Within 10 s, measure light signal in luminometer (LUMinator T) on 'Cell Assay' channel

Relative light unit (RLU) reading is equivalent to somatic cell count ml^{-1}

Milk samples with 300 000 or more somatic cells ml^{-1} are read with good correlation to Direct Microscopic Somatic Cell Count (DMSCC)

Cambridge, UK; Promega Corporation, Madison, US) with a customer designed protocol, Charm Sciences (Malden, MA, US) commercializes the SomaLite, a 2 min single service test for somatic cell count in milk (**Table 9**).

Other Dairy Industry Applications

Other possibilities for the use of ATP measurement include the monitoring of yeast contamination in yoghurt, the determination of bacteria, yeasts and moulds in cheese and the rapid monitoring of micro-

Table 10 Manufacturers of ATP or bacterial bioluminescence kits

Bio-Orbit Oy, PO Box 36, FIN – 20521 Turku, Finland
Tel: +358 2 410 1100 Fax: +358 2 410 1123
bioorbit@bioorbit.com

Biotrace Ltd, The Science Park, Bridgend, CF31 3NA, UK
Tel: +44 (0) 1656 768844 Fax: +44 (0) 1656 768835
www.biotrace.com

Celsis.Lumac, Cambridge Science Park, Milton Rd, Cambridge CB4 4FX, UK
Tel: +44 (0) 1223 426008 Fax: +44 (0) 1223 426003
www.celsis.com

Charm Sciences Inc., 36 Franklin St, Malden, MA 02148–4120, US
Tel: +1 781 322 1523 Fax: +1 781 322 3141 www.charm.com

Promega Corp., 2800 Woods Hollow Rd, Madison, WI 53711-5399, US
Tel: +1 608 274 4330 Fax: +1 608 277 2516 www.promega.com

Valio Ltd, Meijeritie 6, PO Box 10, FIN – 00039 Helsinki, Finland
Tel: +358 10381121 Fax: +358 9 562 5068
www.valio.fi

bial contamination in rinse and process water (**Table 10**).

In the future other commercial applications, such as more specific detection of pathogens, spoilage bacteria and enzymes can be expected, based on bioluminescent immuno- and probe assays, the measurement of luciferase expression in reporter gene studies, bacterial luminescence or a combination of ATP bioluminescence with image analysis.

Conclusions

The microbiological safety and the quality of the dairy products has to be ensured by the application of Hazard Analysis and Critical Control Point (HACCP). ATP bioluminescent assays, mostly rapid and sensitive, can help to achieve this aim by playing an important role in verifying the adequacy of cleaning, the testing of both raw materials and ingredients and the verification of finished products.

See also: Acetobacter. **ATP Bioluminescence**: Application in Meat Industry; Application in Hygiene Monitoring; Application in Beverage Microbiology. **Bacteriophage-based Techniques for Detection of Food-borne Pathogens**. **Biophysical Techniques for Enhancing Microbiological Analysis**: Future Developments. **Electrical Techniques**: Food Spoilage Flora and Total Viable Count (TVC). **Immunomagnetic Particle-based Techniques**: Overview. **Rapid Methods for Food Hygiene Inspection**. **Total Viable Counts**: Pour Plate Technique; Spread Plate Technique; Specific Techniques; MPN; Metabolic Activity Tests; Microscopy. **Ultrasonic Imaging**: Non-destructive Methods to Detect Sterility of Aseptic Packages. **Ultrasonic Standing Waves**.

Further Reading

Bell C, Stallard PA, Brown SE and Standley JTE (1994) ATP-Bioluminescence techniques for assessing the hygienic condition of milk transport tankers. *Int. Dairy J.* 4: 629–640.

Bell C, Bowles CD, Toszeghy MJK and Neaves P (1996) Development of a hygiene standard for raw milk based on the Lumac ATP-bioluminescence method. *Int. Dairy J.* 6: 709–713.

Bossuyt R (1982) A 5 minute ATP platform test for judging the bacteriological quality of raw milk. *Neth. Milk Dairy J.* 36: 355–364.

Campbell AK (Ed.) (1988) *Chemiluminescence. Principles and Applications in Biology and Medicine.* Cambridge: VCH Publishers.

Champiat D and Larpent J-P (Eds) (1993) *Bio-chimiluminescence. Principes et Applications.* Paris: Masson.

Eriksen B and Olsen O (1989) Rapid assessment of the microbial status of bulk milk and raw meat with the new

instrument: BactoFoss. In: *ATP Luminescence. Rapid Methods in Microbiology.* p. 175. Oxford: Blackwell Scientific Publications.

Griffiths MW, McIntyre L, Sully M and Johnson I (1991) Enumeration of bacteria in milk. In: *Bioluminescence and Chemiluminescence, Current Status.* p. 479. Chichester: John Wiley & Sons.

Harding F (Ed.) (1995) *Milk Quality.* London: Blackie Academic & Professional.

Pahuski E, Martin L, Stebnitz K, Priest J and Dimond R (1991) Rapid concentration procedure for microorganisms in raw milk. *J. Food Protect.* 54: 813.

Reybroeck W and Schram E (1995) Improved filtration method to assess bacteriological quality of raw milk based on bioluminescence of adenosine triphosphate. *Neth. Milk Dairy J.* 49: 1–14.

Stanley PE, McCarthy BJ and Smither R (1989) (eds) *ATP Luminescence. Rapid Methods in Microbiology.* Oxford: Blackwell Scientific Publications.

Van Crombrugge J, Waes G and Reybroeck W (1989) The ATP-F test for estimation of bacteriological quality of raw milk. *Neth. Milk Dairy J.* 43: 347–354.

Waes GM and Bossuyt RG (1982) Usefulness of the benzalkon-crystal violet-ATP method for predicting the keeping quality of pasteurized milk. *J. Food Protect.* 45: 928–931.

Application in Hygiene Monitoring

Mansel W Griffiths, Department of Food Science, University of Guelph, Ontario, Canada

Introduction

Bioluminescence is defined as the production of light by a biological process. The bioluminescence reaction most widely studied is that of the firefly. The mechanism by which fireflies emit a flash of light was first analysed and identified by William McElroy in 1947. McElroy found that light emission was the result of an enzyme-catalysed reaction involving adenosine triphosphate (ATP). The reaction can be represented as follows:

$$ATP + D\text{-}luciferin + Mg^{++} \xrightarrow{luciferase} AMP + oxyluciferin + CO_2 + PP_i + light$$

The reaction is stoichiometric and so the amount of light generated is proportional to the concentration of ATP present. All living cells contain ATP because it is the universal energy donor molecule for cellular metabolism, so the amount of ATP present in a sample is related to the number of living cells present. The ATP content can be easily assayed using the luciferase/luciferin reaction wherein the amount of light produced (and hence the amount of ATP present) can be measured using an instrument called a luminometer, of which there are several types.

Sources of ATP

As described above, ATP is essential for all living cells, so in foods ATP is associated with animal or plant cells as well as with any microbial cells that may be present. In addition, many foods contain extracellular ATP associated with the food matrix. For example, in milk ATP is present in the somatic cells derived from the mammary gland, in microbial cells contaminating the milk, and also as free ATP linked to calcium ions in the casein micelles of milk. Thus, by analysis of ATP present on a food contact surface a good indication of its overall cleanliness can be obtained.

To sanitize a food contact surface, it must first be washed to remove any food residues that could act as a nutrient source which might subsequently support the growth of microorganisms. Following washing, the surface is treated with a sanitizing agent to kill the residual microflora. If either process is not done properly, food particles and/or microorganisms can remain on the surface and may compromise the safety and storage life of foods produced afterward. Normal swabbing and plate counting procedures only detect microbial contamination of the surface and may not indicate whether the surface has been properly cleaned. However, ATP bioluminescence detects contamination by food residues as well as microorganisms, and thus is a more reliable indicator of the overall hygienic condition of the area tested. As the luciferase/luciferin reaction is very fast, with results obtained in approximately 2 min rather than the days required by methods involving traditional plate counts, the cleaning crew can respond effectively if deficiencies in the sanitizing process are detected. If the surface is found to have a high ATP level it can be cleaned again before production resumes. The simplicity and rapidity of the ATP bioluminescence technology has led to the development of several commercial systems for hygiene monitoring based on ATP bioluminescence.

Arguably the best illustration of the usefulness of being able to detect food residues as well as microorganisms on surfaces by the ATP bioluminescence assay was provided in a study of cleaning operations in the kitchens of health care institutions. In this study, surfaces of the blades and feeding trays of meat slicers at three kitchens were swabbed before use, immediately after use and after cleaning. The swabs were then analysed by an ATP bioluminescence assay and by plate count. The results indicated that for two of the six samples taken the plate count indicated that the respective surface was clean immediately after use

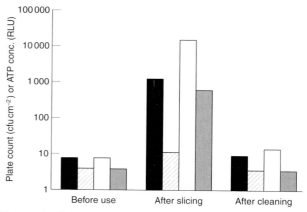

Figure 1 Comparison of results achieved using ATP bioluminescence and plate count to assess the cleanliness of meat slicers in institution kitchens. The results presented are the average from three establishments. Blade ATP, black bar; blade plate count, hatched bar; feed tray ATP, white bar; feed tray plate count, stippled bar. (Data of Seeger and Griffiths, 1994)

but before cleaning, whereas all these surfaces were deemed to be dirty by the ATP bioluminescence assay, which detected ATP in cooked meat residues remaining on the blades and feeding trays (**Fig. 1**).

Methods of ATP Hygiene Monitoring

Several ATP monitoring systems are available (**Table 1**). They are similar in that they involve swabbing a surface, extracting ATP from the swab using a surfactant, and then measuring the ATP present by monitoring the quantity of light emitted after addition of luciferase and luciferin in a buffer containing Mg^{++} ions. The systems differ in the way that the swab sample is presented to the reagents. Several manufacturers have devised disposable swabbing devices in which the whole test can be carried out without the need for pipetting of reagents. These devices also offer the advantage of being less wasteful than kits that require a minimum number of tests to be performed once the reagents are formulated. Each manufacturer also supplies luminometers for use in conjunction

Table 1 Major manufacturers and distributors of commercial ATP bioluminescence hygiene monitoring tests

Instrument/test	Manufacturer
Bio-Orbit	Bio-Orbit Oy, Turku, Finland
Hy-Lite	E. Merck, Darmstadt, Germany
Inspector/System SURE	Celsis Lumac plc, Cambridge, UK
Lightning	IDEXX Labs Inc., Westbrook, ME, USA
Luminator/PocketSwab	Charm Sciences Inc., Malden, MA, USA
Uni-Lite/Uni-Lite Xcel	Biotrace Ltd, Bridgend, UK

with their test system. The instrument readings are generally in arbitrary units called relative light units (RLU), but the values obtained with different luminometers are not necessarily compatible unless they are calibrated against standard concentrations of ATP.

Before hygiene monitoring using ATP bioluminescence is implemented within a processing plant or food service establishment, criteria for the pass/fail cutoff RLU values should be determined by monitoring critical sites within the operation at times when they are known to be clean or dirty. At the same time it should be verified that the correct detergent concentration, contact times and temperature of application are being achieved. During the evaluation period it is prudent to sample surfaces made of different materials, as the nature of the surface material and the sanitation protocol used for each surface may result in variations in the ATP concentration associated with an effective cleaning regime. In this way, realistic target levels can be set.

Many luminometers also come equipped with ports that allow easy transfer of data to computers, and some companies offer software for trend analysis. This allows constant reappraisal of the RLU value that represents a clean surface. The value of trend analysis is illustrated by a study carried out in a typical dairy operation. The ATP bioluminescence hygiene monitoring data (expressed as RLU) were collected over a period of 3 months from a critical control point at a milk filling machine and analysed in retrospect by applying appropriate statistical process control (SPC) tools like cusum charts and individuals charts. The analysis showed that the cusum and individuals charts established a proper trend analysis of the RLU data which was able to provide an advance warning of potential out-of-control (fail) critical control point status. By employing SPC, it is possible to prevent hygiene test failures. Furthermore, if the SPC technique of identifying assignable and unassignable causes of failure were adopted, the total number of failed surfaces should decrease. This would lead to more effective hygiene management and more efficient production.

There have been few independent evaluations of the performance of commercial ATP bioluminescence hygiene monitoring kits, and none in which all the systems have been compared. The most comprehensive comparison was performed under the auspices of *Food Quality* magazine. In this study five systems were compared for sensitivity and repeatability by adding a known volume of bacterial cultures and food liquids (chosen to represent the types of organic residue likely to be found in environmental samples) to swab tips; however, no surface sampling was carried out. The study concluded that potential users of the technology should evaluate systems in their own facilities and that ATP bioluminescence should be used in conjunction with microbiological testing.

Reagents and Instruments

For the ATP bioluminescence hygiene monitoring tests to be reliable, the reagents used in the assay must produce a consistent light signal and be stable over the projected shelf life of the kit. The user must be satisfied that the test performs satisfactorily, and it is the responsibility of the supplier to introduce adequate safeguards to ensure that the product meets specifications. An important constituent of the ATP bioluminescence assay is the extractant used to release the analyte from cells. It is important that the extractant is capable of lysing a broad range of cells in a short period without affecting the activity of the luciferase. Ideally, it should also be able to inactivate ATP hydrolysing enzymes released during lysis of cells. Data on the extractant used by one company has shown that it was 98% efficient in releasing ATP from 71 different bacteria (both Gram-positive and Gram-negative) when compared with a reference method using trichloroacetic acid as the lytic agent.

Perhaps the advance that has had greatest impact on the emergence of ATP bioluminescence as a tool for monitoring hygiene has been the development of constant-light-output reagents. These allow the light generated by the luciferase-luciferin reaction to remain detectable for longer times without decay. This means that the times at which the reagents are added are no longer so important, and these reagents have allowed the use of simple, portable luminometers to detect the signal, so that assays can be performed directly in the plant when and where they are needed.

Sensitivity

Theoretical limits of detection (TDL) for various soil types have been determined for five commercially available test kits (**Table 2**). In a subsequent study comparing three hygiene monitoring systems, a much poorer sensitivity was reported when ATP standard solutions were examined, but the systems had sensitivities comparable to those found in the first study when raw meat juice was analysed.

There is continuing debate as to whether the sensitivity of these ATP bioluminescence hygiene monitoring kits is adequate to assess the sanitary status of food contact surfaces. There are few recommendations as to what constitutes an acceptably clean surface. A maximum microbial load of $40\ cfu\ cm^{-2}$ has been proposed as an acceptable standard. Thus, if an area of $5\ cm \times 5\ cm$ ($25\ cm^2$) is sampled, as recommended by many of the manu-

Table 2 Theoretical detection limits for ATP bioluminescence hygiene monitoring kits

Soil type	Detection limit for most sensitive assay	Detection limit for least sensitive assay
Pure ATP	0.48 fmol	14 fmol
Lactobacillus sp.	1.3×10^3 cfu ml^{-1}	1.2×10^5 cfu ml^{-1}
Pseudomonas sp.	1.3×10^3 cfu ml^{-1}	2.0×10^4 cfu ml^{-1}
Yeast	5.0×10^2 cfu ml^{-1}	5.4×10^4 cfu ml^{-1}
Raw meat juice	1100-fold dilution	59-fold dilution
Cooked meat juice	640 000-fold dilution	43 000-fold dilution
Fresh bean rinse	3300-fold dilution	110-fold dilution
Canned bean juice	95-fold dilution	9.2-fold dilution

(Data of Flowers et al., 1997)

facturers of hygiene monitoring kits, an unacceptable surface would yield more than 1000 microorganisms, which is close to the limit of the most sensitive ATP bioluminescence hygiene monitoring kits. This has created an interest in ways to improve the sensitivity of the ATP bioluminescence reaction.

An ATP recycling reaction that uses an enzyme cascade to amplify low levels of ATP has been proposed as a way of increasing the sensitivity of the luciferase reaction. The AMP generated by the luciferase/luciferin reaction is converted to ADP by the enzyme myokinase and this ADP, in the presence of phosphoenolpyruvate and pyruvate kinase, is subsequently used to regenerate ATP which feeds back into the luciferase-luciferin reaction. The time for the reaction to reach half of the maximum light output is directly related to the log of the ATP concentration, and thus to levels of cleanliness. It is claimed that this system can detect levels of ATP as low as 26 pmol l^{-1}. However, a luminometer that is capable of following reaction kinetics is required to monitor light output, and this may limit its application.

Another method of improving the sensitivity of the ATP bioluminescence reaction is to generate ATP using adenylate kinase, an essential enzyme in virtually all cells. The enzyme catalyses the following reaction:

$$\text{adenylate kinase}$$
$$Mg^{++}{\cdot}ATP + AMP \rightleftharpoons Mg^{++}{\cdot}ADP + ADP$$

ADP can be used as a substrate to drive the reaction to form ATP, which can be subsequently assayed by luciferase/luciferin. A source of pure ADP is necessary because many commercial supplies of this chemical contain low levels of ATP which interfere with the bioluminescence reaction. Adenylate kinase constitutes about 0.1% of cellular protein and has a turnover number of approximately 40 000 min^{-1}. In a bacterial cell the ratio of ATP molecules to adenylate kinase molecules is about 1000:1, and therefore

because of the turnover number, this allows a 40-fold amplification of ATP per minute. By assaying for adenylate kinase from lysed cells rather than ATP, a 10- to 100-fold increase in the sensitivity of the bioluminescence reaction has been recorded if 5 min are allowed for the adenylate kinase reaction. If this reaction time is further increased it may be possible to detect fewer than 10 organisms.

Advantages of ATP Bioluminescence for Hygiene Monitoring

The main advantage of ATP bioluminescence for hygiene monitoring is the speed at which results can be obtained. This 'real-time' testing allows remedial measures to be taken if the cleaning procedure is deemed to be unsatisfactory. The development of single-use devices and portable luminometers makes it easy to test the cleanliness of surfaces directly on the factory floor or in food service premises. The availability of results almost instantaneously generally improves the performance of cleaning crews and provides motivation for them to do a better job at the next cleaning cycle. Hygiene monitoring using ATP bioluminescence can be a useful tool in the training of personnel involved in sanitation. The effectiveness of their efforts to clean plant can be readily demonstrated, and any deficiencies can be remedied quickly.

Limitations of ATP Bioluminescence for Hygiene Monitoring

Sampling Efficiency As with all microbiological testing methods, the ATP technique is only as good as the sampling methodology used. Conventional swabbing procedures may only remove 1–10% of the contaminants from a surface, with poor reproducibility and repeatability. The next advance in hygiene monitoring may relate to the way surfaces are sampled. To try to eliminate the problems associated with sampling, a new instrument has been developed: termed the BioProbe (Hughes Whitlock Ltd., Monmouth, Gwent, UK), it consists of a luminometer attached to a suction cup. Reagents are added directly to the surface to be tested, and the cup is placed over the area. The light emitted from the surface is measured directly, providing an indication of its total cleanliness. Curved surfaces, to which the instrument cannot be attached, can be sampled in the usual way and the swab placed under the cup of the instrument to obtain ATP levels.

Comparison with Plate Counts One of the main limitations of the bioluminescence technique can also be viewed as one of its strengths: that is the inability of the method to differentiate between ATP from microbial and non-microbial sources. When results obtained by the ATP bioluminescence method are compared with plate counts there is agreement in about 60–70% of cases. In most cases of disagreement, the surface is found to be dirty by the ATP assay but clean by plate count. This is due to food residues with low microbial counts present on the surface, and these residues contain ATP which is detected by the bioluminescence assay. In a few cases (< 5%), surfaces fail by the plate count but have low ATP levels; this phenomenon may be due to the presence of spores or stressed cells, both of which contain relatively low concentrations of ATP.

Environmental Effects Because this method is an enzymatic test, the results will be affected by factors that affect the activity of the luciferase reaction, such as pH and temperature. These problems can be largely overcome by adequate buffering of the reaction mixture and by the use of temperature-compensated luminometers. Detergents, sanitizers and other materials used in the sanitation process, some of which may be coloured, may also affect the light reaction.

Detergents used in the food industry have a variable effect on the luciferase reaction. An alkaline chlorinated foam and an acidic foam cleanser have been evaluated for their effects on the ATP bioluminescence signal. In the case of the alkaline foam cleanser, concentrations of less than 0.05% enhanced the light reaction but only by up to a maximum of 15%. The reaction was inhibited (or quenched) by the alkaline foam cleanser at concentrations above 0.05% and this inhibition became significant when the concentration of the cleanser reached 0.2% and above. Similarly, at low concentrations (< 0.25%) of the acidic cleanser the luciferase reaction was enhanced by as much as 20%, but at the higher concentrations (> 0.25%) the ATP bioluminescence reaction was inhibited. This inhibition became significant at concentrations of acidic foam cleanser of 0.4% and above.

This concentration-dependent transition phenomenon was also observed for commercial sodium hypochlorite, household bleach and D-limolene. However, an enhancement effect on the luciferase reaction that did not appear to be dependent on concentration was shown for a quaternary ammonium sanitizer. This study was limited to monitoring the release of ATP from a variety of bacteria associated with food-borne illness and food spoilage. In some instances, it was shown by using a microbial disc assay that cleanser or sanitizer concentrations that were not effective against the bacteria did produce an effect on the bioluminescence signal by either increasing or decreasing the amount of light produced. The authors concluded that this could lead to false positive

Table 3 Effect of commercial cleansers and sanitizers on the ATP bioluminescence reaction using *E. coli* as a source of ATP

Chemical	Concentration (%)	Change in light output (%)	
		Study 1	Study 2
Acid foam cleanser	0.1	+22.5*	ND
	0.2	+8	ND
	0.4	−34*	ND
	1	−92*	ND
Alkaline foam cleanser	0.01	+15*	ND
	0.1	−6	ND
	0.5	−78*	ND
	1	−92*	ND
Quaternary ammonium sanitizer	0.001	ND	0
	0.0025	+10*	ND
	0.003	ND	+1
	0.01	+5	ND
	0.08	+8*	ND
Sodium hypochlorite	0.005	ND	−1
	0.01	+6*	ND
	0.1	−3	ND
	1	−40*	ND

*Significantly different from untreated control ($P \leqslant 0.05$).
ND, not determined.
(Data of Velasquez and Fiertag, 1997; and Green et al., 1998)

or false negative hygiene monitoring results. However, if the cleaning and sanitizing process is conducted properly, with adequate rinsing between steps to dilute the chemical agent, it is unlikely that the residual concentration of chemicals on food contact surfaces would have a significant effect on the bioluminescence reaction.

In a separate study, the effects of chemical sanitizers on *E. coli* cells, chicken blood cells and a pure ATP standard were monitored using the ATP bioluminescence reaction. Neither sodium hypochlorite (30–70 p.p.m.) nor the quaternary ammonium (10–30 p.p.m.) had any effect on the ATP bioluminescence reaction, and this was true regardless of the source of ATP. At these concentrations the results were similar to those previously reported. Lactic acid, trisodium phosphate and trichlosan all significantly inhibited the ATP reaction at concentrations of 0.5% or more, whereas hydrogen peroxide was inhibitory at concentrations of 1% or more. Again, the results were similar regardless of the source of ATP. The effects of cleaners and sanitizers on the ATP bioluminescence reaction are summarized in **Table 3**. These adverse effects can be overcome by the addition of neutralizing agents such as Tween 80 or lecithin to the reaction mixture, as well as by dilution.

Applications in Food Processing Plants

Hygiene monitoring using ATP bioluminescence methods has been applied to many sectors of the food industry including brewing, milk processing, meat

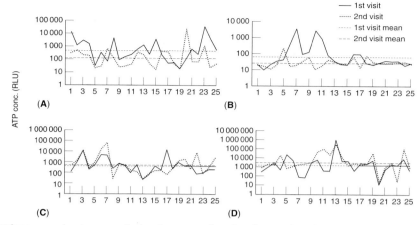

Figure 2 ATP bioluminescence hygiene monitoring to assess the cleanliness of surfaces at four different food handling establishments. (**A**) Meat processing plant. (**B**) Fluid milk plant. (**C**) Vegetable processing plant. (**D**) Retail outlet. Swabs were taken from 25 surfaces and subjected to analysis using the Biotrace Unilite system. (Data of Avant and Griffiths, unpublished)

processing, fresh produce and fruit juice operations. The usefulness of the test is illustrated by data obtained from a meat processing plant, a fluid milk plant, a vegetable processing plant and a retail outlet (**Fig. 2**). The same 25 surfaces were sampled at each of two visits. The results were relayed to the person responsible for cleaning after the first visit and, in the majority of cases, at points where there was evidence of substantial contamination during the first sampling, a significant improvement was noted at the second visit. An ATP bioluminescence hygiene monitoring system was used to evaluate food contact surfaces, including gaskets, pipe fittings, valves, filler parts, and hand-washed items, in four fluid milk plants experiencing shelf life problems. Levels of ATP (measured in RLU) were compared with microbial counts on adjacent sites of equal area. The study concluded that, although limited to use on accessible sites, the hygiene monitoring system was an effective, rapid tool for identifying sources of post-pasteurization contamination in the fluid milk plants evaluated. A similar study also demonstrated the efficacy of the ATP bioluminescence hygiene monitoring test in the identification of inadequately cleaned milk transporters (**Fig. 3**).

An ATP bioluminescence assay for hygiene monitoring in poultry meat production has been evaluated; instead of surfaces, individual carcasses were swabbed, using 232 samples. To obtain the bacterial ATP level and cfu counts, $25 \, cm^2$ poultry breast skin and $10 \, g$ poultry neck skin samples were examined. Samples were taken at various points during the slaughter process. From all samples, ATP was assayed and total viable counts were obtained. The detection limit for the ATP bioluminescence method for mesophilic bacteria was log 3.5–$4.5 \, cfu \, cm^{-2}$ in poultry breast skin, and the overall correlation coefficient

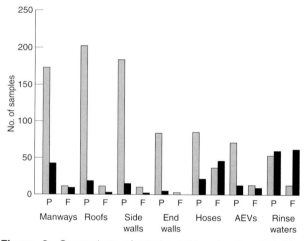

Figure 3 Comparison of results achieved using ATP bioluminescence and plate counts to monitor the efficacy of milk transporter cleaning. Stippled bars, ATP pass; solid bars, ATP fail; P, $< 2500 \, cfu$ per $1000 \, cm^2$; F, $> 2500 \, cfu \, cm^{-2}$. (Data of Bell et al, 1994)

between the two methods was 0.86. In a separate study where ATP bioluminescence measurements were obtained from carcass swabs, it was shown that the bioluminescence assay was more appropriate for identifying and evaluating hygiene at critical control points than plate counts when used in conjunction with a statistically valid sampling protocol.

The ATP bioluminescence assays are also useful for assessing hygiene practices in red meat operations. During an experiment to determine the potential for transfer of *Escherichia coli* O157:H7 from contaminated ground beef to grinding equipment and the inactivation of attached cells during cleaning and sanitizing, samples of beef were consecutively ground in a Hobart meat grinder with stainless steel chips glued to the auger housing. Chips were harvested after grinding, detergent washing with or without manual scrubbing and rinsing, sanitizing in a chlorine or peroxyacetic acid sanitizer, and overnight storage. Approximately 3–4 log cfu cm^{-2} attached to the stainless steel chips after grinding, but after washing and sanitizing in a chlorine or peroxyacetic acid sanitizer, viable bacteria were infrequently recovered by plate count. The recovery rate was reduced by manual scrubbing during the washing step. The scrubbing step also increased the number of passing scores assigned using an ATP bioluminescence assay of total residual soil on the chips sanitized in chlorine. The overall results indicated that plate counts alone may not be a reliable indicator of sanitation efficacy during meat processing.

Hygiene monitoring of ATP is not universally applicable to all sectors of the food industry. Where dry cleaning methods are used, for example in dried milk manufacture and flour milling, food residues are never completely eliminated, so high background ATP readings are always obtained. Also, ATP hygiene monitoring may not be useful during the processing of products that contain little or no ATP, such as sugar: in this case, the only source of ATP would be from microorganisms contaminating inadequately sanitized surfaces, and so only gross contamination would be detected.

When ATP bioluminescence is used during fish and shellfish processing, care must be taken in interpretation of results as these foodstuffs may be contaminated by bacteria and proteins that are naturally luminescent. Their presence will lead to abnormally high light readings and may result in surfaces being assessed as dirtier than they actually are.

As data show that 80–90% of outbreaks of foodborne illness are associated with food service operations, arguably the best application of ATP bioluminescence is to monitor sanitation in institutional or restaurant kitchens. However, the main drawback in applying the technology in these environments is its cost. Although the instruments and reagents are expensive, if they can help to prevent an outbreak that would damage the reputation of the establishment then they can prove to be cost-effective.

Hygiene monitoring by ATP bioluminescence has been adopted by many companies and the suppliers of the technology will continue to develop cheaper, easier and more accurate tests in the future.

See also: **Adenylate Kinase. ATP Bioluminescence**: Application in Meat Industry; Application in Dairy Industry, Application in Beverage Microbiology. **Bacteriophage-based Techniques for Detection of Food-borne Pathogens. Electrical Techniques**: Food Spoilage Flora and Total Viable Count (TVC). **National Legislation, Guidelines & Standards Governing Microbiology**: European Union. **Rapid Methods for Food Hygiene Inspection. Sampling Regimes & Statistical Evaluation of Microbiological Results. Total Viable Counts**: Pour Plate Technique; Spread Plate Technique; Specific Techniques; MPN; Metabolic Activity Tests; Microscopy.

Further Reading

Bautista DA, McIntyre L, Laleye L and Griffiths MW (1992) The application of ATP bioluminescence for the assessment of milk quality and factory hygiene. *Journal of Rapid Methods and Automation in Microbiology* 1: 179–193.

Bell C, Stallard PA, Brown SE and Standley JTE (1994) ATP-bioluminescence techniques for assessing the hygienic condition of milk transport tankers. *International Dairy Journal* 4: 629–640.

Flickinger B (1996) Plant sanitation comes to light: an evaluation of ATP bioluminescence systems for hygiene monitoring. *Food Quality* 2(14): 22–36.

Flowers R, Milo L, Myers E and Curiale MS (1997) An evaluation of five ATP bioluminescence systems. *Food Quality* 3(19): 23–33.

Green TA, Russell SM and Fletcher DL (1998) Effect of chemical sanitizing agents on ATP bioluminescence measurements. *Journal of Food Protection* 61: 1013–1017.

Griffiths MW (1993) Applications of bioluminescence in the dairy industry. *Journal of Dairy Science* 76: 3118–3125.

Griffiths MW (1995) Bioluminescence and the food industry. *Journal of Rapid Methods and Automation in Microbiology* 4: 65–75.

Griffiths MW (1996) The role of ATP bioluminescence in the food industry: new light on old problems. *Food Technology* 50: 62–73.

Griffith CJ, Davidson CA, Peters AC and Fielding LM (1997) Towards a strategic cleaning assessment programme: hygiene monitoring and ATP luminometry, an options appraisal. *Food Science and Technology Today* 11: 15–24.

Hawronskyj JM and Holah J (1997) ATP: a universal hygiene monitor. *Trends in Food Science and Technology* 8: 79–84.

Murphy SC, Kozlowski SM, Bandler DK and Boor KJ (1998) Evaluation of adenosine triphosphate-bioluminescence hygiene monitoring for trouble-shooting fluid milk shelf-life problems. *Journal of Dairy Science* 81: 817–820.

Ogden K (1993) Practical experiences of hygiene control using ATP-bioluminescence. *Journal of the Institute of Brewing* 99: 389–393.

Seeger K and Griffiths MW (1994) ATP bioluminescence for hygiene monitoring in health care institutions. *Journal of Food Protection* 57: 509–512.

Velazquez M and Fiertag JM (1997) Quenching and enhancement effects of ATP extractants, cleansers, and sanitizers on the detection of the ATP bioluminescence signal. *Journal of Food Protection* 60: 799–803.

Application in Beverage Microbiology

Angus Thompson, Scottish Courage Brewing Ltd, Edinburgh, UK

Introduction

Rapid microbiology methods using adenosine triphosphate (ATP) bioluminescence have been developed for quality assurance in the beverage industry since the 1970s. This technique has been applied to hygiene monitoring, the detection of microbial contamination in beverages and even brewing yeast vitality analysis. The advantages and limitations of ATP methods for the analysis of beverages are considered in this article; rapid methods for detecting spoilage microorganisms in beer, carbonated beverages, fruit concentrates and fruit juices are described. In addition, the use of ATP techniques for yeast vitality analysis in the brewing industry is discussed.

The Benefits of Rapid Methods

Growth-dependent ('cultural') microbiology methods have traditionally been used in the beverage industry to assess raw material and product quality, with samples frequently being tested by agar plate methods which take 1–7 days to complete. These methods yield historical information on the status of production processes. They are used for trend plot analysis and show problems needing corrective action. Results from traditional microbiology methods, however, often lag behind processing. Infected materials soon pass through production, causing further problems when analysis is too slow.

In managing production processes, precise knowledge of the current microbial status is much more beneficial than information on what conditions were like a few days ago. Rapid methods are needed for timely information on critical parameters. Microbial problems can be contained and minimized with 'real time' analysis. For example, rapid detection of microbial contamination stops spoilage microorganisms being passed from infected yeast cultures into beer production (fermentation).

The requirement for rapid microbiology methods in the beverage industry has led to scientists developing and exploiting methods that apply ATP bioluminescence. With the introduction of ATP techniques the need to incubate microbiological tests for long periods has decreased. Nevertheless, it should be remembered that rapid tests for microbial contamination wait for faults to occur. Tight process control, not offered by laboratory-based methods, is more useful for ensuring that manufacture is always satisfactory ('right first time'). The process of tunnel pasteurization illustrates this point: rigorous control of heating ensures that packages are commercially sterile after pasteurization, whereas rapid off-line microbiology does not. High-speed bottling lines are able to produce thousands of contaminated packages before detection in a satellite laboratory by rapid methods, so prevention by in-line control loops is much more preferable.

Concentration on quality assurance rather than quality monitoring is particularly important, because raw materials are wasted and disposal costs incurred as a result of beverage processing faults. There is little need for microbial contamination tests with perfect control of processing conditions; however, in an innovative and competitive production environment there is usually a significant risk of process breakdowns, even when best practices are followed. Rapid infection tests are beneficial in the event of process breakdowns. For instance, quick results are required when an antimicrobial filter fails during aseptic packaging and bottles are being held in a warehouse awaiting clearance.

Rapid microbiology methods are particularly useful for the analysis of packaged products that are susceptible to microbial contamination. The microbial stability of a beverage is governed by both its composition and the nature of the production process. Unpasteurized products, such as those packaged in polyethylene terephthalate (PET) bottles, are inherently less stable than pasteurized ones. In addition, drinks supporting the growth of thermoduric spoilage microorganisms such as *Zygosaccharomyces bailii* are often difficult to stabilize.

Choice of Luminometers and Reagents

Choosing luminometers and reagents tends to be a complex process, as the goods available differ in their characteristics and careful consideration is required for selection of the most suitable items. Brewing Research International (BRI) at Nutfield, Redhill, Surrey, UK has produced evaluation reports for a number of ATP bioluminescence systems. These are available to companies who are members of BRI.

Many portable luminometers for swab-based hygiene monitoring are not suitable for beverage analysis. Convenient systems involving unit-dose ('single shot') reagent sticks are available for the analysis of drops of unfiltered liquid, but these do not detect low levels of contamination and can suffer from interference by extracellular (background) ATP which is present in many drinks. Systems that test filtered samples are required to detect low levels of microbial contamination without delay.

Some luminometers accept cuvettes holding reagents from filter tests, while others directly measure light emission from filters. In the case of cuvette-based tests using standard size (47 mm diameter) membrane filters, only about 25% of the membrane washings are normally tested. This is associated with both a loss of sensitivity and a waste of reagents. It is possible to use 'mini' filters (8 mm diameter) that fit into cuvettes and avoid reagent waste; however, it is much easier to filter the large sample volumes required for sensitive analysis through standard size membrane filters. In general, luminometers that directly analyse standard size membranes offer the most sensitive methods of beverage analysis.

Drinks manufacturers tend to use ATP bioluminescence reagents from test kits marketed by specialist suppliers. Kits designed for beer, fruit juice and water analysis are commercially available. These are simple to exploit, since they offer both established methods and suitable (compatible) reagents. **Table 1** offers details of luminometers and reagents that can be used for beverage analysis.

Storage of Reagents

Suppliers normally provide information on the stability and storage of ATP test reagents. Luciferase reagents for quality assurance applications tend to be stabilized for ease of use. Some test kits now yield rehydrated luciferase that is stable at ambient temperature for 24 h and stable for a few days in a refrigerator. Freezing is recommended for long-term storage of rehydrated enzyme reagents; this saves reagent loss when tests are infrequently performed. Non-enzymatic reagents (extractants, nutrient broths and buffer solutions) are usually stored in a refrigerator. Sterile plastic cryovials can be used for the storage of reagents, as they are ATP-free.

Reagent and Luminometer Checks

Today's reagents and luminometers are very reliable when properly handled, but it is still worth checking their performance prior to analysis. Checks reduce the risk of producing spurious results. **Table 2** provides information on ATP bioluminescence reagent checks for beverage analysis.

Brewing Yeast Vitality Analysis

Rapid vitality tests allow brewers to check that yeast cultures in storage vessels are suitable for beer production. The rejection of yeast cultures with low vital-

Table 1 Luminometers and ATP bioluminescence test kits for beverage analysis

Supplier	Address	Luminometer	Test format	Test kit
Bio-Orbit Oy	PO Box 36, FIN-20521, Turku, Finland Tel. +358 2410 1100	1253 system and 1251 system with reagent injectors	Cuvette-based	ATP biomass kit
Biotrace Ltd	Science Park, Bridgend, Mid-Glamorgan, UK Tel. +44 (0) 1656 768835	Uni-Lite system and Uni-Lite Xcel system	Cuvette-based	Bev-Trace kit
Celsis Ltd	Cambridge Science Park, Milton Road, Cambridge, UK Tel. +44 (0) 1223 426008	System Sure Monitor and Biocounters with reagent injectors	Cuvette-based	Beer microbial QM kit Fruit juice test kit Water microbial kit
Charm Sciences Inc.	36 Franklin Street, Malden, Massachusetts, USA Tel. +1 781 322 1523	Luminator K system	Cuvette-based	Microbial quality test kit
Hughes Whitlock Ltd	Singleton Court, Wonastow Rd, Monmouth, Gwent, UK Tel. +44 (0) 1600 715 632	Bioprobe system	Direct analysis of membrane filters	High sensitivity test kit
Millipore Corporation	80 Ashby Road, Bedford, Massachusetts, USA Tel. +1 617 275 9200	MicroStar Rapid Microbial Detection System Rapid fruit juice sterility test luminometer	Direct analysis of membrane filters Cuvette-based	MicroStar reagent kit Fruit juice sterility test reagent kit

Table 2 Luminometer and reagent checks for ATP bioluminescence methods

'Dark cell' test for luminometer performance

1. Insert a blank sample (e.g. empty cuvette) in the luminometer and perform a light measurement
2. Check that the reading is suitably low (e.g. < 20 RLU) before further use
3. Repair the luminometer if necessary (e.g. replace cuvette O ring seal)

Luciferase reagent check

1. Measure the light emitted from an aliquot (e.g. 100 µl) of luciferase reagent
2. Check that the reading is suitably low (e.g. < 30 RLU) before further use
3. Add ATP to the luciferase reagent (e.g. 10 µl of a 1 mmol l^{-1} ATP standard solution)
4. Remeasure the light emission
5. Check the reagent activity by comparing the result with that normally obtained from fresh luciferase reagent supplemented with ATP standard solution
6. Prepare more luciferase reagent if the activity is too low (e.g. 3 times less than normal)

ATPase reagent check

1. Take a completed luciferase check, containing ATP and enzyme that produces a constant light output
2. Add an aliquot of ATPase reagent (e.g. 20 µl) to the reagent check
3. Measure the light emission and then remeasure it after a short time (e.g. 30 s)
4. Check that the light emission is rapidly decreasing before using the ATPase
5. Prepare fresh ATPase if necessary

Non-enzymatic reagent contamination check

1. Mix aliquots of the non-enzymatic reagents (e.g. 100 µl portions of buffer, nutrient broth and extractant)
2. Add an aliquot of pre-checked luciferase reagent (e.g. 100 µl)
3. Measure the light emission and check that it is suitably low (e.g. < 50 RLU for tests with nutrient broth)
4. Test individual reagents for ATP contamination if the result from the mixed reagent check is too high
5. Identify and replace any spoilt reagents

---

ities helps to avoid slower than normal brewery fermentations. In addition, use of vitality results for the calculation of yeast addition (pitching) rates promotes consistency in fermentation performance.

Measurements of cellular ATP or energy charge (EC) can be used to rapidly assess the vitality of brewing yeast, but techniques that do not involve ATP analysis are more frequently exploited in brewery laboratories, such as measurements of yeast acidification power or sterol levels. One study measured concentrations of ATP in brewery yeast cells prior to fermentation and used a pitching rate of 0.2 g ATP per hectolitre of wort for beer production. The yeast cell numbers added at the start of fermentation consequently varied according to the yeast's vitality. It was found that use of ATP measurements for pitching rate determination, rather than traditional biomass (% solids) measurements, improved fermentation control.

In 1977 Hysert and Morrison described a method that measured the EC of brewing yeast cells. These researchers measured cellular concentrations of ATP, adenosine diphosphate (ADP) and adenosine monophosphate (AMP) and then calculated EC values using the following equation:

$$EC = \frac{[ATP] + \frac{1}{2}[ADP]}{[ATP] + [ADP] + [AMP]}$$

The concentrations of ADP and AMP were measured by bioluminescence after conversion to ATP by the reactions described below, catalysed by adenylate kinase (AK) and pyruvate kinase (PK).

$$AMP + ATP \xrightarrow{\text{AK}} 2\,ADP \qquad (1)$$

$$ADP + phosphoenolpyruvate \xrightarrow{\text{PK}} ADP + pyruvic\ acid \qquad (2)$$

It was shown that brewing yeast EC levels fluctuate during beer production. Healthy yeast cells at the start of fermentation have EC values greater than 0.8, while stressed yeast cells at the end of fermentation have EC values between 0.5 and 0.7.

Nowadays ATP technology is seldom applied to vitality analysis in the brewing industry for technical reasons. Obtaining representative samples of yeast can be difficult, because slurries in yeast storage vessels are frequently viscous and not homogenous. In addition, the EC of a yeast sample can easily change after sampling. The turnover of ATP in yeast cells is rapid and EC values soon change in response to a new environment. Technical advantages are offered by in-line yeast health analysers which exploit capacitance, such as the 'Bug Meter' marketed by Aber Instruments (Aberystwyth, Dyfed, UK). These do not have sampling problems and can be used as sensors in pitching control loops. At the moment ATP bioluminescence is rarely used for in-line analysis of brewing yeast.

Prevention of Microbial Contamination

Microbial contamination must be avoided during beverage processing, as it can spoil the quality of products and even cause illness. Food safety laws require manufacturers to ensure that drinks do not contain pathogenic microorganisms. In the UK, beverage producers are legally required to show 'due diligence' in the prevention of food poisoning.

Many companies have successfully used hazard analysis critical control point (HACCP) schemes to prevent the production of contaminated beverages. Rapid ATP techniques have been incorporated into HACCP schemes in the drinks industry, although this application of ATP technology is not universal. The need for ATP methods varies according to the production process and the nature of the critical control points identified by studies. Measurement of parameters other than ATP levels (e.g. detergent strengths, temperatures and contact times during cleaning in place) can ensure that microbial contamination does not occur during processing.

Fortunately pathogens do not proliferate in many beverages. For example, beer does not allow pathogens to grow, because of its content of alcohol, carbon dioxide and antimicrobial hop compounds, and low pH value. Fruit juice tends to be too acidic for growth of food poisoning microorganisms. Beverages are more frequently spoilt by contamination with non-pathogenic microorganisms, producing off flavours, hazes and sediments in packaged products. Methods of ATP analysis are very useful for the rapid detection of both pathogens and spoilage microorganisms in beverages.

Sample Preparation for Contamination Tests

Filtration of samples improves the sensitivity of ATP-based tests for microbial contamination in liquids, because it:

- increases the concentration of microbial cells during analysis
- removes extracellular ATP
- eliminates compounds that quench light emission
- stops bioluminescence inhibitors being passed into tests.

Bright (clear) beverages lend themselves to filtration, but drinks that contain high levels of suspended particles, such as fruit juices with a high pulp content, present a filtration problem. Particular sample treatments are required to analyse beverages containing high levels of suspended particles. Differential centrifugation allows the selective removal of non-microbial particles with a diameter greater than 50 μm. Flocculant yeast cells are lost from samples, however, when centrifugation is used to remove smaller particles.

Samples of bright liquid (e.g. 250 ml volume) can be passed through 0.45 μm pore size, 47 mm diameter, membrane filters. This pore size retains microbial cells and allows rapid filtration. Special types of membrane have also been used for filtration of large volume samples: for instance, a procedure for analysing 2 litre beer samples from PET bottles used positively charged nylon membrane filters (Ultipore, Pall Process Filtration Ltd, Portsmouth, UK). Samples containing suspended particles, such as orange juice, can be double-filtered, with large non-microbial particles being removed by coarse filtration and microorganisms being harvested by subsequent fine filtration.

Many beverages naturally contain non-microbial ATP, which interferes with infection tests if it is not removed. Beer – particularly dark ales and products of high-gravity brewing – contains significant levels of extracellular ATP, released from yeast cells during fermentation and maturation, while fruit juice naturally contains high levels of somatic ATP. Passing sterile water or saline through filters, after the samples have been filtered, helps to remove extracellular ATP. In addition, non-microbial ATP is degraded when filters are immersed in buffered solutions containing ATPase (e.g. calcium-activated ATPase from potatoes) and incubated for a brief period (e.g. 30 min) at ambient temperature. Microbial ATP is not removed while it is contained in cells and not exposed to ATPase. Problems can occur, however, if microbial ATP is rapidly degraded by ATPase reagents after the addition of microbial ATP releasing reagents.

Methods of fruit juice analysis, requiring highly active ATPase reagents to deal with high concentrations of somatic ATP, often involve luciferase reagents which emit flashes of light just after their addition. These reagents need to be dispensed automatically by luminometers fitted with pumps, so that light emissions are measured at the correct time. Luciferase reagents producing a more stable light emission are suitable for testing samples with low levels of background ATP. These can be manually dispensed and are compatible with less complicated luminometers without reagent dispensing pumps.

The physiological state of microorganisms influences results from ATP-based methods, as stressed cells have a lower EC than healthy, metabolically active cells. Heat treatment during final product pasteurization and filtration during sample analysis are stressful for microbial cells. Ensuring that filter membranes do not dry out during analysis helps to decrease interference by stress. Some analysts additionally recommend use of a low vacuum pressure during filtration in order to minimize stress. A brief incubation period after filtration can be used to boost ATP levels in microbial cells and therefore improve test sensitivity. Incubation of filter membranes immersed in a solution containing nutrient broth, buffer solution and ATPase concurrently revives microbial cells and removes extracellular ATP.

Carefully designed sample preparation methods are necessary for accurate results from ATP-based beverage tests. For the prevention of false negative results, test solutions must not contain agents that quench light emission or luciferase inhibitors, plus they must have an appropriate pH value. HEPES buffer (pH 7.75) is often used to adjust the pH value of test solutions. The strength of buffer required depends on the test substance. Acidic, well-buffered drinks such as orange juice are analysed using relatively strong buffers (e.g. $0.25 \, mol \, l^{-1}$ HEPES).

Contamination Analysis after Sample Preparation

It is often worth measuring light emissions from a sample before and after the addition of microbial ATP releasing reagent. Measurement before the addition of ATP extractant is important since it shows whether any high results are due to interfering background ATP. A reading for total ATP is obtained after the addition of microbial ATP releasing reagent. Total and background ATP readings (in relative light units, RLU) can be used to assess product quality, as follows.

FAIL the product if:

$$\frac{total \; ATP \; result}{free \; ATP \; result} > 3$$

and

$$total \; ATP \; result > 100 \; RLU$$

Otherwise PASS the product.

There is no need to perform a background ATP reading whenever a total ATP result is negligible. It is possible to avoid waste of luciferase reagent by performing a total ATP measurement on part of a sample first and then deciding if a background ATP measurement is required.

The background ATP in fruit juice tests varies according to the type of juice. Orange and pineapple juices yield higher levels of free ATP than grapefruit, apple and grape juices. It is more difficult to analyse juices that produce high background ATP results, as contaminants need to be at a higher cell concentration for detection.

RLU readings do not show the exact amount of ATP in a test without calibration, but contaminants can still be detected without calibration, using RLU ratios from total and background ATP tests. Some laboratories prefer not to perform technically demanding calibrations. Use of ATP standard solutions, however, can improve the accuracy of analysis.

At the moment ATP bioluminescence is not used to detect microbial contamination in liquids that naturally contain viable microorganisms, such as beers that possess brewing yeast cells. It is too difficult to detect microbial contaminants in the presence of other microorganisms. Methods that separate contaminants from other microorganisms or techniques that specifically extract ATP from contaminants are required to prevent interference by naturally occurring microorganisms.

Procedures for testing filtered samples vary according to the type of luminometer used and the reagents selected for analysis. **Figure 1** describes a beer sterility test using cuvettes, while **Figure 2** depicts a beer sterility test employing the Bioprobe system from Hughes Whitlock (Monmouth, Gwent, UK), which directly analyses membranes. The methods illustrated in Figures 1 and 2 can also be used for the analysis of carbonated beverages.

Detection Limits

Tests without a growth period, using conventional ATP bioluminescence techniques, do not detect very low levels of microbial contamination, but identify heavily contaminated products in less than 1 h. These tests can be useful for fruit concentrates which tend to contain contaminating microorganisms. Preservative-resistant yeast cells, which do not grow on agar plates, can be detected in fruit concentrates using ATP tests.

For detection, a sample must contain enough microbial ATP to produce a result that is significantly higher than background readings. The number of cells required for detection additionally depends on the type of cells. Yeast cells contain approximately 100 times more ATP per cell than bacterial cells, so tests are able to detect lower numbers of yeast cells than bacterial cells. Test solutions containing more than 100 viable yeast cells or more than 10 000 viable bacterial cells are detected by conventional ATP bio-

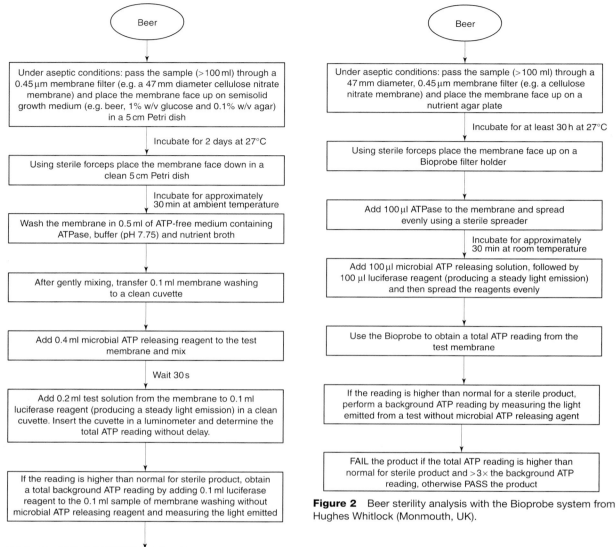

Figure 1 Beer sterility analysis with a cuvette-based system.

Figure 2 Beer sterility analysis with the Bioprobe system from Hughes Whitlock (Monmouth, UK).

luminescence methods. Lower levels of contamination can produce results that are similar to background results and are therefore difficult to detect.

Drink sterility tests using ATP bioluminescence systems generally involve incubation periods, which are necessary for very low levels of infection to grow above the detection threshold. Incubation times can be determined from laboratory trials. In order to detect very low levels of slow-growing lactobacilli in beer (e.g. 2 bacteria per bottle), membrane filtered samples are incubated at 27°C for 2 to 3 days. Sterility testing using ATP techniques is quicker than sterility testing using traditional agar plate methods, although it is still not particularly rapid.

Increases in cell concentrations are efficiently achieved if samples are membrane filtered and then incubated in small (50 mm diameter) Petri dishes containing growth medium. Growth can be promoted by either a nutrient broth or a semisolid medium containing a low level of agar. Enrichments with a liquid medium should involve small aliquots of medium (e.g. 2 ml per membrane), so that cells are not excessively diluted. In tests using semisolid agar, membranes are placed on the medium to yield microcolonies on the upper surfaces of membranes. Semisolid medium, in contrast to liquid medium, does not dilute cells immediately after filtration. Some ATP methods use selective media for the detection of certain problem-causing species; for example, an ATP-based method for beer analysis using Raka-Ray Broth for detection of lactobacilli and Wallerstein Laboratory Broth for the detection of yeast strains has been described.

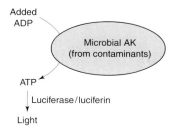

Figure 3 The use of adenylate kinase (AK) to detect microbial contamination.

Methods Exploiting Adenylate Kinase

There has been interest in using adenylate kinase (AK) for rapid beverage sterility tests. In many AK-based tests, ADP is added to samples and converted to ATP by native AK from microbial contaminants; the resultant ATP is detected by bioluminescence (**Fig. 3**).

An extractant is needed to expose the added ADP to AK. The amount of ATP formed depends on both the AK activity and the incubation period. Increasing the latter boosts ATP formation and therefore increases test sensitivity.

Very low levels of contamination can be rapidly detected using AK-based methods, because microbial cells possess enough AK activity to yield significant levels of ATP from added ADP after a short incubation period. For example, a single bacterial cell can produce 1×10^{-15} mol of ATP after 25 min incubation. Methods using AK have a lower detection limit than conventional rapid ATP bioluminescence techniques, because AK increases the ATP concentration in test solutions by as much as a thousandfold.

The extra sensitivity of methods involving AK can be useful for the analysis of beverages that cannot be filtered. In tests on unfiltered samples, however, analysis can be ruined by extracellular AK. Many beverages, such as beer, contain traces of extracellular AK that can result in false positive results. The risk of interference is decreased by filtration, because filtration removes extracellular AK from samples.

Standard ATP bioluminescence methods do not detect dead cells; however, there is concern that techniques involving AK may do so. Research has shown that AK from some microorganisms is thermostable. For instance, AK from *Escherichia coli* can retain activity after boiling for 10 min. Contamination tests using AK are invalidated if there is interference by AK activity from dead cells.

At the moment AK methods are rarely used for quality assurance in the beverage industry, but test kits exploiting AK activity are not under development. The Autotrack system from Biotrace Ltd, launched in 1998, is of interest since it is an on-line, 'real time', microbial detection device which detects down to 1000 cells/ml using AK technology.

Use of the MicroStar System

The MicroStar Rapid Microbial Detection system (Millipore Corporation, Bedford, Massachusetts, USA) was developed during the late 1990s for the detection of very low levels of microbial contamination in liquid samples. It is significantly more sensitive than conventional bioluminescence systems, as it is able to detect a single yeast cell on the surface of a membrane without enrichment. A charge-coupled device (CCD) camera is used by the MicroStar system to measure light emission and a digital image of viable microbes on test membranes is produced. This allows levels of contamination to be measured in terms of colony forming units (cfu) per sample. Other types of luminometer are less accurate at measuring levels of viable cells, since they depend on conversion factors that are affected by both cell size and luciferase activity.

Individual bacterial cells do not contain sufficient ATP for instant detection by the MicroStar system, which detects microcolonies containing more than 50 bacterial cells. Before bacterial analysis, filtered samples are incubated to allow microcolony formation.

The MicroStar system is noteworthy because of its sensitivity and quantitative accuracy, but it is considerably more expensive than many other ATP bioluminescence systems for quality assurance. It is expensive for troubleshooting purposes, but offers much better value when a positive sterility test is required for the release of drinks stored in warehouses. For example, it can be useful for testing, prior to dispatch, the stability of fruit-containing drinks prone to *Zygosaccharomyces bailii* contamination.

Accelerated Forcing Tests

Forcing tests involve incubation of product and process samples to promote growth of microbial contaminants. These tests differ from traditional agar plate tests for the enumeration of microorganisms, as they show whether or not samples contain strains that cause spoilage. For example, forcing shows the ability of lactobacilli to grow in a particular beer, whereas agar plating does not.

Beverage producers perform forcing tests to check the microbial stability of packaged products, but these tests take days and even weeks to complete. In the case of beer, samples are forced at 27°C for up to 3 weeks. An extended incubation period is necessary to guarantee a long product shelf life (e.g. 9 months) and allow the visual detection of slow-growing spoilage microorganisms. With low levels of contamination it takes many generations to produce microbial hazes and sediments.

Manufacturers want to supply fresh beverages, and products like beer are often dispatched before the completion of traditional forcing tests. Poor forcing test results are a sign of poor quality and they often prompt the retrieval of products from trade. It is dangerous for manufacturers not to recall glass bottles containing viable spoilage microorganisms from trade, as vigorous fermentations can cause sealed bottles to explode. The use of quality assurance systems that prevent release of unstable products is of vital importance, because product retrieval is both expensive and damaging to the reputation of suppliers.

Due to its speed and sensitivity, ATP technology can be successfully used in accelerated forcing tests. It can detect invisible increases in levels of microorganisms during the early stages of spoilage. A few spoilage microorganisms (e.g. 5 bacteria) can be detected in a packaged product in less than 3 days, rather than 3 weeks, but an incubation period is essential to allow growth. The growth rate of contaminants governs the time needed for analysis and growth in products designed to be resistant to spoilage tends to be slow. For example, *Pectinatus cerevisiiphilus* has a doubling time of 4 h in some types of beer.

Figure 4 describes an accelerated forcing test that uses ATP bioluminescence to assess the microbial stability of fruit juice samples. One advantage of ATP-based forcing tests is that they are suitable for optically dense liquids, such as orange juice.

Such ATP methods are particularly useful for preventing delays in fruit juice production. They minimize the time that aseptically filled packages of fruit juice must be held in warehouses before release, and decrease the time that fruit concentrates need to be stored before being passed as suitable for processing. In the case of fresh, unpasteurized juices with a short shelf life, rapid product release is essential, so that the products are supplied with an acceptable shelf life.

Conclusion

Rapid methods that use ATP bioluminescence for the detection of microbial contaminants in beverages have the following advantages and limitations.

Advantages

1. Considerably faster than cultural methods such as agar plate techniques or traditional forcing tests.
2. Detect low levels of microorganisms.
3. Show the presence of viable cells.
4. Detect non-culturable microorganisms.
5. Suitable for use at sites without expert microbiologists.

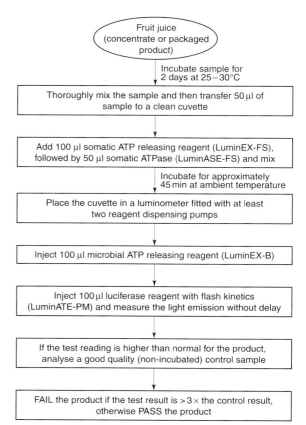

Figure 4 Fruit juice analysis with the Celsis test kit. Fruit juice concentrates are diluted prior to analysis. The degree of dilution is governed by the microbiological specification for the test substance. In the case of concentrates with a specification of less than 1000 cfu g^{-1} a 10^3 dilution is tested.

6. Cost-efficient when decreasing delays in production (e.g. warehouse storage before product release).

Disadvantages

1. Do not identify microbial contaminants.
2. Need an incubation period for the detection of very low levels of infection.
3. Produce poor quantification of microbial contaminants.
4. Do not detect microbial contaminants in samples containing other microorganisms, such as fermentation samples.
5. Destroy a significant volume of beverage when analysis of a large number of samples is necessary.

These characteristics allow people to determine if ATP methods are appropriate for their analytical needs. When in-line process control loops are not able to prevent microbial contamination in beverages, the application of ATP techniques is beneficial because it produces rapid results, allowing quick release of products or detection of spoilage microorganisms.

See also: **Acetobacter**. **ATP Bioluminescence**: Application in Meat Industry; Application in Dairy Industry; Application in Hygiene Monitoring. **Bacteriophage-based Techniques for Detection of Food-borne Pathogens**. **Biophysical Techniques for Enhancing Microbiological Analysis**: Future Developments. **Electrical Techniques**: Food Spoilage Flora and Total Viable Count (TVC); **Immunomagnetic Particle-based Techniques**: Overview. **Rapid Methods for Food Hygiene Inspection**. **Total Viable Counts**: Pour Plate Technique; Spread Plate Technique; Specific Techniques; MPN; Metabolic Activity Tests; Microscopy. **Ultrasonic Imaging**. **Ultrasonic Standing Waves**.

Further Reading

Avis JW and Smith P (1989) The use of ATP bioluminescence for the analysis of beer in polyethylene terephthalate (PET) bottles and associated plant. In: Stannard CJ, Petit SB and Skinner FA (eds) *Rapid Microbiological Methods for Foods, Beverages and Pharmaceuticals*. P. 1. Oxford: Blackwell Scientific.

Boulton CA (1996) A critical assessment of yeast vitality testing. *Ferment* 9: 222–225.

Dowhanick TM and Sobczak J (1994) ATP bioluminescence procedure for viability testing of potential beer spoilage micro-organisms. *J. Am. Soc. Brew. Chem.* 52: 19–23.

Hysert DW and Morrison NM (1977) Studies on ATP, ADP and AMP concentrations in yeast and beer. *J. Am. Brew. Soc. Chem.* 35: 160–167.

Miller R and Galston G (1989) Rapid methods for the detection of yeast and *Lactobacillus* by ATP bioluminescence. *J. Inst. Brew.* 95: 317–319.

Miller LF, Mabee MS, Gress HS and Jangaard NO (1978) An ATP bioluminescence method for the quantification of viable yeast for fermenter pitching. *J. Am. Brew. Soc. Chem.* 36: 59–62.

Van Beurden R and Shepard S (1995) *HACCP and the Lumac Solution: The Meaning of HACCP in Food Safety Management and the Contribution of Lumac*. Lumac bv, PO Box 31101, 6370 AC Landgraaf, The Netherlands.

AUREOBASIDIUM

T Roukas, Department of Food Science and Technology, Aristotle University of Thessaloniki, Greece

Characteristics of the Genus

Aureobasidium is a Deuteromycetes (Fungi Imperfecti) fungus. There are more than 10 species, but *A. pullulans* is the most prevalent in foods.

Species of *Aureobasidium* have pinkish, rapidly expanding colonies which soon become slimy due to production of yeast cells (**Table 1**). Colonies later darken due to the development of chlamydospores. Endoconidia are common. Some species produce extracellular polysaccharides. The main distinguishing character of *Aureobasidium* from related genera is the mode of conidiogenesis – synchronous in *Aureobasidium* and percurrent in *Hormonema*.

A. pullulans shows synchronous conidiation producing 2–14 conidia closely together from lateral and terminal as well as from intercalary cells of young expanding hyphae. Conidial scars are slightly protuberant. Percurrent production of conidia from single butts is observed in a later stage of development. Dense clumps of conidia are seen alongside submerged hyphae by observation of Petri dishes. The vegetative morphology of *A. pullulans* is extremely variable. It can grow as a budding yeast or as a mycelium. The preponderance of any stages in the life cycle can be influenced by nitrogen and carbon sources. Therefore the type of conidiogenesis has been used as a diagnostic feature, rather than general morphology. The species initially shows synchronous conidiation, although the number of conidia produced may be very small. Synchronous conidium initials break through the mother cell wall. Later these loci remain productive; 2–14 conidiogenous loci are present on each expanding hyphal cell.

A. pullulans var. *melanogenum* includes strains of *A. pullulans* that rapidly turn black due to chlamydospore formation. Blackening is a highly variable property which is stimulated when D-galactose, L-sorbose or glucono-delta-lactone is used as sole source of carbon. After repeated subculturing on aromatic compounds *A. pullulans* strains can lose the ability to produce melanin.

Aureobasidium is fairly common in fruits and vegetables. It is a common field fungus. It is a ubiquitous saprophyte from all sorts of moist and decaying environments growing on paint surfaces in food factories. *A. pullulans* has been isolated from a wide range of foods, but only rarely as a cause of spoilage. Many strains have been isolated from the surface of sugar-containing fruits. Particularly overripe fruits are colonized. It commonly occurs on cabbage, strawberries, grapes, green olives, citrus and citrus products. It is also found in shrimp, barley, wheat and flour, oats and nuts. It is prevalent in frozen foods, being the

Table 1 Characteristics of the genus *Aureobasidium* including *A. pullulans*

Characteristics	Aureobasidium	A. pullulans
Morphology		
Colonies on CYA and MEA	Pink or cream, later black	Cream, later grey to black
Colony diameter (cm)	4	2.5–3.5
Mycelium	Hyphae and budding yeast-like cells	Hyphae 2–16 µm in diameter; thick-walled hyphae fall apart into separate cells (arthroconidia)
Reproductive structures	Conidiogenous cells: cylindrical with small denticles directly from the hyphal walls or short lateral protrusions on the hyphae; 2–4 denticles on one cell produce conidia synchronously	Conidiogenous cells: denticles directly from hyphae or small lateral protrusions produce 2–14 blastic conidia synchronously
	Conidia: ellipsoidal (4–6 × 2–3 µm) secondary conidia produced by yeast-like budding of primary conidia	Conidia: yeast-like one-celled (10–16 × 3–6 µm); secondary smaller conidia (7–10 × 3–5 µm)
	Chlamydospores: thick-walled hyphae	Chlamydospores: thick-walled hyphae
Growth temperatures		
Minimum	2°C	–5°C
Optimum	25°C	25°C
Maximum	35°C	35°C
Minimum a_w for growth	Not known	0.90

CYA = Czapek yeast extract agar; MEA = malt extract agar; a_w = water activity.

predominant mould isolated from blueberry, apple and cherry pies. Many years ago it was involved in the spoilage of long-term-stored beef and is still occasionally found in some cheeses.

The most important physiological characteristic which makes *Aureobasidium* significant in food mycology is its ability to withstand the reduced water activity (a_w) in frozen foods as well as the low temperature. *A. pullulans* is a slightly osmophilic species: in vitro the species tolerates raised salt levels (up to 10%). The spores of *A. pullulans* have the ability to germinate and grow at temperatures of –5°C or lower.

In nutritional physiology, *A. pullulans* shows assimilation of glucuronate, soluble starch, lactose, xylose, citrate, melibiose, inositol and alpha-methyl-glucoside and absence of fermentation. Most strains show good or weak gelatin liquefaction.

A. pullulans produces many important enzymes (**Table 2**) and consequently has become an important organism in applied microbiology.

A. pullulans is an industrially important micro-organism because of its capability of producing pullulan, a commercially exploited polysaccharide used in coatings and wrappings potential and as a food ingredient. Other products produced by *A. pullulans* and used as additives in food processes are presented in **Table 3**. *A. pullulans* is a safe microorganism for use as a single-cell protein; it can also be used in controlling and monitoring environmental pollution.

Methods of Detection

Plating Method

Aureobasidium contamination in foods can be detected as shown in **Figure 1**. Identification of the growing colonies on the basis of appearance and microscopy is performed as shown in Table 1. The method evaluates viable propagules.

Immunological Methods

Immunological methods provide a specific and sensitive way to estimate or identify fungi. This method is rapid and easy to use for detection of fungal contamination in foodstuffs. Monoclonal antibodies used in agglutination kits give an agglutination reaction with culture filtrates of *A. pullulans*. With polyclonal antibodies *A. pullulans* gives low reaction. Further evaluations on foodstuffs are necessary to confirm that the commercially available kits are useful in the routine laboratory.

Unacceptable Levels of *Aureobasidium* Species

It is almost impossible to specify the colony (or propagule) counts of *Aureobasidium* which are acceptable in a range of foods. Even very low numbers, e.g. < 1 per 100 g, were a spoilage potential in non-sterile foods distributed through a cold chain, since growth of such fungi may not be inhibited by bacterial development in the product.

Table 2 Enzymes produced by *Aureobasidium pullulans*

Enzyme	Temperature optimum (°C)	pH optimum	Application
Sucrase	35	4.5	Hydrolyses sucrose
Xylanase			Hydrolyses xylan. Causes clarification of fruit pulp and juices
β-Xylosidase	60		
Glucoamylase	80	4.5	Amylolytic ability
Glucoamylase produced by *A. pullulans* A-124	50	5.75	
α-Amylase			Starch or starch hydrolysates can be used for pullulan production and to control the molecular weight of pullulan
β-Galactosidase	45	6.8	Hydrolyses lactose in whey or milk
Pectinolytic enzymes		4.3	In maceration of fruit pulps and for clarification of juices and wines
Polygalacturonase	37	3.8	Hydrolyses sodium polypectate and pectin
Pectin lyase produced by *A. pullulans* LV 10	40	5–7.5	Attacks highly esterified pectins, low-methoxyl pectins, and polygalacturonic acid
Pectin esterase produced by *A. pullulans* AY-037		3–6	Clarification of fruit juices
Xylitol dehydrogenase	25	10–10.5	Oxidizes xylitol to D-xylulose
Laccase	25–35	4.5–6.4	
L-Fucose dehydrogenase	30	9.5	Converts L-fucose to L-fuconic acid
Phosphatase			
Polyamine oxidase			Biological inactivation of polyamines. Component of clinical diagnostic assay kits
L-Rhamnose dehydrogenase			Hydrolyses L-rhamnose
Fructosyl transferase			Hydrolyses sucrose to high-fructose syrup

Table 3 Food additives produced by *Aureobasidium pullulans*

Product	Strains	Application
Pullulan	Most strains isolated	Coating and wrapping agent and as a food ingredient
Erythritol	*Aureobasidium* sp. SN-124A	An artificial sweetener used as food ingredient
Gluconic acid or its salt	*A. pullulans* AHU 9190	
L-Malic acid	*A. pullulans* FERM-P2760	As acidulant
Fructooligosaccharides	Strains producing fructosyl transferase	Non-digestible sweeteners. They have applications in health foods

Importance to the Food Industry

Aureobasidium pullulans was one of the causative moulds for the black spot spoilage of frozen meat transported long distances by sea. It actually penetrates into the tissues of meat incubated at −1°C. This spoilage condition is characterized by uniformly black fungal colonies (3–8 mm diameter) whose hyphae penetrate the superficial layers of the tissue, and although meat presents no health hazard, it is aesthetically unacceptable. Colonies produced by *A. pullulans* are mainly subsurface, with the hyphae spreading along the intercellular junctions, possibly in response to the arid conditions at the frozen meat surface. If the surface remains moist, growth is superficial. The meat can be trimmed, but there is financial loss due to the value of the meat removed and the subsequent downgrading of carcasses and cuts.

A. pullulans is a ubiquitous saprophyte from all sorts of moist and decaying environments, therefore it can be assumed that it is transferred to meat from hides during the carcass-dressing procedure. As colony formation by *A. pullulans* occurs in the subsurface tissue, the spoilage cannot be hindered by wrapping films. So it is important to minimize the extent of contamination from freezing works environments. Since mould growth does not occur when the product is held at a suitably low temperature, the problem may arise when meat is exposed to higher temperatures resulting from defrost cycles in the freezer chest, or mishandling during transport or

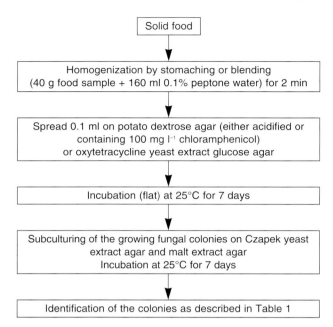

Figure 1 Detection of *Aureobasidium* in foods by the plating method.

storage. Multiple exposures to defrost temperatures can result in spoilage. Nowadays, temperature control is excellent during shipping and distribution. In addition, meat transported under modified atmosphere minimizes the problem.

To control *A. pullulans* as spoilage fungi of vegetables and fruits, modified atmosphere can have positive effects, such as increasing the lag phase of fungal growth, repressing mycelial growth and decreasing spore development. Increasing the carbon dioxide content and/or decreasing the oxygen content of the atmosphere has been shown to be fungistatic on *A. pullulans*.

A. pullulans can cause rotting of healthy fruits of strawberry, producing pectinolytic enzymes. The pectinolytic activity is destroyed by boiling or by sulphiting. Sulphur dioxide treatment also decreases enzyme stability.

Importance to the Consumer

Mycotoxins are not known to be produced by *A. pullulans*. The World Health Organization has classified *A. pullulans* in risk group I, where there is no possibility of infection to either society or laboratory workers.

A. pullulans has been implicated in divergent opportunistic mycoses, such as infection of the mandible after removal of an impacted molar, keratitis, subcutaneous granuloma, localized systemic infection and generalized sepsis in immunocompromised patients. The non-specificity of the clinical pictures demonstrates that the pathogenicity of such strains is low; symptoms vary with the portal of entry and condition of the host.

See also: **Meat and Poultry**: Spoilage of Meat. **Mycotoxins**: Classification.

Further Reading

de Hoog GS and Yurlova NA (1994) Conidiogenesis, nutritional physiology and taxonomy of *Aureobasidium* and *Hormonema*. *Antonie van Leeuwenhoek* 65: 41–54.

Deshpande MS, Rale VB and Lynch JM (1992) *Aureobasidium pullulans* in applied microbiology: a status report. *Enzyme and Microbial Technology* 14: 514–527.

Gill CO, Lowry PD and Di Menna ME (1981) A note on the identities of organisms causing black spot spoilage of meat. *Journal of Applied Bacteriology* 51: 183–187.

Pitt JI and Hocking AD (1997) *Fungi and Food Spoilage*, 2nd edn. London: Blackie Academic & Professional.

Samson RA, Hoekstra ES, Frisvad JC and Filtenborg O (1995) *Introduction to Foodborne Fungi*, 4th edn. Baarn: Centraalbureau voor Schimmelcultures.

BACILLUS

Contents

Introduction

Michael K Dahl, Department of Microbiology,
University of Erlangen, Germany

The genus *Bacillus* is one of the preferentially used organisms for producing metabolites and enzymes by fermentation. This is partly due to the fact that most (but not all) members of the genus are non-pathogenic, excellent protein and metabolite secretors, and easy to cultivate. Non-pathogenic *Bacillus* strains are used both in foods and in industry. The numerous products accepted as safe include enzymes for food and drug processing, as well as foods produced from these strains. The tremendous advances in molecular biology have increased the use of *Bacillus* spp. in heterologous gene expression. One member of the genus, *Bacillus subtilis*, has been especially subjected to intensive microbiological, biochemical and genetic investigations. In the late 1950s, John Spizizen successfully demonstrated the genetic transformation of a particular *B. subtilis* isolate using purified DNA. This laid the foundation for a series of intensive studies of metabolism, gene regulation, bacterial differentiation, chemotaxis and starvation. At present, *B. subtilis*, together with *Escherichia coli*, is one of the best understood prokaryotes, and the complete sequence of the genome of *B. subtilis* is now available, which facilitates further investigations into biological molecular mechanisms. Insights made by analysing molecular mechanisms in *B. subtilis* can easily be transferred to related organisms, thereby helping to advance applied research and food production.

The Genus *Bacillus*

At present the genus *Bacillus* encompasses more than 60 species. It is widespread in nature and can be isolated from food, soil, water and even from eukaryotic organisms. Owing to the enormous genetic diversity of this genus, it is difficult to define it concisely. One of the most important properties for taxonomy is spore formation, since it is easily detectable. However, the fact that sporulation depends on the growing conditions complicates classification by this characteristic. Spores can be detected microscopically and provide a simple characteristic for the endospore-forming family Bacillaceae. In this family, the genus *Bacillus* incorporates many species of Gram-positive, rod-shaped bacteria, which are able to grow under aerobic and facultatively anaerobic conditions and thus differ from *Clostridium* spp. which are strictly anaerobic. However, aerobic endospore-forming bacteria are currently assigned to four genera in the family Bacillaceae. Therefore, the genus *Bacillus* is basically defined by morphological characteristics. Depending on the type of spore formation observed, we distinguish between:

- species producing oval endospores that distend the mother cell
- species producing oval endospores that do not distend the mother cell
- species producing spherical endospores.

Typically, *Bacillus* spores contain dipicolinic acid, but the diversity of spore formation described above makes a classification by this property difficult. Numerical analysis of additional phenotypic features leads to some idea of how the genus *Bacillus* might be reorganized into several genera. There is no generally accepted definition of a prokaryote genus. Nevertheless, it has been recommended that the maximum genetic diversity should not exceed a chromosomal base composition range of 10–12% guanine/cytosine (G+C) content. Phylogenetically, *Bacillus* belongs to the low G+C content group of bacteria, although, depending on the species, the G+C content can vary in the range from 33% (*B. anthracis*) to 69% (*B. thermocatenulatus*). Assuming the G+C content definition is correct, the chromosomal base composition range in *Bacillus* would signify great phylogenetic divergence, leading to the conclusion that *Bacillus* encompasses more than one genus. This assumption is also supported by rRNA sequence analysis, which reveals as much divergence as in the combined families of Enterobacteriaceae and Vibrionaceae. In turn, the size of the genus *Bacillus* complicates the identification of new isolates. Therefore, at least thirty phylogenetic tests must be carried out before grouping an isolate in this genus, including 16S rRNA analysis (**Fig. 1**).

From Genomes to Proteomes

The composition and structural organization of genomes are increasingly important in the classification and subdivision of bacteria into families and genera. Owing to the rapid development of molecular methods and automatic DNA sequencing, a complete genome can be determined within a short period. One of the most famous representatives of the genus *Bacillus*, *B. subtilis*, is used as a model organism in basic research and as a host for recombinant DNA in applied research. In the early 1990s *B. subtilis* was chosen for investigation of its complete genome sequence. This collaborative project involved about 35 groups in Europe, the USA and Japan and was completed in the autumn of 1997. The whole genome is composed of 4 214 814 base pairs (which may vary slightly due to error corrections) harbouring about 4100 open reading frames, which reflect potential protein-encoding genes. The genes present in the genome have been classified according to functional features (**Table 1**). At present, the assignment of genes can be grouped into five categories based on sequence homologies and functional analysis:

- definite assignment of genes due to experimentally identified functions (about 10%)
- probable assignments due to high sequence homologies (about 50%)
- possible assignments due to low sequence homologies (about 15%)
- putative assignments due to weak sequence homologies (about 10%)

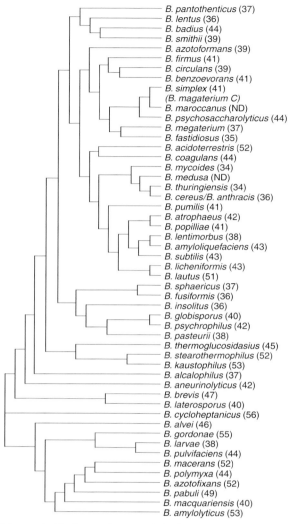

Figure 1 Phylogenetic tree of *Bacillus* spp. according to 16S rRNA analysis. The G+C content is given in parentheses in mole per cent. ND, not determined.

- 15% with no counterpart found in the protein sequence data base.

About half of the genes can be assigned to proteins with a defined or probable function, and half have no clear function. Based on these genome sequence data, a systematic function search programme has been started. The analysis of the protein composition of *B. subtilis* (the proteome) will provide further knowledge of the organism and the genus *Bacillus* in general and lead to the functional assignment of the proteins and the corresponding genes.

Cell Wall Composition

In contrast to the Gram-negative bacteria, the Gram-positive bacteria including *Bacillus* reveal a highly varied peptidoglycan composition and structure.

Table 1 *Bacillus subtilis* genome summary

4099	ORFs are known at present
3044	ORFs have homologues (73.3%)
1326	ORFs contain superfamily assignments (32.3%)
1700	ORFs are assigned to functional categories (41.5%)
578	ORFs have known or homologous three-dimensional structure (14.1%)
540	ORFs have signal peptides (13.2%)
1140	ORFs have at least one transmembrane region (27.8%)
751	ORFs have at least two transmembrane regions (18.3%)
613	ORFs have at least three transmembrane regions (15.0%)
152	ORFs contain more then 20% of low-complexity sequence (3.7%)
185	ORFs contain coiled-coil regions (4.5%)
693	ORFs are all-alpha proteins (16.9%)
181	ORFs are all-beta proteins (4.4%)
1971	ORFs are alpha/beta proteins (48.1%)
114	ORFs are irregular proteins (2.8%)

ORF, open reading frame, including all putative, potential and defined genes.

About a hundred different types have been described. Therefore, cell wall composition is often a useful criterion in taxonomy. The murein sacculus of *Bacillus* consists of peptidoglycan and is composed of up to about thirty layers. The peptidoglycan is a heteropolymer of glycan cross-linked by short peptides. Peptide chains are always composed of alternating L- and D-amino acids. In the genus *Bacillus*, the murein version of most species (with some few exceptions) is of the direct-linked *meso*-diaminopimelic acid type.

Sporulation

Spore formation in *Bacillus* takes place when the cell culture reaches the stationary growth phase. The process of spore formation is an excellent model system for studying the molecular biology of differentiation. During the sporulation process, a vegetative cell (the progenitor) gives rise to two specialized cells differing in cell type both from each other and from the parent cell. Furthermore, in some cases this process is associated with the synthesis of biotechnologically important products such as insect toxins and peptide synthetases creating peptide antibiotics.

The sporulation process is initiated at the end of the exponential growth phase. The development of the endospore formation involves an energy-intensive pathway and requires the production of a complex morphological structure. External (and presumably also internal, however partially unknown) signals force the cell to respond by inhibiting cell division and initiating the sporulation process. Initially a complex signal transduction system is turned on, the phosphorylay, which subsequently, at the end of the signal cascade, activates the transcriptional regulator protein Spo0A through phosphorylation. In contrast to vegetative growth, sporulation gives rise to an asymmetrically positioned septum which partitions the developing cell into compartments of unequal sizes. The smaller part is the forespore, which in its subsequent development exhibits a biochemical composition and structure completely different from the remaining mother cell. During the sporulation process several genes are sequentially activated; this selected gene activation is induced by the communication of mother cell and forespore, by signals transferred across the septum. The subsequent specific transcriptional regulation of spore genes is influenced by the activation of different alternative sigma factors, which confer promoter specificity to the RNA polymerase. In turn, the forespore transforms itself into the endospore, and the mother cell dies by cell lysis.

The sporulation characteristics of various *Bacillus* species are summarized in **Table 2**.

Isolation of Sporulating Bacteria

Spore-formers can be selectively isolated from natural samples after incubation at 80°C for 10 min. This treatment effectively destroys vegetative cells, whereas spores remain viable. The heat-treated probes can be streaked onto plates of medium and further incubated under aerobic conditions at their individual optimal growing temperatures. The colonies obtained are almost exclusively made up of the genus *Bacillus*.

Gene Transfer

Another interesting features of some members of the *Bacillus* genus, which has been well analysed for *B. subtilis*, is the development of natural competence for DNA uptake. Before sporulation initiation, about 10–20% of a cell culture express competence in the post-exponential growth phase under defined growth conditions. Such competent cells efficiently bind, process and internalize available exogenous high-molecular-weight DNA. The DNA can originate either from chromosomal DNA or DNA fragments, which must integrate themselves into the host chromosome in order to survive there, or from plasmid DNA, which can endure and replicate as extrachromosomal DNA in the cytoplasm if it contains a functional origin of replication. Several stages of the DNA transformation process have been described, including binding, fragmentation, uptake, and (in the case of transforming chromosomal DNA) also integration and resolution. In several organisms, including *B. subtilis*, competence has been used to genetically analyse and construct stable and defined mutationally altered strains. The latter can be obtained by allelic exchange based

Table 2 Selected characteristics of representative species of the genus *Bacillus*

(A) Spores oval or cylindrical, facultative aerobes, casein and starch hydrolysis, no swollen sporangia and thin spore wall		
B. coagulans	Thermophiles and acidophiles	Spore position central or terminal
B. acidocaldarius		Spore position terminal
B. licheniformis	Mesophiles	Spore position central
B. cereus		Spore position central
B. anthracis		Spore position central
B. megaterium		Spore position central
B. subtilis		Spore position central
B. thuringiensis	Insect pathogen	Spore position central
Sporangia distinctly swollen, thick spore wall		
B. stearothermophilus	Thermophile	Spore position terminal
B. polymyxa	Mesophiles	Spore position terminal
B. macerans		Spore position terminal
B. circulans		Spore position central or terminal
B. larvae	Insect pathogens	Spore position central or terminal
B. popilliae		Spore position central
(B) Spores spherical, obligate aerobes, casein and starch not hydrolysed		
B. sphaericus	Sporangia swollen	Spore position terminal
B. pasteurii	Sporangia not swollen	Spore position terminal

on homologous recombination, introducing defined mutations, recombinant DNA or foreign genes flanked by DNA fragments with homologies to genomic regions. This important feature of *B. subtilis* makes the strain suitable as a host for genes under regulatory promoter control for industrial use in protein overproduction.

From Starch to Sugar

Many bacilli produce extracellular hydrolytic enzymes essential for the breakdown of polysaccharides or oligosaccharides, nucleic acids, proteins and lipids. The resulting products can be used as carbon sources, nitrogen sources, energy sources and electron donors. However, they also contain hydrolytic enzymes in the cytoplasm, which prepare carbon sources to enter glycolysis by further hydrolysation, phosphorylation and isomerization reactions. The enzymes involved in sugar metabolism are of commercial interest to the food industry and in diagnostic medicine.

Bacillus species are used to manufacture commercially important enzymes (**Table 3**), for example for the production of glucose from corn, wheat or potato starch. The resulting glucose can be attacked by glucose isomerase to produce fructose, which has

a sweeter taste than either glucose or sucrose. This enzymatic process therefore has become increasingly important for the industrial production of sugar from starch, especially as a sweetening agent in soft drinks. In principle, these reactions can be catalysed separately by enzymes which operate sequentially in the conversion reactions. These reactions are composed of three principal steps:

- *Thinning reaction*, in which the starch poly-

Table 3 Examples of commercially produced enzymes from *Bacillus* spp.

Bacillus *species*	Enzyme
B. amyloliquefaciens, B. licheniformis, B. stearothermophilus	Alpha-amylase
B. coagulans	Glucose isomerase
B. stearothermophilus	Glucose kinase
B. stearothermophilus	Glucose-6-phosphate dehydrogenase
B. amyloliquefaciens	Metalloprotease
B. cereus	Phospholipase
B. stearothermophilus	Phosphotransacetylase
B. acidopullulyticus	Pullulanase
B. licheniformis, B. lentus, B. alcalophilus	Serine protease

saccharides are attacked by α-amylase, shortening the chain and reducing viscosity.

- *Saccharification*, which produces glucose from the shortened polysaccharides catalysed by the glucoamylase.
- *Isomerization*, which converts glucose into fructose, catalysed by the glucose isomerase.

The resulting fructose-containing syrup can be used directly to sweeten food products.

Regulatory Aspects Studied in *B. subtilis*

Bacillus detects and responds to environmental changes to survive adverse conditions by adapting to changes in the composition of available nutrients. This is also the case during fermentation in industrial microbial enzyme production, where the metabolization of nutrients and the secretion of end products result in the medium composition continuously changing. These environmental signals subsequently result in a shift of gene expression rates. Therefore, an understanding of signal reception and the molecular mechanisms of cellular response responsible for gene expression is crucial to optimize enzyme production and to cut production expenses. As mentioned above, *B. subtilis* has been chosen as a representative of the genus *Bacillus* for fundamental molecular biological research, and numerous scientific projects deal with aspects of regulatory mechanisms on the transcriptional, translational and enzymatic levels. An important network regulation mechanism on the transcriptional level with a central regulatory component is known as carbon catabolite repression (CCR). The presence of different rapidly metabolized carbohydrates leads to the sequential expression of different specific sugar utilization systems. Glucose and fructose are preferentially metabolized. Therefore, a change in the sugars used during fermentation or food production will affect the concentration and composition of the remaining nutrients and fermentative products, thereby altering the taste of the end product. In *Bacillus* CCR is essentially mediated by the central regulatory component CcpA, after its interaction with HPr (phosphorylated at serine position 46 by an ATP-dependent HPr kinase). When it is phosphorylated at a second site, namely histidine position 15, HPr is one of the central components in the phosphoenolpyruvate-dependent phosphotransferase systems of sugar uptake systems, where it transfers the phosphate from enzyme I to the sugar permease. The component CcpA belongs to the LacI-GalR regulator family. In the mediation of CCR for transcriptional control, it acts as a repressor able to

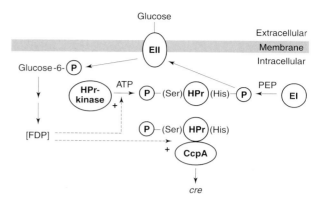

Figure 2 Model of CcpA-dependent mediated carbon catabolite repression in *Bacillus*. ATP, adenosine triphosphate; *cre*, catabolite responsive element (DNA); EI, enzyme I of the phosphoenolpyruvate-dependent phosphotransferase system; EII, specific glucose permease; FDP, fructose diphosphate; His, histidine residue; P, phosphoryl group; PEP, phosphoenolpyruvate; Ser, serine residue.

bind *cis*-active DNA elements (*cre*) (**Fig. 2**) located in a range of 200 base pairs downstream or upstream from the transcriptional start sites of genes or operons under catabolite control.

This newly discovered CCR mechanism differs completely from the mechanism observed in *Escherichia coli*, which presumably exists only in the Enterobacteriaceae. The CcpA protein was detected in eight different species of *Bacillus*, in *Lactobacillus*, *Lactococcus*, *Micrococcus*, *Mycobacterium*, *Staphylococcus*, *Streptococcus* and *Streptomyces*. Therefore, it can be concluded that the CcpA-dependent regulation of CCR is probably widespread in Gram-positive bacteria and a universally used regulatory mechanism in bacteria.

Pathogenesis

Different variants of *Bacillus thuringiensis*, *B. popilliae*, *B. larvae*, *B. cereus*, *B. sphaericus* and other related species are pathogenic to insects. The use of these strains for microbial insect control offers the advantage of being safer than the more toxic chemical control agents. Furthermore, they have relatively slight effects on the ecological balance of the environment. The microbial insecticide comprises spores and crystalline proteins which, when ingested by larvae, cause gut paralysis, probably by upsetting the ionic balance of the gut. The spore survives its passage through the gut, penetrates the weakened midgut wall, and multiplies in the haemolymph. Death results from either intoxication or septicaemia. High selectivity and the absence of harmful side effects on plants, warm-blooded animals or humans give many of the *Bacillus* products an advantage over other insecticides. Several insect-specific pathogens are commercially produced for use as microbial pesticides. *Bacillus thuringiensis* produces insect larvicides. Cultures of sporulated *B. thuringiensis* have been used worldwide to control damage to crops, trees and ornamental plants. During endospore formation, this bacterium produces toxic protein crystals (Bt toxin) that make it a good pesticide, which differs from *B. cereus* (see below). Most of the toxin genes of *B. thuringiensis* are located on conjugative plasmids, which are transmissible by conjugation between *B. thuringiensis* and *B. cereus* under laboratory conditions. The resulting *B. cereus* transconjugants are able to synthesize crystal proteins. Because the dividing line between *B. cereus* and *B. thuringiensis* is so dubious, these organisms could be considered to have changed species to *B. thuringiensis*.

Bacillus popilliae causes a fatal illness called milky disease in Japanese beetle larvae. After ingestion by the larvae, *B. popilliae* germinates in the gut, begins to multiply and invades the haemolymph. After about 10 days a typical milky appearance is observed due to the massive numbers of bacteria.

Bacillus cereus strains are also often pathogenic for insects. They produce phospholipase C, an α-exotoxin which permits the bacteria to pass through the barrier of the intestinal epithelial cells. Subsequent penetration into the haemolymph followed by multiplication kill the insect.

Certain strains of the soil saprophyte *B. cereus* are also pathogenic for humans. Two toxins, one causing diarrhoea and the other provoking vomiting, are produced in starchy foods, custards and dairy products containing the common contaminant *B. cereus*. Another member of the genus *Bacillus* with pathogenic properties is *B. anthracis*. The three toxin genes are located extrachromosomally on a large plasmid. *Bacillus anthracis* causes the animal disease anthrax, which spreads to humans primarily through minor breaks in the skin or mucous membranes. Infection sources can be milk, meat, wool or hairs from infected animals. Cutaneous anthrax first appears as a papule which develops into a vesicle and after 2–6 days into a black eschar. This bacterium produces a potentially lethal toxin and 5–20% of untreated cases are fatal. *Bacillus anthracis* is similar in many respects to *B. cereus* and *B. thuringiensis*. These taxa can be distinguished phenotypically by numerical taxonomy, and there are some phenotypic tests such as sensitivity to penicillin: *B. cereus* possesses a chromosomally encoded β-lactamase, whereas *B. anthracis* is virtually always penicillin-sensitive.

Outlook

In the near future the demand of biotechnological industry for new or improved products will lead to the development of new genetically engineered strains and new isolates from the environment, which will be grouped in the genus *Bacillus*, increasing its enormous phylogenetic diversity. The taxonomy of the genus *Bacillus* is evolving; consequently, definitions for a subdivision of the genus into phylogenetic and phenetic groups must be considered. Furthermore, the genetic information about members of *Bacillus* will dramatically increase. However, our knowledge of chromosomal composition and the genes can only give us a hint about their function when homologous genes and proteins are known. Therefore, biochemical research into protein functions will be a field of increasing significance. In food production, as well as in enzyme production involving bacilli, gene manipulation will lead to strain constructions with new (designed) features allowing an optimized and therefore less expensive production process. This can be achieved by introducing and designing specific metabolic pathways or heterologous enzyme overproduction.

See color Plate 2.

See also: **Bacillus**: *Bacillus cereus*; *Bacillus stearothermophilus*; *Bacillus anthracis*; *Bacillus subtilis*; Detection of Toxins; Detection by Classical Cultural Techniques.

Further Reading

Cliver D (ed.) (1990) *Food Borne Diseases*. San Diego: Academic Press.

Gerhardt P, Murray RGE, Wood WA and Krieg NR (eds) (1994) *Methods for General and Molecular Bacteriology*. Washington: American Society for Microbiology Press.

Glick BR and Pasternak JJ (1998) *Molecular Biotechnology*, 2nd edn. Washington: American Society for Microbiology Press.

Hoch JA and Silhavy TJ (eds) (1995) *Two Component Signal Transduction*. Washington: American Society for Microbiology Press.

Hueck CJ and Hillen W (1995) Catabolite repression in *Bacillus subtilis*: a global regulatory mechanism for the Gram-positive bacteria? *Molecular Microbiology* 15(3): 395–401.

Kunst F, Ogasawara N, Moszer I et al (1997) The complete genome sequence of the Gram-positive bacterium *Bacillus subtilis*. *Nature* (London) 390: 249–256.

Piggot PJ, Moran CP and Youngman P (eds) (1994) *Regulation of Bacterial Differentiation*. Washington: American Society for Microbiology Press.

Sneath PHA (ed.) (1982) *Bergey's Manual of Systematic Bacteriology*. Vol. 2. Baltimore: Williams & Wilkins.

Sonnenshein AL, Hoch JA and Losick R (eds) (1993) Bacillus subtilis *and other Gram-positive Bacteria*. Washington: American Society for Microbiology Press.

Bacillus cereus

Carl A Batt, Department of Food Science, Cornell University, USA

Characteristics of the Species

Bacillus cereus is a diverse species belonging to the larger *Bacillus cereus* group, which also includes *Bacillus mycoides* and *Bacillus thuringiensis*. Distinction between the species is based upon a number of biochemical characteristics; however, distinguishing between the species can be difficult. A number of different schemes have been reported to identify *B. cereus*, thus differentiating it from the other members of the *B. cereus* group. Depending upon which scheme is employed, the overlap between two or all three of the species differs.

The bacterium *B. cereus* is Gram positive and is characterized by its ability to form spores. It has an optimum growth temperature of 28–35°C with a minimum of 4–5°C and a maximum of 48°C. The generation time of the organism is 18–27 min. It grows over a wide pH range of 4.9–9.3 and at salt concentrations of up to 7.5%. The spores are relatively heat-resistant, although the D values tend to be variable. Typically the D_{100} range is approximately 2.2–5.4 min although considerable variation has been observed between different strains. Germination of spores is robust and frequencies of up to 100% have been reported. The germination process is rapid and can occur in some strains within 30 min. Germination requires a number of small molecules including glycine or alanine and purine ribosides.

Bacillus cereus is a common inhabitant of soils and can easily be transmitted into vegetation and subsequently into foods. It is often present in a variety of foods, including dairy products, meats, spices and cereals. In general, foods that are processed by drying or are otherwise subjected to heating can still contain *B. cereus*. Typically, food-borne poisoning involving *B. cereus* results from the consumption of cereal dishes and other predominantly starchy foods. Of all foods, fried rice has been implicated most often in *B. cereus* food-borne illnesses. This is due to the fact that this pathogen is a frequent contaminant of uncooked rice and that *B. cereus* spores can survive the cooking process. Rice cooked, but then held at room temperature, can allow the bacteria to multiply and produce toxin. The subsequent heat treatment during frying is usually not sufficient to kill vegetative cells and certainly insufficient in inactivating the toxin. Thus, when this food product rests at room temperature, the problem is exacerbated by allowing vegetative cells to multiply.

Despite their apparent close phenotypic relationship, members of the *B. cereus* group – *B. cereus*, *B. thuringiensis* and *B. mycoides* – are genotypically diverse, but there are several genes common to all three. Typology using multiple enzyme electrophoresis, carbohydrate profiles, phage, DNA–DNA homology or rRNA suggests differences both within species and between species. Differences within *B. cereus* are perhaps best exemplified by the differences in genome size between isolates. The size can vary by as much as 2 Mb (or over 50% of the genome). Mapping and hybridization analysis reveals that, for example, one *B. cereus* isolate has a small genome (2.4 Mb) and is a subset of a larger *B. cereus* genome and that this genome is conserved as part of the genome of at least four other *B. cereus* strains. Within this conserved genome is at least one virulence factor, phospholipase C (see below). In certain strains, some of this variant genome is 'extrachromosomal' while in other cases it is absent.

The inherent diversity within the *B. cereus* group and the presence of similar, if not identical, toxin genes in a number of other members of the *B. cereus* group, including *B. mycoides* and *B. thuringiensis*, raise the issue of microbiological identity and safety, based solely upon a microbiological name. For example, *B. thuringiensis*, which is commonly used as a biopesticide, is known to carry a number of 'virulence' genes similar, if not identical, to those found in *B. cereus*. These include the haemolysin and the cereolysin genes. Is *B. thuringiensis* a threat to human health? It is probable that it is not especially, as its preparations that are used as biopesticides are typically rendered non-viable. Are all strains of *B. thuringiensis* avirulent? There have been a few limited reports of disease caused by *B. thuringiensis*, although the absolute accuracy of the strain identification may be an issue. Since the discrimination between *B. cereus* and *B. thuringiensis* is typically based upon the presence or the absence of a parasporal crystal, misidentification may arise. As the absolute requirements for virulence are better defined for *B. cereus*, new opportunities for functional identification schemes may be realized.

Virulence in *B. cereus* is a function of a number of different factors. Thus, there are different clinical pictures of the disease (**Table 1**). Two forms predominate; one is an emetic version, while the other is diarrhoeal and is characterized, in part, by abdominal pains. The emetic symptoms develop within 1–5 h

Table 1 Toxins found in the *B. cereus* group

Toxin	Gene	Comments
Haemolysin		A tripartite protein that contains a
BL	*hblA*	haemolysin and two binding
B	*hblC*	proteins. It also has enterotoxin
L_1	*hblD*	activity as demonstrated in a
L_2		rabbit ileal loop assay
Enterotoxin	*bceT*	A single protein whose activity has been established on the basis of a mouse ileal loop assay
Cereolysin		Two genes encoded in a single operon
A	*cerA*	Phospholipase C
B	*cerB*	Sphingomyelinase
Cereulide	Not identified	Small dodecadepsipeptide, produces vacuole response in HEp-2 cells

after the consumption of the contaminated food, while the diarrhoeal may take up to 12 h or more to develop.

The diarrhoeal form of the disease is very similar to *Clostridium perfringens* food poisoning. In general, the symptoms pass, and no further complications arise from *B. cereus*. In a limited number of cases, more severe forms of the disease have been observed in both humans and animals. These more severe forms include bovine mastitis, systemic and pyrogenic infections, gangrene, septic meningitis, lung abscesses and endocarditis. A metastatic bacterial endophthalmitis form has also been described.

There are a number of extracellular enzymes and at least some of these are assumed to be toxins. These extracellular enzymes include proteases, amylases, phospholipases, β-lactamases, haemolysins, and sphingomyelinases. The role of one or more of these enzymes in virulence is difficult to establish because of the absence of appropriate model systems, and isogenic strains specifically deficient in one or more of these enzymes.

The different forms of disease caused by *B. cereus* are presumably a function of the combination of toxins and the health status of the host. In all cases, identifying the toxin and then assigning a functional role to that particular toxin is complicated. There are no perfect model systems for studying either the diarrhoeal or the emetic response. Several model systems have been reported that have varying degrees of authenticity to the actual disease response. They also have varying degrees of difficulty, and usually the degree of difficulty is inversely proportional to the likelihood that a particular assay will truly model the disease state.

The most widely used model to measure compounds for the diarrhoeal response is the rabbit ileal loop assay. In this model, the lower portion of the

intestine of a rabbit is surgically exposed to allow ligation of the ileal region. Multiple regions can be ligated to allow different samples to be tested in a single animal. The sample is injected into one ligated section and the response in terms of fluid accumulation monitored with time. Because injecting material into a ligated ileal loop can cause fluid accumulation without the sample necessarily being toxic, controls are important. The assay is typically qualitative based upon the degree to which the loop is distended due to accumulation of fluid. As with any animal bioassay, local or federal guidelines and requirements may complicate and limit any attempts to apply it.

The toxins that are responsible for the diarrhoeal response have not been firmly established. It is clear that there are a number of toxins including a haemolysin and a cereolysin. The haemolysin designated haemolysin BL comprises three distinct peptides, B, L_1 and L_2. Each of the genes coding for the three components has been cloned and the sequences determined. The B component appears to have the ability to lyse erythrocytes, while the two L components are responsible for binding to the erythrocytes. It is haemolysin BL that is responsible for the discontinuous appearance of the haemolytic pattern that surrounds colonies of *B. cereus*.

A second virulence factor in *B. cereus* is cereolysin, which again is a multicomponent cytotoxic complex. Tandemly arranged genes for phospholipase C and sphingomyelinase are transcribed as an operon. There are in total three phospholipases in *B. cereus* and they hydrolyse phosphatidylinositol. Some, but not all, of these phospholipases are metalloenzymes requiring divalent cations for activity. The sphingomyelinase is also a metalloenzyme that has haemolysin-like activity.

A putative emetic toxin has been isolated and identified. A major difficulty in the identification of the emetic toxin is the lack of a suitable assay for biological activity. The most accepted model system is the monkey, but monkey assays are expensive to carry out owing to the cost of procurement and housing of these animals. Furthermore 'read-out' of the assay is far from exact and the time of onset, as well as the severity of the response, needs to be taken into account. Typically the sample is introduced by a stomach tube, and then the animals observed for approximately 5 hours. A set of six animals is tested and an emetic response in two of the six is considered a positive indication of the toxin. In an effort to find a more amenable model system, it was reported that the adult male suncus (a white footed shrew) was similarly susceptible to emesis. Both the time to emetic response and frequency of episodes is the output of this assay.

The putative emetic toxin is a small dodeca-depsipeptide and has been shown to produce a vacuole response in HEp-2 cells. The involvement of this purified peptide in the emetic response is based upon an in vitro assay, but there is still a need to confirm this using more established assays for the emetic response. The emetic toxin is cyclic comprising a three repeat of D-O-Leu-D-Ala-L-O-Val-L-Ala. Structurally it is related to the ionophore valinomycin and this may suggest its mode of action. No genes coding for cereulide have been identified, but it is likely they will be structurally similar to those involved in cyclic peptide biosynthesis as observed in other *Bacillus* and *Streptomyces* spp. A survey of *B. cereus* reveals that strains of the H-1 serovar were most likely to produce cereulide when compared with any other serotype of this organism.

Recent attempts to develop more facile assays have been reported. As mentioned previously, one such example is an HEp-2 cell assay where the proliferation of the cell line is used as an index of the cytostatic (or emetic) effects of the *B. cereus* toxin. In a survey of *B. cereus* strains, a significant proportion (74%) were enterotoxin producers, but only 5% produced the emetic toxin as measured by this assay. As with any model assay the results need to be confirmed against a 'gold standard'. In this case with the emetic toxin, the gold standard is the monkey assay, and it is rare to see an unequivocal comparison with it made by these assays.

Methods of Detection

Bacillus cereus is difficult to detect, primarily because of the close relationship between it and other members of the *B. cereus* group. Differentiation is usually accomplished by growth on selective media followed by microscopic observation of a parasporal crystal, characteristic of *B. thuringiensis*. *Bacillus mycoides* is characterized by its rapid colony spread, although this trait as well as spore crystal formation in *B. thuringiensis* can be lost upon culture.

Detection of *B. cereus* in foods, typical as for other microorganisms, consists of a series of steps including selective enrichment followed by plating on to selective agar media which contain ingredients to screen for the organism. Homogenates of food are prepared in Butterfield's phosphate buffered water at a 1 : 10 dilution. Direct plate counts can be made using a selective screening agar such as mannitol egg-yolk polymyxin (MYP). The polymyxin is added to suppress the growth of other microorganisms, and *B. cereus* is highly resistant to this antibiotic. Mannitol

is not utilized by most *B. cereus*, and therefore the colonies are pink, as opposed to yellow for mannitol fermenting bacteria. The MYP agar medium contains egg yolk which is a substrate for lecithinase, an enzyme found in *B. cereus*. The precipitate that forms around the colony can easily be distinguished after 24 h at 30°C. In some cases, an additional 24 h is required in order to observe clearly the zone of precipitation.

Where low numbers of *B. cereus* are expected, direct plating may not be suitable. The threshold for direct plating is approximately 10 colony forming units (cfu) g^{-1}. The most probable number (MPN) technique can be used to enumerate bacteria in samples below 10 cfu g^{-1}. The MPN technique for *B. cereus* starts with dilution into trypticase soy-polymyxin broth in triplicate. The tubes are incubated for 48 h at 30°C and dense growth is usually observed. The culture is then streaked onto MYP agar and incubated (as described above). Any presumptive positives must be reconfirmed as *B. cereus*.

Confirmatory Tests

Confirmation of *B. cereus* requires completion of a number of tests (**Table 2**). Unfortunately, there is no single test that can be used to identify *B. cereus* unequivocally. As mentioned previously, the most distinguishing features of *B. cereus* as compared with *B. mycoides* and *B. thuringiensis* are the absence of rhizoid growth and spore crystal respectively. Unfortunately, *B. mycoides* on culture in the laboratory may lose its rhizoid growth and *B. thuringiensis* may lose its ability to form crystals. The basic characteristics of *B. cereus* include large Gram-positive rods with spores that do not swell the sporangium, in addition to its production of lecithinase and failure to ferment mannitol. It produces acid from glucose under anaerobic conditions. Other characteristics of *B. cereus* include: reduction of nitrate to nitrite,

Table 2 Confirmatory tests for the *B. cereus* group including *B. cereus*, *B. thuringiensis* and *B. mycoides*

Confirmation test	B. cereus	B. thuringiensis	B. mycoides
Gram reaction	+	+	+
Catalase	+	+	+
Motility	±	±	−
Nitrate reduction	+	±	+
Tyrosine degradation	+	+	±
Lysozyme resistance	+	+	+
Egg yolk hydrolysis	+	+	+
Glucose utilization (anaerobic)	+	+	+
Voges–Proskauer	+	+	+
Acid from mannitol	−	−	−
Haemolysin	+	+	+

production of acetylmethylcarbinol (Voges–Proskauer positive), degradation of tyrosine and resistance to lysozyme. To assess these various characteristics and to confirm *B. cereus*, additional tests include those detailed below.

Phenol Red Glucose Broth Three millilitres of the broth is inoculated with a loopful of culture and incubated at 35°C for 24 h. Growth is determined by turbidity and anaerobic fermentation of glucose measured by the change in colour from red to yellow.

Nitrate Broth A 3 ml quantity of broth is inoculated with a loop of culture and after 24 h at 35°C, nitrite production is measured by the addition of α-naphthylamine and α-naphthol.

Tyrosine Agar The clearing of the agar medium around a colony streaked out on tyrosine agar indicates utilization.

Lysozyme Broth Nutrient broth supplemented with 0.001% lysozyme can be used to score for lysozyme resistance, a property of *B. cereus* in addition to other bacteria.

Specific Tests

Specific tests to distinguish *B. cereus* from other members of the *B. cereus* group are detailed below.

Motility Motility is measured by stabbing the centre of a tube of semisolid medium and allowing the culture to grow and spread for 18 h at 30°C. Motile bacteria will diffuse out from the stab, forming an opaque growth pattern, while non-motile bacteria do not diffuse out. A second option is to put a loop of culture on a prewet agar plate and observe the spread of bacterial growth beyond the boundaries of the area defined by the loop.

Rhizoid Growth A freshly poured agar plate is inoculated with a loop of an overnight culture and the inoculum is allowed to absorb into the agar. After 48–72 h rhizoid growth is characterized by the production of hair or root-like structures projecting from the inoculated area. Rough colonies should not be confused with typical rhizoid growth of *B. cereus*.

Haemolysin Activity Trypticase soy–sheep blood agar plates are inoculated with an overnight culture and incubated at 35°C for 24 h. Strong haemolytic activity is characterized by a complete zone of haemolysis approximately 2–4 mm around the colony.

Protein Toxin Crystal Formation Nutrient agar slants are inoculated with an overnight culture and left at room temperature for 2–3 days. A smear on a microscope slide is then stained with 0.5% basic fuchsin of TB carbolfuchsin ZN. Toxin crystals from *B. thuringiensis* appear as dark-staining, diamond-shaped objects that are smaller than the spores. They are released from the sporangium upon lysis and therefore unless spore release is observed, the test is inconclusive.

Currently, there are commercial enzyme-linked immunosorbent assays (ELISAs) to detect at least one of the toxins produced by *B. cereus*, and, in a number of studies, this test has proved useful. These assays include the reverse passive latex agglutination (RPLA) enterotoxin assay and a visual immunoassay, the latter is reported to be specific against the diarrhoeal enterotoxin. Detection of this toxin does not, however, resolve all food safety concerns and the assays do not yield equivalent results. It has been documented that the two commercial ELISAs detect either only one component of the haemolysin BL complex or two nontoxic proteins.

Several studies have surveyed the virulence of strains isolated from different sources. For example, 12 *B. cereus* strains isolated from different foods and disease outbreaks were all shown to produce the diarrhoeal enterotoxin. A slightly lower frequency (84–91%) of toxigenic strains was reported from a collection isolated only from food. In another study, only 8 of 11 strains tested produced toxin. One limitation to a number of these studies is the use of commercial kits which may not be accurate in assessing toxigenic potential of *B. cereus* (see below). Since these ELISAs measure the toxin but not its activity, the relevance of these results are not clear. For example, the RPLA test yields a positive result for samples after boiling, a process that inactivates them biologically. Subtle differences in activity or virulence may be missed using this type of analysis.

Toxin production can be variable and dependent upon the growth conditions. For example, the pH and sugars in the growth medium can result in an almost twentyfold difference in toxin production. In these studies, toxin production was measured using the RPLA test which apparently recognizes the haemolysin B component of the BL complex. Under certain conditions, including high glucose concentration, toxin was not produced at detectable levels. Therefore, in assessing the potential for a particular *B. cereus* isolate to cause disease, none of these methods will yield an unequivocal answer. Furthermore, simply isolating *B. cereus* from a food without further assaying its virulence may be a suggestion of risk but without justification.

Regulations

The actual number of cases of food-borne illness involving *B. cereus* is difficult to estimate. In the USA the number of outbreaks reported varies from 6 to 50 per year. The relatively mild symptoms and the short duration of illness contribute to the under-reporting of this food-borne pathogen. In addition, testing for *B. cereus* is not a routine practice in a number of State health laboratories.

In the USA, beyond a general concern about any pathogen in the foods, specific attention has been directed toward the contamination of infant formula with *B. cereus*. The Food and Drug Administration (FDA) has historically expressed concern 'due to levels of *B. cereus* that exceed 1000 &ldqo;colony forming units&rdqo; (cfus) per gram (g) of a powdered infant formula' (see 54 FR 3783, Jan. 26, 1989, and 56 FR 66566, Dec. 24, 1991). Moreover, infant formula is of concern because of the ability of *B. cereus* to replicate rapidly upon rehydration of dried formula. Therefore, recent efforts by the FDA are directed toward reducing the maximum permissible level (M) of *B. cereus* in infant formula to 100 cfu (or MPN) per gram. The FDA will determine compliance with the M values listed below using the *Bacteriological Analytical Manual* (BAM), 8th edn (1995).

Importance to the Food Industry

Bacillus cereus spores are able to survive low-temperature processing, which occurs, for example, in spray drying. Therefore, any food product that is a spray-dried powder is subject to contamination by *B. cereus*. The estimated infectious dose of this pathogen is probably greater than 10^5 and it will not grow in dried ingredients. Problems arise not only with foods that are improperly processed, but, more importantly, with foods that are improperly stored. A compilation of a number of studies reported that the frequency of *B. cereus* positive dairy samples ranged from 4% to 100%. Levels of *B. cereus* ranged from 5 to over 1000 *B. cereus* per gram of sample. As mentioned previously, attention has been given to infant formulae since they are typically composed of spray-dried dairy ingredients. In infant formula *B. cereus* positive samples were found at frequencies of 1.9–100%.

Contamination of dairy products by *B. cereus* presumably originates with the raw milk. Improper cleaning of processing equipment can only contribute to the contamination problem. Thermal processing is not totally effective at killing *B. cereus* spores. Values

of D at 100°C range from 2.2 min to 5.4 min. Removal of spores using processing steps including centrifugation (bactofugation) are very effective at reducing spore loads. Although spray-drying towers are operated at temperatures in the range 150–220°C, rapid cooling of the particles results in their temperature reaching only 40–50°C. Aside from its toxigenic potential, *B. cereus* can cause other problems in foods. The organism causes spoilage which has been termed 'broken cream' or sweet curdling of milk. This is because of its proteolytic activity in the absence of high levels of acid production.

Importance to the Consumer

There are few reports of *B. cereus* intoxication, although certain foods including fried or boiled rice, pasteurized cream, cooked meat, mashed potatoes and vegetable sprouts appear to be common sources of food poisoning. Its ability to sporulate leads to the high frequency of *B. cereus* contamination in dried food products. The species is ubiquitous and hence its isolation from a suspect food associated with a food-borne illness is not strong enough evidence for a causal relationship.

The prototypical *B. cereus* outbreak was reported in 1994 and concerned food poisoning at two child-care centres in Virginia. The initial reports described acute gastrointestinal illness among children and staff at two day-care centres, which were under a single management. The symptoms were reported after the consumption of a catered lunch at these facilities. A total of 67 individuals consumed the lunch and of those 14 or approximately 21% became ill. The 13 individuals at the centres who did not consume the lunch did not become ill. The predominant symptom was nausea, and to a lesser extent abdominal cramps, and diarrhoea. The majority of the cases were in children aged 2.5–5 years. The median onset time of the symptoms was 2 h.

The one dish at the catered lunch that was common to a number of the victims was chicken fried rice. *Bacillus cereus* was isolated from some leftover food (approximately 10^6 cfu g^{-1}) and from the vomitus of one child. Only a single other food (milk) was available for testing and it proved to be negative for *B. cereus*. The rice had been cooked the previous day and cooled at room temperature before refrigeration. The final dish was prepared that day and then stored without refrigeration for approximately 1.5 h.

This incident illustrates the major issues in linking a food-borne illness like *B. cereus* to the consumption of a specific food. Confirmation requires the isolation of the pathogen from the suspected food and then linkage to the incident by epidemiological data. Furthermore, it is widely accepted that contamination levels in excess of 10^5 cfu g^{-1} should be observed in the food to justify a causal relationship.

At least one compelling study was carried out to determine the consequences of consuming B. cereus-contaminated milk. In this study healthy adults consumed pasteurized milk, some of which, after storage, was found to be contaminated with $> 10^8$ B. cereus cells. Only at the highest levels was there any significant correlation to reports of gastrointestinal distress. Below 10^8 there was no significant effect. Diarrhoeal enterotoxin was measured, and although many of the recovered strains produced toxin, the levels in milk, even with high B. cereus numbers, was low.

In general, the consumer can reduce the risk of B. cereus food-borne illness by storing potentially suspect foods at temperatures below 7°C or above 55–60°C. In addition, rapid cooling or heating to reach these temperatures reduces the time the food spends at temperatures that allow B. cereus growth.

See also: **Biochemical and Modern Identification Techniques**: Food-poisoning Organisms. **Direct (and Indirect) Conductimetric/Impedimetric Techniques**: Food-borne Pathogens. **Enzyme Immunoassays**: Overview. **Food Poisoning Outbreaks**.

Further Reading

Agata N, Ohta M, Arakawa Y and Mori M (1995) The bceT gene of Bacillus cereus encodes an enterotoxic protein. Microbiology 141: 983–988.

Agata N, Ohta M, Mori M and Isobe M (1995) A novel dodecadepsipeptide, cereulide, is an emetic toxin of Bacillus cereus. FEMS Microbiology Letters 129: 17–20.

Beecher DJ, Schoeni JL and Wong ACL (1995) Enterotoxic activity of hemolysin BL from Bacillus cereus. Infection and Immunity 63: 4423–4428.

Carlson CR, Caugant DA and Kolsto AB (1994) Genotypic diversity among Bacillus cereus and Bacillus thuringiensis strains. Applied and Environmental Microbiology 60: 1719–1725.

Grannum PE (1994) Bacillus cereus and its toxins. Journal of Applied Bacteriology 76: 61S–66S.

Johnson EA (1990) Bacillus cereus food poisoning. In: Cliver D (ed.) Food Borne Diseases. P. 128. San Diego: Academic Press.

Sutherland AD (1993) Toxin production by Bacillus cereus in dairy products. Journal of Dairy Research 60: 569–574.

Bacillus stearothermophilus

P Kotzekidou, Department of Food Science and Technology, Aristotle University of Thessaloniki, Greece

Characteristics of the Species

Bacillus stearothermophilus is a thermophilic, aerobic, spore-forming bacterium with ellipsoidal spores that distend the sporangium. It is a heterogeneous species in which the distinguishing features are a maximum growth temperature of 65–75°C, a minimum growth temperature of 40°C, and a limited tolerance to acid. The bacterium does not grow at 37°C; its optimum growth is at 55°C. Starch hydrolysis is positive, although some strains do not hydrolyse starch. Hydrolysis of casein and reduction of nitrate to nitrite are variable. Growth in 5% NaCl is scant. The heterogeneity of the species is indicated by the wide range of DNA base composition as well as the diversity of the phenotypic characters (**Table 1**). Minimum pH for growth of B. stearothermophilus is 5.2; the minimum water activity (a_w) for growth at optimum temperature is 0.93.

Table 1. Differential characteristics of Bacillus stearothermophilus

Characteristic	B. stearothermophilus
Morphology	Rods
Width of rod (μm)	0.6–1
Length of rod (μm)	2–3.5
Sporangium swollen	+
Spore shape	Ellipsoidal
Spore position	Terminal or subterminal
Acid from:	
glucose	+
L-arabinose	+/–
D-xylose	+/–
mannitol	+/–
Hydrolysis of:	
starch	+
casein	+/–
Utilization of citrate	+/–
Dihydroxyacetone formation	–
Catalase	+/–
Anaerobic growth	–
Voges–Proskauer test	–
Nitrate reduction	+/–
Growth in NaCl:	
5%	+/–
7%	–
Growth at:	
30°C	–
40°C	+
55°C	+
65°C	+
Gas from glucose	–

Bacillus stearothermophilus is a common inhabitant of soil, hot springs, desert sand, Arctic waters, ocean sediments, food and compost. The incidence of *B. stearothermophilus* in foods is related to the distribution of the microorganism in soil, water and plants. Foods that have been heated or desiccated generally possess an enriched and varied flora of bacterial spores. *Bacillus stearothermophilus* is included in the usual microflora of cocoa bean fermentation as well as of cocoa powder. It is the dominant microorganism of beet sugar, and is isolated from pasteurized and dried skimmed milk.

The incidence of *B. stearothermophilus* spores in canned foods is of particular interest. The spores enter canneries in soil, on raw foods and in ingredients, e.g. spices, sugar, starch and flour. The presence of *B. stearothermophilus* spores in some containers of any given lot of commercially sterile low-acid canned foods may be considered normal. If the food is to be distributed in nontropical regions where temperatures do not exceed about 40°C for significant periods of time, complete eradication of the microorganism is not necessary since it cannot grow at such low ambient temperatures. For tropical conditions, the thermal process must be sufficient to inactivate spores of *B. stearothermophilus* that might otherwise germinate and multiply under these conditions. *Bacillus stearothermophilus* is the typical species responsible for thermophilic flat sour spoilage of low-acid canned foods.

The importance of thermophilic spoilage organisms in the food industry has generated considerable interest in the factors affecting heat resistance, germination and survival of their spores. Because it grows at high temperatures, *B. stearothermophilus* tends to produce heat-resistant spores. However, the genetic variation in moist as well as dry heat resistance between different strains of *B. stearothermophilus* is of considerable magnitude (**Table 2**). The main factors affecting these discrepancies are the composition of the sporulation medium, the sporulation temperature and the chemical state of the bacterial spore, as well as the heating conditions in terms of the water activity, the pH and the ionic environment of the heating medium, the presence of organic substances, the composition of the atmosphere, etc. Under dry conditions *B. stearothermophilus* spores show the greatest increase in heat resistance (Table 2). At high water activity the decimal reduction time at 100°C (D_{100}) of *B. stearothermophilus* spores is no less than 800 min, and under dry conditions the D_{100} is about 1000 min.

In low-acid canned foods D_{120} values of 4–5 min and z values (the temperature increase needed for a tenfold decrease in the D value) of 14–22°C have been reported. Values of D decrease when the pH is reduced from 7.0 to 4.0. Values of z appear to be higher when the medium is acidified, although the difference is not statistically significant. Organic acids and glucono-delta-lactone have the same effect as acidulants in reducing the heat resistance of *B. stearothermophilus* spores. Sodium chloride reduces heat resistance of *B. stearothermophilus* when present at relatively low levels, i.e. less than 0.5 mol l^{-1}.

When a dormant heat-resistant spore is activated and germinates to form a vegetative cell, its heat resistance is lost. Spores frozen at −18°C or freeze-dried exhibit a loss in viability and heat resistance. Heating of spores at sublethal temperatures can result in enhanced heat resistance. Activation of dormant spores by sublethal heating breaks dormancy and increases the ability of spores to germinate and grow under favourable conditions. Heat-shocked spores of *B. stearothermophilus* ATCC 7953 when activated become permeabilized at the outer membrane and become susceptible to lysozyme. When *B. stearothermophilus* spores are plated on conventional media without prior heat shock, commonly less than 10% of the total number of spores germinate; after heat activation 50% germination occurs, and after treatment with 0.5 mol l^{-1} hydrochloric acid almost 100%

Table 2 Moist and dry heat resistance of *B. stearothermophilus* spores

Strain of B. stearothermophilus	Investigated temperature range (°C)	Heating in phosphate buffer (pH 7) or in water		Dry heat	
		D-value (min)	z-value (°C)	D-value (min)	z-value (°C)
NCIB 8923	115–130	D_{120} 5.8	13.0		
NCIB 8919	115–130	D_{120} 5.3	11.0		
NCIB 8924	115–130	D_{120} 1.0	8.9		
ATCC 7953	111–125	D_{121} 2.1	8.5		
ATCC 7953	110–120	D_{118} 10.0	5.7		
ATCC 7953	150–170			D_{160} 0.08	19
NCA 1518	100–160			D_{160} 3.2–27.0	14–22
NCTC 10339	150–180			D_{160} 0.16	26–29

germination results. Spores produced at 65°C are optimally activated after holding at 30°C for 6 h resulting in increased frequency of spore germination. Sublethal heating at 80°C for 10 min may induce dormancy in some strains of *B. stearothermophilus* rather than activation. Optimum germination of spores is a function of temperature, time, pH and suspending medium. After heat treatment maximum recovery of *B. stearothermophilus* spores is obtained at pH 7.0 and decreases as pH falls. Phosphates in the recovery medium result in a progressive decrease in spore recovery, whereas starch improves recovery.

Spores of *B. stearothermophilus* are used as biological indicators for verifying exposure of a product to a sterilizing process. For monitoring steam sterilization, endospores of *B. stearothermophilus* (strains NCTC 10007, NCIB 8157, ATCC 7953) are in current use, particularly for processes performed at 121°C or higher. Biological indicators are available commercially, either as suspensions for inoculating test pieces, or on already inoculated carriers such as filter paper, glass or plastic. After exposure to the sterilization process, the biological indicators are cultured in appropriate media incubated under suitable conditions. Immobilized *B. stearothermophilus* spores are used to monitor the efficacy of a sterilization process, particularly to measure sterilizing values in aseptic processing technologies for viscous liquid foods containing particulates; they can also be used to monitor in-pack sterilization efficacy. The main immobilization matrices are alginate beads or cubes mixed with puréed potatoes, peas or meat, and polyacrylamide gel spheres. Particle dimensions vary between 0.16 cm and 0.5 cm. The estimated z values for immobilized *B. stearothermophilus* spores were 8.5–11.8°C.

Bacillus stearothermophilus produces a wide range of enzymes, many of which are of industrial significance (**Table 3**). Some of them are extracellular, enabling simple recovery from fermentation broths. The microorganism presents a number of advantages for the isolation of intracellular enzymes because its cell yield is generally good. A 400 litre fermentation of *B. stearothermophilus* NCA 1503 yields 5–8 kg of wet cell paste, equivalent to $15 \, g \, l^{-1}$. The majority of enzymes produced are intrinsically thermostable, and this enhanced stability is also exhibited against the action of other protein denaturants such as detergents and organic solvents.

The thermostable enzymes that have found commercial application are essentially intracellular enzymes. Glycerokinase produced from *B. stearothermophilus* NCA 1503 is used as a clinical diagnostic for the assay of serum triacylglycerols. The same strain cloned into *Escherichia coli* produces

lactate dehydrogenase, which is used in a clinical diagnostic kit for the assay of glutamate pyruvate transaminase and glutamate oxaloacetate transaminase. Other diagnostic enzymes produced by *B. stearothermophilus* NCA 1503, including phosphofructokinase, phosphoglycerate kinase and glucose phosphate isomerase, have been used as components of clinical diagnostic assay kits. The strain also produces restriction enzymes for molecular biology, while another strain of *B. stearothermophilus* isolated in China produces thermostable DNA polymerase which is used in polymerase chain reaction and DNA sequencing.

Methods of Detection

Bacillus stearothermophilus possesses greater heat resistance than most other organisms commonly present in foods. This characteristic is advantageous to the examination of foods and ingredients because by controlled heat treatment of samples it is possible to eliminate all organisms except the spores of heat-resistant microorganisms. Further, heat shock or activation is necessary to induce germination of the maximum number of spores. In a standard procedure, 30 min at 100°C or 10 min at 110°C followed by rapid cooling should be used.

Aerobic thermophilic spore-formers can be encountered in heat-shocked samples using dextrose–tryptone agar after incubation at 55°C for 2–3 days. Dehydration of the plates during incubation is minimized by placing the plates in oxygen-permeable bags. *Bacillus stearothermophilus* should be grown preferably in nutrient media supplemented with calcium and iron, as well as with manganese sulphate to promote sporulation (i.e. nutrient agar supplemented per litre with 3 mg of manganese sulphate as well as the following sterile solutions: 10 ml D-glucose 20% w/v, 0.8 ml $CaCl_2$ 5% w/v and 0.8 ml $FeCl_2$ 5% w/v).

When investigating the incidence of 'process-resistant' spores (i.e. spores that will survive the heat treatment of low-acid canned foods that is generally accepted as adequate for elimination of *Clostridium botulinum* spores) in ingredients such as dry sugar, starch, flour or spices, it is convenient to heat suitable portions of the commodities, suspended in brain–heart infusion broth with 1% added starch (pH 7), in a pressure cooker for 4 min at 120°C, followed by rapid cooling. The presence of any surviving thermophilic aerobic spore-formers is demonstrated by incubating the heated samples at 50°C under aerobic conditions.

Samples other than finished products must be handled so that there will be no opportunities for

Table 3 Enzymes produced by *B. stearothermophilus*

Strain of B. stearothermophilus	Enzyme	Temperature optimum (°C)	pH optimum	Thermal stability retained	Application
NCIB 8924	Neutral protease	50	–	100% at 65–70°C	Detergent and leather industry. Food industry in beer and bakery products
KP 1236	Neutral protease	80	7.5	100% at 80°C for 10 min	
503–4	α-Amylase	55–70	4.6–5.1	100% at 70°C for 24 h	Hydrolyses α-1,4 glucosidic linkages in amylose and amylopectin
ATCC 12980	α-Amylase	80	5.5	71% at 85°C for 20 h 95% at 70°C for 2 h	
KP 1064	Pullulanase	60–75	6.0		Splits α-1-6 glucosidase linkages in pullulan to maltotriose
KP 1006	α-Glucosidase	60	6.5	100% at 60°C for 30 min	Hydrolyses α-1,4 and/or α-1,6 linkages in short-chain saccharides
Specific strains	Cyclodextrin glycosyltransferase	60	–		Produces from starch non-reducing cyclodextrins
	Glucose isomerase	55	–	100% at 80°C for 30 min	Production of fructose syrups
H-165	Lipase	75	5.0		Assay of monoacylglycerols
NCIB 11270	Glycerokinase	–	–	50% at 70°C for 5 h	Assay of serum triacylglycerols
NCIB 11270	Glucokinase	–	–		Assay of creatine kinase
ATCC 12980	Leucine dehydrogenase	–	–		Assay of leucine aminopeptidase

spore germination or spore production between the collection of the samples and the start of examination procedures.

Bacillus stearothermophilus includes Gram-positive rods with terminal or subterminal spores which swell the sporangium. It is difficult to identify because of the close relationship between it and other aerobic spore-forming thermophiles. Confirmation of *B. stearothermophilus* requires completion of a number of tests as indicated in Table 1. The bacterium does not grow under anaerobic conditions and is negative on Voges–Proskauer test. Some strains grow in medium containing up to 5% salt. The distinguishing feature of *B. stearothermophilus* compared with other aerobic, thermophilic spore-formers was formerly considered to be starch hydrolysis; however, the isolation of strains unable to hydrolyse starch has restricted the distinguishing features to the temperature growth range and the limited tolerance to acid. Additional tests to confirm *B. stearothermophilus* are described below.

Tests for *B. stearothermophilus*

Acid and Gas Production in Phenol Red Carbohydrate Broth Three millilitres of broth containing 0.5% w/v of the appropriate carbohydrate such as glucose, L-arabinose, D-xylose or mannitol is inoculated with a loopful of culture and incubated at 55°C for 48 h. Acid production is indicated by the change in colour from red to yellow. To observe gas production from glucose an inverted Durham tube is added to a tube of glucose medium and incubated at 55°C for 7 days. Gas bubbles in the Durham tube indicate gas production.

Hydrolysis of Starch A clear zone around a colony streaked out on starch agar after flooding the plate with 95% ethanol indicates hydrolysis.

Hydrolysis of Casein Clearing of the casein around the colony growing on milk agar as streaked inoculum indicates decomposition of casein.

Utilization of Citrate in Phenol Red Medium In a medium containing citrate as the sole carbon source, a red (alkaline) colour indicates its utilization.

Dihydroxyacetone Formation Colonies growing on glycerol agar inoculated by streaking once across the plate are flooded with a mixture of a solution containing copper sulphate and another containing potassium sodium tartrate and sodium hydroxide. A red halo around the growth after 2 h indicates production of dihydroxyacetone.

The incorporation of the majority of the tests indicated in Table 1 into the wells of a microtitre plate facilitates the application of the identification scheme. This miniaturized procedure saves a considerable amount of time in operation, effort in manipulation, materials, labour and space.

Another approach to the identification of _Bacillus_ strains is based on the API 20E and 50CH identification system. These highly standardized, commercially available materials eliminate the problems of interlaboratory variation in media, and improve test reproducibility, but their use is intended primarily for _Bacillus_ species of medical importance.

Regulations

The presence of _B. stearothermophilus_ spores in ingredients for foods other than thermally processed low-acid foods is probably of no significance provided those foods are not held within the thermophilic growth range for many hours. This microorganism has no public health significance.

The National Food Processors Association (NFPA) standard for the total thermophilic spore count in sugar or starch specifies that for the five samples examined there should be a maximum of not more than 150 spores and an average of not more than 125 spores per 10 g of sugar (or starch). The sugar and starch standard may be used as a guide for evaluating other ingredients, keeping in mind the proportion of the other ingredients in the finished product relative to the quantity of sugar or starch used. For canners the NFPA standards for thermophilic flat sour spores (typical species is _B. stearothermophilus_) in sugar or starch specify that for the five samples examined there should be a maximum of not more than 75 spores and an average of not more than 50 spores per 10 g of sugar (or starch).

Importance to the Food Industry

Bacillus stearothermophilus is the typical species responsible for the thermophilic flat sour spoilage of low-acid canned foods. It ferments carbohydrates with the production of short-chain fatty acids that 'sour' the product. Spoilage does not result in gas production and hence there is no swelling of the cans, so the ends of the container remain flat. The species is responsible for the spoilage of low-acid foods such as canned peas, beans, corn and asparagus when they are maintained at a temperature above 43°C for an extended period, or cooling is carried out very slowly, if the food contains viable spores capable of germinating and growing in the product. Since flat sour spoilage does not develop unless the product is at high temperature, proper cooling after thermal processing and avoiding high temperatures during warehouse storage or distribution are essential.

Spores of _B. stearothermophilus_ enter canneries in soil, on raw foods and in ingredients, and their population may increase at any point where a suitable environment exists. For example, equipment such as holding tank blanchers and warm filler bowls may serve as a focal point for the build-up of an excessive population. The spores show exceptional resistance to destruction by heat and chemicals and are therefore difficult to eliminate in a product or in the plant. To minimize spore contamination, control of spore population in ingredients and products entering the plant, as well as the use of sound plant sanitation practices, are suggested.

Dormant _B. stearothermophilus_ spores are of no concern in commercially sterile canned foods destined for storage and distribution where temperatures will not exceed 43°C. However, canned foods in tropical locales and those intended for hot-vend service must not contain thermophilic spores capable of germination and outgrowth in the product in order to be considered commercially sterile.

Importance to the Consumer

The prolonged heating necessary to destroy all _B. stearothermophilus_ spores causing spoilage to low-acid canned foods would impair taste, texture and appearance and lead to loss of nutritional value. It is therefore necessary to store canned foods at temperatures below the minimum required for growth of this microorganism.

The incidence of _B. stearothermophilus_ spores in heat-processed foods may affect the commercial life of the product without presenting a hazard for public health.

Inadequate cooling subsequent to thermal processing is a major contributor to spoilage by _B. stearothermophilus_. Localized warming of sections of stacks of heat-processed foods placed too close to heating appliances is also of importance.

See also: **Heat Treatment of Foods**: Thermal Processing Required for Canning; Spoilage Problems Associated with Canning. **High-Pressure Treatment of Foods**.

Further Reading

American Public Health Association (1992) *Compendium of Methods for the Microbiological Examination of Foods*, 3rd edn. Washington: APHA.

Balows A, Trueper HG, Dworkin M, Harder W and Schleifer KH (eds) (1992) *The Prokaryotes. A Handbook on the Biology of Bacteria: Ecophysiology, Isolation, Identification, Applications*, 2nd edn. Vols 1–4. New York: Springer.

Claus D and Berkeley RCW (1986) Genus *Bacillus* Colin 1872, 174. In: Sneath PHA, Mair NS, Sharpe ME and Holt JG (eds) *Bergey's Manual of Systematic Bacteriology*. Vol. 2, p. 1043. Baltimore: Williams & Wilkins.

Harwood CR (1989) *Bacillus*. New York: Plenum Press.

Sharp RJ, Riley PW and White D (1992) Heterotrophic thermophilic *Bacilli*. In: Kristjansson JK (ed.) *Thermophilic Bacteria*. P. 19. Boca Raton: CRC Press.

Bacillus anthracis

Les Baillie, Biomedical Sciences, DERA, CBD Porton Down, Salisbury, Wiltshire, UK

Characteristics of the Species

Bacillus anthracis, the causative agent of the disease anthrax, is the only obligate pathogen within the genus *Bacillus*. The genus comprises Gram-positive aerobic or facultatively anaerobic spore forming, rod-shaped bacteria. Its ability to form resistant spores accounts for its reported longevity, up to 200 years, and resistance to physical agents, such as heat and chemical disinfectants. The spores will withstand temperatures of 70°C and, depending on the conditions, exposure to acids, alkalis, alcohols, phenolics, quaternary ammonium compounds and surfactants. They are usually destroyed by boiling for 10 min and by dry heat at 140°C for 3 h. They are susceptible to sporicidal agents such as formaldehyde and are inactivated by gamma radiation, an approach that has been used to decontaminate animal hides.

Conditions conducive to germination of *B. anthracis* are not well characterized. Germination is influenced by temperature, pH, moisture and the presence of oxygen and carbon dioxide. Spores will germinate at temperatures of 8–45°C, pH values of 5–9, relative humidity >95% and adequate nutrition. Optimum germination conditions for the Vollum strain of *B. anthracis* have been shown to be 22°C in the presence of the germinant L-alanine.

It is frequently convenient to class *B. anthracis* informally within the '*B. cereus* group' which comprises *B. cereus*, *B. anthracis*, *B thuringiensis* and *B. mycoides* on the basis of phenotypic reactions. However, genetic techniques have provided clear evidence that *B. anthracis* can be reliably distinguished from other members of the bacilli. In practical terms the demonstration of virulence constitutes the principal point of difference between typical strains of *B. anthracis* and those of other anthrax-like organisms.

Although primarily a disease of herbivores, particularly the human food animals, cattle, sheep and goats, the organism can infect humans, frequently with fatal consequences if untreated. In herbivores the disease usually runs a hyperacute course and signs of illness can be absent until shortly before death. At death the blood of the animal generally contains $> 10^8$ bacilli per millilitre.

B. anthracis is regarded as an obligate pathogen; its continued existence in the ecosystem appears to depend on a multiplication phase within an animal host. Spores of anthrax reach the environment either from infected animals and their products or as a consequence of the actions of humans. In the wild it is thought that the release of spores from infected animals plays an important part of the infective cycle; the spores contaminate the soil, healthy animals which graze on contaminated land are exposed to the spores and may subsequently develop infection.

The disease has largely been irradicated from the western world due to mass animal vaccination programmes and the maintenance of stringent veterinary control measures. In other parts of the world where vaccination is not available the organism is still a significant cause of animal mortality and human disease.

In the UK the sudden death of a food animal is investigated by the veterinary authorities, and, if death is due to anthrax, the animal and its products are destroyed. In countries, with less well developed public health systems the meat of an infected animal may be considered too valuable to 'waste' and subsequently the flesh is likely to be consumed or sold on. In Zambia the custom dictates that an animal that has died from unknown causes cannot be disposed of, but it is opened up, shared among relatives and friends and eaten. Efforts to advise local communities on the dangers of such behaviour meet resistance due to the economic loss caused by burying or burning.

Three forms of the disease are recognized in humans: cutaneous, pulmonary and gastrointestinal infection. Development of meningitis is possible in all three forms of anthrax. The gastrointestinal tract and pulmonary forms are regarded as being most frequently fatal due to the fact that they can go

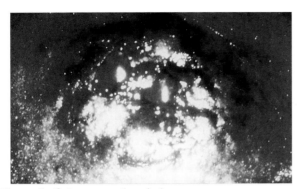

Figure 1 Cutaneous anthrax lesion.

unrecognized until it is too late to instigate effective treatment.

The cutaneous form accounts for the majority of human cases (> 95%). It is generally believed that *B. anthracis* is noninvasive and thus requires a break in the skin to gain access to the body. Infection is normally caused by spores of the organism colonizing cuts or abrasions of the skin (**Fig. 1**). Workers who carry contaminated hides or carcasses on their shoulders are liable to infection on the back of their necks, handlers of other food materials or products tend to be infected on the hands, arms or wrists. Most carbuncular cases recover without treatment but in 20% of cases the infection will progress into a generalized septicaemia which is invariably fatal.

Pulmonary anthrax is caused by the inhalation of spores of *B. anthracis* aerosolized during the processing of contaminated animal products such as hides, wool and hair. The onset of illness is abrupt. The early clinical signs are of a mild respiratory tract infection with mild fever and malaise, but acute symptoms may appear within a few hours with dyspnoea, cyanosis and a fever. Death usually follows within 2–3 days with acute splenomegaly and circulatory collapse as terminal events. This form of infection has an associated mortality rate of > 80%.

Gastrointestinal anthrax is extremely rare and occurs mainly in Africa, the Middle East and central and southern Asia. Where the disease is infrequent or rare in livestock, it is rarely seen in humans. Most cases of intestinal anthrax result from eating insufficiently cooked meat from anthrax-infected animals.

Historically it is estimated that there is one cutaneous case for every 10 infected carcasses butchered and one enteric outbreak for every 30–60 animals eaten. Due to the rareness of the conditions there are no figures for the number of organisms that need to be ingested to cause disease. Although usually fatal there are serological and epidemiological data to suggest that low-grade infection may occur.

There are two clinical forms of gastrointestinal

anthrax which may occur following ingestion of contaminated food or drink:

- Intestinal anthrax: the symptoms include nausea, vomiting, fever, abdominal pain, haematemesis, bloody diarrhoea and massive ascites. Toxaemia and shock develop and death results.
- Oropharyngeal anthrax: the main clinical features are sore throat, dysphagia, fever, regional lymphadenopathy in the neck and toxaemia. Even with treatment, the mortality is about 50%.

It is extremely important that effective treatment is started early as the prognosis is often death. Suspicion of the case being anthrax depends very greatly on awareness and alertness on the part of the physician as to the patient's history and the likelihood that he/she had consumed contaminated food and drink.

The two major virulence factors of *B. anthracis* are the ability to form an antiphagocytic capsule and toxin expression. Both of these factors are carried on different plasmids, the loss of either resulting in a reduction in the virulence of the organism.

The capsule of *B. anthracis* is composed of a polypeptide (poly-D-glutamic acid) which inhibits phagocytosis and opsonization of the bacilli. The genes controlling capsule synthesis, *CapA*, *CapB* and *CapC* are organized in an operon which is located on the plasmid, pXO2. Capsule expression is subject to regulation by CO_2 and bicarbonate via an, as yet, unclear mechanism involving the regulator *atxA*. This regulator also controls the level of expression of the anthrax toxin genes (**Fig. 2**). Why expression of virulence factors should be linked to CO_2 and bicarbonate levels is unclear. It could be that the bacteria 'monitor' the level of these agents in the host as an indication of nutrient availability.

The tripartite anthrax toxin is considered to be the major virulence factor. The three proteins of the exotoxin are protective antigen (PA), lethal factor (LF) and edema factor (EF). The toxins follow the A-B model with the A moiety being the catalytic part and the B moiety being the receptor binding part. PA acts as the B moiety and binds to the cell surface receptor, where LF and EF compete for binding to PA.

EF is an inactive adenyl cyclase which is transported into the target cell by PA. Once in contact with the cytoplasm EF binds calmodulin (a eukaryotic calcium-binding protein) and becomes enzymatically active, converting ATP into cAMP. The resulting effects are the same as those caused by cholera toxin with the affected cells secreting large amounts of fluid.

The contribution of EF to the infective process is ill defined. In general, bacterial toxins which increase cAMP dampen the innate immune responses of phagocytes and there is some evidence that this may

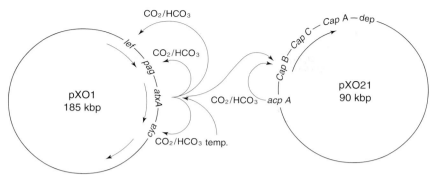

Figure 2 Coordinate regulation of virulence factors. The production of the capsule and anthrax toxin genes are enhanced by CO₂/bicarbonate and temperature. The molecular mechanism of enhanced virulence has not been elucidated.

be true for edema toxin. It is generally considered that the pathological changes seen in infected animals are due to the lethal factor combined with PA. In the only studies directly implicating EF as a virulence factor, mice were found to be killed by lower doses of the lethal toxin when EF was administered simultaneously.

Lethal toxin is the central effector of shock and death from anthrax. Animals injected intravenously with purified lethal toxin succumb in a manner that closely mimics the natural systemic infection. Lethal toxin appears to be a zinc-dependent metalloprotease, but its substrate and mode of action have yet to be defined. It affects most types of eukaryotic cells. Macrophages, which play a key role in combating infection, are particularly sensitive to the toxin. At low levels the toxin appears to interfere with the ability of macrophages to kill bacteria. The toxin also stimulates the production of cytokines within the macrophage. As the level of toxin increases in the blood more cytokines accumulate within the macrophage until the cell is finally lysed. It is proposed that this sudden release of cytokine leads to shock and would explain the rapid death seen in animals.

In addition to the major factors outlined above *B. anthracis* expresses a number of other factors which may contribute to virulence. These 'minor factors' could account for the difference in virulence between strains.

Like many other pathogenic organisms *B. anthracis* produces an S-layer composed of two proteins called Eal and Sap. S-layers are proteinaceous para-crystalline sheaths present on the surface of many Archaebacteria and Eubacteria. S-layers have been found on many bacterial pathogens including *Campylobacter* spp. and *Clostridium* spp. Various functions have been proposed for S-layers, including shape maintenance, molecular sieving or phage fixation. The

S-layer may be a virulence factor, protecting pathogenic bacteria against complement killing.

It has been demonstrated that *B. anthracis* can produce a number of chromosomally encoded extracellular proteases which, like lethal toxin, kill macrophages.

The presence of similar, if not identical, toxin genes in a number of members of the *B. cereus*, *B. thuringiensis* and *B. mycoides* group raises the possibility that these genes may also be present in *B. anthracis*. A homologue to the cereolysin gene of *B. cereus* has been detected in *B. anthracis*. Although functionally inactive in the majority of strains, spontaneously occurring low level activity has been demonstrated. It would not be surprising if homologues to other bacillus virulence factors were not detected.

Detection

Given the scarcity of anthrax in the developed world it is unlikely that many routine diagnostic laboratories would have the experience, or access to the materials required, to identify the organism correctly. The main problem is the differentiation of *B. anthracis* from the phenotypically similar *B. cereus/thuringiensis* group which may also be present in many of the samples examined for anthrax.

Direct detection of the organism in the field is relatively simple in animals that have died suddenly of the disease. At death the blood of an animal generally contains $> 10^8$ bacilli per millilitre. Blood films are dried, fixed immediately by heat or immersion for 1 min in absolute alcohol, and stained with polychrome methylene blue, which after 20 s is washed off. When the slide is dry, it is examined for characteristic deep blue, square-ended bacilli surrounded by a well-demarcated pink capsule (McFadyean's

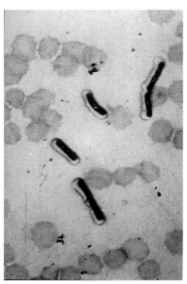

Figure 3 Capsule stain.

reaction) (**Fig. 3**). It should be noted that in some animal species such as pigs the terminal bacteraemia is limited and the bacilli are unlikely to be seen in McFadyean-stained blood smears.

Antigen-based direct detection methods have been developed which are more sensitive than staining. A highly specific immunochromatographic assay has been developed utilizing a monoclonal capture antibody to the anthrax toxin component, PA. This assay can detect as little as $25\,\mathrm{ng\,ml^{-1}}$ of PA and can be performed in a few minutes without the need for special reagents. This test could be used in addition to staining to screen animal blood and tissue and confirm the presence, or absence, of the organism.

DNA-based detection using the polymerase chain reaction (PCR) has the potential to detect a single organism in a sample. PCR methodologies have been used successfully to detect the presence of *B. anthracis* in environmental samples. Once the problem of PCR inhibitors in blood and animal tissue have been overcome it should be possible to detect the organism in animal samples.

Unless there is an index of suspicion it is unlikely that animal products would be routinely examined for the presence of *B. anthracis*. In cases where contamination with *B. anthracis* is suspected the World Health Organization in their *Comprehensive and Practical Guidelines on Anthrax* propose the isolation protocol shown in **Figure 4**.

The sensitivity limit of this technique is approximately five spores per gram of starting material. It should be borne in mind that the number of bacteria isolated very much depends on the distribution of the organism within the sample.

The PLET agar described in the method is a semi-

selective medium for *B. anthracis* which contains polymyxin (30 000 units l^{-1}), lysozyme (300 000 units l^{-1}), EDTA ($0.3\,\mathrm{g\,l^{-1}}$) and thallous acetate ($0.04\,\mathrm{g\,l^{-1}}$).

Once colonies have been isolated further testing is required to confirm their identity (**Table 1**). Many saprobic species of aerobic spore-forming bacilli are hard to distinguish from *B. anthracis* except on the basis of pathogenicity. The most commonly encountered are *B. cereus/thuringiensis/mycoides*, *B. sublilis* and *B. licheniformis*.

The preliminary tests shown in Table 1 are those routinely used by the Anthrax Section, CAMR, Porton Down, Salisbury, UK and allow the presumptive identification of an isolate as *B. anthracis*.

- *Lack of motility*: log phase cultures of the organism grown in nutrient broth at 22 and 30°C are examined for motile organisms by phase contrast microscopy. Unlike the other closely related bacilli, *B. anthracis* is non-motile.
- *Lack of haemolysis*: when cultured on 7% defibrinated horse blood agar colonies of *B. anthracis* are large, opaque, white and have a very rough surface and an irregular edge; they are normally non-haemolytic although the occurrence of haemolytic colonies has been reported.
- *Sensitivity to diagnostic gamma phage*: sensitivity to *B. anthracis* specific phage is determined by spreading 200 μl of a log phase culture over the surface of a blood agar plate. After incubation for 1 h at 37°C, 20 μl of *B. anthracis* specific gamma phage suspension is spotted on the plate. After overnight incubation at 37°C the plate is examined for plaques. On rare occasions phage-negative *B. anthracis* and phage-positive *B. cereus* may be encountered.
- *Sensitivity to penicillin*: the test organism is subcultured to a blood agar plate; a 10 unit penicillin disk is spotted on the culture and the plate is incubated overnight at 37°C. *B. anthracis* is sensitive to penicillin whereas *B. cereus* is resistant. Very rarely penicillin-resistant *B. anthracis* isolates are encountered.

Commercially available biochemical screening systems such as API 50CHB (bioMerieux, France) and Biolog (Biolog Inc., Hayward, USA) have been evaluated for their ability to identify *B. anthracis* (**Fig. 5**). These systems offer the advantage of being easy to use and show promise as simple, first line, one-step screening tests for the presumptive identification of *B. anthracis*.

The tests described above are called presumptive tests as other strains of bacilli can give similar reactions to *B. anthracis*. The demonstration of virulence constitutes the principal point of difference between

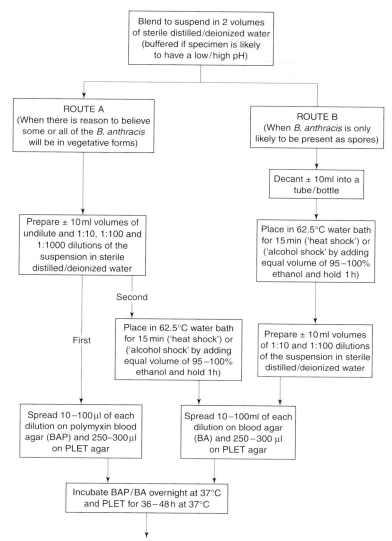

Figure 4 WHO protocol for the isolation of anthrax.

Table 1 Detection and identification methods for *B. anthracis*

Direct
  Microscopy (McFadyean's stain)
  Antigen detection
  Polymerase chain reaction
Preliminary tests
  Lack of motility
  Lack of haemolysis
  Sensitivity to diagnostic gamma phage
  Sensitivity to penicillin
  Commercial Biochemical kits: API 50CHB, Biolog
Confirmatory tests (specialist lab)
  Virulence in animals – guinea pig
  Capsule formation – McFadyean's stain
  Toxin detection – immunoassays
  Virulence gene detection – polymerase chain reaction

typical stains of *B. anthracis* and those of other anthrax-like organisms.

Traditionally the guinea pig has been the model used to demonstrate virulence. The animal is injected with the sample, and if it dies, the cause of death is confirmed by isolation of *B anthracis* from the blood. Although this traditional technique is sensitive it is likely to be replaced by more sensitive in vitro tests.

Virulent isolates of *B. anthracis* produce both a capsule and exotoxins. Detection of capsule formation is relatively simple. Capsule-forming organisms when grown on medium containing bicarbonate and in the presence of CO_2 produce colonies that are raised and mucoid in appearance whereas non-capsule forming organisms produce flat, dull colonies. In addition the presence of the capsule can be confirmed by McFadyean stain.

Detection of active toxin production is not as

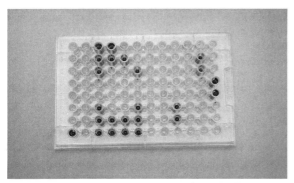

Figure 5 Biolog plate showing positive profile for anthrax.

straightforward and requires either an animal system, a tissue culture assay using toxin-sensitive cell lines or an immunological technique such as an enzyme-linked immunosorbent assay.

The PCR allows detection of the genes encoding the virulence factors without the need for their expression. Specific DNA primers have been developed for the detection of capsule and toxin genes. The primer pair GAG GTA GAA GGA TAT ACG GT and TCC TAA CAC TAA CGA AGT CG produce a 596 bp amplicon from the PA gene whereas the primers CTG AGC CAT TAA TCG ATA TG and TCC CAC TTA CGT AAT CTG CG produce an 846 bp capsule gene specific product.

Primers have been developed specific to the genome of the organism allowing the detection of atypical, non-virulent strains of *B. anthracis*.

Regulations

Most countries have regulations concerning the handling and disposal of infected food animals and their products. Concerns over the importation of contaminated animal products into the UK at the beginning of the twentieth century led the government to set up disinfection stations to treat all animal hair and leather goods.

The United Kingdom Anthrax Order 1991 prescribes the steps that should be taken to deal with an animal that has, or is suspected of having anthrax. This measure calls for the infected animals and its products such as milk to be destroyed thus removing them from the food chain.

The World Health Organization has produced detailed comprehensive and practical guidelines on anthrax detailing best practices on all aspects of the disease.

In many areas where anthrax is endemic, particularly Africa, the problem is not the lack of regulations but rather the will and the means to enforce them in the face of local customs.

Importance to the Food Industry

The number of reported cases of food-borne illness involving *B. anthracis* is extremely small compared to other traditional food-poisoning organisms. To date there has never been a documented case in the United States. In countries with well-developed veterinary and public health systems infected animals will be identified and removed from the food chain. In countries where such systems are not in place there is potential for contaminated animals and their products to be processed and consumed. A survey of animals in a slaughterhouse in eastern Nigeria revealed that 5% of cattle and 3.3% of sheep were positive for anthrax. These infected animals not only pose a risk to the people consuming the meat but an occupational risk to workers exposed to the carcasses. In the same survey it was found that 13% of butchers/skinners had acquired cutaneous anthrax. Slaughterhouse waste in the form of offal for animal feed, and slurry discharged into the environment, represents a further source of potential infection. A study of uncut anthrax-contaminated slaughterhouse waste showed that viable anthrax could still be recovered after the offal had been heat treated for 30 min at 130°C.

The use of bone charcoal by the food industry in the production of sugar products presents an avenue for anthrax contamination. The bones are normally obtained from areas of the world in which anthrax is endemic and on occasions *B. anthracis* has been isolated. For this reason the bones must be sterilized, usually by gamma irradiation, prior to use.

Importance to the Consumer

Due to the scarcity of the disease there are few published records of human infection. As one would expect the cases that are published mainly originate from Africa, the Middle East and central and southern Asia. Figures for human anthrax in China showed that of 593 recorded cases, 384 were linked to the dismembering and processing of infected animals and only 192 cases were due to the consumption of contaminated meat. In neighbouring Korea sporadic outbreaks of human anthrax have been reported. From 1992 to 1995, three outbreaks occurred, a total of 43 cases, all linked to the consumption of contaminated beef or bovine brain and liver.

An outbreak in India was centred on an infected sheep. Of the five individuals who skinned and cut up its meat for human consumption, four developed fatal anthrax meningitis. Another person who wrapped the meat in a cloth and carried it home on his head developed a malignant pustule on his forehead and went on to develop meningitis. It is noteworthy that a large number of people who cooked, or ate, the cooked meat of the dead sheep remained well.

See also: **Bacillus**: Bacillus cereus. **Bacteria**: Bacterial Endospores; Classification of the Bacteria – Traditional; Classification of the Bacteria – Phylogenetic Approach. **Nucleic Acid-based Assays**: Overview.

Further Reading

Anthrax Order (1991) Statutory Instruments 1991. No. 1824, Animals. London HMSO.

George S, Mathai D, Balraj V, Lalitha MK and John TJ (1994) An outbreak of anthrax meningoencephalitis. *Trans. R. Soc. Trop. Med. Hyg.* 88: 206–207.

Okolo MI (1985) Studies on anthrax in food animals and persons occupationally exposed to the zoonoses in Eastern Nigeria. *Int. J. Zoonoses* 12: 276–282.

Reiddinger O and Strauch D (1978) Some hygienic problems in the production of meat and bone meal from slaughterhouse offal and animal carcasses. *Ann. Inst. Super. Sanita*, 14: 213–219.

Turnbull PCB (ed.) (1996) Proceedings of the International Workshop on Anthrax, Winchester, UK, 1995. *Salisbury Medical Bulletin*, Special Suppl. no. 87.

World Health Organization (1997) Comprehensive and Practical Guidelines on Anthrax, The World Health Veterinary Public Health Anthrax Working Group, WHO/Zoon/97. Geneva: World Health Organization.

Bacillus subtilis

Michael K Dahl, Department of Microbiology, University of Erlangen, Germany

Besides *Escherichia coli*, *Bacillus subtilis* is one of the best-understood prokaryotes. The DNA sequence of the entire genome of *B. subtilis* has been available since autumn 1997. This analysis of genomic sequence data yields more insights facilitating further investigations concerning biological molecular mechanisms. The organism has been studied intensively, including its metabolism and the uptake of metabolites, gene regulation, replication, sporulation, chromosomal septation, chemotaxis and cell wall composition. Due to the non-pathogenecity of *B. subtilis*, the strain (and related organisms) as well as genetically modified variants of the organism are used in industrial fermentation, or as a host for plasmids carrying recombinant DNA for overproduction of encoded heterologous proteins. Genetic data will

result in more in-depth knowledge of such major cell processes and enable *B. subtilis* and closely related organisms to be investigated more fully.

Characteristics of the Species

B. subtilis is an aerobic, endospore-forming, rod-shaped, motile, flagella-containing bacterium, commonly found in soil and in association with plants. In the presence of sucrose it produces carbohydrate capsules composed of levan and inulin. An extracellular levansucrase is involved in this synthesis. The original isolate of the Marburg strain, designated 168 in 1936, is also termed *B. subtilis* 1A1 according to the *Bacillus* Genetic Stock Center (BGSC, Ohio State University, Dept. of Biochemistry, Columbus, OH 43210, US) and is referred to as the wild-type strain. It has been classified as a bacterium with a low guanine and cytosine (G+C) content of the chromosome. This was verified by the genome-sequencing project, in which the G+C content was determined to be exactly 43.5%. In addition to the general characteristics described for the genus *Bacillus*, *B. subtilis* is able to develop a natural competence for DNA uptake (see below). Another important characteristic is its sporulation, which enables it to survive in unfavourable living conditions and endure long starvation periods.

The name *B. subtilis* refers to another important feature of the strain: subtilisin. Subtilisin E is an extracellular alkaline serine protease encoded by the *aprE* gene, expressed simultaneously with the early sporulation genes. The fact that subtilisin E is found extracellularly reflects the fact that some proteins synthesized in the cytoplasm of *B. subtilis* are preferentially exported to the environment by a specific secretion mechanism.

B. subtilis has a single 4 214 814 bp length chromosome, originally divided into 360°. Since the chromosome has been completely sequenced, gene positions are now preferentially described in bp position numbers, where position 0 is determined as the origin of replication *oriC* (the replication terminator *terC* is located at 2.017 kb). Correction of the DNA sequence and annotation of the complete genome sequence is an ongoing task. At present, about half of the genes can be assigned to proteins with a defined or probable function and half of the open reading frames, about 2800, have no clear or biochemically identified function. However, the sequence data give indications about some general features of the genome. For example, the chromosome contains an average gene density of 1 gene per 1.03 kb, and 9.3% of the genome are prophage(-like) regions. The overall coding is covered by 87.8% of the chromosome containing

12.2% intergenic regions. In the light of these genome sequence data, a systematic function search programme has been started. The analysis of the protein composition of *B. subtilis* (the proteom) will provide further knowledge of the organism and the genus *Bacillus* in general and lead to functional assignment of the proteins and corresponding genes.

Safety Aspects of *B. subtilis*

No case demonstrating invasive properties of *B. subtilis* has been described up to now. In general, this strain is considered an opportunistic microorganism with no pathogenic potential to humans. However, *B. subtilis* is virtually ubiquitous and it is therefore inevitable that sometimes it may be found in association with other microorganisms in infected humans. In a few cases, *B. subtilis* was found to be associated with drug abusers or severely debilitated patients. Nevertheless, there is no evidence of any pathogenic potential of *B. subtilis* to humans in general.

B. subtilis has been isolated in some cases of food poisoning, but the number of episodes is low (for example, lamb and chicken have been incriminated in food-poisoning episodes). The organism demonstrates the production of a highly heat-stable toxin, which may be similar to the vomiting-type toxin produced by *B. cereus*. Although these findings may suggest that *B. subtilis* is toxic, the reports dealing with these finding are speculative and in no cases were confirmatory toxicological studies conducted. The association of *B. subtilis* with some cases of food poisoning may be due in part to misclassification of *B. cereus* (a well-established cause of food poisoning). In the UK in the period 1975–1986 only four episodes were reported. In one example people suffered from severe vomiting and diarrhoea after eating vanilla slices from the same bakery. *Staphylococcus aureus* was isolated in large numbers from vanilla slices from the same batch as those giving rise to symptoms, as well as from faecal specimens from affected persons. In addition, both *B. cereus* and *B. subtilis* were isolated from the slices. Thus there are few examples of *B. subtilis* strains as confirmed causes of food poisoning.

Even recombinant *B. subtilis* strains are classified as safe. Permission to produce enzymes from recombinant *B. subtilis* strains has been given in several states. It must be concluded that the strain is a safe host for the production of harmless products.

Detection and Growth of *B. subtilis* and Storage of Strains

Growth Conditions

The soil bacterium *B. subtilis* is quite hardy, growing in a temperature range of 25–43°C, usually under aerobic conditions at pH 5.5–8.5. However, it has also been reported that *B. subtilis* can be classified as an obligate anaerobe. The standard growing condition is at 37°C with agitation. Under these conditions and in a rich growth medium, *B. subtilis* shows a generation time of approximately 20–30 min. Both Luria Bertani (LB; containing 1% tryptone, 0.5% yeast extract, and 0.5% NaCl, adjusted with NaOH to pH 7.4) or nutrient broth (NB; 0.8% nutrient broth, 0.4% NaCl) media can serve as rich media. Alternatively, minimal media containing well-defined compositions of nutrients can be used. The most common of these is Spizzen minimal medium, a buffer composed of $14 \, \text{g} \, \text{l}^{-1}$ K_2HPO_4 and $6 \, \text{g} \, \text{l}^{-1}$ KH_2PO_4, $2 \, \text{g} \, \text{l}^{-1}$ $(NH_4)_2SO_4$ and $1 \, \text{g} \, \text{l}^{-1}$ sodium citrate. After autoclaving, filter-sterilized glucose is added to a final concentration of $25 \, \text{mmol} \, \text{l}^{-1}$ and, if necessary, auxotrophic requirements. Instead of glucose, several laboratories use 0.4% K-glutamate or 0.5% succinate, respectively. It has also been reported that the addition of $2 \, \mu\text{g} \, \text{ml}^{-1}$ of ferric citrate essentially supports growth in minimal medium. The *B. subtilis* wild-type isolate 168 Marburg, however, harbours a tryptophan auxotrophy (*trpC2*). Therefore, minimal medium must be supplemented with tryptophan, usually at a final concentration of $50 \, \mu\text{g} \, \text{ml}^{-1}$.

Supplements in Special Growth Conditions

Many of the *B. subtilis* derivatives are auxotrophic mutants and require nutrient supplements for optimal growth on minimal medium. Some mutants are unable to grow on specific carbon sources such as sucrose, trehalose, glucose, mannitol and fructose. These mutants can be tested in minimal medium in which glucose has been exchanged by the relevant carbon source. Amino acid auxotrophs require addition of the appropriate amino acid to the minimal medium at a final concentration of $50 \, \mu\text{g} \, \text{ml}^{-1}$. Purine auxotrophs require supplements at a final concentration of $20 \, \mu\text{g} \, \text{ml}^{-1}$, whereas pyrimidine auxotrophs require $50 \, \mu\text{g} \, \text{ml}^{-1}$.

In rich media, some mutants (*cysA* or *thyA/B*) exhibit improved growth based on the addition of cysteine or thymine. However, *dal* or *ctrA* mutants have an absolute requirement even in rich media, for D-alanine and cytidine, respectively.

Antibiotic Concentrations

B. subtilis is often used as a host for extra-chromosomal, self-replicating DNA, such as plasmids. Furthermore, *B. subtilis* is also used as a host for foreign DNA or defined DNA constructs by integration of the DNA into the chromosome (see below). In both cases, the selective markers listed in **Table 1** can be used.

Sporulation

Spore formation takes place when the stationary growth phase is reached under aerobic conditions. The development of endospores is an energy-intensive pathway and results in the production of different complex morphological structures. The sporulation process is divided into seven distinct steps (**Fig. 1**) and is probably best investigated in *B. subtilis*. The process usually takes about 6–8 h under optimized laboratory conditions. The regulation of these sporulation steps, sequential expression of different gene classes and regulation of different biochemical reactions in the forespore and the mother cell involve complex networks of regulation machinery, which are not completely understood and which have not been identified in all their details up to now.

Even standard rich medium yields only small quantities of spores unless Mn^{++} ions in $\mu mol\,l^{-1}$ concentrations are added. Iron and calcium ions also improve the morphology and heat resistance of these spores. The best spore yield is obtained using an optimized sporulation medium based on nutrient agar, as described by Schaeffer in 1965. The so-called Schaeffer sporulation medium (SSM) contains 0.8% NB, 0.025% $MgSO_4 \cdot 7H_2O$ and 0.1% KCl. After autoclaving, the medium is supplemented with 1 ml each of the following sterile stock solutions: 1 mmol l^{-1} $FeSO_4 \cdot 7H_2O$ (filter-sterilized), 10 mmol l^{-1} $MnCl_2 \cdot 4H_2O$ and 1 mol l^{-1} $CaCl_2 \cdot 2H_2O$.

The resulting spores are heat-resistant at 80°C. This feature makes it possible to detect and classify spore-formers such as *B. subtilis*.

Table 1 Selective markers in *Bacillus subtilis*

Selective marker (antibiotic)	Concentration ($\mu g\,ml^{-1}$)	Dissolved in
Chloramphenicol	5	95% ethanol
Erythromycin	1	95% ethanol
Kanamycin	30	H_2O
Lincomycin	25	50% ethanol
Neomycin	5–7	H_2O
Phleomycin	0.5	H_2O
Spectinomycin	100	H_2O
Tetracycline	20	50% ethanol

Long-term Storage

After *B. subtilis* strains have produced endospores they can be stored easily in cool conditions, or even at room temperature for many years. Storage for decades has already been reported, and presumably spores can survive for centuries. There are two common storage conditions: culture slopes containing a slant of sporulation agar and paper filters.

Culture Slopes For sporulation agar SSM is commonly used, supplemented with 23 g l^{-1} agar. When the medium reaches a temperature of about 55°C the liquid should be poured into small screw-capped tubes, as shown in **Figure 2**. After the agar has cooled and set, cells can be streaked on the agar surface and incubated overnight at 37°C, leaving the screw cap slightly open for aerobic conditions. Usually the cells become thick, often rugose and turn brown when sporulated. The screw caps of culture slopes con-

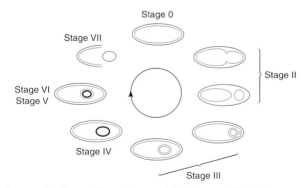

Figure 1 Sporulation stages of *Bacillus subtilis*. The seven stages based on cytological changes are numbered from 0 to VII. Stage 0 refers to the vegetative cell. Stage I is no longer used to define a morphological and distinct state in sporulation. Therefore, sporulating cells immediately enter stage II where the appearance of a forespore septum can be observed. In stage III the mother cell engulfs the forespore. Cortex formation occurs in stage IV, when the developing spore is refractile. Stage V represents inner spore coat protein deposition on the surface of the outer forespore membrane. In stage VI, the outer spore coat is deposited on the surface of the inner spore coat. Stage VII shows release of the mature spore and lysis of the mother cell.

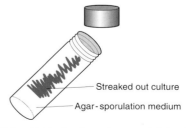

Figure 2 Screw-capped culture slope containing *Bacillus subtilis* spores for long-term storage.

taining spores can subsequently be closed and stored under cool conditions or at room temperature.

Filters   For this procedure SSM is used. A few drops of concentrated spore suspension (see above) are placed on a small stack of sterile pieces of paper or filters. The paper can be stored in sterile aluminium foil for long-term storage or used for shipping the bacterial material.

Some asporogenous strains of *B. subtilis* require special care (including citric acid cycle and several *spo* mutants). They can form translucent colonies and will lyse in a few days. For long-term storage these strains must be taken from a logarithmically growing culture and glycerol or dimethyl sulphoxide added. Under these conditions the cultures can be frozen at −70°C.

Recovery of Cultures from Spores

Spores from *B. subtilis* can be streaked on rich medium agar plates (as for single colonies) and incubated at an appropriate temperature. Overnight incubation of the agar plate at 30–37°C usually results in vegetative cells and single colonies.

By withdrawing the papers with sterile forceps and dropping them on to the surface of an agar plate, cultures can be recovered from the filter papers. Additionally, some drops of liquid medium to wet the paper will allow rapid outgrowth. Incubation of the agar plate at 30–37°C overnight usually results in a dense patch of cells. Before proceeding to use these strains it is recommended that the patch of cells is restreaked on another agar plate to obtain single colonies that can be further tested for their individual phenotypes, if present.

Natural Competence for DNA Uptake

Many strains of *B. subtilis* are inherently capable of binding DNA fragments and transporting the fragments to an intracellular location. This does not seem to require sequence-specific binding sites. This feature was discovered by Spizizen in 1958 and since then has been commonly used to manipulate this organism genetically. Shortly after logarithmic growth stops and before sporulation begins, about 10–20% of a *B. subtilis* cell culture develops natural competence for DNA uptake. *B. subtilis* can be transformed with either plasmid DNA or DNA fragments, later flanked by homologous sequences (of at least 100 bp), enabling the fragment to integrate into the chromosome by double homologous recombination. However, the development of natural competence for DNA uptake requires special growing conditions. Previously the minimal medium of the culture was exchanged at the beginning of the stationary phase. Newer protocols

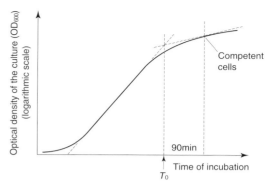

Figure 3   Determination of the end of the logarithmic growth phase, defined as T_0. Ninety minutes after T_0, 10–20% of the cell culture develop natural competence for DNA uptake.

use a much faster, one-medium procedure, where the cell culture is grown at 37°C in MGE (minimum glucose glutamate) medium: Spizizen minimal medium (see above) containing 2% glucose, 0.4% K-glutamate, 2 µg ml⁻¹ ferric citrate, 0.025% yeast extract, 2 mmol l⁻¹ MgSO₄, 1 mmol l⁻¹ CaCl₂ and, if necessary, auxotrophic requirements. The optical density of the growing culture is monitored at 600 nm over the time period (usually every 30 min), resulting in a typical growth curve (**Fig. 3**), which allows determination of the zero time point (T_0). T_0 is defined as the time point when the switch between logarithmic growth and stationary phase takes place (Fig. 3). The cells develop competence for DNA uptake 90 min after T_0 and can be used directly for transformation.

Competent cells can also be frozen in the presence of 15% glycerol and stored for several months before use. However, the freezing procedure reduces transformation efficiency. In contrast to *E. coli* frozen competent cells, the solution containing *B. subtilis* competent cells must be thawed quickly.

Importance to the Food Industry

B. subtilis can secrete various proteins, which fall into two categories:

- The truly soluble exoproteins, mainly exoenzymes secreted directly into the growth medium, for example degradative enzymes such as α-amylases and proteases.
- Proteins comprising heterogeneous groups of proteins that remain associated with the cell wall after export, including autolytic enzymes, peptidoglycan-associated proteins and substrate-binding proteins of ABC-transporters (ABC: ATP-binding cassette).

The genes for these extracellular proteins are of particular interest for biotechnology and the food industry, as are the secretion features and protein export mechanisms. Extracellular proteins offer several advantages. Easy recovery, purification and potential for extremely high yields are important considerations in the cost-effectiveness of the protein production process. However, a major disadvantage of *B. subtilis* is the prevalence of extracellular proteases that could degrade foreign proteins. Therefore, protease-free *B. subtilis* strains have been constructed for use in overproduction and secretion of important proteins.

In principle, secretion requires two main components: the necessary export machinery and the information inherent in the protein to be exported, which this machinery recognizes as an export signal. The export machinery in *B. subtilis* is not as well characterized as in *E. coli*, but consists of several components, mostly designated Sec-proteins (encoded by *sec* genes). The molecular mechanism is not completely understood and numerous investigations on that mechanism are still in process. However, it is known that the exported precursor proteins must be kept in a so-called export competent state during and after translation, mediated by chaperones. Therefore, the second component, the export signal in the protein, is usually present in the N-termini of exoprotein precursors. The signal peptide (or signal sequence) which has some salient features comprises this part. In *B. subtilis* the average length of the signal peptide is 29 amino acids. The N-terminal region contains 2–7 positively charged amino acids. This positively charged N-terminus is assumed to interact with the negatively charged lipid layer of the cell wall. A hydrophobic core region composed of 17 residues (average length) follows the N-terminal region. Residues such as glycine or proline that were predicted to favour a β-turn found in the majority of signal peptides define the hydrophobic core termination. Glycine and proline are positioned at −5 to −7 from the C-terminus of the signal peptide. Position +1 is defined for the N-terminal amino acid of the mature protein, also reflecting the cleavage of the signal peptide after translocation. The supposed β-turn is followed in over 90% of all cases by alanine at position −1 which defines the cleavage site of the signal peptide. For the secretion mechanism it was proposed that the hydrophobic core region of the signal peptide might span the cell membrane to facilitate translocation initiation.

B. subtilis is able to secrete foreign proteins harbouring the signal peptide for secretion; this indicates that *B. subtilis* signal peptides are interchangeable and that several heterologous proteins can be secreted, albeit with varying efficiency. Thus, signal peptides are sufficient to target heterologous proteins to the secretion machinery. However, the varying efficiency suggests that the mature protein sequence might also play a role in export. This suggestion is supported by the few exceptions known, in which the fusion of a mature protein to the signal sequence of *B. subtilis* did not result in successful export of the heterologous and recombinant protein.

Strain Constructions

To construct strains with special, defined and individual features or carrying genes of other organisms for heterologous gene expression, two categories of DNA can be introduced into *B. subtilis* by the transformation procedure (see below): either extrachromosomal plasmid DNA capable of self-replication or plasmid DNA incapable of replication or linear DNA fragments which must both integrate into the chromosome to remain in the organism. In this last process circular DNA can integrate by a single homologous recombination, whereas linear fragments must integrate via a double homologous recombination (**Fig. 4**).

Transformation

In some special cases, when using mutants of *B. subtilis* with a negative effect on the development of natural competence for DNA uptake, it is necessary to transform those strains using protoplast or electroporation methods. Nevertheless, the principle of a defined strain construction of *B. subtilis* remains the same, regardless of the method used to introduce the recombinant DNA.

When using *B. subtilis* strains with natural competence for DNA uptake, it is sufficient to incubate the strains with the DNA for 30 min at 37°C under permanent agitation for aeration. Even the slightest traces of detergent remaining from a DNA preparation should be removed, otherwise the cells could lyse. Subsequently, especially when recombinant strains are to be selected on an antibiotic medium, the cells must be incubated for another 30–45 min for an outgrowth in which selective gene marker expression can be started. A mixture of yeast extract and casamino acids must be added to the cell-DNA suspension to achieve this and in some cases small concentrations of the selective marker (usually 1/10 of the concentration used for selection) to induce gene expression. After outgrowth the cells are spread on selective agar plates (Table 1) and incubated overnight at 37°C. It has been reported that better homologous recombination results can be achieved by incubating the strain overnight at 43°C. Usually, on agar plates incubated overnight, colonies occur with the appro-

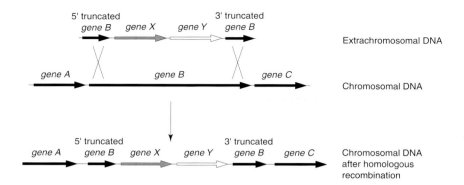

Figure 4 Homologous recombination in the *Bacillus subtilis* chromosome. Foreign or manipulated DNA can be stably integrated by this mechanism into the chromosome at defined sites. *Gene X* may represent the gene of interest, which should be expressed in *B. subtilis*; *gene Y* may represent the gene encoding for the selective marker or vice versa. *Gene B* denotes the locus of integration into the chromosome; *genes A* and *C* are the flanking regions. As a result, the introduction of the recombinant DNA (*genes X* and *Y*) by double homologous recombination leads to the disruption of the target gene (*gene B*).

priate resistance and a stable integration of the gene introduced (Fig. 4).

A second method for introducing recombinant DNA is self-replicating extrachromosomal DNA: the plasmids. Several plasmids are known to be useful for *B. subtilis* transformation. The competence for plasmid and chromosomal DNA uptake occurs at the same growth stage and the kinetics for both types of DNA are similar. However, plasmid transformation is inefficient, requiring a DNA input equivalent to about 1000 plasmids to achieve a single plasmid-containing transformant. This unsatisfactory value is explained by the fact that stable plasmid-containing transformants only result when the donor plasmid is a trimer or in higher multimeric forms. Because monomers make up more than 90% of a purified plasmid preparation, this is the apparent explanation. To solve this problem and to obtain higher transformant percentages it is usually sufficient to increase the plasmid DNA concentration to the microgram range during the transformation process.

B. subtilis Plasmids

Several plasmids can be successfully transferred into *B. subtilis* and are maintained in a stable form in this species. Many of these plasmids were originally identified in *S. aureus* as small extrachromosomal elements that specify resistance to different drugs, as listed in Table 1. They are maintained at copy numbers of about 10 or more per chromosome (up to 200 copies were reported). Also, plasmids from strains of *B. subtilis*, *B. cereus* and *Streptococcus* have been identified as being able to replicate in *B. subtilis*. When the plasmids contain a selective marker gene, encoding mostly for drug resistance, and the plasmid-harbouring cells are always grown on selective media,

the plasmids usually remain stable in the cell. However, cloning large DNA fragments into some of the plasmids can decrease both copy number and stability. Plasmid instability is a major problem associated with the successful cloning of genes in *B. subtilis*. Two types of instability have been determined: segregational instability, involving the loss of the entire plasmid correlated with the loss of the selective marker/drug resistance and structural instability that may arise by deletion, insertion or rearrangement of the plasmid DNA. *E. coli* plasmids cannot replicate in *B. subtilis*. Therefore, these plasmids can only persist in *B. subtilis* by single recombination in a target gene of the chromosome or by double homologous recombination, as described for transformation with chromosomal DNA (see above), if the corresponding homologous DNA sequences are present on the plasmid.

In addition to *B. subtilis* vectors, new plasmids have been constructed by achieving combinations or joining them to *E. coli* plasmids to generate shuttle vectors, replicating on both organisms (or generally in two different organisms). They contain two origins of replication, one for *E. coli* and a second for *B. subtilis*. Because of the low plasmid DNA concentration which is usually obtained from plasmid preparation with *B. subtilis* as a host, shuttle plasmids are used for plasmid preparation in *E. coli* as a host for high DNA concentration yields. However, in some cases, heterologous gene expression led to problems, especially when products of *B. subtilis* genes have toxic effects in *E. coli*. This may result in plasmid instability or failure to obtain transformants in *E. coli*.

See color Plates 3 and 4.

See also: **Bacillus**: Introduction. **Escherichia coli**: *Escherichia coli*.

Further Reading

Dubnau D (1997) Binding and transport of transforming DNA by *Bacillus subtilis*: the role of type-IV pilin-like proteins – a review. *Gene* 192: 191–198.

Dubnau D, Hahn J, Roggiani M, Piazza F and Weinrauch Y (1994) Two-component regulators and genetic competence in *Bacillus subtilis*. *Res. Microbiol.* 145: 403–411.

Errington J (1993) *Bacillus subtilis* sporulation: regulation of gene expression and control of morphogenesis. *Microbiol. Rev.* 57: 1–33.

Freudl R (1992) Protein secretion in Gram-positive bacteria. *J. Biotechnol.* 23: 231–240.

Grossman AD (1995) Genetic networks controlling the initiation of sporulation and the development of genetic competence in *Bacillus subtilis*. *Annu. Rev. Genet.* 29: 477–508.

Harwood CR (ed.) (1989) *Bacillus*. Biotechnology Handbooks 2. New York: Plenum Press.

Kunst F, Ogasawara N, Moszer I et al (1997) The complete genome sequence of the Gram-positive bacterium *Bacillus subtilis*. *Nature (Lond.)* 390: 249–256.

Meijer WJ, Wisman GB, Terpstra P et al (1998) Rolling-circle plasmids from *Bacillus subtilis*: complete nucleotide sequences and analyses of genes of pTA1015, pTA1040, pTA1050 and pTA1060, and comparisons with related plasmids from Gram-positive bacteria. *FEMS Microbiol. Rev.* 21: 337–368.

Moszer I (1998) The complete genome of *Bacillus subtilis*: from sequence annotation to data management and analysis. *FEBS Lett.* 430: 28–36.

Piggot PJ, Moran CP Jr and Youngman P (eds) (1994) *Regulation of Bacterial Differentiation*. Washington, DC: American Society for Microbiology Press.

Sonnenshein AL, Hoch JA and Losick R (eds) (1993) *Bacillus subtilis and other Gram-positive bacteria*. Washington, DC: American Society for Microbiology Press.

Wong SL (1995) Advances in the use of *Bacillus subtilis* for the expression and secretion of heterologous proteins. *Curr. Opin. Biotechnol.* 6: 517–522.

Detection of Toxins

S H Beattie and **A G Williams**, Hannah Research Institute, Ayr, UK

Introduction

Bacteria from the genus *Bacillus* occur widely within the environment and are frequently detected both in raw materials used in the food industry and in food products at the point of sale. Although the majority of *Bacillus* spp. are non-pathogenic, *Bacillus cereus* is a recognized food-borne enteropathogen and causative agent of food poisoning in humans. Other species in the genus that have been implicated in food poisoning outbreaks include *B. subtilis*, *B. licheniformis*, *B. pumilus*, *B. brevis* and *B. thuringiensis*. The problems associated with *Bacillus* spp. are exacerbated as their spores resist – or are activated during – food processing, and in addition psychrotrophic strains are capable of growth in milk and food products correctly stored at refrigeration temperatures.

Food-borne Illness

Illness Caused by *Bacillus cereus*

Bacillus cereus is associated with two distinct food-borne gastrointestinal disorders, the diarrhoeal and emetic syndromes. The diarrhoeal illness was first described following an outbreak of food poisoning in a Norwegian hospital in the 1940s, although earlier unconfirmed reports described outbreaks with a similar aetiology. The syndrome is typified, in the absence of fever, by abdominal discomfort, profuse watery diarrhoea, rectal tenesmus, and on some occasions nausea that rarely produces vomiting. The illness is usually associated with the consumption of one of a diverse range of proteinaceous foods that include milk products, cooked meats, sauces and desserts. Onset of the symptoms occurs some 8–16 h after consumption of the contaminated food, and this delay is indicative of subsequent bacterial growth and toxin formation in the small intestine. The inactivation of preformed toxin in contaminated foods by digestive enzymes and gastric pH reduces the effects of its ingestion. The symptoms normally resolve within 12–24 h without the need for medical intervention, although there are reports of more severe symptoms developing in specific population groups.

The emetic syndrome is caused by a preformed toxin produced as a consequence of the growth of toxigenic strains of *B. cereus* in the food. Farinaceous foods are most commonly associated with the emetic illness. This disease is more common than the diarrhoeal form in Japan, whereas in North America and European countries the diarrhoeal syndrome is more prevalent. The onset of the emetic syndrome occurs within 1–5 h of consumption of the contaminated food, and the symptoms (which include malaise, nausea, vomiting and occasionally diarrhoea), persist for 6–24 h. The diarrhoea is most probably caused by the concomitant synthesis of enterotoxin in some emetic strains. Although the symptoms of the emetic illness are generally regarded as being relatively mild, there is a published case report describing the progress of the disease in an Italian teenage boy; his subsequent death was attributed to liver failure induced by the emetic toxin produced by a strain of *B. cereus* isolated from a pasta sauce that he had consumed.

Incidence of *B. cereus*-mediated Food-borne Illness   Although *B. cereus* is now recognized as an important cause of food-borne illness in humans, it is difficult to ascribe definitive figures to the number of outbreaks caused by the microorganism because of inherent inadequacies in existing reporting procedures. It is recognized that the official statistics for all food-borne disease outbreaks underestimate the true extent of the problem, and may only represent 10% of the number of cases that actually occur. The short duration and nature of the illness caused by *B. cereus* limit medical diagnosis and accurate recording of incidents, with the result that the full extent of the *B. cereus* problem is not known.

Analysis of food poisoning statistics collected in North America during the decades commencing in 1970 and 1980 led reviewers to conclude that *B. cereus* was a relatively unimportant food poisoning agent. Only 3.1% and 6.9% of the cases of bacterial food-borne diseases in the USA and Canada respectively were caused by the microorganism; this represented only 1–2% of the total number of cases recorded. There are, however, geographical differences in the incidence of outbreaks and number of cases attributable to *B. cereus*. Data obtained in several studies over various periods between 1960 and 1992, in Europe, Japan and North America, indicate that 1–22% of outbreaks and 0.7–33% of food poisoning cases could be attributed to *B. cereus*. In both Norway and the Netherlands, where more detailed surveillance of food-borne illness has been undertaken, *B. cereus* has emerged as the most frequently identified bacterial food-borne pathogen.

Bacillus cereus occurs widely in raw and processed foods. The microorganism is ubiquitous in nature and it seems impossible to exclude its spores from the food chain. Strains of *B. cereus* will grow over a wide pH and temperature range and at salt concentrations up to 7.5%. The generation time of the organism under optimum conditions is approximately 20 min. It is, therefore, evident that the organism will be able to grow in foods that are improperly prepared or subjected to temperature abuse during storage. Should *B. cereus* growth occur, the potential exists for food poisoning to ensue. Foods particularly associated with *B. cereus* include dairy produce, meat products, spices and cereals. Food containing more than 10^4–10^5 cells or spores per gram may not be safe for consumption as the infectious dose has been calculated to vary from 10^5–10^8 cells or spores per gram. The variation in the estimated infectious dose reflects the large interstrain differences in the amount of toxin produced and the inherent variability in the susceptibility of the population at large. It has been suggested that repeated exposure to low levels of *B. cereus* in foods, especially milk, may lead to a partial protective immunity developing.

Illness Associated with Other *Bacillus* Species

Bacillus species other than *B. cereus* are present in a range of food products and on occasions have been isolated from food samples implicated in outbreaks of food poisoning. The species most frequently isolated in such cases are *B. subtilis* and *B. licheniformis*, although *B. pumilus*, *B. brevis* and *B. thuringiensis* have been associated with a smaller number of outbreaks. A number of incidents of intestinal anthrax have been caused by the consumption of meat from animals infected with *Bacillus anthracis*. In addition, some strains of *B. mycoides*, *B. circulans*, *B. lentus*, *B. polymyxa* and *B. carotarum* produce extracellular components that cross-react with antibodies raised against the enterotoxin of *B. cereus* for use in commercially available kits for toxin detection. This implies some form of structural relatedness. A strain of *B. brevis* involved in a food poisoning incident was also able to produce a heat-labile enterotoxin.

Bacillus subtilis and *B. licheniformis* are widely distributed in the environment and have been implicated in incidents of food-borne illness. *Bacillus subtilis* has been identified as a causative agent of food-borne disease in the UK (49 episodes with more than 175 cases between 1975 and 1986), Australia and New Zealand (14 incidents) and Canada. Ingestion of contaminated food was characterized by a peppery or burning sensation in the mouth; the onset of symptoms, which typically include diarrhoea and vomiting, can occur within a very short period although the median incubation period is 2.5 h (range 10 min to 14 h). Other symptoms can include abdominal pain, nausea and pyrexia; the duration of the episode is 1.5–8 h. Incriminated foods include meat, seafood, pastry dishes and rice. The levels of *B. subtilis* in these implicated food vehicles was in the range 10^5–10^9 colony forming units (cfu) per gram.

There have been detailed descriptions of food poisoning outbreaks caused by *B. licheniformis* in North America and in the UK where 24 episodes (218 cases) were recorded between 1975 and 1986. Cooked meats and vegetables were the principal food vehicle from which large numbers (> 10^6 cfu g^{-1}) of the causative organism could be isolated. As a result of infection *B. licheniformis* may dominate the intestinal bacterial population. The median incubation period prior to onset is about 8 h (range 2–14 h). The most common symptom is diarrhoea, although vomiting and abdominal pain have been reported to occur in about 50% of cases; nausea, pyrexia and headaches are not characteristic of *B. licheniformis*-mediated food poisoning. The duration of the illness is approximately 6–24 h.

Although *B. pumilus* is closely related to *B. licheniformis* and *B. subtilis*, there are few reports implicating this species as a food-borne pathogen. In five incidents reported in England and Wales during the period 1975–1986, symptoms of diarrhoea and vomiting developed after varying periods (0.25–11 h) following the consumption of food contaminated with large numbers (10^6–10^7 cfu g^{-1}) of this microorganism. Foods implicated included meat products, canned fruit juice and cheese sandwich.

An outbreak of food poisoning involving *B. brevis* has been reported, and in other incidents the microorganism has been isolated from the suspected food vehicle and the faeces of a patient. The mean incubation time in the *B. brevis*-mediated outbreak was 4 h and the symptoms reported were nausea, vomiting and abdominal pain. A heat-labile enterotoxin was responsible for the illness.

Bacillus thuringiensis is closely related to *B. cereus* and, although widely used as an insecticide, it has the potential to be pathogenic to humans. Isolates of *B. thuringiensis*, belonging to the H serotypes *kurstaki* and *neoleonensis*, recovered from food products such as milk, pitta bread and pasta were shown to be enterotoxigenic. Faecal isolates from an outbreak of gastroenteritis in a chronic care institution in Canada were identified as enterotoxin-producing strains of *B. thuringiensis*. In addition, the microorganism has also been shown to induce food-borne illness in human volunteers. Recent reappraisal studies, using specific molecular probes based on variable regions of 16S rRNA, have indicated that causative strains from food poisoning incidents that had been initially identified by phenotypic characteristics as *B. cereus* were in fact stains of *B. thuringiensis*. The potential for enterotoxin-producing strains of *B. thuringiensis* to cause food poisoning should not be overlooked in diagnostic laboratories, especially as preparations of the microorganism are used widely to control insect pests in many countries. Cases of *B. cereus*-mediated diarrhoeal outbreaks resulting from the consumption of raw and improperly cooked vegetables have been recorded. In view of the phenotypic relatedness of *B. cereus* and *B. thuringiensis* it is possible that some of these incidents, and other outbreaks, may have been caused by *B. thuringiensis*. The actual incidence of *B. thuringiensis*-mediated food-borne illness may therefore be greater than reported figures currently indicate. Some characteristics of food-borne illness associated with *Bacillus* spp. are summarized in **Table 1**.

Bacillus cereus Diarrhoeal Syndrome

Bacillus cereus Enterotoxins

Bacillus cereus diarrhoeal enterotoxin is produced during the logarithmic stage of growth. The enterotoxin interacts with the membranes of epithelial cells in the ileum, and causes a type of food poisoning that is almost identical to that of *Clostridium perfringens*. Both species produce toxins damaging to the membrane, but with different modes of action. *Clostridium perfringens* requires Ca^{++} ions in order to bind to target cells, and thus cause leakage. Conversely, *Bacillus cereus* enterotoxin is inhibited from causing cell leakage if Ca^{++} ions are present. *Bacillus cereus* enterotoxin is about a hundred times more toxic to human epithelial cells than the *C. perfringens* toxin.

The diarrhoeal syndrome associated with *B. cereus* is considered to be caused by viable cells or spores, rather than preformed toxin, because the time between consumption of incriminated food to onset of the symptoms is too long for the disease to be an intoxication. *Bacillus cereus* is capable of growth and enterotoxin production under anaerobic conditions, and is therefore capable of forming enterotoxin in the ileum.

Nutrient availability appears to be important in the production of diarrhoeogenic toxin. In uncontrolled batch fermentations high levels of sugar did not support toxin formation, whereas starch enhanced toxin production. However, in controlled fermentations where pH was regulated, sugar and starch neither enhanced nor repressed toxin formation, indicting that the repression occurring in the presence of high sugar levels was due to the accompanying pH fall rather than to the sugar itself. Water activity has a significant effect on growth and toxin production of *B. cereus*. Low pH inhibits enterotoxin formation, and outside the range pH 5–10 a rapid loss in activity occurs. The diarrhoeal enterotoxin is unstable over a wide range of conditions, with ionic strength being particularly critical. However, enterotoxin stability is greater after heating in milk than in cell-free culture supernatants.

The amount of toxin produced by different strains of *B. cereus* varies considerably. It has been shown that 60–70% of strains isolated from milk products are able to produce diarrhoeal enterotoxin; however, the number of strains that are able to produce sufficient enterotoxin to constitute a health risk is probably limited. It is also unlikely that diarrhoeogenic toxin production will occur in dairy products that are maintained in the cold chain. Nevertheless, the presence of the organism still constitutes a potential hazard to the consumer, particularly since *B. cereus* is able to grow well at 37°C and under low oxygen

Table 1 Characteristics of food-borne illness caused by *Bacillus* spp.

	B. cereus emetic syndrome	B. cereus diarrhoeal syndrome		B. thuringiensis	B. subtilis	B. licheniformis	B. pumilus	B. brevis
		Type I	Type II					
Symptoms	Malaise, nausea, vomiting	Abdominal pain, watery diarrhoea	Gastroenteritis	As for B. cereus diarrhoeal syndrome	Vomiting, diarrhoea, nausea, abdominal pain, pyrexia, headaches	Diarrhoea, vomiting, abdominal pain	Diarrhoea, vomiting, nausea	Nausea, vomiting, abdominal pain
Implicated food vehicle	Farinaceous rice, pasta, noodles, pastry	Proteinaceous dairy products, meats, sauces, desserts	Proteinaceous dairy products, meats, sauces, desserts	As B. cereus and unwashed sprayed vegetables	Meat, seafood, pastry, rice	Cooked meat and vegetable dishes	Meat products	Meat
Infective dose (cfu g^{-1})	10^5–10^8	$>10^4$	$>10^4$	$>10^4$	$>10^5$–10^9	$>10^6$	$>10^6$	$>10^8$
Incubation period (h)	0.5–5	8–16	8–16	8–16	<1–14	2–14	<1–11	1–9.5 (av.4)
Duration of illness (h)	6–24	12–24	12–24	12–24	1.5–8	6–24	Unknown	Unknown
Nature of toxin	Heat-stable, cyclic dodecadepsi-peptide, ingested in food	Haemolytic, dermonecrotic, heat-labile complex of 3 peptides. Formed *in situ* in gut	Non-haemolytic heat-labile complex of 3 peptides	Enterotoxin, structure not determined but reacts with antibodies to B. cereus enterotoxin	Unknown	Unknown, but culture supernatants of some strains react with antibodies to B. cereus enterotoxin	Unknown, but culture supernatants of some strains react with antibody in Oxoid RPLA assay for B. cereus enterotoxin detection	Heat-labile enterotoxin
Potential detection method	No kit / Cytotoxicity, boar spermatozoa motility, emesis in primates and *Suncus* sp.	Oxoid BCET-RPLA / Rabbit ligated ileal loop, cytotoxicity, gel diffusion	Tecra VIA kit / Cytotoxicity	Oxoid/Tecra kits for B. cereus enterotoxin / Cytotoxicity	Not detected by kits for B. cereus enterotoxin / Cytotoxicity	Some strains Tecra VIA/Oxoid BCET kit for B. cereus enterotoxin detection / Cytotoxicity	Some strains Oxoid BCET-RPLA kit for B. cereus enterotoxin detection / Cytotoxicity	Cytotoxicity

Table 2 Some characteristics of toxins formed by *Bacillus cereus*

Toxin	Molecular mass (kDa)	Characteristics
Enterotoxins		
Haemolysin BL	B 37.8	Haemolytic, heat-labile, tripartite enterototoxin
	L_1 38.5	
	L_2 43.2	
Non-haemolytic	39	Heat labile
	45	
	105	
Enterotoxin T	41	Encoded by the *bceT* gene
Emetic	1.2	Heat stable
		Stable to proteolysis
		Stable over a range of pH (2–11)
Haemolysin		
Cereolysin	56	Thiol activated, heat labile, mouse lethality
Haemolysin BL	See above	
Sphingomyelinase	34	Stable metalloenzyme, haemolytic, only hydrolyses sphingomyelin
Haemolysin II	30	Heat labile, sensitive to proteolytic enzymes
Phospholipase C		
Phosphatidylinositol hydrolase	34	Non-metalloenzyme, specifically hydrolyses phosphatidylinositol (PI) and PI-glycan-containing membrane anchors
Sphingomyelinase	See above	
Phosphatidylcholine hydrolase		Stable metalloenzyme (Zn^{++}, Ca^{++}), hydrolyses phosphatidylcholine, phosphatidylethanolamine and phosphatidylserine

concentrations, conditions typical of the gastrointestinal ecosystem.

Structure of the Enterotoxin There has been considerable debate concerning the structure and molecular mass of *B. cereus* diarrhoeogenic toxins. Three different enterotoxins have now been characterized: two tripartite enterotoxin complexes and a single protein enterotoxin (**Table 2**).

Haemolysin BL One of the enterotoxin complexes is haemolysin BL, which is haemolytic, cytotoxic and dermonecrotic, causes vascular permeability changes, and has been shown to cause fluid accumulation in the ligated rabbit ileal loop. Haemolysin BL is made up of three components: B, L_1 and L_2, with molecular masses of 37.8 kDa, 38.5 kDa and 43.2 kDa respectively. The individual components of haemolysin BL do not possess the enterotoxic activities separately, and all three components are required for maximal activity. The B protein component binds haemolysin BL to the target cell; L_1 and L_2 components have lytic functions. The L_2 component of haemolysin BL interacts with the antibody component of the Oxoid BCET-RPLA toxin detection kit (Oxoid, Unipath, Basingstoke, UK).

Non-haemolytic Enterotoxin Complex The causative strain (0075-95) of a large outbreak of *B. cereus* diarrhoeal food poisoning in Norway in 1995 was shown to produce a different, non-haemolytic tripartite enterotoxin complex. This second complex, referred to as the non-haemolytic enterotoxin, comprises three proteins (39 kDa, 45 kDa and 105 kDa), which are nontoxic individually but are cytotoxic in combination. The 45 kDa and 105 kDa proteins react with a commercially available visual immunoassay (Tecra) (Tecra Diagnostics, Batley, UK), but the 45 kDa protein is considerably more reactive than the 105 kDa component. This strain of *B. cereus*, although a food-borne pathogen, reacted negatively when tested with the Oxoid BCET-RPLA kit.

Enterotoxin T The *bceT* gene of *B. cereus* encodes an enterotoxin protein with the characteristics of the diarrhoeal toxin, known as enterotoxin T. Enterotoxin T has a molecular mass of 41 kDa; it has not been implicated in any outbreaks of food poisoning to date, but approximately 43% of randomly selected *B. cereus* strains isolated from different food products possessed the *bceT* gene. However, the *BceT* gene was absent from 5 out of 7 *B. cereus* strains that had been involved in food poisoning incidents.

There is evidence to suggest that more than one enterotoxin may be produced by a single strain of *B. cereus*. Many strains of *B. cereus* have been found to react with both the Oxoid and Tecra detection kits. This indicates that some strains are able to produce both tripartite enterotoxin complexes.

Bacillus cereus Emetic Syndrome

The emetic syndrome was first characterized in the UK following several incidents associated with the consumption of rice from Chinese restaurants and take-away outlets. The emetic syndrome is an intoxication as opposed to an infection; it has a rapid onset of up to 5 h after consumption of incriminated foodstuff. The symptoms of the illness are vomiting and nausea, with accompanying diarrhoea in about 30% of cases. The syndrome is not associated with fever. The emetic toxin of *B. cereus* causes similar symptoms to *Staphylococcus aureus* toxin.

The emetic toxin of *B. cereus* is a cyclic dodeca-depsipeptide named cereulide, which is structurally related to valinomycin. Cereulide has a molecular mass of 1.2 kDa, and was originally believed to be a breakdown product of a lipid component in the growth medium. However, the molecule is now known to consist of a ring structure of three repeating tetrameric units containing amino and oxyacids (D-O-Leu-D-Ala-L-O-Val-L-Val). The toxin molecule is very stable to heat, extremes of pH and proteolysis with trypsin. Emetic toxin formation is generally associated with H-1 serovars of *B. cereus*, and occurs after spore formation. The emetic toxin causes swelling of the mitochondria, and uncoupling of mitochondrial oxidative phosphorylation in HEp-2 cells. In higher animals the toxin mode of action is through binding to the 5-hydroxytryptamine receptor and stimulation of the vagus afferent nerve.

Factors Affecting Emetic Toxin Formation

Emetic toxin production is affected by the composition of the culture medium. Milk and rice-based media support effective emetic toxin production, whereas the toxin is not detectable after growth on brain–heart infusion (BHI) broth or tryptone–soya broth. Factors controlling emetic toxin formation have not been determined, although it has been observed that optimal emetic toxin production occurs after 20 h incubation at 30°C in batch cultures. The toxin is also detectable in non-sporulating chemostat cultures grown at a dilution rate of 0.2 h on a whey protein medium at pH 7 at 30°C.

Other *B. cereus* Toxins

In addition to producing food poisoning toxins, *B. cereus* also produces other toxic substances. The characteristics of these compounds are described in Table 2. They include phospholipase C and haemolysins. One of the phospholipases, sphingomyelinase, is also a haemolysin. Phospholipases, along with proteinases and lipases, are the degradative enzymes responsible for the off flavours and defects associated with the growth of *B. cereus* in milk. All of the toxins of *B. cereus*, with the possible exception of the emetic toxin, are produced during the exponential phase of the life cycle.

The haemolysins of *B. cereus* consists of sphingo-myelinase, cereolysin, cereolysin AB, haemolysin II, haemolysin III, haemolysin BL, and a 'cereolysin-like' haemolysin. Several of the extracellular haemolysins, including haemolysin BL, are considered to be virulence factors.

Toxins of Other *Bacillus* Species

Although other *Bacillus* species have been associated with outbreaks of food-borne disease, there is no definitive information on the identity of the toxins formed. Culture supernatants of some isolates of *B. circulans*, *B. lentus*, *B. mycoides* and *B. thuringiensis* are cytotoxic to Chinese hamster ovary (CHO) cells, although the activity is lost on heating, suggesting that like *B. cereus* and *B. brevis*, the component is heat-labile enterotoxin. Culture supernatants of some strains of *B. carotarum*, *B. circulans*, *B. lentus*, *B. licheniformis*, *B. mycoides*, *B. pumilus*, *B. polymyxa* and *B. thuringiensis* react positively with the antibody present in the Oxoid *B. cereus* enterotoxin detection kit. Strains belonging to the species *B. thuringiensis*, *B. circulans*, *B. lentus*, *B. licheniformis* and *B. thuringiensis* were, however, positive with antibodies supplied in the Tecra kit for *B. cereus* enterotoxin detection. This implies that extracellular components produced by these species have some structural similarity to components present in the *B. cereus* tripartite haemolytic and non-haemolytic complexes respectively.

Detection of *B. cereus* Toxins

In Vivo

Detection of *B. cereus* toxins has traditionally involved the use of studies in vivo (**Table 3**). Methods for detection of the diarrhoeal enterotoxins have included the rabbit or guinea pig ileal loop test, vascular permeability testing, dermonecrotic tests on guinea pigs, mouse lethality testing and rhesus monkey feeding trials. Rhesus monkey and *Suncus murinus* feeding trials, and mouse and *Suncus murinus* lethality testing, are suitable for determining the presence of the emetic toxin. However, studies in vivo are expensive; they require highly trained, licensed staff to perform them; and to many people they are morally unacceptable. Therefore alternative in vitro methods have been developed for use in diagnostic and research laboratories.

Table 3 Detection of *B. cereus* emetic and diarrhoeal enterotoxins

Assay	Toxin detected	Mode of action
In vivo		
Rhesus monkey feeding trials	Enterotoxin and emetic toxin	Monkeys fed rice culture slurry, and are observed for symptoms of the syndromes
		Non-specific
Suncus murinus feeding trials	Emetic	Oral and intraperitoneal injection resulting in emesis
Mouse lethality	Enterotoxin and emetic toxin	Mice intravenously injected with *B. cereus* culture supernatants
Suncus murinus lethality	Emetic toxin	*Suncus murinus* intravenously injected with *B. cereus* culture supernatants
Rabbit or guinea pig ileal loop test	Enterotoxin	Fluid accumulation in ligated ileal loops
Vascular permeability testing	Enterotoxin	Intradermal injection causes changes in vascular permeability (oedema and haemorrhage)
Dermonecrotic tests on guinea pigs	Enterotoxin	Skin cell death
In vitro		
Antibody:		
BCET-RPLA (Oxoid)	Haemolysin BL enterotoxin	Reverse passive latex agglutination technique
	Other *Bacillus* spp.	Detects the L_2 component of haemolysin BL
BDE-VIA (Tecra)	Enterotoxin – non-haemolytic enterotoxin	ELISA detects six different proteins, including 45 kDa and 105 kDa components
	Other *Bacillus* spp.	
Cytotoxicity:		
Visual	Enterotoxin and emetic toxin	Microscopic detection of visual changes
		Emetic – vacuolization of HEp-2 cells
Metabolic assessment (MTT)	Enterotoxin and emetic toxin	Proliferation assay
Lactate dehydrogenase release	Enterotoxin and emetic toxin	Measurement of LDH leakage from lysed cells
Disruption of monolayer		Crystal violet
Diffusion:		
Gel diffusion assay	Haemolysin BL	Haemolysin BL causes a discontinuous haemolysis pattern on blood agar plates
Fluorescent immunodot assay	Enterotoxin	Substrate gel system, measured by fluorescence
Microslide immunodiffusion assay	Enterotoxin	Detected by a line of identity on immunodiffusion assay
Other:		
Motility of boar spermatozoa	Emetic	Loss of motility of boar spermatozoa
PCR	Haemolysin BL	Amplification of DNA to look for haemolysin BL gene

In Vitro

In vitro assay methods for the detection of *B. cereus* food poisoning toxins include the application of antibody-based reactions (BCET-RPLA and Tecra BDE), cell cytotoxicity and various diffusion techniques (e.g. microslide immunodiffusion, disc diffusion and gel diffusion assays) (see Table 3).

Cell Cytotoxicity   Cell culture techniques have been used for the detection of both diarrhoeal toxin and emetic toxin of *B. cereus* (see Table 3). Cultured cell lines used include HeLa, HEp-2, Vero, McCoy and CHO cells. Different approaches have been used for the detection of cytotoxicity. Initially the presence of toxin was detected by the microscopic monitoring of any morphological response by cells in the presence of toxin; however, such methods were subjective. Detection methods have been improved by the measurement of specific cellular responses in the presence of the toxin. These methods include an assessment of the metabolic status of the cells using the tetrazolium salt   3-(4,5,-dimethylthiazol-2-yl)-2,5-diphenyltetrazolium bromide (MTT), and detection of lactate dehydrogenase release from damaged cells.

A cell culture assay for the detection of the emetic toxin was developed following the observation that culture supernatant fluids from 87% of *B. cereus* strains isolated from emetic syndrome outbreaks caused vacuoles to appear in HEp-2 cells. The emetic toxin affects the proliferation of cells, and this has been exploited in cytotoxicity assays. Cereulide causes vacuole formation in HEp-2 cells; the vacuolation factor is thought to be the emetic toxin itself. Electron microscopy has revealed that the apparent vacuoles are swollen mitochondria. Oxygen consumption rate was found to increase in the vacuolated HEp-2 cells; the toxin appeared to be acting as an uncoupler of oxidative phosphorylation in the mitochondria.

Different cell lines respond in different ways to the effects of emetic toxin. For instance, Chinese hamster ovary cells have been found to be as sensitive as HEp-2 cells to emetic toxin, but instead of forming vacuoles, the CHO cells become spherical with granulation of the cell contents. In all cell lines cell multiplication was arrested in the presence of the emetic toxin.

Other cell culture assays developed for emetic toxin detection include a cell proliferation assay which measures total metabolic activity of cultured cells, monitoring acid formation by HEp-2 cells induced by B. cereus emetic toxin, monitoring of amino acid uptake, and staining reactions with crystal violet (Table 3). Cell cytotoxicity methods can be used to detect the presence of enterotoxin and emetic toxins in culture supernatants and incriminated food samples. The methods are semi-quantitative and can be used to establish the toxigenic potential of isolates.

Immunological Methods   Two commercially available *in vitro* immunoassay kits have been developed to detect B. cereus diarrhoeal enterotoxins. These kits are the B. cereus enterotoxin reverse passive latex agglutination (BCET-RPLA) assay (Oxoid), and the *Bacillus* diarrhoeal enterotoxin visual immunoassay (BDE-VIA) (Tecra) (Table 3).

These two test kits are antibody based. The Oxoid reverse passive latex agglutination assay uses latex particles to amplify the antibody : antigen reaction. The latex particles are coated with antibody to detect a specific antigen; the antibody in this protocol has been raised against the L_2 component of the haemolysin BL enterotoxin.

The Tecra BDE VIA kit is a sandwich enzyme-linked immunosorbent assay (ELISA) in which the antibody is absorbed onto the solid phase, and the antigenic sample (enterotoxin) is added to complex with the antibody. Unbound antigen is removed by washing, and an enzyme-labelled conjugate which binds to the antigen is added. Excess conjugate is removed by washing, and the complex detected colorimetrically by the enzyme-mediated release of a chromophore from a specific exogenous substrate.

The two commercial test kits detect different antigens. While the Oxoid kit detects the L_2 component of haemolysin BL, the Tecra kit has been shown to react with six different proteins, including at least one from the non-haemolytic enterotoxin complex. In the past, several studies have compared the detection sensitivities of Oxoid, Tecra and cell cytotoxicity assays. However, it is now recognized that two distinct enterotoxin complexes are produced by B. cereus, and that the kits detect components in the different enterotoxins. Thus, the haemolytic enterotoxin is detectable with the Oxoid RPLA assay and the non-

haemolytic complex with the Tecra VIA. The apparent differences in efficacies therefore reflect the differences in the proportion of B. cereus strains that produce the different enterotoxins. The toxic potential of an unknown isolate should therefore be established by both methods. Cell cytotoxicity assays have the advantage of detecting toxic effect, rather than specific enterotoxins. Therefore, cell cytotoxicity assays can be used to detect all of the toxins produced, including the emetic toxin, for which there are currently no commercial test kits available. However, cytotoxicity assays are subject to interference from any other toxic metabolites that may be present in the samples assayed.

Other In Vitro Methods   In addition to the commercial test kits and cell cytotoxicity assays, several other in vitro methods have been developed. These methods include a gel diffusion assay, a fluorescent immunodot assay and a microslide immunodiffusion test (see Table 3). The emetic toxin adversely affects the motility of boar spermatozoa, and an assay system monitoring spermatozoa activity has been developed for the detection of the emetic toxin. A DNA probe has been designed for the detection of haemolysin BL using polymerase chain reaction (PCR) amplification. Further development work may result in the commercialization of one or more of these detection methods.

Components in culture supernatants of strains of *Bacillus* species, other than B. cereus, also react positively in the Oxoid BECT-RPLA and Tecra BDE VIA assays, and induce cytotoxic effects in CHO cells. Strains of B. thuringiensis, B. mycoides, B. circulans, B. lentus, B. polymyxa, B. carotarum and B. licheniformis produced putative enterotoxins that reacted positively with the Oxoid antibody preparation, causing latex particle agglutination. Bacillus thuringiensis, B. circulans, B. lentus and B. licheniformis strains, however, produced a moiety that resembled a component of the non-haemolytic complex, as a positive reaction has been obtained using the Tecra kit. Culture supernatants from strains of B. brevis, B. circulans, B. lentus, B. licheniformis and B. subtilis were cytotoxic when tested with a CHO cell line. Until more specific protocols are developed, methodologies developed for the detection of B. cereus toxins may be adapted for use in the detection of toxins produced by other species of *Bacillus*.

See also: **Bacillus**: Detection by Classical Cultural Techniques. **Biochemical and Modern Identification Techniques**: Introduction. **Sampling Regimes & Statistical Evaluation of Microbiological Results**.

Further Reading

Agata N, Ohta M, Arakawa Y and Mori M (1995a) The bceT gene of *Bacillus cereus* encodes an enterotoxin protein. *Microbiology* 141: 983–988.

Andersson MA, Mikkola R, Helin J, Andersson MC and Salkinoja-Salonen M (1998) A novel sensitive bioassay for detection of *Bacillus cereus* emetic toxin and related depsipeptide ionophores. *Applied and Environmental Microbiology* 64: 1338–1343.

Beattie SH and Williams AG (1999) Detection of toxigenic strains of *Bacillus cereus* and other *Bacillus* spp. with an improved cytotoxicity assay. *Letters in Applied Microbiology* 28: 221–225.

Beecher DJ and Wong ACL (1994a) Improved purification and characterisation of haemolysin BL, a haemolytic dermonecrotic vascular permeability factor from *Bacillus cereus*. *Infection and Immunity* 62: 980–986.

Beecher DJ, Schoen JL and Wong ACL (1995) Enterotoxic activity of haemolysin BL from *Bacillus cereus*. *Infection and Immunity* 63: 4423–4428.

Granum PE and Lund T (1997) Mini Review: *Bacillus cereus* and its food poisoning toxins. *FEMS Microbiology Letters* 157: 223–228.

Granum PE, Andersson A, Gayther C, te Giffel M, Larsen H, Lund T and O'Sullivan K (1996) Evidence for a further enterotoxin complex produced by *Bacillus cereus*. *FEMS Microbiology Letters* 141: 145–149.

Griffiths MW (1995) Foodborne illness caused by *Bacillus* species other than *B. cereus* and their importance to the dairy industry. *Bulletin of the International Dairy Federation* 302: 3–6.

Isobe M, Ishikawa T, Suwan S, Agata N and Ohta M (1995) Synthesis and activity of cereulide, a cyclic dodeca-depsipeptide ionophore as emetic toxin from *Bacillus cereus*. *Bioorganic and Medical Chemistry Letters* 5: 2855–2858.

Jackson SG (1989) Development of a fluorescent immuno-nodot assay for *Bacillus cereus* enterotoxin. *Journal of Immunological Methods* 120: 215–220.

Kramer JM and Gilbert RT (1989) *Bacillus cereus* and other *Bacillus* species. In: *Foodborne Bacterial Pathogens*, Doyle MP (ed.). P. 21–70. New York: Marcel Dekker.

Lund T and Granum PE (1996) Characterisation of a non-haemolytic enterotoxin complex from *Bacillus cereus* isolated after a foodborne outbreak. *FEMS Microbiology Letters* 141: 151–156.

Notermans S and Batt CA (1998) A risk assessment approach for food-borne *Bacillus cereus* and its toxins. *Journal of Applied Microbiology* 84: 51S–61S.

Schultz FJ and Smith JL (1994) *Bacillus*: Recent advances in *Bacillus cereus* food poisoning research. In: Foodborne disease handbook, diseases caused by bacteria. Vol. 1. Hui YH, Gorham JR, Murrell KD and Cliver DO (eds). New York: Marcel Dekker.

Detection by Classical Cultural Techniques

Ian Jenson, Gist-brocades Australia Pty Ltd, Moorebank, NSW, Australia

The genus *Bacillus* consists of a rather heterogenous group of Gram-positive endospore-forming rods that grow aerobically and usually produce catalase. A number of species have recently been moved to new genera. The only species of interest to food microbiologists in these new genera is *Alicyclobacillus acidoterrestris*. *Bacillus* spp. are easily detected on a wide range of media. The species most likely to be found in food generally have simple nutritional requirements and can be grown on media such as nutrient agar. All species grow aerobically and some are facultative anaerobes. Colonial morphology of *Bacillus* species is often highly suggestive diagnostically, but considerable variation may be observed, even within a single species. Colonies are usually translucent to opaque and white to cream coloured. Most species do not produce pigments.

Range of Media and Applications

The most general test which is performed for members of the genus *Bacillus* is for 'aerobic mesophilic spore-formers' or 'thermophilic flat sour spore-formers'. More specific tests may be performed for spore-formers which are classed as 'aciduric flat sour spore-formers' or 'rope spores'. The only established methods for a *Bacillus* species of interest to food microbiologists are for *Bacillus cereus*. Psychrotolerant strains of *B. cereus* have been assigned to a new species, *B. weihenstephanensis*. Strains of this species may also produce toxins and will be identified as *B. cereus* in the standard microbiological tests. The other species of general interest to food microbiologists are *B. licheniformis* and *B. subtilis* (**Table 1**).

Table 1 *Bacillus* species of interest in food microbiology

Species	Reason
Bacillus cereus	Human illness, bitty cream
Bacillus subtilis	Rope in bakery products, human illness, production of some fermented soy products such as natto and kinema
Bacillus licheniformis	Rope in bakery products, human illness, flat sour defect in canned foods
Bacillus stearothermophilus	Thermophilic flat sour defect of canned foods
Bacillus polymyxa	Flat sour defect of canned foods
Bacillus macerans	Flat sour defect of canned foods
Bacillus coagulans	Aciduric flat sour defect of canned foods
Bacillus smithii	Aciduric flat sour defect of canned foods

Table 2 Media used for detection of *Bacillus* species

Species/Functional group	Food	Media
Bacillus cereus	Cereals, spices, milk products, legumes, food associated with characteristic illness	Mannitol–egg yolk–polymixin agar; polymixin–egg yolk–mannitol–bromothymol blue agar; tryptone–soy–polymixin broth
Rope spores	Flour, bread, bakery ingredients	Dextrose–tryptone agar
Mesophilic aerobic spore-formers	Starch, dried fruits, vegetables, dried milk, spices, cereals, low acid (pH > 4.6) canned foods	Dextrose–Tryptone agar, tryptone–glucose extract agar
Thermophilic flat sour spores	Sugar, starch, flour, spices	Dextrose–tryptone agar
Aciduric flat sour spore-formers	Tomato products, dairy products	Dextrose–tryptone agar, thermoacidurans agar

The medium chosen depends on the purpose of the examination and the level of sensitivity required (**Table 2**). If the purpose is to determine whether potential spoilage organisms are present, then media are used that support a wide range of *Bacillus* as well as non-*Bacillus* species. When testing for the presence of food-borne pathogens, selective media are employed. In most cases plating techniques will give sufficient sensitivity but sometimes enrichment tests for *B. cereus* are performed. Additionally, in some cases it is desirable to test for the presence of vegetative organisms and at other times a test for spores is required.

Bacillus Species and Choice of Media

A wide range of *Bacillus* species may be involved in the mesophilic spoilage of low acid canned foods. Flat sour spoilage is the result of acid production with little or no gas production; product pH is decreased but the can is not distended. *B. coagulans* and *B. circulans* are largely responsible. Swollen cans may result from spoilage with *B. subtilis*, *B. pumilus* or *B. macerans*.

Low-acid foods are more likely to be spoiled by thermophilic bacilli such as *B. stearothermophilus*. This species, *B. coagulans* and the closely related *B. smithii* are responsible for flat sours. *B. subtilis* may produce some gas and swelling of the can.

High-acid foods are not as susceptible to spoilage by bacilli. An exception are the flat sours of tomato products that are frequently associated with *B. coagulans*.

Rope in bread and bakery products is due to the growth of *B. subtilis* or *B. licheniformis* which hydrolyse starch to produce estery odours and characteristic stickiness. In extreme cases, the bread structure breaks down resulting in strands of spoiled material that can be removed from the bread surface.

Alicyclobacillus acidoterrestris is a thermophile which has been associated with spoilage of fruit juices and the production of taints. The extent of problems caused by this or other organisms, which have received little attention, is not known.

Bacillus cereus is a food-borne pathogen which has been associated with consumption of a wide range of foods. It can be responsible for both emetic and diarrhoeal syndromes. The organism can grow rapidly in some foods, and for this reason, enrichment techniques able to detect very low numbers of organisms have been developed. Dried milk products which are to be used in infant foods are an example of foods which might be tested by such procedures. *B. cereus* can be responsible for the breakdown of fat in cream which results in a flakey appearance when added to a hot beverage. This is referred to as 'bitty' cream.

Bacillus Species and Sample Type

Samples of spoiled canned food or ingredients for canned food may be tested for a number of groups of *Bacillus* species, depending on the likely temperature of product storage and also the level of product acidity. Thermophilic spoilage should be considered if the product may be stored for long periods above 43°C. Low-acid foods have a pH above 4.6, whereas high-acid foods have a pH below 4.6. Samples of ingredients, such as sugar, spices and starch, may be tested for *Bacillus* species.

Rope spores may be tested for in bread, flour or bakery ingredients. It is relevant to test raw materials for the presence of spores that might survive processing, germinate and cause product spoilage. Once a product is actively spoiling it is more relevant to test for appropriate vegetative organisms, though in already spoiled products spores are, once again, likely to be present.

Raw foods such as rice, flour, raw milk and spices are recognized sources of *B. cereus*. Prepared meat, bakery, egg, lentil and rice products have been associated with outbreaks. Levels of $10^5\,g^{-1}$ or more are usually found in food associated with illness.

Table 3 Procedures for detecting *Bacillus* species in foods

	Sample	Dilution	Heating	Plating	Incubation
Mesophilic spores	50 g food	10^{-1} dilution in 0.1% peptone then 100 ml TGE agar	80°C for 30 min	5 plates	35°C for 48 h
Thermophilic flat sour	20 g sugar	Water up to 100 ml	100°C for 5 min	2 ml in each of 5 plates DTA	50–55°C for 48, 72 h
Thermophilic flat sour	20 g starch	Water up to 100 ml then 100 ml DTA	100°C for 3 min then 108°C for 10 min	5 plates with 2% water agar overlay	50–55°C for 48, 72 h
Aciduric flat sour	Liquefied tomato products or milk		88°C for 5 min	1 ml in each of 2 DTA and 2 TAA plates	55°C for 48 h
Aciduric flat sour	10 g non-fat dried milk	0.02N sodium hydroxide up to 100 ml	108°C for 5 min	2 ml in each of 10 plates DTA	55°C for 48 h
Aciduric flat sour	20 ml cream	Special diluent up to 100 ml	108°C for 5 min	2 ml in each of 5 plates DTA	55°C for 48 h
Rope spores	20 g	Butterfield's diluent up to 100 ml then DTA	94 ± 2°C for 15 min	Add tetrazolium salts and pour 5 plates	35°C for 24, 48, 72 h
Alicyclobacillus acidoterrestris			80°C for 10 min	0.2 ml spread on Orange serum agar	44–46°C for 48 h
Bacillus cereus (direct plating)	10 g–50 g	Serial dilution in Butterfield's diluent or 0.1% peptone solution		0.1 ml spread on MYP or PEMBA	30°C or 37°C for 24 h, optionally followed by 24 h at room temperature
Bacillus cereus (MPN)	10 g	Serial dilution in Butterfield's diluent or 0.1% peptone solution		Tryptone soy polymixin broth *then* MYP or PEMBA	48 h at 30–37°C 24 h at 30–37°C

Formulations

All media can be produced using standard laboratory techniques. Commercially available dehydrated media may be used in many cases. Formulations can be found in the appendix.

Detailed Procedures

The procedures for different types of *Bacillus* have a number of features in common (**Table 3**). Samples are diluted and then frequently are heated to destroy vegetative cells. Heating may occur either in the diluent or in the agar. After plates are poured or inoculated they are incubated and colonies are counted to provide a count per gram of the original sample. Comments on the general aspects of these procedures are given below before providing details on each procedure.

General Notes

Sample Size A minimum sample size of 10 g should be taken in an attempt to ensure that the sample is representative. Some authorities recommend testing samples of up to 50 g to ensure that a representative sample is tested.

The quantity of sample inoculated into media is frequently large. Many authorities recommend the testing of up to 1 g of product. For instance, 10 ml of a 10^{-1} dilution is added to 100 ml agar and distributed over five Petri dishes. Obviously dilutions need to be made if the number of organisms is expected to be large. However, it should be noted that the practice of plating a sample over several plates will result in methods which are both sensitive and able to accommodate highly contaminated samples.

Heating Samples being tested for the presence of spores are heated to destroy vegetative organisms and encourage spore germination. It is generally accepted that 80°C for 10 min is sufficient to destroy vegetative cells. Spore germination is necessary if an accurate count of spores is to be obtained. The spores of some *Bacillus* species are more heat resistant than others and this feature is used in some methods to make them more selective.

When samples are heated there are several points which need to be watched. Some spores can germinate very quickly, therefore it has been recommended that the period between preparing the first dilution and heating is less than 10 min, preferably at as low a temperature as possible. The heating and cooling periods should also be as short as possible. A small sample should be heated in a sealed tube (to prevent contamination from the waterbath or evaporation of the sample) and a pilot tube used to measure the temperature of the sample. The level of the waterbath must be above the level of the sample in the tube. Samples should be agitated during the heating and cooling stages. When a temperature above 100°C is required this is most conveniently achieved in an autoclave. For instance, 108°C is equivalent to applying a pressure of 5 psi (34.5 kPa).

Some methods require the sample to be added to agar before the heat treatment step. In these cases the agar is maintained at around 50°C. Once the sample is added the agar is quickly heated. After the required time, the agar is cooled as quickly as possible, taking care not to cause the agar to gel. After a short period of equilibration at 45°C, the plates are poured.

Incubation A temperature of 30–37°C is used for mesophiles and 50–55°C for thermophiles. At the higher temperatures the plates should be sealed in plastic bags or containers containing water so that the plates will not dry out. It is usual to examine the plates during the required incubation period to ensure that the plates do not become overgrown with large or spreading colonies and that acid reactions do not revert to alkaline by continued incubation.

Mesophilic Aerobic Spore-formers

The method given here is that of the American Public Health Association. Usually 0.1–10 ml of the initial dilution is inoculated into the tryptone–glucose extract (TGE) agar. All colonies appearing on the plates are counted. If 10 ml is used to inoculate TGE agar then the sum of the number of colonies on five plates can be expressed as the number of mesophilic aerobic spore-formers per gram of the original sample.

Some authorities suggest that the sample need only be heated for 10 min at 80°C and suggest an incubation temperature of 30°C.

Thermophilic Flat Sour Spore-formers

The methods here are those of the National Food Processors Association (USA) and the method for sugar, additionally has approval as an AOAC Official method (972.45).

The method for sugar allows either solid or liquid sugar to be tested. If the liquid product is tested, the amount added to the initial dilution is adjusted to contain 20 g dry sugar equivalent. After heating to 100°C, 2 ml of the heated sugar solution is added to each of five Petri dishes before adding dextrose–tryptone agar (DTA). In the AOAC procedure the plates are incubated at 55°C for 35–48 h.

The method for starch requires 10 ml of the starch suspension to be added to DTA and boiled for 3 min to gelatinize the starch before proceeding to heat the suspension further. After pouring the agar into plates and allowing it to gel, a thin layer of 2% water agar is overlayed to prevent the spread of some organisms across the surface of the agar.

Typical colonies are round, 1–5 mm in diameter with an opaque central spot and a yellow halo in the agar. This halo may be missing. Subsurface colonies are compact and may be pinpoint in size. It may be necessary to isolate subsurface colonies by streaking onto the surface of fresh DTA to confirm their typical appearance.

Aciduric Flat Sour Spore-formers

The methods given here are those of the American Public Health Association. Tomato products and other liquid products, dry products such as non-fat dry milk and cream are tested by different procedures.

Tomato and milk products are tested by plating onto DTA and Thermoacidurans agar. Raw tomatoes and similar tomato products may need to be blended so that spore tests can be performed on a liquid product. *B. coagulans* colonies appear slightly moist, slightly convex and pale yellow on the surface of DTA. Subsurface colonies appear as compact yellow to orange colonies 1 mm or more in diameter with fluffy edges. On Thermoacidurans agar this organism will produce large colonies, which are white to cream in colour.

Non-fat dry milk is suspended in 0.2M sodium hydroxide before being heated; 2 ml of the heated suspension is added to each Petri dish before adding DTA. Incubation conditions and expected colonial morphology are as given above.

Cream is suspended in a special diluent and heated. The suspension has high viscosity. It is recommended that the DTA is poured into Petri dishes and the cream suspension added before the agar sets. Incubation conditions and expected colonial morphology is given above.

Rope Spores

The method given here is that of the American Association of Cereal Chemists (method 42–20). Volumes of 10 ml and 1 ml of the first dilution are added to molten DTA. The flasks should reach 94°C within 5 min and are maintained at this temperature for 15 min. After cooling, 1 ml tetrazolium salts solution is added before pouring the plates. After 24 h, subsurface colonies with a yellow halo are drawn to the surface of the agar with a sterile inoculating needle. After a further 24 h the plates are inspected for the presence of typical colonies. Typical colonies are grey–white, moist and blister-like at first and may become drier and wrinkled with age. The colonies have a stringy consistency when touched with an inoculating needle. If any further subsurface colonies have appeared they are treated and inspected as for those appearing at 24 h. The total count of typical colonies over the five plates is used to calculate the number of rope spores per gram.

Alicyclobacillus acidoterrestris

Only tentative methods have been proposed for this organism. A presence/absence test can be performed after incubating a sample at 44°C for 48 h, if desired. Tentative identification can be made by Gram stain which reveals Gram-positive rods with terminal to subterminal spores which are slightly swollen.

Bacillus cereus

Two direct plating methods and one enrichment method are commonly used to detect *B. cereus* in foods. Two methods are commonly used to confirm the identity of presumptive *B. cereus* detected by these procedures. Most regulatory authorities use the mannitol-egg yolk-polymixin (MYP) agar procedure (for example, AOAC method 980.31 and 983.26) but there is increasing support for the use of polymixin–egg yolk–mannitol-bromothymol blue agar (PEMBA).

In the direct plating procedure, dilutions of the food under test are made in either Butterfield's diluent (AOAC) or 0.1% peptone solution. Incubation conditions vary between 30°C and 37°C for 24–48 h, sometimes at 25°C for the second day. If the longer incubation time is used, the plates should be examined at 24 h to avoid problems due to overgrowth. Typical colonies on MYP are crenate to fimbriate, 3–6 mm in diameter with a ground glass surface surrounded by a zone of precipitate and pink agar. On PEMBA typical colonies are similar but 3–5 mm diameter and surrounded by a zone of precipitate and turquoise to peacock blue agar.

In the enrichment procedure, dilutions of the food under test are made as in the direct plating procedure and inoculated in tryptone–soy–polymixin broth. If it is desired to enumerate low levels of *B. cereus* in a food then the enrichment is configured as an MPN test. The broth is incubated at 30°C for 48 h before plating onto MYP or PEMBA and incubating according to the requirements of the standard method being followed.

Presumptively positive colonies may be confirmed by either biochemical/physiological identification or the use of a staining technique (if PEMBA was used). It is considered by many that the characteristics of *B. cereus* are so distinctive that the staining technique of Holbrook and Anderson is sufficient to confirm *B. cereus* isolated on PEMBA or other media containing low concentrations of nitrogen that encourage sporulation. Confirmatory tests to differentiate *B. cereus* from most other *Bacillus* species include Gram stain,

anaerobic glucose fermentation, nitrate reduction, Voges–Proskauer, tyrosine decomposition, lysozyme sensitivity, mannitol fermentation and egg yolk reaction (**Table 4**). To differentiate *B. cereus* from other closely related species (*B. anthracis*, *B. mycoides*, *B. thuringiensis*) it is necessary to determine motility, rhizoidal growth, haemolysis or toxin crystal production. Serological methods have been described, which are able to differentiate *B. cereus* from other species and are also of use in epidemiological typing, but the antisera are not widely available.

Presumptive *B. cereus* colonies are grown on nutrient agar or tryptone–soy broth for 18–24 h at 30°C. If nutrient agar is used, a colony is suspended in a small volume of Butterfield's diluent to produce a turbid suspension. Confirmatory tests are performed as detailed below.

Gram Stain   *B. cereus* will appear as large Gram-positive rods in short to long chains; spores are ellipsoidal, in a central to subterminal position and do not swell the cell.

Anaerobic Glucose Fermentation   After inoculating phenol red dextrose broth with a small inoculum and incubating in an anaerobe jar for 24 h at 37°C, acid production is indicated by a change in the indicator from red to yellow.

Nitrate Reduction   After inoculating nitrate broth with a small inoculum and incubating at 37°C for 24 h, 0.25 ml of each of nitrite reagents A and B is mixed and added. A orange colour developing within 10 min indicates a positive reaction.

Voges–Proskauer Inoculate Voges–Proskauer medium and incubate at 37°C for 48 h. Transfer 1 ml to an empty test tube and add 0.2 ml of 40% potassium hydroxide, 0.6 ml of α-naphthol and a few crystals of creatine. If the solution turns pink within 1 h the reaction is considered positive.

Tyrosine Decomposition   Inoculate the surface of the slope and incubate at 37°C for 48 h. Examine for clearing of the agar around the growth which indicates tyrosine decomposition. Incubate for a further 24 h and examine again, if necessary.

Lysozyme Sensitivity   Inoculate nutrient broth containing lysozyme and a control nutrient broth with a small inoculum and incubate at 37°C for 24–48 h. Record strain as sensitive if no growth occurs in broth containing lysozyme.

Mannitol Fermentation   If it was not possible to

Table 4 Identification of *Bacillus* species of public health interest

	B. cereus	B. mycoides	B. thuringiensis	B. anthracis	B. subtilis	B. licheniformis
Cell diameter > 1.0 µm	+[a]	+	+	+	−	−
Anaerobic glucose fermentation	+	+	+	+	−	+
Nitrate reduction	+	+	+	+	+	+
Voges–Proskauer	+	+	d	+	+	+
Tyrosine decomposition	+	d	d	d	−	−
Lysozyme sensitivity	+	+	+	+	d	d
Mannitol fermentation	−	−	−	−	+	+
Egg yolk reaction	+	d	d	+	−	−
Motility	+	−	+	−		
Rhizoid growth	−	+	−			
Haemolysis	+	+	+	−		
Toxin crystals	−	−	+	−		

[a]+, 90% or more of strains are positive; −, 90% or more of strains are negative; d, 11–89% of strains are positive.
Source: Claus and Berkeley (1986).

record mannitol fermentation from the primary isolation plate, inoculate the strain onto MYP or PEMBA and incubate at 37°C for 24 h. The agar will become pink or blue around the growth indicating a lack of mannitol fermentation.

Egg Yolk Reaction   If it was not possible to record egg yolk reaction from the primary isolation plate, inoculate the strain onto MYP or PEMBA and incubate at 37°C for 24 h. A white precipitate around the growth indicates a positive egg yolk reaction.

Holbrook and Anderson Stain   Smears may be produced from the centre of a 24 h colony or the edge of a 48 h colony growing on PEMBA. Smears are air dried and fixed with minimal heating. Stain with malachite green over a boiling waterbath for 2 min. After washing the slide and blotting it dry, stain with Sudan black for 15 min. Then rinse the slide in xylol for 5 s and blot dry before staining with safranin for 20 s. *Bacillus cereus* will appear 4–5 µm long and 1.0–1.5 µm wide with square ends. Lipid globules, staining black, are present in vegetative cells and spores, staining green, are ellipsoidal, central to subterminal in position and do not swell the sporangium.

Motility   Stab inoculate BC motility medium and incubate at 30°C for 24 h and examine for diffuse growth away from the stab, indicating that the strain is motile.

Rhizoid Growth   Inoculate the centre of a predried nutrient agar plate with a loopful of inoculum in one spot and allow to dry. Inoculate the plate right side up at 30°C for 24 h. Rhizoid growth is indicated by root or hair-like structures growing from the centre of the colony.

Haemolysis   Inoculate the sheep blood agar plate with a loopful of inoculum in one spot and allow to dry. Incubate at 30°C for 24 h and examine for a zone of complete haemolysis around the colony.

Toxin Crystal Production   After allowing a culture on a nutrient agar slope to grow for 24 h at 30°C, hold at room temperature for 2–3 days. Smears are air dried and fixed with minimal heating. Flood smear with methanol for 30 s and drain. Dry slide in burner flame. Flood with basic fuchsin. Heat until steam rises and remove heat. Repeat heating after 1–2 min. After a further 30 s, pour off stain and rinse well. Dry slide without blotting and examine for dark-coloured, tetragonal crystals which have been liberated from lysed sporangia. It may be necessary to allow more time for spores to lyse.

Bacillus subtilis and Bacillus licheniformis

These species are sometimes implicated in cases of food poisoning but no standard methods exist. They are able to grow on MYP or PEMBA but are able to ferment mannitol. For example, on PEMBA these species produce flat colonies about 3 mm in diameter and green to grey–green in colour. They do not produce an egg yolk precipitate. These species may be identified using the confirmatory tests specified for *Bacillus cereus* and the reactions in Table 4.

Alicyclobacillus acidoterrestris

The numbers of cells and spores of *A. acidoterrestris* are generally very low in foods, so even spoiled foods will need to be heated and enriched prior to plating. Liquid products such as fruit juices can be heated (80°C for 10 min) without dilution. Heated samples should be incubated for 48 h at 44–46°C and then plated onto orange serum agar (OSA). *A. acidoterrestris* is able to grow at pH values less than 4.0 but not at neutral pH values.

Advantages and Limitations

Mesophilic Aerobic Spore-formers, Flat Sour Spore-formers, Aciduric Flat Sour Spore-formers

The procedures outlined here are considered standard methods but it is possible that other methods are more applicable to certain foods and certain situations. Incubating canned food and observing for signs of spoilage is both easier and more sensitive than microbiological tests. The tests are therefore most relevant to raw materials.

Rope Spores

The result of the Rope Spore test is highly dependent on the heating procedure used and the subjective analysis of colony types.

It is widely acknowledged that this test does not correlate with the development of rope in bakery products. Bakery products receive different heat treatments to the one used in this test. Also, some spores will germinate and grow more slowly than others in bakery products. Strains vary in their amylase activities and their ability to produce odours and stickiness in product. An actual baking test, though qualitative, is the most predictive for the development of rope in products.

Bacillus cereus

The MYP medium was considered a significant advantage over earlier media because it combined selective (polymixin B) and differential (mannitol, egg yolk) features into one agar and gave a quantitative recovery of the target organism. Care needs to be taken to examine the plates after 24 h incubation because the mannitol fermentation reaction can become positive as other mannitol positive organisms grow on the plate. Also the plate can become overgrown, making colonies difficult to count and egg yolk reactions difficult to read. *B. cereus* does not sporulate well on this agar, making the confirmatory microscopy test of little value. Closely related species will be indistinguishable from *B. cereus* on this agar.

PEMBA was developed from the KG medium which, in turn, was developed from the MYP medium. It allows *B. cereus* to sporulate after 24 h, provides more buffering to assist in reading mannitol fermentation reactions and contains sodium pyruvate to improve the reading of the egg yolk reaction. There is less growth of competitive organisms on PEMBA when testing most foods, which makes the reactions easier to read. Egg yolk reactions can sometimes be difficult to detect and some *B. cereus* strains may appear to be negative. Some closely related species will be indistinguishable on this agar.

Both media can only be stored at 4°C for 4 days after pouring as egg yolk reaction become less intense with storage.

Collaborative Evaluations/Validations

AOAC Collaborative evaluations of the MYP agar method, the tryptone–soy–polymixin broth MPN method and the biochemical confirmatory tests, both for differentiation from unrelated *Bacillus* species and closely related *Bacillus* species have been performed. The MYP agar method was considered to be preferable to the MPN method for counting high numbers of *B. cereus*. The MPN method was suitable for counting low numbers of *B. cereus* but had a higher standard deviation both within and between laboratories. The between-laboratory standard deviation for the plating method was $0.1–0.2 \log_{10}$ and was $0.48–0.55 \log_{10}$ for the MPN method. The collaborative studies show that the methods are very reliable for differentiating *B. cereus* from other bacilli.

Holbrook and Anderson's validation of the PEMBA method and confirmatory staining was very thorough. They used a number of *B. cereus* strains as well as closely related and other species of *Bacillus* in their tests and showed that the strains gave typical egg yolk and staining reactions in nearly all cases. There were no problematic egg yolk reactions with PEMBA as there were with MYP. They showed equivalent recovery of *B. cereus* on MYP and PEMBA.

A study sponsored by the European Commission and published as an appendix to the International Standards Organisation method for enumeration of *B. cereus* has shown that MYP and PEMBA give equivalent quantitative results. In both media, the variation within a laboratory would result in an expectation that 95% of results on duplicate samples would be within $0.3 \log_{10}$ of each other and between laboratories would result in an expectation that 95% of the results on duplicate samples would be within $0.5 \log_{10}$ of each other.

See also: **Bacillus**: Bacillus cereus. **Direct (and Indirect) Conductimetric/Impedimetric Techniques**: Food-borne Pathogens. **Food Poisoning Outbreaks**. **National Legislation, Guidelines & Standards Governing Microbiology**: European Union; Japan. **Sampling Regimes & Statistical Evaluation of Microbiological Results**.

Further Reading

Claus D and Berkeley RCW (1986) Genus *Bacillus* Cohn 1872. In: Sneath PHA, Mair NS, Sharpe ME and Holt JG (eds) *Bergey's Manual of Systematic Bacteriology*, vol. 2. Baltimore: Williams & Wilkins.

Harmon SM (1982) New method for differentiating members of the *Bacillus cereus* group: collaborative study. *Journal of the Association of Official Analytical Chemists* 65: 1134–1139.

Holbrook R and Anderson JM (1980) An improved selective and diagnostic medium for the isolation and enumeration of *Bacillus cereus* in foods. *Canadian Journal of Microbiology* 26: 753–759.

Jensen N (1999) *Alicyclobacillus* – a new challenge for the food industry. *Food Australia* 51: 33–36.

Jenson I and Moir CJ (1997) *Bacillus cereus* and Other *Bacillus* Species. In: Hocking AD, Arnold G, Jenson I, Newton K and Sutherland P (eds) *Foodborne Microorganisms of Public Health Significance*, 5th edn, p. 379. Sydney, Australia; Australian Institute of Food Science and Technology (NSW Branch) Food Microbiology Group.

Kramer JM and Gilbert RJ (1989) *Bacillus cereus* and Other *Bacillus* Species. In: Doyle MP (ed.) *Foodborne Bacterial Pathogens*. New York: Marcel Dekker.

Lancette GA and Harmon SM (1980) Enumeration and confirmation of *Bacillus cereus* in foods: collaborative study. *Journal of the Association of Official Analytical Chemists* 63: 581–586.

Lechner S, Mayr R, Francis KP et al (1998) *Bacillus weihenstephanensis* sp. nov. is a new psychrotolerant species of the *Bacillus cereus* group. *International Journal of Systematic Bacteriology* 48: 1373–1382.

Parry JM, Turnbull PCB and Gibson JR (1983) *A Colour Atlas of Bacillus Species*. London: Wolfe Medical.

Pettipher GL, Osmundson ME and Murphy JM (1997) Methods for the detection and enumeration of *Alicyclobacillus acidoterrestris* and investigation of growth and production of taint in fruit juice and fruit juice-containing drinks. *Letters in Applied Microbiology* 24: 185–189.

Vanderzant C and Splittstoesser DF (eds) (1992) *Compendium of Methods for the Microbiological Examination of Foods*, 3rd edn. Washington: American Public Health Association.

van Netten P and Kramer M (1992) Media for the detection and enumeration of *Bacillus cereus* in foods: a review. *International Journal of Food Microbiology* 17: 85–99.

Appendix Formulations

Diuents/Solutions

Butterfield's Phosphate
Stock Solution

Potassium dihydrogen phosphate	34.0 g
Distilled water	500 ml

Adjust pH to 7.2 with approximately 175 ml of 1N NaOH
Adjust final volume with distilled water to 1000 ml
Store refrigerated

Diluent

Stock solution	1.25 ml
Distilled water to	1000 ml

Dispense and sterilize by autoclaving at 121°C for 15 min.

Peptone Diluent

Bacteriological peptone	1.0 g
Distilled water to	1000 ml

Dispense and sterilize by autoclaving at 121°C for 15 min.

Cream Diluent

Gum tragacanth	1.0 g
Gum arabic	1.0 g
Water	100 ml

Autoclave for 20 min at 121°C.

Tetrazolium Salts

2,3,5-Triphenyl-tetrazolium chloride	10.0 g
Water to	100 ml

Sterilize by membrane filtration through a 0.2 μm filter.

Nitrite Reagents
Reagent A

Sulphanilic acid	8.0 g
5N Acetic acid	1000 ml

Reagent B

α-Naphthol	2.5 g
5N Acetic acid	1000 mL

Holbrook and Anderson Stain
The following solutions are required:

5% w/v malachite green
0.3% Sudan black B in 70% ethanol
Xylol
0.5% w/v safranin

Toxin Crystal Stain
0.5 g basic fuchsin dissolved in 20 ml ethanol then made up to 100 ml with water.

Media for Enumeration

Mannitol–Egg Yolk–Polymixin (MYP) Agar

Beef extract	1.0 g
Peptone	10.0 g
D-Mannitol	10.0 g
Sodium chloride	10.0 g
Phenol red	0.025 g
Agar	12–18 g
Water	940 ml

Adjust pH so that it will be 7.1 ± 0.2 at 25°C after autoclaving at 121°C for 15 min. Add 10 ml of filter-sterilized polymixin B sulphate solution (10 000 units ml^{-1}) and 50 ml of 50% egg yolk emulsion per 940 ml of the basal agar.

Polymixin–Egg Yolk–Mannitol–Bromothymol Blue Agar (PEMBA)

Tryptone	1.0 g
D-Mannitol	10.0 g
Sodium pyruvate	10.0 g
Magnesium sulphate heptahydrate	0.1 g
Sodium chloride	2.0 g
Disodium hydrogen phosphate	2.5 g
Bromothymol blue	0.12 g
Agar	12–18 g
Water	940 ml

Adjust pH so that it will be 7.2 ± 0.2 at 25°C after autoclaving at 121°C for 15 min. Add 10 ml of filter-sterilized polymixin B sulphate solution (1000 units ml^{-1}) and 50 ml of 50% egg yolk emulsion per 940 ml of the basal agar.

Tryptone–Soy–Polymixin Broth

Tryptone	34.0 g
Soya peptone	6.0 g
Sodium chloride	10.0 g
Glucose	5.0 g
Dipotassium hydrogen phosphate	5.0 g
Water	1000 ml

Adjust pH so that it will be 7.3 ± 0.2 at 25°C after autoclaving at 121°C for 15 min. Add 1 ml of filter-sterilized polymixin B sulphate solution (1000 units ml^{-1}) per 100 ml broth.

Dextrose–Tryptone Agar (DTA)

Tryptone	10.0 g
Dextrose	10.0 g
Agar	12–18 g
Bromocresol purple	0.04 g
Water	1000 ml

Adjust pH so that it will be 6.7 ± 0.2 at 25°C after autoclaving at 121°C for 15 min.

Thermoacidurans Agar (TAA)

Yeast extract	5.0 g
Proteose peptone	5.0 g
Dextrose	5.0 g
Dipotassium phosphate	4.0 g
Agar	20.0 g

Adjust pH so that it will be 5.0 ± 0.2 after autoclaving at 121°C for 15 min.

Tryptone–Glucose Extract Agar (TGE)

Beef extract	3.0 g
Tryptone	5.0 g
Dextrose	1.0 g
Agar	15.0 g

Adjust pH so that will be 7.0 ± 0.2 after autoclaving at 121°C for 15 min.

Orange Serum Agar

Orange serum	200 ml
Yeast extract	3 g
Tryptone	10 g
Dextrose	4 g
Dipotassium phosphate	2.5 g
Agar	17 g
Water to	1000 ml

Adjust pH to give 5.5 ± 0.2 after autoclaving at 121°C for 15 min

Media for Confirmation

Phenol Red Glucose Broth

Proteose peptone no. 3	10.0 g
Beef extract	1.0 g
Sodium chloride	5.0 g
Phenol red	0.018 g
Dextrose	5.0 g
Water to	1 litre

Dispense in 3 ml quantities in small test tubes. Autoclave for 10 min at 121°C. Final pH should be 7.4 ± 0.1

Nitrate Broth

Beef extract	3.0 g
Peptone	5.0 g
Potassium nitrate	1.0 g
Distilled water to	1 litre

Adjust pH to 7.0 ± 0.1 and dispense 5 ml quantities into small test tubes. Autoclave 15 min at 121°C.

Modified VP Medium

Proteose peptone	7.0 g
Dextrose	5.0 g
Sodium chloride	5.0 g
Water to	1000 ml

Adjust to give a pH of 6.5 ± 0.1 after autoclaving and dispense 5 ml quantities into small tubes. Autoclave for 10 min at 121°C.

Tyrosine Agar

Prepare nutrient agar and after autoclaving, add 10 ml of water containing 0.5 g of L-tyrosine (sterilized by autoclaving at 121°C for 15 min) to each 100 ml of nutrient agar. Dispense into slopes in sterile bottles.

The tyrosine will not dissolve and must be evenly suspended throughout the agar.

Nutrient Broth with Lysozyme

Dissolve 0.1 g lysozyme hydrochloride in 100 ml water and sterilize through a 0.2 μm membrane filter. Add 1 ml of this solution to 99 ml nutrient broth. Dispense 2.5 ml volumes into small sterile tubes.

Nutrient Broth/Agar

Beef extract	3.0 g
Peptone	5.0 g
Agar (if required)	15.0 g
Water to	1000 ml

Adjust pH to give 6.8 ± 0.1 after autoclaving at 121°C for 15 min.

BC Motility Medium

Trypticase	10.0 g
Yeast extract	2.5 g
Dextrose	2.5 g
Disodium hydrogen phosphate	2.5 g
Agar	3.0 g
Water to	1000 ml

Adjust pH to give 7.4 ± 0.2 after autoclaving. Dispense into tubes and autoclave at 121°C for 10 min.

Sheep Blood Agar

Trypticase	15.0 g
Phytone peptone	5.0 g
Sodium chloride	5.0 g
Agar	15.0 g
Water to	1000 ml

Adjust pH to give 7.0 ± 0.2 after autoclaving. Autoclave at 121°C for 15 min. Cool to 48°C and add 5 ml sterile defibrinated sheep blood per 100 ml medium and dispense into Petri dishes.

BACTERIA

Contents

The Bacterial Cell

Robert W Lovitt and **Chris J Wright**, Department of Chemical and Biological Process Engineering, University of Wales Swansea, UK

Introduction

The kingdom of bacteria is an extremely diverse group of microorganisms and can be found in any environment where liquid water is present. There are over 5000 recognized species of bacteria, distinguished by structural and biochemical characteristics. However, they share a basic cellular organization. This article describes the basic components of the cell and their function. Much of the study of bacteria is restricted to a relatively few well-worked organisms; these include the Gram-negative enteric bacteria as exemplified by *Escherichia coli* and *Salmonella*, the pseudomonads. The study of Gram-positive bacteria is dominated by *Bacillus*, *Clostridium* and the lactic acid bacteria including *Streptococcus* and *Staphylococcus*.

The cell is the basic unit of a living system and as such the understanding of its structure is of prime importance to their growth and survival in the environment. The structure allows them to compete for food, survive hostile environments and occupy specific niches within the environment. By definition the distinction between prokaryotes and eukaryotes is best seen at the level of cellular organization.

The basic composition of a typical bacterial cell is shown in **Table 1** and illustrates the average composition of *E. coli*. Table 1 shows that over 95% of the mass is made up of macromolecules. It also shows the estimated number of molecules of specific components. The composition of bacterial cells is never constant; it is highly dynamic responding to changes in the environment. The types of molecules produced and the proportions of these components are very much dependent on interaction with the environmental conditions in which the organism is growing and the control systems programmed by the genetic material within the cell. The composition of the cell is constantly changing as the cellular material is turned over.

Types of Morphology

Of all living cells the bacteria are the smallest and most rapidly growing. There are a number of clearly discernible morphological types and these are shown in **Figure 1**. Most of the common bacteria are simple cocci or rods (bacillus) or spirals or curved forms. However, more complex forms exist. Cocci may occur in pairs, tetrads and sarcina forms or as chains or grape-like forms. The bacillus form may easily be mistaken for a coccus when the rods are very short, e.g. coccobacillus. They also can occur as long spindly

Table 1 Molecular composition of a typical bacterial cell

Component	Mass ($\times 10^{-13}$g)	Percentage of total mass	Molecular weight	Molecules per cell
Entire cell	15	100		
Water	12	80	18	4×10^{10}
Dry weight	3	20		
Protein				
Ribosomal	0.22	1.5	4×10^4	3.3×10^5
Non-ribosomal	1.5	10	5×10^4	1.8×10^6
RNA				
Ribosomal 16S	0.15	1	6×10^5	1.5×10^4
Ribosomal 23S	0.30	2	1.2×10^6	1.5×10^4
tRNA	0.15	1	2.5×10^4	3.5×10^5
mRNA	0.15	1	10^6	9×10^5
DNA	0.15	1	4.5×10^9	2
Polysaccharides	0.15	1	1.8×10^2	5×10^7
Lipids	0.15	1	10^3	9×10^6
Small molecules	0.08	0.5	4×10^2	1.2×10^7

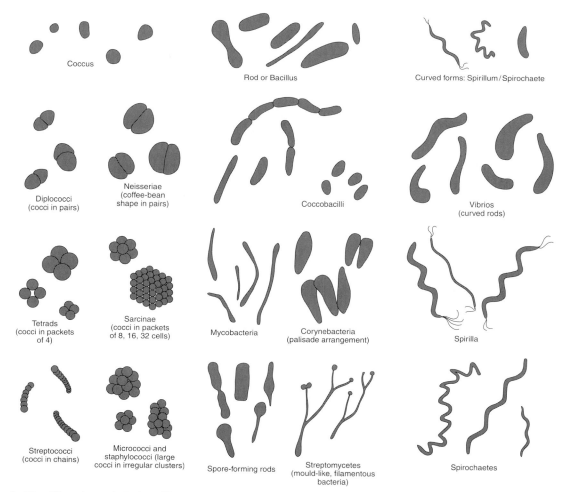

Figure 1 The different morphologies of bacterial cells.

Table 2   Size and composition of various parts of bacterial cells

Part	Size range	Comments
Slime layer	5–500 nm	
Microcapsule		Complex hydrated materials that vary in composition, mainly carbohydrate
Capsule		but can contain significant amounts of protein. Often responsible for the
Slime		main antigenic properties of the cell.
Cell wall	10–20 nm	20% cell dry weight
Gram-positive		Mainly a mixed polymer of muramic acid, peptides techoic acid and
		polysaccharides
Gram-negative		Have a multilayered wall structure consisting of an asymmetric outer
		membrane that is semi-permeable. There is a thin muramic acid layer
		and the space between the cell membrane and the outer membrane, the
		periplasm consists mainly of proteins in solution
Outer membrane		
Periplasm		
Cell membrane	10–20 nm	5–10% cell dry weight, 50% protein, 30% lipid 20% carbohydrate
		A lipid bilayer; the main semi-permeable barrier of the cell; the membrane
		also contains linked electron transport systems which are coupled to
		energy generation and selective transport processes for ions and organic
		materials
Flagellum	0.1–10 000 nm	This largely protein structure arises from the membrane and is responsible
		for motility. Rotation of the flagellum is coupled to proton flux across the
		cell membrane
Pili	0.2–2000 nm	Protein structures protruding from the envelope. They function to attach
		cells to surfaces
Inclusions		
Spores	0.5–2 μm	Specialized resistant cellular structures that are formed in adverse
		conditions
Storage granules	0.05–2 μm	Consist of polysaccharide, lipid, polyhydroxybutyrate and sulphur
Chromatophores	50–100 nm	Specialized cell membrane invaginations that contain photosynthetic
		apparatus
Ribosomes	10–30 nm	Organelles for protein synthesis; consist of RNA and protein and make up
		about 20% of the dry wt of cells. Their concentration is a function of
		growth rate
Nuclear material		Poorly aggregated materials but can occupy up to 50% of the cell volume.
		Consists of DNA duplex and can make up to 3% of the cell dry weight
Cytoplasm		Made up of proteins, mostly in the form of enzymes

or fat distorted forms such mycobacteria and the corynebacteria that often takes the form of Chinese characters. Finally, rods can differentiate with the formation of spores or may assume mould-like filamentous structures as with the Streptomycetes. The curved spiral forms can be found in the form of vibrios or curved bacteria. Spirilla can also be found and they may be flagellated. Long spiral forms are exemplified by the Spirochaetes.

Table 2 shows the size of cellular forms and other structures found within the cells. These structures are described in more detail below.

Environmental Influences on Morphology

Although a great diversity of morphology can be found in bacteria, the types of morphological forms observed depends very much on the environmental conditions in which the cells are grown. The growth rate or physiological state and the physical environment can influence the shape, colour, size and motility of the cells. Under certain conditions the cells

may differentiate, e.g. sporulation or the formation of aerial hyphae.

Organization of the Prokaryotic Cell

The organization of the prokaryotic cell bears little relation to the eukaryotic cell. Prokaryotic cells contain no organelles bound by membranes. The genetic material is never organized into complex structures such as chromosomes. They contain no endoplasmic reticulum, Golgi apparatus, mitochondria, chloroplasts, microtubules, or a membrane-bound nucleus. The bacterial cell consists of important macromolecular structures as shown in **Figure 2**. The envelope which encloses the cell comprises a series of complex substructures: the cell membrane, the cell wall and sometimes (in Gram-negative organisms) an outer membrane. In some organisms there is also a well-defined region between the outer membrane and cytoplasmic membrane called the periplasm. In the interior of the cell, the cytoplasm or cytosol is densely

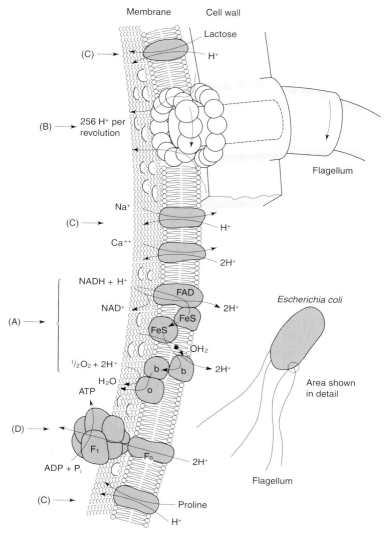

Figure 2 The structure and activities of the cytoplasmic membrane involving proton transfer.

packed with ribosomes and the nuclear region. In some organisms other discernible bodies can be found and are normally associated with storage. Developing spores can also be found. Bacteria also possess a number of important surface structures. Capsules, flagella and pili are commonly present. The following sections review the chemistry and function of these structures.

Structure and Chemistry of the Bacterial Envelope

The structures found within the cell envelope represent the solution to problems that the cell encounters in its natural environment. The strength of the cell wall infers resistance to high osmotic pressure and resistance to phage and enzymatic attack. Its selectivity combats organic poison and antibiotic action. In addition, it functions in the selective acqui-

sition of nutrients and in the survival of changing environments.

Cell Membrane All bacteria possess a cell membrane, which usually consists of phospholipids and many types of protein. Over 200 proteins have been identified in *E. coli* membrane preparations. Typically about 70% of the weight of a membrane consists of protein. Sterols are usually absent but other analogous lipid terpenoid-derived materials can be present. At or near the optimum growth temperature the membrane is in a fluid state. Individual lipids can exchange places with one another but considerable order exists within the membrane especially around proteins. The proteins in the membrane are capable of moving through, or rotating within the membrane.

The membrane functions as an osmotic barrier

modified by the presence of complex transport systems and energy generation systems. Figure 2 shows an idealized structure. Zone A in Figure 2 illustrates electron transport processes that are used to drive energy generation via ATPase (zone D) and the transport of ions and sugars (zone C). For a cell to be alive it is thought that the membrane must be energized and intact. The energization of the membrane normally means that it maintains a potential difference between the inside and the outside of the membrane. This can take the form of a pH gradient (usually slightly more alkaline on the inside of the cell) and an electrical potential. These are maintained by the pumping of protons and other ions across the membrane. This is achieved by the harnessing of the redox process that alternatively reduces and oxidizes electron carriers which straddle the membrane.

Because of the complexity of the membrane structure it is not surprising that materials that disrupt the cell membrane can have catastrophic consequences for the bacterial cell. Some of the most common food preservatives are thought to act on cell wall structure and function, for example, fatty acids (acetic acid) and parahydroxybenzoic acids, alcohols and other solvents, detergents and mineral acids and alkalis. Temperature also has a significant effect on the composition of the lipids and it has been shown that lowering the temperature will freeze the membrane and stop membrane function.

Cell Wall One of the main features of bacterial cells is their extreme toughness and their resistance to mechanical stress. Much of their remarkable strength can be attributed to the cell wall. The cell wall not only prevents the cell from bursting but also protects the delicate cell membrane from chemicals that could cause its disruption. The organization of the cell envelope can be considerably different between bacteria. Indeed one of the fundamental distinctions between different types of bacteria is made on the basis of the wall structure. The Gram stain functions to distinguish between these marked differences in structure. **Figure 3** shows a comparison of the cell structures of Gram-positive and Gram-negative bacteria and their dimensions.

Gram-positive Cell Wall The Gram-positive cell wall consists mainly of a thick layer of murein or peptidoglycan that is interspersed with techoic and techuronic acids. These layers are laid down upon one another and wrap around the cell forming a sacculus. This determines the overall shape of the cell. The precise structure of the murein layer is difficult to visualize. However, the two basic structures are represented by the chemical composition. The peptidoglycan consists of an alternating sugar backbone of N-acetylglucosamine and N-acetylmuramic acid that form very long chains. The chains are cross-linked with small bridging tetra-peptides. The precise composition of the peptide bridge is to some extent species dependent. The type and proportions of the peptide can be used to distinguish certain groups of bacteria.

The techoic acids found in Gram-positive bacteria are also species dependent. The basic structure of techoic acid comprises smaller repeating units of sugars, glycerol and amino acids that are linked via phosphodiester bonds.

Apart from the mechanical strength of the cell wall, the surface may also act as powerful ion-exchange or chelation systems for sequestering ions from the environment. The techoic and techuronic acids have been implicated in this. It has been demonstrated that under phosphate limitation the levels of techoic acid will decrease within the cell wall but techuronic acid levels increase so the capacity of the cell walls for binding magnesium is almost unchanged. The higher techuronic acid levels may aid the sequestration of magnesium and other compounds. Thus the precise composition of the cell wall is very much dependent on the environmental condition in which the organism is grown. In general, the cell is the hydrophilic, inhibiting the movement of hydrophobic materials that may seriously disrupt the environment.

The high strength of the cell wall means that bacteria are capable of tolerating hypotonic solutions. However, Gram-positive bacteria are normally susceptible to lysozyme (muramidase) which disrupts the cell wall to such an extent that the cells burst in hypotonic environments. Lysozyme is found in many body fluids but notably in teardrops where contaminating bacteria are lysed. It is possible to create wall-less cells or sphaeroplasts if treated with lysozyme in a hypertonic environment such as 0.5M sucrose.

Gram-negative Cell Walls Gram-negative organisms have a more complex cell envelope than Gram-positive organisms. They have developed a different approach to protecting the membrane. As shown in Figure 3, the Gram-negative cell wall contains relatively small quantities of murein in a thin layer. However, there is an additional layer, the outer membrane that is built upon the murein layer. The outer membrane is chemically distinct from the cell membrane in that it is chemically resistant and highly asymmetric. The bi-layer structure on the inner side is very similar to a normal cell membrane, but the outward-facing side of the outer membrane is made up of a unique material, lipopolysaccharide (LPS).

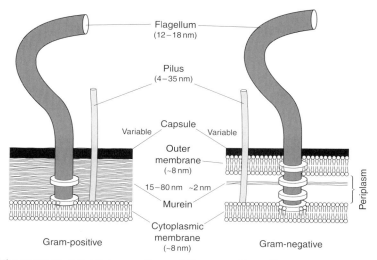

Figure 3   Comparison of the structure of the Gram-positive and Gram-negative cell envelopes.

One of its unique properties is the ability to exclude hydrophobic compounds. There are three parts to the LPS structure: (1) Lipid A, which anchors the structure to the membrane, consists of fatty acids slightly shorter than those typically found in cell membranes. (2) Core carbohydrates, connected to Lipid A, consist principally of ketodeoxytonic acid, octonoic acid and heptose. (3) Connected to the core carbohydrate is the O antigen that consists of up to 40 sugars. These hydrophilic carbohydrate chains cover the surface of the cell. The Gram-negative outer membrane therefore represents an effective barrier to both hydrophobic and hydrophilic materials. To allow materials through the envelope, several proteins or porins straddle the membrane and allow passive diffusion of low molecular weight compounds. In addition, there are specific protein molecules which translocate specific compounds.

The outer membrane structure confers a greater resistance to antibiotics than the Gram-positive system. The Gram-negative structure is highly reactive when introduced into animals. The components of the outer wall are often toxic causing fever and activating the immunological system. The outer O antigen can be used for the identification of species and variants of important food poisoning bacteria such as the salmonellae.

The outer membrane is fixed to the rest of the cell by covalent bonding via the outer membrane lipoprotein and by weaker bonds between the outer membrane and the cell wall proteins and proteins in the cell membrane.

The Periplasm   The cell membrane and the outer membrane create a compartment between them called the periplasm. Although deceptively small when seen in electron micrographs of the cell, the periplasmic space can be up to 40% of the membrane. Within the periplasm lies the cell wall and a whole range of proteins which either bind important materials or act as enzymes to hydrolyse materials into more utilizable forms. The space may also contain detoxifying enzymes, such as penicillinase. Many of the enzymes and proteins within the cell can easily be released from the periplasm by osmotic shock.

Acid-fast Cell Walls   A few significant bacteria have a further development of the cell wall structure in that they contain large quantities of waxy materials. These are complex long-chained hydrocarbons substituted with sugars and other materials. One of the most common types is the mycolic acid found in mycobacteria, which can have a carbon backbone of up to 90 carbons. This unique wall structure forms the basis of the acid-fast tests. When a dye is introduced into the cells, for example by heating, it cannot be removed by dilute hydrochloric acid as with most other bacteria, and so these bacteria are said to be acid fast. The wall structure is typically Gram-positive containing murein, polysaccharides and lipids in addition to the waxy materials. The waxy coat makes them resistant to many poisonous chemicals and to white blood cells. Another important consequence is that they are very slow growing with a doubling time well over 24 h.

Crystalline Surface Layers   Crystalline surface layers have also been found in some bacteria and represent another way to organize the cell envelope. The surface consists of a protein layer in the form of a crystal and is sometimes referred to as the S-layer and is located on the outermost layers of cells. It represents an additional layer to either the Gram-positive or -negative cell wall architecture and can occur in layers several molecules deep. An S-layer is made up of a single kind of protein which sometimes has carbohydrates attached. The function of the S-layer is not clearly understood but it does afford protection against phagocytosis. It may also serve to protect against phages and may aid the bacteria in adhesion to surfaces.

Other Cell Surface Structures

Capsules Many bacterial cells, both Gram-positive and Gram-negative, secrete a hydrophilic slime layer usually constructed from high-molecular-weight polysaccharides. This layer is termed the capsule (Fig. 3) and can extend a distance many times that of the cell diameter. The polysaccharides may be either heteropolymeric or homopolymeric, for example dextrin (poly-glucose) in the capsule of *Leuconostoc mesenteroides*. The formation of the capsule depends on the cell's environment and its secretion is not essential. Capsule-forming bacteria will grow under laboratory conditions without forming a slime layer. The capsule, however, functions to aid cell survival in a variety of environments. It protects the cell from physical and chemical attacks such as those found when food surfaces or preparation equipment are cleaned. The 'stickiness' of the capsule promotes cell adhesion to surfaces, a survival advantage. In addition the capsule protects the cell from phagocytosis. The 'slipperiness' of the capsule hinders the uptake of the bacteria by phagocytic cells. Many pathogenic bacteria are able to travel unchallenged through the bloodstream to the target organ. Well-known capsule-forming bacteria include *Streptococcus pneumoniae*, *Haemophilus influenzae* and species of meningococci.

Flagella Some bacteria are motile by means of flagella rotation. A flagellum is a helical filament that is rotated by a 'motor' located at its base in the bacterial cell envelope. The filament imparts movement by rotation, not by bending which is the case for eukaryotic flagella. Bacteria can be differentiated by the different arrangements of their flagella. Some species have a single polar flagellum, for example some *Pseudomonas* species; others have multiple polar flagella. When flagella are located all over the bacterial cell envelope this is termed peritrichous flagella. *E. coli* has approximately 10 peritrichous flagella.

There are three component parts to a flagellum, the extending filament, the hook and the basal body (Figs 2 and 3). The hook attaches the filament to the basal body that acts as the motor that rotates the flagellum. The filament can be up to 10 µm in length. Each filament consists of several thousand units of the protein flagellin, an extremely rigid protein. Single molecules of flagellin aggregate spontaneously to produce the characteristic structure of the flagellum

filament. The filament is formed by constant distal growth. The hook is a short curved structure that acts as a universal joint holding the filament in the basal body. The hook is wider than the filament and has a constant length. Like the filament it is an aggregation of a single protein called hook protein. The basal body consists of at least 15 proteins that form a rod structure with four rings. These four rings anchor the flagellum, yet allow it to rotate. It is unknown how the basal body is held in the cell envelope, however each ring of the basal body is seen to correspond to the layers of the Gram-negative cell boundary (Fig. 3).

The precise mechanism that rotates the basal body is unknown, however it is known to be linked to membrane potential. The flagellum rotation mechanism is efficient, requiring only the transport of about 1000 protons per turn. The flagellum motor exhibits chemotaxis responding to attractive or repulsive chemical stimuli. The concentration of the chemical dictates which direction the flagellum rotates, and thus which direction the bacterium swims, and also for how long.

Pili Pili or fimbriae are protein structures that extend from the bacterial cell envelope for a distance up to 2 µm (Fig. 3). They function to attach the cells to surfaces. *E. coli* cells can have up to 300 of these organelles. They are constructed from structural proteins, called pilins, arranged in a helix to form a straight cylinder. Some pili contain more than one type of pili. Pili can have other proteins located at their tip responsible for attachment specificity. When these proteins promote the adhesion of bacteria in a host–pathogen relationship they are termed adhesins. For example the adhesins of pili found on the cell envelope of *Neisseria gonorrhoeae* are responsible for the binding to specific receptors found on the urinary or genital tracts, in this case believed to be glycoproteins. Pili termed type 1 or common type are involved with the attachment of cells to substrates such as eukaryotic cells. Sex pili, as their name suggests, are involved in the conjugation of bacterial cells, promoting the initial joining of mating pairs.

Cellular Contents and Inclusions

The Cytosol The bacterial cell envelope contains the cytosol which is essentially a highly concentrated solution housing regions of biosynthesis, energy production and genetic information. Its major component is protein but it also contains all the compounds involved in the cell's metabolic functioning. There are no internal lipid membrane barriers thus all the metabolite products pass very quickly to sites of macromolecule formation. This organizational

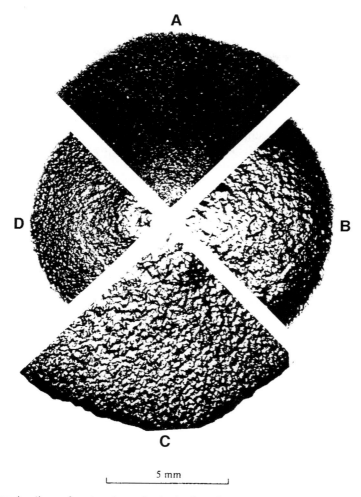

Figure 4 Photographs showing the surface topology of colonies from four strains of *Bacillus cereus* grown on tryptone–soy agar at 30°C. Strain numbers: (**A**) NCTC 9680; (**B**) NCTC 9947; (**C**) C19; (**D**) D11.

feature is cited in the high metabolism and growth rates of bacterial populations. The extent of cytosol organization is debated, however it is clear that the self-assembly of protein aggregates infer order to this region. Cytosol order can be seen in the formation of organelles that are specific functional regions of the cytosol. These can be enclosed by protein to form a diffusion barrier holding for example gas, as in the gas vesicles of aquatic bacteria conferring buoyancy to the cell. Chlorobium vesicles are protein-bound organelles that enclose photosynthetic pigments in the cytosol of photosynthetic green bacteria. Important regions that can be differentiated within the cytosol and are essential to bacterial survival are now discussed.

Polysomes The cytosol can be packed with ribosomes, organelles responsible for the translation of messenger RNA (mRNA). A polysome is the name of the structure formed when several ribosomes trans-

verse the same mRNA molecule. In an actively growing bacterium up to 90% of ribosomes are bound to a polysome structure. It is thought that in bacteria there is no specialization of ribosomes to synthesize specific proteins. The integrated cytosol of the bacterium allows transcription, mRNA synthesis and translation to occur simultaneously. In eukaryotes these processes take place in the nucleus and endoplasmic reticulum. The prokaryotic system allows faster adjustment of gene expression enhancing survival in changing environments. The ribosome organelles are complexes of RNA and protein. The ribosomes of *E. coli* consist of 62% RNA and 38% protein. Recent studies have shown that ribosomal structure has been conserved throughout the prokaryotes. Bacterial ribosomes are complex highly asymmetric structures constructed from two subunits, 70S and 30S, designated according to their different centrifugal separation.

Nucleoid In bacteria, an amorphous region is seen to be distinct from the cytosol; this is the location of the cell's DNA. The region, termed the nucleoid, is an undulating irregular shape. Bacteria can have several nucleoids depending on their growth rate. Real-time images of actively growing bacteria have shown that nucleoids simply 'pull apart' and divide without the complexities seen in eukaryotic cell division. The genetic material of bacteria consists of a single covalently linked ring-shaped molecule. Its length (10^6 nm) is many times that of the bacterium. To be housed in the nucleoid it is consequently very thin (3 nm). The dense packaging of the DNA molecules is achieved by their supercoiling, which is thought to be induced by the ionic environment in the nucleoid region.

Storage Granules Other regions can be differentiated from the cytosol and these are generally responsible for storage. Bacteria contain storage granules that function to supply compounds when they are limiting in the environment. For example, E. coli has glycogen granules, about 50 nm in diameter which accumulate when carbon is in excess and compounds containing nitrogen are growth limiting. These storage granules disappear when external carbon becomes limiting and the glycogen is used as a carbon source. Many bacteria store carbon in the form of glycogen but other carbon-rich compounds can be used, such as poly-β-hydroxyalkane that is accumulated by pseudomonads. Other elements are stored by prokaryotes in granules. Certain bacteria are able to store phosphate and sulphur as polyphosphate and elemental sulphur, respectively. Some inclusion bodies are formed within the cytosol to perform highly specialized functions. For example some bacteria form iron deposits enabling them to respond to magnetic fields.

Endospores A few bacterial species are able to form endospores within the vegetative cell. *Bacillus* and *Clostridium* species are important spore formers that pose extreme problems to public health and the food industry. Spores are structures that can survive extremes of chemical and physical attack in harsh environments. They can remain viable for centuries to germinate in a favourable environment. Cleaning regimes within the food industry and food preservation methods must kill spores or risk contamination by spore-formers such as *Clostridium botulinum*. When nutrients become limiting the bacterial population begins to form endospores. The nucleoid divides and the cell splits unequally to produce a small forespore containing a copy of the cell's DNA and a mother cell. The forespore is then engulfed by the mother cell, which further refines

the forespore. Finally, the cortex and spore coat are thickened, prior to lysis and release of the endospore. The formation of endospores requires a high degree of cellular chemical and morphological differentiation. The endospore is substantially different from the mother cell. It is smaller with lower water content, a thicker wall and a much higher amount of calcium dipicolinate. The function of the latter is unknown.

Structural Changes During Cell Division

Typically, bacteria divide by binary fission into two equal daughter cells. Others have more complex patterns that may involve more unequal division. Indeed the basis of the many different cell morphologies is thought to be due to division occurring only in one plane to yield sheets. Division along three perpendicular planes results cubical packets. Even more intricate patterns can be seen in developmental cycles, for example actinomycetes.

Cell division requires that the cell partitions the cytosol contents by invagination of the cell envelope; only then can the daughter cells separate. Cell division therefore proceeds by a series of steps. E. coli and other Gram-negative rods divide by making a ring-shaped furrow around the mid-cell region. This invagination curves inwards until two hemispherical caps are formed. A further unfolding of the membrane and murein layers occurs to form a septum. The cells, although now separate, may stay linked together for some time again giving rise to a characteristic morphology.

In some Gram-positive bacteria cell division occurs without constriction at the girth. In these cases a septum forms tangentially to the surface and grows inwards. Whatever the system of division the cell surface components must double in thickness at the septum site. The cells may be held together by incompletely separated cell walls or membranes or by extra-cellular polysaccharide.

Bacterial Colony Formation and Characteristics

One of the main methods of distinguishing one bacterium from another is the type of colony that it forms on agar. The bacterial colony represents growth of bacteria under heterogeneous conditions that are encountered on an agar plate. When cells are inoculated onto an agar plate there is rapid growth, however this soon becomes limited by diffusion of nutrients in agar gel or from the gas phase above the colony. Many bacterial colonial forms can be recognized by characteristic colour and spatial patterns that imply that the morphogenic structure can

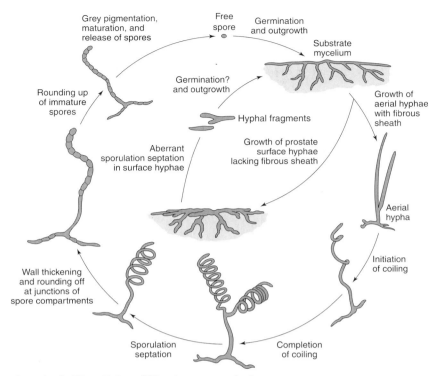

Figure 5   The life cycle and cell differentiation of *Streptomyces coelicolor*.

be generated from a single cell. **Figure 4** shows the surface topology of four strains of *Bacillus cereus*. It is quite clear that the colonies formed have different textures and sizes. The basis of these differences is the subtle response of cells to the environment that they are in. Detailed analysis of colonies has shown that depending on their position the cells will have different physiological and structural characteristics. For example individual microbe cells may be long on the edge of the colonies whereas they are short on the interior. The enzyme composition of cells will also vary with their position in the colony.

Thus the architecture of a developing colony is complex and depends not only on intrinsic factors, including shape of individual organisms, size range, method of reproduction, the production of extracellular molecules and motility, but also on extrinsic factors, such as diffusion of gases and nutrients into the colony.

Cellular Differentiation

A further extension to cellular activity of a bacterial colony formation is the response of cells to the environment causing cells to differentiate and form new structures. The formation of resistant endospores is one good example; another is the life cycle of the *Caulobacter*. One form is a vibrio-shaped cell with a single prostheca or stalk which has a localized stick region called a holdfast; once attached to a surface it

can release a motile swarmer cell. The swarmer cell does not divide, but with time develops a stalk and settles on the surface.

The actinomycetes are prokaryotes with a growth habit that in some respects resembles that of the fungi. **Figure 5** shows the life cycle of *Streptomyces ceolicolor*. Starting from the germinating spore, a mycelium is formed that grows over and into the agar surface. In some cases aerial hyphae develop which initially coil and then begin to form spores that on maturation are released to restart the cycle.

Conclusions

The structure and function of bacterial cells are intimately related. The efficiency of this relationship has meant that bacterial species exist in a vast range of environments, overcoming most of the problems that are present in an environmental niche. Sometimes their survival is useful but often it is to the annoyance of the food microbiologist and frequently compromises public health.

See also: **Bacteria**: Bacterial Endospores; Classification of the Bacteria – Traditional; Classification of the Bacteria – Phylogenetic Approach.

Further Reading

Dawes IW and Sutherland IW (1992) *Microbial Physiology*, 2nd edn. Oxford: Blackwell Scientific.

Neidhardt F, Ingraham JL, Low B, Magasanik B, Schaechter M and Umbarger HE (eds) (1996) *Escherichia coli* and *Salmonella*: Cellular and Molecular Biology, 2nd edition. ISBN 1-55581-084-5. Washington: ASM Press.

Neidhardt, F, Ingraham JL and Schaechter M (1990) *Physiology of the Bacterial Cell: A Molecular Approach*. Sunderland: Sinauer.

Truper HG (ed.) (1991) The Prokaryotes. A Handbook on the biology of bacteria: ecophysiology, isolation, identification and applications. Second edition. ISBN 0-387-97258-7. Berlin: Springer-Verlag.

Bacterial Endospores

Grahame W Gould, Bedford, UK

Thirteen genera of bacteria form endospores. Endospores are spores that are formed *within* their 'mother cells' in contrast to the cysts and exospores of other microorganisms. The spore-forming genera that are best known and most important in food microbiology are the aerobic and facultatively anaerobic *Bacillus* and the strictly anaerobic *Clostridium*. Their importance derives firstly from the extreme resistance of their endospores to heat and to other stresses such as ionizing and non-ionizing radiation, biocidal chemicals and enzymes. They are also more resistant than vegetative cells to new and emerging preservation techniques including the application of high hydrostatic pressure and high-voltage electric discharges. Secondly, some spore-formers cause food poisoning, which may be mild (*Bacillus subtilis* and *Bacillus licheniformis*), severe (*Bacillus cereus*, *Clostridium perfringens*) or life-threatening (*Clostridium botulinum*).

Spore Formation

Endospores are formed generally in response to environmental conditions that are unfavourable for the continued growth of vegetative cells. The most common stimulus in nature is probably the exhaustion of nutrients. The sequence of biochemical and morphological changes that occur during spore formation and the genetics of their control have been extensively studied. The major morphological changes are summarized in **Figure 1**. The earliest event after the cessation of vegetative cell growth and division is *asymmetric septation*: a membrane forms towards the end of the mother cell to create two compartments of unequal size. The next stage is *engulfment*, during which the larger compartment grows around the smaller one. This small compartment will later become the central protoplast or 'core' of the spore, but at this stage it is separated from the environment by three membranes, the middle one of which is topologically and functionally back-to-front. It is thought that this arrangement, which is unique in prokaryotic cells, leads to interference with the normal transport of ions and low-molecular-weight nutrients, etc., so that the osmolality in the enclosed 'forespore' compartment falls, leading to loss of water from it by simple osmosis. At the same time, the pH within the forespore compartment reduces by about 1. As these changes are occurring, metabolic activity falls, i.e. dormancy is imposed. Next, calcium and dipicolinic acid (**Fig. 2**) accumulate and are deposited in the forespore protoplast at levels up to about 10% of the dry weight of some spores, accompanying further dehydration; this leads to an increase in refractive index which causes spores to appear bright when viewed microscopically using positive phase-contrast optics. At the same time as these changes are occurring, peptidoglycan is laid down between the two opposing forespore membranes to form the growing cortex. Finally, the mainly proteinaceous coat layers are deposited outside the back-to-front outer membrane of the forespore. The mother cell then usually lyses and releases the fully formed spore. The heat resistance of the developing spore increases most dramatically during the later stages of dehydration of the protoplast and the synthesis of the cortex around it, during stages V to VI (**Fig. 3**).

The structure of the mature spore is therefore characterized by a type of compartmentalization which is unusual in prokaryotic cells (**Fig. 4**). The central protoplast contains a complete genome, ribosomes, cytoplasmic enzymes, etc., but dormant and stabilized in an environment relatively low in water and with high levels of calcium dipicolnate (CaDPA), and other components such as spore-specific small acid-soluble proteins (SASPs) that bind to and stabilize spore DNA (see below). The protoplast is surrounded by a membrane outside of which is the cortex. The cortex is made up of a thin layer of mother cell-type peptidoglycan adjacent to the forespore membrane and a thicker layer of spore-specific peptidoglycan around it. The peptidoglycan in spores is loosely cross-linked and lacks the amino acid cross-links between adjacent peptide chains which are common in some vegetative cell peptidoglycans (**Fig. 5**). Outside this is the

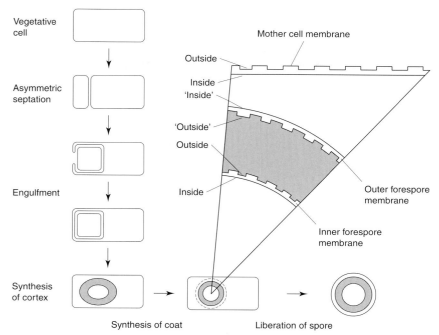

Figure 1 Major morphological changes occurring during the formation of bacterial endospores, especially regarding the disposition of the membranes and the formation of the cortex. The vegetative cell wall is omitted for clarity.

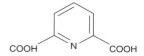

Figure 2 Structure of dipicolinic acid (pyridine-2,6-dicarboxylic acid).

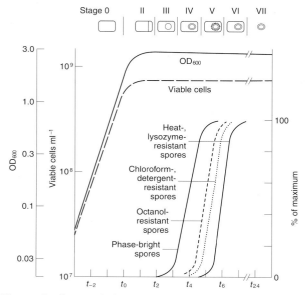

Figure 3 Stages during the sporulation of *Bacillus subtilis* related to the onset of resistance to heat and other agents. OD = optical density; *t* values indicate hours, at 37°C, following the end of vegetative growth (stage 7). (From Freeze and Heinze, pp. 102–172 in Hurst and Gould, 1983.)

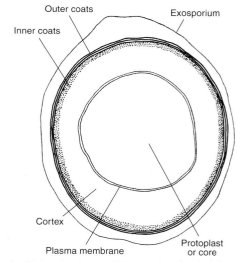

Figure 4 Major structures within a mature bacterial endospore.

reversed-polarity membrane, or at least its remnants, and then the coats.

Dormancy and Longevity of Endospores

The degree of dormancy of spores is so deep that if unheated (unactivated) they have virtually undetectable metabolism. Their levels of ATP, NADH and NADPH are less than one-thousandth of the levels in vegetative cells. The spore protoplast contains many enzymes and the substrates on which they act, but in the spore no such action occurs. However, while most

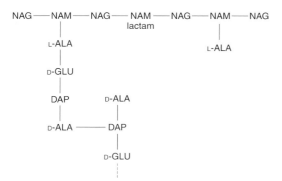

Figure 5 Generalized structure of spore-type peptidoglycans. NAG, *N*-acetyl glucosamine; NAM, *N*-acetyl muramic acid; L-ALA, L-alanine; D-GLU, D-glutamic acid; DAP, diaminopimelic acid; D-ALA, D-alanine.

Table 1 Heat resistances of bacterial endospores (*Bacillus* spp.) relative to the resistances of their parent vegetative cells. Adapted from the data of Warth (1978) *Journal of Applied Bacteriology* 134: 699–705

Species	Heat resistance (°C)[a]		Additional resistance of spores (°C)
	Vegetative cell	*Spore*	
B. sphaericus	47	88	+ 41
B. megaterium	47	89	+ 42
B. caldolyticus	72	115	+ 43
B. cereus type-T	48	92	+ 44
B. licheniformis	54	99	+ 45
B. stearothermophilus	71	118	+ 47
B. brevis	55	106	+ 51
B. subtilis	57	111	+ 54

[a]Heat resistance defined as the temperature at which the decimal reduction time (D) of a heated suspension is 10 min.

of the spore's enzymes are dormant, some of them remain outside the general dormancy and resistance mechanisms. For instance, enzymes involved in the initial stages of germination (see below), such as the receptors that interact first with specific germinants, must remain reactive to these compounds, and the enzymes that catalyse the subsequent reactions leading to loss of general dormancy remain uninhibited in the otherwise dormant cell.

The longevity of endospores is impressive, and estimates continue to grow. Spores of *Bacillus anthracis* grown by Pasteur were found to be viable after more than 60 years. Spores of thermophiles have been isolated from some of the earliest canned foods after well over 100 years of storage. Dried soil on the roots of herbarium plants yielded viable spores after 320 years. *Thermoactinomyces* spores were isolated from stratified lake sediments deposited over 7000 years ago. Some reports of spore isolation from pre-Cambrian rock and coal seams are generally suspected to have suffered from contamination during the isolation techniques. However, recent claims of the isolation of *Bacillus sphaericus* from the abdominal cavities of bees trapped in Dominican amber that was 25–40 million years old were supported by rigorous techniques and persuasive genetic analysis and comparison with extant organisms. Thus, dormancy and survival of endospores in the wet state for thousands of years and in the dry state possibly for millions of years must be seriously considered.

Endospore Heat Resistance

While the heat resistances of endospores of different types of bacteria vary widely, the most resistant have enormous tolerances. Some strains of the anaerobe *Desulfotomaculum nigrificans* have wet heat D values (the time for a tenfold reduction in numbers) 121°C as high as 55 min. The spores that are most resistant to wet heat are those of the anaerobic thermophile

Clostridium thermosaccharolyticum, with reported D values at 121°C of nearly 5 h. On the other hand, spores of some psychrotrophic strains such as C. botulinum type E have D values as low as a few minutes at 80°C. Resistance to dry heat is much greater than to wet heat. Values for D as high as 3.5 min at 160°C have been reported. For this reason, sterilization systems employing superheated (dry) steam rather than saturated (wet) steam require temperatures about 50°C higher in order to have similar efficacy.

At least four distinct elements contribute to the heat resistance of a particular type of spore. The first is the intrinsic heat resistance of the organism and its components, as they exist in the unprotected vegetative cell. For instance, vegetative cells of thermophiles such as C. thermosaccharolyticum are far more heat-tolerant than those of mesophiles such as B. cereus, and these are more tolerant than psychrotrophs such as C. botulinum type E. The second component is the additional resistance imposed on its constituents by the mechanisms operating in the spore. The additional resistance commonly adds about 40–55°C, so that spores are normally about 40–55°C more heat-resistant than the vegetative cells from which they were formed. When a number of types of spores are considered, their heat resistances therefore have a general relationship with the heat resistances of the corresponding vegetative cells (**Table 1**). However, the correlation is not exact. Some spores add on considerably more resistance to the basic vegetative resistance than others, and the additional resistance is not correlated with the absolute resistance of the vegetative cells or spores of a particular strain. The third component depends on the ERH, a_w (water activated) and osmolality of the environment at the time of heating. Generally, equilibrium at low ERH (equilibrium relative humidity), reduction of a_w or the addition of solutes raises heat resistance

Table 2 Increase in the heat resistance of bacterial endospores at low water activities (a_w)

Procedure used to reduce a_w	Microorganism	Heat resistance[a] of spores at a_w			
		0.9	0.7	0.5	0.3
Equilibration in the absence of solutes	B. stearothermophilus	3.2	73	180	235
	B. megaterium	5.3	67	2700	1340
	C. botulinum type E	0.4	10^3	8×10^3	10^5
Addition of glycerol	B. subtilis	1.3	5.5	19	71
	B. stearothermophilus	1.2	8.4	47	108

[a]Heat resistance is expressed as the increase relative to that at an a_w of approximately 1.0.

and, while the magnitude of this effect varies greatly for different spore types and means of a_w reduction, the increases can be very large – up to 10^5-fold or so for spores of C. botulinum type E equilibrated at an ERH of 30% (**Table 2**). Fourthly, spores produced by sporulation at higher temperatures generally have higher heat resistances than those of the same strain sporulated at lower temperatures.

A low water content (and, it has recently been suggested, possibly a glassy state in the spore protoplast) is widely agreed to be the major factor in protecting components of the protoplast from denaturation by heat, and may help to protect spore DNA from oxidative damage. However, the main mechanism for protection of DNA in spores derives from its saturation with small spore-specific acid-soluble proteins. These are synthesized early in the sporulation process, then remain bound to the chromosome of the dormant spore with a stoichiometry of about 1 SASP molecule per 5 base pairs. They are degraded within the first few minutes of germination (see below). The contribution of SASPs to heat resistance has been best illustrated by studies of SASP-deficient mutants. While these remain far more resistant than vegetative cells, they are significantly more wet heat-sensitive than spores of the wild-type. Under dry heating conditions, SASP-deficient spores are much more heat-sensitive, and much more mutable, than wild-type spores, most probably owing to the high levels of depurination that occur in the absence of SASPs.

Heat resistance of spores is usually estimated by counting the numbers of survivors following heat treatment, i.e. the numbers able to germinate, grow and produce a visible colony in a bacteriological medium. There is a major potential pitfall here, because when some types of spore are heated they are unable to form a colony (i.e. they are scored as 'dead') not because their basic heat resistance mechanism has been overcome, but because they have an unusually heat-sensitive germination mechanism. Although scored as dead, such spores therefore contain a fully

viable protoplast that is unable to escape. The best-known examples of this phenomenon are some spores of C. perfringens and psychrotrophic strains of C. botulinum. With these clostridia, the apparently heat-inactivated spores can be shown to be alive because they can be recovered by bypassing the inactivated germination system; this can be done by the addition of lysozyme, which passes through the leaky spore coat to hydrolyse its peptidoglycan substrate in the cortex beneath. Most types of spores do not have coats that are naturally permeable to lysozyme, but treatment with reagents that rupture disulphide bonds in coat proteins will make them sufficiently leaky for lysozyme to pass through and cause germination.

Spore Germination and Outgrowth

Dormancy, longevity and resistance are all essential attributes of spores that underpin their role as effective survival structures. However, at the same time as being metabolically inactive while the environment is adverse, the spore must also remain able to sense the environment so as to be capable of rapidly germinating with the return of suitable conditions, such as will occur in many foods. Germination is characterized by a series of degradative events triggered by specific germinants which lead to loss of typical spore properties. It can be triggered by a range of different 'nutrient germinants' that are highly specific for different types of spores (e.g. amino acids such as L-alanine, ribosides such as inosine and adenosine, sugars such as glucose and fructose, cations such as K^+, alone or in synergistic combinations). It can also be triggered by certain non-nutrient chemicals such as long-chain alkyl monoamines (e.g. dodecylamine), by the spore-specific chelate calcium dipicolinate, by some peptidoglycan-degrading enzymes (e.g. lysozyme, spore-derived lysins) if the coat is first made permeable, and by some physical processes (e.g. the application of high hydrostatic pressure, physical abrasion).

Germination of spores of some, but not all types, is more rapid and complete if the spores are first 'activated'. The most common way to activate spores is by sublethal heat treatment. Activation leads to rising colony counts at the commencement of heat-inactivation curves of some types of spores. Germination of some types of spores following activation can be very rapid, i.e. with 90% or so of a population germinating within 1–2 min following the addition of the correct germinants. However, in many cases germination is much slower and also usually incomplete, so that attempts to control spores in foods by the addition of germinants prior to brief incubation then mild heating to inactivate the germinated forms (akin to the old technique of 'Tyndallization'), have met with failure.

Although the complete sequence of events that occur during germination have not been elucidated, the essential elements have been worked out from genetic analysis, mainly in *B. subtilis* and biochemical analysis, mainly in *B. megaterium*. Following the initial interaction of a nutrient germinant with its receptor, a number of changes in spore germination enzymes occur. Binding of the germinant causes a conformational change in a receptor, exposing a proteolytic site on it. The active protease then cleaves the pro-form of a peptidoglycan lytic enzyme – the germination-specific lytic enzyme (GLSE). The pro-form in the dormant spore is immobilized by being bound to peptidoglycan and is inactive, but the cleaved-off portion is mobile and active and begins to hydrolyse peptidoglycan in the cortex of the spore. As this occurs, the cortex peptidoglycan becomes depolymerized and water moves into the protoplast, which swells greatly. Some of the soluble components of the protoplast cytoplasm, including calcium, dipicolinic acid, potassium and other cations are excreted. Later, some amino acids generated by hydrolysis of SASPs are lost from the spore and appear in the surrounding medium, and fragments of hydrolysed peptidoglycan are excreted. Some of these components remain lost to the spore. Others (e.g. K^+ and amino acids) may be reabsorbed as membrane transport functions recommence and the germinated spore begins to metabolize.

Early in this sequence, heat resistance and the other resistance properties that are characteristic of the ungerminated spore are rapidly lost. The hydration of the protoplast and the loss of dry matter from the spore which occurs at the same time result in a large fall in refractive index. This causes the phase-bright spores to become phase-dark, so that germination can be easily monitored microscopically. Alternatively, the absorbance of suspensions of spores falls during germination, sometimes by up to 60% or so, and offers another quick and easy way to monitor the process.

Having germinated, the spore then 'outgrows'. This involves partial lysis and shedding of the remnants of the spore coat as the new vegetative cell grows out. Most food preservatives that inhibit spores do so at this stage, i.e. after germination has occurred, rather than inhibiting germination itself. For instance, the bacteriocin nisin, which is used to prevent growth from spores in some cheeses and to prevent the growth of thermophiles from spores in some canned foods, stops growth immediately the spore has completed the rapid degradative reactions of germination. Lysozyme, which is used to prevent the spoilage of some cheeses by *Clostridium tyrobutyricum*, lyses vegetative cells of this organism once the spore coat has been shed at the commencement of outgrowth.

Conclusion

All bacterial endospores are extremely dormant and all have high, but very wide-ranging, resistances to heat and other physical and chemical agents and enzymes. Heat resistance is commonly up to 10^5 times greater than that of the corresponding vegetative cell. Spores add between 40°C and 55°C additional resistance to the intrinsic resistance of their contents. Spore dormancy and resistance mechanisms are multicomponent. They include a low water content in the central protoplast, high levels of calcium dipicolinate in the protoplast, immobilization of small molecules, and protection of DNA by small acid-soluble proteins, whose binding to DNA is promoted by dehydration. A key factor in the dormancy and resistance mechanisms of spores is the unusual compartmentalization that occurs during sporulation, resulting in a central protoplast surrounded by the peptidoglycan cortex and coats. The coats of spores are not necessary for the maintenance of resistance and dormancy, but the cortex is. It somehow maintains the low water status in the protoplast that it encloses, but by means that remain uncertain.

Although spores of some types may be made to germinate rapidly and lose their resistance properties, the deliberate use of germination as a spore control procedure in foods has not been successful because of the reluctance of populations of naturally occurring spores to germinate quickly and completely. New preservation techniques may eventually offer alternative spore control procedures. For instance, high hydrostatic pressures cause spores to germinate, though insufficiently at present to form the basis of new sterilization processes.

See also: **Bacillus**: Introduction. **Clostridium**: Introduction; *Clostridium botulinum*. **Heat Treatment of Foods**: Ultra-high Temperature (UHT) Treatments; Principles of Pasteurization. **High-pressure Treatment of Foods**.

Further Reading

Berkeley RCW and Ali N (1994) Classification and identification of endospore-dorming bacteria. *Journal of Applied Bacteriology Symposium Supplement* 76: 1S–8S.

Errington J (1993) *Bacillus subtilis* sporulation: regulation of gene expression and control. *Microbiological Reviews* 57: 1–33.

Foster SJ (1994) The role and regulation of cell wall structural dynamics during differentiation of endospore-forming bacteria. *Journal of Applied Bacteriology Symposium Supplement* 76: 25S–39S.

Hurst A and Gould GW (eds) (1983) *The Bacterial Spore*, vol. 2. London: Academic Press.

Johnstone K (1994) The trigger mechanism of spore germination: current concepts. *Journal of Applied Bacteriology Symposium Supplement* 76: 17S–24S.

Marquis RE (1998) Bacterial spores – resistance, dormancy and water status. In: Reid DJ (ed.) *The Properties of Water in Foods ISOPOW VI*. Pp. 486–504. London: Blackie.

Marquis RE, Sim J and Shin SY (1994) Molecular mechanisms of resistance to heat and oxidative damage. *Journal of Bacteriology Symposium Supplement* 76: 40S–48S.

Moir A, Kemp EH, Robinson C and Corfe BM (1994) The genetic analysis of bacterial spore germination. *Journal of Bacteriology Symposium Supplement* 76: 9S–16S.

Setlow P (1994) Mechanisms which contribute to the long-term survival of spores of *Bacillus* species. *Journal of Bacteriology Symposium Supplement* 76: 49S–60S.

Setlow P (1995) Mechanisms for the prevention of damage to DNA in spores of *Bacillus* species. *Annual Reviews of Microbiology* 49: 29–54.

Classification of the Bacteria – Traditional

Vilma Morata de Ambrosini, Carlos Horacio Gusils, Silvia Nelina Gonzalez and **Guillermo Oliver**, Centro de Referencia para Lactobacilos and Universidad Nacional de Tucumán, Argentina

What is a bacterium? Generally the dictionary defines bacterium–bacteria as very small living things, some of which cause disease. From the microbiological point of view this definition is inadequate as it does not indicate that the bacteria have principally a beneficial role in nature. The term bacterium is derived from the Greek word *bakterion* meaning a small stick. Bacterial cells are either round, rod-shaped or spiral in form. The Latinized Greek word *kokkos*, from which cocci/coccus is derived, means a berry and it is used to indicate a round shape. Bacteria with a rod-shape are known as bacilli/bacillus (Latin: *bacillus*, a stick). Bacterial cells able to grow in a twisted, helical shape are termed vibrios, spirilla or spirochaetes (Greek: *speira*, coil).

Bacteria are prokaryotic cells, which lack a clearly defined membrane-bound nucleus. Prokaryote is a term derived from two Greek words: *pro*, meaning before and *karyon*, a kernel. They have characteristic shapes and staining properties that can be examined from a film stained using the methods of Gram, Schaeffer–Fulton or Ziehl–Neelsen.

According to traditional classification the most commonly used bacterial characteristics are: morphology, nutritional requirements, cell wall chemistry, cellular inclusions and stored products, capsule chemistry, pigments, ability to use different energy sources, fermentation subproducts, gas requirements, temperature and pH tolerance, antibiotic sensitivity, pathogenicity, symbiotic relations, immunological characteristics and habitat. Some of these characteristics might be useful to identify a particular group and the number of characters to be considered should be assessed in order to obtain a correct classification for this group. Although this is a casual way of classification without being systematic, this has been very useful in certain groups of bacteria.

Gram staining, which mainly depends on cell wall composition, is a widely used distinguishing characteristic. Most bacteria can be grouped into Gram-positive or Gram-negative bacilli and cocci. Special cell structures and cell morphology (spiral, sheathed, etc.) are widely used to classify bacteria.

Organisms need energy which they obtain in various ways. Organisms that are able to use the energy of sunlight are called phototrophic. Other microorganisms termed chemotrophic are able to transform chemical energy (organic and inorganic molecules) into specific energy molecules (ATP). According to the carbon source (C) used, these energy transformation processes occur in two nutritional types of microorganisms: heterotrophic microorganisms obtain their energy from organic compounds and autotrophic microorganisms from carbon dioxide (CO_2).

Microorganisms vary in their need for, or tolerance of, oxygen. Those which lack a respiratory system are called anaerobes. There are two types of anaerobes: aerotolerant which can tolerate oxygen and grow in its presence even though they cannot use it; obligate or strict anaerobes are killed by oxygen. Microorganisms that grow when in the presence of oxygen are called aerobes, of which there are three groups: obligate aerobes, which require oxygen; facultative aerobes, which do not require oxygen but grow better under aerobic conditions; and microaerophilic aerobes which need small amounts of oxygen below atmospheric level. **Table 1** shows bacteria–oxygen relations.

Bacteria can be classified into five groups according to the temperature range over which growth is possible. Psychrophiles, grow best at a relatively low temperature (range 0–20°C). Some microbiologists refer to bacteria that have an optimum temperature of 15–20°C with a maximum of 35°C as facultative psychrophiles. Mesophiles grow best at temperatures

Table 1 Oxygen–bacteria relations

Microorganisms	Effect of oxygen
Aerobes	
Obligate	Required
Facultative	Growth better with oxygen
Microaerophilic	Required but at low levels
Anaerobes	
Aerotolerant	Growth better without oxygen
Obligate or strict	Lethal

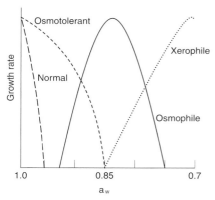

Figure 1 Growth rate of bacteria of different solute tolerances.

of 20–45°C, thermophiles grow at, or tolerate, high temperatures (25–70°C) and extreme thermophiles have very high temperature optima (about 80°C).

Each living cell has a pH range within which growth is possible. Bacteria that grow at low pH values (range 1–5) are classified as acidophiles. Some microorganisms are considered alkalinophilic because they have pH optima as high as 10–11. In most natural environments the pH is in the range 5–9 and microorganisms with optima in this range are most common and are called neutrophiles.

All organisms need water to live. In microbiology, water availability is expressed as a physical term: water activity (a_w). Numerically a_w varies between 0 and 1 but the lowest value compatible with microorganism life is 0.65.

$$a_w = \frac{\text{relative humidity}}{100}(\%)$$

Normal bacteria can survive at a_w values of 1–0.95. There exist soluble compounds that have high affinity to water, such as salts and sugars. Bacteria which can survive high solute concentrations (a_w 1–0.85) are called facultative osmophiles (osmotolerant). Microorganisms generally inhibited at both higher and lower concentrations of solutes are known as osmophiles and those able to live only in very dry environments are called xerophiles or obligate osmophiles (a_w 0.85–0.65). Organisms that can live in environments with high salt concentrations are called halophilic bacteria. The growth rates of bacteria of different solute tolerances are shown in **Figure 1**.

Taxonomy comprises three interrelated topics: classification, nomenclature and identification. Classification is the disposition of organisms into taxa on the basis of relationships and several levels or ranks are used. Nomenclature is the special assignment of names to taxa in agreement with international instructions. Identification is the process of determining that a new isolate belongs to one of the established taxonomic groups.

Classification of prokaryotes requires knowledge of their characteristics obtained by experimental and observation techniques. Biochemical and genetic

properties are necessary in addition to morphological features for an adequate description of a taxon.

Before the development of taxonomic methods based on molecular biology, traditional classification characterized bacteria as fully as possible. The arrangement was carried out in agreement with the opinion of the systematists. Several initial decisions survive to the present day, even under scrutiny from computer-assisted methodologies.

The nomenclature of the different kinds of living organisms is either by informal names or scientific names of taxonomic groups (pl. taxa, s. taxon). Informal or vernacular groups are defined by common descriptive names. Examples of such groups are: lactic acid bacteria, methane-oxidizing bacteria, etc. The scientific names, regulated by the International Code of Bacterial Nomenclature, also called the Revised Code, are latinized forms of easily recognized words and possess definite positions in the taxonomic registry. The Revised Code has three main purposes for scientific names: stability, unambiguity and necessity.

Bacteria, like some other microorganisms that are classified only on properties shown in artificial cultures, have 'type strains' as the reference specimen for the name. The type strain is by definition authentic but it may not be entirely typical. In work on both classification and identification, the type strains are very important to establish comparisons. These strains are preserved in culture collections from which they are available for microbiological studies.

The species name has two parts, i.e. it is binominal or a binomen. The first is the genus name, always written in italics and with a capital initial letter, which is a latinized substantive of any gender. The second is a latinized adjective used to describe the genus. This specific epithet, written in italics with lower case letters, is in agreement with the gender of the genus name. A trinominal name can result by adding an extra subspecific character when subspecies names are used. Above the genus level most names are plural latinized adjectives in the feminine gender. Family names end in -aceae and Order names all end in -ales.

Identification and classification objectives are not identical. An identification description can be carried out only after that group has first been classified. It is based on a pattern of properties shown by all the members of the group, which other groups do not possess. The properties used in identification are often different from those used in the classification of the group. Generally, the characteristics chosen for an identification plan should be easily determinable, whereas those used for classification may be quite difficult to determine such as DNA homology. Genera and species identification might be based not only on a few tests, but rather on the pattern given by a whole battery of tests. The members of the family Lactobacillaceae represent one example of this. To alleviate the need for inoculating large numbers of tube media, some rapid multitest systems have been devised and are commercially available (such as API or biological systems). Each manufacturer provides charts, tables, codes and characterization profiles for use with the particular system.

There is currently renewal of interest in bacterial taxonomy stimulated by the recognition of novel groups of bacteria that do not fit comfortably into current systematic schemes and by new understanding of the taxonomic utility and phylogenetic significance of molecular and genetic data.

Classical and Numerical Taxonomy

Most taxonomic work with bacteria consists of examining individual strains of bacteria. However, the entities that can be classified may be of various forms – strains, species, genera – for which no common term is available. These entities, t in number, are called operational taxonomic units (OTUs). A character is defined as any property that can vary between OTUs. The values it can assume are character states. Thus, 'length of bacillus' is a character and '1.0 μm' is one of its states. It is obviously important to compare the same character in different organisms.

For classical and numerical taxonomy, the characters should cover a range of properties: morphological, physiological, biochemical. The test reactions and character states are coded for numerical analysis into positive and negative form. The resulting table, therefore, contains entries '+' and '−' (or 1 and 0 which are more convenient for computation). In classical taxonomy all characters do not have the same weight. This is the principal difference between classical and numerical taxonomy. The question arises as to what weight should be given to each character relative to the rest. The philosophy of numerical tax-

onomy derives from the opinions of the eighteenth century botanist, Adanson, and therefore numerical taxonomies are sometimes referred to as Adansonian.

The result is a tree-like diagram or dendogram (more precisely phenogram, because it expresses phenetic relationships), in which the tightest bunches of twigs represent clusters of very similar OTUs.

Nucleic Acids in Bacterial Taxonomy

Historically, classification of bacteria has been based on similarities in phenotypic characteristics. The first unique feature of DNA that was recognized as having taxonomic importance was the mole percent of guanine plus cytosine content (mol% G+C). Among the bacteria, the mol% G+C values range from ca. 25–75 and the value is constant for a given organism. Closely related bacteria have similar mol% G+C values. However, it is important to recognize that two organisms that have similar mol% G+C values are not necessarily closely related; this is because the mol% G+C values do not take into account the linear arrangement of the nucleotides in the DNA.

Concept of a Bacterial Genus and Higher Taxa

Comparisons of RNA cistrons, by rRNA homology experiments and by 16S oligonucleotide catalogue similarities, are providing data from which a more unifying phylogenetic concept for higher bacterial taxa is possible. The establishment of several genera has been proposed on the basis of rRNA homology results. For example, the 16S rRNA oligonucleotide similarity values have contributed greatly to the present taxonomic classification of the methanogenic bacteria and to the establishment of division IV Mendosicutes in the kingdom Procaryotae.

Genetic Methods in Bacterial Taxonomy

In the last two decades it has become clear that the genetic complement of a bacterial cell lies not only in the main chromosome but, in many cases, also in extrachromosomal elements such as plasmids, transposons and lysogenic or temperate phages. All these elements carry genetic material capable of phenotypic expression. The contribution that such extrachromosomal entities make to a particular bacterial phenotype, either by direct expression or interaction with the chromosomal DNA of the cell, is only just beginning to be understood. Plasmids have been observed in virtually every bacterial genus examined. Many plasmids detected by physical screening methods are not known to code for any phenotypic trait in the host bacterium. They are called cryptic plasmids.

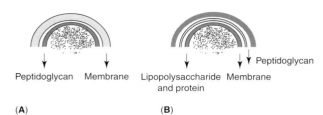

Figure 2   Schematic diagrams of cell walls of (**A**) Gram-positive and (**B**) Gram-negative bacteria.

Serology in Bacterial Taxonomy

Serological techniques depend on the ability of the chemical constituents of bacterial cells to behave as antigens eliciting production of antibodies in vertebrates. Enterobacterial common antigen (ECA) is typically found in all members of the Enterobacteriaceae. It is localized in the outer membrane and contains a linear chain of polysaccharides composed of three different amino sugars. It demonstrates a strong cross-reaction among the species studied.

Serological studies of value in bacterial taxonomy can be divided into two broad classes: (1) those concerned with detecting differences or similarities between bacteria on the basis of their cell surface and associated antigenic complement (e.g. flagella, pili, cell walls, cytoplasmic membranes, capsules and slime layers) and (2) the use of antisera raised against purified enzymes to assess structural similarities between homologous proteins from different bacteria.

Chemical Bacterial Taxonomy

Chemotaxonomy

Cell wall composition   The cell wall envelops prokaryotic cells and because of its rigid structure it gives the cell its shape and protects the plasmic membrane from rupture by differences in osmotic pressure. Eukaryotic cells which posses a membrane-bound nucleus, generally lack a cell wall. The cell wall also allows classification of eubacteria by the traditional Gram-staining technique into two large groups: Gram-positive and Gram-negative bacteria. Gram-positive bacteria retain iodine-fixed crystal violet in the presence of alcohol, whereas Gram-negative bacteria are stained with a contrast dye (safranin). There are variations in this reaction due to the stage of development, colouring time, etc. A clear distinction between the two groups is based on the chemical composition of the cell wall and its ultrastructure. Observation by electron microscopy reveals a great difference between the two groups (**Fig. 2**). Cell walls of Gram-positive eubacteria are 30–80 nm wide and are quite homogeneous, whereas cell walls of Gram-negative eubacteria are thinner and less homogeneous with an external trilayer membrane structure similar to the cytoplasmic membrane. The chemical composition of the cell walls is quite different. Peptidoglycan is the main component of cell walls of Gram-positive bacteria (never under 30% of total cell wall weight). Polysaccharides and/or teichoic or teichuronic acids are present to a lesser extent and lipids are very scarce. In Gram-negative cells, peptidoglycan constitutes about 10% of the cell wall weight and major components are lipolysaccharides, phospholipids, proteins and lipoproteins.

Acid-fast bacteria such as *Mycobacterium*, *Nocardia* and *Corynebacterium*, which cannot be included in either Gram group, contain large quantities of lipids in their cell walls, particularly mycolic acid. This characteristic is used in the Ziehl–Neelsen staining technique and because of this property the bacteria are referred to as acid–alcohol resistant.

The chemical composition of cell walls of cyanobacteria is similar to Gram-negative cell walls; cell wall width is somewhat greater and there is a larger number of cross-linkages.

Peptidoglycan, murein or mucoprotein are common constituents of Gram-positive and -negative bacteria. Murein is a polymer which contains glycan strands that are cross-linked through short peptides. The glycan strand comprises two glucosamine derivatives: N-acetylglucosamine and N-acetylmuramic acid, β-1,4-cross-linked. The peptide molecule binds to N-acetylmuramic acid and has both L- and D-amino acids. Classification according to Schleifer and Kandler (1972) is based on peptidoglycan peptide characteristics and proposes that the bacteria are assigned a letter, a number and a Greek letter.

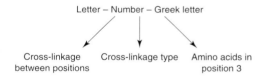

Variation within the peptidoglycan is much wider in Gram-positive than in Gram-negative eubacteria and is mainly observed in the peptides. The glycan strand presents minor variations, as for example in mycobacteria and nocardia where the N-acetyl group is oxidatively linked to N-glycolyl. The peptidoglycan peptide both in the stem peptide and the interpeptide bond show structure and composition variations among different species, but they are stable and meet the requirements for being a taxonomic marker. There are no mutations known which modify their structure in one step. Schleifer and Kandler (1972) have classified mureins as a function of their primary structure, thus defining peptidoglycan types. This has proved to be extremely valuable information from a taxonomic point of view which has revealed differences between organisms that could not be confirmed by qualitative analysis. In the particular case of some Gram-positive bacteria such as actinomycetes and coryneforms determination of cell wall composition is an essential requirement for classification and identification.

In Gram-positive eubacteria murein-linked cell wall polysaccharides appear as neutral polysaccharides and/or teichoic acids. Generally, little is known about the polysaccharides of these bacteria except for some better-studied groups like streptococci, in which serological typing has been carried out. Other polysaccharides of taxonomic interest are arabinogalactans and arabinomannans which bind to mycolic acids of acid-fast bacteria.

Teichoic acids have only been found in a limited number of Gram-positive eubacteria such as actinomycetes, bacilli, brevibacteria, certain lactobacilli, listeria and staphylococci. In the latter, composition, structure and serology have been used to classify them and to distinguish them from micrococci, which lack this polymer.

Mycolic acids are 3-hydroxy acids with long alkyl branches in position 2, which are only present in *Caseobacter*, *Corynebacterium*, *Mycobacterium*, *Nocardia* and *Rhodococcus*. They are highly valuable chemotaxonomic markers for classification and identification of these genera.

Lipopolysaccharides (LPS), present in Gram-negative bacteria, make up the chemical base for the O or surface antigen, which is the most important of the three antigens used for serological classification of Enterobacteria; the other antigens are H (flagellar) and K (capsular). Antigens are composed of three covalently linked segments: side chain, core and lipid A. O-specific side chains consist of up to 30 repeating units and each one contains 3–6 sugar residues. The structure of polysaccharide cores may be useful for classification to genus level as in the case of *Salmonella* and *Shigella*. Differentiation within the same species (infrasubspecific taxon) can be obtained by chemotyping (quantitative composition of the polysaccharide molecule) and serotyping of the O-specific chains. Within the genus *Escherichia* there are variations in the LPS core. Lipid A could be of great significance in the classification of the Rhodospirillaceae.

Archaebacteria exhibit at least five morphological types of cell envelope, which are very useful for identification and classification.

Major Taxa of Bacteria

The following is proposed as an arrangement of higher taxa, which can serve during this time of taxonomic transition. It involves some amendments of rank and new names. This arrangement continues to recognize the absence or presence and nature of cell walls as determinative at the highest level.

The Kingdom Procaryotae

These are single cells or simple associations of similar cells forming a kingdom defined by cellular, not organismal, properties. The nucleoplasm is never separated from the cytoplasm by a unit membrane system. Four major categories or divisions are described on a phenotypic basis. Microorganisms included in the three first divisions are known as eubacteria.

Division I. Gracillicutes These are prokaryotes that have a complex (Gram-negative type) cell wall profile consisting of an outer membrane and an inner, thin peptidoglycan layer (which contains muramic acids) and a variable complement of other components outside or between these layers. The dry weight of the cell wall comprises about 11–22% lipid and < 10% peptidoglycan.

- Class I. Scotobacteria: non-photosynthetic Gram-negative bacteria
- Class II. Anoxyphotobacteria: phototrophic anaerobic bacteria
- Class III. Oxyphotobacteria: phototrophic aerobic bacteria

A total of 16 groups are included within these classes based on phenotypic characteristics such as morphology, motility, oxygen metabolism, trophic category and kind of life (free or parasitic).

Division II. Firmicutes These are prokaryotes with cell wall profile of Gram-positive type; reaction with Gram's stain is generally, but not always, positive. The dry weight of the cell wall usually contains < 4% lipid and > 30% peptidoglycan.

- Class I. Firmibacteria: cocci forming spores and non-sporing bacilli
- Class II. Thallobacteria: actinomycetes and related organisms. Sporogenous bacteria

A total of 13 groups are included within these classes based on phenotypic characteristics such as morphology, motility, oxygen metabolism, sporulation ability, capacity to synthesize antibiotics, trophic category and kind of life (free or parasitic).

Division III. Tenericutes These prokaryotes which lack a cell wall are commonly called the mycoplasmas. They are enclosed by a unit membrane, the plasma membrane.

- Class I. Mollicutes

Only one group is included within this class based on the phenotypic characters of pleomorphism, requirement of sterols for growth, gliding motility, oxygen metabolism, appearance as 'fried egg' on agar cultures.

Division IV. Mendosicutes These are prokaryotes with unusual walls that give evidence of an earlier phylogenetic origin from ancestral forms other than the groups included in Division I (Gracillicutes) and Division II (Firmicutes).

- Class I. Archaeobacteria

Five groups are included in this class on the basis of their superficial structures. The members are ecologically and metabolically diverse and live in somewhat extreme environments.

See also: **Bacteria**: The Bacterial Cell; Classification of the Bacteria – Phylogenetic Approach. **Nucleic Acid-based Assays**: Overview.

Further Reading

Balows A (ed.) (1991) *The Prokaryotes. A Handbook on the Biology of Bacteria: Ecophysiology, Isolation, Identification, Applications* (2nd ed.). New York: Springer.

Schleifer KH and Kandler O (1972) Peptidoglycan types of bacterial cell walls and their taxonomic implications. *Bacteriological Reviews* 36: 311–340.

Schleifer KH and Stackebrandt E (1983) *Annual Review of Microbiology* 37: 143–187.

Sneath PHA, Mair NS, Sharpe ME and Holt JG (1986) *Bergey's Manual of Systematic Bacteriology*. Baltimore: Williams & Wilkins.

Classification of the Bacteria – Phylogenetic Approach

E Stackebrandt, DSMZ – German Collection of Microorganisms and Cell Cultures, Brunswick, Germany

Introduction

Those who have chosen systematics, classification and taxonomy as a research topic consider it to be an exciting and important discipline. For others it is a boring subject which, through changes in names of microbial taxa, causes confusion in daily work. Changing the affiliation of species, genera, families and higher taxa is an inherent part of classifying the prokaryotes, which can be explained by the desire of taxonomists to create a system that reflects the natural relationship (phylogeny) between the taxa. Indeed, since the 1970s there has been an avalanche of

changes, and new insights into the phylogeny of prokaryotic species may have caused individual species to be affiliated within a short time span to several different genera. Nevertheless, of all taxonomic schemes of microorganisms the taxonomic system of prokaryotic species is the most developed in that it strictly follows the natural relatedness among the species. The cornerstone of a natural classification above the species level was the recognition that all the traits of an excellent phylogenetic marker are present in ribosomal (r) RNA and the genes coding for rRNA.

Principles of 16S rDNA Identification

Phylogenetic markers have to be universally present among all representatives of a given group of phylogenetically related organisms to allow them to be studied. In order to cover the complete range of taxa the most useful marker molecules are those that are ubiquitous: among these are the genes coding for rRNA and for proteins involved in basic biochemical reactions. These marker molecules evolved from a common ancestor and are homologous. This property can be derived from their sequences and their constraint in function during evolution. Another prerequisite is genetic stability: analysis of molecules involved in lateral gene transfer would disturb any phylogenetic conclusions. Size is an important factor, as small molecules carry too little evolutionary information and large molecules are difficult to sequence routinely.

The most powerful and most extensively used phylogenetic marker molecule is the 16S ribosomal RNA and the genes coding for this molecule (rDNA). The advantages of working with this molecule are obvious. As part of the protein-translating apparatus, the ribosomes must have been present in the evolutionary earliest prokaryotic cell, the components of the ubiquitously occurring ribosomes have not changed their function, and the presence of mostly multiple genes coding for rRNA makes horizontal transfer of these genes highly unlikely. This is in contrast to many protein-coding genes, which seem to have undergone significant horizontal gene transfer during early evolution of life forms. More than 12 000 sequences of 16S rDNA are available for prokaryotic strains from public data bases such as the Ribosomal Database Project (RDP) and ARBOR (ARB). The 23S rDNA has been analysed less frequently because of the double number of nucleotides (1550 compared with 3000 nucleotides). Prominent features of rRNA molecules are:

- The high degree of conservatism in sequence. The molecules consist of regions of varying degree of conservatism, depending upon their function within the ribosome: while highly important parts show almost no differences among most unrelated organisms, moderately and highly variable regions are scattered throughout the molecule. Even humans and *Escherichia coli* still share about 60% 16S rDNA sequence similarity, while the other 40% of the sequence has sufficient differences to permit phylogenetic definition down to the level of genera and species.

- The capacity of the 16S rRNA molecule to form highly developed secondary, tertiary and quaternary structures by short- and long-distance intermolecular interactions of inverse complementary sequence stretches. Many of these helices and the linking loop regions show significant variability in length and sequence. The secondary structure facilitates the alignment of sequences needed for a meaningful analysis.

Laboratory Procedures

Sequence Determination

Amplification of rRNA genes by the polymerase chain reaction (PCR) provides easy access to sequenceable material. The conserved regions which are scattered over the rRNA molecules or genes (rDNA) serve as target sites of oligonucleotide primers (usually 14–20 bases in length) needed for amplification and sequencing. Thus, a set of no more than 10 primers is sufficient to analyse a wide spectrum of phylogenetically diverse organisms. In principle a few bacterial cells are sufficient to perform the analysis. Thus even uncultured bacteria or microbial communities from natural samples are accessible to phylogenetic analysis. In the latter case the resulting mixture of rDNA fragments with different primary structures can be separated by cloning. The amplified DNA fragments are then sequenced directly by the chain termination method. Automated DNA sequence analysis is most conveniently carried out as linear PCR cycle sequencing reaction.

Sequence Alignment

Given the high content of invariant and conserved positions or regions along the sequence of rDNA, alignment of 16S rDNA is a straightforward procedure. Within the variable and highly variable regions with a high degree of length variation it is often difficult or even impossible to recognize homology from primary structure similarity. The alignment of these regions can in many cases be improved by taking into account the predicted higher order structure.

Phylogenetic Analyses and Phylogenetic Trees

The phylogenetic relationships of organisms based on comparative sequence analyses can be graphically presented. Two formats of graphic representation are generally used. Radial trees resemble 'botanical' trees; the distances between two nodes (organisms) are measured by the sum of edges between the nodes. Dendrograms (see Figs. 1 and 2) arrange the organisms in a fork-like fashion; only the horizontal components of connecting lines are summed to read the distances.

Treeing Methods Two major types of tree inferring approaches are commonly used: pairwise distance, and maximum parsimony. For the application of distance methods a matrix of pairwise dissimilarity values is calculated from the sequence alignment. Based on distance matrices, phylogenetic trees are preferentially reconstructed applying additive tree methods. These methods seek the tree for which distances expected from topology and branch lengths are most similar to those calculated from present-day sequences. A disadvantage of distance methods is that only overall dissimilarity values are used and all information about individual sequence positions is disregarded. Maximum parsimony methods use information of the individual positions of aligned sequences directly. The underlying model of evolution assumes that contemporary sequences were derived from their ancestors through the minimum number of changes. These methods seek for the most parsimonious trees among all possible tree topologies by determining the sum of changes that must have occurred to give the sequences in the alignment.

The significance of the relative branching order in a phylogenetic tree can be tested by resampling techniques such as the 'bootstrap' method. This approach randomly resamples alignment positions and generates trees (between 100 and 1000). The more often an individual branching point is resampled, the higher the value that defines a branching point to be monophyletic.

Application of Results

It should be noted in the context of the use of 16S rDNA (or any other nucleic acid) in taxonomy that a phylogenetic tree of a gene unravels the evolution of that particular gene but not necessarily the evolution of the genome of the organism. The more similar the topologies of phylogenetic trees of different genes, larger is the fraction of the genome that evolved along the same evolutionary path. Although important to an understanding of the evolution of an organism, a phylogenetic tree can rarely be used alone to decide

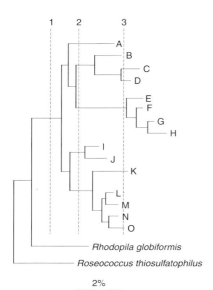

Figure 1   Dendrogram of 16S rDNA relatedness between hypothetical isolates A to O for which nothing is known other than the primary structure of this molecule. The vertical dashed lines 1 to 3 indicate some of the several possible levels of similarity at which taxa could be separated, illustrating that in the absence of phenotypic data, genera and species taxa cannot be defined meaningfully at this level of information. The bar represents 2 nucleotide substitutions per 100 nucleotides.

on the phylogenetic rank of a novel organism. This statement is explained as follows.

The phylogenetic branching pattern such as the one shown in **Figure 1** indicates that the organisms A to O are members of a single monophyletic line of descent which is separated from species of *Rhodopila* and *Roseococcus* and any other member of the alpha subclass of the class Proteobacteria. Most branching points are supported by high bootstrap values indicating their statistical significance. Isolates for which no other information is available other than their phylogenetic position form clusters at different levels of relationship, but no obvious hints are given from the branching pattern about how to interpret these clusters.

If one assumes that lineages that are separated by 16S rDNA differences of more than 2.5% define species (see below), are then all lineages A to O part of the same family, or the same genus (vertical line 1), or should five taxa be separated along vertical line 2 (separating lineages A, B–D, E–H, I–J and K–O), or should the delineation of taxa be done at line 3? The difficulty in the interpretation of phylogenetic patterns is partly caused by the presence of different branch lengths of the lineages. These are due to differences in evolutionary tempo and mode which are probably different in different groups of prokaryotic taxa. However, even if organisms evolve isochronally (at the same rate), the place of an organism relative to its

phylogenetic neighbours gives no clue to the physiological properties used for the description of a species. Also, as the phylogenetic position of an organism may change slightly as more phylogenetic neighbours are included in the analysis, the decision about the rank must await analysis of chemotaxonomic, biochemical, morphological and other more traditional taxonomic properties. The main advantage of knowing the phylogenetic position of a strain relative to its phylogenetic neighbours is the provision of a stable underlying structure needed for a modern approach to the polyphasic classification of genera and species. The situation is different in the description of higher taxa, i.e. above the level of genera. Here, for the description of families, orders, classes and the like, common phenotypic properties are often not known and clusters of phylogenetically similar ranks can be defined by common properties at the level of 16S rDNA and, in the future, of properties of other genes.

Links with 'Traditional' Bacteriology

Analysis of 16S rDNA has become a standard method in bacterial identification and classification. Although, as explained below, the sequence of this gene is too conserved to define a prokaryotic species, its analysis speeds up the identification process. The availability of the 16S rDNA sequence in combination with a huge data base containing more than 12 000 sequences from more than 90% of all described genera provides an advantage unmatched by any other rapid screening method. Of the species-rich genera, such as *Streptomyces*, *Bacillus* and *Clostridium*, and of large families such as Enterobacteriaceae, Pasteurellaceae and Pseudomonadaceae, all or nearly all species have by now been subjected to 16S rDNA analysis, allowing a reliable phylogenetic placement of novel isolates. The phylogenetic position of a new isolate next to its nearest phylogenetic neighbours immediately indicates whether the isolate falls within the radius of members of a described genus, or whether it forms a separate branch outside the boundaries of a genus. This information is extremely useful in selecting the identification strategy to be used. In the past time-consuming experiments were needed to obtain information on superficial resemblances to known species; now analysis of 16S rDNA guides the taxonomist immediately to the required tests. A branching point within the genus would concentrate on the characters distinguishing species, while a branching point outside the radiation of a genus would aim firstly at the investigation of genus-specific properties and secondly at characteristics defining a new species.

Identification turns into classification when an isolate shares only moderate 16S rDNA homologies with described species and common phenotypic properties are not shared. In such a case the presence of a novel taxon is indicated. However, in contrast to identification, classification is a subjective matter, and although they may be working with the same objective information, different taxonomists may come to different conclusions regarding the depth and breadth of a new taxon (see Fig. 1). No classification system can claim to reflect the natural situation, because prokaryotes and lower eukaryotes do not reveal the nature of their relationships. Thus, a validly described family containing two genera will remain taxonomically valid even if a different research group separates this family into two families each containing a single genus. It is the user of taxonomy who must be convinced that either the new system works better in identification, or makes better sense from the overall biological point of view – if the user sees no practical advantage in working with a new classification system, the system will simply not be used.

Rather than discussing problems involved in the interpretation of 16S rDNA dendrograms theoretically, an actual example can be used. The dendrogram depicted in Figure 1, used to outline some of the problems involved in the interpretation of phylogenetic data, reflects the situation seen among the acetic acid bacteria. For decades, the biotechnologically important bacteria involved in the oxidation of ethanol to acetic acid were classified in the genera *Acetobacter* and *Gluconobacter*. Reclassification started with more detailed chemotaxonomic and phylogenetic analyses. The genus *Acidomonas* was established for *Acetobacter methanolica* strains growing on methanol. Subsequent chemotaxonomic analysis of ubiquinone pointed towards the heterogeneity of members of Acetobacteraceae. While some species of *Acetobacter* contain Q-9, other species of this genus, as well as those of *Acidomonas* and *Gluconobacter*, possessed Q-10. Consequently, two *Acetobacter* subgenera were described to embrace species with the two different ubiquinone types (subgenus *Acetobacter* for the Q-9 species, subgenus *Gluconacetobacter* for the Q-10 species). Phylogenetic analysis then revealed the incoherence of the genus *Acetobacter* in that the subgenus *Gluconacetobacter* clustered separately from members of the subgenus *Acetobacter*. Subsequently the subgenus *Gluconacetobacter* was elevated to generic rank.

As shown in **Figure 2**, the four genera form a phylogenetically coherent identity which confirms their placement in the family Acetobacteraceae. All branching points are supported by high bootstrap values, indicating the statistical significance of the branching order of lineages. Members of the four genera can be differentiated from each other not only by phenotypic properties but also by a set of a few

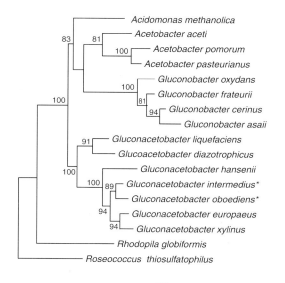

Figure 2 Dendrogram of 16S rDNA showing the phylogenetic position of members of the four genera of Acetobacteraceae. Numbers at branching points refer to bootstrap values. The bar represents 2 nucleotide substitutions per 100 nucleotides. Organisms marked with an asterisk have been described as members of *Acetobacter*.

unique 16S rDNA nucleotides (see Fig. 1). In combination these nucleotides are characteristic for members of a given taxon. These 'signature' nucleotides are qualitative measures for the phylogenetic clustering of organisms. The more isolated the phylogenetic position of a taxon, the more distinctive the set of signature nucleotides. Except for *Gluconobacter* spp. which form a closely related cluster in the 16S rDNA dendrogram (Fig. 2), signatures are not very pronounced among the highly related members of the Acetobacteraceae. Signatures of 16S rDNA have been described for the domains Archaea and Bacteria, and for the majority of main lines of descent within the latter domain. Often, signature nucleotides are located in the discriminating regions of oligonucleotides which are part of the *in situ* detection of organisms in the environment (**Table 1**).

The dendrogram in Figure 2 depicts the bifurcation that separates the genera *Acidomonas*, *Acetobacter* and *Gluconobacter* from the genus *Gluconacetobacter*. Within *Acetobacter*, *A. aceti* stands isolated, while within *Gluconacetobacter* two clusters are formed that separate *G. liquefaciens* and *G. diazotrophicus* from the other *Gluconacetobacter* species. The bifurcation points of these two clusters are as low as the one that separates *Acetobacter* and *Gluconobacter*. A typical question asked in the interpretation of such situation is: do the separate species – *A. aceti* on the one side and *G. liquefaciens* and *G. diazotrophicus* on the other – represent novel genera?

Table 1 Selection of 16S rDNA signature nucleotides which define the four genera of the family Acetobacteraceae. Unique information is printed in bold

Position	Acidomonas	Acetobacter	Gluconobacter	Gluconacetobacter
241–285	C-G	**U-A**	C-G	C-G
379–384	C-G	**G-C**	C-G	C-G
995	Base present	**Base missing**	**Base missing**	Base present
1027–1034	Base pair present	Base pair present	**Base pair missing**	Base pair present
1028–1033	Base pair present	Base pair present	**Base pair missing**	Base pair present
1116–1184	U-G	U-G	U-G	**C-G**
1117–1183	G-U	**A-U**	G-U	G-U
1118–1155	U-A	U-A	U-A	**C-G**
1310–1327	C-G	C-G	**U-A**	C-G
1426–1474	**U-A**	U-G	**C-G**	U-G

This question cannot be answered without a closer look at the phenotypic data. A survey of these properties indicates that differences in metabolic properties are not sufficiently obvious among the species of *Acetobacter* and *Gluconacetobacter* to support the establishment of novel genera. At the molecular level *G. liquefaciens* and *G. diazotrophicus* differ from the other four species of the genus in a few 16S rDNA signatures (position 155–166, C-G versus U-A; position 442–492, A-U versus G-C; position 591–648, U-A versus G-C; position 591–648, U-A versus G-C; and position 999–1041, G-U versus C-A). More detailed analysis of chemotaxonomic markers such as fatty acids, polyamines, polar lipids and the like is needed to decide the further taxonomic fate of the two genera.

Limitations of This Approach

As in prokaryotic microbiology a natural entity 'species' cannot be recognized as a group of strains that is genetically well separated from its phylogenetic neighbours, the taxon 'species' is defined by a pragmatic, polyphasic approach. This includes the recognition of genomic and phenotypic similarities and dissimilarities among strains, followed by the decision as to which of these strains should be affiliated to the same species. In this process DNA-DNA reassociation plays an important role in that phenetically similar strains sharing values of about 70% similarity should be considered to be members of the same species. Organisms that share 70% or more DNA similarity will share at least 96% DNA sequence identity at the genome level.

The primary structure of the 16S rDNA is highly conserved and species sharing 70% or more DNA similarity share usually more than 97% sequence identity. The 3% (or 45 nucleotides) differences are not evenly scattered along the primary structure of the molecule but are concentrated mainly in certain hypervariable regions. Recent evolutionary progress, expressed in the diverging primary structure of genes

and consequently in changes in phenotype, may not be shown directly at the level of the conservative rRNA genes. Calibration of the 16S rRNA clock reveal approximately 2–4% fixed substitutions per 100 million years. It appears plausible that within this epoch significant progress can be made in the speciation of prokaryotic species. Thus, two species, defined according to the present species definition, may possess very similar, if not identical, 16S rDNA primary structures. To complicate the interpretation of molecular data in taxonomy even more: the correlation of the two phylogenetic parameters, DNA similarity and 16S rDNA homology, is not linear. Comparative data have demonstrated that either method is strong in that area of relationships where the other method fails to reliably depict relationships. Sequence analysis is superior from the level of domains (starting at about 55% homology) to moderately related species, i.e. below 97.5% similarity. Above this value, DNA reassociation values were found to be either low (around 20–40%), or as high as 100%. Several groups of organisms have been identified which share almost identical 16S rDNA sequence but hybridize significantly lower than 70%, thus representing individual species. However, below sequence homology values of about 97.5% it is unlikely that two organisms share more than 60–70% DNA similarity, hence are related at the species level.

The range of methods in identification and classification has been extended significantly through the introduction of molecular techniques. Molecular analyses can be performed at breathtaking speed through the application of PCR technologies, automated sequencing and automated molecular characterization, pulsed field electrophoresis and oligonucleotide probing. At this stage of knowledge it appears unlikely that the definition of a species will change in such a way that the polyphasic approach is replaced by analysis of a single molecule. On the other hand, taxonomists have witnessed the tremendous impact of molecular approaches on the classification of taxa between the ranks of domains and genera. We should therefore not exclude the possibility that in the near future the rank of species may also be defined predominantly on the basis of molecular sequences.

See also: **Acetobacter**. **Bacteria**: Classification of the Bacteria – Traditional. **Gluconobacter**. **Nucleic Acid-based Assays**: Overview. **PCR-based Commercial Tests for Pathogens**.

Further Reading

Goodfellow M and O'Donnell AG (1993) *Handbook of New Bacterial Systematics*. London: Academic Press.

Li WH and Graur D (1991) *Fundamentals of Molecular Evolution*. Sunderland: Sinauer Associates.

Maidak BL, Larsen N, McCaughey NJ et al (1994) The Ribosomal Database Project. *Nucleic Acids Research* 22: 3485–3487.

Olsen GJ, Woese CR and Overbeek R (1994) The winds of (evolutionary) changes: breathing new life into microbiology. *Journal of Bacteriology* 178: 1–6.

Stackebrandt E (1992) Unifying phylogeny and phenotypic properties. In: Balows A et al (eds) *The Prokaryotes, A Handbook on the Biology of Bacteria: Ecophysiology, Isolation, Identification, Applications*, 2nd edn. New York: Springer.

Stackebrandt E and Goebel BM (1994) A place for DNA-DNA reassociation and 16S rRNA sequence analysis in the present species definition in bacteriology. *International Journal of Systematic Bacteriology* 44: 846–849.

Stackebrandt E and Rainey FA (1995) Partial and complete 16S rDNA sequences, their use in generation of 16S rDNA phylogenetic trees and their implications in molecular ecological studies. In: Akkermans ADL et al (eds) *Molecular Microbial Ecology Manual*. Amsterdam: Kluwer.

Stackebrandt E, Rainey FA and Ward-Rainey NL (1997) Proposal for a new hierarchic classification system, *Actinobacteria* classis nov. *International Journal of Systematic Bacteriology* 47: 479–491.

Vandamme P, Pot B, Gillis M, De Vos P, Kersters K and Swings J (1996) Polyphasic taxonomy, a consensus approach to bacterial systematics. *Microbiological Reviews* 60: 407–438.

Woese CR (1987) Bacterial evolution. *Microbiological Reviews* 51: 221–271.

Woese CR, Gutell R, Gupta R and Noller H (1983) Detailed analysis of the higher order structure of 16S like ribosomal ribonucleic acids. *Microbiological Reviews* 47: 621–669.

Zuckerkandl E and Pauling L (1965) Molecules as documents of evolutionary history. *Journal of Theoretical Biology* 8: 357–366.

Bacterial Adhesion *see* **Polymer Technologies for Control of Bacterial Adhesion**.

BACTERIOCINS

Contents
Potential in food preservation
Nisin

Potential in Food Preservation

Triona O'Keeffe and **Colin Hill**, Department of Microbiology and National Food Biotechnology Centre, University College Cork, Ireland

The following review of antimicrobial factors produced by lactic acid bacteria focuses primarily on bacteriocins and their use in food for both preservation and food safety. Nisin, the best-known and most widely used bacteriocin, is discussed with reference to its use in various food systems including dairy, meat and canning systems. Other bacteriocins currently in use are also discussed along with their potential uses and those of newly emerging bacteriocins. Advantages and disadvantages of bac-

teriocins as food additives are considered in addition to the optimization of production for large-scale purification and economic addition to foods. Interaction of bacteriocins with other inhibitory factors is outlined as part of the 'hurdle' approach to food preservation where bacteriocins can provide an additional barrier to the growth of unwanted microorganisms.

Antimicrobial Factors Produced by Lactic Acid Bacteria

The lactic acid bacteria (LAB) – including the genera *Lactobacillus, Lactococcus, Leuconostoc* and *Pediococcus* – have long been used in fermentations to preserve the nutritive qualities of various foods. The major function of a starter culture is the production of lactic acid at a suitable rate to ensure a consistent and successful fermentation. Other functions include the production of flavour compounds such as diacetyl and CO_2 from citrate by mesophilic cultures, and acetaldehyde from lactose by thermophilic cultures; acting as a source of proteolytic enzymes during growth in milk and ripening of many cheeses; and finally, contributing to the preservation of the fermented product as a consequence of a number of inhibitory metabolites produced by the lactic cultures.

The major metabolite of lactic acid bacteria is lactic acid. The production of this organic acid is responsible for the associated drop in pH, which may be sufficient to antagonize many microorganisms. In addition to a direct effect on the pH, the undissociated form of the molecule can cause the collapse of the electrochemical proton gradient of susceptible bacteria, leading to bacteriostasis and eventual death. Outside of its use in food fermentations, the main application of lactic acid in the food industry is in the decontamination of meat and poultry carcasses. Acetic and propionic acids are also produced in small amounts by lactic acid bacteria. They act in a similar manner to lactic acid and are widely used as food additives; however, they are not usually derived from LAB fermentations for this purpose. They do play an important antimicrobial role in some fermented foods, and it is known that acetic acid has a synergistic antimicrobial effect when present with lactic acid.

Diacetyl and acetaldehyde, as well as imparting aroma and flavour to cultured dairy products, also have an antimicrobial effect. Acetaldehyde can inhibit cell division in *Escherichia coli*, and diacetyl inhibits yeasts and Gram-negative and Gram-positive bacteria. However, the use of the diacetyl as a food preservative is precluded owing to its intense aroma and the large amounts required for preservation.

Hydrogen peroxide is also produced by a large number of lactic acid bacteria lacking the enzyme catalase – in particular by *Lactobacillus* spp. – and inhibits other microorganisms such as *Staphylococcus aureus* and *Pseudomonas* spp. The lactoperoxidase system is a naturally occurring antimicrobial system in milk in which the antagonistic hypothiocyanite radical is generated by the action of lactoperoxidase on thiocyanate and peroxide. This system has been successfully used to extend the shelf life of raw milk and cottage cheese and to inhibit pathogens in raw and processed milk products. The potential of H_2O_2 produced by lactic acid bacteria for food preservation may be limited by the oxidizing nature of the molecule, and free radicals produced may have profound effects on the sensory quality, causing rancidity of fats and oils and discoloration reactions.

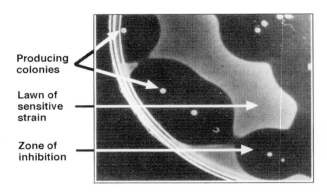

Figure 1 A lawn of sensitive bacteria, with zones of inhibition surrounding bacteriocin-producing colonies.

Reuterin (β-hydroxypropanaldehyde) is an inhibitory compound produced by *Lactobacillus reuterii* under anaerobic conditions in the presence of glycerol. Reuterin has a very wide spectrum of activity including Gram-positive and Gram-negative bacteria, yeasts, fungi, and protozoa. Organisms of public health significance that are inhibited by reuterin include species of *Salmonella*, *Shigella*, *Clostridium*, *Staphylococcus*, *Listeria*, *Candida* and *Trypanosoma*.

The focus of this review is on another category of inhibitory molecules, the bacteriocins. These antimicrobial peptides are produced by many bacterial species, but of particular interest to the food industry are those produced by members of the lactic acid bacteria, since they enjoy 'generally recognized as safe' (GRAS) status and thus have the potential to be used as preservatives in food.

Bacteriocins

Bacteriocins are ribosomally synthesized, extracellularly released bioactive peptides or peptide complexes which have a bactericidal or bacteriostatic effect on other (usually closed related) species. In all cases the producer cell exhibits specific immunity to the action of its own bacteriocin. Bacteriocin-producing strains can be readily identified in a deferred antagonism assay, in which colonies of the putative producer are overlaid with a bacterial lawn of a sensitive strain. After further incubation, zones of inhibition are visible in the sensitive lawn (**Fig. 1**). The term 'bacteriocin' was originally coined in 1953 specifically to define protein antibiotics of the colicin type (produced by *Escherichia coli*), but is now accepted to include peptide inhibitors from any genus. They are generally considered to act at the cytoplasmic membrane and dissipate the proton motive force through the formation of pores in the phospholipid bilayer.

Table 1 Bacteriocins characterized from lactic acid bacteria

Bacteriocin	Producer	Inhibitory spectrum[a]	Size (aa)	Food source[b]
Class I: Lantibiotics				
Nisin (A and Z)	Lactococcus lactis	Broad	34	Milk
Lacticin 481	Lactococcus lactis	Broad	27	
Lactocin S	Lactococcus sake	Broad	37	Fermented sausage
Carnocin U149	Carnobacterium pisicola	Broad	35–37	Fish
Variacin	Micrococcus varians	Broad	25	Meat fermentations
Class II: Non-lantibiotic, small, heat-stable				
Lactococcin A	Lactococcus lactis	Narrow	54	Cheese
Lactococcin B	Lactococcus lactis	Narrow	47	Cheese
Lactococcin M	Lactococcus lactis	Narrow	48	Cheese
Lactacin F	Lactobacillus johnsonii	Narrow	57, 48	
Mesenterocin 52B	Leuconostoc mesenteroides	Narrow	32	Raw milk
Curvaticin FS47	Lactobacillus curvatus	Medium	31	Ground beef
Pediocin-like bacteriocins:				
Sakacin A	Lactobacillus sake	Medium	41	Meat
Sakacin P	Lactobacillus sake	Medium	41	Fermented sausage
Carnobacteriocin A, B	Carnobacterium piscicola	Medium	53, 48	Meat
Pediocin AcH/PA-1	Pediococcus acidilactici	Medium	44	Meat
Leucocin A-UAL-187	Leuconostoc gelidum	Medium	37	Meat
Enterocin 1146/A	Enterococcus faecium	Medium	47	Fermented meat/milk
Piscicolin 126	Carnobacterium piscicola	Medium	44	Ham
Mesenterocin Y105	Leuconostoc mesenteroides	Medium	37	Goat's milk
Class III: Large, heat-labile				
Helveticin J	Lactobacillus helveticus	Narrow	333	

[a]Narrow spectrum indicates bacteriocins that only affect the producer genus; medium spectrum indicates bacteriocins that affect the producer genus and members of one or two other genera.
[b]Where known.

A number of schemes for differentiating between different bacteriocin types have been proposed, but it is generally agreed that at least three definable classes exist; **Table 1** comprises a non-exhaustive list of some of the bacteriocins characterized to date. Class I encompasses the small, post-translationally modified, broad host range lantibiotics, of which nisin is the best-known example. Class II includes the small, heat-stable unmodified peptides, while class III contains larger, heat-labile molecules such as helveticin J.

The increasing demand for high-quality 'safe' foods that are not extensively processed has created a niche for natural food preservatives. The ideal natural food preservative should fulfil the following criteria:

- acceptably low toxicity
- stability to processing and storage
- efficacy at low concentration
- economic viability
- no medicinal use
- no deleterious effect on the food.

While most bacteriocins fulfil all these criteria, to date nisin is the only bacteriocin to be commercially exploited on a large scale, having gained Food and Drug Administration (FDA) approval in the USA in 1988, although it had been in use in Europe for some time (the World Health Organization approved the use of nisin in 1961). Its success has stimulated further research targeted towards identifying new bacteriocins from lactic acid bacteria which could be potentially used in a similar manner. Many bacteriocins have now been characterized that exhibit antibacterial activity against a range of pathogenic and food spoilage bacteria. It is to be expected that bacteriocins and bacteriocin-producing lactic acid bacteria (used as starters or protective cultures) will find many roles in both fermented and nonfermented foods as a means of improving food quality, naturalness and safety.

The best-studied bacteriocin is nisin, but this is but one of many different inhibitory peptides produced by members of the LAB. Nisin is discussed below in some detail as an example of what has been achieved

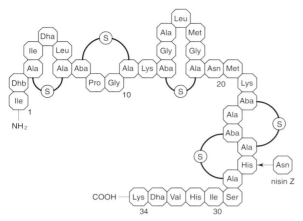

Figure 2 The mature nisin molecule including post-translational modifications. Dha, dehydroalanine; Dhb, dehydrobutyrine; Ala-S-Ala, lanthionine; Aba-S-Aba, β-methyl-lanthionine.

with this molecule, and what possibilities may exist for the exploitation of other peptide inhibitors.

Nisin

Nisin, a bacteriocin produced by certain strains of *Lactococcus lactis* subsp. *lactis*, is inhibitory to a wide range of Gram-positive bacteria, including strains or species of streptococci, staphylococci, lactobacilli, micrococci, *Listeria* and most spore-forming species of *Clostridium* and *Bacillus*. Gram-negative bacteria, yeasts and fungi are not normally affected. Nisin is a lantibiotic (class I, Table 1), indicating that it has undergone significant post-translational modification (the mature active nisin molecule is shown in **Fig. 2**). Nisin has a bactericidal effect on sensitive cells, effecting a rapid death within a minute of addition. The target of nisin and related molecules has been shown to be the cytoplasmic membrane of sensitive cells. Adsorption does not appear to involve specific protein receptors (as is the case for colicins), but involves direct interaction between the nisin molecule and the membrane. The action of nisin proceeds through insertion and pore formation, leading to a rapid and aspecific efflux of low-molecular-weight compounds and the depolarization of the membrane. Membrane insertion relies upon a charged membrane, and does not occur in non-energized liposomes. It is currently mooted that the membrane-associated nisin molecules aggregate to form amphiphilic pores, through which the low-molecular-weight compounds can exit. The affected cell is unable to derive energy with which to synthesize macromolecules such as proteins or nucleic acids, and is rapidly killed. It has been observed that Gram-negative cells, normally insensitive to the action of nisin, can be sensitized by the addition of chelating agents which disrupt the integrity of the outer membrane and allow the bacteriocin access to the cytoplasmic membrane.

Nisin was initially proposed for use as an antibiotic but its relatively narrow antibacterial spectrum, sensitivity to digestive proteases and instability at physiological pH rendered it unsuitable. However, nisin possesses many features that make it ideal for use as a food preservative. The form of nisin most commonly used by food processors is manufactured by Aplin & Barret in Dorset, England, and is marketed as Nisaplin, which has a high and consistent activity. It is currently recognized as a safe food preservative in approximately 50 countries. The physicochemical properties of nisin dictate that stability, solubility and activity are greatest at low pH. It is used mainly to prevent the outgrowth of spores in a wide variety of foods, including processed cheese and canned vegetables (see below). Some species of spore-formers are more sensitive than others. *Bacillus stearothermophilus* is more sensitive than *B. cereus*, *B. megaterium* or *B. polymyxa*. It is effective in preventing the outgrowth of *Clostridium botulinum* types A, B and E, but the proteolytic types are more resistant than the non-proteolytic ones. Sensitivity increases with lower pH, increased temperature, length of heat shocking and lower spore load. Nisin is sporicidal rather than sporistatic.

Nisin has found many applications in the food industry, only a few of which are presented here.

Dairy Products Nisin is used in pasteurized, processed cheese products to prevent outgrowth of spores such as those of *Clostridium tyrobutyricum* that may survive heat treatments as high as 85–105°C. Use of nisin allows these products to be formulated with high moisture levels and low NaCl and phosphate contents, and also allows them to be stored outside chill cabinets without risk of spoilage. The level of nisin used depends on food composition, likely spore load, required shelf life and temperatures likely to be encountered during storage.

Nisin is also used to extend the shelf life of dairy desserts which cannot be fully sterilized without damaging appearance, taste or texture. Nisin can significantly increase the limited shelf life of such pasteurized products.

Nisin is added to milk in the Middle East where shelf-life problems occur owing to the warm climate, the necessity to transport milk over long distances and poor refrigeration facilities. It can double the shelf life at chilled, ambient and elevated temperatures and prevent outgrowth of thermophilic heat-resistant spores that can survive pasteurization. It can also be used in canned evaporated milk.

Canned Foods Nisin may also be added to canned foods at levels of 100–200 IU g^{-1} to control thermophilic spore-formers such as *Bacillus stearothermophilus* and *Clostridium thermo-saccharolyticum* which may survive and grow in canned foods stored at high temperatures. It also allows a reduction in heat processing required without compromising food safety. It is used in canned potatoes, peas, mushrooms, soups, and cereal puddings. It increased activity at acid pH levels makes it ideally suitable in low pH foods such as canned tomatoes, to inhibit acid-tolerant spoilage flora such as *B. macerans* and *C. pasteurianum*.

Meat Concern about the toxicological safety of nitrite used in cured meat has led to investigation into the use of nisin to allow a reduction in nitrite levels. However, uneconomically high levels are required to achieve good control of *Clostridium botulinum*, perhaps as a consequence of nisin binding to meat particles, uneven distribution, poor solubility in meat systems, or possibly interference in activity by meat phospholipids.

Wine The insensitivity of yeasts to nisin allows its use to control spoilage lactic acid bacteria in beer or wine. It can maintain its activity during fermentation without any effect on growth and fermentative performance of brewing yeast strains and with no deleterious effect on taste. It can therefore be used to reduce pasteurization regimens and to increase shelf life of beers. It has similar applications in wine except for those that require a desirable malolactic fermentation. However, nisin-resistant bacterial starter cultures such as resistant strains of *Leuconostoc oenos*, in conjunction with nisin, can be used to actually control the malolactic fermentation. Nisin may also be used to reduce the amount of sulphur dioxide used in winemaking to control bacterial spoilage.

Other Applications Another application of nisin is the control of bacterial contamination in the baking industry in high-moisture, hot baked flour products and liquid egg. It is also useful to control the growth of Gram-positive contaminants in fermentations that depend on Gram-negative microorganisms or fungi, e.g. single-cell protein, organic acids, polysaccharide, amino acid or vitamin production.

Pediocin-like Bacteriocins

Pediocin-like bacteriocins are members of the class II bacteriocins (Table 1), a group of bacteriocins in which there is considerable commercial interest. They are small, heat-resistant peptides that are not post-translationally modified to the same extent as the class

I bacteriocins, apart from the cleavage of a leader sequence from a double glycine site upon export of the bacteriocin from the cell, and the presence of disulphide bridges in some molecules. All of the pediocins share certain features, including a seven amino acid conserved region in the N-terminal of the active peptide (-Tyr-Gly-Asn-Gly-Val-Xaa-Cys-). Perhaps the best-known is pediocin PA-1, which is produced by *Pediococcus acidilactici*. A commercial formulation has been introduced under the trade name ALTA. Pediococci are important in the fermentation of vegetables and meat for both acid production and flavour development. The pediocin-like bacteriocins (which are also produced by genera other than the pediococci, see Table 1) are active against other lactic acid bacteria but are particularly effective against *Listeria monocytogenes*, a food-borne pathogen of increasing concern to the food industry. *Listeria* may be found in raw milk, dairy products, vegetables and meat products and can grow under conditions such as refrigeration temperatures (growth has been reported at temperatures as low as –1°C), high salt concentrations (up to 10%), low pH (pH 5.0), and high temperatures (44°C). Pediocin PA-1 has been observed to inhibit *Listeria* in dairy products such as cottage cheese, ice cream, and reconstituted dry milk. It has also been demonstrated as a biocontrol agent on meat systems. *In situ* production in dry fermented sausage inhibits *L. monocytogenes* throughout fermentation and drying, possibly owing to a combination of the reduction in pH and bacteriocin production. *Pediococcus acidilactici* is also used as a low-level inoculum in reduced-nitrite bacon to prevent the outgrowth of *Clostridium botulinum* spores and subsequent toxin production.

Other 'pediocin-like' bacteriocins include sakacins A and P (*Lactobacillus sake*), leucocin A (*Leuconostoc gelidum*), and enterocin A (*Enterococcus faecium*). The relative insensitivity of starter lactic acid bacteria to some of the above, e.g. enterocin A, suggests a potential role in food fermentations where normal starter activity is required, through addition at the start or else *in situ* production during the fermentation. One example of this is a strain of *Enterococcus faecalis* isolated from natural whey cultures utilized as starter in the manufacture of mozzarella cheese from water-buffalo milk. It inhibits *Listeria monocytogenes* but not other useful LAB. Others, such as leucocin A whose producer was isolated from meat, show promise in the preservation of vacuum-packed meat. Such meat is stored at low temperatures in anaerobic conditions, possibly with added organic acids. However, spoilage bacteria of meat are psychrotrophic, facultatively anaerobic, and acid tolerant; it is therefore necessary to control them by other means such as bacteriocinogenic LAB which must be able to compete with the relatively high indigenous microbial loads of raw meat. Leucocin A has the advantage over nisin in that it is stable in meat and can be produced during chilled storage without undesirable organoleptic changes under aerobic or anaerobic conditions for extended periods. It prevents spoilage due to sulphide production by *Lactobacillus sake*.

Potential Uses for Other Bacteriocins

As more and more bacteriocin producers are being isolated and characterized, usually from food environments, the potential for their use increases. *Lactobacillus plantarum* produces plantaricins S and T in the Spanish-style green olive fermentation. They are active against a number of natural competitors and spoilage microorganisms such as propionibacteria and clostridia, resulting in a more reliable product without completely eliminating the indigenous microflora and detracting from the quality of the final product. A number of enterococci were isolated from Argentinian milk samples and milk products which produce proteinaceous compounds that inhibit *Vibrio cholerae*. This implies that they may already play an important role in natural preservation of foods, especially in regions where cholera is epidemic or endemic. Similarly, many traditional African foods are fermented by lactic acid bacteria before consumption. Naturally occurring bacteriocin-producing strains in these products may have the potential to improve the quality and shelf life of other African fermented foods which are often plagued by problems such as inconsistent quality, hygienic risks and premature spoilage.

Other bacteriocins which have been isolated from food environments include plantaricin F, from chilled processed channel catfish; acidocin B, produced by *Lactobacillus acidophilus* with a narrow spectrum of activity which includes *Clostridum sporogenes* and a narrow range of other lactobacilli; and salivaricin B, produced by *Lactobacillus salivarus* with a very broad host range including *Listeria monocytogenes*, *Bacillus cereus*, *Brochothrix thermosphacta*, *Enterococcus faecalis* and many lactobacilli, which may have a more widespread application. Another recently identified bacteriocin with a broad host range similar to that of nisin is lacticin 3147, produced by a strain of *Lactococcus lactis*. Production and immunity are plasmid-encoded traits and can be conjugally transferred to commercial starters. Several diverse applications have been suggested for this bacteriocin, ranging from a veterinary use in cattle teat seals to inhibit mastitic streptococci, to the manufacture of cheeses with improved safety and quality attributes.

The ability to eliminate non-starter lactic acid bacteria (NSLAB) during Cheddar cheese ripening may allow further investigation of the role which NSLAB play in flavour development. Since lacticin 3147 is also an effective inhibitor of many Gram-positive food pathogens and spoilage microorganisms, these starters may provide a very useful means of controlling the proliferation of undesirable microorganisms during Cheddar cheese manufacture.

An alternative application of bacteriocins other than the preservation and protection of food is bacteriocin-induced starter lysis to accelerate ripening of Cheddar cheese, particularly where starters have low autolysis levels. A strain of *Lactococcus lactis* producing lactococcins A, B and M causes lysis of susceptible cells, so it could be used as an adjunct culture to accelerate the lysis of starter lactococci whose slow autolysis might otherwise produce a bitter cheese.

Microgard (Wesman Foods Inc., USA) is commercially produced from grade A skim milk fermented by a strain of *Propionibacterium shermanii*, and has a wide antimicrobial spectrum including some Gram-negative bacteria, yeasts and fungi. This product is added to 30% of the cottage cheese produced in the USA as an inhibitor against psychrotrophic spoilage bacteria. It is available as a liquid concentrate, spray-dried or freeze-dried preparation. It is added to a variety of dairy products such as cottage cheese and yoghurt and a nondairy version is also available for use in meat and bakery goods. The inhibitory activity almost certainly depends primarily on the presence of propionic acid, but there has also been a role proposed for a bacteriocin-like protein produced during the fermentation. This use of milk fermented by a bacteriocin producer as an ingredient in milk-based foods may be a useful approach for introducing bacteriocins into foods at little cost.

Advantages and Disadvantages of Bacteriocins as Food Additives

One of the advantages associated with the use of bacteriocins in food is that these molecules can be said to be normal constituents of the human and animal diet, in that meat and dairy systems are particularly rich sources of bacteriocinogenic lactic acid bacteria. Bacteriocins are proteinaceous in nature and would therefore be expected to be inactivated by proteases of gastric or pancreatic origin during passage through the gastrointestinal tract. Therefore, such bacteriocins, if used in foods, should not alter the digestive tract ecology or result in risks related to the use of common antibiotics.

In addition, most bacteriocins have good thermostability and so can survive the thermal processing cycle of foods. Others can work at both low pH and low temperature and could therefore be useful in acid foods and cold-processed or cold-stored products. Bacteriocins may also have applications in minimally processed refrigerated foods, e.g. vacuum and modified atmosphere-packaged refrigerated meats and ready-to-eat meals, which lack the multiple barriers or hurdles to the growth of pathogenic and spoilage bacteria formerly conferred by traditional preservation techniques. In addition, the genetics of the better-known bacteriocins are well characterized and thus it is possible that the genes encoding bacteriocin production and immunity could be transferred to non-producing starter strains for *in situ* production. This is particularly true for bacteriocins whose genes are located on naturally transmissible elements, like nisin (conjugal transposon) and lacticin 3147 (conjugal plasmid).

One possible drawback to the use of bacteriocins in foods is that they are hydrophobic molecules which may partition to the organic fat phase within a food matrix. It has been suggested by many researchers that the use of bacteriocins will be limited by this fact. However, while most bacteriocins are indeed very hydrophobic, they are relatively small molecules and so can easily diffuse into the water phase of food products. Nonetheless, binding to food surfaces and poor activity are often observed when bacteriocin-producing strains are added to food systems. The food specific environment may have other drawbacks such as poor solubility or uneven distribution of the bacteriocin molecules, sensitivity to food enzymes, and the negative impact of high levels of salt or other added ingredients affecting the production or activity of the bacteriocin. In addition to this, spontaneously bacteriocin-resistant mutants of target strains may arise. Nisin-resistant mutants of *Listeria monocytogenes* appear at frequencies of 10^{-6} to 10^{-8}. However, in properly processed food such high levels should not be encountered. While these disadvantages have been identified by many scientists, in practice bacteriocins have been shown to be effective in a number of food systems, including full-fat cheeses and meats.

Another problem to overcome is the reluctance of industry to incorporate new methodologies over old tried and tested ones, particularly if they have to embark on the substantial and expensive programmes of toxicological testing that may be necessary for the introduction of a new antimicrobial as a purified additive. While it is unclear how many detailed toxicity trials have been performed to date, no evidence of bacteriocin toxicity has been reported. Toxicity studies for nisin were carried out using amounts far in excess of the amount that would be used in food,

with no ill-effects. Nisin is rapidly inactivated in the intestine by digestive enzymes and is undetectable in human saliva 10 minutes after consumption. There was no evidence of sensitization and no evidence of any cross resistance that might affect the therapeutic effect of antibiotics. The use of any new food ingredient has to undergo strict regulatory considerations. However, in the case of biologically derived macromolecules with well-understood pathways of digestion and metabolism, such as proteins, they may be determined to be safe for consumption by utilizing available knowledge of their structure, biological activity, digestibility, and biological and compositional factors. In the case of bacteriocins, safety assessment may require characterization of the substance as completely as possible, description of the preparation, proposed use and proportion in food, knowledge of the effect in the food and the metabolic fate in the gastrointestinal tract and also perhaps an environmental impact assessment. There was earlier concern that the use of bacteriocins such as nisin might hide the use of poor-quality materials or poor manufacturing practice, but this fear is unfounded because bacteriocins have a relatively low antimicrobial activity, and efficacy is dependent on a low microbial load.

Production of Bacteriocins

Economically reliable processes have to be developed for bacteriocin production. Optimizing production and enhancing stability and activity are necessary for such an economic breakthrough. In addition, bacteriocin-producing starter cultures may prevent the growth of spoilage and pathogenic microorganisms more efficiently in food and feed products if the production *in situ* and stability of active bacteriocin are increased. Fermentations should be based on cheap substrates, and a suitable and low-cost downstream processing strategy devised to produce bacteriocins for direct use as food biopreservative, or biomass as starter culture for *in situ* bacteriocin production. Most bacteriocins are produced during the active growth phase but often there is a sharp decrease in activity at the end of the log phase due to protein degradation, adsorption to cell surface, protein aggregation, complex formation, etc. Maximum bacteriocin amounts can be harvested by collecting it immediately when activity peaks or by using conditions that minimize adsorption, e.g. low pH. Bacteriocin could be removed from the fermentation either batchwise or continuously by such adsorbants. Bacteriocin production seems to be stimulated by stress factors such as low temperature and competing microorganisms. A lower specific growth rate may lower acid production thus making for less successful competition and necessitating other factors such as bacteriocin production for competition. Production may be controlled by a two-component regulatory mechanism through signal transduction, a cell–cell communication system known as quorum sensing. Molecular techniques may allow one strain to produce a number of bacteriocins, thus increasing the spectrum of bacteria sensitive to that strain. Protein engineering may increase activity, stability and host range. It is possible to make nisin synthetically and the sequence has been altered to examine the specific role of different amino acids. However, such engineering must overcome regulatory hurdles. An ideal protocol for bacteriocin production would be one which is applicable for large-scale purification having low production and recovery costs, leading to a bacteriocin yield greater than 50% and purity greater than 90%.

Future Prospects for Bacteriocins in Food

Bacteriocins should not be seen as a primary means of food preservation. Rather, they can contribute to the 'hurdle' approach to food preservation and safety whereby a number of barriers, both intrinsic and extrinsic, act as 'hurdles' for microorganisms to overcome in the food. Many bacteriocins have been observed to be more stable and effective at acid pH, higher temperatures (important in the case of temperature abuse) or lower temperatures (important for refrigerated foods). Bacteriocins serve as bactericidal barriers which can help to reduce the levels of contaminating bacteria while biostatic measures such as modified atmosphere packaging or water reduction can prevent the remaining population from growing. They can also be used in conjunction with other antibacterial factors. The animal-derived protein, lysozyme, lyses many Gram-positive cells and is used to prevent gas formation in some cheeses. Lysozyme and nisin can act synergistically to inactivate cells of *Listeria monocytogenes*. Bacteriocin activity can also be enhanced by the effect of chelators. Plant-derived antimicrobials, some of which are already in use such as benzoic, sorbic, acetic, and citric acids, and also phenolic compounds and essential oils such as those found in garlic, can be used in food preservation but have disadvantages in that the food may not be considered 'natural' or the level required for inhibition may introduce too strong a flavour to the food. Bacteriocins may replace or permit a lower level of such inhibitors. Some bacteriocins (e.g. lactocin S) have a much slower rate of killing than others. Combining fast- and slow-acting bacteriocins in a food may allow it to be 'safe' for longer. Alternatively, bacteriocins could be used as a remedial measure in

conjunction with rapid detection methods to remove contamination, e.g. in brewing. The future for bacteriocins does not lie in discovering or engineering the perfect bacteriocin for all applications; instead, it is more practical to imagine specific bacteriocins for specific tasks. Because of their relatively large size, bacteriocins may be considered a finite population of macromolecular inhibitors, so the relative amount required to inactivate target cells may depend on the population of cells that may be present. This and the other factors already discussed that limit bacteriocin effectiveness on foods illustrate that bacteriocins alone cannot be depended on to ensure the safety of a particular food.

It is likely that the types of food and the application will be determined by practicalities such as the economics of bacteriocin production on an industrial scale. Fermented foods provide an obvious application. The bacteriocin could be introduced to the product at little or no cost through the use of a bacteriocin-producing starter bacterium. This would ensure even distribution of the inhibitor through the food, and the reduced pH of fermented foods would allow maximal activity of most bacteriocins. Also, LAB are categorized as GRAS for producing fermented foods and so regulatory considerations can be overcome. Addition of partially purified bacteriocin preparations to foods may be less cost-effective, but has been accomplished with nisin. Perhaps a more readily acceptable means of addition of bacteriocin will mirror the current use of Microgard (derived from a milk fermentate) as an ingredient in milk-based foods as a way of introducing bacteriocins at little cost. Genetic analysis of bacteriocin operons will continue to pave the way for bacteriocin applications, since the ability to over-produce the inhibitor will certainly have an impact on the cost-effectiveness. Further research into the prevalence of natural bacteriocinogenic strains in retail foodstuffs in conjunction with toxicological studies may provide an even stronger case for the safe use of bacteriocins in the food chain.

See also: **Bacteriocins**: Nisin. **Cheese**: Microbiology of Cheese-making and Maturation. *Clostridium*: *Clostridium tyrobutyricum*. **Fermented Foods**: Origins and Applications. *Lactobacillus*: Introduction. *Lactococcus*: Introduction; *Lactococcus lactis* Sub-species *lactis* and *cremoris*. *Leuconostoc*. **Meat and Poultry**: Spoilage of Cooked Meats and Meat Products. *Pediococcus*. **Wines**: Microbiology of Wine-making.

Further Reading

Abee T, Krockel L and Hill C (1995) Bacteriocins: modes of action and potentials in food preservation and control of food poisoning. *International Journal of Food Microbiology* 28 (2): 169–185.

De Vuyst L and Vandamme EJ (1994) Nisin, a lantibiotic produced by *Lactococcus lactis* subsp. *lactis*: properties, biosynthesis, fermentation and applications. In: De Vuyst L and Vandamme EJ (eds) *Bacteriocins of Lactic Acid Bacteria*. Glasgow: Blackie.

Gould GW (1996) Industry perspectives on the use of natural antimicrobials and inhibitors for food applications. *Journal of Food Protection* (supplement) 82–86.

Hill C (1995) Bacteriocins: natural antimicrobials from microorganisms. In: Gould G (ed.) *New Methods of Food Preservation*. Glasgow: Blackie.

Montville TJ, Winkowski K and Ludescher RD (1995) Models and mechanisms for bacteriocin action and application. *International Dairy Journal* 5: 797–814.

Morgan S, Ross RP and Hill C (1997) Increasing starter cell lysis in cheddar cheese using a bacteriocin-producing adjunct. *Journal of Dairy Science* 80: 1–10.

Muriana PM (1996) Bacteriocins for control of *Listeria* spp. in food. *Journal of Food Protection* (supplement) 54–63.

Piard JC and Desmazeaud M (1992) Inhibiting factors produced by lactic acid bacteria. 2. Bacteriocins and other antibacterial substances. *Lait* 72: 113–142.

Rodriguez-Gomez JM (1996) Applications of nisin in the food industry. *Alimentaria* 34 (271): 93–97.

Salih MA and Sandine WE (1990) Inhibitory effects of Microgard™ on yogurt and cottage cheese spoilage organisms. *Journal of Dairy Science* 73: 887–893.

Vandenbergh PA (1993) Lactic acid bacteria, their metabolic products and interference with microbial growth. *FEMS Microbiology Reviews* 12: 221–238.

Nisin

E Alison Davies and **Joss Delves-Broughton**, Technical Services and Research Department, Aplin & Barrett Ltd (Cultor Food Science), Dorset, UK

Introduction

Nisin is a natural, toxicologically safe, antibacterial food preservative. It is regarded as natural because it is a polypeptide produced by certain strains of the food-grade lactic acid bacterium *Lactococcus lactis* subsp. *lactis*, (hereafter referred to as *L. lactis*), during fermentation. Nisin exhibits antimicrobial activity towards a wide range of Gram-positive vegetative bacteria, and is particularly effective against bacterial spores. It shows little or no activity against Gram-negative bacteria, yeasts or moulds.

History

Nisin was discovered in the late 1920s and early 1930s, when problems arose during cheese-making. Batches of milk starter culture used in the process had become contaminated with a nisin-producing strain of *L. lactis* (then called *Streptococcus lactis*), and as a result of nisin's inhibitory properties, the development of the cheese was detrimentally affected. Nisin was named accordingly, from group N (streptococcus) Inhibitory Substance. Subsequently, it was shown to have antimicrobial activity against a wide range of Gram-positive bacteria, particularly spore-formers, but not against Gram-negative bacteria, yeasts or fungi.

Initial research on nisin focused on its potential therapeutic qualities, for medical and veterinary uses. At that time it was found to be unsuitable for such purposes, mainly because of its limited antibacterial spectrum and its low solubility and instability in body fluids. The potential use of nisin as a food preservative was first suggested in 1951 by Hirsch, who demonstrated that clostridial gas formation in cheese could be prevented by the use of nisin-producing starter cultures. Subsequently, numerous other applications of nisin were identified and in 1969, nisin was approved for use as an antimicrobial in food by the Joint Food and Agriculture Organization/World Health Organization Committee on Food Additives.

The suitability of nisin as a food preservative arises from the following characteristics: it is nontoxic; the producer strains of *L. lactis* are regarded as safe (food-grade); it is not, at present, used clinically; there is no apparent cross-resistance in bacteria that might affect antibiotic therapeutics; it is digested immediately and it is heat-stable at a low pH. Since 1953, nisin has been sold as a commercial preparation under the trade name Nisaplin by Aplin & Barrett Ltd (UK), and is currently permitted as a food additive (labelled 234) in over 50 countries. The activity or potency of a nisin preparation is expressed in terms of International Units (IU): 1 g of pure nisin is usually equivalent to 40×10^6 IU and 1 g of Nisaplin is equivalent to 1×10^6 IU. The assay method in most common use for actively measuring nisin levels in foods is nisin bioassay. This involves measuring zones of inhibition in agar seeded with the test organism *Micrococcus luteus*.

The principal commercial applications of nisin are in foods and beverages which, by their nature, are pasteurized but not fully sterilized. Examples of such foods are processed cheese (including spreads), milk (plain and flavoured), clotted cream, dairy desserts, ice cream mixes, liquid egg and hot-baked flour products such as crumpets and potato cakes. In warm

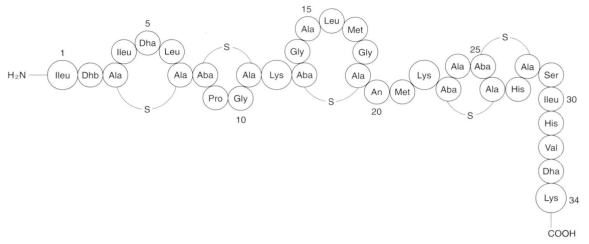

Figure 1 The structure of nisin. Aba: aminobutyric acid; Dha: dehydroalanine; Dhb: dehydrobutyrine (β-methyldehydroalanine); Ala–S–Ala: lanthionine; Aba–S–Ala: β-methyllanthionine.

climates, nisin is used in canned products to prevent spoilage by thermophilic, heat-resistant spore-formers. Other applications include alcoholic beverages (e.g. beer and wine) and salad dressings, in which nisin controls spoilage by lactic acid bacteria, and natural cheeses such as ripened cheese and soft white fresh cheeses, to control *Listeria monocytogenes*.

Structure and Biosynthesis

In 1971, Gross and Morell elucidated the complete structure of the nisin molecule (**Fig. 1**). Nisin was at that time a novel oligopeptide, but subsequently a number of similar bacteriocins have been identified and characterized. Although nisin is, as yet, the only commercially accepted bacteriocin for food preservation, most of the lactic acid bacteria that produce these similar bacteriocins can also be used commercially in starter cultures.

Nisin belongs to a group of bacteriocins collectively known as *lantibiotics*. Lantibiotics are produced by Gram-positive bacteria of different genera, e.g. *Lactococcus* (nisin, lacticin 481), *Lactobacillus* (lactocin S), *Staphylococcus* (Pep 5, epidermin, gallidermin), *Streptococcus* (streptococcin A-FF22, salivaricin A), *Bacillus* (subtilin, mersacidin), *Carnobacterium* (carnocin U149), *Streptomyces* (duramycin) and *Actinoplanes* (actagardine). Like nisin, the other lantibiotics are effective against a range of Gram-positive bacteria. On the basis of their different ring structure, charge and biological activity, the lantibiotics are classified into two subgroups: Type A, lantibiotics of the nisin type; and Type B, lantibiotics of the duramycin type. Established Type A lantibiotics include Pep 5, epidermin, gallidermin, subtilin, mersacidin and actagardine.

Lantibiotics are relatively small polycyclic polypeptides, nisin consisting of 34 amino acids (3354 daltons). They are so named because they contain, besides protein amino acids, the unusual amino acids lanthionine and/or β-methyllanthionine, both of which form interchain thioether bridges. Nisin also contains another two unusual amino acids, dehydroalanine and dehydrobutyrine. In total, nisin has two dehydroalanine (Dha), one dehydrobutyrine (Dhb), one lanthionine and four β-methyllanthionine residues. Dha and Dhb arise from the dehydration of serine and threonine respectively, and the condensation of Dha or Dhb with cysteine generates thioether bonds and the amino acids lanthionine and β-methyllanthionine respectively. Subtilin is a natural analogue of nisin. They each contain the same number of dehydro-residues and lanthionine rings, with conserved locations of the Dha residues and rings. However, there are 12 amino acid differences, and nisin has 34 residues whereas subtilin has 32.

Lanthionine is known to introduce a high level of hydrophobicity, and a high proportion of basic amino acids gives nisin a net positive charge. Nisin can form dimers or even oligomers, which possibly arise through a reaction between the dehydroamino acids and amino groups of two or more nisin molecules. In aqueous solution, nisin is most soluble at pH 2. At a high pH, the presence of nucleophiles makes Dha and Dhb susceptible to modification, which may explain the decreased solubility and instability of nisin under basic conditions. Using NMR analysis, it has been shown that nisin exists in a rigid three-dimensional structure due to the constraints imposed by the five thioether rings.

Nisin preparations have been resolved into five polypeptides (nisins A–E), but nisins B–E are thought to be degradation products of nisin A. It is not, as yet, clear which components of the nisin molecule are essential for activity. Apart from chemically derived modifications, nisin A variations can arise through changes in DNA sequence. Nisin-like molecules with different activity spectra are produced by different strains of *L. lactis*. This phenomenon is due to minor differences in amino acid sequence. For example, nisin Z is identical to nisin A except for a substitution of Asn for His as amino acid residue 27. This amino acid change is a result of a single nucleic acid substitution.

Nisin is initially synthesized ribosomally as a precursor peptide, which is then enzymatically cleaved (to give pronisin) and post-translationally modified to generate the mature lantibiotic. The prepronisin structural gene has been cloned and sequenced, and has been designated *spaN* and *nisA* by different workers. The primary transcript of prepronisin consists of an N-terminal leader peptide of 23 amino acids, followed by a C-terminal propeptide of 34 amino acids, from which the lantibiotic is matured.

Nisin biosynthesis genes are encoded by a novel conjugative transposon (70 kbp), generally thought to be located on the chromosome as opposed to plasmid-mediated. Sucrose fermentation, nisin immunity, conjugal transfer factors, *N*-(5-carboxyethyl)-ornithine synthase and bacteriophage resistance determinants have all been linked with nisin production. The nisin genes are organized into an operon-like structure, with the functions of genes *nis* A, B, T, C, I, P, R, K, F, E, G having been identified (**Fig. 2**):

- *nisA* gene: encodes nisA, the prepronisin structural protein
- *nisB* and *nisC* genes: encode the enzymes needed for the modification of the lantibiotic precursor peptides
- nisT: involved in the transport of (precursor) nisin molecules across the cytoplasmic membrane
- *nisI* gene: encodes a putative lipoprotein, involved in immunity to nisin
- *nisP* gene: encodes a subtilisin-like serine protease, involved in cleavage of the leader peptide sequence from the final precursor peptide
- *nisR* gene: encodes a positive regulatory protein needed for the activation of expression of the *nis* genes
- *nisK* gene: encodes another regulatory protein, histidine kinase
- *nisF*, *nisE* and *nisG* genes: also thought to be involved in immunity to nisin.

Mode of Action and Antimicrobial Effect

The primary target of nisin in sensitive Gram-positive cells is the cytoplasmic membrane. It was originally thought that nisin acted as a cationic surface-active detergent, due to the strong adsorption of nisin to cells, causing leakage of cellular material and subsequent lysis. However, it is now thought that membrane disruption is due to the incorporation of nisin into the membrane, with subsequent ion channel or pore formation. Adsorption does not appear to involve specific protein receptors, but the interaction between nisin and the membrane is perhaps facilitated by negatively-charged cell wall components. It is thought that nisin takes the form of oligomers inside the membrane, but it is not known whether it inserts into the membrane as a monomer and then self-assembles into an oligomer to form the pore, or whether the aggregation event precedes membrane binding or insertion. It is possible that nisin may interact with integral constituents of the membrane such as proteins or membrane-bound cell wall precursors.

Compounds of low molecular weight effuse through the nisin-induced membrane pores, causing dissipation of the membrane potential and the pH gradient and resulting in the collapse of the proton motive force (p.m.f.). It is now believed that depletion of the p.m.f. is the common mechanistic action of bacteriocins from lactic acid bacteria. The dehydro residues and thioether rings of nisin are thought to play a role in its activity against vegetative bacteria, but their action against spores and vegetative cells may differ. Gram-negative cells are normally protected from nisin by the presence of an outer membrane in their cell walls. However, when the outer membrane is weakened, Gram-negative cells also show sensitivity to nisin, indicating that their cytoplasmic membranes are susceptible. Weakening of the Gram-negative outer membrane can be achieved by chelating agents such as EDTA. Chelating agents bind divalent ions which serve to stabilize the lipopolysaccharide layer of the outer membrane.

The action of nisin against spores is predominantly sporostatic as opposed to sporicidal. Nisin affects the post-germination stages of spore development. It inhibits pre-emergent swelling, and thus the outgrowth and formation of vegetative cells. It is thought that the active sites in spores are membrane-bound sulphydryl groups present in newly germinated spores, reactivity occurring with the nisin dehydro groups. Nisin's success as a sporostatic agent in the preservation of heat-processed foods is dependent on an effective level of nisin being sustained throughout the shelf life of the food. However, progressive heat

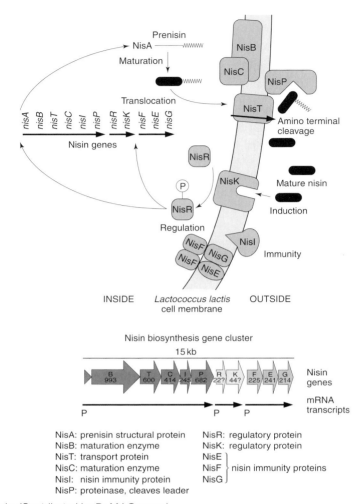

Figure 2 Nisin biosynthesis. (Contributed by Dr MJ Gasson.)

damaging of spores results in their increased sensitivity to nisin.

The nisin sensitivity of Gram-positive vegetative cells and spores varies considerably, even between strains of the same species. Susceptible Gram-positive bacteria include lactic acid bacteria (often associated with food spoilage), the food-borne pathogen *Listeria monocytogenes*, and the spore-forming genera *Bacillus* and *Clostridium*, both of which can cause pathogenic and spoilage problems. Nisin's effectiveness is concentration-dependent, in terms of both the amount of nisin added and the number of spores or vegetative cells that need to be inhibited or killed. Its action against vegetative cells can be either bactericidal or bacteriostatic, depending on a number of factors including nisin concentration, bacterial population size, physiological state of the bacteria and the conditions of growth. Bactericidal effects are enhanced under optimal growth conditions and when the bacteria are in an energized state. In contrast, bacteriostatic effects are enhanced when nisin forms part

of a multi-preservation system in which conditions are non-optimal and other inhibitory factors are exerted (hurdle technology). The fact that different conditions are required for nisin application to ensure either bacterial destruction or inhibition must be taken into consideration.

Solubility and Stability

The commercial preparation Nisaplin contains approximately 2.5% nisin, the remainder consisting of residual milk solids from the fermentation process and NaCl. Nisaplin is an extremely stable product, showing no loss of activity over a 2-year period providing it is stored under dry conditions, in the dark, and at temperatures below 25°C. Nisin shows increased solubility in an acid environment and becomes less soluble as the pH increases. However, due to the low level of Nisaplin used in food preservation, solubility does not present a problem. Nisin solutions are most stable to autoclaving (121°C for

15 min) in the pH range 3.0–3.5 (< 10% activity loss). pH values below and above this range cause a marked decrease in activity especially those furthest removed from the range (> 90% activity loss at pH 1 or 7). Losses of activity at pasteurization temperatures are, however, significantly less (< 20% during standard processed cheese manufacture at pH 5.6–5.8). Food components can also protect nisin during heat processing as compared to a buffer system.

The stability of nisin in a food system during storage is dependent upon three factors – incubation temperature, length of storage and pH. Greatest nisin retention occurs at lower temperatures. For instance, the manufacture of a pasteurized processed cheese spread (85–105°C for 5–10 min at pH 5.6–5.8) results in an initial 15–20% nisin loss, nisin retention after 30 weeks' storage being approximately 80% at 20°C, 60% at 25°C and 40% at 30°C. Thus, a higher level of nisin addition will be required if storage at unusually high ambient temperatures is intended. Residual nisin levels in canned foods after heat processing can be as low as 2%, but the fact that heat-resistant thermophilic spores are highly nisin-sensitive and heat-damaged spores have increased sensitivity to nisin means that extremely low levels of residual nisin can be effective. Pre-acidification of the brine used in canned vegetables, to pH 4 with citric acid, also improves nisin retention, with minimal effect on the pH of the final product after heat processing.

Applications

Processed Cheese Products

Nisin has become established as a most effective preservative in pasteurized processed cheese products, including block cheese, slices, spreads, sauces and dips. This is because typical heat processing (85–105°C for 5–10 min) of the raw cheese during melting does not eliminate spores. Without the addition of nisin, the composition of the pasteurized processed cheese would favour the outgrowth of the spores. Spore-formers associated with processed cheese include *Clostridium butyricum*, *C. tyrobutyricum*, *C. sporogenes* and *C. botulinum*. Spore outgrowth of the first three species listed may result in spoilage due to the production of gas, off odours and liquefaction of the cheese, whereas *C. botulinum* more seriously produces a potentially fatal toxin. The level of nisin required to inhibit the outgrowth of spores in processed cheese and other products is dependent on a number of factors: the level of clostridial spores present, the composition of the food, e.g. NaCl, disodium phosphate, pH and moisture content; the shelf life required and the temperature of storage. Generally, levels of nisin used to control non-botulinal

spoilage in processed cheese vary from 6.25 to 12.5 mg kg^{-1}. For anti-botulinum protection, the level required is 12.5 mg kg^{-1} or higher.

Other Pasteurized Dairy Products

Other pasteurized dairy products, such as chilled desserts, cannot be subjected to full sterilization without damaging their organoleptic qualities, and are thus sometimes preserved with nisin to extend their shelf life. For example, tests on chocolate dairy dessert demonstrated a 20-day increase in shelf life with 3.75 mg kg^{-1} of nisin added at 7°C. Similarly, canned evaporated milk has an extended shelf life with added nisin.

The addition of nisin to milk is permitted in some countries due to shelf-life problems associated with the climate (high ambient temperatures), long-distance transport and inadequate refrigeration. The use of nisin at levels of 0.75–1.25 mg l^{-1} has been demonstrated to more than double the shelf life of the product. However, it is not permitted in the UK and other countries with temperate climates. The addition of nisin to high-heat-treated flavoured milk has also been shown to extend shelf life.

Pasteurized Liquid Egg Products

Pasteurized liquid egg products can comprise the whole egg, the egg yolk or the egg white. Heat treatment applied (62–65°C for 2–3 min) kills salmonella but not all bacterial spores and vegetative cells. Many of the surviving bacteria are psychrotrophic, and so pasteurized liquid egg products usually have a limited shelf life. In a trial conducted with pasteurized liquid whole egg, nisin (5 mg l^{-1}) caused a significant increase in refrigerated shelf life, of between 6–11 days and 17–20 days. Nisin also protected the egg from the growth of the psychrotrophic *Bacillus cereus*.

High-moisture Hotplate Bakery Products

Typical products include crumpets and potato cakes. They are flour-based, have a high moisture content (48–56%), are non-acid (pH 6–8) and are lightly cooked on a hotplate during manufacture. Sold at ambient temperature, they are traditionally toasted before consumption. The flour used in the manufacture of these products invariably contains a low number of *Bacillus* spores, and conditions are ideal for outgrowth, so it is not surprising that outbreaks of food poisoning linked to *B. cereus* toxins have been reported.

These bakery products have a short shelf life (3–5 days), but in a recent survey conducted in the UK very high levels of *Bacillus* (10^8 cfu g^{-1}) were detected in potato cakes well within the sell-by date. The popularity of crumpets in Australia and the risk of *B.*

cereus food poisoning, accentuated by high ambient temperatures, led to factory trials incorporating nisin. The addition of nisin to crumpet batter at concentrations of $3.75\,mg\,kg^{-1}$ and above effectively inhibited the growth of *B. cereus*, resulting in safe levels. This resulted in regulations in Australia allowing the use of nisin in such high-moisture hotplate products.

Canned Foods

Nisin is used in canned foods principally for the control of thermophilic spoilage. It is mandatory in the UK that low-acid canned foods (pH > 4.5) should receive a minimum heat process ($F_0 = 3$) to ensure the destruction of *Clostridium botulinum* spores. The survival of heat-resistant spores of the thermophiles *Bacillus stearothermophilus* and *C. thermosaccharolyticum* during this process are responsible for spoilage, particularly under warm conditions. The bacterial spoilage of high-acid foods (pH < 4.5) is restricted to non-pathogenic, heat-resistant, aciduric, spore-forming bacterial species such as *C. pasteurianum*, *B. macerans* and *B. coagulans*. Spoilage resulting from the growth of all these bacteria can be effectively controlled by nisin. Addition levels are generally between 2.5 and $5.0\,mg\,kg^{-1}$ and product examples include canned vegetables, soups, coconut milk and cereal puddings (such as rice, semolina and tapioca).

Meat and Fish Products

In relation to processed meat products, nisin has been considered as an alternative preservative system to that of nitrite, due to concerns about toxicological safety. However, various studies have shown that nisin does not perform at its full potential in meat systems. Results generally indicate that nisin is only effective at high levels, i.e. $12.5\,mg\,kg^{-1}$ and above. Proposed reasons for the inadequacy of a nisin preservative system in meats include: poor solubility in meat systems; binding of nisin onto meat particles and surfaces; uneven distribution; and possible interference with nisin's mode of action by phospholipids. However, both modified atmosphere and vacuum packaging in combination with nisin have shown more promising results.

Relatively few studies have been carried out on the use of nisin as a preservative of fish and shellfish. However, the potential hazard of botulism from chilled fish packed under vacuum or modified gas atmospheres prompted a trial application of nisin by spray to fillets of cod, herring and smoked mackerel inoculated with *Clostridium botulinum* type E spores. Toxin production was delayed by 5 days compared with the control at 10°C, but only by half a day at

26°C. Recent research into the application of nisin to canned lobster meat, to control *Listeria monocytogenes*, has been very positive. Heat processing of canned lobster, which is retailed frozen, can only be achieved by heating at 60°C for 5 min without undesirable product shrinkage occurring. Such heat processing results in a 2 log reduction of *L. monocytogenes*, whereas the addition of nisin to the brine at $25\,mg\,l^{-1}$ increases the reduction by 5–6 logs.

Natural Cheese

Nisin can be used to prevent blowing in some hard and semihard ripened cheeses such as Emmental and Gouda. This is caused by contamination with the anaerobic spore-formers *Clostridium butyricum* and *C. tyrobutyricum*, usually from a milk source when the cow has been fed with silage. The bacteria convert lactic acid into butyric acid, which causes the off flavour and aroma of the cheese. The formation of H_2 and CO_2 gas during ripening also results in the development of too many large holes in the cheese. Cheeses can also be contaminated with *Lactobacillus* spp., causing off flavours and gas production, and with the food poisoning pathogens *Listeria monocytogenes* and *Staphylococcus aureus*, all of which are susceptible to nisin.

The use of nisin is an attractive alternative to other agents, including sodium nitrate, which has become increasingly unpopular and usually only works against specific microorganisms (e.g. *Clostridium*). However, nisin-resistant starter cultures must be used in conjunction with nisin to ensure successful development of the cheese. The use of nisin producing starter cultures, to manufacture cheese with significant levels of nisin, is also being investigated. At present no existing nisin-producing starters have the flavour-generating, eye-forming and acidifying activities and the bacteriophage resistance which are suitable for the manufacture of most cheese types. However, a nisin-producing starter culture for Gouda production has been developed using the food-grade genetic transfer technique of conjugation. During production, clostridial blowing and *S. aureus* growth were both inhibited over the whole period of ripening.

Soft white fresh cheeses (e.g. ricotta, paneer) do not require starter cultures, being alternatively coagulated by direct acidification, calcium chloride or rennet, and in these cheeses nisin very effectively controls the growth of *Listeria monocytogenes*. Shelf-life analysis of ricotta in an inoculated trial demonstrated that the addition of $2.5\,mg\,l^{-1}$ nisin to the milk pre-production could effectively inhibit the growth of *L. monocytogenes* at 6–8°C for at least 8 weeks. Ricotta made without the addition of nisin contained unsafe levels of the organism within 1–2 weeks of incubation.

Yoghurt

The addition of nisin to stirred yoghurt post-production has an inhibitory effect on the starter culture (a mixture of *Lactobacillus delbrueckii* subsp. *bulgaricus* and *Streptococcus thermophilus* strains), thereby preventing subsequent over-acidification of the yoghurt. Thus an increase in shelf life is attained by maintaining the flavour of the yoghurt (less sour).

Salad Dressings

The development of salad dressings with reduced acidity gives improved flavour and protects the added ingredients. However, raising the pH (from 3.8 to 4.2) can make salad dressings prone to lactic acid bacterial spoilage during ambient storage. Such growth has been successfully controlled by the addition of nisin at levels of 2.5–5 mg l^{-1}.

Alcoholic Beverages

Nisin has a potential role in the production of alcoholic beverages. It has been demonstrated that nisin is effective in controlling spoilage by lactic acid bacteria such as *Lactobacillus*, *Pediococcus* and *Leuconostoc* at a level of 0.25–2.5 mg l^{-1} in both beer and wine. Yeasts are completely unaffected by nisin, which allows its addition during the fermentation. Identified applications of nisin in the brewing and wine industry include: its addition to fermenters to prevent or control contamination; increasing the shelf life of unpasteurized beers; reduction of pasteurization regimes; and washing pitching yeast to eliminate contaminating bacteria (as an alternative method to acid washing, which affects yeast viability). Formerly, nisin could not be used during wine fermentations that depend on malolactic acid fermentation. However, this problem has been overcome by developing nisin-resistant strains of *Leuconostoc oenos*, that can grow and maintain malolactic fermentation in the presence of nisin. In the production of fruit brandies, the addition of nisin reduces the growth of competitive lactic acid bacteria and directly favours the growth of the fermenting yeast, to increase alcohol content by at least 10%.

Antagonistic Factors

Various factors can detrimentally affect nisin's action. In non- or minimally heat-processed foods, proteolytic enzymes of microbial, plant or animal origin can degrade nisin during the shelf life of the food. The extent of degradation is dependent on the length and temperature of storage and on pH. Thus the likely retention of nisin during shelf life is a factor dictating nisin addition levels.

There is evidence that both fats and proteins in food can interfere with nisin action. As nisin is predominantly hydrophobic, it is thought that it binds onto certain food particles, thus becoming unavailable for antibacterial action. This is particularly important in meat systems, in which nisin is thought to bind onto phospholipids, resulting in a much lower efficacy than in other products.

Generally, nisin works best in liquid and homogenous, rather than solid and heterogenous, foods. Certain food additives have been shown to be antagonistic to nisin. For example, nisin is degraded in the presence of titanium dioxide (a whitener) or sodium metabisulphite (an antioxidant, bleaching agent and broad-spectrum antimicrobial agent).

Some bacterial species such as *Listeria monocytogenes*, can offer resistance to nisin – that is, strains with acquired nisin resistance can arise in the presence of sublethal nisin concentrations. However, this phenomenon does not occur in all strains and the frequency of isolation of these nisin resistant cells is very low (approximately 10^{-6} to 10^{-8}). External factors such as temperature, pH and salt influence the frequency of nisin resistance in *L. monocytogenes*. At reduced temperature (10°C), pH (5.5) and salt concentration (0.5%), nisin resistance is eliminated. Thus only in cases of high-level contamination with high storage temperatures allowing rapid growth, or long shelf lives at low temperatures (under suitable conditions) and/or low levels of nisin, could nisin-resistant mutants of *L. monocytogenes* potentially arise. The nisin resistance mechanism in *L. monocytogenes* is thought to be associated with adaptation of the cell envelope, to prevent the incorporation of nisin into the cytoplasmic membrane. Recent research has indicated the possible involvement of both the cytoplasmic membrane and the cell wall. It has also been reported that some bacterial species, including *Lactobacillus plantarum*, *Streptococcus thermophilus* and *Bacillus cereus* produce an enzyme *nisinase*, which specifically deactivates nisin. However, it has been shown that nisinase is not produced by *Listeria monocytogenes*.

See also: **Bacillus**: *Bacillus cereus*. **Bacteriocins**: Potential in Food Preservation. **Cheese**: Microbiology of Cheese-making and Maturation; Role of Specific Groups of Bacteria. **Clostridium**: *Clostridium tyrobutyricum*; *Clostridium botulinum*. **Eggs**: Microbiology of Egg Products. **Fermented Milks**: Yoghurt. **Fish**: Spoilage of Fish. **Genetics of Microorganisms**: Bacteria. **Heat Treatment of Foods**: Spoilage Problems Associated with Canning; Principles of Pasteurization. **Lactococcus**: *Lactococcus lactis* Sub-species *lactis* and *cremoris*. **Listeria**: *Listeria monocytogenes*. **Meat and Poultry**: Spoilage of Cooked Meats and Meat Products. **Natural Antimicrobial**

Systems: Preservative Effects During Storage. **Preservatives**: Classification and Properties. **Starter Cultures**: Cultures Employed in Cheese-making. **Wines**: Microbiology of Wine-making.

Further Reading

Delves-Broughton J (1990) Nisin and its uses as a food preservative. *Food Technology* 44: 100–117.

Delves-Broughton J and Gasson MJ (1994) Nisin. In: Dillon VM and Board RG (eds) *Natural Antimicrobial Systems and Food Preservation*. P. 99. Wallingford: CAB International.

De Vuyst L and Vandamme EJ (1994) Nisin, a lantibiotic produced by *Lactococcus lactis* subsp. *lactis*: properties, biosynthesis, fermentation and applications. In: De Vuyst L and Vandamme EJ (eds) *Bacteriocins of Lactic Acid Bacteria: Microbiology, Genetics and Applications*. P. 151. London and Glasgow: Blackie Academic & Professional.

Fowler GG and Gasson MJ (1991) Antibiotics – nisin. In: Russell NJ and Gould GW (eds) *Food Preservatives*. P. 135. London and Glasow: Blackie Academic & Professional.

Hurst A (1981) Nisin. *Advances in Applied Microbiology* 27: 85–123.

Hurst A and Hoover DG (1993) Nisin. In: Davidson PM and Branen AL (eds) *Antimicrobials in Foods*. P. 369. New York: Marcel Dekker.

BACTEROIDES AND *PREVOTELLA*

Harry J Flint and **Colin S Stewart**, Rowett Research Institute, Aberdeen, UK

Synopsis

Bacteroides and *Prevotella* spp. are the most numerous groups of bacteria found in the human colon and in the hind gut and rumen of farm animals. They are obligate, or strict, anaerobes whose activities play an important role in the breakdown and conversion of food and feed components. In the human colon, *Bacteroides* spp. are thought to make a very significant contribution to the metabolism of plant and host-derived polysaccharides, while *Prevotella* spp. contribute to the utilization of plant material in the rumen. Both genera play important roles in protein metabolism, and potentially also in the metabolism of carcinogens and xenobiotics.

Characteristics and Classification

Obligately anaerobic bacteria belonging to the CFB (*Cytophaga*, *Bacteroides*, *Flavobacterium*) phylum are among the most abundant cultured isolates obtained from gut samples. Most isolates from human faeces belong to the genus *Bacteroides*, with the commonest species being those of the *B. fragilis* group: *B. eggerthii*, *B. fragilis*, *B. ovatus*, *B. thetaiotaomicron*, *B. uniformis* and *B. vulgatus*. *Bacteroides* is widely reported to be the most abundant genus isolated from human faeces by anaerobic culture techniques, with numbers often in the range 10^{10}–10^{11} per gram. Many species previously classified as *Bacteroides* have now been placed in new genera within the CFB phylum, e.g. *Prevotella* which includes oral species such as *P. melaninogenica* and *P. oralis* and major rumen species such as *P. ruminicola*. Many more species formerly classified as *Bacteroides* are now recognized as being quite unrelated. These include some significant representatives of the normal flora of the human and animal gut such as *Mitsuokella multiacidus*, *Megamonas hypermegas*, *Fibrobacter succinogenes* and *Ruminobacter amylophilus*.

Bacteroides/Prevotella spp. are Gram-negative, non-sporing bacteria that show pleomorphic morphology, ranging from short cocco-rods to long, irregular rod-shaped cells. The best-defined group are the human colonic *Bacteroides* spp., which are bile resistant, saccharolytic and produce acetate and succinate as major fermentation end products. They possess sphingolipids and a mixture of long-chain fatty acids with predominantly straight-chain saturated, anteisomethyl branched and isomethyl branched acids. *Meso*-diaminopimelic acid is present in the peptidoglycan layer and menaquinones (mainly MK10 and MK11) are present within the cell. *Bacteroides* spp. have a DNA %G+C composition of 39–48% (**Table 1**).

Bacteroides strains fail to take up aminoglycoside antibiotics, making them intrinsically resistant, and many strains produce β-lactamases. Many strains show transferable resistance to clindamycin and to tetracyclines, and these resistances are generally found to be transmitted on chromosomally located conjugative transposons. They are generally sensitive to the antibiotic metronidazole which is activated by anaerobic conditions, but some resistant strains have been reported.

The composition of *Bacteroides* lipolysaccharide differs from that of other Gram-negative bacteria and

Table 1 Characteristics of some *Bacteroides* and *Prevotella* species and their occurrence. Data from Holdeman, Cato and Moore (1977), Holdemann, Kelley and Moore (1986) and Shah (1992)

	B. fragilis	B. vulgatus	B. distasonis	B. ovatus	B. thetaiotaomicron	B. uniformis	B. eggertii	P. oris	P. oralis	P. ruminicola/bryantii/brevis	P. melaninogenicus	P. bivius	P. intermedius
Products from PYG	SAppa (ibivl)	SAp (ivibl)	SAppa (ivibl)	SAppa (ibivl)	SAppa (ivibl)	Sapl (ivib)	SAp (ivibl)	SA (pibiv)	SAf (l)	Saf (pivib)	SA (fivibl)	SA iv(libf)	SA iv (libpf)
Esculin hydrolysed	+	+	+	+	+	+	+	+	+	+	-+	-	-
Indole produced	-	-	-	+	+	+	+	-	-	-	-	-	+
Gelatin digested	-w	+	-w	d	-w	-	-	+-	+	+	+	+	+
Acid produced from													
Arabinose	-	+	-	+	+	+w	+	+-	-	+	-+	-	-
Cellobiose	-w	-	+	+	+	+w	-w	+w	+	+	-	-	-
Glucose	+	+	+	+	+	+	+	+	+	+	+	+	+
Inositol	-	-	-	-	-	-	-	-	-	-	-	-	-
Lactose	+	+	+	+	+	+	+	+	+	+	+	+	-
Maltose	+	+	+	+	+	+w	+	+	+	+	+	+	+
Melibiose	+w	+w	+	+	+	+w	-	d	+	+	-+	-	-
Salicin	-	-	+	+	-+	+-	+w	+	+	+/-	-	-	-
Starch	+	+	+	+	+	+	+w	+	+	+	+	+	+w
Sucrose	+	+	+	+	+	+w	-	+	+	+	+	-	+
Trehalose	-	-	+	+	+	-	-	+	-	-	-	-	-
Xylose	+	+	+	+	+	-	+	+	-	+/-	-	-	-
G + C (%)	41-44	40-42	43-45	39-43	40-43	45-48	44-46	42-46	43	39-52	36-40	40	41-44
Normal flora													
Human intestine	√	√	√	√	√	√	√	-	√	√	√	-	-
Rumen	-	-	-	-	-	-	-	-	-	√	-	-	-
Animal intestine	Chicken, pig	Chicken, pig, rat	Pig, rat	Chicken	Rat	-	-	Chicken	-	Chicken, pig	-	-	-
Other sites						Urogenital tract		Mouth	Mouth	-	Mouth, Urogenital tract	Urogenital tract, Mouth	-

* Products from PYG (peptone-yeast extract-glucose) broth cultures. Capital letters (for monobasic acids) = >1m mol acid/100 ml culture, small letters, <1m mol acid/100 ml culture. A = acetic, B = butyric, f = formic, iB = isobutyric, iV = isovaleric, P = propionic, Pa = phenylacetic, L = lactic, S = succinic. Products in brackets may or may not be detected. Sugars, + = 90% or more strains produce final pH below 5.5; = 90% or more strains produce final pH above 5.7; w = weak reaction, final pH 5.5–5.7. d = 11–89% of strains positive. Where two reactions are given, the first is more common.

Table 2 Peptone–trypticase–yeast medium for cultivation of *Bacteroides* spp.

Petone	0.5 g
Trypticase	0.5 g
Yeast extract	1.0 g
Glucose	0.5 g
Resazurin solution[a]	0.4 ml
Salts solution[b]	4.0 ml
Distilled water	100 ml
Add after boiling:	
Haemin solution[c]	1.0 ml
Vitamin K_1[d]	0.02 ml
Cysteine HCl.H_2O	0.05 g

[a] Resazurin solution: 25 mg resazurin dissolved in 100 ml distilled water.
[b] Salts solution (per litre): 1 g K_2HPO_4; 1 g KH_2PO_4; 10 g $NaHCO_3$; 2 g NaCl; 0.2 g anhydrous $MgSO_4$; 0.2 g anhydrous $CaCl_2$.
[c] Haemin solution: dissolve 50 mg haemin in 1 ml 1 N NaOH and make up to 100 ml in distilled H_2O. Autoclave.
[d] Vitamin K_1 solution: 0.15 ml Vitamin K_1 dissolved in 30 ml 95% ethanol. Store at 4°C for up to 1 month.

the lipopolysaccharide of *B. fragilis* involved in soft tissue infections is less toxic than that of enterobacteria to mammalian cells.

Light microscopic examination of negatively stained *Bacteroides* reveals the presence in some strains of extracellular capsules, which in some cases exceed the diameter of the cell. Two extracellular structures, an electron-dense layer and a fibrous network, can be seen by electron microscopic examination of *B. fragilis* grown in vitro. Pilus-like structures have also been seen in this species. Capsular polysaccharides of *B. fragilis* contain L-fucose, D-galactose, D- and L-quinovosamine, D-glucosamine and galacturonic acid. At least some of the aminosugars are N-acetylated. The composition of the extracellular polysaccharide (EPS) of rumen *Prevotella* strains has also been studied; all strains studied possess galactose, glucose and mannose in varying proportions, and fucose is present in the EPS of most strains. *B. fragilis* produces both catalase and superoxide dismutase (SOD), whereas other species may possess only SOD or neither enzyme.

Cultivation

Most human colonic *Bacteroides* species are only moderately oxygen sensitive, but must be grown in prereduced media under anaerobic conditions. A variety of media can be used for routine cultivation, one of which is described in **Table 2**.

Media are prepared by boiling three times and bubbling with oxygen-free CO_2 before dispensing into sealed tubes or bottles and autoclaving. To maintain strictly anaerobic conditions, inoculation of liquid media should be performed in an anaerobic glove box, on the bench by syringe injection through suitable rubber septa, or using screw-cap Hungate style tubes together with a gassing hook system. For most human colonic *Bacteroides* strains strict anaerobiosis is not essential at all times and colonies can be successfully transferred between agar plates in air, for example, provided exposure times are kept short (less than 1 hour) and are followed by anaerobic growth. Gas Pac jars can also be used successfully to grow the less oxygen-sensitive species.

Wilkins–Chalgren agar and Brucella blood agar have been used for total counts. Partially selective isolation of human colonic *Bacteroides* can be achieved relying on their bile resistance using media incorporating $20 \, g \, l^{-1}$ oxgal, with aesculin as added carbohydrate energy source, and $100 \, \mu g \, ml^{-1}$ gentamicin (Bacteroides bile esculin agar). Media containing kanamycin and vancomycin (KVLB) can also be used to select against facultative anaerobes.

Bacteroides and *Prevotella* species are not always easily differentiated on nutritional and morphological criteria, and this is particularly true for the less studied animal isolates. Molecular methods using ribosomal RNA specific primers and probes are becoming increasingly important and oligonucleotides specific for *Bacteroides/Prevotella* have been reported.

Action on Food Components in the Human Gastrointestinal Tract and Interactions with the Host

Bacteroides and related obligate anaerobes proliferate in regions of the gut where transit times are slow enough, and energy sources of dietary and host origin are sufficient, to allow the development of an anaerobic fermentation. Facultative anaerobes play an important role in the establishment of an anaerobic flora by rapidly using up most of the available oxygen. Nevertheless oxygen is constantly entering the system from the gut wall and from digesta, creating microenvironments with differing oxygen concentrations.

Polysaccharide Breakdown

More than 50% of the plant polysaccharides ingested by the host are degraded during passage through the colon. Several *Bacteroides* species are thought to play important roles in the fermentation of plant polysaccharides of dietary origin. Polysaccharides fermented by some *Bacteroides* strains in pure culture include amylose, amylopectin, xylan, pectin, polygalacturonic acid, alginate and guar gum, locust bean gum and arabinogalactan. Most strains of *B. ovatus* and *B. thetaiotaomicron* and some *B. fragilis* strains ferment many of these, although only *B. ovatus* strains ferment xylan (**Table 3**). Although other species of

Table 3 Fermentation of plant and mucin-derived polysaccharides by human colonic *Bacteroides* spp. Data from Salyers et al (1977).

	B. fragilis	B. thetaiotao-micron	B. ovatus	B. eggertii	B. vulgatus	B. distasonis	Other Bacteroides
Strains examined	53	22	24	6	22	11	50
Amylose	53	22	16	6	22	0	17
Dextran	4	22	16	6	22	0	18
Xylan	11	0	24	0	1	0	21
Polygalacturonate	17	20	24	6	6	0	0
Pectin	17	22	23	0	9	0	19
Gum tragacanth	2	0	19	0	0	0	0
Locust bean gum	0	0	4	0	1	0	17
Guar gum	0	0	4	0	1	0	17
Alginate	0	0	10	0	0	0	0
Laminarin	0	10	1	0	1	11	34
Arabinogalactan	0	22	15	0	16	0	44
Hyaluronate	0	22	24	0	0	0	23
Heparin	12	22	24	6		0	0
Chondroitin Sulphate	2	22	23	0	0	0	21
Ovomucoid	0	22	17	0	0	0	0
Bovine mucin	2	0	0	0	0	0	1

human colonic bacteria are able to ferment some plant polysaccharides, few show such a wide range of substrate fermenting activity as these *Bacteroides* species. Most strains of *B. thetaiotaomicron* and *B. ovatus* species are also capable of fermenting a wide range of host-derived polysaccharides including mucopolysaccharides such as hyaluronate, heparin, chondroitin sulphate and ovomucoid, although most are unable to ferment pig or bovine mucin. Degradation of mucins in the colon, which has potentially important consequences for intestinal health as the mucin layer is thought to provide a protective barrier against infection and damage, is likely to involve combined action together with other components of the microflora including *Ruminococcus* spp. Many of the mucopolysaccharides contain uronic acids as major components and do not appear to be fermented by other groups of human colonic bacteria. It is suggested that fermentation of uronic acids such as D-glucuronate and D-galacturonate may be very uncommon outside the *Bacteroides/Prevotella* grouping.

Glycosidases

In addition to polysaccharide fermentation, *Bacteroides* utilize a wide range of soluble sugars as energy sources. Their glycosidase activities may be important in the conversion of a wide variety of plant glycosides in the colon. In some cases these may yield toxic products, as in the case of cyanogenic glycosides, whereas in other cases the products may inhibit other bacteria such as *Escherichia coli*. A potentially significant role is in the enterohepatic circulation of toxic compounds, whereby diet-derived toxins converted by the liver into glucuronide derivatives are reconverted by bacterial enzymes in the large intestine to release the toxin. This type of circulation also applies to certain drugs such as morphine, thus complicating the problem of achieving the correct dosage.

Other potentially important interactions between colonic *Bacteroides* and the host include the possibility that bacterial activities can influence the state of glycosylation of the cells lining the intestine. This was suggested by recent observations on the effects of *B. thetaiotaomicron* strains introduced into the digestive tract of gnotobiotic (germ-free) mice.

Protein Metabolism

The *B. fragilis* group appear to be the most numerous proteolytic bacteria in the human colon, and utilize trypsin and chymotrypsin particularly well. Their proteases are largely cell bound, at least in growing cells. It has been suggested that *Bacteroides* proteinases could cause damage to mucosal membranes in the gut. Products of peptide and amino acid fermentation by colonic bacteria have many possible effects. Accumulation of ammonia can be cytopathic for colonic epithelial cells, whereas amines such as putrescine affect colonic cell growth and differentiation. Breakdown of aromatic amino acids yields a range of phenolic compounds.

Metabolism of Xenobiotics and Carcinogens

Other aspects of the conversion of dietary components may have important consequences for the host. Possible correlations between a diet rich in meat, a high population of *Bacteroides* spp. and a high incidence of colon cancer have been postulated but the validity of this link is still unclear.

Azo compounds are present in the diet as food colours and as pharmaceuticals. Azoreductases present in colonic bacteria, including *Bacteroides* spp., cleave these compounds, releasing aromatic amines. In some cases, there is concern that these products may be carcinogenic, and this has led to restrictions on the numbers of compounds permitted for use in food. In other cases, such as the anticolitis drug salazopyrin, the azoreductase activity results in the release of an antibiotic (sulphapyridine) and an anti-inflammatory (aminosalicylic acid) in the colon.

The nitroreductases of *Bacteroides* and other gut bacteria act on a wide range of substituted nitrophenols. For example, nitropyrenes, common in urban air particles, can be reduced to potent mutagens. In addition to generating mutagenic compounds (in particular a glyceryl ether lipid) *Bacteroides* can also act on some nitro-compounds to reduce their mutagenicity.

The bile is an important route for the excretion of glutathione conjugates of xenobiotics by the xenobiotic metabolizing enzyme system. The colonic flora is presented with cysteine conjugates as a result of the metabolism of glutathione in the liver, bile and small intestine. The enzyme C-S lyase (β lyase), produced by *Bacteroides*, *Fusobacterium*, *Eubacterium* and other species converts cysteine conjugates into a thiol metabolite, pyruvic acid and ammonia.

Bile Salt Metabolism

Bacteroides probably play an important role in bile salt metabolism in the gut, and may be beneficially affected by the products. Most colonic *Bacteroides* isolates produce cholglycine hydrolase, the enzyme responsible for the first step in bile salt metabolism, releasing free cholic acid and glycine. A significant proportion of strains produce 7-dehydrolase and hydroxysteroid dehydrogenases. Desoxycholate stimulates growth of *Bacteroides* strains, possibly a selective advantage for *Bacteroides* in vivo, since this compound suppresses growth of some other bacteria.

Pathogenicity

Strains of *Bacteroides*, particularly *B. fragilis*, are reported as opportunistic pathogens involved in a variety of deep infections and abcesses. *B. fragilis* strains isolated from calves and lambs with diarrhoea have been shown to produce a heat-labile enterotoxin of M_r 19 500 whose action has been demonstrated in tests using rabbit ileal loops. There has been no suggestion that *Bacteroides* strains can spread as food-borne human pathogens and their obligately anaerobic nature makes this improbable.

Bacteriocins and Bacteriophages

Bacteriocin production by *Bacteroides* spp. was first reported in the mid-1950s in France and occurs in around one quarter of faecal strains of *Bacteroides*. As with other species, the ecological role of bacteriocin production by *Bacteroides* is not clear. Although bacteriocin-sensitive strains outnumber producers in the gut, it remains possible that bacteriocin production is a factor in the survival of the producer strains. Bacteriophages have been reported that infect human colonic *Bacteroides* and rumen *Prevotella*.

Importance in Agriculture and Food Production

Prevotella species, related to colonic *Bacteroides*, are the largest single bacterial group reported from the rumen of cattle and sheep under most dietary regimes. These organisms are highly diverse, and the single species *P. ruminicola* has been reclassified into four new species, *P. ruminicola*, *P. bryantii*, *P. brevis* and *P. albensis*. Most species utilize starch as a growth substrate, and several species or strain groupings contribute to the utilization of plant cell-wall material by ruminants through their ability to hydrolyse xylans and pectins and to utilize breakdown products from plant cell-wall degradation. This group is considered to have a particularly important role in the metabolism of protein and peptides, since many strains are actively proteolytic and possess a characteristic dipeptidyl peptidase activity that is readily detectable in rumen contents. Although the host animal benefits from post-ruminal utilization of microbial cell protein, breakdown of dietary protein by ruminal bacteria can cause serious loss of amino acid nitrogen as ammonia. *Prevotella* and *Bacteroides* species are also reported to be significant members of the hind gut microflora of pigs.

See also: **Bacteriocins**: Potential in Food Preservation. **Microflora of the Intestine**: The Natural Microflora of Humans.

Further Reading

Benno Y, Endo K, Mizutani T et al (1989) Comparison of fecal microflora of elderly persons in rural and urban areas of Japan. *Appl. Environ. Microbiol.* 55: 1100–1105.

Bry L, Falk PG, Midvedt T and Gordon JI (1996) A model of host–microbial interactions in an open mammalian ecosystem. *Science* 273: 1380–1383.

Levett PN (1990) *Anaerobic Bacteria*. Milton Keynes: Open University Press.

Levett PN (ed.) (1991) *Anaerobic Microbiology: A Practical Approach*. Oxford: IRL Press.

Hill MJ (ed.) (1995) *Role of Gut Bacteria in Human Toxicology and Pharmacology*. London: Taylor and Francis.

Hobson PN and Stewart CS (eds) (1997) *The Rumen Microbial Ecosystem*, 2nd edn. London: Chapman Hall.

Holdemann LV, Cato EP and Moore WEC (1977) *Anaerobe Laboratory Manual*, 4th edn. Blacksburg, VA: Virginia Polytechnic Institute.

Holdemann LV, Cato EP and Moore WEC (1993) *Anaerobe Lab Manual Update*. Blacksburg, VA: Virginia Polytechnic Institute.

Holdeman LV, Kelley RW and Moore WEC (1986) Genus 1. Bacteroides Castellani and Chalmers 1919, 959 AL. In: Kreig NR and Holt JG (eds) *Bergey's Manual of Systematic Bacteriology*. Vol. 1, 604. Baltimore, Williams and Wilkins.

Hoskins L (1993) Mucin degradation in the human gastrointestinal tract and its significance to enteric microbial ecology. *Eur. J. Gastroenterol. Hepatol.* 5: 205–213.

Gibson GF and MacFarlane GT (eds) (1995) *Human Colonic Microbiology*. Boca Raton: CRC Press.

Mackie RI and White BA (eds) (1996) *Gastrointestinal Microbiology*, vol 1. London: Chapman & Hall Microbiology series.

Mackie RI, White BA and Isaacson RE (eds) *Gastrointestinal Microbiology*, vol 2. London: Chapman & Hall Microbiology series.

Moore WEC and Moore LH (1995) Intestinal floras of populations that have a high risk of colon cancer. *Appl. Environ. Microbiol.* 61: 3202–3207.

Salyers AA (1984) *Bacteroides* of the human lower intestinal tract. *Annu. Rev. Microbiol.* 38:293–313.

Salyers AA, Vercellotti JR, West SEH and Wilkins TD (1977) Fermentation of mucin and plant polysaccharides by strains of *Bacteroides* from the human colon. *Appl. Environ. Microbiol.* 33: 319–322.

Shah HN (1992) The genus *Bacteroides* and related taxa In: Balows A, Trüper AG, Dworkin M, Harder W and Schleifer KH (eds) *The Prokaryotes*. Vol. IV, p. 3593. New York: Springer-Verlag.

BACTERIOPHAGE-BASED TECHNIQUES FOR DETECTION OF FOOD-BORNE PATHOGENS

Richard J Mole and **Vinod K Dhir**, Biotec Laboratories Ltd, Ipswich, Suffolk, UK

Stephen P Denyer, Department of Pharmacy, The University of Brighton, UK

Gordon S A B Stewart (dec), Department of Pharmaceutical Sciences, The University of Nottingham, UK

Introduction

The drive toward the rapid detection of bacterial pathogens has focused on novel technologies that are poised to supersede conventional culture-based methods. Bacteriophages (bacterial viruses) possess features, such as specificity and rapid growth, which make them ideal agents for the rapid detection of bacteria. Several methodologies that have taken advantage of these attributes include: expression of bacteriophage-encoded bioluminescent genes in specific target cells (*lux*-bacteriophage); detection of bacteria by the intracellular replication of specific bacteriophages (termed 'phage amplification'); and the use of bacteriophage enzymes that specifically lyse target bacteria, with the concomitant release of ATP (termed 'lysin-release ATP-bioluminescence'). At present these methodologies are not commercially available within the food industry, but their potential for providing the basis for future rapid detection systems is significant (**Table 1**).

Bacteriophage Characteristics

Bacteriophages (often called phages) are obligate intracellular viral parasites that infect bacteria. In their native form they are metabolically inactive, comprising nucleic acid (usually double-stranded DNA) surrounded by a protein coat called a capsid, and, in some instances, a lipid envelope. Native bacteriophage particles are capable of existing for long periods of time between liberation from the parent cell and contact with new host cells. In the presence of a susceptible cell, the phage becomes attached to the cellular surface and the viral DNA enters the cell (**Fig. 1**). Subsequently, the intracellular bacteriophage exploits the host's metabolic machinery to achieve the replication of the viral genome and capsid. The regulation of the replicative cycle requires the expression of phage genes, some of which encode for the protein sub-units that constitute the capsid. Association of the replicated viral genome with the capsid proteins results in the formation of mature phage particles, which are usually liberated from the infected

Table 1 Perceived advantages and disadvantages of bacteriophage-based detection of pathogenic bacteria

Methodology	Advantages	Disadvantages
lux-bacteriophage	Rapid (1–2 h)	Genetically modified phage
	Specific	Equipment-orientated
	Sensitive	Host range of phage
	Quantitative	Food matrix may interfere with detection sensitivity
	Only detects viable cells	
Phage amplification	Rapid (4–10 h)	Host range of phage
	Specific	
	Sensitive	
	Quantitative	
	Simple	
	Only viable cells detected	
Lysin-release ATP-bioluminescence	Extremely rapid (10 min)	Equipment-orientated
	Specific	Low sensitivity
	Quantitative	
	Compatible with current ATP assay	
	Only viable cells detected	

cell by lysis. This event is normally achieved by phage-induced breakdown of the bacterial peptidoglycan layer, resulting in loss of the structural integrity of the cell wall, causing the cell to rupture and release progeny bacteriophages. This process is often mediated by a two-component system, comprising a peptidoglycan digesting enzyme (lysin) and a transport holin protein, which facilitates the passage of the lysin across the membrane to the cell wall.

Applicability of Bacteriophages for the Detection of Bacterial Pathogens

Although bacteriophages were discovered at the beginning of the twentieth century, their usefulness for the detection of bacterial cells has largely been overlooked. Instead, attention has focused primarily on their use as molecular biology tools and in phage-typing schemes for bacterial identification. The latter application highlights one of the key attributes – specificity – that renders bacteriophages suitable for use in the detection of bacterial pathogens.

It is possible to differentiate bacterial strains by their pattern of susceptibility to a panel of bacteriophages – a property that is exploited in phage typing. This property arises because of restrictions in the ability of bacteriophages, which can infect only a certain range of bacterial hosts – usually members of the same genus – which are capable of sustaining the replication of the phage. For instance, the *Campylobacter* bacteriophage targets *Campylobacter* cells, irrespective of the presence of other bacteria, due to the inherent specificity encoded by the bacteriophage.

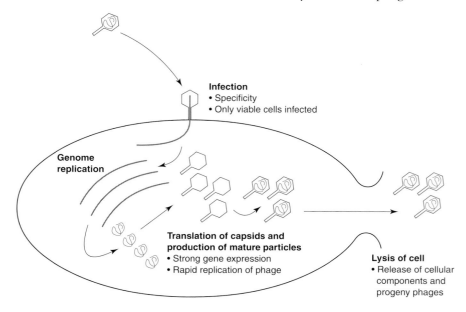

Figure 1 Replication of a lytic bacteriophage and the physiological characteristics that make bacteriophages suited to the detection of bacterial pathogens.

The identification of bacterial pathogens by cell culture often requires confirmatory tests to be performed following presumptive identification. For example, confirmation of the identity of *Campylobacter* species after growth in a selective environment requires Gram staining, visualization of motility and oxidase testing. The detection of *Salmonella* species requires several overnight incubation steps in enrichment media for unequivocal identification. However, the specificity of bacteriophages for target genera obviates the need for laborious biochemical, serological and morphological confirmatory testing.

Bacteriophages are reliant on the metabolic machinery of their host cells, hence successful replication can only be sustained in metabolically active cells. Dead cells are generally of little concern to public health – therefore the inability of bacteriophages to infect such cells can be used as a viability indictor.

Traditional culture techniques rely on the exponential doubling of bacterial cells, whereas the geometric replication of bacteriophage particles enables their detection within a shorter time frame. For example, bacterial growth over 30 min will result in a doubling of cell numbers (assuming a doubling time of 30 min) whereas the replication of bacteriophages would generate a twenty- to hundredfold increase in phage particles. The liberation of progeny phage can be visualized more rapidly than cell growth, either by plaque formation on lawns of susceptible bacteria (visualization is possible after 4 h on lawns of *Salmonella*) or by lysis of liquid cultures.

The rapidity of phage replication requires tightly regulated gene expression within the cell, with strong promotion of bacteriophage gene transcription and translation. Gene products from infecting phages can be detected rapidly (usually in < 30 min) post-infection, and hence can be used as an indication of bacterial presence.

The *lux*-bacteriophage

Concept

An innovation in the use of bacteriophages for the detection of cells involved bacteriophage-carrying genes which produced 'visible' products within infected cells. For instance, the *luxAB* genes, that code the enzyme responsible for bioluminescence in bacteria (luciferase) have been cloned into the genomes of several bacteriophages. In the production of light in this reaction, aldehyde (R.CHO) is converted to carboxylic acid (R.COOH). The reaction also requires both the reducing agent flavin mononucleotide ($FMNH_2$ and O_2). (**Equation 1**):

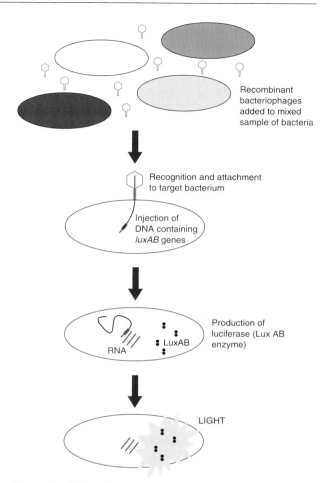

Figure 2 Schematic representation of the steps involved in cell detection using recombinant bacteriophages carrying the bioluminescent *luxAB* genes.

$$FMNH_2 + R.CHO + O_2 \xrightarrow{\text{luciferase(luxAB)}}$$
$$FMN + R.COOH + H_2O + LIGHT$$

(**Equation 1**)

Once a recombinant bacteriophage has recognized and infected a target cell, bacteriophage DNA enters the cytoplasm and there is sequential transcription and translation of the bacteriophage genome. If the *luxAB* genes are placed downstream of a strong promoter (e.g. a promoter regulating the production of the bacteriophage structural components), expression of the LuxAB enzyme (luciferase) will follow (**Fig. 2**). Hence infected cells, with the luciferase enzyme present, will produce light (on the addition of exogenous aldehyde), which can be detected by sensitive light cameras, luminometers or on photographic film. Uninfected cells will not produce luciferase, and remain dark. Viability is essential for light production, because dead cells will not permit phage infection or luciferase gene expression. In addition, $FMNH_2$ essential for light production is present only in meta-

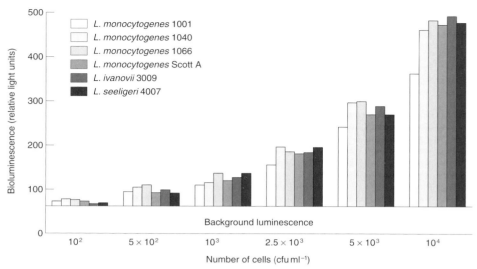

Figure 3 Detection of low numbers of *Listeria* species using the bioluminescent bacteriophage A511::*luxAB*. Cultures were diluted to the concentrations shown, before infection with 3×10^8 A511::*luxAB* particles per millilitre. Bioluminescence was measured after 130 min incubation at 20°C. Background bioluminescence (approximately 60 relative light units) is also shown. (Reproduced with permission from MJ Loessner, CED Rees, GSAB Stewart and S Scherer (1996).)

bolically active cells. The rapidity and sensitivity of both the *lux*-phage-based technology and modern light detection equipment provide the basis for a test capable of detecting single cells within an hour.

Listeria lux-phage

A *luxAB* cassette (from *Vibrio harveyi*, a marine bioluminescent bacterium) has been inserted into the genome of the virulent *Listeria* phage, A511, downstream of the major capsid protein gene, *cps*. A strong promoter governs this area of the genome, resulting in the expression of these genes in the later stages of the bacteriophage replicative cycle – luciferase expression and light production occur 20 min post infection. The presence of the *luxAB* genes within the genome has no detrimental effect on bacteriophage replication and propagation.

Bacteriophage A511 is capable of infecting 95% of *Listeria monocytogenes* serovars 1/2 and 4 which are responsible for most, if not all, cases of human listeriosis. Hence the recombinant phage A511::*luxAB* will cover the majority of clinically important *Listeria* isolates, permitting their rapid detection.

Detection of Listeria Cells using lux-phages

The A511::*luxAB* phage has been used successfully for the detection of several *Listeria* strains (**Fig. 3**). The maximum sensitivity of bacterial detection using *lux*-phages was determined as approximately 100 cells per millilitre (10^2 cfu ml^{-1}), below which the light output of infected cells was indistinguishable from the background bioluminescence. Broth enrichment

of the sample prior to infection with phages improved the ability of the test to detect low numbers of bacterial cells. For example, *L. monocytogenes* was detected in food samples seeded at 0.1 cfu g^{-1} (cabbage), 1 cfu g^{-1} (milk) and 10 cfu g^{-1} (Camembert cheese). The variability in the cell detection limits observed in different foods is thought to be a reflection of the complexity of the food matrices and the level of competitive microflora expected within the food.

The presence of *Listeria* can be confirmed by the production of light from samples using *lux*-phage technology (including the broth enrichment step) after 24 h. In contrast, conventional techniques take up to 4 days for presumptive listerial identification, with the possible need for further confirmatory testing. A phage-based (A511::*luxAB*) most probable number (MPN) method has been described, which achieves reliable and rapid detection of listerial cells (**Table 2**). Dilutions of cell suspensions were incubated for 20 or 44 h in a manner similar to conventional MPN techniques, prior to infection with *lux*-phage. Emitted light, rather than culture turbidity, was taken as an indication of cell growth. Estimations of the cell numbers in the original samples were then made using conventional MPN statistical tables.

Detection of Salmonella Cells using lux-phages

An MPN technique using bioluminescent phage P22 (carrying *luxAB* genes), capable of infecting salmonella cells, has been developed. The technique gives a direct representation of the presence of *salmonella*, with no apparent interference by competitive microflora. In addition, direct correlation between the

Table 2 Enumeration of *Listeria monocytogenes* Scott A from foods by the luciferase reporter phage-based MPN method (Reproduced with permission from MJ Loessner, M Rudolf and S Scherer (1997).)

Food sample	Approx. initial contamination rate (cells g⁻¹)	Luciferase phage-based MPN assay (cells g⁻¹) after incubation for:	
		20 h	44 h
Minced meat (hamburgers)	1.6×10^2	1.1×10^2	2.0×10^2
	1.6×10^3	4.3×10^4	4.6×10^4
Shrimp	8.0×10^1	6.0×10^1	6.0×10^1
	2.0×10^3	1.5×10^3	1.5×10^3
Ricotta cheese	8.0×10^1	2.3×10^2	2.3×10^2
	2.0×10^3	2.3×10^3	2.4×10^3
	1.6×10^4	1.5×10^4	1.5×10^4
Lettuce (sliced iceberg)	8.0×10^1	1.5×10^2	1.5×10^2
	2.0×10^3	4.3×10^3	4.3×10^3
	1.6×10^4	4.6×10^4	4.6×10^4

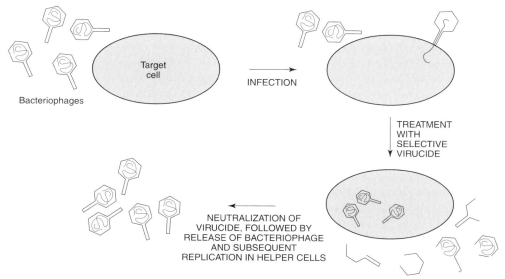

Figure 4 Schematic representation of the steps taken in a phage amplification assay.

number of *S. typhimurium* cells and enumeration by the *lux*-phage-based MPN technique has been established.

Phage Amplification

Concept

The 'Phage Amplification' technique for detecting bacteria relies on two key characteristics: the specificity of bacteriophages to target cells, and the ability of a potent virucidal agent to rapidly inactivate free (extracellular) phages whilst remaining non-destructive to bacterial cells. These attributes permit the detection of specific groups of bacteria on the basis of their ability to produce progeny phage particles.

Target bacteria are exposed to a bacteriophage reagent during an initial infection period: this permits recognition, attachment and infection of the host cells by the bacteriophages (**Fig. 4**). This is followed by treatment with the selective virucide, resulting in the destruction of all exogenous bacteriophage particles, without damage to the target cells. Phage particles that have infected target cells are shielded from the virucidal process and continue replication, eventually yielding progeny phages. Hence, any phages present after the virucidal treatment will have originated from infected cells. The virucide is neutralized after a 3-min contact time with bacteriophages, so that phages released from infected cells are not inactivated by residual virucide. Progeny phage particles are not immediately detectable, but if helper bacteria are added (typically a phage-susceptible, non-pathogenic variant), amplification of the phage signal can be achieved. For instance, plaques may be seen in the lawns of helper bacteria, signifying the presence of bacteriophages and hence of target cells that were infected by the bacteriophages. The number of plaques present should correlate with the number of cells in the original sample. Alternatively, the presence of progeny bacteriophages may be assessed by the lysis of liquid cultures of helper bacteria, resulting in a semi-quantitative estimation of target cell populations.

An assay may be tailored to the detection of specific bacterial genera by the choice of bacteriophage used in the assay. For instance, if a *Campylobacter* phage were used, only *Campylobacter* cells would be detected. The differentiation of dead cells from viable cells is possible with phage amplification, because the bacteriophages require viable cells for replication. The rapidity of detection by phage amplification is a major benefit compared to conventional cell replication-based techniques (**Fig. 5**). The visualization of plaques in lawns of *Salmonella* and *Pseudomonas* is possible within 4 h, permitting detection of these pathogens within a working day. Phage Amplification has been successfully applied to the specific detection of *Campylobacter*, *Pseudomonas*, *Salmonella*, *Staphylococcus* and *Mycobacterium* cells, the latter in a diagnostic test for tuberculosis.

Detection of *Campylobacter* by Phage Amplification

Campylobacter jejuni cells have been detected by Phage Amplification with a level of sensitivity of approximately 1 in 12 cells (**Fig. 6**), equating to a limit of detection of 120 cells per millilitre in pure culture. The presence of high numbers of competitive microflora, as may be encountered in a food environment, has a limited effect on the assay sensitivity, with only a marginal decline in detection efficiency (**Fig. 7**). Initial work involving the application of technology to the food matrix, in particular the detection of *Campylobacter* cells in chicken meat, has been extremely encouraging.

An advantage of phage amplification technology is its adaptability. For instance, an integrated assay is capable of detecting either *Salmonella* or *Campylobacter* cells, depending on the choice of helper bacteria and the growth conditions employed for plaque formation (**Fig. 7**). This technique is likely to reflect

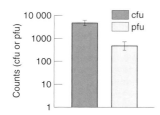

Figure 6 Detection of *Campylobacter jejuni* ATCC 35922 with bacteriophage φ5 by Phage Amplification in 100 μl of growth medium. Dilutions in growth medium of overnight *C. jejuni* ATCC 35922 cultures were assayed by conventional plate culture (cfu) and by Phage Amplification (pfu). Bacteriophage φ5 was incubated with 100 μl of cell suspension for 20 min at 42°C prior to a 3-min virucidal treatment and neutralization step. *C. jejuni* ATCC 35922 was used as the helper bacterium to form lawns in the agar overlays. (cfu:pfu ratio = 12.2 (SEM ± 1.7), *n* = 21).

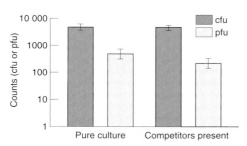

Figure 7 Detection of *Campylobacter jejuni* ATCC 35922 and *Salmonella typhimurium* DB in a combined *Campylobacter* and *Salmonella* Phage Amplification assay. A dilution series of *C. jejuni* ATCC 35922 and *S. typhimurium* DB was prepared in growth medium. Cells were detected by either conventional plate culture (cfu) or phage amplification (pfu). 100 μl of mixed cell suspension was incubated with 10 μl of combined *Campylobacter* phage (φ5) and *Salmonella* phage (Felix-01). After virucidal treatment (3 min) and neutralization, either *C. jejuni* ATCC 35922 or *S. typhimurium* DB helper cells were added and incubated in conditions appropriate to permit amplification of each bacteriophage, and hence the determination of the number of individual target bacteria. (*Campylobacter* – cfu : pfu ratio = 33.4 (SEM ± 2.0), *n* = 2. *Salmonella* – cfu : pfu ratio = 21.8 (SEM ± 6.3), *n* = 2.)

the needs of the food industry for the detection and differentiation of several bacterial pathogens concurrently even amongst competitor organisms (**Fig. 8**).

Bacteriophage Lysin-release ATP-bioluminescence

Concept

Successful phage replication requires a strategy for the release of mature particles from the host cell. This is normally achieved by the phage-induced degradation of the peptidoglycan layer of the bacterial cell wall, the layer which maintains the structural integrity of the cell. Recently, a bacteriophage gene encoding a cell wall degrading enzyme (lysin) has been cloned from a *Listeria* phage. Unexpectedly, this enzyme demonstrated specificity for the peptido-

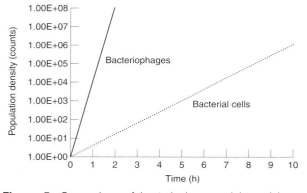

Figure 5 Comparison of bacteriophage and bacterial replication rates.

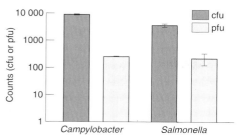

Figure 8 Detection of *Campylobacter jejuni* ATCC 35922 by Phage Amplification assay in mixed populations, with bacteriophage φ5. Dilutions of *C. jejuni* ATCC 35922 were cultured overnight in media containing: *Pseudomonas fluorescens* (approx. 10^8 cfu ml^{-1}), *Escherichia coli* (approx. 10^9 cfu ml^{-1}), *Citrobacter freundii* (approx. 10^9 per ml) and *Salmonella typhimurium* DB (approx. 10^9 cfu ml^{-1}). The cultures were assayed by conventional plate culture (cfu) and by Phage Amplification (pfu). Bacteriophage φ5 was incubated with 100 µl of cell suspension for 20 min at 42°C prior to a 3-min virucidal treatment and neutralization step. *C. jejuni* ATCC 35922 was used as the helper bacterium and was grown in agar overlays on selective medium. (Pure culture – cfu : pfu ratio = 12.2 (SEM ± 1.8), n = 6. Competitors present – cfu : pfu ratio = 82.1 (SEM ± 52.7), n = 6.)

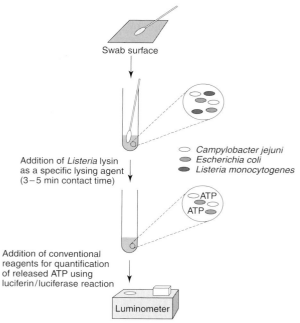

Figure 9 Schematic representation of the proposed lysin-release ATP-bioluminescence assay.

glycan from *Listeria* cell walls, and was incapable of digesting the peptidoglycan from other bacterial genera (except two strains of *Bacillus*). Hence, this lysin provides the capability of specifically lysing *Listeria* cells within a sample, while other bacterial cells remain intact.

This finding may be exploited, to enhance the detection of bacteria using ATP-bioluminescence, by adding specificity to an otherwise nondiscriminatory test for microbial contamination. Conventional ATP-bioluminescence relies on a nonselective releasing agent that lyses all the cells present in a sample, leading to the release of their ATP. This rise in ATP can be monitored using the firefly luciferase/luciferin system, which produces light at the expense of ATP (**Equation 2**). In the presence of the luciferin substrate and O$_2$, the luciferase enzyme converts the energy stored in ATP into light:

$$\text{Luciferin} + \text{ATP} + \text{O}_2 \xrightarrow{\text{Luciferase}}$$
$$\text{Oxyluciferin} + \text{AMP} + \text{pyrophosphate} \quad \text{(Equation 2)}$$
$$+ \text{LIGHT}$$

Hence light production (usually detected by a luminometer) signifies the presence of bacteria in a sample, but gives no information about the identity of the bacteria present. If lysin were used instead of the nonselective releasing agent, the release of ATP (and hence light) could be attributable to a specific genus – for instance, *Listeria* if the *Listeria* lysin were employed (**Fig. 9**).

This combined technology can yield results quickly, mirroring the rapidity of conventional ATP-bioluminescence systems, with detection times within

10 min. The sensitivity of detection is limited by the ability to detect rises in ATP concentration, a problem that is inherent in the detection of very low levels of light. However, recent advances in ATP-detection technology will directly enhance the sensitivity of the lysin-release ATP-bioluminescence assay. Conventional ATP-bioluminescence is already widely established in the food industry, so implementation of the combined technology will be straightforward.

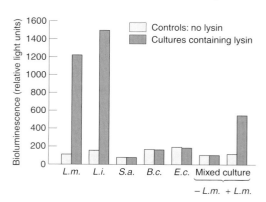

Figure 10 Lysin-release ATP-bioluminescence detection of *Listeria* species. Detection of released ATP using a firefly luciferase–luciferin assay after specific lysis of *Listeria* cells with lysin HPL118. *L.m.*: *Listeria monocytogenes*; *L.i.*: *Listeria ivanovii*; *S.a.*: *Staphylococcus aureus*; *B.c.*: *Bacillus cereus*; *E.c.*: *Escherichia coli*. Mixed cultures contained *S. aureus*, *B. cereus*, and *E. coli*. Bacterial cultures were diluted to approximately 10^7 cfu ml^{-1}, mixed with 5 U of lysin, incubated for 10 min at room temperature and assayed for released ATP. Controls had no lysin added. (Reproduced with permission from GSAB Stewart, MJ Loessner and S Scherer (1996).)

Detection of *Listeria* using Lysin-release ATP-bioluminescence

A combined lysin-release ATP-bioluminescence reaction demonstrates the specificity of lysin for *Listeria* cells (**Fig. 10**). *Listeria*-seeded samples exhibited increased production of light as a consequence of the release of ATP from *Listeria* cells, whereas samples lacking *Listeria* did not show any significant increase in light production. The technique coped with the presence of competitive microflora, permitting the differentiation of *Listeria*-positive samples from samples lacking *Listeria* cells.

See color Plates 5 and 6.

See also: **ATP Bioluminescence**: Application in Meat Industry; Application in Dairy Industry; Application in Hygiene Monitoring; Application in Beverage Microbiology. *Campylobacter*: Detection by Cultural and Modern Techniques. **Enrichment Serology**: An Enhanced Cultural Technique for Detection of Foodborne Pathogens. *Listeria*: Detection by Classical Cultural Techniques; *Listeria monocytogenes*; *Listeria monocytogenes* – Detection by Chemiluminescent DNA Hybridization. **Molecular Biology – in Microbiological Analysis**. *Mycobacterium*. *Salmonella*: Introduction. **Nucleic Acid-based Assays**: Overview. *Salmonella*: Introduction; Detection by Classical Cultural Techniques.

Further Reading

Ackermann HW and DuBow MS (1987) *Viruses of Prokaryotes*. Vol. 1. Boca Raton: CRC Press.

Loessner MJ, Rees CED, Stewart GSAB and Scherer S (1996) Construction of luciferase reporter bacteriophage A511::*luxAB* for rapid and sensitive detection of viable *Listeria* cells. *Applied and Environmental Microbiology* 62 (4): 1133–1140.

Loessner MJ, Rudolf M and Scherer S (1997) Evaluation of luciferase reporter bacteriophage A511::*luxAB* for detection of *Listeria monocytogenes* in contaminated foods. *Applied and Environmental Microbiology* 63 (8): 2961–2965.

Graham J (1996) Timely test spots TB in hours. *New Scientist*, No. 2043, 17th March.

Stewart GSAB (1997) Challenging food microbiology from a molecular perspective. *Microbiology* 143: 2099–2108.

Stewart GSAB, Jassim SAA, Denyer SP et al (1998) The specific and sensitive detection of bacterial pathogens within four hours using phage amplification. *Journal of Applied Microbiology* 84(5): 777–783.

Stewart GSAB, Loessner MJ and Scherer S (1996) The bacterial *lux* gene bioluminescent bio-sensor revisited. *ASM News* 62 (6): 297–301.

Turpin PE, Maycroft KA, Bedford J, Rowland CL and Wellington EMH (1993) A rapid luminescent-phage based MPN method for the enumeration of *Salmonella typhimurium* in environmental samples. *Letters in Applied Microbiology* 16: 24–27.

Young RY (1992) Bacteriophage lysis: Mechanism and Regulation. *Microbiological Reviews* 56 (3): 430–481.

Beer *see* **Lager**.

Benzoic Acid *see* **Preservatives: Permitted Preservatives – Benzoic Acid**.

Beverage Microbiology *see* **ATP Bioluminescence: Application in Beverage Microbiology**.

BIFIDOBACTERIUM

Dallas G Hoover, Department of Animal and Food Sciences, University of Delaware, Newark, USA

Introduction

Bifidobacteria were once largely unknown by people working in the area of food science and technology, but since the mid-1980s there has been a revival of interest due to the expanded use of bifidobacteria as additives in products that are now marketed as nutriceuticals or functional foods. Consumer interest in nutritional health and wellbeing is the driving force for application of these anaerobic gut bacteria for use as probiotic cultures. Subsequently, there have been studies to evaluate bifidobacteria in foods and to study the physiological response of people fed bifidobacteria and supplements that enhance proliferation of resident bifidobacteria (prebiotics). This article describes the genus *Bifidobacterium* and reviews current knowledge of these organisms used in our food supply.

Historical Perspective

Discovery

Bifidobacteria have generated attention from people interested in host–bacterium relationships in humans. This was true at the time of the discovery of bifidobacteria by Henri Tissier in 1900 from the faeces of newborn infants. Tissier called his Gram-positive, curved and bifurcated (clefted, X- or Y-shaped) rod-like cells *Bacillus bifidus communis*. (Tissier's original isolate is now referred to as *Bifidobacterium bifidum* Ti.) Soon afterwards, his colleague at the Institut Pasteur, Nobel Prize laureate biologist Elie Metchnikoff, incorporated Tissier's bacilli into his theories of vigour and long life. Although there were earlier reports of fermented milks with implied health benefits, Metchnikoff was the first to put the subject on a scientific basis. Metchnikoff spoke and published on his theories of sound health and longevity from the ingestion of lactobacilli and other bacteria present in such foods as yoghurt, kefir and sour milk. His work and statements led to a 20-year public demand for sour-milk products. Metchnikoff developed and perpetuated the theory that not only does the intestinal microbiota control the outcome of infection by enteric pathogens, but it also regulates the natural chronic toxaemia which plays a major role in ageing and mortality.

General interest in Metchnikoff's bacteriotherapy greatly diminished with the outbreak of World War I and Metchnikoff's death at the age of 71; however, studies on the use of lactic acid cultures in dietary regimen continued through the century and are now as popular as ever. The discovery of bifidobacteria in high numbers in healthy breast-fed infants and the fermentative/acidulating nature of bifidobacteria have long implied a beneficial relationship in human nutrition and gastrointestinal health. None the less, most of the studies in this century on nutrition therapy (or beneficial host–bacterium relationships) have focused on yoghurt cultures and lactobacilli such as *Lactobacillus acidophilus*. This is due in part to past practice, as well as the reputation of bifidobacteria as being difficult to work with and maintain. However, bifidobacteria, streptococci, enterococci, yeasts and other microorganisms have now attracted considerable attention for probiotic application. Not only have humans been evaluated for beneficial effect from the consumption of probiotic cultures, but domestic livestock and other animals as well.

Taxonomy

Over this century, the bifidobacteria have been assigned to at least eleven different genera. These have included genus names that could be anticipated, such as *Bacillus*, *Tisseria*, *Lactobacillus* and *Bifidobacterium*. Other assigned genera names were *Bacteroides*, *Bacterium*, *Nocardia*, *Actinomyces*, *Actinobacterium*, *Cohnistreptothrix* and *Corynebacterium*. Species names have varied accordingly but in the early years usually included some base form of bifid, such as *bifidus*, *bifidum*, *bifida* and *parabifidus*. *Bifidus* means cleft or divided in Latin (the characteristic split ends of the cells are evident when nutrition is restricted). The genus *Bifidobacterium* (originally from Orla-Jensen in 1924 and described in the 9th edition of *Bergey's Manual*) is phenotypically and morphologically outlined in **Table 1**.

Bifidobacteria were classified in the order Actinomycetales; a group characterized by the formation of branching filaments. This property can be described as a fungal appearance. Actinomycetes, which are well distributed in nature, were separated into two large subgroups: the very numerous oxidative forms that are commonly found in soil, and the fermentative types that are primarily found in the natural cavities of humans and animals. It is to this latter subgroup that the bifidobacteria belonged. *Bifidobacterium* was taxonomically separated from other actinomycetes (such as *Streptomyces* and *Nocardia*) by cell-wall type; the bifidobacteria were designated as having a type VIII cell wall (relatively high concentrations of ornithine).

Table 1 Genus description: *Bifidobacterium* (from Scardovi 1986)

Rods of various shapes: Short, regular, thin cells, slightly bifurcated club-shaped elements in star-like aggregates or disposed in 'V' or 'palisade' arrangements.

Gram-positive: Usually catalase-negative, non-sporeforming, non-motile; colonies on agar smooth, convex, entire edges, cream to white, glistening and of soft consistency.

Anaerobic: Some species aerotolerant (degree depends on species and culture medium); optimum growth temperature 37–41°C (minimum 25–28°C; maximum 43–45°C); optimum pH for initial growth 6.5–7.0, no growth at pH 4.5–5.0 or 8.0–8.5.

Saccharoclastic: Acetic and lactic acids are formed primarily in the molar ratio of 3:2; carbon dioxide is not produced; glucose is degraded exclusively and characteristically by the fructose 6-phosphate shunt (fructose 6-phosphoketolase cleaves fructose 6-phosphate).

Habitat: Intestines of humans, various animals and honeybees; also found in sewage and human clinical material.

On agar plates, colonies of bifidobacteria closely resemble those of lactic acid bacteria (especially lactobacilli). Bifidobacteria can easily be confused with lactobacilli and are often referred to as a member of the lactic acid bacteria; however, bifidobacteria are not closely related to any of the traditional lactic acid bacteria used in the production of fermented foods. For example, compared to lactobacilli, bifidobacteria are not as acid-tolerant, nor can their growth be termed 'facultative anaerobic'. Indeed, bifidobacteria do produce lactic acid from the fermentation of carbohydrates, but acetic acid is normally produced in equal or higher amounts than lactic acid and the catabolic pathway used is distinct from the homofermentative and heterofermentative pathways used by lactic acid bacteria. A key determinative assay to distinguish bifidobacteria from lactobacilli is the fructose 6-phosphate phosphoketolase assay. This enzyme splits the hexose phosphate to erythrose 4-phosphate and acetyl phosphate; bifidobacteria have this enzyme, lactobacilli do not. Another important distinction between the two genera is that the mean mol% G+C of DNA for *Lactobacillus* is approximately 37%, for *Bifidobacterium* the mean is about 58%.

As with other bacteria, technical improvements in identification protocols and expansion of information in microbial systematics have steadily increased the number of defined species in the genus. *Bergey's Manual of Systematic Bacteriology* (1986) identifies 24 species of *Bifidobacterium* (**Table 2**), of which the types considered primarily human in origin are the species: *bifidum, longum, infantis, breve, adolescentis, angulatum, catenulatum, pseudocatenulatum* and *dentium*. Most of these species predominate in the human colon and subsequently can be found in faeces and sewage. All the species associated with humans can ferment lactose, an important characteristic when considering the application of bifidobacteria in dairy products and as probiotic cultures with intended effectiveness in easing the discomfort of lactose malabsorption.

To secure probiotic benefits for human consumers, it is normally assumed that the most beneficial host–bacterium relationship would be dependent on the utilization of cultures isolated from humans. Isolated from animal faeces, *B. animalis* has been used as a human probiotic culture primarily due to its resistance to viability loss with refrigerated storage, but its use in foods appears to be diminishing because of its source. With limited application, the species of bifidobacteria commonly associated with animals have not been studied in as much detail as the human types. Taxonomically, the three species of *Bifidobacterium* associated with honeybees (*asteroides, coryneforme* and *indicum*) have relatively little in common genetically and phenotypically with any of the other species in the genus.

Not all bifidobacteria can be considered GRAS (generally regarded as safe) for use in foods. Isolates of *B. dentium* are potentially pathogenic. Strains can be isolated from human dental caries and the oral cavity, faeces of the human adult, the human vagina, abscesses and the appendix. Strains of *B. dentium* have also been named *B. appendicitis* and *Actinomyces eriksonii*. Although pathogenic, members of *B. dentium* are not considered highly infectious nor virulent in comparison to many common bacterial pathogens.

Intestinal Ecology

Approximately 10^{14} microorganisms populate the human gastrointestinal tract. This is over ten times the total number of human cells in the body. It has been estimated that up to 450 different species of microorganism reside in the human gut. Most of these organisms are located in the lower portion of the small intestine and the colon. The stomach and the upper intestine possess gastric acid, bile salts and a highly propulsive motility to keep the concentrations and diversity of the microbiota low. Along the length of the small intestine the microbiota gradually increases. With healthy conditions the population of bacteria in the upper intestine is generally less than 10^5 organisms per millilitre of contents. The middle of the small intestine is a transitional zone between the sparse populations of the upper intestine and the luxuriant levels found in the large intestine. The ileum contains approximately 10^7 bacterial cells per millilitre. Most of the intestinal lactobacilli reside here. Once past the ileocaecal valve, the intestinal population of the microbiota increases dramatically. The total concentration of bacteria in the large intestine approaches the theoretical limit that can fit into a given mass, approximately 10^{11}–10^{12} organisms per millilitre. Bifidobacteria are most prevalent in the large intestine, especially in the area of the caecum.

Given the large amount of microorganisms in residence, the human colon is a very active bioreactor. The microbiota of the colon is mostly anaerobic (about $1000 : 1$, anaerobes : aerobic or facultative bacteria). The large intestine can be described in three sections: the right ascending colon, the transverse (middle) colon, and the left descending colon. The ascending colon receives its contents from the small intestine via the ileocaecal valve. The right colon features active fermentation with high bacterial growth rates; the concentration of total short-chain fatty acid content is about $127 \, \mathrm{mmol \, l^{-1}}$ and the pH is 5.4–5.9. As the intestinal contents move towards elimination from

Table 2 Description of the 24 species of *Bifidobacterium* (consolidated from Scardovi (1986))

B. adolescentis	Predominates in faeces of human adults, occurs frequently in sewage; ferments pentoses; four biovars (*a–d*) vary in fermentation of sorbitol and mannitol; cannot be distinguished phenotypically from *B. dentium*, analysis of transaldolase isozymes necessary.
B. angulatum	Characteristically in 'V' (angular) or palisade forms, branching absent; more sensitive to oxygen than most bifidobacteria; isolated from human faeces and sewage; most strains do not ferment sorbitol and could be confused with *B. globosum*, *B. pseudolongum* and sorbitol-negative strains of *B. pseudocatenulatum* from calf faeces.
B. animalis	Isolated from faeces of calf, sheep, rat, chicken, rabbit and guinea pig, and sewage; phenotypically very similar to *B. longum* but inactive toward melezitose; DNA unrelated to any other species; two bivars *a* and *b*; can be distinguished from 'human' species by the absence of gluconate fermentation.
B. asteroides	Found in the intestine of the western and the asiatic honeybees; lactose-negative; growth in static fluid generally adheres to the glass walls leaving the liquid clear; hydrogen peroxide vigorously decomposed when grown in 90% air + 10% carbon dioxide (necessary for aerobic growth); strains contain high number of plasmids; 50% DNA homology to *B. choerinum*.
B. bifidum	Type species of genus; highly variable in appearance; serovar *a* predominates in faeces of human adults, *b* predominates in that of infants; contains strains once identified as *Lactobacillus bifidus* var. *pennsylvanicus* (György).
B. boum	From bovine rumen and pig faeces; cell morphology varies greatly; can grow in 90% air + 10% CO_2 becoming catalase-positive; nearly 70% DNA-related to *B. thermophilum*; distinction from *B. thermophilum* and *B. choerinum* by transaldolase electrophoresis or with polyacrylamide gel electrophoresis of proteins; lactose-negative.
B. breve	Represents the shortest and thinnest rods among bifids found in the human intestine; also found in human vagina and clinical material; most related to *B. infantis* and *B. longum* (40–60% DNA homology).
B. catenulatum	Cells form chains; found in faeces of human adult and sewage; carbon dioxide has no effect on oxygen sensitivity; distinguishable from *B. adolescentis* and *B. pseudocatenulatum* based on ability to ferment melezitose and lack of starch utilization; 55 (T_m) is lowest mol% G + C of DNA in the bifidobacteria.
B. choerinum	Found in faeces of piglets; mol% G + C of DNA is 66.3 (T_m); if not distinguishable from *B. thermophilum* and *B. boum* by sugar fermentation patterns, then transaldolase electrophoresis or polyacrylamide gel electrophoretic patterns of soluble proteins can be used.
B. coryneforme	Isolated from intestine of European honeybees; lactose-negative; CO_2 does not influence growth; grows poorly on TPY (trypticase–phytone–yeast extract medium) but very well on deMan, Rogosa and Sharpe medium; 60% DNA-relatedness to *B. indicum*.
B. cuniculi	Found in faeces of adult rabbit; highly anaerobic, CO_2 has no effect on growth; lactose, ribose and raffinose are not fermented which distinguishes from *B. globosum*, *B. pseudolongum* and *B. animalis*, and also the morphologically different *B. magnum*.
B. dentium	Morphological similarity to *B. infantis*; isolated from human dental caries and human abscesses, considered to have pathogenic potential; also found in faeces of human adult and human vagina; some DNA-relatedness to *B. adolescentis*, CO_2 does not affect growth; requires riboflavin and pantothenate for growth.
B. globosum	Anaerobically grown cells are short, coccoid or almost spherical; found in faeces of pig, suckling calf, rabbit and lamb, and rumen of cattle, occasionally sewage; displays anaerobic aerotolerance (in the presence of 10% CO_2); most closely related to *B. pseudolongum*; harbours high-molecular-weight plasmids.
B. indicum	From the intestine of honeybees; CO_2 required for aerobic growth; unrelated by DNA homology to any other species in the genus; lactose-negative.
B. infantis	Pentose-negative forms predominate in faeces of breast-fed infants, also found in human vagina; closely related to *B. longum*, can be differentiated on the basis that members of *B. longum* ferment both arabinose and melezitose.
B. longum	Can form very elongated and relatively thin cells; two biovars are defined, biovar *a* is found in adult humans and biovar *b* is found in infants and is mannose-negative; only species isolated from humans that usually harbours a large variety of plasmids.
B. magnum	From faeces of rabbit; cells are usually long and thick, and occur in aggregates; species is the most acid-tolerant of the bifidobacteria, originally optimum pH for growth is 5.3–5.5, growth is slow at 5.0–5.9, no growth at 4.2 or 7.0; DNA unrelated to any other species.
B. minimum	Small cells of bifidobacteria isolated from sewage or wastewater; few strains studied; no DNA-relatedness to other species; distinct polyacrylamide gel electrophoresis protein pattern; Lys-Ser interpeptide bridge of peptidoglycan unique among bifidobacteria.
B. pseudocatenulatum	Abundant in sewage, in faeces of infants and suckling calves; cell morphology is extremely variable and shows highly diverse traits according to strain and origin; DNA-related to *B. catenulatum* but DNA G + C content different by 3 mol%; riboflavin, pantothenate and nicotinic acid required for growth.
B. pseudolongum	Faeces of chicken, cattle, rat and mice; four biovars recognized on the basis of different fermentative patterns of mannose, lactose, cellobiose and melezitose; most similar to *B. globosum*, distinguished by G + C mol% of DNA and DNA homology patterns.

Table 2 Description of the 24 species of *Bifidobacterium* (consolidated from Scardovi (1986))—continued

B. pullorum	Faeces of chicken; requires nicotinic acid, pyridoxine, thiamin, folic acid, *p*-aminobenzoic acid and Tween 80 for growth; does not ferment lactose and starch; acetic and lactic acids produced in 3.5 : 1 ratio but unlike any other species of bifidobacteria, the isomeric type of lactic acid formed is DL; the mol% G + C DNA is 67.4 (T_m), highest in the genus; no DNA-relatedness to any other species.
B. subtile	From sewage and wastewater; optimum temperature for growth is 34–35.5°C, markedly lower than other species; lactose is not fermented; few strains have been studied; unrelated by DNA homology to any other species in the genus.
B. suis	Found only in faeces of piglets; riboflavin the only growth factor required; unrelated in DNA homology to any other species in the genus; can be distinguished from other species found in pig by ability to ferment arabinose and xylose and inability to ferment starch.
B. thermophilum	Faeces of pig, piglet, chicken, calf, rumen of cattle, and sewage; can grow at 46.5°C and survive 60°C for 30 min; four biovars have been defined; only DNA homology relatedness in bifidobacteria is with *B. longum* (27–80%).

the body, nutrients are depleted and bacterial activity slows. In the transverse colon, total short-chain fatty acid content is about $117 \, \text{mmol} \, \text{l}^{-1}$ and the pH is approximately 6.2. In the left colon, there is little carbohydrate fermentation; the end products of protein fermentation (phenols, indoles and ammonia) are relatively high. Total short-chain fatty acid concentration is about $90 \, \text{mmol} \, \text{l}^{-1}$ and the pH is 6.6–6.9. Thus the microbiota is capable of fermenting carbohydrates and proteins while metabolizing a wide range of compounds such as bile acids, fats and drugs.

Bifidobacteria thrive in this environment. *Bifidobacterium* species can be isolated from faeces of humans at any age. At birth, bifidobacteria are one of the first groups to establish themselves in the intestinal tract and usually are the largest group represented in infants. For breast-fed babies, levels of 10^{10}–10^{11} per gram of faeces are common. It is generally believed that during the birth process, bifidobacteria residing in the mother's vagina and faeces act as an oral inoculum for the developing intestinal microbiota of the newborn infant. Bottle-fed babies normally have 1-$\log_{10}$ count less bifidobacteria present in faecal samples than breast-fed babies, and bottle-fed infants generally have higher levels of enterobacteriaceae, streptococci and anaerobes other than bifidobacteria. Bifidobacteria constitute up to 90–99% of the intestinal flora in healthy breast-fed infants, whereas lactococci, enterococci and coliforms represent less than 1% of the population; bacteroides, clostridia and other organisms are absent. Such findings suggest a health advantage to breast-feeding due in part to the establishment and maintenance of high numbers of acidulating bifidobacteria in the gut.

The relationship between breast-feeding and high intestinal levels of bifidobacteria led to the belief that bifidobacteria require a growth factor present only in human milk, but this has been shown not to be the case. Apparently bifidobacteria grow better in human milk than bovine milk because of a lower protein content and a diminished buffering capacity, so that many infant formula manufacturers now adjust the protein and mineral profile to more closely approximate that of human milk.

With the change of diet and the ageing process following infancy the level of bifidobacteria declines so that Bacteroidaceae predominate in the adult gut, with eubacteria, bifidobacteria and Peptococcaceae represented in that order. In the elderly, bifidobacteria continue to decline with the increase in the faecal populations of coliforms, enterococci, lactobacilli and *Clostridium perfringens*.

The microbiota in the human colon varies significantly between individuals. This variation not only involves the species present, but also the fermentation capacity and metabolic product profile. The ability of the intestinal microbiota of an individual to ferment different carbohydrates depends on past diet and the species of bacteria present. These bacteria affect digestion and absorption, their metabolic products provide nutrients and affect the wellbeing of the host.

In healthy adults, the intestinal microbiota is fairly stable; however, in infants it is not particularly stable and is susceptible to fluctuations caused by small disturbances of diet or common childhood diseases. At any age, the equilibrium of the human intestinal ecosystem can be altered due to stress, diet, disease and drugs (i.e. antibiotics).

Prebiotics

Diet will affect the microorganisms of the intestinal tract. Some dietary fibres increase stool output and colonic content turnover, resulting in increased bacterial turnover and growth. These substances include cellulose, pectins, vegetable mucous substances, microbial and dietary polysaccharides, oligosaccharides, scleroproteins and Maillard products. In the case of some of these fibres, the increased bulk of bacterial cells is the major component of the increase in weight of the stool.

The term, prebiotic, is often used to describe use of a component intentionally added to the diet for desirable health benefits linked to stimulation of metabolism and proliferation of desirable gut bacteria while preferably inhibiting or minimizing the growth of undesirable varieties. Prebiotics are included in the segment of products known as functional foods or nutriceuticals, that is, foods that can prevent and treat diseases. Regarding prebiotics for bifidobacteria (e.g. bifidus growth factors), the earliest studies centred on the effects of human milk on gut bacteria. The list of compounds that have been examined and used as prebiotic compounds for specific growth enhancement of resident intestinal bifidobacteria include *N*-acetylglucosamine, glucosamine, galactosamine, human and bovine casein digestates, lactoserum of bovine milk, porcine gastric mucin, yeast extract, liver extracts, colostrums of various milks, milk glycoproteins, lactoferrin, lactulose, lactitol, carrots, chitin, raffinose, stachyose, inulin, Jerusalem artichoke flour, tri- and pentasaccharides of dextran, neosugar, fructooligosaccharides and galactooligosaccharides. Effects from ingestion of these prebiotic compounds vary and efficacy has been debated.

In Japan, oligosaccharides are one of the most popular functional food components. These physiologically functional oligosaccharides are the short-chain polysaccharides called fructooligosaccharides, galactooligosaccharides and soybean oligosaccharides. The two requirements for their use are that they are not digestible by human digestive enzymes and they are preferentially metabolized by bifidobacteria in the large intestine. An advantage in using prebiotics (oligosaccharides) instead of probiotics (ingestion of viable cultures of bifidobacteria) to elevate and maintain populations of colonic *Bifidobacterium* is that prebiotic compounds can easily be added to foods as a stable ingredient whereas the delivery of viable bifidobacteria in food products is difficult given the rigors of food processing and storage (e.g. exposure to low pH, oxygen, heat and cold).

Probiotics and Implied Health Benefits of Bifidobacteria

For a culture to be considered a viable candidate for use as a dietary adjunct, it must be a normal inhabitant of the intestinal tract, survive passage through the upper digestive tract, be capable of surviving and growing in the intestine, produce beneficial effects when in the intestine, and maintain viability and activity in the carrier food before consumption. Most bacteria are killed after ingestion by the severe acid conditions in the stomach and the bile juice that is released into the duodenum. Once in the intestine, only a limited number of bacteria can reside there. Indigenous bacteria tend to eliminate transient or exogenous species spontaneously.

Bifidobacteria have been shown to activate immunological, antibacterial and antitumour effects in animals even though bifidobacteria demonstrate low antigenicity compared to other intestinal bacteria. Also, the metabolic activities of bifidobacteria do not result in the production of ammonia or other detrimental compounds such as putrescine, cadaverine, indole, skatole, hydrogen sulphide, phenols, cresols, aglycones, tyramine, tryptamine, or histidine, and do not reduce nitrate to form nitrite (that can lead to the formation of nitrosamines). Such compounds are foul smelling and more importantly are toxic or potentially carcinogenic. Putrefactive bacteria such as the clostridia, coliforms and enterococci contribute many of these noxious compounds.

Regardless as to whether or not gut bifidobacterial numbers are increased by prebiotics, probiotics, or both, it is widely accepted that elevation and maintenance of bifidobacterial populations in the intestinal tract relative to other bacterial populations is a desirable circumstance. Several benefits are implied since bifidobacteria produce acetic and lactic acids from the fermentation of carbohydrates that lowers faecal pH. The increased level of acidity and greater numbers of bifidobacteria reduce the levels of undesirable bacteria which results in the reduction of toxic metabolites and detrimental enzymes. This leads to a number of beneficial situations which are outlined in **Table 3**.

Table 3 Benefits attributed to bifidobacteria

Stabilization of intestinal microbiota/resistance to enteric diseases
Prevention of pathogenic and autogenous diarrhoea/treatment of some diarrhoeas
Reduction of toxic metabolites and detrimental enzymes related to ageing process
Deconjugation of bile salts
Prevention of constipation/stimulation of peristaltic movement/controls mucin at intestinal surface
Protection of liver function
Reduction of serum cholesterol
Reduction of blood pressure
Induction of cell-mediated immunity
Antitumorigenic activity
Production of nutrients and vitamins
Improvement of lactose-tolerance to milk products
Important role in infant nutrition
Prevention of vaginal yeast infections
Degradation of nitrosamines/metabolism of ammonium ions
Aid in absorption of calcium
Intestinal recolonization following antibiotic treatment, chemotherapy or radiation treatment

It has been shown that children with high numbers of bifidobacteria resist some enteric infections very effectively. In fact, the feeding of dairy products containing bifidobacteria has been used to treat these infections in Japanese children with success. Regular supplementation of the infant diet with bifidobacteria can be used to maintain normal intestinal conditions; it can also be used in conjunction with antibiotic therapy to correct abnormal conditions, such as intractable diarrhoea.

Compared to children and adults, the elderly have lower counts of indigenous bifidobacteria. With this decline there is a corresponding increase in the population of *Clostridium perfringens* detected in the elderly. *C. perfringens* is a pathogenic bacterium that produces toxins and volatile amines. Adults fed foods containing high numbers of bifidobacteria over a 5-week period demonstrate a significant decrease in clostridia with an increase in bifidobacteria. Also, elderly patients suffering from bowel obstruction respond favourably to treatment with yoghurt containing bifidobacteria. The presence of high numbers of bifidobacteria in the infant and adult colon seems to be desirable and can be influenced by dietary supplementation. Bifidobacteria are known to exhibit inhibitory effects on many pathogenic organisms, both in vivo and in vitro, in addition to *C. perfringens*; these include other clostridia, *Salmonella*, *Shigella*, *Bacillus cereus*, *Staphylococcus aureus*, *Campylobacter jejuni*, and the pathogenic yeast, *Candida albicans*.

Bifid-amended Foods and Beverages

In the US before the 1980s, the use of bifidobacteria in foods was limited to a few products intended for therapeutic treatment. Among the earliest products was a bifidus milk developed by Mayer in the 1940s for use in treatment of infants afflicted with nutritional deficiencies. By the 1960s enough evidence had been accumulated to show it was possible to modify intestinal biota with *Bifidobacterium bifidum*. In the 1970s, Japan produced its first bifidus product, a fermented milk containing *B. longum* and *Streptococcus thermophilus* (in 1971). Bifidus yoghurt followed in 1979. Growth of bifidus foods and bifidus growth factor supplements continues to this day in Japan with other countries of the world following suit. Products that have been formulated with viable bifidobacteria and/or bifidus growth supplements include fermented and non-fermented milks, buttermilk, yoghurt, sour cream, dips and spreads, ice-cream, powdered milk, infant formula, cookies, fruit juices and frozen deserts. Bifidus growth factors (e.g. neosugar, fructooligosaccharides) are available at health food stores together with gel caplets and liquids containing bifidobacteria that are often in combination with *Lactobacillus acidophilus*.

Enumeration and Isolation Methods

Maintenance of anaerobic conditions is essential when culturing bifidobacteria. Accordingly, bifidobacteria require reducing agents in culture media for optimum growth (i.e. ascorbic acid, thioglycolate or cysteine). Cysteine and cystine are considered essential amino acids for growth. Normally, ammonium salts can serve as the sole source of nitrogen. Iron (both oxidation forms), magnesium and manganese are necessary trace elements. Bifidobacteria of human origin usually require a full complement of the B vitamins for optimal growth that can be supplied by yeast extract, even though some human strains of bifidobacteria can synthesize relatively large amounts of vitamins B_6 (pyridoxine), B_9 (folic acid) and B_{12} (cyanocobalamin). Most strains of *Bifidobacterium* can utilize glucose, galactose, lactose, lactulose, oligosaccharides, products of starch hydrolysis, bicarbonates and carbon dioxide as carbon sources.

Complex growth media are favoured for optimal propagation of bifidobacteria. For pure-culture growth, common commercial media such as deMan, Rogosa and Sharpe (MRS) broth and reinforced clostridial medium (RCM) normally work very well. There are numerous examples of selective media that have been developed for bifidobacteria. Many of the older media were designed to select for bifidobacteria from faecal material. More recent media have been devised to select bifidobacteria from fermented dairy foods. In yoghurts and fermented milks the difficulty is in distinguishing bifidobacteria from probiotic lactobacilli and lactic acid bacteria used as starter cultures.

Due to the varied physiological requirements of the different species in *Bifidobacterium*, it is nearly always the case that no single selective medium permits growth of all types of bifidobacteria while preventing the growth of other genera. A case in point is the use of antibiotics to select out for bifidobacteria in samples of mixed microbiota. Bifidobacteria are known to be resistant to nalidixic acid, polymyxin B, kanamycin, paromomycin and neomycin. Therefore, these antibiotics have been incorporated into various selective solid media to inhibit colony formation by yoghurt bacteria and *Lactobacillus acidophilus*; however, natural antibiotic resistances do occur, some types of bifidobacteria do display sensitivities to these compounds and individual variation among strains is not uncommon. As a result, other confirming tests must be used. For example, colony morphology and the use of oligosaccharide- or arabinose-containing agars have accompanied the use of selective agars containing antibiotics and selective inhibitors such as lithium chloride, sodium azide and propionic acid.

Newer methods of molecular biology have been developed and applied to the detection and identification of bifidobacteria. Rapid identification of bifidobacteria by gas chromatographic analysis of membrane fatty acid composition, DNA probe technology employing polymerase chain reaction amplification, and base sequence studies of DNA encoding for ribosomal RNA have been adapted. However, the equipment and technology of these newer molecular methods are limited to laboratories possessing adequate resources and expertise. These procedures are currently quite costly to perform on a regular basis.

See also: **Biochemical and Modern Identification Techniques**: Microfloras of Fermented Foods. **Fermented Milks**: Yoghurt. *Lactobacillus*: *Lactobacillus acidophilus*. **Microflora of the Intestine**: The Natural Microflora of Humans; Biology of Bifidobacteria; Detection and Enumeration of Probiotic Cultures. **Probiotic Bacteria**: Detection and Estimation in Fermented and Non-fermented Dairy Products.

Further Reading

Ballongue J (1993) Bifidobacteria and probiotic action. In: Salminen S and von Wright A (eds) *Lactic Acid Bacteria.* P. 357. New York: Marcel Dekker.

Edwards C (1993) Interactions between nutrition and the intestinal microflora. *Proceedings of the Nutrition Society* 52: 375–382.

Fuller R (1992) Problems and prospects. In: Fuller R (ed.) *Probiotics, The Scientific Basis.* P. 378. London: Chapman & Hall.

Heine W, Mohr C and Wutzke KD (1992) Host–microflora correlations in infant nutrition. *Progress in Food and Nutrition Science* 16: 181–187.

Hoover DG (1993) Bifidobacteria: activity and potential benefits. *Food Technology* 47(6): 120–124.

Hughes DB and Hoover DG (1991) Bifidobacteria: their potential for use in American dairy products. *Food Technology* 45(4): 74–83.

Ishibashi N and Shimamura S (1993) Bifidobacteria: research and development in Japan. *Food Technology* 47(6): 126–136.

Mitsuoka T (1990) *A Profile of Intestinal Bacteria.* Tokyo: Yakult Honsha.

Scardovi V (1986) Genus *Bifidobacterium.* In: Sneath PHA (ed.) *Bergey's Manual of Systematic Bacteriology.* Vol. II, p. 1418. Baltimore: Williams & Wilkins.

Shah NP (1997) Isolation and enumeration of bifidobacteria in fermented milk products: a review. *Milchwissenschaft* 52(2): 72–76.

Tomomatsu H (1994) Health effects of oligosaccharides. *Food Technology* 48(10): 61–65.

BIOCHEMICAL AND MODERN IDENTIFICATION TECHNIQUES

Contents

Introduction
Food Spoilage Flora (i.e. Yeasts and Moulds)
Food-poisoning Organisms
Enterobacteriaceae, Coliforms and *Escherichia coli*
Microfloras of Fermented Foods

Introduction

Daniel Y C Fung, Department of Animal Sciences and Industry, Kansas State University, USA

In the past 15 years applied microbiologists have developed and tested a large number of biochemical identification techniques and modern techniques within the discipline entitled Rapid Method and Automation in Microbiology. This field of study has been defined as dynamic areas of study that address the utilization of microbiological, chemical, biochemical, biophysical, immunological and serological methods for the study of improving isolation, early detection, characterization and enumeration of microorganisms and their products in clinical, food, industrial and environmental samples. Clinical microbiologists started to utilize these techniques in the early 1960s and in the past 10 years food microbiologists have accelerated their involvements in this area (**Fig. 1**). This introductory article provides an overview of the developments of this field and sets the stage for more detailed discussions on practical applications of some of these methods and procedures in food spoilage flora, food poisoning organisms, Enterobacteriaceae,

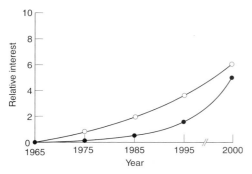

Figure 1 Relative interest in rapid methods among medical microbiologists (○) and food microbiologists (●). (Fung 1995)

coliforms and *Escherichia coli* and microfloras of fermented foods.

There are five major areas of developments in this field: (1) improvements in sampling and sample preparation; (2) alternative methods for viable cell count procedures; (3) instruments for estimation of microbial population and biomass; (4) miniaturized microbiological techniques and (5) novel and modern techniques. Each development has a definite influence on the total discussion of the following articles.

Improvements in Sample Preparation

The stomacher is a very successful instrument designed more than 25 years ago by Antony Sharpe to massage food samples in a sterile bag. The food is placed in the sterile disposable plastic bag to which appropriate sterile diluents are added. The bag with the food is placed in the open chamber. After the chamber is closed, the bag is massaged by two paddles for a suitable time period, usually 1–5 min. There is no contact between the instrument and the sample. During massaging microorganisms are dislodged into the diluent for further microbiological investigation. Recently, a new instrument called the pulsifier has been developed which can dislodge bacteria from food by high speed pulsification of food and diluent in a bag in the instrument. An evaluation of the pulsifier showed that the stomacher and pulsifier provided similar bacterial counts of paired studies of 96 samples. However, the pulsifier provided much clearer liquid samples which are advantageous for further microbiological manipulations, such as measurement of ATP, immunological tests, and polymerase chain reaction procedures.

Another development in sample preparation is to have instruments which can dispense a desired amount of liquid automatically into a vessel for blending of solid or liquid food. An instrument called Diluflo can accurately dispense from 0.1 ml to 100 ml into a bag or container with samples already in place. Furthermore, the instrument dispenses proportionally the amount of liquid in relation to the weight of the food sample. For example if a 1:10 dilution of a food is required, a sample of food is placed into the vessel (e.g. 9 g) and automatically the Diluflo will deliver 81 ml of sterile dilution into the vessel thus making exact manual weighing of the food sample and exact application of sterile diluent unnecessary and saving considerable amount of operation time. The instrument can be programmed to make 1:10, 1:50, 1:100 or other dilution factors.

Alternative Methods for Viable Cell Count Procedure

The conventional viable cell count or standard plate count method has been in use for more than a century. It involves preparing food into a slurry and then serially (1:10 series) diluting it to a final desired concentration of somewhere between 1:100 to 1:1 000 000 depending on the estimated concentration of microbial population. Then the diluted liquids are accurately pipetted into a sterile Petri dish (usually 0.1 or 1 ml) and then melted nutrient agar (48°C) is poured into the Petri dish. After solidification of the agar, the Petri dishes are then placed into the incubator at the desired temperature, e.g. 32°C, 35°C, or other temperatures for microorganisms, to grow to visible colonies, usually 24–48 h before counting the number of colonies and converting the number in counts per millilitre or per gram of the food. Although this time-honoured procedure is practised all over the world, it is time-consuming, labour intensive and wasteful of glassware and large numbers of disposable items, such as plastic pipettes and Petri dishes. Several ingenious methods have been developed to make the viable cell count more efficient, automatic and cost effective. These new methods were first designed to perform total viable cell counts but more recently due to improvements of media development and additional tests these methods can also detect and enumerate pathogens such as *Salmonella*, *E. coli* O157:H7, and other pathogens.

The Spiral Plating system (Spiral Biotech, Bethesda, MD) can spread a liquid sample on the surface of nutrient agar in a Petri dish automatically in a spiral shape (the Archimedes spiral) with a concentration gradient starting at the centre and decreasing as the spiral progresses outward on the rotating plate. The volume of liquid deposited at any segment of the agar plate is known. After the liquid containing microorganisms is spread, the agar plate is incubated overnight at an appropriate temperature for the colonies to develop. The colonies developed along the spiral pathway can be counted either manually or electronically. New versions of the original Spiral Plater

can automatically perform all the functions, including picking up a sample with a stylus, spreading the sample on the agar, lifting the stylus away from the plate and then rinsing and sterilizing the stylus for another sample. This system has been in use for more than 20 years in the food industry with excellent results.

The Isogrid System (QA Lab, San Diego, CA) consists of a square filter with hydrophobic grids printed on the filter to form 1600 squares for each filter. Food samples are weighed, blended and enzyme treated before passage through the membrane filter containing the grids. The filter is then placed on agar containing a suitable nutrient for growth of bacteria, yeast, moulds, faecal coliforms, *E. coli*, *Salmonella*, etc. The hydrophobic grids prevent colonies from growing further than the grids; thus all colonies have a square shape and are easily counted either manually or electronically. Both the Spiral Plating method and the Isogrid method require much less dilution of food sample compared with the conventional method. Usually only a 1:10 dilution of the food is necessary before application of the sample to the Spiral Plating system or the Isogrid system.

Rehydratable nutrients are embedded into films in the Petrifilm system (3M Co., St Paul, MN) which is about the size and thickness of a plastic credit card. An analyst can lift up the plastic cover of the unit and then apply 1 ml of liquid sample (with or without dilution) into the rehydratable nutrient gel and then replace the cover. The thin units (up to 10 can be stacked together) are then placed into the incubator at suitable temperature for 24 or 48 h for microbial growth. After incubation the colonies are counted directly through the clear plastic cover. The film can be kept as a permanent record of the microbial sample. Besides total count this system has films for coliform count, *E. coli* count, yeast and mould count and others. Simplicity and ease of operation along with long shelf life (1 year or more in cold storage) and smallness of the units have made this a very attractive system for small microbiological laboratories.

Another convenient viable count system is the Redigel (also marketed by 3M). This system consists of sterile nutrients with a pectin gel in a tube and no traditional agar. The tube is ready to be used at any time and no heat is needed to 'melt' agar. A 1-ml food sample is first pipetted into the tube. After mixing, the entire content is poured into a special Petri dish previously coated with calcium. When the liquid comes in contact with the calcium, a calcium-pectate gel is formed and the complex swells to resemble conventional agar. After incubation the colonies can be counted exactly as in the conventional standard plate count method. This system also has units for total count, coliform count, etc. similar to the Petri-film system.

A comprehensive analysis of all four methods against the conventional method on seven different foods (20 samples each) showed that these newer systems and the conventional method were highly comparable and exhibited a high degree of accuracy and agreement ($r = 0.95+$). Other methods such as Simplate, etc. are also being developed and tested. The aim is to find the easiest, fastest, and most automated systems for making the conventional viable count method more efficient and less time consuming in both operation and reading of results.

Instruments for Estimation of Microbial Populations and Biomass

Counting viable colonies is only one way to monitor growth of microorganisms in our food and the environment. A variety of chemical, physical, and biochemical methods have been used to study microbial populations and measure biomass. Some of these methods can be used to rapidly estimate viable cell numbers in food, water and other specimens since these methods can be measured within seconds or minutes whereas viable cell counts need hours to days to measure. In order to make use of these methods one must establish linear correlation between these parameters with viable cell numbers as a population of microbial cells grow or die. Thus we need to obtain standard curves of parameters such as dry weight of cell, protein contents, DNA or RNA concentrations, cellular components, adenosine triphosphate (ATP) level, detection time of electrical impedance or conductance, generation of heat, radioactive CO_2 etc. against viable cell count of a microbial population. By knowing the relationship one can then estimate the viable cell count by matching the units being measured. Theoretically, these methods can detect as little as one viable cell in the sample if the incubation period is long enough (days or weeks). On the practical side, the limit is usually 10^3–10^5 cells per millilitre.

All living things utilize ATP. In the presence of a firefly enzyme system (luciferase and luciferin system), oxygen, and magnesium ions, ATP will facilitate the reaction to generate light. The amount of light generated by this reaction is proportional to the amount of ATP in the sample. So the relative light units can be used to estimate the biomass of the cells in a sample. Using this principle, many researchers have used ATP to estimate bacterial cell number in meat, wine, fish and other foods.

One of the major problems is the presence of non-bacterial ATP in the food sample. In this situation one must then remove non-bacterial ATP either by filtering out the bacterial cells or extract the non-bacterial ATP, destroy the ATP and then extract bacterial ATP and monitor the bacterial ATP. Another problem is that different microbes have different amounts of ATP. For example a yeast cell has 50 times more ATP than a bacterial cell. Also the same organism may have a different amount of ATP at different stages of the growth cycle. Thus ATP is not a very good method for estimating actual number of bacteria in a food sample without a lot of sample manipulation. Currently the trend is to use ATP to monitor the hygiene of the food preparation environment. The theory is that if a certain level of ATP is found on the surface of a cutting board then the board is not clean. This does not take into account where the ATP came from, since food particles, blood, dirt and microbes are all not desirable in the food preparation environment. There are more than ten commercial companies producing ATP kits for the rapid monitoring of ATP on surfaces within minutes. These kits can greatly assist food companies in their sanitation programmes because in a very short time a team of cleaners can decide if they have performed the work properly or not. A clean surface for food preparation should have very little or no ATP. Companies marketing ATP kits include IDEXX, (Westbrook, ME), Lumac (Landgraaf, Netherlands), Biotrace (Plainsboro, NJ), Charm Science (Malden, MA) and New Horizon (Columbus, MD). Other sections in this encyclopedia also describe the use of ATP for applied microbiology.

As microorganisms grow and metabolize nutrients, large molecules change to smaller molecules in a liquid system and cause a change in electrical conductivity and resistance. By measuring the changes in electrical impedance, capacitance and conductance, the number of microorganisms in the liquid can be estimated, because the larger the number of microorganisms in the liquid the faster the change in these parameters, which can be measured by sensitive instruments. The Bactometer (bioMerieux Vitek, Inc., Hazelwood, MO) is designed to measure impedance changes in a food sample and is fully automated with the capability of handling 64 samples or more at any one time. As microorganisms metabolize and grow the impedance of the liquid will be changed and when the cells reach about one million per millilitre there will be a distinct change in the impedance curve. This is the 'detection time' of the sample in this system. A food sample having a large initial microbial population will cause the impedance curve to change earlier (shorter detection time) than a sample with smaller initial microbial population. The detection time is inversely proportional to the initial population, thus by knowing the relationship between microbial population and detection time one can use the detection time (e.g. 4 h) to estimate the initial population of the food (e.g. 1×10^6 per gram in the food). The Malthus Instrument (Crawley, UK) works by measuring the conductance of the fluid and generates conductance curves similar to the impedance curve of Bactometer. These instruments have been used to monitor the microbial quality of brewing liquids, milk, seafood, meat, etc. The Bactometer has been used to determine the shelf-life potential of pasteurized whole milk. Besides estimating bacterial numbers in the food these systems can be used to screen for food-borne spoilage and pathogenic organisms such as *Salmonella*, coliforms and yeasts.

An instrument called the 'Omnispec bioactivity monitor system' (Wescor, Inc. Logan, UT) is a tristimulus reflectance colorimeter that monitors dye pigmentation changes mediated by microbial activities in liquid foods. The instrument can monitor colour and hue changes from the bottom of optically clear growth vessels of different sizes without disturbing the sample, making it a unique non-destructive monitoring system. By using a microtitre plate containing 96 wells (about 0.4 ml per well) almost 400 samples can be studied simultaneously making it a very useful tool for studying large numbers of variables in microbiological investigations. In the author's laboratory Omnispec has been a very valuable tool to study the effects of a variety of chemicals (antimicrobial, enzymes such as oxyrase, etc.) on a large number of bacteria (e.g. *Listeria monocytogenes*, *Salmonella*, *Enterobacter*, *E. coli* O157:H7, *Yersinia*, *Hafnia*) automatically.

The catalase test is another rapid method to estimate microbial population of certain foods. Catalase is a very reactive enzyme and provides results in a matter of seconds. Microorganisms can be classified as catalase-positive or catalase-negative organisms. Both groups are important in food microbiology; however, under aerobic cold storage conditions (such as meat, poultry, fish, etc. in the refrigerators) catalase-positive organisms, such as *Pseudomonas*, *Micrococcus* and *Staphylococcus*, predominate. By measuring the catalase activities of these food one can estimate the bacterial populations therein. Catalase activity can also be used as an index of the cleanliness of meat-processing areas. A 5×5 cm area is swabbed with a cotton swab which is then placed in a tube containing hydrogen peroxide. If the surface is contaminated with meat, blood, aerobic microorganisms, etc. gas (molecular oxygen) will be generated by the reaction of catalase and catalase-like enzymes with hydrogen peroxide. The amount of gas is proportional to the degree of contamination. Yet another exciting use of catalase activity is to monitor how well foods,

such as chicken and fish, are cooked. Catalase is heat sensitive and when food is well cooked to 71°C catalase activities will be destroyed. This is a rapid test since it takes only a few seconds to measure the reaction.

Miniaturized Microbiological Techniques

Biochemical testing methods have been used in applied microbiology to differentiate groups of microorganisms for almost 150 years. Microorganisms can metabolize a great variety of organic materials and can generate acidic, basic, and neutral end products with or without the production of gas or coloured compounds from these reactions. By ingenious design of growth media in solid or liquid forms microbiologists have been able to use this information to identify and characterize closely related bacteria into genera and species. A typical set of biochemical tests for the differentiation of the family Enterobacteriaceae would include indole, methyl red, Voges–Proskauer, Simmons' citrate, hydrogen sulphide, urea, KCN, motility, gelatin, lysine decarboxylase, arginine dihydrolase, ornithine decarboxylase, phenylalanine deaminase, malonate, gas from glucose, fermentation of glucose, lactose, sucrose, mannitol, dulcitol, salicin, adonital, inositol, sorbitol, arabinose, raffinose and rhamnose. By growing pure cultures in these media for a period time, usually 24–48 h, and by observing changes of colour of the liquid from red to yellow (or other pH indicator colours), or typical reactions after addition of reagents one can identify unknown cultures using a variety of diagnostic schemes matching the biochemical data of the unknown with well established profiles of known cultures. This is the basis of classical identification methods using the *Bergey's Manual of Determinative Bacteriology* as the guide. This type of procedure has been used for more than 100 years and has served bacteriology well. However, the procedure is time-consuming, labour intensive, and uses large amount of culture media, chemicals, glass ware, tubes, cap, bottles, Petri dishes and incubator space. In addition, a microbiologist has to be very skilful in interpreting the results and making subtle judgements on the accuracy of the tests. A slight shade of colour difference may mean a test is interpreted as positive or negative. In this type of diagnostic scheme one error in judgement can easily result in a wrong identification. There is a definite need to improve the conventional method of identification of unknown cultures using biochemical tests.

About 30 years ago this author initiated a systematic approach to miniaturize all biochemical tests for the identification of bacteria from foods and labelled this set of tests as miniaturized microbiological techniques. In this system the volume of reagents and media was reduced from 5–10 ml to about 0.2 ml for microbiological testing in microtitre plates. The basic components of the miniaturized system are the microtitre plates for test cultures (8 × 12 multiwell configuration), a multiple inoculation device and containers to house solid media (large Petri dishes) and liquid media (another series of microtitre plates). The procedure involves placing liquid cultures (pure cultures) to be studied into sterile wells of a microtitre plate to form a 'master plate'. Each microtitre plate can hold up to 96 different cultures, 48 duplicate cultures or various combinations as desired. The cultures are then transferred by a sterile multipoint inoculator (96 needles protruding from a template) to solid or liquid media. Sterilization of the inoculator is by alcohol flaming. Each transfer represents 96 separate inoculations in the conventional method. After incubation at an appropriate temperature, the growth of cultures on sold or liquid media can be observed and recorded, and the data analysed. These miniaturized procedures save a considerable amount of time in operation, effort in manipulation, materials, labour and space. These methods have been used for the study of large numbers of bacterial and yeast isolates from foods and developed many bacteriological media and procedures. Many useful microbiological media were discovered through this line of research. For example, an aniline blue *Candida albicans* medium was developed and marketed by DIFCO under the name Candida Isolation Agar. Some excellent agars for *E. coli* O157:H7, *E. coli*, *Yersinia enterocolitica*, etc. are being developed and studied. The progression of development of miniaturized microbiological techniques and modern rapid methods is depicted in **Table 1**.

At around the time when the author was working on miniaturization of microbiological techniques in late 1960s to 1970s an important trend in diagnostic microbiology started to unfold. This was the commercialization of miniaturized diagnostic kits in Europe and USA. These kits can be characterized as agar-based kits, dehydrated media-based kits and paper-impregnated media-based kits. This section gives an introduction to how these systems came into being. More information about these kits and their applications is provided in later articles.

Agar-based Kits   The R/B and Enterotube II systems are the two prime examples and are among the oldest commercial diagnostic kits. The R/B system is similar to the familiar TSI tube for differentiating *Enterobacteriacae* except that it has eight different reactions in two tubes. After inoculating the pure culture into

Table 1   Major developments (Fung 1992)

Diagnostic tests for liquid, semi-solid and solid media
1. Large tubes: One reaction per tube
2. Large tubes: multiple reactions per tube
3. Small tubes: one reaction per tube
4. Small tubes: multiple reactions per tube
5. Wells in a tray of different configurations:
   (a) One type of reaction for many organisms per tray
   (b) Many types of reactions for one type of organism per tray
   (c) Many types of reactions for a few organisms per tray
6. Diagnostics kits

Inoculations into diagnostic tests
1. One inoculation per tube
2. Multiple inoculations (manually or by instruments)
   (a) Liquid in a tray
   (b) Solid in agar
   (c) Agar in multiple compartments
   (d) Liquid dispensing to multiple wells

Automated instruments for monitoring
1. Cell mass
2. Cell components
3. Cell metabolites
4. Cell activities

Development of serological and immunological techniques
1. Immunoblotting
2. Electrophoresis
3. Radioimmunosorbant assay
4. Enzyme-linked immunosorbent assay

Development of genetic techniques
1. DNA probes, RNA probes
2. Polymerase chain reaction
3. RiboPrinting, random amplified polymorphic DNA
4. Ligase chain reaction, Q-beta replicase

Concepts involving the living cell
1. Living cell versus dead cell
2. Growing cell versus non-growing cell
3. Meaning of viable cell count
4. Correlation between total count and other parameters
5. Amplification of cells
6. Concentration of cells
7. Signal versus background
8. Sensitivity versus detection line

a larger tube and a smaller tube the tubes are incubated for 24 h and then the reactions are recorded and compared to colour charts of known cultures for identification. There is a lot of colour bleeding in this system which makes it very difficult to interpret the data.

The Enterotube II system has 12 separate compartments in a cigar-shaped plastic tube with a sterile needle placed through all 12 chambers. After removing the caps from both ends the sterile inoculation needle is used to touch a pure colony grown on agar. The needle is then slowly pulled through all 12 chambers in one motion. This deposits the culture in the 12 agars in the respective chambers. After incubation appropriate reagents are added into the chambers and the colour reactions are recorded. The data are then fed into a data sheet and a number is generated from blocks of three reactions based on the positive reactions. The numerical score is then transformed into a code. A code book is used to identify the unknown culture. This procedure is used by most other kits systems to be discussed. These agar-based systems are easy to use but have a short shelf life (a few months) and there are problems with dehydration and occasional contamination.

Dehydrated Media-based Kits   The best example is the API system. In 20 small chambers housed in a long strip 20 different media are placed in the chambers and dehydrated. A pure culture is first suspended in a sterile liquid and then aliquots are carefully placed into each chamber. Some tests require an oil overlay to ensure anaerobic condition. After overnight incubation, reagents are added to some chambers and the colour of the tubes are recorded. As described for Enterotube II system a code for the unknown culture is generated and compared with a code book for identification. API has the largest database of all the kits and has become *de facto* the standard diagnostic test kit for Enterobacteriaceae.

Biolog is a dehydrated medium system using 95 carbon sources with one positive control in the microtitre plate. An unknown culture is first homogenized in liquid and then the liquid culture is placed in all 96 wells using a multichannel pipetter. The design of the system is such that when an organism utilizes a particular carbon source the liquid will turn blue. Thus there is only one colour to read in these wells. An analyst can match the positive growth pattern of the unknown culture with the pattern of a known culture for identification. A better way is to put the microtitre plate with growth results into a specially designed instrument which can automatically match the pattern of the unknown culture with patterns of known cultures in the data bank for identification. This is an ambitious system designed to identify hundreds of clinical, food and environmental cultures.

The crystal system is also a dehydrated medium kit. In this system 30 different biochemical substrates are dehydrated at the tip of small plastic rods. A culture suspension is first placed into the trough of the bottom unit. A unit with 30 small rods each with a different dehydrated medium at the tip is placed firmly into the bottom unit with the culture. The unit is then incubated. After growth and reaction the colour of the tips of the rods indicates positive or negative reactions. The unit is placed into a reader to register the reaction pattern which is then matched with the database of known culture patterns for identification.

The RapidID system is also based on dehydrated medium housed in small chambers. The difference between RapidID system and API is that chambers are not inoculated individually but rather the ingenious design receives 1 ml of culture suspension in a trough. By tilting the trough perpendicular to the openings of the small chambers with the media in one movement 10 chambers can be inoculated simultaneously thus saving much time and labour compared with the API inoculation procedure.

One of the earliest and most automated dehydrated medium systems is the Vitek system. This system comprises a plastic card (about the size of a credit card) with 30 different dehydrated media placed in tiny wells connected to each other by a series of microtubes in the card. A pure culture is first suspended in liquid and then by vacuum the liquid is introduced into the 30 wells in the card. The card is then placed into an incubator unit. At regular intervals the card is scanned and identification is done automatically by computer. This system has been used in hospital environments for more than 20 years.

The major advantage of the dehydrated medium-based kits is long shelf life (1.5 years) in refrigerated storage.

Paper-impregnated Medium-based Kits   The two kits in this category are the MicroID and Minitek systems. The MicroID system has 15 separate chambers, 10 of which have one paper disc containing one reaction medium. The other five chambers have one paper disc in the bottom portion of the chamber and another paper disc in the top portion of the chamber for secondary reaction. A liquid culture is introduced to the opening of each of the 15 chambers; the liquid drops to the bottom of the chamber and wets the paper disc with medium. After 4 h incubation one reagent is added to the first chamber and the unit is rotated through 90° so that the liquid from the five chambers will come into contact with the discs at the top portion of these chambers for secondary reaction. The colour of the five discs in the top part of the chamber are read as well as the 10 paper discs of the other 10 chambers. Identification of the unknown is similar to other systems by finding the code number and comparing with the code book. This was the first 4 h test from the time the analyst picked the colonies from the agar plate. This is possible because this system utilized pre-formed enzymes in the cultures for the reactions.

The Minitek system is more flexible than the other kits. The manufacturer sells 36 different substrates on paper discs contained in small tubes. These tubes can be dispensed in an instrument and then the discs with substrates are dropped into the wells of a 10-unit multiwell plastic plate. After the paper discs are in place, an automatic pipetter with the liquid culture is used to inoculate the culture in all the chambers.

Usually, 20 paper discs in two 10-well plates are used for identification of enterics. After inoculation of the culture some wells are filled with mineral oil to keep the test under anaerobic conditions. After incubation, identification is done by first reading positive and negative test results and the code generated is matched with a code book. It should be emphasized that the code book of one system cannot be used to identify unknowns from another system. Also the biochemical results of one system cannot be transposed to biochemical results of another system. The same organism may give a positive result of a test in one system but a negative result in another system because of the amount of chemicals used by different systems.

These paper-impregnated medium kits also have long shelf life of 1.5 years. These kits are based on biochemical changes due to enzymes in the cells. These methods have found the greatest use in clinical microbiology. Many systems now include bacterial isolates from food sources and put the information in the database for identification of food isolates. There are many other diagnostic kits made by different countries throughout the world but the basic principles are the same as the three types of kits described here.

An example of the variety of methods developed and tested to identify one family of bacteria, the Enterobacteriaceae, is given in **Table 2**. Many charts for the detection of other pathogens are described in the literature.

Refinements of Novel Methods

This section describes new developments in immunology and genetic techniques.

Immunology

Antibody and antigen reactions for diagnostic microbiology have been used for more than 50 years in clinical sciences, food science, biological sciences and related sciences. A variety of formats have been used such as agglutination tests, precipitation tests, haemagglutination, single gel diffusion, double gel diffusion, microslide diffusion, latex bead agglutination, etc. The most popular format in terms of commercial kits is the enzyme-linked immunosorbent assay (ELISA). In the Organon Teknika (Durham, NC) system two monoclonal antibodies specific for *Salmonella* detection are used; one for capturing the organism and the other for reporting the captured antigen. The first antibody is fixed in a solid base such as a microtitre well. A suspected sample containing *Salmonella* is then added to the well. If *Salmonella* is present it is captured by the antibody. The second antibody labelled with an enzyme is then added to the well. It reacts with the captured *Salmonella* and after

addition of the appropriate substrate a colour reaction occurs. The colour reaction can be detected visually or by use of colorimeter. A series of washing steps are involved in this type of ELISA test. Another system which utilizes monoclonal antibodies is the Assurance EIA test marketed by BioControl (Bothell, WA). The Tecra system (International BioProducts, Redmond, WA) was developed in Australia and uses polyclonal antibodies to detect *Salmonella*. Such ELISA test kits have been developed for *Listeria*, *E. coli*, *Campylobacter*, etc. Many companies provide a host of polyclonal and monoclonal antibodies for a variety of diagnostic tests, including some food pathogens.

In the VIDAS system (bioMerieux Vitek, Hazelwood, MO) all intermediate steps are automated. Other completely automated ELISA systems include Tecra OPUS (International BioProduct, Redmond, WA), Bio-tek Instruments (Highland Park, VT), and Automated EIA Processor (BioControl, Bothell, WA). In recent years a series of 'lateral migration' ELISA tests have been developed. After overnight pre-enrichment, an analyst only needs to add a drop of the suspect liquid (boiled or unboiled) into the first well of the unit. If a suspect culture (e.g. *E. coli* O157) is present an antibody will react with the antigen and form a complex which will migrate laterally to another part of the unit where another specific antibody is fixed to capture the target organism (e.g. *E. coli* O157). A coloured particle is attached to the first antibody; thus reaction is reported as a colour band in the unit. Excess antibodies will continue to migrate to a region where they will be captured by another antibody and form a visible complex. This is the test control to ensure the system is performing properly. The entire reaction takes only about 10 min, making this type of test very rapid indeed. Currently REVEL (Neogen, Lansing, MI) and VIP (BioControl, Bothell, WA) are two popular systems for rapid detection of *Salmonella* and *E. coli* by lateral migration technology. It should be emphasized that with this type of test a negative result would allow the food products, such as ground beef, to be released for shipment. However, when the test is positive the conventional approved methods must be used to confirm the identity of the culture.

The UNIQUE system (Tecra system, Roseville, Australia) for *Salmonella* is another method of using immunocapture technology. In this system a dip stick with antibody against *Salmonella* is applied to the pre-enriched broth. The antibody captures the *Salmonella*, if present. This charged dip stick is then placed in a fresh enrichment broth and the cells are allowed to multiply for a few hours. After the second enrichment step, the dip stick with a much larger population of *Salmonella* attached to it will be subject

Table 2 Miniaturized biochemical assays: Enterobacteriaceae (Kalamaki, Price and Fung 1997)

Type of kit	Supplier	Limit of detection	% Correctly identified	% Total errors	Sensitivity	Specificity	% Agreement	Total time	Cost per assay
API20E	bioMérieux Vitek, Hazelwood, MO	Pure colony	77[a] 95.6[b] 78.7[c] 95.2[d]	1.6 (a)	NR[e]	NR	NR	21 h	$4.17
Enterotube II	Roche, Basel, Switzerland	Pure colony	NR	NR	NR	NR	97	18–24 h	NR
MicroID[f]	REMEL	Pure colony	NR	NR	NR	NR	97 98.8 (Salmonella) 97.7 (E. coli) 88.1 (Other enterics)	4 h	NR
MUCAP Test	Biolife, Italy	Pure colony	NR	NR	100	80 90.1	NR	NR	NR
Rambach Agar[g]	Technogram, France	Pure colony	NR	NR	91 88	100 76	NR	NR	NR
RapIDonE	Innovative Diagnostic Systems Inc.	Pure colony	NR	4.6	NR	NR	92.1	4 h	NR
Salmonella-strip	LabM, UK	Pure colony	NR	NR	100	99	NR	NR	NR
SM-ID[c]	bioMérieux Vitek, Hazelwood, MO	Pure colony	NR	NR	93	37	NR	NR	NR
Spectrum-10	ABL Austin Biological Laboratories	Pure colony	NR	NR	NR	NR	91	18–24 h	NR
Vitek GNI[f]	bioMérieux Vitek, Hazelwood, MO	Pure colony	NR	4.40	NR	NR	84.5[a] 92.8[b]	NR	NR

[a]After initial incubation.
[b]After additional biochemical tests were performed as directed by the manufacturer.
[c]After 24 h of incubation.
[d]After 48 h of incubation.
[e]NR, not reported.
[f]AOAC final action.
[g]Selective agar.

to further ELISA procedures. The entire test is housed in a convenient plastic self-contained unit. A similar system is developed for *Listeria monocytogenes* by the same company. This type of self-contained unit is very useful for the smaller laboratory where automated systems may not be practical for routine use.

Immunomagnetic capture technology is another exciting development in applied microbiology. In this system paramagnetic beads are coated with antibodies designed to capture target pathogens such as *Salmonella*. The beads are placed in a liquid culture and the antibodies capture any *Salmonella* present. Then a powerful magnet is applied to the side of the glass container to localize all the paramagnetic beads with captured target pathogens, thus greatly concentrating the population from the liquid. The rest of the liquid is discarded and the tube washed to remove compounds that are not needed. The beads are released from the side of the glass tube by removing the magnetic field.

Further microbiological processes are performed to identify the target pathogen. This procedure saves at least one day in most pathogen detection systems. Vicam (Somerville, MA) and Dynal (Oslo, Norway) are two systems using this technology.

Motility enrichment is another way to rapidly screen for motile organisms such as *Salmonella*, *Listeria*, etc. A motility flash system has been developed that can presumptively detect the presence of *Salmonella* in food in about 16 h. Confirmation takes another 24 h with this system. BioControl (Bothell, WA) have marketed a 1–2 test system for *Salmonella* which utilizes motility as a form of selection. The food is first pre-enriched for 24 h in lactose broth and then 0.1 ml is inoculated into one of the chambers in the L-shaped system. The chamber contains selective enrichment liquid medium. There is also a small hole connecting the liquid chamber with the rest of the system which contains a soft agar for *Salmonella* to migrate. An opening on the top of the second chamber allows the introduction of a drop of polyvalent anti-H antibody for reaction with flagella of *Salmonella*. If the sample contains *Salmonella* from the lower side of the unit, they will migrate through the hole and up the agar column. When the antibody meets the *Salmonella* a visible 'immunoband' forms. The presence of the immunoband indicates that the original food sample contained *Salmonella*. Further confirmation tests are necessary for final identification. Stimulation of the growth of pathogens in these systems will shorten the detection time. The author's laboratory has developed a variety of procedures utilizing an enzyme named oxyrase to stimulate the growth of pathogens such as *Listeria monocytogenes*, *Campylobacter*, *E. coli*, etc. in the pre-enrichment or enrichment stage so that the cells reach 10^6 per millilitre rapidly for secondary detection such as ELISA or other technologies. Oxyrase can also be used to stimulate the growth of starter cultures in the fermentation of a variety of food products such as buttermilk, yoghurt, bread dough, sausages, beer and wine.

Genetic Methods for Identification

DNA and RNA probes have been used for more than 15 years in the diagnostic field. At first the target was DNA of pathogens such as *Salmonella* and used radioactive compounds to report the hybridization. More recently the target has been RNA and the reporting system is a probe with enzyme attached to change the colour of substrates for reporting the hybridization. The reasons are that in a bacterial cell there is only one copy of DNA but up to 10 000 copies of RNA thus by probing RNA these systems will be more sensitive and enzyme systems are far more user friendly than radioactive materials for reporting the hybridization reaction. For more than 10 years, Genetrak (Hopkinton, MA) has been marketing DNA and RNA for the detection of *Salmonella* and *Listeria monocytogenes*.

Polymerase chain reaction (PCR) systems are the latest development in DNA amplification technology and have recently gained much attention in food microbiology. Originally the procedures were highly complicated and a very clean environment was needed to perform the test. Recently, much research has been directed at simplifying the procedure for laboratory analysts. Qualicon (Wilmington, DE) is marketing BAX screening system which utilizes pre-packaged tubes for PCR tests of pre-enriched sample for pathogens such as *Salmonella* and *E. coli*. All the reagents necessary for PCR are in the tube (primers, buffer, $MgCl_2$, TAQ, and nucleotides). The target DNA, if present in the pre-enriched sample will be subjected to the PCR procedure automatically in the thermal cycler. The cycle consists of heating the liquid to 95°C for a few seconds or minutes to cause the DNA to unfold, then lowering the temperature to about 50°C for the primers (oligonucleotides for specific sequence of bases of the target pathogen) to anneal to the target sections of the half DNA with another specific primer attached to the opposite region of the other half DNA. The temperature is then raised to 72°C for the enzyme TAQ to complete the polymerization of the half DNAs to complete DNA by depositing complementary nucleotides to the unfolded DNA. After the completion of polymerization one original DNA becomes two identical DNA pieces. The cycle repeats and the number of DNA will increase exponentially. Depending on the speed of each cycle one piece of DNA can be amplified to 1×10^6 pieces in about 2 h. The PCR products can then be detected by electrophoresis, dot blotting, Southern blotting or ELISA type hybridization. In the BAX system electrophoresis is used to detect PCR products for *Salmonella* and *E. coli* O157. A new system named Probelia developed by Pasteur Institute is introduced by BioControl in the USA for effective PCR test for *Salmonella* and *Listeria monocytogenes*. There are a number of differences between BAX and Probelia systems: (1) In Probelia the nucleotides used are adenine, uracil, guanine and cytosine instead of adenine, thymine, guanine and cytosine. (2) A special enzyme uracil-D-glycolase is in the system which can destroy all Probelia PCR products from a previous run; thus for a new run there will be no contaminants before the start of another new sequence of PCR cycles. (3) There is an internal control in the same tube with all the other ingredients for PCR. (4) The PCR products are detected by an ELISA type hybridization procedure. These systems are now being introduced into food laboratories and will be very useful when all the technical details are solved for common food laboratories.

Qualicon also markets a Ribotyping system which can track the origins of several pathogens in food plants and other environments. This is especially important for epidemiological work in food-borne outbreaks. In this system a pure culture must be isolated from a suspected sample. DNA from the culture is then extracted and digested by special enzymes into fragments. These fragments are subjected to electrophoresis for separation and then the fragments are loaded on a membrane and the membrane is processed. A highly sensitive photo system is used to record patterns of the fragments and the data are processed through sophisticated computer systems and a riboprint pattern of the culture is obtained. The pattern is then matched with the database to identify the culture. It is important to know that the same

Table 3 Predictions of food microbiological developments (Fung 1995, presented at the American Society for Microbiology Annual Meeting)

1. Viable cell counts will still be used
   (a) Early sensing of viable colonies on agar, 3–4 h
   (b) Electronic sensing of viable cells under microscope, 2–3 h
   (c) Improvement of vital stains to count living cells
   (d) Early sensing of MPN
2. Real-time monitoring of hygiene will be in place
   (a) ATP
   (b) Catalase
   (c) Sensor for biological materials
3. PCR, ribotyping, genetic tests will become reality in food laboratories
4. ELISA and immunological tests will be completely automated and widely used
5. Dipstick technology will provide rapid answers
6. Biosensors will be in place in HACCP programmes
7. Instant detection of target pathogens will be possible by computer generated matrix in response to particular characteristics of pathogens
8. Effective separation; concentration of target cells will greatly assist in rapid identification
9. Microbiological alert system will be in food packages
10. Consumer will have rapid alert kits for pathogens at home

organism can have many different patterns. For example *Salmonella* has 97 RiboPrint Patters, *Listeria* has 80, *E. coli* has 65 and *Staphylococcus* has 252 patterns. Thus when there is an outbreak of *Salmonella*, for example, it is possible to trace the exact origin of the contamination by matching patterns of the culture causing the outbreak versus the source. Finding a culture of *Salmonella* in a certain food is not enough to pin-point the source of this culture to the outbreak but if the RiboPrint of the culture matches exactly with the culture that caused the sickness then it is more reliable to identify the source of the problem. This process is completely automated once the pure culture is introduced into the RiboPrint instrument. In about 8 h eight samples can be processed simultaneously. Also every two hours a new set of eight samples can be introduced to the instrument. This instrument won the 1997 Institute of Food Technologists Industrial Award for the excellence of the process and the potential impact on tracing foodborne pathogens.

Other techniques of this type of work include the random amplified polymorphic DNA (RAPD) method, pulsed-field electrophoresis, multiplex RAPD, etc.

It is not possible to mention all the new and useful methods, suffice to say that there are many chemical, biochemical and physical methods that can be used to identify microorganisms such as gas liquid chromatography, GC mass spectrometry, fatty acid profile, protein profiles, pyrolysis, calorimetry, etc.

In conclusion, this article has described a variety of methods that are designed to improve current methods, explore new ideas and develop new concepts and technologists for the improvement of applied microbiology. This field will certainly grow, and many food microbiologists will find these new methods very useful in their routine work in the immediate future. Many methods described here are already being used by applied microbiologists nationally and internationally.

Table 3 (compiled in 1995 by the author) lists some predictions of food microbiological developments. As we move into the twenty-first century many of the predictions have become realities. The future is bright for this field of endeavour for promoting food safety and protecting the health of consumers nationally and internationally.

See also: **Biochemical and Modern Identification Techniques**: Food-poisoning Organisms.

Further Reading

Adams MR and Hope CFA (1989) *Rapid Methods in Food Microbiology*. Amsterdam: Elsevier.

Bourgeois CM, Leveau JY and Fung DYC (1995) *Microbiological Control for Foods and Agricultural Products*. New York: VCH Publishers.

Chain VS and Fung DYC (1991) Comparative analysis of Redigel, Petrifilm, Isogrid, and Spiral Plating System and the Standard Plate Count method for the evaluation of mesophiles from selected foods. *Journal of Food Protection* 54: 208–211.

Doyle MPO, Beuchat LR and Montville TJ (1997) *Food Microbiology: Fundamentals and Frontiers*. Washington, DC: ASM Press.

Feng P (1997) Impact of molecular biology and the detection of foodborne pathogens. *Molecular Biotechnology* 7: 267–278.

Fung DYC (1992) Historical development of rapid methods and automation in microbiology. *Journal of Rapid Methods and Automation in Microbiology* 1: 1–14.

Fung DYC (1995) What's needed in rapid detection of foodborne pathogens. *Food Technology* 49(6): 64–67.

Fung DYC (1997) Overview of rapid methods of microbiological analysis. In: Tortorello MC and Gendel SM (eds) *Food Microbiological Analysis: New Technologies*. New York: Marcel Dekker.

Fung DYC and Kraft AA (1970) A rapid and simple method for the detection and isolation of *Salmonella* from mixed cultures and poultry products. *Poultry Science* 49: 46–54.

Fung DYC and Mathews RF (eds) (1991) *Instrumental Methods for Quality Assurance in Foods*. New York: Marcel Dekker.

Fung DYC, Sharpe AN, Hart BC and Liu Y (1998) The Pulsfier: a new instrument for preparing food suspension for microbiological analysis. *Journal of Rapid Methods and Automation in Microbiology* 6(1): 43–50.

Fung DYC, Yu LSL, Niroomand F and Tuitemwong K (1994) Novel methods to stimulate growth of food pathogens by oxyrase and related membrane fractions. In: Spencer RC, Wright EP and Newsome SWB (eds) *Rapid Methods and Automation in Microbiology.* Andover, UK: Intercept.

Kalamaki M, Price RJ and Fung DYC (1997) Rapid methods for identifying seafood microbial pathogens and toxins. *Journal of Rapid Methods and Automation in Microbiology* 5: 87–137.

Mossel DAA, Corry JEL, Struijk CB and Baird RM (1995) *Essentials of the Microbiology of Foods.* New York: John Wiley.

Oslon WP (ed.) (1996) *Automated Microbial Identification and Quantitation: Technologies for the 200s.* Buffalo Grove, Il: Interpharm Press.

Patel PD (1994) *Rapid Analysis Techniques in Food Microbiology.* New York: Chapman & Hall.

Swaminathan B and Feng P (1994) Rapid detection of food-borne pathogenic bacteria. *Reviews in Microbiology* 48: 401–426.

Tortorello ML and Gendel SM (1997) *Food Microbiology Analysis: New Technolgies.* New York: Marcel Dekker.

Food Spoilage Flora (Yeasts and Moulds)

George G Khachatourians, Department of Applied Microbiology and Food Science, University of Saskatchewan, Saskatoon, Canada

Dilip K Arora, Department of Botany, Banaras Hindu University, Varanasi, India

Microbiological contamination of foods is a global problem, and a significant amount of expense is being incurred as a result of such contamination and spoilage of foods. To avoid the spoilage of food quality, early detection of these fungi are important. Recent advancement in biotechnology is rapidly altering the diagnostic procedures used in the identification of food fungi. Biochemical identification assays have been miniaturized and automated making them faster, economical, reliable and less time-consuming. However, these identification techniques and tools need modification and still greater challenges exist in rapid identification of fungi and yeasts from food due to complexity and variety, and their interference with detection procedures. Methods to sequester target fungi from the interfering food compounds are needed for efficient rapid identification of fungi when they are present at very low levels. Comparative evaluation of the various protocols in terms of cost, sensitivity,

specificity, speed and reproducibility need to be undertaken so that the true applicability of these methods can be determined. In future, molecular methods will find increasing applications in food-borne identification systems that can significantly enhance our ability to detect cultural, non-cultural and non-viable moulds; however, it is unlikely that these techniques can totally replace conventional methods.

Biochemical identification of fungi is based on chemical characteristics. The significance of biochemical identification may depend on taxonomic levels, the evaluation of chemical constituents, and the organism's chemical and physiological activities. Biochemical techniques have been used extensively in the identification of bacteria and yeasts, but their extensive practical applicability in the identification of filamentous fungi is still lacking. This situation is, however, greatly improved by the introduction of polymerase chain reaction (PCR) technology. Chemical characterizations of ascomycetous yeasts and other filamentous food spoilage fungal flora have been mainly based on physiological and chemical tests, such as cell-wall biopolymers, quantitative profiles of sterols, total fatty acids, pyrolysis of cells, phospholipids etc. Other biochemical tests that have been used for identification include immunology, isozymes and protein profiling; however, generally the most common modern detection methods in food mycology are based on immunoassays and nucleic acid hybridization-based technology. This article outlines some of the important biochemical characters used in the identification of yeast and filamentous fungi, and discusses the merits and demerits of biochemical markers, the immunoassay, isozymes and automated systems over molecular detection techniques.

Biochemical Diagnostic Markers

Fatty Acids, Proteins and Isozymes

Although the fatty acid (FA) composition of lipids in fungi is well established, the importance of FA profiles in the identification of fungi at this stage is not well defined and still remains unclear. Many research papers and reviews have been published on the value of FA profiles to differentiate between fungi. The presence and absence of $\omega 3$ and $\omega 6$ of FAs and their relative amounts (C_{16} and C_{18} FAs) are important in the identification of food-borne organisms. FA profiles with other phenotypic characters have been used successfully by various yeast taxonomists to differentiate a number of taxa belonging to *Schizosaccharomyces* and *Nadasonia*. However, detailed lipid analyses are needed and should include lipid separation into neutral, glycolipid and phospholipid fractions. Each of these fractions can further be separated and the FA composition of each subfraction should be determined in order to understand the species composition of the test fungi.

Promising results regarding the use of FAs for identification of filamentous fungi have been reported. With the aid of FA profiles it is possible to differentiate between various species of *Aspergillus*, *Mucor* and *Penicillium*. The cellular FA composition of several strains of yeasts, especially associated with wine spoilage (*Torulaspora delbreuckii* and *Zygosaccharomyces bailli*), has been repeatedly examined as a criterion to differentiate species. The strains of *Saccharomyces cerevisiae* and other wine-associated yeast species can be differentiated by using capillary gas chromatography which is an easy, quick and cheap method. This method has been applied to determine the causes of 'suck' fermentation in a South African food and beverages industry. Methods based on statistical quality-controlled chart utilizing cellular FA composition and morphology as criteria, were successfully applied to monitor the fungal contaminants in the bioprotein pilot plants in South Africa. In the case of *Rhodosporidium*, FA profiles can be used for the rapid differentiation between species. Similarly, *Saccharomyces* may be identified by FA profiles determination. The use of FA and sterol profiles (FAST-profiles) test has recently been described for fungi to define species and intraspecific variation. A total of 20 fatty acids and seven sterols were used to determine the identity of a batch of 1740 fungal isolates collected from Finland.

Electrophoresis of fungal cellular proteins significantly contributed in the identification of fungal isolates, the recognition of mutants, mating types, *formae speciales* and determination of specific heterogeneity. Usually whole cell protein has been used to detect species or strains commonly inhabiting the foods and feeds. Reliability of the results depends on the approach and resolution of specific detection techniques. The protein profile of the same fungus may vary depending on the growth and metabolic conditions of fungi. Detection of common moulds from contaminated foods using protein-profiling has potential difficulties, as there is no objective comparative measurements of relatedness of protein profiles. The detection protocols based on protein profiling needs simplification, standardization and automation.

Isozyme electrophoresis is a very useful technique that has many applications in food mycology. This biochemical tool is frequently used in fungal taxonomy, systematics, population genetics, spread of fungal plant pathogens and tracing of ploidy level in fungi. Detection of isozymes allows a genetic inter-

pretation of variations in alleles and loci. A different enzyme locus exhibits variations at various taxonomic levels. Isozyme data have been used in the identification of several fungi, including most commonly occurring food-borne fungi such as *Penicillium* and *Aspergillus*. Though isozyme bands are genetic markers, useful to detect and identify fungi, the application of this technique is not very useful to detect fungi from contaminated food and feeds. Several reviews and books on isozyme techniques are available.

Isozymes are molecular forms of an enzyme, which have similar and often identical, enzymatic properties, but have different amino acid sequences. Because different amino acids create net charge differences for isozymes, they can be detected during electrophoretic separation. As a result, isozymes can be used to assess fungal isolates based on different alleles of a single gene locus (allozymes), multiple loci coding for a single enzyme, and post-translationally modified isozymes. Polyacrylamide gel electrophoresis (PAGE) or starch gel electrophoresis and isoelectric focusing gels, can be used to separate isozymes which are then used for 'fingerprinting' of fungi and yeasts. Today's electrophoresis systems allow both a large number of enzymes to be detected from a single or several fungal samples, and the detection of allozymes and isozymes coded by different loci as long as a compatible buffer system is available. The use of isozyme analysis has been described for the identification of plant-pathogenic fungi. This approach can also be used for the examination of food-contaminating fungi. In brief, starch or polyacrylamide based gel is boiled and poured into a mould to form the gel. After the gel cools, depending on the gel system, a sample containing the enzymes is brought into contact, and a defined current for voltage, amperage and time, depending on the buffer used, is applied. After electrophoresis, the gel can be processed and tested for the particular enzyme activity. Examples of assays are slicing of the gel and assay for particular activity, staining of the gel for transformation of a chromogenic substrate (e.g. *o*-nitrophenylated sugars), and overlaying of the gel with a gel-substrate which can be cleared by enzymatic activity (e.g. casein, albumin, starch etc.). With automated systems many more gels can be run in a single day under identical conditions.

As with any technique, isozyme analysis has its strengths and weaknesses. When starch gel electrophoresis is used, the technique is relatively inexpensive compared to either immunological tests or techniques involving the PCR. With isozymes, a large number of staining systems is available thus allowing the comparison of numerous genetic loci coding for enzymes from many metabolic pathways. This has the advantage of allowing us to draw conclusions about the genetic variability. Further, isozyme analysis detects only significant differences in enzyme structure, most often those which are at the species level. Finally, isozyme analysis staining systems are usually quite specific and generally detect a single enzyme of known identity on a gel slice. This is in contrast to the numerous bands obtained from a general protein stain. However, difficulties arise when analysing some fungi, which are difficult to grow, or populations that may not represent a single genotype. The amount of material required for isozyme analysis may vary from organism to organism. Time requirements may be another disadvantage of isozyme analysis. Although the electrophoretic test can be conducted quite rapidly, it often takes several days, or even weeks, to isolate and grow the organisms.

Fungal Volatile and Secondary Metabolites

Secondary metabolites are outward-directed differentiation products of normal cellular metabolism that may function as chemical signals between organisms or species. Different analytical methods can be applied for the separation and detection of secondary metabolites. Some of these methods involve the use of thin layer chromatography, gas chromatography, high performance liquid chromatography (HPLC), micellar capillary electrophoresis, flow injection electrospray mass spectrometry, ultraviolet diode array detection and nuclear magnetic resonance detection. There are different opinions on the use of secondary metabolites in characterization and identification of filamentous fungi. Some workers believe that secondary metabolites are strain-specific, whereas others think that they are very sensitive to growth and environmental factors, and therefore could not be considered for diagnostic purposes. However, in some of the few genera of filamentous fungi, secondary metabolites have been shown to be very reliable and act as highly diagnostic characters. For example, secondary metabolites have been particularly effective in identifying some of the most common food-spoilage fungi such as species of *Penicillium*, *Aspergillus* and *Fusarium*. The food-borne terverticilliate penicillia are very difficult to characterize by using traditional characters, but the secondary metabolites of many of these species have given a very clear identification system of this most complex species group. Several closely related species of *Penicillium* can be separated using secondary metabolites by using diode array detection or flow injection analysis electrospray mass spectrometry. There are several pitfalls in using secondary metabolites for identification of food-borne fungi. (1) Based on characterization of these metab-

olites, only highly specialized taxonomists are able to make a correct identification. (2) No simplified diagnostic procedure has been developed for fungi which commonly contaminate foods. (3) Most fungal species have the ability to produce secondary metabolites, but in some cases a particular taxon or isolate needs specific stimuli to initiate the accumulation of some of these metabolites. (4) Several types of secondary metabolites may remain unnoticed because of inefficient extraction procedures or low analytical sensitivity or low reproducibility under different growth and metabolic conditions.

Fungi often develop characteristic odours due to the production of unique combinations of volatile metabolites like alcohols, ketones, esters, terpenes and other hydrocarbons. However, volatile metabolites have only rarely been used for identification purposes. Recent studies have shown that toxic and non-toxic isolates of both *Aspergillus* and *Fusarium* species can be distinguished from each other based on their production of sesquiterpenes. Similarly, it has been demonstrated that a large number of *Penicillium* species can also be classified based on profiles of volatile metabolites. Based on volatile profiles, *Penicillium roqueforti* and *P. commune* can be easily identified; the latter species is most frequently a contaminant of cheese.

Immunological Techniques

Immunological techniques have become and will remain an important and widespread technology in the detection and identification of filamentous fungi and yeasts in food. This is particularly possible because of the availability of monoclonal antibody (MAb) technology, which has revolutionized the development process in detection, and diagnosis of organisms. Like many other microorganisms, fungi produce a number of antigens, which can be used for identification with desired specificity. It is possible to raise isolate-, species- and genus-specific antibodies that are very sensitive and target-specific. However, raising monoclonal antibodies from the infected food materials creates problems in extraction of fungal antigens. The development of in vitro technology whereby antibody genes are expressed in bacteria will make the process of production of monoclonal antibodies simpler and more widely accessible. Though there is overlap with both medical and plant pathology, within the past decade immunological diagnosis of food-borne fungi has resulted in several advancements resulting in the characterization of fungal antigens in an effort to characterize some of the important immunodominant sites. Several antigenic sugars and proteins have been characterized in some of the common food-borne fungi such as *Aspergillus*,

Botrytis, Cladosporium, Fusarium, Geotrichum, Monascus, Mucor, Penicillium and *Rhizopus*. However, the rapid detection of common food-spoilage flora in foods (e.g. *Aspergillus, Penicillium* and *Fusarium* which produce mycotoxins and spoils foods) using immunological techniques is not well developed.

The modern detection assays in food mycology are based on recognition of fungal cell wall- and cell surface-associated or extracellular polysaccharides and several classes of protein antigens of the cell surface by specific immunoassays. Up to a dozen different antigens can be isolated from any one species. These antigens can be used to obtain immune sera. Thermostable extracellular polysaccharides (EPS) of fungi are made of mannose, galactose, glucose, fucose and occasionally glucoronic acid. These EPS are released to the growth medium by many species; however, the EPS yield of fungi can vary depending on species, a number of variables and culture conditions. The EPS or cell surface proteins being antigenic to animals could be used to produce polyclonal immunoglobulin (Ig)G antibodies of rabbits to be specifically and sensitively used in a number of immunoassays. Commercially available kits include: mould latex agglutination test (Holland Biotechnology, B. V. Leiden, The Netherlands) and Pastorex *Aspergillus* Test (PAT; Eco-Bio Diagnostic Pasteur, Genk, Belgium).

The methodology for the use of immunological techniques depends on the test methodologies and instruments. In addition, such identification techniques need trained human resources. The methods included in this group of tests are mould latex agglutination test, enzyme-linked immunosorbent assay (ELISA), radioimmunoassay (RIA), enzyme immunoassay (EIA), for which in some cases commercial kits are available. The use of ELISA to detect food-spoilage fungal flora basically depends on the particular objective of the investigator, i.e. the level of sensitivity desired, application of technique in pure culture or food materials, the need for quantification and use of polyclonal or monoclonal antibodies as reagents. The indirect and double-antibody sandwich ELISA has been widely used for the identification and detection of fungi in food samples. Sandwich ELISA has some advantages over other techniques as chlorophyll and other interfering components from the plant food materials can be easily removed. Interpretation of ELISA is usually straightforward, although appropriate controls must be included to avoid false positive or negative results. Multiwell ELISA formats for detection of mould from food are available commercially; they are very rapid and less time consuming. Detailed descriptions of commercially

available ELISA kits can be found in the literature. Though immunofluorescence (IF) is considered to be more sensitive than ELISA and other tests, IF tests are difficult to standardize and interpretation of results is often difficult because of contamination of food samples with large numbers of non-target microorganisms may interfere with the assay unless specific antibodies are used.

Evaluation of Commercial Techniques and Range of Tests Available

The unique physiological properties of yeasts and filamentous fungi have made them relatively difficult to test and identify using commercial kits. Commercially available kits for yeast identification have been developed primarily for the use of clinical microbiologists, and are not always useful for food-borne spoilage yeasts and filamentous fungi due to the absence of an appropriate database. However, in recent years, several biotechnological companies have enhanced their databases for food-borne yeasts, and made the diagnostic techniques automated using computer-controlled readers and interpretation of data. Most of these systems are expensive but can be afforded by the larger food companies/institutions. A number of commercialized systems (**Table 1**) are becoming key elements in the determination of yeasts. These techniques are convenient and useful for identification of isolates. However, it still takes 1–3 days to obtain the results. Newer methodologies or kits should be possible as similar systems are being developed around human infectious fungi and yeasts. Fuller surveys of these methods and sources of materials are available in the literature.

Among all commercially available kits perhaps the API 20C, API 50CHB (which is designed for *Bacillus* spp., but can be successfully used for yeast and fungi) and API YEAST-IDENT (Analytab Products, Plainview, NY, USA) have been used for identification of a wide range of yeasts and fungi. To perform the test, a heavy suspension of yeast or fungal spores is prepared and inoculated into the API test system as per the instructions of the manufacturer. The resulting biochemical reactions are read after 2–3 days and used for identification.

The BIOLOG identification system (Biolog Inc., Hayward CA, USA) is a standardized, computer-linked semi-automated technology for the identification of yeasts and fungi. The Biolog system has evolved from conventional methods for identification by introducing a number of co-metabolism tests and many assimilation and oxidation assay techniques not usually common in conventional methods. This test incorporates a wide range of substrates and a redox

Table 1 Characteristics of some commercial identification systems

Principle	System	Method[a]	Number of tests	Number of species in database	Days	Accuracy of results (%)
Growth based	API 20C	M	20	42	3	99
	ATB 32 ID	M/A	32	63	2	91
	AutoMicrobic	M	30	62	1	83–97
	Microring YT	M	6	18	2	53
	Minitek	M	12	28	3	53
	Quantum II	A	20	34	1	82–86
	Uni-Yeast-Tek	M	15	42	2	40
Enzyme based	MicroScan	M/A	27	42	1/6	85–96
	YeastIdent	M	20	42	1/6	55–60

[a]M, manual; A, automated.

dye, tetrazolium violet (TZV), as an indicator of substrate utilization. During cellular metabolic activity of the test substrate NADH is formed and in order for it to be reoxidized electrons passing through the electron transport chain (ETC) cause irreversible reduction of TZV to formazan which is purple. Because the TZV functions independently of any ETC it will accept electrons irrespective of metabolism of many of some 95 substrates, e.g. amino acids and other carbon or nitrogen sources. Thus an extensive list of substrates can be used. The formation of purple formazan can be read by eye or in a micro-plate reader with a filter cut-off of 600 nm. The results are then compared with the Biolog database for yeasts. The specificity and sensitivity of the test system depends on growth and metabolism of yeast or fungal species. Recently, the Biolog system was evaluated for the identification of 21 species (72 strains) of yeasts of foods and wine origin. Using these tests several yeasts species such as *Saccharomyces cerevisiae*, *Debaryomyces hansenii*, *Yarrowia lipolytica*, *Kluyveromyces marxianus*, *Koeckera apiculata*, *Dekkera bruxellensis*, *Schizosaccharomyces pombe*, *Zygosaccharomyces bailii* and *Z. rouxii* were identified correctly 50% of the time and *Pichia membranaefaciens* 20% of the time. Another study also tested 46 strains of yeasts representing 14 species using the automated Biolog and ATB32C systems. Both systems correctly identified many species, with Biolog correctly identifying 38 of the 46 strains, and the ATB correctly identified 30 out of 46 strains. BioMeriex 32 C strips were also tested for the identification of many food-borne yeasts, and most yeast isolates were identified with a probability of 95% or greater.

The Biolog system protocol is simple and includes the following advantages. (1) The strain of interest is cultivated on a simple agar medium (available from Biolog). (2) Cells are removed from the surface of the agar and suspended in sterile water at specified density. (3) Cell suspension is inoculated into each of the 96 wells of the Biolog MicroPlate. (4) The MicroPlate is then incubated at 26°C for 24–72 h until a sufficient metabolic pattern is found. For species identification, the MicroPlate must be read with the Biolog MicroStation Reader. Currently, 267 species of yeast have been identified by the Biolog System. Biolog has kits for the identification of yeasts and filamentous fungi, however, recently Agriculture and Agri-food Canada, and Biolog are in a process of developing an effective identification for green mould disease fungus on mushrooms. Traditional identification based on morphological criteria can not differentiate at the strain or pathovar level. Rapid and accurate identification of the strains of green moulds (*Trichoderma*) occurring on commercial mushrooms are essential since not all species of *Trichoderma* that colonize compost cause disease.

Molecular Techniques

Molecular techniques for the identification of food-associated fungi and yeasts are versatile because of: (1) the ability of these tests to recognize genomic differences; (2) the speedier application of methods such as DNA fingerprinting, chromosome karyotyping, protein electrophoretic patterns and fatty acid profiles in identification; and (3) the ability to consider ecological or processing source in tracking these agents. As yeasts and fungi are eukaryotic organisms, their genomic DNA, in addition to chromosomal and mitochondrial DNA, can have plasmids, killer factors, and Ty-elements. These DNAs because of their unique sequence, size, topological (single, double, covalently closed, linear etc.) features should avail themselves for analyses. The molecular methods and tools used for the identification of these DNAs is described in other books and publications but will be given briefly here.

Electrophoretic Karyotyping

Compared to classical karyotyping of fungal chromosomes by cytochemical methods, electrophoretic karyotyping (EK) uses pulsed-field gel electrophoresis, where chromosomes can migrate through a series of reorienting arrangements in an electric field to eventually align according to their sizes and numbers. Pulsed field gel electrophoresis (PFGE) was introduced in 1984 and further developed to be a powerful technique for studying the electrophoretic karyotype of fungi. Chromosomes of many industrially important yeasts and fungi have been characterized in this manner. Further newer data indicate that various isolates or strains of certain species show distinct EK 'fingerprint' differences. Contour-clamped homogenous electric field (CHEF) gel electrophoresis was used to separate intact, chromosome-size DNA of different species of *Saccharomyces* and *Zygosaccharomyces*. Strains of the same *Saccharomyces* species had similar electrophoretic karyotypes. Furthermore, differences between individual chromosomal bands can be performed to show strain-specific chromosome length polymorphism (CLP). The karyotype differences of strains beyond chromosome length polymorphism can also be established. In this case DNA–DNA hybridization techniques will test conspecificity, by using individual chromosomes as templates to hybridize against gene specific or randomly primed probes. This method has been used to study genetic diversity of a given yeast species and species identification and taxonomy. Some of the chromosome specific DNA probes used in the identification of yeasts and fungi are given in **Table 2**.

RFLP and DNA Amplification Techniques

For restriction fragment length polymorphic (RFLP) analyses, fungal DNA is isolated and subjected to cleavage by DNA restriction enzyme(s). There are over 600 such enzymes, which under defined conditions cut double-stranded DNA at specific nucleotide sequences. The resulting fragments, depending on their size are separated electrophoretically on agarose gels. Such DNA fragments when stained by ethidium bromide show differences in their size and banding patterns, and therefore show polymorphisms. DNA from such gels is transferred on to suitable membranes by electroblotting and hybridized by (radio-isotopically or non-isotopically labelled) probes and viewed after an exposure to x-ray films.

RFLP patterns of many yeasts and fungi have been established and are useful in unequivocal discrimination and identification. However, such analy-

ses require a substantial amount of time (several days). An alternative to this is to use PCR technique and employ random amplified polymorphic DNA (RAPD) analysis. This method requires a minimum quantity of fungal DNA that would result from the formation of a number of DNA sequence amplifications from one set of primers of arbitrary nucleotide sequence. The product of amplification would generate DNA bands in an electrophoretic gel, which should vary in size and sequence. This type of analysis has been useful in the RAPD analysis of wine yeasts.

The PCR, developed by Cetus (Emeryville, California) scientists in 1985, is a powerful in vitro method for amplifying a segment of DNA that lies between two regions of known sequence, defined by a set of primers. The basic steps of PCR are denaturation of the DNA, annealing the primers to complementary sequences and extension of the annealed primer with *Thermus aquaticus*. Taq-DNA polymerase can survive extended incubation at 95°C, therefore, all components can be added at the start of the reaction without any further replenishment. Together these steps are referred to as a cycle. Also, since annealing and extension can be carried out at elevated temperatures, mispriming is reduced thus resulting in improvements in the specificity and yield of the amplification reaction.

There are many choices for the amplification technique. **Table 3** shows a comparison of these. The instrumentation and tools needed for the technical tasks are shown in **Table 4**. The reader is referred to other books on the subject. Compared to the original use of PCR, i.e. to amplify segments of DNA located between two specific primer hybridization sites, a single-sided PCR method has been developed that initially requires specification of only one primer hybridization side. The second site is then defined by the ligation-based addition of a unique DNA linker. This method, referred to as ligase chain reaction (LCR), allows for exponential amplification of any fragment of DNA. Another modification of the basic PCR has led to the development of anchored PCR. By using anchored PCR it is possible to amplify full-length mRNA when only a small amount of the sequence information is available. PCR-based tech-

Table 2 Chromosome specific DNA probes for identification

Probe DNA	Marker characteristic
Chromosomal	β-tubulin
	rRNA, ITS

Table 3 DNA amplification techniques for identification

Method	Requirements		
PCR	Temperature cycling	2 primers	Taq polymerase
LCR	Temperature cycling	4 primers	Taq polymerase+ DNA ligase
GapLCR	Temperature cycling	4 primers	Taq polymerase

Table 4 Enzymological aspects of rDNA work with pure cultures

Enzyme	Function	Purpose
Cell lysis	Lysozyme, cellulase	Removes cell walls
Proteases	Pronase, protease K	Removes proteins
RNAses	RNase H	Elimination of RNA
Nick translation	E. coli DNA polymerase I	Synthesize DNA
Klenow fragment		
T4 DNA polymerase		
Copy DNA	Reverse transcriptase	(cDNA) from mRNA
Process DNA	Nuclease Bal31	
Exonuclease digestion		
Mung-bean nuclease, restriction endonucleases	Endonuclease cuts	
DNA methylases	Methylated bases	
Phosphatases	Remove 5'-phosphate	
Ligases	Polynucleotide ligase	Join ends of DNA or RNA
Polymerase chain reaction	Taq polymerase	Produce oligonucleotides

niques such as random amplified polymorphic DNA (RAPD), amplified fragment length polymorphism (AFLP), DNA amplification fingerprinting (DAF) and random amplified microsatellite (RAMS) PCR have been developed that are used for identifying DNA markers.

Critical Evaluation of the Techniques

Assessment of the extent of mycological contamination of food ingredients and processed foods is an essential part of any quality assurance or quality control programme in the food industry. Likewise, enumeration of desirable and undesirable yeasts associated with fermented foods and beverages is important if high quality products are to be routinely produced. Although in some instances the counts of total fungal load in a commodity is needed, in others, for example, regulatory situations, exact knowledge of not only counts of viable moulds (e.g. in spices, dried vegetables, human and animal food, frozen and fresh vegetables and meat etc.) but also the presence of mycotoxin-producing fungi especially when combinations of their toxins is suspected, is necessary for quality assurance and public health safety. It is in the last context that critical evaluation of techniques with a rapid and exact knowledge would be valued. Moreover, in commodities that have been damaged or deteriorated to the point of inability to recover viable fungal cells, the molecular methodology becomes the sole source of evaluation.

The molecular methods based on DNA although still brand new have changed our ability for detection and identification of a wide variety of fungi and yeasts in foods. The PCR-based methods can amplify a limited amount of a nucleic acid with a high degree of sensitivity. Finally, there are methods for the removal of the interfering material within foods that can affect these amplification test systems to add power to the detection of fungi in food science.

Molecular over Biochemical

With either biochemical kits or molecular techniques based on antibodies there are some differences in the detection of the presence of fungi in foods. Antibody methods for the detection of antigens rely on the principle that the presence of such antigens always relates to the presence of yeasts or moulds, to which the antibodies bind. Therefore, the accuracy of diagnosis and identification requires that: (1) the EPS is always made by the fungal agents but not other agents or cells (animal, microbial or plant species); (2) the antigen or the epitope is made in sufficient quantities to be detected by the antibodies; (3) the antigen is accessible and stable within the food's composition, processing and the environment; and (4) the antibody for the test system has high avidity and specificity. For example, with EPS detection by immunological techniques some cross-reacting immunological reactions occur. Although the EPS ELISA of some yeasts is specific it is not so with basidiomycetous yeasts. The EPS of *Zygosaccharomyces bailii* could be detected in a highly specific competitive ELISA but not in a sandwich ELISA or in a latex agglutination test. The cell surface specific antibodies can be used, e.g. rapid detection of infectious wild yeasts in brewer's and baker's yeast, in certain instances for classification and a reasonably rapid and reliable method of identification of *Saccharomyces* and *Zygosaccharomyces* species. With the exception of capsulated species, in which cell wall antigens are masked, yeast cells are readily agglutinated by specific antiserum. Some antigens are present in many ascomycetous yeasts and in some basidiomycetous yeasts whereas other antigens display more general and species specificity.

ELISA tests for *Aspergillus* and *Penicillium* spp.

have been shown to be quick, reliable and sensitive in testing of 161 food samples. The use has been reported of latex agglutination test, i.e. latex particles sensitized with IgG specific for EPS produced by *Apergillus* and *Penicillium*. This test system was collaboratively tried by nine laboratories and compared to conventional methods (colony counts) for the detection of fungi in food samples. Eight of the nine laboratories were able to detect 5–15 ng ml^{-1} of purified EPS. Fair correlation was shown between colony counts and latex agglutination titres for cereals, spices and animal feed, but there was no correlation for fruit juices, and walnuts gave false positive results. It is concluded that the latex agglutination test is a rapid (c. 10–15 min to read results), simple and reliable quantitative method for the detection of *Penicillium* and *Aspergillus* in cereals, spices and animal feeds. The Mould Reveal Kit (MRK, Eco-Bio, Genk), a rat monoclonal antibody against *Aspergillus* galactomannan coated onto latex beads has been used and compared to HBT mould latex agglutination test kit (Holland Biotechnology, Leiden, The Netherlands) an EPS induced polyclonal antibody test. Both kits are used for the rapid screening of foodstuffs for mould contamination. The HBT kit was negative for several fungi and results were not as sensitive. The sensitized latex beads detect this EPS at a 15 ng ml^{-1} after 5 min incubation (**Table 5**). Of 35 common foodborne fungi tested, 27 gave positive reactions in the latex agglutination test, including all 16 *Aspergillus* and *Penicillium* spp. tested (**Table 6**). The Mould Reveal Kit was shown to be faster than the latex agglutination test. It took 2–5 min for the MRK to produce macroscopically visible agglutination. The test is simple and semi-quantitative.

Advantages/Limitations of Biochemical Over Other Techniques

Although the assumption can be that binding of antibody to an antigen is similar to that of nucleic acids, DNA-based diagnostics are based on principles of greater specificity and authenticity. In contrast to

Table 6 Screening of culture filtrates (1 : 10 dilutions) of different fungi with the mould reveal kit

Positive cultures	*Aspergillus clavatus, A. candidus, A. flavus, A. fumigatus, A. niger, A. ochraceus, A. versicolor, Eurotium repens, E. herbariorum, Emericella nidulans, Neosartorya fischeri, Penicillium brevicompactum, P. camemberti, P. chrysogenum, P. digitatum, P. expansum, P. roqueforti, P. viridicatum, P. glabrum, P. spinulosum. Talaromyces sp., Eremascus albus, Paecilomyces variotii, Trichoderma sp., Wallemia sebi, Cladosporium cladosporioides, C. herbarum, Aureobasidium pullulans, Botrytis cinerea*
Negative cultures	*Alternaria sp., Epicoccum sp., Scopulariopsis sp., Geotrichum candidum, Fusarium solani, F. oxysporum, Mucor racemosus, M. plumbeusi, Syncephalastrum racemosum, Rhizopus oryzae*

immunodiagnostics, DNA has an advantage for not only specificity of binding, but also amplification and authentication by sequencing. However, the assumption to be satisfied is that, such DNA from yeasts or fungi will be present intact within or outside the microbial agent (a cell or a spore) in partially preserved form, interfering substances would be absent, and the sequence used for diagnostics would be sufficiently distinct from the foods.

Both sets of fungal identification tests rely on genetic and chemical taxonomic diversity of the species. The advantage of molecular and biochemical tests are that there are standard commercial kits for performing simple tests and complex procedures for the isolation and amplification of DNA from most yeasts and fungi. DNA of high purity from chromosomes, plasmids and mitochondria can be extracted and purified from proteins and other cellular constituents. A number of commercial plant or fungal DNA purification kits which utilize silica-based resins

Table 5 The sensitivity or the mould reveal kit towards fungal ethanol precipitates or culture filtrates (exoantigens) and purified fungal polysaccharides

Sensitivity[a] (ng ml^{-1})	Species	
	Exoantigens	Polysaccharides
15	*Aspergillus fumigatus*	*A. fumigatus, P. digitatum*
50	*Penicillium digitatum*	*C. cladosporioides*
100	*P. marneffei*	*Botrytis tulipae*
250	*Wallemia sebi*	
500	*Cladosporium herbarium*	
1000 or more	*C. cladosporioides, Trichoderma viride, Fusarium solani, Cryptococcus neoformans*	*Candida albicans, Saccharomyces cerevisiae*

[a]Hexose concentrations.

and anion exchangers are now available. Finally, the greatest advantage here is the detection of non-viable and non-culturable fungal agents. For non-culturable (but viable) agents, the demonstration of messenger or ribosomal RNA by reverse transcription will be an important indicator of live fungal agents.

In many cases the PCR detection system can identify desired gene sequences quickly, with high specificity and in large volumes. Reports indicate that amplification of DNA sequences from a number of regions can be used for identification and differentiation of yeast and fungi. The variable region of the 5′ end of the large nuclear ribosomal DNA (rDNA) (28S), internal transcribed spacer (ITS) sequences of ribosomal DNA, intron splice sites and random amplified micro-satellite sequences (RAMS) have been used for detection and identification purposes. As for sequencing, the best analytical tool available for this is mass spectrometry. A mass spectrometer works by vaporizing the DNA and then accelerating the molecules through a vacuum chamber with the help of an electric field. Tiny differences in the time that it takes the fragments to reach the detector reveal small differences in their mass and hence their sequence. This technique is known as MALDI-TOF and it has the capability of analysing hundreds of DNA samples within a few minutes. Although originally used over a decade ago for protein analysis it was not available for DNA analysis until 1993 when various matrices were developed that would work with DNA fragments as long as 100 base pairs. However, for practical sequencing MALDI-TOF would have to work with DNA fragments much longer than the current 100 base-pair capacity. At the present time new matrices are being studied that could extend MALDI-TOF to reach 1000 bases and if this works then this technique would be a major breakthrough for high-throughput sequencing.

In spite of its power, DNA amplification methods although overcoming the sensitivity limitations of direct DNA probe assays and immunological assays, contain an inherent limitation and problem of carry-over contamination of amplification products. In a typical PCR amplification reaction with nanomolar concentrations of reagents and products, one can estimate 1012 molecules per 0.1 millilitre of amplification products to be present. A carryover of just under 10 copies of product in 10^{15} l (femtolitre) can generate false positive results and hence the need for extreme care. To further prevent such problems, post-amplification sterilization of amplification reaction products through UV-irradiation, use of uracil DNA glycosylase, addition of psoralens and copper bis(1,10-phenanthroline) are encouraged. Finally, proper negative controls, internal and external amp-

lification controls and a DNA extraction control should be prerequisites for a reliable identification and detection system. It is important to realize that despite the strengths or weaknesses of any one test system, the powerful approaches of molecular or biochemical diagnostics are here to complement the conventional techniques. It should be interesting to see how the future developments shape the 'modern' techniques of today.

See also: **ATP Bioluminescence**: Application in Beverage Microbiology. **Biochemical and Modern Identification Techniques**: Introduction; Food-poisoning Organisms. **Fungi**: Food-borne Fungi – Estimation by Classical Cultural Techniques. **Mycotoxins**: Immunological Techniques for Detection and Analysis.

Further Reading

Bonde MR, Micales JA and Paterson GL (1993) The use of isozyme analysis for identification of plant-pathogenic fungi. *Plant Disease* 77: 961–968.

Bridge PD, Arora DK, Reddy CA and Elander RP (eds) (1998) *Applications of PCR in Mycology*. Wallingford, UK: CAB International.

Deak T and Beauchat LR (1996) *Handbook of Food Spoilage Yeasts*. Boca Raton, FL: CRC Press.

deRuiter GA, Hoopman T, van-der Lugt AW, Notermans SHW and Nout MJR (1992) Immunochemical detection of Mucorales species in foods. In: Samson RA, Hocking AD, Pitt JI and King AD (eds) *Modern Methods in Food Mycology*, p. 213. Amsterdam, Netherlands: Elsevier Science.

Frisvad JC, Bridge PD and Arora DK (eds) (1998) *Chemical Fungal Taxonomy*. New York: Marcel Dekker.

Frisvad JC, Thrane U and Filtenborg O (1998) Role and use of secondary metabolites in fungal taxonomy. In: Frisvad JC, Bridge PD and Arora DK (eds) *Chemical Fungal Taxonomy*. New York: Marcel Dekker.

Hocking AD, Fleet GH, Praphailong W and Baird L (1994) *Assessment of Some Commercially Available Automated and Manual Systems for Identification of Food-borne Yeasts*. Third International Workshop on Standardization of Methods for the Mycological Examination of Foods, p. 16.

Khachatourians GG and Woytowich AL (1999) Genetic Engineering Part 1: Principles and Applications. In: Francis FJ (ed.) *The Wiley Encyclopedia of Food Science and Technology*, 2nd edn. New York: John Wiley.

Kock JLF and Botha A (1998) Fatty acids in fungal taxonomy. In: Frisvad JC, Bridge PD and Arora DK (eds) *Chemical Fungal Taxonomy*. New York: Marcel Dekker.

Koshinsky HA and Khachatourians GG (1995) Cloning restriction fragment length polymorphism and karyotyping technology. In: Hui YH and Khachatourians GG (eds) *Food Biotechnology: Microorganisms*, New York: John Wiley.

Koshinsky HA and Khachatourians GG (1994) Mycotoxicoses: the Effects of Mycotoxin Combinations. In: Hui Y, Gorham JR, Murrey KD and Cliver DO (eds) *Foodborne Disease Handbook. Diseases Caused by Viruses, Parasites and Fungi*, vol. 2, p. 463. New York: Marcel Dekker.

Land GA, McGinnis MR and Salkin IF (1991) Evaluation of commercial kits and systems for the rapid identification and biotyping of yeasts. In: Vaheri A, Tilton RC and Balows A (eds) *Rapid Methods and Automation in Microbiology and Immunology*. Berlin: Springer-Verlag.

Middelhoven WJ and Notermans S (1993) Immuno-assay techniques for detecting yeasts in foods. *International Journal of Food Microbiology* 19: 53–62.

Muller MM and Hallaskela A-M (1998) A chemotaxonomic method based on FAST-profiles for the determination of phenotypic diversity of spruce needles endophytic fungi. *Mycological Research* 102: 1190–1197.

Muller MM, Kantola R and Kitunen V (1994) Combining sterol and fatty acid profiles for the characterization of fungi. *Mycological Research* 98: 593–603.

Notermans SHW and Kamphuis HJ (1992) Detection of Fungi in Foods by Latex Agglutination: A Collaborative Study. In: Samson RA, Hocking AD, Pitt JI and King AD (eds) *Modern Methods in Food Mycology*, p. 205. Amsterdam, Netherlands: Elsevier Science Publishers.

Praphailong W, Van Gestel M, Fleet GH and Heard GM (1997) Evaluation of the Biolog system for the identification of food and beverage yeasts. *Letters in Applied Microbiology* 24: 455–459.

Robison BJ (1995) Use of commercially available ELISA kits for detection of foodborne pathogens. *Methods in Molecular Biology* 46: 123–131.

Stager CE and Davis JR (1992) Automated systems for identification of microorganisms. *Clinical Microbiology Reviews* 5: 302–327.

Stynen DL, Meulemans A and Braendlin GN (1992) Characteristics of a Latex Agglutination Test Based on Monoclonal Antibodies for the Detection of Fungal Antigens in Food. In: Samson RA, Hocking AD, Pitt JI and King AD (eds) *Modern Methods in Food Mycology*, p. 213. Amsterdam, Netherlands: Elsevier Science.

Torok T and King AD (1991) Comparative study on the identification of food borne yeasts. *Applied and Environmental Microbiology* 57: 1207–1212.

Torok TD, Rockhold T and King AD (1993) Use of electrophoretic karyotying and DNA–DNA hybridization in yeast identification. *International Journal of Food Microbiology* 19: 63–80.

van der Horst M, Samson RA and Kerman H (1992) Comparison of Two Commercial Kits to Detect Moulds by Latex Agglutination. In Samson RA, Hocking AD, Pitt JI and King AD *Modern Methods in Food Mycology*, p. 241. Amsterdam, Netherlands: Elsevier Science Publishers.

Food-poisoning Organisms

Daniel Y C Fung, Kansas State University, Manhattan, USA

The classical scheme of identification of bacteria by biochemical methods depends on whether a pure culture of the organism of interest can grow in an agar plate, an agar slant, a broth, a paper strip, or other supportive material containing specialized growth promoters or inhibitors in the presence of a fermentable or degradable compound, resulting in the medium changing colour, development of gas, development of fluorescent compound and other manifestation of metabolic activities. If the behaviour of known cultures in these media is known, an unknown culture can be matched with these characteristics and an analyst can make an identification of the unknown culture. This process is tedious, time-consuming and requires a lot of labour, materials, time and energy to perform the tests. In addition the skill of the analyst in interpreting the reactions and arriving at a correct judgement makes this process subjective and often unreliable. To make a complete identification, a great many tests can be done, as shown in **Table 1**.

To improve on the classical methods of biochemical identification several developments have been made and refined in recent years. Collectively these methods are considered as modern biochemical identification techniques.

Table 1   Information needed on food-borne pathogens

Morphology under magnification and on agar plates
Gram reaction and special staining properties
Biochemical activity profile and special enzyme systems
Pigment production, bioluminescence, chemiluminescence and fluorescent compound production
Nutritional and growth factor requirements
Temperature and pH requirements and tolerance
Fermentation products, metabolites and toxin production
Antibiotic sensitivity pattern (antibiogram)
Gas requirements and tolerance
Genetic profile: DNA/RNA sequences and fingerprinting
Pathogenicity to animals and humans
Serology and phage typing
Cell wall, cell membrane and cellular components
Growth rate constant and generation time
Ecological niche and survival ability
Motility and spore formation
Extracellular and intracellular products
Response to electromagnetic fields, light, sound and radiation
Resistance to organic dyes and special compounds
Impedence, conductance and capacitance characteristics
Others

From Fung (1995) with permission.

Miniaturization of Biochemical Methods

The author has miniaturized a large number of conventional biochemical tests and used the microtitre plate as the vessel for growth of the pure cultures. A multipoint inoculator has been used to expedite the efficiency of transfer of large numbers of cultures simultaneously. With the sterile inoculator he was able to transfer large numbers of cultures from a master plate containing up to 96 individual pure cultures into a variety of liquid media housed in separate microtitre plates or large Petri dishes containing different types of agars. With these miniaturized methods he and his colleagues studied large numbers of bacterial and yeast isolates over the years with excellent results. These miniaturized systems are flexible and can be applied to many groups of microbes.

Diagnostics Kits

The advantages of miniaturization and multitest units led to the commercialization of diagnostic kits in the 1970s. The two main types of diagnostic kits are agar-based and dehydrated media-based. In these systems the pure cultures grow in a variety of solid or liquid media, changing colour, or gas formation or utilizing their enzymes to change the colour of the substrates. Diagnostic charts can be used to identify unknown cultures or the numerical manuals of computer-assisted systems. It should be emphasized that most of these systems were first designed to identify the family Enterobacteriaceae. Later some systems branched out to identify other organisms, such as the non-fermentors, lactics, yeast, etc. The next article of this section will concentrate on modern biochemical identification of Enterobacteriaceae, coliforms, and *Escherichia coli*. Some overlapping of information between this article and the next article will occur. The following are synopses of how diagnostic kits operate and the range of organisms tested. As far as possible, information concerning comparative analysis of these kits with conventional methods will be made (also see the Further Reading section).

Agar-based Diagnostic Kits

Although several agar-based multimedia diagnostic kits have appeared on the market over the years, the only system still available is the Enterotube system (Roche Diagnostic, Nutley, NJ). The Enterotube II is a self-contained, compartmented plastic tube containing 12 different conventional media and an enclosed inoculating wire, which is threaded through the entire unit. This system permits 15 standard biochemical tests to be inoculated and performed from a single bacterial colony. Reagents are added to the indole test and Voges–Proskauer test before colour

reactions and gas formation are read. **Table 2** shows the colour reactions of the tests. Since such a table exists for every diagnostic kit described in this chapter, this table will serve as a model for other kits. Similar tables will not be repeated.

After reading the reactions, each result is given a score according to the system. After all the scores are added an identification (ID) value in the form of a five-digit number will be generated. Other systems (to be described) may have seven- or 10-digit numbers. From the code book the organism can be identified. This procedure is repeated for most other systems and will not be described again. The Enterotube II was developed for identifying Enterobacteriaceae only. This system received a 94% accuracy rating compared with the conventional method. It is worth mentioning that most comparative analyses of diagnostic kits were done in the 1980s. The consensus of opinion was that, when a system is 90–95% correlated with the conventional method, it is rated as good. When the value dropped to 85% the system is marginally acceptable and any value below that is not acceptable.

A similar unit, the Oxi/ferm Tube, was designed for Gram-negative non-fermenters.

Advantages of Enterotube II include rapidity and ease of inoculation, that inoculum suspension is not required and a single colony can be used for identification.

Table 2 Reactions of biochemical tests for Enterotube II

Test		Positive reaction	Negative reaction
GLU	Glucose utilization	Yellow	Red
Gas	Gas production	Wax lifted	Wax not lifted
LYS	Lysine decarboxylase	Purple	Yellow
ORN	Ornithine decarboxylase	Purple	Yellow
H_2S	H_2S production	Black	Beige
IND	**Indole formation**	**Pink-red**	**Colourless**
ADON	Adonitol fermentation	Yellow	Red
LAC	Lactose fermentation	Yellow	Red
ARAB	Arabinose fermentation	Yellow	Red
SORB	Sorbitol fermentation	Yellow	Red
VP	**Voges–Proskauer**	**Red in 20 min**	
DUL	Dulcitol fermentation	Yellow or pale yellow	Green
PA	Phenylalanine deaminase	Black to smoky grey	Green
UREA	Urease	Red-purple	Yellow
CIT	Citrate utilization	Deep blue	Green

Tests in **bold** require addition of reagents.
Indole test: Add 1–2 drops of Kovac's reagent through plastic film of H_2S/indole compartment using a needle.
Voges–Proskauer: Add 2 drops of 20% KOH solution containing 0.3% creatine and 3 drops of 5% α-napthol in absolute ethyl alcohol. Read colour within 20 min.
From Thippareddi and Fung (1998) with permission.

Disadvantages include that it is only useful for Enterobacteriaceae, it is difficult to stack in the incubator, and it has a short shelf life.

Dehydrated Media Diagnostic Kits

Concurrently with the development of miniaturized microbiological methods, various diagnostic kits using miniaturized concepts were developed in the early 1970s. Dehydrated media kits have the advantage of much longer shelf life than agar-based media (18 months versus a few months). Those currently used in clinical, environmental, industrial and food microbiology will be discussed in the following sections. Of course, many similar systems are available worldwide: the systems discussed here have been well tested and used in the US and Europe.

API (bioMérieux Vitek, Hazelwood, MO) This is probably the most popular system for diagnostic bacteriology in the world, especially for Enterobacteriaceae. The API 20E system is miniaturized microtube system which has 20 small wells designed to perform 23 standard biochemical tests from isolated colonies of bacteria on plating medium. The system has procedures for same-day and 18–24-h identification of Enterobacteriaceae. It consists of microtubes containing dehydrated substrates. The substrates are reconstituted by adding a bacterial suspension into each of the 20 wells; some of the wells are filled with mineral oil to create anaerobic conditions. The unit is then incubated so that the organisms react with the contents of the tubes and read when the indicator systems are affected by the metabolites or added reagents – generally after 18–24 h incubation at 35–37°C. After all the reactions are read and recorded in a data sheet a number code is generated and the code can be matched with the code book for identification. The strip can also be read in a reader and the results interpreted by a computer.

This system has been evaluated by a vast number of investigators worldwide and currently is considered almost the standard test for identification of Enterobacteriaceae. Using API 20E bacteria were identified to the genus and species level with 90.2% and 93% accuracy.

Using similar formats, other organisms can be identified, such as API NET for Gram-negative non-Enterobacteriaceae, API CAMPY for *Campylobacter* spp., API Staph-Ident for staphylococci and micrococci, API 20A for anaerobes, API Listeria for *Listeria*, API 50 CH for *Lactobacillus*, API 20C for yeasts, Rapid Strep for streptococci, and Rapid Coryn for *Corynebacterium*.

Advantages include being the most complete system commercially available for the identification of Enterobacteriaceae and an excellent database. Disadvantages include that it is difficult and time-consuming to inoculate, problems in handling and stacking of tray and lids due to flexible plastic materials, and that a competent microbiologist is needed to read and interpret the colour changes.

MicroID (Organon Teknika, Durham, NC) This is the first system that provides results in 4 h: the system measures enzyme activities and not growth of the culture. It consists of a moulded polystyrene tray containing 15 reaction chambers and a hinged cover. The first five reaction chambers contain a single combination substrate/detection disc with upper and lower discs in the same trough. The remaining 10 reaction chambers each contain a single combination substrate/detection disc. Discs contain all substrate and detection reagents required to perform the indicated biochemical tests (except for the Voges–Proskauer test). The surface of the tray is covered with clear polypropylene tape to prevent spillage and also for reading the reactions.

A few cultures from a Gram-negative isolation agar plate are first mixed into a liquid form and 0.2 ml of the liquid is then introduced into each of the 15 wells. The unit is then incubated at 35°C for 4 h, after which time two drops of 20% KOH are added into the VP well. The unit is then rotated 90° so that the liquid from the lower part of the first five wells comes into contact with the upper discs of the same chamber for final reactions. The reactions of these five tests are read from the upper discs. Then the reactions from the remaining 10 discs are read. Again, a number is generated in the data sheet and the code number is matched, with codes in the code book for identification. MicroID has achieved 93.5% and 97% accuracy levels.

A similar format with different substrates was available to identify *Listeria* spp.

Advantages of the system include high accuracy, speed of reaction (4 h) and convenience – it is self-contained, easy to use, requires only one reagent addition and has a long shelf life. A disadvantage is that a competent microbiologist is needed to read the colour reaction.

Minitek (Becton Dickinson Microbiology Systems, Cockeyville, MD) This is a flexible system. The unit contains 10 wells. Two units (20 wells) can be used to identify one culture. The system supplies 36 different substrates, thus the user can choose which test to perform. First, paper discs containing individual substrates are applied to the wells, one disc per well. A liquid culture is prepared and applied to each well using an automatic application gun (about 0.2 ml per well). Some wells will be filled with mineral oil to create an anaerobic environment. The unit is then incubated overnight at 35°C. After incubation the colour reactions are read and identification is made with the aid of a code book.

Minitek achieved a 97% accuracy rate. Because the system is flexible different combinations can be used to study organisms other than Enterobacteriaceae.

On the one hand, the advantage of the system is versatility and flexibility, but this may be a disadvantage when no code book is available for organisms other than Enterobacteriaceae. The construction of the unit is sturdy. Disadvantages include that the various components of the total system are handled excessively in preparation and operation. Again skill is needed to read borderline reactions in the discs.

BBL Crystal (Becton Dickinson Microbiology Systems, Cockeysville, MD) Going from the extreme of having a lot of manipulation in the Minitek system, Becton Dickinson then developed a system that has very little manipulation – the BBL Crystal system. In one system (Enteric/Non-fermentor ID Kit) both enteric and non-fermenters can be identified. It is important to ensure that the unit is marked correctly as to whether an oxidase-positive (non-fermenter) or oxidase-negative (fermenter) pure culture is to be analysed.

The system is easy to use. On one panel, 30 dried biochemical substrates are housed and a companion unit (base) is used for the liquid sample. The liquid culture (ca. 2 ml) is carefully poured into the trough of the base. Then the upper unit containing the 30 tests is simply snapped into the base such that the culture interacts with the 30 substrates. The unit is then incubated at 35°C overnight. After incubation the unit is introduced into a Crystal light box to record reactions and for identification using a 10-digit system. Identification can also be made using a computer.

Since this is a relatively new system, no extensive comparative analysis has been made.

In addition, there is also a Rapid Stool/Enteric ID kit for stool samples.

Advantages include sturdy panels, ease of operation and computer-assisted identification. Very few disadvantages are noted.

RapidID One System   (Remel, Lenexa, KS) This is a miniaturized unit housed in an ingenious chamber. On one side of the chamber there is a trough where a liquid culture can be introduced. Then the unit can be tilted slowly at a 45° angle forward: the liquid will flow into individual wells, each containing a separate substrate. Thus, inoculation into 20 wells can be made

in one motion. This is more convenient than the API system where the analyst needs to insert 20 drops of liquids into 20 miniaturized wells. After incubation, the colour reactions can be read after 4 h incubation and the cultures identified. The forerunner of the RapidID enteric system is the Spectrum 10 system. The Spectrum 10 is rated as 91% accurate. Using the basic design, Remel markets strips for Enterobacteriaceae, non-fermenters, yeasts, anaerobes, streptococci, *Leuconostoc*, *Pediococcus*, *Listeria*, *Neisseria*, *Haemophilus* and urinary tract bacteria. Identification of various anaerobes to the genus level using the anaerobic RapidID system ranges from 83% to 97% accuracy and to the genus level, 76–97%.

Advantages include results in 4 h, clear chromogenic reactions and one-step inoculation. A disadvantage is the skill required to read the colour changes.

ATB (bioMérieux Vitek, Inc., Hazelwood, MO) This is a 32-carbon assimilation test system. The culture is first made into a solution and then the liquid is introduced to the unit. After incubation (4–24 h depending on the culture) the tests can be read manually or automatically. Test strips are available for anaerobes, staphylococci, micrococci, yeast, Enterobacteriaceae, streptococci and Gram-negative bacilli. The automatic reader can also read API 20 and API 50 test strips.

Biolog System (Biolog, Hayward, CA) This is a miniaturized system utilizing the microtitre plate format for growth of bacteria in various liquid media. In contrast to the miniaturized system developed by the author about 30 years ago, 95 different carbon sources are used in the microtitre plate; one well, containing a rich growth medium, is used as the positive control. An unknown culture is suspended in a liquid medium and the liquid aliquots are injected into the 96 wells. The plate is then incubated overnight at 35°C and after incubation the colour of the wells is examined. The advantage of this system is that the colour is either clear (no reaction) or blue (as a result of reduction of the dye in the medium). The pattern of blue wells will indicate the identity of the unknown culture. Using the human eye to interpret these data would be tedious and unreliable. The company developed an automatic reader to provide instant identification of the unknown culture by matching the profile of the known cultures in the data bank against the profiles of the unknown cultures. This is indeed a simple system to use and interpretation of the results is easy. The system is exceedingly ambitious in that it is designed to identify all bacterial cultures! Currently it has an identification database for 1118 species/groups of microorganisms. It suffers by having a limited database for identification of organisms. Sometimes non-typical isolates are not identifiable. This is not a negative comment on this system because once the databank reaches a sufficient level it can be a powerful system to identify isolates from all kinds of environments.

Vitek System   (bioMérieux Vitek, Hazelwood, MO) This is possibly the ultimate automated system for identification of microorganisms from clinical, environmental, industrial and food samples. The system has its origin in the Viking Mission during the early stages of the US space programme a few decades ago. The heart of the system is a plastic cord with 30 tiny wells containing selective media and specialized substrates designed to discriminate bacterial taxa by the growth pattern and kinetics of the unknown culture in media in the 30 wells. A pure culture is first suspended in a liquid and liquid is introduced into the card (the size of an ordinary credit card) by pneumatic pressure, such that all 30 wells will be filled with an aliquot of the culture. The card is then placed into the incubator. Up to 240 cards can be inserted into one large unit. More units can be tested at the same time if more incubators are connected to the system. The instrument periodically scans each card and determines the kinetics of the growth of the organism in each well and then determines the identity of the unknown culture. For typical cultures the identification can be completed in 2 h. Other bacterial cultures can be identified in about 18 h.

The system has been evaluated extensively over the years and constantly received over 90% accuracy compared with the conventional method: the author's group rated it 99% accurate compared with the conventional method. **Table 3** shows the performance of the Vitek system on various organisms. Overall, its performance is exceptionally good. This method is truly automated and provides reliable data in a short time for most samples. It can identify Gram-negative, Gram-positive, yeast, *Bacillus*, anaerobes, non-fermenters, *Neisseria/Haemophilus* and other classes of

organisms. It constantly receives high correlations with conventional methods of identification of unknown microbial cultures.

In addition to its high cost the system has had no negative comments from participants in the International Workshop on Rapid Methods and Automation in Microbiology held at Kansas State University annually since 1981.

Range of Food Applications

The methods and systems described previously are designed for the identification of pure cultures obtained from clinical, food, industrial and environmental samples. Almost all foods are potential sources of contamination of pathogenic microorganisms. Thus, all microbiological methods are designed to enrich, isolate, enumerate, characterize and identify the unknown culture in question. The results of the diagnostic tests are only as valuable as the purity of the culture. If there is a mixed culture in the primary isolation, all the valuable identification capabilities of these systems will be meaningless. Thus, for food microbiologists it is essential that all food samples must be properly prepared before either directly plating the sample on selective agars or enriching the foods in pre-enrichment and enrichment liquid media and isolating pathogens on appropriate agar plates. The development of excellent primary isolation agars for selectively isolating the target organism has assisted in selecting the isolates for further identification by one or more of the diagnostic systems described. Announcements of new selective agars for specific food-borne pathogens appear almost weekly. It is up to the user to decide which agar to test or use.

Another related development of identification of food-poisoning organisms by modern biochemical techniques is the variety of screening tests on the market. Tests for pathogens such as ELISA tests, DNA probes, polymerase chain reaction tests, dipstick techniques and motility tests for pathogens are considered to be screening tests. Negative screening tests would allow the food processors to ship their products to the market but a positive screening test will necessitate an embargo of the product and a confirmation test to be done on the suspected food. This procedure involves conventional methods as well as some of the diagnostic kits mentioned in this article.

What are the bacterial food pathogens facing us these days? The list is long but worth reiterating: *Salmonella* spp., *Staphylococcus aureus*, *Clostridium perfringens*, *C. botulinum*, *Campylobacter jejuni*, *Escherichia coli* O157:H7, *Yersinia enterocolitica*, *Shigella* spp., *Vibrio*, *Aeromonas* and *Plesiomonas*,

Table 3   Vitek identification data

Organism	Correct (%)	False-negative rate (%)	False-positive rate (%)
Edwardsiella	86.1	1.06	0
Enterobacter	97.2	0.21	0
Escherichia coli	97	0.45	0
Hafnia	65	2.19	0
Salmonella	96.7	1.03	0
Yersinia	93.1	0.53	0

From Kalamaki et al (1997) with permission.

Table 4 Salmonella spp. screening and identification test kits

Test kit	Supplier	Assay principle	Limit of detection	Sensitivity	Specificity	Agreement (%)	False-negative rate (%)	False-positive rate (%)	Total time	Cost per assay
1-2 Test[1]	BioControl Systems	Motility enrichment/immunodiffusion	>10^2 cfu ml^-1	100%	100%	96% 99% 94.4%	5.2% <1% 2.3% 0-5.38%	0% 1.4% <1%	36-58 h	$6.97
Assurance Salmonella[1]	BioControl Systems	ELISA	10^4-10^6 cfu ml^-1	98%	96.2%	97.2 95.6%	1.8% 2.3%	0%	48 h	$3.23
Bactigen	Wampole Laboratories	Latex agglutination	Pure colony	81.6% 96%	67%	NR	18.4% 2.3%	33%	3-5 min for test only	NR
BacTrace	KPL	ELISA	10^5 cfu ml^-1	89.2%	NR	NR	14% 8%	53.8% 5.5% 52.1%	42-52 h	NR
EIAFoss	Foss Electric	Immuno-fluorescence	10^5 cfu ml^-1	NR	NR	NR	NR	<2% 0%	<24 h	$6.00
Equate[3]	Binax	ELISA	NR	NR	NR	98%	4.5%	11.54%	49 h	NR
Gene-Trak colorimetric[1]	Gene-Trak	Nucleic acid hybridization	10^7-10^8 cfu ml^-1	99% 100%	93% 100%		2.5% 13.4% 0%	1.4% 4.9% 0%	46.5 h	$7.60
Gene-Trak isotopic	Gene-Trak	Nucleic acid hybridization	10^6-10^7 cfu ml^-1	90.8%	NR	NR	9.2%	NR	46 h	NR
Locate	Rhone Poulenc	ELISA	10^5-10^6 cfu ml^-1	81.30%	97%	NR	18.7%	3%	45 h	NR
Malthus 2000[2]	Malthus Instruments	Automated	NR	96.5%	91.8%	NR	7.5%	8.2%	Variable	NR
Microelisa[3]	Dynatech	ELISA	10^5 cfu ml^-1	99.2%	99%	NR	NR	NR	48 h	NR
Microscreen	Neogen	Immuno-precipitation	10^4-10^7 cfu ml^-1	NR	98%	NR	<1%	<2%	21 h	$7.60
Microscreen	Mercia Diagnostics	Latex agglutination	Pure colony	89.5% 96%	67% 96%	NR	10.5% 4%	18.4% 4%	3-5 min for test only	NR
Path-Stik[5]	Integrated Biosolutions	Immuno-chromatography	5 x 10^5-5 x 10^6 cfu ml^-1	91.4% 93%	98.6% 96.4%	97.5%	5.2%	4.8%	38-56 h	$7.00
Salmonella enteritidis screen/verify[4]	Vicam	Immunomagnetic separation/plating/latex agglutination	0.2-2.6 cfu g^-1	NR	100%	100%	0%	0%	22 h	$7.40
Salmonella latex test	Unipath	Latex agglutination	NR	NR	NR	NR	NR	NR	NR	NR
Salmonella rapid test	Unipath	Motility enrichment/biochemical detection	NR	98%	100.00%	99%	2%	0%	42 h	$7.28

Table 4 *Salmonella* spp. screening and identification test kits (Continued)

Salmonella screen/verify[4]	Vicam	Immunomagnetic separation/plating/latex agglutination	0.4–11.1 cfu g⁻¹	100%	100%	100%	0%	0%	22 h	$6.90
Salmonella Tek[2]	Organon Teknika	ELISA	10^5 cfu ml⁻¹	96.5%	89.3% 97.6%	97.6%	0.4% 1.6% 15.4% 8.4%	6% 3.8% 43.2%	48 h	$2.10
Salmonella unique	TECRA	Immuno-enrichment/EIA	1.5×10^4–6.5×10^4 cfu ml⁻¹	97%	80%	NR	6%	20% <0.5%	<22 h	$8.00
Salmonella VIA[2]	TECRA	ELISA	30 cfu g⁻¹ or 10^5 cfu ml⁻¹	94% 86%	98% 86.5%	99.4% 96.7% 96% 95%	1.4% 5.8% 0%	4.1% 2.22% 2.5% 5%	42–48 h	$3.80
Spectate	May & Baker Diagnostics	Latex agglutination	10^7 cfu ml⁻¹	66.30%	83%	NR	7.3%	17%	3–5 min for test only	
Vidas	bioMérieux Vitek	Immuno-fluorescence	10^5–10^6 cfu ml⁻¹ 1.8×10^6 cfu ml⁻¹	84–99%	93–98%	92.9% 97%	1–16% 0.6% 11%	7% 1.8% 0%	24–48 h	$4.50
Wellcolex color *Salmonella*[6]	Wellcome Diagnostics	Latex agglutination	10^6 cfu ml⁻¹	99% 98.4% 88.2%	100% 98%	100% correlation 98%	97.5%[7] 100%	63%[8] 78%	3 min for test only	

NR, not reported.
[1] AOAC first action.
[2] AOAC final action.
[3] No longer available.
[4] AOAC approved.
[5] AOAC certificate of performance.
[6] Serotype identification.
[7] Negative predictive value.
[8] Positive predictive value.
From Kalamaki et al (1997) with permission.

Bacillus cereus, Listeria monocytogenes and others. The biochemical techniques can identify most of these organisms in a laboratory setting but most commercial diagnostic kits are designed for specific groups of organisms. Thus, knowledge of the basic principles of diagnostic microbiology is essential for all food microbiologists, regardless of whether one uses the diagnostic kits described. Which diagnostic system is best for the identification of a particular food-poisoning organism is the subject of much debate. **Table 4** illustrates the numerous identification systems and screening systems for *Salmonella* alone. A similar chart is available for *Listeria*, Enterobacteriaceae, *Campylobacter*, *E. coli*, *Staphylococcus aureus*, *Vibrio cholerae* and *V. vulnificus*.

In conclusion, modern biochemical identification techniques are more convenient modes of improving the conventional biochemical techniques. They make sample operation, inoculation, incubation, reading, data collection and interpretation of data for diagnostic purposes more convenient than the conventional methods. To make a quantum jump in the identification of food-poisoning organisms one has to look to polymerase chain reaction and biosensor technologies to obtain real-time rapid identification.

See color Plate 7.

See also: **Biochemical and Modern Identification Techniques**: Introduction; Enterobacteriaceae, Coliforms and *E. coli*.

Further Reading

Celig DM and Schreckenberger PC (1991) Clinical evaluation of the RapID-ANA II panel for identification of anaerobic bacteria. *J. Clin. Microbiol.* 29: 457–462.

Fung DYC (1995) What's needed in rapid detection of foodborne pathogens. *Food Tech.* 49: 64–67.

Fung DYC (1998) *Handbook for Rapid Methods and Automation in Microbiology in Microbiology Workshop.* Kansas State University, Manhattan, KS: Department of Animal Sciences and Industry.

Fung DYC, Goldschmidt MC and Cox NA (1984) Evaluation of bacterial diagnostic kits and systems at an instructional workshop. *J. Food Protec.* 47: 68–73.

Goosh WM III and Hill GA (1982) Comparison of MicroID and API 20E in rapid identification of Enterobacteriaceae. *J. Clin. Microbiol.* 15: 885–890.

Kalamaki M, Price RJ and Fung DYC (1997) Rapid methods for identifying seafood microbial pathogens and toxins. *J. Rapid Methods Automation Microbiol.* 5: 87–137.

Russell SM, Cox NA, Bailey JS and Fung DYC (1997) Miniaturized biochemical procedures for identification of bacteria. *J. Rapid Methods Automation Microbiol.* 5: 169–178.

Thippareddi H and Fung DYC (1998) Laboratory Manual Section. In: Fung DYC (ed.) *Handbook for Rapid Methods and Automation in Microbiology Workshop.* Kansas State University, Manhattan, KS: Department of Animal Sciences and Industry.

Enterobacteriaceae, Coliforms and *Escherichia coli*

R R Beumer and **M C te Giffel**, Wageningen Agricultural University, Laboratory of Food Microbiology, Bomenweg, Wageningen, The Netherlands

A G E Griffioen, Wageningen, The Netherlands

Introduction

The identification of microorganisms with conventional methods involves a great amount of materials and work. Therefore, many rapid identification systems have been developed. These techniques are based on dehydrated substrates ('dry systems') or on ready-to-use media ('wet systems'), placed in cupules or tubes. The shelf life varies from a few months to approximately 1.5 years.

There are a number of conventional and commercially available automated and non-automated systems to identify Gram-negative bacilli. Except for reference testing, conventional macrotube biochemical tests for bacterial identification have been replaced by commercial systems, because the classical methods are too expensive, slow and unwieldy for routine use in the microbiological laboratory. Some of the systems are restricted to one genus (i.e. API*Listeria*), others can be applied for large groups (BBL® Crystal™ Grampositive Identification System). These systems range from visual interpretation of miniaturized biochemical panels with computerized taxonomic databases to semi-automated or automated systems that can interpret and analyse results in a matter of hours.

Systems for determination of Enterobacteriaceae are more available than identification systems for other microorganisms. As most of these systems were initially developed for application in clinical microbiology, bacteria from animal, food, feed and/or environmental sources may be less commonly tested and incorporated in the identification databases of the system. The diversity of these microorganisms, originating from various sources, may cause problems for the identification systems, since bacterial strains of the same species may vary slightly in their biochemical reactions.

Though results obtained with various identification

systems for Enterobacteriaceae have been described in the literature, worldwide most tests for the biochemical identification of Enterobacteriaceae (in medical, industrial and research laboratories) are performed with the miniaturized systems API 20E, BBL® Enterotube™ and BBL® Crystal™ Enteric/Nonfermenter ID system. In this article these systems are described in detail, and the results obtained with the systems in identifying test strains of Enterobacteriaceae are discussed.

Brief Principles and Types of Commercially Available Tests

Some of the commercially available tests, e.g. VITEK (bioMérieux, Marcy-l'Etoile, France) or BIOLOG (Biolog Inc., Hayward, California), can only be used in combination with expensive equipment, so their use is restricted to laboratories testing large numbers of strains. Other tests (e.g. Hy-enterotest, Hy-laboratories, Israel) are not available worldwide, or can only be applied for determination at the genus level.

The API 20E (bioMérieux, Marcy-l'Etoile, France), BBL® Enterotube™ and BBL® Crystal™ (Becton Dickinson and Company, Maryland, USA) tests do not need high investments in apparatus, are user-friendly, and are well-known in laboratories for clinical, veterinary and food microbiology.

Principle of the API 20E Identification System for Enterobacteriaceae

API 20E is a standardized identification system for Enterobacteriaceae and other non-fastidious Gram-negative rods which uses 23 miniaturized biochemical tests and a database. The API 20E strip consists of 20 microtubes containing dehydrated substrates. These tests are inoculated with a bacterial suspension which reconstitutes the media. During incubation, metabolism or metabolite produces colour changes that are either spontaneous or revealed by the addition of reagents.

The reactions are analysed according to the interpretation table, the analytical profile index (a code book) or the APILAB software.

Principle of the BBL® Enterotube™ for the Identification of Enterobacteriaceae

The Enterotube™ and its computer coding and identification system are specially designed for the identification of Enterobacteriaceae, i.e. aerobic, Gram-negative rods, which are oxidase-negative. The Enterotube™ consists of 12 compartments in a row, filled with ready-to-use media ('wet' system), with which 15 reactions can be performed.

Principle of the BBL® Crystal™ Enteric/ Nonfermenter ID System for Enterobacteriaceae

The BBL® Crystal™ Enteric/Nonfermenter (E/NF) identification system is a miniaturized identification method employing modified conventional and chromogenic substrates. The kit comprises lids, bases and inoculum fluid tubes. The lid contains 30 dehydrated substrates on the tips of plastic prongs. The base has 30 reaction wells. The test inoculum is prepared with the inoculum fluid and is used to fill all 30 wells in the base. When the lid is aligned with the base and snapped in place, the test inoculum rehydrates the dried substrates and initiates test reactions.

After an incubation period, the wells are examined for colour changes, resulting from metabolic activity of microorganisms. The resulting pattern of the 30 reactions is converted into a ten digit profile number that is the basis for identification.

Detailed Protocols for the Biochemical Identification of Enterobacteriaceae

The test protocols given here are according to the manufacturer's instructions.

API 20E Identification System for Enterobacteriaceae

Materials Provided   The API 20E kit consists of: 25 API 20E strips, 25 incubation boxes, 25 report sheets, one clip seal and one instruction manual. To use the API 20E, the following materials are necessary (also available from the manufacturer): suspension medium or sterile water (5 ml), reagent kit, zinc reagent, mineral oil, pipettes, API 20E profile index or identification software and an ampoule rack. The following general laboratory equipment is also required: incubator (35–37°C), refrigerator, Bunsen burner and marker pen.

For the determination of fermentative or oxidative metabolism and motility, API OF-medium and API M-medium might be necessary.

Directions for Use
Preparation of the Strip

- Prepare an incubation box, tray and lid, and distribute about 5 ml of water into the honeycombed wells of the tray to create a humid chamber.
- Record the strain reference on the elongated tab of the tray.
- Place the strip in the tray.
- Perform the oxidase test on an identical colony. Only use oxidase-negative colonies for the biochemical determination of Enterobacteriaceae.

Preparation of the Inoculum

- Open an ampoule of suspension medium or sterile water without additives.
- With the aid of a pipette, remove a single well-isolated colony from an isolation plate.
- Carefully emulsify to achieve a homogenous bacterial suspension.

Inoculation of the Strip

- With the same pipette, fill both the tube and cupule of tests CIT (citrate utilization), VP (acetoin production, Voges–Proskauer reaction) and GEL (gelatinase), with the bacterial suspension.
- Fill only the tubes (and not the cupules) of the other tests.
- Create anaerobiosis in the tests ADH (arginine dihydrolase), LDC (lysine decarboxylase), ODC (ornithine decarboxylase), URE (urease) and H_2S by overlaying with mineral oil.
- Close the incubation box and incubate at 35–37°C for 18–24 h.

Reading of the Strip

- After 18–24 h at 35–37°C, read the strip by referring to the interpretation table.
- Record all spontaneous reactions on the record sheet.
- If the glucose reaction is positive and/or three tests or more are positive: reveal the results which require the addition of reagents: VP, TDA (tryptophane desaminase), IND (indole production) and NO_2.
- Add the reagents required and record the results on the report sheet.
- If the glucose reaction is negative and the number of positive tests is less than or equal to two, do not add reagents.

Identification

- Using the identification table: compare the results recorded on the report sheet with those given in the table.
- With the analytical profile index or the identification software: the pattern of the reactions obtained must be coded in a numerical profile.

On the report sheet, the tests are separated into groups of three, and a number (4, 2 or 1) is indicated for each. By adding the numbers corresponding to positive reactions within each group, a 7-digit profile number is obtained for the 20 tests of the API 20E strip.

In some cases, the 7-digit profile is not discriminatory enough and additional tests should be carried out: reduction of nitrates to nitrites (NO_2),

reduction of nitrites to N_2 gas, motility, growth on MacConkey agar medium, oxidation of glucose and fermentation of glucose. The technical assistance service can be consulted for any unlisted profile.

Disposal After use, all ampoules, strips and incubation boxes must be autoclaved, incinerated, or immersed.

BBL® Enterotube™ for the Identification of Enterobacteriaceae

Materials Provided Enterotubes™ are delivered in boxes containing 5 or 25 units, 5 or 25 report sheets and an instruction manual.

Directions for Use

Preparation of the Tube Remove both caps from the tube. The tip of the inoculation needle is under the white cap. Without flaming the needle pick a well-isolated colony directly on the top of the needle. Do not puncture the agar.

Inoculation of the Tube

- Inoculate the tube by first twisting the needle, then withdrawing it through all the compartments using a turning motion. Reinsert the needle (without sterilizing) into the tube until the notch on the needle is aligned with the opening of the tube. The tip of the needle should be seen in the citrate compartment.
- Break the needle at the notch by bending. The portion of the needle remaining in the tube maintains the anaerobic conditions necessary for true fermentation of glucose, decarboxylation of lysine and ornithine and the detection of gas production. Replace both caps.
- With the broken off part of the needle punch holes through the foil, covering the air inlets of the last eight compartments (adonitol, lactose, arabinose, sorbitol, Voges–Proskauer, dulcitol/PA (phenylalanine desaminase), urea and citrate) in order to support aerobic growth in these compartments.
- Incubate the tube at 35–37°C, standing it upright if possible in the test-tube support with its glucose compartment pointing upward, or lay the tube on its flat surface.

Reading of the Tube

- Interpret all the reactions with the exception of indole and Voges–Proskauer, in comparison with a reference picture or a not inoculated tube. Record the reactions on the interpretation pad. The reaction is negative if the compartment remains unchanged (exceptions: indole and Voges–Proskauer).

- Perform the indole test and the Voges–Proskauer test by adding the Kovacs' reagent into the H$_2$S/indole compartment (directly under the plastic film) and α-naphthol and potassium hydroxide (through the air inlet) into the Voges–Proskauer compartment.

In order to identify an isolate, the numbers for the positive reactions are written down on the interpretation pad. The circled numbers for 5 × 3 reactions are added together, resulting in a five-digit number (ID value). The number thus obtained, is then compared to the Computer Coding and Identification system (available at Becton Dickinson) which results in the identification of a microorganism.

If further tests are required, or if the purity of culture has to be checked, an inoculum from the incubated tube can be taken and applied to a suitable medium or broth for subcultivation as follows:

- The inoculation needle is drawn out with sterile forceps and streaked on a plate.
- Bacterial substance is extracted from a positive compartment with a sterile loop after the plastic film has been removed.

Disposal of the Tube   After reading, the Enterotube™ must be disposed of by autoclaving.

BBL® Crystal™ Enteric/Nonfermenter ID System for Enterobacteriaceae

Materials Provided   The BBL® Crystal™ kit consists of 20 lids, 20 bases, 20 tubes with inoculum fluid, two incubation trays and one colour reaction card and a results pad.

To use the kit the following materials are required (but not provided): sterile cotton swabs, incubator (35–37°C, 40–50% humidity), a BBL® Crystal™ light box, the BBL Crystal™ ID system electronic codebook, non-selective culture plates (e.g. tryptone–soy agar) and reagents to perform the indole test and the oxidase test (BBL DMACA Indole and BBL Oxidase reagent dropper).

Directions for Use
Preparation of the Panels

- Remove the lids from the pouch. Discard the dust cover and desiccant. Once removed from the pouch, the lids should be used within 1 h.
- Take a tube with inoculum fluid, and label with the number of the strain to be tested. Using aseptic technique, pick one well-isolated large (ca. 2–3 mm) colony (or 4–5 smaller colonies of the same morphology) with the tip of a sterile cotton swab or a wooden applicator stick from a blood plate such as Trypticase® Soy Agar with 5% sheep blood or MacConkey agar (or any suitable non-selective medium).

- Suspend the colony material in the inoculum fluid.
- Recap the tube and vortex for approximately 10–15 s.
- Take a base, and mark the number of the strain on the side wall.
- Pour the entire contents of the inoculum into the target area of the base.
- Hold base in both hands and roll the inoculum gently along the tracks until all of the wells are filled. Roll back any excess fluid to the target area and place the base on a bench top.
- Align the lid so that the labelled end of the lid is on top of the target area of the base.
- Push down until a slight resistance is felt. Place thumb on edge of lid towards middle of panel on each side and push downwards simultaneously until the lid snaps into place (listen for two 'clicks').
- Place inoculated panels in incubation trays. All panels should be incubated upside down in a non-CO$_2$ incubator with 40–60% humidity. Trays should not be stacked more than two high during incubation. The incubation time is 18–20 h at 35–37°C.

Reading of the Panels

- After the recommended period of incubation, remove the panels from the incubator. All panels should be read upside-down using the BBL® Crystal™ light box. Refer to the colour reaction chart for an interpretation of the reactions. Use the BBL® Crystal E/NF results pad to record reactions.
- Each test that is positive is given a value of 4, 2 or 1, corresponding to the row where the test is located. A value of 0 (zero) is given to any negative result. The numbers resulting from each positive reaction in each column are then added together, resulting in a 10-digit number: the BBL®-Crystal™ profile number.
- This number, and the off-line tests for indole and oxidase tests, should be entered on a personal computer in which the BBL® Crystal™ ID system electronic codebook has been installed, to obtain the identification.

Disposal   After use, all infectious materials including plates, cotton swabs, inoculum tubes, filter papers used for oxidase or indole tests and BBL® Crystal™ panels must be autoclaved prior to disposal or incinerated.

Comparison of the Tests

The biochemical reactions used in the identification systems can be classified according to the reaction patterns in the following groups: decarboxylation reactions, hydrolysis, oxidation and/or fermentation reactions, production of characteristic substances and 'other reactions'. The biochemical reactions used for each of the three identification systems are given in **Table 1**.

In our laboratory, the three tests were used for the identification of 60 bacteria belonging to the family of the Enterobacteriaceae. Before inoculation of the tests all strains were investigated for Gram-reaction (negative), specific growth on violet red bile agar (VRBG, Oxoid CM 485), fermentation of glucose (positive) and oxidase reaction (negative).

As a criterion for a 'good identification' 90% confidence was chosen. With this requirement the BBL® Crystal™ system identified 82% of the strains correctly, the BBL® Enterotube™ 78% and the API 20E test 67%. The manufacturer of API 20E uses, apart from the confidence percentage, the so-called T-index. This value varies from 0 to 1 and is inversely proportional to the number of atypical reactions. In the enclosed instruction manual only little practical information is given about the use of the T-index. According to the manufacturer, apart from a percentage of identification of at least 90%, percentages below 90% with a T-index between 0.5 and 1.0 are also acceptable. If the T-index is included in the identification with API 20E, the percentage of 'good identifications' increases to 85%.

For identification on the genus level the tests performed only slightly better: API 20E 88%, BBL® Crystal™ 87% and the BBL® Enterotube™ 83%.

It is interesting to know which reactions often gave problems in reading the results. **Table 2** presents nine reactions which often led to false results.

Part of the false reactions can be explained by a short incubation time and/or a (too) low level of inoculation. In some cases extending the incubation period to several hours (4–8 h) gave better results. Especially the reading of the decarboxylation reactions gave false results. This is caused by the design of the BBL® Crystal™ system: a micro-aerobic environment is created resulting in lysine-positive reactions for some bacteria, in contrast with conventional lysine tests. The computer program (Crystal™ Electronic

Table 2 Atypical test results (percentage) obtained with 60 confirmed strains of Enterobacteriaceae) using three identification systems for Enterobacteriaceae (API 20E, BBL® Enterotube™ and BBL® Crystal™)

Biochemical test	API 20E	BBL® Enterotube™	BBL® Crystal™
Arabinose	5	2	2
Citrate	22	0	2
H$_2$S	5	7	–
Indole	5	12	–
Inositol	17	–	13
Lysine	17	3	32
Ornithine	7	10	–
Sorbitol	5	2	3
Urea	0	7	12

–, this biochemical reaction is not part of the system.

Table 1 Biochemical reactions used in three identification systems for Enterobacteriaceae (API 20E, BBL® Enterotube™ and BBL® Crystal™)

Type of reaction	API 20E	Enterotube™	BBL® Crystal™
Decarboxylation	Lysine, ornithine	Lysine, ornithine	Lysine
Hydrolysis	Arginine p-n-p[a]-galactoside, urea	Urea	Arginine, aesculin, p-n-p-acetyl-glucosaminide, p-n-p-arabinoside, p-n-p-bisphosphate, p-n-p-β-galactoside, p-n-p-α-β-glucoside, p-n-p-β-glucuronide, p-n-p-γ-L-glutamyl-*p*-n-anilide, p-n-p-phosphate, p-n-p-phosphorylcholine, p-n-p-xyloside, proline *p*-n-anilide, urea
Oxidation fermentation	Amygdalin, arabinose, citrate, glucose, inositol, mannitol, melibiose, rhamnose, sorbitol, sucrose	Adonitol, arabinose, citrate, dulcitol, glucose, lactose, sorbitol	Arabinose, adonitol, citrate, galactose, inositol, malonate, mannitol, mannose, melibiose, rhamnose, sorbitol, sucrose
Production	Acetoin, H$_2$S, indole, nitrite	Gas (glucose), H$_2$S, pyruvic acid, indole	–
Other reactions	Gelatin, (degradation), tryptophan, (deamination)	–	Tetrazolium (reduction), p-n-DL-p-alanine (oxidative) deamination), glycine (degradation)

[a]p-n-p = ortho-nitro-phenyl.

Codebook) takes this into account. The citrate conversion of the API 20E system gave 22% false negative reactions. This can be explained by the short incubation time. The traditionally used citrate reaction (with Simmons citrate agar) is usually incubated for 1–2 days and, if necessary, even longer.

The Enterotube™ showed no false negative citrate reactions, probably caused by the high inoculum level used: fresh colony and first compartment to be inoculated.

The atypical results in the indole production of the Enterotube™ (3.3% false positive and 8.3% false negative) were caused by reading faults. The red colour after addition of the indole reagent was often difficult to observe. There is no explanation for the high number of false positive inositol reactions in BBL® Crystal™ and API 20E.

The inoculation of the BBL® Crystal™ test and the Enterotube™ is very user-friendly and quick. The API system is less user-friendly, because the inoculation is time consuming. While inoculating, air bubbles are easily trapped in the cupules and some cupules need to be covered with mineral oil.

Before reading, the tests reagents must be added to API 20E and BBL® Enterotube™, whereas in the BBL® Crystal™ system no reagents have to be added. Since test results for the three systems are comparable, user-friendliness and price will eventually determine the choice.

See color Plate 8.

See also: **Biochemical and Modern Identification Techniques**: Introduction; Food-poisoning Organisms. **Enterobacteriaceae, Coliforms and *E. coli***: Classical and Modern Methods for Detection/Enumeration. **Enzyme Immunoassays**: Overview. **Hydrophobic Grid Membrane Filtration Techniques (HGMF)**. **Immunomagnetic Particle-based Techniques**: Overview. **National Legislation, Guidelines & Standards Governing Microbiology**: European Union; Japan. **Nucleic Acid-based Assays**: Overview. **Petrifilm – An Enhanced Cultural Technique**. **Rapid Methods for Food Hygiene Inspection**. **Sampling Regimes & Statistical Evaluation of Microbiological Results**. **Water Quality Assessment**: Modern Microbiological Techniques.

Further Reading

Griffioen AGE and Beumer RR (1995) Identificatie van Enterobacteriaceae met 6 systemen. *Voedingsmiddelentechnologie* 18: 18–23.

Holmes B, Willcox WR, Lapage SP and Malnick H (1977) Test reproducibility of the API (20E), Enterotube™, and Pathotec systems. *Journal of Clinical Pathology* 30: 381–387.

Holmes B, Costas M, Thaker T and Stevens M (1994) Evaluation of two BBL® Crystal™ systems for identification of some clinically important Gram-negative bacteria. *Journal of Clinical Microbiology* 32: 2221–2224.

Klingler JM, Stowe RP, Obenhuber DC et al (1992) Evaluation of the Biolog automated microbial identification system. *Applied and Environmental Microbiology* 58: 2089–2092.

Micklewright IJ and Sartory DP (1995) Evaluation of the BBL® Crystal™ Enteric/Nonfermenter kit for the identification of water-derived environmental Enterobacteriaceae. *Letters in Applied Microbiology* 21: 160–163.

Monnet D, Lafay D, Desmonceaux M et al (1994) Evaluation of a semi-automated 24-hour commercial system for identification of Enterobacteriaceae and other Gram-negative bacteria. *European Journal of Clinical Microbiology and Infectious Diseases* 13: 424–430.

Microfloras of Fermented Foods

Jyoti Prakash Tamang, Microbiology Research Laboratory, Department of Botany, Sikkim Government College, India

Wilhelm H Holzapfel, Institute of Hygiene and Toxicology, Federal Research Centre for Nutrition, Karlsruhe, Germany

Fermented foods are produced by the action of microorganisms, either spontaneously or by adding starter cultures, which modify the substrates biochemically and organoleptically into edible products, and are thus generally palatable, safe and nutritious. Filamentous fungi, yeasts and bacteria, mostly lactic acid bacteria (LAB) constitute the microfloras associated with traditional fermented foods (**Table 1**) which are

Table 1 Microfloras associated with fermented foods

Type	Microorganisms
Filamentous fungi	*Absidia, Actinomucor, Amylomyces, Aspergillus, Monascus, Mucor, Neurospora, Penicillium* and *Rhizopus*
Yeasts	*Candida, Debaryomyces, Geotrichum, Hansenula, Kluyveromyces, Pichia, Saccharomyces, Saccharomycopsis, Torulopsis, Trichosporon* and *Zygosaccharomyces*
Bacteria	Species of lactic acid bacteria: *Carnobacterium, Enterococcus, Lactobacillus, Lactococcus, Leuconostoc, Oenococcus, Pediococcus, Streptococcus, Tetragenococcus, Vagococcus* and *Weissella*. Other genera: *Acetobacter, Bacillus, Bifidobacterium, Brevibacterium, Citrobacter, Klebsiella* and *Propionibacterium*

present in or on the ingredients, utensils or environment, and are selected through adaptation to the substrate. The majority of the fermented foods involving filamentous fungi are produced in the East and Southeast Asia. In the Indian subcontinent, mostly due to wide variation in agro-climatic conditions and diverse forms of dietary culture of different ethnic groups of people, in most cases all three major types of microorganism are associated with traditional fermented foods. In Africa, Europe and America, fermented products are prepared exclusively using bacteria (mostly LAB) or bacteria–yeasts mixed cultures; moulds seem to be little used.

Uses of Microorganisms in Fermented Food

Microorganisms bring about some transformation of the substrates during food fermentation: (a) enrichment of diet with improved flavour and texture; (b) preservation of foods; (c) enrichment of nutritional value; (d) destruction of undesirable components; (e) improvement to digestibility; and (f) probiotic function.

Enrichment of Diet

Biotransformation of bland vegetable protein into meat-flavoured amino acids sauces and pastes by mould-fermentation (*Rhizopus* spp., *Aspergillus* spp.) is common in Japanese miso and shoyu, Chinese soy-sauce and Indonesian tauco. In ang-kak, a traditional fermented rice food of Southeast Asia, *Monascus purpureus* produces a purple–red water-soluble colour in the product which is used in colouring meats and rice wine. Halophilic microorganisms such as *Pediococcus halophilus*, contribute flavour and quality to fermented fish products of Southeast Asia.

Biological Preservation

Biological preservation implies a significant approach to improve the microbiological safety of foods without refrigeration by lactic acid fermentation. Species of lactic acid bacteria during fermentation produce lactic acid, the characteristic fermentative product, which reduces pH to a level where putrefactive, pathogenic and toxinogenic bacteria are inhibited. Thus, the LAB can act as a 'biopreservative'. LAB in food fermentation show special promise for implementation as protective cultures which do not pose any health risk, and thus are designated as 'GRAS' (generally recognized as safe) organisms due

to their antibacterial properties (**Table 2**). Common fermented vegetable products preserved by lactic acid fermentation are: gundruk and sinki in the Himalayan regions of India, Nepal and Bhutan; kimchi in Korea; sauerkraut in Germany and Switzerland; and sunki in Japan.

Bioenrichment of Nutritional Value

Biological enrichment of food substrates by fermentation which enhances nutritive value of raw material has high significance for developing countries where the majority of the people cannot afford to have commercially available expensive fortified nutritive foods. In tempeh, a traditional fermented soybean food of Indonesia, the levels of vitamins such as niacin, nicotinamide, riboflavin and pyridoxine are increased by *Rhizopus oligosporus*, whereas cyanocobalamin is synthesized by non-pathogenic strains of *Klebsiella pneumoniae* and *Citrobacter freundii* during fermentation. Essential amino acids and mineral contents are increased in kinema, a traditional soybean food of the Himalayas, fermented by *Bacillus subtilis* by which the nutritive value of the product is enriched.

Destruction of Undesirable Components

Microorganisms associated with the fermented foods produce desirable amounts of enzymes which may degrade unsatisfactory or antinutritive compounds and thereby convert the substrates into consumable products with enhanced flavour and aroma. Bitter varieties of cassava (*Manihot esculenta*) tubers, the main staple crop in West Africa contain the cyanogenic glycoside linamarin which can be detoxified by species of *Leuconostoc*, *Lactobacillus* and *Streptococcus* in gari, a fermented cassava product. Cassava tubers are thereby rendered safe to eat.

Improved Digestibility

In yoghurt and cheese, lactic acid bacteria largely convert the lactose into more digestible lactate, proteins into free amino acids imparting digestibility to

Table 2 Antimicrobial properties of lactic acid bacteria

Product	Main target organisms
Lactic acid	Putrefactive and Gram-negative bacteria, some fungi
Acetic acid	Putrefactive bacteria, clostridia, some yeasts and fungi
Hydrogen peroxide	Pathogens and spoilage organisms, especially in protein-rich foods
Lactoperoxidase	Pathogens and spoilage organisms (milk and dairy products)
Diacetyl	Gram-negative bacteria
Bacteriocins → nisin	Spore-forming Gram-positive bacteria

the product. People suffering from lactose-intolerance may benefit from the lactic fermentation of milk and by the fermentative reduction of the lactose content through the action of the β-galactonidase enzyme.

Probiotic Function

Some strains of LAB, primarily species of *Lactobacillus*, *Enterococcus* and even species of *Bifidobacterium* (which are not true lactics due to higher (55–57 mol%) G+C contents than true lactics) are used as probiotic adjuncts and as biotherapeutic agents as protection against diarrhoea, stimulation of the immune system, alleviation of lactose intolerance symptoms, reduction of enzymes implicated in carcinogenesis, and reduction of serum cholesterol. Probiotic cultures are thought to provide substantial health benefits by stabilizing the gastrointestinal tract.

Biochemical Identification

Isolation, enrichment, purification, characterization and identification of dominant microorganisms involved in fermented foods are important aspects of microbial systematic characterization. Proper identification and authentic nomenclature ensure the quality control and normalized production of fermented foods. Biochemical identification is mostly based on: carbon and nitrogen sources, energy sources, sugar fermentation, secondary metabolites formation and enzymes and toxins production. The following steps are required to determine the types of microflora associated with fermented foods.

1. Information on traditional processing methods such as temperature of incubation and fermentation period should be sought from the local people.
2. Dominant microorganisms must be isolated from samples, collected from different places.
3. Each of the isolates must be tested by producing native product in the traditional way.
4. Proven producing strains must be identified by phenotypic characteristics such as colony and cell morphology, Gram staining, physiological and biochemical tests, and an authentic nomenclature at least up to generic name should be assigned.
5. Unknown strains of isolates should be further identified to species level using genotypic identification methods such as DNA base composition, DNA hybridization and ribosomal RNA sequencing and chemotaxonomical tools such as cell wall, cellular fatty acids and isoprenoid quinones.
6. Identified strains may be further studied on their enzymatic profiles, antimicrobial activities, toxicity, vitamin formation, biogenic amines formation and other biochemical activities, to determine their specific roles in particular fermented product; this may help for improvement of the native product.
7. Identified cultures should be deposited in a culture collection for preservation and documentation.

Automated or Commercialized Identification Systems

Conventional approaches for phenotypic studies on microorganisms are cumbersome, time-intensive and difficult to apply in routine fermentation studies. Several semi-automated, new systems have been developed in the first instance for phenotypic typing of Gram-negative bacteria, and especially those for clinical samples. However, some of these systems have been improved and adapted for application to lactic acid bacteria. They are most valuable for studying the dynamics and domination of LAB during fermentation of food substrates.

API system

One of the widely used modern biochemical identification methods for prompt sugar fermentation test of microorganisms is the API system. The API system of bioMérieux (API System – SA, France) is a miniaturized biochemical kit for the phenotypic identification of different groups of bacteria and even of yeasts. Of social interest for studying LAB involved in food fermentation, is the 'API-50 CHL' which enables the determination of the fermentative ability of 49 different carbohydrates by an isolated strain. In this way, the working load during routine identification studies, is drastically reduced. The system is standardized, and every step is exactly defined, e.g. for inoculation of each of 50 wells in the 'strip' with a cell suspension, using a Pasteur pipette, and after growth under exactly defined conditions. The incubation temperatures (e.g. 26, 30 or 37°C) are selected according to the product and the typical environmental conditions of fermentation. Fermentation reactions are recorded after 3, 6, 24 and 48 h and assimilation reactions after 1, 2, 4 and 8 days, also noting the intensity of the reaction on a scale from 0 (negative) to 5 (strongly positive). The system has some drawbacks for LAB. The AutoMicrobic system of bioMérieux makes use of a computer-driven optical reader and specific cards for generating a biochemical profile. Using the 'Profile Index' may, however, lead to erroneous identification, when no additional tests are performed. More reliable results are obtained when isolated strains are preliminarily grouped according to hetero- and homofermentation, the lactic acid isomers (DL, L(+) or D(−)) produced from glucose, cell morphology and some key physiological tests such as growth at different temperatures, NaCl concentrations and pH values. For confirmation of *Lactobacillus plantarum* and *Lactobacillus pentosus* strains, the presence of *meso*-diaminopimelic acid (DAP) in the cell wall should be determined. Hydrolysis (6 N HCl) of the cell biomass from 5 ml growth in de Man Rogasa and Sharp (MRS) broth followed by thin-layer chromatography may be used as a routine approach.

Rapid API-ZYM System

The study of the enzyme profiles of LAB involved in food fermentation is enabled by another development of bioMérieux, called the Microenzyme 'Rapid API-ZYM System' and has found applications (amongst others) for LAB involved in different fermented foods. The relative activities of 19 hydrolytic enzymes may be determined semi-quantitively. The kit contains 19 chromogenic substrates which have been dried on absorbent microcupules. The intensity of enzymes such as proteinase, peptidase, arylamidase and esterase–lipase activities may thus be determined. The system has special value for studying cultures in food fermentation.

Biolog System

Biolog System (Biolog, USA), or BioSys-inova (Sweden) is a simple, automated and high-resolution system for subtyping bacterial strains phenotypically below the species level. It has excellent potential for studying bacterial succession and dynamics even on the strain level. It uses 96 microwell plates containing a basal medium with 94 different carbon sources with two additional control wells. A redox indicator enables the rapid detection of the microbial activity in each well. After incubation for 4–24 h, a pattern of 'active' wells is obtained yielding a metabolic 'fingerprint'. The strain is identified by comparison with patterns of the reference database. A turbidometer and a computer-aided Micro-Plate Reader, together with the appropriate software, are also supplied by the company, and simplify the final identification. The Biolog System provides an extended database for 57 LAB species (representing the genera *Lactobacillus*, *Lactococcus*, *Leuconostoc* and *Pediococcus*). Both the automated and the manual systems make use of the database.

Immunofluorescent System

Other modern developments such as immunofluorescent or immunomagnetic separation and isolation procedures have thus far been mainly developed and applied for clinical strains. However, they offer extremely elegant and highly precise typing methods for studying microbial communities during food fermentation. Monoclonal as well as polyclonal antibodies are being used. Labelling using commercial immunoassay kits is mainly based on enzymes (e.g. peroxidase or alkaline phosphatase), and such compounds that participate in a luminescent reaction (e.g. acridinium ester, isoluminal derivatives).

See also: **Electrical Techniques**: Lactics and other Bacteria. **Probiotic Bacteria**: Detection and Estimation in Fermented and Non-fermented Dairy Products.

Further Reading

Aurora G, Lee BH and Lamoureux M (1990) Characterization of enzyme profiles of *Lactobacillus casei* species by a rapid API-ZYM system. *Journal of Dairy Science* 73: 264–273.

Stiles ME and Holzapfel WH (1997) Lactic acid bacteria of foods and their current taxonomy. *International Journal of Food Microbiology* 36: 1–29.

Tamang JP (1998) Role of microorganisms in traditional fermented foods. *Indian Food Industry* 17: 162–167.

Vandamme P, Pot B, Gillis M et al (1996) Polyphasic taxonomy, a consensus approach to bacterial systematics. *Microbiological Reviews* 60: 407–438.

BIOFILMS

Brigitte Carpentier, National Veterinary and Food Research Centre, Maisons-Alfort, France

O Cerf, Alfort Veterinary School, Maisons-Alfort, France

Introduction

The study of biofilms began relatively recently, notably in the context of the food industry, and many aspects are still under exploration. This article deals only with biofilms on chemically inert surfaces. It is intended to present the current state of knowledge and debate, and unresolved questions are presented as such.

One topic of debate concerns the definition of the term 'biofilm'. According to some authors, biofilm bacteria are embedded in a matrix of extracellular polymeric substances (EPS). According to others, bacteria form a biofilm even if EPS are absent or below a detectable level. The characteristic of biofilm bacteria of most importance to the food hygienist is their high resistance to antimicrobial agents. Increased resistance is acquired within a few hours following the adhesion of the bacteria, and before any embedment in an EPS matrix can occur. Therefore, the definition of a biofilm for the food hygienist could be 'a microbial community, adhering to a surface and frequently embedded in an extracellular matrix'.

A second area of debate concerns the architecture of biofilms and the factors which influence it. On surfaces common in the food industry, biofilms frequently do not form a continuous film, but rather distant microcolonies, which adhere onto surfaces (**Fig. 1**). In other situations, biofilms may be large mushroom-shaped microbial clusters, separated by voids or water channels that can be observed under scanning confocal laser microscope. Dense confluent biofilms are also described. The latter two biofilm structures are unlikely to be found on surfaces that are regularly cleaned and disinfected, but rather on immersed surfaces that are not subjected to frequent hygiene operations (e.g. heat exchangers in power plants, bioreactors, catheters).

Biofilm formation is a natural phenomenon which occurs wherever there are microorganisms and surfaces – whether only slightly wet or immersed, and whether surrounded by a high or a low level of nutrients (even the low concentration of nutrients in ultra-pure water is sufficient to support biofilm formation). This explains why biofilms are easy to produce in laboratory conditions, and many techniques have been published. But laboratory findings may not apply to field biofilms, as explained below. The contamination of slightly wet surfaces is a main concern of only the food hygienist. For this reason the literature on such biofilms is sparse, and the application to food science of the knowledge acquired from studies on immersed biofilms is necessary.

Biofilms can build up on any surfaces in the food industry. Among the most contaminated surfaces are floors, drains, and some conveyor belts. Less contaminated surfaces are walls, ceilings, and the inside and outside of equipment. Recesses and crevices, where cleaning is difficult, are most prone to biofilm accumulation, leading to an increasing awareness of the need for equipment and premises to be designed so as to enable good hygiene. After cleaning, the maximum reported biofilm density is around 10^7 cfu cm^{-2}, but greater contamination is likely to be found in places inaccessible to cleaning.

Biofilm Formation

Adhesion

Adhesion is a physico-chemical process of interaction,

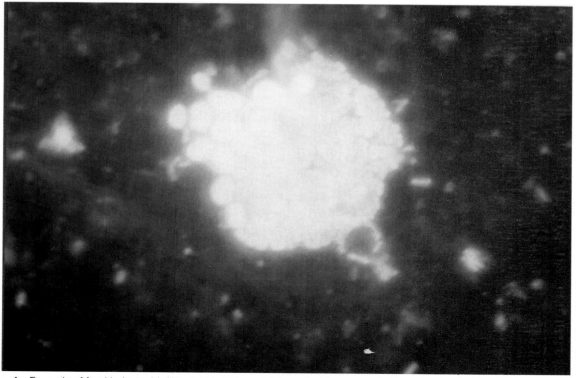

Figure 1 Example of food industry biofilm, developed on stainless steel from a meat site and stained with acridine orange.

between molecules that are situated in the outermost layer of the inert surface and of the microorganisms, and molecules of the surrounding fluid. Interaction occurs due to three types of force, that combine into free interfacial energy: van der Waals' forces, which are attractive; and electron acceptor/electron donor interactions and electrostatic interactions – these can be either repulsive or attractive. In aqueous phase of foods, in which the ionic strength is high, electrostatic interactions are negligible. In water, in which the ionic strength is weak, for example in ducts, electrostatic interactions are not negligible and they limit adhesion to a variable extent, because both the surface of the microorganisms and the inert surface are generally negatively charged.

As soon as a solid material is placed in a liquid, in a matter of seconds, soluble molecules in the liquid concentrate on the surface of the solid and form a 'conditioning film'. Microorganisms need more time to adhere. Consequently, the surface to which microorganisms stick is conditioned. In the food industry, conditioning results from the adhesion of molecules of food and/or cleaning agents and disinfectants. Work surfaces close to each other tend to become similar in terms of free energy depending on the type of food being processed and on the cleaning and disinfection operations. **Figure 2** shows that the water contact angles (one of the values used to calculate surface free energy) of different floor materials in a pastry site are no longer different for each material from the third

week onwards. This would be consistent with hygiene operations causing the spread of residual food and cleansing agents, and hence the coating of floor surfaces. However, it is not known whether the conditioning of a given work area is attributable only to the treated food and hygiene operations, or whether it is also dependent on the soiling of the predominating material within the work area (in terms of the area of coverage and the density of the remaining soiling). For example, a hydrophobic material may show a high affinity for the fat components of the food being processed, whereas a hydrophilic material may have a high affinity for other components. If the hydrophobic material predominates, it is likely that the conditioning film of all the surface materials in the work area will contain a high amount of fat.

The surface properties of bacteria depend on the growth conditions: pH, temperature, culture medium composition, growth phase, etc. It is frequently observed that starved cells or those in the stationary phase adhere better to surfaces than do cells in the exponential growth phase.

In our opinion, the most important discovery of the 1990s is that the bacteria which adhere to inert surfaces are phenotypically profoundly different from planktonic bacteria. Several works report that adhesion triggers the expression of a σ factor that derepresses a large number of genes. For example, adhesion can trigger the production of several enzymes that produce exopolysaccharides and in *Vibrio parahaemolyticus*, the *laf* gene is derepressed, leading to the production of lateral flagella.

Colonization

After adhesion, growth of the adherent microbial cells leads to the colonization of the surface. Exopolysaccharide production has often been shown to be more abundant after bacterial adhesion. The architecture of the biofilm is dependent on different factors. It has recently been demonstrated that regulation of gene expression in response to changes in all density ('quorum sensing') by *Pseudomonas aeruginosa* was involved in biofilm differentiation. A *lasI* mutant that cannot produce the signal molecule (an acylated homoserine lactone) formed a flat, thin and uniform biofilm, whereas in the biofilm of the wild strain, the cells are in a loose confederation with considerable space in-between. These signals have been detected in natural biofilms. Factors including bacterial strains, nutrient availability, hydrodynamic conditions in flowing systems, and other environmental conditions also influence the amount of biofilm produced and its architecture.

The time taken to achieve maximum population density and a stable state can be long – it can be

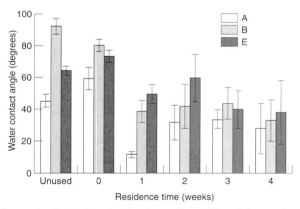

Figure 2 Contact angle of water on three different floor materials A, B and E in a pastry site, as a function of residence time of materials in the site. A and B = vitrified extended tiles. E = resin-based material. Number of repetitions = 30. Error bars represent standard deviation. Unused = unused material. Residence time = corresponds to floor materials installed in the site and submitted once to one hygiene operation. (Mettler E and Carpentier B, GNEVA, France (1998) Variations over time of microbial load and physico-chemical properties of floor materials after cleaning in food industry premises. *Journal of Food Protection* 61: 57–65. Reprinted with permission from *Journal of Food Protection*. Copyright held by the International Association of Milk, Food and Environmental Sanitarians, Inc.)

expressed in days, weeks or even months. For example, biofilm accumulation, assessed by ATP measurements on surfaces exposed to ground water that did not contain a disinfectant, was shown not to have reached steady state after 4 months, although heterotrophic plate counts stopped increasing after 10 days. This indicates that culturable bacteria represent only a fraction of the active biomass ($\leq 10\%$). On floors in a meat industry, culturable populations reached a steady state after 4 weeks.

The composition of biofilm on the floors of different premises used for food processing, after hygiene operations, is reported in **Table 1**. In the meat and milk sites the genera *Pseudomonas* and *Staphylococcus* predominate, whereas yeasts and *Corynebacterium* predominate in the pastry site. The genera *Pseudomonas* and *Staphylococcus* are very often reported to colonize inert surfaces in the food industry. In addition, some *Pseudomonas* strains have been shown to help surface colonization by *Listeria monocytogenes*, which alone has poor colonizing properties.

Properties of Biofilms

Properties Linked to EPS

The biofilm matrix contains water, extracellular polymeric substances (EPS), nucleic acids and entrapped substances, the precise composition depending on the environment. The biochemistry of EPS themselves is not well known. The main components appear to be exopolysaccharides and exoproteins. The latter do not seem to have been studied much to date. Methods for detecting *in situ* minor differences in small amounts of exopolysaccharides are not available, and it is hypothesized that in the majority of environments the polysaccharides of planktonic and of biofilm bacteria have the same composition but different physical properties.

Table 1 Microflora isolated after growth on nonselective culture media from floor materials at three food industry sites after hygiene operations in descending order of predominance (after results from Mettler E and Carpentier B (1998) Variations over time of microbial load and physico-chemical properties of floor materials after cleaning in food industry premises. *Journal of Food Protection* 61: 57–65)

Meat site	Pastry site	Milk site
Pseudomonas	Yeasts	Pseudomonas
Staphylococcus	Corynebacterium	Staphylococcus
Enterobacter	Leuconostoc	
Flavobacterium	Pseudomonas	
Yeasts	Staphylococcus	
Kluyvera	Aerococcus	
	Achromobacter	
	Acinetobacter	
	Moulds	

Role of EPS in Microbial Adhesion and in Maintaining the Structure of the Biofilm As discussed above, it is not clear whether exopolysaccharides are necessary for adhesion to occur. It has been shown that non-polysaccharide-producing mutants were able to adhere. However, it is important to distinguish between thin, and possibly adhesive, polymer layers on the cell surface and the extensive extracellular network of polymers (EPS) frequently present in biofilms. EPS seem to be necessary for the aggregation of microbes and microcolony formation. It is often reported that polysaccharides are required to effectively retain the bacteria within the biofilm, but it has been shown that an increase in polysaccharide production is not always linked with improvement in bacterial attachment. It has been observed recently that the biofilms obtained with the *Pseudomonas aeruginosa* mutant *lasI* described previously and those obtained with the wild-type strain showed similar uronic acid and total polysaccharide content. Despite this, the mutant biofilm was easily detached by the detergent sodium dodecyl sulfate (SDS), whereas SDS had no detectable effect on the wild-type strain biofilm.

EPS Entrapment Capacity Exopolysaccharides can trap nutrients from the bulk liquid or those produced in the biofilm by cells or by cell leakage. Some exopolysaccharides can bind cations, toxic metallic ions and other substances coming into contact with the biofilm. This entrapment capacity is essential in aquatic environments or in bioreactors, because the ions and molecules are easily metabolized, helping to purify contaminated water.

EPS and Desiccation Another possible role of exopolysaccharides is desiccation resistance, because many of them form a gel with a high capacity for retaining water. It has been shown that reducing the availability of water enhanced exopolysaccharide production. This could be one of the reasons for the survival of Gram-negative bacteria such as *Pseudomonas* on food industry surfaces that are intermittently dry. In fact, Gram-negative cells are usually less resistant to desiccation (but more resistant to disinfection) than those of Gram-positive bacteria.

Adaptation of Biofilm Bacteria to Environmental Stress

In the food industry, biofilms have to survive stressful conditions because of factors such as an unsteady nutrient supply, chemical shock and desiccation. For example, floors are one of the main reservoirs of *Listeria monocytogenes*, floors in dairy plants are usually acidic, and it is possible that adaptation to acid occurs in the biofilms of the surface of the floor. It has been shown that suspended cells of acid-adapted *L. monocytogenes* can survive a 90-minute contact time in a solution with pH 3. The induction of the acid-tolerance response also protects *L. monocytogenes* against the effects of other environmental stresses and probably also those of some chemicals used in hygiene operations.

Presence of Active but Non-culturable Cells

A proportion of biofilm microorganisms is not culturable on the media classically used to perform cfu counts, as is the case with microorganisms from other environments. Some non-culturable cells can show activity, for example respiration as revealed by a capacity to reduce CTC (5-cyano-2,3-ditolyl tetrazolium chloride) (see **Fig. 3**). Active cells whose viability (classically defined as the ability to multiply) is not demonstrated are frequently called 'viable but non-culturable cells', abbreviated to VBNC. The true status of VBNC is still being discussed, to determine for example whether they are cells which have simply not been provided with appropriate conditions to support culture, and whether the VBNC state needs a factor and a signal to facilitate resuscitation?

Strength of Biofilm Adhesion

Little is known about the mechanisms governing the strength of adhesion of biofilm cells. As stated above, the role of exopolysaccharides in attachment is not fully understood. The strength of adhesion increases

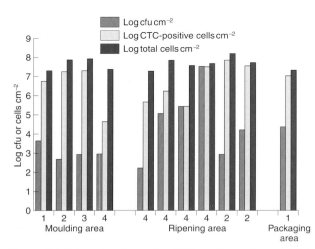

Figure 3 Colony forming units, CTC-positive cells and total count of bacterial cells isolated from surfaces in three areas of a cheese-making plant. (1) Conveyor belts. (2) Equipment. (3) Trolley wheels. (4) Floor. (Courtesy of M Alliot, Laboratoire Soredab, La Boissière-Ecole, France.)

with biofilm age, and is dependent on the nature of the substratum. Erosion of the biofilm is easily achieved: pouring liquid onto a contaminated surface will detach microorganisms, but the number of detached cells is negligible compared to the biofilm population, so no decrease in the population can be detected after a simple rinse. It is impossible to detach all the microorganisms adhering to a surface without killing them. For example, the action of a 50-bar water jet was shown to cause the detachment of about 90% of the biofilm population of ceramic tiles, whether formed in the lab or in the food industry. It is therefore difficult to measure surface contamination accurately and to determine the identity of all the surface microorganisms. Furthermore, it is not known whether the proportion of detachable cells is the same before and after hygiene operations. Consequently, it is not possible to ascertain the accuracy of the measurement of the efficacy of hygiene operations.

Resistance to Cleaning Agents

Neither neutral surfactants nor acidic products are more effective than water at detaching a biofilm from a surface. Alkaline products facilitate cell detachment, but only to a small extent. Several commercial alkaline products were shown to detach 10–90% of the bacterial cells of a laboratory-grown biofilm whose initial population was 3×10^7 cfu cm^{-2}. However, a chlorinated alkaline product did not significantly reduce the number of adhering cells in a biofilm naturally formed on new tiles, placed on the floor of a cheese factory for 4 weeks.

Resistance to Antimicrobial Agents

The major property of biofilms of concern to food hygienists is their high resistance to nearly all antimicrobial agents. This is illustrated in **Figure 4**, which demonstrates that the antimicrobial resistance of *Listeria monocytogenes* increases with biofilm age: single cells adherent to a glass slide were obtained after 4 hours' incubation in culture broth, and adherent microcolony cells were obtained after 14 days' incubation. However, laboratory results probably underestimate the real resistance of natural biofilm because adaptation probably occurs in response to environmental stress. For example, division by 10^3 of the culturable population of a laboratory biofilm (10^7 cfu cm^{-2} *Pseudomonas fluorescens* grown on tiles) could be obtained by the chemical action of a hygiene operation. Yet only a division by 10 of a natural biofilm (10^4 cfu cm^{-2} on the same tiles, formed in a cheese-making site during 4 weeks) was obtained by the same chemical treatment.

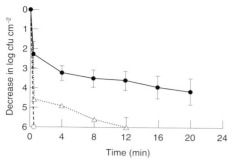

Figure 4 The decrease in log cfu cm^{-2} of adherent micro-colonies, adherent single cells, and planktonic cells of *Listeria monocytogenes* caused by 800 p.p.m. benzalkonium chloride. ○–○; planktonic cells; ●–● adherent microcolonies; △–△ adherent single cells. (Frank JF and Koffi RA, University of Georgia, USA (1990) Surface-adherent growth of *Listeria monocytogenes* is associated with increased resistance to surface sanitizers and heat. *Journal of Food Protection* 53: 550–554. Reprinted with permission from *Journal of Food Protection*. Copyright held by the International Association of Milk, Food and Environmental Sanitations, Inc.)

Increased resistance occurs soon after adhesion, before detectable EPS production, and vanishes when the cells are detached and suspended in a liquid. The increase in resistance depends on the nature of the disinfectant. Surface-active disinfectants (e.g. quaternary ammonium compounds, amphoteric agents) have a markedly reduced efficacy on biofilms compared to that on suspended cells. However it has been shown that phenol has the same efficacy in relation to *Pseudomonas aeruginosa*, whether the cells were suspended or within a biofilm (**Table 2**).

The nature of the biofilm substratum has a strong influence on biocide efficacy in laboratory conditions. For example, stainless steel appears to be more easily disinfected than do many other materials, including aluminium and polymers. However, as stated before, food industry surfaces are conditioned and different surfaces tend to be conditioned in the same way in a given workshop. Such conditioning can decrease the influence of materials on biocide efficacy.

The hypothesis that the EPS matrix constitutes a barrier to diffusion, that would limit the penetration of biocides within the biofilm, is now sometimes rejected because it is known that the architecture of many biofilms allows biocide flow. However, EPS can react chemically with some antimicrobial agents. For example, exoproteins and food protein residues entrapped in a biofilm or adsorbed onto a contaminated surface react with chlorine, leading to a decrease in free chlorine concentration and hence to a decreased disinfection efficacy of the chlorinated solution. This is of primary importance, because members of the family of chlorinated alkaline products are widely used as cleaning and disinfection agents in the food industry.

The starvation of deep-lying cells in the bottom layers of biofilms has often been assumed to explain the high resistance of biofilm bacteria. This assumption is based on the following facts: (1) bacteria within the bottom layer or inside a cluster do not receive sufficient nutrients or O$_2$, because these are consumed by bacteria in the top layers and (2) suspended cells grown at a sub-optimal growth rate are more resistant than their equivalents grown at the optimum rate. However, several works show that contrary to the properties of suspended cells, biofilm bacteria are less resistant to disinfectants when grown in diluted media (e.g. tryptone soya broth) than when grown in the same medium at full concentration.

In conclusion, the physiological modification of adherent bacterial cells is now the main hypothesis used to explain biofilm resistance to antimicrobial agents. A recent work showed that such a modification could explain the antibiotic resistance of sessile *Escherichia coli*. It has been observed that the porin protein OmpF was under-expressed in immobilized *E. coli* that displayed increased resistance to the β-lactam antibiotic latamoxef. This porin, which constitutes a nonspecific diffusion channel of the outer membrane, is a possible route of entry of the β-lactam antibiotic to the cell.

Table 2 The efficacy of disinfectants against biofilm or planktonic *Pseudomonas aeruginosa*. Results are reported as the ratio of MBC of biofilm to MBC of planktonic cells, where MBC = minimal bactericidal concentration resulting in division by 10^5 of the initial population in 5 min at 20°C. (Calculation after results from Ntsama-Essomba C, Bouttier S, Ramaldes M and Fourniat J (1995) Influence de la nature chimique des désinfectants sur leur activité vis-à-vis de biofilms de *Peusodomonas aeruginosa* obterus en conditions dynamiques. In: Bellon-Fontaine MN and Fourniat J (eds) *Adhésion des Micro-organismes aux Surfaces*. P. 282–294. Paris: Lavoisier Tec & Doc)

Quaternary ammonium compounds		Amphoteric surfactant	Oxidizing agents		Phenolic compounds	
Cetrimide	Benzalkonium chloride	Tegol	Peracetic acid	Sodium hypochlorite	Phenol	Ortho-cresol
<400	100	25	4	5	1	4

Reducing Biofilm Build-up

Dryness

The best method for limiting biofilm development is to maintain dryness of the surfaces and atmosphere. All available means of achieving dryness should be used, and the stagnation of water should be prevented. The slope of floors and gutters should be > 1.5% to allow adequate flow of the water used for hygiene operations. The number of traps and siphons should be adequate.

Surface Texture

Surfaces should be made from materials which are not porous. Thus concrete, or materials containing a high proportion of cement, are not recommended. Wood has a bad reputation because of its high porosity. Nevertheless, provided the necessary hygiene precautions are well-understood and implemented, wood is used in some instances because of its other characteristics. In particular, wooden chopping boards and blocks are used in domestic kitchens and in butchers' shops, because wood limits the sliding of knives and resulting accidents, and in cheese-ripening rooms, because wood harbours the microflora needed for ripening.

Materials should exhibit and retain low roughness and, of course, should have no pits or crevices. Scrutiny under a binocular lens with side illumination is necessary to check for the absence of pits in floor materials – indeed, resin-based floors frequently have holes which are capable of harbouring bacteria and yet cannot be cleaned. The arithmetical mean roughness (R_a) and/or peak-to-valley height (R_z) are usually used to characterize roughness. But recent work on the cleanability of floor materials showed that another parameter, reduced valley depth (R_{VK}), should be taken into account. R_{VK} was better correlated with cleanability when calculated with a cutoff value of 0.8 mm rather than 2.5 mm, indicating that the gross topographic irregularities of floor materials were not related to their cleanability. However the parameter R_{VK} is not useful for every type of material – for example it is not useful for materials containing irregularly placed holes and pits. However, the tests available for characterizing the cleanability of materials (involving soiling, cleaning, disinfection, and the assessment of residual contamination) are not very discriminating. They result in only two, or possibly three, classes of materials.

Hygienic Design

The need to avoid crevices and recesses in surface materials, and the importance of water drainage, have already been stressed. The same requirements should be applied to the inside of equipment. They are sometimes not obvious to mechanical engineers, and are even contrary to their traditions. Therefore, recommendations were prepared and published by the European Hygienic Equipment Design Group. These are being standardized at international and continental level (ISO, 3A in the USA, CEN in the European Union). The recommendations encompass, for example, global design, material and surface finishing, the welding of stainless steel, the design of valves and joints and methods for measuring cleanability.

Surface Modification

The 'antimicrobial material' concept has been under investigation in the field of medical sciences since the early 1980s, in the context of the prevention of infection related to the materials used for catheters and prostheses. Some antimicrobial materials have been proposed to the food industry, for example paints and PVC floors that contain a fungicide, but there are no published results showing their efficacy on a long-term basis.

The copolymers of ethylene oxide and propylene oxide, marketed under the brand names Pluronic or Superonic, have proved highly effective at preventing the microbial colonization of hydrophobic surfaces in the laboratory. These substance have been incorporated in hygiene chemicals and the products used to prevent biofilm accumulation in cooling towers, but their real efficacy in field conditions is difficult to assess.

Some new concepts in surface treatment seem effective at the laboratory level, for example the adsorption of nisin or the incorporation of catalysts for the degradation of peroxides and persulphates. This is a development that needs further exploration.

Eliminating Biofilm: Cleaning and Disinfection

As stated before, biofilms are difficult to eliminate, either by cleaning or by disinfection. Cleaning (removal) can be achieved by mechanical action, using brushes or high-pressure water jets on open surfaces, or high velocity of cleaning solution to increase shear stress in ducts. The maximal effects are achieved rapidly, and prolonging such treatment brings no improvement. Disinfection is necessary for the inactivation of some of the microorganisms that have not been removed by cleaning. Despite the high resistance of biofilm cells to antimicrobial agents, disinfection is an important step within the hygienic operations. Indeed, it must be conducted frequently because it is likely that pathogenic microorganisms introduced into the workshop accidentally will not have the high resistance of bacteria of the old and adapted biofilms that constitute the resident microflora.

Disinfection can be achieved by some cleaning agents – for example, caustic soda is commonly used and is efficient at removing organic matter and inactivates Gram-negative bacteria, but is practically ineffective against Gram-positive bacteria. However, some disinfectants, such as chlorine-containing solutions and hydrogen peroxide, have been shown to remove biofilms in the laboratory.

It is now acknowledged that laboratory methods for estimating the disinfectant activity of chemicals on planktonic bacteria cannot be extrapolated to biofilms. Furthermore, attempts to produce biofilms in the laboratory, under simulated field conditions, have not been totally successful to date.

See also: **Good Manufacturing Practice**. **Polymer Technologies for Control of Bacterial Adhesion**. **Process Hygiene**: Designing for Hygienic Operation; Overall Approach to Hygienic Processing; Modern Systems of Plant Cleaning; Risk and Control of Aerial Contamination; Testing of Disinfectants.

Further Reading

Bellon-Fontaine MN, Mozes N, van der Mei HC et al (1990) A comparison of thermodynamic approaches to predict the adhesion of dairy microorganisms to solid substrata. *Cell Biophysics* 17: 93–106.

Boulangé-Peterman L (1996) Processes of bioadhesion on stainless steel surfaces and cleanability: a review with special reference to the food industry. *Biofouling* 10: 275–300.

Bossier P and Verstraete W (1996) Triggers for microbial aggregation in activated sludge. *Applied Microbiology and Biotechnology* 45: 1–6.

Bradshaw DJ (1995) Metabolic responses in biofilms. *Microbial Ecology in Health and Disease* 8: 313–316.

Carpentier B and Cerf O (1993) Biofilms and their consequences, with particular reference to hygiene in the food industry. *Journal of Applied Bacteriology* 75: 499–511.

Christensen BE (1989) The role of extracellular polysaccharides in biofilms. *Journal of Biotechnology* 10: 181–202.

Costerton JW, Lewandowski Z, Caldwell D, Korber D and Lappin-Scott HM (1993) Microbial biofilms. *Annual Review in Microbiology* 49: 711–745.

Czechowski MH and Banner M (1992) Control of biofilms in breweries through cleaning and sanitizing. *MBAA Technical Quarterly* 29: 86–88.

Davies DG, Parsek MR, Pearson JP, Iglewski BH, Costerton JW and Greenberg EP (1998) The involvement of cell-to-cell signals in the development of a bacterial biofilm. *Science* 280: 295–298.

Flint SH, Bremer PJ and Brooks JD (1997) Biofilms in dairy manufacturing plant – description, current concerns and methods of control. *Biofouling* 11: 81–87.

Marshall KC (1992) Biofilms: an overview of bacterial adhesion, activity, and control at surfaces. *ASM News* 58: 202–207.

Mattila-Sandholm T and Wirtanen G (1992) Biofilm formation in the industry: a review. *Food Reviews International* 8: 573–603.

Stewart PS (1996) Theoretical aspects of antibiotic diffusion into microbial biofilms. *Antimicrobial Agents and Chemotherapy* 40: 2517–2522.

Sutherland IW (1995) Biofilm-specific polysaccharides – Do they exist? In: Wimpenny J, Handley P, Gilbert P and Lappin-Scott H (eds) *The Life and Death of Biofilm*. P. 103. Cardiff: Bioline.

BIOPHYSICAL TECHNIQUES FOR ENHANCING MICROBIOLOGICAL ANALYSIS

Andrew D Goater and **Ron Pethig**, Institute of Molecular and Biomolecular Electronics, University of Wales, Bangor, UK

By using microelectrode structures, various forms of electric fields, such as non-uniform, rotating and travelling wave, can be imposed on particles of sizes ranging from proteins and viruses to microorganisms and cells. Each type of particle responds to the forces exerted on them in a unique way, allowing for their controlled and selective manipulation as well as their characterization. Moreover, particles of the same type but of different viability can be distinguished in a simple, reliable manner. The principles that govern the way in which bioparticles respond to these various field types are described with examples of current and potential biotechnological applications.

Basic Concepts

The induced motion or orientation of bioparticles in electrical fields has been observed for over 100 years. Until comparatively recently, only particle motion or *phoresis*, induced by DC electric fields was studied. From the generic idea of electrophoresis, a whole new branch of novel electrokinetic manipulation methods of bioparticles has arisen, simply by taking advantage of another dimension, the particle response to the frequency of the applied field.

Innate Electrical Properties of Bioparticles

In order to understand the interactions of a particle with an electric field one must first consider the innate electrical properties of that particle. The important passive electrical properties of a bioparticle, such as a cell or microorganism, are its effective conductivity and electrical capacitance (i.e. dielectric permittivity) as well as its surrounding electrical double layer. A generalized bioparticle suspended in an aqueous solution (weak electrolyte) is represented in **Figure 1** with the relative distribution of innate charges, both bound and free. Many of the molecules that make up biological particles possess ionizable surface chemical groups such as COOH or NH_2. The ionizable head groups of lipids in the plasma membrane are one such example and because of these the particle possesses a net surface charge. An electrostatic potential due to these charges will be present around the particle, the effect of which decreases to that of the bulk medium with increasing distance from the particle. Ions of opposite charge, counter-ions, to those on the surface will be attracted towards the particle by this electrostatic potential. Together, the bound surface charges and the surrounding counter-ion atmosphere, shown as the cation dense region in Figure 1, form what is termed an electrical double layer.

Application of a DC Field to Particles

On the application of a DC electric field across the bioparticle, all the charges, bound and free, in the system will be attracted to the electrode of opposite polarity (**Fig. 2**). If the solution is more or less neutral only relatively small concentrations of H^+ and OH^- will be present, ions such as Na^+ and Cl^- will carry the bulk of the current. Those ions associated with the electrical double layer will respond to the field to form an asymmetric distribution around the particle, the new equilibrium of which is established by the magnitude of the electric field and the opposing ionic concentration diffusion gradient, which tends to restore the random, symmetrical distribution. Any motion of the particle towards the electrodes in a DC field is due to the net surface charge. Human erythrocytes, for example, in a standard saline solution under the influence of a DC field of $1\,V\,cm^{-1}$ migrate towards the anode at around $1\,\mu m\,s^{-1}$. Particle separation is therefore possible due to differences in their mobility in an electric field, which may be due to their size, mass or charge.

Whereas bound charges and polar molecules in the system may orientate in the field, free charge carriers (e.g. ions) will migrate towards the electrodes, that is unless they encounter a material with different electrical properties. Ions encountering the plasma membrane, will be prevented from free motion towards the electrodes by this membrane if it is intact. The membranes of viable cells are only semi-(selectively) permeable to ions and non-lipid soluble molecules (i.e. are relatively non-conducting). The conductivity of the cell membrane tends to be around $10^{-7}\,S\,m^{-1}$, some 10^7 times less conductive than that of the interior which can be as high as $1\,S\,m^{-1}$. For particles the size of erythrocytes, then within about a microsecond after the application of an electric field, the ions will have fully built up at the particle boundary forming an aggregation of interfacial charges.

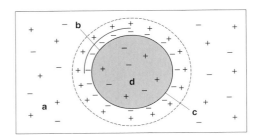

Figure 1 The relative distribution of charge for a suspended particle. A simplified cell (solid circle) suspended in an aqueous medium at neutral pH showing the relative distributions of charge, both free and bound. Approximate conductivity ($\sigma = S\,m^{-1}$) and relative permittivity (ε_r where air = 1) of the bulk solution (a) $\sigma = 10^{-4}\,\varepsilon_r = 80$, cell wall (where present) (b) $\sigma = 10^{-2}\,\varepsilon_r = 60$, membrane (c) $\sigma = 10^{-7}\,\varepsilon_r = 3$, and interior (d) $\sigma = 10^{-1}\,\varepsilon_r = 70$ for a typical viable cell.

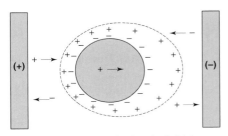

Figure 2 Application of a DC electric field to a suspended particle. On the application of a DC electric field to a cell in aqueous solution, charges will experience a force towards the oppositely charged electrode. Ions in the bulk solution are free to migrate to the electrodes, whereas charges associated with the *electrical double layer* are restricted and show a distortion or polarization.

Importantly these induced charges are not uniformly distributed over the bioparticle surface, forming predominantly on the sides of the particle facing the electrodes. These charges and the distorted electrical double layer lend to the particle the properties of an electrical dipole moment, m. This dipole moment is in the order of 2.5×10^5 debye units (D) for a cell of 5 μm diameter (c.f. 1.84 debye for a water molecule); the cellular dipole moment is therefore described as macroscopic, although the magnitude of the induced charge is still only a fraction, around 0.1%, of the net surface charge carried by cells and microorganisms.

Application of an AC Field to Particles

If we now consider the application of an alternating field to a particle, we see that various phenomena occur over different frequency ranges of applied field. Starting close to the DC condition, with a field that reverses direction a few times a second, the particle motion is dominated by electrophoretic forces. The particle may follow reversals of the field electrophoretically for frequencies up to a few hundred hertz, where reversals of the field take less than a few milliseconds. Because of the particle's inertia, this electrophoretic motion becomes vanishingly small for frequencies above around 1 kHz.

Other mechanisms can respond to field reversals of much higher frequencies such as the dynamic behaviour of the electrical double-layer distortion or polarization around cells. This can follow changes in field direction that take as little as a few microseconds. Any faster than this (i.e. frequencies > 50 kHz) then the counter-ion cloud around cells does not have time to distort. Like the fall off in the electrophoretic motion with increasing field frequency, the decrease in response of the double layer to the changing field occurs gradually over a range of frequencies. This is termed a dielectric dispersion.

Interfacial polarizations are even more responsive to changing field directions and for subcellular-sized particles can take as little as tens of nanoseconds to respond to a reversal in field direction, they can therefore exert their influence up to frequencies of 50 MHz and beyond. This is still nowhere near as responsive as small polar molecules such as water to alternating fields. A measure of the ability of molecules in a material to align in an electric field is given by the relative permittivity of that material, which for bulk water molecules at 20°C in an alternating field less than 500 MHz has a value of 80. At frequencies above 100 GHz the relative permittivity of water falls to that typical of non-polar molecules, around 4. A similar fall in permittivity is seen above about 50 kHz on the freezing of water, because the molecules of the liquid become restricted in a solid lattice and can no longer rotate so freely to align with the field.

On cell death, membrane integrity is lost, it becomes permeable to ions and its conductivity increases by a factor of about 10^4 with the cell contents freely exchanging material with the external medium. This transition in the properties of the membrane shows up as a large change in the polarizability of the cell in an electric field. Other causes for particles having different polarizabilities include differences in their morphologies or structural architecture, which may be associated with the cells belonging to different species, different stages of differentiation or physiological state. Two such particles, that differ in polarizability, are shown in **Figure 3** subjected to an alternating homogenous field created between two parallel electrodes. The direction of the dipole moment formed by the interfacial charges is shown to depend on the relative polarizabilities of the particle compared with the medium.

Particle Motion in Inhomogeneous AC Electric Fields: Dielectrophoresis

Homogeneous AC electric fields do not induce motions in electrically neutral particles, due to equal forces acting on both sides of the polarized particle. If the particle carries a net charge, it will oscillate back and forth as a result of electrophoresis. As the frequency increases these translational oscillations become vanishingly small. Net translational motion is possible, however, if instead the field is inhomogeneous (**Fig. 4**). To distinguish this force from electrophoresis, Herbert Pohl adopted the term dielectrophoresis (DEP) from the term dielectric which is used to describe liquid and solid materials of low

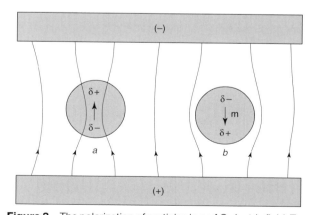

Figure 3 The polarization of particles in an AC electric field. Two particles in an aqueous medium between two parallel electrodes. Particle a is more and particle b is less electrically polarizable than the surrounding fluid. Electrical charges are induced on the surfaces of both particles, to produce induced dipole moments m.

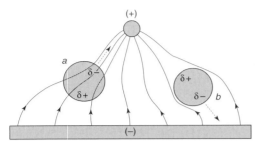

Figure 4 Polarization of particles in non-uniform AC electric fields. Two particles of different polarizability in a non-uniform (inhomogeneous) electric field. The highly polarizable particle *a* experiences a positive DEP force directing it toward the high-field region near the pin electrode, while the weakly polarizable particle *b* is directed away from the high-field region by a negative DEP force.

conductivity. For example, an intact membrane is a dielectric material characterized by having a conductivity 10^{16} times smaller than copper and a dielectric permittivity 3 times that of air. Examples of some particles investigated with DEP are given in **Table 1**.

The DEP Force as a Function of Medium Conductivity

Figure 4 shows that the polarity of the force exerted on the particle depends on the polarity of the induced dipole moment, which in turn is determined by the relative polarizability of the particle and the medium. As a consequence, by altering the polarizability of the medium one can control the direction of motion of a particle. This principle can be exploited to gain particle separations by choosing a suspending medium with an intermediate polarizability, that is between the polarizabilities of two particles in the mixture, so that each particle type will be under the influence of a DEP force of different polarity. Selective manipulation using the DEP force has been used to enable sep-

arations of various interspecific mixtures such as between some Gram-positive and Gram-negative bacteria, as well as the intraspecific separation of live and dead cells, or cancerous from normal cells. Examples of separations demonstrated are listed in **Table 2**, together with the appropriate medium polarizabilities (conductivity) and field frequency. The DEP force imparted on a particle by an electrical field is also proportional to a number of other factors; the particle size, shape and the magnitude and degree of non-uniformity of the applied electric field.

The electrode geometry is very important in maximizing the forces on the particles. For example, small and sharply pointed electrodes create strong field gradients, and therefore large DEP forces. Microelectrodes and the relatively low conductivity required for these separations both have the advantage of reducing heat production at the electrodes, and electrolysis. Fabricated using standard photolithographic techniques, they typically take the form of thin 0.1 µm layers of gold on chromium, evaporated on glass (microscope slide size) substrates. In one design, the interdigitated castellated electrodes (**Fig. 5**), through their geometry, provide an efficient means of repeating regions of high and low field gradient, which, when fabricated over large areas, provide the means of large-scale separations of particles. **Figure 6** illustrates the local cell separation between the electrode castellations.

Separation of particles under positive and negative DEP can be achieved either by gravity or fluid flow over the electrodes, which selectively removes the less-immobilized particles under the influence of negative DEP and enables their subsequent collection. Those cells still held, under positive DEP, can be released by turning off the field and collecting in a similar manner.

Table 1 Examples of particles investigated by non-uniform AC electric fields (DEP)

Particle type		Examples
Acellular	Virus	Trapping of single virion herpes simplex type 1
Prokaryotes	Bacteria	Characterization and separation of bacteria
Eukaryotes	Protozoa	Differentiation between normal and *Plasmodium falciparum*-infected erythrocytes
	Yeast	Batch separation of viable and non-viable (heat treated) *Saccharomyces cerevisiae*
	Plant cells	Batch separation of plant cells from mixture containing yeast and bacteria.
Mammalian cells	Cell lines	MDA231 human breast cancer cell separation from erythrocytes and T-lymphocytes
	Lymphocyte	Removal and collection of human leukaemic cells from blood
Other particles	Proteins	Collection of proteins, e.g. avidin 68 kDa and ribonuclease A 13.7 kDa.
	DNA	Separation of different sizes of DNA (9–48 kb) using positive DEP with field flow fractionation
	Liposomes	Alignment of cell size liposomes for subsequent electrofusion
	Artificial nanoparticles	Separation of latex beads of diameter 93 nm, with differing surface charge

Table 2 Values of suspending medium conductivity and voltage frequency used to dielectrophoretically separate cell mixtures

Cell mixture		Conductivity (mS m⁻¹)	Frequency (kHz)	Released cell
Escherichia coli (Gram − ve)	Micrococcus luteus (Gram + ve)	55	100	E. coli
Erythrocyte	M. luteus	10	10	Erythrocyte
Non-viable yeast	Viable yeast	1	10 MHz	Non-viable
Blood cells	Leukaemic cells	10	80	Blood cells
Human peripheral blood	Breast cancer cells	10	80	Erythrocyte
Bone marrow	Peripheral blood	1	5	CD34+ subpopulation

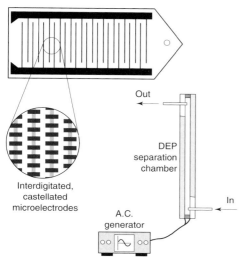

Figure 5 A typical DEP separation chamber consisting of two sealed glass plates with microelectrode arrays fabricated on their inner surface, and inlet and outlet ports for the passage of cell mixtures and suspending fluids. The interdigitated, castellated, electrode design enables cells to be separated locally under the influences of negative and positive dielectrophoresis.

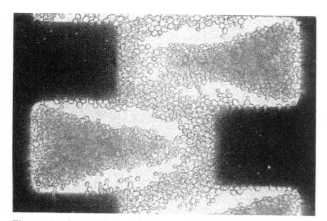

Figure 6 Separation of viable and non-viable cells by DEP. By applying a 4 MHz signal to a cell suspension on castellated interdigitated electrodes, healthy and non-viable cells can be separated. Non-viable cells stained by a dye, experience a negative force and collect into loosely held triangular formations in regions of low electric field strength. The unstained viable cells, experience a positive dielectrophoretic force and collect in chains between opposite castellations.

Separation chambers based on this mechanism are usually composed of two electrode arrays sandwiching a thin layer of fluid. Thin chambers are used because the DEP force decays with distance in a near exponential manner, and an effective DEP force is considered to extend no further than 300 µm from the plane of the microelectrode. Despite this possible limitation, separations of more than 10⁴ cells per second can be achieved by using quite simple equipment.

The DEP Force as a Function of Field Frequency

The polarizability of a particle also changes as a function of the frequency of the applied field. A single particle may therefore exhibit both positive and negative dielectrophoresis as its polarizability changes over a frequency range, for a constant medium conductivity. A typical DEP frequency spectrum illustrating such a transition is shown for a live yeast cell in **Figure 7**. Also represented is the DEP spectra for a dead yeast cell, which only experiences a change in the polarity of DEP force for frequencies greater than a few megahertz at a conductivity of 8 mS m⁻¹. DEP spectra are obtained by measuring the particle motion in a chamber with, for example, polynomial type electrodes (**Fig. 8**) energized with sinusoidal voltages, with 180° phase difference between adjacent electrodes.

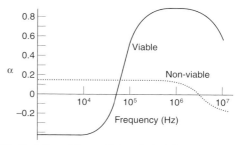

Figure 7 Variation of the particle polarizability α as a function of the frequency of the applied electric field for viable and nonviable yeast cells in a suspending medium of 8 mS m⁻¹.

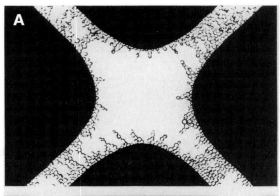

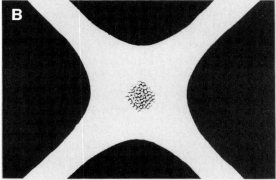

Figure 8 Examples of positive (**A**) and negative (**B**) dielectrophoretic collection of yeast cells (*Saccharomyces cerevisiae*). Positioning of cells in the centre of a polynomial array by negative dielectrophoresis is convenient prior to electrorotation analysis using the same electrodes with the appropriate electrical connections.

Levitation of Particles

Contact with the electrode induced by positive DEP may impinge on subsequent removal of the particle (e.g. by fluid flow or gravitational forces). The attractive or repulsive forces on the particles by DEP so far described are for interactions where both the particle and electrode are in the same plane, the particle resting on the substrate. These forces can also be applied to particles to make them levitate above the substrate, either from the result of an attractive high field region presented above the particle in the form of an electrode probe or by the repelling action of interdigitated electrodes on the plane of the glass, where the particle can be confined in a stable position above the electrodes. Particle levitation can be combined with other techniques, for example field flow fractionation (FFF), whereby particles levitated to different heights (up to 100 μm and above) are exposed to different rates of fluid flow. Negative DEP forces can also be exerted simultaneously from above and below to trap particles in a '3D field cage'.

Cell Handling for Electrofusion

Another application for DEP is the manipulation of cells prior to electrofusion. Attractive interactions between the induced dipoles of adjacent cells can result in the formation of chains of cells (pearl chains) of variable length. Close cell contact, followed by a high field strength DC pulse(s) of kV cm^{-1} and ms duration can lead to cell fusion of two to several thousand cells, so that giant cells can be formed as well as hybrid cells with two nuclei.

Are Cells Damaged?

To induce cell fusion, or indeed electrical breakdown of the cell membrane, a field strength of at least ten times more than is typically used in DEP separations is required. Hybrid cells from electrofusion are viable, which suggests that cells having undergone exposure to normal DEP forces are not damaged. Further evidence includes the exclusion of trypan blue from dielectrophoretically separated erythrocytes and the successful culture of various cell types including yeast cells and CD34+ cells.

The fluid flow during a DEP separation procedure produces a maximum shear stress on the cell of around 3 dyn cm^{-2}. T-lymphocytes and erythrocytes have been reported to be able to withstand a shear stress some 50 and 500 times this value, respectively. Therefore almost insignificant levels of shear stress are experienced by these cells in DEP chambers.

The conductivity of suspending medium used is normally much below that of a normal physiological medium, however, as long as the osmolarity is of the right value, osmotically sensitive cells can be investigated. This is achieved by additives such as sucrose at 280 mM, which has little effect on the conductivity. An alternative approach has been to use sub-micrometre electrodes which minimize heating effects enabling the use of normal physiological strength media.

DEP: Concluding Remarks

The method is non-invasive and does not require the use of antibodies or cell surface antigens or other labelling, although in some applications the use of specific markers or dielectric labels may be an advantage. DEP can be employed at either the single-cell or multicell (more than 10^4 cells per second) level, and it has already been demonstrated for a variety of applications, notably: the purification of cell cultures by DEP separation of non-viable or contaminating species; the isolation or enrichment of cell subpopulations; and the rapid isolation of toxic microorganisms in water and food.

Manipulation of sub-micrometre particles such as single virions of Herpes simplex virus (type-1) both in enveloped and in capsid form, gives an indication of the potential for sub-micrometre applications, such as the study of single virion–bacterium interactions or virus harvesting. Rapid biopolymer (DNA or protein) fractionation has also been described in a method termed DEP chromatography.

Particle Motion in Rotating Electric Fields: Electrorotation

Whereas conventional DEP utilizes stationary fields, two closely related techniques utilize moving fields, more specifically either of rotating or travelling wave form. The investigation of particle motion in these moving fields has led to the development of some different applications. Inducing cellular spin by subjecting the cell to a uniform (homogeneous) rotating electrical field is termed electrorotation (ROT). Applications of ROT include the real-time assessment of viability of individual cells and their characterization.

A uniform rotating electric field can be generated by energizing four electrodes with sinusoidal voltages, with 90° phase difference between adjacent electrodes. Creation of the dipole moment in a particle takes a characteristic time to reach its maximum value, equally when the field changes direction the dipole will respond and decay at a rate determined in part by the passive electric properties of the particle and suspending medium that appertain to the frequency of the applied voltage. Torque resulting in cellular spin is induced by the interaction between the rotating electric field and the remnant dipole. As illustrated in **Figure 9**, the torque created can result in spin of the particle in the opposite direction to the field as well as in the same direction as the field (not shown).

For a given particle, there is a unique rotation rate for each frequency of applied voltage. This variation

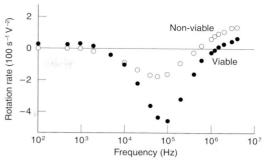

Figure 10 ROT spectra of live and dead *Cryptosporidium parvum* oocysts. Viability was confirmed with the fluorogenic vital dyes 4′,6-diamidino-2-phenylindole (DAPI) and propidium iodide (PI).

in rotation rate is shown in **Figure 10** for a viable and non-viable oocyst of *Cryptosporidium parvum* suspended in a $5 \, \mu S \, cm^{-1}$ solution, whose viability had been confirmed using a fluorogenic vital dye technique. Although the field may be rotating at rates greater than $10^7 \, s^{-1}$, the induced particle rotation rate which is dependent on the square of the field strength remains measurable by the human eye. Depending on the frequency, typical rotation rates observed are between -3 and $+1.5$ rotations per second for a viable *C. parvum* oocyst subjected to a rotating field of around $10 \, kV \, m^{-1}$, with negative rotation rates indicating antifield rotation of the particle. There is a frequency (around 800 kHz for this conductivity) in the ROT spectra of Figure 10 where the viable and non-viable oocysts rotate in opposite directions, providing a convenient, single frequency, viability check on individual oocysts.

After concentration from a sample, particles for observation in a ROT chamber (which can be manufactured on a reusable glass slide or as a cheap 'use once – throw away' device) only require a few washes followed by resuspension in a known conductivity medium. Analysis by ROT observation of a sample using a normal microscope can require less than 15 min preparation. Although the particle suspension may require a purification step to avoid particle–debris interactions, ROT to date has found many applications, both with biological and synthetic particles (**Table 3**).

As well as the rapid (a few seconds per cell) straightforward assessment of the viability of individual cells, the viability of larger numbers of cells (e.g. 30 cells of diameter 5 μm in a field of view at a magnification of 400) can also be assessed simultaneously. To assist the analyst, automatic measurement of the rotation rate for a full spectrum is also possible. A full frequency ROT spectrum, which can be thought of as a 'fingerprint' for heterogeneous particles like oocysts and cells, provides information not only about the via-

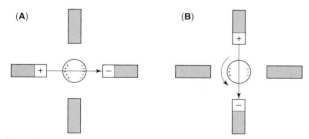

Figure 9 Generation of particle torque in a ROT chamber. In a stationary field (**A**) the induced dipole moment for a particle that is less polarizable than the medium is directed against the field. On turning the field in a clockwise direction (**B**), the field interacts with the decaying charges to produce a torque on the particle. In the example shown the resultant spin of the particle opposes the direction of the moving field, this is termed *anti-field electrorotation*. Conversely for a particle that is more polarizable than the surrounding medium the torque induced results in a spin in the same direction as the field or *co-field rotation* (not shown).

Table 3 Examples of particles investigated by rotating electric fields (ROT)

Particle Type		Examples
Acellular	Virus	Virus erythrocyte interactions
Prokaryotes	Bacteria	Biocide treatment of bacterial biofilms
Eukaryotes	Protozoa	*Cryptosporidium* spp. oocysts
	Yeast	*Saccharomyces cerevisiae* comparison of wild type/ vacuole deficient mutant
	Algae	*Neurospora* slime
	Plant cells	Barley mesophyll protoplasts
	Insect cell line	Effect of osmotic and mechanical stresses and enzymatic digestion on IPLB-Sf cell line of the fall armyworm (*Spodoptera frugiperda*, Lepidoptera)
Mammalian cells	Cell lines	MDA-231 human breast cancer cells
	Lymphocyte	Influence of membrane events and nucleus
	Erythrocyte	Erythrocytes parasitized by *Plasmodium falciparum*
	Platelet	Influence of activators
Other particles	Liposomes	Liposomes with 1 to 11 bilayers
	Latex bead	Effect of surface conductance

bility of the particle, but also the conductivity and permittivity of the various 'compartments' within its structure. After ROT analysis, as with DEP, the particle remains intact and unchanged, and because ROT is a noninvasive method the particle can be subjected to further holistic or destructive analytical methods.

A variety of particle types, including cells, protozoan cysts and bacteria can be investigated by this technique. By probing a common difference between all dead and viable cells, namely membrane integrity, ROT is applicable to many particles. Potential applications also include distinguishing between subtypes or strains of bacteria, whose surface or membrane properties differ, for example the rapid diagnosis of the causitive agents of food poisoning to direct appropriate action.

Particle Motion in Travelling Wave Electric Fields

Like ROT, a third AC electrokinetic technique also uses moving fields, instead of rotating they are in the form of linear travelling waves, made simply by applying AC voltages in phase sequence to a linear array of electrodes. At low frequencies ($< 100\,Hz$) translational motion is induced in the particles by largely electrophoretic forces, associated with surface charge characteristics. To overcome problems such as erroneous particle trajectories and motion caused by

the convection of suspending fluid, higher frequencies are more commonly utilized at which DEP forces have the strongest effect on the motions of particles.

Unlike DEP, the motion of particles by travelling wave dielectrophoresis (TWD) is achieved in a stationary supporting fluid; without the need for fluid flow there is no dilution of particle density. Indeed, the concentration of particles without centrifugation may be important for certain delicate particles which may be distorted or damaged. Loss of particles through adhesion to the container is avoided as TWD can manipulate particles without contact with the substrate, a small distance above an array of electrodes. Indeed, for successful translational motion of particles by TWD, the particle must be under conditions of negative DEP (or negligible positive DEP). Examples of particles investigated by travelling wave electric fields is given in **Table 4**.

Selective retention or transportation of subpopulations from a suspension is one application of TWD. In this way target organisms can be separated from benign background cells. Separation of yeast cells using TWD has been demonstrated, both by retaining viable cells at 5 MHz and moving non-viable cells, and at a higher conductivity by moving viable cells while retaining non-viables. The TWD response of a particle can be predicted by examination of the respective DEP and ROT spectra; the sense and magnitude of the ROT indicating the direction and magnitude of the TWD force on the particle in the travelling wave.

Electrode geometries also influence the resultant TWD force, the optimum electrode gap is found to be similar to the effective particle size. Particles can be made to move over lines of electrodes (of the appropriate geometry, spacing width and voltage) or for more convenient viewing, in the gap between the tips of many rows of electrodes as shown in **Figure 11**.

Unless spiral electrode geometries are utilized, whose area can be increased simply by adding further helical turns, there are limits to the size of planar

Table 4 Examples of particles investigated by travelling wave electric fields (TWD)

Particle type		Examples
Eukaryotes	Protozoa	*Cryptosporidium parvum* oocysts
	Yeast	*Saccharomyces cerevisiae*
	Plant cells	Membrane covered pine polls
Mammalian cells	Blood cells	Separation of components of whole blood
Other particles	Artificial spheres	Cellulose spheres

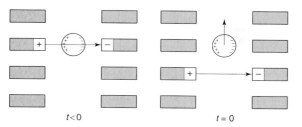

Figure 11 Schematic representation of the travelling field effect for a particle less polarizable than the medium ($\alpha_p < \alpha_m$). In the instant ($t < 0$) before voltage switching between electrodes occurs the dipole moment induced is opposed to the direction of the field. On switching the electrode voltages at time $t = 0$, the interaction between the remnant dipole and the field induces a translational force in the particle in the opposite direction to the travelling field.

monolayer electrode arrays as there are special limitations for the connections to the individually addressed electrodes. Multilayer electrode fabrications only require four connections (for quadrature phases) to energize one or more TWD arrays of any length. Theoretically, multilayer TWD devices can be built up (like a model railway track) so that particles may be taken to many investigative units such as ROT chambers, in a single integrated device. Separation, manipulation and characterization of particles in a single device, a sort of 'laboratory-on-a-chip device' has been proposed.

Conclusions

Detection, enumeration and characterization of low numbers of microorganisms in foods and water by methods that do not require a culture stage would be advantageous. The rapid concentration, separation and identification of microbes (and their viability) already present in samples would avoid testing delays due to slow bacterial growth phases. The novel dielectric methods of DEP, ROT and TWD may offer a solution.

Although some applications have already been demonstrated on bench-scale experiments, further data must be collected on the electrokinetic responses of a wider range of bioparticles. These include non-target particles such as plant spores and non-pathogenic protozoan cysts, so the specificity of the tests can be optimized. Particles from different sources must also be investigated as this may alter the electrokinetic response.

For these technologies to proceed and compete with current techniques they must be more specific, sensitive, reliable, rapid or more competitively priced.

One simple way of achieving many of these requirements is to make the tests fully automated by integrating them onto a single disposable device. Following the introduction of microelectrodes into this field of study using photolithography and associated semiconductor micro-fabrication technologies, and more recently the development of multilayer fabrication techniques, dielectrophoresis, electrorotation and travelling wave dielectrophoresis have all developed into techniques that can be incorporated onto a single 'bioprocessor-chip' device.

See also: **Adenylate Kinase. Biosensors**: Scope in Microbiological Analysis. **Flow Cytometry. Polymer Technologies for Control of Bacterial Adhesion. Rapid Methods for Food Hygiene Inspection. Sampling Regimes & Statistical Evaluation of Microbiological Results. Ultrasonic Imaging. Viruses**: Introduction; Detection. **Water Quality Assessment**: Modern Microbiological Techniques.

Further Reading

Arnold WM and Zimmermann U (1988) Electrorotation: development of a technique for dielectric measurements on individual cells and particles. *Journal of Electrostatics* 21: 151–191.

Fuhr G, Zimmermann U and Shirley SG (1996) Cell Motion in Time-varying Fields: Principles and Potential. In: Zimmerman U and Neil GA (eds) *Electromanipulation of Cells*. Boca Raton: CRC Press.

Jones TB (1995) *Electromechanics of Particles*. Cambridge: Cambridge University Press.

Lynch PT and Davey MR (eds) (1996) *Electrical Manipulation of Cells*. New York: Chapman & Hall.

Markx GH, Huang Y, Zhou X-F and Pethig R (1994) Dielectrophoretic characterisation and separation of micro-organisms. *Microbiology* 140: 585–591.

Markx GH, Talary MS and Pethig R (1994) Separation of viable and non-viable yeast using dielectrophoresis. *Journal of Biotechnology* 32: 29–37.

Pethig R (1979) *Dielectric and Electronic Properties of Biological Materials*. Chichester: Wiley.

Pethig R (1991) Application of AC electrical fields to the manipulation and characterisation of cells. In: Karube I (ed.) Automation in Biotechnology, p. 159. Amsterdam: Elsevier.

Pethig R (1996) Dielectrophoresis: Using inhomogeneous AC electrical fields to separate and manipulate cells. *Critical Reviews in Biotechnology* 16: 331–348.

Phol HA (1978) *Dielectrophoresis*. Cambridge: Cambridge University Press.

BIOSENSORS

Scope in Microbiological Analysis

Millicent C Goldschmidt, Department of Basic Sciences, The University of Texas Health Center at Houston, USA

Introduction

In many respects the growth of biosensor technology mimics the beginnings of a bacterial growth curve. The instrumentation is just emerging from a long lag period of 20–30 years and will soon be in the explosive potential growth phase. Market sales projections for the US alone are predicted to reach over $40 million by the year 2000 for agricultural and environmental groups, almost $100 million for medical and consumer groups and over $800 million for industrial and other areas.

Biosensor technology is already changing the manner in which diagnostic methodologies are being performed in diverse areas of science. As demands for more sensitive, accurate and faster procedures increase in almost every field, biosensors will become more commonplace and will be universally accepted as approved methods by regulating agencies.

Although the main body of literature dealing with biosensors and foods is fairly recent, publications are also reaching the end of the lag period and are poised for an exponential explosion of their own. The US Department of Commerce National Technical Information Service has biosensor citations. If one has computer access to the National Academy of Medicine's Medline searches (now free), general or specific references can be obtained.

The first portion of this chapter will deal with definitions of the various types of biosensors and their instrumentation. Biosensors that can be used for monitoring microbial contamination of foods and the environment will follow. Instrumentation that could be applied to on-line and flow injection analysis (FIA) monitoring as well as spot analyses that could be performed in real time will also be discussed. Limitations and future directions conclude the chapter.

Biosensor Definitions

The bio in biosensor refers to the biological sensing portion or sensor. It can either be a specific group of molecules or somewhat general in nature, depending on the degree of specificity required by the specific problem facing the investigator. The transducer measures the changes that occur when the sensor couples with its analyte or target and converts them to a digital readout. These changes must also reflect some relationship between the intensity of the signal and the concentration of the analyte. Again, the degree of sensitivity is determined by the type of transducer ultimately employed. The transducer is in close contact or coupled to the sensor. The analyte is the substance that is to be detected. Again, it can be specific as a distinct compound or more general as a related group of substances. As will be seen, one of the important aspects of biosensor usage is the adaptability and versatility of the various systems relating to many different areas of microbiology, foods, food products and pathogen detection.

The Sensor

As mentioned above, the sensor can be a selective biorecognition moiety such as a polyclonal or monoclonal antibody or a single-stranded DNA molecule. It can also be more general, as a tester microorganism which responds to many different toxic compounds by a decrease in one or more metabolic activities. Enzymes have also found frequent use in biosensors responding again both to specific and nonspecific substrates. **Table 1** lists typical sensors which have been used to detect target analytes. Details on several of these follow.

Immunoglobulin Sensors Immunoglobulins have the advantage of being able to recognize their target antigens (analytes) in a mixture of organisms such as is found in foods. This can be contrasted to the activity or extreme specificity of other types of sensors or transducers. Polyclonal antibodies have been used to identify various *Salmonella*, staphylococci, staphylococcal enterotoxin B as well as atrazine and other herbicides in drinking water. Monoclonal antibodies as sensors have been used to identify botulinum toxin A and Mojave toxin. The above examples indicate the specific reactions conferred by antibody–antigen couplings. Although there appears to be no published research using antibodies to bacterial ribosomal RNA or DNA as sensors, these antibodies are available and could rapidly detect the presence and concentration

Table 1 Biosensing entities

Sensor	Analyte detected	Recognition signal
Enzyme immunoassay fluorescein-labelled	Mojave toxin	Voltage change (light-addressable)
Immunoglobulins	*Salmonella typhimurium*	Piezoelectric resonance changes
Hexacyanoferrate III	Fructose and ascorbate in fresh foods	Current changes
Acetylcholinesterase (FITC[a]-tagged)	Insecticides	Fibreoptic (fluorescence)
Enzymes	Milk, fruit juice, wine on-line monitoring, lactose and galactose analysis	Voltage change in enzyme electrode, thermal
Antibodies	Protein A in foods, protein interactions, atrazine, herbicides, staphylococcal enterotoxin B	Piezoelectric, resonance changes
Nucleoside phosphorylase + xanthine oxidase	Phosphates in food products	Hydrogen peroxide, uric acid, current changes
Invertase, maltose, saccharose, glucose oxidase	Sucrose in syrups and jams	Current changes
Monoclonal antibodies	Plant viruses, viral epitopes, botulinum toxin	Polarize light angle changes, evanescent waves
Immunoglobulins/NAD-Redox	*Staphylococcus*-bearing protein A, *Salmonella*	Current changes
Glucose oxidase	Glucose, immunoassays, glucose in molasses	Current changes
β-Galactosidase and glucoamylase	Lactose in raw milk	Current changes
Single-stranded DNA binding protein	DNA in processed pharmaceuticals	Voltage change when captured
Luciferase + NAD-dependent enzymes	Sorbitol, ethanol, oxaloacetate	Luminescence changes
Polyclonal and monoclonal antibodies	Botulinum toxin A	Fibreoptic (fluorescence) temperature changes
Enzyme/amperometric	Atrazine, chlorocresol, phenols in ointments	Current changes, colorimetry
Enzyme/thermistor	Cholesterol, glucose, lactose	Heat emission changes
DNA hybridization, antibody–antigen reaction	DNA antibody, antigens	Current changes, optical (evanescence, fluorescence)
Tyrosinase and peroxidase	Phenol and peroxide in pharmaceuticals	Current changes
Bis methylacridinum nitrate and luminol	DNA, ATP	Luminescence
β-Lactamase	Penicillins	Temperature changes
Luminol	ATP	Luminescence
Calorimetric	Heavy metal ions, immunoassays, glucose	Exothermic responses
Glucoamylase and α-amylase	ATP, NAD+, starch, disaccharides	Current changes
Lactate oxidase	Lactate	Current changes
Nitrate-sensing electrode	Nitrates in nitrifying biofilms	Current changes
Frog bladder membranes	Sodium ion-channel blockers such as tetrodotoxins, saxitoxin	Sodium electrode peak inhibition

[a]Fluorescein isothiocyanate.

of generic bacteria in foods, pharmaceuticals or sterile liquids. Observing a heightened reaction with time could indicate bacterial growth. The addition of a vital stain such as acridine orange and a spectrophotometric transducer could distinguish viable from non-viable cells.

Antigens and Other Compounds as Sensors   Antigen sensors are mainly used to detect antibodies. A slaughterhouse might wish to test cattle for immunoglobulins to certain pathogens as an indicator of disease. Single-stranded DNA-binding proteins have been used to capture DNA in processed pharmaceuticals as indications of contamination. Avidin can capture biotin-labelled compounds. Small peptides synthesized on automated peptide synthesizer instru-

mentation have been used to mimic antibody-binding sites as receptor molecules or affinity ligands. These synthetic antibodies have functioned well as sensors to detect much larger protein analytes. They hold much possibility for wide use in biosensors. Lectin saccharides, hormone receptors, hormones and leukocytes are among the sensors that have been used.

Enzyme Sensors   When an enzyme reacts with its proper substrate (analyte), a measurable change occurs. Enzymes have been used for many years in basic and applied areas. Over 400 enzymes can be purchased from various chemical and pharmaceutical companies. Many enzymes such as glucose oxidase, horseradish peroxidase and alkaline phosphatase are well understood and well characterized. These three have been used as markers for immunological reactions, including enzyme-linked immunoassays (ELISA) and other enzyme-liquid immunoassays (EIA, EI). As will be seen from Table 1 and in the section on transducers, many of these enzymes – glucose oxidase in particular – have been coupled to electrodes and to other biosensor instrumentation.

Glucose oxidase was one of the first enzymes to be used in biosensors. Some of the impetus in the biosensor field in this area was prompted by the search for rapid methods to detect glucose concentrations in diabetics, with a view to eventually developing an implantable device that would automatically add insulin in response to glucose blood levels.

The Biotechnology Centre at Cranfield, UK, has developed a knife-type biosensor incorporating glucose-sensing enzymes to measure the concentration of glucose in animal muscle tissues as an indicator of meat freshness. The lower the concentration, the less fresh the meat (due to microbial utilization of the glucose). An indicator of fish freshness using enzymes to detect hypoxanthine has been reported and fish freshness with an enzyme system was also determined. The use of enzymes for biosensor detection of other carbohydrates, including galactose oxidase, β-galactosidase, invertase and glucoamylase has been reviewed, including detection of alcohol, lactate, pyruvic acid, choline, cholesterol and biological purines including inosine by enzyme electrodes. Fish freshness can also be determined by the degradation of ATP to inosine and eventually hypoxanthine.

Enzyme sensors have been commonly employed to measure milk, fruit juice, wine, phosphates in foods, sucrose in syrups and jams and aspartame. Table 1 lists several other enzyme sensors and the analytes they detect which are important in the food industry.

Microbial Cells as Sensors The use of microorganisms as sensors demonstrates the range of sensitivities that can be employed in biosensors. Whole cells of *Alteromonas putrefaciens* were used to determine the freshness of tuna based on the presence of glucose. The fresher the fish, the higher the glucose content and the greater the metabolic activity of the bacteria. Short-chain free fatty acids in milk (e.g. butyric acid using *Arthrobacter nicotianae* were determined in an amperometric biosensor with a 3 min response time. Microbial biosensors for glucose estimations as well as monitoring fermentation processes have been reported. *Clostridium acidiurici* was used to detect serine using a potentiometric NH$_3$ gas sensor. A *Proteus* sp. was used to detect L-cysteine. *Pseudomonas* cells created a specific biosensor to identify and quantify L-proline. A microbial electrode was used to study biological oxygen demand. Molecular

biologists have aided biosensor technology by constructing specialized bacteria for various purposes. Of particular interest is the transfer of the *lux* AB genes which encode for bioluminescence (luciferase) into the genome of several bacteria. As a result of this fusion with the *flic* genes of *Escherichia coli*, luminescence could be induced in the presence of as little as 1 µg ml^{-1} aluminium but not copper, iron or nickel. The metal responsit *smt* region of *Synechococcus*, a cyanobacterium, was fused to the *lux* cDABE genes of *Vibrio fischeri*. These transformed cells of *Synechococcus* luminesced in the presence of 0.5–4 µmol l^{-1} ZnCl$_2$ as well as trace levels of CuSO$_4$ and CdCl$_2$. Conversely, the addition of antimicrobial agents (and probably other inhibitory compounds) can extinguish light production in *E. coli*. Since the genetics of *E. coli* has been extensively studied and understood, this organism has been the bacterium of choice in many instances. However, the insertion of these same *lux* genes, when fused next to the genes encoding mercury detoxification (*mer* genes) in *Serratia marcescens*, has resulted in light emission in the presence of mercury in a quantitative manner.

In a different experiment a lysogenic strain 6Y5027 of *E. coli* was used to detect mutagens such as aflatoxins. The mutagens caused phage induction which killed the bacteria and decreased respiration. A recombinant *E. coli* was used to detect a wide spectrum of organophosphorous pesticides. *Hansenula anomala* has been used in a flow-through system to measure organic pollution. *Rhodococcus* cells and their enzyme extracts were used to determine phenol, cresol, benzoate and 2-methyl-4-chlorophenol, and nitrifying bacteria were used to determine creatinine. *Trichosporon cutaneum* was immobilized on the electrodes of a disposable amperometric biosensor to determine biochemical oxygen demand (BOD) of treated and untreated waste water from municipal sewage and industrial effluents. Response time was 7–20 min compared to the conventional 5 day methodology. **Table 2** lists many more whole cells used in the detection of widely varied substances including sulphur dioxide in orange juice, glucose, lactic acid in milk and L-aspartate. Thus, one can see the widespread use of whole organisms in biosensor methodology. Various tissue and organs have also been used, with relatively minor applications to the scope of this article. A tissue biosensor of a frog bladder was used to measure sodium ion channel blockers such as tetrodotoxin and saxitoxin (shellfish poisoning). One assay took 5 min and was a modified-flow system. A microbial cyanide biosensor has been described using *Saccharomyces cerevisiae* to monitor the presence of cyanide in river water. Oxygen electrodes monitored the yeast's respiration in a flow-

Table 2 Biosensors using whole cells

Cell type	Analyte detected	Recognition signal
Sarcina flava	Gutamine, NH_3	Voltage change
Thiobacillus thioxidans	SO_2 in orange juice foods	Oxygen production
Nocardia erythropolis	Cholesterol	Current changes
Hansenula anomala	Glucose, organic pollution	Voltage change, current changes
Alcaligenes eutrophus + Lux genes	Heavy metal resistance	Optical detection
Bacterium cadaveris	L-Aspartate	Voltage changes
Escherichia coli (lysogenic strain)	Mutagens, aflatoxin, etc.	Phage induction
E. coli	Organophosphorus pesticides, lipoic acid reduction, glutamic acid, neurotoxins	Voltage changes
Human leukocytes	Egg allergens	Voltage changes
E. coli containing luxAB or mer-lux genes	Environmental monitoring for aluminium, mercury	Luminescence (luciferase)
Pseudomonas aminovorans	Trimethylamine levels in fish tissue	Oxygen uptake
Pseudomonas spp.	L-Proline	Current changes
Acetobacter pasteurianus	Lactic acid in yoghurt, milk	Current changes
Carboxydotrophic bacteria	Carbon monoxide	Oxidation/reduction
Gluconobacter suboxydans	Ethyl alcohol	Voltage changes
Trichosporon brassicae	Ethyl alcohol	Oxidation/reduction
T. cutaneum	Phenol, biochemical oxygen demand	Oxidation/reduction, current changes
Clostridium acidiurici	Serine, NH_3	Voltage changes
Alteromonas putrefaciens	Glucose (freshness of tuna)	Current decreases
Rhodococcus spp.	Phenol, cresol, benzoate	Current changes
Clostridium butyricum	BOD, organic matter	Current changes
Hansenula polymorpha	Formaldehyde	Voltage changes
Saccharomyces cerevisiae	Nystatin, cyanide	Current changes
E. coli containing luciferase	Biologically active antimicrobials	Light emission changes
Arthrobacter nicotianae	Short-chain fatty acids in milk	Current changes
Synechococcus (smt-lux transcriptional fusion)	Heavy metals: $ZnCl_2$, $CuSO_4$, $CdCl_2$	Luminescence

BOD = Biochemical oxygen demand.

through system. Linearity was observed over the range of 0.3–150 μm cyanide.

The Transducer

The transducer portion of a biosensor senses a signal change when sensor and analyte have coupled or reacted together and converts it into a digital readout. The type of transducer that is ultimately employed depends on the type of analyte to be detected, the type of sensor available and the required degree of accuracy and speed of the reactions. The transducer is usually coupled to the sensor or is in close proximity, although the signal can be sent to a receiver located at a distance from the reaction. Biosensors with fairly short reaction times are usually employed under these circumstances. The time allotted to detect a measurable level or threshold of analyte may be more lengthy. The signal ultimately generated by the transducer must be related to the concentration, presence or absence of the analyte. Transducers can be as simple as a pH or Clark oxygen electrode or as complex as automated instrumentation based on the charge in the angle of light refraction from a mirrored surface

(surface plasmon resonance). **Table 3** lists various transducing elements and some of the instrumentation involved. More details follow.

Electrochemical Transducers The combination of sensors with electrodes has resulted in biosensors which function most efficiently. Depending on the stability of the sensor, electrode-based biosensors are accurate and easily used. They were among the first to be employed. Electrochemical transducers include potentiometric, amperometric, piezoelectric, capacitative and conductive instrumentation. The latter two are not used as often as the others. The electric signal can be digitized, recorded accordingly and the data stored in a computer for future analysis or reports.

Potentiometric Transducers Potentiometric transducers are among the most popular types of biosensors. When the sensor and analyte react on the surface or in proximity to a potentiometric transducer, a membrane potential (charge) develops that can be correlated with analyte concentration. At a steady current therefore, the resultant change in voltage (potential) can easily be measured. The transducer

Table 3 Transducing elements used in biosensor instrumentation

Transducing element	Instrumentation	Measurement
Electrochemical		
Potentiometric	pH and ion-sensitive electrodes, gas sensors (CO_2, NH_3), semi-conductors	Change in potential, (e.g. voltage at constant current)
Amperometric	Clark oxygen electrodes, carbon electrodes + benzoquinone mediators, channelling system, silicone technology	Current change in constant voltage
Piezoelectric and surface acoustic wave	Horizonal polarized surface acoustic waves, quartz or sapphire crystals and oscillating electrical field	Megaherz changes
Capacitance and conductance	Impedance types: combination with amperometrics, pH-sensitive field-effect transistors	Dielectric effects, current changes
Optical		
Visible light	Luminometer, spectrophotometer,	Photon emission/change in spectra
Luminescence	fluorometer, light-addressable sensor	
Fluorescence	CCD camera, silicon microchips	
Fibreoptic: optical fibres in contact with instrument	Luminometer, spectrophotometer, fluorometer	Photon emission
Surface plasmon resonance, evanescent wave field detection	Spectrophotometer (refractometry), evanescent wave field detection	Change in reflected light angle, detection in evanescent fields
Other transducer systems		
Calometric, other thermal	Calorimeters, infrared detectors, thermistors	Change in emission or absorption of heat

CCD = Charge-coupled device.

can vary from a simple hand-held disposable probe to automated analysers and pH meters. Probably the simplest involve the classic glass pH electrode. As electrodes have developed to a finer degree through the years, gas and ion-sensing potentiometric electrodes have been exploited as transducers in biosensors. More recently, solid-state transducers have been used employing semi-conductors.

Most sensors can be combined with potentiometric electrode transducers. In particular, antibodies and enzymes are among those most frequently used in relation to the food industry. One of the particularly fascinating aspects of biosensor technology is reflected in the fact that the limits of application are only bounded by human ingenuity. The methods of attaching the sensor to the electrode usually include some type of encapsulation. This can vary from a collodion layer and a nylon mesh to other types of films. If a solenoid valve is added to the transducer, the sensor can first be affixed to magnetic particles and easily removed from the transducer to be exchanged or reactivated. Silicone, rubber membranes, various polymers with pre-impregnated sensors such as antibodies or enzymes have also been used. The simplest method of attachment appears to be covalently binding or cross-linking the sensor to the transducer. The same techniques used in immunoassay procedures to anchor antibodies to glass or plastic surfaces are applicable here (pH, glutaraldehyde, tosylation, etc.). The solid-state transducers do not need an internal

reference solution as do the classic liquid-filled transducers. The latter can only sense a single analyte while the former can be formulated using chips and gates to sense several analytes simultaneously. The solid-state transducers can also be miniaturized. Among the enzyme sensors coupled to this type of transducer, glucose oxidase, other peroxidases and urease have been used. Maltose and starch can be determined by first passing the specimens through an amyloglucosidase bed and then detecting the liberated glucose by glucose oxidase potentiometry. A batchwise detector has been developed using a recombinant *E. coli* to detect various organophosphorus pesticides. **Tables 1–4** reflect the use of potentiometric transducers by the various indicated recognition signals. Light-addressable potentiometric transducers have a special silicon wafer which is illuminated from the back side by a light-emitting diode. The light causes a photocurrent to flow which depends on the pH. A chemical reaction on the surface that changes the pH affects this photocurrent. Where a potential is applied externally to maintain a constant current, a change in this potential results and is recorded. It reflects the reaction occurring on the surface of the wafer.

Amperometric Transducers Amperometric transducers measure the change in current at a fixed potential (or voltage) between working and reference electrodes, usually of Ag/AgCl. Occasionally three electrodes are used. Platinum appears to be one of

Table 4 Flow injection and on-line systems employing biosensors

Transducing element methodology	Analyte detection	Recognition signal
Amperometric	Hypoxanthine/fish	Current change
Potentiometric	Aspartame	Voltage change
Potentiometric	Pesticides, organophosphates, insecticides	Voltage change
Amperometric	Creatinine, ammonia, urea	Current change
Amperometric	Glucose	Current change
Amperometric	*Salmonella* in foods	Current change
Amperometric	Penicillin	Current change
Amperometric	Glucose and cellobiose	Current change
Amperometric	Phenols, dopamine, BOD	Current change
Amperometric	Aspartame	Current change
Amperometric	Vitamin C in foodstuffs	Current change
Amperometric	Starch	Current change
Piezoelectric and acoustic wave	Drugs of abuse, pesticide analyses, atrazine, organophosphorus compounds	Change in resonance (MHz) spectra
Optical/photometric	Phenols, dopamine	Change in absorbance spectra
Optical/photometric	H_2O_2, glucose, lactose	Change in absorbance spectra
Optical/spectrophotometric	Lysine, cadaverine (ammonium)	Change in spectra of indicator dye
Optical/surface plasmon resonance, evanescent wave	Antibody–antigen reactions, mycotoxin a peptides	Change in reflected light angle
Optical/fibreoptics	Penicillin	Fluorescence
Optical/fibreoptics/amperometric	Glucose	Fluorescence
Optical/fibreoptics	Glucose, fructose, sorbitol, gluconolactone	Fluorescence
Calorimetric/thermistor	Penicillin, glucose, urea, lactose	Temperature changes
Calorimetric/thermistor	Immunoglobulins	Temperature changes
Calorimetric	Penicillin	Temperature changes
Nitrate-detecting ion electrode	Nitrifying organisms in a biofilm	Changes in nitrate ion responses

BOD = Biochemical oxygen demand.

the metals of choice for electrodes. Newer ones use impregnated carbon fibres. The first enzyme electrode was developed in 1962. A Clark oxygen electrode was modified by adding an enzyme layer and using a platinum electrode which responded linearly with a change in current during the production of hydrogen peroxide when the affixed enzyme, glucose oxidase, reacted with glucose in the presence of oxygen at a constant potential. Later, other electron acceptors such as NADH or hexacyanoferrate III were added to the amperometric configuration.

There are other newer and more detailed third-generation amperometric transducers. Organic conductors in contact with enzyme sites have been reported. Lactate oxidase was used to measure soluble L-lactate in yoghurt and buttermilk using screen-printed sensors made under industrial conditions. Accuracy needed to be improved. An amperometric enzyme-channelling immunosensor has been developed using a disposable polymer-modified carbon electrode with two enzymes. One is bound to the electrode and co-immobilized with an antibody while the other one is free and amplifies the signal.

As can be seen, a wide variety of sensor entities can be employed. These include whole cells, glucose-sensing enzymes, detection of phosphates in food products by phosphatases, detection of sucrose in syrups and jams by enzymes such as invertase, mut-arase and saccharase and the detection of phenols in ointments. Organophosphorus and carbamate pesticides have been determined by inhibition of acetylcholinesterase activity. Monoclonal and polyclonal antibodies have been used to detect *Salmonella* in foods as well as staphylococcal cells containing protein A. A multivalent amperometric immunosensor system based on silicone technology in which the capture molecule is streptavidin covalently immobilized on silica. Miniaturized needle-type biosensors that detect glucose amperometrically have been reported.

Conductance and Capacitative Transducers
Conductance is measured in ohms and is the reciprocal of resistance (or current/voltage). Capacitance is measured in farads as a reflection of the dielectric changes (using semi-conductors) that occur when the voltage varies with time and produces current changes proportional to the rate of voltage changes. Staphylococcal enterotoxin B has been detected by immunospecies immobilized on the surface of silanized SiO_2. An enzyme biosensor has been developed to determine penicillin concentrations using conductometric planar electrodes and pH-sensitive field-effect transistors. Capacitance has been used with an acetylcholine receptor to detect *Crotalus* snake toxin. Since impedance techniques and instrumentation have been

accepted and used in the food industries, perhaps this type of transducer will become more applicable in the future.

Piezoelectric and Acoustic Wave Transducers The piezoelectric transducer can be either a sapphire or quartz A/T cut crystal transducer which is electrically stimulated to oscillate or resonate. The sensor (such as an immunoglobulin, enzyme or single-stranded DNA molecule) is fixed to the surface of the crystal and a frequency or resonance value (in megaherz) is established. When the sensor and analyte combine, there is an increase in mass and a subsequent shift in the oscillation frequency which can be related to the concentration of the analyte. Horizontal polarized surface acoustic waves (HP-SAW) were first applied in an immunosensor at 345 MHz. Piezoelectric immunobiosensors were used for atrazine herbicides in drinking water. As little as one part per billion was detected with polyclonal or monoclonal antibodies as sensors. An immuno-piezoelectric transducer was used to detect staphylococcal enterotoxin B and compared favourably with a capacitative biosensor.

An immunological system has been used in which one side of the crystal was first coated with staphylococcal protein A: a resonant frequency of 10 MHz was established. This method was subsequently used for an immunobiosensor. Protein G silanized to the crystal was used for stable immobilization of ligand-bound compounds. The addition of immunoglobulin G decreased the resonant frequency at a ratio of 1 MHz to each 10 ng of added immunoglobulin. Metal ions (Cu and Ni) were selectively absorbed from solution over a wide range of concentrations. A piezoimmunosensor was developed to detect *Salmonella typhimurium*.

Piezoelectric A/T-cut crystals coated with a special film sensitive to a pH change near its isoelectric point have been used to monitor bacterial growth and metabolic rates: pH changes of 0.001 unit could be detected. A bulk acoustic wave ammonia biosensor has been used to determine the lag time as well as the specific growth rate of *Proteus vulgaris* and the influence of various temperatures on these parameters.

Optical Transducers The use of optical instrumentation in various analyses has long been accepted. It is no surprise, therefore, that the field of optics has been successfully adapted to biosensor technology for over 30 years. Optical transducers range from simple absorbance, luminescence, reflectance and fluorescence determinations to the use of more complex optical fibres in various procedures.

A charge-coupled device (CCD) camera has been used to perform DNA hybridization on microchips and nucleic acid hybridization has been detected on the surface of a CCD device.

Optical Transducers without Fibreoptics Spectrophotomers, fluorometers and luminometers have been used singly or in conjunction with other transducers. Glucose oxidase was immobilized on a polyaniline polymer. When glucose reacted with this electrode, the optical absorption spectra changed and a linear response of up to 2.2 mmol l^{-1} glucose with a 2.5-4 min response time was observed. The *lux* (luciferase) genes have been transferred into *Escherichia coli*, *Serratia marcescens* and other bacteria and used as sensors with various transducers to detect various antimicrobial substances by production or inhibition of luminescence.

The optical detection of DNA by biosensor technology has been reported using bis-methylacridinium nitrate and luminol in an inexpensive luminometer. The affinity sensor (IAsys) has been evaluated based on detection with an evanescent field within a few hundred nanometers from the sensor surface. An evanescent wave immunosensor has also been used to determine botulinum toxins. The IAsys was used with immobilized antibody to carboxypeptidase in studying quantitative characteristics of protein interactions.

Fibre optic Transducers The use of optical glass or plastic fibres allows remote measurement to be made and recorded. The combined transducer complexes of bioluminescent or chemiluminescent enzymes and dehydrogenases bound to optic fibres have been used in the rapid detection of ATP, NADH or H_2O_2. Chemiluminescent optical biosensors can measure reactive oxygen species. A fibreoptic evanescent wave immunosensor detects protein A production by *Staphylococcus aureus* in foods. A 40 mV argon-ion laser (488 nm) was used here and antibodies to protein A were adsorbed on the optical fibre. A single optical fibre (100 μm in diameter) was used as a pH sensor. An evanescent wave fibreoptic biosensor detects endotoxin from *E. coli* using immobilized polymyxin B on fibres. An automated optical biosensor system has been described based on fluorescence detection of 16 *mer* oligonucleotides in DNA hybridization assays. Insecticides have been studied via the optical detection of luminescence in a fibreoptic biosensor.

NADH has been determined with fibreoptic transducers and there have been reports on the microdetermination of sorbitol, ethanol and oxaloacetate in this system. Another hybrid-transducing system has been described in which a pH indicator (bromopyrogallol) is incorporated into polymer membranes fixed to fibreoptic bundles. The change in calorimetric and refractive indices quantitate the reaction. In addition, these hybrids help overcome, to an extent, the attenuations that occur during light propagation along the fibres. A fully automated fibreoptic spectrofluorimeter has been described which could operate up to 18 biosensors from a remote centre. Of course, the use of antibody–antigen reactions on fibreoptic transducers soon followed. The antigens or the antibodies could be bound to the optic bundle to naked fibres, tapered fibres (evanescent wave), along the fibres or at the cut surface ends of the fibres. Tapered optical probes were used to develop a fluoroimmunoassay for the F1 antigen of *Yersinia pestis* and the protective antigen of *Bacillus anthracis*. The rapid detection of *Clostridium botulinum* toxin A was reported using a sandwich immunoassay employing rhodamine-labelled polyclonal antitoxin which produced a fluorescent signal. With 1 min, toxin concentrations as little as $5 \, ng \, ml^{-1}$ were detected. Botulinum neurotoxins were reported using fibreoptic biosensors.

A competitive immunoassay was used to immobilize herbicide triazine derivatives on the surface of fibreoptic transducers. Fibreoptics have been used to identify polymerase chain reaction (PCR) products in detecting *Listeria*. An optical biosensor has been developed in which an ethylene vinyl acetate polymer was used as a controlled-release system to deliver fluorescent reagents to the optical fibre. Continuous and stable measurements were provided over a lengthy period. Many optical instruments utilize fluorescence either directly incorporated with the sensor or in solution, separated only by a membrane. A fibreoptic immuno-biosensor was developed to detect staphylococcal enterotoxin B in ham as well as in urine and serum.

Surface Plasmon Resonance and Other Evanescent Wave Transducers

If the base of an optical prism is coated with or coupled to a thin semi-transparent metal layer (such as gold), it is possible to excite an electromagnetic wave, called a surface plasmon, along the surface when incident polarized light at a certain defined critical angle is beamed at the prism. Reactions occurring at or very near the metal surface cause a change in the refractive index. Thus, when an antibody–antigen or DNA–DNA coupling has occurred, there is a change in mass and the system then moves out of this optical resonance. The resultant angle of light reflected from the prism is changed and can be used both to quantitate as well as to indicate that a reaction has occurred at (or coupled with) the metal-coated layer, producing a sensorgram. Since the reflected angle is all that is needed, it is therefore possible to measure native reactions that occur

without additional reagents or markers. This is a very important advantage! This technique has been termed biospecific interactions analysis or BIAcore. BIAcore AB in Sweden markets an instrument (the BIAcore) which utilizes this technology and has made it possible to conduct very sensitive tests.

Several other instruments are reported such as the IAsys and BIOS-1 which uses recognition on an optical grating surface. As is necessary with all new technology, comparisons with accepted gold standard techniques such as ELISA, hapten–antibody binding and DNA–DNA hybridization have been performed and were found to exhibit excellent correlations. The BIAcore instrument also appears to have overcome several inherent biosensor problems through the use of replaceable sensor chips containing the ligands, an integrated flow-through liquid-handling system for the transport of the samples (analytes) and sensor reactivation after use. Kinetic studies of BIAcore reactions have been discussed in relation to the mass transfer rates of one species in a flowing solution reacting with an immobilized component in the hydrogel. The BIAcore instrument has been described determining kinetic association and dissociation rate constants at different temperatures. The technique of polymerase-induced elongation on nucleotide hybridization holds real promise for sensitive, accurate and rapid determination for the investigation of foods, food products and their biological safety as well as environmental concerns. DNA–DNA hybridization of 10 fmol of a 97 base target could be detected in less than 5 min. DNA probes used in PCR reactions could easily be utilized with this instrumentation. An octamer probe was used to analyse oligonucleotide affinities. Again, no additional reagents were needed to indicate that a reaction had occurred. Antibodies have been used to measure small specific antigenic regions in the form of peptides which were specific epitopes from one of the envelope glycoproteins of the HIV-1 virus, and the results correlated with conventional ELISA tests and in fact had an extended response range. This instrument was used to study the monoclonal antibody reactions with human vaccinia and polioviruses and two plant viruses (cowpea mosaic virus and tobacco mosaic virus). This technology was more advantageous than the conventional methods.

In a study of tetanus toxoid, dissociation and association rates were evaluated as well as affinity constants of IgG, IgM and Fab fragments. Surface plasmon resonance was used to detect and measure antibody–antigen affinity and kinetics. Protein interactions using an Fab fragment of an anti-paraquat monoclonal antibody in the BIAcore have been studied. The amino acid moieties of an antibody that react with the antigen actually compose only a minor portion of the whole immunoglobulin molecule. The rest of the molecule could actually cause some steric hindrance. Therefore, the use of synthetic peptides which mimic the antibody-binding site should (and do) give more accurate results. Similarly, peptides can be used as ligands to combine with antibody analytes. Peptides and super antigens have been discussed reacting with major histocompatibility classes. Since several bacterial toxins, such as staphylococcal enterotoxin B, are super antigens, this methodology is applicable. Although most literature dealing with surface plasmon resonance is presently biomedical in nature, applications to rapid detection of food pathogens, products and the environment are bound to follow. A fibreoptic sensor utilizing surface plasmon resonance has been discussed: a section of the cladding surrounding the optic fibre was removed and a layer of reflecting silver was symmetrically deposited on the core, creating the sensing element. This eliminates the need for a coupling prism as there is a fixed angle of incidence with modulated wavelength measurements. Samples of high-fructose corn syrup were diluted. The resultant SPR (surface plasmon resonance) spectra were in good agreement with able refractometer values. BIAcore AB has developed a similar system, the BIAcore probe with a gold interface on the optic fibre. Detecting molecules such as antibodies could identify target analytes such as antigens within minutes.

Thermal Transducers Measurements of heat absorption or production have been used for many years to assess the varied activities of microorganisms. Older instruments were relatively insensitive, measuring only gross temperature changes. However, newer instrumentation can measure very small changes (10^{-5}°C) within minutes. The two main types of thermal transducers involve the use of either thermistors or thermopiles. In a thermistor, energy flow is measured and the voltage across a semi-conductor varies as the current is increased. Therefore, a relatively small rise in temperature results in a relatively large change in resistance. Thermopiles, on the other hand, are merely arrays of thermocouples which measure voltage changes.

Very sensitive thermistor-based biosensors have been used in bioprocess monitoring and environmental control and the development of an integrated thermal biosensor for the simultaneous determination of multiple analytes has been developed. A thermal transducer using a thermopile composed of strips of silicone/aluminum integrated on 5 μm silicone membrane has been described. Glucose oxidase, urease and penicillinase enzymes were immobilized on the back side of the thermopile in an FIA system. Genetically prepared enzyme conjugates have been used in lactose and galactose analyses. The on-line production of penicillin V using an enzyme (penicillinase) coupled with a thermistor was monitored. The values compared well to those obtained from high-pressure liquid chromatography. A thermodynamic analysis of antigen–antibody binding was conducted using biosensor measurements on a BIAcore at different temperatures. A general enzyme thermistor based on specific reversible immobilization using antibody–antigen interactions was discussed. Perkin-Elmer (Norwalk, CT) markets a differential scanning calorimetric (DSC-7) robotic system (DSC-7) with a 48-position carousel. Gilson (Worthington, OH) sells a microscal flow microcalorimeter. Although neither instrument is a biosensor, the addition of sensing molecules such as enzymes, antibodies and antigens could probably convert these instruments into thermal biosensors without too much difficulty. The Gilson instrument is probably the more readily adaptable of the two.

Flow Injection Analysis and On-line Systems Employing Biosensors

It is important to be able quickly and continuously (more or less) to monitor on-line various aspects of microbial growth, production of valuable end products (vitamins, amino acids and glucose) as well as rapidly determine the sterility of finished products. FIA methodology has been elaborated in great detail in several reviews. The discussion here will focus on the use of biosensor systems which can be uniquely adapted to many different analyses employing FIA and other on-line or flow-through systems. This will serve to highlight biosensor capabilities. FIA has been mentioned earlier in this article. As can be seen from Table 4, biosensor technology has been easily adapted into FIA procedures. A few of these approaches will be discussed to demonstrate biosensor versatility.

As early as 1983, optical fluorescence sensors were reported for the continuous measurement of chemical concentrations in biological systems using glucose oxidase amperometrically and in combination with fibreoptics. An on-line glucose sensor was developed for fermentation monitoring. Similarly an optoelectronic sensor employing penicillinase to generate a measurable pH change when reacted with penicillin was used. Different biosensors were used for on-line analysis of this antibiotic. Antibody-coated piezoelectric crystals have been used for continuous gasphase analyses. A flow-through genosensor using DNA fragments on silicone was reported. Immobilized acetylcholinesterase has been used to detect organophosphorus and carbamate insecticides. An optical biosensor for lysine based on lysine decarboxylase and an optical transducer as well as optical biosensors for biological oxygen demand studies have been described. When lysine reacts with lysine decarboxylase, cadaverine is produced. When the cadaverine ion transports into a membrane it is coupled with the transport of a proton (of an indicator dye) out of the membrane; thus, a measurable spectral change occurs in the indicator dye and is detected optically in the flow-through system.

The development of the BIAcore instrument, with its system for the flow of analyte samples, presents an excellent example of an adaptable and sensitive on-line biosensor. It is easy to use and regenerates a powerful biospecific interaction analysis involving antibodies and antigens; it is adaptable for the identification of specific proteins in mixtures. If a heating unit could be inserted just in front of an FIA injecting unit to produce single-stranded RNA, DNA or DNA (or RNA) fragments, then PCR or other similar probes could be used to detect and identify quickly and accurately target microorganisms using the BIAcore instrumentation.

A miniaturized enzyme column integrated with a micro-electrochemical biosensing agent attached to a flow cell for the flow injection detection of glucose has been described. Glucose oxidase was the enzyme. The amperometric micro-electrode was fabricated and integrated on a silicone wafer by various micromachining techniques. The biosensor was then connected to a conventional FIA system and demonstrated a linear relationship between current and glucose concentration in the range of 1–25 mmol l^{-1}. Various sensors for environmental analyses, including water quality and heavy metal screening, have been reported. Glucose determinations, amperometric immunosensors and microbial sensors for water pollutants have been discussed. A nitrate biosensor was used to determine the structure and function of a nitrifying biofilm in a trickling filter of an aquaculture water recirculation system. In a study of bioprocess monitoring, either permeabilized cells of *Zymomonas mobilis* or oxido-reductase enzymes isolated from this organism were used as a model for a fibreoptic-fluoro biosensor that could be integrated into an FIA system. NADP(H) was oxidized or reduced during the enzymatic reactions depending on the substrate and operating conditions. Fluorescence intensity of the NADP(H) was measured at 450 nm after excitation at 360 nm.

An amperometric biosensor was constructed to determine the concentration of galactose in on-line yeast fermentation broths using galactose oxidase. More than 900 samples were measured in 6 weeks.

A correlation coefficient of 0.991 compared to the standard UV method was found. A potentiometric flow-through system was used to detect organophosphorus pesticides employing recombinant *E. coli*. The response time was 20 min. A flow-through system containing a piezoelectric immunosensor was used to determine atrazine concentrations. The measurement of biological parameters during fermentation processing were discussed, as was the role of biosensors in the monitoring and control of manufacturing processes related to food analyses. A flow-through biosensor, the origen (Igen International, Gaithersburg, MD) was used to detect *E. coli* O157 and *Salmonella typhimurium* added to foods and fomites. This instrument combines immunomagnetic separation with electrochemiluminescence. It was possible to detect 100–1000 per millilitre of the bacteria.

A new type of biosensor was developed using an optical flow cell and detection of H_2O_2 production. The peroxidase was immobilized on an upper layer of controlled pore glass in the 1 mm path of the flow cell. The bottom layer contained an ion-exchange resin where the reaction product was temporarily retained. In this biosensor, the hydrogen peroxide produced reacted with 4-amino phenazone and absorbance was measured. The flow cell could be placed in a spectrophotometer. It could also be connected to an FIA system. A bolus of NaCl flushed the coloured product from the flow cell and regenerated the system for next sample injection. An FIA biosensor system was developed to determine aspartame. The aspartame was first cleaved to phenylalanine by pronase and then an L-amino acid oxidase, immobilized on an amperometric transducer, reacted with the phenylalanine. Each assay took 4 min. When dietary food products were tested for aspartame concentrations, the results agreed quite well with the manufacturer's data.

A microprocessor-controlled FIA system was utilized to analyse the ascorbic acid content in a range of foodstuffs. This system was based on an amperometric detection of a wall-jet electrode coupled with an ascorbate oxidase-packed bed. A robotic auto-sampler-diluter further automated the system. Concentrations of 1–200 mg ml^{-1} gave a linear response in 4 min with a correlation of 0.98 when compared to existing methods.

As mentioned above, thermal transducers have been used in FIA systems. The Microscal microflow calorimeter marketed by Gilson could easily be adapted for FIA analysis. Surface plasmon resonance instrumentation was used with different flow cells on a single biosensor which was interfaced with a matrix-assisted laser desorption/ionization (MALDI) time-of-flight mass spectrometer looking at an immunoassay for myotoxin a peptide.

Caveats

Thus far, a rosy picture has been painted concerning the utility and value of biosensor technology. But as with a rose, there are thorns which one must recognize in order to avoid a major or minor (scientific) bleed. Some of the main problems are summarized in the following points.

Stability

Instability is inherent, to different degrees, in all biological molecules. Many authors report sensor stabilities ranging from a few days or months to (rarely) a year. Antibodies in particular appear to lose activity after constant regeneration. Enzymes appear to be somewhat more stable depending on the buffering capacity or pH of the system. DNA molecules, DNA fragments and peptides appear to be fairly stable. Disposable sensors would solve this problem.

Shelf Life

Refrigeration and even freezing of sensors when not in use have contributed to their longevity. If sensors can be lyophilized on transducers by companies which manufacture biosensors, their shelf life would be greatly enhanced since they would not become activated until used.

Regeneration of Activity

Once a test has been completed, the sensor must be regenerated or reactivated for the next sample. In the case of antibodies, the antigen must be uncoupled. With enzymes, the reactive products must be removed. Piezoelectric surface mass must be restored. Interesting solutions have been reported, ranging from magnetic particles to the use of staphylococcal protein A which couples to the FC end of an antibody. The antibody can be easily uncoupled from this protein by a brief acid rinse. Plug-in duplicate modules (containing sensors, transducers, or both) could be used if regeneration times constitute a serious bottleneck. Flow injection and on-line analyses are dependent upon a quick purge of the system. The BIAcore instrument has appeared to solve many of these problems.

Acceptance of Techniques by Regulatory Agencies and Potential Users

As mentioned throughout this chapter, biosensor technology must be compared to existing, accepted methodologies (the gold standard). So far, the reported results have had high coefficients of agreement. In many cases, the range, speed and accuracy of biosensors have been superior to other commonly used methods. That sufficient data must be accumulated to prove the efficacy of these biosensor procedures almost goes without saying.

The National Institute of Standards and Technology in Gaithersburg, MD, in conjunction with more than six companies, has become involved in a consortium to study and eliminate obstacles to biosensor acceptance and use.

Assessment of Future Use of Biosensors

All indications point toward biosensors as the great wave of the future methodology in many diverse disciplines, including the rapid detection and characterization of microorganisms and their reactions. With the aid of molecular biologists, geneticists and other scientists and engineers, new and ingenious sensors and transducer methodologies will be created. Biosensors will eventually greatly shorten the time needed to detect the presence of food-related pathogens, toxins and environmental pollutants. They will provide real-time kinetics in the study of immunological and nucleic acid interactions and have already done so on a limited basis. On-line biosensors will improve present capabilities for rapid monitoring of fermentations, food-processing and pharmaceutical procedures. In short, biosensors will eventually change the manner in which we perform our rapid methodologies. The fun and attendant creative exhilaration has only just begun. Prepare for the explosion which is certain to come.

See also: **Enzyme Immunoassays**: Overview. **Nucleic Acid-based Assays**: Overview.

Further Reading

Billard P and DuBow MS (1998) Bioluminescence-based assays for detection of bacteria and chemicals in clinical laboratories. *Clinical Biochemistry* 31(1): 1–14.

Goldschmidt MC (1993) Biosensors: Blessing or bane. *Rapid Methods and Automation in Microbiology* 2: 9–15.

Hall EAH (1991) *Biosensors*. Prentice Hall Advanced Reference Series, Engineering: New Jersey.

Kaufmann JM (1992) Enzyme electrode biosensors: theory and applications. *Bioanalytical Applications of Enzymes* 36: 63–113.

Luong JH et al (1997) Developments and applications of biosensors in food analysis. *Trends in Biotechnology* 15(9): 367–377.

Lowe CR (1999) Chemoselective biosensors. *Current Opinion in Chemical Biology* 3(1): 106–111.

Malmqvist M (1999) BIAcore: an affinity biosensor system

for characterization of biomolecular interactions. *Biochemical Society Transactions* 27(2): 335–40.

Mulchandani A and Bassi AS (1995) Principles and applications for bioprocess monitoring and control. *Critical Reviews in Biotechnology* 15(2): 105–124.

Van Regenmortel MH et al (1998) Measurement of antigen–antibody interactions with biosensors. *Molecular Recognition* 11(1–6): 163–167.

Wang J (1998) DNA biosensors based on peptide nucleic acid (PNA) recognition layers – A review. *Biosensors and bioelectronics* 13(7–8): 757–762.

Bio-Yoghurt *see* **Fermented Milks**: Yoghurt.

BOTRYTIS

M D Alur, Food Technology Division, Bhabha Atomic Research Centre, Mumbai, India

This article briefly describes the characteristics of the genus *Botrytis* and its related species. Methods of enumeration and detection in food materials are given. The importance of this genus in relation to spoilage of fruits and vegetables as well as its usefulness in the preparation of wine of superior quality are discussed.

Characteristics of the Genus and Species

The genus *Botrytis* belongs to the class Fungi Imperfecti (Deuteromycetes) which contains all those fungi that do not have a sexual (perfect) stage. The conidial stages of these fungi are similar to those of Ascomycetes. The genus *Botrytis* belongs to Moniliaceae, the largest form-family of the class Fungi Imperfecti. The family includes all imperfect fungi which produce conidia on unorganized hyaline conidiophores or directly on the somatic hyphae. Most species are saprobic, but many are well-known plant pathogens and others are human pathogens. In *Botrytis*, large oval or spherical conidia are produced at the tips of erect conidiophores which are simple or branched. The conidia are not in chains but in head-like formations and are attached singly on sterigmata. They may be hyaline or, in some species, brightly coloured. Many species are known to have teleomorphs in the genus *Botryotinia*. The organisms include many plant pathogenic species, e.g. *B. allii*, *B. cinerea* and *B. fabae*. The classification of Fungi Imperfecti is based on the secondary fruiting forms and other external traits and serves exclusively the practical aims of naming and identification.

Growth Characteristics

Fungi require free oxygen for growth. Most fungi can grow over a wide range of pH (2–8.5) but some are favoured by an acid pH. Fungi in general can utilize foods ranging from simple to complex substrates. Most fungi possess hydrolytic enzymes such as amylases, pectinase, proteinase and lipase.

The growth of fungi is slow compared to that of bacteria and yeast. Most fungi are mesophilic, growing between 25 and 31°C. In general, most fungi require less available moisture than yeasts and bacteria. The minimum water activity (a_w) for spore germination has been found to be as low as 0.62 for some fungi and as high as 0.93 for others, e.g. *Mucor*, *Rhizopus* and *Botrytis*.

Sorbic acid, propionates and acetates specifically act as fungicides in nature.

Cultural Characteristics

The gross appearance of a mould growing on food is enough to indicate its genus. Definite zones of growth distinguish some moulds, e.g. *Aspergillus niger* pigments red in the mycelium; purple, yellow, brown and grey-black are characteristics, as are the pigments of masses of asexual spores; green, blue-green, yellow, orange, pink, lavender, brown, grey, black etc. These colours are characteristics of masses of asexual spores of *A. niger*. The reverse side of a mould on an agar plate appears opalescent blue-black or greenish black in colour.

Distinguishing Characteristics of the Genus *Botrytis*

1. Woolly pale, dirty brown septate mycelium
2. Fairly long, stiff conidiophores which are branched irregularly at the end and bear conidia in grape-like bunches
3. Small, ovate conidia
4. Dirty-green sclerotia

Morphological distinguishing characteristics of some species of the genus *Botrytis* are given in **Table 1**.

Methods of Enumeration and Detection

Fungi are frequently encountered actively growing contaminants in and on various commodities including foods, inadequately cleaned food-processing equipment and food storage facilities.

Moulds can initiate growth over a wide pH range, from below pH 2 to above pH 9. The temperature range for growth of most moulds is likewise broad (5–35°C). Many food-borne moulds can grow in foods with a_w of 0.85 or below. Most fungi of importance in foods are aerobes.

Traditionally, acidified media have been used to enumerate moulds in foods. Such media are now recognized as inferior to antibiotic-supplemented media that have been formulated to suppress bacterial colony development. However, antibiotic-supplemented plate count agar or dichloran rose bengal agar are recommended as general-purpose media for enumerating mould by the dilution plating procedure. The antibiotics most frequently employed in media are chlorotetracycline (Aureomycin), chloramphenicol, oxytetracycline, gentamicin and kanamycin ($100 \mu g ml^{-1}$). Inoculated plates should be incubated in an upright position at 22–25°C for 5 days before colonies are counted.

Acidified Media

Potato dextrose agar (PDA) acidified with sterile 10% tartaric acid to a pH of 3.5 is the medium of choice. The acidification of media makes it selective for fungi. The medium is acidified after sterilization at temperature of 47–50°C and 15–20 ml is poured into Petri dishes (9 cm diameter) and then dried overnight at room temperature.

Preferred Antibiotic Method

Plate count agar or DRBC (Dichloran rose Bengal chloramphenicol) agar containing $100 \mu g ml^{-1}$ chloramphenicol is recommended. The chloramphenicol can be added to the medium before sterilization. Petri plates 9 cm in diameter are filled with 15–20 ml of the medium and then dried overnight at room temperature (21–25°C). Store DRBC plates in the dark to avoid photogeneration of inhibitors from the rose bengal dye. Inoculate 0.1 ml of appropriate decimal dilutions in duplicate on the solidified agar and spread over the entire surface using a sterile bent rod. Incubate plates for 5 days at 22–25°C. Count plates containing 30–150 colonies. Report counts as colony-forming units per gram of sample.

Staining Method

The use of fluorescent materials on the stage has made everyone familiar with their macroscopic possibilities as differentiating agents. Because it is visible to the naked eye at very low dilutions, fluorescein is useful for tracing the source of contamination of water

Table 1 Morphological distinguishing characteristics of some species of the genus *Botrytis*

Characters	Species of the genus Botrytis			
	B. pyramidalis	B. bassiana	B. terrestris	B. cinerea
Turf	White	White	Grey	Grey
Colony appearance	White	White	White to grey	Grey-green to brown-black
Conidiophores	Long septate, dichotomously branched	Erect, white, unbranched; 2–3 μm in diameter	Erect, ascending septate, branched 2–3.5 μm in diameter; 50–200 μm high	Erect, unbranched, septate, 11–23 μm thick towards the tips
Conidia	Ovoid, 2–3 μm. The tips of the branchlet are swollen and on the vesicle occur 3–6 short branches; swollen, club-shaped on their tips carrying phialides. Conidia single on the phialides, long, egg-shaped, round at the tips of the baseline. Papilla 5–7 μm wide	Globose on lateral phialides: 2–3 μm on the sides of the conidiophores in heads 2–3 μm in diameter	Small 2.5–3 × 3–4 μm. Uniform 2.5–3 × 3–4 μm hyaline to light grey. Clusters of conidia separate easily. Conidia produced on the ends of the branches form compact triangular clusters	Large 9–12 × 6.5–10 μm; hyaline with several projections at the tip. Conidia ovoid or elliptical to globose; apiculate at the base 9–12 × 4.5–10 μm

supplies. In mycology, acridine orange has been used to distinguish dead and living materials of fungi. A combination of this material with potassium hydroxide has been valuable in the microscopic diagnosis of fungal infections of skin and hair using UV microscopy. A visible fluorescing reaction product will be seen as a positive reaction, as in the direct test.

When viewed under UV light the indirect reaction will make a spore appear larger since two coats of antibodies have been applied. The use of fluorescent antibody techniques for the identification of fungi has been reported for the genus *Botrytis* and other genera.

The direct epifluorescence technique (DEFT) could be employed by producing an antibody 'cocktail' from *B. cinerea* and other moulds such as *Alternaria alternata*, *Fusarium solani*, *Rhizopus stolonifer* and *Mucor pyriformis*. Samples of tomato paste were mixed with the antibodies plus fluorescent dye and then examined by fluorescent microscopy. The fluorescent characteristic of mould-contaminated commodities has also been used to detect the presence of mould optically by measuring fluorescence at 442 and 607 nm.

Immunological Method

Recently, research efforts have shifted to the detection of moulds in foods using immunological methods, especially the enzyme-linked immunosorbent assay (ELISA). This method is based on the production of antibodies to moulds and the use of these antibodies to bind to mould antigens present in foods. Antibodies to *Alternaria*, *Aspergillus*, *Botrytis*, *Cladosporium*, *Fusarium*, *Geotrichum*, *Mucor*, *Penicillium* and *Rhizopus* have been produced and used to detect these moulds in bread, cereals, cottage cheese, fruits, nuts, spices and other foods. ELISA could detect *Alternaria*, *Cladosporium*, *Fusarium* and *Geotrichum*, to the genus level. Moulds and related families such as *Aspergillus* and *Penicillium* or *Mucor* and *Rhizopus* cross-reacted since they share common antigens. When a mould mixture of *Alternaria alternata*, *B. cinerea*, *F. solani*, *M. piriformis* and *R. stolonifer* was used to produce antibodies, about 50% cross-reactivities occurred between all genera. There was no cross-reactivity with yeast. The mould antigens were heat-stable and water-soluble. ELISA corrrelated with the amount of mould added to food at detection levels as low as 100 μg to ng per gram of food. It was found that moulds with high aflatoxin B, patulin and penicillinic acid contents had high levels of antigens to the *Penicillium*/*Aspergillus* ELISA. The immunodominant fraction of the mould antisera is being determined so that rapid test kits can be developed. A 10–20 min latex agglutination test for the detection of antigens to *Penicillium* and *Aspergillus* sp. was used to determine if these moulds were in nuts and spices. Research is continuing on the immunological detection of moulds in foods since the method has promise for the development of rapid methods.

Fluorescence Microscopy

Fluorescence microscopy has primarily been used for the enumeration of single cells and cell clumps. However, procedures have been developed in which microorganisms in diluted food homogenate are collected on membrane filters, stain beads, fluorescent dyes (acridine orange or aniline blue) and then counted with a microscope equipped with an epifluorescent illumination (DEFT). The transfer of membrane to agar media following filtration permits microcolonies to be counted. Advantages of the DEFT method are that it is rapid, it may permit the differentiation of live from dead cells and it permits the detection of low numbers of organisms.

For rapid viable counts on selective media, homogenize 10 g food in 90 ml 0.1% peptone water; filter through 5 μm millipore filter. Filter through 0.4 μm Nucleopore polycarbonate membranes. Transfer the membrane top-side-up to the desired agar media, incubate media for 3–6 h at 30°C. Overlay the membrane with 2 ml of stain for 2 min before applying vacuum. The stain consisted of 0.025% acridine orange and 0.025% Tinopal AN in 0.1 mol l^{-1} citrate-NaOH buffer, at pH 6.6. Rinse the membrane under vacuum with 2.5 ml 0.1 mol l^{-1} citrate-NaOH buffer, pH 3.0 and 2.5 ml 95% ethanol.

Air-dry and then mount on a slide for examination under a fluorescent microscope using non-fluorescent immersion oil. Count orange fluorescent cell clumps and single cells that are present in a number of random fields. Calculate the count in the original sample by multiplying the average count per field by the number of fields in a 25 mm membrane and then dividing by the millilitres of sample that were filtered.

Method of Enumeration of *Botrytis cinerea*

Th medium outlined in **Table 2** is a selective medium to distinguish between *B. allii* and *B. cinerea* on onion. The former species is restricted by sorbose.

Importance to the Food Industry

Physiological injuries caused by environmental stress can predispose fruits and vegetables to post-harvest diseases. For example, exposure of tropical produce to low temperatures induces chilling injury and enhanced microbial rotting, e.g. *Botrytis* in tomato.

Table 2 Method of enumeration of *Botrytis cinerea*

Components	Concentration
Potassium chloride (KCl)	1.0 g
Potassium dihydrogen phosphate (KH$_2$PO$_4$)	1.5 g
Sodium nitrate (NaNO$_3$)	3.0 g
Magnesium sulphate (MgSO$_4$)	0.5 g
Casein hydrolysate	5.0 g
Yeast extract	3.0 g
Glycerol	5.0 g
L-Sorbose	2.5 g
Agar	20 g
Tap water	1 l

Chilling injury is known to increase the susceptibility of the surface of tomatoes to infection by *B. cinerea*.

Several types of rot caused by the species of the genus *Botrytis* are described here.

Neck Rot

Neck rot is caused by *B. aclada* (*B. allii*) which is more common in stored onions.

Noble Rot

Noble rot is a rot of grapes caused by *B. cinerea*. On ripe or over-ripe white grapes the mould penetrates the grape skins and leads to loss of water. Must from such grapes is more concentrated than normal and is used for making high-quality sweet wines. The same mould on unripe black grapes renders it useless for wine-making.

Chocolate Spot

Chocolate spot is a disease of beans (particularly broad beans, *Vicia faba*) caused by *B. fabae*. Rounded chocolate-brown spots develop on all parts of the plant. In severe cases, the entire plant may be blackened and killed. Infection is usually initiated by conidia.

Soft Rot (of Fruits and Vegetables)

This a type of rot caused by pectinolytic organisms in which fruits or vegetables degenerate into a wet slimy mass. In acidic fruits (e.g. tomatoes, apples) soft rot is caused by pectinolytic fungi (e.g. species of *Botrytis*).

Grey Mould (Botrytis)

Grey mould is a disease caused by *B. cinerea* which can affect a wide range of plants as well as stored fruits and vegetables (e.g. beans, peas, raspberries and strawberries). The disease is encouraged by cool damp conditions and is characterized by a grey, fluffy, surface mould overlaying a soft, commonly brown rot. *B. cinerea* frequently rots pears in storage and may colonize stems. Stem infections may grow into the fruit and completely colonize it.

Blight of Tulips

B. tulipae causes a serious blight of tulips while *B. peoniae* causes great damage to peonies.

Advantages of *Botrytis cinerea* in the Food Industry

Some European sweet wines undergo microbial transformation. Before picking, the grapes become spontaneously infected with a fungus, *B. cinerea*. This infection causes water loss (thus increasing sugar content) and destruction of malic acid (thus decreasing the acidity of the grapes). Certain favourable changes in flavour and colour occur. The must from infected grapes is fermented by glucophilic yeasts, yielding a sweet dessert-type wine.

French dry sherry is of interest because it is made from grapes which have a high sugar content as a result of being dried out by an infecting grey mould, *B. cinerea*. They therefore, yield a wine with high alcohol content.

The production of some of the most prestigious sweet white wines is based on the use of overripe grapes and the development of noble rot. The development of *B. cinerea* on ripe grapes is favoured under special climatic conditions when periods of high humidity and sunshine alternate. According to the degree of infection one distinguishes the berries, filled with juice with a brownish-violet skin colour, and the berries which are wrinkled, dehydrated and covered with a mycelium and conidophores. Harvest is carried out with several selections. One collects only those berries or bunches of grapes which have reached the threshold of the rot stage. The grape must contain 260–400 g l^{-1} sugar, be high in glycerol concentrations and be deacidified by a decrease in tartaric acid.

Citric acid, a product of primary metabolism, is also produced in substantial amounts by *B. cinerea*.

Control of *Botrytis*

Antifungal agents used against *Botrytis* infections include Benomyl, dicarboximides, dichlofluanid, dicloran, thiram and zineb.

Irradiation with gamma rays was first tested as a means of controlling *B. cinerea* in order to prevent spoilage of strawberries. This method extended the period for which strawberries might be displayed on the greengrocer's shelf. Summer-grown strawberries appear to be able to cope with the doses of 2–3 kGy that are necessary to kill the spores of the fungi causing rotting.

See also: **Enzyme Immunoassays**: Overview. **Fungi**: Classification of the Deuteromycetes. **Microscopy**:

Transmission Electron Microscopy. **Spoilage Problems**: Problems caused by Fungi. **Total Viable Counts**: Microscopy. **Wines**: Microbiology of Wine-making.

Further Reading

Booth C (ed.) (1971) *Methods in Microbiology*. London: Academic Press.

Coley-Smith JR, Verhoeff K and Jawid WR (eds) (1980) *The Biology of* Botrytis. London: Academic Press.

Frazier WC (1958) *Food Microbiology*. New York: McGraw-Hill.

Gilman JC (1959) *A Manual of Soil Fungi*, 2nd edn. Iowa, US: Iowa State University Press.

Murray DR (1990) *Biology and Food Irradiation*. John Wiley & Sons Inc., New York.

Rehm HJ and Reed R (eds) (1983) *Biotechnology*. Vol. 5. *Food and Feed Production with Microorganisms*. Florida: Verlag Chemie.

Stanier RV, Ingraham JL, Wheelis ML and Painter PR (1993) *General Microbiology*, 5th edn. London: Macmillan.

Van der Zant C and Splittstoesser DF (eds) (1992) *Compendium of Methods for the Microbiological Examination of Foods*. Washington, DC: American Public Health Association.

BOVINE SPONGIFORM ENCEPHALOPATHY (BSE)

D M Taylor and **R A Somerville**, Neuropathogenesis Unit, Institute for Animal Health, Edinburgh, UK

Introduction

Bovine spongiform encephalopathy (BSE) belongs to a group of transmissible degenerative encephalopathies (TDE) listed in **Table 1**. These fatal diseases are caused by unconventional but uncharacterized transmissible agents which have many unusual properties, including a relatively high degree of resistance to inactivation. A principal feature of the TDE is that PrP protein becomes conformationally (but not chemically) modified as a consequence of infection. This protein can be found in a variety of tissues but is present at the highest levels within the central nervous system (CNS). The disease-specific form of the protein (designated PrP^{Sc}) resists normal degradation in the host, and accumulates as pathological deposits within the CNS; this is usually accompanied by vacuolar lesions in neurons (**Fig. 1**), which is why these diseases are often described as spongiform encephalopathies. The PrP^{Sc} protein appears to be associated specifically with TDE but there is a continuing debate as to whether it is (a) the infectious agent per se as suggested by the protein-only ('prion') hypothesis, (b) a component of the agent as proposed by the 'virino' hypothesis, or (c) simply a pathological product of infection.

BSE in Britain

Bovine spongiform encephalopthy was observed initially in only a few English cattle in 1986 but later became a major epidemic in Britain which peaked in 1992; it had affected more than 172 000 cattle by July

Table 1 Transmissible degenerative encephalopathies

Disease	Affected species
Scrapie	Sheep, goats, moufflon
Transmissible mink encephalopathy	Mink
Chronic wasting disease	Elk, mule-deer
Bovine spongiform encephalopathy	Cattle, captive exotic ruminants
Feline spongiform encephalopathy	Domestic cats, captive exotic felids
Creutzfeldt–Jakob disease	Humans
New variant Creutzfeldt–Jakob disease	Humans
Gerstmann–Sträussler–Scheinker disease	Humans
Kuru	Humans
Fatal familial insomnia	Humans

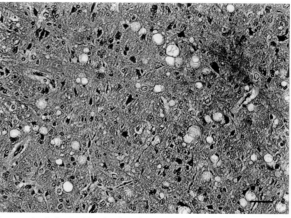

Figure 1 Infected brain showing spongiform encephalopathy.

1998. Epidemiological studies in Britain demonstrated a probable association between feeding calves with diets containing meat and bone meal (MBM), and their later tendency to develop BSE. Consequently, a ban on feeding ruminant-derived protein to ruminants was introduced in 1988 in Britain, and resulted in a downturn in the incidence of BSE from 1993 onwards. The delayed effect was a simply a reflection of the average incubation period for BSE which is around 5 years, and supported the hypothesis that MBM had been the source of the infectious agent. The suspect MBM had been manufactured by the rendering industry from animal tissues obtained mainly from abattoirs, and would have included sheep tissues infected with the transmissible agent that causes the sheep disease, scrapie, which is endemic in Britain. The only detectable lesions in BSE-affected cattle are confined to the central nervous system (CNS), the principal lesion being neuronal vacuolation which is similar to that observed in the brains of sheep with scrapie. Although this tends to support the idea that BSE was caused by the presence of scrapie agent in MBM, the existence of a previously unrecognized scrapie-like disease of bovines cannot be formally excluded. The hypothesis that infected MBM was the source of the BSE problem was supported further by studies which showed that BSE and scrapie agents could survive processes used within the European Union to manufacture MBM. Despite the British ban in 1988 on feeding ruminant-derived proteins to ruminants, by August 1987 BSE had occurred in more than 33 000 cattle born after the feed ban. The majority of cases occurred in cattle born shortly after the ban was imposed, because there had been no attempt to remove and destroy any ruminant-derived MBM already in existence at the time of the ban. However, a significant number of cases were detected in cattle born well beyond the time of the feed ban, and it was subsequently observed that the legally required exclusion of potentially BSE-infected specified bovine offals (as described later) from the animal food chain had not been observed conscientiously. Epidemiological evidence shows that BSE infectivity probably cross contaminated cattle diets produced in feed mills that were processing ruminant-derived proteins to be fed to pigs and poultry. One study has suggested that an additional factor that may have contributed to the emergence of BSE in cattle born long after the feed ban is the low incidence of BSE in cattle born to BSE-infected dams. This particular study did not provide any direct evidence of infection being passed from mother to offspring, and the question of inheritance of genetic susceptibility was raised. However, other studies have indicated that there are no particular genotypes of

cattle that have an enhanced susceptibility to BSE. Nevertheless, even if true maternal transmission did occur at the reported level, this would only delay, but not prevent, eradication of BSE.

BSE-related Diseases

In Animals

Since the emergence of BSE in Britain, a previously unrecorded TDE, described as feline spongiform encephalopathy (FSE), has been observed in British domestic cats. By August 1998 the total number reported was 85 in Great Britain, and one case in Northern Ireland. There have also been single cases in Liechtenstein and Norway. Within the same time frame, novel BSE-like diseases have been observed in captive exotic felids and ruminants that were born in Britain, and were fed bovine carcasses and MBM, respectively (**Table 2**).

In contrast to the variety of strains of agent that can be recovered from sheep with scrapie, BSE appears to be caused by a single strain. This conclusion is based upon the consistent pattern of incubation periods in a panel of five strains of inbred mice injected with BSE-infected cattle brains obtained from varied locations at different times throughout the epidemic. The patterns of severity of spongiform encephalopathy in different areas of the brains of these mice are also extremely consistent. Samples from FSE-infected cats, and from two of the affected captive ruminant species (kudu and nyala) have produced identical results, which indicates that these diseases were all caused by the BSE agent, since no strain of agent with the same properties has ever been recovered from scrapie-infected sheep. Furthermore, the BSE agent has been found to retain this characteristic strain type in mice after experimental passage through goats, pigs or sheep. Sheep are susceptible to BSE by experimental oral challenge, and the ensuing clinical and neurohistopathological features are indistinguishable from those of scrapie. Because some British sheep were fed ruminant-derived proteins until 1988, it is possible that BSE has been masquerading

Table 2 British-born exotic species affected by BSE

Ungulates	Felids
Ankole	Cheetah
Arabian oryx	Ocelot
Bison	Puma
Eland	Tiger
Gemsbok	
Kudu[a]	
Nyala[a]	
Scimitar-horned oryx	

[a] Species from which transmission studies have been carried out.

as scrapie in sheep. In contrast to cattle with BSE, the spleen (and probably other tissues) becomes infected in sheep with BSE. If the placenta becomes infected, this could provide a mechanism whereby BSE could pass from generation to generation in sheep by perinatal contact with this infected tissue after birth. Such a mechanism is considered to account for the perpetuation of scrapie in sheep, in which the placenta has been shown to be infected.

In Humans

Although there is no evidence that the scrapie agent has ever infected humans, it now appears that the BSE agent probably has. By August 1998 a total of 27 cases of a new variant form of Creutzfeldt–Jakob disease (CJD) had been observed in Britain but not elsewhere, apart from one case in France. These cases are distinguishable from classical CJD because of (a) a much younger age distribution, (b) different clinical symptoms, and (c) different pathological lesions in the brain. In the absence of any other apparent explanation, it was considered that these cases might be attributable to the incorporation of BSE-infected CNS tissue in human foodstuff until this was prohibited in 1989 in Britain. Transmission studies in mice have now shown that the strain characteristics of the agent that causes the new form of the human disease are exactly the same as those of BSE agent, and are unlike those of any other agent characterized to date.

Protection of Human and Animal Health

In 1988, BSE was made a notifiable disease in Britain. Cattle suspected of having the disease are required to be slaughtered, and the carcasses destroyed. Incineration is the routine method of destruction, although landfill was used to some extent in the earlier days when there was insufficient capacity to incinerate all such carcasses.

A ban on feeding ruminant-derived proteins to ruminants was introduced in Britain in 1988, and this was followed in 1994 by a European Community ban on feeding mammalian-derived proteins to ruminants. As a result of studies showing the ineffectiveness of many rendering procedures for inactivating BSE and scrapie agents, the rules for rendering within the EC were changed in April 1997. The only procedure now permitted for producing MBM for animal consumption is a process that involves exposure of the raw materials to steam under pressure at 133°C for 20 min. As a consequence of the occurrence of the new variant form of CJD in humans, and its potential association with BSE, the incorporation of meat and bone meal into the diets of any species of farmed animal has been prohibited in Britain since 1996.

Although epidemiological studies have failed to demonstrate any enhanced risk of humans developing CJD through dietary or occupational exposure to scrapie agent, it was considered prudent to exclude potentially BSE-infected tissues from human and animal foodstuff. These regulations were introduced in 1989 and 1990 respectively, and the selected tissues were designated as specified bovine offals (SBO). These represented the tissues that were likely to contain the highest levels of infectivity, based upon what was known about scrapie in sheep: namely brain, spinal cord, spleen, tonsil, thymus and intestine. However, it was shown later that the only tissues containing detectable infectivity in cattle with naturally acquired BSE are brain, spinal cord and retina. This suggests that the pathogenesis of BSE is different from scrapie. At the end of 1995 a ban was introduced in Britain on the use of meat recovered mechanically from bovine vertebral columns as human food. In 1996, regulations were introduced which require that human food derived from bovines will come only from cattle under the age of 30 months, at which time there is likely to be very little infectivity in BSE-infected cattle.

In view of the possibility that sheep might have become infected with the BSE agent, and that the disease could be clinically and neurohisto-pathologically indistinguishable from scrapie, it is now a statutory requirement in the UK that sheep heads, spinal cords and spleens are not incorporated into animal or human foodstuff.

BSE in Europe

The disease has occurred in a number of cattle born in other European countries. The number of cases identified in the indigenous cattle populations of these countries by July 1998 are shown in **Table 3**. It seems likely that some, if not all, of these cases have arisen through feeding MBM exported from Britain, before such exportation was prohibited. The international marketing arrangements for trading MBM make it impossible to determine the ultimate destination of all MBM exported from Britain. Nevertheless, the

Table 3 BSE in indigenous cattle populations by July 1998

Country	No. of cases
England, Scotland and Wales	172 468
Northern Ireland	1 770
Belgium	4
France	38
Luxembourg	1
Portugal	133
Republic of Ireland	298
The Netherlands	2
Switzerland	274

strain of the agent responsible for the Swiss BSE epidemic has been found to be identical to that of the unique strain associated with the British epidemic. Considering that the British rendering industry was particularly successful in exporting MBM to Europe after the 1988 British ban on feeding ruminant-derived proteins to ruminants, it is entirely possible that any European country that imported MBM during this period could have received MBM of British origin. The key point is whether that MBM was used to feed cattle. In addition to the cases that have arisen through the importation of British MBM, there is the question of the cases that should have been observed from the exportation of adult cattle from Britain to Europe for breeding purposes between 1985 and 1990. The number of cases of BSE recorded in mainland Europe is well below the number that can be calculated to have been developing the disease by the time they were exported, quite apart from indigenous cattle that acquired their disease through the consumption of meat and bone meal. This has never been explained satisfactorily.

Diagnosis

At present there is no fully validated, practical, preclinical diagnostic test for BSE or other TDEs. There is no classical immune response or other host reaction to disease and normal serological tests are therefore not applicable. Initial diagnosis of BSE depends on the observation of clinical signs of disease, which include incoordination, increased fear, increased startle response and decreased rumination. Postmortem confirmation of diagnosis of TDEs traditionally relies on histopathological examination of the brain where vacuolation (spongiform change), neuronal loss and a reactive astrocytosis can be observed to differing degrees. In BSE, vacuolation in the mesencephalon, medulla and pons is particularly prominent. Examination of the medulla has been found to be a reliable means of confirming diagnosis. A significant number of clinically suspect cattle are not confirmed as BSE-positive cases. At the height of the UK epidemic about 10% of cases were found not to be BSE on neuropathological examination, but this ratio has risen to around 20% as the epidemic has waned.

Since the association of abnormal, protease-resistant forms of the protein PrP with the TDEs was discovered, its detection has been a potentially valuable diagnostic tool. This is a host-encoded protein which is found in brain and other tissues. In TDEs it is found in an altered form (PrP^{Sc}), distinguished biochemically from the normal form (PrP^c) by its sedimentation from detergent-treated tissue extracts

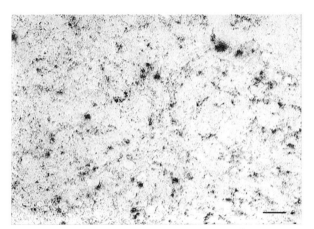

Figure 2 Immunocytochemical staining of PrP^{Sc} deposits in brain.

and its partial resistance to protease digestion. Fibrillar structures termed scrapie-associated fibrils (SAF) can also be observed by negative stain electron microscopy in the pellets of detergent extracts. Deposits of PrP^{Sc} can be observed by immunohistochemistry in infected brain (**Fig. 2**). In some TDEs, although not in BSE, amyloid plaques consisting of PrP^{Sc} can be observed microscopically in the brain.

The basis of the deposition of PrP^{Sc} is considered to be the conversion of the normal form of the protein (PrP^C) into PrP^{Sc}. Operationally, the normal form is defined by its solubility in detergents and its susceptibility to proteases, while PrP^{Sc} is defined by its sedimentation and partial proteolytic resistance. The two forms differ in their tertiary structure. The structure of PrP^{Sc} may be associated with aggregation of the protein leading to its amyloid-like deposition as fibrils and plaques.

In experimental mouse models of TDEs the sequence of deposition of PrP^{Sc} can be studied in brain, spleen and other tissues throughout the incubation period of the disease. These data show that in some cases PrP^{Sc} can be detected soon after infection, but in other cases PrP^{Sc} is detected later. In a mouse model of BSE, PrP^{Sc} was not detected in spleen until late in the incubation period, and only in some animals. When and where PrP^{Sc} was detected was controlled by the strain of TDE agent, the route of infection, and the PrP genotype of the host. These data have implications for the diagnostic potential of PrP^{Sc} detection. The PrP^{Sc} protein was found in the brains of all sheep affected by natural scrapie and cattle affected by BSE. However, although PrP^{Sc} was detected in sheep spleens, it was not detected in any cattle spleens. Furthermore, with the exception of a report of the detection of PrP^{Sc} in blood from one familial form of CJD, PrP^{Sc} has not been detected in blood or other body fluids in any TDE. However, there are recent reports of the detection of PrP^{Sc} in tonsils of sheep infected with scrapie (using immunohistochemical techniques) well before the onset of clinical disease.

At present, therefore, detection of PrP^{Sc} is valid as an adjunct or as an alternative to postmortem neuropathological diagnosis, particularly when the tissue has become autolytic. It may be detected by biochemical analysis, immunohistochemical procedures or by electron microscopy as SAF. Sensitivity of detection of PrP^{Sc} is much lower than the detection of infectivity by bioassay. Its use in preclinical diagnosis remains questionable. It may be possible in some models of the disease to use the detection of PrP^{Sc} in a preclinical diagnostic test but only when factors such as infecting agent, route of infection and host genotype are constant, and if and when PrP^{Sc} deposition in the test tissue (such as tonsil) has been characterized. However, in other models of the disease such a test may be impossible, if accessible organs or body fluids are not affected by the disease and do not accumulate PrP^{Sc}.

Tissues and body fluids from TDE-infected and uninfected animals have been compared to determine if any differences in protein composition or in metabolites can be detected which could then be used diagnostically. In cerebrospinal fluid (CSF) a protein designated 14-3-3 has been detected in elevated amounts in patients with CJD. Other brain proteins, e.g. tau, have been found in similar circumstances. The 14-3-3 protein has also been found in clinically affected animals with TDE, including animals with clinical signs of BSE. There are no data yet to show whether preclinical detection of elevated levels of this protein is possible. However, since its appearance probably arises from its release into the CSF from affected cells in the brain, it is unlikely that it will be detected much before the first clinical signs of disease are diagnosed. In humans, elevated levels of 14-3-3 are also found in a few other neurological conditions so its specificity is not complete. Nevertheless a test for 14-3-3 on biopsy samples of CSF may be of value in supporting a TDE diagnosis ante mortem.

A study of the electrochemical properties of urine showed elevated levels of some metabolites in BSE-infected compared with uninfected cattle. The value of this finding for diagnosis in preclinical animals has yet to be demonstrated, but again will be of limited use if these metabolic changes are only associated with significant neurodegeneration.

Overall there are few candidate approaches available for in vivo diagnosis. This is perhaps not surprising given the pathogenesis of the disease. Typically after peripheral infection of a TDE agent infectivity first replicates in lymphoid organs before passing to and replicating in the CNS later in the incubation period. However, in BSE-infected cattle no infectivity has been found in peripheral organs (apart from the distal ileum of experimentally infected cattle). There is no histopathological sign of infection in peripheral organs and therefore little reason to predict altered levels of metabolites or other molecular markers which might aid diagnosis. In the brain, there is pathological damage which increases progressively from the time that it first becomes infected until clinical disease becomes manifest, presumably due to the pathological lesions. Molecular or other consequences of pathological change cannot therefore usually be expected to appear until late in the infection, and close to clinical disease, as is the case so far.

At present, the diagnosis of TDEs in food-animal species is dependent upon one or more of the following procedures:

- microscopic examination of stained sections of fixed brain tissue for spongiform encephalopathy
- electron microscopy examination of negatively stained detergent extract of brain tissue for SAF
- microscopic examination of immunocytochemically stained sections of fixed brain tissue for PrP^{Sc}
- immunoblotting samples of brain tissue for PrP^{Sc}.

See color Plates 9 and 10.

See also: **National Legislation, Guidelines & Standards Governing Microbiology**: European Union.

Further Reading

Allen IV (1993) *Spongiform Encephalopathies*. Edinburgh: Churchill Livingstone.

Baker HF and Ridley RM (eds) (1996) *Prion Diseases*. Totowa: Humana Press.

Bradley R and Marchant B (eds) (1994) *Transmissible Spongiform Encephalopathies*. Working Document for the European Commission F.11.3 – JC/0003. Brussels: EC.

Bruce ME et al (1994) Transmissions to mice indicate that 'New Variant' CJD is caused by the BSE agent. *Nature* 389: 498–501.

Collee JG and Bradley R (1997) BSE: a decade on – part 1. *Lancet* 349: 636–641.

Collee JG and Bradley R (1997) BSE: a decade on – part 2. *Lancet* 349: 715–721.

Court L and Dodet B (eds) (1996) *Transmissible Subacute Spongiform Encephalopathies: Prion Diseases*. Proceedings of the Third International Symposium on Transmissible Spongiform Encephalopathies: Prion Diseases, 18–20 March 1996, Paris. Paris: Elsevier.

Spongiform Encephalopathy Advisory Committee (1994) *Transmissible Spongiform Encephalopathies: A Summary of Present Knowledge and Research*. London: HMSO.

Taylor DM (1996) Bovine spongiform encephalopathy – the beginning of the end? *British Veterinary Journal* 152: 501–518.

Taylor DM and Woodgate SL (1997) BSE: the causal role of ruminant-derived protein in cattle diets. *Revue Scientifique et Technique of the Office International des Epizooties* 16: 187–198.

Wilesmith JW et al (1988) Bovine spongiform encephalopathy: epidemiological studies. *Veterinary Record* 123: 638–644.

BREAD

Contents
Bread from Wheat Flour
Sourdough Bread

Bread from Wheat Flour

Rekha S Singhal and **Pushpa R Kulkarni**, University Department of Chemical Technology, University of Mumbai Matunga, Mumbai, India

The discovery of leavened bread is generally attributed to ancient Egyptians. Up to about 3000 BC, bread-making consisted of crushing the wheat grains between stones followed by wetting to form a dough and then further baking. With experience, it was realized that dough left undisturbed for some period of time before baking gave a better texture and improved digestibility. The importance of yeast in bread-baking was then recognized, and earlier attempts – though unsuccessful – were made with yeast and barley meal. The empirical knowledge acquired through the ages was then transmitted to the succeeding civilizations, and today the science behind the so-called art has been unveiled to a major extent. Progress continues to be made in various aspects of bread-baking, the benefit of which is ultimately passed on to mankind.

Outline of the Bread-making Process

Bread is principally produced from wheat flour, water, salt and yeast. Specialized products can also be made using numerous ingredients such as fats/oils, sugars, milk powder, eggs, honey, syrups, fruits, spices and other flavouring substances. The influence of some of these ingredients on bread quality is outlined in **Table 1**. For instance, fat mellows baked products as it spreads as a film between the protein and starch components of the flour. Fats with low melting and solidification are better in this respect than very hard fats. **Figure 1** gives an outline of the three different commercial bread-making processes. The advantages and disadvantages of each process are summarized in **Table 2**. The continuous mix process developed from the observation that oxidative and enzymatic changes during fermentation can be replaced by intensive mixing of the dough at high speed. Initially, the Do-Maker and Amflow processes were developed; these used liquid preferment. The difference in the two methods was that the liquid preferment in the Do-Maker process consisted of all the ingredients except flour and shortening, while in the Amflow process some proportion of the flour (10–50%) was contained in the liquid preferment. The Charleywood bread process, introduced by the British Baking Industries Research Association in England in 1961, substitutes mechanical and chemical development for biological maturation by intensive dough-mixing in a batch high-speed mixer for 3–5 min with controlled energy input. This process is promising for the production of bread from wheats which have low gluten strength.

The steps involved in bread-baking serve an important function as many physico-chemical changes occur at each step (**Table 3**). Each of the wheat flour constituents plays an important role in bread-making, and these are summarized in **Table 4**. Many additives

Table 1 Effect of bakery ingredients on bread quality

Ingredient	Effect on quality
Fat	Mellows the baked product by making the crumb finer and silkier; increases volume; aids retention of freshness; makes the crust soft and pliable
Salt	Makes the crumb uniform; improves slicing properties; slows down water imbibition and swelling properties; shortens the gluten; improves gas retention in the dough and resistance to extension
Egg	Aids in emulsification (due to lecithin), mellowing (due to fat) and binding; also affects browning, crumb colour, taste and nutrition value
Sugar	Enhances fermentation and browning; improves dough stability, elasticity and shortness; mellows the baked product
Water	Binds the flour constituents; important for yeast development; serves as a leavening agent due to evaporation during baking

(A)

Add all ingredients

↓

Mix to optimum development

↓ Ferment 100 min

Punch

↓ Ferment 55 min

Divide

↓ Intermediate proof 25 min

Mould and pan

↓ Proof 55 min

Bake

(B)

Mix part of flour, part of water, yeast, and
yeast food to a loose dough (not developed)

↓ Ferment 3–5 h

Dough mix   { Add other ingredients
              and mix to optimum
↓ Floor time 40 min   development

Divide

↓ Intermediate proof 20 min

Mould and pan

↓ Proof 55 min

Bake

(C)

Liquid brew (may include some flour)

Remaining flour
Other dry ingredients

↓

Dough incorporator

↓

Dough pump

↓

Developer

↓

Pan

↓ Proof 90 min

Bake

Figure 1 Outline of (**A**) straight dough-baking process; (**B**) sponge-and-dough baking process and (**C**) continuous baking process.

Table 2 Comparison of different bread-making procedures

Bread-making procedure	Comments
Straight-dough system	Gives a chewier bread; has a coarse cell structure and less flavour; product quality is sensitive to the timing between individual process steps
Sponge-and-dough procedure	Gives a soft bread; has a fine cell structure; has well-developed flavour; is tolerant to time and other conditions
Continuous bread-making procedure	Economical; requires fewer personnel and less time to produce the same amount of bread

are used in bread-making as they perform special functions. **Table 5** summarizes these additives and their functions.

Bread-processing from wheat flour essentially consists of three basic operations: mixing or dough formation, fermentation and baking.

Mixing with Dough Development

Generally low-protein flours (< 12%) require more mixing time. Mixing time is associated with the glutenin portion of the flour. Adding reducing agents such as cysteine and sodium bisulphite shortens the mixing time by breaking disulphide bonds in the glutenin proteins. This decreases the size of the proteins which in turn aid in its rapid hydration. pH alters the charge on the proteins and hence alters the mixing time. Lower pH shortens the mixing time and higher pH lengthens it. Overmixing produces a wet mixing dough due to shear thinning. Fast-acting oxidants used in bread-processing induce a rapid breakdown.

Fermentation

Incorporation of bakers' yeast into dough brings about its fermentation, which is manifested as an alteration in the dough rheology. This is clear from the fact that it is the yeast and not the products of yeast fermentation that brings about rheological changes. This can be clearly seen from the spread ratio, which is defined as the ratio of the width to the height of the dough. Dough having viscous-flow properties has a higher spread ratio, and a dough that is elastic in nature has a smaller ratio.

Baking

The biochemical activities that continue in the dough through the proof period are arrested when responsible enzymic and microbial systems are inactivated. Stabilization of an otherwise unstable colloidal system is achieved. This is due to starch gelatinization (65°C) and drastic alterations in the gluten which make the bread stronger and more elastic. During baking, the inside of the loaf does not exceed 100°C. However, the crust attains a temperature of 110–150°C, and sometimes about 150–200°C. **Figure 2** shows how the direct and indirect baking changes are affected by temperature.

Other Types of Bread

Bread can also be made from rye flour, either alone or in various combinations with wheat flour. However, unlike conventional bread made from wheat flour which uses yeast for fermentation, rye breads use sourdough starter cultures for fermentation. A portion of the sourdough from an earlier batch can

Table 3   Outline of the different stages in the bread-making process and the physico-chemical changes involved therein

Process	Physico-chemical changes
Mixing for dough development	• Causes rapid hydration of the flour particles with hydrogen or hydrophobic bonding or both • Gives a cohesive and partially elastic character to the dough resistant to extension • Permits air (50% of the possible) to be incorporated
Fermentation	• Rapid consumption of oxygen by yeast, making the process anaerobic • Major products are CO_2 and ethanol • A decrease in pH from 6.0 to 5.0 and saturation of CO_2 in the aqueous phase • Leavening of the system takes place
Punching or remixing the dough	• Subdivides the larger gas cells into many more smaller cells • Creation of newer gas cells • Brings yeast cells and fermentable sugars together
Moulding of the dough followed by proofing at 85% relative humidity for 55–60 min	• Aligns the protein fibrils to give a proper dough
Baking	• Loss of surface moisture causes formation of crust • Oven-spring or increase in dough size takes place due to increase in the volume of the gases on heating, increase in the activity of the yeast followed by its destruction • Vaporization of the other materials, e.g. water–ethanol mixture • Browning of the crust due to Maillard reaction and caramelization • Formation of a wide variety of carbonyl substances which endow the baked product with a desirable organoleptic profile

Table 4   Role of wheat flour constituents in bread quality

Constituent	Impact on quality
Starch	Furnishes fermentable sugars; makes itself amenable to a strong union with gluten adhesive; absorbs water from gluten by gelatinization, causing the film to set and become rigid; affects loaf volume and crumb structure due to the coherence attributed to the partially gelatinized state; dilutes the gluten to an optimum level; provides a bread structure permeable to gas so that the baked bread does not collapse on cooling
Gluten	Slows down the rate of diffusion of CO_2 from the dough, resulting in better texture
Lipids	Polar lipids, especially glycolipids, are particularly important in increasing the bread volume
Pentosans	
Water-soluble	Important in producing an optimum loaf volume, presumably by increasing the viscosity of the aqueous phase; decreases dough development time and also dough stability
Water-insoluble	Impairs loaf volume by impairing oven-spring and crumb grain; produces bread with higher moisture content; has a tendency to decrease staling of bread

serve as a starter for a subsequent batch of rye bread. Sourdough starters are available commercially and contain 2.0×10^7 to 9.0×10^{11} bacteria per gram and 1.7×10^5 to 8×10^6 yeast cells per gram. The bacteria belong mostly to the group of heterofermentative lactobacilli, sometimes up to nine different species are found in a single sample of sourdough starter culture. Yeast species include *Pichia saitoi*, *Saccharomyces cerevisiae*, *Candida krusei* and *Torulopsis holmii*. Some species such as *S. inusitus* and *S. exiguus* are considered typical of sourdough because they can grow in the presence of lactic acid bacteria.

Physical Forms of Yeast Employed

The production of bakers' yeast (*S. cerevisiae*) exclusively for bread dates back to the late 1800s. Initially it was produced using grain mashes which were subsequently replaced by inexpensive molasses. Yeast is capable of aerobic respiration where it utilizes the carbohydrates efficiently, as shown by the following equation:

$$C_6H_{12}O_6 + 6O_2 \rightarrow 6CO_2 + 6H_2O + \text{energy}$$

The first major development in bread-making occurred with the introduction and universal adoption of bakers' compressed yeast. Later, technological developments gave the baker products such as active dry yeast and high active dry yeast (instant yeast). All three physical forms of yeast are presently used in commercial practices.

Compressed Yeast

The most commonly used method of manufacture of compressed yeast involves the use of a molasses–ammonia process in which carefully selected yeast strains are seeded in a dilute solution of molasses, mineral salts and ammonia, and allowed to grow with constant aeration using sterile air. This aeration inhibits fermentation of the carbohydrate materials

Table 5 Role of additives in bread-making

Compound	Principal function
Emulsifiers	
Lecithin and hydroxylated lecithin	Dispersing agent
Monoglycerides and diglycerides and their propylene glycol derivative	Crumb softener
Polysorbate 60, ethoxylated monoglycerides and diglycerides	Volume improver and dough strengthener
Diacetyl tartaric acid ester of monoglycerides and diglycerides	Crumb softener, volume improver and dough strengthener
Calcium and sodium stearoyl lactylate and lactylic stearate	Dough strengthener, crumb softener, volume improver and processing tolerance
Oxidizing agents	
Bromates and iodates of potassium and calcium	Oxidation of -SH groups in the gluten to improve the gas retention of the dough
Azodicarbonamide	Fast-acting substitute for iodate
Calcium peroxide	Strengthens gluten, increases absorption
Reducing agents	
L-Cysteine, ascorbic acid	Dispersal of gluten to reduce mixing requirements
Enzymes	
Diastase from malt preparations	Provides fermentable carbohydrates and stimulates gas production in doughs
Protease of microbial (Aspergillus niger) or plant (papain from papya or bromelain from pineapple) origin	Improves dough machinability and extensibility
Soybean lipoxidase	Oxidizes carotenoids in wheat to whiten the crumb
Fermentation accelerators	
Sulphate, phosphate or chloride of ammonium	Nitrogen source for yeast metabolism
Acidulants and buffers	
Calcium hydrogen phosphate, vinegar, lactic acid	Acid salts and acids lower dough pH; also work as flavour additives
Calcium salts such as sulphate, lactate, carbonate and phosphate	Buffer salts control pH; source of calcium to strengthen gluten

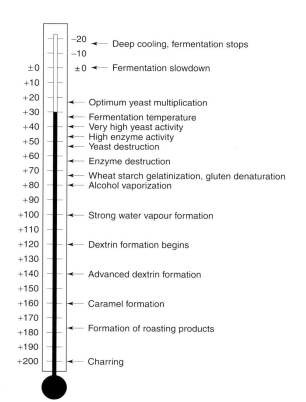

Figure 2 Indirect and direct baking changes affected by temperature (°C).

and leads to their efficient utilization. Careful control of the process parameters such as temperature, pH, ammonia and mineral supply and other factors influencing yeast growth is exercised. Some molasses are not satisfactory for yeast production due to insufficient concentration of certain nutrients, or the presence of substances inhibitory to the yeast growth, such as fatty acids, nitrites or various insecticides and herbicides. Other carbon sources such as hydrolysed starch and hydrolysed lactose from cheese whey have also been evaluated for the manufacture of bakers' yeast.

After production of its biomass, the yeast is then separated from the solution by filtration or centrifugation, washed, compressed and packaged in blocks of desired weight.

Active Dry Yeast

Strains of *S. cerevisiae* which are resistant to drying and also show an acceptable baking activity are used for the production of active dry yeast. Such strains have a very high activity in the wet state, but are less active in the dried form on an equal solid basis due to heat applied during drying. Dry yeast has lower moisture content and therefore can be stored at room temperature for a longer time. Transportation costs are also reduced due to the removal of a large volume of water during drying. **Figure 3** outlines the steps involved in the manufacture of active dry yeast.

The yeast strain selected for the manufacture of active dry yeast should not contain much nitrogen (maximum 6.5% of dry weight basis) and also it should have a bud index of less than 1% since otherwise loss of activity during drying would result. This is possible by monitoring the wort feed schedule and the aeration pattern. A trehalose concentration of 12% or greater (solid basis) is suitable for the production of active dry yeast. Methyl cellulose or carboxymethyl cellulose have been suggested as suitable swelling agents at 1–2% of the yeast solids. Emulsifiers such as monoglycerides, lecithin, glycerol polyesters or sorbitan monostearate are sometimes added before drying to facilitate wetting under usage conditions. This also lightens the colour of the product. The storage stability of such products can also be increased by adding 0.1% butylated hydroxyanisole to the basis of yeast solids. Rehydration of active dry yeast requires special procedures. A ratio of 4:1 water:yeast, and temperature 40–45°C are ideal for such a purpose. Active dry yeast is generally sold packaged under nitrogen atmosphere for home baking through retailers. It is extensively used by institutions and areas where it is too expensive to distribute refrigerated compressed yeast.

High Active Dry Yeast

This has been an important milestone in the bakers' yeast industry, and has an advantage of not requiring a rehydration procedure. It can be directly added to flour and other dry ingredients. The strains for this product have been developed by protoplast fusion, classical hybridization or mutation. Its commercial production parallels that of active dry yeast to the point of pressing the yeast cream to obtain the cake yeast. The cake yeast is then extruded through a finer mesh, followed by drying in fluid-bed airlift dryers. These dryers take advantage of the cooling effect of moisture evaporation; they use air inlet temperature of 65–70°C, and permit drying in a shorter time. The particle size is about 0.2–0.5 mm in diameter and 1–2 mm in length, and has a high porosity. Due to its large surface area, it is unstable in air and hence requires nitrogen or vacuum packaging.

Contaminants and Storage of Bakers' Yeast

Since the fermentation conditions cannot always be kept sterile, bakers' yeast, whether compressed or active dry yeast, always contains contaminating organisms, in particular lactic acid-producing organisms (homofermentative *Lactobacillus* or heterofermentative *Leuconostoc*) and *Escherichia coli*. Since active dry yeast is made by drying compressed yeast, the counts are lower, but essentially the same. This is substantiated by the fact that yeast is in itself an excellent growth medium for bacterial contaminants. The specifications for bakers' yeast require the absence of *Salmonella*, coliforms must be < 1000 per gram and *E. coli* must be < 100 per gram.

Storage of manufactured (compressed) yeast is of immense importance to the baker. Yeasts can survive for long periods without deterioration if the temperature is close to that of the freezing point of water. Storage at 0°C, 13.3°C and 22.2°C gives a shelf life of about 2–3 months, 2 weeks and 1 week respectively. Storage of compressed yeast above 10°C results in the metabolism of storage carbohydrates with concomitant production of heat. Active dry yeast loses 7% activity per month if improperly stored, and 1% per month under nitrogen or vacuum packing. Once the pack has been opened, it should be used quickly and should then be protected from high humidity in the atmosphere.

Important Characteristics of *Saccharomyces*

Saccharomyces is one of the 33 genera of yeast, and has about seven species. Although approximately 50 morphological and physiological criteria are used to differentiate and characterize yeast species, it is not uncommon to find certain yeast strains that are difficult to group correctly. Industrial yeast strains include top- and bottom-fermenting beer yeast, bakers' yeast, distillery yeast, wine yeast and food yeast. Distillery, wine and bakery yeast do not show any clear-cut differentiation, but are not interchangeable.

Production strains have originated from wild

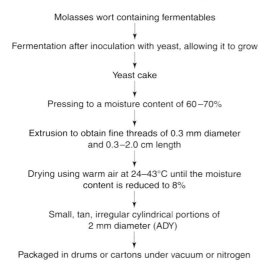

Molasses wort containing fermentables

↓

Fermentation after inoculation with yeast, allowing it to grow

↓

Yeast cake

↓

Pressing to a moisture content of 60–70%

↓

Extrusion to obtain fine threads of 0.3 mm diameter and 0.3–2.0 cm length

↓

Drying using warm air at 24–43°C until the moisture content is reduced to 8%

↓

Small, tan, irregular cylindrical portions of 2 mm diameter (ADY)

↓

Packaged in drums or cartons under vacuum or nitrogen

Figure 3 Flowsheet for the manufacture of active dry yeast (ADY).

strains by genetic processes. The major fermentation characteristics of industrial strains of bakers' yeast are given in **Table 6**.

Selection of Yeast Strains

The introduction of compressed yeast towards the end of the 18th century heralded the development of the baking industry. Earlier bakers used an inoculum consisting of a mixture of spent brewers' yeast, wild yeast and some bacterial contaminants. This led to erratic fermentations and baked products of inferior quality.

Currently, selected strains of *Saccharomyces cerevisiae* with special osmotic properties are available to the baker for leavening of the dough. While strains for lean, straight and sponge dough formulations require yeasts that have lesser osmotolerance, sweet dough strains are generally selected for osmotolerance in addition to good fermentative ability in high sugar environments. Strains having a strong maltose fermenting capability are selected to leaven lean doughs.

Yeasts grown to contain a high nitrogen content are not suitable for preparing dried yeast. A nitrogen content of 6.5% on a dry basis is suggested to be suitable for the same. Under these conditions, trehalose and acid-soluble glycogen increase rapidly; these are responsible for stabilizing the cell membrane and providing energy for cell maintenance respectively. Similarly, buds, being heat-sensitive, should not be present above 1% for the manufacture of dried yeast.

Biochemical Action of Yeast in Dough

Yeast ferments the carbohydrates to ethanol and carbon dioxide anaerobically as per the following reaction:

$$C_6H_{12}O_6 + Yeast \rightarrow 2C_2H_5OH + 2CO_2 + 27 \text{ calories}$$

100 parts	51.1 parts	48.9 parts
Glucose	Ethanol	Carbon dioxide

Table 6 Major fermentation characteristics of commercially used bakers' yeast

Characteristic	
Raw material	Flour, sugar
Fermentation	
Time (h)	1–3
Temperature (°C)	30–35
pH	4.7–5.2
Final ethanol concentration (% w/v)	2–3
Number of cells (10^6 per ml)	
Start	275
End	300
Gas production rate	18–35
(mmol ethanol h^{-1} g^{-1} yeast solids)	

Starch is susceptible to fermentation after amylolytic splitting into glucose and then passing through the pathway shown in **Figure 4**. The enzymes catalysing the reactions in Figure 4 are all found in yeast. Various other side reactions that can lead to the formation of glycerol, organic acids (lactic, succinic, formic, glyceric), higher alcohols, esters, acetaldehyde, ammonia and many others also take place, but seldom account for more than 6% of the total fermented sugars in normal fermentation. The agents or factors responsible for conversion of sugar into alcohol and carbon dioxide are referred to as zymase and are a mixture of several enzymes, coenzymes and inorganic salts.

Yeast solids utilize about 0.77–3.0 g sugar $h^{-1}g^{-1}$. On this basis it is calculated that 3 h fermentation with yeast (containing 29% solids) and 55 min proofing should consume 1.75–6.82 g sugar compared with estimated values of 3.5 g, and produce the corresponding value of carbon dioxide.

The CO_2 produced is partially lost to the atmosphere. Only about 45% of the CO_2 produced during fermentation is retained in the bread dough. CO_2 is retained either as a gas within the gas cells, or it exists in a dissolved state in the aqueous phase. Some of it combines with water to form carbonic acid. Being weak, very slightly ionized and unstable, it contributes only minimally to the lowering of the pH of the dough. The lactic acid and acetic acid bacteria present in the flour convert sugar to lactic acid, and ethanol into acetic acid respectively. Lactic acid has a measurable effect in reducing the pH of the dough, while acetic acid is of lesser significance. Yeast foods such as ammonium sulphate and ammonium chloride also influence the pH of the dough.

Susceptibility of Bread to Microbial Spoilage

Mouldiness and ropiness, commonly termed as mould and rope, are the chief types of microbial spoilage of bread. Although the baking temperatures employed are sufficient to kill all mould spores, the spores can be introduced during air cooling, handling or from wrappers. Growth is commonly observed between the slices of the bread, or in the creases of the loaf.

Contaminated slicing machines, long cooling time, wrapping of the bread when warm and storage in a warm humid place all favour mould spoilage. Counteractive measures include filtering the air and irradiating the room with ultraviolet rays, adequate cooling before wrapping, ultraviolet irradiation of the surface of the loaf and the slicing knife and using amtimycotic food additives such as sodium or calcium propionate at 0.1–0.3%.

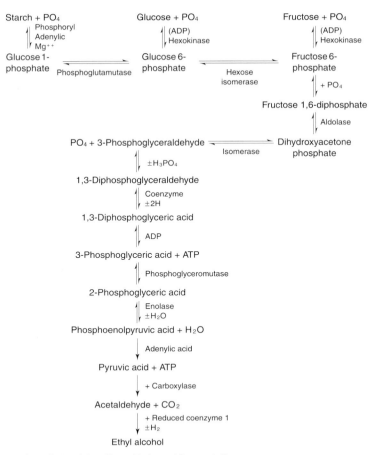

Figure 4 Pathway for conversion of starch to ethanol in bread fermentation.

Ropiness is generally observed during hot weather in home-baked rather than commercially baked breads. The bacteria responsible for ropiness withstand the temperature of baking and capsulation of the bacillus together with the enzymatic hydrolysis of starch and flour gluten encourage rope formation. Ropiness is favoured by heavy contamination of the dough with spores, contamination from slicing knives, slow cooling of the bread after baking, relatively high pH of the bread, and storage in a warm humid atmosphere. Using flour which has a low bacterial spore count (< 20 per 100 g flour), adequate cleansing or sanitizing of the equipment, prompt cooling after baking, storage at cool temperatures and use of 0.1–0.3% calcium or sodium propionate are recommended to overcome these problems.

The different types of microbial spoilage in bread, the causative organisms and observed manifestations of spoilage are given in **Table 7**.

See also: **Bread**: Sourdough bread. **Saccharomyces**: Introduction; *Saccharomyces cerevisiae*. **Spoilage of Plant Products**: Cereals and Cereal Flours. **Starter Cultures**: Uses in the Food Industry. **Yeasts**: Production and Commercial Uses.

Table 7 Microbial spoilage of bread and its observed manifestations

Type of spoilage/causative organism	Observed manifestation
Mould spoilage	
Rhizopus nigricans	White cottony mycelium and black dots of sporangia
Penicillium expansum or *P. stoloniferum*	Green spores
Aspergillus niger	Greenish or purplish-brown to black conidial heads and yellow pigment diffusing into the bread
Monilia (Neurospora) sitophila	Pink conidia giving pink or reddish growth
Endomycopsis fibuliger and *Trichosporon variable*	Chalky bread, white chalk-like spots
Bacterial spoilage	
Bacillus subtilis, B. licheniformis	Ropiness that is yellow to brown in colour, and soft to stick to touch; unpleasant odour
Serratia marcescens	Bloody bread, caused by the release of pigment

Further Reading

Athnasios AK (1996) Yeasts. In: Elvers R, Haekins S, Russey W and Schulz G (eds) *Ullmann's Encyclopedia of Industrial Chemistry*. Vol. A28, p. 461. Weinheim, Germany: VCH Verlagsgesellschaft.

Burrows S (1970) Baker's yeast. In: Rose AH and Harrison JS (eds) *The Yeasts*. Vol. 3, p. 349. London: Academic Press.

Frazier WC (1967) Contamintion, preservation and spoilage of cereals and cereal products. In: *Food Microbiology*, 2nd edn. New York: McGraw-Hill.

Hoseney RC (1994) Yeast leavened products. In: *Principles of Cereal Science and Technology*, 2nd edn. St Paul, MN: American Association of Cereal Chemists.

Nagodawithana TW (1986) Yeasts: their role in modified cereal fermentations. In: Pomeranz Y (ed.) *Advances in Cereal Science and Technology*. St Paul, MN: American Association of Cereal Chemists.

Pomeranz Y (1968) Relation between chemical composition and bread-making potentialities of wheat flour. In: *Advances in Food Research*. Vol. 16, p. 335. New York: Academic Press.

Pomeranz Y (1987) *Modern Cereal Science and Technology*. New York: VCH Publishers.

Pyler EJ (1979) Dough fermentation. In: *Baking Science and Technology*. Vol. 2. Chicago IL: Siebel.

Spicher G and Pomeranz Y (1985) Bread and other baked products. In: Campbell T, Pfefferkorn R and Rounsaville JE (eds) *Ullmann's Encyclopedia of Industrial Chemistry*. Vol. A4, p. 331. Weinheim, Germany: VCH Verlagsgesellschaft.

Sourdough Bread

Brian J B Wood, Department of Bioscience and Biotechnology, University of Strathclyde, Glasgow, Scotland

Introduction

Sourdough bread-making is an ancient craft, which is currently undergoing a revival of interest in developed countries. The technology and microbiology of the constituent processes are examined, and the diversity of the processes is illustrated. Connections with other traditional ways of fermenting grains and legume seeds are noted.

History

The origins of the making of all breads are so ancient that everything said about them must be pure speculation. I suggest that the products now known as sourdough breads are more ancient than breads made with the aid of added yeast. In support of this view

I offer the following evidence: (1) The sourdough fermentation will start spontaneously if a mixture of flour and water is left in a warm place for a few hours, and satisfactory bread can be made from such a ferment; and (2) Many traditional fermentations of maize, cassava and other starchy substrates in primitive societies use processes very similar to those employed in sourdough production, even though the product is more often akin to a porridge or gruel rather than a bread. It would be plausible to suggest that the production of such a porridge was the original process, out of which the production of bread would develop fairly easily.

In India, several related products are made by fermentation of a mixture of rice and a pulse (legume seed), ground or milled to various degrees of fineness. The fermentation is spontaneous, and dominated by lactic acid bacteria – indeed, no yeasts are present. Despite this important difference from sourdough breads, the mixture, after the addition of water to form a batter, undergoes fermentation in which there is some leavening. The leavening is due to the formation of CO_2, resulting from the heterofermentative metabolism of sugars by some of the lactic acid bacteria present in the batter. Normally the batter is left to ferment overnight, then cooked by steaming to make a soft, moist, spongy cake (idli). A thinner batter is fried to make a kind of pancake (dosa). There are several other variants on the theme, depending upon the choice of legume seed, how fine or coarse the grind of the rice and the legume, the method of cooking, etc. For the present purpose I draw attention to the similarities in concept and use between this family of foods and the sourdough batters, which are used in pancake production as well as in bread-making.

Bread production in Old Testament times probably used sourdough technology, particularly if rye or primitive barley (such as that still cultivated as bere barley in the Orkney Islands), were significant components of the dough mixture. The excess yeast produced in beer-brewing, however, provided an alternative way of leavening wheaten breads, and the baking process could be speeded up by using the brewers' yeast – this technology is the direct ancestor of the modern baking industry. Nevertheless, sourdough breads still play a significant part in the market in much of Europe (particularly Scandinavia, Germany and eastern Europe), in the former Soviet Union and in parts of the Middle East.

In the USA, sourdough bread was vital to the pioneers travelling west across the vast plains, mountains and deserts in slow-moving wagon parties, with no means of preserving yeast for baking. As will be explained, sourdough bread starters are relatively easy to conserve, and if all else failed, another starter could be prepared overnight from flour and water. The sourdough was used for bread and also for the breakfast pancakes. It was so important a part of the survival kit of the pioneers and of the adventurers seeking gold in areas such as Alaska and the Yukon that they became known as 'sourdoughs', as featured in the poems of Robert Service (1957) and the novels of Jack London.

In modern America, sourdough bread is usually associated with San Francisco, California, where the tradition and practice of sourdough bread production survived in numerous small craft bakeries in the century after the Californian gold rush. It has re-emerged in the 1980s and 1990s to become big business, with 'San Francisco sourdough bread' on sale at airports throughout the USA. Comparable processes survived in other countries until the middle of the twentieth century, but have now completely disappeared where modern, large-scale bread production has displaced most of the craft baking. For example, there was a process called the 'Parisian barm' process in Glasgow and the west of Scotland, using a starter which had the consistency of a thin cream and which was sometimes used in conjunction with pressed bakers' yeast, to give a good flavour with the speed advantage conferred by the pressed yeast. The last bakery to use this process ceased production in the early 1970s.

In some cases, bakers have used a type of sourdough technology without realizing that they were doing so. In one case the bakers were using a mixture of whole grains and kibbled (crushed) grains as part of the mix for Vogel Formula and Turkestan loaves. These components were soaked in water overnight to ensure the full uptake of water by the grain, and the high temperature in the bakery ensured that by the following morning the mixture was fermenting vigorously. This imparted a subtle but delicious lactic flavour to the finished bread, but the bakers had not understood the role of this fermentation in producing the unique flavour of their products until it was pointed out to them. This particular process has now been lost – the bakery was taken over and closed by a food conglomerate.

Pattern of Consumption

Above, I have tried to show that the sourdough process is probably the original type of bread-making. It would be easy for a consumer in the English-speaking world to assume that sourdough bread has been superseded in all but a few specialist cases, such as San Francisco sourdough bread and certain Jewish rye breads, by yeast-leavened bread produced with the aid of compressed or dried pure-culture yeast. A

preference for the blander yeast-leavened bread was exported rather successfully to the Colonies during the colonial period, and seems to have survived well, even this long after independence was achieved by the former colonies. Thus the expanding interest in the San Francisco bread is seen as a rather new phenomenon. In fact Scandinavia, Germany, the Low Countries, eastern Europe, and the countries of the former Soviet Union all have thriving baking industries based on sourdough technology. To a British visitor, the variety of breads on offer in these countries can seem most bewildering – or stimulating, if forewarned and interested in this subject. A walk through the pedestrianized central area of the lovely north Westphalian town of Detmold (home of the principal German centre for baking research, the Bündesforschungsanstalt für Getreide- und Kartoffelverarbeitung, Institut für Backerietechnologie) on a fine Saturday morning offers the sights and delicious smells of a myriad of inviting breads, in such profusion that one wonders how a profit is made by all the bakers.

In Scotland, on the other hand, the traditional Parisian barm seems to have completely disappeared from baking, but it was still made by at least one baker of the traditional 'morning rolls' until fairly recently – although incorporating pressed yeast to aid the final proving of the dough. Some Polish Jewish bakers, driven from home by the Nazi invasion of their homeland, settled in Glasgow and set up to bake dense, close-textured rye bread, which they still sell, and which is popular in health-food and wholefood shops throughout the region. At least two small craft bakeries which have opened in the Glasgow area more recently offer sourdough bread. I suggest that three markets are operating here: (1) people displaced from eastern Europe, traditional Jews and possibly others whose indigenous culture embraced sourdough bread; (2) patrons of health-food and wholefood stores, who perceive sourdough breads as part of a desirable lifestyle; and (3) new consumers who are eating a widening range of foods, may not even realize that the tasty filled roll purchased from the fancy delicatessen in Milngavie (an up-market commuter town serving Glasgow) is made with sourdough bread, but like the 'continental' taste which the bread imparts, and which is probably associated with continental holidays. The first two of these markets are probably mature, or even declining slightly, but the third is ripe for development in a more targeted manner than has occurred so far. The remarkable expansion of the American market for San Francisco-style sourdough bread in recent years supplies a model for this development, and anecdotal evidence suggests that equivalent changes in consumption are taking place in other countries, e.g. Australia.

Raw Materials and Methods of Production

Clearly, an ancient process will have many variations – the following sections present an outline of the basic process, to which variations can be linked, and then draw analogies with other processes. The raw materials are flour and water. Much sourdough bread, particularly in continental Europe, is made from rye flour – alone or in blends with wheat flour – but the American market has a significant fraction devoted to wheat-flour sourdough, often using white flour. Breads containing rye are darker than wheat-only breads, and in some cases (pumpernickel and the very dark 'black bread') are very highly coloured.

The Basic, Traditional Technology

Starter

A fresh starter is easily prepared by mixing flour and water and leaving the mixture in a warm place overnight. By the following morning the yeasts and bacteria present in the flour will have grown in number very greatly and the mixture will possess a sour, alcoholic odour. If a fairly stiff flour–water mixture (a dough or batter) had been prepared, it would have expanded in volume as CO_2 was released by the ethanolic fermentation of sugars present in the flour, but prevented from escaping by the stiffness of the dough. If a pourable cream had been prepared, it would be covered with a head of froth. Although the baker imposes no control over the types of microbe which can develop, the conditions so favour yeasts and lactic acid bacteria that they dominate the fermentation quite rapidly, and are in complete control by the next morning.

Maintenance of the Starter

The starter is mixed with the other ingredients to make a batch of bread. If a dough was originally prepared, a portion of it is removed before or after the first phase of dough fermentation, and is set aside as the starter for the next batch of bread. This process is repeated every time a new batch of dough is prepared. If the original starter was prepared as a cream, then part of the cream will be mixed with the other ingredients comprising a batch of bread dough, and the remaining cream will be mixed with flour and water to make up another batch of starter. Again, this procedure will be repeated with the preparation of every batch of bread. The process of repeatedly making a fresh batch of starter is often referred to as 'rebuilding' the starter.

The inoculation of each new batch of starter with an actively fermenting brew of organisms results in fermentation beginning much more rapidly than would otherwise be the case. It also has the very important effect of selecting for organisms particularly fitted to survive in the flour–water mixture being used. The yeasts and lactic acid bacteria grow synergistically, and this process, ensuring constant reselection, is very similar to the 'back slopping' found in some traditional beer-brewing and other liquid fermentations. This results in the emergence of a very stable consortium of organisms, which can survive as a regularly-used ferment for very long periods of time: some are claimed to be over a century old. However, a price must be paid for these advantages and it is expressed in the form of labour. Rebuilding is a labour-intensive process which does not lend itself to much automation – in the case of some San Francisco processes, the starter is apparently rebuilt very 8 h, irrespective of bread production schedules. However, methods are being developed which are intended to reduce the demands of the most traditional procedures (see below).

Methods which Combine Selected Lactic Acid Bacteria with Bakers' Yeast

Recently, some companies which provide starter cultures for food fermentations have identified a niche market for lactic acid bacteria cultures, which can be used in association with regular bakers' yeast. The bacteria are reported as isolates from traditional sourdoughs, and include *Lactobacillus brevis* (heterofermentative) and *L. delbrueckii* and *L. plantarum* (both homofermentative). The starters are grown up in a mixture of flour and water, then added to the main batch of bread dough – up to a 30% addition if wheat bread is being made, but up to 40% for rye bread. It is perhaps surprising that the suppliers report that these additions do not inhibit the bakers' yeast, which is normally regarded as being a little sensitive to acidic conditions.

From the consumer angle the principal effect will be on the flavour – the changes and metabolites produced by the bacteria, including the increased acidity, enhance the flavour of the bread. In addition, the decreased pH tends to inhibit amylase activity, which is claimed to improve the texture of the bread. This effect is particularly important in rye bread, in which the starch plays an important part in stabilizing the spongy structure of the raised dough – a task carried out almost exclusively by the proteins in wheat bread. The lactic fermentation is also claimed to increase the shelf life of the bread and to reduce the need for the additives which are used in much modern bread-making.

Some recent reports divide sourdough processes into three categories: type I processes involve the traditional doughs, with regular rebuilding of a stable consortium of yeast(s) and lactic acid bacteria; type II processes are of the type discussed above in this section, in which the leavening action is provided by regular bakers' yeast and the lactic acid bacteria are propagated separately, to provide enhanced flavour and other modifications; and type III processes involve the use of dried starters.

German Processes

The term 'German processes' is used for convenience here, because the processes have been most thoroughly documented (by workers in Detmold) in the context of German bread-making and because some are named in association with German cities, e.g. the Detmolder and Berliner processes. Studies by Spicher and his associates in Detmold emphasize that the processes involve multi-species fermentation, in contrast to the two-species system described for the San Francisco process. Some sources suggest that it is the practice to restart ferments from flour and water (without the addition of existing sour ferment) fairly frequently – this could influence the number of species participating in the process. However, Spicher states that 'The development of commercial sourdough starter cultures over several decades has led to a certain natural selection of organisms that predominate in the fermentation'. Spicher also states that attempts to replace these natural associations of organisms with a combination of pure cultures has not proved satisfactory, although other workers have asserted the contrary claim. This apparent difference may simply reflect the very different ways of conducting, and ingredients used in, fermentations for different bread types.

The simplest processes recognized in Spicher's classification are single-stage sours with the addition of yeast to ensure good leavening of the dough – these are comparable to processes such as the Parisian barm process, formerly used for morning rolls in Glasgow. In two-stage and multi-stage processes, the flora which develops is adequate to ensure good leavening without the further addition of yeast.

Spicher describes three single-stage processes. In the Detmold process, the complete souring of the initial mixture requires about 15–24 h fermentation, the temperature range is 20–28°C and the amount of starter (as a proportion of the total dough) ranges from 2% at an ambient temperature of 27–28°C to 20% at 20–23°C. The sourdough mix comprises 55.6% rye flour and 44.4% water. When acidification is complete, this sour is blended with wheat and rye flours, water and yeast to make the final dough for proving and baking. A typical final mix would comprise 29.5% sourdough, 25.3% rye flour, 17.9% wheat flour and 27.4% added water (additional to that in the sour rye mix). This would give a final proportion of 70% rye flour to 30% wheat flour. Pressed yeast is added to the mixture at the rate of 1.0–1.5% by weight. The same rate of yeast addition is used for the Berlin process, in which a higher temperature and a softer mixture shorten the souring process to 3–4 h. The sourdough mix contains 9.5% starter, 47.6% rye flour and 42.9% water. After acidification, this is mixed with the other ingredients in the proportions 47.7% sourdough, 17.0% rye flour, 17.0% wheat flour and 18.2% water, retaining the overall 70 : 30 ratio for rye to wheat. In the limited information on these two processes in English, there is no mention of the amount of salt added to the final mixture, although some salt would be essential for the proper flavour of the bread.

In the Manheim (or salt–sour) process, on the other hand, the salt level is carefully specified. Salt is essential to the process, and its relatively high level typically extends the souring process to 48 h, although it can enable extension to as long as 80 h without loss of quality. The initial souring mixture comprises 9.0% starter, 45.0% rye flour, 45.0% water and 0.9% salt. After fermentation at 30–35°C initially and 15–20°C after 48 h, the sourdough (32.0%) is blended with 26.0% rye flour, 17.3% wheat flour and 24.8% water. Yeast has to be added at 2.5–3.5% by weight to the final dough mixture, in order to overcome the inhibitory effect of the elevated salt content (0.29% of the final dough). The composition prescribed above retains the 70 : 30 rye : wheat flour ratio.

In the two-stage and multi-stage German processes, the dough is built up by the successive addition of rye flour and water to the initial sourdough mixture. This approach seems to permit more development of the yeasts in the consortium, thus requiring the addition of less, or even no, pressed yeast to the final dough. Lonner's group has studied a Swedish rye sourdough process, using an initial mixture containing equal weights of rye meal and water. This starter (9.5%) was mixed with 47.6% rye flour and 42.9% water to make a final dough. This seems to correspond to Swedish practice in rye bread production, although experience would suggest that a mixture devoid of wheat gluten would rise very poorly, and give a very heavy loaf.

Wheat Sourdough Breads

Although there is a tendency to associate the words 'sourdough' and 'rye', not least because of the improved leavening imparted to the weak rye doughs by the souring, there is substantial production of wheat sourdough, the best-known being the San Francisco-type. Workers from the US Dept. of Agriculture's Western Regional Research Laboratory carried out extensive investigations into this class of product, at about the time that its market began to expand, and their investigations shed much light on the microbiology of the process.

The starter is maintained in an active state by rebuilding every 8 h, although it can be refrigerated for several days without loss of quality. Rebuilding involves mixing 100 parts by weight of an existing sponge with 100 parts of high-gluten wheat flour and 48–52 parts of water. The mixture is left to develop at bakery temperature. For baking, a mixture comprising 9.2% of the ripe sponge, 45.7% plain white bread-making flour (regular patent wheat flour in American terminology), 44.2% water and 0.9% salt is blended and after proving for 7 h is baked. At this stage, the initial pH of 5.2–5.3 has declined to 3.9.

A very similar type of process underlies the production of the Indian nan bread.

Festival and other Sweet Breads based on Sourdough Technology

The reference to 'festival' highlights the fact that these sweet products are often associated with celebrations. The best example of this is the Italian panettone, which has strong associations with the Christmas and Easter festivities, and which has a number of variants, involving the addition of things such as sultanas, chocolate chips and orange peel. It is available in the USA and in most places where there is a significant Italian population. There are other local products which conform to the same general type, for example North American 'friendship cake'.

Panettone is prepared according to a recipe along the following general lines. Strong, white wheat flour (300 g), water (200 ml) and mother sponge starter (30–59 g) are blended to form a batter or sponge, which ferments at 30–35°C for 1–2 h. After removal of a portion of the mixture to form the next mother sponge, the rest is mixed with sugar (150–200 g), salt (15–20 g), malt (5–10 g), butter or oil (180–200 g), egg yolks (200 g), sultanas (250–300 g), candied citrus fruit peel (100–150 g) and strong white flour (200 g). The mixture is divided into portions of about 500 g and left to ferment for 2–6 h at 25–35°C, with a 'knock back' (remoulding of the dough) at some stage of the process. It is then placed into baking tins and allowed to prove for a final 60–90 min, then baked. This information is taken from Campbell-Platt: other authors differ on points of detail, but this is a very representative recipe.

Kvass (Kvas, Kwas)

This is a traditional drink of eastern Europe and the countries of the former Soviet Union. Its method of production has sufficient connections with that for sourdough breads to justify a mention here.

Kvass is a refreshing, sharp, somewhat effervescent drink, which is slightly alcoholic. It can be made from rye, wheat, or rye bread. The ingredient is combined with hot water and the extract is fermented, according to some sources with the aid of malted barley – this would be particularly relevant when using cereal grains. The inoculum can be yeast or a portion of sourdough bread starter. The fermentation is a mixed yeast–lactic fermentation, in order to give the required tart, refreshing taste. When rye bread is the substrate it may be rusked (heated in an oven until thoroughly dry, a process which helps to degrade the starches in the bread), then crumbled and mixed with water and starter. Once a fermentation has been established it can be operated by a kind of fed-batch process, in which a portion of the liquid is poured off for consumption, then water and bread are added to the fermentation, restoring it to the original volume. The drink is fairly alcoholic, possibly containing up to 4% ethanol, depending on how the fermentation is operated.

At its best, kvass is a golden-brown liquid with a pleasing nutty flavour. An attempt to place its production on a proper scientific basis was made in Poland some years ago, but commercial production still seems to rely on small-scale local artisans, and its microbiological quality must be very dubious. Its production and distribution is reminiscent of that of sorghum beer (pito, Bantu beer, etc.) in Africa.

Microbiology and Biochemistry

The sourdough breads discussed in this article are characterized by a somewhat acidic taste, but even the conventional mass-produced Chorleywood-process loaf has a small flavour contribution from the lactic acid bacteria which are always present in the yeast and the raw materials. The longer the process time of yeasted breads, the more scope there is for lactic acid bacteria to grow in the dough and make a significant contribution to the flavour of the bread. Information about sourdough bread microbiology and biochemistry must therefore have some relevance to the processes which occur during yeast-bread production.

Yeasts

In many bread-making processes, some pressed yeast (*Saccharomyces cerevisiae*) is added after the initial souring is complete, and even when not specifically added it will sometimes be present in the ferment. However, the acidic conditions are somewhat inhibitory for this yeast – the dough pH will drop to, or below, 3.8–3.9 during the fermentation. Thus selective pressures, particularly in sourdough starters which have been maintained through many cycles of rebuilding, will favour yeasts better able to survive these conditions. Chief among these is *Candida milleri* (also referred to, particularly in earlier work, as *S. exiguus* or *Torulopsis holmii*). This yeast is particularly interesting because of its inability to ferment maltose, the principal sugar in cereal flour, although the amylases present in the grain are activated by the addition of water to the flour. In this respect, the yeast is quite different from *S. cerevisiae*, which avidly consumes maltose. This inability on the part of *C. milleri* is the key to the stability of sourdough mother sponges, because the yeast depends upon the lactic acid bacteria, which metabolize maltose (by the maltose phosphorylase pathway) and excrete glucose which the yeast can use. In return, the yeast provides vitamins of the B group and other substances which are essential or stimulatory for the growth of the lactic acid bacteria. Among these there may be a peptide which has been isolated from hot-water extracts of bakers' yeast, but which is absent from commercial microbiological yeast extracts.

C. milleri is undoubtedly a major participant in sourdough starters from widely dispersed areas, but its dominance in the San Francisco family of sourdough breads, which are probably the most intensively studied among these breads, may have given it a somewhat excessive importance. For example, some workers in Poland have reported the presence of *T. candida* and *C. krusei* alongside it in commercial starters. Swedish studies reported *S. delbrueckii* as the only yeast in a sponge prepared from only rye meal.

Lactic Acid Bacteria

The most frequently reported lactic acid bacterium in a variety of sourdoughs, first discovered in San Francisco sourdoughs, is *Lactobacillus sanfrancisco*. However, various reports suggest a more diverse population of lactic acid bacteria. *L. sanfrancisco* is a bacterium which is very well adapted to life in sourdough batters. It metabolizes maltose by the maltose phosphorylase pathway and forms a very stable partnership with the yeast *Candida milleri*, flourishing with a ratio of about 1:100 yeast cells : bacteria. The bacterium originally isolated from

San Francisco metabolized only maltose and was unable to use other sugars, but other *L. sanfrancisco* isolates, from German and Italian bakeries, were shown to be more diverse – about 25% utilizing only maltose and the rest utilizing maltose and a variety of other sugars. *L. sanfrancisco* is a heterofermentative organism, but seems to produce relatively low levels of acetic acid in the dough. This low level is desirable, because low levels are largely driven off during baking, but too much acetic acid can give a vinegary taste and smell to the finished bread. Acetic acid is useful in fermentations of this type because it inhibits the germination of *Bacillus* spores, and so helps to prevent the development of their ropy spoilage of bread. This is rarely encountered in modern mass-produced breads made from flour with a very low spore count, but is still encountered in whole-grain flours, and might be expected to be more prevalent in rye flour.

Other bacteria have been isolated from bread doughs and are claimed to have desirable effects on the flour and other organoleptic qualities of bread. These include *L. delbrueckii* and *L. plantarum* (homofermentative, i.e. produce lactic acid as the main or only product of hexose sugar fermentation) and *L. brevis* (heterofermentative, producing lactic and acetic acids, ethanol and CO_2 from hexoses). Work on Swedish rye dough also revealed *Pediococcus pentosaceus* as a frequently-occurring bacterium. *L. pontis* and *L. bavaricus* have been reported to occur in sourdough starters from Germany and Italy.

Technological Effects of Sourdough Fermentation

The principal effects can be largely attributed to the acidity resulting from acid production by the lactic acid bacteria. These are particularly important in rye breads, because rye flour is deficient in the proteins which give wheat-flour doughs their remarkable elasticity and extendability. The starches therefore assume a greater rheological importance in rye breads compared to wheat breads. The low pH caused by the lactic acid bacteria inhibits the cereal amylases, reducing degradation of the starch and so improving the dough's properties. The low pH also improves the swelling of pentosans, which is also significant for rye, but not for wheat, doughs. Improvements attributable to lactic acid bacteria, which are beneficial to both rye and wheat dough, include: modification of the dough's water-binding capacity; improvements in the nutritional value of the bread; and prevention of malfermentation and spoilage. However, the benefits of sourdough fermentation are not exclusively due to acidification. For example, it is suggested that lactic acid bacteria may assist in preventing the microbial spoilage of bread, through the production of bacteriocins.

Adding acetic and lactic acids to a dough will not entirely reproduce the changes in dough rheology which occur through a sourdough fermentation, although the effects are considerably more marked than those in dough acidified to the correct pH with a mineral acid. The effects most noticeable to the consumer include, as well as the development of the taste and aroma characteristic of sourdough bread, improvements to the crumb elasticity and bread volume and delay in the onset of staling. The low pH also improves the activity of phytases present in the flour, and the reduction in phytic acid concentration reduces the binding of essential metal nutrients. It also enhances the availability of the phosphate stored in the phytic acid, although it is doubtful if this is significant for a consumer eating a reasonably varied diet. However there is increasing concern that even the American diet is only marginally adequate in terms of trace metals, and so a reduction in metal-binding by phytic acid may be significant. There is some evidence that some of the lactic acid bacteria may be able to hydrolyse phytate, but more work on this is needed.

Conclusion

Although sourdough fermentation is probably the most ancient way of improving the edibility of cereal grains, it has not yet been completely superseded by more technological procedures. There is some encouraging evidence to suggest that there is a substantial revival of interest in its applications, and that it can be usefully combined with modern engineering principles.

See also: **Bacteriocins**: Potential in Food Preservation. **Bread**: Bread from Wheat Flour. *Candida*: Introduction. **Ecology of Bacteria and Fungi in Foods**: Influence of Redox Potential and pH. **Fermentation (Industrial)**: Basic Considerations; Control of Fermentation Conditions. **Fermented Foods**: Origins and Applications; Fermentations of the Far East; Beverages from Sorghum and Millet. **Lactobacillus**: Introduction: *Lactobacillus brevis*. **Metabolic Pathways**: Release of Energy (Anaerobic). **Saccharomyces**: Introduction: *Saccharomyces cerevisiae*. **Starter Cultures**: Uses in the Food Industry; Importance of Selected Genera. **Torulopsis**. **Yeasts**: Production and commercial uses. **Yersinia**: Introduction.

Further Reading

Berg RW, Sandine WE and Anderson AW (1981) Identification of a growth stimulant for *Lactobacillus sanfrancisco*. *Applied and Environmental Microbiology* 42: 786–788.

Hammes WP and Gänzle MG (1998) Sourdough breads and related products. In: Wood BJB (ed.) *The Microbiology of Fermented Foods*, 2nd edn. P. 199. London: Blackie.

Jenson I (1998) Bread and baker's yeast. In: Wood BJB (ed) *The Microbiology of Fermented Foods*, 2nd edn. P. 172. London: Blackie.

Oura E, Suomalainen H and Viskari R (1982) Breadmaking. In: Rose AH (ed.) *Economic Microbiology*. Vol. 7, p. 87. London: Academic Press.

Platt G Campbell (1987) *Fermented Foods of the World, A Dictionary and Guide*. London: Butterworths.

Service R (1957) *Songs of a Sourdough*. Reset edition. London: Ernest Benn, Toronto: McGraw-Hill Ryerson.

Spicher G and Pomeranz V (1985) Bread and other baked products. In: *Ullmann's Encyclopedia of Industrial Chemistry*, 5th edn. Vol. A4, p. 331. Weinheim: VCH Verlagsgesellshaft.

Steinkraus KH (1996) *Handbook of Indigenous Fermented Foods*, 2nd edn. P. 149. New York: Marcel Dekker.

Sugihara TF (1985) Microbiology of breadmaking. In: Wood BJB (ed.) *Microbiology of Fermented Foods*. Vol. 1, p. 249. London: Elsevier Applied Science.

Wood BJB (1993) Sourdough bread. In: Macrae R, Robinson RK and Sadler MJ (eds) *Encyclopedia of Food Science, Food Technology and Nutrition*. Vol. 1, p. 479. London: Academic Press.

BRETTANOMYCES

Juan Jimenez, **Manuel Fidalgo** and **Marcos Alguacil**, Departamento de Genética, Facultad de Ciencias, Universidad de Málaga, Spain

General Characteristics of the Species

Yeasts of the genera *Brettanomyces* were isolated in 1904 from the late fermentation of English 'stock beer', but it was not until 1940 that strains of these yeasts were examined in sufficient detail to classify them. In 1960, some strains of *Brettanomyces* were observed to form ascospores. These ascosporogenous strains were transferred to the genus *Dekkera*.

The number of species differentiated in the genera *Brettanomyces* and *Dekkera* differs from one study to another (up to 18 *Brettanomyces* spp. and four *Dekkera* spp. in the early 1970s) but could be reduced to four in the case of *Brettanomyces* and to two for *Dekkera* strains. In this scheme, two teleomorphic *Dekkera* species, *D. anomala* (CBS77-type strain) and *D. bruxellensis* (CBS74-type strain: synonym *D. intermedia*) and four anamorphic *Brettanomyces* spp., *B. anomalus* (CBS77-type strain: synonym *B. claussenii*), *B. bruxellensis* (CBS72-type strain: synonym *B. abstinens*, *B. custersii*, *B. intermedius*, *B. lambicus*), *B. custersianus* (CBS4805-type strain) and *B. naardenensis* (CBS 6042-type strain) are recognized. However, while *Brettanomyces* and *Dekkera* are genealogically distinct, it is evident from studies at the level of DNA divergence that these organisms are phylogenetically very closely related.

A large number of morphological and physiological studies and phylogenetic relationships deduced from DNA analysis indicate that *Brettanomyces* and *Dekkera* are the same organism. The two genera differ only in that the latter shows alternation of generations with the formation of asci and ascospores. This is similar to *Saccharomyces* in which sexual and asexual strains can be isolated from their natural environments. Taking these observations into consideration, four *Brettanomyces* (*Dekkera*) spp. represented by *B. (D.) anomalus*, *B. (D.) bruxellensis*, *B. (D.) custersianus* and *B. (D.) naardenensis* represent a plausible classification of all these strains.

For simplicity and industrial relevance, further descriptions of these yeasts will be mainly focused on *Brettanomyces* strains, but all the characteristics described below are common to both genera.

Brettanomyces strains are routinely grown on malt agar medium (2% malt extract, 2% glucose, 0.1% peptone, 2% agar) and can tolerate a wide range of incubation temperatures (from 20 to 37°C). Rich media currently used for other yeasts such as YPD (2% glucose, 1% peptone, 0.5% yeast extract, 2% agar) can be employed. Vegetative proliferation of *Brettanomyces* cells occurs by budding. In exponentially growing cultures these cells may be spheroidal, but frequent ogival or cylindrical cells may also be observed (**Fig. 1**); average sizes are $2.0–6.5 \times 4.0–22.0\,\mu m$. Very elongated filamentous cells are also observed when these strains are cultured on corn meal agar. Under these conditions pseudomycelium is abundantly produced during both aerobic and anaerobic growth. Formation of pseudohyphae is also induced by addition of 0.25% (v/v) isoamyl alcohol to the growing media. The cell cycle of *Brettanomyces* yeasts follows the conventional cycle found in most of the budding yeast species in which the G1 phase is predominant. In these cells a single nucleus at the G1 phase of the cell cycle is usually found (**Fig. 2**).

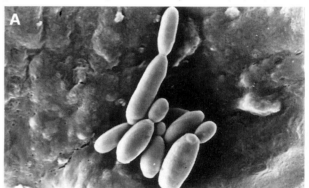

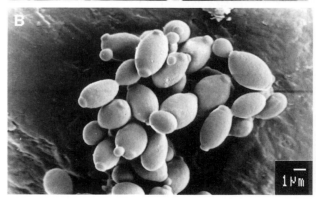

Figure 1 Scanning electron micrographs of *Brettanomyces* cells showing different cell morphologies usually found in these species.

Physiological and Nutritional Properties

The genus *Brettanomyces* is primarily characterized by physiological and metabolic properties. In malt

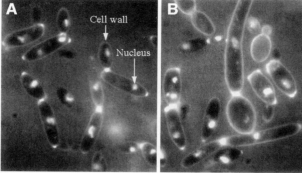

Figure 2 Fluorescent micrographs of (**A**) 4′,6 diamino-2-phenylindole (or DAP) (nuclear DNA) and (**B**) Calcofluor (cell-wall) staining of ethanol-fixed *Brettanomyces* cells.

extract *Brettanomyces* yeasts grow slowly and produce a characteristic aroma. All species are fermentative and fermentation is usually stimulated by molecular oxygen. *Brettanomyces* are markedly auxoheterotrophic in that they require an external vitamin source. All strains examined have a basic requirement for both thiamin and biotin. Nitrate assimilation varies among the species but all utilize ethylamine hydrochloride as nitrogen source. Nutritional and vitamin supplementations required for these yeasts to grow may impose important ecological restrictions. *Brettanomyces* can be found in honey and three exudates, but fermenting products and their associated environs are generally an accepted habitat for these yeasts.

Differences in fermentative and assimilatory properties are summarized in **Table 1**. The genus is rather restricted in its utilization of carbon compounds. For instance, none of the species metabolizes L-sorbose,

melibiose, soluble starch, D-xylose, L-arabinose, D-arabinose, L-rhamnose, erythritol, galactitol, D-mannitol, citric acid or inositol. The utilization of D-ribose, ribitol, D-glucitol, succinic acid and lactic acid is frequently variable for these species.

The yeast *Brettanomyces* shows the Custer effect in that, under strictly anaerobic conditions, in the presence of glucose, CO_2 production is negligible. The Custer effect is provoked by the reduction of NAD^+ in the conversion of acetaldehyde to acetic acid. The addition of acetaldehyde and other organic hydrogen acceptors alleviates the Custer effect by restoring the redox balance. The addition of oxygen also restores CO_2 production, but it has a different effect on the glycolysis.

Under aerobic conditions, *Brettanomyces* are strongly acidogenic, producing large amounts of acetic acid when grown on glucose. Production of acetic acid results from oxidation of ethanol rather than via pyruvate. Accumulation of acetic acid could result from low-level activity of Tricarboxylic acid (TCA) cycle enzymes as well as from the imbalance in reduced/oxidized states of the coenzyme involved in oxidation of ethanol to acetic acid.

Like *Saccharomyces*, these yeasts also generate petite colonies which have lost the mitochondrial DNA genome. However, petite mutants of *Brettanomyces* are able to maintain a oxidative pathway. This respiratory pathway is resistant to cyanide and azide and could account for 50% of the respiratory capacity. The presence of this respiration pathway constitutes a compensation mechanism for the reduction in acetaldehyde dehydrogenase activity. This alternative pathway may be a fundamental element

Table 1 Salient characteristics of species in the genus *Brettanomyces*

Fermentation

	G	Ga	Su	Ma	Ce	Tr	La	Ra	Mz	α-G
B. anomalus	+	+	+	v	+	v	+	v	v	v
B. custersianus	+	−	−	−	−	s	−	−	−	−
B. bruxellensis	+	−	+	+	−	+	−	−	+	+
B. lambicus	+	v	+	+	−	+	−	−	+	+
B. intermedius	+	+	+	+	+	+	−	v	+	+
B. custersii	+	s	+	+	+	+	−	s	+	+
B. claussenii	+	+	+	+	+	+	+	+	+	+

Assimilation

	Ga	Su	Ma	Ce	Tr	La	Ra	Mz	α-G	Et	NO₃
B. anomalus	+	+	+	+	+	+	s	v	v	+	+
B. custersianus	v	w	v	−	+	−	−	−	−	+	−
B. bruxellensis	−	+	+	−	+	−	w	+	+	+	v
B. lambicus	s	+	+	−	+	−	−	+	+	+	+
B. intermedius	+	+	+	+	+	−	v	+	+	+	v
B. custersii	+	+	+	+	+	s	s	+	+	+	+
B. claussenii	+	+	+	+	+	+	+	+	+	+	+

G = Glucose; Ga = galactose; Su = sucrose; Ma = maltose; Tr = trehalose; Ce = cellobiose; La = lactose; Ra = raffinose; Mz = melezitose; α-G = α-methyl-D-glucoside; Et = ethylamine hydrochloride; NO₃ = potassium nitrate; s = slow; v = variable; w = weak.

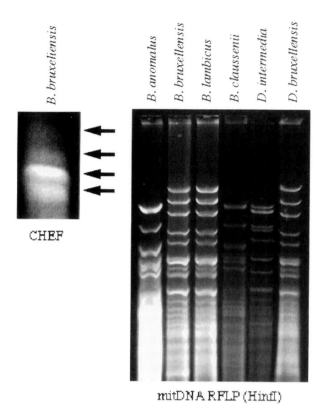

CHEF

mitDNA RFLP (HinfI)

Figure 3 Genomic properties of *Brettanomyces/Dekkera* strains. EtBr-stained DNA of the electrophoretic karyotype (contour-clamped homogeneous electric field, or CHEF) of a *Brettanomyces* strain (left); four bands are usually distinguished in these species. Restriction fragment length polymorphism (RFLP) patterns of mitDNA (EtBr-stained) observed by electrophoresis in agarose gel after digestion of DNA samples with *HinfI* (right), following the method described in the text.

of the Custer effect (transient inhibition of alcoholic fermentation as a result of anaerobic conditions) characteristic of these strains.

Genomic Analysis

Electrophoretic karyotypes, mitochondrial DNA restriction patterns and DNA content per cell represent simple approaches to define the characteristic structure of the entire genome of yeast species. The presence of one to five large chromosomes (larger than 1000 kb) is the most frequent situation found in members of the ascomycetous yeasts. *Saccharomyces* is the exception, in which many chromosomal bands, some of which are of relatively low molecular weight (less than 500 kb) are observed. The molecular karyotype of *Brettanomyces/Dekkera*-type strains usually shows four large chromosomes (**Fig. 3**). In these strains, the relative amount of DNA is higher in some bands than in others (Fig. 3). According to flow cytometry analysis, the amount of DNA per cell ranges from

0.9 to 1.8-fold that of haploid *Saccharomyces cerevisiae* cells. Overall, cytometry analysis and molecular karyotyping data suggest different degrees of aneuploidy in *Brettanomyces/Dekkera*.

Mitochondrial genomes from yeasts in the *Brettanomyces/Dekkera* group vary in size from 28 to 101 kb: *B. naardenensis* (41.7), *B. custersianus* (28.5), *B. custersii* (101.1), *B. anomalus* (57.7), *D. bruxellensis* (85.0), *D. intermedia* (73.2). Genomes smaller than 42 kb have a similar gene order whereas the largest mitochondrial DNAs are rearranged related to the smaller molecules. A rapid, simple method based on mitochondrial DNA restriction analysis yields characteristic mitochondrial DNA patterns in these yeasts (Fig. 3). Using this method, different patterns can be obtained (restriction fragment length polymorphism or RFLP) in *Dekkera/Brettanomyces* strains which are closely related, indicating that the mitDNA is polymorphic in these yeast strains; however, in most cases synonym strains share identical mitDNA RFLP patterns (Fig. 3). This method provides a simple and efficient technique for the rapid characterization of *Brettanomyces/Dekkera* yeasts (the experimental procedure is described below).

Simple Procedure for RFLP Analysis of mitDNA in *Brettanomyces* Yeasts

The mitDNA in *Brettanomyces* is biased in its nucleotide composition (A + T enriched), yielding large DNA molecules when digested with four-cutter restriction enzymes that contain G and C in their target sequences (*AluI*, *HinfI* and *RsaI* are frequently used). These large molecules are easily distinguishable from the completely digested nuclear DNA in agarose gel electrophoresis (Fig. 3), making of this technique a powerful and simple method for direct analysis of polymorphism in mitDNA.

1. A flask containing 10–25 ml of YPD medium is inoculated with cells of the desired strain. The culture is incubated until the early stationary phase is reached.
2. Cells of 5 ml of this culture are collected by centrifugation, washed twice and suspended in 500 μl sorbitol 1 mol l^{-1}, EDTA 0.1 mol l^{-1}, pH 7.5 containing 4 U zymoliase 20 T. The cells are incubated for 1 h at 37°C.
3. Harvest the suspension, remove the supernatant and add 500 μl Tris-HCl 50 mmol l^{-1}, EDTA 20 mmol l^{-1}, pH 7.5 to the pellet. Add 50 μl sodium dodecyl sulfate 10% and incubate for 30 min at 65°C.
4. Add 200 μl potassium acetate 5 mol l^{-1}, vortex and incubate in ice. After 30 min incubation, centrifuge

the suspension and remove 650 µl of the supernatant (containing nucleic acids).

5. Add 640 µl isopropanol (ice-cooled), centrifuge and wash the pellet twice with 70% ethanol (ice-cooled). Let the pellet dry at room temperature.

6. Resuspend the pellet in 20–50 µl Tris-HCl 10 mmol l^{-1}, EDTA 1 mmol l^{-1} pH 8. Then 15 µl of this solution is used for DNA digestion with the appropriate four-cutter restriction enzyme to determine mitDNA restriction patterns by electrophoresis in agarose minigels.

Implications in Fermented Beverages and Foods

Wines

Brettanomyces and its ascosporogenous sexual state *Dekkera* have been well documented as some of the more troublesome yeasts that grow in fruit juice and wines. These microorganisms (and certain lactic acid bacteria) are responsible for the unpleasant odour and taste known as mousiness, and cause economic losses that run into the hundreds of thousands of dollars. The habitat for both yeasts has been classically ascribed to barrel-ageing red wine, but *Brettanomyces/Dekkera* yeasts have been isolated from many other fermented beverages such as beer, cider and a diversity of white wines, including fino sherry wines. Various sensory descriptors have been used to characterize *Brettanomyces*-tainted fermentations, but sensory interpretations vary widely, not only when red and white wines are compared, but also within either group. These range from spicy to medicinal. The origin of these sensory properties is due to the production of the volatile ethylphenols 4-ethyl phenol (from *p*-coumaric acid) and 4-ethyl guaiacol (from ferulic acid), the former being found at higher concentrations and a useful sensory marker of the presence of *Brettanomyces*. Formation of mousy components associated with *Brettanomyces*-infected wines is probably due to amino acid derivatives, 2-acetyl-1,4,5,6-tetrahydropyridine, but the fault found in phenolic red wines caused by *Brettanomyces* is more frequent than the classical mousy taint attributed to this yeast genus. The mechanism of biosynthesis of ethylphenols is due to the sequential activity of a cinnamate decarboxylase, which transforms certain cinnamic acids into the correspondent vinylphenols, and a vinylphenol reductase which catalyses the reduction of these compounds into ethyl-phenols (**Fig. 4**). Additional fatty acid metabolites that potentially contribute to the sensory profile of tainted wines include isobutyric, isovaleric and 2-methylbutyric acids.

Although growth of most yeasts is significantly affected by increased pressures, *Brettanomyces* spp. have been isolated in sparkling wine fermentations.

Lambic Beer

In some cases, *Brettanomyces* spp. are involved in specific fermentations. For instance, *Saccharomyces* and *Brettanomyces* yeasts are both essential organisms to brew lambic-style beer. *Brettanomyces* mainly contributes to the flavour profile of the finished product. The involvement of these yeasts in desired characteristic flavours of certain red wines has also been proposed, but *Brettanomyces/Dekkera* strains have not been isolated from them, and consequently, these sensory properties cannot necessarily be attributed to these yeasts.

Fruit Juices and Soft Drinks

Essentially, fruit juices and soft drinks are high-acid, low-pH products that contain substantial concentrations of fermentable sugars. As a consequence, they become selective environments for yeast growth. *Brettanomyces* spp. are among the yeasts most frequently associated with the spoilage of soft drinks and fruit juice.

Cider

Some cider factories support populations of *Brettanomyces* spp. and only rigid microbiological control of the bottling plant prevents this yeast from infecting the finished product.

Pickles

Subsurface yeasts have been associated with a gaseous fermentation and a type of spoilage known as bloater or hollow cucumber formation. *Torulopsis caroliniana*, *Hansenula subpelliculosa*, *Zygosaccharomyces* spp. and *Brettanomyces* spp. have been related with this gaseous-type spoilage of commercially prepared cucumber pickles.

On the basis of extensive studies on occurrence of yeasts in cucumber fermentations, yeasts occur in the following approximate order of frequency: *B. versatilis*, *H. subpelliculosa*, *T. caroliniana*, *T. holmii*, *S. rosei*, *S. halomembranis*, *S. elegans*, *S. delbrueckii*, *B. sphaericus* and *H. anomala*. These yeasts were found in brines containing from 4% to 18% salt by weight and ranging in pH value from 3.1 to 4.8.

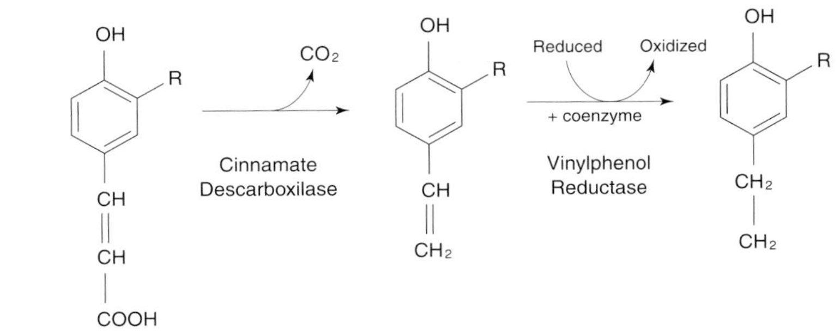

R = H: *p*-coumaric acid   4-vinylphenol   4-ethylphenol
R = OCH₃: ferulic acid   4-vinylguaiacol   4-ethylguaiacol

Figure 4   Biosynthesis of ethylphenols by *Brettanomyces* yeasts.

Methods of Detection

Based on Physiological Properties

The time frame for the development of maximal cell number and subsequent decline of *Brettanomyces/Dekkera* yeasts probably depends on wine chemistry and other environmental factors. Programmes monitoring the presence of *Dekkera* yeasts are essential. The monitoring method of choice generally involves collection of microbes by membrane filtration and subsequent culture, using appropriate media. However, monitoring and controlling these species are difficult, partly because of the difficulties encountered in routine isolation and identification at the genus level. These microorganisms grow very slowly in current yeast media, and the rapid growth of other yeasts and bacteria makes the isolation step difficult. Attempts to isolate minor populations of slowly growing *Brettanomyces/Dekkera* yeasts from mixed flora include cycloheximide as a selective inhibitor (10–50 mg l^{-1}) to impede the growth of numerically superior microbes such as *Saccharomyces* sp. (cycloheximide-sensitive). However, cycloheximide resistance is also common to *Hansenioaspora* or *Kloeckera*, for instance, which are part of the normal flora of wine fermentations. Species of *Brettanomyces/Dekkera* are strongly acidogenic, producing high amounts of acetic acid from the oxidation of ethanol. Thus, in addition to the cycloheximide-resistance character, the production of acetic acid (it is able to solubilize $CaCO_3$ 2% in agar media over time) has been used as a diagnostic test for the occurrence of *Brettanomyces* and *Dekkera* yeasts (**Fig. 5**). Sorbic acid exhibits inhibitory activity towards fermentative yeasts. This preservative inhibits *Saccharomyces*, but it exhibits little activity toward *Brettanomyces/Dekkera* strains, and consequently is also useful for diagnostic purposes.

Detection of Intact *Dekkera/Brettanomyces* Yeast Cells by Nested PCR

The method described above takes too long for identification, and in fact alternatives such as enzyme-linked immunosorbent assays (ELISA) have been proposed. The polymerase chain reaction (PCR) is another reliable, sensitive and rapid method capable of widespread application in processes such as wine production. PCR allows the specific detection of DNA sequences, and thus it is ideally suited to the detection of contaminating microorganisms.

The following nested PCR method can select and specifically identify those *Brettanomyces/Dekkera* strains usually documented as spoilage microorganisms. The great advantage of this method is that PCR works with intact *Brettanomyces/Dekkera* cells, and samples of microorganisms can be directly used without the need for any DNA purification step.

Wine is a chemically complex beverage that influences the enzymatic amplification of DNA. For this reason, the efficient and sensitive detection of *Brettanomyces/Dekkera* cells in wines requires the previous removal of the wine by centrifugation of the testing sample. Alternatively, samples could be filtered and the retained microorganisms suspended in water before testing. As an example, 1.5 ml samples of wine are collected, the samples are centrifuged and the pellet suspended in 100μl of sterile water. Aliquots (10μl) of the suspension can be used in each reaction mixture as follows.

Sequences of the primers are:

- BD1:   AGAAGTTGAACGGCCGCATTTGCAC (sense)
- BD2:   AGGATTGTTGACACTCTCGCGGAGG (antisense)
- BD3: CGGCATATCGAAGACAG (sense)
- BD4: CATCCTCGCCATACAAC (antisense)

Primers BD1 and BD2 are used in the first reaction,

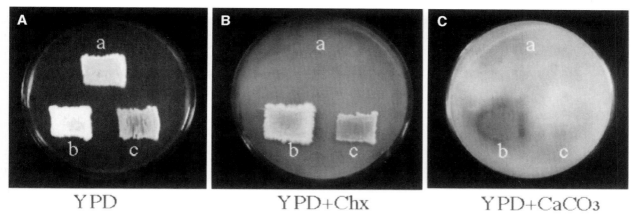

Figure 5 Three yeast strains (**A**: *Saccharomyces*; **B**: *Brettanomyces*: **C**: *Hansenioaspora*) isolated from fermented beverages growing in YPD agar, YPD agar with cycloheximide (testing for resistance to cycloheximide) and YPD agar with $CaCO_3$ (to assay for acetic acid production by solubilization of $CaCO_3$). This is a standard procedure for the diagnosis of *Brettanomyces/Dekkera* strains, which are acidogenic and cycloheximide-resistant.

and BD3 and BD4 are internal primers to be used in a nested PCR reaction. The first reaction mixture is preheated for 5 min at 95°C, and then 30 cycles are carried out, each of which consists of a denaturation step (30 s, 95°C), an annealing step (30 s, 55°C) and an extension step (1 min, 72°C). After the last cycle, extension should be continued for a further 10 min to allow the reaction to go to completion.

To initiate the second reaction 1 µl of this PCR-amplified DNA is used (BD3 and BD4 are used as nested primers), and the reaction is achieved at identical conditions to the first PCR, except that 65°C is used as the annealing temperature.

The first reaction renders a PCR fragment of 471 bp in length, and the second one of 327 bp. The amplified DNA products can be detected by simple electrophoresis in agarose gel (EtBr staining). When the level of contamination is high the corresponding DNA band can be observed from the first PCR reaction; if the level is low, the DNA product is observed in the second reaction. The nested PCR makes possible the detection of fewer than 10 intact cells of *Dekkera* strains in the original sample with no need for DNA purification steps.

Selective Media and Methods for Control

The genus displays a rather strict habitat specificity and has only been recovered from industrial products or equipment. All strains tested failed to develop on 50% (w/w) glucose–yeast extract agar but a number of different media have been used successfully. Yeast can be isolated on universal beer agar with 25 mg l^{-1} gentamicin and 50 mg l^{-1} oxytetracycline and 0.5% calcium carbonate (UBAC), incubated at 28°C for 5–6 days. Cycloheximide-resistant yeasts can be determined on UBAC containing 10 mg l^{-1} cycloheximide

(actidione), incubated at 28°C for 5–6 days. In lambic beer it is quite selective for *Brettanomyces*. For the isolation of these yeasts from wines, the following *Brettanomyces/Dekkera*-specific medium (DHSA) has been used: 5 g l^{-1} yeast extract, 5 g l^{-1} bacto-peptone, 5 g l^{-1} threalose, 45 g l^{-1} saccharose, 0.5 g l^{-1} monopotassium phosphate, 0.125 g l^{-1} potassium chloride, 0.125 g l^{-1} magnesium sulphate (7 H$_2$O), 0.0025 g l^{-1} ferric sulphate, 0.0025 g l^{-1} manganese sulphate (H$_2$O), 20 g l^{-1} bacto-agar, 0.030 g l^{-1} green of bromocresol, 0.05 g l^{-1} cycloheximide and 0.25 g l^{-1} sorbic acid. The pH is adjusted to 4.8 and autoclaved (10 min, 120°C). Penicillin and gentamicin are usually added just before seeding. Also biphenyl (0.010 g l^{-1}) may be added to culture media to inhibit mould growth. *Brettanomyces/Dekkera*-selective media containing 0.0024 l bacto-glycerol, 48.3 g wort agar and 0.05 g l^{-1} cycloheximide may be used for general purposes.

These yeasts are relatively sensitive to the effect of molecular sulphur dioxide (0.8 mg l^{-1}), therefore SO$_2$ may be useful in controlling the growth of this yeast. However, they can survive in the inside of barrels where the irregular and difficult-to-clean surface confers to these organisms an apparent resistance to the common preservative sulphur dioxide. Thus, once established in wood cooperage, elimination is extremely difficult.

Passive adherence of these yeasts to insects, predominantly to adult *Drosophila* flies, is the main mechanism of dispersal and transmission throughout fermenting juice and wines. The difficulty of eliminating *Brettanomyces* together with their rapid dissemination make these species among the more problematic yeasts.

See also: **Cider (Hard Cider)**. **PCR-based Commercial Tests for Pathogens**. **_Saccharomyces_**: Introduction; _Saccharomyces cerevisiae_. **Wines**: Microbiology of Winemaking; Specific Aspects of Oenology.

Further Reading

Boekhout T, Kurtzman CP, O'Donnell K and Smith MT (1994) Phylogeny of the genera _Haseniaspora_ (anamorph _Kloeckera_), _Dekkera_ (anamorph _Brettanomyces_), and _Eeniella_ as inferred from partial 26S ribosomal DNA nucleotide sequences. _Int. J. Syst. Bacteriol._ 44: 781–786.

Ciani M and Ferraro L (1997) Role of oxygen on acetic acid production by _Brettanomyces/Dekkera_ in winemaking. _J. Sci. Food. Agric._ 75: 489–495.

Fugelsang KC, Osborn MM and Muller CJ (1993) _Brettanomyces_ and _Dekkera_. Implications in Wine Making. In: Gump BH (ed.) _Beer and Wine Production, Analysis, Characterization, and Technological Advances._ P. 110. Washington, DC: American Chemical Society.

Ibeas I, Lozano I, Perdigones F and Jimenez J (1996) Detection of _Dekkera_ and _Brettanomyces_ yeasts in sherry wines by the nested polymerase reaction method. _Appl. Environ. Microbiol._ 62: 998–1003.

Kuniyuki AH, Rous C and Sanderson JL (1984) Enzyme-linked immunosorbent assay (ELISA) detection of _Brettanomyces_ contaminants in wine production. _Am. J. Enol. Vitic._ 35: 143–145.

BREVIBACTERIUM

Bart Weimer, Center for Microbe Detection and Physiology, Utah State University, Logan, USA

Introduction

Brevibacterium linens, originally known as IX, was changed to _Bacterium linens_ in 1910. In 1953, the name was changed back to _Brevibacterium linens_. Interest in finding a taxonomic niche for this organism was sparked again in the United States and Japan during the 1970s. This line of research continues, but today _Brevibacterium_ is recognized as a genus in its own right. The preoccupation with taxonomy has left a void in knowledge about industrially important characteristics of _B. linens_.

Currently, the genus _Brevibacterium_ is limited to _B. linens_ (the type strain is ATCC 9172), _Brevibacterium epidermidis_, _Brevibacterium iodinum_ and _Brevibacterium casei_, with _species incertae sedis_ including _Brevibacterium incertum_, _Brevibacterium acetylicum_, _Brevibacterium oxidans_, _Brevibacterium halotolerans_, _Brevibacterium frigoritolerans_ and _Brevibacterium rufescens_. Many investigators suggest that this genus should be transferred to the genus _Arthrobacter_ (especially _Arthrobacter globiformis_) because of the coryneform morphology of _Brevibacterium_. However, pigmentation, cell-wall composition and DNA–DNA hybridization analysis show _Brevibacterium_ is distinct from _Arthrobacter_. Formerly, _Brevibacterium_ was a repository for organisms that had no other place in the taxonomic scheme. Thus, many of the species in this genus are unrelated in their physiological and biochemical attributes. Many investigators suggest that _B. linens_ be placed in its own genus, separate from the other members of its current group. This discussion focuses on _B. linens_ because it is the most notable member for industrial purposes. However, in recent years _Brevibacterium lactofermentum_ has been studied extensively.

In the 1980s, the primary interest in this organism turned from taxonomic to biochemical. Most interest in this organism has centred around catabolism of amino acids (AA), particularly aromatic AA, and proteases. Early research indicates that methanethiol and other compounds produced by _B. linens_ are important in preservation and flavour development of surface-ripened cheeses. Albert et al (1994) published an excellent review of the early work done in Europe and the United States dating back to Duclaux in 1893. Subsequently, Hosono and Tokita published extensively on the metabolism of brevibacteria.

Pure Culture Characteristics

Brevibacterium is of interest industrially because it produces various products such as AA (especially glutamic acid and lysine) and enzymes important to cheese ripening. Hence, pure culture characteristics are described in terms of their use in cheese. _B. linens_ is a non-motile, non-spore-forming, non-acid-fast, Gram-positive obligate aerobe with a growth temperature range of 8–37°C and an optimum of 21–23°C. Strains isolated from human skin have a higher optimum growth temperature of 37°C. It produces rods in singlets, pairs or short chains ranging from 0.6 to 2.51 μm. With time, about 2 days, the rods are replaced with cocci about 0.6–1 μm. Rods predominate in the exponential phase and change to cocci in the stationary phase (**Fig. 1**). The cellular morphology change is associated with methionine concentration, growth medium pH, growth temperature, and aeration. *B. linens* also reduces nitrates to nitrites, and is urease and oxidase-negative, and lipase-, catalase-, litmus milk- and DNase-positive.

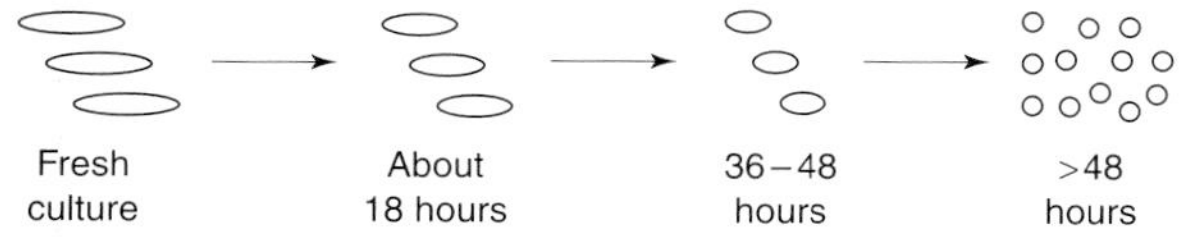

Figure 1 Cellular morphology change during incubation.

When grown on nutrient agar, colonies are opaque, small (0.5–1 mm in diameter) and convex, with a shiny, smooth surface. After 4–7 days of incubation the colonies become large, 2–4 mm in diameter. During growth via aerobic respiration, this organism produces a cell membrane-associated carotenoid pigment (**Fig. 2**). The colour formation of *B. linens* is highly variable and is manipulated by varying the oxygen, salt and light concentrations during cheese ripening. This may account for the different colony coloration in pure culture. White- or cream-coloured colonies, are not as fastidious about their vitamin or nitrogen requirements as orange colonies. Cheesemaking strains in pure culture can be manipulated to stop producing pigment. Cream-coloured colonies have been isolated from sources other than cheese, such as marine fish and human skin. Orange strains isolated from marine fish are remotely related to the industrially important *B. linens* strains.

B. linens is heat labile, resistant to drying, and survives carbohydrate starvation for at least 6 weeks. Cellular polysaccharide content remained constant after 56 days of starvation as did the basal rate of respiration (0.03% $^{14}CO_2 h^{-1}$). The low endogenous metabolism of *B. linens* is attributed to their ability to survive starvation. Salt (NaCl) tolerance is widely variable among strains and ranges from 0 to 20% with an average of 5%.

B. linens grows in a wide pH range, starting at pH 5.5 and continuing to 9.5 with the optimum being 7.0. As the salt concentration increases, the ability of the organism to grow at lower pH decreases. This pH dependence is overcome in surface-ripened cheese by growing a succession of organisms; yeasts initiate growth and raise the pH, and brevibacteria subsequently replace the yeast to produce bases in the cheese.

As reported by many investigators, *B. linens* is sensitive to antibiotics commonly used to treat mastitis, but is resistant to many of the common antibiotics (methicillin, nafcillin, cloxacillin, oxacillin, furadantin and nalidixic acid).

B. linens (ATCC 9175 and an unnumbered strain) produce a bacteriocin called linecin. This bacteriocin inhibits only other strains of *B. linens* (ATCC 8377 and 9172), but is similar to rosecins produced by *Microbacterium roseus* (ATCC 186 and EFO 3764). Culture filtrates of *B. linens* (ATCC 9174 and 9175) are inhibitory to *Clostridium botulinum*, *Staphylococcus aureus* and *Bacillus cereus*. The inhibitory substances are produced in a variety of growth media, with soybean meal broth being best. These compounds also have biochemical activity in animal models. The physiological role in animals is unclear and further work is needed to demonstrate the interaction between these bacteria and animal hosts.

Tryptone glucose extract broth, trypticase broth, and beef infusion broth do not support the production of bacteriostatic substances. Bacteriostatic compounds are soluble in polar solvents, heat stable (121°C for 25 min), pH stable (pH 2–12 for 20 h) and are associated with the yellow to red pigment. *B. linens* (ATCC 9172 and 8377) also produce antifungal agents effective against *Penicillium expansum*, *Aspergillus niger*, *Rhizopus nigricans* and *Mucor racemosus*. This activity is associated with production of methanethiol (CH_3SH). Concentrations in the range of 0.33–25 p.p.m. of methanethiol inhibit fungal spore germination and injure actively growing yeast cultures. *B. linens* produce methanethiol in concentrations of 2 p.p.m. in 60 h on a medium containing methionine.

Cheese Surface Growth Characteristics

Brevibacterium linens grows in association with salt-tolerant yeast on the surfaces of Limburger-style cheeses. The pioneer organisms, yeasts, show noticeable growth after 2–4 days. The yeasts are from a variety of species including: *Mycoderma*, *Debaryo-*

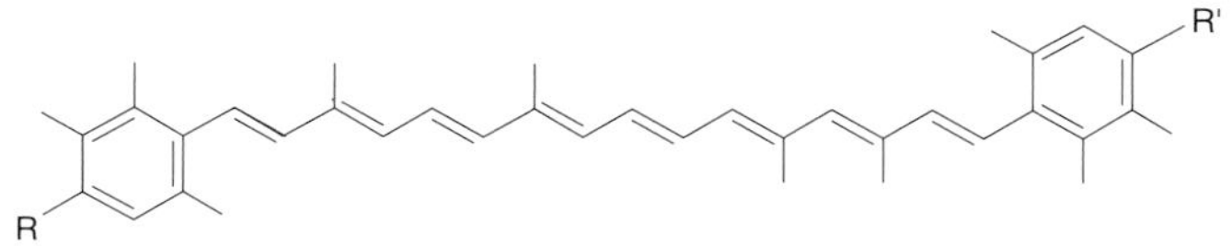

Figure 2 Carotenoid pigment found in brevibacteria. R,R′ = –H is isorenieratene; R = H, R′ = –OH is 3-hydroxyisorenieratene; R,R′ = –OH is 3,3′-dihydroxyisorenieratene.

myces and *Trichosporon*, depending on the cheese type. *B. linens* start to grow as the pH rises to about 5.9–6.5 because the yeasts use lactic acid. After 5–10 days, *B. linens* dominates the surface microflora of the cheese.

B. linens depends on these yeasts by raising the pH of the curd and produces growth factors and vitamins. The vitamin requirements of *B. linens* are strain dependent and vary from no requirement to specific requirements. The yeasts produce pantothenic acid, riboflavin, niacin and biotin that are needed for growth by *B. linens*. In pure culture, yeast extract is the only additive that increases both cell number and growth rate.

Studies on flavour development in Limburger cheese showed that *Candida mycoderma* and *Debaryomyces kloeckeri* are the pioneer organisms (10^7 yeast per gram of cheese) and they are gradually replaced by *B. linens*. Micrococci are also in the normal succession of this cheese. The yeasts produce volatile fatty acids, H_2S, carbonyl compounds, and large quantities of citric acid and n-butyric acid, leading some researchers to conclude that end products produced by the yeasts contribute to the flavour of Limburger cheese. The yeast end products are limited when compared to production of the same compounds by *B. linens*. *B. linens* is more proteolytic than either yeast in the same time period (48 h).

Micrococci are the organisms most often confused with *B. linens*. A number of media exist for the isolation of *B. linens* from the surface of cheese. LGCS agar is commonly used to isolate *B. linens* from the surfaces of cheese and is based on the lack of acid production from glucose. This medium includes beef extract, peptone, glucose, NaCl, $CaCO_3$, agar and pimafucine (to inhibit yeast). The recommended incubation time of 5 days at 25°C with daily observation for a clearing zone under or around the colony shows acid production and, therefore, not *B. linens*. Micrococci, lactococci, *Staphylococcus aureus* and coliforms produce distinct clearing zones on the selective agar and lactobacilli do not grow.

Biochemical Characteristics

Carotenoid Pigment

The primary pigment of *B. linens* is a yellow to orange aromatic carotenoid, 3,3′-dihydroxy-isorenieratene. The pigment has two minor forms corresponding to the monohydrate (3-hydroxyisorenieratene) and the hydrocarbon (isorenieratene). Molecular studies have shown that the pigments associate with the cell membrane and not the cell wall (Fig. 1).

Many investigators report the pigment production

of *B. linens* is related to cultural conditions. Colour production in *B. linens* is dependent on the dissolved oxygen and L-methionine concentration in the growth medium. Colony colour difference between plates incubated in air opposed to oxygen is also common. Cultures incubated in light increase in colour compared to those incubated in the dark. The organisms must be grown in light from the first day of growth for colour to develop. Colonies will not become coloured if the culture is started in the dark even if it is moved to light. Addition of 4% (w/v) NaCl to the medium decreases colour production by the colonies. However, drying the agar and adding 3% (w/v) NaCl and calcium carbonate to the growth medium increases pigmentation, and after 6 weeks of growth the colour disappears. Temperature has little effect on pigment production.

The biochemical characteristics of *B. linens* have received a great deal of attention in recent years. Research in Europe has produced more information about the biochemical capabilities of *B. linens* and more recently, attention has focused on aromatic amino acid and methionine metabolism to produce cheese flavour compounds.

Substrate Utilization

Substrates used by *B. linens* are divided into four classes: carbohydrate, organic acid, alcohol and amino acids. Utilization characteristics are dependent on strain and are used in numerical taxonomic studies to assign a taxonomic location. Over 90% of cheese strains require an organic nitrogen source, with the major source being glutamic acid and sometimes ammonia.

Carbohydrate Use *B. linens* is unusual in its use of sugar. *B. linens* does not convert sugars into their corresponding acids for cellular metabolism, rather it produces bases. Brevibacteria use a number of carbohydrates for carbon and energy sources (**Table 1**).

The best carbon and energy source for cell growth is lactate with acetate. Saccharose is variable in its use, whereas *B. linens* does not utilize sucrose or

Table 1 Possible carbon or energy sources for *Brevibacterium linens*

L-Arabinose	Acetate	Glucose
Xylose	Salicin	Fructose
Rhamnose	Glycerol	Erythritol
Adonitol	Indole	Hydrogen sulphide
Mannose	Mannitol	Dextrin
Galactose	Dulcitol	Raffinose
Sorbose	Sorbitol	Cellobiose
Inositol	Saccharose	Methyl gluconoside
Trehalose	Arbutin	Acetyl methyl carbinol
Maltose	Aesculin	Lactate

lactose. Glucose and glycerol are good for cell growth, but may decrease endogenous respiration. Neither salt concentration nor pH influence sugar use.

Organic Acid Use   *B. linens* metabolizes propionate, valerate, carpronate, heptanoate, 4-aminobutyrate and 4-hydroxybenzoate; but not citrate. This is interesting since the yeast that grow before *B. linens* on surface-ripened cheeses produce large amounts of citrate.

Alcohol Use   The ability of *B. linens* to utilize alcohols decreases with increasing molecular size (**Table 2**). Brevibacteria usually do not metabolize hexyl and heptyl alcohols.

Amino Acid Use   AAs are used extensively by *B. linens*, with many having inducible or repressible transport systems. Culture conditions, including carbon and energy source, growth temperature, salt and growth stage modulate transport of serine and aromatic amino acids.

  B. linens is considered indole-negative, indicating that it lacks lyases needed to split the C–C bond of the ring in aromatic amino acids. Catabolism of tryptophan results in an accumulation of anthranilic acid at optimum pH whereas at the pH of Cheddar cheese, tryptophan use stops. Use of phenylalanine and tyrosine results in phenylacetic acid and unknown non-growth metabolites, respectively. These two amino acids are also not used at pH 5.2.

Metabolic End Products

This area of study is of great interest in recent years since brevibacteria produce many compounds and enzymes important in cheese ripening and the AA fermentation industry. They have many capabilities that are associated with protein utilization, and are industrially important.

Amino Acid Production   Seven free AAs (Lys, Glu, Pro, Ala, Val, Leu and Phe) appear in greatest quantity when *B. linens* is used to ripen Trappist-type cheese. Glutamic acid is found in the greatest quantity followed by proline and leucine. The concentration of proline decreases as glutamic acid increases in the curd over time. It has been suggested that proline is converted to glutamic acid. Use of these bacteria for production of amino acids is common, with *Brevibacterium lactofermentum* use prevalent.

Amino Acid Decarboxylation   Decarboxylation of AAs releases CO_2 and yields the corresponding amine with the requirement of pyridoxal phosphate. The optimum pH is in the range 5–7 at 30°C for the decarboxylation of lysine, leucine and glutamic acid. Decarboxylation of AAs has been investigated in relation to flavour production by *B. linens*. *B. linens* was found to produce the greatest quantity of base when whey and 0.5% (w/v) acid-hydrolysed casein were used as substrates. Decarboxylation occurs between pH 4 and 8 with lysine being utilized most actively. None of the decarboxylases released CO_2 at 20°C. Glucose increases decarboxylation, but a corresponding decrease in pH does not occur. Amines resulting from decarboxylation reactions by *B. linens* are cadaverine (Lys), monomethyl amine (Ala), isoamylamine (Leu), γ-aminobutyric acid (Glu) and tyramine (Tyr), monoethylamine, dimethylamine, triethylamine and ammonia.

Amino Acid Transamination   Aminotransferases, with the coenzyme pyridoxal phosphate, are responsible for α-keto acid production during AA interconversion. These reactions are important in cheese-ripening and AA anabolism, but the exact mechanism is unclear.

Amino Acid Deamination   *B. linens* produce ammonia in substantial quantities from serine, glutamine and asparagine, and in minor amounts from threonine, arginine, alanine, glutamic acid, lysine and glycine. Ammonia production correlates with a decrease of lactate in cheese. Subsequently, it was found that *B. linens* produces small amounts of ammonia in the early stages of ripening but later produces large amounts of ammonia.

Oxidation–Reduction Reactions   These reactions have not been shown to be important in the metabolism of *B. linens*, and it seems likely that oxidases are important, since *B. linens* is catalase-positive, an ammonia producer and an obligate aerobe. Dehydrogenases, which use NAD^+ to accept electrons, convert L-amino acids, with addition of water, into their corresponding α-keto acids, ammonia and NADH.

Volatile Products   Investigations of the production of volatile fatty acids (VFA) by brevibacteria have focused on whole milk, butter fat, milk fat, carbohydrates and individual AAs as substrates. Many studies have demonstrated that brevibacteria asso-

Table 2   Alcohol usage by *Brevibacterium linens*

Alcohol	Acid produced
Ethanol	Acetic acid
Amyl alcohols	Valeric acid
Butyl alcohols	Butyric acid
Propanol	Propionic acid

ciated with Limburger cheese produce these compounds which are acidic, neutral or alkaline and produce typical flavours associated with cheese.

VFA production by *B. linens* from AAs is dependent on the medium with the best medium being whey plus acid-hydrolysed casein or whey plus glycine. Glycine, alanine, glutamic acid, leucine, aspartic acid, asparagine, methionine and cystine are metabolized to acetic acid primarily, whereas alanine, asparagine and cystine also yield caproic acid but leucine is converted into isovaleric acid.

VFA formation from carbohydrate sources has no influence with lactose. Galactose and glucose play important roles in the formation of VFA. Glucose has the greatest influence on VFA production; after 3 days of incubation at 21°C VFA concentration peaked. The optimum pH range for VFA production is 7 and 8 for glucose and galactose, respectively.

Acetic acid, n-butyric acid and caproic acid were the primary VFAs when the base medium was supplemented with butterfat. With this substrate incubation for 4 days at 21°C was required for peak production at pH 7. In whole milk, *B. linens* produces acetic acid, isovaleric acid and caproic acid. *B. linens* produces almost twice the amount of VFA produced by the yeast associated with Limburger cheese.

The primary volatile carbonyl compounds produced by *B. linens* are acetone, formaldehyde and 2-pentanone in whole milk. The volatile carbonyl compounds produced from AAs, carbohydrates and milk fat are acetaldehyde and acetone from any AA except glycine, tyrosine and methionine. Formaldehyde is produced from glycine, leucine, aspartic acid, tyrosine and 2-pentanone is produced from glutamic acid.

Acidic carbonyl compounds are derived from fatty acids and are direct precursors for methyl ketones. Glucose yields formaldehyde, acetaldehyde and 2-pentanone whereas pyruvic acid is converted into acetaldehyde. Casein and fat yield more volatile carbonyl compounds more than do carbohydrate sources. n-Butyric acid is the original fatty acid for acetone with the intermediate being α-ketobutyric acid. However, casein and milk fat are more important in volatile carbonyl compound production by *B. linens* than is glucose or pyruvic acid.

Methanethiol Production   Production of alkylthiols, specifically methanethiol or methyl mercaptan (CH₃SH), has been the subject of great interest in recent years. A putrid aroma arises with the appearance of the reddish colour in surface-ripened cheeses and in pure cultures of *B. linens*. Production of volatile sulphur compounds is strain variable within brevibacteria and isolates from human skin also produce

CH₃SH. Methionine is linked to growth of coryneform bacteria. Studies using [³⁵S]methionine demonstrated that it is converted into a number of sulphur compounds, including hydrogen sulphide, disulphides and mercaptans (especially methanethiol) by the bacteria on Trappist-type cheese, but not *B. linens* specifically.

Routes of CH₃SH production are demethiolation, deamination or transamination and demethiolation followed by deamination in brevibacteria. However direct conversion of methionine into CH₃SH is also possible via L-methionine-γ-lyase. Methanethiol production, and the capacity of the culture to produce methanethiol, depend on the dissolved oxygen concentration (optimum 25%), culture age (optimum 25 h), temperature (optimum 30°C) and pH (optimum 8–9). Glucose inhibits CH₃ formation and favours cell growth. AAs, other than methionine, have no effect on the production of CH₃SH. Lactate favours both cell growth and methanethiol production. Repression of methanethiol production by glucose is connected to the coenzyme pyridoxal phosphate or transport enzymes at the cell surface, because CH₃SH production is associated with intracellular concentrations of L-methionine.

Milk fat-coated microcapsules of *B. linens* containing cystine and methionine, showed that cells remain viable after 10 days of anaerobic conditions inside the microcapsules. Cells inside the microcapsules convert methionine into CH₃SH, both anaerobically and aerobically, but under aerobic conditions, the production rate is three times greater. However, these results do not agree. Some strains of *B. linens* (ATCC 8377, B11, B12, and B13) produce *S*-methylthioacetate and 2,3,4-trithiopentane in addition. *B. linens* in association with *Micrococcus* stains produce more *S*-methylthioacetate than pure cultures of *B. linens*.

Major Enzymes of Proteolysis and Lipolysis

The characterization of proteolytic and lipolytic enzymes in *Brevibacterium linens* is an old field of interest but there has been renewed interest because of the ability of this organism to accelerate cheese ripening. *B. linens* is active in the proteolysis and lipolysis of Limburger cheese since protein and fat breakdown products are abundant. From 1959 to 1970 about one paper per year was published, but in the 1970s a number of investigators published important work describing extracellular proteases of *B. linens*. Recently a single protease from *B. linens* ATCC 9174 was isolated and characterized.

Proteolytic Enzymes

Brevibacteria are proteolytic due to an extracellular protease and multiple aminopeptidases. There have been numerous reports of the proteolytic activity of *B. linens* with gelatin, casein, milk and paracasein as substrates and a variety of detection methods. The protease activity of *B. linens* is unusual because the enzyme activity curve cycles during the incubation time. The optimum incubation time for total cell density is 6 days, but the optimum incubation time for enzyme activity is 1 day with a rapid decrease in enzyme activity after 2 days. The optimum pH is 7 for proteolysis and neither glucose nor oxygen affects proteolysis in cheese. Glucose favours growth, but hinders production of extracellular proteases, and produces a difference in enzyme activities in preparations after 2 days of growth compared to preparations after 6–8 days of growth. Peptone, yeast extract, NaCl and K_2HPO_4 supplemented with casein increase protease activity. Cultures incubated at 20°C have the greatest enzyme activity after 24 h, but at 25°C, the maximum enzyme activity is delayed to 48 h. Activity cycles over time, but not because of temperature. The pure extracellular protease has optimum activity at pH 7.0 and 25°C and is sensitive to heat above 40°C. The best substrate for the extracellular protease is casein, although it shows some activity toward haemoglobin and albumin.

B. linens contains an intracellular protease that degrades casein at an optimum pH and temperature of 7.9 and 45°C. This intracellular protease is inhibited by reducing agents, metal chelating agents, mercury and *p*-hydroxymercuric benzoic acid. In addition, *B. linens*, isolated from Trappist-type cheese, contain polypeptidase activity.

Aminopeptidase activity is high and varies with growth conditions and medium components. The aminopeptidase is more heat stable than the protease, and has activity in a wide range of pH and temperature. When stored at 0–20°C the aminopeptidase is stable for one year at pH 8.0. The enzyme is specific for L-leucine, but activity is influenced by AA residues at the C-terminus with hydrolysis of dipeptides. The enzyme is composed of two subunits with positive cooperation, with subunit molecular weights of $48\,000 \pm 3000$ Da. Aminopeptidase is activated by cobalt and requires a minimum pre-incubation period of 1 h at 20°C. Inhibitory substances include (listed in order of their effectiveness): heavy metals, metal-complexing reagents and reducing agents. Aminopeptidase activity decreases, unexpectedly, with cadmium; this is unique to this enzyme. Some AAs also inhibit activity (arranged in order of their inhibition strength): histidine and serine, glutamic acid and cystine. Methanol, ethanol, propanol, butanol and amyl alcohols are also inhibitory.

Additional Uses of Brevibacteria

Brevibacteria are used on many industrial processes, despite the scant amount of publicly available information. Some specific applications of these bacteria include: insecticide degradation, amino acid production, waste-water treatment, vitamin production, polyacrylamide production, bronchodilator production and antimutagen production. Many of these applications are described in patents held by chemical companies. A recent use of these bacteria is in the dairy industry for the addition of flavour compounds, specifically in Cheddar cheese.

Genetic studies and engineering of brevibacteria are limited and this is limiting future engineering prospects. Most studies use *B. lactofermentum* (now known as *Corynebacterium glutamicum*). As more industrially important uses develop for these bacteria more detailed genetic studies will appear. The genome of *B. lactofermentum* is estimated to be about 3.0 Mb, whereas *B. linens* is about 3.1 Mb. Plasmid studies indicate that the plasmid pool varies with strain and many are cryptic, with sizes ranging from 3.4 to 6.8 kb. Efforts are underway to produce cloning vectors for these bacteria. Many of these plasmids are greater than 5.0 kb and use antibiotic selection markers.

Conclusion

Brevibacteria are important in many industrial processes. Despite the lack of detailed information surrounding their metabolism, more uses are being found. As these uses increase, more detailed information will appear. Use of the organisms in the food and chemical industry is leading the way for industrial applications. Many unique biochemical pathways found in these bacteria make them interesting to study.

See also: **Cheese**: Microbiology of Cheese-making and Maturation; Role of Specific Groups of Bacteria; Microflora of White-brined Cheeses.

Further Reading

Albert J, Long H and Hammer B (1944) Classification of the organisms important in dairy products. IV. *Bacterium linens*. Bulletin no. 328, Agricultural Experiment Station Iowa State College.

Archer D (1989) Biology of *Corynebacterium glutamicum*: a Molecular Approach. In: Hershberger A (ed.) *Genetic and Molecular Biology of Industrial Microorganisms*. P. 27. Washington, DC: ASM Press.

Crombach W (1974) Relationships among coryneform bacteria from soil. Cheese and sea fish. *Antonie van Leeuwenhoek* 40: 361.

Dias B and Weimer B (1998) Purification and characterization of methionine γ-lyase from *Brevibacterium linens* BL2. *Appl. Environ. Microbiol.* 64: 3327.

Ferchichi M, Hemme D and Nardi M (1987) Na⁺-stimulated transport of L-methionine in *Brevibacterium linens* CNRZ918. *Appl. Environ. Microbiol.* 53: 2159.

Foissy H (1978) Some properties of aminopeptidases from *Brevibacterium linens*. *FEMS Microbiol. Lett.* 3: 207.

Jones D (1978) An evaluation of the contributions of numerical taxonomic studies to the classification of coryneform bacteria. In: Bousfield IJ and Calley AG (eds) *Coryneform Bacteria*. P. 33. London: Academic Press.

Tokita F and Hosono A (1972) Studies on the extracellular protease produced by *Brevibacterium linens*. I. Production and some properties of the extracellular protease. *Jap. J. Zootech. Sci.* 43: 39.

Brewer's Yeast see *Saccharomyces*: Brewer's Yeast.

BROCHOTHRIX

Richard A Holley, Department of Food Science, University of Manitoba, Winnipeg, Canada

Introduction

Brochothrix thermosphacta is translated to mean loop filaments sensitive to heat, and aptly describes this bacterium. The organism was originally included in the genus *Microbacterium* but because it was not particularly thermotolerant, had a DNA base composition (mol% G + C = 36) lower than 58–64% of other members of the genus, did not have an operational tricarboxylic acid (TCA) cycle and contained *meso*-diaminopimelic (*m*-DAP) in the peptidoglycan, it was moved from this genus and tentatively placed in the family Lactobacillaceae. Recently it has been shown that it more closely resembles *Listeria* since catalase activity and cytochromes are present in both genera (**Table 1**) and there is 16S rRNA oligonucleotide sequence homology. *Brochothrix* and *Listeria* are included in the *Clostridium–Lactobacillus–Bacillus* supercluster of taxa at present.

Currently, the genus *Brochothrix* contains two species, *B. thermosphacta* and *B. campestris*, which are biochemically similar. Both are indigenous to the farm environment and can be found in soil and on grass, but only *B. thermosphacta* has been found associated with animal and food microflora. Isolation of the latter organism has frequently been made from hogs and pork carcasses as well as from beef, lamb, poultry, fish and a variety of other foods (frozen vegetables, tomato salad and dairy products). Isolations have also been made from processing equipment and animal faeces.

B. thermosphacta has drawn considerable attention because it frequently causes early spoilage of fresh and cured meats. This is due in part to its ability to tolerate high concentrations of salt and grow at both low water activity (a_w) and low temperature in the presence of little oxygen (>0.2%). Nonetheless, the

Table 1 Features distinguishing *Brochothrix* from other Gram-positive non-sporing rods

Feature	Brochothrix	Listeria	Lactobacillus	Carnobacterium	Kurthia	Erysipelothrix
Rod diameter (μm)	0.6–0.8[a]	0.4–0.5	0.5–1.6	0.5–0.7	0.7–0.9[a]	0.2–0.5
Facultatively anaerobic or microaerophilic	+	+	+	+	−	+
Catalase	+	+	−	−	+	−
Motility	−	+[b]	−	±[c]	+	−
Growth at 37°C	−[d]	+	+	±[c]	±[c]	+
Growth on STAA agar	+	−	−	−	−	−
Peptidoglycan diamino acid	*m*-DAP	*m*-DAP	*m*-DAP, lysine, ornithine	*m*-DAP	Lysine	Lysine

[a]Pleomorphic.
[b]At 20–25°C.
[c]Species dependent.
[d]Occasional strain grows.
m-DAP, *meso*-diaminopimelic acid.
Adapted from Dodd and Dainty (1992), with permission

exact range of the natural habitat of this organism and *B. campestris* has not been fully characterized. This article focuses on *B. thermosphacta* and where information is available on *B. campestris*, it is presented.

Brochothrix thermosphacta
Characteristics

Brochothrix thermosphacta is a Gram-positive filamentous rod measuring 0.6–0.8 μm in diameter and 1–2 μm long. Cells occur individually, in chains or in characteristic long filaments that often fold into loops or knots. In older cultures, coccoid forms are found which yield rod-shaped cells upon subculture. Cells do not form spores, do not have capsules and are non-motile. The organism is facultatively anaerobic and produces non-pigmented colonies. Catalase activity and cytochromes are present. However, tests for catalase should be conducted using cells grown on specified media (e.g. all purpose tween, APT, a commercial medium, Difco or RBL) within the optimal temperature range for the organism (20–25°C). Cells cultivated at higher temperature or on other media may lose their catalase activity. *Brochothrix* is a psychrotroph and will grow at 0 to 30°C, but over 30°C growth seldom occurs. They are non-haemolytic and non-pathogenic to humans. *Brochothrix* are thermosensitive and it is generally agreed that they do not survive exposure at 63°C for 5 min. The *D* value at 55°C is 0.1 min and the *Z* value has been calculated to be 8°C. Fermentation of glucose mainly gives rise to L(+)-lactic acid but small amounts of acetic and propionic acids have been detected. Ethanol can be formed anaerobically in glucose-limited continuous culture. Under aerobic conditions glucose is metabolized to acetoin, diacetyl, plus acetic, isobutyric and isovaleric acids as well as a number of other branched chain fatty acids and alcohols. Fatty acid residues are formed from amino acids and not by lipolysis. Several of these products are organoleptically unpleasant, yielding sour, acidic, malty, musty, sickly sweet or sweaty odours and explain why *B. thermosphacta* contributes to substantially shortened food product shelf life. Acetoin is only produced aerobically and precursors are glucose, glycerol or ribose. Interestingly, indole and H_2S are not produced.

The organisms are methyl red and Voges–Proskauer-positive and reduce both potassium tellurite and tetrazolium salts at 0.01% (w/v). Added citrate cannot be utilized. Enzymes of the TCA cycle are almost completely absent when cells are grown in a complex medium; however, in chemically defined media these enzymes may be active enough to provide substrates for synthesis but not active enough to yield energy. The organism forms acid weakly but no gas from a number of carbohydrates (arabinose, cellobiose, dulcitol, glucose, inositol, lactose, maltose, mannitol, sucrose and xylose). Organic growth factors (biotin, cysteine, lipoate, nicotinate, pantothenate, *p*-aminobenzoate and thiamine) are required for both aerobic and anaerobic growth in glucose–mineral salts medium. Pyruvate, acetate, propionate and citrate (as mentioned), cannot be used as sole sources of carbon. The cell wall peptidoglycan is directly cross-linked by *m*-DAP. Cellular content of long-chain fatty acids is characteristic and consists mainly of the straight chain saturated iso- and anteiso-methyl-branched chain types. *B. thermosphacta* may be distinguished from *Listeria* spp. by its greater content of (anteiso-$C_{15:0}$) 12-methyl tetradecanoic acid (41–70%) compared with the 22–31% present in *Listeria*. The major respiratory quinones present in both genera are menaquinones and are not useful in differentiation.

Brochothrix contains a glycerol esterase but this lipase attacks short-chain fatty acids within the temperature range of 35–37°C and it has no activity at 20°C. Tributyrin and Tween 60 are utilized as substrates but not other tweens or beef fat. Lecithinase was present in just over half the strains tested. *Brochothrix* are essentially non-proteolytic and cannot attack either casein or gelatin. On meat its activities are largely confined to exposed or cut surfaces. The organisms are unable to hydrolyse arginine and have no effect on the meat protein myoglobin. Nitrate is not reduced to nitrite by these organisms.

Brochothrix is capable of growth over a pH range of 5.0–9.0 (optimum pH 7.0). All strains can grow in 6.5% NaCl and some grow in 10% NaCl. Under aerobic conditions these organisms grow in substrates with a_w of 0.96–0.94 at 20–25°C. Under anaerobic conditions growth is more restricted by low temperature, pH and low a_w. Nitrite is slightly more inhibitory toward *Brochothrix* than lactobacilli but the former can grow in up to 100 p.p.m. nitrite at ⩾pH 5.5 and 5°C, aerobically in the presence of 2–4% NaCl. Except for pH, these conditions approximate the average composition of cured meat products and the manner in which they are often stored. In the absence of oxygen, or if the nitrite concentration is doubled to 200 p.p.m. at this pH, growth is inhibited. The inclusion of CO_2 in growth atmospheres is not inhibitory to *Brochothrix* until concentrations reach 50% provided oxygen is present. Low concentrations of oxygen have no effect on growth rate until they fall below 0.2%.

Comparison of *Brochothrix* Species

The two species of *Brochothrix* can be distinguished on the basis of several biochemical differences. *B. campestris* does not grow in the presence of 8% NaCl within 2 days or in the presence of 0.5% potassium tellurite, both characteristics possessed by *B. thermosphacta*. In contrast, *B. campestris* produces acid from rhamnose and hydrolyses hippurate whereas *B. thermosphacta* does not. The end products of glucose metabolism by *B. thermosphacta* have been intensely studied because of their impact on meat spoilage, but those produced by *B. campestris* (which is not known to be present in food) have not yet been documented. *B. campestris* has been shown to produce a bacteriocin, brochocin-C, which was active against *B. thermosphacta*, a variety of lactobacilli, *Listeria* spp. and other Gram-positive bacteria. *B. thermosphacta* is not known to produce bacteriocins, but more study is needed.

Although little work has been done on the serology of *Brochothrix* spp., investigations of bacteriophage specificity among isolates of *B. thermosphacta* from beef have been conducted. The 14 different phage lysotypes that were identified showed intra-genic specificity with some indication that support for further speciation of isolates from this genus may occur in the future. Simultaneous taxonomic work based on esterase gel electophoresis supports this contention.

Isolation and Enumeration

Normally present in meat and meat products stored aerobically or vacuum packed at chill temperatures, *B. thermosphacta* is usually detected in such samples without enrichment. This organism may be recovered from stored meats if present by directly plating swabs of meat surfaces or suitable dilutions of macerated meat in 0.1% (w/v) peptone directly onto suitable media such as glycerol nutrient agar. The latter is prepared by dissolving: 20 g peptone; 2 g yeast extract; 15 g glycerol; 1 g K_2HPO_4; 1 g $MgSO_4.7H_2O$, and 13 g agar in 1 l distilled water and adjusting the pH to 7.0. The medium is autoclaved at 121°C for 15 min. This medium will allow growth of a variety of other bacteria as well (*Kurthia* spp., pseudomonads, staphylococci and lactobacilli). The direct selective isolation of *Brochothrix* on STAA medium is the procedure of choice and normally enrichment is not believed necessary. STAA medium is prepared as for glycerol nutrient agar but when, after autoclaving, the sterile liquid reaches 50°C the following solutions, prepared with sterile distilled water, are added: streptomycin sulphate to a final concentration of 500 µg ml^{-1}; actidione to 50 µm ml^{-1}; and thallous acetate to 50 µm ml^{-1}. After these additions, the liquid

is mixed well and dispensed in Petri plates and solidified. These can be stored for up to 2 weeks at 4°C before use. Appropriate sample dilutions are spread on the agar surface and plates are incubated at 20–25°C for 2–3 days. Almost all colonies that develop (whitish colour, 1–4 mm in diameter) are *Brochothrix* but some pseudomonads, if present in the sample, will grow on this medium. The latter may be detected by their positive-oxidase reaction following flooding of the plate with a fresh 1% solution of tetramethyl-*p*-phenylenediamine dihydrochloride. Oxidase-positive colonies become blue whereas the oxidase-negative *Brochothrix* remain uncoloured. The selectivity of STAA is based on the use of a high concentration of streptomycin sulphate which inhibits many Gram-negative and some Gram-positive bacteria, especially the coryneform bacteria that morphologically resemble *Brochothrix* spp. Thallous acetate and actidone inhibit practically all yeasts as well as many aerobic and facultatively anaerobic bacteria but not all lactobacilli and streptococci are inhibited by the 0.005% thallous acetate present in STAA. Many are inhibited by the presence of streptomycin. Nonetheless, STAA is not perfectly selective and difficulty can be encountered with faecal samples where *Brochothrix* are present in low numbers relative to other organisms. Normally bacilli, coryneforms, lactobacilli and streptococci do not grow on STAA and growth on STAA is used as a confirmatory test for *Brochothrix*. To further improve selectivity some success has been obtained by the addition of nalidixic acid (5 µg ml^{-1}) and oxacillin (5 µg ml^{-1}) to the original STAA medium. This formulation has been used to isolate both *Brochothrix* species from soil and grass. In another modification of STAA for recovery of *Brochothrix* spp. from meat and meat products, blood agar base (Merck) was used following supplementation with (per litre): 2 g yeast extract; 1 g K_2HPO_4; 0.8 g $MgSO_4.7H_2O$; 0.35 g Na_2CO_3; 10 g inositol; plus 10 ml of a 0.3% solution of neutral red as indicator. After pH adjustment to 7.0, autoclaving and cooling to 50°C, 0.5 g l^{-1} of filter-sterilized streptomycin sulphate is added. Streptomycin is the major selective agent and *Brochothrix* spp. produce acid from inositol to give pink colonies.

It is not known to what extent the incorporation of inhibitors including antibiotics in media for the direct recovery of *Brochothrix* from food and environmental samples may have on stressed or injured organisms. This is particularly true of thallous acetate where more study is needed. The finding that 1 of 25 strains of *Brochothrix* was sensitive to the presence of streptomycin in STAA suggests that the selectivity of this medium may restrict the isolation of some members of the genus.

Alternative Rapid Detection of *Brochothrix*

Based on 16S rRNA sequencing data, polymerase chain reaction (PCR) primers and probes were developed to yield a method capable of genus-specific identification of *Brochothrix*. Although the method was initially successful, it was of limited value because detection sensitivity was not as good as that obtainable by the STAA medium when used with food samples. When present, staphylococci reduced sensitivity further.

International Guideline for *Brochothrix* Enumeration

The Nordic Committee on Food Analysis (NMKL) completed a controlled multi-laboratory, blinded study on the use of STAA for recovery of *Brochothrix* strains in the presence of the natural microflora isolated from food samples. The repeatability and reproducibility of the method were good but the number of false positives was higher than desirable. The Committee recommended that STAA should be incubated for better defined periods at a more precise temperature and specified 48 ± 3 h at 25 ± 1°C. In addition to the test for oxidase, a catalase test was deemed necessary when lactobacilli were suspected of being present in samples. They also noted from other work that actidione did not improve STAA selectivity and suggested it not be included in the medium formulation. The thus modified STAA medium and procedure for the recovery of *Brochothrix* spp. from food was adopted as an official method by NMKL.

Importance to the Food Industry

Since *Brochothrix* spp. are non-pathogenic to humans, these organisms are of importance because they cause premature spoilage of meat and meat products by virtue of their production of objectionable odours in refrigerated products packaged with residual concentrations of oxygen greater than 0.2%. This can occur even though they may not be the dominant population of bacteria present in samples. Once levels of about $5 \log_{10}$ cfu g^{-1} or cm^{-2} are reached, sensory evidence of their presence can lead to product rejection. They do not cause discoloration of meat pigments.

Brochothrix spp. are natural contaminants on food animal carcasses and inevitably find their way into meat processing plants where they can be isolated from equipment surfaces. They do not survive the thermal process normally used for cooked products but recontaminate these during packaging operations. *Brochothrix* spp. are more of a problem on cured meats which have higher pH (6.2–6.5) than fresh meats (pH 5.3–5.5) and because cured meats are often stored at higher temperature during retail distribution and display (< 9°C). Provided good oxygen barrier films for vacuum packages are used (having an O_2 permeability of < 15 cm^{-3} m^{-2} day^{-1} atm at 23°C and 75% RH) such as polyvinylidene chloride (PVDC)-based films, *Brochothrix* spp. will not cause problems. Improvements in O_2 barrier film materials and reduced costs for their production mean that this group of organisms will be of lesser importance to the food industry in the future even though product shelf-life extensions to 60 days are the current benchmark.

B. thermosphacta can form the dominant portion of the microflora on refrigerated meat and meat products when stored in air, under vacuum or on meat of normal pH stored under high O_2-modified atmosphere. However, they are a minor part of the microflora of these products when stored under 100% CO_2 or when CO_2–N_2 mixed atmospheres are used for packaging meat products. These observations are related to the greater ability of this group of organisms to grow at lower pH in the presence of O_2. *Brochothrix* spp. are innocuous when O_2 is absent from the packaging atmosphere and behave in a manner similar to homofermentative lactic acid bacteria under these conditions, producing mainly lactic acid.

In the presence of measurable O_2, growth of *Brochothrix* is unaffected by the presence of other organisms and malodorous metabolic products are generated. *Brochothrix* spp. do not grow anaerobically at pH < 6.0 and this is the reason that they are infrequently identified as a problem in fresh meats of normal pH that are vacuum packaged with suitable O_2 barrier films. They are present in dry fermented sausage but the pH after initial fermentation is sufficiently low (< 5.3) to retard *Brochothrix* development. Numbers are usually significantly lower than $5 \log_{10}$ cfu g^{-1}.

Historically this group of organisms has been a continual problem in refrigerated retail-ready British fresh sausage where $\leqslant 450$ p.p.m. SO_2 is permitted as a preservative. Since O_2-permeable film is traditionally used for packaging to maintain meat pigment colour and because *Brochothrix* can grow in the presence of up to 1000 p.p.m. SO_2, their influence on product shelf-life is significant. Sulphite keeps the normally dominant pseudomonads in check, providing opportunity for growth and spoilage by *Brochothrix*. Improved meat plant sanitation and handling practices can have a major impact on reducing the prevalence of these organisms.

Brochothrix spp. can dominate in meat packages where products are preserved under high O_2-modified atmosphere but they are inhibited in packaging systems where $\geqslant 50\%$ CO_2 is used with very low or no residual O_2. At 10°C they are overgrown by lactobacilli, particularly in reduced O_2 environments where packaging films of low permeability are used. It may be anticipated that *Brochothrix* will become of even greater importance in determination of fresh meat shelf life where meats are prepared for centralized distribution as retail-ready cuts. In these systems, which are gaining industry popularity, retail-ready cuts are prepared at a central cutting facility where they are packaged on trays in traditional high gas-permeable films. Packages are grouped together and overwrapped with a high barrier film and back-flushed with CO_2 or N_2 to achieve very low residual O_2 in the 'master package'. Alternative systems use high O_2-modified atmospheres in master packages, ostensibly to maintain meat pigment stability, particularly with beef. However, very low residual O_2 (< 300 p.p.m.) also fosters colour stability and delays microbial growth. From the central site, meat is distributed to retail stores where it is displayed in the primary package upon its removal from the master pack without additional cutting or manipulation. Provided there is good temperature control during master package storage (-1.5 ± 0.5°C) fresh meat products can be held for 3 weeks before retail display and achieve the same retail display shelf-life as freshly cut meat. These systems, particularly the master packages containing high O_2-modified atmosphere, provide almost ideal conditions for growth and spoilage by *Brochothrix* spp., if present.

It is also of interest that in the US permission has been granted for the irradiation of red meats as well as poultry. *Brochothrix* may be able to establish themselves as the dominant organisms in meats preserved in this manner because they are about 10 times more resistant to irradiation than the pseudomonads, particularly if high permeability films are mandated for packaging irradiated meats.

See also: **Listeria**: Introduction; Detection by Classical Cultural Techniques. **Meat and Poultry**: Spoilage of Meat; Spoilage of Cooked Meats and Meat Products. **Total Viable Counts**: Spread Plate Technique.

Further Reading

Borch E, Kant-Muermans M-L and Blixt B (1996) Bacterial spoilage of meat and cured meat products. *International Journal of Food Microbiology* 33: 103–120.

Dodd CER and Dainty RH (1992) Identification of *Brochothrix* by intracellular and surface biochemical composition. In: Board RG, Jones D and Skinner IA (eds) *Identification Methods in Applied and Environmental Microbiology*. Pp. 297–30. London: Blackwell Scientific.

Egan AF and Grau FH (1981) Environmental conditions and the role of *Brochothrix thermosphacta* in the spoilage of fresh and processed meat. In: Roberts TA, Hobbs G, Christian JHB and Skovgaard N (eds) *Psychrotrophic Microorganisms in Spoilage and Pathogenicity*. P. 211. London: Academic Press.

Feresu SB and Jones D (1988) Taxonomic studies on *Brochothrix, Erysipelothrix, Listeria* and atypical lactobacilli. *Journal of General Microbiology* 134: 1165–1183.

Gardner GA (1981) *Brochothrix thermosphacta* (*Microbacterium thermosphactum*) in the spoilage of meats: a review. In: Roberts TA, Hobbs G, Christian JHB and Skovgaard N (eds) *Psychrotrophic Microorganisms in Spoilage and Pathogenicity*. P. 139. London: Academic Press.

Holzapfel WH (1992) Culture media for non-sporulating Gram-positive food spoilage bacteria. *International Journal of Food Microbiology* 17: 113–133.

Kandler O and Weiss N (1986) Regular, nonsporing Gram-positive rods. In: Sneath PHA, Mair NS, Sharpe ME, and Holt JG (eds), *Bergey's Manual of Systematic Bacteriology*. 2: 1208–1253. Baltimore: Williams and Wilkins.

Peterz M (1992) Evaluation of method for enumeration of *Brochothrix thermosphacta* in foods. *Journal of AOAC International* 75: 303–306.

Skovgaard N (1985) *Brochothrix thermosphacta*: comments on its taxonomy, ecology and isolation. *International Journal of Food Microbiology* 2: 71–79.

BRUCELLA

Contents
Characteristics
Problems with Dairy Products

Characteristics

J Theron and **T E Cloete**, Department of Microbiology and Plant Pathology, Faculty of Biological and Agricultural Sciences, University of Pretoria, South Africa

Introduction

The bacterial genus *Brucella* is a genetically homologous group containing six species designated primarily on the basis of host specificity. Brucellae are Gram-negative facultative intracellular pathogens which cause infectious disease of the genitourinary tract of sheep, goats, pigs, cattle, dogs and other animals, and of the reticuloendothelial system of humans. Humans usually acquire the disease through ingestion of contaminated livestock products; through contact with infected animals (e.g. among shepherds, farmers and veterinarians); and through inhalation of infectious aerosols (e.g. by workers in abattoirs and microbiology laboratories). Brucellosis (Malta fever, undulant fever) is therefore primarily a contagious disease of animals that is also transmittable to humans (zoonosis). This has made brucellosis a problem of both public health and economic significance, not only because of direct or indirect transmission from infected animals to humans, with consequent illness, physical incapacity and loss of personnel, but also because it causes diminution of much needed foodstuffs, especially animal proteins, which are essential for human health and wellbeing.

Brucella Species

Brucellae are facultative intracellular bacteria that can infect many species of animals as well as humans. The genome of *Brucella* contains two chromosomes of 2.1 and 1.5 Mb, respectively. Both replicons encode essential metabolic and replicative functions and therefore are chromosomes and not plasmids. Based on DNA–DNA hybridization studies, the genus *Brucella* is a highly homogeneous group with all members showing greater than 95% DNA homology, thus classifying *Brucella* as a monospecific genus. Six nomen species are recognized within the genus *Brucella*: *B. abortus*, *B. melitensis*, *B. suis*, *B. ovis*, *B. canis* and

B. neotomae. This classification is mainly based on the difference in pathogenicity and in host preference. Within the respective species, seven biovars are recognized for *B. abortus*, three for *B. melitensis* and five for *B. suis*. The other species have not been differentiated into biovars, although variants do exist. The species and biovars are currently distinguished by differential tests based on phenotypic characterization of lipopolysaccharide antigens, phage typing, dye sensitivity, CO_2 requirement, H_2S production, and metabolic properties.

B. abortus primarily infects cattle but is transmitted to buffaloes, camels, deer, dogs, horses, sheep and humans. *B. melitensis* causes a highly contagious disease in sheep and goats although cattle can also be infected. It is highly infectious in humans. *B. suis* covers a wider host range than most other *Brucella* species. Biovars 1 and 3 infect swine primarily, biovar 2 causes infection in European wild hares, biovar 4 is responsible for infection in reindeer and wild caribou and biovar 5 was isolated from rodents in Russia. All biovars can be transmitted to humans with the possible exception of biovar 2. *B. canis* infects dogs, but is occasionally transmitted to humans, causing a mild type of brucellosis. *B. ovis* primarily infects sheep, but goats are susceptible to the disease by experimental infection. *B. neotomae* is only known to infect the desert wood rat under natural conditions, and no other cases have been reported. Of all the species, *B. melitensis* occurs most frequently in the general population and it is the most pathogenic and invasive species of *Brucella*, followed, in order by *B. suis* and *B. abortus*. More recently, isolations of previously unidentified species of *Brucella* have been reported from marine mammals. The strains isolated from marine animals form a separate group and have been unofficially designated *B. maris*. At least two subdivisions of this strain can be distinguished, corresponding approximately to strains isolated from cetaceans and seals, respectively.

Morphology and Physiology

Brucellae are small non-motile, non-sporing Gram-negative capnophilic coccobacilli. Although coccobacillary forms are predominant, cocci and longer rods (0.5–0.7 µm by 0.6–1.5 µm), occurring either singly, in pairs or short chains may be observed. Resting stages are not known. Most *Brucella* strains are slow-growing fastidious organisms on primary isolation and grow poorly on nutrient media unless supplemented with 5–10% serum or blood. Growth occurs aerobically and many strains require increased (i.e. 5–10%) carbon dioxide for optimal growth, but no growth occurs under strict anaerobic conditions. The optimal growth temperature is 37°C but growth occurs within a temperature range of 10–40°C. The optimal pH range is 6.6–7.4

The biochemical characteristics of brucellae are summarized in **Table 1**. The brucellae use carbohydrates but do not produce detectable amounts of either acid or gas. Some biotypes produce H_2S gas, and some are susceptible to the growth-inhibiting effect of the dyes, basic fuchsin and thionin. The growth of many strains is improved by calcium pantothenate and *meso*-erythritol.

Importance to the Food Industry

Epidemiology of the Disease

Worldwide, brucellosis remains a major source of disease in humans and domesticated animals. Epi-

Table 1 General characteristics of the genus *Brucella*

Haemin (X factor)	Not required
NAD (V factor)	Not required
Catalase	+
Oxidase	+ (except *B. neotomae* and *B. ovis*)
Urease	Variable
H_2S	Variable
Nitrate reduction	+ (except *B. ovis*)
Methyl red	–
Voges–Proskauer	–
Indole	–
Hugh and Leifson's O/F medium	Variable
Litmus milk	No change or may render it alkaline
Release of *o*-nitrophenol from ONPG O-Nitrophenyl β-D-Galactopyranoside	–
Citrate	–
Gelatin liquefaction	–

demiological studies have shown that the risk of transmission of the disease to: other animals and humans is closely related to the greatly expanded international and national trade in live animals, animal products and animal feedstuffs; methods of processing milk for butter, cheese and other products; standards of animal and personal hygiene; the growth of urbanization, coupled with the increased numbers of domesticated or half-wild animals living in close association with humans in cities which exposes more people to zoonoses; tourism and other movements of people; and new systems of animal farming which may lead to changes in the ecology that disseminate and increase animal reservoirs of zoonoses.

The reported incidence and prevalence of the disease vary widely from country to country and in different regions within a country. With the exception of countries where it has been eradicated, bovine brucellosis caused mainly by *B. abortus* is the most widespread form and is prevalent in South America as well as in developing regions in Africa, Asia and Australasia. Sheep and goats and their products remain the main source of human infection. In humans, ovine and caprine brucellosis caused by *B. melitensis* has a limited geographic distribution, but remains a significant problem in the Mediterranean basin of Europe, western Asia, and parts of Africa and Latin America. *B. melitensis* in cattle has emerged as an important problem in some southern European countries, Israel, Kuwait and Saudi Arabia.

Although few recent outbreaks of disease by *B. suis* biovar 4 have been reported, foci of the infection persist in the Arctic regions of North America and Russia. *B. ovis* infection appears to be distributed in all major sheep-raising countries, but *B. ovis* has not been demonstrated to cause overt disease in humans. *B. canis* can cause disease in humans, notably in dog handlers, laboratory workers and children with infected pet dogs. However, this is rare even in countries where the infection is common in dogs. *B. canis* infection has been microbiologically confirmed in a number of countries, such as the United States, Mexico, Argentina, Spain, China, Japan and Tunisia. The recent isolation of distinctive *Brucella* strains from marine animals extends the ecologic range of the genus, and possible pathogenicity and zoonotic potential.

Mechanism of Entry into Food and Transfer to Humans

Although brucellosis is a notifiable disease in many countries, the disease is often unrecognized and unreported, and the true incidence of brucellosis is therefore not known. The number of infections that occur each year may be 10–25 times higher than the reported figure. The reported incidence in endemic-disease areas varies widely, from less than 0.01 to more than 200 per 100 000 population. Despite being primarily a contagious disease of domesticated animals, humans contract the disease through various means. In the more susceptible species of domestic animals (cattle, goats, sheep and pigs), enormous multiplication of brucellae occurs in the uterus in the latter part of pregnancy and, to a lesser extent, in the mammary glands during lactation. Abortion leads to massive contamination of the environment. Aborted and infected placental material may contain up to 10^{13} bacteria per gram. On the other hand, excretion of brucellae in the milk, usually in much smaller numbers, may continue for years. Thus, milk and dairy products are frequently the source of food-borne brucellosis. Moreover, brucellae are intracellular parasites of the reticuloendothelial system and penetrate the whole body of the infected host. *B. melitensis* and *B. suis* transmit more easily to humans than *B. abortus*. The usual portal of entry in humans is by mouth, either directly by consuming infected dairy and meat products, or indirectly through contact with hands contaminated during work.

Direct contact with infected animals and animal products, such as dung, urine, uterine discharge, and abortion products, also influences risk to the population. Hence, individuals recognized to be at increased risk include shepherds, farmers, veterinarians, butchers and meat packers. Nomadic herdsmen in developing countries are continually exposed to brucellosis through their daily contact with animals. Caring for newborn animals in the family dwelling greatly increases the risk of human infection. In addition, water used by cattle and for drinking and bathing by humans serves as another source of infection. Arctic dwellers are at increased risk of infection, since both wild and semi-domesticated reindeer supply the native population with milk, meat and clothing and brucellosis is enzootic in these animals.

In addition to the intimate contact that rural inhabitants have with their animals, national or local dietary customs and habits also contribute to the transmission of brucellosis to human populations. Well-documented examples of dietary practices that expose both rural and urban populations to food contaminated with *Brucella* are the habit of the Mongolians of drinking airig (fermented mares' milk), the Eskimos of eating bone marrow and uncooked liver and kidneys from freshly killed reindeer, the Sudanese of eating raw liver and other offal with spices (umfitfit or Marrara), and the Peruvians of consuming fresh cheese made from unpasteurized goats' milk.

Inhalation of infectious aerosols is a recognized occupational hazard in workers in abattoirs and microbiology laboratories. Accidental self-inoculation or corneal contamination with the vaccine strains of *B. abortus* strain 19 and *B. melitensis* strain Rev 1 has been reported. Although examples of human-to-human transmission by blood transfusions, bone marrow transplantation or sexual contact have been occasionally reported, they are seemingly insignificant. Neonatal brucellosis has also been reported, suggesting the possibility of transplacental transmission during pregnancy or the time of delivery.

Fate of Bacteria During Processing and Storage

Although brucellae are resistant to environmental stress, they are rapidly killed by high temperatures, such as those used in pasteurization and for cooking (smoking) processed meats. Therefore, meats are rarely implicated in outbreaks of brucellosis, because cooking is usually sufficient to destroy the *Brucella* organisms. They are very sensitive to sunlight, being killed in a few hours, and also quite sensitive to acid conditions (pH 4 or below). However, highly relevant to the transmission of brucellae by food products is the extent of survival of these organisms in food. It appears that brucellae are a group of sturdy organisms which can survive prolonged periods in milk and dairy products as well as in raw meat products, but not in smoked (heated) products. *B. abortus* survived for 75 days in a salted Italian sausage. Brucellae can survive in ham during pickling in brine for 21 days, and in salted and otherwise cured meats for 150 days, but not after the smoking process. *B. suis* has also been shown to survive for 21 days in pork under refrigeration and *B. abortus* in sausages for 175 days. The organisms can survive in tap water for 10 days at 25°C, or 57 days at 8°C, but in the presence of organic matter, such as cattle urine or faeces, soil and, even lake water at 25°C, they can survive for more than 2 years.

Pathogenicity and Symptomatology

Brucellae probably enter the body through small abrasions of the skin and directly through the mucosa of the oropharynx or conjunctiva. They are carried to the liver, spleen, kidneys and bone marrow by the lymphatics and circulation, where they enter and replicate in fixed macrophages and parenchymal host cells. Spread from these foci of infection to other organs and tissues may occur by septicaemic dissemination. Pathogenesis depends on strain virulence and the status of the host immune response.

The onset of brucellosis can be acute in approximately one-half of the cases and insidious in the remainder. The incubation period in humans is highly variable. The symptoms usually appear from 2 to 8 weeks after infection, but acute cases average 10–14 days. The clinical features of the disease are also variable and may range from a mild 'flu-like disease' to a prolonged incapacitating illness. Human brucellosis is typically characterized by fever, malaise, headache, weakness, profuse sweating, chills, arthralgia, depression, weight loss and generalized aching. The fever pattern can be intermittent or irregular and of variable duration, leading to the term undulant fever. The case fatality rate without treatment is less than 2%, but is higher for *B. melitensis* infections. In rare cases, extreme pyrexia can cause death. Complications of *Brucella* infection occur in 10–15% of patients, frequently affecting the skeletal and genitourinary systems. Rarely, neurologic and cardiac complications may occur. Neurologic manifestations include meningitis, encephalitis, brain abscess and psychosis, and cardiovascular manifestations include endocarditis, myocarditis and pericarditis. Infective endocarditis accounts for the majority of brucellosis-related deaths.

Part or all of the original syndrome may reappear as relapses, especially on re-exposure. Relapses are common over a 2- to 4-month convalescent period. Recurrent episodes are generally shorter in duration than the primary attack. Chronic brucellosis is diagnosed on the basis of a history of brucellosis and the clinical symptoms of weakness, malaise and emotional disturbances.

Detection Methods

Cultivation

Cultural isolation of brucellae constitutes the definitive diagnosis of brucellosis. Since brucellosis is one of the most easily acquired laboratory infections, work should only be carried out under containment level 3 conditions. Detection of *Brucella* consists of a series of steps including selective enrichment followed by plating on to selective agar media which contain ingredients to screen for the organism.

Primary isolation of the organisms from lymph-node and bone-marrow aspirates and blood may be enhanced using tryptose broth, brain heart infusion broth, or *Brucella* broth in biphasic bottles. Fibrous clots, exudates and tissues are aseptically ground and the resulting material is inoculated onto trypticase soy agar containing 5% sheep blood agar, *Brucella* agar with 5% serum, or serum dextrose agar. If contamination of the sample by other microorganisms is a strong possibility, selective media should be employed for primary isolation, e.g. Farrel's agar medium supplemented with antibiotics (bacitracin, polymyxin B, nalidixic acid, vancomycin, cyclohexamide, and

nystatin). However, the growth of brucellae may be significantly retarded by selective media. The plates are incubated at 37°C in an atmosphere of 5–10% CO_2. After 48–72 h of incubation, smooth *Brucella* isolates produce circular, convex colonies, 1–3 mm in diameter, with a smooth glistening surface. Rough *Brucella* isolates produce colonies of similar size and shape, but of a more opaque off-white colour and often with a granular surface. Plates must be incubated for a minimum of 4 weeks before being discarded as negative.

Once an isolate has been identified as a *Brucella* culture, the colonies can be transferred to tryptose agar slants for further characterization. The species and biovars may subsequently be identified by tests based on agglutinin absorption assay, phage typing, dye sensitivity, CO_2 requirement, H_2S production, and metabolic properties. In all cases, when typing organisms suspected to be *Brucella*, it is essential to include in the tests at least the *Brucella* reference strains *B. abortus* 544, *B. melitensis* 16M and *B. suis* 1330 as a check on media and methods. Automated blood culture systems may also be used, but allowance should be made for the relatively slow growth of the organism. In contrast, the use of commercial identification kits can be particularly problematic as *Brucella* has been misidentified as *Moraxella phenylpyruvica*.

Serological Identification

As mentioned, brucellae can present itself on culture with a smooth or rough colony morphology, but some present a mucoid phenotype. *B. ovis* and *B. canis* occur normally in the rough form, whereas the other four species are usually isolated in the smooth form. It is possible for smooth colonies to spontaneously become rough and some rough *Brucella* can revert to the smooth morphology. Smooth strains are often markedly more pathogenic than the rough variants. Coupled to the rough versus smooth morphology is the composition of the lipopolysaccharide (LPS) molecule of *Brucella*. Smooth organisms have an LPS molecule containing a polysaccharide 'O' chain consisting of a homopolymer of 4-formamido-α-D-4,6-dideoxymannose. The structure of the LPS of rough strains is basically similar to that of the smooth LPS except that the O chain is either absent or reduced to a few residues. The O chain plays a central role in the serological diagnosis of brucellosis since it is an immunodominant antigen and most diagnostic serological tests are based on the detection of antibodies to the O chain.

Smooth strains of *Brucella* are agglutinated with unabsorbed antisera to smooth *Brucella* (commercially available). However, cross-reactions occur with other Gram-negative bacteria such as *Francisella tularensis*, *Escherichia hermanni* and *Escherichia coli* O157, *Salmonella* O30, *Stenotrophomonas maltophilia*, *Vibrio cholerae* O:1 and *Yersinia enterocolitica* O:9. Rough strains cross-react when unabsorbed anti-rough *Brucella* sera are used, but no cross-reactions occur when these organisms are agglutinated with unabsorbed anti-smooth sera.

Agglutination in monospecific antiserum is performed by preparing a dense suspension of the organism to be tested in 0.5% phenolized saline and heated at 60°C for 1 h. A drop of the suspension is added to a drop of each monospecific antiserum and, the solution is mixed. Agglutination should occur within 1 min with one of the sera. Control cultures should be used for this test and these should agglutinate their respective homologous sera within 1 min, without agglutinating in the other sera. Results from the previously mentioned typing tests will identify almost all *Brucella* strains as a particular biotype, but phage typing and oxidative-metabolic tests are required to identify occasional strains.

The diagnosis of human brucellosis is based on clinical suspicion, epidemiological evidence, and positive culture or serology. Since the protean manifestations of the disease, especially in the chronic stage, can be misleading, and brucellae grow rather slowly in vitro so that primary isolation can be delayed, the preliminary diagnosis often depends on the results of serologic tests. Immunoglobulin (Ig)G and IgA antibodies appear to be the most useful indicators of active infection. *Brucella* enzyme-linked immunosorbent assay (ELISA) is currently widely used for serologic diagnosis of the disease in humans and other species. It is the most sensitive and specific of the *Brucella* serologic tests and has successfully been used for diagnosis of acute and chronic brucellosis as well as neurobrucellosis. Polymerase chain reaction (PCR) assays with random or selected primers have given promising results, but standardization and further evaluation are needed.

Treatment

The purpose of chemotherapy for brucellosis in humans is to control the illness promptly and to prevent complications and relapses. Treatment of brucellosis is initiated on the basis of clinical evidence and serological results, since culture confirmations are delayed and infrequent. Several antimicrobial agents and regimens have been used in the treatment of brucellosis, but the most effective, least toxic chemotherapy for human brucellosis is still disputed. The intracellular location of the microorganisms makes it refractory to the action of many antibiotics. Most

authorities consider doxycycline to be the most effective single drug for uncomplicated brucellosis, but the rates of relapse with single-drug therapy are in the range 5–40%. Therefore, combination therapy is generally recommended.

The treatment recommended by the World Health Organization for acute brucellosis in both adult men and non-pregnant adult women is rifampicin 600–900 mg day^{-1} and doxycycline 200 mg day^{-1} given orally for a minimum of 6 weeks. This combination is convenient, nontoxic and highly effective, and relapses are infrequent after its use. Trimethoprim-sulphamethoxazole (TMP-SMZ), alone or in combination with rifampicin or gentamicin, is also useful for treating pregnant women or patients intolerant to tetracyclines. Rifampicin, or alternatively cotrimoxazole, has been recommended for uncomplicated disease in children. Both are associated with a high relapse rate if used singly, and best results are achieved by using them in combination.

Doxycycline in combination with TMP-SMZ and rifampicin has been used successfully for treatment of brucella meningitis. In brucellosis of the nervous system, an effective regimen has been a prolonged course of TMP-SMZ plus rifampin and a brief course of corticosteroids. Brucellar valvular endocarditis is usually rapidly fatal, but a few cures have been reported. These are based on early aggressive combination chemotherapy with rifampicin, a tetracycline and an aminoglycoside.

Control and Prevention of Brucellosis

In view of the high incidence and wide distribution of brucellosis in both humans and animals, various control measures against brucellosis have been conducted in many countries. However, the prevention of brucellosis is dependent on the eradication or control of the disease in animal hosts, the exercise of hygienic precautions to limit exposure to infection through occupational activities, and the effective heating of potentially contaminated foods. Thus, meat of all kinds, especially from unknown sources or from contaminated areas, must be cooked thoroughly. Vaccination now has only a small role in the prevention of human disease, although in the past, various preparations have been used, including the live attenuated *B. abortus* strains 19-BA and 104M, the phenol-insoluble peptidoglycan vaccine and the polysaccharide-protein vaccine. All of these vaccines had limited efficacy. The live vaccines were associated with potentially serious reactogenicity and have provoked unacceptable reactions in individuals sensitized by previous exposure to *Brucella* or if inadvertently administered by subcutaneous rather than percutaneous injection. Ribosomal proteins such as L7/L12 stimulate protective responses to *Brucella*. They therefore appear to have potential as candidate vaccine components.

Other, more practical, preventative measures have been recommended and include the following. Farmers and workers in abattoirs, packing plants and butcher shops must be educated as to the nature of the disease and the risk in the handling of carcasses or products of potentially infected animals. Such occupational exposure can furthermore be minimized by wearing impermeable clothing, rubber boots, gloves and particularly face masks, for respiratory and eye protection, and by practising good personal hygiene. Since brucellae are able to survive in the environment in soil, water, urine and manure for periods of 1 day to several weeks depending on the temperature, contaminated areas should be disinfected. Care should be taken in the handling and disposal of placenta, discharges and fetus from an aborted animal. Decontamination of utensils and clothing requires exposure to 1% phenolic soap or chloramine for 30 min. The area contaminated by abortion products may be disinfected by 20% chlorine solution or washed with slaked lime. The general public should also be educated not to drink untreated milk or eat products made from unpasteurized or otherwise untreated milk. Livestock should be serologically tested to identify infected animals and such animals should subsequently be eliminated by segregation and/or slaughter. In areas of high prevalence, young goats and sheep should be immunized with live attenuated vaccines such as the *B. melitensis* strains Rev I, whereas cattle should be vaccinated with *B. abortus* strain 19. Newer vaccines, such as the *B. abortus* rough strain RB51 can serve as an effective vaccine to prevent infection from exposure to virulent strains of *B. abortus*. *B. melitensis*, *B. suis* and *B. ovis* in various animal strains including cattle and swine.

See also: **Biochemical and modern identification techniques**: Food Poisoning Organisms. **Enzyme Immunoassays**: Overview. **Fermented Milks**: Products of Eastern Europe and Asia. **Milk and Milk Products**: Microbiology of Liquid Milk; Microbiology of Dried Milk Products; Microbiology of Cream and Butter.

Further Reading

Alton GG, Jones LM, Angus RD and Verger JM (1988) *Techniques for the Brucellosis Laboratory.* Paris: Institut National de la Recherche Agronomique.

Corbel MJ (1997) Vaccines against bacterial zoonoses. *Journal of Medical Microbiology* 46: 267–269.

Corbel MJ (1997) Brucellosis: an overview. *Emerging Infectious Diseases* 3: 213–221.

Hall WH (1990) Modern chemotherapy for brucellosis in humans. *Reviews of Infectious Diseases* 12: 1060–1099.

Jimenez de Bagues MP, Elzer PH, Jones SM et al (1994) Vaccination with *Brucella abortus* rough mutant RB51 protects BALB/c mice against virulent strains of *Brucella abortus*, *Brucella melitensis* and *Brucella ovis*. *Infection and Immunity* 62: 4990–4996.

Lord VR, Schurig GG, Cherwonogrodzky JW, Marcano MJ and Melendez GE (1998) Field study of vaccination of cattle with *Brucella abortus* strains RB51 and 19 under high and low disease prevalence. *Amrican Journal of Veterinary Research* 59: 1016–1020.

Matyas Z and Fujikura T (1983) Brucellosis as a world problem. *Developments in Biological Standardization* 56: 3–20.

Moyer NP, Holcomb LA and Hausler WJ (1992) *Brucella*. In: Balows A, Hausler WJ Jr, Hermann KL, Isenberg HD and Shadomy HD (eds) *Manual of Clinical Microbiology*. P. 457. Washington, DC: ASM Press.

Ross HM, Foster G, Jabans KL and MacMillan AP (1996) Isolation of *Brucella* from seals and small cetaceans. *Veterinary Record* 138: 587–589.

Young EJ (1991) Serologic diagnosis of human brucellosis: analysis of 214 cases by agglutination tests and review of the literature. *Reviews of Infectious Diseases* 13: 359–372.

Problems with Dairy Products

Photis Papademas, Department of Food Science and Technology, University of Reading, UK

Sir David Bruce (Australian bacteriologist and physician: 1855–1931) was the first to discover the microorganism that had caused the death of a man suffering from Malta fever, back in the 1880s. This microorganism was later called *Brucella melitensis*, and it is regarded as the first species of the genus ever to be isolated. The fact that the same agent was later isolated from caprine and ovine milks reflects the zoonotic character of the disease, widely known as brucellosis.

Since the pioneer work by Bruce, five other species have been isolated and identified from animals: *B. abortus* in cattle, *B. suis* in swine, *B. ovis* in sheep, *B. neotomae* in rats and *B. canis* in dogs.

All *Brucella* species are pathogenic for humans and animals, but the most important ones are *B. melitensis*, *B. suis* and *B. abortus* in order of descending pathogenicity. The abundance of *B. melitensis* in dairy products makes it both the most important economically and the most hazardous to health.

Factors Affecting Survival and Growth in Milk and Milk Products

In liquid milk and in milk products, such as cheese, a number of factors have to be satisfied in order for *Brucella* spp. to survive and cause the disease in humans.

Storage Temperature

The survival of *Brucella* spp. has been reported over a range of temperatures from –40°C to 37°C; the general trend is that as storage temperature increases survival decreases.

Sodium Chloride

An increased sodium chloride content in milk products may prove inhibitory to the growth of *Brucella*. *B. abortus* survived in unsalted butter for 13 months whereas the survival time of the same species was approximately halved when the salt content of the butter was increased to 2.3%.

However, survival of the microorganism also depends on the storage temperature. The survival time of *B. melitensis* in sheep's milk cheese stored in 27% sodium chloride brine and at 11–14°C was 45 days. Another example reflecting the effect of storage temperature on the survival of *Brucella* species in foodstuffs, even if the sodium chloride is increased, is the case of Domiati cheese. At a sodium chloride concentration of 7.60–7.66% and storage at 18–22°C, the microorganism survived for 12 days. The survival time was doubled when the cheese was stored at 2–4°C and the salt concentration ranged between 7.60 and 8.99%.

Fat and Water Content

A high fat content may have a protective effect on the survival of *Brucella* spp.; data on cheeses support this view.

Brucella survive for shorter periods of time in cheeses with a low water content – 6 days survival in Gruyère compared to 57 days survival in a soft cheese such as Camembert. Generally, mature cheeses kept in storage for empirically determined periods before marketing are safer than fresh ones.

Two important reasons for the above observations are, first, the fact that, during the production of hard cheeses, a cooking step is included in the manufacturing procedure (**Table 1**) which helps to eliminate *Brucella* at the beginning of the process; and second, contamination levels are reduced during ripening through the combined effect of reduced water activity and pH.

pH

B. abortus survived in a model system using sterilized milk and lactic acid for 34 days at a pH range of 5.0–5.8, but when the pH dropped to 3.9, the survival period decreased to only 2 days. The apparent effect of pH on *Brucella* is concealed in milk products which are much more complex food systems; factors such as fat and water content seem to have a greater influence than pH over the survival time of the pathogen.

Eradication/Control of *Brucella* in Foods

Pasteurization

The numbers of *Brucella* in raw milk vary between infected animals depending on the physiological status of the animal and the route of infection; 12–44% of infected cows and up to 60% of infected goats will excrete infective *Brucella*. Numbers such as 50 000–500 000 cfu ml^{-1} of raw milk have been reported but, in 55% of the samples, the count was less than 1000 cfu ml^{-1}.

Table 1 Behaviour of *Brucella abortus* during cheese manufacture, storage and ripening

Cheese	Cooking temp. (°C)	Time (min)	cfu^{-1} ml (milk)[a]	After 24 h	Storage time (days)	Survival time (days)
Hard						
Emmental	52	50	10 000	794	57	6
Gruyère	48	25	10 000	1259	57	6
Semihard						
Tilsiter	43	15	10 000	1585	57	15
Soft						
Muenster	38	10	10 000	1585	57	> 57
Camembert	32	60	10 000	1995	57	> 57

[a] Artificially inoculated.

The most effective way of eradicating *Brucella* in milk is by pasteurization or sterilization before marketing or any further processing. *Brucella* in milk have a D value at 65.6°C of 6–12 seconds and a z value between 4.4 and 5.6°C. *B. abortus* was killed in artificially infected fresh milk (10^6 cfu ml^{-1}) when heat-treated in a high temperature short time (HTST) simulator at 65.8°C for 5 s.

There are no quantitative data on the infected dose for humans.

Control

The risk of food-borne brucellosis can be controlled when two demands are complied with: first, strict hygiene and process control during dairy product manufacture, and second, the application of control measures at farm level.

The phosphatase test can be used to confirm the correct pasteurization of milk used for cheese production and special vigilance is required when cheese is produced in areas where brucellosis is endemic, and/or applied technology does not give a high degree of certainty with respect to the elimination of *Brucella* from the product before marketing.

An effective hazard analysis critical control point (HACCP) plan is crucial on a production line to ensure the delivery of a safe product to the consumer, and to avoid economic losses and possible litigation.

At farm level, the most successful means of eradication are elimination by test-and-slaughter and by vaccination (*B. abortus* 45/20 and strain 19 vaccines are used in cattle). In the Republic of Ireland, *B. abortus* strain 45/20 vaccine has been used with success, when administered to cattle 9-months-old or above, whether pregnant or not.

Great improvements in animal brucellosis control has been achieved in Cyprus by the test-and-slaughter method where, during the brucellosis control campaign (1973–1985), the percentage of infection in the total sheep and goat population dropped from 1.06 in 1973 to 0.004 in 1985, making Cyprus the only country in the Mediterranean region free of animal brucellosis.

Continuous surveillance is required in brucellosis-free herds to ensure that the status is maintained. This control, as in other countries, was also achieved in Cyprus, by the use of the milk ring test (MRT) in dairy cattle and the allergic skin test (AST) in sheep and goats.

The principle of the MRT is the agglutination of the *Brucella* organisms if antibodies are present in the milk sample. To 1 ml of well-mixed milk, 50 µl of MRT-stained (haematoxylin) *Brucella* antigen is added and, after thorough mixing, the sample is incubated for 1 h and 18–20 h at 37°C and 4°C respect-

ively. The formation of a deep purple coloured ring on the surface of the milk sample is regarded as a positive result.

The AST is primarily used in sheep and goats. The efficiency of the test is based on the hypersensitivity to *Brucella* antigens acquired by animals during the course of infection or following vaccination. Animals are injected intradermally into the upper or lower eyelid with a *Brucella* allergen (0.1 ml) and a positive result will produce a swelling in the injected eyelid which is readily apparent in comparison with the other eye. The palpebral route is considered to be the most convenient inoculation procedure when testing sheep and goats.

Positive results from the MRT and AST must be carefully interpreted, and veterinary services in Cyprus carried out confirmatory tests, including the Rose Bengal plate test for dairy cattle – a spot agglutination test using a Rose Bengal-stained *Brucella* antigen buffered to a pH of 3.6–4.0. Since the Rose Bengal plate test is considered to be oversensitive to the cattle vaccine strain 19, it is recommended that its use should be limited to identifying non-immunized flocks. The complement fixation test for both cattle and sheep/goats involves warm fixation at 37°C for 30 min, between test serum, antigen and complement. The end point of the reaction is taken as the last dilution showing appreciable fixation.

Additional measures to be taken include adherence to sanitation and disinfection procedures in farms, and also monitoring the transport of animals to brucellosis-free herds from areas where the infection exists.

Brucellosis in Humans

Some of the factors that will determine whether a pathogen will invade the host are the virulence of the organism, the immune status of the host, the infecting inoculum and the route of inoculation.

Epidemiology

Countries affected by the problem of brucellosis (*B. melitensis*) in animals and subsequently in humans stretch from the southern (Tunisia) to the eastern edges of the Mediterranean basin (where Israel, Lebanon, Egypt and Syria report an annual total of more than 90 000 cases of human brucellosis) through central Asia, where sheep are often affected. Other countries, such as France, Greece, Italy, Spain and Portugal, account for most cases of human brucellosis in Europe.

Another endemic area is South America where the goat is almost exclusively affected. Argentina has a prevalence rate of 5% caprine brucellosis

Table 2 Outbreaks of human brucellosis associated with the consumption of milk and/or milk products

Country and state	Year	Pathogen	Number of cases	Fatalities	Type of food	Factors
US: Colorado	1973	*Brucella melitensis*	3	0	Mexican soft cheese	Raw milk suspected
US: Texas	1983	*B. melitensis*	31	1	Mexican soft cheese	Raw milk suspected
US: Texas	1985	*B. melitensis*	9	0	Mexican soft cheese	Raw milk suspected
France	1985	*B. melitensis*	5	0	Goat's or sheep's cheese and caprine milk	Unripened cheese and raw milk suspected
Israel	1988	*B. melitensis*	498	ND	Goat's and sheep's cheese and milk	Raw milk suspected
Saudi Arabia	1990–1991 (3-month period)	*B. melitensis* and *B. abortus*	90	0	Raw milk	Raw milk and occupation-acquired
UK	1992–1994	*B. melitensis* (9 cases) *B. abortus*, *Brucella* spp. (35 cases)	44	0	ND	Infection acquired abroad (22 cases)
Malta	1995	*Brucella* spp.	135	1	Soft cheese	Raw ovine and caprine milk
UK	1995	*B. melitensis* (5 cases)	9	0	Caprine milk and/or cheese	Raw caprine milk suspected

ND, no data.

(characterized by abortion in females and occasionally orchitis in males). Mexico, Brazil, Colombia, Ecuador and Peru are countries in the Americas that are also affected by human brucellosis.

Brucellosis incidents in countries where the disease is endemic are highest in spring and summer and coincide with increased abortion, parturition and milk production of animals such as sheep and goats.

Although *B. abortus*, which in the past accounted for most cases in northern Europe, has a much greater geographical distribution that *B. melitensis*, it does not pose a potential threat to humans since the risk of transmitting *B. abortus* via bovine milk is low. In the unlikely event of infection, the symptoms are mild or subclinical. On the other hand, *B. melitensis* can be dangerous to humans (**Table 2**).

Transmission

The transmission of *Brucella* from infected animals to humans can be direct or indirect. Direct transmission is common in groups of people who come to contact with infected animals, i.e. sheep and goat owners, veterinarians or people living near farms. These groups are more vulnerable to infection during the period when goats and sheep are giving birth due to the huge discharge of bacteria; this discharge may continue for up to 3 months after birth. Aerosols of contaminated dust can also cause infection, since bacteria can remain viable in dust for up to 10 weeks.

Generally, the major route of infection is through the mucous membranes of the upper respiratory tract, the oropharynx and the conjunctiva. Introducing *Brucella* into the conjunctival sac infects laboratory animals, and this route of infection is suspected to be the cause of laboratory- and abattoir-associated brucellosis.

During the winter, *Brucella* can survive for 10 weeks in soil, 7 weeks in faeces and for up to 25 weeks in urine. Goat manure, which is used as a fertilizer, may be regarded as a potential vehicle for the indirect transmission of brucellosis to humans.

B. melitensis can also be directly transmitted to humans by consumption of raw milk and fresh cheese, as many of the *Brucella* present in milk can end up in the cheese. Infected cheese can cause urban outbreaks, so that controlling and eradicating the pathogen from milk is of primary importance.

Incubation Period

Due to the differences in virulence of *Brucella* species, different routes of infection and variations in the quantity of infecting agent, the incubation period varies. A period of 3–4 weeks is about the average for the onset of symptoms.

Clinical Significance and Symptoms

The characteristic of acute brucellosis is an intermittent fever, in the range of 38–41°C, but hyperpyrexia, i.e. a substantially higher temperature, is a significant cause of death. At onset of brucellosis, other symptoms include malaise, back pain, weight loss, anorexia and, especially during the evening, chills and sweats.

Brucella localizes within tissues of the body that are rich in elements of the reticuloendothelial system, and infection involves lymph nodes, spleen, liver and bone. The majority of patients suffering from infection by *B. melitensis* have enlargement of the liver, spleen and superficial lymph nodes. In the case of *B. abortus* infection, fewer cases of spleen enlargement were reported and the number decreased further with regard to perceptible liver enlargement.

The most frequent complication in humans is osteomyelitis due to localization of the disease in the bones. Neurological involvement in the form of meningoencephalitis is rare, accounting for only 5% of brucellosis cases reported.

Treatment

As *Brucella* is an intracellular bacterium and relatively inaccessible to antibiotics, prolonged treatment is advised. Treatment has to be effective in acute illness to alleviate the symptoms and prevent complications and/or relapse. In humans, a combination of tetracycline with streptomycin for the first 2–3 weeks of therapy is advised, and the tetracycline is continued for a further 4–6 weeks.

See also: **Cheese**: In the Market Place; Microbiology of Cheese-making and Maturation; Mould-ripened Varieties; Role of Specific Groups of Bacteria; Microflora of White-brined Cheeses. **Food Poisoning Outbreaks**. **Hazard Appraisal (HACCP)**: The Overall Concept. **Heat Treatment of Foods**: Principles of Pasteurization. **Milk and Milk Products**: Microbiology of Liquid Milk; Microbiology of Cream and Butter.

Further Reading

Alton GG (1991) *Brucella melitensis*. In: Frank J (ed.) *Networking in Brucellosis Research*. P. 205. Tokyo, Japan: United Nations University.

Alton GG, Jones LM and Pietz DE (1975) *Laboratory Techniques in Brucellosis*, 2nd edn. Monograph series no. 55. Geneva: World Health Organization.

Corbel MJ (1997) Brucellosis: an overview. *Emerging Infectious Diseases* 3 (2): 213–221.

Corbel MJ and MacMillan AP (1998) Brucellosis. In: Collier L, Balows A and Sussman M (eds) *Topley and Wilson's Microbiology and Microbial Infections*. Vol. 3, p. 819. Avon: Bath Press.

Joint FAO/WHO Expert Committee on Brucellosis (1986) Technical report series 740, 6th report. Geneva: World Health Organization.

Roberts TA, Baird-Parker AC and Tompkin RB (1996) *Characteristics of Microbial Pathogens*. Vol. 5, p. 36. London: Blackie Academic and Professional.

Young EJ (1983) Human brucellosis. *Reviews of Infectious Diseases* 5 (5): 821–841.

Young EJ and Corbel MJ (1989) *Brucellosis: Clinical and Laboratory Aspects*. Boca Raton, FL: CRC Press.

Burkholderia cocovenenans see ***Pseudomonas***: *Burkholderia cocovenenans*.

Butter see **Milk and Milk Products**: Microbiology of Cream and Butter.

BYSSOCHLAMYS

P Kotzekidou, Department of Food Science and Technology, Aristotle University of Thessaloniki, Greece

Characteristics of the Genus

Byssochlamys is an Ascomycetes fungus. It is the teleomorph state of certain species of *Paecilomyces*. Two species significant in food mycology are included in the genus: *B. fulva* and *B. nivea*. Distinction between the species is based mainly upon morphological characteristics.

Byssochlamys is an Ascomycetes genus characterized by the absence of ascocarp. Asci are spherical to oval, borne in open clusters, composed of loose wefts of hyaline, thin, twisted hyphae. Each ascus contains eight ascospores. The anamorph state produces reproductive structures (penicilli) borne from surface hyphae or long trailing aerial hyphae and asexual spores (conidia). The characteristics of the two species of the genus are presented in **Table 1**.

Byssochlamys is fairly common in soil, particularly in vineyards, orchards and fields where fruits are grown. It contaminates fruits and other vegetation on contact with soil, before delivery to the processing plant. *Byssochlamys* is almost uniquely associated with food spoilage, and in particular with the spoilage of heat-processed acid foods. Its ascospores persist in a dormant yet viable state in the soil for extended periods of time, and contaminate fruits harvested from or close to the ground. Since these ascospores are able to survive routine heat pasteurization treatments applied to many fruit products, spoilage may occur due to post-pasteurization germination and subsequent outgrowth. The species *B. nivea* is commonly detected in raw milk (as a contaminant with soil and silage feeding) and its ascospores, which are resistant to normal pasteurization, may occur in cream cheese and fermented milk.

The most important physiological characteristic which makes *Byssochlamys* significant in food mycology is the heat resistance of its ascospores. It is assumed that the rather thick cell wall with a distinct electron-transparent layer between the outer cell wall and the cytoplasmic membrane protects the ascospores against heating. Intrinsic heat resistance of ascospores varies markedly between strains of the same species and with heating conditions (**Table 2**). The nature of the heating medium – pH, soluble solids and organic acid content – influences the sensitivity of ascospores to elevated temperatures. Ascospores are more susceptible to heat if the pH is low and/or if preservatives such as organic acids (especially fumaric, lactic and acetic acids) or SO_2 are present. On the other hand, high levels of sugar and sodium chloride have a protective effect. The protective mechanism is due in part to establishment of an osmotic pressure differential between the heating medium and ascospores, which favours heat resistance.

For *B. fulva* a D value between 1 and 12 min at 90°C and a z value of 6–7°C are practical working values. *B. nivea* ascospores are marginally less heat-resistant than those of *B. fulva*. A D value at 88°C of 0.75–0.8 min, with z values ranging from 4.0 to 6.1°C, is of practical importance.

Attempts have been made to inactivate *Byssochlamys* ascospores in fruit juices by the combined effect of high pressure and temperature as well as by ionizing radiation. Estimated values for an effective pasteurizing process are reported in **Table 3**.

Another physiological characteristic which makes *B. fulva* and *B. nivea* outstanding spoilage fungi in canned, bottled or cartoned fruit products is their ability to grow at very low oxygen tensions, producing CO_2. A small amount of oxygen contained in the head space of a jar or bottle, or the slow leakage of oxygen through a package such as a Tetra-Brik, can provide sufficient oxygen for these fungi to grow. The production of gas may cause swelling and spoilage of the product. *Byssochlamys* spp. are particularly tolerant of conditions of elevated carbon dioxide. *B. fulva* and *B. nivea* are capable of growth in atmospheres containing up to 60% carbon dioxide with less than 0.5% oxygen.

The species of the genus *Byssochlamys* produce pectolytic enzymes which are responsible for the degradation of pectic substances and the maceration of plant tissues. Pectolytic enzymes from *B. fulva* and *B. nivea* have been implicated in the softening and breakdown of canned fruits, such as apricots, strawberries, grapes, cherries and apples. Ascospores of the fungus, which may be present on the raw fruit, survive the heat of can processing. Germination and limited growth occur and pectolytic enzymes are produced before the O_2 within the can is exhausted. All fruits within a can are typically affected but several months of storage may elapse before all have softened completely. Off odours and a slightly sour taste may develop and gas production may occur. The pectolytic enzymes produced vary between strains of the same species. *B. fulva* produces a variety of pectolytic enzymes, including polygalacturonase, pectinesterase, polymethylgalacturonase and pectate lyase. *B. nivea* produces endopolygalacturonase and endopolymethylgalacturonase. Pectolytic enzymes are particularly tolerant to heat and show a definite bimodal heat stability. Stability to heating is minimal at 50–80°C and increases at about 100°C. Pectolytic enzymes which survive during thermal processing of canned fruits cause significant softening of the fruit tissues during post-process storage.

Some strains of *B. fulva* and *B. nivea* may produce patulin. Its production is affected by head space in glass jars of heat-processed fruit juice, controlled atmospheres, temperature, water activity (a_w) and preservatives. Very low levels of patulin can be produced under atmospheres containing less than 0.5% O_2 and 20–60% CO_2. Minimum a_w values for patulin production are 0.92 at 21°C and 0.87 at both 30 and 37°C. However, there is no evidence of patulin production in commercially processed fruit juices. Some *B. fulva* strains produce byssochlamic acid only under aerobic conditions; partial or complete exclusion of air prevents its formation. Byssochlamic acid can be produced from a wide variety of sugars at various concentrations and at a pH range of 2.5–8.0. Both mycotoxins are important because of their potent biological effects (**Table 4**). They can conceivably gain entrance to the human food chain via fruits and fruit juices.

Methods of Detection

Plating Techniques

Because of the low incidence of the genus *Byssochlamys* in foods, it is important that relatively large samples are analysed for its detection. In some liquid fruit products centrifugation may be necessary to concentrate the ascospores. Laboratory pasteurization of

Table 1 Characteristics of the genus *Byssochlamys*, including *B. fulva* and *B. nivea*

Characteristics	B. fulva	B. nivea
Anamorph	*Paecilomyces fulvus*	*Paecilomyces niveus*
Morphology		
Colonies on CYA and MEA	Buff to brown	White to cream
Colony diameter	6–9 cm	7–9 cm
Mycelium	Septate	Septate
Reproductive structures		
Anamorph	Penicilli on short stipes	Penicilli on short stipes
	Phialides flask-shaped (12–20 μm long), narrowing gradually	Phialides cylindrical (12–20 μm long), gradually tapering
	Conidia: cylindrical (7–10 μm long)	Conidia: ellipsoidal to pyriform (3–6 μm long)
	Chlamydospores: no	Chlamydospores: spherical to pyriform (7–10 μm long)
Teleomorph	Asci: without distinct wall, spherical (9–12 μm diameter)	Asci: without distinct wall, spherical (8–11 μm diameter)
	Ascospores: ellipsoidal (5–6.5 × 3–4 μm)	Ascospores: ellipsoidal (4–5.5 × 2.5–3.5 μm)
Growth temperatures		
Minimum	10°C	10°C
Optimum	30–35°C	30–35°C
Maximum	45°C	40°C
Minimum a_w for growth		0.87
Ascospore maturing time at:		
25°C	Occasionally	10–14 days
30°C	7–12 days	7–10 days
37°C	10–14 days	Rarely

CYA, Czapek yeast extract agar; MEA, malt extract agar; a_w, water activity.

Table 2 Heat resistance of *Byssochlamys fulva* and *B. nivea* ascospores

Species	Heating medium	Heat resistance
Byssochlamys fulva	Glucose (16° Brix) tartaric acid (0.033 mol l⁻¹), pH 3.6 and 5.0	90°C, 1.2–46 min 3 $\log_{10}$ inactivation time
	Tomato juice	90°C, 8.1 min 1 $\log_{10}$ inactivation time
	Grape juice	$D_{87.8°C}$, 11.3 min
Byssochlamys nivea	Grape juice	88°C, survived 60 min
	Apple juice	99°C, survived in juice containing 4.7% sucrose
	Cream (10% w/w fat)	$D_{92°C}$, 1.6–19 s
	Tomato juice	90°C, 1.5 min 1 $\log_{10}$ inactivation time

Table 3 Inactivation of *Byssochlamys* ascospores by high pressure and ionizing radiation

Species	Pressure	Ionizing radiation
Byssochlamys fulva	300–600 MPa at 10–60°C	> 7 kGy for pasteurization
Byssochlamys nivea	⩾ 700 MPa at ⩾ 70°C	

fruit juices, pulps and concentrates aids to the selective isolation of the heat resistant fungi. An inactivation of vegetative cells of fungi and bacteria, as well as less heat-resistant spores, is also achieved. Simultaneously, a heat activation of ascospores of heat resistant moulds to germinate is obtained. The composition of the heating medium influences the rate and extent of activation. However, the achievement of maximal activation is species dependent. In heat processed foods since ascospores may be stressed by the heating process highly acidic media are not recommended for heat activation or detection. In low-acid foods that

Table 4 Effects of mycotoxins produced by *Byssochlamys* spp.

Mycotoxin	Origin	Toxic effects
Byssochlamic acid	*Byssochlamys fulva*	Inhibition of some essential enzymes; hepatotoxic and haemorrhagic effects
Patulin	*Byssochlamys fulva* *Byssochlamys nivea*	General toxicity and possible carcinogenicity

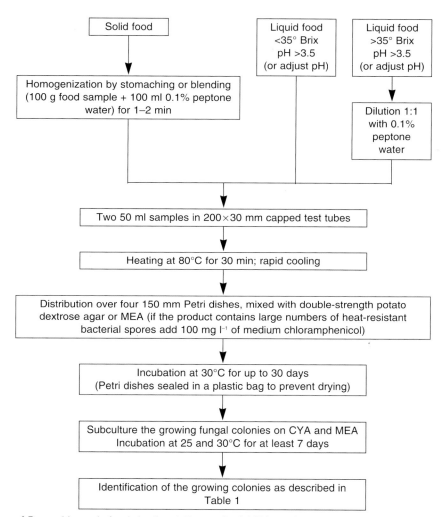

Figure 1 Detection of *Byssochlamys* in foods by the plating method. MEA = malt extract agar; CYA = Czapek yeast extract agar.

are heavily contaminated with bacterial spores, the addition of antibiotics (chloramphenicol) to the plating medium is required to inhibit heat resistant bacterial spores.

Byssochlamys can be detected and enumerated by two methods: the plating method adapted for larger samples (an overview is given in **Fig. 1**) and the direct incubation method (**Fig. 2**).

The plating method is recommended for solid and liquid foods, such as fruits and products containing pieces of fruits, whereas the direct incubation method is suitable for semisolid foods, such as homogenates and fruit pulps. The disadvantage of the plating method is that aerial contamination during plating may be a problem. An indication of contamination is the occurrence of green *Penicillium* colonies or colonies of common *Aspergillus* spp. such as *A. flavus* and *A. niger*. To minimize the problem, use of a laminar flow hood is recommended. Alternatively, the direct incubation method can be used, as risk of contamination from the air is avoided and loss of moisture is minimized. A disadvantage of the direct incubation method is that colonies growing in flasks must be picked and cultivated in suitable media.

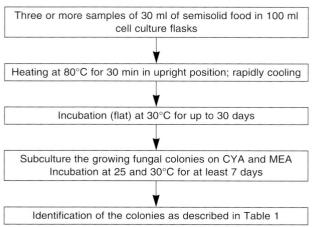

Figure 2 Detection of *Byssochlamys* in semisolid foods by the direct incubation method. CYA = Czapek yeast extract agar; MEA = malt extract agar.

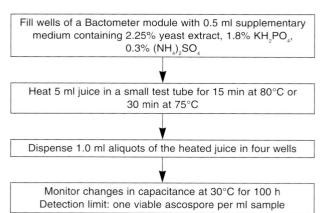

Figure 3 Detection of *Byssochlamys* in fruit juices by impedimetry and conductimetry.

However, subculturing of the colonies is also recommended for identification of fungi detected by the plating method.

Impedimetry and Conductimetry

Since detecting fungi of the genus *Byssochlamys* by plating techniques is laborious and time-consuming, requiring incubation for at least 7 days and ideally up to 30 days, impedance monitoring has been suggested as a useful tool for rapid detection. An impedimetric method for *Byssochlamys* detection in fruit juices is described in **Figure 3** using a Bactometer (Vitek Systems, UK). This instrument is capable of simultaneously monitoring 256 samples distributed in 16 disposable modules of 16 wells. The instrument monitors changes in impedance and its two components, conductance and capacitance, over time. Incubation temperature is 30°C. Fruit juice is pasteurized at 80°C for 15 min or at 75°C for 30 min before being dispensed into the wells. Undiluted fruit juice concentrates give poor capacitance changes even

when the fungi grow in the concentrate. Therefore diluting concentrates before analysis is recommended. In samples contaminated with spore-forming bacteria, antibiotics generally used in mycological media, e.g. 50 p.p.m. chloramphenicol and 50 p.p.m. chlortetracycline, may be added. This has little or no effect on curve quality and detection time.

Impedimetry and conductimetry are effective rapid methods when used under well-defined conditions in a particular kind of food. They can only be used on a broader scale with considerable developmental studies.

Unacceptable Levels of *Byssochlamys* spp.

The acceptable level of contamination of a raw material with ascospores of the genus *Byssochlamys* depends on the type of product into which the material will be incorporated and the heat process to which it will be subjected. If it will be incorporated into frozen desserts such as ice creams and ice confections, or short-life chilled desserts such as fruit salads, cakes and yoghurts, there is no need to set a specification. Products which are at risk from spoilage by *Byssochlamys* ascospores are shelf-stable products which receive a relatively light process (such as conventional or UHT pasteurization) and do not contain preservatives such as sorbate or benzoate. A count of 5 ascospores per 100 g or ml of product at a stage just before the retort or heat exchanger indicates a serious problem. For UHT-processed fruit juice blends without preservatives, even a lower level of contamination is unacceptable.

In Australia, practical experience has shown that the most common spoilage problems caused by *Byssochlamys* are associated with passionfruit juice or pulp. A contamination level of less than 2 spores per 100 ml gives a negligible spoilage rate in most finished products. Contamination levels of 2–5 spores per 100 ml are marginal, and more than 5 spores per 100 ml are unacceptable. However, for some products, such as UHT-processed fruit juice blends (preservative-free) containing a high proportion of passionfruit juice, the specification of one manufacturer requires that *Byssochlamys* spores are absent from a 100 ml sample taken from each 200 l drum of raw material.

Importance to the Food Industry

Byssochlamys spp. produce ascospores that frequently show high heat resistance and survive the thermal processes given to some fruit products. Germination of ascospores results in growth of the fungi on fruits and fruit products, producing pectic enzymes which cause complete breakdown of texture in fruits and off flavour development. Some *Byssochlamys* spp.

produce patulin and byssochlamic acid and therefore may constitute a public health hazard.

B. fulva and *B. nivea* have been implicated in spoilage of strawberries, blackberries, apricots, grapes, plums and apples in cans and bottles, blended juices (especially those containing passionfruit) and fruit gel baby foods. The soil acts as the primary reservoir for *Byssochlamys* ascospores and fruits which come in contact with soil directly or from rain splash are susceptible to contamination. The number of ascospores on fruits is low – less than one per gram. *B. nivea* appears to be a less common problem in foods than *B. fulva*. *Byssochlamys* spp., although only occurring sporadically, are a continuing problem to the food industry. Ascospores can survive heat treatments normally applied to hot-packed canned fruit products and subsequently grow under reduced oxygen. Spoilage in cans is evidenced by growth of fungi where small amounts of oxygen remain in the container head space. Pasteurization temperatures applied to canned-acid foods may stimulate spore activation, thus resulting in post-pasteurization germination and subsequent outgrowth. To overcome the problem in canned fruits and fruit juices, washing fruit before canning or juice extraction, rejecting difficult-to-clean wrinkled fruit and screening products for heat-resistant ascospores is suggested.

B. nivea ascospores may be present in raw milk when contamination with soil occurs. They can survive the pasteurization processes applied to milk and cream. The fungus can occasionally cause spoilage in heat-processed cheeses such as cream cheese, in the case of prolonged storage and inadequate cooling. Rarely, it causes spoilage in UHT dairy products. To ensure that only 1 out of 10^6 packs of cream cheese (500 g packages) produced is infected, a heat treatment time of 24 s at 92°C is required. Problems caused from *B. nivea* in packaged ravioli can be overcome by packing in an atmosphere of 60% CO_2, 39.4% N_2 and 0.05% O_2.

See also: **Heat Treatment of Foods**: Spoilage Problems Associated with Canning. **Milk and Milk Products**: Microbiology of Liquid Milk. **Spoilage Problems**: Problems caused by Fungi.

Further Reading

Beuchat LR and Pitt JI (1992) Detection and enumeration of heat-resistant molds. In: Vanderzant C and Splittstoesser DF (eds) *Compendium of Methods for the Microbiological Examination of Foods*, 3rd edn. Washington, DC: American Public Health Association.

King AD, Pitt JI, Beuchat LR and Corry JEL (1986) *Methods for the Mycological Examination of Food*. New York: Plenum Press.

Pitt JI and Hocking AD (1997) *Fungi and Food Spoilage*, 2nd edn. London: Blackie Academic & Professional.

Samson RA (1974) *Paecilomyces* and some allied Hyphomycetes. *Studies in Mycology* 6: 1–119.

Samson RA, Hocking AD, Pitt JI and King AD (eds) (1992) *Modern Methods in Food Mycology*. Amsterdam: Elsevier.

Cakes *see* **Confectionary Products**: Cakes and Pastries.

CAMPYLOBACTER

Contents
Introduction
Detection by Cultural and Modern Techniques
Detection by Latex Agglutination Techniques

Introduction

M T Rowe and **R H Madden**, Department of Agriculture for Northern Ireland and Queen's University of Belfast, Northern Ireland

This article provides a synopsis of the current state of knowledge on those aspects of the genus *Campylobacter* which the authors consider relevant. These include taxonomy, general physiology, ecology, pathogenicity, typing of the genus, methods of control and viable but non-culturable forms. It should be appreciated that because of the relatively recent recognition of the importance of members of the genus as food-poisoning agents and developments in molecular methods, the taxonomy of the genus, in particular, is evolving rapidly.

Introduction

Theodor Escherich in the 1880s made the first recorded observation of spiral bacteria in the faeces of patients with infantile diarrhoea but he was unsuccessful in culturing them and regarded them as non-pathogenic. The first isolation of *Campylobacter* spp. was achieved by King in 1957 when he successfully isolated 'vibrios' from blood samples of humans with diarrhoea. A major advance in the culture of these organisms was made by Martin Skirrow who developed a selective medium that obviated the need for a laborious filtration stage, making possible the routine isolation of these organisms. Currently *Campylobacter* spp. are isolated more frequently than

Salmonella spp. in human gastroenteritis patients. In these cases *C. jejuni* is the major species responsible, with the incidence of *C. coli* being approximately 10%, but this figure may be higher depending on the geographical location.

The incubation period for *Campylobacter* enteritis is usually 1–7 days but can extend to 10 days. The onset of symptoms is characterized by fever with confusion or delirium and general malaise followed by severe abdominal cramping which in turn is followed by profuse diarrhoea that becomes watery, often containing blood and mucus. Although the diarrhoea usually lasts for 2–7 days, abdominal pain and cramping can persist for up to 3 months. Systemic infection is uncommon but complications such as Guillain–Barré syndrome (GBS) and reactive arthritides (reactive arthritis and Reiter's syndrome) can arise. Fortunately, most *Campylobacter* enteritis cases are self-limiting and fatality rates are low. The infective dose can be as low as 5–800 organisms with the attack rate correlating with increasing dose.

Guillain–Barré Syndrome

This disorder is rare and affects the peripheral nerves of the body. It can vary greatly in severity from the mildest cases where clinical treatment is not sought to a complete paralysis that brings the patient close to death. Lipopolysaccharides isolated from *C. jejuni* strains implicated in GBS have regions homologous to the human gangliosides GM1 and GD1b. The molecular mimicry between these regions may lead to autoimmunity and GBS. Miller–Fisher syndrome, similar to GBS, is an acute neuropathic disorder characterized by paralysis of the eye muscle, absence of reflexes and facial weakness.

Reactive Arthritides (Reactive Arthritis and Reiter's Syndrome)

Both syndromes are characterized by sterile inflammation of joints from infections originating from non-articular sites and are mediated by T cells. This may be triggered by enteritis involving *Campylobacter* spp. or other pathogens such as *Salmonella* or *Yersinia*. Although viable organisms are not present, bacterial antigens are probably transported to the joints within phagocytic cells.

Taxonomy

The *Campylobacter*-like organisms are members of the family Spirillaceae, which comprises Gram-negative curved rods and consists of the related genera *Anerobiospirillum*, *Arcobacter*, *Campylobacter* and *Helicobacter*. The unusual physiology of *Campylobacter* spp. led to them being discovered as a significant cause of food-borne disease only relatively recently. In 1984 *Bergey's Manual* listed five *Campylobacter* species; however, in 1996 a probability matrix for the identification of campylobacters (and related organisms) was published which listed eight more, with several subspecies also described.

The use of DNA hybridization studies led to a major reorganization of campylobacters in 1991 with the creation of the genus *Arcobacter* to which two species of *Campylobacter* were assigned. At that time 11 species of *Campylobacter* were recognized and both *C. jejuni* and *C. coli* were listed in the genus. It is probable that more campylobacters are awaiting discovery since techniques based on isolating and amplifying DNA from normal and pathological samples of the gastrointestinal tracts of animals and humans indicate the presence of species which have not yet been isolated in pure cultures. There is even a possibility that campylobacters are present in humans as commensal organisms.

Differentiating between the two species most implicated in food poisoning is usually simple since *C. jejuni* can hydrolyse hippuric acid whilst *C. coli* cannot. Thus the hippurate hydrolysis test is of great importance in food microbiology. Care must be exercised in its application, however, as inadequate buffering of the reaction mixture can lead to false-negative results. There have also been reports of hippurate-negative *C. jejuni* isolates. Further investigations have shown the gene required for hippurate hydrolysis to be present in such isolates but not to be transcribed. Thus genetic probes can help to identify these species correctly.

General Physiology

Campylobacters have a distinctive morphology, being slender, spirally curved rods 0.2–0.5 µm wide and 0.5–5 µm long. Species are highly motile by means of a single polar flagellum at one or both ends, giving rise to a characteristic corkscrew-like motion. The principal distinguishing feature of the physiology of this genus is that they are microaerophilic with a respiratory type of metabolism. Thus oxygen is required for energy production but can only be tolerated at levels below normal atmospheric pressure. This property was partly responsible for the genus remaining undetected until relatively recently as it could tolerate neither fully aerobic nor anaerobic conditions, i.e. those normally employed to isolate organisms from animals and humans.

Optimal oxygen concentrations have been quoted as being 3–6% but media supplements can be used to allow growth at higher concentrations, for example FBP (ferrous sulphate, sodium metabisulphite and sodium pyruvate) increases aerotolerance, allowing growth at oxygen levels of 15–20%. The size and state of the inoculum will also dictate whether growth in synthetic media takes place with a heavy inoculum usually advised to ensure growth. Elevated levels of CO_2 are also recommended with levels of 2–10% having been cited, and the growth of some species is dependent on the presence of hydrogen. Latterly it has been recommended that hydrogen is present in the atmosphere used to incubate any clinical samples.

Despite their sensitivity to oxygen, *C. jejuni* and *C. coli* possess catalase, oxidase and superoxide dismutase activity; however, these enzymes appear to give limited protection from hydrogen peroxide and superoxide ions as shown by the increased growth resulting from the addition of FBP supplement which destroys these compounds. Blood also encourages good growth and contains both catalase and superoxide dismutase.

Campylobacters are chemo-organotrophs which neither ferment nor oxidize carbohydrates. Instead, energy is derived from either amino acids (aspartate and glutamate can be utilized) or tricarboxylic acid (TCA) cycle intermediates. The amino acids are deaminated to provide TCA intermediates for subsequent oxidation but no complex molecules, such as proteins, are utilized.

In terms of growth temperature, campylobacters are mesophilic, as would be expected from their association with warm-blooded creatures. Growth temperatures range from about 25 to 45.5°C. The latter temperature can be referred to as thermophilic in medical circles and hence organisms growing at 42°C are sometimes grouped under the general term of thermophilic campylobacters. None of the genus are however true thermophiles.

Ecology

The normal habitats of *Campylobacter* spp. are selected niches (intestinal tract, reproductive organs and oral cavity) of warm-blooded animals. For those organisms related to gastroenteritis the normal habitat is the lower part of the gastrointestinal tract. In this environment the organisms are exposed to controlled temperatures in the range 37–41°C and hence the inability of campylobacters to grow below 30°C is of no consequence. The low oxygen tensions found in the lumen of the gut mean that campylobacters do not require protective mechanisms to counter the toxic effects of atmospheric levels of oxygen, whilst the high nutrient levels are conducive to the proliferation of these highly fastidious organisms.

Campylobacters do not persist in the environment due to their limited defences against oxygen, relatively high minimum growth temperature and complex nutritional requirements. Their spread is therefore most likely to be by oral–faecal contamination. In the case of foodstuffs they are relatively sensitive to heat, hence normal cooking will kill them and transmission will therefore be due to underprocessing or raw–cooked contamination.

Species Other than *C. jejuni* and *C. coli*

Campylobacter upsaliensis

This species can be carried by healthy puppies and kittens. In a survey of ribotypes isolated from humans and dogs, the human strains were found to possess a unique 16S ribotype and carried plasmids more frequently than did canine strains. Besides human enteritis, strains can cause abortion, bacteraemia and haemolytic–uraemic syndrome. Transmission is much more likely to be via contact with domestic pets rather than food-borne.

Campylobacter sputorum

C. sputorum has been isolated from dairy cows and calves. Although this species has not been associated with human enteritis, it has been implicated in periodontitis. Three biovars have been proposed and their characteristic biochemical tests are as follows:

- *C. sputorum* biovar *sputorum*: catalase-negative
- *C. sputorum* biovar *faecalis*: catalase-positive
- *C. sputorum* biovar *paraureolyticus*: urease-positive.

If this proposal is adopted then this would require the unification of biovars *sputorum* and *bubulus* into biovar *sputorum* and the biovar name *bubulus* would be discarded.

Campylobacter concisus

C. concisus has mainly been associated with periodontitis but has been isolated from stool samples of children suffering from enteritis. However, in one study of children with and without enteritis there was no significant difference in isolation rates of *C. concisus* between the groups. Specific primers based on 23S rDNA have been developed for this species.

Campylobacter fetus

This species has two subspecies: *fetus* and *venerealis*. Both are primarily associated with animals. Although better known as a cause of sporadic abortion in cattle, they are also a rare cause of human disease. The presence in the blood stream of *C. fetus* subsp. *fetus* may be associated with cancer. Primers common to both subspecies have been identified.

Campylobacter gracilis

C. gracilis may be involved in periodontal disease and has been shown to cause infections in dental implants. No conclusive evidence exists that it is a cause of human enteritis.

Campylobacter helveticus

This species has been isolated from the faeces of domestic cats and dogs. No conclusive evidence exists that it is involved in human disease. A species-specific recombinant DNA probe has been developed to help determine its pathogenicity.

Campylobacter hyointestinalis

Two subspecies exist for this species: *hyointestinalis* and *lawsonii*. The subspecies may be differentiated by phenotypic tests, whole-cell protein and macro-restriction profiling. Although this species was first considered to be a pathogen of pigs, it has now been established as an occasional human pathogen.

Campylobacter lari

This species has been isolated from mussels and oysters. It is presumed that the shellfish become contaminated by the faeces of gulls feeding in the growing waters. In one survey relatively extensive DNA polymorphisms were found within a single batch of shellfish, indicating a high degree of genetic diversity within the species. It has only been infrequently associated with human enteritis and bacteraemia but there is one published case where it was deemed to have caused chronic diarrhoea and bacteraemia in a neonate. Reactive arthritis may be a complication following enteritis. Unique polymerase chain reaction (PCR) primers for *C. lari* have been identified based on a 23S rDNA sequence.

Pathogenicity

Campylobacter spp. can express virulence either directly, by invasion of the epithelial cells of the gut and releasing toxin or indirectly, by inducing an inflammatory response. Like many pathogens, the pathological changes can be multifactorial in nature, with a combination of determinants being involved. The main factors described are motility, adhesion, invasion, iron acquisition and toxin production.

C. jejuni has a polar flagellum (at one or both ends of the cell) and is capable of rapid motility which, when combined with a spiral morphology, gives the organism a selective advantage in penetrating and colonizing the thick viscous mucus barrier of intestinal cells. *Campylobacter* spp. also exhibit chemotaxis regulated by the *cheY* gene. The flagella are highly immunogenic and can undergo both phase and antigenic variations which help them to evade the immune response of the host. Two flagella genes have been identified: *flaA* and *flaB* which are in a tandem chromosomal arrangement separated by a short intervening sequence. The *flaA* and *flaB* genes are of approximately equal size (1.7 kbp) with predicted molecular masses of 59 588 and 59 909 respectively. Flagella, in particular flagella with type A flagellin, do appear to be necessary for invasion and internalization but flagella may or may not act as adhesion factors.

Carbohydrate moieties, probably a glycoprotein, have been shown to be important for adhesion since pre-treatment with L-fucose and D-mannose inhibits adherence to INT 407 epithelial monolayers. A variety of outer membrane proteins have also been described that bind to eukaryotic cells. *Campylobacter* spp. may also possess fimbriae (4–7 nm in width) whose synthesis is enhanced by bile salts. Non-fimbriated mutants, however, are still able to adhere to and invade INT 407 cells and colonize ferrets but with ameliorated disease symptoms, which suggests some role in virulence.

Although evidence of epithelial cell invasion in vivo is sparse host cell invasion, which occurs within a very short time period, has been observed experimentally in macaque monkeys and in the colon of

patients. In addition to epithelial cell invasion the organism may overcome the gut barrier by translocation (passing between cells) possibly via M cells. Certainly *Campylobacter* spp. have been observed to associate preferentially to intercellular junctions. The ability to invade is strain-dependent. Using Hep-2 cells, no correlation was found between invasiveness and the type of symptoms observed, showing that host factors such as immune status are important. *Campylobacter* spp. do not exhibit a positive Sereny test and show variability in their ability to invade a range of tissue cell lines, although invasion has been shown to be more efficient when cells of human origin are used.

Campylobacter spp. require iron for normal cell division since in the absence of the metal ion cells elongate, lack septa and become filamentous. *C. jejuni* and *C. coli* do not produce siderophores (iron chelating agents) but are capable of utilizing siderophores produced by other microorganisms, including enterobactin and ferrichrome. A transport system encoded by the *ceu* operon may be involved in this process. Most strains of *C. jejuni* and *C. coli* produce a cell-associated haemolysin and possess a mechanism to transport haemolysis products into the cell and release iron from the complexes. *C. jejuni* also elaborates the iron storage protein ferritin which helps protect the bacterium against iron overload which may result in iron-catalysed damage to cellular components. The organism also elaborates an iron-containing superoxide dismutase (SOD) which provides protection against oxidative stress during invasion. *C. jejuni* possesses a *fur* gene which, if similar to *fur* genes in other bacteria, is involved in the synthesis of SODs and outer membrane siderophore receptors as well as regulating other genes involved in pathogenesis.

The reported frequency of enterotoxin elaboration amongst isolates varies greatly (**Table 1**). Differences in methodology and in parameters, such as the number of passages, strain storage and culture conditions and polymyxin B treatment, probably contribute towards this observed variation. *Campylobacter* appear to produce several types of cytotoxins, including a Shiga-like toxin (SLT), a cytolethal distending toxin (CLDT), *C. jejuni* toxin (CJT; similar to *Vibrio cholerae* toxin) and heat-labile toxin (similar to *Escherichia coli* LT toxin). Some of the cytotoxins which have a molecular weight range of 50 000–70 000 are as yet poorly characterized but on tissue culture cell lines the effects include cell rounding with nuclear condensation, loss of cell monolayer adhesiveness and cell death within 24–48 h. *Campylobacter* produce a cytotoxin (SLT) similar to Shiga toxin, as evidenced by its neutralization by Shiga toxin antibody and by a monoclonal antibody against the

Table 1 Incidence of enterotoxin production among strains of *Campylobacter jejuni* and *C. coli*

Number and origin of strains	Enterotoxin-positive (%)	Method of detection
25 clinical, Belgium	100	CHO
62 clinical, US	94	GM₁ ELISA[a]
22 clinical, South Africa	77	Y-1
44 clinical, Belgium	75	CHO
32 clinical, Mexico	65	CHO, RILT
80 clinical, Algeria	65	CHO
12 clinical, diverse origin	50	CHO, GM₁ ELISA, RILT
316 clinical, Canada	48	CHO, Y-1
44 clinical, Costa Rica	47	Y-1
372 clinical, diverse origin	45	CHO
39 clinical, US	36	GM₁ ELISA
202 clinical, Sweden	32/19[b]	CHO, GM₁ ELISA[b]
22 clinical, India	32	CHO, GM₁ ELISA
22 clinical, US	0	CHO, GM₁ ELISA
15 clinical, US	0	CHO, GM₁ ELISA
47 carriers, India	12	CHO, GM₁ ELISA
30 carriers, Algeria	60	CHO
6 carriers, Mexico	16	CHO, RILT
8 carriers, Mexico	0	CHO
77 carriers, India	0	CHO, GM₁ ELISA, RILT

CHO, Elongation of Chinese hamster ovary cells; GM₁ ELISA, ELISA with ganglioside GM₁ as the solid phase and antiserum against enterotoxin CT or LT as primary antiserum; Y-1, rounding of mouse adrenal tumour cells; RILT, fluid accumulation in the rat ileal loop test.
[a]The primary antiserum used with homologous serum against enterotoxin of *C. jejuni*.
[b]Different results were obtained with the two methods. Adapted from Wassenaar (1997).

B subunit of *E. coli* SLT-1. The titre produced by *Campylobacter* is however < 1000-fold less than that produced by *Shigella dysenteriae* or *E. coli* O157. In chinese hamster ovary cells the CLDT of *Campylobacter* caused accumulation of F-actin assemblies which resembled actin stress fibres and this was accompanied by a block in cell division which in vivo would result in cytokinesis. Strains of *C. jejuni* have been shown to produce significantly more CLDT than *C. coli*. Published research on *C. jejuni* toxin currently presents a confusing picture, particularly as regards its pathogenesis, chemical structure and genetic basis. It is however closely related to cholera toxin and LT toxin of *E. coli* since it can be neutralized by antibodies raised against the latter two toxins and also by the fact that it binds strongly to the ganglioside GM1 on target cells. Evidence from molecular studies suggests that all *C. jejuni* strains possess the toxin gene but that it is not always expressed. A further toxin, proteinaceous in nature, has been shown to be elaborated by *C. jejuni* and *C. coli* which causes elongation of Vero and Hep-2 cells and rounding of CHO cells.

Typing of *Campylobacter* spp.

The identification of specific subspecies of pathogens is essential for effective epidemiology. However, biotyping of campylobacters is restricted by their limited range of biochemical activities, although schemes were devised and applied. Serotyping has also been developed and the method of Penner, based on heat-stable antigens, has been most widely applied. However, this method has been limited by the availability of antisera and is generally only used by public health laboratories which have facilities to raise their own antisera. Coincidentally, the increasing importance of *Campylobacter* spp. as food-poisoning pathogens occurred at a time when methods of genotyping were becoming more widely available and, in fact, the genomic sequence of *C. jejuni* has been recently published.

The lack of an established biotyping scheme for *C. jejuni* led to genotyping methods being applied and, as methods evolved, or were invented, they have been applied to investigate specific aspects of the epidemiology of this species. As with the development of all tools, the ultimate aim of a given investigation will determine which genotyping method is selected. Thus a UK Department of Health investigation into appropriate methods of subtyping campylobacters aimed to define the most appropriate methods for use in clinical laboratories. The group concluded that a two-stage process was best, with biotyping used to screen incoming isolates into groups (using biochemical reactions and Penner serotyping), then pulse-field gel electrophoresis (PFGE) being applied to define the specific subtype.

PFGE is based on the entire genome of the target species being released from cells held in a gel and then digested by a rare-cutting restriction enzyme *in situ*. The very large fragments of DNA require the application of an oscillating electrical field to migrate through the agarose gel and the subtyping is based on the pattern of DNA fragments which have their molecular weight estimated from a comparison with a molecular-weight standard included in each gel. The latter analysis can take place using commercial software which analyses images of the gel patterns and can subsequently produce similarity dendrograms based on numerical taxonomy principles.

However, PFGE is a relatively slow process: the electrophoresis takes about 24 h to complete, and for a full analysis it is recommended to use digestion with more than one restriction enzyme. Hence, analytical methods aimed at specific regions of the genome have been applied, such as ribotyping and PCR restriction fragment length polymorphism (PCR-RFLP). PCR-RFLP typing has been applied based on ribosomal

Table 2　Overview of typing methods applied to *Campylobacter* spp.

Method	Advantages	Disadvantages
Biotyping	Simple, cheap, quick	Low discrimination
Phage typing	Relatively simple, reasonable discrimination	Phage sets not widely available
Serotyping	Relatively simple, discrimination can be good	Antisera not widely available
Flagellin typing	Fast and uses widely available reagents	Interlaboratory comparison of types difficult, long-term stability of types in question
Pulsed-field gel electrophoresis	Highly discriminatory	Slow, specific equipment required, interlab comparison of types difficult
Automated ribotyping	Fast, intermediate level of discrimination	High capital and running costs
Amplified fragment length polymorphism analysis	Fast, level of discrimination can be defined by primers	Needs expensive capital equipment, method still undergoing development

RNA genes and also targeting the flagellin gene, of which there are two copies, *fla* A and *fla* B. Whilst individual groups report successes with this method, the long-term stability of flagellin types has been called into question.

An overview of some benefits and drawbacks of the typing methods applied to *Campylobacter* spp. is presented in **Table 2**.

Methods of Control

Pets, water and improperly handled and cooked foods account for most cases of *Campylobacter* enteritis. Untreated water and unpasteurized milk have been responsible for those outbreaks with large numbers of associated cases. However, undercooked poultry products have mainly been responsible for the large numbers of sporadic cases of campylobacteriosis. It is probably in this latter area where adequate control strategies still require most research effort. Since *C. jejuni* is often found in high numbers ($> 10^4$ cfu) per processed carcass, perhaps the best approach is to concentrate on devising methods which will ensure that the birds arrive at the processing plant with significantly reduced *Campylobacter* contamination. This focuses attention on the rearing conditions. *Campylobacter* is rarely found in poultry feed or the hatchery environment and generally colonizes the chicks only after the second or third week. The most likely vectors are flies, wild birds, rodents or contaminated water. Three main strategies are currently being employed: drinking water quality, vaccination and competitive exclusion. Certainly some authors contend that disinfection of drinking water is likely to have the greatest impact on the prevalence of *Campylobacter* spp. Passive immunization of chicks has resulted in reduced colonization by the organism but the cost-effectiveness of this approach still has to be determined. Competitive exclusion involves the administration early in the chick's life of a cocktail of organisms that prevents subsequent colonization of the bird when challenged with *Campylobacter* spp. A three-strain mixture comprising *Klebsiella pneumoniae*, *Citrobacter diversus* and *Escherichia coli* ($O13:H^-$) provided 43–100% (average 78%) protection.

Viable but Non-culturable Forms

Campylobacter cells, in common with other genera such as *Vibrio*, *Salmonella* and *Shigella*, have been shown to metamorphose into a viable but non-culturable (VNC) state when subjected to unfavourable conditions such as would be experienced in water, which generally has a low nutrient status. With *Campylobacter* the cells transform from a motile spiral form to a coccoid VNC form which is incapable of cell division in normal media entirely suitable for the normal culturable form.

If the VNC form of *Campylobacter* is capable of initiating an infection in humans or colonizing the gut of domestic animals and poultry or indirectly via food contact surfaces, then contaminated water must pose a risk. There is still much controversy over the infectivity of VNC *Campylobacter* cells. It must be stated that such a phenomenon has been found with *Vibrio cholerae* and other related enteric water-borne pathogenic bacteria. Such authors suggest that the VNC form is a degenerative state and that there is a continuum of physiological states, with one extreme being highly culturable and the other dead cells. The VNC state is between these but tending towards the latter state. Certainly more research is needed to elucidate the role, if any, of the VNC form of *Campylobacter* in the transmission of disease and colonization of domestic animals and birds.

See also: **Campylobacter**: Detection by Cultural and Modern Techniques; Detection by Latex Agglutination Techniques. **Food Poisoning Outbreaks**. **Milk and Milk Products**: Microbiology of Liquid Milk.

Further Reading

Ketley JM (1997) Pathogenesis of enteric infection by *Campylobacter*. *Microbiology* 143: 5–21.

On SLW (1988) In vitro genotypic variation of *Campylobacter coli* documented by pulsed-field gel electrophoretic DNA profiling: implications for epidemiological studies. *FEMS Microbiology Letters* 165: 341–346.

On SLW, Neilsen EM, Engberg J and Madsen M (1998) Validity of *Sma*I-defined genotypes of *Campylobacter jejuni* examined by *Sal*I, *Kpn*I and *Bam*HI polymorphisms: evidence of identical clones infecting humans, poultry and cattle. *Epidemiology and Infection* 120: 231–237.

Skirrow MB (1994) Diseases due to *Campylobacter*, *Helicobacter* and related bacteria. *Journal of Comparative Pathology* 111: 113–149.

Smith JL (1996) Determinants that may be involved in virulence and disease in *Campylobacter jejuni*. *Journal of Food Safety* 16: 105–139.

Wassenaar TM (1997) Toxin production by *Campylobacter* spp. *Clinical Microbiology Reviews* 10: 466–476.

Detection by Cultural and Modern Techniques

Janet E L Corry, Division of Food Animal Science, Department of Clinical Veterinary Science, University of Bristol, UK

Introduction

A recent sharp increase in the number of *Campylobacter* infections has focused the attention of public health services and the food industry – particularly the supermarket chains – on the need for methods of detection of these microorganisms.

There are about 20 species and subspecies of *Campylobacter*. Most reported cases of human campylobacter diarrhoea in the UK are caused by *C. jejuni* subsp. *jejuni* and *C. coli*. Other species that have been reported to cause gastrointestinal illness in humans are *C. lari*, *C. jejuni*, subsp. *doylei*, *C. fetus* subsp. *fetus*, *C. upsaliensis*, *C. hyointestinalis* subsp. *hyointestinalis*, *C. concisus*, *C. sputorum* biovar, *sputorum*, *C. sputorum* biovar. *paraureolyticus* and *C. curvus*. The last, along with several others, also causes periodontal (mouth) infections. Many *Campylobacter* species can also cause systemic infections in humans and animals, including septicaemia, abortion, meningitis and abscesses.

Members of two other closely related genera, *Arcobacter* and *Helicobacter*, originally included in the genus *Campylobacter*, cause gastroenteritis in humans. These are *Arcobacter butzleri*, *A. cryaero-* *philus*, *Helicobacter pullorum*, *H. fennelliae*, *H. heilmanii*, *H. cinaedi* and *H. canis*. The helicobacters are closely related to *H. pylori*, which colonizes the human stomach wall and is an important cause of gastritis and duodenal and peptic ulcers, and has also been implicated in gastric cancer in humans. This article discusses methods for the isolation of the most important species of *Campylobacter*, *Arcobacter* and *Helicobacter* (referred to as 'campylobacteria'). Methods for *H. pylori*, although it may be transmitted in food, are not considered.

Campylobacter Gastroenteritis

The most common symptoms of gastroenteritis caused by *Campylobacter jejuni* are diarrhoea, sometimes bloody, and abdominal pain. The infectious dose is low – 500 is most frequently quoted for *C. jejuni*. The incubation period is 2–11 days, with the symptoms lasting for up to 3 weeks. As with many gastrointestinal infections, there may be complications after the infection appears to have cleared up. These include arthritis and, most serious, Guillain–Barré (GB) syndrome, which has an annual incidence of 1–2 per 100 000 and is associated with recent campylobacter infection. The syndrome is a nervous disease which causes paralysis. Usually patients make a full recovery, but 5–10% of patients die and 15–20% are left with significant residual nerve damage.

In developing countries most victims of campylobacter diarrhoea are children, and the older population develops resistance to subsequent infection. In the UK and similar countries most cases occur in young children and young adults. Young men catering for themselves for the first time are especially likely to be infected.

Sources of Infection

Campylobacteraceae (including *C. jejuni*) can be found sometimes in large numbers in the intestinal contents of many wild and domestic animals and birds, where they usually cause few or no symptoms. However, *C. upsaliensis* and *C. helveticus*, as well as *C. jejuni*, sometimes cause diarrhoea in cats and dogs (especially kittens and puppies). Lambs and calves sometimes suffer from diarrhoea, usually due to *C. fetus* subsp. *fetus* or *C. jejuni*. Humans can therefore be infected directly from these animals, and outbreaks have been reported, particularly amongst parties of children visiting farms and coming into contact with calves and lambs. There have also been large outbreaks from raw milk and from untreated water supplies. Milk can be contaminated by cow faeces (or sometimes via campylobacter mastitis). Surface water (e.g. in streams and lakes) is often a rich source of

Table 1   Percentage of various species of Campylobacter isolated from human cases of diarrhoea in Wales and northwest England (source: PHLS Campylobacter Reference Laboratory, 1999)

Campylobacter species	Percentage of total number of infections
C. jejuni subsp. jejuni	93
C. coli	6.5
C. lari	0.5
C. jejuni subsp. doylei	A few cases
C. fetus subsp. fetus	A few cases
C. upsaliensis	A few cases

campylobacters, due to faecal contamination from wild and domestic animals and birds.

Campylobacter infections differ from those caused by *Salmonella* in that they mostly appear to be sporadic cases rather than outbreaks involving two or more people. This means that the sources are more difficult to determine and also that cases are less likely to be reported and investigated. Investigations have also been hampered by the fact that not all strains isolated from animals or the environment are pathogenic to humans, and there is no easy method of determining pathogenicity. In addition, there has been no convenient typing system, like that for salmonellas, that could be used in epidemiological investigations when trying to trace causes of infection. Progress is at last being made in typing. In the UK, the Public Health Laboratory Service (PHLS) Central Public Health Laboratory set up a specialist laboratory to speciate, serotype and phage-type isolates from human cases, using pulsed field gel electrophoresis (PFGE) where needed to investigate possible outbreaks. To date this specialist laboratory only examines isolates from Wales and the northwest of England. Few clinical laboratories even determine the species of campylobacter isolated. The species most commonly isolated from this area are listed in **Table 1**. The vehicles of infection in approximate order of frequency are:

- undercooked chicken meat
- raw or poorly pasteurized milk
- contaminated water
- bird-pecked bottled milk
- cross-contaminated ready-to-eat foods
- puppies and kittens

Detection of Campylobacteria

Campylobacteria are Gram-negative, spiral-shaped bacteria, motile with one or more polar flagella, and are oxidase-positive. They have a reputation for being difficult to grow and for needing complex media. In fact, most will grow in or on quite ordinary media

such as nutrient broth or agar, provided the atmosphere and humidity are appropriate. *Campylobacter jejuni*, *C. coli*, *C. lari* and *C. upsaliensis* are commonly known as thermophilic campylobacters because they are able to grow at 43–45°C, but not at 30°C or below. *Campylobacter jejuni* and other campylobacters require a microaerobic atmosphere containing 5–7% oxygen and about 10% carbon dioxide, and their growth, particularly on solid media, is assisted by the use of substances that neutralize toxic oxygen derivatives. The most commonly used of these are blood, whole or lysed, 'FBP' – a mixture of ferrous sulphate (F), sodium metabisulphite (B) and sodium pyruvate (P) – and charcoal or haematin plus BP. A few (e.g. *Helicobacter pullorum*) require about 10% hydrogen in addition to the standard microaerobic atmosphere, while *Arcobacter* spp. can grow aerobically or microaerobically. The classical campylobacters also grow much better on the surface of solid media if it is not too dry. This aspect does not appear to have been investigated for the newer strains. Many of the 'new' campylobacteria are unable to grow at 42°C or 43°C, or in the presence of some of the antibiotics commonly used in selective media devised for the classical thermophilic strains. Biochemically campylobacteria are relatively inert, making routine identification using traditional methods, e.g. sugar fermentation, difficult.

Selective Media

Bearing in mind that campylobacters only really came to widespread attention after 1977, the number of different media that have been devised for their isolation is amazing. However, the number of selective agents used is relatively small, and most media differ only in terms of the concentrations of selective agents and/or their combinations, basal medium and oxygen-quenching agents. Cycloheximide, amphotericin or nystatin are used to inhibit fungal competitors, rifampicin is active against both Gram-positive and Gram-negative bacteria, while Gram-positive bacteria are commonly inhibited by vancomycin or bacitracin in combination with cefoperazone. Polymyxin B or E (colistin) inhibits most Gram-negative rod-shaped bacteria, except *Proteus* spp., for which trimethoprim is added. Alternatively, one of the most popular media uses deoxycholate.

Table 2 summarizes the constituents of some of the most important plating media. Butzler and co-workers have devised various media. The earlier versions contained bacitracin and novobiocin with polymyxin B or polymyxin E (colistin). Later the bacitracin and novobiocin were replaced by rifampicin and cefoperazone (or the novobiocin was replaced by vancomycin and cephazolin). All con-

Table 2 Inhibitors and other agents used in various plating media for campylobacters (concentrations in mg per litre)

Medium	Cephalosporins	Trimethoprim	Polymyxins*	Vancomycin	Rifampicin	Other agents
Butzler 1973	–	–	10 000 (B)	–	–	5 novobiocin 25 000 bacitracin
Lauwers 1978	15 cephalothin then 15 cephalexin	–	10 000 (E)	–	–	Sheep blood
Skirrow 1977	–	5	2 500 (B)	10	–	Lysed horse blood
Campy-BAP 1978	15 cephalothin	5	2 500 (B)	10	–	Sheep blood
Preston 1982	–	10	5 000 (B)	–	10	Lysed horse blood
Butzler (Virion) 1983	15 cefoperazone (30 – 1986)	–	10 000 (E)	–	10	Sheep blood
mCCD 1984 (Aspinall 1993 mod. CAT agar)	32 cefoperazone (8 cefoperazone, 4 teicoplanin)	–	–	–	–	Haemin FeSO$_4$ 1% deoxycholate, charcoal
Karmali 1986	32 cefoperazone	–	–	20	–	Haemin Charcoal Pyruvate
Goossens semisolid 1989	30 cefoperazone	50	–	–	–	–
Exeter 1986/91	15 cefoperazone	10	4 mg (B or E)	–	10	Lysed horse blood

* International units per litre.

tained sheep blood except the medium of Goossens, which is semisolid and uses only cefoperazone in combination with a high level of trimethoprim as selective agents. This medium relies on the superior ability of campylobacters to swarm or swim, which reduces the need for selective agents. The plates are inoculated with small volumes of neat faeces near the edge of 50 mm diameter plates. Campylobacters swarm just below the surface, thus apparently avoiding the need for oxygen-quenching compounds. Subcultures are made from the edge of the swarming zone. Having no added blood or other anti-oxygen system, this medium has the advantage of being relatively simple and cheap to make. (According to the authors it can be used with a candle jar, avoiding the need to generate a microaerobic atmosphere by use of bottled gas or sachets.) It is therefore particularly useful where budgets are tight, but needs to be prepared not more than 4 days in advance and stored chilled and in the dark. However, similar precautions are advisable for all campylobacter media.

The medium of Skirrow (e.g. Oxoid CM 331 with 5% lysed horse blood and SR69) is still widely used. It contains blood and replaces some of the polymyxin B used in the Butzler media with trimethoprim and vancomycin, while the Campy-BAP medium developed by Blaser and colleagues in the USA is basically Skirrow medium with added cephalothin.

The media from the Preston group – Oxoid CM67, SR117, SR84; CCD (charcoal, cephazolin, later modified to cefoperazone); deoxycholate agar (mCCDA: e.g. Oxoid blood-free selective agar plus SR125); CAT (cefoperazone, amphotericin, teicoplanin) agar (Oxoid blood-free selective agar base plus Oxoid SR 174) – are the only ones that offer a logical rationale for their development. All these media have nutrient agar as a base. Preston medium has a similar formulation to that of Skirrow (lysed horse blood, but double the concentration of trimethoprim and polymyxin B, rifampicin instead of vancomycin because of its wider spectrum of antibacterial activity, and amphotericin to suppress fungi). CCD agar is one of only a few that contain no blood. Charcoal, pyruvate and ferrous sulphate are used as oxygen quenchers and casein hydrolysate is added to support the growth of *C. lari* strains. Selective agents were reduced to cephazolin (later modified to cefaperozone) and deoxycholate. The medium normally also now contains amphotericin which enables it to be incubated at 37°C rather than 42°C. It is now known as mCCD (modified CCD) agar. Cefoperazone–amphotericin–teicoplanin (CAT) agar is a modification of mCCD agar, devised in order to isolate *C. upsaliensis* in addition to the classical thermophilic strains. Some strains of *C. upsaliensis* are sensitive to 32 mg l^{-1} of cefoperazone used in mCCD agar and many other campylobacter selective media. CAT agar uses 8 mg l^{-1} of cefoperozone and 4 mg l^{-1} of teicoplanin, resulting in a medium that is slightly less selective but grows a wider variety of campylobacters. Another charcoal-containing medium was devised by Karmali: this contains vancomycin and 32 mg l^{-1} of cefoperazone (e.g. Oxoid CM908 and SR139).

Membrane Filtration Method

The membrane filtration method has been found useful by a number of workers because it avoids the use of selective media – and the possibility of failing to find campylobacteria sensitive to the inhibitors in selective isolation media. A 47 mm, 0.45 µm or

0.65 μm pore cellulose triacetate membrane filter is laid on the surface of an agar plate (usually blood agar, but it could be selective medium) and a small volume of faecal or other suspension is dispensed onto the filter, taking care not to allow it to spill over the edge. The plate is incubated aerobically face up for 30–60 min before the filter is removed and incubation continued under microaerobic atmosphere. Campylobacteria appear to be able to penetrate the membrane while other bacteria cannot. However, numbers around $10^5 \, ml^{-1}$ of suspension seem to be needed before they can be detected using this method. When the technique is carefully carried out, very few contaminants are observed, even when using nonselective blood agar.

Enrichment Media

Table 3 summarizes the most commonly used liquid media. Many of the solid media have been adapted for use as enrichment media (e.g. Preston broth, mCCD broth, Exeter broth). Others have been devised specifically as enrichment media (e.g. VTP FBP and the medium of Park and Sanders and of Hunt and Radle). However, even in the last three, the selective and oxygen-quenching agents differ little from those used in many of the solid media. Liquid media are sometimes used to isolate campylobacters from faeces, but more often are used for examining food or water, where these microorganisms are likely to be present in low numbers. The medium of Bolton (LAB-M, LAB135, plus X131) is satisfactory for isolating thermophilic campylobacters from foods. Adding CAT supplement (Oxoid SR174) instead of X131 allows it to be used to isolate arcobacters.

Atmosphere for Enrichment There appears to be no consensus on this. The simplest method, which appears to work quite well, is to use bottles with a small air space and a lid tightly closed. This is particularly useful if supplements are to be added during incubation (see below). An alternative is to use bottles with loose lids in a microaerobic atmosphere. Other systems that have been used are conical flasks with side-arms, or plastic bags that are flushed with microaerobic gas mixture three times before starting incubation.

Isolation of Thermophilic Campylobacters from Foods

As thermophilic campylobacters do not grow at temperatures below 32°C, there is little likelihood of their multiplying in foods, except in the hottest kitchen. The possibility of there being sublethally damaged organisms present is much more real. Various methods have therefore been devised in order to resuscitate damaged campylobacters. Temperatures of 42–43°C have been observed by several workers to be inhibitory. Various antibiotics have also been found to be toxic to damaged campylobacters (e.g. polymyxin, rifampicin and deoxycholate). Methods of enrichment for organisms from foods therefore generally involve incubating at 37°C for 4 h or 6 h, or even 4 h at 31–32°C, followed by 37°C for 2 h before transfer to 42°C or 43°C. **Table 4** summarizes the International Standards Organization (ISO) method.

Comparison of Isolation Methods Comparative studies have been made of three different enrichment media (Doyle and Roman, Park and Sanders, and

Table 3 Inhibitors and other agents used in various enrichment media (concentrations in mg per litre)

	Cephalosporins	Trimethoprim	Polymyxin B*	Vancomycin	Rifampicin	Other agents/methods
Preston 1982	–	10	5 000	–	10	LHB, FBP
Doyle and Roman 1982	–	5	20 000	15	–	LHB, cysteine, succinate
VTP FBP	–	7.5	5 000	15	–	LHB, FBP
mCCD	32 cefoperazone					
Park and Sanders (1991)	32 cefoperazone (add after 4 h)	10	–	10	–	LHB, pyruvate, citrate 4 h 32°C 2 h 37°C then 42°C (shaking)
Exeter	15 cefoperazone	10	4 mg	–	10	LHB, FBP a/b added after 4 h at 37°C
Hunt and Radle 1992	15 cefoperazone + 15 after 3 h	12.5	–	10	–	LHB, FBP 3 h at 32°C, 2 h 37°C then 42°C
Bolton (Lab-M, LAB135)	20 cefoperazone	–	–	20	–	LHB, haemin α-ketoglutamate, BP

a/b, antibiotics; B, sodium metabisulphate; F, ferrous sulphate; LHB, lysed horse blood; P, sodium pyruvate.
* International units per litre.

Hunt and Radle) and three plating media (Campy-BAP, mCCDA and Campy-Cefex) for the isolation of campylobacters from 50 samples of naturally contaminated chicken. Ten samples were positive by direct plating and 49 samples were positive by one or a combination of enrichment and plating media. The Park and Sanders and Hunt and Radle enrichment broths detected similar numbers of positive samples (40 and 43 respectively), while Doyle and Roman gave only 23. Overall Park and Sanders broth in combination with mCCDA gave the best results. In a collaborative trial, it was found that Park and Sanders broth incubated statically, aerobically was best for detecting low numbers (2 cells per 10 g) of campylobacters in naturally contaminated frozen chicken. The Steele and McDermott method of plating from enrichment broth was as effective as selective agars such as mCCDA.

Isolation of Campylobacters from Faeces

There are surprisingly few published comparisons of plating media, and **Table 5** summarizes the results. On the whole mCCDA and Karmali perform well, although in cases of human diarrhoea, the Steele and McDermott method was reported to give many more isolates than mCCDA of *Campylobacter* spp. other than the classical thermophilic strains.

Table 4 International Standards Organization method for isolating thermophilic campylobacters (ISO 10272, 1995)

1.	Enrich in Preston broth (18 h 42°C) or Park and Sanders broth 32°C 4 h, 37°C 2 h, 42°C 40–42 h (if possibly stressed)
2.	Use microaerobic atmosphere, e.g. 5% O_2; 10% CO_2; 85% N_2
3.	Solid media: Karmali Modified Butzler (or Skirrow, mCCDA or Preston) at 42°C, inspecting after 48 h, 72 h and if necessary up to 5 days

Table 5 Comparisons of selective plating media for thermophilic campylobacters

1.	Pure cultures + human and animal faeces Butzler not good for *C. coli* and *C. lari* Productivity: Preston > Blaser > Skirrow > Blaser > Campy-BAP > Butzler. Selectivity: Preston > Butzler > Skirrow > Blaser > Campy-BAP
2.	Seagull faeces. Preston, Skirrow, Butzier and Blaser – all similar except Butzler not good for *C. coli* and *C. lari*.
3.	Human faeces. All media similar except mCCDA had fewer contaminants
4.	*Campylobacter jejuni* only – pure cultures and spiked faeces. mCCDA > Karmali > Skirrow > Preston > Butzler > Blaser Wang. Karmali preferred on basis of appearance of colonies and selectivity
5.	Human diarrhoea stools mCCDA at 42°C compared with membrane filter method Many more isolates with membrane filter – *Campylobacter* spp., *C. upsaliensis, C. jejuni* subsp. *doylei*

Taking Environmental Samples

Campylobacters are sensitive to exposure to oxygen and to drying, so the optimum method of taking these samples is to put swabs straight into enrichment media on site.

Isolating Less Common Campylobacteria

It seems probable that the importance of strains other than the 'classical' thermophilic campylobacters (*C. jejuni* subsp. *jejuni, C. coli* and *C. lari*) has been underestimated because the media and methods used are frequently unsuitable for isolating them. In particular, some clinical laboratories apparently are still using 42°C for isolation of campylobacteria from faeces. This situation is changing: CAT agar was designed to select for strains of *C. upsaliensis* and has also proved useful, together with an enrichment broth containing the same combination of selective antibiotics, for isolating *Arcobacter butzleri* from chicken and *Campylobacter sputorum* biovar. *paraureolyticus* from cows. A comparison of CAT agar, mCCDA, Karmali agar and the semi-solid medium (SSM) of Goossens for growth of some of the non-thermophilic campylobacters and arcobacters indicated that none of these media was suitable for all strains.

A comparison of various combinations of the Steele and McDermott membrane filter method and a selective medium for isolating *C. concisus* from human faeces is summarized in **Table 6**. The total number of positive results obtained by any method was 116 and the most effective combination of methods was the Steele and McDermott method with a 0.65 μm pore size membrane in combination with selective medium (blood agar with 10 mg l^{-1} nalidixic acid and 10 mg l^{-1} vancomycin).

Several methods have been used to isolate *A. butzleri*. One of these was a microaerobic method at 25°C for poultry carcass swabs, with enrichment in

Table 6 Sensitivities of different culture methods for *C. concisus* (from Van Etterijck et al, 1996)

Culture method	No. of positive cultures	Sensitivity[a] (%)
Filter (0.45 μm pore size)	59	51
Filter (0.65 μm pore size)	62	53
Selective medium[b]	51	44
Filter (0.45 μm and 0.65 μm pore size)	81	70
Filter (0.45 μm pore size) and selective medium	97	84
Filter (0.65 μm pore size) and selective medium	99	85

[a]Sensitivity was calculated on the basis of the total number of positive cultures (116) obtained by the use of the three methods combined.
[b]Blood agar + 10 mg l^{-1} nalidixic acid + 10 mg l^{-1} vancomycin.

a medium containing bile salts, cefoperazone ($7.5 \, \text{mg} \, l^{-1}$), polymyxin E and 5-fluorouracil, and plating by the Steele and McDermott method onto Karmali agar base with the same selective agents, minus bilt salts. Other workers, examining poultry rinses, incubated at 30°C, microaerobically for the enrichment and aerobically for the plating medium. The enrichment culture was filtered and the filtrate plated onto mCCDA. Another method used a modified leptospira enrichment broth with $100 \, \text{mg} \, l^{-1}$ 5-fluorouracil to isolate *Arcobacter* spp. from pork. Aerobic incubation at 30°C for 9 days was followed by plating onto a modified *Yersinia* selective agar (cephsulodin-irgasan-novobiocin supplemented with 200 mg 5-fluorouracil), cephalothin–vancomycin–amphotericin (CVA) agar and by the Steele and McDermott method (0.45 µm pore membrane filter) onto blood agar. Incubation was at 30°C for 48 h microaerobically. All the isolates tested were identified as *A. butzleri*. The CVA and modified *Yersinia* selective agars were considered to be the best plating media. Another reported method involved the use of an enrichment broth with piperacillin $75 \, \text{mg} \, l^{-1}$; cefoperazone $32 \, \text{mg} \, l^{-1}$; trimethoprim $20 \, \text{mg} \, l^{-1}$; cyclohex-imide $100 \, \text{mg} \, l^{-1}$. The plating medium was semisolid with the same antibiotics. Incubation was at 24°C aerobically, incubating the enrichment medium for 48 h and the plating medium for 48–72 h. Arcobacters were isolated from beef and pork at low prevalence (< 5%) and from 24% of poultry samples. All were identified as *A. butzleri* or '*A. butzleri*-like'. It seems that *A. butzleri* can readily be isolated by any of a number of methods.

Studies on chicken carcass washes have shown that enrichment of equal volumes of carcass wash with double-strength CAT broth (30°C aerobically for 48 h) followed by plating onto CAT agar and 5% sheep blood agar (both by the Steele and McDermott membrane filter method) yielded *A. butzleri* from all of the 25 carcasses examined. A lower proportion yielded *A. cryaerophilus* and/or *A. skirrowii*. *Arcobacter cryaerophilus* was detected on both CAT agar and blood agar, but *A. skirrowii* was only detected on blood agar. *Campylobacter jejuni* could be detected by direct plating onto mCCDA or onto blood agar by the Steele and McDermott method, but could not be detected after enrichment in CAT broth, owing to overgrowth by *A. butzleri*. *Helicobacter pullorum* was isolated from 9 of 15 carcasses, but only by direct plating onto blood agar using the Steele and McDermott method. The plating media were incubated at 30°C microaerobically in an atmosphere that included hydrogen (obtained by two-thirds evacuation of the jar and refilling with anaerobic gas mixture, omitting the catalyst). Incubation of plates was extended for up to 7 days. *A. butzleri* is a robust organism that grows more rapidly than other arcobacters, while the *A. cryaerophilus*, *A. skirrowii* and *H. pullorum* colonies grow more slowly and are frequently only isolated on non-selective blood agar.

Rapid Detection Methods

There are numerous reports of polymerase chain reaction (PCR) methods for detecting campylobacters. For routine purposes these have several disadvantages: the possibility of detecting non-viable cells; the lack of an easy method of determining numbers; and most important, the lack of an isolate for further study – unless the original sample is examined by traditional means. However, in combination with a preliminary enrichment step, and for purposes where mostly negative results are expected, a PCR technique or other method such as an enzyme-linked immunoassay (ELISA) might be appropriate. Methods utilizing PCR ribotyping and DNA dot-blot techniques can be extremely useful for identification of pure culture isolates, however, particularly for less common species.

Future Developments

There is controversy over the 'viable but non-culturable' (VNC) concept. Improvements in methods of isolation of sublethally damaged campylobacters are likely at least to reduce the incidence of this phenomenon. An interesting development in this field is Oxyrase, a bacterial membrane preparation, which has been demonstrated to reduce levels of oxygen in culture media, aiding the multiplication of anaerobes. To date it has not been found any better than other oxygen scavengers for campylobacteria.

Conclusion

There are a number of reliable methods for isolating the classical thermophilic campylobacters from faeces. A satisfactory method for foods is to use the Bolton enrichment broth, with a resuscitation step (e.g. 4 h at 37°C before moving to 42°C) if the bacteria might be sublethally damaged, and then plating onto mCCDA. If a wider variety of campylobacters is sought, the membrane filter plating technique onto nonselective blood agar can be used with a hydrogen-containing microaerobic atmosphere, preceded by enrichment at 37°C in a Bolton medium with CAT selective supplement (Oxoid SR174). For species such as *Helicobacter pullorum*, direct plating onto blood agar using a membrane filter, or a PCR detection method, are the only options that seem satisfactory.

See also: **Campylobacter**: Introduction. **Food Poisoning Outbreaks**. **PCR-based Commercial Tests for Pathogens**.

Further Reading

Adler H and Spady G (1997) The use of microbial membranes to achieve anaerobiosis. *Journal of Rapid Methods and Automation in Microbiology* 5: 1–12.

Aspinall ST, Wareing DRA, Hayward PG and Hutchinson DN (1996) A comparison of a new campylobacter selective medium (CAT) with membrane filtration for the isolation of thermophilic campylobacters including *Campylobacter upsaliensis*. *Journal of Applied Bacteriology* 80: 645–650.

Aspinall ST, Wareing DRA, Hayward PG and Hutchinson DN (1993) Selective medium for thermophilic campylobacters including *Campylobacter upsaliensis*. *Journal of Clinical Pathology* 46: 829–831.

Aspinall ST, Wareing DRA, Hayward PG and Hutchinson DN (1996) A comparison of a new campylobacter selective medium (CAT) with membrane filtration for the isolation of thermophilic campylobacters including *Campylobacter upsaliensis*. *Journal of Applied Bacteriology* 80: 645–650.

Atabay HI and Corry JEL (1997) The isolation and prevalence of campylobacters from dairy cattle using a variety of methods. *Journal of Applied Microbiology* 84: 733–740.

Atabay HI, Corry JEL and On SLW (1998) Diversity and prevalence of *Arcobacter* spp. in broiler chickens. *Journal of Applied Microbiology* 84: 1007–1016.

Atabay HI, Corry JEL and Post DE (1996) Comparison of the productivity of a variety of selective media for *Campylobacter* and *Arcobacter* species. In: Newell DG, Ketley J and Feldman RA (eds) *Campylobacter VIII. Proceedings of the 8th International Workshop on Campylobacters, Helicobacters and Related Organisms*. P. 19. New York: Plenum.

Collins CI, Wesley IV and Murano EA (1996) Detection of *Arcobacter* spp. in ground pork by modified plating methods. *Journal of Food Protection* 59: 448–452.

Corry JEL and Atabay HI (1997) Comparison of the productivity of cefoperazone amphotericin teicoplanin (CAT) agar and modified charcoal cefoperazone deoxycholate (mCCD) agar for various strains of *Campylobacter*, *Arcobacter* and *Helicobacter pullorum*. *International Journal of Food Microbiology* 38: 201–209.

Corry JEL, Post DE, Colin P and Laisney MK (1995) Culture media for the isolation of campylobacters. In: Corry JEL, Curtis GDW and Baird RM (eds) *Culture Media for Food Microbiology. Progress in Industrial Microbiology* 34: 129–162 also *International Journal of Food Microbiology* (1995) 26: 43–76.

Goossens H, Vlaes L, Galand I, Van Den Borre C and Butzler JP (1989) Semisolid blood-free selective motility medium for the isolation of campylobacters from stool specimens. *Journal of Clinical Microbiology* 27: 1077–1080.

Humphrey T, Mason M and Martin K (1995) The isolation of *Campylobacter jejuni* from contaminated surfaces and its survival in diluents. *International Journal of Food Microbiology* 26: 295–303.

Hunt JM and Abeyta C (1995) Campylobacter. In: Food and Drug Administration, *Bacteriological Analytical Manual* 8th edn. p. 7.01. Arlington: AOAC.

International Standards Organization (1995) *Methods for microbiological examination of food and animal feeding stuffs* 17. Detection of thermotolerant *Campylobacter* ISO 10272 (BS 5763 part 17, 1996).

Lammerding AM, Harris JE, Lior H, Woodward DE, Cole L and Muckle CA (1996) Isolation method for recovery of *Arcobacter butzleri* from fresh poultry and poultry products. In: Newell DG, Ketley J and Feldman RA (eds) *Campylobacter VIII. Proceedings of the 8th International Workshop on Campylobacters, Helicobacters and Related Organisms*. New York: Plenum.

On S (1996) Identification methods for campylobacters, helicobacters and related organisms. *Clinical Microbiology Reviews* 9: 405–422.

Skirrow MB (1977) Campylobacter enteritis: a new disease. *British Medical Journal* 2: 9–11.

Steele TW and McDermott SW (1984) The use of membrane filters applied directly to the surface of agar plates for the isolation of *Campylobacter jejuni* from faeces. *Pathology* 16: 263–265.

Van Erterijck R, Breynaert J, Revets H et al (1996) Isolation of *Campylobacter concisus* from faeces of children with and without diarrhoea. *Journal of Clinical Microbiology* 34: 2304–2306.

Detection by Latex Agglutination Techniques

Wilma C Hazeleger and **Rijkelt R Beumer**, Wageningen University and Research Centre, Department of Food Technology and Nutritional Sciences, Wageningen, The Netherlands

Introduction

The traditional methods for the isolation and confirmation of food-borne pathogens usually consist of (pre)enrichment followed by isolation on selective media, subculturing of suspect colonies on non-selective media after which confirmation of isolates takes place. Traditionally, these confirmation tests involve examination of morphology, Gram reaction and biochemical tests which take several days. In some cases, especially in medical microbiology, this procedure might take too long and therefore several rapid tests have been developed to shorten the time needed for confirmation of pathogens. The latex agglutination test is easy to perform and the result is known rapidly. In clinical laboratories, latex agglutination assays are frequently used for the identification or confirmation of pathogenic bacteria or the toxins produced by pathogens such as *Clostridium difficile* and *Vibrio cholerae* and also for viruses such as Rotavirus.

Since the mid 1980s, latex agglutination tests have also been developed to facilitate the confirmation of food-borne pathogens such as *Staphylococcus aureus*, *Salmonella*, *Yersinia* and *Campylobacter*. For entero-pathogenic *Campylobacter* spp., three different latex tests are commercially available. According to the manufacturers, these assays can be used for the confirmation of suspect colonies and, in one case, detection of campylobacters in enrichment broth or directly in faeces.

In this article, latex agglutination assays and test protocols for detection of *Campylobacter* are described in detail. Furthermore, the use of the different tests in the detection and confirmation of *Campylobacter* is discussed. It is concluded that the latex agglutination tests can be used successfully for the quick confirmation of colonies after microscopical examination. All tests perform well but there are differences in the time needed to carry out the tests. If a latex assay is performed directly on faeces or on enrichment broth, great care has to be taken with respect to the interpretation of the results.

Brief Principles and Types of Commercial Tests Available

The isolation and identification of *Campylobacter jejuni* involve many steps. Usually, *Campylobacter* is isolated from food and environmental samples via selective enrichment in broth followed by isolation on selective agar plates. In most cases, isolation from faeces takes place directly on selective agar plates due to the expected high numbers of *Campylobacter* present in stool samples. After cultivation on solid media, colonies are subcultured, examined for morphology (small curved bacilli) and motility (+), Gram stained (–) and finally confirmed via biochemical reactions such as oxidase (+), catalase (+), glucose, lactose and sucrose utilization (–) and growth at 25°C (–) and 42°C (+). If it is necessary to make a distinction between *C. jejuni*, *C. coli* and *C. lari*, additional tests such as antibiotic resistance and hippurate hydrolysis can be performed. To shorten the time needed for confirmation, several rapid tests have been developed to facilitate detection of campylobacters such as immunological and DNA methods. The latex agglutination assay is an immunological test.

The basic principle of this test is that polystyrene latex particles with a diameter of 0.8–1.0 μm are coated with rabbit immunoglobulins raised against antigen preparations from selected strains. When these antibody-labelled latex particles are mixed with a suspension containing antigens such as whole cell bacteria or cell wall parts, such as flagella, a sensitive and specific immunochemical reaction takes place

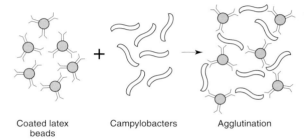

Coated latex beads   Campylobacters   Agglutination

Figure 1 Principle of a latex agglutination reaction. When latex beads, coated with antibodies raised against *Campylobacter*, are mixed with a suitable amount of campylobacters, agglutination will occur. The aggregates formed in this manner are visible to the naked eye.

Table 1 Reaction of three different latex agglutination tests with *Campylobacter* and *Helicobacter* species. Data provided by the manufacturers

Strain	BBL Campyslide	Meritec Campy	Microscreen
C. jejuni	+	+	+
C. coli	+	+	+
C. lari	+	+	+
C. fetus	+	–	+/–
C. upsaliensis	a	a	+
H. pylori	a	–	variable

+, clearly visible agglutination; +/–; weak agglutination; –, no visible agglutination; a, no information provided.

causing the latex particles to agglutinate into aggregates that are macroscopically visible (**Fig. 1**).

For confirmation of enteropathogenic campylobacters, three commercial latex agglutination tests are on the market: BBL Campyslide (BBL Microbiology Systems, Cockeysville, USA), Meritec Campy (Meridian Diagnostics, Cincinnati, USA) and Microscreen Campylobacter (Mercia Diagnostics, Guildford, UK). All tests consist of latex beads coated with rabbit antibodies against *C. jejuni* strains. Additionally, the particles of Meritec Campy and BBL Campyslide contain antibodies against *C. coli*, *C. lari* and the latter also against *C. fetus* strains. All tests detect *C. jejuni*, *C. coli* and *C. lari*. Some kits also show additional agglutination with other *Campylobacter* species and *Helicobacter pylori* (**Table 1**). No agglutination is observed with the closely related *Arcobacter* (non-published results).

Detailed Protocols of Tests and Their Points of Application in the Cultural Techniques for *Campylobacter*

Tests (Protocols According to the Manufacturers' Instructions)

BBL Campyslide This can be applied for the confirmatory genus-level identification of selected campylobacters (*C. jejuni*, *C. lari*, *C. coli* and *C. fetus* subsp. *fetus*) from culture.

Materials

- Extraction reagent (dilute solution of acetic acid)
- Neutralization reagent (dilute solution of NaOH with a pH indicator)
- Reactive latex (anti-*Campylobacter* spp. rabbit antibody-coated latex suspension in buffer)
- Control latex (normal rabbit immunoglobulin-coated latex suspension)
- Positive antigen control (extracted antigens of *Campylobacter* species)
- Plastic stirrers
- Glass test slide.

For the positive control, positive antigen control is mixed with reactive latex and with control latex. Distinct agglutination with the reactive latex should occur, whereas no agglutination greater than a slight graininess should be observed with the control latex.

For the negative control, extraction reagent is first mixed with neutralization reagent and then with the reactive and control latex. Neither of the latex reagents should produce agglutination other than a slight graininess.

Test Protocol. Colony material must be thoroughly resuspended with extraction reagent in a small test tube to achieve a slightly turbid homogeneous suspension. After incubation for 20–30 min neutralization reagent is added. After thorough mixing, the colour of the suspension should be mid to deep purple. Two drops of the mixture are applied to two separate circles on the test slide. Reactive latex is added to one circle and control latex to the other. Suspensions are mixed with plastic stirrers and the slide must be rocked for 3–4 min by hand or by using a mechanical rotator (100 r.p.m.). Then the slide can be examined under intense light. When weak agglutination occurs, an additional rotation of 6 min is recommended. The test is negative if agglutination occurs with neither the reactive nor the control latex. A positive test shows distinct visible agglutination with the reactive test but not with the control latex. When the control latex reagent shows agglutination, the test is non-interpretable. In this case, a heavy amount of inoculum may be the cause and it is recommended that the test is repeated with a lighter organism suspension.

Meritec-Campy This is a latex agglutination test for the confirmatory identification to the genus level of *C. jejuni*, *C. coli* and *C. lari* from culture.

Materials

- Latex detection reagent (rabbit antiserum to common antigens of selected *Campylobacter* species bound to latex particles suspended in buffer)
- Extraction reagent (dilute solution of hydrochloric acid)
- Neutralization reagent (glycine buffer)
- Positive antigen control reagent (neutralized acid extract of appropriate *Campylobacter* organisms in buffer)
- Test slide.

For the positive control, positive control reagent is mixed with latex detection reagent. Definite agglutination should be observed.

For the negative control, extraction reagent is mixed with neutralization reagent after which latex detection reagent is added. No agglutination should be observed.

Test Protocol One to six colonies are resuspended in extraction reagent and mixed with a wooden stick to dissociate all visible clumps of the inoculum. No incubation time is required for this step. Neutralization reagent and subsequently latex detection reagent are added to the extract. The contents must be mixed thoroughly and the slide placed on a rotator for 5 min (100–110 r.p.m.) after which the agglutination reaction can be observed under a high intensity light. A positive test is indicated when the latex detection reagent clearly agglutinates with the test specimen and no agglutination occurs in the negative control. When no agglutination with the latex detection reagent occurs, the test is negative. If an extremely weak agglutination reaction occurs, the procedure can be repeated with a larger initial inoculum.

Microscreen Campylobacter This is an assay for the identification of enteropathogenic campylobacters in enrichment broth, on selective plates or in acute phase, diarrhoeal stool samples.

Materials

- Test latex reagent (latex particles coated with rabbit immunoglobulins raised against antigen preparations from selected *C. jejuni* serotypes)
- Control latex reagent
- Positive control
- Sample diluent
- Disposable test slides.

For the positive control, positive control suspension is mixed with test latex reagent and control test reagent. Easily discernible agglutination of the test latex reagents, with no significant agglutination of the control latex, indicates normal reagent function.

No negative control test is described in this assay.

Test Protocol Preparation of the sample is dependent on the source of the isolates to be examined. For identification from agar plates: several colonies are mixed in sample diluent to form an even suspension. For identification from faeces, a fresh faecal sample must be mixed with sample diluent and coarse particles must be allowed to sediment. The supernatant can be used for the latex test. Material from enrichment broth can be used directly. For the latex test, the bacterial suspension which is obtained in either of the above mentioned ways, can be applied on two adjacent ovals on the slide. Control latex reagent is added to one oval and test latex reagent to the other. After thorough mixing, the slide must be rocked gently from side to side for 2 min after which the test can be read. Agglutination of the test latex combined with no agglutination of the control latex is an indication that campylobacters were present. If both test and control latex show no agglutination, *Campylobacter* was not present in sufficient numbers to be detected by the test. A specimen which causes the control latex reagent to agglutinate shows nonspecific agglutination and cannot be interpreted.

Comparison of the Test Protocols

A short overview of the three different tests is given in **Table 2**. The principles of the tests are the same and all tests are easy to perform and do not require the purchase of expensive devices. For the BBL Campyslide and the Meritec-Campy tests, extraction and neutralization steps are necessary. Since an incubation step of 20–30 min is essential for the extraction of antigens in the BBL Campyslide test, this assay will take considerably more time to perform than the two

others. In all tests, several samples can be tested simultaneously.

Points of Application

Latex agglutination tests are often used for the confirmation of campylobacters isolated from food, stool or environmental samples. In these cases, after isolation on solid media, suspect colonies are examined microscopically. If the cells show the characteristic morphology of campylobacters (typical spiral shape and rapid darting motility) latex tests can be applied instead of the biochemical confirmation steps. In this way the time needed for confirmation is shortened by one to two days. In the Microscreen Campylobacter test, it is claimed that it is also possible to use the latex agglutination assay directly on faeces or on material from enrichment broth. However, in these cases many false negatives may occur probably due to low amounts of campylobacters present in this material. Also, especially with faeces, the high density of the material may cause nonspecific agglutination resulting in non-interpretable tests. It must be kept in mind that non-culturable coccoid forms also show positive reaction with the latext tests. Even culture filtrates are able to aggregate the latex beads which is an indication that not only whole cells but also cell components and dead cells play a role in the agglutination reaction.

Regulations, Guidelines, Directives

At the present time, the use of latex tests for the confirmation of *Campylobacter* is not described in ISO regulations or validated by the AOAC. The Dutch guideline (NEN 6269), however, allows latex agglutination tests as a substitute for the biochemical confirmation reactions.

Table 2 Comparison of basic steps in the procedures of three latex agglutination assays for *Campylobacter* according to the manufacturers' instructions

	BBL Campyslide	*Meritec-Campy*	*Microscreen*
Detection	Colonies	Colonies	Colonies, enrichment broth, faeces
Sample preparation/extraction	Mix with extraction reagent	Mix with extraction reagent	Mix with sample diluent
Incubation period	20–30 min	No	No
Neutralization necessary	Yes	Yes	No
Positive control provided	Yes	Yes	Yes
Negative control	Extraction reagent + neutralization reagent + reactive and control latex	Extraction reagent + neutralization reagent + latex detection reagent	Not described
Incubation time with latex beads	3–4 min	5 min	2 min
Estimated time for total assay	30–40 min	10 min	5 min

Table 3 Detection limits (in colony forming units per ml) of three different latex agglutination tests

Strain	BBL Campyslide	Meritec-Campy	Microscreen
C. jejuni	$9 \times 10^6 - 10^8$ (2)	$9 \times 10^6 -$ 5×10^8 (2)	$10^3 - 7 \times 10^5$ (1) $2 \times 10^6 - 10^8$ (2) 10^2 (3)
C. coli	5×10^7 (2)	5×10^7 (2)	10^7 (2) 10^5 (3)
C. lari	7×10^6 (2) na (3)	7×10^6 (2)	7×10^5 (2)
C. upsaliensis	nt	nt	10^5 (3)
H. pylori	nt	nt	10^5 (3)

na, no agglutination observed; nt, not tested; (1) Baggerman and Koster (1992); (2) Hazeleger et al (1992); (3) Sutcliffe et al (1991).

Table 4 Sensitivity and specificity of three latex agglutination tests used for confirmation of colonies (data provided by the manufacturers)

	BBL Campyslide	Meritec-Campy	Microscreen
Sensitivity	98.7%	99.1%	100%
Specificity	99.1%	100%	100%
Positive predictive value	99.3%	100%	a
Negative predictive value	98.2%	95.2%	a
Accuracy	98.8%	99.2%	a

a, no information provided by the manufacturer.

Detection Limits

No apparent differences in detection limits are reported between the three tests. The minimal amount of cells needed for a positive latex agglutination test varies between different strains. In general the detection limit is $10^5 - 10^8$ colony forming units (cfu) per millilitre (**Table 3**). However, some authors report much lower detection limits. This could be due to the fact that the non-culturable coccoid form of *Campylobacter* and culture filtrate also agglutinate with the latex tests. When many of these coccoid cells or cell components are present in the culture to be tested, an underestimation of the amount of cells will be made if this culture is plated on solid media. Another explanation for conflicting results can be the fact that the different latex tests may vary in sensitivity between one kit and another due to the amount and type of antibodies used, which cannot easily be standardized. Occasionally, longer incubation of the test latex with the test suspension enhances the sensitivity of the tests. However, it must be taken into account that waiting too long before reading the test, may cause drying out of the material which could wrongly be taken for agglutination.

Advantages and Limitations Between Latex Tests and Other Techniques

Although latex assays have not, so far, been included in international regulations for confirmation of campylobacters, these tests can be used successfully in the process of isolation and identification. The reported sensitivity and specificity data provided by the manufacturers do not indicate large differences between the tests (**Table 4**). A positive test will give a quick indication that *Campylobacter* was present in food, environmental samples or in faeces.

Occasionally false positive reactions have been reported with other Gram-negative bacteria such as *Pseudomonas aeruginosa*, *Proteus mirabilis* and *Aci-netobacter calcoaceticus*. However, this should not be a problem as long as the test is used for confirmation after microscopical examination of the suspect colonies. The mentioned microorganisms can be easily discriminated from *Campylobacter* with respect to their morphology.

If the test is used directly on enrichment broth, much care must be taken with respect to the interpretation of the results. In these cases, a negative result of a latex agglutination assay is not evidence that *Campylobacter* was not present in the enrichment broth. The amount of campylobacters in these broths is often below the relatively high minimal detection level. On the other hand, elements of the enrichment broth may inhibit the immunological reaction. Even if the latex assay is positive, it should be taken into account that no prior microscopical examination has been done and that other microorganisms occasionally react with the coated latex beads. The same applies to the use of a latex agglutination assay directly on stool specimens. Due to the high amount of bacteria and debris, false positive or false negative reactions are likely to occur.

With the application of latex agglutination assays, no distinction can be made between *C. jejuni*, *C. coli* and *C. lari*. However, this should not be a problem since these strains are all considered to be pathogenic. If further characterization is desirable, additional tests can be carried out to distinguish between the entero-pathogenic strains. If a rapid test is needed earlier in the isolation and/or identification process, DNA methods such as polymerase chain reaction methods are preferable. The latter are highly specific, have much lower detection limits and can, therefore, be applied to enrichment broth. However, these methods require specific experimental experience and investment in expensive equipment. Another disadvantage of DNA methods is the fact that the microorganism is not isolated and, as a result of that, is not available for further examination.

In conclusion, the latex agglutination assays provide a quick and accurate way of confirmation of colonies in the traditional isolation procedure of *Campylobacter jejuni*, *Campylobacter coli* and *Campylobacter lari*. Furthermore, the method is easy to perform, does not require any expensive equipment and the results are quickly available. However, these tests are less suitable for the detection of campylobacters directly in faeces or in enrichment broth.

See also: **Campylobacter**: Introduction; Detection by Cultural and Modern Techniques. **Escherichia coli**: Detection by Latex Agglutination Techniques. **Salmonella**: Detection by Latex Agglutination Techniques.

Further Reading

Baggerman WI and Koster T (1992) A comparison of enrichment and membrane filtration methods for the isolation of *Campylobacter* from fresh and frozen foods. *Food Microbiol.* 9: 87–94.

Barbour WM and Tice G (1997) Genetic and Immunologic Techniques for Detecting Foodborne Pathogens and Toxins. In: Doyle MP, Beuchat LR and Montville TJ (eds) *Food Microbiology, Fundamentals and Frontiers.* P. 710. Washington DC: ASM Press.

Goossens H and Butzler J-P (1992) Isolation and Identification of *Campylobacter* spp. In: Nachamkin I, Blaser MJ and Tompkins LS (eds) Campylobacter jejuni, *Current Status and Future Trends.* P. 93. Washington DC: ASM Press.

Hazeleger WC, Beumer RR and Rombouts FM (1992) The use of latex agglutination tests for determining *Campylobacter* species. *Lett. Appl. Microbiol.* 14: 181–184.

Hodinka RL and Gilligan PH (1988) Evaluation of the Campyslide agglutination test for confirmatory identification of selected *Campylobacter* species. *J. Clin. Microbiol.* 26: 47–49.

Nachamkin I and Barbagallo S (1990) Culture confirmation of *Campylobacter* spp. by latex agglutination. *J. Clin. Microbiol.* 28: 817–818.

Phillips CA (1995) Incidence, epidemiology and prevention of foodborne *Campylobacter* species. Review. *Trends Food Sci. Technol.* 6: 83–87.

Smibert RM and Krieg NR (1994) Phenotypic characterization. In: Gerhardt P, Murrey RGE, Wood WA and Krieg NR (eds) *Methods for General and Molecular Bacteriology.* P. 640. Washington DC: ASM Press.

Sutcliffe EM, Jones DM and Pearson AD (1991) Latex agglutination for the detection of *Campylobacter* species in water. *Lett. Appl. Microbiol.* 12: 72–74.

CANDIDA

Contents
Introduction
Yarrowia (Candida) lipolytica

Introduction

Rolf K Hommel, Cell Technologie Leipzig, Germany

Peter Ahnert, Department of Biochemistry, The Ohio State University, Columbus, USA

Imperfect yeast summarized in the genus *Candida* does not display any sexual state of growth and reproduction. Many species are anamorphs of perfect species (teleomorphs) belonging to a variety of genera. The perfect state of most *Candida* is not known. The genus *Candida* is heterogeneous; it contains all imperfect ascomycetous species that are not classified otherwise. Representatives of the 163 *Candida* species are wide spread in natural and artificial habitats. The genus covers the most important human (and animal) pathogen, *Candida albicans*, and also many yeasts used in the production of various foods and feeds for thousands of years. Their high biochemical potency makes some *Candida* useful for biotechnological processes.

Characteristics of the Genus *Candida*

General taxonomical properties of *Candida* yeasts may be summarized as follows. Cells appear in different forms: globose, ellipsoidal, cylindrical or elongated, and occasionally ogival, triangular or lunate. Pseudohyphae and non-septate true mycelium may be formed. Dimorphism, the alternate occurrence of unicellular and hyphal and/or pseudohyphal phases occur in many species (e.g. *C. albicans*). The reproduction proceeds by holoblastic budding. The wall is ascomycetous and two-layered. Arthroconidia and ballistoconidia are not formed. Endospores, i.e. vegetative cells formed inside other cells, also occur in some *Candida*, mostly in long-standing cultures. *Candida vulgaris* (syn. *Candida tropicalis*) is the taxonomic type species. The physiological and biochemical properties reflect the heterogeneity of the genus; sugars may be fermented, nitrate may be assimilated, pellicles may be formed in liquid media. Extracellular starch-like compounds are not produced. Inositol is assimilated by some species, urease is not produced, gelatin may be liquefied. The reaction with diazonium blue B is negative. Xylose, rhamnose and fucose are not present in cell hydrolysates. Ubiquinones Q_8, Q_9 or Q_{10} may be present.

The assimilation of inositol may be positive or negative. Most inositol-positive strains form pseudomycelia. For convenience and ease of identification, *Candida* is divided into physiological groups based on this latter property (**Table 1**). This division is not known to be backed genetically. Some species are highly variable and therefore occur in several groups. The lack of a sexual phase is less suitable for genetic analysis and manipulation. However, stable fusants of interspecific and intergeneric hybridization are possible, for example *C. tropicalis* and *Saccharomycopsis lipolytica* or *C. boidinii* and *C. tropicalis*.

Physiological Properties

As diverse as are the natural habitats of *Candida* yeasts, so are their adaptations in a wide range of physiological properties. As for most yeasts, the majority of *Candida* are mesophilic, growing well at temperatures of 25°C–30°C, with extremes of 0–48°C. *C. austromarina*, *C. psychrophila* and *C. scottii* (*Leucosporidium scottii*) are obligate psychrophilic yeasts. The obligate thermophilic *C. slooffii*, *C. pintolopesii* and *C. bovina* are no longer classified as *Candida*.

Candida, like other yeasts, do not photosynthesize or fix nitrogen. They generally cannot grow anaerobically. Some strains survive and reproduce under microaerophilic conditions. Respiratory metabolism is preferred in the presence of oxygen. Only a few former *Torulopsis* species prefer fermentative metabolism under this condition.

Some species, such as *C. apicola*, *C. bombicola*, *C. famata*, *C. magnoliae* and *C. lactis-condensi*, are osmotolerant. *C. glucosophila* is osmophilic. Their natural habitats normally impose a continuous osmotic stress, usually accompanied by low levels of nitrogen. One response to low water activity is the intracellular accumulation of polyhydroxy alcohol (glycerol).

Conventional carbon sources metabolized by all wild types are glucose, mannose and fructose. Nicotinic acid is required by most *Candida*. Some species requite pyridoxine. *C. utilis*, one of the most useful species, does not require biotin; it accumulates dulcitol and lipids (oleogenous yeast) intracellularly and has the broadest S-metabolism spectrum among yeast.

In the presence of energy (carbohydrate and nitrogen) it utilizes inorganic sulphate, sulphite, thiosulphate, sulphur-containing amino acids, sulphide and taurine. The ability of *C. utilis* to grow well on spent sulphite liquor, rich in pentoses, is used in the Waldorf process, the oldest industrial process to produce fodder yeast. Biomass is also produced on the basis of whey and whey products (*C. utilis* and *C. krusei*), ethanol and acetic acid (*C. utilis*) and methanol (*C. boidinii*). Pentoses (xylose and cellobiose) are fermented by *Candida*. Fermentation of xylose by *C. shehatae* is of biotechnological importance. *C. blankii* uses pentoses as substrates for single-cell protein (SCP, bagasse) production.

C. tropicalis, *C. intermediata* and *C. maltosa* are well-studied examples for SCP production using paraffin oil or wax. Many *Candida* species are able to grow on n-alkanes as sole source of carbon, which is important for bioremediation. The degradation proceeds via a complex regulated metabolic system, involving cytochrome P-450 in the primary step. *C. maltosa*, like other 'alkane *Candida*', has a high biotechnological potential for the production of: hydrophobic intermediates, enzymes usable as biocatalysts, muconic acid, amino acids from racemic substrates (enantio-selective) and heterologous proteins.

Most methanol-utilizing yeasts are *Pichia* or *Candida*. In the presence of methoxy groups from lignin and pectin, from which methanol may be liberated on hydrolysis of methyl esters, the number of methylotrophic *C. boidinii* and *C. sonorensis* is enlarged. Methanol assimilation is accompanied by radical morphological and metabolic changes, such as packing of the cytoplasm with microbodies containing alcohol (methanol) oxidase.

Candida yeasts are used to produce a variety of biotechnologically interesting compounds like glycerol, higher alcohols, organic acids, esters, diacetyl, aldehydes, ketones, acids, long-chain dicarboxylic acids and xylitol. Other products are nicotinic acid, biotin and D-β-hydroxyisobutyric acid. Some strains extracellularly accumulate sophorosides when grown on n-alkanes, alkenes, fatty acids, esters or triglycerides.

Extracellular enzymes, like pectinases, β-glucosidases and lipases, are of commercial interest. *C. cylindracea* lipase, which is commercially available, has been well studied and used for enzymatic synthesis in non-aqueous phases or at interphases for the synthesis of odorous and other chemically important substances. The lipase from *C. rugosa*, displaying a high stereoselectivity and enantiopreference, is used for the hydrolysis of milk fat.

Table 1 Differentiation of *Candida* into physiological groups based on inositol assimilation[a]

Group	Assimilatory capabilities									Example of species
	Inositol	Nitrate	Erythritol	Raffinose	Maltose	Growth at 40°C	Cellobiose	Melezitose	Trehalose	
I	+									C. blankii, C. castrensis, C. santajacobensis
II	–	+								C. etchellsii, C. lactis-condensi, C. magnoliae, C. nitratophila, C. utilis, C. vartiovaarae, C. versatilis
III	–	–	+	+						C. entomophila, C. insectorum
IV	–	–	+	–	+					C. atlantica, C. atmosphaerica, C. sophiae-reginae
V	–	–	+	–	–					C. fermenticarens, C. lipolytica
VI	–	–	–			+				C. albicans, C. glabrata, C. inconspicua, C. lusitaniea, C. tropicalis
VII	–	–	–	+		–	+			C. gropengiesseri, C. intermedia, C. tenuis
VIII	–	–	–	+		–	–			C. apicola, C. milleri, C. stellata
IX	–	–	–	–		–	+	+		C. maltosa, C. oregonensis, C. pulcherrima, C. reukaufii, C. sake
X	–	–	–	–		–	+	–		C. dendrica, C. zeylanoides
XI	–	–	–	–		–	–		+	C. catenulata, C. parapsilosis
XII	–	–	–	–		–	–		–	C. cylindracea, C. sorboxylosa, C. vini

[a] This grouping is not based on known genetic relatedness for species within these groups, made soley for convenience and ease in identification. Species not included: C. austromarina and C. psychrophila which do not grow at 25°C; C. famata (var. famata; var. flareri) shows an extreme physiological variability; C. glucosophila is highly osmophilic and does not grow in media without addition of sugar. Some others also omitted appear in at least two groups, like C. bombicola, C. famata, C. guilliermondii, C. krusei, C. rugosa, C. shehatae, C. tenuis, C. viswanathii, C. zeylanoides.

Habitats

The wide taxonomic boundaries of the genus *Candida* result in a vast variety of settled habitats. Being heterotrophic, they depend on other organisms to convert available substrates into usable forms. Plants are common hosts: leaf surfaces, slime fluxes, nectaries and nectar of flowers, flower petals and other flower parts, skin of fruit, decaying fruit (preferably damaged), stems and plant-associated habitats, including soil. Leaf surfaces are populated with nitrogen-fixing bacteria or producers of sugary compounds, all providing nutrients for yeasts. Many tropical fruits from Africa and South America display a consistent colonization with *Candida* and *Rhodotorula*. In Japan, *C. famata* preferentially colonizes fruit surfaces.

Candida boidinii, for example, is associated with tanning solutions containing sugars, nitrogenous compounds and mineral salts (pH 4.0–5.9). *C. sonorensis* strains, isolated from rotting cactus tissue, metabolize methanol liberated by bacteria.

Nectaries have a high sugar and low nitrogen content and are settled by the fermentative *C. pulcherrima* and *C. reukaufii* (nectar and bumblebee nests). A non-spoiling association of *Serratia plymuthica* and *C. guilliermondii* is involved in pollination of a commercial fig variety. Fallen green figs are settled by *C. fructus*. *C. sorboxylosa* has been isolated from souring figs. Different former *Torulopsis*, now *Candida*, are spoilers of berries and currants. Another example is *C. krusei*, isolated from decaying oranges.

Candida yeasts are found in bark beetles and other borers. *C. nitratophila*, for example, is present in bark beetles and their larvae. *C. tenuis* settles on many coniferous trees and species of beetles and is isolated from cactus roots. *C. oregonensis*, *C. shehatae* and *C. nitratophila* are associated with ambrosia beetles, their larvae, or their borings.

Insects serve as vectors of yeast distribution (*Drosophila* species, bees, bumblebees). Studies in California revealed that *Candida* species represent the majority of yeast isolates found in collected nectar and pollen. Xerotolerant yeasts predominate in association with bees: *C. apicola* and *C. magnoliae* in the crops of honey bees, and *C. apicola* and *C. bombicola* in nests of bumblebees. Various former *Torulopsis* spp. are intracellular symbionts of insects. *C. krusei* and *C. sonorensis* are closely associated with *Drosophila* species.

Candida are not permanently resident in soil. Their number there depends on the amount of available nutrients. Surface layers are preferred because *Candida* are aerobic or microaerophilic. *Candida* are present in soils of many regions. Plant-associated yeasts reach the ground, washed off by rain or along with falling fruit. There they survive the winter and are transported back at the beginning of summer (wind, insects).

As normal inhabitants *Candida* are present in 'natural' and polluted waters (rivers, lakes, pulp mill basins, sewage plants, etc.) and sediments. In marine environments the number of *Candida* increases with decay of marine plants, kelp and plankton. From these sources *C. famata*, *C. guilliermondii*, *C. tropicalis* and *C. parapsilosis* have been isolated. Other species like *C. glabrata* and *C. parapsilosis* are often isolated from seafood, *C. inconspicua* and *C. parapsilosis* from fish, *C. stellata*, *C. sake* and *C. parapsilosis* from oysters. *C. krusei* and *C. valida* occurred together with other yeasts especially in polluted sediments. The presence of the 'C. krusei complex' may indicate sewage pollution. The pathogenic *C. albicans* is suggested as one indicator for general pollution. The higher the pollution with domestic sewage, the higher the cell counts of pathogenic *Candida* in seafood such as oysters and mussels. Oil pollution results in a large increase of *C. lipolytica*, *C. guilliermondii*, *C. tropicalis* and *C. maltosa*.

Pathogenic Yeasts of the Genus *Candida*

Yeasts of the genus *Candida* are the most numerous, widespread, versatile and troublesome. Among those known to cause systemic infections in humans, eight belong to this genus: *C. albicans*, *C. glabrata*, *C. guilliermondii*, *C. krusei*, *C. lusitaniae*, *C. parapsilosis*, *C. tropicalis* and *C. viswanathii*. Two others, *C. kefyi* and *C. norvegensis*, have been reclassified as *Kluyveromyces marxianus* and *Pichia norvegensis*, respectively.

The most important, clinically relevant *C. albicans* is endogenous in the oral, gastrointestinal and urogenital tracts of humans. It is a ubiquitous pathogen of most warm-blooded animals like poultry, pigs, cattle, dogs, various primates and wild animals. Approximately 40–60% of the adult human population harbour this yeast, which may be considered as an obligate, harmless and often symptomless commensal. Its host-free occurrence is rare. Infections are in general not transmitted through food. Infections arise only in damaged or compromised hosts.

In infected tissues the mycelial forms, which tend to be less susceptible to antibiotics and more pathogenic, dominate. Hyphae formation may be a transitional response to high temperatures and neutral pH values. *C. albicans* possesses a circular mitochondrial genome of approximately 40 kb on which subunits of complex I in the electron transport chain are encoded, three others have been mapped to the nuclear genome. The mitochondrial genome encodes subunits 1 and 2 of cytochrome oxidase. Hybridization probes and polymerase chain reaction (PCR) primers exist for the mitochondrial genome and chromosomes 1 to 7 and R. Additionally, many probes have unknown or multiple locations. Some fragments of chromosome R, containing the rDNA, vary in size. The karyotype of *C. albicans* is variable, making it unsuitable for systematic studies and detection. With *C. albicans* and *C. glabrata*, parasexual genetic systems undergoing mitotic segregation were established by protoplast fusion.

The intestinal population, which is responsible for the primary attack, acts via three mechanisms: (1) production of acetaldehyde from sugars via ethanol; (2) disruption of the intestinal lining, allowing large immunogens to enter; (3) release of a large number of antigenic and toxic substances. Adherence to epithelial host tissue plays a key role in the transition from colonization to invasive candidiasis. Adherence involves adhesins expressed on germ tubes but only to a small degree on blastospores or yeast-phase cells. Adhesins bind to host fibrinogen, fibronectin, laminin and collagen. They interact lectin-like with host glycosides. Adhesins are associated with the phosphomannoproteins which constitute antigenic factors for serotyping. Antibodies generated against germ tube specific adhesins, allow discrimination of invasive candidiasis from colonization. Syndromes produced by *C. albicans* may occur singly or in combination depending on the nature of the infection (superficial, locally invasive or deep): vulvovaginitis, dermatitis, cystitis, fever, hepatic dysfunction, chronic fatigue, depression, mental confusion, etc. This yeast accounts for over 90% of systemic or deep infections. Here, and in all other infections, the deepness of the illness depends on the physical status of the host; chronic superficial candidiasis is often associated with malnutrition, pregnancy and diseases with impaired immune system (e.g. diabetes). Acquired immunodeficiency syndrome (AIDS) increases the occurrence of, for example, oral candidiasis. Deep-seated or systemic candidiasis of *C. albicans*, including two or more organs, is frequently iatogenic as result of hyperalimentation, broad-spectrum antibiotics, immunosuppressive or antineoplastic therapies.

The number of deep or systemic infections with *C. tropicalis* tends to increase and the prognosis for a disseminated infection may become more serious than with *C. albicans*. *C. tropicalis* may be isolated from organically enriched soil and from aquatic habitats. It causes the syndromes peritonitis and endocarditis. Most syndromes associated with *Candida* infections are disseminated. However, definite syndromes are associated with: *C. glabrata* (pyelonephritis, osteomyelitis), *C. guilliermondii* (endocarditis), *C. parapsilosis*, (endocarditis, endophthalmitis), or *C. viswanathii* (meningitis). In addition, *C. maltosa* has been isolated from immunocompromised patients or animals. The virulence is strain, but not species, specific. Some further *Candida* yeasts are associated with transient fungaemia, e.g. *C. famata*, *C. heamulonii*, *C. krusei*, *C. lipolytica* and *C. rugosa*. Often affected by transient and superficial infections are housewives, fruit canners, workers handling fish or having wet hands for long periods of time. Systemic avian candidiasis may demand killing of millions of hens as happened in Taiwan and Hong Kong.

Methods of Detection and Identification

The isolation and identification of *Candida* to the species level is difficult since they are widely distributed, very variable, change physiology with growth conditions, and are usually associated with other yeasts and moulds. The taxonomic classification does not provide a generally valid criterion aiding in isolation and identification.

Methods are based on standard assays for carbohydrate assimilation and fermentation, morphology, reproduction characteristics, genotype and antigenotype. Most common, glucose containing, nonselective media for yeast separation, cultivation and enumeration may be used: dextrose agar (pH 6.9), dextrose broth (pH 7.2), Sabouraud medium, dextrose tryptone agar, malt extract medium or plate count agar. Acidification of these media, preferably with lactic acid or tartaric and citric acid (10%, final pH = 3.5), or addition of antibiotics, such as cycloheximide, streptomycin, chloramphenicol and gentamycin, increase the selectivity (lactobacteria and other yeasts are inhibited). Czapek Dox broth tests the acceptance of nitrate as sole nitrogen source (Table 1). Microscopic enumeration, discriminating between bacteria and yeast, is hampered by yeast clusters.

The sample source determines subsequent conditions. Isolates from foods with high sugar content prefer to grow on sugar-rich media, highly osmophilic strains require low water activity. Isolates from (fermenting) brined foods prefer acidified dextrose agar over malt agar. Colony appearance differentiates film-forming surface yeasts from fermentative subsurface ones; the former are generally dull and very rough. Reduction of tartaric acid to 3% allows growth of osmotolerant yeasts. Temperature tolerance increases with increasing osmotic pressure. Two incubation temperatures, 25°C and 30–32°C, should be chosen. Incubation times are generally in the range of 3–5 days and must be extended for osmotolerant and osmophilic yeasts to 5–10 days and 14–28 days, respectively.

Candida are lipolytic, excreting lipases which sometimes cause flavour changes. Lipase activity is detected by single- or double-layer agar methods, based on the liberation of free unsaturated fatty acids from triglycerides and colour changes of Victoria blue B. *Candida* lipases are non-lipolytic with the substrate tributyrin.

The pathogens *C. albicans* and *C. tropicalis* are selectively differentiated as smooth, brownish-black, round colonies on ABY (acid bismuth yeast) agar. In conjunction with BiGGY (bismuth sulphite-glucose-glycine-yeast) agar, *C. albicans* is differentiated further: brown to black colonies without pigment diffusion or sheen. For chlamydospore production by this species a modified Czapek Dox agar (pH 6.8) is available. Most pathogenic *Candida* are detectable by automated test systems. Molecular approaches for identification and typing are: gas liquid chromatography, pyrolysis mass spectrometry, protein fingerprinting, lipopolysaccharide profiles. Electrophoretic karyotyping may discriminate species but not strains. PCR, the recent method of choice for identification, is applicable for clinical samples such as *C. albicans*.

Serotyping detects and identifies *Candida*. Some antigenic markers are widely shared, others are very specific. By using combinations of antisera, serotypes can be identified, most commonly as serotypes A and B. Identification to the level of species and strains is possible using combinations of monovalent antisera or monoclonal antibodies. Serotyping is used for detecting *Candida*, distinguishing invasive candidiasis from colonization, and tracking of candidiasis outbreaks in healthcare and agriculture.

Serotyping procedures can be fast, sensitive and accurate but antigen presentation varies with growth conditions, cell cycle and mutagenesis. Structural analysis of the antigens, use of commercially available monoclonal antibodies, and the use of techniques like Western blotting alleviate these problems.

Serotype A contains antigens absent in serotype B but in one study *Candida* A reacted most strongly if grown at pH 4.5 and least at pH 2. *Candida* B did not react when grown at acidic pH but did react when grown at pH 7–9. *Candida* A are often more susceptible to drug treatment (flucytosine). *Candida* antigens are classified as factor 1, factor 4 and so on. They are cell-wall mannoproteins which can be branched and phosphorylated. Cross-reactivities are shown in **Table 2**.

Importance to the Food Industry

Fermentation of food as a method of preservation has been used widely in the Far East (Japan, China, Indonesia), India and Africa for thousands of years and involves *Candida*. These processes impart a moderately acidic flavour to the foods and form small quantities of ethanol. Natural microbial consortia involved are mixtures of fermentative yeasts and, mainly, lactic acid bacteria. In sourdough bread, *Candida milleri* and lactic acid bacteria are quasi-symbiotic. Fermented alcoholic beverages like saké, mead, teekvass (Russia), kefir (Caucasus region), kumiss (Asia) or leven (Egypt) are made with the participation of *Candida*.

Candida are involved in the microbial consortia producing various Indian fermented foods and drinks. *C. tropicalis* and *C. guilliermondii* occur in kanji fermentations. *C. krusei* is present in the starter for the rice-based chhang, from which arrack is distilled. The same species is present in phool warries and even in the microflora of fermented food based on milk, fruit or vegetables. *C. vartiovaarae, C. krusei, C. famata, C. prapsilosis* and other species are found in many legume-based fermented foods.

In fermentations of olives, brined cucumber, cured meat, fruit wine, etc. from other regions, *Candida* play an important role, contributing to the flavour and increasing the content of proteins, vitamins, etc.

Japanese fermented food is characterized by its high salt content. In soy sauce fermentation *Candida* help to ferment the decomposed proteins, starch and lipid, made by the Koji enzymes. Fermentation proceeds in the presence of 18% NaCl. Strains detected in young soy mash are *C. famata* and *C. polymorpha*, producing considerable amounts of glycerol. In old mash *C. versatilis*, for example, is found, producing various additional polyols. Some strains are able to ferment galactose, sucrose or maltose (*C. versatilis, C. etchellsii*) in high saline media. These consortia produce the rich aroma of soy sauces in the second stage of fermentation. *C. versatilis* and *C. etchellsii* are involved as minor components in miso fermentation at 5–13% salt. The longer the fermentation phase (weeks to months), the better the flavour. *C. versatilis* is isolated from the inner part of cheeses, *C. sake* and *C. intermedia* from the surface. *Candida* dominate the surface of goat cheese.

Among the well-balanced microbial populations of cocoa fermenting yeasts, *C. catenulata, C. famata, C. holmii, C. krusei, C. parapsilosis* and *C. zeylanoides* form a dominant group during the first 1–2 days. *C. famata* or *C. krusei* are present from the flower to ripe fruit. *C. krusei* may utilize ethanol, formed under microaerobic conditions by other yeasts. The

Table 2 Cross-reactivity of *Candida* by serotyping. *Candida* species in rows were shown to share antigens with organisms in columns

	C. glabrata	C. guilliermondii	C. krusei	C. parapsilosis	C. tropicalis
C. albicans A	+			+	+
C. albicans B	+			+	+
C. krusei I					
C. parapsilosis		+	+		
C. tropicalis					

Table 2 – continued

	Escherichia coli	Pityrosporum ovale	Pichia delftensis	Pichia zaruensis	Salmonella ssp.
C. albicans A		+			
C. albicans B		+			
C. krusei I			+	+	
C. parapsilosis					
C. tropicalis	+				+

pectinolytic activities of *C. zeylanoides* (polygalacturonidase), *C. famata* (pectin methylesterase), and others are important in this fermentation.

Some species are frequently associated with wine from specific regions (e.g. *C. sake*, *C. vini*, *C. versatilis*), others are fortuitous. *C. famata* belongs to the yeasts dominating the surface of wine grapes. In the first stages of primary fermentation, *Candida* originate from grapes and must. In the secondary fermentation of sherry-style wine, surface film-forming yeasts form important flavour components (aldehydes, other oxidation products). Spoilage of other wines by film-forming *Candida* during storage usually presents a cosmetic problem. In wineries, film-forming *Candida* belong to non-occasional yeasts. *Candida*, with predominantly oxidative metabolism, are present where grape must (pH 3.5, sugar concentrations up to 20%, w/v) is formed.

The public health significance of yeasts in foods is considered minimal or negligible by most authorities. However, the food quality will be lowered and chances for bacterial and mould settlement increase. Food spoiling by *Candida* proceeds as surface growth (e.g. *C. parapsilosis*, *C. zeylanoides*). Dramatic effects may accompany this growth, such as swelling and explosion of packages (CO_2 production) or gross alterations in food appearance. Glycerol, higher alcohols, organic acids, esters or diacetyl may be formed and affect flavour. Under more oxidative conditions aldehydes, ketones and acids are produced. The conversion of phenylalanine into phenethyl-alcohol gives a distinctive aroma (cheese). Organic sulphides and H_2S from sulphur-containing amino acids result in a foul smell. The action of extracellular enzymes like pectinases, lipases or proteases alters texture. The metabolization of nitrate or nitrite, added for meat conservation, supports bacterial spoiling. **Table 3** summarizes examples of typical *Candida* spoilers in foods. When both red and poultry meats (e.g. turkey carcasses) are kept frozen (–5 to –10°C), the initially

low cell counts of *C. zeylanoides* among the microbial flora increase from 5% to >90%. In nearly all processed meat (salted, vacuum packed, stored refrigerated at –4–7°C) *Candida* species develop well. Treatment of poultry meats with antibiotics stimulates yeast growth. After long-term storage under refrigerated conditions, cell counts of *Candida* on fruit, damaged fruit or fruit juices increased. However, *C. krusei*, the most common *Candida* spoiler, showed decreased viability at temperatures below 0°C. Lipolytic *Candida* are involved in primary or secondary spoiling of dairy products. The main effects of spoiling are the results of the fermentation of lactose (and sucrose), hydrolysis of casein and fat, and the ability to grow at storage temperatures; surface slime with or without pigmentation may be formed on cream cheese. The additional utilization of lactic acid increases the pH and results, for example, in cheese overripening. Typical results of *Candida* spoilage in yoghurt are yeasty and bitter off-flavours, and gassy or frothy texture. Containers start swelling at cell counts above 10^5 or 10^6 cells per gram. Similar effects may be observed for mayonnaise and salad dressings. Species of *Candida* are involved in softening of brined vegetables, like sauerkraut or pickles. Fermentation or formation of pellicles and sediments are indicators of *Candida* spoiling in nonalcoholic beverages.

The great variability of biochemical potency and sensitivity of *Candida* yeasts, including their special adaptations, make it difficult to select effective preservation agents or methods. Chemical and physical factors determine spoilage allowing *Candida* to grow and survive. Heat treatment is more effective than refrigeration. Ascospore-forming yeasts are more heat resistant, especially *C. krusei*. Heat resistance in fruit juices depends on the type of juice, its concentration, and the presence of preservatives and antioxidants. Addition of sucrose to juices reduces the effect of heat inactivation. At low water activity the range for

Plate 1 *Aspergillus flavus* growth on maize. (Photograph courtesy of Patrick Dowd.) See also entry **Aspergillus: *Aspergillus flavus***.

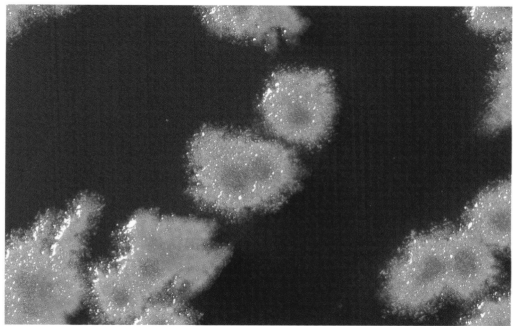

Plate 2 (*above*) *Bacillus licheniformis* 6346 colonies. (With permission from Public Health Laboratory Service, London, UK.) See also entry **Bacillus: Introduction**.

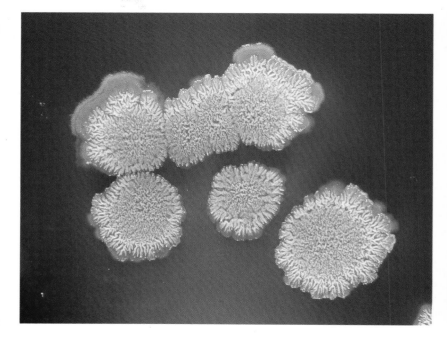

Plate 3 (*right*) Colonies of *Bacillus subtilis*. (With permission from Custom Medical Stock Photo.) See also entry **Bacillus: Bacillus subtilis**.

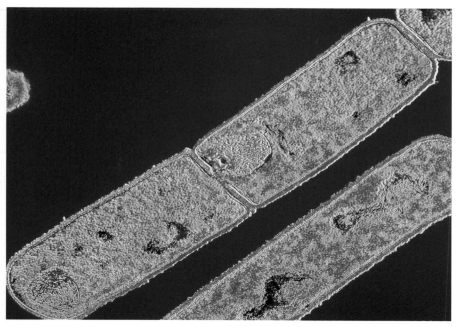

Plate 4 False-colour transmission electron micrograph of *Bacillus subtilis* undergoing a cellular division. Here, in this late phase of replication, the cell wall (at centre, pink) separating the two new sister cells has almost completely formed. A membranous structure is seen in each cell; bacteria do not contain a nucleus, and the DNA genetic material is scattered (light pink on yellow). *B. subtilis*, or the Hay Bacillus, is an aerobic rod-shaped Gram-positive bacteria. It is sometimes pathogenic on humans, causing severe eye infections (iridocyclitis and panopthalmitis). Magnification: ×7400 at 6×7cm size. (With permission from Science Photo Library.) See also entry **Bacillus: Bacillus subtilis**.

Plate 5 (a) (*below left*) Bacteriophage T4 on cell. **(b)** (*below right*) Enhanced *E. coli* and T4 phage. (With permission from Custom Medical Stock Photo.) See also entry **Bacteriophage-based Techniques for Detection of Foodborne Pathogens**.

(a)

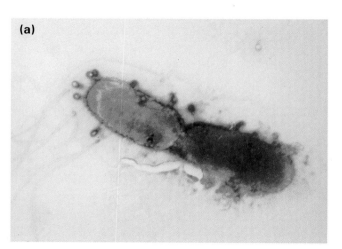

(b)

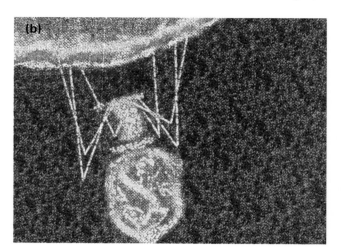

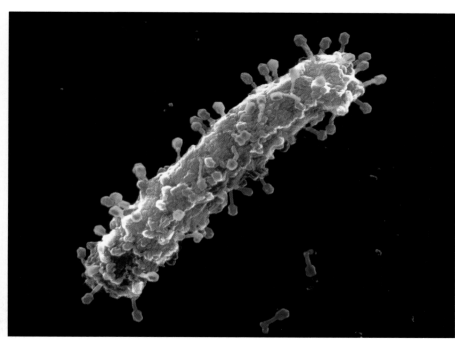

Plate 6 (*left*) Coloured scanning electron micrograph of T-bacteriophage viruses attacking a bacterial cell of *Escherichia coli*. Dozens of blue-coloured viruses are seen around the cell, each having a head and tail. T-bacteriophages are specific parasites of *E. coli* bacteria. The virus attaches itself to the cell wall of the *E. coli* cell using its tail. The elongated tail is a contractile sheath which acts like a syringe to squirt the contents of the head, the DNA genetic material, into the host cell. Viral DNA commandeers the genetic machinery of the cell, forcing it to reproduce more bacteriophages. Magnification: ×18,000 at 6×6cm size. (With permission from Science Photo Library.) See also entry **Bacteriophage-based Techniques for Detection of Foodborne Pathogens**.

Plate 7 (*right*) An example of an API strip from bioMérieux. Different biochemical changes may (+) or may not (−) take place in each tubule, and the pattern of changes can be used for identifying bacteria or yeasts to genus and species level. (With permission from Dr Pradip Patel, Leatherhead Food RA, Surrey, UK.) See also entry **Biochemical Identification Techniques – Modern Techniques: Food Poisoning Organisms**.

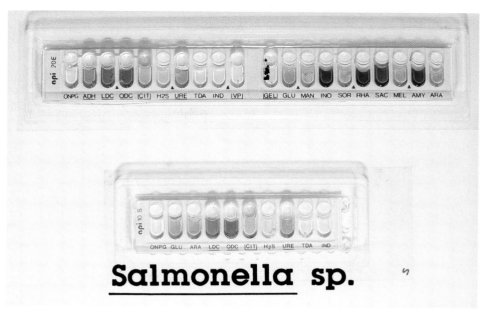

Salmonella sp.

Plate 8 (*left*) Detection and enumeration of coliforms and identification of *E. coli* from food products at 37°C using the selective chromogenic medium Coli ID (bioMérieux). (With permission from bioMérieux.) See also entry **Biochemical Identification Techniques – Modern Techniques: Enterobacteriaceae, Coliforms and E. coli**.

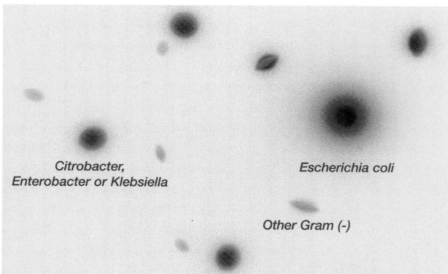

Citrobacter, Enterobacter or Klebsiella

Escherichia coli

Other Gram (-)

Plate 9 (*right*) Photomicrograph of histological section showing single and multiple vacuoles within the cytoplasm of neurons (nerve cells) in the brain of a bull with BSE. Some vacuoles distend the neurons. (With permission from M.J. Stack, Electron Microscopy Unit, Veterinary Laboratories Agency, Surrey, UK.) See also entry **Bovine Spongiform Encephalopathy (BSE)**.

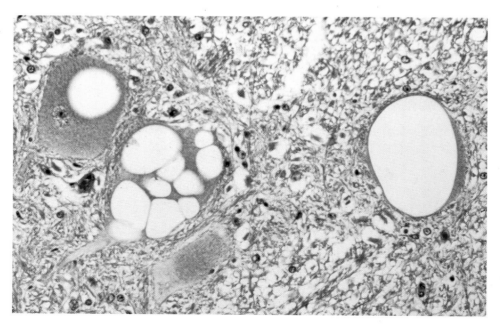

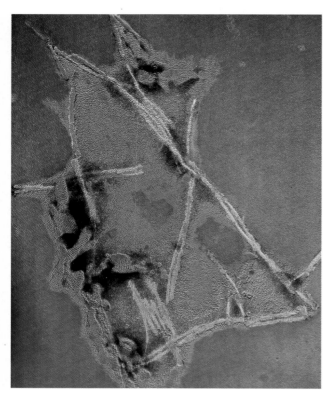

Plate 10 (*left*) Coloured transmission electron micrograph of prion fibrils in the brain of a cow infected with BSE (bovine spongiform encephalopathy) or "mad cow" disease. Prions are virus-like organisms made up of a prion protein. These elongated fibrils (orange) are believed to be aggregations of the protein that makes up the infectious prion. Prions attack nerve cells producing neurodegenerative brain disease. "Mad cow" symptoms include glazed eyes and uncontrollable body tremor. Prions cause BSE in cattle, scrapie in sheep and goats, and Creutzfeldt-Jakob disease in humans. Negatively stained. Magnification: ×62,000 at 6×4.5cm size. (With permission from Science Photo Library.) See also entry **Bovine Spongiform Encephalopathy (BSE)**.

Plate 11 (*below*) Coloured scanning electron micrograph of the fungus *Candida albicans*, cause of human thrush. Groups of rounded yeast-like spores are seen connected by long filaments, known as hyphae. These hyphae spread into a large fungal network, known as a mycelium, and produce more spores. *Candida albicans* causes the infection known as candidiasis. This affects moist mucous membranes of the body, such as skin folds, mouth, respiratory tract and vagina. Oral and vaginal conditions are known as thrush. Genital thrush is sexually transmitted. Magnification: ×3000 at 6×4.5cm size. (With permission from Science Photo Library.) See also entry **Candida: Introduction**.

Plate 12 (below left) Stilton cheese. The milk is ripened initially with lactic acid bacteria, and the mould grows internally during maturation; enzymes from mould promote flavour development. (With permission from J. D. Owens, Department of Food Science and Technology, University of Reading, UK.) See also entry **Cheese: Mould-ripened Varieties**.

Plate 13 (*below right*) An oocyst of *Cryptosporidium parvum*. Magnification: ×1000. (Photograph courtesy of R. Girdwood, Scottish Parasite Diagnostic Laboratory, Glasgow.) See also entry **Cryptosporidium**.

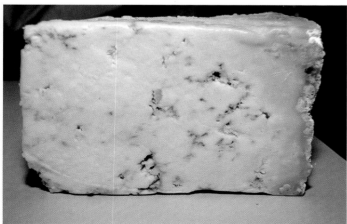

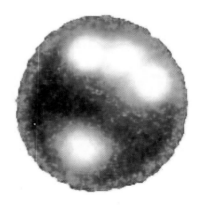

Table 3 Examples of spoiling *Candida* species frequently isolated from foods

Food from which Candida *yeast isolated*	Examples of Candida *spp. characterized*
Meat	
Red meat	C. zeylanoides
Poultry meat	C. zeylanoides, Candida subsp.
Processed meat	
Frankfurter, ham	Candida subsp.
Preferred in high salt	C. rugosa, C. lipolytica, C. zeylanoides
(>4%) bacon	C. famata
Low salt bacon	C. tropicalis, C. krusei,
Khundi, smoked meat (Nigeria)	C. albicans, C. famata
Seafood	
Shellfish, oysters, quahogs, mussels, crabs	C. albicans, C. parapsilosis, C. sake, C. stellata, C. tropicalis, C. glabrata, C. inconspicua
Dairy products	
Cheese	C. famata
Retail cheese	C. lipolytica, C. famata (same regions in Egypt: C. albicans, C. tropicalis, C. parapsilosis)
Yoghurt	C. famata, C. versatilis, C. lusitaniae
Fruits and vegetables	C. haemulonii, C. sake, C. famata, C. krusei, C. stellata
Fruit juices, soft drinks,	C. tropicalis, C. sake, C. apicola, C. magnoliae, C. krusei
Concentrated juices	C. magnoliae, C. krusei
Nonalcoholic beverages	C. inconspicua, C. famata, C. krusei, C. vini
Food with high sugar content	Osmotolerant Candida spp.: C. mogii, C. apicola, C. bombicola, C. lactis-condensi
Sugar	C. apicola
Syrups/molasses	C. valida
Chocolate syrup	C. etchellsii, C. versatilis
Jam	C. cantarelli
Beer	C. utilis
Mayonnaise, salad dressings, mixed salads, tomato sauces	C. sake, C. stellata, C. zeylanoides, C. krusei

growth of osmotolerant yeasts (e.g. *C. lactis-condensi*) is shifted towards higher temperatures.

Conventional acidification by addition of citrate and lactate encourages yeast growth in and on meat. The combination of sorbic acid, acetic acid and benzoates on one hand, and citric acid and lactic acid on the other, may reveal synergistic inhibitory effects. For example, the increased temperature tolerance of xerophilic yeasts is not accompanied by a noticeably affected tolerance towards benzoates. Highly acidic foods, like mayonnaise, do not support yeast growth; the addition of fruits and other ingredients lowers the acetic acid content and introduces metabolizable substrates, allowing *Candida* to grow.

See Color Plate 11.

See also: **Bread**: Sourdough bread. ***Candida***: *Yarrowia (Candida) lipolytica*. **Cocoa and Coffee Fermentations**. **Ecology of Bacteria and Fungi in Foods**: Influence of Available Water; Influence of Temperature; Influence of Redox Potential and pH. **Fermentation (Industrial)**: Production of Organic Acids; Production of Oils and Fatty Acids; Colours/Flavours Derived by Fermentation. **Fermented Foods**: Fermentations of the Far East. **Fish**: Spoilage of Fish. **Molecular Biology – in Microbiological Analysis. PCR-based Commercial Tests for Pathogens. Single Cell Protein**: Yeasts and Bacteria. **Spoilage Problems**: Problems caused by Fungi. ***Torulopsis*. Yeasts**: Production and Commercial Uses. **Fermentation (Industrial)**: Production of Oils and Fatty Acids.

Further Reading

Cygler M, Grochulski P and Schrag JD (1995) Structural determinants defining common stereoselectivity of lipases towards secondary alcohols. *Canadian Journal of Microbiology* 41: 289–296.

Gatfield IL (1997) Biotechnological production of flavor-active lactones. *Advances in Biochemical Engineering/ Biotechnology* 55: 221–238.

Gow NA and Gooday GW (1987) Cytological aspects of dimorphism in *Candida albicans*. *Critical Reviews in Microbiology* 15: 73–78.

Jeffries TW and Kurtzman CP (1994) Strain selection, taxonomy, and genetics of xylose-fermenting yeasts. *Enzyme Microbial Technology* 16: 922–932.

Mauersberger S, Ohkuma M, Schunck WH and Takagi M (1996) *Candida maltosa*. In: Wolf K (ed.) *Nonconventional Yeasts in Biotechnology*. P. 411. Berlin: Springer.

McCullough MJ, Ross BC and Reade PC (1996) *Candida albicans*: a review of its history, taxonomy, epidemiology, virulence attributes, and methods of strain differentiation. *International Journal of Oral and Maxillofacial Surgery* 25: 136–144.

Odds FC (1987) *Candida* infections: an overview. *Critical Reviews in Microbiology* 15: 1–5.

Pfaller MA (1992) The use of molecular techniques for epidemiologic typing of *Candida* species. *Current Topics in Medical Mycology* 4: 43–63.

Pla J, Gil C, Monteoliva L et al (1996) Understanding *Candida albicans* at the molecular level. *Yeast* 12: 1677–1702.

Ratledge C (1994) Microbial conversions of agro-waste materials to high-valued oils and fats. *Institute for Chemical Engineering Symposium Series* 137: 25–33.

Samson RA, Hocking AD, Pitt JI and King AD (eds) (1992) *Modern Methods in Food Mycology*. Amsterdam: Elsevier.

Spencer JFT and Spencer DM (eds) (1997) *Yeasts in Natural and Artificial Habitats*. Berlin: Springer.

Sudbery PE (1994) The non-*Saccharomyces* yeasts. *Yeast* 10: 1707–1726.

Sullivan DJ, Henman MC, Moran GP et al (1996) Molecular genetic approaches to identification, epidemiology and taxonomy of non-albicans *Candida* species. *Journal of Medical Microbiology* 44: 399–408.

Towner KJ and Cockayne A (1993) *Molecular Methods for Microbial Identification and Typing*. London: Chapman & Hall.

Yarrowia (Candida) lipolytica

G M Heard and **G H Fleet**, Department of Food Science and Technology, The University of New South Wales, Sydney, Australia

Characteristics of the Genus

Yarrowia lipolytica is an ascomycetous yeast that is notable for its ability to degrade lipids and proteins and grow on *n*-paraffins as sole carbon source. Early studies reported the isolation of this species from a corn-processing plant, margarine, olives, petroleum products, soil and the cornea of the human eye. Today, it is commonly isolated from a range of food products, especially dairy products. Since early researchers did not find a sexual state for their isolates, the species was classified in the genus, *Candida*. With the discovery of compatible mating types, the formation of ascospores was observed in the late 1960s and the species was subsequently described within the genus *Endomycopsis* as *Endomycopsis lipolytica*. Taxonomic problems associated with the naming of the genus *Endomycopsis* caused it to be changed to *Saccharomycopsis* in the 1970s and the species became known as *Saccharomycopsis lipolytica*. However, *S.*

lipolytica was observed to be significantly different from other species in the genus, especially in the morphological features of its ascospores and its ability to produce coenzyme Q-9. Based on these differences, a new genus *Yarrowia* was created, and the most recent taxonomic classification describes *Candida lipolytica* as *Yarrowia lipolytica*. It is the only species within the genus. Phylogenetic trees constructed from sequence data of the large 26S ribosomal DNA subunit and the small ribosomal RNA subunit have confirmed the unique taxonomic status of this species. The G+C composition (mol% 49–51) of its DNA is also unusually high for an ascomycetous yeast. Pulsed-field electrophoresis of the DNA has given 4–6 chromosomal bands of size 2–5 Mb.

Yarrowia lipolytica grows on malt extract agar giving tannish-white colonies that can be soft, smooth and glistening to dull, tough, old and wrinkled, depending on strain. Asexual growth occurs by multilateral budding on a narrow base. The budding cells are spheroidal, ellipsoidal or elongate and occur singly as pairs or small clusters. The organism is characteristically dimorphic and, in addition to yeast cells, produces abundant pseudomycelium and branched, septate true mycelium (hyphae) 3–5 μm in width and up to several millimetres in length. The septa have a single central micropore. The proportion of yeast to mycelial growth depends on the strain and the cultural conditions. Heat stress and growth in the presence of N-acetylglucosamine induce transition from yeast to mycelial growth, whereas culture on hydrocarbons tends to give more yeast cells.

The species is heterothalic and mating types A and B have been described. Sexual reproduction with the formation of ascospores is induced by the culture of diploid cells on YM agar or V8 agar. Citrate stimulates sporulation. Asci are usually produced on hyphal cells as short side branches or at the end of hyphae. They are spherical to oval and contain one to four ascospores that are quite variable in shape. Depending on strain, the ascospores can be spheroidal, hat-shaped and saucer-like, and can have a smooth to rough surface, with or without a ledge.

Other key features of the species are its: inability to ferment sugars; assimilation of a range of alcohols, polyols and organic acids; production of urease, coenzyme Q-9 and extracellular lipases and proteases; and inability to assimilate nitrate (**Table 1**).

Physiological and Biochemical Properties

As outlined in Table 1 and discussed in a later section, *Yarrowia lipolytica* exhibits a range of properties of industrial or biotechnological significance. With the goal of optimizing the commercial value of these prop-

Table 1. Significant taxonomic and growth properties of *Yarrowia lipolytica*

Taxonomic		Growth
Multipolar budding	Assimilation of:	Temperature 4–37°C
Pseudohyphae	glucose, mannitol, glucitol,	pH 2.5–8.0
Septate hyphae	glycerol, erythritol, lactate,	NaCl 0–15%; a_w 0.90
Asci from hyphal cells	succinate, citrate,	Sucrose 0–50%, a_w 0.94
Hat-shaped/spheroidal/irregular ascospores	hexadecane, *N*-acetylglucosamine, ethanol,	Sorbate 1000 mg l^{-1}, pH 5.0
Sugars not fermented		Benzoate 1200 mg l^{-1}, pH 5.0
Urease production		
Lipolytic, proteolytic		
Co-Q9		
G+C (mol %) 49–51		

erties, substantial progress has been made in understanding the molecular biology of this organism. Nevertheless, there is a need for more basic information about the growth and metabolic behaviour of this species and about the effects of environmental factors on these properties.

Y. lipolytica is non-fermentative and derives its energy by the aerobic assimilation of carbon substrates. With the exception of glucose, it is unable to assimilate most hexose and pentose sugars but is well known for its capacity to grow on various hydrocarbons (e.g. n-hexadecane, n-decane), fatty acids (from lipid degradation), and various alcohols, polyols and organic acids. Induction of the glyoxylate pathway is necessary for the utilization of these substrates and isocitrate lyase, the key enzyme of this cycle has been well studied. However, more specific enzymes are also needed for the metabolism of these substrates and include cytochrome P-450 monooxygenase for the hydroxylation of n-alkanes, acyl-CoA synthetase for the utilization of fatty acids and alcohol dehydrogenase for the metabolism of alcohols. The production of an extracellular emulsifier, liposan, probably assists in the assimilation of hydrocarbons.

The occurrence, growth and biochemical activity of *Y. lipolytica* in foods is very much linked to its strong production of extracellular proteases and lipases. The biochemistry and genetics of the production and excretion of proteases have been extensively studied. At least four different proteases have been identified, and include three acid proteases (pH optima in the range 3.1–4.2), an alkaline protease (pH optimum 9.0) and a neutral protease. In contrast, the lipolytic activity of *Y. lipolytica* is less studied but appears to be a complex of cell bound and extracellular enzymes that preferentially act at the 1 and 3 positions of the glyceride. Other extracellular or wall-bound enzymes produced by this yeast are RNase, acid phosphatase and α-mannosidase.

Factors limiting the growth of *Y. lipolytica* are listed in Table 1. This species is capable of good growth at temperatures as low as 4–5°C, and we have recorded growth rates of 0.06^{h-1} in yeast nitrogen base–glucose medium (pH 3.0–7.0) at 5°C. Lag phases of about 1 day were generally observed. The upper limit for growth is 35–37°C. Growth in yoghurt, milk and cheese stored at 4–10°C has been reported. Strong growth occurs in the range pH 3.0–8.0. Growth is weak at pH 2.0 and absent at pH 1.5 and 9.0. *Yarrowia lipolytica* has not been regarded as a salt-tolerant or sugar-tolerant yeast but all of our isolates grow in the presence of 12.5% (w/v) NaCl (a_w 0.92) and one isolate grew at 15% (w/v) NaCl (a_w 0.90). Salt tolerance is greatest at pH 5.0–7.0 and decreases at pH 3.0, but all strains were able to grow in the presence of 10% (w/v) NaCl at pH 3.0. The salt-tolerant property of this species explains, in part, its frequent isolation from cheeses. Growth occurs in the presence of 50% (w/v) sucrose (a_w 0.94) (pH 3.0–7.0) but not at 60% (w/v) sucrose. There has been one report of this species growing at a_w 0.89 (10% NaCl + 10% sucrose).

Y. lipolytica has never been recognized as a member of the group of preservative-resistant yeasts such as *Zygosaccharomyces bailii* or *Pichia membranifaciens*. This view needs revision as all of our isolates and those reported by Guerzoni and co-workers grow in the presence of 1000–1200 mg l^{-1} benzoate or sorbate at pH 5.0–7.0. At pH 3.0, growth occurred in the presence of 250 mg l^{-1} benzoate. These properties equate with the resistance of *Z. bailii* to these preservatives. However, it is susceptible to oleoresins from cinnamon, clove and thyme and there is a synergistic effect of these compounds with other stresses such as heat. *Y. lipolytica* is inhibited by cinnamon (50 μg ml^{-1}), clove (500 μg ml^{-1}) and thyme (500 μg ml^{-1}) oleoresins but is resistant to the oils from garlic and onion. Heating increases the sensitivity of the species to the oleoresins; for example, thermally stressed cells are inhibited by 25 μg ml^{-1} cinnamon oleoresin.

Enumeration and Identification

Yarrowia lipolytica grows on the wide range of media generally used for the isolation and enumeration of yeasts. These include YM medium, malt extract agar, glucose–yeast extract agar and tryptone glucose, yeast extract agar. Bacterial growth on these media may be restricted by the addition of antibiotics such as chloramphenicol or oxytetracycline ($100\,\mu g\,l^{-1}$) or acidification to pH 3.5. However, acidification may prevent the growth of sublethally injured yeast cells. For the isolation of *Y. lipolytica* from cheeses that contain abundant mould populations, (e.g. blue-veined cheese, Camembert), biphenyl ($50\,mg\,l^{-1}$) can be incorporated into the medium as an antimicrobial agent. Another medium developed for the isolation of *Y. lipolytica* from cheeses includes the addition of tyrosine (10 mM) to the nutrient base. Metabolism of the tyrosine results in the production of melanin pigment that causes the colony to be surrounded by a brown coloration.

Differential and selective media have been developed for enumeration of food-borne yeasts by adding dyes at specific concentrations to a nutrient agar base. *Yarrowia lipolytica* is selectively isolated from other yeasts on crystal violet (CV) agar (crystal violet 1 : 1000 final dilution) and malachite green (MG) agar (1 : 250 000 final dilution). On both agars, species of *Saccharomyces* are inhibited whereas *Y. lipolytica* appears as white, coarsely folded colonies. The basis of selectivity is not understood. *Yarrowia lipolytica* weakly fluoresces at a wavelength of 366 nm, when grown on a medium containing aniline blue (1 : 10 000 dilution) and either chlortetracycline HCl ($100\,\mu g\,l^{-1}$) or chloramphenicol ($100\,\mu g\,l^{-1}$). This property may be used for selective isolation of the species from foods, although other species, including *Candida albicans*, also fluoresce under these conditions. The intensity of fluorescence may be used as an indicator to differentiate between the species.

Identification of *Yarrowia lipolytica* follows the morphological, physiological and biochemical tests described in *The Yeasts, a Taxonomic Study*, 4th edition. The main characteristics of the species are listed in Table 1. Production of pseudo- and true hyphae, formation of asci on hyphal cells containing round to hat-shaped spores are all factors assisting recognition of the species. Microscopic analysis and sporulation tests provide rapid presumptive diagnosis of the species. The strong lipolytic and proteolytic reactions are key features of this species. Its lipolytic activity is seen by plating onto a nutrient agar medium into which butter fat, olive oil or beef tallow has been incorporated. Lipolysis is indicated by strong clear zones surrounding yeast growth. Proteolytic activity is similarly revealed but the yeast is cultivated on a medium into which skim milk or gelatin has been incorporated. Several commercially available yeast identification kits give reproducible and reliable identification of this species. These include the API 20C and ID 32C systems (bioMérieux) and the Biolog YT plate system (Biolog Inc., California, USA).

A polymerase chain reaction–restriction fragment length polymorphism (PCR–RFLP) method for targeting single-stranded ribosomal DNA (ss rDNA) has been used to differentiate *Y. lipolytica* from species of *Saccharomyces*, *Zygosaccharomyces* and *Candida*. A 1200 bp internal fragment of the ss rDNA is amplified and characteristic fragmentation patterns are generated by digestion with restriction endonucleases. The primers for DNA amplification were designed on the conserved region at the beginning and middle of the gene encoding ss rDNA for *Candida albicans* (primers: 5′-GTC TCA AAG ATT AAG CCA, TG-3′ and 5′-TAA GAA CGG CCA TGC ACC AC-3′). The PCR product was digested using the restriction enzymes *Mse*I, *Ava*II, *Taq*I, *Scr*FI, *Hha*I, *Sau*3AI, *Msp*I and *Cfo*I. *Yarrowia lipolytica* was only differentiated from the other species by the *Mse*I digest.

Alternatively, PCR fingerprinting primers for the microsatellite sequences may be used to identify *Y. lipolytica*. Microsatellite sequences occur repeatedly and randomly within the genome of yeasts and are used to provide a fingerprint of individual species. Primers consisting of the oligonucleotides GAC_5 and GTG_5 are synthesized and used in a PCR reaction to generate PCR fragments of differing lengths that are visualized using gel electrophoresis and staining techniques. A database of species based on the profiles from PCR patterns can be used to identify and type yeast isolates. Random amplified polymorphic DNA (RADP) assays can also be applied to differentiate *Y. lipolytica* from other species. Randomly applied primers are used to generate a pattern of DNA fragments. Patterns are visualized and compared as described above.

Significance in Foods and Beverages

Yarrowia lipolytica is significant in the food and beverage industries because it (1) causes product spoilage, (2) is a potential source of food ingredients and processing aids, (3) can be produced as biomass for feedstock, and (4) can be used to process food waste (**Table 2**). There has been one report associating *Y. lipolytica* with human disease. It has been isolated from hospital patients diagnosed with persistent fungaemia and from patients suffering from sinusitis. Fungaemia can develop from the use of intravascular agents, such as a catheter. It was concluded that the

Table 2 Significance of *Yarrowia lipolytica* in the food and beverage industries

Food spoilage	Dairy products: cheese, cream, butter, yoghurt, milk.
	Meat products: fresh meats, ground meats, cured and fermented
	Seafood, fish silage
	Mayonnaise-based salads
Source of food ingredients and processing aids	Citric acid
	Peach flavouring agent, γ-decalactone
	Biotransformation of monoterpenes to flavour compounds
	Cheese flavour product
	Emulsifier, liposan
Starter culture and source of enzymes	Maturation culture and source of proteases and lipases for cheese and sausage fermentations
Biomass (single-cell protein) production for feedstocks	Growth on grains, vegetable by-products, hydrocarbons
Processing of food waste	Reduction of BOD[a] and COD[b] of vegetable processing waters

[a] BOD, biological oxygen demand;
[b] COD, chemical oxygen demand.

species was a weakly virulent, opportunistic pathogen but there are no reports linking such clinical problems with food consumption.

Spoilage of Foods and Beverages

Y. lipolytica is an oxidative species and, unlike many other food spoilage yeasts, is not implicated in the fermentative spoilage of foods such as sugar syrups and fruit juices. Because of its oxidative nature, it often causes the spoilage of foods as a result of mycelial-like growth on the surface. Its production of lipolytic and proteolytic enzymes selects for its proliferation in foods of high protein and fat content. These foods include dairy products, seafoods and meats, but the species has also been isolated from processed vegetables such as dressed salads.

Y. lipolytica is frequently associated with the spoilage of dairy products including milk, kefir and cheeses, and may also be found in yoghurts. It is capable of strong growth in milk, giving final populations of 10^7–10^8 cfu ml^{-1}. Although unable to utilize the lactose in milk, it metabolizes other components such as protein, fat and the organic acids lactic, citric, pyruvic and acetic. Major end products of this growth include free amino acids, such as leucine, glutamic acid, valine, phenylalanine and arginine, that are indicative of protein degradation, and glycerol and free fatty acids, including oleic, caprylic, lauric, caproic, myrystic and palmitic, that indicate lipid breakdown.

In yoghurts, diffusion of oxygen through packaging material permits the growth of the species to populations of 10^6–10^7 cfu g^{-1} during storage at 4–12°C for up to 20 days. It occurs in many cheeses at populations up to 10^6–10^8 cfu g^{-1} and its growth may be accompanied by off-odours and softening of the cheese curd. It also metabolizes tyrosine to produce melanin pigments that result in brown discoloration of cheese surfaces.

Y. lipolytica has been isolated together with a range of other yeasts and coryneform bacteria from the surface of brick or red smear cheeses. It is frequently isolated from surface and interior locations of soft, mould-ripened cheeses, such as the Camembert and blue-veined varieties, where it can be the predominant yeast species. It appears to be a natural component of the secondary or maturation flora of these products, and its population may be as high as 10^6–10^8 cfu g^{-1}. However, its precise contribution to quality of these cheeses is not clear.

There is very little quantitative information on the occurrence of yeasts in butter, cream and oils. However, lipolytic species, such as *Y. lipolytica* can grow on the surface of butter and in cream, resulting in the production of off-flavours. *Y. lipolytica* has been isolated in mixed yeast populations in vegetable and fish oils, where it occurred at up to 40% of the total yeast population.

Y. lipolytica is associated with spoilage of meat products, including seafood, poultry and red meat products. Spoilage of seafoods is generally attributed to the growth of bacteria, but there have been several reports implicating *Y. lipolytica* in the spoilage of fish, fish silage, crab, lobster and mussels. Mycelial like growth is observed on the outer surface of these products during refrigerated storage and is accompanied by the production of ammoniacal odours, volatile bases and general putrefaction. The ability of *Y. lipolytica* to survive at water activities as low as a_w 0.89 and migrate to the lipid fraction of the food gives it a competitive advantage over bacteria in the salty, seafood environment.

The microflora of other meat products also includes *Y. lipolytica* but, generally, there are only qualitative reports of its occurrence. It has occurred at 3.1% of the total yeast population on fresh, processed poultry carcasses and up to 14% of the yeast population on spoiled carcasses. In cooked meat products, it can represent up to 40% of the total yeast population. *Y. lipolytica* tolerates the low water activity and pH conditions of processed meat products such as cured meats and fermented products. In unfermented sausages it tolerates added SO_2 at levels of approximately 450 µg g^{-1}. In other processed meat products, including fermented sausages, the species must survive the presence of preservatives such as sorbic and citric acids and lactic and acetic acids produced by lactic acid bacteria. It can constitute more than 20% of the yeast population of fermented sausages.

In recent years, Y. lipolytica has been isolated as part of the microflora of vegetable salads, specifically, low pH, dressed salads. These salads generally undergo fermentative spoilage by yeasts such as species of Saccharomyces. The role of Y. lipolytica in the salad environment is not yet defined but it may metabolize the lipid components of the dressing.

Source of Food Ingredients and Processing Aids

Y. lipolytica exhibits a range of biochemical properties of potential value to food processing (Table 2). Citric acid is widely used in the food and pharmaceutical industries as an acidulant and flavouring agent. Commercially, it is produced by fermentation processes using Aspergillus niger, but Y. lipolytica is known to excrete high concentrations of this acid and is considered as an alternative organism for its production. Numerous studies have examined the biochemistry and kinetics of citric acid production by Y. lipolytica and, depending on conditions, production levels of $15–78 \, g \, l^{-1}$ have been reported. The amino acids lysine and methionine are also overproduced by this yeast and may offer commercial prospects.

The production of specific food flavouring agents by Y. lipolytica is another area of commercial interest. The peach flavouring agent, γ-decalactone, has significant value to the food industry and is readily produced by Y. lipolytica from the bioconversion of ricinoleic acid in castor oil. Several patents covering the process have been approved and production levels up to $9.4 \, g \, l^{-1}$ by Y. lipolytica have been reported. Y. lipolytica has been used for the biotransformation of low priced, commercially available monoterpenes, such as limonene and piperitone to highly priced flavour and fragrance chemicals. The process, conducted in fermentation culture, involves the conversion of limonene into perillic acid (50% conversion) and piperitone into 7-hydroxypiperitone (8% conversion).

As mentioned above, Y. lipolytica produces an emulsifier, liposan, that exhibits good emulsification activity over the pH range 2–5 and has potential use in food processing. Liposan is approximately 85% carbohydrate and 17% protein and may be used to stabilize oil-in-water emulsions with commercial vegetable oils. It is stable at temperatures up to 70°C. Production yields of liposan are not reported.

The strong proteolytic and lipolytic activities of Y. lipolytica have attracted significant commercial interest in the dairy industry. A number of patents exist for the use of Y. lipolytica as a novel starter culture in cheese production. As a starter culture to assist in flavour development during maturation, the species has the ability to compete with other yeasts present in the cheese, it is compatible with the lactic acid bacteria starters used in cheese production and it produces the appropriate proteolytic enzymes for flavour enhancement. There could be similar applications of this yeast in the production of fermented meat sausages. Moreover, isolated proteases or lipases from the yeasts could be added directly to cheese or sausages to enhance flavour development. A process for the production of a cheese-flavoured substance involving Y. lipolytica lipase and proteinase activity has been patented. It is based on a mixed fermentation with Y. lipolytica and two lactic acid bacteria, Lactococcus lactis and Lactobacillus casei.

Production of Biomass for Feedstocks and Processing Food Waste

Several reports in the literature describe the use of Y. lipolytica to produce biomass for feedstocks. A number of patents describe a process for silage production. It has been grown on soybean meal, crude oils, wastes of olive oil processing, soapstocks and vegetable material to produce single cell protein. Plant leaf waste from vegetables such as turnip, mustard, radish and cauliflower plants have been fermented by a mixed culture of yeasts, including Y. lipolytica, to produce animal feeds, high in protein and vitamins. The biological oxygen demand of the wastes was also reduced after the fermentation.

See also: **Cheese**: Mould-ripened Varieties. **Fish**: Spoilage of Fish. **Meat and Poultry**: Spoilage of Meat. ***Pichia pastoris***. **Spoilage Problems**: Problems Caused by Fungi. **Starter Cultures**: Cultures Employed in Cheesemaking. ***Zygosaccharomyces***.

Further Reading

Barth G and Waber H (1985) Improvement of sporulation in the yeast *Yarrowia lipolytica*. *Antonie van Leeuwenhoek* 51: 167–177.

Barns SM, Lane D, Sogin ML, Bibeau C and Weisburg WG (1991) Evolutionary relationships among pathogenic *Candida* species and relatives. *Journal of Bacteriology* 173: 2550–2255.

Barth G and Gaillordin C (1997) Physiology and genetics of the dimorphic fungus, *Yarrowia lipolytica*. *FEMS Microbiology Reviews* 19: 219–237.

Casarégola S, Feynerol C, Diez M, Fournier C, Gaillardin C (1997) Genomic organization of the yeast *Yarrowia lipolytica*. *Chromosoma* 106: 380–390.

Dillon VM and Board RG (1991) A review. Yeasts associated with red meats. *Journal of Applied Bacteriology* 71: 93–108.

Fleet GH (1990) Yeasts in dairy products – a review. *Journal of Applied Bacteriology* 68: 199–211.

Guerzoni ME, Lanciotti R and Marchetti R (1993) Survey of the physiological properties of the most frequent yeasts associated with commercial chilled foods. *International Journal of Food Microbiology* 17: 329–341.

Kurtzman CP (1998) *Yarrowia* van der Walt & von Arx. In: Kurtzman CP and Fell JW (eds) *The Yeasts, a Taxonomic Study*, 4th edn. P. 420, Amsterdam: Elsevier Science.

Ogrydziac DM (1993) Yeast extracellular proteases. *Critical Reviews in Biotechnology* 13: 1–55.

Pagot Y, Endrizzi A, Nicaud JM and Belin JM (1997) Utilisation of an autotrophic strain of yeast *Yarrowia lipolytica* to improve Y-decalactone production yields. *Letters in Applied Microbiology* 25: 113–116.

Praphailong W and Fleet GH (1997) The effect of sodium chloride, sucrose, sorbate and benzoate on the growth of food spoilage yeasts. *Food Microbiology* 14: 459–468.

Praphailong W, Van Gestel M, Fleet GH and Heard GM (1997) Evaluation of the Biolog system for the identification of food borne yeasts. *Letters in Applied Microbiology* 24: 455–459.

Rane KD and Sims KA (1993) Production of citric acid by *Candida lipolytica* Y 1095: effect of glucose concentration on yield and productivity. *Enzyme and Microbial Technology* 15: 646–651.

Rodriguez C and Dominguez A (1984) The growth characteristics of *Saccharomyces lipolytica*: morphology and induction of mycelium formation. *Canadian Journal of Microbiology* 30: 605–612.

Roostita R and Fleet GH (1996) Growth of yeasts in milk and associated changes to milk composition. *International Journal of Food Microbiology* 31: 205–219.

Roostita R and Fleet GH (1996) The occurrence and growth of yeasts in Camembert and blue-veined cheeses. *International Journal of Food Microbiology* 28: 393–404.

van der Vossen Jos MBM and Hofstra H (1996) DNA based typing, identification and detection systems for food spoilage microorganisms: development and implementation. *International Journal of Food Microbiology* 33: 35–49.

Van der Walt JP and von Arx JA (1980) The yeast genus *Yarrowia* gen. nov. *Antonie van Leeuwenhoek* 46: 517–521.

Walsh TJ, Salkin IF, Dixon DM and Hurd J (1989) Clinical, microbiological, and experimental animal studies of *Candida lipolytica*. *Journal of Clinical Microbiology* 27: 927–931.

CANNING *see* **HEAT TREATMENT OF FOODS**: Thermal Processing Required for Canning; Spoilage Problems Associated with Canning.

CATERING INDUSTRY *see* **PROCESS HYGIENE**: Hygiene in the Catering Industry.

CELLULOMONAS

M I Rajoka, National Institute for Biotechnology and Genetic Engineering (NIBGE), Faisalabad, Pakistan

K A Malik, Pakistan Agricultural Research Council, Islamabad, Pakistan

Characteristics of the Genus

The genus *Cellulomonas* contains a heterogeneous collection of cellulose-decomposing bacteria principally isolated from soil materials. The genus *Cellulomonas* along with *Jonesia*, *Oerskovia*, and *Promicromonospora* has recently been assigned to a new family Cellulomonadaceae. Phylogenetically the family belongs to the order Actinomycetales. A combination of chemotaxonomic and morphological properties differentiate members of Cellulomonadaceae from related taxa. Despite their apparent close phenotypic relationship, members of the genus *Cellulomonas* are genotypically different though they have several common genes. Distinction between species is based on a number of morphological and biochemical characteristics for which different schemes have been reported to identify species of *Cellulomonas*. Strains of the genus produce cellulases and hemicellulases which have several industrial applications. Intergeneric hybrids have been constructed between *Cellulomonas* and *Zymomonas* or multiple genes coding for different enzymes have been cloned in *Z. mobils* and *Saccharomyces cerevisiae* to extend their substrate consumption efficiency and produce valuable biochemicals from inexpensive substrates.

Initially 27 putative species were collected in the genus but later on in the *Seventh Bergey's Manual of Determinative Bacteriology*, only 10 species were recognized; the rest were found to be synonymous with these ten. The 10 species were distinguished from each other by motility, nitrate reduction, ammonia production, chromogenesis and, in the case of *C. fimi*, by the fermentation of xylose and arabinose. They were regarded as Gram-negative rods; a portion of rods is arranged at an angle to each other to give V-formations. However, only six authentic cultures were available and *C. biazotea* was regarded as the type species. Since cellulose decomposition is a feature of many other soil bacteria and fungi, therefore, this genus received only academic importance as it had no economic utility. Recently, efforts have been made to utilize cellulosic crop residues, major agro-industrial wastes and environmental pollutants, for production of value-added products. Therefore, renewed interest has developed in isolation, identification and application of *Cellulomonas* spp. in the production of glucose from such wastes. Recently, they have been isolated from rumen, activated sludge and cellulose-enriched environments.

In the eighth edition of *Bergey's Manual*, only one species was recognized. It was suggested that minor variation in chromogenesis, motility, and nitrate reductase activity, did not permit recognition of 10 species. *Cellulomonas flavigena* and not *C. biazotea* was recognized as a type species.

Later, the data from DNA–DNA homology and biochemical reaction studies recognized seven species of the genus *Cellulomonas*. Because of a substantial degree of similarity *Cellulomonas* was included in the family Corynebacteriaceae. Neighbouring groups are defined by *Arthrobacter*, *Renibacterium*, *Micrococcus*, *Stomatococcus*, *Dermatophilus*, *Brevibacterium* and *Microbacterium* and related genera. In the second volume of *Bergey's Manual of Systematic Bacteriology*, six species, namely *Cellulomonas biazotea*, *C. cellasea*, *C. fimi*, *C. flavigena*. *C. gelida* and *C. uda*, have been recognized. Distinction was based on a number of biochemical and chemataxonomic characteristics, namely peptidoglycan type, fatty acid composition, and other conventional tests namely morphology, cultural conditions for good growth, biochemical reactions exhibited by these species and cell wall composition with respect to carbohydrates. Gram reaction in these species can some times be difficult to determine; some Gram positive strains may lose colour and may appear Gram negative.

Some authors have proposed a scheme for identification of these six species of *Cellulomonas* which can be used to tentatively separate these species. According to this scheme, *C. biazotea* can be separated because it is the only species which grows on raffinose. Among the others, *C. fimi* is lysine and ornithine positive, all other species being negative. *C.*

fimi, *C. flavigena*, and *C. uda* are o-nitrophenyl-β-D-galactopyranoside (ONPG) positive whereas *C. gelida* and *C. cellasea* are negative. *C. gelida* produces acids from glucose whereas *C. cellasea* does not and can be separated. Among *C. fimi*, *C. flavigena* and *C. uda*, *C. fimi* does not produce nitrate reductase. *C. flavigena* does not produce acid from lactose and can be separated from *C. uda*. According to this classification, *Cellulomonas* spp. are usually Gram-negative during the first 24 h of growth, thereafter, Gram-positive or Gram-variable staining is obtained on cells grown under optimum growth conditions. In several old cultures, Gram reaction is usually negative and cells are pleomorphic in nature and are motile.

Recently, 280 physiological characters have been used in a numerical taxonomic study of 604 strains to differentiate the genera *Arthrobacter*, *Aureobacterium*, *Brevibacterium*, *Cellulomonas*, *Clavibacter*, *Corynebacterium*, *Curtobacterium*, *Microbacterium*, *Nocardia*, *Nocardioides*, *Rhodococcus*, *Terrabacter* and *Tsukamurella* by cluster analysis. A high degree of similarity between the genera *Aureobacterium*, *Cellulomonas*, *Clavibacter*, *Curtobacterium* and *Microbacterium* has been found in phylogenetic-based studies and could be confirmed phenotypically. Bacteria belonging to the genus *Corynebacterium*, including *Corynebacterium glutamicum*, and eight other species of this genus are distinctly different from *Cellulomonas* spp.

Creation of a distinctly different genus based on cellulolysis was not well accepted but further studies using modern taxonomic methods, such as molecular techniques using electrophoresis, composition with respect to peptidoglycans, fatty acid and carbohydrate profiles of cell wall, biochemical properties, DNA–DNA homology, 16S and 5S rRNA analyses, have supported the earlier conclusions on the species of *Cellulomonas*. Now the well-recognized species are *C. biazotea*, *C. cellasea*, *C. cellulans*, *C. fimi*, *C. flavigena*, *C. fermentans*, *C. gelida* and *C. uda*. They have an optimum temperature for growth of 28–33°C. They grow over a wide pH range (pH 5.5–7.8). The generation time varies between 2 and 4 h. They require nitrogen and vitamins for good growth and cellulase production. Under optimum growth conditions, cell-wall composition, peptidoglycan structure, menaquinone composition and fatty acid profiles are unique properties and can be used to separate this genus from other related genera. All species are usually nitrate reductase and catalase positive, and have yellow chromogenesis.

In the ninth edition of *Bergey's Manual of Determinative Bacteriology*, only seven species have been recognized. For their differentiation, seven tests have been proposed. The above-mentioned tests can be easily performed in all microbiology laboratories and both can be combined for confirmatory tests.

Strains of *Cellulomonas* produce hydrolases to consume carbohydrates namely starch, xylan and cellulose. Some of these enzymes are multifunctional, possess a multidomain structure and can be induced by many substrates. They consist of a number of extracellular and intracellular enzymes produced by different species of *Cellulomonas*. Extracellular and intracellular cellulases and xylanases produced by *Cellulomonas* spp. are presented in **Table 1**. It has been suggested that cellulases of *Cellulomonas* spp. operate by a lysozyme-type reaction mechanism. An endoglucanase and an exoglucanase of *C. fimi*, like cellulases of other bacteria and fungi, consist of a catalytic and a cellulose binding domain. Each of these enzyme components can be truncated by a serine protease to separate the catalytic and cellulose-binding domains. Cellulases of *C. fimi*, *C. uda*, *C. flavigena* and *C. biazotea* have been purified to homogeneity level. The enzymes have been extensively studied for their biochemical and kinetic parameters and active site residues involved in catalysis have been identified. These active site residues can be chemically modified to suit industrial applications.

In *Cellulomonas* strain CS1–1, *C. biazotea* and their mutants, production of α-L-arabinofuranosidases (EC 3.2.1.55), β(1,4)-mananases (EC 3.2.1.78), α-D-(1,4)-galactoside galactohydrolase (EC 3.2.1.22) showing activity towards galactomannan, hemicellulase A, hemicellulase B, α- and β-glycosidases has been studied in greater detail compared with other *Cellulomonas* spp. Some species produce cellobiase activity to produce glucose from cellobiose but those which lack this enzyme, possess cellobiose phosphorylase (EC 2.4.1.20) to consume cellobiose. Cellobiose dehydrogenase (EC 1.1.99.18),

glycerol dehydrogenase (EC 1.1.1.6) and L-amino oxidase (EC 1.4.3.2) are also produced by some species of *Cellulomonas*.

Since strains of *Cellulomonas* are inhabitants of soil and decaying cellulosic substrates, they can easily be transmitted into foods. They have been found in a variety of foods, including meats, liquor of spiced olives, starchy foods and raw cow's milk. Production of toxins by *Cellulomonas* spp. is still not a well-characterized phenomenon. Therefore toxicity due to their fermentation products is not clear. Some species of *Cellulomonas* produce amylases and may contaminate starchy foods like cooked rice if stored at room temperature. They may release polysaccharides and glucose which may be used by pathogenic organisms to grow. The subsequent heat treatment, usually frying, is usually not sufficient to kill vegetative cells. Recently one species, namely, *Cellulomonas hominis* sp. nov. has been isolated from clinical specimens. Its role in pathogenicity is not well characterized. Its characterization has been based on 16S rRNA genes, phenotypic and molecular data compared with related species but its taxonomic status is to be determined by DNA–DNA hybridization studies.

Methods of Detection

Strains of *Cellulomonas* resemble many of their close relatives in the family Cellulomonaceae. Differentiation is usually done by growth on filter paper, Walseth cellulose, Sigmacell 100, or cottonwool followed by microscopic observations for mobility. The strains of *Cellulomonas* leave a yellow colony print on the surface of a cellulose medium.

Detection of *Cellulomonas* spp. in food and related specimens, as in other organisms, demands a series of tests including selective enrichment onto cellulose

Table 1 Production of cellulases and xylanases studied in different strains of the genus *Cellulomonas*

Enzyme studied	EC number	Organisms extensively studied	Comments
Cellobiohydrolase	3.2.1.91	*C. biazotea*, *C. fimi*, *C. flavigena*, *C. uda*, *Cellulomonas* CS1–1	Intra- and Extracellular production
Endo-1,4-β-D-glucanase	3.2.1.4	*C. biazotea*, *C. cellasea*, *C. fimi*, *C. flavigena*, *C. uda*, *Cellulomonas* CS1–1	Extracellular production
Filter paper activity	—	*C. biazotea*, *C. cellasea*, *C. fimi*, *C. flavigena*, *C. uda*, *Cellulomonas* CS1–1	Produced extracellularly, measured as filter paper activity (FPase)
1,4-β-Glucosidase	3.2.1.21	*C. biazotea*, *C. fimi*, *C. cellasea*, *C. flavigena*, *C. uda*, *Cellulomonas* CS1–1	Produced intracellularly
Endo-1, 4-β-xylanase	3.2.1.8	*C. biazotea*, *C. cellasea*, *C. flavigena*, *C. fimi*, *C. uda*, *Cellulomonas* CS1–1	Produced extracellularly
1,4-β-xylosidase	3.2.1.37	*C. biazotea*, *C. cellasea*, *C. fimi*, *C. favigena*, *C. uda*, *Cellulomonas* CS1–1, *Cellulomonas* sp. NCIM	Produced intracellularly

medium, followed by plating onto cellulose agar medium made in basal salts medium (see later) containing yeast extract. Homogenates of foods are prepared in Butterfield's phosphate-buffered water at a 1:1 or 1:10 dilution or direct plate counts can be made using the selective medium. Penicillin is added to suppress fungal growth. All *Cellulomonas* spp. liberate extracellular filter paper activity, called FPase activity, which can produce a cellulose hydrolysing zone around single colonies. Some organisms also produce cell-associated cellobiohydrolase activity which can also help in clearing cellulose around single colonies. The plates are incubated at 30°C for up to 14 days for confirmation of test. In some cases, 21–30 days may be required to clear cellulose.

When a low number of strains of the genus *Cellulomonas* is expected, direct plating may not be suitable. The threshold for direct plating is a $10\,\text{cfu}\,\text{g}^{-1}$ sample. The most probable number (MNP) technique can be useful to enumerate bacteria. The MNP technique for *Cellulomonas* spp. starts with dilution of cellulose-grown cultures in triplicate. The tubes are incubated at 30°C for 72–96 h for dense growth and then plated onto cellulose agar plates for viable counts. The culture is streaked on the surface of a cellulose–agar plate to any assumptive positives. Colonies of different shape and sizes (which clear cellulose), appearing simultaneously, or those appearing at different time intervals are picked and streaked on fresh salt cellulose agar medium. Confirmation of *Cellulomonas* spp. requires completion of a number of tests (**Table 2**). Unfortunately there is no single test for unequivocal determination of *Cellulomonas* spp. They produce acetic and lactic acids from glucose under aerobic conditions (*C. biazotea*, *C. cellasea*, *C. flavigena*, *C. cellulans* and *C. gelida*) or acetic acid, lactic acid, succinic acid and ethanol under anaerobic conditions (*C. fimi*, *C. uda* and *C. fermentans*).

Other characteristics of *Cellulomonas* spp. include reduction of nitrate to nitrite, and non-production of acetylmethylcarbinol (Voges–Proskauer negative). To assess various characteristics to confirm these species, additional tests are required.

For these tests, the strains are maintained on (preferably) Dubos salts-agar medium containing $(\text{g}\,\text{l}^{-1})$ $K_2HPO_4.7H_2O$ 0.5, KCl 1.0, $NaNO_3$ 0.5, $MgSO_4$ 0.5, $FeSO_4.7H_2O$ 0.1, yeast extract 2, Sigmacell 100 10 and agar 25 g (DYEA) plates and slants. Then basal liquid medium is dispensed as 10 ml into test tubes (16 × 150 mm) and autoclaved at 15 pounds pressure for 15 min. After cooling to ambient temperature, the test carbohydrates are added aseptically to a concentration of 1% (w/v). The carbohydrates or other carbon sources are filter-sterilized with bacteriological filters (0.2 μm) if soluble or autoclaved

if not soluble. Fresh cultures of each organism are prepared in Dubos salts yeast extract–glucose medium. After overnight growth, cells are harvested aseptically by centrifugation (15 000 g, 30 min), washed twice with 0.89% NaCl solution, and resuspended in saline and their absorbance measured at 610 nm. The absorbance is then brought to McFarland tube no. 4 by dilution or by concentration. A 1 ml sample of cells is then aseptically resuspended in DYE liquid medium containing test carbohydrates at 1% final carbohydrate concentration (w/v). The cells are grown statically at 30°C. The test tubes are individually examined after each day up to 30 days of incubation. The whole experiment may be repeated three or four times for declaring negative test as compared with type cultures namely *C. biazotea* (ATCC 486, DSM 20112, NCIB 8077), *C. cellasea* (ATCC 487, DSM 20118, NCIB 8073), *C. fimi* (ATCC 484, DSM 20113, NCIB 8980), *C. flavigena* (ATCC 482, DSM 20109, NCIB 8073), *C. gelida* (ATCC 488, DSM 20111, NCIB 8076), *C. fermentans* (DSM 3133) and *C. uda* (ATCC 491, DSM 20107, NCIB 8200) available in *The Prokaryotes* (2nd edition).

Motility Motility is usually measured by stabbing the centre of a tube of semisolid growth-supporting medium and allowing the culture to grow. The cultures are incubated for 24 h at 30°C. Motile bacteria diffuse out from the stab, forming an opaque growth pattern, whereas non-motile bacteria do not diffuse out.

Hydrolysis of o-Nitrophenyl-β-D-Galactopyranoside (ONPG) Cells are streaked on the surface of DYEA plates containing ONPG in the medium and incubated at 30°C. After 3–28 days growth, the plates are spread with carbonate buffer, $200\,\text{mmol}\,\text{l}^{-1}$ (pH 8.5). The colonies are screened by measuring the diameter of the yellow halo surrounding the colonies.

Nitrate Broth A 1 ml sample of grown cultures is inoculated separately in 10 ml culture medium in test tubes and incubated at 30°C. After 24 h, nitrite production is measured by addition of α-naphthylamine and α-naphthol.

Hydrolysis of Aesculin Hydrolysis of aesculin is determined in DYE–aesculin–ferric ammonium citrate–agar medium by streaking the cells from a single colony and examining after each day of incubation. Aesculin-hydrolysing organisms produce β-glucosidase (located in the periplasmic fraction of the cell) which hydrolyses aesculin to liberate aesculitin which, in turn, reacts with ferric ions to give a black precipitate. The hydrolysing efficiency can be quan-

Table 2 Biochemical and morphological tests for identification of species of the genus *Cellulomonas*

Confirmation test	Species						
	Biazotea	cellasea	cellulans	fimi	flavigena	gelida	uda
Gram reaction	+	+	+	+	+	+	+
Catalase	+	+	−	+	+	+	
Motility	+	+	+	+	+	+	
Nitrate reduction	+	+	+	−	+	+	+
Voges–Proskauer	−	−	−	−	−	−	−
Methyl red	−	−	−	−	−	−	−
ONPG	+	−	+	+	+	−	+
Glucose utilization (anaerobic)	−	−	+	−	−	+	+
Aesculin	+	+	+	+	+	+	+
Urease	+	+	+	−	−	−	+
Ammonia from peptone	−	−	+	−	+	+	−
Lactose	+	+	−	+	−	+	+
Ribose	−	−	+	−	+	−	−
Raffinose	+	−	−	−	−	−	−
Proline	−	+	+	−	−	−	−
Acid from glucose	+	+	+	+	+	+	+
Arabinose	+	+	+	+	−	+	−
Fructose	+	+	+	+	+	+	+
Xylose	+	+	+	+	+	+	+
Maltose	+	+	+	+	+	+	+
Lysine decarboxylase	−	−	−	−	−	−	+
Arginine dihydrolase	−	−	−	−	−	−	+
Ornithine decarboxylase	−	−	−	−	+	−	+
(L)-Lactate	+	+	+	+	−	−	−
Acid from dextrin	−	−	DN	+	+	+	+

ONPG, *o*-nitrophenyl-β-D-galactopyranoside.
DN, not determined.

tified by measuring the zone of blackening (in mm) around single colonies. Differences in values indicate the genetic variability among species.

Hydrolysis of Carboxymethyl Cellulose (CMC) CMC is added to DYEA medium and 10 μl of cells of equal density are poured on the surface of the plate. After growth at 30°C, CMC hydrolysis is visualized as a yellow zone when stained with Congo red followed by treatment with NaCl. Differences in the diameter of the yellow zone indicate the genetic variability among species.

Hydrolysis of Sigmacell 100 Crystalline Cellulose Hydrolysis of Sigmacell 100 is visualized by streaking the cells on the surface of a plate containing Sigmacell 100 in the top layer of DYEA and DYEA in the base layer. After incubation, the streaked area is examined every day to see the zone of Sigmacell clearance. The diameter of the zone of clearance around single colonies indicates the organisms ability to decompose cellulose. Differences in values indicate the genetic variability among species.

Acid Production from Carbohydrates A 0.1 ml sample of grown culture is inoculated into 10 ml of culture medium in a test tube containing nutrient broth supplemented with test saccharides (glucose, D-ribose, raffinose, lactose, dextrin etc.) with an inverted Durham tube. After 24 h, phenol red is added which is red at neutral pH but changes to yellow at acidic pH (pH 6.8). The appearance of gas in the Durham tube, indicates the fermentative ability of the organism.

Utilization of L(+)-Lactate A 0.1 ml sample of grown culture is inoculated into 10 ml of culture medium in a test tube containing nutrient broth supplemented with L(+)-lactic acid. The appearance of dense growth indicates the utilization of L(+)-lactate in the fermentation medium.

Indole Production Test A 0.1 ml sample of culture is inoculated into trypticase soy broth, and incubated at 30°C. Then 10 drops of Kovac's reagent are added into dense growth and the tube is agitated gently. The appearance of a brick-red colour indicates the production of indole from tryptophan in the medium.

Utilization of Proline A 0.1 ml sample of grown culture is transferred to 10 ml of culture medium in a test tube containing nutrient broth supplemented with proline. After 24 h, proline is consumed by the organism and dense growth is visible.

Utilization of Lysine and Ornithine A 0.1 ml sample of grown culture is transferred into 10 ml of culture medium in a test tube containing nutrient broth supplemented with lysine or ornithine. After 24 h, the compounds are consumed by the organism and dense growth in each case is visible.

Applications of *Cellulomonas* spp.

Almost all species of *Cellulomonas* have been screened to produce cellulases, namely endo-β-glucanase, 1,4-β-D-glucan glucohydrolase, exocellobiohydrolase, β-1,4-D-glucosidase, endo-β-xylanases and β-xylosidase, for the utilization of agro-industrial wastes. These enzymes find application in the textile, paper and pulp industries. All cellulases and xylanases produced by *Cellulomonas* spp. have been purified to homogeneity and multiple isomers have been obtained. The cellulose-binding domain of endo-β-1,4-glucanase produced by *C. fimi* has been used to purify human interleukin-2 for clinical applications. The enzyme preparations can be used for obtaining stable botanical extracts for food/medical applications.

Food Applications

Cellulases and xylanases produced by *Cellulomonas* spp. have not been used commercially but, like fungal enzymes, they can be applied in baked foods, fruit processing, preparation of dehydrated vegetables and food products, preparation of essential oils and flavours for the food industry, and starch processing. Concentrated enzyme preparations have been used in making jams, baby food, foods for invalids and fruit juices. Hemicellulase-rich cellulases are implicated in degumming of coffee extracts and making instant coffee. Endoglucanase has been used to remove turbidity of glucan present in wines and beers. In a well-balanced enzyme preparation, cellulosic materials are hydrolysed to monomeric sugars which may be used for production of food-grade products namely, lactic acid, acetic acid and xylitol.

Cellulomonas fimi, *C. biazotea*, *C. flavigena* and *C. uda* have been used alone or in combination with yeasts or bacteria for the production of single cell protein, to enhance the digestibility and nutritive values of poultry and animal feeds. In South Korea, farmers add inoculum of *Cellulomonas* spp. to rice straw or wheat straw to improve taste, digestibility and protein content. The organisms have been found to have a balanced level of all amino acids. This has been observed in other species of *Cellulomonas* and with *C. biazotea* grown alone or with *Saccharomyces carlsbergensis* in combined saccharification and fer-

mentation studies (**Table 3**). These organisms may serve as a good source of amino acids in amino acid-deficient poultry/animal diets. Cellulaseless enzyme preparations rich in xylanases and glycosidases are useful in the reduction of chlorine and chlorine dioxide consumption in the paper and pulp industry to abate environmental pollution due to tetrachloro compounds of lignin precursors present in waste waters. Endoglucanases are used in oil drilling where guar gum or CMC is used for improved oil recovery. They are also used for plant cell wall destruction to produce plant protoplasts for genetic manipulation for use in the production of transgenic cereal crops with improved grain productivity and protein content.

Cellulomonas spp. have been considered as potential candidates for waste disposal, composting of grasses, pith, dried palm oil effluents and shredded newspapers Mixed cultures of *Cellulomonas* spp., *Desulfovibrio vulgaris* and *Methanosarcina barkerii* were found to be highly efficient at producing methane from cellulose and xylan. A mixed culture of *Cellulomonas* spp. and *Rhodopseudomonas capsulata* has been used to produce hydrogen as a source of energy. Similarly, a mixed culture of *Cellulomonas* mutant and *Azospirillum brasilense* has been used to enhance the nitrogen-fixing ability of the latter.

The cellulosic hydrolysates have been used to produce ethanol using *Saccharomyces cerevisiae*, a thermotolerant yeast strain of *Kluyveromyces marxianus* or *Zymomonas mobilis*. Strains of *Cellulomonas* possess multiple genes for the production of cellulases and xylanases. These genes have been

Table 3 Amino acid analyses of *Cellulomonas biazotea* (C), *Saccharomyces carlsbergensis* (Y) alone and in combination (C + Y)

Amino acid (g per 100 g dry cells)	C	Y	C + Y
Aspartic acid	2.92	2.59	2.87
Threonine	1.20	1.30	1.25
Glutamate	2.69	2.29	2.45
Serine	1.80	1.20	1.60
Proline	1.85	1.80	1.60
Glutamine	0.99	0.77	0.89
Alanine	2.05	1.97	0.50
Tyrosine	0.39	0.80	0.50
Valine	1.68	1.13	1.28
Methionine	0.71	0.20	0.48
Isoleucine	0.85	1.05	0.88
Leucine	2.03	1.80	1.90
Threonine	1.45	1.20	1.30
Phenylalanine	1.00	0.81	0.76
Lysine	0.99	1.61	1.45
Histidine	0.34	1.30	0.81
Arginine	0.92	0.66	0.73
Cysteine/cystine and others	3.14	2.52	3.26

cloned in *Escherichia coli*, *S. cerevisiae* or *Zymomonas mobilis* for the purification of the individual components so as to understand the structure–function–relationship of cellulases, to hyperproduce one enzyme component in greater quantity, and to enhance the substrate-consumption efficiency of noncellulolytic organisms or to produce ethanol from cellulosic wastes. The rDNA methodology also directly aids in the development of more active cellulases. The application of cellulases through improvement and/or hyperproduction of specific components can permit them to be mixed optimally for routine and new industrial applications.

See also: **Single Cell Protein**: Yeasts and Bacteria.

Further Reading

Cappuccino JA and Sherman A (1987) *Microbiology: A Laboratory Manual.* Menlo Park, California: Benjamin/Cummings.

Chaudhary P, Kumar NN and Deobagkar DN (1997) The glucanases of *Cellulomonas. Biotechnology Advances* 15: 315–331.

Elliston KO, Yablonsky MD and Eveleigh DE (1991) Cellulase – insights through recombinant DNA approaches source. *ACS Symposium Series* 460: 290–300.

Holt JG, Krieg NR, Sneath PHA, Staley JT and Williams ST (1995) Irregular, nonsporing Gram-positive rods. *Bergey's Manual of Determinative Bacteriology,* 9th edn, pp. 571–596. Baltimore, Maryland: Williams and Wilkins.

Kompffer P, Seiler H and Dott W (1993) Numerical classification of coryneform bacteria and related taxa. *Journal of General and Applied Microbiology* 39: 135–214.

Kumar NN and Deobagkar DN (1995) Presence of multiple amylases encoded by independent genes in *Cellulomonas* sp. NCIM2353. *Biotechnology Letters* 17: 797–802.

McHan F and Cox NA (1987) Differentiation of *Cellulomonas* species using biochemical tests. *Letters in Applied Microbiology* 4: 33–36.

Rajoka MI, Bashir A, Hussain M-RA et al (1998) Cloning and expression of bgl genes in *Escherichia coli* and *Saccharomyces cerevisiae* using shuttle vector pYES2.0. *Folia Microbiologica* 43(2): 129–135.

Rickard PAD, Rajoka MI and Ide JA (1981) The glycosidases of *Cellulomonas. Biotechnology Letters* 3(9): 487–492.

Siddiqui KS, Rashid MH, Durrani IS, Ghauri TM and Rajoka MI (1997) Purification and characterization of an intracellular β-glucosidase from *Cellulomonas biazotea. World Journal of Microbiology and Biotechnology* 13: 245–247.

Siddiqui KS, Azhar MJ, Rashid MH and Rajoka MI (1997) Stability and identification of the active-site residues of carboxymethylcellulases from *Aspergillus niger* and *Cellulomonas biazotea. Folia Microbiologica* 42: 312–318.

Stackebrandt E and Prauser H (1991) Assignment of the genera *Cellulomonas, Oerskovia, Promicromonospora* and *Jonesia* to Cellulomonadaceae fam nov. *Systematic and Applied Microbiology* 14: 261–265.

Stackebrandt E and Prauser H (1992) The family Cellulomonadaceae. In: Balows A, Trüper MG, Dworkin M, Harder W and Schleifer KH (eds) *The Prokaryotes.* P. 1323. New York: Springer.

Wilson DB (1993) Structure–function relationships in cellulase genes. *ACS Symposium Series* 516: 243–250.

CEREALS *see* **SPOILAGE OF PLANT PRODUCTS**: Cereals and Cereal Flours.

CENTRIFUGATION *see* **PHYSICAL REMOVAL OF MICROFLORAS**: Centrifugation.

CHEESE

Contents

In the Marketplace

A Y Tamime, Scottish Agricultural College, Auchincruive, Ayr, UK

Evolution of Cheese-making

The manufacture of cheese is probably the most effective way of preserving milk – a highly nutritious but perishable foodstuff. Cheese-making converts milk into a product which is not likely to deteriorate in a short period of time.

The exact origin of cheese-making are difficult to establish, but from archaeological evidence, cheese was first produced around 7000 BC. Modern cheese-making could have evolved in two main stages. The first was the production of sour or fermented milks, including yoghurt, through the ability of lactic acid bacteria to grow in milk and to produce acid, to reduce the pH to the isoelectric point of caseins, at which they coagulate. The second was the removal of whey from the broken acid milk gel, – this concentrates the curd into a form which can be either consumed fresh or stored for future use. The acid whey may be consumed immediately as a pleasant and refreshing drink. The addition of salt helped to preserve the concentrated curd and also improved its flavour. Hence it was probably realized that by further dehydrating the curd and preserving it in brine, its shelf life would be greatly extended. This fermented milk product was later known as 'pickled cheese', and some examples are Feta, Domiati, Akkawi and Halloumi, which are still manufactured in many countries in the Middle East.

The next step in the evolution of cheese-making was probably the precipitation of the caseins by proteolytic enzymes, such as the rennet which originated from the animal stomachs used as bags to store the milk. However, plant-derived coagulating enzymes from fig trees and thistles appear to have been used widely by the Romans. Rennetted curd is very different from acid curd: the former is more robust, and the majority of cheeses nowadays are made using coagulating enzymes.

Cheese Names and Nomenclature

At present, several thousand generic names are applied to cheeses throughout the world, but many may belong to the same variety of cheese. The exact origin of cheese names can be attributed to: the region and/or town(s) where the cheese is made; religious institutons; type of milk (cow, goat, sheep or buffalo); borrowed or made-up names; the shape, appearance, method of ripening or type of cheese, or the type of additives used.

Over the last few decades, national and international bodies have been involved in preparing or improving the existing specifications for cheese. According to the FAO/WHO (Food Standards Programme – Codex Alimentarius Commission), international standards for 35 individual cheese varieties have been accepted by many government bodies. These varieties include Cheddar, Cheshire, Emmental, Gruyère, Gouda, Edam, Danbo, Fynbo, Brie, Camembert, Provolone, Cottage cheese, blue-veined varieties and extra-hard cheese for grating.

Definition of Cheese

In the latest report of the Codex Alimentarius Commission, the revised definition of cheese allows, for example, the use of ultrafiltered milk in cheese-making. The definition is as follows:

Cheese is the fresh or mature solid or semi-solid product obtained:

- by coagulating milk, skimmed milk, partly skimmed milk, cream, whey cream, or buttermilk, or any combination of these materials, through the action of rennet or other suitable coagulating agents, and by partially draining the whey resulting from such coagulation;

or

- by processing techniques involving coagulation of milk and/or materials obtained from milk which give an end product which has similar physical, chemical and organoleptic characteristics as the product defined under above.

Concentration of the milk by ultra-filtration before cheese-making will not affect the whey protein-to-casein ratio in the retentate. However the above definition of cheese prohibits the addition of whey proteins to the cheese milk. The specific designation of each variety of cheese includes details of chemical analyses, e.g. the permitted level of fat (expressed as minimum percentage of milk fat in dry matter (FDM) and the moisture content (expressed as maximum percentage of water). Some other analytical attributes of cheese which are of importance, but not covered in the cheese regulations, are the salt percentage (calculated as per cent salt in water), pH, and the percentage of moisture in fat-free cheese (MFFC).

Classification of Cheese

The FAO/WHO definitions used for the classification of cheese are:

- Cured or ripened cheese
- Mould-cured or mould-ripened cheese
- Uncured, unripened or fresh cheese.

However, there are also FAO/WHO standards for other types of cheese, these being whey cheese, processed cheese and spreadable processed cheese.

Cheese products made by non-traditional methods should not be overlooked. These are known as 'analogues', 'imitation', 'substitute', 'artificial', 'extruded', 'synthetic' and/or 'filled'.

World Production and Marketing of Cheese

Data on the world production of cheese show that in 1995 around 85% of cheese was manufactured in Europe, Oceania and North America, which are the major producers of cow's milk (see **Table 1**). Hence it is safe to assume that most of the cheeses produced in the world are made from cow's milk, and the rest are made from goat's, sheep's and buffalo's milk.

In 1970 the world production of cheese was 7.7 million, and by 1995 production had almost doubled. Table 1 shows the trends in cheese production in different parts of the world. In the early 1990s, cheese output dropped in eastern Europe, although it appeared to stabilize towards the end of the decade. However, during the same period, increased cheese production in the most important production areas compensated for the cut-backs in eastern Europe and the former Soviet Union and cheese production also increased in the rest of the world. However, outputs are primarily used for domestic consumption, except in some countries in South America.

Until the mid-1990s the world market for cheese

Table 1 World production figures of all types of cheeses (thousand tonnes)[a]

	1970	1980	1990	1995
Africa	294	364	466	510
America (North and Central)	1574	2471	3519	3943
America (South)	377	462	535	658
Asia	478	661	803	876
Europe	4768[b]	7165[b]	8904[b]	7391[c]
Oceania	177	254	306	421
World	7668	11 799	14 533	13 799

[a] After FAO (1981, 1996).
[b] Data include production figures in the former USSR.
[c] Production figure for former USSR was not reported.

increased each year, but since then the market has not experienced overall growth. In 1996 the world market was dominated by the EU countries (50% followed by New Zealand (16%), Australia (11%), Switzerland (5%), USA (3%) and the rest of the world (14%). Dramatic increases in imports to the former Soviet Union and eastern Europe were reported due to a drop in milk production, but the trend subsequently stabilized. Imports to the Far East also increased, while those to the Middle East, particularly to Iran, Iraq and Egypt, and to Africa are reported to have declined by about 25%.

Cheese-making Processes and Methods of Manufacture

Milk as a Raw Material Milk from different species of mammal (cow, goat, sheep or buffalo) can be used for the production of cheese. There are major differences in the chemical composition of these milks, and consequently the quality of the cheese depends on the type of milk used. The lactose in milk is the energy source for the starter microflora; the milk protein plays an important role in the formation of the coagulum, and hence the physcal characteristics of the product; and the milk fat provides the cheese with its feel and contributes its flavour. The flavour of cheese is mainly produced by comlex biochemical reactions (the hydrolysis of fat and protein), initiated by microbial activity, but the flavour of the cheese milk, which varies from species to species, also affects that of the end product.

Preliminary Treatment of the Milk The milk is handled differently, depending on the type of cheese being produced. For example: the milk fat content is adjusted for the manufacture of low-fat cheeses; the casein-to-fat ratios standardized for the production of cheeses with consistent organoleptic properties; cellular matter and other contaminants may be removed either by using cloth/metal filters or by centrifugal clarification; ad spore-formers may be removed or controlled by bactofugation, microfiltration or the addition of nitrates or lysozyme.

Normally the cheese milk is not homogenized. Heat treatment of the milk may either be governed by traditional practice (i.e. raw milk is used) or be mandatory, depending on existing statutory standards. Once the milk is ready for processing, the initial manufacturing stages (i.e. ripening, coagulation and cutting of the curd) are more or less similar for most cheese varieties. However the way in which the curd is handled, including the type of starter culture used, is dependent on the type of cheese being produced. **Figure 1** illustrates the major differences in the manufacture of a wide range of cheese varieties.

Starter Cultures A wide range of microorganisms is used as starter culture, which may consist of a mixture of many species or a few strains of the same species. Starter cultures are selected on the basis of many criteria, including the specific blend of microflora reflecting traditional cheese-making methods, the ability to produce acid from lactose and the ability to produce metabolites that provide the flavour profile

of a given cheese variety. **Table 2** shows some different starter cultures used in the cheese industry. However the precise combination of the starter microflora is dependent not only on the cheese variety but also on the inclination of the cheese maker.

Coagulants These are enzymes known as peptide hydrolases (proteinases), because of their activity on the casein micelles during the coagulation of the milk. Their hydrolytic activity is divided into three phases: the primary phase is the destabilization of κ-casein, which leads to micellar aggregates and the formation of the coagulum; the secondary phase is the partial hydrolysis of the protein during the cheese-making process; and the tertiary phase is significant in the development of flavour and texture during the maturation period. About 10% of the added coagulant is retained in the curd, whilst the rest is lost in the whey.

The coagulants used in the cheese industry can be classified into four groups: animal (calf, lamb, kid, cow, pig); microbial (e.g. *Mucor miehei*, *M. pusillus*, *Endothia parasitica*); plant (e.g. *Carica papaya*, *Ficus carica*); and recombinant or genetically engineered chymosin.

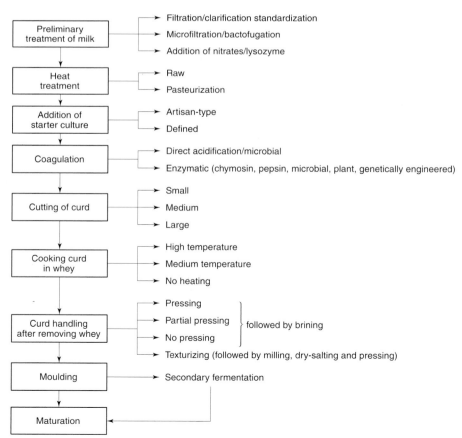

Figure 1 A generalized scheme illustrating the main stages in the manufacture of most varieties of cheese.

Table 2 The application of starter cultures and associated organisms in cheese-making

Microorganisms	Cheese variety	Organisms present
Lactic acid bacteria	Parmesan	5, 7, 8, 9, 11 or 17
Mesophilic	Romano	5, 7, 11
1. Lactococcus lactis subsp. lactis	Sbrinz Sapsago	18
2. L. lactis biovar diacetylactis	Emmental, Gruyère	5–9, 14–16
3. L. lactis subsp. cremoris	Cheddar, Cheshire	1, 3 +[a] 2, 4, or 12[b]
4. Leuconostoc mesenteroides spp. cremoris	Graviera	1, 3 + 5, 9 or 5, 7
Thermophilic	Roncal, Idiazabal	17
5. Stretococcus thermophilus	Kefalotori	5, 7
6. Lactobacillus delbrueckii subsp. delbrueckii	Manchego	1–4
7. L. delbrueckii subsp. bulgaricus	Kasseri, Majorero	17
8. L. delbrueckii subsp. lactis	Colby, Caerphilly	1, 3
9. L. helveticus	Kashkaval	17
10. L. paracasei subsp. paracasei	Mozzarella	5, 7
11. L. plantarum	Provolone	5, 7, 8, 9
Miscellaneous bacteria	Gouda, Edam	1–4
12. Enterococcus faecium or durans	Blue mould (Roquefort, Stilton, Danablu,	1, 3, 20 + 2, 4[c]
13. Brevibacterium linens	Gorgonzola, Mycella)	
14. Propionibacterium freudenreichii	White mould (Camembert, Brie, Pont l'Eveque,	1, 3, 19, + 2, 4[c]
15. P. globosum	Coulommier)	
16. P. shermanii	Surface-ripened (Limberger, Brick, Munster,	1–4, 14, 21
17. Indigenous microflora	Tilsiter)	1–4, 14, 21
18. Sour whey		
Moulds		
19. Penicillium camemberti		
20. P. roqueforti		
21. Geotrichum candidum		

[a] For the production of flavour in the cheese; this results in the formation of gas holes in the texture of the product.
[b] For the production of modified Cheddar.
[c] The addition of such microorganisms is optional.
Data compiled from Robinson (1990, 1995).

Extra-hard Cheese Varieties

A number of extra-hard cheese varieties are produced in Mediterranean Europe, Switzerland and Argentina, e.g. Asiago, Parmesan/Grana-types, Romano-types, Caciocavallo, Montasio, Reggianto and Sbrinz. The averae moisture content is between 25 and 30 g per 100 g and the texture is granular. The cheese is usually matured for 2 years or more, but some varieties can be consumed after a few months as fresh cheese (i.e. containing more moisture than matured cheese). The flavour of the matured cheese may be strong and harsh or delicate and faintly aromatic. The cheese is normally used for grating.

Most types of extra-hard cheese are made from raw cow's or sheep's milk, in some instances the milk being partly skimmed (e.g. Asiago d'Allevo). Parmesan/Grana-tyes of cheese are traditionally made as follows. Raw milk is left in shallow vats overnight, and the following day part of the cream is removed before mixing it with fresh full-fat milk. The milk is processed in a copper vat, allowing the indigenous microflora to ripen the milk before coagulation using crude extracts of calf vells. The curd is cut into small pieces, cooked to a temperature of approximately 55°C, and when sufficiently firm is scooped into a cloth drawn across the bottom of the vat. The curd mass is then lifted from the whey, allowing the free whey to drain, and transferred to moulds. Wooden boards are placed on the moulds and lightly pressed. Frequent turning of the mould assists whey drainage. The cloth is then removed, and full pressure applied overnight. The following day the cheese is brined, and later matured at ambient temperature or at 16–18°C and 85% humidity, for 2 years. In some factories, whey starter cultures have been used, consisting of a mixture of thermophilic lactic acid bacteria (see Table 2).

Hard Cheeses (Close-textured)

Cheeses in this group are amongst the most popular varieties, and are produced worldwide, e.g. Cheddar, British territorials, Cantal, Friesian Clove and Leiden, Greek Graviera, Idiazabal, Kefalotori, Machengo, Ras, Roncal. In general, cow's milk is used, but certain varieties (e.g. Manchego, Serina, Roncal and Kefalotori) are made from sheep's milk.

The basic manufacturing processes of hard-pressed cheeses have much in common. Mesophilic, homo-fermentative lactic acid bacteria (i.e. no gas producers) are usually used to ripen the milk, but are sometimes replaced or accompanied by *Lactococcus lactis* biovar *diacetylactis*, *Leuconostoc mesenteroides* subsp. *cremoris*, *Streptococcus thermophilus*, *Lactobacillus helveticus* or *L. delbrueckii* subsp. *bulgaricus*. After coagulation, the curd is cut into small pieces and scalded to 35–40°C, and the whey is removed. The curd is then texturized, milled, dry-salted, moulded and pressed under considerable pressure, to produce a cheese containing 30–45 g per 100 g moisture. There may be between 1.5 and 3 g per 100 g residual salt in the cheese.

Hard and Semi-hard Cheeses (Open-textured or with Eyeholes)

Cheeses that may contain large numbers of small eyes include Kefalograviera, Esrom, Havarti, Maribo and Tilsiter. The first of these is made in Greece from a mixture of cow's and sheep's milk (60:40), with the occasional inclusion of goat's milk (<20%). Some of these varieties can be surface-ripened with bacteria (see below).

Those cheeses with visible 'eyes' – holes in the texture of the product – are hard or semi-hard in texture.

The moisture content of these cheeses may lie between 36 and 48 g per 100 g. The holes containing CO_2) may originate from either heterofermentative mesophilic lactic acid bacteria or the use of *Propionibacterium* species. The holes resulting from the former are rather small and very sparse (e.g. in Dutch-type cheeses), while propionic acid bacteria produce large holes (e.g. in Emmental and Gruyère) during the early stages of maturation, when the cheese is stored at about 25°C for a few weeks after brining.

The cooking temperature of the curds and whey differs according to the type of cheese. Some varieties are scalded to 55°C (e.g. Emmental, Beaufort), while others are maintained at about 35°C (e.g. Samsoe, Gouda, Edam, Jarlsberg). During the manufacture of the latter type of cheese, part of the whey is removed and replaced with water at either the same temperature or higher, so that the curd/whey mixture is slightly heated. In some instances, after the pre-pressing and pressing stages, the cheese is dipped in hot whey to assist in further 'plasticizing' the curd texture, prior to brining and maturation.

Swiss-type varieties contain less moisture than Gouda cheese, and this is achieved by cutting the coagulum into smaller pieces, scalding to a higher temperature and pressing at higher pressures. It is important that the curd is pre-pressed while still submerged in the whey, to eliminate air bubbles, the final pressure being applied after the whey is removed. In the traditional process, each cheese vat will yield a cheese 'wheel' of about 80 kg, and the vats are made of copper alloy. However in a mechanized system, pressing tables are used to produce rindless blocks of cheese, each of approximately 80 kg (see **Fig. 2**).

Care should be taken during the manufacture of these cheeses to suppress the development of spores of *Clostridium* species. Clostridial growth in these low-acid cheeses can be minimized by the addition of nitrate or lysozyme to the milk, or by bactofugation or microfiltration of the cheese milk.

Semi-hard Cheeses

Cheese varieties designated as 'semi-hard' are largely those with a moisture content of approximately 45 g per 100 g. Their texture may be crumbly and/or soft. Some examples include Lancashire, Caerphilly and Wensleydale (British) and Colby and Monteray Jack (American). Arguably, Gouda and Edam could be classified as semihard cheeses if the visible eyes were not taken into account. Other cheeses in this group are Bara (Edam-like, made in Argentina), Caciotta (Italy) and Majorero (Spain), which are made from cow's, sheep's or goat's milk. Defined mesophilic lactic acid bacteria are used as starter cultures, but in the traditional Italian and Spanish cheese varieties the desirable acidity during manufacturing is provided by the indigenous microflora of the milk, predominantly species of *Lactococcus*, *Lactobacillus* and *Enterococcus*.

Brined Cheeses

In the Mediterranean and Balkan regions, brined cheeses are made from sheep's and/or goat's milk, and may contain approximately 50 g per 100 g moisture, which could justify their classification as 'semihard'. However, preservation of the cheese in brine, to prevent spoilage in subtropical conditions, results in a salt content of approximately 5 g per 100 g, which influences the characteristics of the cheese. Some of the most popular varieties are Feta, Teleme, Domiati and Bulgarian white. Some closely related varieties are preserved in brine for much shorter durations than those listed, and sometimes thermophilic lactic acid bacteria are used to ripen the milk. Normal practice is to preserve the pressed curd in brine, but during the manufacture of Domiati cheese, the salt is added to the cheese milk before the addition of the coagulant. This traditional process is still practised by the dairy industry in Egypt.

Feta-type cheeses made from ultrafiltered cow's milk have been developed in Denmark. Depending on

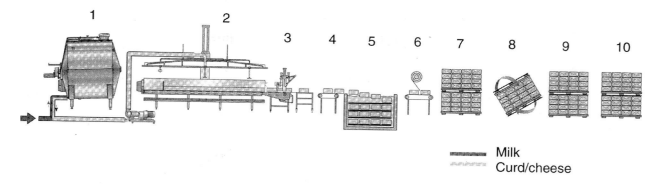

Emmental: 1, cheese vat; 2, press vat for total pressing of the curd; 3, unloading and cutting device; 4, conveyor; 5, brining; 6, wrapping in film and cartoning; 7, palletized cheese in 'green' cheese store; 8, turning the cheese; 9, fermenting store; 10, maturation store.

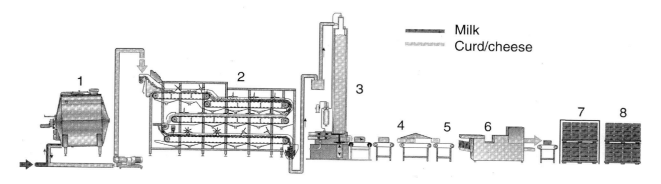

Cheddar: 1, cheese vat; 2, Alf-O-Matic cheddaring machine; 3, block former; 4, vacuum sealing; 5, weighing; 6, carton packer; 7, palletizer: 8. maturation store.

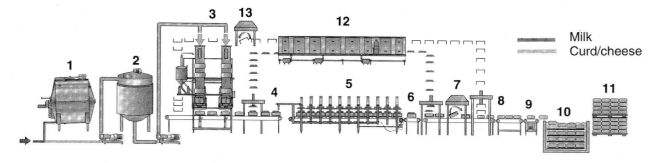

Gouda: 1, cheese vat; 2, buffer tank; 3, Casomatic pre-pressing machine; 4, lidding; 5, conveyor press; 6, de-lidding; 7, mould turning: 8. mould emptying: 9. weighing: 10. brining: 11. maturation store: 12. mould and lid washing: 13. mould turning.

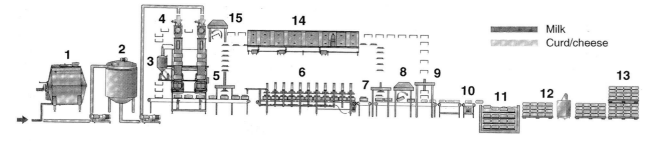

Tilsiter: 1→3, similar to Gouda; 4, rotary strainer; 5→11, similar to Gouda (i.e. 4→10); 12, fermenting store with smearing machine; 13, maturation store; 14, moulds and lids washing; 15, mould turning.

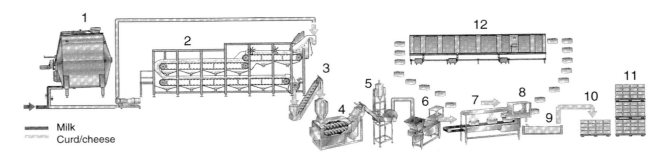

Mozzarella: 1, cheese vat; 2, Alf-O-Matic cheddaring machine; 3, screw conveyor; 4, cooker/stretcher; 5, dry salting; 6, multi-moulding; 7, hardening tunnel; 8, de-moulding; 9, brining; 10, palletizing; 11, maturation store; 12, mould washing.

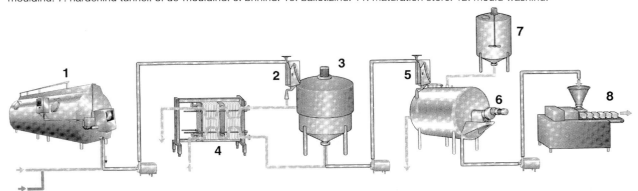

Cottage: 1, cheese vat; 2, whey strainer; 3, cooling and washing tank; 4, plate heat exchanger (PHE); 5, water drainer; 6, creamer tank; 7, dressing tank; 8, filling machine.

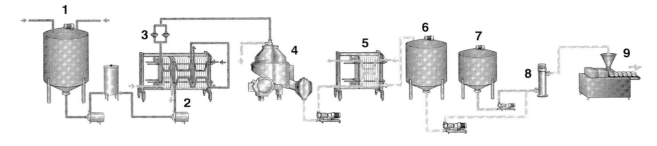

Quarg: 1, ripening tank; 2, PHE for thermization; 3, filter system; 4, nozzle separator; 5, plate heat exchanger (PHE); 6, intermediate tank; 7, cream tank; 8, dynamic mixer; 9, filling machine.

Figure 2 Flowcharts showing the mechanized production of different cheeses. (Reproduced by courtesy of Tetra Pak (Processing Systems Division) A/B, Sweden.)

the level of concentration of the milk solids, one of two cheese varieties can be made. A threefold concentration yields a product known as 'structured Feta' that has textural properties similar to those of a traditional variety. A five-fold concentration needs no removal of whey during the manufacture of the cheese, which is a very close-textured product known as 'Cast Feta'. The label 'Feta' is applicable only to cheese originating from Greece.

Mould-ripened Cheese

Mould-ripened cheeses fall into two broad categories: those varieties with the white mould *Penicillium camemberti* growing on the surface, and those with the blue or greenish mould *Penicillium roqueforti* growing internally in the structure of the cheese. Cheeses that fall into either of these categories may be hard, semi-hard or semi-soft. Mould-ripened varieties are predominantly of European origin. Some examples of blue-veined cheeses are Stilton, Gorgonzola, Mycella, Danablu, Bleu d'Auvergne, Edel-pilkäse, Bleu des Causses, Cabrales and Roquefort. The most popular white-mould varieties are Camembert and Brie. Blue and white moulds can be used together, e.g. in Bavarian blue and Lymeswold. The mould strain may be chosen in terms of its lipolytic and proteolytic activities, which can ultimately affect flavour development in the cheese. Also, additional microorganisms such as blue moulds and/or *Brevibacterium linens* may grow on the surface of traditionally made white-mould cheeses.

Mould-ripened cheeses are made from cow's, goat's or sheep's milk. For blue-veined cheese, the block of cheese is produced in the conventional way, followed by brining and/or dry-salting after pressing. The mould may be added to the milk before cheesemaking, or alternatively skewers are dipped into a solution containing mould spores before piercing the block of cheese. Piercing is essential, to allow the entry of air. During Roquefort-making, the blocks of cheese are pierced in the maturation caves, where the mould is introduced into the cheese from the environment; however yeasts can be also introduced, and can play a role in flavour development. Similarly, *Penicillium camemberti* spores can be either added to the milk or sprayed onto the cheese during the primary maturation phase. Traditionally during the production of Camembert and Brie, the removal of whey occurs as the curd is ladled by hand into open-ended moulds with drainage holes. The industrialized process involves the partial removal of why before moulding, to speed up the curd handling stage(s).

Both white- and blue-mould cheeses are matured at approximately 15°C and 85% relative humidity, to provide optimal conditions for the spores to germinate and grow. The standard of hygiene is important in the maturation rooms, particularly for white-mould varieties, to prevent the growth of undesirable species on the surface of the cheese, e.g. as black spots.

Italian-style Cheeses

During the production of a number of cheeses made from cow's, sheep's or buffalo's milk, the curd is heated at about 55°C in water or brine, followed by stretching. This process causes changes in the physical characteristics of the casein to produce plastic curd, which is slightly fibrous or, in extreme cases, pliable enough to produce braided cheeses. Typical examples are Mozzarella, Kashkaval, Halloumi, Provolone and Ostiepok. These include varieties made from full- or low-fat milk. Thermophilic starter cultures are normally used. Traditionally, after the curd has been stretched and formed into blocks, the cheese is brined. However the mechanized process uses combined dry-

salting and brining for the manufacture of, for example, Mozzarella. The moisture content of these cheeses may range from 40 to 60 g per 100 g, which places them into the categories of semi-hard and/or semi-soft varieties.

The residual galactose content in mozzarella cheese can influence the colour of the cheese melt on pizza. A brownish melt of mozzarella is associated with a high retention of galactose in the cheese, and the choice of galactose-positive variants of *L. delbrueckii* subsp. *bulgaricus* and *S. thermophilus* could minimize this effect.

Bacterial Surface-ripened Cheeses

Well-known cheeses such as Brick, Esrom, Époisses, Limburger, Livarot, Mont d'Or, Munster, Reblochon, Saint-Nectaire, Taleggio and Tilsiter have distinctive flavour characteristics that originate from a surface-growing microflora. The moisture content of these cheeses may lie between 40 and 50 g per 100 g, and hence these products may be classified as semi-hard and/or semi-soft.

The dominant bacterial growth on the surface has been identified as *Brevibacterium linens*, but salt-tolerant yeasts, along with *Geotrichum candidum*, also become established on the surface of the cheese. Species of *Micrococcus*, *Arthrobacter* and *Caseobacter* have also been isolated from bacterial surface-ripened cheeses, especially in the absence of competition from *Penicillium camemberti*. The numbers of these microorganisms can be very high. They contribute to flavour development through varied activities such as alcohol production, utilization of lactic acid and the release of proteolytic and lipolytic enzymes. *B. linens* may also be present on the surface of low-moisture cheeses, e.g. traditional Emmental, but its concentration is reduced by wiping the surface of the cheese with brine or, in other varieties, allowing the rind to dry.

After the brining stage, the block is initially matured – for example in Brick cheese, at 15°C for 2 weeks at a humidity of 90%. Afterwards, the cheese is transferred to a drying room, waxed and finally matured for 2–3 months at < 10°C. Frequent turning of the cheese in the maturation rooms, and wiping the surface with a cloth soaked in brine, encourages the growth of the surface microflora and their spread from cheese to cheese.

Soft Cheeses

The forerunners of the soft cheeses we know today were varieties with a moisture content of between 55 and 80 g per 100 g. Some modern examples are

Cottage, Karish, Kopanisti, Queso Blanco, Quarg and Fromage Frais. In general, mesophilic or thermophilic lactic acid bacteria are used as starter cultures. The traditional varieties of cheese rely on the acidification of milk, as do fermented milk products like yoghurt. This is followed by drainage, using earthenware vessels, animal skins, wooden barrels or cloth bags to concentrate the coagulate. Thus, the method of manufacture is completely different from the conventional stages of cheese-making (see Fig. 1). Technically, some of these cheese varieties (e.g. Quarg or Fromage Frais) should be classified as concentrated fermented milk products, rather than cheese. Cottage and Quarg cheeses are both made from skimmed milk, but there are differences in their manufacture. In Cottage cheese-making, once the gel is firm at pH 4.5 it is cut, allowed to heal for 30 min, cooked in the whey to 50–55°C, cooled, washed, creamed and packaged. In the manufacture of Quarg, the fermented skimmed milk is thermized, concentrated in a special centrifuge, blended with cream and/or fruits and packaged. These products have a short shelf life, and extreme care should be exercised during production to minimize contamination.

Compositional Quality and Industrial Processes for Cheese-making

Milk from different species of mammals is used for the manufacture of cheese – hence the gross chemical composition of the products will vary (see **Table 3**). In extra hard varieties of cheese, the moisture content is < 30 g per 100 g and these cheeses require up to 2 years for maturation. Fresh cheeses, i.e. those with a moisture content of > 60 g per 100 g, require no maturation and can be consumed directly after manufacture.

In response to the centralization of cheese production in many countries, mechanized systems have been developed which ensure retention of the characteristics of the traditional varieties of cheese. Fig. 2 provides some examples of factory production systems for Emmental, Cheddar, Gouda, Tilsiter, mozzarella, cottage and quarg cheese. Note that the equipment used for the production of cheeses from ultrafiltered milk is not shown in Figure 2.

See also: **Arthrobacter**. **Brevibacterium**. **Cheese**: Microbiology of Cheese-making and Maturation; Mould-ripened Varieties; Role of Specific Groups of Bacteria; Microflora of White-brined Cheeses. **Fermentation (Industrial)**: Basic Considerations. **Fermented Milks**: Range of Products. **Geotrichum**. **Lactobacillus**: Introduction. **Lactococcus**: *Lactococcus lactis* sub-species *lactis* and *cremoris*. **Micrococcus**. **Milk and Milk**

Table 3 Typical averages of the gross chemical composition (g per 100 g) of different cheeses

Cheese variety	Moisture	Protein	Fat	Salt	FDM[a]
Parmesan	29.6	35.6	25.8	1.5	32
Grana Padano	30.0	35.0	28.0	–[b]	32
Sbrinz	28.0	31.0	34.0	4.5	47
Emmental	35.7	28.7	29.7	1.5	45
Gruyère	33.2	29.8	32.3	–	45
Cheddar	36.3	25.4	32.2	1.8	50
Manchego	37.0	23.0	34.0	–	54
Edam	41.9	24.8	28.3	1.9	45
Gouda	36.4	25.5	29.2	2.1	45
Mozzarella	60.1	19.9	16.1	–	–
Provalone	39.6	36.3	28.9	–	–
Munster	52.1	21.6	22.6	2.6	45
Brie	45.5	22.6	27.9	2.8	50
Camembert	52.0	21.0	22.3	2.2	45
Edilpilzkase	42.8	21.1	29.8	3.5	50
Danablu	< 48.0	21.0	26.0	4.5	50
Gorgonzola	42.4	19.4	31.2	–	–
Roquefort	39.4	21.5	30.6	4.6	–
Limberger	51.7	22.4	19.7	3.0	40
Brick	40.0	22.0	30.0	2.1	50
Tilsiter	40.6	26.3	27.7	1.8	45
Cottage	78.5	12.3	4.3	1.0	–
Quarg (skimmed)	81.3	13.5	0.3	–	–
(full-fat)	73.5	11.1	11.4	–	40
Kopanisti	60.0	16.5	19.5	3.0	–
Queso Blanco	55.0	23.0	15.0	2.5	33

[a]FDM: fat in dry matter expressed in (%).
[b]Data not reported.
Data compiled from Souci et al (1994) and Robinson (1995).

Products: Microbiology of Liquid Milk. **Preservatives**: Traditional Preservatives – Sodium Chloride. **Starter Cultures**: Cultures Employed in Cheese-making. **Streptococcus**: Introduction; *Streptococcus thermophilus*.

Further Reading

Anonymous (1989) *Dairy Legislation in the EEC Member Countries*. Kiel: Federal Dairy Research Centre.

FAO (1981) *Production Yearbook*. Vol. 34. Rome: Food and Agriculture Organization of the United Nations.

FAO (1996) *Production Yearbook* Vol. 49. Rome: Food and Agriculture Organization of the United Nations.

FAO/WHO (1990) *Codex Alimentarius – Abridged Version*. Rome: Food and Agriculture Organization of the United Nations.

Fox PF (ed) (1993) *Cheese: Chemistry, Physics and Microbiology*, 2nd edn. Vols 1 and 2. London: Chapman & Hall.

IDF (1981) *Catalogue of Cheeses*. Document 141. Brussels: International Dairy Federation.

IDF (1997) *IDF The World Market for Cheese*. Document 326. Brussels: International Dairy Federation.

Johnson ME (1997) *Cheesemaking*. London: Chapman & Hall.

Masui K and Yamada T (1996) *French Cheeses*. New York: D K Publishing.

Robinson RK (ed) (1990) *Dairy Microbiology – The Microbiology of Milk Products*, 2nd edn. Vol. 2. London: Elsevier Applied Science.

Robinson RK (ed) (1993) *Modern Dairy Technology – Advances in Milk Products*, 2nd edn. Vol. 2. London: Elsevier Applied Science.

Robinson RK (ed) (1995) *A Colour Guide to Cheese and Fermented Milks*. London: Chapman & Hall.

Robinson RK and Tamime AY (eds) (1991) *Feta and Related Cheeses*. Chichester: Ellis Horwood.

Robinson RK and Wilbey RA (1998) *Scott's Cheesemaking Practice*, 3rd edn. London: Chapman & Hall.

Souci SW, Fachmann W and Kraut H (1994) *Food Composition and Nutrition Tables*. Stuttgart: Medpharm GmbH Scientific Publishers.

Microbiology of Cheese-making and Maturation

Nana Y Farkye, Dairy Science Department, California Polytechnic State University, San Luis Obispo, USA

Introduction

Cheese-making is the conversion of milk from its fluid state into a semisolid mass. The process involves the coagulation of milk proteins and acidification of the coagulated proteins by the action of starter bacteria. After manufacture and during ripening, the activities of the starter bacteria, secondary (adjunct) starter and non-starter microorganisms present influence the maturation process, to give each cheese its distinctive characteristics. The maturation process involves a series of complex biochemical reactions including glycolysis – the fermentation of lactose to lactic acid and the metabolism of lactate; proteolysis – the hydrolysis of cheese proteins (primarily casein) to peptides and amino acids; and lipolysis – hydrolysis of fat (triglycerides) into free fatty acids that are further broken down in flavourful compounds.

Typical Composition of Cheese Milk

There are over 400 cheese varieties, manufactured from cow, goat, sheep, or buffalo milk. However, worldwide, most cheese is produced from cow's milk. The typical chemical composition of cow's milk is given in **Table 1**. The chemical composition (casein to fat ratio) of milk influences cheese composition and determines the legal identity of the cheese.

Table 1 Approximate chemical composition of bovine milk.

Component	Mean (%)	Range (%)
Water	87.3	86.1–89.4
Lactose	4.7	4.5–5.0
Fat	3.9	3.3–4.7
Protein	3.2	2.9–3.5
Casein	2.6	2.4–2.8
Whey protein	0.6	0.5–0.7
Salts	0.7	0.6–0.9

Microbiological Quality of Milk

Cheese may be manufactured from raw or pasteurized milk. In the USA and many other countries, cheese made from raw milk must be held back for at least 60 days before consumption. As a result, it is preferable to use pasteurized milk for cheese manufacture – pasteurization destroys pathogenic organisms in the raw milk. Milk is legally pasteurized when given a minimum heat treatment of 63°C for 30 min (low temperature long time – LTLT process) or 72°C for 15 s (high temperature short time – HTST process). In some countries, cheese milk is given a heat treatment called thermization (65°C for 15 s) to inactivate psychrotrophic bacteria. Thermization also eliminates most yeasts and coliforms, which otherwise will cause excessive gas openings and off flavours in cheese. Milk that has been given thermization is either used directly for cheese-making or pasteurized before use.

Although milk in the mammary gland is sterile, it immediately becomes contaminated as it leaves the udder, by various bacteria present outside the udder or in the surrounding environment. Therefore, milk may contain a few to millions of microorganisms of different genera. Organisms isolated from raw milk include pathogenic bacteria (e.g. *Salmonella*, *Listeria*, enteropathogenic *Escherichia coli*), *Pseudomonas*, *Enterobacter*, *Klebsiella*, *Alcaligenes*, *Acinetobacter*, *Microbacterium*, *Flavobacterium*, *Bacillus*, *Lactococcus*, *Lactobacillus*, *Propionibacterium*, coliforms, yeasts and many others.

Due to the differences in the microbiological quality of milk, many countries have instituted standards that classify milk according to microbiological quality. The differences in microbiological quality influence cheese quality, whether the milk is pasteurized or not. The microbiological limits of raw milk sold in the US is given in **Table 2**.

Principles of Cheese-making

The basic principle of cheese-making involves the coagulation of milk proteins (primarily casein) by one of:

Table 2 Microbiological classification of raw milk sold in the US

Class	Standard plate counts (cfu ml^{-1})	Coliforms (cfu ml^{-1})	Direct microscopic count (cells ml^{-1})
Raw Grade A			
Sold direct to consumers	15 000 (maximum)		
Pasteurized before sale	50 000 (maximum)	750 (maximum)	
Manufacturing milk			
1st quality			⩽ 500 000
2nd quality			⩽ 1 000 000
Undergrade			> 1 000 000

- Addition of the milk-clotting enzyme, rennet. This method is used in the manufacture of most cheeses.
- Direct acidification of milk to pH 4.6, or *in situ* production of acid by starter bacteria. This method is used for the manufacture of cottage cheese.
- Heat/acid coagulation, in which hot milk (80–90°C) is acidified to pH 4.6–5.3. This method is used for the manufacture of ricotta, quark, paneer and queso blanco.

Traditionally, most rennet-coagulated cheeses are ripened or matured before consumption whereas acid- and heat/acid-coagulated cheeses are consumed fresh (unripened). During the manufacture of rennet-coagulated cheeses, the rate of acidification, syneresis (expulsion of whey from curd) and salt addition are important and unique to individual varieties. These steps are dependent on the types of starter used for manufacture and the non-starter bacteria present.

Starter Bacteria, Starter Adjuncts and Non-starter Bacteria in Cheese

Starter Bacteria and Starter Adjuncts

Starter bacteria are primarily lactic acid bacteria, added to milk for cheese-making. Their primary purpose is to ferment lactose to produce acid. A list of lactic acid bacteria used as starters for cheese-making is given in **Table 3**. Some microorganisms are added as starter adjuncts or secondary starters to achieve a specific function (e.g. flavour, textural attributes) in certain varieties. Others are added because they produce antimicrobial substances called bacteriocins, that inhibit other organisms that may be present in cheese. Microorganisms that are traditionally usually used as secondary starters are listed in **Table 4**. Also present in cheese are non-starter lactic acid bacteria (NSLAB). These are organisms that arrive in cheese either by surviving pasteurization or from post-pasteurization contamination.

Lactic acid bacteria used as cheese starter are classified into two groups – mesophilic or thermophilic – depending on their ability to produce acid at different temperatures. Lactic acid bacteria are also classified

Table 3 Classification of lactic acid bacteria used in cheese-making

Species of lactic acid bacteria	Cheese type manufactured with species
Mesophilic (growth optimum at 20–40°C)	
Homofermentative	
Lactococcus lactis subsp. *cremoris*	Cheddar-type, Gouda-type, blue mould-type Limburger, Brie, Camembert, cottage, cream
Lactococcus lactis subsp. *lactis*	
Heterofermentative	
Lactococcus lactis subsp. *lactis*, biovar *diacetylactis*	Gouda, Dutch-type, blue mould-type, cottage,* cream
Leuconostoc mesenteroides subsp. *cremoris*	
Thermophilic (growth optimum at 30–55°C)	
Homofermentative	
Streptococcus thermophilus	Emmental/Swiss-type, mozzarella, Romano, Parmesan, Provolone, other Italian types
Lactobacillus helveticus	
Lactobacillus delbrueckii subsp. *bulgaricus*	

* The heterofermentive species are used to culture the cream dressing to impart flavour and to extend shelf-life by inhibiting psychotropic bacteria.

Table 4 Microorganisms used as secondary starters in cheese-making

Microorganism	Cheese type manufactured with species
Propionibacterium freudenreichii var. *shermanii*	Swiss-type (e.g. Emmental, Gruyère)
Brevibacterium linens	Limburger
Penicillium roqueforti	Blue mould-types (e.g. Roquefort, blue Stilton)
Penicillium camemberti	White mould-types (e.g. Camembert)
Penicillium candidum	Brie

as homofermentative or heterofermentative. Homofermentative organisms ferment glucose to produce lactic acid exclusively. Heterofermentative lactic acid bacteria ferment glucose to acetic acid, CO_2 and/or ethanol in addition to lactic acid.

For cheese manufacture, pure cultures of individual species or mixtures of species of lactic acid bacteria are propagated in special media. These cultures are added to cheese milk at levels ranging from 0.02 to 5.0%, to ferment lactose to produce lactic acid. The viable cell population used for cheese-making ranges from 10^5 to 10^7 colony forming units (cfu) per millilitre, although concentrated starters, for direct in-vat inoculation, may contain as many as 4×10^{11} cfu ml^{-1}. When frozen concentrated starters are used, the levels added are usually low (< 0.01%). In some traditional cheese-making processes, indigenous microflora in the milk are relied on for acid production, without the use of starter bacteria. The amount of starter added to cheese milk depends on the cheese type, the medium in which the organism was grown, conditions of propagation, and the rate of acidification desired.

Over the duration of the cheese-making (usually < 5 h for most varieties) the starter population increases a hundred times to about 10^9 cfu g^{-1} in the finished cheese. The cooking temperature used during cheese-making influences starter activity. Cheeses made with mesophilic starters are cooked at lower temperatures than those made with thermophilic starters. However, starter activity in the finished cheese is dependent largely on the concentration of salt in the moisture and on the elimination of lactose. High salt concentrations (salt-in-moisture > 6%) inhibit the activity of most starter lactic acid bacteria.

Cultures of lactic acid bacteria used as starters for Cheddar and related types of cheese consist primarily of single or mixed strains of *Lactococcus lactis* subspp. *cremoris* or *lactis*, although some manufacturers include *L. lactis* subsp. *lactis* biovar *diacetylactis* or *Leuconostoc* spp. Swiss-type (e.g. Emmental, Gruyère) and related cheeses with holes (eyes) are manufactured using very low levels (0.12–0.2%) of mixtures of *Streptococcus thermophilus* and *Lactobacillus delbruekii* subsp. *bulgaricus*, *Lactobacillus helveticus* or *Lactobacillus lactis* as primary starter. Mesophilic lactococci may also be included as primary starter. *Propionibacterium freudenreichii* var. *shermanii* is added as secondary starter, for eye formation and flavour development. The levels of propionic acid bacteria added to cheese milk are low. However, when the cheese is transferred to a warm (18–25°C) room during ripening, the cell densities of propionic acid bacteria in the cheese may reach 10^8–10^9 cfu g^{-1}.

Lactococcus lactis subspp. *cremoris* or *lactis* are used in conjunction with *Leuconostoc mesenteroides* subsp. *cremoris* and/or *Lactococcus lactis* subsp. *lactis* biovar *diacetylactis* as starters for Gouda and Dutch-type cheeses. Starters used for Mozzarella and other Italian varieties comprise a combination (usually 1:1 ratio) ratio of *Streptococcus thermophilus* and *Lactobacillus delbruekii* subsp. *bulgaricus*. The ratio of the two organisms varies depending on the characteristics desired in the finished cheese.

Starter bacteria used in blue-veined and other mould-ripened cheeses (e.g. blue Stilton, Roquefort, Camembert) are primarily *Lactococcus lactis* subspp. *lactis* or *cremoris*. Mould (*Penicillium roqueforti*, *P. gorgonzola*, *P. glaucum* or *P. camemberti*) is added as secondary starter, depending on the variety being manufactured. In blue-veined mould varieties, air (O_2) is let into the cheese by piercing with needles during ripening to allow internal mould growth and to give the cheese a distinctive blue, veiny appearance. Mould colour varies from blue (*P. roqueforti*) through blue-green (*P. gorgonzola*) to green (*P. glaucum*). In surface mould-ripened cheeses, e.g. Camembert, white mould (*P. camemberti*) is added to the milk or sprayed onto the finished cheese. In bacterial surface-ripened cheeses such as Limburger and Brick, *Lactococcus lactis* subspp. *cremoris* or *lactis* are the primary starters. Surface smear organism, *Brevibacterium lineus* is brushed on the surface of the cheese from a whey or broth culture or from an older cheese with a well-developed smear.

Non-starter Bacteria

The predominant non-starter lactic acid bacteria (NSLAB) in cheese are lactobacilli (*Lactobacillus casei* subspp. *casei*, *pseudoplantarum* or *rhamnosus*, *L. brevis*), and are *Pediococcus* species (e.g. *Pediococcus pentosaceus*). Other non-starter, non-lactic acid bacteria are *Micrococcus* species. *Micrococcus* spp. are components of adventitious flora. They are aerobic, oxidative and nonfermentative. Extensive studies on Cheddar cheese show that as starter numbers decline during the early stages of ripening, the population of NSLAB increases, and is dependent on the rate of cooling of the cheese after pressing. Typical cell densities of lactobacilli in cheese range from 10^4 to 10^6 cfu g^{-1} after the first few days to about 10^7 to 10^8 cfu g^{-1} within a few weeks post-manufacture. Thereafter, the levels remain constant throughout the rest of the ripening period. The rate of decline of starter numbers depends on the elimination of lactose (the primary energy source), inhibition by salt, autolysis, ripening temperature and redox potentials in the cheese. In addition to NSLAB, moulds (e.g. *Geotrichum candidum*) and yeasts (e.g. *Kluyveromyces marxianus* var. *lactis*, *Saccharomyces cerevisiae* and *Debaryomyces hansenii*) are present in mould-ripened cheeses. Non-starter organisms found in the surface smear are yeasts, *G. candidum*, *Brevibacterium linens* and *Micrococcus* species.

Microbiological Changes during Cheese-making

The principal action of starter bacteria during cheese-making is the metabolism of lactose, to produce lactic acid. The lactic acid lowers the pH and increases the acidity in cheese during manufacture. In addition, it has other important functions during the manufacture and ripening of cheese. These include:

- aiding the coagulation of milk and influencing the activities of residual coagulant and of plasmin and other enzymes that aid in the cheese-ripening process
- aiding the solubilization of colloidal calcium phosphate – which in turn affects cheese texture and other characteristics
- aiding whey expulsion (syneresis)
- stimulating the growth of symbiotic organisms present
- inhibiting the growth of contaminating microorganisms during manufacture and ripening, thereby ensuring safety and extending the shelf life of the cheese.

The metabolic pathways used by lactic acid bacteria for acid production differ between organisms. Lactic acid bacteria use two systems to transport lactose into the cell: either the phosphoenolpyruvate-dependent phosphotransferase system (PTS) or the lactose permease system. The lactococci (*Lactococcus lactis* subspp. *cremoris*, *lactis* and *lactis* biovar *diacetylactis*) use the PTS for the assimilation of lactose (as lactose phosphate) from milk and from the serum phase of cheese. The lactose is translocated to yield a lactose phosphate, which is hydrolysed into glucose and galactose 6-phosphate. The glucose is then fermented via the Embden–Meyerhof–Parnas (glycolytic) pathway to lactic acid, and the galactose 6-phosphate is fermented via the tagatose 6-phosphate pathway to lactic acid.

In *Leuconostoc* species, lactose assimilated by a permease system is hydrolysed by β-galactosidase into glucose and galactose. The glucose is metabolized via the phosphoketolase pathway to lactic acid, ethanol and CO_2. The galactose moiety is metabolized via the Leloir pathway, in which galactose is converted into glucose-1-phosphate, then glucose-6-phosphate which follows the Emden–Mayerhof–Parnas pathway to lactic acid. *Streptococcus thermophilus*, *Lactobacillus helveticus* and *Lactobacillus delbrueckii* subsp. *bulgaricus* also transport lactose via a permease system into the cell, where it is hydrolysed into glucose and galactose. Only the glucose moiety is subsequently metabolized by *S. thermophilus* and *L. delbruekii* subsp. *bulgaricus* – the galactose is expelled back into the medium. There is some evidence to suggest that *L. helveticus* ferments galactose by the Leloir pathway to produce lactic acid. The end products of lactose fermentation by the various organisms are summarized in **Table 5**.

Microbiological Changes During Cheese Maturation

Metabolism of Lactic and Citric Acid

The concentration of lactate (lactic acid) varies between cheese varieties. Levels in Camembert, Swiss cheese, Cheddar and Romano are 1.0, 1.4, 1.5 and 1.7% respectively. The fate of lactate in cheese depends on the types of microorganism present. In Swiss-type cheeses, propionic acid bacteria metabolize lactic acid in the pH range 5.0–5.3 at 18–25°C, to give propionic acid, acetic acid and CO_2.

$$3CH_3.CHOH.COOH \longrightarrow 2CH_3.CH_2.COOH + CH_3.COOH + CO_2 + H_2O$$

The CO_2 generated accumulates in the cheese and is responsible for the characteristic holes (called 'eyes') in the cheese. In Cheddar and Dutch-type cheeses, L(+)-lactate is isomerized by NSLAB (e.g. pediococci and some lactobacilli) to D(−)-lactate, resulting in a racemic mixture of both isomers in the cheese. D(−)-Lactate reacts with calcium to form an insoluble calcium salt, that crystallizes and appears as undesirable white specks in the cheese.

Bovine milk contains approx. 8 mmol citrate. Although most of the citrate in milk is soluble and is lost in the whey during manufacture, Cheddar cheese contains 0.2–0.5% citrate. Not all lactic acid bacteria metabolize citrate. *Lactococcus lactis* subspp. *cremoris* and *lactis*, *Streptococcus thermophilus* and thermophilic lactobacilli do not metabolize citrate. Citrate is metabolized by *Lactococcus lactis* subsp. *lactis* biovar *diacetylactis*, *Leuconostoc* species and mesophilic lactobacilli such as *Lactobacillus casei* and *L. plantarum*. Citrate is transported via a pH-dependent inducible permease into the cell, where it is converted to oxaloacetate and acetate. The oxaloacetate is decarboxylated to give pyruvate, which is oxidized to diacetyl and CO_2.

$$2CH_3.CO.COOH + O_2 \longrightarrow CH_3.CO.CO.CH_3 + 2CO_2 + H_2O$$

Diacetyl can be converted to acetoin and thence to 2,3-butylene glycol and 2-butanone. Diacetyl and

Table 5 Summary of end products of lactose fermentation

Organism	Principal end products (mole of product per mole of lactose used)	Isomer of lactic acid
Lactococci	4 lactic acid	L(+)
Leuconostoc	2 lactic acid + 2 ethanol + 2 CO_2	D(−)
Streptococcus thermophilus	2 lactic acid	L(+)
Lactobacillus delbrueckii subsp. bulgaricus	2 lactic acid	D(−)
Lactobacillus helveticus	4 lactic acid	DL

acetate contribute to the flavour and aroma of Dutch-type cheeses, in which *Lactococcus lactis* subsp. *diacetylactis* and *Leuconostoc* species are used as starters. The production of diacetyl increases as pH decreases. The CO_2 produced from citrate metabolism is responsible for the characteristic eyes in Dutch-type cheeses. The production of CO_2 in Cheddar-type cheeses leads to a textural defect called openness – therefore lactic acid bacteria that metabolize citrate are not normally used in the manufacture of Cheddar and related types. Although diacetyl is an important contributor to the flavour and aroma of cottage cheese, the excessive activity of citrate-metabolizing organisms during cottage-cheese manufacture leads to curd flotation.

Lipolysis

Although many lactic acid bacteria possess esterolytic and lipolytic enzymes, they are only weakly lipolytic towards milk fat. Therefore lipolysis in bacteria-ripened cheeses like Cheddar, Dutch-type and Swiss-type cheeses is very low. In cheeses such as Provolone, pregastric lipases (from kid or lamb) or microbial lipases are used to give the typical 'piccante' flavour. Lipolysis due to microbial enzyme activity is greatest in mould-ripened cheeses. *Penicillium camemberti* secretes one extracellular lipase, whereas *P. roqueforti* secretes two types of extracellular lipase (acid and alkaline), which catalyse the hydrolysis of fat in cheese to produce free fatty acids. The free fatty acids are further oxidized to form β-ketoacids, that are decarboxylated to give methyl ketones (mainly 2-nonanone and 2-pentanone). The methyl ketones are reduced further to secondary alcohols.

Proteolysis

Lactic acid bacteria are nutritionally fastidious, requiring an exogenous source of nutrients, including free amino acids, for growth. The concentration of free amino acids in milk is low, so starter bacteria must hydrolyse milk proteins to give the free amino acids needed for growth. Starter bacteria possess proteinases (proteases) – enzymes that hydrolate peptide bonds in proteins and peptidases – enzymes that hydrolyze peptide bonds in small peptides (e.g. dipeptides, tripeptides).

In cheese, the sequence of proteolytic events that leads to the formation of free amino acids is:

1. Initial hydrolysis of milk proteins (primarily α_s- and β-caseins) by residual milk-clotting enzymes or plasmin, to give large polypeptides.
2. Hydrolysis of the large polypeptides by microbial proteases and peptidases, to small peptides.
3. Breakdown of the small peptides into amino acids by microbial peptidases.

Lactic acid bacteria possess extracellular proteases, that are associated with the cell wall. These proteases hydrolyse casein or casein-derived peptides into small peptides, that are transported into the cell. Research shows that peptides containing up to six amino acid residues can be transported into the cell. Lactic acid bacteria that lack extracellular proteases and grow slowly in milk are characterized as proteinase-negative (Prt−). The cell wall-associated proteases are classified into two groups, P_I and P_{III}-types. P_I-type proteases preferentially hydrolyse β-casein, whereas P_{III}-type proteases have broad specificities and hydrolyse both α_s- and β-caseins. Generally, lactococci producing P_{III}-type proteases are thought to produce less bitter-tasting peptides from casein hydrolysis and in cheese.

Lactic acid bacteria also possess various peptidases (endopeptidases, aminopeptidases, dipeptidases, tripeptidases and proline-specific exopeptidases), that have different peptide bond specificities (**Table 6**). No carboxypeptidase activity has been detected in lactococci. The collective action of the various starter peptidases results in the production of free amino acids in cheese.

Organisms used as secondary starters also possess proteolytic enzymes that contribute to cheese-ripening. *Penicillium roqueforti* and *P. camemberti* produce aspartyl proteases and metalloproteases, that have similar specificities with regard to α_s- and β-caseins. In addition, *Penicillium* species possess both aminopeptidases and carboxypeptidases. Yeasts (e.g. *Debaryomyces*, *Kluyveromyces* and *Saccharomyces*) that develop on the surface of soft cheeses, and *Geo-*

Table 6 Peptidases present in various lactic acid bacteria species

Enzyme	Other name/Abbreviation	Bond Cleaved
Aminopeptidase P		X↓Pro-Y-Z. . .
Proline iminopeptidase		Pro↓X-Y-Z. . .
Proline iminodipeptidase	prolinase (PIP)	Pro↓X
Imidodipeptidase	prolidase	X↓Pro
X-Prolyl dipeptidylaminopeptidase	PepX	X-Pro↓Y-Z
Pyrolidonyl carboxylyl peptidase	PCP	pGlu↓Y-Z
Aminopeptidase A	PepA	Asp(Glu)↓Y-Z
Aminopeptidase C	PepC	X↓Y-Z
Aminopeptidase N	PepN	X↓Y-Z
Dipeptidase	DIP	X↓Y
Tripeptidase	TRP	X↓Y-Z
Endopeptidase	PepO	..W-X↓Y-Z..

trichum candidum that grows on the surface of Camembert made from raw milk, also produce intracellular proteolytic enzymes that contribute to proteolysis in cheese. Because of the high proteolytic activity of *Penicillium* species, the levels of proteolysis in mould-ripened cheeses are higher than in bacteria-ripened varieties such as Cheddar-type, Gouda-type and Swiss-type cheeses.

The microorganisms in cheese also break down the amino acids produced by proteolysis, to give numerous compounds that contribute to the flavour and aroma of cheese. Microbial enzymes involved in the catabolism of amino acids include deaminases, decarboxylases and transaminases.

These catalyse the deamination, decarboxylation and transamination of amino acids to produce, respectively, α-ketoacids, amines and other amino acids, which are then converted into aldehydes, alcohols and acids. For example, in blue-veined cheeses, arginine is converted to ornithine and citrulline by *P. roqueforti*. Biogenic amines such as tyramine, histamine, tryptamine, putrescine and cadaverine are produced by the decarboxylation of amino acids in cheese. *Brevibacterium linens* and other coryneform bacteria produce an enzyme, demethiolase, which produces methanethiol directly from methionine.

Defects of Bacterial and Fungal Origin

Undesirable microorganisms in cheese may lead to defects in its appearance, flavour or texture. Undesirable organisms are present either as a result of using raw milk with poor microbial quality or due to contamination during cheese manufacture, handling or packaging. Pathogenic organisms (e.g. *Staphylococcus aureus*, *Listeria monocytogenes*) can survive and grow in cheese. This underlies the importance of pasteurizing cheese milk, in order to inactivate pathogenic organisms, and of adhering strictly to good manufacturing practices during cheese-making – although many countries have no microbial standards

for cheese. Microorganisms causing defects in cheese include some lactic acid bacteria, coliforms, psychrotrophs, spore-forming bacteria, coryneforms, yeasts and moulds.

Defects Caused by Lactic Acid Bacteria

The predominance of heterofermentative lactic acid bacteria (e.g. *Lactobacillus brevis*, *L. casei* subsp. *pseudoplantarum*) in close-textured cheese such as Cheddar results in an open texture due to gas production. Pediococci and lactobacilli convert L(+)-lactate to D(−)-lactate, which reacts with calcium to form insoluble calcium lactate crystals, that appear as undesirable white specks in ripened Cheddar. In brine-salted cheeses, such as Dutch- and Swiss-type cheeses, contaminating salt-tolerant lactobacilli metabolize amino acids to give undesirable phenolic and putrid H_2S-like flavours during ripening. Studies show that concentrations of gas-producing lactobacilli of > 10^3 cfu ml^{-1} in brine are detrimental to cheese quality. Soft body defect in mozzarella has been attributed to the proteolytic activity of mesophilic lactobacilli (e.g. *L. casei* subsp. *casei*). In Swiss-type cheeses, some strains of *L. delbrueckii* subsp. *bulgaricus* produce a pink discoloration, and pink spots have been attributed to the growth of pigmented strains of propionibacteria. The presence of faecal streptococci (e.g. *Enterococcus durans* and *E. faecalis*), that occur most frequently in cheeses made from raw milk, causes an undesirable flavour and high levels of amines (e.g. histamine and tyramine).

Defects Caused by Psychrotrophic Bacteria

Psychrotrophic bacteria cause defects particularly in fresh soft cheeses (e.g. cottage), the principal causative organisms belonging to the genera *Pseudomonas*, *Aeromonas* and *Acinetobacter*. The most common defects are surface discoloration, off odours and off flavours. Thermostable lipolytic enzymes produced by psychrotrophic bacteria in raw milk survive cheese-making, causing rancidity in the cheese. Also, thermostable proteases from psychrotrophic bacteria may cause proteolysis, leading to bitterness in cheese.

Defects Caused by Coliform Bacteria

The presence of coliforms in cheese is an indication of poor sanitation, because the coliform bacteria present in raw milk are killed by pasteurization. Coliforms grow rapidly in cheese during the first few days of storage. The metabolites of coliforms include lactic acid, acetic acid, formic acid, succinic acid, ethanol, 2,3-butyleneglycol, H_2 and CO_2. The production of H_2 and CO_2 results in early gas blowing of the cheese. In retailed prepacked cheese, coliform concentrations of approximately 10^7 cfu g^{-1} result in a gassy defect and swelling of the plastic packaging.

Defects Caused by Spore-forming Bacteria

The main source of clostridial contamination of milk is silage. The spores survive the pasteurization of milk and germinate in cheese, causing late gas blowing and the development of off flavours in many varieties (mostly Swiss- and Dutch-type cheeses). *Clostridium tyrobutyricum* is the major spore-forming organism that causes late gas blowing. Other causative organisms are *C. butyricum* and *C. sporogenes*. These organisms metabolize lactate (or glucose) to produce butyric acid, acetic acid, CO_2 and H_2 gas. The accumulation of the CO_2 and H_2 causes the late gas blowing. This defect is prevented by the removal of spores from the milk, by bactofugation or microfiltration before cheese-making or by the addition of lysozyme or nitrate to the cheese milk. Research shows that butyric acid fermentation and late gas blowing may occur in cheese made from milk containing 5–10 spores per litre.

Defects Caused by Coryneform Bacteria, Yeasts and Moulds

The presence of large numbers of coryneform bacteria and yeasts on cheese surfaces results in a slimy rind, discoloured appearance and undesirable flavours. The growth on cheese of moulds (other than those added to mould-ripened cheeses) is a common and recurrent problem. The most common moulds found on cheese are *Penicillium* species. Other moulds that occur include *Aspergillus*, *Alternaria*, *Cladosporium* and *Fusarium*. Undesirable mould growth in cheese results in discoloration, poor appearance and off flavours. Furthermore, some moulds produce mycotoxins that may pose health risks to consumers. Airtight packaging can prevent undesirable mould growth.

See also: **Cheese**: In the Marketplace; Mould-ripened Varieties; Role of Specific Groups of Bacteria; Microflora of White-brined Cheeses. **Lactobacillus**: Introduction. **Lactococcus**: Introduction. **Milk and Milk Products**: Microbiology of Liquid Milk. **Penicillium**: *Penicillium* in Food Production. **Spoilage Problems**: Problems caused by Bacteria. **Starter Cultures**: Cultures Employed in Cheese-making. **Yeasts**: Production and Commercial Uses.

Further Reading

Fox PF (ed.) (1933) *Cheese: Chemistry, Physics and Microbiology.* Vols. 1 and 2. London: Chapman & Hall.

Gilliland SE (ed.) (1985) *Bacterial Starter Cultures for Foods.* Boca Ratan: CRC Press.

Gripon JC, Monnet V, Lamberet G and Desmazeud MJ (1991) Microbial enzymes in cheese ripening. In: Fox PF (ed.) *Food Enzymology.* London: Elsevier Applied Science.

Malin EL and Tunick MH (ed.) (1995) In: *Chemistry of Structure-Functional Relationships in Cheese.* New York: Plenum Press.

Law BA (ed.) (1997) *Microbiology and Biochemistry of Cheese and Fermented Milk.* London: Chapman & Hall.

Mould-ripened Varieties

A W Nichol, Charles Sturt University, NSW, Australia

Introduction

This article describes the manufacture of internally and externally mould-ripened cheeses, and the processes involved in their maturation. The roles of starter organisms and of the moulds *Penicillium roqueforti* and *P. camemberti* are described, in particular in relation to the degradation of casein and milk fat during the maturation process and to the production of the flavour and texture profile typical of these cheeses.

Mould-ripened cheeses are of two major types: those ripened using moulds which grow on the surface of the cheese, and those ripened by moulds which grow internally. The best known of the surface-ripened cheeses are Camembert and Brie, generally ripened by *P. camemberti* and *Geotrichum candidum* respectively. The internally mould-ripened cheeses are best represented by Danish blue, Roquefort, Bleu d'Auvergne, Stilton and Gorgonzola. The major organism used for ripening these cheeses is *P. roqueforti*.

History

The history of each of the major blue cheeses is interesting not only for its own sake but because it reveals key characteristics of the cheeses, which are relevant to their manufacture.

True Roquefort has its origins in the French village of Roquefort-sur-Soulzon, which is located in an area of limestone caves called Combalou. The cheese is made from sheep milk, and the curd is prepared in the usual way using a mesophilic starter and rennet. Once the curd is hooped, it is spiked to allow air into the interior, then stored in the limestone caves where the atmosphere is both cool and humid. Cracks in the limestone act as natural filters and allow the circulation of fresh air with the correct temperature and relative humidity for optimal mould growth. Roquefort cheese is believed to predate the Roman occupation of Gaul: its history is thus deeply rooted in the past.

Gorgonzola also is known to have been produced for at least a thousand years, in the Valassima region of northern Italy, on the river Po. It is made from cow's milk, which has been set with rennet after the addition of lactic acid bacteria and a culture of *P. roqueforti*. However, the strain of mould used for the production of Gorgonzola differs from that used in Roquefort production. It is more green in colour than the mould used for other blue cheeses, and has more proteolytic activity. The traditional method of manufacture involves the use of two curds, prepared for evening and morning milk. By using these curds some mechanical openness is provided, and this allows the mould to grow after spiking by permitting additional air exchange. The cheese was matured originally in the caves of Valassima, in a cool, moist environment.

Stilton is one of the strongest-flavoured of the blue cheeses. In relative times, its origins are recent. The earliest records of its production are dated 1720 and are attributed to Daniel Defoe, who reported enjoying it at the village of Stilton in Cambridgeshire. It is made from high-fat milk and ripened in cool, moist cellars. Mechanical openness is achieved by breaking the cheddared curd into large pieces, and also by spiking. The strong flavour of Stilton suggests that a highly lipolytic strain of *P. roqueforti* is used. An interesting historical variant of the blue cheeses is Dorset blue or Blue Vinny. This cheese was made from partially skimmed milk on farms in the Dorset region, in a manner similar to Stilton. It is said that the mould was obtained from horses' bridles, dipped into the milk by the cheese maker.

Danish blue, although one of the best-known of the blue cheeses, appears to have been developed as recently as the 1950s. It is probably an imitation of cheeses such as Bleu d'Auvergne, which may itself have begun as a cow's-milk imitation of Roquefort. It is a high-fat cheese, in some cases very high-fat (up to 60% fat in dry matter).

Common features of the production of all these cheeses include: incubation of the curd to give a rela-tively low initial pH; attempts to create openness in the structure of the curd; piercing of the curd to allow further gas exchange; and maturation in an environment with low temperature and high humidity.

Brie and Camembert are the best-known but by no means the only cheeses which are ripened by moulds growing on their surfaces. Brie is the more ancient of the two types. The first authenticated historical reference to Brie dates from the eighth century, when Charlemagne's secretary, Eginhard, recorded that the king sampled a Brie cheese at a village 20 miles east of Meaux, and was so delighted with it that he ordered two batches each year. Brie cheeses conforming to AOC (appellation d'origine contrôlée) regulations are between 25 and 37 cm in diameter.

The apparent similarities between Brie and Camembert are explained by the history of Camembert. Marie Fontaine, who was born in Roiville in 1761, was assisted in developing Camembert by a young priest, Abbé Gorbeut, a native of Brie. She adapted the Brie method to take into account the smaller cheese hoops used in the area. The first Camembert cheeses were 11 cm in diameter. The cheeses were probably blue-green in colour, because it was not until 1910 that Roget Lepetit replaced the blue mould *P. glaucum* (now *P. roqueforti*) with the white mould *P. candidum* (now *P. camemberti*). Camembert de Normandie retains its original dimensions (10.5–11 cm diameter). The smaller size of Camembert compared to Brie allows for a much shorter maturation period (affinage). Camembert may be ready for the market in 3 weeks, whereas Brie normally requires 8 weeks.

Many other surface mould-ripened cheeses, less well-known than Brie and Camembert, are produced in France. These include Saint-Nectaire, Neufchâtel, Gaperon, and many goat's milk cheeses. Most of these are ripened by the action of *P. camemberti* and *Geotrichum candidum*, but other moulds may participate as well. *Sporotrichum aureum* and *Mucor* species are reported to contribute, for example, to the ripening of Saint-Nectaire.

Manufacture of Blue Cheeses

Probably the major factor influencing the manufacture of blue cheeses is the need to allow O_2 to enter the interior of the cheese and to allow CO_2 out. This is often rather appropriately known as allowing the cheese to breathe. This need arises because the mould is a typical eukaryote, and requires O_2 to act as a terminal electron acceptor, for the production of energy and growth. O_2 supply is not generally a problem for surface mould-ripened cheeses, but for internally mould-ripened cheeses, considerable care is necessary for the creation of an environment in which gas exchange can occur. Steps aimed at creating this environment include the following:

- *Pre-treatment of milk:* when a high level of lipolysis is desired, homogenization not only activates endogenous lipase in the milk but, more importantly, creates a porous curd. Homogenized milk curds are less dense than curd from non-homogenized milk due to the incorporation of air.

- *Control of the development of acidity:* if a mesophilic culture is used for the development of acid, it is usually *Lactococcus lactis* subsp. *lactis*. For Gorgonzola-style cheeses, thermophiles are commonly used – generally mixtures of *Lactobacillus delbrueckii* subsp. *bulgaricus* and *Streptococcus salivarius* subsp. *thermophilus*. The aim is to create a relatively acid environment that will give a short, crumbly-textured curd with considerable mechanical openness. A pH of 4.7–4.9 at salting has been suggested, but in the author's experience, stronger acidities are helpful (pH 4.5–4.7).

- *Incorporation of gas producers:* the use of mixed cultures containing *Leuconostoc* species helps to create openness in the curd, by the formation of eyes due to the fermentation of citrate, with the release of CO_2.

- *Adequate drainage of the curd:* well-drained curd gives a lower moisture content in the matured cheese than does curd that has been inadequately drained. Mould growth in the inadequately-drained curd is usually poor, and results in cheeses with a lower final pH (about 5.8) than cheeses made from adequately drained curd (pH 6.5–7.1). More importantly, the cheese made from inadequately drained curd has flavour defects such as sourness. However, the moisture in mould-ripened cheeses must be maintained at an optimal level. Too low a level will result in the poor distribution and activation of enzymes released from the mould, and defective maturation. This is the major reason why the maintenance of high humidity is necessary in the maturation of mould-ripened cheeses, especially those with a natural rind.

- *Spiking at the correct stage:* spiking, to allow the exchange of CO_2 with O_2, is usually carried out 1–4 weeks after salting. Spiking early results in rapid maturation, but if it is too early the spike-holes may collapse. If the cheese is spiked too late, slow maturation and the presence of competing organisms such as yeasts in the spike-holes may result in the holes being blocked and hence poor gas exchange. This may be overcome to some extent by piercing through a wax or plastic coating.

A typical flow sheet for the production of cheese made in the Danish blue-style is shown in **Figure 1**. There are many variations on this procedure. Although it is generally agreed that the addition of the mould is best carried out by its addition to the milk, it can also be done at the hooping stage. Salting can be achieved by rubbing the exterior, by brining, or by mixing the curd with salt, but there is little evidence that brining gives a superior product.

Manufacture of Surface-ripened Cheeses

Generally, surface-ripened cheeses are manufactured in smaller batches than are internally mould-ripened cheeses. This is because the development of acidity may be so rapid that in larger batches, the time needed for casting the whole batch of cheese curd into hoops may lead to significant differences in pH within the batch. This leads to differences in mould growth and hence quality.

In the manufacture of traditional or Normandie-style Camembert, the pH at hooping is similar to that for internally mould-ripened cheese, i.e. 4.6–4.7. This pH is achieved by using a mixed mesophilic culture, of the Flora Danica or similar type, if pasteurized milk is used. Although Camembert and Brie meeting AOC criteria must be dry-salted, most factory-produced cheeses made in this style are brine-salted. *Affinage*, ripening of the cheese, sees the initial growth of a layer of lactose-fermenting yeasts on the surface of the cheese. *Debaryomyces hansenii* predominates, but *Kluyveromyces* also contributes. This is followed by growth of the moulds *Geotrichum candidum* and *Penicillium camemberti*. A balanced growth of these moulds leads to high quality cheese. Excessive growth of *P. camemberti* can lead to bitterness, due to the formation of bitter hydrophobic peptides from β-casein. Camembert matures in about 4 weeks following hooping, and then has a pH of 7–8.

Cheese Maturation

Mould Growth

The growth of *Penicillium roqueforti* in experimental loaves of blue cheese has been investigated using electron microscopy (Washam et al, 1997). The internal structure of the cheese was shown to be very porous, with the mould growing in cracks and crevices. At day 5 most, but not all, the conidia had germinated. The cheeses were pierced at about this stage. Fully germinated conidia were seen after 2 weeks. At 3 weeks the growth of the mycelia was dense and supported a large number of spores. By 6 weeks the mycelia had partly degenerated. This process continued until the cheese was mature, after approxi-

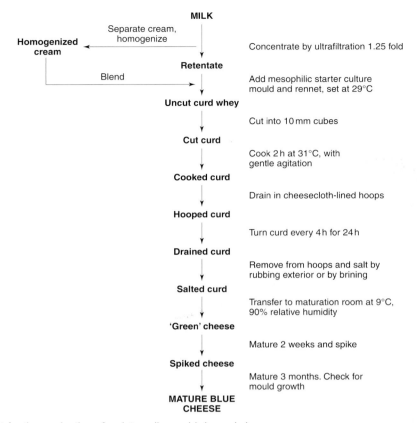

Figure 1 Flow sheet for the production of an internally mould-ripened cheese.

mately 9 weeks. Detachment of conidio from the conidiophores characterized this stage.

The physical changes in the cheese and the growth of mould within it are paralleled by chemical changes associated with the mould growth and metabolism. One of the most marked features of this process is the dramatic change in pH during maturation. **Figure 2** typifies these changes. During the first week after salting, pH continues to drop due to the continued activity of lactic acid bacteria. At piercing, mould

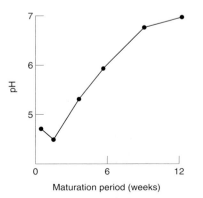

Figure 2 pH changes during the ripening of a Gorgonzola-style cheese.

growth commences and a steady rise in pH takes place, which peaks at about 10 weeks. Final pH values are generally in the range 6.6–6.9.

Mould Metabolism

The changes in pH shown in Figure 2 are associated with two major metabolic events. The first of these is the consumption of lactic acid, which is the principal carbon source for mould. Although the literature contains few reports of studies on lactic acid consumption by *P. roqueforti*, the depletion of lactic acid due to the growth of *P. camemberti* in Camembert is well-documented.

As the mould begins to grow, it produces a number of extracellular proteases as well as intracellular or mycelial proteases. It would appear that in the early phases of growth the proteolysis of casein, principally β-casein, is largely due to extracellular proteases. The sharp increase in proteolysis which occurs at about week 10 of maturation is suggestive of the sudden release of intracellular proteases, due to mycelial degradation. Proteolysis is a feature of the maturation of most cheeses, the enzymes involved being a combination of endogenous milk protease (plasmin), added enzymes (chymosin and pepsin) and microbial proteases, especially those derived from lactobacilli. The activity of residual chymosin is directed principally towards α_{s1} casein, while plasmin and mould proteases are relatively specific for β-casein degradation. Electrophoretic studies of proteolysis in a range of Australian cheeses showed very extensive degradation of β-casein to γ-casein in Gorgonzola-style cheese. Degradation of α_{s1} casein to its derivatives, α_{s1-1}, α_{s1-2} and α_{s1-3} caseins, was also apparent. As would be expected, similar patterns are also seen in Brie- and Camembert-style cheeses. Although other proteolytic enzymes are present in cheese, it is the mould and its enzymes which are responsible for the bulk of protein degradation in blue cheeses. This can be shown by growing mould in curd which is otherwise aseptic. The pattern of casein breakdown is very similar to that seen in normally matured cheese.

Gripon (1987) reports that there are two major extracellular proteases: a metalloprotease, which is activated by cobalt, zinc or manganese; and an aspartate protease, so called because it appears to have an aspartyl residue at its active site. The aspartate protease appears to be active in early maturation because it has a low pH optimum, while the metallo enzyme is active later in maturation, as reflected by its pH optimum of 6.0. Enzymic activity is inhibited by NaCl at concentrations higher than 0.5%. For this reason, proteolytic activity is highest in the centre of the cheese and lowest near the rind. This characteristic allows the use of salt to control proteolysis during ripening.

Towards the end of maturation, peptides, amino acids and other forms of non-protein N_2 accumulate in the cheese. The decarboxylation of amino acids gives rise to amines, such as tyramine. Deamination produces acids, which in turn may produce aldehydes, alcohols and other flavour components. Whether these substances are produced by the mould and its enzymes, or by bacteria such as *Brevibacterium linens*, is not clear. The production of ammonia at this stage of maturation is, however, very clear to anyone who has entered a cellar where maturation is in progress. Ammonia and other amines contribute very significantly to the rise in pH which occurs during curd maturation. The degradation of casein to basic nonprotein derivatives is thus the second major metabolic event which contributes to pH rise.

Proteolysis contributes to not only the characteristic flavour of the cheese but also, and perhaps more importantly, to its body and texture. The short, crumbly texture of the low-pH curd changes to a creamy texture, the creaminess depending on the degree of proteolysis. This is extensive in cheeses such as Gorgonzola, Brie and Camembert, which have a rich creamy texture. If proteolysis becomes too exten-

sive the cheese becomes liquid and rank in odour, due to the presence of excessive amounts of amines.

As well as producing enzymes which break down proteins, *P. roqueforti* produces enzymes which attack and degrade milk fat. The characteristic peppery flavour of Danish blue, Stilton and similar cheeses is primarily due to this process. *P. roqueforti* produces an inducible lipase, which requires the presence of amino acids, as an N_2 source, and of milk fat for production. The enzyme is relatively heat-stable and causes the release of 20–30% of the fatty acids from milk-fat triglyceride during the course of maturation. Homogenization assists this process markedly. Milk lipase is only marginally involved in the release of fatty acids, as shown by the fact that cheese from unpasteurized homogenized milk produces only marginally more free fatty acid over a 24-week period than does cheese from pasteurized homogenized milk.

Methylketones produced by the oxidation of fatty acids play an important role in determining the flavour of blue cheese. The most important are 2-nonanone and 2-heptanone. There is evidence that these compounds are formed by β-oxidation of fatty acids. A likely metabolic pathway is shown in **Figure 3**. Similar lipases and oxidative enzymes occur in *P. camemberti* and in *Geotrichum candidum*. The latter

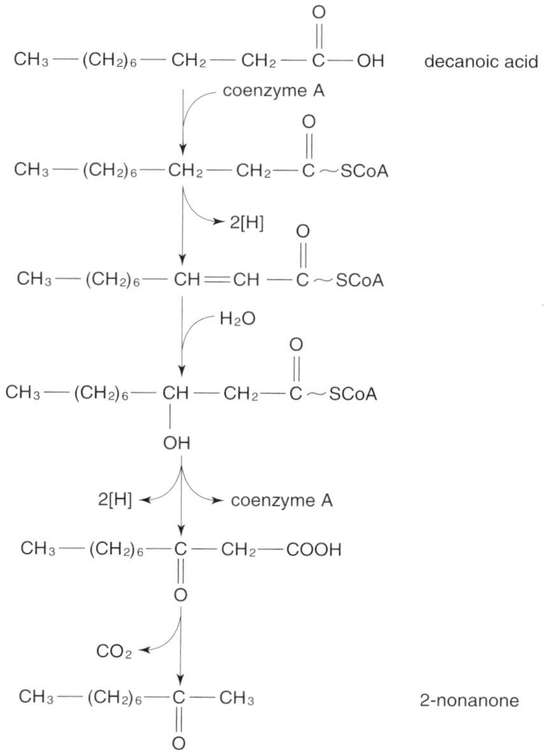

Figure 3 Possible pathway for the production of 2-nonanone in mould-ripened cheese.

produces a lipase that is relatively specific for the hydrolysis of triglycerides containing oleic acid, which occurs in high concentration in Camembert cheese.

Contribution of Other Organisms to Maturation

Mould-ripened cheeses often display considerable diversity with respect to the organisms they contain. For example, Roquefort is reported to contain 94 strains of *Lactococcus* and 49 strains of *Leuconostoc*. In our own work (unpublished) we have identified the following organisms from the crust of Gorgonzola-style cheeses: the moulds *Penicillium citrovorum*, *P. brevicompactum* and *Geotrichum candidum*; the yeasts *Yarrowia lipolytica* and *Candida catenulata*; and the bacteria *Micrococcus luteus*, *Arthrobacter* spp. and *Brevibacterium linens*. There is no doubt that these organisms contribute to the maturation process, especially *B. linens*, which is particularly active in the degradation of amino acids, with the release of volatile sulphur-containing compounds. The extent of the contribution of these organisms relative to that of *P. roqueforti*, however, is hard to determine.

Defects in Mould-ripened Cheeses

The most serious defect in blue cheese production is failure of the mould to grow or, in a less extreme situation, poor mould growth. In the real world this defect is almost always caused by closure of the spike-holes too soon after spiking, and/or the texture of the cheese being insufficiently open. The effect is the same in either situation: insufficient O_2 penetrates to the interior of the cheese, and the mould fails to grow. This problem arises due to many different causes. Some of the most important are:

- Inadequate development of acidity in spring milk, due to the inhibition of starter organism growth by the lactoperoxidase system. This leads to a curd with a texture that is too soft to maintain openness.
- Use of late-lactation milk. Late-lactation milk retains moisture in the curd, and the soft curd tends to collapse into the spike-holes. In extreme situations, collapse of the crust also occurs.
- Pushing of yeasts into the interior of the cheese at spiking. Very often the yeasts will grow rapidly, and block the spike-holes.
- Inadequate drainage has already been mentioned as a cause of poor mould growth.

Interestingly, the addition of too little mould (too few mould spores) to the milk at or before rennetting is rarely a cause of poor mould growth. Inoculation rates of 5×10^6 spores per litre of milk have been recommended, but inoculation rates of less than one

tenth of this figure usually give adequate mould growth.

Poor mould growth is associated with defects in flavour, texture and body. The body of the cheese will be short and crumbly, because proteolysis is poor or absent. The texture may appear chalky. The flavour will be sour, due to an insufficient rise in pH. If no mould growth has occurred the pH may drop to around 4.0, especially if excessive amounts of whey have been trapped in the curd.

Browning reactions are a problem in many mould-ripened cheeses, especially Gorgonzola-style cheeses. The pigment which causes browning is a melanin-like substance produced by the action of yeasts, especially *Yarrowia lipolytica*, which contains the enzyme tyrosinase. Recent evidence suggests that the causative yeasts are most likely to grow in curd produced from late-lactation milk. The higher water activity in late-lactation curds provides a favourable environment for the growth of these yeasts.

Common defects in Camembert and Brie include textural defects, due to an initial pH (at hooping) which is too high or too low, and browning reactions, which are almost always associated with the presence of high levels of free tyrosine and tyrosinase-containing yeasts. The growth of inappropriate mould on these cheeses (e.g. *Penicillium roqueforti*, *Mucor* and *Oidium* spp.) also causes important defects.

Raw-milk versus Pasteurized-milk Cheeses

There is little doubt that cheeses made from raw milk have more complexity and individuality than do cheeses made from pasteurized milk. This is especially true for mould-ripened cheeses, because the maturation or ripening process is heavily dependent on an appropriate succession of microbiological flora. However in the New World, most soft cheeses are made from pasteurized milk that has been heat-treated for 15 s or longer at 72°C. Many of the microorganisms destroyed by this procedure must be replaced, by the addition of mixed cultures containing lactic acid bacteria, micrococci, yeasts, etc. Moulds are generally added at the same time but separately. Soft cheeses with high pH and short maturation period are especially susceptible to contamination by *Listeria monocytogenes*. This organism can grow at maturation-room temperatures, but is destroyed by pasteurization. Listeriosis, the disease caused by this organism, is a serious condition that can result in abortions in humans. It is also associated with a high mortality rate in the immunocompromised. There is continuing argument about whether recent occurrences of the disease were associated with raw-milk cheeses or with the contamination of milk and milk products by the organism after pasteurization. Evidence points to both sources of infection. In Europe, codes of practice emphasizing cleanliness and sanitation, both on the farm and in the factory, safeguard against the disease to some extent, as does a recognition within society of the potential hazards of raw-milk products. The development of starter organisms possessing activity against *Listeria* may provide another defence against the hazards posed by this organism.

See color Plate 12.

See also: **Cheese**: Microbiology of Cheese-making and Maturation; Role of Specific Groups of Bacteria. **Geotrichum**. **Lactobacillus**: *Lactobacillus bulgaricus*. **Lactococcus**: Introduction. **Listeria**: *Listeria monocytogenes*. **Penicillium**: Introduction; *Penicillium* in Food Production. **Starter Cultures**: Cultures Employed in Cheese-making. **Streptococcus**: *Streptococcus thermophilus*.

Further Reading

Androuet P (1983) *Guide du Fromage*, English edn, revised. P. 184. Wiltshire: Aidan Ellis.

Choisy C, Gueguen M, Lenoir J, Schmidt JL and Tourneur C (1986) The ripening of cheese, microbiological aspects. In: Eck A (ed.) *Cheesemaking, Science and Technology*. P. 259. New York: Lavoisier.

Coghill D (1979) The ripening of blue-vein cheese: a review. *Aust. J. Dairy Technol.* 34: 2–75.

Eckhof-Stork N (1976) *World Atlas of Cheese*, English edn. Pp. 59, 88. New York: Paddington Press.

Galloway J (1995) Production of soft cheese. *J. Soc. Dairy Technol.* 48: 36–43.

Gripon JC (1987) Mould ripened cheeses. In: Fox PF (ed.) *Cheese Chemistry, Physics and Microbiology*. Vol. 2, p. 121. Elsevier, London.

Hewidi M and Fox PF (1984) Ripening of Blue cheese: characterisation of proteolysis. *Milchwissenschaft* 39: 198–201.

Kinsella JE and Hwang DH (1976) Enzymes of *P. roqueforti* involved in the biosynthesis of cheese flavour. *Crit. Rev. Food Science & Nutrition* 6: 191–228.

Morris HA (1981) Blue veined cheeses. In: *Pfizer Cheese Monographs*. Vol. 7, p. 1. New York: Pfizer.

Nichol AW, Harden TJ and Tuckett H (1996) Browning reactions in mould ripened cheeses. *Food Australia* 48: 136–138.

Washam CJ, Kerr TJ and Todd RL (1979) Scanning electron microscopy of Blue cheese: mould growth during maturation. *J. Dairy Sci.* 62: 1384–1389.

Role of Specific Groups of Bacteria

M El Soda, Department of Dairy Science and Technology, Faculty of Agriculture, Alexandria University, Egypt

Propionibacterium

Propionibacterium species are characterized by being Gram-positive, catalase-positive, non-spore-forming, non-motile, facultative anaerobes. They are usually pleomorphic, diphtheroid (i.e. resembling *Corynebacterium diphtheriae*) or club-shaped, with one end rounded and the other tapered or pointed. Individual cells may be coccoid, elongated, bifid or branched. They occur singly or in pairs, clumps, short chains or various other configurations.

The genus *Propionibacterium* includes two distinct groups of microorganisms: the acne or cutaneous propionibacteria, that form a major part of the skin flora of humans, and the dairy or classical propionibacteria, that have traditionally been isolated from dairy products, particularly cheese. The latter group includes four species: *Propionibacterium freudenreichii* subspp. *freudenreichii* and *shermanii*, *P. jensenii*, *P. thoenii* and *P. acidi-propionici*. The economic value of the propionibacteria of dairy origin derives from their important role in eye formation and flavour development in Swiss-type cheeses. Dairy propionibacteria also have industrial applications outside the cheese industry:

Propionic Acid Production   Propionic acid and its salts are used in the food industry as antifungal agents. A large part of the world's production of propionic acid (> 120 000 t) is obtained from the petrochemical industry. However, production involving fermentation processes using propionibacteria has been described, and will probably increase in the near future due to increasing consumer demand for natural and biological products.

Production of Vitamin B12   *P. freudenreichii* strains have been specifically selected for their high yields of vitamin B12. Yields of 19–23 mg l^{-1} were reported in a two-stage process (a primary anaerobic stage followed by a secondary aerobic phase).

Propionibacteria as Probiotics   There is clear evidence that propionibacteria have probiotic (a mono or mixed culture of microorganisms which when applied to animal or human affect the host beneficially) effects, based on their production of beneficial metabolites (e.g. vitamin B12) and antimicrobial compounds such as propionic acid and bacteriocins. Cells of *P. freudenreichii* subsp. *shermanii* were reported to also exhibit antimutagenic activity. In probiotic food products, propionibacteria are usually combined with lactic acid bacteria and/or bifidobacteria.

Metabolic Activity during Eye Formation

The total number of cheese varieties reported in the literature is 400–1200. Although the basic steps in cheese-making are to a great extent similar, cheeses come in different shapes and colours, have different consistencies and develop different flavours. One of the major factors leading to this enormous variety of cheese is the nature of the various microorganisms involved in the cheese-making process and the ripening of the cheese.

One group of bacteria, including special heterofermentative species of *Lactococcus* and *Leuconostoc* in addition to *P. freudenreichii* subsp. *shermanii*, liberate CO_2 during the fermentation of lactose, citrate or lactate. This has led to the development of a distinct group of cheeses known as 'cheeses with eyes'. Propionibacteria are used in the production of these so-called Swiss cheeses. There are several types (**Table 1**), differing in terms of the size of the cheese and the number of holes.

The number and activity of propionibacteria are controlled to a great extent by the production process and the physical properties of the curd. In Swiss-type cheeses, lactose is metabolized to lactic acid by the thermophilic streptococcus *S. thermophilus* and the thermophilic lactobacilli *L. helveticus* and *L. delbrueckii* subspp. *bulgaricus* and *lactis*. *S. thermophilus* metabolizes lactose to L(+)-lactic acid, using the glucose moiety. Galactose is then fermented to a mixture of D(+)- and L(−)-lactic acid in the presence of *L. helveticus*. Lactose catabolism begins during

Table 1 Cheeses with eyes produced by propionibacteria

Cheese variety	Country of origin	Weight (kg)
Appenzeller	Switzerland	6–8
Beaufort	France	14–70
Comté	France	38–40
Danbo	Denmark	6
Elbo	Denmark	6
Emmental	Switzerland	60–130
Emmental français	France	45–100
Fynbo	Denmark	7
Gruyère	Switzerland, France	20–45
Herregardsost	Sweden	12–18
Jarlsberg	Norway	10
Maasdamer	Netherlands	12–16
Samsoe	Denmark	14
Svecia	Sweden	12–16
Tybo	Denmark	3

processing in the cheese vat. After 4–6 h of moulding, the sugar is entirely hydrolysed. Propionic acid fermentation is initiated by a rise in the curd temperature to 18–25°C. At these temperatures propionibacteria levels reach up to 10^9 cfu per gram of cheese. Hotroom curing takes 5–7 weeks, during which L(+)-lactate is metabolized preferentially by propionibacteria compared to the D(−)-isomer. As a result of the fermentation of L(+)- and later D(−)-lactate, propionic acid, acetic acid and CO_2 are produced according to the following pathways:

- Lactate is oxidized, in the presence of a flavoprotein as H_2 acceptor, to pyruvate:

$$\text{Lactate} \xrightarrow{-2[H]} \text{Pyruvate}$$

- Pyruvate accepts a carboxyl group from methylmalonyl-CoA by a transcarboxylase reaction leading to the formation of oxaloacetate and propionyl-CoA:

$$\text{Pyruvate} \xrightarrow[\text{Methylmalonyl-CoA}]{+[COOH]} \text{Oxaloacetate} + \text{Propionyl-CoA}$$

- Propionyl-CoA reacts with succinate to produce succinyl-CoA and propionate, in the presence of a CoA transferase. Succinate results from the reduction of oxaloacetate to fumarate, which is then reduced to succinate):

$$\text{Oxaloacetate} \longrightarrow \text{Malate} \longrightarrow \text{Fumarate}$$
$$\longrightarrow \text{Succinate}$$

$$\text{Propionyl-CoA} + \text{Succinate} \xrightarrow{\text{CoA transferase}}$$
$$\text{Succinyl-CoA} + \text{Propionate}$$

- In a reaction catalysed by an isomerase, methylmalonyl-CoA is obtained from succinyl-CoA, to complete the cycle:

$$\text{Succinyl-CoA} \xrightarrow{\text{Isomerase}} \text{Methylmalonyl-CoA}$$

- Part of the pyruvate resulting from the oxidation of lactate is converted to acetyl-CoA and CO_2 by the action of pyruvate dehydrogenase:

$$\text{Pyruvate} \xrightarrow{\text{NAD, CoA}} \text{Acetyl-CoA} + \text{NADH} + CO_2$$

- Acetyl-CoA is then converted to acetate.

$$\text{Acetyl-CoA} \longrightarrow \text{Acetyl-P} \longrightarrow \text{Acetate} + Pi$$

The CO_2 generated is responsible for the development of eyes. The texture of the cheese and the temperature at which the propionic acid fermentation takes place play a key role in the process.

The steps in eye formation in Swiss-type cheeses can be summarized as follows:

- CO_2 diffusion occurs before propionic acid fermentation begins, some CO_2 being produced from the hydrolysis of lactose.
- Most of the CO_2 needed for eye formation is produced by the action of the propionic acid bacteria on lactate.
- A critical gas pressure is reached, at which the gas forms a small bubble, or becomes part of another bubble in a favourable part of the cheese. Gas generated nearby moves to the initial eye, which expands.
- The number and size of the eyes depend on the pressure and the rate of diffusion of the CO_2 produced in the cheese matrix. If gas production is too slow, saturation does not occur and few or no eyes are obtained. The resultant cheese is described as 'blind'. In an 'overset' cheese, an excessive number of small eyes are produced because of rapid generation of CO_2. However, excessively rapid gas production causes breaking of the cheese structure, and the formation of very large holes is observed.

Brevibacterium linens

Brevibacterium linens, which is the type species of the genus *Brevibacterium*, is Gram-positive with both rod and coccoid forms. Cells of older cultures (3–7 days) are composed of coccoid cells, while cells in the exponential phase are characterized by their irregular rod shapes. *B. linens* is an obligate aerobe that does not produce acid from lactose. The microorganism grows well at neutral pH. Growth also occurs in the pH range 6.5–8.5, and in NaCl concentrations up to 15%. *B. linens* strains produce colonies which are yellow to deep orange-red, on a variety of media. *B. linens* is a major component of surface-ripened cheeses (**Table 2**).

Surface-ripened cheeses can be defined as varieties with desirable microbial growth on the surface, that plays a key role in the development of the characteristic flavour of the cheese. Surface-ripened cheeses can be differentiated according to the types of

Table 2   Varieties of surface-ripened cheese

Cheese variety	Country of origin	Average weight
Appenzeller	Switzerland	6–8 kg
Beaufort	France	20–60 kg
Brick	USA	2.5 kg
Époisses	France	4.5 kg
Limburger	Belgium, Germany	200 g–1 kg
Livarot	France	300–500 g
Mont d'or	France	200 g–3 kg
Muenster	Germany	500 g–1 kg
Pont L'Évêque	France	350 g
Reblochon	France	240–500 g
Ridder	Norway	2 kg
Romadur	Germany	80–180 g
Saint-Nectaire	France	800 g–1.5 kg
Saint-Paulin	France	1.5–2 kg
Serra da Estrela	Portugal	1.5–2 kg
Taleggio	Italy	2 kg
Tilsiter	Germany	1.5–2 kg
Trappist	Germany	1.5–2.7 kg

microorganism growing on their surface, into cheeses with mould and those with yeasts and bacteria. In the latter, surface-ripening is the result of the symbiotic growth of the bacteria and yeasts.

Yeasts are present in higher concentrations during the earlier stages of the ripening process, because they can develop at rather low temperatures and at relatively high humidities. They can also tolerate the low pH and high NaCl concentration at the cheese surface. The yeast flora is composed mainly of *Debaryomyces*, *Candida* and *Torulopsis*, and it plays a key role in the transformation of the environment on the cheese surface The yeast flora uses lactic acid as a carbon source, transforming it to H_2O and CO_2. As a result the pH of the cheese surface is considerably increased from close to 5.0 to about 5.9. The yeasts also stimulate the growth of *B. linens* and of micrococci, through the synthesis of vitamins including riboflavin, niacin and pantothenic acid. The yeast flora disappears after 1–20 days, giving way to the micrococci and *B. linens*.

The micrococci isolated from surface-ripened cheeses have been identified as *Micrococcus caseolyticus* and *M. freudenreichii*. It is believed that micrococci play a role in the proteolysis of cheese and in flavour development. *B. linens*, along with microorganisms of the genus *Arthrobacter*, forms the predominant flora of the smear of surface-ripened cheeses. Through their various metabolic activities, these microorganisms cause changes in the texture of the cheese and play a key role in the development of its characteristic flavour.

Action of *Brevibacterium* during the Maturation of Smear-coated Cheeses

B. linens strains give the smear its distinctive orange or orange-brown colour, reflecting their ability to synthesize orange pigments. Pigment formation seems to be light-dependent, because some strains do not synthesize pigments in the dark. The presence of O_2, the age of the culture and the composition of the media also play a role in pigment formation.

In contrast to many cheese-related microorganisms, *B. linens* exhibits a wide range of protein-, peptide- and amino acid-degrading enzymes. Indeed, both intracellular and extracellular proteinase activities have been detected in *B. linens*, indicating that the extracellular proteolytic system can hydrolyse cheese proteins from the first days of ripening. Hydrolysis continues after the death of the cells, due to the release of their intracellular proteinases. The resulting peptides are then degraded by the various extracellular aminopeptidases, to amino acids. Intracellular aminopeptidases and dipeptidases play a similar role after cell autolysis.

B. linens possesses the ability to decarboxylate a wide range of amino acids including lysine, leucine and glutamic tyrosine and serine. As a result of this action, volatile and non-volatile amines, which play an important role in cheese flavour, are produced. The deamination of several amino acids, including phenylalanine, tryptophan, histidine, serine and glutamine leads to the formation of ammonia, which is also an important player in the flavour and aroma of smear-coated cheeses. Ammonia production also raises the pH, leading to a softer cheese body. Volatile sulphur compounds, resulting from the degradation of methionine through demethiolase activity, also make a significant contribution to the flavour characteristics of smear-coated cheeses. **Figure 1** summarizes the possible role of the different enzymes produced by *B. linens* in protein degradation during surface-ripening.

B. linens also produces lipolytic enzymes: extracellular lipase, as well as extracellular, cell-bound and intracellular esterases have been detected in various strains. It is believed that the lipolytic activities of *B. linens* and other surface microflora make a significant contribution to lipolysis in varieties of cheese such as Brick, Port-Salut and Limburger, in which fatty acid levels in the range 700–4000 mg per kg of cheese have been reported.

The compounds responsible for the typical flavour of surface-ripened cheeses, which are produced on the surface, diffuse into the interior until equilibrium is reached.

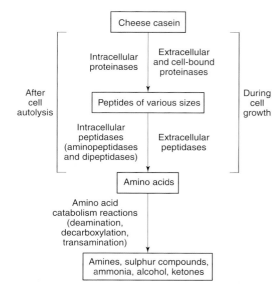

Figure 1 Degradation of casein and formation of flavour compounds by *Brevibacterium linens*.

See also: **Brevibacterium**. *Candida*: Introduction; *Yarrowia (Candida) lipolytica*. **Cheese**: In the Market-place; Microbiology of Cheese-making and Maturation; Mould-ripened Varieties. **Debaryomyces**. **Fermentation (Industrial)**: Basic Considerations; **Fermented Milks**: Range of Products; **Lactobacillus**: Introduction; **Lactococcus**: Introduction; *Lactococcus lactis* sub-species *lactis* and *cremoris*; **Micrococcus**. **Process Hygiene**: Designing for hygienic operation; **Propionibacterium**. **Streptococcus**: Introduction; *Streptococcus thermophilus*; **Yeasts**: Production and Commercial Uses.

Further Reading

Boyaval P and Desmazeaud MJ (1983) Le point des connaissances sur *Brevibacterium linens*. *Lait* 63: 187–216.

Boyaval P and Cow C (1995) Production of Propionic acid. *Lait* 75: 453–462.

Eck A and Gilles JC (eds) (1997) *Le fromage de la Science à l'Assurance-qualité*, 3rd edn. Paris: Lavoisier techniques & documentation.

Fox PF (ed.) (1993) *Cheese: Chemistry, Physics and Microbiology*, 2nd edn. Vols. 1 and 2. London: Chapman & Hall.

Gautier M, Lortal S, Boyaval P et al (1993) Les bactéries propioniques laitières. *Lait* 73: 257–263.

Hemme D, Bouillanne C, Métro F and Desmazeaud MJ (1982) Microbial Catabolism of amino acids during cheese ripening. *Sciences Des Aliments* 2: 113–123.

Hettinga DH and Reinbold GW (1972) The propionic-acid bacteria – A review II. Metabolism. *J. Milk Food Technol.* 35: 358–372.

Kosikowski FV and Mistry VV (1997) *Cheese and Fermented Milk Foods*, 3rd edn. Vol I. Westport: FV Kosikowski LLC.

Langsrud T and Reinbold GW (1973) Flavor development and microbiology of Swiss cheese – A Review III.

Ripening and flavor production. *Journal of Milk and Food Technology* 36: 593–609.

Martley FG and Crow VL (1996) Open texture in cheese: the contributions of gas production by microorganisms and cheese manufacturing practices. *Journal of Dairy Research* 63: 489–507.

Mocquot G (1979) Swiss-type cheese. *Journal of Dairy Research* 46: 133–160.

Robinson RK (ed.) (1995) *A Colour Guide to Cheese and Fermented Milks.* London: Chapman & Hall.

Wood HG (1989) Metabolic cycles in the fermentation by propionic acid bacteria. *Current Topics in Cellular Regulation* 18: 255–287.

Microflora of White-brined Cheeses

Barbaros H Özer, The University of Harran, Faculty of Agriculture, Department of Food Science and Technology, Sanliurfa, Turkey

Introduction

The manufacture of pickled cheeses dates back thousands of years. Pickled cheeses are classified as soft or semihard cheeses, which are whitish in colour and preserved and stored in brine. They are particularly popular in the Balkans, the Middle East, the Mediterranean region and eastern Europe. They are well-suited to both the climatic and the economic conditions of these regions, but the processing methods vary from one region to another. The nomenclature and descriptions of some pickled cheese varieties are given in **Table 1**.

For centuries, the production of pickled cheeses was limited to small-scale units, which made standardization of the properties and composition of these varieties difficult. However, the introduction of new technologies such as membrane processing has made the large-scale production of brined cheeses possible and, in recent years, pickled cheeses have been gaining in popularity in the international dairy market.

Sheep's milk is preferred in the manufacture of pickled cheese because it gives a white colour, but blends of milk are also used. The main objection to goat's milk in the manufacture of white-brined cheeses is that it produces a hard, dry cheese which is atypical of this type of cheese. Cow's milk is not ideal because it gives a yellowish colour and a characteristic 'cowy' odour to the cheese.

The pickled white cheeses of most industrial importance are Feta, Halloumi and Teleme, which are ripened in brine, and Domiati, which is produced by the coagulation of previously-salted milk. The technology and microbiology of the first three of these varieties are discussed in this article.

Technology of White-brined Cheeses

Feta Cheese

Feta cheese is traditionally produced from sheep's milk. Mixtures of sheep's and goat's milks also give Feta cheese of good quality, as long as the proportion of goat's milk is <20%. Traditionally, the rennet for Feta cheese-making was produced by the cheese makers themselves from the abomasa of lambs and kids slaughtered before weaning, but at present traditional rennets and commercial coagulants are being used together in the ratio 1:3 (traditional: commercial). Pasteurized milk is used in the manufacture of traditional Feta cheese. Yoghurt starters (blends of *Streptococcus thermophilus* and *Lactobacillus delbrueckii* subsp. *bulgaricus*) are traditionally preferred by cheese makers, especially in Greece.

The initial acidity of the milk for Feta cheese should be <0.25% lactic acid, or the pH should be >6.5. Before the addition of the rennet, the milk is standardized to give a casein-to-fat ratio of 1:1 and then either pasteurized at 72°C for 15 s (HTST) or heat-treated at 65–68°C for 30 min (Holder method). Calcium chloride (0.02% v/v) is used with heat-treated milk to restore the calcium balance after heating. Coagulation is achieved at 32°C in 45–50 min. Lactic starters are usually added at 32–34°C, 15–30 min before rennetting.

When coagulation is complete, the curd is cut into cubes (2–3 cm^3) and, in order to increase cohesion, it is left to stand in the whey for about 5–10 min. Afterwards, the cubes are transferred into a mould and drainage of the excess whey takes place. During drainage, at 14–16°C, the moulds are turned upside down every 2–3 h. When the desired amount of whey has been removed, the fresh cheese is taken out of the mould and cut into blocks. The blocks are dry-salted by sprinkling granular salt onto all their surfaces, and this process is repeated three to four times over several days. The salted blocks are left on the cheese table for about 10–15 days, to allow the formation of a thick skin on the exposed surfaces. Then the cheese blocks are washed with water or brine and placed in a barrel. Brine solution containing 6–8% NaCl is added and the barrels are left for maturation at 14–16°C, until a pH of 4.4–4.6 is reached. The final salt content of the cheese is about 3.0%.

The technology of the manufacture of Feta cheese as described above is the traditional one. Over the past few decades, many advances have taken place in the cheese industry. Mechanization and automation in cheese-making have enabled much progress in the marketing of Feta-type cheeses at an international

Table 1 Nomenclature and description of some white-brined cheese varieties

Name	Country of origin	Milk used for manufacture	Comments
Akawi	Lebanon, Syria, the Czech Republic	Cow's, sheep's and goat's	Fresh white cheese, kept in brine for a few days
Beli Sir U Kirskama	Former Yugoslavia	Cow's, sheep's or mixture	White cheese, kept in brine for 10–20 h then ripened in vacuum-packed foil
Beyaz peynir (Salamura, Edirne peyniri)	Turkey	Cow's, sheep's or mixture (pasteurized)	Curd is pressed, cheese blocks are kept in brine overnight and then stored in tins with the brine. It is ripened for about 6 months before consumption.
Bjalo Salamureno Sirene	Bulgaria	Sheep's	Semihard; salty and acidic flavour. Mixture of *Lactobacillus lactis* subsp. *lactis* and *L. casei* is used as starter.
Brinza	Israel, Russia, the Czech Republic	Cow's, sheep's, goat's or mixture	Average composition of sheep's milk cheese is moisture 60%, fat 20%, protein 14%, ash 2.3% and salt 3.5–3.7%
Feta	Greece, Denmark, UK	Cow's, sheep's, goat's or mixture	
Halloumi	Cyprus, Lebanon	Sheep's	
Domiati	Egypt	Cow's, buffalo's	Previously-salted milk is rennetted and then processed into cheese.
Kefalotyri	Greece, Syria	Goat's, sheep's	Hard cheese; high dry-matter content; firm texture with small gas holes; strong, piquant and salty flavour
Teleme	Greece, Turkey, Bulgaria	Goat's, sheep's, cow's, buffalo's or mixture	

level. Modern technologies have replaced old-fashioned techniques in the processing of milk into cheese, involving ripening, coagulation, cutting the coagulum and drainage. Details of the industrial production of structured Feta cheese are summarized in **Figure 1**.

Halloumi Cheese

Halloumi is a firm pickled cheese usually made from sheep's milk or sheep's milk mixed with goat's milk, to give a blend containing 5.2–5.3% fat. However, in recent years the use of cow's milk in the manufacture of Halloumi has become more popular, being more abundant and relatively cheaper than sheep's or goat's milk. The compositional differences between cow's milk and sheep's milk make modifications to the traditional cheese-making procedure inevitable.

In the traditional Halloumi-making process no heat treatment is applied, but in larger-scale operations, a low temperature short time heat treatment can be applied. Starter culture is not used in this product. Coagulation is achieved by means of rennet, added at a sufficient level to give a firm curd in 30–45 min at 30–34°C. After the coagulum has set, the curd is cut into cubes (1–2 cm³). This stage is particularly important, in order to ensure the sliceable texture of the Halloumi cheese. The temperature of the curd is then raised to 38–42°C, while it is being constantly stirred. Once the desired temperature has been achieved, the curd is stirred for a further 10 min to

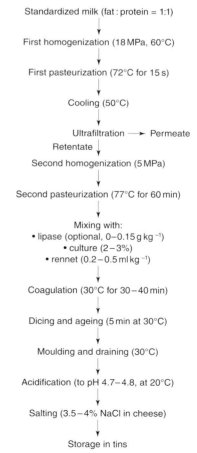

Standardized milk (fat : protein = 1:1)
↓
First homogenization (18 MPa, 60°C)
↓
First pasteurization (72°C for 15 s)
↓
Cooling (50°C)
↓
Ultrafiltration → Permeate
Retentate ↓
Second homogenization (5 MPa)
↓
Second pasteurization (77°C for 60 min)
↓
Mixing with:
• lipase (optional, 0–0.15 g kg⁻¹)
• culture (2–3%)
• rennet (0.2–0.5 ml kg⁻¹)
↓
Coagulation (30°C for 30–40 min)
↓
Dicing and ageing (5 min at 30°C)
↓
Moulding and draining (30°C)
↓
Acidification (to pH 4.7–4.8, at 20°C)
↓
Salting (3.5–4% NaCl in cheese)
↓
Storage in tins

Figure 1 Industrial production of structured Feta cheese.

increase its firmness, and then allowed to settle. Afterwards, the curd is compressed to remove whey, and the compressed curd is cut into large blocks (approx. 10 kg). Further drainage is achieved by pressing. The curd is then cut into small blocks (approx. 10 × 15 × 5 cm³) and these are cooked in the boiling whey (collected during the pressing of the curd) for about 40–80 min. Afterwards, the curd is cooled to a suitable temperature (generally ambient temperature) for salting. Dry salt (5%) is sprinkled onto the cheese blocks, and the blocks are left for some 24 h. Salt is then added with pieces of dried mint leaves (*Mentha viridis*), to give the finished cheese attractive appearance and to improve the flavour. Finally, the Halloumi cheese is stored in brine for ripening. The method of making Halloumi cheese from sheep's milk is outlined in **Figure 2**.

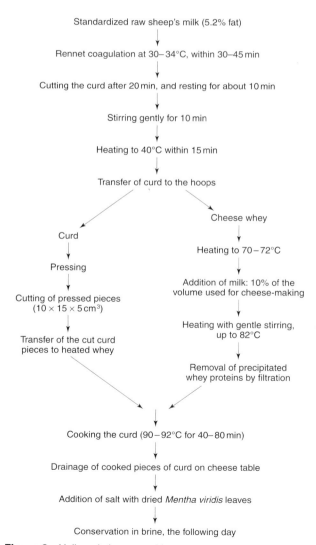

Standardized raw sheep's milk (5.2% fat)

Rennet coagulation at 30–34°C, within 30–45 min

Cutting the curd after 20 min, and resting for about 10 min

Stirring gently for 10 min

Heating to 40°C within 15 min

Transfer of curd to the hoops

Curd   Cheese whey

Pressing

Cutting of pressed pieces (10 × 15 × 5 cm³)

Transfer of the cut curd pieces to heated whey

Heating to 70–72°C

Addition of milk: 10% of the volume used for cheese-making

Heating with gentle stirring, up to 82°C

Removal of precipitated whey proteins by filtration

Cooking the curd (90–92°C for 40–80 min)

Drainage of cooked pieces of curd on cheese table

Addition of salt with dried *Mentha viridis* leaves

Conservation in brine, the following day

Figure 2   Halloumi cheese-making.

Teleme Cheese

Teleme cheese is a representative variety of soft white cheese made from cow, sheep, buffalo or goat's milk or a mixture of two or more of these. It is ripened and kept in brine until delivered to the consumer. Apart from marginal differences between the methods of curd drainage and salting of Feta and Teleme cheeses, they resemble each other and Teleme is recognized as a Feta-type cheese. In making Teleme cheese, the curd is drained by the application of pressure. Salt is not sprinkled onto the surface of the cheese but, instead, the cheese blocks are soaked in dense brine for some hours.

In the manufacture of Teleme cheese, the milk is pasteurized at 72°C for 15 s, and then cooled to a suitable temperature (32°C) for the addition of the starter culture (usually a yoghurt starter). Alternatively, a pure culture of *Lactococcus lactis* subsp. *lactis* and *Lactobacillus delbrueckii* subsp. *bulgaricus* at a ratio of 1:3 and a dosage rate of 0.5% v/v produces good results. Following the addition of calcium chloride (0.025%), chlorophyll (for decoloration) and, if wished, lipase, the milk is rennetted at 32°C. After about an hour, the coagulum is cut into cubes (1–2 cm³) and transferred to moulds lined with cheesecloth, which facilitates drainage. After drainage is complete, the curd is cut into blocks which are placed in a dense brine solution (18% brine, 15–16°C, 20 h). Following removal from the brine, the pH of the cheese blocks is allowed to drop to 4.8 at 16–18°C in ripening rooms and granular salt is sprinkled onto the surface of the cheese. When the pH has reached 4.8, the cheese blocks are placed in a tin, covered with brine (7–8% NaCl) and stored in a cold room (3–4°C) for about 1–2 months (**Fig. 3**).

Starter Cultures used in the Manufacture of White-brined Cheeses

The production of white-brined cheese depends on the fermentation of lactose by lactic acid bacteria, to form primarily lactic acid. The selection, maintenance and use of starter cultures are perhaps the most important aspects of cheese-making, particularly in the context of a modern, mechanized process for which predictability and consistency are essential. The use of starter culture leads to: inhibition of the growth of undesirable bacteria; the regulation of coagulum formation, by aiding the action of rennet; improved drainage of whey; the production of characteristic flavours and aromas; and the controlled development of acid.

The fact that white-brined cheese is often traditionally produced without any starter at all is one of the causes of its frequently indifferent quality. Many

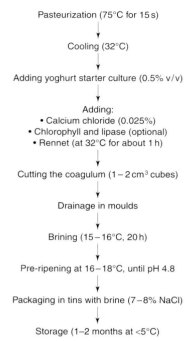

Pasteurization (75°C for 15 s)

↓

Cooling (32°C)

↓

Adding yoghurt starter culture (0.5% v/v)

↓

Adding:
• Calcium chloride (0.025%)
• Chlorophyll and lipase (optional)
• Rennet (at 32°C for about 1 h)

↓

Cutting the coagulum (1–2 cm³ cubes)

↓

Drainage in moulds

↓

Brining (15–16°C, 20 h)

↓

Pre-ripening at 16–18°C, until pH 4.8

↓

Packaging in tins with brine (7–8% NaCl)

↓

Storage (1–2 months at <5°C)

Figure 3 Teleme cheese-making.

Table 2 Combinations of starter cultures used in the production of white brined cheeses

Starter cultures and ratios	Type of cheese
Yoghurt starters	Feta
Lactococcus lactis subsp *lactis* + *Lactobacillus casei* + *Leuconostoc mesenteroides* subsp. *cremoris* (3 : 1 : 1)	Feta
Lactobacillus lactis subsp. *lactis* + *Lactobacillus casei* + *Enterococcus durans* + *Leuconostoc mesenteroides* subsp. *cremoris* (6 : 2 : 1 : 1)	Feta
Lactococcus lactis subsp. *lactis* + *Lactobacillus casei* + *Enterococcus durans* (6 : 2 : 2)	Feta
Lactococcus lactis subsp *lactis* + *Lactobacillus cremoris* (1 : 1)	Feta
Lactobacillus delbrueckii subsp. *bulgaricus* + *Lactococcus lactis* subsp. *lactis* (3 : 1)	
Yoghurt (shock culture)	Feta
Lactococcus lactis subsp. *lactis* + *Lactobacillus casei* (3 : 1)	White-brined cheeses
Lactococcus lactis subspp. *lactis* + *cremoris* + *lactis* biovar *diacetylactis* (1 : 1 : 2)	White-brined cheeses
Lactobacillus helveticum + *Streptococcus* spp.	White-brined cheeses
Lactococcus lactis subsp. *cremoris* + *Enterococcus durans*	Feta
Lactobacillus delbrueckii subsp. *bulgaricus* + *Lactococcus lactis* subsp. *lactis*	Feta
Streptococcus thermophilus + *Lactococcus lactis* subsp. *lactis* + *Lactobacillus casei* + *Lactobacillus delbrueckii* subsp. *bulgaricus*	Feta
Pediococcus pentosaceus	Teleme
Lactococcus lactis subsp. *lactis* + *Lactobacillus delbrueckii* subsp. *bulgaricus* (1 : 3)	Teleme
Enterococcus faecalis + *Leuconostoc* spp. + *Lactobacillus plantarum*	Osetinskii cheese
Lactococcus lactis subspp. *lactis* + *lactis* biovar *diacetylactis* + *Streptococcus paracitrovorous*	Imeretinskii cheese
Kefir grains (*Lactococcus lactis* subsp. *lactis* + *Lactobacillus caucasicus*)	White-brined cheeses
Enterococcus faecalis biovars *liquefaciens* + *zymogenes* (1 : 9)	Unripened white cheeses
Homofermentative lactic acid bacteria	Brinza cheese
Lactococcus lactis subsp. *cremoris*	Halloumi cheese
Lactococcus lactis subsp. *lactis* + *Lactobacillus casei* + *Lactobacillus plantarum*	Turkish white cheese
Lactococcus lactis subspp. *lactis* + *lactis* biovar *diacetylactis* + *Lactobacillus casei*	Turkish white cheese

species of starter microorganisms have been used in the manufacture of white-brined cheeses, and many combinations have been tested, to investigate the best match of composition and properties with the characteristics of specific varieties of cheese. As mentioned earlier, with the exception of a few varieties, white-brined cheeses have a limited market and so in varieties other than these, either starter culture is not used at all or no research has been carried out. The combinations of starter cultures used in the production of the major white-brined cheese varieties are listed in **Table 2**.

To obtain cheese with a good consistency and flavour, the selection of the starter culture is of particular importance. For example, different varieties of cheese may demand a high tolerance of salt, proteolytic activity and/or acid-forming capacity. Some strains of *Lactococcus lactis* subspp. *lactis*, *cremoris* and *lactis* biovar *diacetylactis* are salt-tolerant, and the thermophilic starter culture comprising *Lactobacillus delbrueckii* subsp. *bulgaricus* and *Streptococcus thermophilus* may show a degree of salt resistance.

A wide range of inoculation rates for starter cultures has been proposed, depending on the type of starter culture used: 0.1–0.2% of a mixture of *Lactococcus lactis* subsp. *lactis* and *Lactobacillus casei* is satisfactory for Feta cheese, while an inoculation rate of 1–2% of *Streptococcus thermophilus* and *Lactobacillus delbrueckii* subsp. *bulgaricus* is optimal for Feta cheese-making using a thermophilic culture.

A new approach is the use of hydrolysed starters (starter with added rennet powder) instead of the normal culture, to accelerate the ripening time of the cheese by up to 50%. The suggested inoculation rate for a hydrolysed starter is 0.5–0.6% for white-brined cheeses. In the manufacture of Teleme cheese, a combination of starter bacteria and yeasts such as *Pichia membranaefaciens*, *Hansenula anomala*, *Debaryo-*

myces hansenii and *Kluyveromyces marxianus var. lactus* is recommended, in order to improve the aroma and flavour.

Composition of Cheese in Relation to Starter Activity

Acid Development and Proteolysis A crumbly body is required in Feta cheese, and so a strongly acid-producing starter culture should be used. A starter culture with a 1:1 ratio of streptococci and lactobacilli (1% v/v) is able to convert enough lactose to lactic acid to create a crumbly body in Feta cheese. In general, salt-tolerant strains (tolerant to 6–7% NaCl) have a high proteolytic activity and acid-forming capacity. Hydrolysed cultures containing rennet powder cause rapid proteolysis, but the soluble N_2 level depends on the culture used. Yoghurt cultures are inhibited at refrigeration temperature, so fresh cheeses made with yoghurt cultures have higher pH values than those made with mesophilic cultures, but the pH values of cheeses made with mesophilic cultures do increase during storage for up to 6 months.

The highest ripening coefficient in Teleme cheese was obtained when a mixed starter culture of *Lactococcus lactis* subsp. *lactis* and *Lactobacillus casei* subsp. *casei* was used. The next highest was obtained with *Lactococcus lactis* subsp. *lactis* + *Streptococcus thermophilus*, or with *Lactococcus lactis* subspp. *lactis* + *cremoris*. The yield of cheese is normally not influenced by the type of starter culture used.

Texture and Aroma The white-brined cheeses made with thermophilic starters suffer from a lack of characteristic aroma and flavour. Textural problems may also be pronounced: a fragile structure and a bitter taste are sometimes quoted as the main drawbacks to using yoghurt cultures. However, the combination of yoghurt culture with *Enterococcus durans* eliminates these problems to a great extent and yields a cheese with a firm but spreadable consistency and a pronounced aroma. It should be noted, however, that the use of *Enterococcus* species in starter cultures is not permitted in many countries. The use of salt-tolerant starters in the production of white-brined cheese, together with ripening in 18–20% brine, produces cheeses with an elastic texture. The combination of *Lactococcus lactis* subspp. *cremoris* and *lactis* is agreed to be the best combination of starters as far as flavour and aroma are concerned.

Non-starter Organisms in White-brined Cheeses

Two aspects of the microbiology of brined cheeses should be considered: the growth of pathogens, and the prevention of microbial deterioration of the product during storage over its anticipated shelf life.

The microbiological quality of cheese is closely related to the method of manufacture and, as unpasteurized milk is widely used in the manufacture of white-brined cheeses, the initial microbiological load of the milk determines the quality of the final product. On the positive side, non-starter bacteria may contribute to the formation of flavour in cheese and, by affecting its levels of soluble N_2 and peptides, and the pH of pickled cheese, they can alter the sensory and textural properties of cheeses. On the other hand, non-starter bacteria can cause spoilage.

In Feta cheese, the total counts of aerobes, lactic acid bacteria and proteolytic bacteria reach maximum levels at approximately 5, 20 and 45 days respectively, and a decrease is observed at 3 or 4 months. Thereafter, the number of these microorganisms remains unchanged. Enterococci (*Enterococcus faecalis*, *E. faecium*, *E. durans*), *Leuconostoc* species (*L. mesenteroides* subsp. *dextranicum*) and lactobacilli (*Lactobacillus plantarum*, *L. casei*, *L. brevis*) are the predominant genera in fresh Feta cheese, the salt-resistant enterococci becoming more dominant later.

Psychrotrophs tend to increase during the first few weeks of storage, and then their numbers fluctuate. *Pseudomonas*, *Aeromonas* and *Acinetobacter* are among the genera of psychrotrophs most frequently found in white-brined cheeses. Coliforms are often present in high numbers during the early stages of ripening, due to the use of unpasteurized milk and/or poor sanitary conditions during cheese-making. However, coliforms are soon reduced to negligible levels under the usual conditions for the ripening and storage of white-brined cheese.

Certain acute human diseases and conditions (e.g. gastroenteritis, encephalitis, meningitis, abortion) are associated with foods contaminated with *Yersinia enterocolitica* and *Listeria monocytogenes*. Dairy products prepared from pasteurized milk should ordinarily not pose a health problem, because the pasteurization process should be adequate to destroy both pathogens. The survival of *Y. enterocolitica* in cheese is dependent on the rate of development of acid and the final pH of the product: if acid production is slow and the final pH is > 4.5, *Y. enterocolitica* can survive for up to 30 days, but with rapid acid development and low pH it can survive for only 4 days. *L. monocytogenes* is more acid-tolerant than *Y. enterocolitica*, and can remain active in pickled cheese for up to 90 days in pH 4.3 and a salt concentration of 6%. *Staphylococcus aureus* is another pathogen which can survive in brined cheese, especially in the presence of yeasts: even at low pH and high salt concentrations, mutual stimulation between yeasts and *S. aureus* is evident.

Micrococci that can cause spoilage include *Micrococcus cascolyticus*, *M. freudenreichii* and *M. varians*. These can grow actively during the production of cheese, but the sensitivity of the group to acid and salt means that their overall number declines during ripening.

Yeasts are present at low levels in brined cheese, and may have an important role in the formation of flavour. They can also enhance proteolysis and, as mentioned earlier, they are recommended for inclusion in the starter culture for the manufacture of Teleme cheese.

The growth of moulds on white-brined cheese is quite common. Unless they are capable of producing mycotoxins, they do not carry any potential health risk for humans, but the aroma, flavour and appearance of the cheese may be changed. The genera *Penicillium*, *Mucor*, *Aspergillus*, *Cladosporium* and *Fusarium* have been isolated from Teleme and Feta cheeses, and there is concern that some species, including *P. cyclopium*, *P. viridicatum*, *A. flavus* and *A. ochraceus*, are able to produce mycotoxins. In addition, aflatoxins may pass into cheese from brine, and penetrate it as deeply as 15–20 mm from the surface. Therefore washing the surface of cheese may not be sufficient to remove aflatoxins. However, aflatoxin production depends on the storage temperature, and at temperatures of 5–10°C, it is synthesized at only low levels.

Effects of Brine on Microorganisms

The main functions of NaCl are: to control the total solids content of cheese, by syneresis; to give cheese a characteristic flavour; and to reduce the water activity, a_w, of cheese. The inhibitory effect of NaCl on the activity of starter bacteria is species- and strain-specific, and is undoubtedly due to its influence on a_w. The effect of NaCl on the lactic acid bacteria results in a reduction in both their multiplication and biological activity. A good example of the latter effect is the observation that increased salt concentrations cause a decrease in the production of diacetyl, acetoin and CO_2 by *Lactococcus lactis* subsp. *diacetylactis* and by *Leuconostoc* species.

Defects in White-brined Cheeses

In general, high concentrations of brine affect the texture of cheese, causing softness, but a high salt content also inhibits the growth and activity of starter cultures. At the same time, acid-tolerant halophilic microorganisms may find suitable conditions for

growth and, as a result of their metabolic activity in the cheese, the blowing of tins may be observed.

'Early blowing' is one of the major problems in the production of white-brined cheeses, with coliforms and/or yeasts being primarily responsible. *Klebsiella aerogenes* and *Aerobacter aerogenes*, which are salt-tolerant, are both able to produce gas and cause holes in the cheese. 'Late blowing' is caused by *Clostridium tyrobutyricum* and *C. butyricum*, but is not normally observed in white cheeses due to the inhibitory effect of salt on butyric acid bacteria, as long as the salt concentration in the brine is adequate.

A slimy brine is sometimes observed during the storage of white-pickled cheeses, and this is caused by ropy strains of *Lactobacillus plantarum* and/or *L. casei*. These bacteria are inhibited at low pH (< 4.0) and high salt content (> 8%).

See also: **Cheese**: In the Market-place; Microbiology of Cheese-making and Maturation. **Clostridium**: *Clostridium tyrobutyricum*. **Debaromyces**. **Enterococcus**. **Fermented Milks**: Yoghurt. **Hansenula**. **Klebsiella**. **Lactobacillus**: Introduction; *Lactobacillus bulgaricus*. *Lactobacillus casei*. **Lactococcus**: Introduction; *Lactococcus lactis* Sub-species *lactis* and *cremoris*. **Listeria**: Introduction; *Listeria monocytogenes*. **Micrococcus**. **Milk and Milk Products**: Microbiology of Liquid Milk. **Pichia pastoris**. **Preservatives**: Traditional Preservatives – Sodium Chloride. **Saccharomyces**: Introduction. **Starter Cultures**: Uses in the Food Industry; Importance of Selected Genera; Cultures Employed in Cheese-making. **Streptococcus**: Introduction; *Streptococcus thermophilus*. **Yeasts**: Production and Commercial Uses. **Yersinia**: Introduction; *Yersinia enterocolitica*.

Further Reading

Abd El-Salam MH (1987) Domiati and Feta Type Cheeses. In: Fox PF (ed.) *Cheese: Chemistry, Physics and Microbiology*. Vol. 2, pp. 46–49. UK: Elsevier Applied Science.

Anifantakis EM (1991) Traditional Feta Cheese. In: Robinson RK and Tamime AY (eds) *Feta and Related Cheeses*, pp. 276–309. London: Ellis Harwood.

Anifantakis EM (1991) *Greek Cheeses. A Tradition of Centuries* p. 95. Athens: National Dairy Committee of Greece.

Caric M (1987) Mediterranean Cheese Varieties: Ripened Cheese Varieties Native to Balkan Countries. In: Fox PF (ed.) *Cheese: Chemistry, Physics and Microbiology*. Vol. 2, pp. 257–275. UK: Elsevier Applied Science.

Haddadin MYS (1986) Microbiology of White-Brined Cheeses. In: Robinson RK (ed.) *Development of Food Microbiology*. Vol. 2, pp. 67–89. UK: Elsevier Applied Science.

Kehagias C, Koulouris S, Samona A, Malliou S and Koumoutsos G (1995) Effect of various starters on the quality of cheese in brine. *Food Microbiology* 12: 413–419.

Nunez M, Medina M and Gaya P (1989) Ewe's milk cheese: technology, microbiology and chemistry. A review. *Journal of Dairy Research* 56: 303–321.

Robinson RK (1991) Halloumi Cheese – the Product and its Manufacture. In: Robinson RK and Tamime AY (eds) *Feta and Related Cheeses*, pp. 144–159. London: Ellis Harwood.

Tamime AY and Kirkegaard J (1991) Manufacture of Feta Cheese In: Robinson RK and Tamime AY (eds) *Feta and Related Cheeses*, pp. 70–143. London: Ellis Harwood.

Tzanetakis U, Vafapoulou A and Tzanetaki E (1989) The quality of white-brined cheeses from goat's milk made with different starters. *Food Microbiology* 12: 55–63.

CHEMILUMINESCENT DNA HYBRIDIZATION see *LISTERIA*: *Listeria monocytogenes* – Detection by Chemiluminescent DNA Hybridization.

CHILLED STORAGE OF FOODS

Contents

Principles
Use of Modified-atmosphere Packaging
Packaging with Antimicrobial Properties

Principles

Brian P F Day, Campden & Chorleywood Food Research Association, Chipping Campden, Gloucestershire, UK

Introduction

Foods are chilled to retard or prevent both perceptible deterioration and the growth of microorganisms: it was recognized as early as 1936 that chilled storage greatly retarded or prevented the growth of food-borne pathogens. The following definition of chilled foods, issued by the Institute of Food Science and Technology (UK), is widely accepted: 'Perishable foods, which to extend the time during which they remain wholesome, are kept within specified ranges of temperatures above −1°C and below +8°C.' In recent years, chilled foods have become increasingly popular and major retailers have devoted more shelf space to them, often at the expense of frozen and ambient-stable foods.

Many factors have contributed to the commercial success of chilled foods in the modern marketplace:

- Consumers perceive chilled foods as fresh and wholesome, and the nearest alternative to home-made versions which could be prepared from fresh ingredients. A perception of healthy eating is not usually associated with heat-processed, ambient-stable foods.
- The transparent packaging of chilled foods encourages consumers to view them as fresh and ready-to-eat, e.g. bagged mixed salads, sandwiches, cream cakes, pastries.
- Consumer demand for convenience: many chilled foods are ready-to-eat or can be heated in their containers in microwave ovens, much more quickly than their frozen equivalents.
- Retailers have embraced chilled foods, because they enhance their image and provide opportunities for the development of a wide range of new added-value products.

The success of chilled foods, particularly in the UK, has been facilitated by efficient chilled distribution and retail display systems. An extremely wide range of food types is chilled. Traditionally this has included fresh meats, fish, poultry and dairy products, but it now also includes cooked, cured and processed meats, fish and poultry; ready meals; sandwiches and other combination products; fresh pasta products; cooked and dressed vegetable products; and fresh, prepared fruit and vegetables.

Microfloras

All chilled foods are likely to contain a wide variety of microorganisms, present either as part of the normal food microflora or due to the contamination of the food during processing or handling. The spoilage of chilled foods is a complex phenomenon involving physical, chemical, biochemical and microbiological changes. This article is concerned only with microbiological changes. Chilled foods may also be considered microbiologically unacceptable due to pathogenic microorganisms. Spoilage microorganisms can cause an unacceptable change in the characteristics of a food product, whereas the growth of pathogens can render it unsafe to eat. A number of pathogenic bacteria are capable of causing food-borne illness. These can be categorized as either infective organisms, which invade the body (e.g. *Salmonella* species, *Yersinia enterocolitica*, *Listeria monocytogenes*, *Clostridium perfringens*, *Aeromonas hydrophila*, *Escherichia coli* O157:H7 and *Campylobacter* species), or organisms which produce a toxin within the food (e.g. *Clostridium botulinum*, *Bacillus cereus* and *Staphylococcus aureus*). Chilled foods may appear fit to eat despite the presence of large numbers of pathogens or their toxins. Hence a lack of visible spoilage cannot be used as an indicator of food safety. In food manufacture, food safety must take precedence over matters relating to other aspects of quality.

Effect of Temperature

The effect of the temperature of chilled foods on microorganisms is either to totally inhibit growth, or to increase the lag phase and reduce subsequent growth. As the minimum growth temperature of each type of microorganism is approached, the microorganisms often become more susceptible to the effects of other preservative features of the food, such as acidity (pH) and water activity (a_w). Although microbial growth does not occur at temperatures below the minimum for growth, the microorganisms may nevertheless survive, and grow if the temperature rises. Appropriate temperatures must therefore be maintained throughout the shelf life of the food product.

Table 1 shows the minimum growth conditions of selected microorganisms associated with chilled foods. Many pathogenic bacteria are unable to grow at temperatures < 4°C. These include the most common food poisoning bacteria: *Salmonella*, *Campylobacter* species, *Escherichia coli* O157:H7, *Clostridium perfringens* and *Staphylococcus aureus*. However, some psychrotrophic pathogenic bacteria such as *Listeria monocytogenes*, *Yersinia enterocolitica* and *Aeromonas hydrophila* are capable of growth at temperatures as low as –2°C. Also some spore-forming and toxin-producing pathogens, including strains of *Clostridium botulinum* and *Bacillus cereus*, are capable of growth at 3–5°C.

Spoilage

The microbiological spoilage of chilled foods depends on the types of microorganism present, the nature of the food substrate and the temperature. Different spoilage microorganisms utilize different substrates,

and also different metabolic pathways, which can often be switched in response to the available nutrients and conditions. Overt food spoilage occurs when the perceptible characteristics of a product, such as visual appearance, flavour, odour and texture, become unacceptable to the consumer. Compounds causing off odours and flavours are mainly volatile, and are often produced by the microbial metabolism of protein. Souring, another frequently encountered type of spoilage, is caused by acid-producing microorganisms. Textural changes can be caused by either the presence of certain microorganisms themselves, their products (e.g. polysaccharides causing ropiness in milk and dairy products), or the effects of their by-products (e.g. enzymes causing soft rot of fruit and vegetables and gas causing holes in hard cheese). Visual spoilage of microbial origin can take a variety of forms, such as discoloration, pigmentation, gas disruption, surface growth and cloudiness. **Table 2** lists some of the microbiological spoilage problems associated with chilled foods.

Safety

Certain general principles may be applied to the microbial safety of chilled foods:

- The microbiological status of all raw materials should be known, and only materials of high quality should be used.
- All stages of processing should be defined, monitored and controlled.
- The temperatures and times of chilled storage, transport and display should be controlled throughout, including raw materials, and preferably should extend beyond to the home.
- Attention must be given to hygiene throughout the entire process, to ensure that microbial contamination is minimized.

These objectives may be achieved through the application of a quality assurance scheme, for example using hazard analysis critical control points (HACCP). The better education of all involved in chilled food manufacture, distribution and retail sale, and improved consumer education in hygiene and temperature control, would be of great benefit in maintaining the quality and assuring the safety of chilled foods.

Legislative Controls

In response to the expansion of the market for chilled food, legislation and associated controls have been introduced in order to minimize the potential threats to public health posed by microbial growth. Legislation specifically directed at chilled foods imposes

Table 1 Minimum growth conditions of selected spoilage and food poisoning microorganisms. (NB Growth and survival limits assume other factors to be optimal. Interactions between factors are likely to alter these values considerably.)

Type of microorganism	Minimum pH for growth	Minimum a_w for growth	Anaerobic growth	Minimum growth temperature (to nearest °C)
Aeromonas hydrophila	4.0	0.98*	Yes	0
Bacillus cereus	4.4	0.91	Yes	4
Clostridium botulinum (proteolytic strains, types A, B and F)	4.8	0.94	Yes	10
Clostridium botulinum (non-proteolytic type E strain)	4.8	0.97	Yes	3
Clostridium botulinum (non-proteolytic strains, types B and F)	4.6	0.94	Yes	3
Clostridium perfringens	5.5	0.93	Yes	5
Enterobacter aerogenes	4.4	0.94	Yes	2
Escherichia coli	4.4	0.9	Yes	4
Lactobacillus species	3.8	0.94	Yes	4
Listeria monocytogenes	4.4	0.92	Yes	0
Micrococcus species	5.6	0.9	No	4
Moulds	< 2.0	0.6	No	− 10
Pseudomonas species	5.5	0.97	No	−4
Salmonella	3.8	0.92	Yes	4
Staphylococcus aureus	4.0	0.83	Yes	8
Vibrio parahaemolyticus	4.8	0.94	Yes	5
Yeasts	1–5.0	0.8	Yes	−5
Yersinia enterocolitica	4.5	0.96*	Yes	−2

* A_w value calculated from % salt concentration.

Table 2 Examples of microbiological spoilage problems associated with chilled foods.

Problem	Cause	Food	Microorganism
Off odours and flavours			
Nitrogenous	Hydrogen sulphide, ammonia, trimethylamine	Fresh meat and dairy products Heat-processed meat	Pseudomonas spp. Acinetobacter spp. Moraxella spp. Clostridium spp.
Souring	Acids: lactic, acetic, butyric	Meat, dairy products, fruits and vegetables	Lactobacillus spp. Streptococcus spp. Brochothrix thermosphacta Bacillus spp. Coliforms
Earthy	Geosmin	Mushrooms, water, meat	Actinomycetes Cyanobacteria Streptomycetes
Fruity	Lipase or esterase actvity	Milk	Pseudomonas fragi
Potato	2-methoxy-3-isopropyl pyrazine	Surface-ripened cheese	Pseudomonas spp.
Textural spoilage			
Slime	Polysaccharides	Meats	Pseudomonas fragi
Ropiness	Polysaccharides	Milk	Alcaligenes spp.
Bittiness	Polysaccharides	Milk	Bacillus cereus
Rotting	Enzyme activity: pectinases, cellulases, xylanases	Fruits and vegetables	Erwinia spp. Clostridium spp. Yeasts Moulds
Visual spoilage			
Bubbles	Gas formation	Cottage cheese, coleslaw	Coliforms
Holes	Gas formation	Hard cheese	Coliforms
Discoloration	Fluorescent pigment Black pigment Surface growth	Meat, eggs Dairy products Potato salad	Pseudomonas spp. Pseudomonas nigrifaciens Pichia membranaefaciens

temperature controls which are stricter than those applicable to lower-risk frozen and ambient-stable foods.

General Food Hygiene

In the UK, chilled foods must comply with the Food Safety (General Food Hygiene) Regulations, SI 1995 No. 1763. These regulations implement European Community Council Directive 93/43/EEC on the hygiene of foodstuffs, excluding the requirements which relate to temperature control (see below). The regulations apply at all stages of food production (except primary production), but do not apply to those products of animal origin covered by specific hygiene regulations. In addition to requiring all aspects of food handling and production to be carried out in a hygienic manner, the regulations place a number of obligations on food businesses, as follows:

- To identify steps in the activities of the food business which are critical in terms of food safety, and to ensure that controls are in place, maintained and reviewed (e.g. by using an HACCP approach). (However, the regulations do not require a documented hazard analysis system.)
- To meet requirements with respect to rooms where foodstuffs are prepared, treated or processed.
- To use potable water wherever necessary, to ensure that foodstuffs are not contaminated.
- To prohibit anyone known or suspected to be suffering from, or to be a carrier of, a disease likely to be transmitted through food, from working in any food handling area where there is a possibility of contaminating the food with pathogenic microorganisms.
- To ensure that all food handlers are supervised and instructed and/or trained in food hygiene matters, appropriate to their work activities.

These regulations introduced the new concept of voluntary industry guidelines to good hygiene practice. These provide detailed guidance on complying with the regulations, and relate to specific sectors, e.g. catering, vending. The guidelines have promoted cooperation between the food industry and enforcers, and a greater uniformity in compliance with the regulations.

Temperature

In the UK, chilled foods must also comply with the Food Safety (Temperature Control) Regulations, SI 1995 No. 2200. For England and Wales, these regulations require foods to be stored at or below 8°C if likely to support the growth of pathogenic microorganisms or the production of toxins, although there is flexibility allowing certain food businesses to

increase the temperature, if this is justified on the basis of a scientific safety assessment. In Scotland, however, chilled food must be kept in a refrigerator, refrigerating chamber or a cool ventilated place, subject to exemptions. The regulations also contain a hot-holding requirement. For England and Wales, food which has been cooked or reheated, and is for service or on display for sale, and must be kept hot to control the growth of microorganisms or the formation of toxins, must not be stored at a temperature < 63°C, unless justified by a scientific assessment. Scotland has additional reheating requirements. The regulations also contain a general overall temperature requirement, that any raw materials, ingredients, intermediate products or finished products likely to support the growth of pathogenic microorganisms or the formation of toxins must not be kept at temperatures posing a risk to public health.

Labelling

The Food Labelling Regulations, SI 1996 No. 1499, include the requirement that foods which, from the microbiological point of view, are highly perishable (and so could constitute an immediate health risk after a short period of time) must be labelled with an appropriate use-by date and any storage conditions which need to be observed. In addition, any particular conditions regarding storage or use after the package has been opened must be given. The regulations also require that foods with extended durability by virtue of packaging in authorized gases must be marked 'packaged in a protective atmosphere'.

Refrigeration

Obviously, refrigeration is essential for the production, storage and distribution of chilled foods, but the range of refrigeration equipment required is less readily apparent. Two distinct applications of refrigeration are involved: the chilling operation, in which food is cooled from ambient temperature or from cooking temperature (e.g. > 70°C), and chilled storage, at a closely controlled temperature of between −1°C and +8°C. Chilling equipment and chilled storage equipment are quite different in terms of their requirements and their design. Although some chilling equipment may be used for chilled storage, storage equipment is not designed to cool products – only to maintain their temperature. Refrigerated transport for chilled food distribution is a special case of storage, and such equipment should not be expected to provide rapid cooling.

Chilling Equipment

The rate at which heat can be extracted during chilling is dependent on many factors:

- the size and shape of the pack or container affects the rate of transfer of heat to the cooling air (or, in some cases, water)
- temperature and speed of cooling air
- weight, density, water content, specific heat capacity, thermal conductivity and latent heat content of food
- initial food temperature.

The factors leading to rapid cooling also lead to the rapid loss of moisture from unpackaged foods, and this needs to be considered. The thinner packs, higher airspeeds and lower air temperatures which facilitate more rapid chilling also entail higher operating costs, so equipment design has to be a compromise in terms of the best overall operating system. Consequently, a range of equipment is available for different applications, and selection from it must be dependent on the particular operation planned.

Air-blast or cryogenic cooling chambers or tunnels are used for most prepared foods. Water immersion (hydrocooling) is used for some vegetables, but for fresh, leafy produce, vacuum coolers may be more appropriate. Some fresh produce with a relatively long storage life may be cooled in storage chambers, although the rate of cooling is often increased by means of enhanced air circulation. The types of cooling system available are considered below.

Air-blast Coolers

Air-blast coolers or chillers operate by passing cold air at high velocity over food. They cool food rapidly, and are necessary for highly perishable products. The movement of air leads to increased heat transfer, which in turn results in higher rates of chilling. An air velocity of $2–4\,m\,s^{-1}$ and an air temperature of about $-4°C$ are necessary when chilling thick (> 5 cm) products. The correct pattern of air flow around the stacks of food products is important, to minimize chilling time. Also, the relative humidity must be maintained at high levels to prevent the undue drying of unwrapped moist foods.

Cold Rooms

Cold rooms are used to maintain food at chilled temperatures, but are not suitable for rapid chilling. Chilled air is normally delivered at the top of the room, flowing downwards through the stacked product. The warmed air returns directly to the refrigeration unit. In general, the rate of chilling is very low, due mainly to inadequate air circulation, but it may be increased by using extra fans.

Cryogenic Chillers

The use of direct-contact refrigerants, such as liquid N_2 and CO_2 snow, has grown in popularity in recent years. These cryogens cause very rapid chilling, but control is necessary to prevent freezing. Cryogenic chillers are particularly useful when it is desirable to reduce the temperature rapidly e.g. for chilling hot cooked products.

Hydrocoolers

Chilled water, either sprayed down within a chamber or in an immersion tank, provides very rapid cooling with no risk of freezing. It is normally applicable only to fresh fruits and vegetables, which can withstand water immersion, but may also be applied to vacuum-packed prepared foods. The water in such systems is normally recirculated, so strict hygiene is necessary. A degree of hydrocooling can be combined with normal cleaning operations in the case of items such as root vegetables.

Vacuum Coolers

Vacuum coolers are highly specialized and expensive, and well-suited to the rapid cooling of prepacked leafy vegetables. They apply low pressure to wet produce in a sealed chamber, achieving cooling through the evaporation of water. The produce is usually placed on pallets or trolleys and processed in batches of several tonnes, with cooling times of about 15–30 min. Vacuum cooling must be controlled to prevent freezing. Every 5°C reduction in the temperature of the produce results in a 1% loss of weight, and so some crops are sprayed with water before cooling, to minimize this loss.

Ice-slush and Ice-bank Chilling

In these methods of chilling, the produce is held in containers and the cooling air is blown through the stacks. The air is cooled by close contact with water, which is cooled either by melting ice (ice-slush chilling) or by a bank of refrigerated tubes (ice-bank chilling). Most crops can be cooled by these methods. Ice-bank systems are highly successful, and have been specifically designed for both precooling and short-term storage. The benefits are that the cooling rate is better compared to a conventional cold room, high humidities are maintained because fine droplets of cold water are carried in the air, and there is no danger of freezing.

Chilled Storage

Chilled storage equipment is available in a wide variety of sizes, ranging from the smallest, e.g. the absorption-cycle refrigerator used in caravans and boats, through larger domestic and commercial refrigerated storage cabinets, small walk-in stores and, finally, stores large enough to be served by fork-lift trucks handling pallets or bins and accommodating thousands of tonnes of produce. Some refrigerated fruit stores have a controlled atmosphere, in which low levels of O_2 and high levels of CO_2 can be maintained so as to enhance the storage life of respiring fresh produce. Precooked chilled foods should not be stored with any other products.

For most chilled food preparation and short-term storage, walk-in stores are appropriate. These can be constructed and designed integral to a building, but more often are modular units sited within the overall structure. Modular stores are available as self-contained units with volumes of about $2\,m^3$ to $30\,m^3$ or more. The condensers of the refrigeration equipment may be mounted above or alongside the store, or may be remotely sited if there would otherwise be insufficient ventilation to remove the heat. Banks of multi-compartment stores can accommodate chilled, frozen and fresh produce. Many arrangements are possible in order to meet requirements, an example of a modular store constructed in a building with a low ceiling height being illustrated in **Figure 1**.

The design of larger stores must take into account the methods of moving and handling the products, which dictate features such as store height, requirement for a fixed racking system for pallet or carton storage, and store size. Large stores are usually designed for specific applications but tend to be used for different purposes subsequently. Therefore flexible designs are desirable, for example if should be possible to subdivide large stores without major structural alteration. It is essential to consider air flow in the context of every likely store loading pattern, and to ensure that there is sufficient fan power to provide even temperatures. The design of large stores should also ensure the protection of vehicles and products during loading and unloading operations. Protection against the sun and rain is essential, and for the more critical operations, temperature-controlled loading bays may be necessary.

Chilled Transport

The transport of chilled foods must be viewed as part of an operation achieving the movement of chilled goods from one fixed storage area to another: the operation comprises a chain of events, of which the movement of goods in a road vehicle, freight container, rail wagon, ship or aircraft is only a part. Temperature maintenance throughout the chain is essential for success – the finest transport equipment cannot compensate for poor handling during loading, or inadequate cooling of the product.

The term 'chilled transport' may itself be misleading, 'temperature-controlled transport' being more accurate. For example, in cold winter conditions it may be necessary to heat chilled foods in order to prevent damage due to freezing, and quite moderate temperatures can produce irreversible damage to many fresh tropical fruits due to chilling. In general, transport equipment is designed to maintain temperature and not to achieve cooling. Foodstuffs can be cooled to some extent during transport, but this is a slow and non-uniform method of cooling and is unreliable. Therefore precooled foodstuffs should be loaded under temperature-controlled conditions whenever possible.

The range of refrigerated transport equipment is wide, as is that of the needs for transport. At its simplest, the equipment could be an insulated box for containing water ice; at its most complex, it might be an intermodal freight container with integral refrigeration machinery, capable of maintaining frozen or chilled goods at any temperature between $-25°C$ and $+30°C$ in ambient temperatures of $-20°$ to $+50°C$. The most common refrigerated transport is the road vehicle, designed for either local deliveries or long-distance bulk distribution.

Control of the temperature of chilled foods is more difficult to achieve than that of frozen foods. To ensure uniformity of temperature in a load of chilled foodstuffs, relatively high rates of continuous air circulation and fine control of the temperature are necessary: careful stowage within the vehicle may be needed to achieve these.

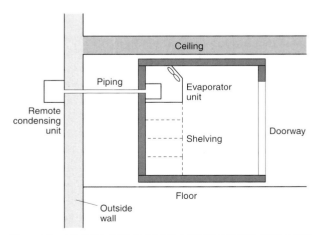

Figure 1 Modular cold room for accommodation with a low ceiling height.

Road Vehicles

Refrigerated road vehicles fall into two basic categories: large semitrailers, with refrigeration units which can be run independently of the tractor unit, and rigid-bodied vehicles, whose refrigeration units may be independent or may be driven by the vehicle engine. Semitrailers are used for long-distance or bulk movements, generally with a small number of destinations. Journey times may range from 2 h for supermarket distribution to several days for the transportation of fresh produce. A typical semitrailer is shown in **Figure 2**. Increasingly, multipurpose multi-compartment vehicles are being produced, which can carry frozen, chilled and fresh produce simultaneously in different compartments. Rigid-bodied vehicles range in size from large vehicles, functionally similar to semitrailers, to small delivery vehicles used for multiple deliveries of chilled foods to corner shops. Delivery vehicles may need walk-in access, for the selection of products from fixed shelving, and may have to cope with their doors being opened frequently.

Intermodal Freight Containers

Intermodal freight containers with integral refrigeration machinery are widely used for the long-distance transport of fresh fruit and vegetables and chilled meat. Their journey times of up to 6 weeks necessitate highly developed refrigeration and control systems, and they are capable of operating over a wide range of conditions. They are normally used only for point-to-point international transport involving a substantial sea journey, although the lease of such a container can be a very convenient way of obtaining a temporary chilled storage facility.

Chilled Display Cabinets

The display cabinets used in retail premises fall into two distinct groups. Most common are the vertical, multi-deck cabinets for the display and self-service retailing of packaged chilled foods. There are also the delicatessen or 'serve-over' cabinets, for foods which are normally not packaged and may be cut prior to serving.

Multi-deck cabinets have a refrigeration evaporator in the base. This may either be supplied from a self-contained condensing unit or, in larger installations, be piped to a central store cabinet behind or under the display area. Fans blow cooled air forwards from behind the shelves and also downwards in an 'air curtain' from the top at the front of the cabinet. Warmed air is returned through a grille at the base of the cabinet. Modern multi-deck cabinets are designed to maintain food at temperatures of 5°C or below, but not to chill food: hence they should not be loaded with warm food. Correct product loading is important in achieving proper air circulation.

In serve-over display cabinets the food is displayed on a base, over which cold air flows, normally behind a glass front. Air from a rear evaporator may be gravity-fed or fan-assisted, but because much of the food in these cabinets is not wrapped, excessive air flow must be avoided to prevent dehydration and weight loss. For the same reason, these cabinets are usually used for display only while sales are in progress, other cabinets being used for overnight storage.

See also: **Chilled Storage of Foods**: Use of Modified Atmosphere Packaging; Packaging with Antimicrobial Properties. **Hazard Appraisal (HACCP)**: The Overall Concept. **National Legislation, Guidelines & Standards Governing Microbiology**: Canada; European Union; Japan. **Packaging of Foods**. **Process Hygiene**: Involvement of Regulatory Bodies.

Further Reading

Advisory Committee on the Microbiological Safety of Food (1992) *Report on Vacuum Packaging and Associated Processes*. London: HMSO.

Betts GD (ed.) (1996) *A Code of Practice for the Manufacture of Vacuum and Modified Atmosphere Packaged Chilled Foods. Guideline 11*. Chipping Campden: Campden & Chorleywood Food Research Association.

Campden & Chorleywood Food Research Association (1997) *Evaluation of Shelf-life of Chilled Foods. Technical Manual 28*. Chipping Campden: Campden & Chorleywood Food Research Association.

Chilled Food Association (1997) *Guidelines for Good Hygienic Practice in the Manufacture of Chilled Foods*. 3rd edn. London: Chilled Food Association.

Dennis C and Stringer M (eds) (1992) *Chilled Foods: a Comprehensive Guide*. Chichester: Ellis Horwood.

Department of Health (1993) *Assured Safe Catering: A Management System for Hazard Analysis*. London: HMSO.

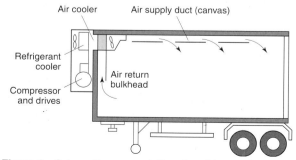

Air cooler   Air supply duct (canvas)

Refrigerant cooler

Air return bulkhead

Compressor and drives

Figure 2 Schematic representation of a refrigerated semitrailer. Arrows indicate air flow.

Gormley R (1995) *Chilled Foods: Some Pointers for Success*. Dublin: National Food Centre.

Hui Y (ed.) (1991) *Encyclopaedia of Food Science and Technology*. New York: J Wiley.

Institute of Food Science & Technology (1991) *Food and Drink – Good Manufacturing Practice: a Guide to its Responsible Management*, 3rd edn. London: Institute of Food Science and Technology.

Use of Modified-atmosphere Packaging

R E O'Connor-Shaw, Food Microbiology Consultant, Birkdale, Queensland, Australia

V G Reyes, Food Science Australia, Victoria, Australia

Introduction

Modified-atmosphere packaging (MAP) extends the shelf life of chilled foods in various ways, depending upon the product. For example, in the case of rapidly respiring foods, such as fruits and vegetables, the primary function of MAP is to retard physiological deteriorative processes, such as respiration; with regard to meat, MAP is used to inhibit microbial spoilage. Consequently, the optimum gas regimes for different foods are product-specific. Package atmospheres comprise air modified in one of several ways: by vacuum packaging, passively by the product, by gas flushing or by 'active' packaging.

A major concern associated with the use of MAP for chilled foods is that the low O_2 concentrations involved may inhibit the growth of aerobic spoilage microorganisms but permit that of anaerobic, non-spoilage pathogens, which may reach high numbers during the extended shelf life of these products. Psychrotrophic pathogens of concern include non-proteolytic strains of *Clostridium botulinum* and *Listeria monocytogenes*. Food safety guidelines for the use of MAP for chilled foods must therefore, be developed and implemented. Low temperature storage and limiting shelf life so the food is eaten before it becomes toxic are two important control measures for ensuring the microbiological safety of these foods.

Types of Modified-atmosphere Packaging

MAP can be defined as packaging which encloses a food along with an atmosphere with a different composition from that of air (normally comprising about 0.03% CO_2, 0.93% argon, 20.95% O_2 and 78.09% N_2). Four packaging techniques are encompassed by this definition: vacuum packaging (VP), passive MAP, gas-flush MAP, and MAP by active packaging.

Vacuum packaging involves the removal of air from a package without replacement by another gas. A 'partial' vacuum is generally recommended for respiring products such as fruit and vegetables, to avoid anaerobic conditions, which may be conducive to anaerobic fermentation. Non-respiring products, such as processed foods and fresh meats, are packaged under 'full' vacuum. Vacuum skin-packaging (VSP) is a form of vacuum packaging which uses a full vacuum and produces a package with the shape of the product. It is marketed under the trademarks Darfresh (Cryovac) and Intact (Trigon).

Modified atmospheres can be created within a package either passively by the commodity, by gas flushing, or by using active packaging materials such as gas absorbers or generators. Passive MAP relies on the selective permeability of the packaging materials to different gases and on product respiration, and is traditionally used with fresh and minimally processed fruits and vegetables. For example, with tomatoes, which respire rapidly, the desired gas concentrations are attained within 1–2 days of packaging. Gas-flush MAP involves the establishment of a specified gas composition within the package in a single stage during the packaging operation, by flushing with the selected gas mixture before sealing. Depending on the desired level of residual O_2, a vacuum operation may be needed prior to gas flushing. The instant modification of the package atmosphere is the main advantage of gas-flush MAP over the relatively slow passive process. MAP can also be achieved by using active packaging materials or inserts, which absorb or generate specific gases. The desired gas compositions can be achieved within several hours of sealing.

The gases used in MAP applications are usually CO_2, O_2 and N_2. The role of CO_2 is primarily to inhibit the growth and metabolism of microorganisms. CO_2 is both water- and lipid-soluble, and so may dissolve in the product to some extent. N_2 is an inert gas with low solubility in water, and so is often used as a filler. High N_2 levels, i.e. > 20%, are generally used to avoid the collapse of modified-atmosphere packages containing high levels of CO_2. **Table 1** gives a summary of the recommended gas regimes for various foods.

Fresh and Minimally Processed Fruits and Vegetables

In general, the recommended atmospheres for fresh, unprocessed fruits and vegetables are CO_2: 0–10% and O_2: 1–5%. However, some fruits, including cherries, blackberries, blueberries, figs, raspberries and strawberries, have a maximum shelf life in CO_2: 10–20% and O_2: 3–10%. Minimally processed fruits and vegetables respire, and so require O_2 to maintain their natural respiratory activities. Optimal storage

Table 1 Recommended gas regimes for modified-atmosphere packaging of various foods

Product	CO_2 (%)	O_2 (%)	N_2 (%)
Whole fruits and vegetables	0–5	1–5	90–99
Minimally processed fruits and vegetables	0–15	2–5	80–90
Fresh meat (retail, short shelf life)	15–25	75–80	1–10
Fresh meat (long shelf life)	50–100	0	
Cured/cooked meats	20–100	<0.5	0–80
Fresh seafood	20–80	0–10	10–80
Precooked foods	20–80	0	20–90
Bakery products	60–100	0	0–40

atmospheres, however, aim to prevent oxidative reactions which can lead to the development of physiological disorders including enzymatic browning and lignification (e.g. 'white blush'). MAP, if not properly managed, can create a risk to public health because a low O_2 concentration and increased shelf life may permit the growth of anaerobic pathogens. To prevent this, the packaging material for minimally processed produce should be permeable, to ensure that O_2 levels within the package do not fall below 2%.

Fresh Meat

Four different MAP concepts are available for fresh meat, depending on the shelf life requirements of the target market:

- Vacuum packaging and vacuum skin-packaging: uses high-barrier packaging materials.
- High-O_2 packaging, using materials with a high O_2 barrier. High-O_2 MAP, currently favoured in western Europe, results in the retention of the red colour of meat but fosters the oxidation of fat. Atmospheres of O_2: 40–60%, CO_2: 20–40% and the balance of N_2 are employed. The ratio of meat volume to modified atmosphere volume is about 1 : 1.2. This form of MAP gives a short shelf life.
- High-CO_2 packaging: if meat colour is not a major consideration, CO_2 concentration >40% can be used in the absence of O_2, resulting in a long shelf life. A form of high-CO_2 packaging is in 'master packing' in high CO_2 concentrations. This involves the opening of the impermeable master pack to expose the gas-permeable material used for the primary package: this admits air, which reoxygenates the meat for the purposes of appearance. The permeability of the primary package to O_2 should be >9000 cm^3 m^{-2} per kPa per day. This form of packaging is designed to give a long shelf life.

Seafood

High-CO_2 MAP can delay the spoilage of fresh seafood, and may protect the quality of the product. CO_2 concentrations of 20–100% have been recommended. The optimum CO_2 concentration is dependent on the initial microbial population, the biology of the seafood, the gas : seafood ratio in terms of volume, and the packaging method. CO_2 concentrations of 40–60% are usually used commercially. MAP is more effective if used in conjunction with low-temperature storage, the addition of sorbates and low-dose irradiation.

Soft Bakery Products

The shelf life of soft bakery products, e.g. breads, rolls, cakes and muffins, is generally limited by staling, moisture loss, and/or mould growth. Staling involves the crystallization of starch and leads to textural hardening, which may or may not be inhibited by high CO_2 levels. The recommended gas composition for MAP for soft bakery products is CO_2: 0–100%, N_2: 20–100%. The presence of 20–30% N_2 is recommended in order to prevent package collapse.

Active Packaging

Substances which interact with the surface of the food as well as with the atmosphere in the package can be used either alternatively to or supplementary to vacuum packaging and/or gas flushing. These 'active' materials are either combined with the packaging material or contained within package inserts (sachets). Their functions include O_2 removal, CO_2 removal/release, water removal, gas indication, antimicrobial action, preservative release and aroma release. Most of the techniques used in the manufacture of these materials originated in Japan, where some had been used for many years before introduction into North America, Europe and Australia.

Oxygen Absorbers

O_2 absorbers consist of a powder of finely divided reduced iron. A typical sachet contains 1 g of iron, which will react with 300 cm^3 of O_2. These sachets can remove O_2 from the package interior to a level of <0.01%. In addition, the presence of an O_2 absorber obviates the effects of any O_2 entering as a consequence of permeation or transmission after the package is sealed. Some commercially available O_2 absorbers are: Ageless (Mitsubishi Gas Chemical Co., Japan), KF-Pack (Toppan Printing Co., Japan) and FreshPax (Multiform Dessicants Inc.). In general, the recommended O_2 transmission rate for a food package containing an O_2 absorber is a maximum of 20 cm m^{-2}

per kPa per day. For O_2 absorbers to be fully effective, the integrity of the package seal must be unbroken.

O_2 absorbers, available in a range of sachet sizes for some time, have more recently been incorporated into packaging materials. One version combines the O_2 absorber with a paper/polymer laminate, which consequently doubles as a product support board and an O_2 absorber, and is used with products such as baked goods, fresh pasta and processed meat. An O_2-absorbing closure liner for bottle and jar caps, TRI-SO_2RB, is available from Tri-Seal International Inc. (New York, USA). The O_2-scavenging agent is branded Amosorb, from Amoco Chemical Co. (Chicago, USA).

Carbon Dioxide Absorbers/Generators

A CO_2-generating or CO_2-scavenging system may be incorporated into packaging material or a sachet. Some O_2 absorbers also generate CO_2, the volume of which is equal to that of the absorbed O_2. Ageless-E is an example of an O_2 and CO_2 absorber, and Ageless-G is an example of an O_2 absorber and CO_2 generator.

Antimicrobial Agents

Spoilage due to microbial growth strictly on the food surface can be limited by packaging material or a sachet which emits an antimicrobial agent, e.g. an antioxidant. The major commercial system of this type is an ethanol emitter, sold under the trade names Ethicap or Antimold 102 (Freund Industries Co., Japan). Ethicap is in the form of a sachet that releases ethanol vapour, which settles on the surface of the food, preventing the growth of moulds, yeasts and bacteria. Trace amounts of vanilla or other flavouring are usually added, to mask the odour of the alcohol. The reported disadvantage of this system is ethanol absorption by the product. However, heating the product prior to consumption would eliminate most of the ethanol, and so Ethicap may be most useful as a preservative for 'brown and serve' products.

Indicator Systems

Another active packaging approach involves placing a time–temperature indicator (TTI) system on the package, to warn the consumer if a particular time–temperature regime has been exceeded and the product is spoiled or possibly dangerous to eat. The first such products on the market were Fresh-Scan and Fresh-Check (LifeLines Technology, NJ), MonitorMark (3M) and I-point (I-point Biotechnologies AB, Malmo, Sweden), followed later by Vitsab (Telatemp Corp., California, USA). Vitsab is an enzyme-based indicator in the form of tags, which are used on large loads rather than retail packages

and have been used as indicators of product shelf life.

Ethylene Absorbers

Many companies make ethylene scavengers based on potassium permanganate ($KMnO_4$), activated carbon, and activated earth. Ethylene scavengers are used in packages of fruit and vegetables. They are available as sachets for use inside the package, or can be integrated into the packaging material. However, some of the claims regarding the ethylene-absorbing capacity of packaging materials are poorly documented and controversial.

Packaging Materials and Equipment

The selection of MAP materials for specific applications depends on a variety of factors, including:

- permeability of packaging material to O_2, CO_2, and water vapour
- changes in film permeability to O_2, CO_2 and water vapour caused by changes in storage temperature and relative humidity (RH)
- effective surface area of final package
- integrity of seal
- resistance of packaging material to mechanical abuse
- suitability for packaging machinery (e.g. vertical form–fill–seal machine)
- cost of packaging material
- peelability characteristics (i.e. ease of opening)
- anti-fog properties, especially for chilled products
- clarity of packaging material.

The permeability of packaging materials is conventionally measured in terms of O_2 transmission rate (OTR), in $cm^3 m^{-2}$ per day and water vapour transmission rate (WVTR), in $g m^{-2}$ per day. Permeability specifications include the temperature and RH at which the measurements are carried out.

For the packaging of respiring products, such as minimally processed vegetables, film made of uncoated oriented polypropylene (PP) and pouches made of ethylene-vinyl acetate (EVA) and polyethylene blends have been used. The O_2 permeability of the pouches lies in the range 2000–6000 $cm^3 m^{-2}$ per day (23°C, 70% RH). Non-respiring products, such as processed foods, fresh meats, and bakery products, are normally packaged in materials with a low gas permeability, the two most common types being those laminated with PVDC (polyvinylidene chloride) and those coextruded with EVOH (ethylene-vinyl alcohol).

Influence of Modified Atmospheres on Spoilage

Whole and Minimally Processed Fruit and Vegetables

MAP of whole fruits and vegetables extends their shelf life by reducing respiration, ripening rate and water loss, and by inhibiting ethylene-induced effects such as the rapid ripening of many fruits. Concentrations of O_2 of 2–10% reduce the rate of respiration, inhibit the oxidation of phenolic compounds and the subsequent development of brown discoloration, and increase the resistance of the product to invasion by plant pathogens, due to a delay in the onset of senescence. High levels of CO_2 also reduce the respiration rate. Concentrations of CO_2 of 1–2% reduce tissue sensitivity to ethylene, and at concentrations $\geqslant 10\%$, the growth of many spoilage microorganisms is inhibited. Many factors influence the precise gain in shelf life resulting from MAP: these include the variety of fruit or vegetable, its initial quality and its physiological age.

The range of minimally processed fruits and vegetables, e.g. salads, trimmed vegetables, and pineapple cylinders, is expanding and diversifying: in 1996, the fresh cut produce industry experienced a 70–90% annual growth rate. The appeal of these products is related to their natural image and to convenience. However, minimal processing results in increased perishability, because cutting enhances the rates of respiration and ripening of most products. Minimal processing also increases microbiological spoilage, through the transfer of the skin microflora to the flesh, where microorganisms can grow rapidly due to exposure to nutrient-laden juices. Several studies have shown that the extension of the shelf life of these products achieved by MAP is due to its effect on the physiological state of the product, rather than on the growth of spoilage microorganisms. For example, appropriate atmospheres can enhance the shelf life of chilled microbiologically sterile, diced cantaloupe flesh to >28 days by retarding physiological deterioration.

Fresh Red Meats and Poultry

The two major forms of deterioration of fresh red meats are loss of the bright red pigment which consumers associate with freshness, and microbial spoilage. The colour of muscle tissue is determined by the state of oxidation of the muscle pigment, myoglobin. The deoxygenated form is dull purple, whereas the oxygenated form, oxymyoglobin, has the preferred bright red colour. Myoglobin is slowly oxidized to metmyoglobin, which has an undesirable, dull brown colour, and is more susceptible than oxymyoglobin to oxidation.

When O_2 is present in non-limiting amounts, the principal spoilage microorganisms in meat are pseudomonads, which produce malodorous, putrid compounds from amino acids. The pH of meat is usually low, in the range 5.4–5.8, resulting from the postmortem glycolytic conversion of glycogen to lactic acid. However, some meat, described as dark, firm and dry (DFD), has a much higher pH, >6.0, reflecting the depletion of glycogen in the muscles of live animals, due to stress. In meat with pH < 5.8 and in O_2 concentrations which limit growth, microbial populations are dominated by lactobacilli. However, in any O_2-containing pockets, aerobic microorganisms grow and metmyoglobin is formed. In O_2 concentrations which limit growth, in meat with pH > 5.8, enterobacteria, *Brochothrix thermosphacta* and *Shewanella putrefaciens* prevail. Enterobacteria cause putrid spoilage, *B. thermosphacta* causes sour aromatic odours, and *S. putrefaciens* causes putrid, sulphurous odours.

Fully effective vacuum packaging inhibits metmyoglobin formation and the growth of aerobic spoilage microorganisms. In high-O_2 MAP ($CO_2 : O_2 : N_2$ 25% : 60% : 15%), the O_2 maintains the muscle pigments in the desirable oxymyoglobin state, and the CO_2 concentrations of >20% approximately halve the growth rates of aerobic bacteria. In low-O_2 MAP ($CO_2 : O_2 : N_2$ 50–90% : 1–10% : 10–40%) high CO_2 and low O_2 levels inhibit the spoilage bacteria. MAP using 100% CO_2 suppresses the growth of all spoilage microorganisms at < 0°C. At > 0°C, *B. thermosphacta* can become a problem if it forms a substantial proportion of the microflora. Storage under anoxic conditions prevents the oxidative deterioration of myoglobin.

Fish

Fish spoilage can be divided into four phases. In phase 1, the fish loses its fresh, often sweet, flavour to become bland due to the action of enzymes intrinsic to its flesh. The end of phase 1 is known as the 'end of high-quality shelf life'. During phases 2 and 3, the numbers of bacteria and their end products increase, and phase 4 involves putrid spoilage. MAP has little effect on the deteriorative processes of phase 1.

MAP exerts a much smaller effect on the microbiological spoilage of fish, and hence on its shelf life, than is the case with other foods. There are several reasons for this. One of the major osmoregulatory compounds in fish is trimethylamine oxide (TMAO). Gram-negative bacteria, including the important spoilage bacterium *Shewanella putrefaciens*, use TMAO as an alternative electron acceptor to O_2.

In low O_2 concentrations, spoilage bacteria reduce TMAO to trimethylamine (TMA), causing undesirable amine odours. This type of spoilage is mainly associated with fish from cold or temperate waters. *S. putrefaciens* also produces H_2S, from sulphur-containing amino acids. Another important spoilage bacterium, *Photobacterium phosphoreum*, is highly resistant to CO_2 and has an increased growth rate under anaerobic conditions. Psychrotrophic bacteria are naturally found on fish from cold or temperate waters, and consequently are preadapted for growth on the harvested fish during chilled storage. The pH of fish flesh is near neutral, which is conducive to microbial growth. Also, the flesh has a low carbohydrate content, and so bacterial proteolytic spoilage commences immediately after harvesting.

Bakery and Pasta Products

Microbial spoilage, chemical spoilage (e.g. rancidity), and physical spoilage (e.g. moisture loss, staling) all occur in bakery and pasta products, although microbial spoilage is often the factor which limits shelf life. The type of spoilage which occurs is dependent on many factors, but particularly on moisture content and water activity (a_w), the latter being the major factor which influences microbial growth. The most important microbiological problem in high- and intermediate-moisture bakery products is mould spoilage and bacterial spoilage is also a major problem limiting the shelf life of fresh and precooked pasta products, and bakery products with high a_w (e.g. crumpets and bagels).

Gas mixtures used to extend the shelf life of bakery products contain 50–100% CO_2, with any balance made up by N_2: the optimum blend varies according to the product. Appropriate modified atmospheres can extend mould-free shelf lives by up to 3 months at room temperature. Recently, fresh pasta packaged in CO_2–N_2 atmospheres containing $\geqslant 20\%$ CO_2 has been introduced. This has a shelf life of $\geqslant 14$ days at $\leqslant 4°C$, and a longer shelf life (> 30 days) is possible if the product is pasteurized prior to MAP. However, a major concern associated with these products is their microbiological safety.

Influence of Carbon Dioxide and Oxygen on Microbial Growth

Carbon Dioxide

High concentrations of CO_2 may reduce the growth rate of microorganisms. They may also inhibit growth by extending the lag phase, although this is also controversial: there is evidence that the duration of the lag phase depends on the physiological state of the bacteria at the time of the application of CO_2, but further investigation is required. The inhibitory effect of the CO_2 disappears once the microorganism is withdrawn from it and placed in air: the residual inhibitory effects of CO_2 which have been observed in foods have been attributed by some authors to changes in the composition of the food microflora, caused by CO_2. Oxidative Gram-negative bacteria and moulds are more sensitive to high CO_2 concentrations than are Gram-positive bacteria, lactic acid bacteria being most resistant to CO_2.

Several other factors also influence the effect of CO_2 on microbial growth. These include:

- temperature, which affects the solubility of CO_2 in water (solubility increases as the temperature decreases)
- pH: greater inhibition of microbial growth occurs with decreasing pH
- CO_2 concentration: inhibition increases with increasing CO_2 concentration
- presence of O_2.

The mechanism by which CO_2 exerts its effect on microbial growth is not well understood.

Oxygen

There is a continuum of O_2 tolerance amongst microorganisms, from strict anaerobes to strict aerobes. Obligate aerobes generate most of their energy by oxidative phosphorylation, using O_2 as the terminal electron acceptor in the process. Obligate anaerobes generate energy without using molecular O_2, demonstrating a degree of adverse sensitivity to O_2 which renders them unable to grow in normal air pressure. Obligate aerobes, e.g. *Pseudomonas fluorescens* and *P. putida*, will grow in atmospheres containing as little as 0.2–0.5% O_2, although maximum populations are smaller in these atmospheres compared to those grown in air.

Strict anaerobes are killed in air because of the toxic effect of the superoxide radical. All aerotolerant microorganisms contain the enzyme superoxide dismutase, which converts the radical to H_2O_2 and O_2.

Influence of Modified Atmospheres on Pathogenic Microorganisms

The low O_2 concentrations used in MAP may inhibit the growth of aerobic spoilage microorganisms while permitting the growth of anaerobic non-spoilage pathogens, which may grow to high numbers during the extended shelf life associated with MAP. Low-temperature storage is often the sole method for maintaining microbial safety, but the difficulty of ensuring temperatures of $\leqslant 5°C$ during distribution, retailing and home storage leads to concern. The psychrotrophic pathogens of concern in this context include *Clostridium botulinum*, *Listeria monocytogenes*, *Aeromonas hydrophila* and *Yersinia enterocolitica*. Clostridia are anaerobes and the others are facultative anaerobes, the highest mortality rates being associated with *C. botulinum* and *L. monocytogenes*.

Clostridium botulinum

The non-proteolytic strains of *C. botulinum* types B, E and F are of concern in chilled foods especially those packaged in low O_2 concentrations and which have an extended shelf life. In most chilled foods, spoilage precedes toxigenesis, which prevents the food being eaten. However there are exceptions. In chilled cod and whiting fillets packaged in 100% CO_2, toxin production occurred before organoleptic rejection. Toxin production by *C. botulinum* type E has been described in three types of fish prior to spoilage, at O_2 levels of 2% and 4%, indicating that the inclusion of O_2 in a modified atmosphere does not guarantee the absence of *C. botulinum*.

The popularity of a class of foods known as REPFEDs (Refrigerated Processed Foods of Extended Durability), e.g. pasteurized pasta, has led to many studies looking at the temperatures, pH and O_2 concentrations which limit the growth of non-proteolytic *C. botulinum* strains. These studies have shown that growth and toxin production can occur at temperatures as low as 3–3.2°C. However, the probability of growth of a single vegetative cell at low temperatures is very low, e.g. 1 in 10^4 cells at 4°C in 5 days, because a long lag phase precedes growth. In optimal growth conditions at 20°C, at least 2×10^4 and $> 2 \times 10^5$ spores are needed to produce growth at PO_2 values of between 1.22 and 1.62 kPa.

Listeria monocytogenes

There is particular concern about the growth of *L. monocytogenes* in fruits and vegetables, because contamination can easily occur while the plant is growing. The first food-borne outbreak of disease attributed to *L. monocytogenes* involved cabbage (in coleslaw).

The effects of modified atmospheres on the growth of two *L. monocytogenes* strains on fresh asparagus spears, broccoli and cauliflower florets, stored at 4°C, have been studied. The atmospheres were different for each vegetable and contained 3–10% CO_2 and 11–18% O_2, with N_2 providing the balance. The most important finding was that although modified atmospheres extended the shelf life of all the vegetables by 7 days, they had no effect on the growth of *L. monocytogenes*. At end of shelf life, populations of

L. monocytogenes on asparagus were significantly higher in a modified atmosphere compared with storage in air. The microbiological safety of broccoli and cauliflower was not compromised by storage in a modified atmosphere, because these vegetables did not support the growth of *L. monocytogenes*. Thus the safety risk associated with MAP of fruits or vegetables must be assessed on an individual basis.

The growth rate of *L. monocytogenes* also varies in other categories of foods depending on the gaseous composition of the storage atmosphere. For example, its rate of growth on cod was similar whether stored in air or in modified atmospheres containing either $CO_2 : N_2 : O_2$ 60% : 40% : 0% or 40% : 30% : 30%. On trout, however, *L. monocytogenes* grew more rapidly in air than in atmospheres containing $CO_2 : N_2$ at 60% : 40% or 80% : 20%. Variable results regarding the growth and survival of *L. monocytogenes* in different meats and poultry have also been reported.

Advantages and Disadvantages of Modified-Atmosphere Packaging

Advantages

- Extended shelf life: product can be distributed over greater distances and stockpiled during periods of reduced demand, reducing production costs.
- Cost: processing, storage and distribution costs are lower than those associated with frozen products.
- Extended shelf life is achieved without the addition of preservatives or overt processing: hence foods are perceived as 'natural': a valuable selling-point.
- Convenience: products require little or no preparation by the consumer: minimal processing, i.e. peeling, trimming and cutting, is done by the processor.
- Packaging: attractive presentation can encourage brand loyalty in consumers.

Disadvantages

- Different atmospheres are required for different products. Optimal atmospheric storage conditions for many products have not yet been determined.
- Difficulty in maintaining low temperatures during product distribution. Chilled storage is crucial for the premium quality and microbiological safety. Consumer education must emphasize that chilled foods in MAP are a distinct class of foods, and that the storage conditions specified on packs are mandatory in terms of safety.
- Microbiological safety: further investigation of the growth of pathogens and their production of toxins is warranted, particularly with regard to the abuse of temperature and atmosphere.

- Package integrity: the loss of the integrity of the package and/or its seal is a common and serious problem for the food industry. Loss can be due to a variety of reasons including improper sealing, puncture, holes, and other defects in the packaging materials.

Future Developments

The use of other preservative techniques in conjunction with chilled, low-O_2–high-CO_2 storage, e.g. the application of the hurdle concept, would enhance the antimicrobial efficacy of MAP. The hurdle concept involves the use of preservation techniques in combination to prevent microbial growth; the less extreme use of individual treatments minimizes product damage.

The development of models to predict the microbial safety and shelf life of chilled foods with MAP will also be a significant advance. Mathematical models which take into account the effects of E'_0, CO_2 and O_2 concentrations, temperature and pH on the growth of spoilage and pathogenic microorganisms in food are urgently required.

In addition, food safety guidelines must be developed and implemented by the food industry. Storage temperatures and use-by dates should be prominently displayed on product labels. Ideally, temperature–time indicators, taking into the account the likely development of unacceptable numbers of pathogens or spoilage microorganisms, should be included in packaging to safeguard consumers from the effects of temperature abuse.

See also: **Brochothrix**. **Clostridium**: *Clostridium botulinum*. **Ecology of Bacteria and Fungi in Foods**: Influence of Redox Potential and pH. **Fish**: Spoilage of Fish. **Listeria**: *Listeria monocytogenes*. **Meat and Poultry**: Spoilage of Meat. **Metabolic Pathways**: Release of Energy (Aerobic); Release of Energy (Anaerobic). **Predictive Microbiology & Food Safety**. **Shewenella**.

Further Reading

Brody A (1989) Modified atmosphere/vacuum packaging of meat. In: Brody A (ed.) *Controlled/Modified Atmosphere/Vacuum Packaging of Foods*. P. 17. Trumbull: Food & Nutrition Press.

Davies AR (1997) Modified atmosphere packaging of fish and fish products. In: Hall, GM (ed.) *Fish Processing Technology*, 2nd edn. P. 200. London: Blackie Academic & Professional.

Davies A and Board R (eds) (1998) *The Microbiology of Meat and Poultry*. London: Blackie Academic & Professional.

Dixon NM and Kell DB (1989) A review. The inhibition by CO_2 of the growth and metabolism of microorganisms. *Journal of Applied Bacteriology* 67: 109–136.

Eyles MJ, Moir CJ and Davey JA (1993) The effects of modified atmospheres on the growth of psychrotrophic pseudomonads on a surface in a model system. *International Journal of Food Microbiology* 20: 97–107.

Farber JM (1991) Microbiological aspects of modified atmosphere packaging technology – A review. *Journal of Food Protection* 54: 58–70.

Farber JM and Dodds KL (1995) *Principles of Modified Atmosphere and Sous Vide Product Packaging*. Lancaster: Technomic Publishing.

Kader AA, Zagory D and Kerbel EL (1989) Modified atmosphere packaging of fruits and vegetables. *Critical Reviews in Food Science and Nutrition* 28: 1–30.

O'Connor-Shaw RE (1998) Shelf life and safety of minimally processed fruit and vegetables. In: Ghazala S (ed.) *Sous Vide and Cook-Chill Processing for the Food Industry*. P. 165. Gaithersburg: Aspen Publishers.

Peck MW (1997) *Clostridium botulinum* and the safety of refrigerated processed foods of extended durability. *Trends in Food Science & Technology* 8: 186–192.

Rooney ML (1995) *Active food packaging*. London: Blackie Academic & Professional.

Sea Fish Industry Authority (1985) *Guidelines for the handling of fish packed in controlled atmosphere*. Edinburgh: Sea Fish Industry Authority.

Stammen K, Gerdes D, and Caposo F (1990) Modified atmosphere packaging of seafood. *Food Science and Nutrition* 29: 301–331.

Packaging with Antimicrobial Properties

David Collins-Thompson and **Cheng-An Hwang**, Nestlé Research and Development Center, New Milford, Connecticut, USA

Introduction

Food products undergo numerous physical, chemical and microbial changes during storage. These changes, such as moisture loss, oxidation and microbial growth, can lead to the rejection of the product due to deterioration and spoilage. Suitable packaging can slow down these changes, and hence extend the shelf life of the food.

Food packaging primarily provides food products with mechanical protection, but it can also act as a barrier to oxygen or moisture – the terms 'active packaging', 'interactive packaging' and 'intelligent packaging' can be used to describe such materials. Examples of active packaging include: susceptors and reflectors in microwaveable food packages, which enhance the appearance of the product after cooking; films for fresh fruits and vegetables, which absorb ethylene to delay ripening; and modified-atmosphere packaging, which delays the oxidation of lipids and inhibits the growth of aerobic microorganisms in food products. The term 'antimicrobial packaging' covers any packaging techniques used to control microbial growth. These include packaging materials and edible films and coatings that contain antimicrobial agents, and also techniques which modify the atmosphere within packs.

Principles of Antimicrobial Packaging

The main cause of spoilage of many refrigerated foods is microbial growth on the product surface. The application of antimicrobial agents to packaging can create an environment inside the package which delays or even prevents the growth of microorganisms on the product surface and hence may lead to an extension of the shelf life and/or the improved safety of the product. For example, an extension of the shelf life of fresh fruits and vegetables can be obtained by using an antifungal agent to coat the boxes used for shipping them. Alternatively, O_2 absorbers and generators of CO_2 or alcohol vapour can be placed inside food packages to create an environment which inhibits microbial growth.

In recent years, antimicrobial packaging has attracted much attention from the food industry, because of increasing consumer demand for minimally processed, preservative-free products. Reflecting this demand, the preservative agents must be applied to packaging in such a way that only low levels of preservative come into contact with the food.

The antimicrobial effect of a packaging material is dependent on the release of a preservative and its deposition on the surface of the product. This can be achieved by incorporating the antimicrobial agent either into the packaging material, to be released into the space surrounding the food, or into an edible film or coating, applied by dipping or spraying onto the food. The film/coating technique is considered to be more effective, although more complicated to apply.

Packaging Materials

The antimicrobial agents that are generally recognized as safe (GRAS) and are used with foods in the US are listed in Title 21 of the Code of Federal Regulations (CFR), Part 172 (food additives permitted for direct addition to food for human consumption), Part 182 (substances generally recognized as safe), and Part 184 (direct food substances affirmed as generally recognized as safe). They include: benzoic acid, butylated

hydroxyanisole, butylated hydroxytoluene, calcium sorbate, diacetyl, ethyl alcohol, hydrogen peroxide, natamycin (pimaricin), nisin preparation, potassium nitrate, potassium sorbate, propionic acid, propylparaben, sodium acetate, sodium diacetate, sodium lactate, sodium nitrate, sodium nitrite, sodium propionate, sodium sorbate and sorbic acid. These GRAS antimicrobial agents may be incorporated into materials used for packaging, films or coatings, in accordance with regulations prescribed by the US Food and Drug Administration. They may or may not be used for food products regulated by the US Department of Agriculture.

The range of packaging materials which incorporate antimicrobial agents includes polyethylene film coated with sorbic acid, to prevent the growth of mould in foods. A second example is film containing acidic groups bound in a polymer matrix, used to wrap raw meat: contact between the film and the moist surface of the meat creates an acidic environment which inhibits microbial growth on the surface. Some packaging materials have a modified polymer surface that has antimicrobial activity. These materials are produced by treating natural and synthetic polyamides, polyureas, polyhydrazides, polyurethanes and copolymers with selective ultraviolet (UV) light, electron irradiation or chemical reducing agents. The antimicrobial activity is due to the presence of functional groups such as amides or hydrazines on the surface of the material.

Microban Product Co. in the US has designed an antimicrobial additive, Microban, for use with polymers. Microban is introduced into the empty spaces of the polymer matrix and released as needed to control the growth of microbial cells that come into contact with the polymer: eukaryotic cells are unaffected. Microban can be incorporated into the materials used for making toys, polyethylene wrapping films, polypropylene chopping boards, work surfaces, waste bins and shipping crates. Another antimicrobial additive is Surfacine (Surfacine Consumer Products LLC). In this application, a three-dimensional network is fixed onto a surface and impregnated with tiny particles of silver halide, forming a silver halide–polymer complex. The silver is transferred directly into microbial cells that come into contact with the coating, leading to a toxic accumulation that results in cell death. The accumulated silver is not toxic to neighbouring cells because it remains with the dead microorganism, but the halide reservoir within the polymeric network replenishes the silver at the surface, so that the antimicrobial activity of the coating is maintained (**Fig. 1**).

New antimicrobial packaging materials are being developed continually. Many of them exploit natural

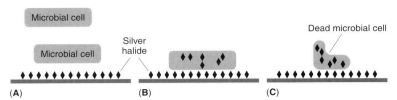

Figure 1 Surfacine antimicrobial packaging. (**A**) Polymer surface impregnated with silver halide. (**B**) Silver halide transferred into microbial cells which come into contact with the coating. (**C**) Silver halide accumulates inside cells, causing death. (Adapted from Surfacine® – Smart Polymer Technology at Work. Surfacine Consumer Products LLC.)

agents, for example lysozyme from egg white, used with polyvinyl alcohol, nylon or cellulose acetate to control common food-borne microorganisms. Pediocin, a natural antimicrobial agent produced by *Pediococcus* species, has been used to coat cellulose casings and powder bags and has been shown to inhibit the growth of inoculated *Listeria* species in chilled meat and poultry. A similar compound, nisin, may also be appropriate for such an application. A packaging material designed to release vaporized allyl isothiocyanate, a compound which occurs in mustard seeds, has been shown to have a general antimicrobial effect.

Edible Films and Coatings

Antimicrobial agents incorporated into edible films or coatings are released onto the surface of the food to control microbial growth. Such coatings can also serve as a barrier to moisture and oxygen. Other additives can also be incorporated, to contribute to the preservation of the colour, flavour, and texture of the food. Edible coatings have become popular in the food industry because they produce less waste, cost less and offer protection after the package has been opened.

Alginates, fats, gums and starch-based coatings have been used to extend the shelf life of frozen meat, poultry and seafood. Their application to food can be as simple as dipping the product into a solution containing the antimicrobial agent, or spraying the solution onto the product. Antimicrobial agents commonly used for this purpose are benzoic acid, sodium benzoate, sorbic acid, potassium sorbate and propionic acid.

Polysaccharides, proteins, waxes and oils are also commonly used in films and coatings materials, either alone or as components of composite materials (**Table 1**). The choice of material for a film or coating is largely dependent on its required function. For example, waxes constitute an effective barrier to water.

Table 1 Materials commonly used for edible films and coatings

Polysaccharides	Proteins	Waxes and oils
Cellulose and derivatives	Corn zein	Beeswax
Starch and derivatives	Wheat gluten	Carnauba wax
Pectin	Milk protein	Candelilla wax
Carrageenan	Casein/whey protein	Paraffin wax
Gum arabic	Soya protein	Polyethylene wax
Chitosan	Collagen	Fatty acids
Alginate	Gelatin	Monoglycerides
Gellan gum	Peanut protein	Shellac
	Keatin	Acylglycerols

Polysaccharide Films and Coatings

Polysaccharides have been used mainly with agricultural products. They are nontoxic and widely available. They are also selectively permeable to CO_2 and O_2, so can be used to retard the respiration and ripening of many fruits and vegetables by limiting the availability of O_2.

Polysaccharide-based coatings, e.g. pectin, are mainly used for processed fruit and nut products. Several water-soluble, composite coatings are made commercially from cellulose, the most abundant polysaccharide in nature, e.g. carboxymethyl cellulose with sucrose–fatty acid esters. Derivatives of cellulose, such as hydroxypropyl methylcellulose, form tough and flexible water-soluble films. Chitosan, a form of chitin derived from exoskeletons such as crab shells, exhibits antifungal activity and can be used in coatings as a natural antimicrobial agent . For example, a film containing methylcellulose, chitosan and 4% sodium benzoate or potassium sorbate has been used for controlling moulds.

Protein Coating

Proteins are suitable for coating many fruits and vegetables, and may be derived from maize, wheat, soybeans or gelatin. Protein-derived coatings provide good barriers to O_2 and CO_2 but not to water. Zein, an alcohol-soluble protein fraction from corn, is classified as a substance generally recognized as safe (GRAS) by the US Food and Drug Administration.

Films made from zein, by a solvent evaporation process, are tough, glossy and grease-resistant. A zein coating made of vegetable oils, glycerin, citric acid and antioxidants is used for products such as nuts, to prevent rancidity. A popular source of protein coatings is gelatin, which is derived from collagen and can be extruded to form casing for meat products. When formed from aqueous solutions, it produces clear, flexible, strong and oxygen-impermeable films. Collagen treated with potassium sorbate has been used in the production of fermented sausage, to control spoilage during chilled storage.

Wax and Oil Coatings

Waxes and oils used as coatings include beeswax, carnauba wax, paraffin wax, polyethylene wax, shellac, fatty acids and monoglycerides. Their commercial use is mainly for fruits and vegetables, including citrus fruits, apples, tomatoes and cucumbers. Many of the fatty acids used in this application are derived from vegetable oils.

Sachet Technology

Modified-atmosphere packaging (MAP) and controlled-atmosphere packaging (CAP) have been used to preserve food products for many years. Both involve the introduction of a gas mixture, e.g. N_2 and CO_2 into a food pack to create conditions which limit chemical, physical and microbiological changes in foods. MAP is commercially successful, but has the disadvantage of repeated gas evacuation and flushing being needed to achieve the desired atmosphere in the pack. Also, this atmosphere often changes during distribution and storage.

A newer approach to controlling the atmosphere inside the pack is the use of sachet technology. This involves placing a sachet of gas absorber or generator into a package to create and maintain the required atmosphere inside it. This technique is easy to apply during manufacture, and sachets of O_2 absorbers and generators of CO_2 or ethanol vapour are widely used in food packaging.

O₂ Absorbers

Oxygen is essential for both oxidation and microbial growth, and so can lead to the deterioration and spoilage of foods. Therefore the depletion of O_2 from a package can significantly increase the shelf life of the product. O_2 absorbers remove O_2 chemically and are used, for example, with chilled pasta and meat products. They mainly prevent the growth of moulds, which are unable to grow if the O_2 level falls below 0.1%, and which grow slowly in comparison with bacteria.

A commonly used O_2 absorber uses finely divided iron, which is oxidized to ferric hydroxide. Two examples of commercially available sachets of O_2 absorbers are Ageless (Mitsubishi Gas Chemical Co.) and FreshPax (Multisorb Technologies Inc.).

Another technique for depleting the O_2 from a food pack is the use of packaging film which incorporates oxidizing chemicals or enzymes that bind O_2, e.g. glucose oxidase, alcohol oxidase, organic chelators. However these are not widely used for chilled food products, being enzyme-dependent and therefore temperature-sensitive.

Alcohol Vapour

Alcohol is a well-known antimicrobial substance. It can be entrapped in silica gel, which is then enclosed in a permeable sachet. The alcohol vaporizes and is deposited on the food surface, where it will inhibit the growth of many microorganisms. Alcohol-vapour sachets are used for extending the shelf life of chilled bakery goods such as pizza bases, by the control of mould growth. Ethicap, manufactured by Freund Industrial Co., contains silicon dioxide powder impregnated with food-grade alcohol and water. When placed in a package, the sachet absorbs moisture and releases alcohol vapour, which is taken up by the food product, where it exerts antimicrobial effects.

Applications

Fruits and Vegetables

High proportions of fruits and vegetables are lost after harvesting, due to deterioration and spoilage. Preservation is usually attempted by storage at a low temperature in a low-O_2 atmosphere, or by treatment with an antimicrobial agent. Typical modified atmospheres used for packaging fruits and vegetables comprise 3–5% CO_2, 2–5% O_2 and 90% N_2. Antimicrobial agents, such as food-grade fungicides, can be applied as an edible film or coating by spraying, or dipping into, a solution of antimicrobial agent in water. After dipping, light rinsing with fresh water removes some of the unabsorbed agent but leaves sufficient on the surface to exert a preservative effect. Dipping in fungicide and fungicide-impregnated wrapper have been used to delay the decay of peaches and nectarines during storage. Fungicides commonly used for fruits include benomyl, imazalil, and thiabendazole.

A technique currently under development for controlling the post-harvest loss of fruits and vegetables is the incorporation into an edible film or coating of antagonistic microorganisms, which would inhibit the growth of spoilage microorganisms. An example is the use of the yeast *Debaryomyces hansenii* to control the growth of *Bacillus subtilis* on peaches and of *Penicillium digitatum*, *P. italicum* and *Geotrichum candidum* on grapefruit and lemons.

Meat and Poultry

Modified-atmosphere and vacuum packaging are widely used for red meat. A modified atmosphere of 40–50% O_2 and 20% CO_2 is used for packaging raw meat, to maintain its pink colour, to suppress microbial growth and to extend its shelf life under refrigeration. Dipping in potassium sorbate reduces the total number of viable bacteria on meat surfaces at refrigerated temperatures. In seafood packaging, O_2 is needed to prevent the growth of anaerobic microorganisms indigenous to water: an example of modified-atmosphere packaging for fish is 80% CO_2 and 20% O_2.

Processed Foods

Antimicrobial packaging is used mainly in conjunction with modified atmospheres for processed foods. O_2 absorbers in particular are used to extend shelf lives, for example of cheese, fresh pasta, powdered drinks, bakery products, nuts, processed meat and dehydrated meat products. For example, pizza can be packaged in a tray of plastic with a high impermeability to O_2, using an atmosphere of 50% CO_2 and 50% N_2 to extend its refrigerated shelf life. Delicatessen products such as patés can be kept in good condition for up to 21 days in modified-atmosphere packaging containing 20–50% CO_2 balanced with N_2.

Regulations and Control

In the US, the laws governing substances that may be used in foods and in packaging materials are stated in Parts 170–189 of Title 21 of the Code of Federal Regulations (CFR). The law is enforced by the Food and Drug Administration (FDA). Part 170 (Food Additives) of Title 21 of the CFR indicates that any component of packaging materials that might migrate into foods is treated as a food additive, including antimicrobial agents incorporated into materials used either for packaging or for films or coatings. The use of substances other than those listed in the CFR must be approved by the FDA, and it is the responsibility of the manufacturer of the packaging material to ensure that it complies with the FDA regulations. In the US, meat and poultry products are regulated by the Department of Agriculture (USDA) and only food additives specifically permitted by the USDA may be used in some meat and poultry products. When a substance is permitted by the FDA or USDA for use in foods or packaging materials, the amount used

must not exceed the absolute minimum required to accomplish the intended effect.

Future Perspectives

Current trends suggest that in due course, packaging will generally incorporate antimicrobial agents and sealing systems will continue to improve.

One challenge to be overcome is the limitation of the applicability of antimicrobial agents to food surfaces. Only a few agents, e.g. ethanol, are capable of penetrating food surfaces. The combination of antimicrobial herbal extracts with a vaporizing system or within the packaging film might be possible. Many of these herbs contain volatile oils, that could be released when the packaging material is handled by the consumer. It might be possible to include inert material carrying an antimicrobial agent within the food. The release of the agent could be controlled by the characteristics of the packaging film or atmosphere. A packaging material may also be designed to release antimicrobial agents to sanitize the hands of a food handler, to minimize cross contamination. One can visualize this system being applied to raw products, especially meat and poultry. The focus of packaging in the past has been on the colour, size and integrity of the package. A greater emphasis on safety features associated with the addition of antimicrobial agents is perhaps the next area for development in packaging technology.

See also: **Chilled Storage of Foods**: Principles; Use of Modified-atmosphere Packaging. **Meat and Poultry**: Spoilage of Meat. Spoilage of Cooked Meats and Meat Products. **National Legislation, Guidelines & Standards Governing Microbiology**: Canada; European Union; Japan. **Packaging of Foods. Polymer Technologies for Control of Bacterial Adhesion.**

Further Reading

Bakker M and Eckroth D (1986) *The Wiley Encyclopedia of Packaging Technology*. New York: John Wiley.

Davidson PM and Branen AL (1993) *Antimicrobials in Foods*. New York: Marcel Dekker.

Griffin RC (1985) *Principles of Package Development*. Westport: AVI Publishing.

Jay JM (1996) *Modern Food Microbiology*, 5th edn. New York: Chapman & Hall.

Krochta JM, Baldwin EA and Nisperos-Carriedo M (1994) *Edible Coatings and Films to Improve Food Quality*. Lancaster: Technomic Publishing.

Rooney ML (1995) *Active Food Packaging*. New York: Blackie Academic & Professional.

US Food and Drug Administration (1997) *Code of Federal Regulations Title 21–Food and Drug, Parts 170 to 199*. Washington: Office of the Federal Register National Archives and Records Administration.

CIDER (HARD CIDER)

B Jarvis, Ross Biosciences Ltd, Ross-on-Wye, Herefordshire, UK

Introduction

Cider (cyder, US: hard cider) is an alcoholic beverage produced by the fermentation of apple juice; a related product (perry) is produced by the fermentation of pear juice. Cider has been produced for more than 2000 years in temperate areas of the world where apple trees flourish. The largest volume of cider is produced in England; but it is also produced in Argentina, Australia, Austria, Belgium, Canada, China, Finland, France (Normandy and Brittany), Germany, Ireland, New Zealand, South Africa, (northern) Spain, Sweden, Switzerland and the USA. Traditional cider-making in most countries is based on farmhouse production (originally for personal consumption), but commercial cider production has become increasingly important in many countries.

In England, commercial cider-making started during the late nineteenth century, although there is evidence of farmhouse cider having been sold commercially since the early eighteenth century. Total cider production in England in 1900 is estimated at 0.25×10^6 hl, of which about 0.025×10^6 hl was produced commercially. In 1996, the total production of cider was about 5.5×10^6 hl, of which 5.2×10^6 hl was produced by commercial companies. In 1996, total cider production in Europe was about 8.5×10^6 hl.

Cider Production

Ciders are made by fermentation of the juice of apples, often with added pear juice. The juice may be either fresh or reconstituted from concentrate. In England, France and Spain most cider is produced from the juice of special cultivars of 'cider' apples, referred to as bittersweet, bitter-sharp, sweet or sharp depending on the relative levels of tannins and acids. Ciders are also made from the juice of culinary or dessert apples. The alcohol content of cider made only from juice generally ranges up to about 6.5% alcohol by volume (abv), depending on the sugar content of the apple juice. In many countries, *chaptalization* (i.e. addition of fermentation sugars) is practised widely, in years when juice sugars are low, to increase the total fermentable sugar such that the fermentation product may contain up to 12% abv. Such 'strong' ciders are blended to produce commercial ciders, generally within the range 1.2–8.5% abv.

Preparation of the Cider Juice

The fruit is transported from the orchards to the cider mill, where it is washed and then milled using various types of equipment such as a knife mill. The milled fruit is pressed using either batch or continuous presses, and the pomace from the first pressing may be treated with water to maximize the yield of sugar and tannins. In some processes, the milled fruit may be liquefied by treatment with pectolytic and amylolytic enzymes, before centrifugation to separate the juice from residual solids. The juice is normally treated with SO_2 gas or sodium metabisulphite to a level of $100–200 \mu g \, l^{-1}$ and allowed to stand for about 24 h prior to use. If a clear juice is required, the cloudy pressed juice may be treated with pectinases and amylases; enzyme treatment is normal if the juice is to be concentrated for storage purposes. Concentrated juices are generally prepared in a multistage evaporator, and may have volatile aromas added back. The concentrate, at about 72° Brix, can be stored for 2 or 3 years at refrigeration temperatures without serious loss of quality. The spent apple pomace is used for the extraction of pectin (provided that enzyme pre-treatment has not been used), as cattle feed or as a soil conditioner.

Cider Fermentation

The juice is transferred to fermentation vats, where yeast nutrients such as ammonium phosphate, ammonium carbonate and pantothenic acid are added, together with glucose syrup and the appropriate yeast culture. Fermentation is allowed to proceed at 15–25°C until all the fermentable sugars have been used – usually taking about 3–8 weeks, depending on the temperature. The raw cider is sometimes chilled, to facilitate flocculation of the yeast, before being racked off from the lees and transferred to other vats for maturation. The maturation process can take up to 2 months, but the cider is often stored for > 1 year before further processing.

Final Preparation

The strong cider base (up to 12% abv) is centrifuged and/or fined to remove solids, and the resultant bright cider is blended to give an appropriate level of alcohol (usually 3.5–8.5% abv). At this stage, sweetener, colour and other ingredients may be added and the acid–sweetness balance adjusted, according to the organoleptic style of cider required. The blended cider is carbonated and packaged into bottles, cans, kegs or barrels for distribution and sale. Processes involved in the preparation of certain special ciders are discussed later.

Fermented cider and perry may be distilled to produce apple-based liquors (e.g. Calvados). The addition of distilled liquor to a product is not permitted if duty is to be levied on the basis of cider. However, blends of cider and distilled cider liquor may be sold as intermediate products (i.e. > 15% abv), e.g. Cider Royale.

The Microbiology of Apple Juice and Cider

The fermentation of apple juice to cider can occur naturally, through the metabolic activity of the yeasts and bacteria present on the fruit at harvest. These are transferred into the apple juice on pressing. Other microorganisms, from the milling and pressing equipment and the general environment, can also contaminate the juice at this stage. Examples of typical juice-associated organisms are shown in **Table 1**, together with an indication of their susceptibility to SO_2 and acid. Unless such organisms are inhibited, for example by the use of SO_2, a mixed fermentation occurs: this causes significant variations in organoleptic characteristics between batches, even if the composition of the apple juice remains constant.

Commercially, the preferred approach for the production of cider is control of the indigenous and adventitious microorganisms, followed by deliberate inoculation with a selected strain of yeast. However, even in this situation, the transfer of fermented juice into different maturation and storage vessels may result in a secondary fermentation, by microorganisms which occur naturally in the traditional oak vats which are frequently used. These organisms may produce beneficial or detrimental changes in the chemical and organoleptic properties of the final cider.

The Role of SO_2 in Apple Juice and Cider

The use of SO_2 as a preservative in cider-making is controlled by legislation in most countries. The maximum level permitted in the final product in Europe is $200 \, mg \, l^{-1}$ but different limits may be applied elsewhere. The addition of SO_2 to apple juice results in the formation of so-called sulphite addition compounds, through binding to carbonyl compounds. When dissolved in water, SO_2 or its salts produce a mixture of 'molecular SO_2', bisulphite and sulphite ions, the equilibrium of which is pH-dependent (**Fig. 1**). The antimicrobial activity of SO_2 is believed to be due to the 'molecular SO_2' moiety of the SO_2 which remains unbound (the so-called 'free' SO_2). Less SO_2

Table 1 Typical microbial contaminants of freshly pressed apple juice

Typical species	SO_2 sensitivity – (insensitive) ++++ (very sensitive)	Ability to grow in juice – (unable to grow) ++++ (grows well)
Yeasts		
Saccharomyces cerevisiae var. cerevisae	± or –	++++
Saccharomyces cerevisiae var. uvarum	± or –	++++
Saccharomyces cerevisiae var. carlsbergensis	± or –	++++
Saccharomycodes ludwigii	–	++++
Kloeckera apiculata	+++	++++
Candida pulcherrima	++++	++++
Pichia spp.	++++	++++
Torulopsis famata	++	++++
Rhodotorula spp.	++++	++++
Moulds		
Penicillium spp.	++	++++
Aspergillus spp.	++++	++++
Paecilomyces varioti	+	++++
Byssochlamys fulva	–	++++
Cladosporium spp.	++++	++
Botrytis spp.	+	++++
Bacteria		
Acetobacter xylinum	++	++++
Pseudomonas spp.	++++	–
Escherichia coli	++++	– (some strains ±)
Salmonella spp.	++++	–
Micrococcus spp.	++++	–
Bacillus spp.	– (spores)	–
Clostridium spp.	– (spores)	–

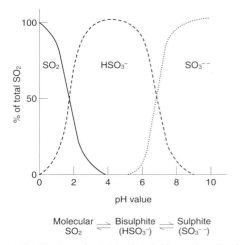

Figure 1 Distribution of sulphite, bisulphite and molecular SO_2 as a function of pH in aqueous solution. (Reproduced with permission from Hammond SM and Carr JG (1976).)

is needed in juices of high acidity: for instance, $15 \, mg \, l^{-1}$ of free SO_2 at pH 3.0 has the same antimicrobial effect as $150 \, mg \, l^{-1}$ at pH 4.0.

The binding of SO_2 is dependent on the nature of the carbonyl compounds present in the juice. Naturally occurring compounds which bind SO_2 include glucose, xylose and xylosone. If the fruit has undergone some degree of rotting, other binding compounds will include 2,5-dioxogluconic acid and 5-oxofructose (2,5-D-threo-hexodiulose). Such juices require increased additions of SO_2 if wild yeasts and other microorganisms are to be controlled effectively. The addition of SO_2 to fermenting juice results in rapid combination with acetaldehyde, pyruvate, and α-oxoglutarate, produced by the fermenting yeasts. Consequently, all additions of SO_2 must be completed immediately after pressing the juice, although provided initial fermentation is inhibited, further additions, to give the desired level of free SO_2, can be made during the following 24 h. Recent studies have shown that the presence of sulphite-binding compounds in fermented cider is dependent upon: the quality of the original fruit; the type of apple juice (i.e. cider, dessert or culinary juice) and whether pectinases were used for clarification; the strain of yeast and its ability to produce sulphite compounds; the fermentation conditions; and the extent to which yeast nutrients have been added.

Fermentation Yeasts

In traditional farmhouse cider-making, especially when the juice is not sulphite-treated, the indigenous yeasts of importance in fermentation include *Candida* spp., *Kloeckera apiculata* and *Saccharomyces* spp. Generally the *Candida* spp. and *Kloeckera* spp. die out within the first few days, but may be of importance in the initial fermentation. When the juice has been treated with sulphite, the fermentation process is carried out primarily by strains of *Saccharomyces* spp. especially *S. cerevisiae* vars. *cerevisiae*, *bayanus*, *capensis*, *carlsbergensis* and *uvarum*.

In commercial practice, specific strains are added to the sulphite-treated juice as a pure culture. The starter culture is prepared in the laboratory from freeze-dried or liquid-nitrogen-frozen cultures, which are resuscitated and then cultivated through increasing volumes of a suitable culture medium, to give an inoculum for use in a starter propagation plant. The nature of the cultivation medium varies, but is often based on sterile apple juice supplemented with appropriate nitrogenous substrates (e.g. yeast extracts) and with vitamins such as pantothenate and thiamin. Increasingly, commercially produced dried or frozen yeast cell preparations are used, either for direct vat inoculation or as inocula for the yeast propagation

Table 2 Desirable characteristics of yeasts for cider-making (modified from B Jarvis, MJ Forster and WP Kinsella (1995))

Attribute	Objective
Produces polygalacturonase	Hydrolyses soluble pectin
High vitality and viability, producing consistently high fermentation rate	Strong fermentation characteristics
Resistant to SO_2 and low pH	Competes well with 'wild' yeasts
Resistant to high OG and ethanol	Good commercial characteristics
Ferments to 'dryness'	Efficient utilization of sugars
Does not produce excessive foam	Avoids loss of product by frothing
Strongly flocculating	Ensures good racking-off
Minimal production of SO_2	Avoids excessive levels of SO_2
Minimal production of SO_2-binding compounds	Minimizes binding of SO_2
Non-producer of H_2S and acetic acid	Avoids undesirable metabolites
Compatible with malolactic bacteria	Important for malolactic fermentation
Good production of aroma compounds, organic acids and glycerol	Important for flavour production

plant. The condition of the yeast at pitching is critically important – the culture must have both high viability and high vitality if cider fermentation at high original gravity (OG) is to be effective. The ideal attributes of a cider yeast are summarized in **Table 2**.

After inoculation the starter yeasts, together with those SO_2-resistant wild yeasts selected naturally from the juice, increase in number from an initial level of about $10^5 \, cfu \, ml^{-1}$ to $5 \times 10^6 - 5 \times 10^7 \, cfu \, ml^{-1}$. Following an initial aerobic growth phase, the resulting O_2 limitation and high carbohydrate levels in the media trigger the onset of the anaerobic fermentation process. Fermentation typically takes some 3–8 weeks to proceed to dryness (i.e. specific gravity 0.990–1.000), at which time all the fermentable sugars have been converted to alcohol, CO_2 and other metabolites.

In controlled fermentations, a maximum temperature of 25°C will generally be permitted, although fermentations controlled at or below 16°C are common in some countries. Because of the exothermic nature of the fermentation process, temperatures of 30°C or above can be attained during periods of high ambient temperature. In Australia, it is common for temperatures as high as 35–40°C to occur in the vat, in the absence of a cooling facility. Generally, temperatures > 25°C are considered undesirable, because during rapid fermentation many desirable flavour compounds are not produced, some undesirable flavours are produced and alcohols and other metabolites may be lost by evaporation. In addition the activity of the desirable yeast strain may be

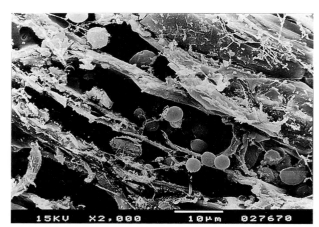

Figure 2 Electron micrograph of a section 1.2 cm below the surface of an oak wood block suspended in fermented cider for 10 weeks, showing colonies and individual microbial cells within the structure of the wood. (Reproduced with permission from Swaffield CH, Scott JA and Jarvis B (1997).)

inhibited, leading to 'stuck' fermentations and the growth of undesirable thermoduric yeasts and spoilage bacteria. Stuck fermentations can sometimes be restarted by the addition of nitrogen (10–50 mg l^{-1}), usually as ammonium sulphate or di-ammonium phosphate with thiamin (0.1–0.2 mg l^{-1}) or a yeast cell preparation.

At the end of fermentation, the yeast cells should flocculate and settle to the bottom of the vat – this process may be aided by chilling the cider in-vat. A certain amount of cell autolysis occurs, liberating cell constituents into the cider. The raw cider is racked off the lees as a cloudy product and transferred to storage vats for maturation. In some plants, the cider may be centrifuged or rough-filtered at this time. If the cider is left too long on the lees, the extent of autolysis may become excessive. This leads to an increase in nitrogenous materials which act as substrates for subsequent undesirable microbial growth and the development of off flavours, such as those caused by mercaptans, in the product.

Maturation and Secondary Fermentation

Traditionally, cider vats are made of wood (usually oak). The wood acts as a reservoir of microorganisms such as yeasts, lactic acid bacteria and acetic acid bacteria, which are important in the secondary fermentation of cider (**Fig. 2**). Modern processes using sterilizable stainless steel vats for fermentation and maturation lack the native microflora. If secondary fermentation is required, it is necessary either to inoculate the vats with a culture of malolactic organisms suitable for cider (NB malolactic cultures sold for wine are unsuitable for cider making!) or to use a process of 'back slopping', in which part of an earlier

batch of matured cider is used as an inoculum (with all the inherent risks of such action). The maturation vats are filled with racked-off cider and provided with an 'over-blanket' of CO_2 or otherwise sealed to prevent the ingress of air, which would stimulate the growth of undesirable film-forming yeasts (e.g. *Brettanomyces* spp., *Pichia membranaefaciens*, *Candida mycoderma*) and aerobic bacteria (e.g. *Acetobacter xylinum*).

During the maturation process, the growth of lactic acid bacteria (e.g. *Lactobacillus pastorianus* var. *quinicus*, *L. mali*, *L. plantarum*, *Leuconostoc mesenteroides* and other species, *Pediococcus* spp.) can occur extensively, especially if wooden vats are used. The 'malolactic' fermentation (MLF) results in the conversion of malic acid to lactic acid and also produces secondary metabolites. The MLF reduces the acidity of the cider and imparts subtle flavour changes which improve the flavour of the product. However, in certain circumstances, metabolites of the lactic acid bacteria may damage the flavour and result in spoilage, e.g. the excessive production of diacetyl (and its vicinal-diketone precursors), the butterscotch-like taste of which can be detected in cider at a threshold level of about 0.6 mg l^{-1}.

In ciders made without sulphite, such as the farmhouse ciders of the Basque region of Spain, it is common for malolactic fermentation to occur concurrently with the yeast fermentation. This leads to very complex flavour development and, since the lactic acid bacteria also metabolize some of the sugar, to reduced alcohol levels.

Spoilage and Pathogenic Microorganisms in Cider

Bacterial pathogens such as *Salmonella* spp., *Escherichia coli* and *Staphylococcus aureus* may occasionally occur in apple juice, being derived from the orchard soil, farm and processing equipment or human sources. Outbreaks of food poisoning have occurred due to *E. coli* O157:H7 strains in freshly pressed non-pasteurized apple juice (usually known in the USA as 'cider'). The acidity of both apple juice and cider normally prevents the growth of pathogens, which survive for only a few hours. However, the specific strains involved in food poisoning have a greater tolerance to acid and can survive for up to 30 days at 20°C in apple juice. These strains are destroyed by normal pasteurization conditions and do not survive in fermenting cider for more than 3 days, due to the interaction between alcohol and acidity. The presence of bacterial endospores from species of *Bacillus* and *Clostridium* may be indicative of poor plant hygiene. They can survive for long periods and are frequently found in cider but, because of its low pH value, they do not create a spoilage threat.

The juice from unsound fruits and juice contaminated within the pressing plant may show extensive contamination by microfungi, such as *Penicillium expansum*, *P. crustosum*, *Aspergillus niger*, *A. nidulans*, *A. fumigatus*, *Paecilomyces varioti*, *Byssochlamys fulva*, *Monascus ruber*, *Phialophora mustea*, and species of *Alternaria*, *Cladosporium*, *Botrytis*, *Oosporidium* and *Fusarium*. None of these are of particular concern in cider-making, except that spores of heat-resistant species (e.g. *Byssochlamys* spp.) can survive pasteurization and can grow in cider if it is not adequately carbonated.

The growth of *Penicillium expansum* on, or in, apples leads to the occurrence of the mycotoxin patulin in the apple juice. Many countries have imposed a guideline limit of $50\,\mu g\,l^{-1}$ for patulin. At high levels, patulin inhibits the yeasts used as starter cultures, but they rapidly develop resistance and metabolize the patulin within a few days, into a number of compounds including ascladiol. Patulin would not, therefore, be expected to occur in cider unless patulin-contaminated juice were added to sweeten the fermented cider.

The role of organisms such as *Brettanomyces* spp. and *Acetobacter xylinum* in the spoilage of ciders during the latter stages of fermentation and maturation is mentioned above. Of equal concern is the yeast *Saccharomycodes ludwigii*, which is often resistant to SO_2 levels as high as $1000-1500\,mg\,l^{-1}$. *S. ludwigii* can grow slowly during all the stages of fermentation and maturation, and is an indigenous contaminant of cider-making premises. Its presence in bulk stocks of cider does not cause an overt problem. However, if it is able to contaminate 'bright' cider at bottling, its growth will result in a butyric flavour and the presence of flaky particles which spoil the appearance of the product. Although the organism is sensitive to pasteurization processes it is not unknown for it to contaminate products at the packaging stage, either as a low-level contaminant of clean but nonsterile bottles or directly from the packaging plant and its environment. Clumps of the organisms may also survive if it is present in unfiltered cider at the time of pasteurization.

Environmental contamination of final products can also result from other yeasts such as *Saccharomyces cerevisiae* vars. *cerevisiae*, *bailii* and *uvarum*. These will metabolize any residual or added sugar, to generate further alcohol and, more importantly, to increase the concentration of CO_2. Strains of these organisms are frequently resistant to SO_2. In bottles of cider inoculated with such fermentative organisms, carbonation pressures up to 9 bar have been recorded.

To avoid any risk of burst bottles, it is essential to maintain an adequate level of free SO_2 in the final product, particularly in multi-serve containers which may be opened and then stored containing a reduced volume of cider. Alternatively a second preservative such as benzoic or sorbic acid can be used, where permitted by legislation. This precaution is not necessary for product packaged in single-serve cans and bottles, which receive a terminal pasteurization process after filling.

Special Secondary Fermentation Processes

Traditional 'Conditioned' Draught Cider This results from a live secondary fermentation process. After filling the barrels, a small quantity of fermentable carbohydrate is added to the cider, followed by an inoculum of active alcohol-resistant yeasts. The subsequent growth is accompanied by low-level fermentation, during which sufficient CO_2 is generated to produce a pétillant cider, together with a haze of yeast cells. Such products have a short shelf life in the barrel, of about 4–6 weeks.

Double Fermented Cider This is initially fermented to an alcohol content which is lower than normal (e.g. 6% abv), by restricting the amount of chaptalization sugar added or, sometimes, by stopping the fermentation process by chilling. The liquor is racked off immediately and is either sterile-filtered or pasteurized prior to transfer to a second sterile fermentation vat. Additional sugar or apple juice is added and a secondary fermentation is induced following inoculation with an alcohol-tolerant strain of *Saccharomyces* spp. Such a process permits the development of very complex flavours in the cider.

Sparkling Ciders These are normally prepared nowadays by artificial carbonation to a level of 3.5–4 vol. CO_2. Traditional sparkling ciders are prepared according to the 'Méthode champenoise'. After bright filtration of the fully fermented dry cider, it is put into bottles containing a small amount of sugar and an appropriate 'Champagne' yeast culture. The bottles are corked, wired and laid on their sides for the fermentation process, which will last 1–2 months at 5–18°C. Following this stage the bottles are placed in special racks with the neck in a downward position. The bottles are gently shaken each day to move the deposit down towards the cork, a process which can take up to 2 months. The disgorging process involves careful removal of the cork and yeast floc without loss of any liquid – sometimes the neck of the bottle is frozen to aid this process. The disgorged product is then topped up using a syrup of alcohol, cider and sugar prior to final corking, wiring and labelling. It is not difficult to understand why this process is rarely used nowadays!

Biochemical Changes during Cider-making

The chemical composition of cider is dependent on the composition of the apple juice, the nature of the fermentation yeasts, microbial contaminants and their metabolites, and any additives used in the final product.

Composition of Cider Apple Juice

Apple juice is a mixture of sugars (primarily fructose, glucose and sucrose), oligosaccharides and polysaccharides (e.g. starch), together with: malic, quinic and citromalic acids, tannins (i.e. polyphenols), amides and other nitrogenous compounds, soluble pectin, vitamin C, minerals and a diverse range of esters which give the juice a typical apple-like aroma (e.g. ethyl- and methyl-iso-valerate). The relative proportions are dependent on: the variety of apple; the environmental and cultural conditions under which it was grown; the state of maturity of the fruit at the time of pressing; the extent of physical and biological damage (e.g. rotting due to mould); and, to a lesser extent, the efficiency with which the juice was pressed from the fruit.

The treatment of fresh juice with SO_2 is also important in the prevention of enzymic and non-enzymic browning reactions of the polyphenols, and SO_2 complexes carbonyl compounds, to form stable hydroxy-sulphonic acids. If the apples contain a high proportion of mould rots, appreciable amounts of carbonyls such as 2,5-dioxogluconic acid and 2,5-D-threo-hexodiulose will occur.

Products of the Fermentation Process

The primary objective of fermentation is the production of ethyl alcohol. The biochemical pathways which govern this process are well-recognized (Embden–Meyerhof–Parnas pathway). Various intermediate metabolites can be converted to form a diverse range of other end products, including glycerol (up to 0.5%). Diacetyl and acetaldehyde may also occur, particularly if the process is inhibited by excess sulphite and/or uncontrolled lactic fermentation occurs. Other metabolic pathways may operate simultaneously, with the formation of long- and short-chain fatty acids, esters, lactones, etc. Methanol is produced in small quantities ($10–100\,mg\,l^{-1}$) as a result of demethylation of the pectin present in the juice.

The tannins in cider change significantly during fermentation, for instance chlorogenic, caffeic and p-coumaryl quinic acids are reduced, with the formation of dihydroshikimic acid and ethyl catechol. The most important nitrogenous compounds in apple juice are the amino acids asparagine, aspartic acid, glutamine and glutamic acid; smaller amounts of proline and 4-hydroxymethylproline also occur. Aromatic amino acids are virtually absent from apple juice. With the exception of proline and 4-hydroxymethylproline, the amino acids are largely assimilated by the yeasts during fermentation. However, leaving the cider on the lees for an appreciable length of time significantly increases the amino nitrogen content as a consequence of the release of cell constituents during yeast autolysis.

Inorganic compounds in cider are mostly derived from the fruit, and depend on the conditions prevailing in the orchard. Their levels do not change significantly during fermentation. Trace quantities of iron and copper may occur naturally. The presence of larger quantities results in significant black or green discoloration, due to the formation of iron or copper tannates, and flavour deterioration.

Changes during Cider Maturation

Maturation results in further changes in the composition of the cider, but their nature is only partly understood. The primary effect of the malolactic fermentation is the conversion of malic acid into lactic acid which, being a weaker acid, results in a reduction in the apparent acidity. Much of the lactic acid is then esterified, with the formation of ethyl, butyl and propyl lactates. This removes harshness and gives a more balanced, smoother flavour. Other desirable flavour changes arising from the malolactic fermentation include the production of small quantities of diacetyl, which gives a butterscotch flavour to the cider – although as noted above, excessive levels of diacetyl are undesirable.

Some strains of lactic acid bacteria also produce excessive quantities of acetic acid if residual sugar is present in the maturing cider. Sulphur aromas and flavours resulting from yeast autolysis are generally lost during maturation, although unpleasant sulphur compounds such as mercaptans may be produced if the cider is infected by film yeasts. Acetic acid may be formed either from the uncontrolled growth of heterofermentative lactic acid bacteria or, more commonly, from the growth of strains of *Acetobacter* spp. Butyric flavours are generally caused by the growth of *Saccharomycodes ludwigii* and mousy flavours (believed to be due to 1,4,5,6-tetrahydro-2-aceto-pyridine and related compounds) are generally ascribed to the growth of film yeasts such as *Brettanomyces*

Table 3 Some key flavour compounds in cider (modified from B Jarvis, MJ Forster and WP Kinsella (1995))

Group of compounds	Examples of important flavour metabolites[a]
Alcohols	**Ethanol**; **propan-1-ol**; **butanol-1-ol**; **iso-pentan-1-ol**; **heptan-1-ol**; **hexan-1-ol**; **2-** and **3-methylbutan-1-ol**; **2-phenylethanol**
Organic acids	**Malic**; **lactic**; *butyric*; *acetic*; **hexanoic**; nonanoic; octanoic; succinic
Aldehydes	**Acetaldehyde**; **benzaldehyde**; butylaldehyde; **hexanal**; nonanal
Carbonyls	Pyruvate; decalactone; decan-2-one
Esters	*Amyl*, *butyl* and *ethyl* acetates; **ethyl** and **butyl lactate**; diethyl succinate; **ethyl benzoate**; **ethyl hexanoate**; **ethyl guiacol**; **ethyl-2-** and **ethyl-3-methylbutyrate**; **ethyl octanoate**; **ethyl octenoate**; **ethyl decanoate**; **ethyl dodecanoate**
Sulphur compounds	*Methanediol*; *ethanthiol*; *methyl thioacetate*; *dimethyl-disulphide*; *ethyl-methyl-disulphide*; *diethyl-disulphide*
Others	*Diacetyl*; *1,4,5,6-tetrahydro-2-acetopyridine*

[a]Compounds in *italics* are generally considered undesirable when more than traces are present; compounds in **bold** are essential flavour constituents.

spp. **Table 3** illustrates some of the key flavour compounds found in cider.

See also: **Aceterobacter**. **Escherichia coli O157**: *Escherichia coli 0157:H7*. **Fermentation (Industrial)**: Basic Considerations; Media for Industrial Fermentations; Control of Fermentation Conditions. **Lactobacillus**: Introduction. **Packaging of Foods**. **Penicillium**: Introduction; *Penicillium* in Food Production. **Preservatives**: Permitted Preservatives – Sulphur Dioxide. **Saccharomyces**: Introduction; *Saccharomyces cerevisiae*; *Saccharomyces*: Brewer's Yeast. **Spoilage Problems**: Problems caused by Bacteria. **Starter Cultures**: Uses in the Food Industry; Importance of Selected Genera. **Wines**: Microbiology of Wine-making; The Malolactic Fermentation. **Yeasts**: Production and Commercial Uses.

Further Reading

Beech FW (1972) English Cider making: Technology, Microbiology and Biochemistry. In: Hockenhull DJD (ed.) *Progress in Industrial Microbiology*, Vol 11, pp. 133–213. Edinburgh: Churchill Livingstone.

Beech FW and Davenport RR (1983) New Prospects and Problems in the Beverage Industry. In: Roberts TA and Skinner FA (eds) *Food Microbiology: Advances and Prospects*. (S.A.B. Symposium Series No. 11), pp. 241–256.

Charley VLS (1949) *The Principles and Practice of Cider-Making*. London: Leonard Hill Ltd.

Dinsdale MW, Lloyd D and Jarvis B (1995) Yeast vitality during cider fermentation: two approaches to the measurement of membrane potential. *J. Inst. Brew.* 101, 453–458.

Hammond SM and Carr JG (1976) The antimicrobial activity of SO_2 – with particular reference to fermented and non-fermented fruit juices. In Skinner FA and Hugo WB (eds) *Inhibition and Inactivation of Vegetative Microbes*. S.A.B. Symposium Series No. 5, pp. 89–110. London: Academic Press.

Jarvis B (1996) Cider, perry, fruit wines and other alcoholic fruit beverages. In: Arthey D and Ashurst PR (eds) *Fruit Processing*. London: Blackie Academic and Professional.

Jarvis B, Forster MJ and Kinsella WP (1995) Factors affecting the development of cider flavour. In: Board RG, Jones D and Jarvis B (eds) *Microbial Fermentations: Beverages, Foods and Feeds*. S.A.B. Symposium Series No. 24, *J. Appl. Bacteriol. Symposium Supplement* 79: 5s–18s.

Lea AGH (1995) Cidermaking. In: Lea AGH and Piggott JR (eds) *Fermented Beverage Production*. P. 66. London: Blackie Academic and Professional.

Swaffield CH, Scott JA and Jarvis B (1997) Observations on the microbial ecology of traditional alcoholic cider storage vats. *Food Microbiology* 14: 353–361.

Williams RR (ed.) (1991) *Cider and Juice Apples: Growing and Processing*. Bristol: University of Bristol.

Citric Acid *see* **Fermentation (Industrial)**: Production of Organic Acids.

Citrobacter *see* **Salmonella**: Detection by Enzyme Immunoassays.

CLOSTRIDIUM

Contents
Introduction
Clostridium perfringens
Detection of Enterotoxins of Clostridium perfringens
Clostridium acetobutylicum
Clostridium tyrobutyricum
Clostridium botulinum
Detection of Neurotoxins of Clostridium botulinum

Introduction

Hans P Blaschek, Department of Food Science and Human Nutrition, University of Illinois, Urbana, USA

Characteristics of the genus *Clostridium*

The genus *Clostridium* contains physiologically and genetically diverse species involved in the production of toxins as well as acids and solvents. The broad range of mol % G+C values together with 16S rRNA cataloguing demonstrates a high degree of phylogenetic heterogeneity within the genus *Clostridium*. Cato and Stackebrandt indicated that the genus does not comprise a phylogenetically coherent taxon. From an evolutionary standpoint, members of this genus appear to have evolved during ancient times, perhaps during the anaerobic phase of evolution.

Species within the genus *Clostridium* have both medical and industrial significance. The genus *Clostridium* was originally described in 1880. Because of the observed heterogeneity, it should not be surprising that the genus is quite large, on the order of 100 species. The clostridia are ubiquitous and are commonly found in the soil, marine sediments and animal and plant products. They are typically found in the intestinal tract of humans and in the wounds of soft tissue infections of humans and animals. In order to be included within this genus, the isolate must be anaerobic or microaerophilic, able to produce an endospore-forming rod, Gram-positive or Gram variable and be unable to carry out dissimilatory sulphate reduction. There is considerable interspecies variability in these observed criteria, and intraspecies Gram-stain variability appears in some cases to be related to the age of the culture. Most clostridia are motile via peritrichous flagella; however, for some species such as *C. perfringens*, motility has not been observed.

Clostridial spores appear to be produced only under anaerobic conditions and in some species, sporulation occurs with considerable difficulty. In certain cases, special media have been formulated to promote sporulation. Depending on the species, endospores may occur in a central, subterminal or terminal position and, because of the slender nature of the mother cell sporangium, cells containing spores appear swollen, unlike the thicker bacilli. The location of the endospores within the cell has important taxonomic value. The heat resistance of clostridial spores is also a function of the species; however because of their foodborne association and pathogenic nature, the greatest concern is with the spores produced by *C. botulinum* and *C. perfringens*. These spores may be able to survive routine cooking procedures and germinate and resume vegetative growth once nutritionally and environmentally appropriate conditions return. Because of the spore-forming capability of this foodborne pathogen, *C. botulinum* is regarded as the 'target microorganism' for the development of appropriate time/temperature heat treatment relationships for canned food products.

Species within the genus *Clostridium* produce a wide diversity of exoproteins, many of which function as virulence factors. Some of these proteins are antigenic in nature and some have associated enzyme activity. An overview of the major and minor antigens produced by *C. perfringens* is given in the article on *C. perfringens*. **Table 1** lists a representative group of clostridial species which cause various diseases. The most important species with respect to human disease include *C. botulinum*, *C. perfringens*, *C. tetani* and

Table 1 Species of *Clostridium* involved in causing diseases

Species	Diseases
C. perfringens	Food poisoning, gas gangrene, necrotic enteritis, minor wound infection
C. tetani	Tetanus
C. botulinum	Botulism food poisoning
C. difficile	Pseudomembraneous enterocolitis
C. novyi	Gas gangrene
C. histolyticum	Gas gangrene
C. septicum	Gas gangrene

C. difficile. The role of toxins produced by these species in causing disease has been well characterized.

From the standpoint of metabolism, there appears to be a delineation between clostridia that are principally saccharolytic and those described as proteolytic. Genetic studies have demonstrated that strains falling into these two groups are unrelated with respect to DNA similarity. Following growth on carbohydrates, the clostridia usually produce mixtures of alcohols and organic acids. The clostridia use the Embden–Meyerhof–Parnas pathway for breakdown of monosaccharides. Although carbohydrates appear to be the preferred carbon source, metabolism of alcohols, amino acids and other organic compounds may also occur. Purines and pyrimidines have also been shown to be fermented by various species of clostridia. The industrial utility of the clostridia is enhanced by their ability to degrade and utilize a diverse group of polysaccharides. Various species of clostridia are able to degrade polymers such as cellulose, starch and pectin and produce useful products such as acids and solvents. The acetone–butanol–ethanol (ABE) fermentation using *C. acetobutylicum* growing on starch or molasses dominates the history of clostridial fermentations.

For most clostridial species, growth occurs most rapidly between pH 6.5 and 7.0 and at a temperature of 30–37°C, although some species, such as *C. perfringens* have very rapid growth (generation times as low as 10 min) at temperatures of 40–45°C. There are also a number of thermophilic clostridia which are able to grow up to a maximum temperature of 80°C. Because of the industrial potential of hydrolytic enzymes such as amylase, pullalanase and glucoamylase recovered from the thermophilic clostridia, these microbes have recently been the subject of intensive investigation. A list of representative thermophilic clostridial species is presented in **Table 2**.

With respect to the development of genetic systems (gene transfer, shuttle vectors, etc.) for the clostridia, the model species have been the food-borne pathogen, *C. perfringens* and the solventogenic clostridia, which include principally *C. acetobutylicum* and *C. beijerinckii*. Most of the early plasmid work initiated 20

years ago was carried out with *C. perfringens*, whereas the industrial significance of strains which are able to produce acetone and butanol has resulted in a renewed emphasis on genetic systems development in *C. acetobutylicum* and *C. beijerinckii*. The molecular tools developed over the last 20 years are now being used to investigate the mechanism of toxin production in the pathogenic clostridia (e.g. *C. perfringens* and *C. botulinum*) and to understand the molecular basis for acid and solvent production, (e.g. *C. acetobutylicum* and *C. beijerinckii*). With respect to the latter, the goal is to improve the fermentation characteristics of these industrially significant species.

Selected Clostridial Species

Clostridium perfringens

C. perfringens has been described as the most ubiquitous pathogenic bacterium in our environment. This anaerobic, Gram-positive bacterium is an inhabitant of the soil and the intestinal tract of both humans and animals. It produces as many as 12 biologically active toxins. Although primarily associated with food-borne disease, it is also responsible for causing gas gangrene, lamb dysentery, necrotic enteritis and minor wound infection.

Clostridium botulinum

Botulism food poisoning is caused by the consumption of food containing heat-labile neurotoxin produced by *C. botulinum*. *C. botulinum* was first isolated in 1895 by E. van Ermangen from salted ham. The causative microorganism was named *Bacillus botulinus* (from the Latin 'botulus' meaning sausage). It is described as an intradietic intoxication in which the exotoxin is produced by the microorganism during growth on the food. The types of *C. botulinum* are identified by neutralization of their toxins by the antitoxin. There are seven recognized antigenic types of *C. botulinum*, A–G (**Table 3**). In addition to toxin production, the types are differentiated on the basis of their ability to produce proteolytic enzymes. The production of proteolytic enzymes by *C. botulinum* when present on food results in a putrid, unpleasant odour which can be a useful deterrent to consumption. Although strains of *C. botulinum* are variably proteolytic, they are always saccharolytic, and are able to ferment glucose with the production of energy as well as acid and gas. Most outbreaks (ca. 72%) of botulism have been traced to home canned foods and vegetables in particular. These outbreaks have been traced to foods that have been improperly handled or insufficiently heated to destroy spores. In the United States, *C. botulinum* types A and B are most common, whereas in Europe, meat products

Table 2 Optimal growth temperatures of thermophilic *Clostridium* species

Species	Optimal growth temperature (°C)
C. thermoaceticum	58
C. thermosulfurogenes	60
C. thermocellum	60
C. fervidus	68
C. thermosuccinogenes	58–72
C. thermobutyricum	57
C. stercorarium	65

Table 3 Occurrence of *C. botulinum* types

Location	Type	Proteolytic	Disease	Antitoxin
Western USA, Canada	A	+	Human	Specific
Europe, Eastern USA	B	+/−	Human	Specific
Widespread	Cα	−	Paralysis of birds	X-react Cα and Cβ
Widespread	Cβ	−	Forage poisoning	X-react with D
South Africa, Russia, USA	D	−	Cattle	X-react with Cβ
Northern hemisphere	E	−	Human	Specific
Japan	F	+	Human	Specific
Argentina	G	+	No disease	Unknown

have frequently served as the vehicle and botulinal food poisoning is due primarily to type B strains. *C. botulinum* is a strictly anaerobic, Gram-positive rod which produces heat-stable spores which are located subterminally on the mother cell sporangium. The microorganism is motile via peritrichous flagella. The neurotoxins produced by *C. botulinum* and *C. tetani* comprise the most potent group of bacterial toxins known. The toxins act by inhibiting the release of neurotransmitters from presynaptic nerve terminals inducing a flaccid paralysis (*C. botulinum*) or a spastic paralysis (*C. tetani*). Although the symptoms induced by the toxins appear dramatically different, the toxins have similarities in their structures and modes of action.

Detection of *C. botulinum* Neurotoxins The botulinum neurotoxins are simple proteins composed of only amino acids. These toxins are among the most toxic substances known. Ingestion of as little as 1–2 μg toxin may prove fatal. Although the toxins produced by *C. botulinum* have all been purified and characterized, type A neurotoxin is best characterized and was the first to be purified. The complete covalent structure of the proteolytically processed, fully active type A neurotoxin has been determined. In addition to being a neurotoxin, haemagglutinin activity is normally associated with type A toxin. Haemagglutinin is believed to stabilize the toxin in the gut. The toxin molecule that is produced by a toxigenic culture is referred to as a progenitor toxin and consists of a toxic and an atoxic component. The progenitor toxin is the precursor of the more toxic derivative toxin. The progenitor toxins can be converted into the derivative form by the action of proteases in the digestive tract of the host or via the direct action of proteolytic enzymes associated with the microorganism. Unlike staphylococcal toxins, botulinal toxins are heat-sensitive proteins. They are destroyed by boiling for 10 min. Therefore, a food can be rendered nontoxic by heating, although the cooking of a suspect food is not considered a worthwhile risk. On the other hand, consumption of low levels of spores by a healthy adult will apparently do no harm. Tryptophan has been

shown to be required for toxin production together with carbon dioxide.

Typically, one portion of the food to be examined is set aside and examined for the presence of the microorganism, and the other portion is used in toxicity testing. Food samples containing suspended solids are centrifuged and the supernatant fluid examined for toxin. Solid food is extracted with an equal volume of gel-phosphate buffer. The macerated food sample is centrifuged under refrigeration and the supernatant used for assay of the toxin. Food samples containing toxins of non-proteolytic *C. botulinum* may require trypsin activation in order to be detected. In this case, the trypsin-treated preparation is incubated for 1 h with gentle agitation.

The mouse lethality assay was the first method employed for detection of toxins produced by food-borne pathogens, and although still used for assay of botulinal toxins, its use has become more limited with the advent of alternative assays. The approach when using the mouse lethality assay is quite straightforward. Pairs of mice are injected intraperitoneally (i.p.) with trypsin-treated and untreated preparations. A portion of untreated supernatant fluid or culture is heated for 10 min at 100°C. All injected mice are observed for 3 days for symptoms of botulism or death. If, after 3 days, all mice except those receiving the heated preparation have died, the toxicity test should be repeated using higher dilutions of supernatant fluids or cultures. This approach allows determination of the minimum lethal dose (MLD) as an estimate of the amount of toxin present. From these data, the MLD per millilitre can be calculated. The precision of the mouse lethality assay for estimating the activity of *C. botulinum* toxin has been shown to be of the order of ±5%. Protocols for typing of the toxin involves rehydrating antitoxins with sterile physiological saline. Antisera can be obtained from the Centers for Disease Control, Atlanta, GA or from the FDA, Washington, DC. Monovalent antitoxins to the various types are employed. Mice are injected with the respective monovalent antitoxins 30–60 min before challenge with toxic samples. A pair of unprotected mice (no injection of antitoxin) is injected with the toxic sample as a control. Mice are observed for 48 h for symptoms of botulism and to record deaths.

Additional approaches for the detection of botulinal toxins include gel diffusion, specifically electro-immunodiffusion, which has a reported sensitivity of 5 mouse LD_{50} per 0.1 ml and the polymerase chain reaction (PCR) which has been applied for detection of *C. botulinum* types A to E toxin genes with a reported sensitivity of 10 fg. Another approach is the evanescent wave immunosensor to detect type B botulinum toxin. The sensor detects fluorescently tagged, toxin-bound antibody. The enzyme-linked immunosorbent assay (ELISA) system has also been successfully used for the detection of *C. botulinum* toxins. For type A *C. botulinum* toxin, a double-sandwich ELISA detected 50–100 mouse LD_{50} of type A and less than 100 mouse LD_{50} of type E. A double-sandwich ELISA using alkaline phosphatase was able to detect 1 mouse LD_{50} of type G toxin. *C. botulinum* toxin type A was detected at a level of 9 mouse LD_{50} per millilitre when using a monoclonal antibody.

The Solventogenic Clostridia: *C. acetobutylicum* and *C. beijerinckii*

The fermentation of carbohydrates to acetone, butanol and ethanol (ABE) by the solventogenic clostridia is well known. For an overview of developments in the genetic manipulation of the solventogenic clostridia for biotechnology applications, the reader is referred to the further reading list. Currently, this value-added fermentation process is attractive for several economic and environmental reasons. Prominent among the economic factors is the current surplus of agricultural wastes or by-products that can be utilized as inexpensive fermentation substrates. Examples include mycotoxin-contaminated corn which is unsuitable for use as animal feed and 10% solids light corn steep liquor (CSL) which is a low value by-product of the corn wet milling industry.

It has been suggested that the instability of certain solventogenic genes (*ctfAB*, *aad*, *adc*) may be the cause of strain degeneration in *C. acetobutylicum*. Specifically, the genes for butanol and acetone formation in *C. acetobutylicum* ATCC 824 were found to reside on a large 210 kb (pSOL1) plasmid whose loss leads to degeneration of this strain. Eight genes concerned with solventogenic fermentation in *C. beijerinckii* 8052 were found at three different locations on the genome. In *C. beijerinckii* 8052, genomic mapping studies suggest that the *ctfA* gene is localized on the chromosome and is co-located next to the acetoacetate decarboxylase gene. An examination of the effects of added acetate on culture stability and solvent production by *C. beijerinckii* showed that one of the effects may be to stabilize the solventogenic

genes, and thereby prevent strain degeneration. In order to examine this hypothesis, further genetic analysis of the solventogenic genes will need to be carried out.

The application of amplified fragment length polymorphism (AFLP) technology as an important tool in bacterial taxonomy has been demonstrated and protocols have been adapted for use with bacteria in general and clostridial species in particular. AFLP is based on selective PCR amplification of restriction fragments from a total genomic DNA digest. The technique involves restriction of the DNA and ligation of oligonucleotide adapters, selective amplification of restriction fragments, and gel analysis of amplified fragments. DNA sequencing gels are able to identify length differences as small as one nucleotide. AFLP technology is proposed as a means to identify and characterize subtle, genomic-level changes which occur in the hyper-butanol producing *C. beijerinckii* BA101 mutant which was produced using chemical mutagenesis. Once the polymorphic loci have been identified, the genetic markers generated from AFLP can be further used to clone and sequence these fragments. Differences observed for the *C. beijerinckii* BA101 strain at the sequence level can be directly compared to the parent strain. Determination of the genomic alterations responsible for the physiology associated with the *C. beijerinckii* BA101 hyper-butanol phenotype will ultimately lead to the development of a strategy for engineering a strain of *C. beijerinckii* with enhanced solvent-producing characteristics for industrial applications.

The genome of *C. acetobutylicum* has been sequenced by Genome Therapeutics Corporation (Waltham, MA). The size of the *C. acetobutylicum* genome was found to be 4.11 Mb, with an overall G+C ratio of 29.2%. There is an expectation for 4200 genes and analysis of the sequence has revealed similarity, although not necessarily functionality, to a number of antibiotic-resistant genes, clostridial toxin genes and various substrate hydrolytic genes. It is expected that analysis of the chromosome sequence will provide important information regarding the phylogenetic relatedness of the solvent-producing clostridia.

Recommended Methods of Detection and Enumeration in Foods

The clostridia generally can be isolated on nutritionally complex media which are appropriate for the cultivation of anaerobes. This may, for example, include blood agar and cooked meat medium. Tryptone–glucose–yeast extract (TGY) medium is easy to prepare and can meet the nutritional requirements of many different species of clostridia. The media should be reduced, normally by the addition of L-cysteine or sodium thioglycollate. In order to selectively recover clostridia from the soil or intestinal contents, it is useful to heat the sample at 80°C for 10 min. This process destroys most vegetative cells and allows the spores to predominate. It has also been shown to be useful for the recovery and regeneration of solvent-producing clostridia such as *C. acetobutylicum* and *C. beijerinckii*.

Methods for detection and enumeration of *C. perfringens* are found in a separate article. Although not as fastidious as *C. perfringens*, the nutritional growth needs of *C. botulinum* are complex, and include a number of amino acids, B vitamins and minerals. Routinely, *C. botulinum* is cultivated in brain–heart infusion or cooked meat medium. Although many foods satisfy the nutritional requirements for growth, not all provide anaerobic conditions. Growth in foods can be restricted if the product is of low pH, has low a_w, a high concentration of salt or an inhibitory concentration of a preservative such as sodium nitrite. A food may contain viable cells of *C. botulinum*, and yet it may not cause disease. For this reason the focus is primarily on detection of the neurotoxin (see above). However, because of the heat lability of *C. botulinum* neurotoxin, processed foods should be examined for the presence of viable cells as well as toxin.

The detection of viable *C. botulinum* typically involves enrichment. Cooked meat medium or trypticase–peptone–glucose–yeast extract (TPGY) is inoculated with 1–2 g solid or 1–2 ml liquid food and incubated. If the organism is suspected of being a nonproteolytic strain, TPGY containing trypsin should be used. After 7 days' incubation, the culture is examined for gas production, turbidity and digestion of meat particles. The culture is also examined microscopically. A typical cell shows distention of the mother cell sporangium due to the presence of the spore which results in a bulging or swollen appearance. If enrichment results in no growth after 7 days, the sample may be incubated for an additional 10 days to detect injured cells or spores. Pure cultures of *C. botulinum* are isolated by pre-treatment of the sample with either absolute alcohol or heat treatment (typically 80°C for 10 min). Heat or ethanol-treated cultures may be streaked on to anaerobic egg yolk agar in order to obtain distinct and separate colonies. The selection of typical *C. botulinum* colonies involves using a sterile transfer loop in order to inoculate each isolated colony into TPGY or cooked meat medium broth. Cultures are incubated for 7 days as described above and tested for toxin production. Repeated serial transfer through enrichment media

may help to increase the cell numbers enough to permit pure colony isolation.

C. botulinum and *C. perfringens* are particularly important species in the food industry because of their ability to produce heat-stable spores and their ability to grow rapidly under anaerobic conditions. Although normally producing only a mild form of food poisoning, *C. perfringens* is of particular concern to the food service industry in those cases where food is prepared in advance, reheated and held on steam tables. It is primarily problematic because of its ubiquitous nature and rapid growth rate given appropriate nutritional and environmental conditions. Because of the devastating nature of botulism food-borne illness, minimum heating times for ensuring the safety of canned foods have been developed with the *C. botulinum* microorganism in mind.

See also: **Clostridium**: *Clostridium perfringens*; Detection of Enterotoxins of *C. perfringens*; *Clostridium acetobutylicum*; *Clostridium tyrobutyricum*; *Clostridium botulinum*; Detection of Neurotoxins of *C. botulinum*.

Further Reading

Andreesen JR, Bahl H and Gottschalk G (1989) Introduction to the Physiology and Biochemistry of the Genus *Clostridium*. In: Minton NP and Clarke DJ (eds) *Clostridia*. P. 27. New York: Plenum Press.

Blaschek HP and White BA (1995) Genetic systems development in the clostridia. *FEMS Microbiol. Rev.* 17: 349–356.

Cato EP, George WL and Finegold SM (1986) Genus *Clostridium*. In: Sneath PHA, Mair NS, Sharpe ME and Holt JG (eds) *Bergey's Manual of Systematic Bacteriology*. P. 1141. Baltimore: Williams and Wilkins.

Hauschild A (1989) *Clostridium botulinum*. In: Doyle MP (ed.) *Foodborne Bacterial Pathogens*. P. 112. New York: Marcel Dekker.

Jay JM (1996) *Modern Food Microbiology*, 5th edn. P. 220. New York: Chapman & Hall.

Johnson JL and Chen J-S (1995) Taxonomic relationships among strains of *Clostridium acetobutylicum* and other phenotypically similar organisms. In: Durre P, Minton NP, Papoutsakis ET and Woods DR (eds) *Solventogenic Clostridia. FEMS Microbiol. Rev.* 17: 233–240.

Kautter DA, Solomon HM and Rhodehamel EJ (1992) *Bacteriological Analytical Manual*, 7th edn. P. 215. Arlington, VA: AOAC International.

Morris JG (1993) History and future potential of the clostridia in biotechnology. In: Woods DR (ed.) *The Clostridia and Biotechnology*. P. 1. Stoneham, MA: Butterworth–Heinemann.

Steinhart CE, Doyle ME and Cochrane BA (1996) *Food Safety 1996*. P. 404. New York: Marcel Dekker.

Sugiyama H (1990) In: Cliver DO (ed.) *Foodborne Diseases*, p. 108. San Diego: Academic Press.

Wrigley DM (1994) In: Hui YH, Gorham JR, Murrell KD and Cliver DO (eds) *Foodborne Disease Handbook: Diseases Caused by Bacteria.* Vol. 1, p. 97. New York: Marcel Dekker.

Clostridium perfringens

Hans P Blaschek, Department of Food Science and Human Nutrition, University of Illinois, Urbana, USA

Clostridium perfringens Food Poisoning

Food poisoning caused by *C. perfringens* was suggested as early as 1895 by Klein. The nature of the food poisoning was recognized after World War II due to the precooking and hoarding of meat rations. In 1953, a number of outbreaks occurred in Great Britain; the foods involved and the microorganism responsible were identified. It is estimated that *C. perfringens* accounts for approximately 10% of the food-borne disease outbreaks in the United States. *C. perfringens* has also been described as the third most common cause of bacterial food-borne disease after *Staphylococcus aureus* and *Salmonella* spp.

C. perfringens food poisoning is typically associated with banquet or cafeteria-style settings, during which bulk foods are prepared for a large number of people. The US Centers for Disease Control has reported that there are about 10 000 cases of *C. perfringens* food poisoning each year, with approximately 100 cases associated with each outbreak. *C. perfringens* food poisoning occurs because foods (particularly meat or poultry) are improperly treated by subjection to long, slow cooking followed by non-refrigerated storage and inadequate reheating procedures. The greater involvement of meats and poultry may be due to the higher incidence of *C. perfringens* food-poisoning strains in these foods and the nutritionally fastidious nature of this microorganism. *C. perfringens* requires at least 13 amino acids as well as vitamins and nucleotides for growth. Up to 24% of veal, pork, and beef samples have been shown to contain heat-resistant *C. perfringens* endospores.

Since *C. perfringens* is widely distributed in nature, it must be accepted that it will be present in many foods and cannot be eradicated from our food supply. Prevention of *C. perfringens* food poisoning must be concerned with the control of outgrowth or germination of spores and the subsequent multiplication of vegetative cells in cooked foods. *C. perfringens* food poisoning is prevented by rapid and adequate cooling and reheating to prevent growth of the microorganism.

In 1993, outbreaks of *C. perfringens* were reported in Ohio and Virginia, during which nearly 300 individuals became ill. One of the largest outbreaks ever reported was that following a banquet in New York City attended by 1800 people, over 900 of whom developed symptoms of *C. perfringens* gastroenteritis. Most outbreaks of *C. perfringens* food poisoning have been associated with commercially prepared food in restaurants and institutions. Outbreaks at home are less common and more likely to go unreported.

In addition to causing food poisoning, *C. perfringens* is also responsible for gas gangrene, necrotic enteritis, lamb dysentery and minor wound infections. Recent evidence suggests that the microorganism has been implicated in sporadic cases of diarrhoea, antibiotic-associated diarrhoea, and diarrhoea in chronic care facilities. It is an important cause of food-borne illness because of its widespread occurrence. It has been described as the most ubiquitous pathogenic bacterium in our environment and is commonly found in the soil, marine and fresh water sediments and in the intestinal tract of humans and animals. Because it is an inhabitant of the animal intestinal tract, it can easily contaminate ground beef and ground poultry during processing.

Characteristics of Clostridium perfringens

C. perfringens is a Gram-positive rod, non-motile and encapsulated. It is typically 2–4 µm long by 0.8–1.5 µm wide with blunt ends. In foods or other complex media, the bacilli may appear shorter and fatter. Hydrogen sulphide is produced and most strains produce a 'stormy fermentation of milk' reaction which involves the rapid formation of a firm, tight clot of casein which is torn by gas bubbles and rises to the surface. Although the microorganism is not regarded as strictly anaerobic, a reduced oxidation reduction potential of ca. −45 mV is necessary for rapid growth. Sporulation occurs with difficulty and special media such as Duncan–Strong or Ellners have been developed which induce sporulation. Spores are rarely observed in smears from foods. However, when formed, they are subterminal and oval. The optimum growth temperature for *C. perfringens* is 37–45°C, whereas the growth range is between 20 and 50°C. *C. perfringens* is sensitive to low temperature storage. Vegetative cells, but not spores, inoculated into meat are slowly inactivated when held at 1, 5, 10 or 15°C. The lowest temperature for growth is 15°C, and there is no growth at 55°C. The optimum pH for growth is 6.0–7.0. When grown at the optimum temperature, the generation times can be as short as 7–10 min, which makes *C. perfringens* one of the most rapidly growing microbes. The consequences of this are obvious. Given the appropriate environmental conditions, *C. perfringens* is able to proliferate rapidly to a high cell population.

The spores of *C. perfringens* differ widely in their resistance to heat, and heat-sensitive and heat-resistant strains have been observed. Heat-resistant spores have been shown to survive 100°C for 5 h. Also, cooked meat exerts a protective effect and enhances the heat resistance of spores. Several studies have shown that spores of *C. perfringens* may survive routine cooking procedures. At 50°C, an interesting phenomenon called the 'Phoenix effect' occurs. Most vegetative cells introduced as inoculum perish in the first few hours; however, the survivors start multiplying at their maximum rate and continue to do so for several hours. By taking advantage of the high temperature tolerance, *C. perfringens* can be isolated from mixed cultures.

Classification

Twelve soluble antigens have been detected in *Clostridium perfringens* culture filtrates, all of which are protein in nature; some are well-known enzymes such as collagenase, proteinase, lecithinase, hyaluronidase and deoxyribonuclease. *C. perfringens* is divided into five types (A to E) on the basis of the production of four major necrotizing or lethal toxins: alpha, beta, epsilon and iota (**Table 1**). All strains produce alpha toxin (lecithinase or phospholipase C). Alpha toxin is a multifunctional metalloenzyme which is responsible for the cytotoxicity, necrosis and haemolysis observed in gas gangrene caused by *C. perfringens*. The alpha toxin gene has been cloned and sequenced. Comparison of the amino acid sequences of *C. perfringens* alpha toxin and the phospholipase C derived from *Bacillus cereus* revealed extensive homologies and the presence of 65 identical amino acid residues. *C. perfringens* alpha toxin attacks membrane phosphorylcholine associated with intestinal villus cells,

but does not appear to play an important role in human food poisoning. Beta-toxin is produced during vegetative growth of *C. perfringens* and appears to be associated with an intestinal disease called necrotic enteritis. This disease has had a history of affecting poorly nourished individuals in postwar Germany and New Guinea. Epsilon toxin is associated with gastrointestinal diseases in livestock. Iota toxin is normally associated with type E strains and causes necrosis. The methodology used for typing of *C. perfringens* strains involves using specific antisera and examining neutralization by injecting the toxin–antiserum mixture into mice or into the skin of guinea pigs. The food poisoning strains belong to type A and produce relatively heat-resistant spores. In addition to these toxins, *C. perfringens* produces an enterotoxin which is a spore-specific protein (i.e. its production occurs together with that of sporulation). There is a high correlation between enterotoxin production by *C. perfringens* strains and their ability to cause food poisoning. The administration of 8–10 mg of purified enterotoxin to healthy adults has been shown to cause food poisoning.

Clinical Features of Disease

Symptoms of *C. perfringens* food poisoning are characterized by severe diarrhoea and lower abdominal cramps. Normally there is no vomiting, fever, nausea or headache. The incubation period is usually 8–24 h before the onset of symptoms. Symptoms generally abate within 12–24 h. Fatalities mainly occur among debilitated persons (i.e. the elderly) and average less than one per year. Diagnosis of *C. perfringens* food poisoning is confirmed by isolating *C. perfringens* with the same serotype from the faeces of patients and the implicated food. The detection of enterotoxin in faeces aids in confirmation of the disease.

Mechanism of Intoxication

The mechanism of intravital (in vivo) intoxication by *C. perfringens* involves the ingestion of 10^6–10^7 living cells per gram of food. Because *C. perfringens* is able to grow rapidly under optimum conditions, it is able to reach the threshold level in only a few hours. *C. perfringens* enterotoxin is produced in the large intestine during sporulation of the microorganism and is released upon lysis of the sporangia. The function of the enterotoxin during sporulation is not yet understood. *C. perfringens* enterotoxin has a molecular weight of 36 kD, an isoelectric point of 4.3, and is heat sensitive (i.e. it is destroyed after heating at 60°C for 10 min).

The relationship between enterotoxin production

Table 1 Distribution of major lethal toxins among the types of *C. perfringens*

| Type | Disease | Toxin | | | |
		Alpha	Beta	Epsilon	Iota
A	Food poisoning, gas gangrene	+	–	–	–
B	Lamb dysentery	+	+	+	–
C	Necrotic enteritis, enterotoxaemia of sheep, lambs, piglets	+	+	–	–
D	Enterotoxaemia of sheep, goats, cattle	+	–	+	–
E	Pathogenicity doubtful	+	–	–	+

and sporulation in *C. perfringens* was demonstrated by the use of mutants with an altered ability to sporulate. When the mutants reverted to sporulation, enterotoxin production was demonstrated. The peak for toxin production is just before lysis of the cell sporangium and the toxin is released with the spores. In culture media, enterotoxin is produced only where endospore formation is permitted. Enterotoxin can be detected in cells about 3 h after inoculation of vegetative cells into media that encouraged sporulation. Enterotoxin accumulates intracellularly, and because of limited solubility, is able to form inclusion bodies. Most heat-resistant strains that sporulate well in Duncan–Strong medium produce high concentrations of enterotoxin. Heat-sensitive spore-formers do not sporulate as well. When spores are formed, the toxin can be detected outside the cell in the culture filtrate after about 10 h. The toxin production peak coincides with the release of free mature spores from the sporangia. Although cells sporulate readily in the intestinal tract, sporulation in cooked foods is poor. The ingestion of preformed enterotoxin in food as would be the case in an intradietic intoxication is not normally an issue with *C. perfringens*, as the food is considered to be unpalatable in those rare cases when sporulation and enterotoxin can be demonstrated.

In the small bowel, enterotoxin has been shown to bind to a brush border membrane receptor of intestinal epithelial cells, which then induces a calcium ion-dependent breakdown of permeability, resulting in the loss of low-molecular-weight metabolites and ions. This loss alters intracellular metabolic function and eventually results in cell death. *C. perfringens* enterotoxin may act as a superantigen, and specifically stimulates human lymphocytes. Therefore, pathogenesis of *C. perfringens* food poisoning may involve a massive release of inflammatory factors via its reaction with a large proportion of T lymphocytes. The *C. perfringens* enterotoxin gene has been cloned and the amino acid composition of the protein determined. The cloned gene has been a useful tool for the epidemiological screening of *C. perfringens* strains isolated from food poisoning outbreaks. Comparative studies suggested that hybridization with an enterotoxin gene probe was more reliable than an immunologically based assay for detecting enterotoxigenic *C. perfringens* strains.

Factors leading to *C. perfringens* food poisoning are fairly clear cut. Inadequate cooking temperatures will allow the survival of spores of *C. perfringens*. The danger exists in the prolonged cooling of cooked meat containing small numbers of surviving spores. These spores are able to germinate and grow rapidly at holding temperatures around 45–50°C, as may occur during the malfunction of a steam table. Meat and poultry dishes with histories of storage at warm temperatures (i.e. below 60°C) for at least two hours after cooking are common factors in almost all outbreaks due to *C. perfringens*.

Detection and Enumeration

Because *Clostridium perfringens* vegetative cells are sensitive to cold temperatures, food samples should be examined as quickly as possible. Since confirmation of *C. perfringens* food poisoning depends on the detection of a large number of cells in the implicated food, cold storage of samples may result in a false negative confirmation. In order to minimize cell death during transport and storage, it is recommended that food samples be mixed 1 : 1 (wt/vol) with 20% glycerol and stored on solid CO_2 or at −60°C. Implicated food samples are aseptically transferred to a sterile container and a suitable diluent (normally peptone fluid) is added to bring about a 1 : 10 dilution. The food is subsequently stomached in order to bring about uniform homogenization of the sample. Serial dilutions are prepared over a range of 10^{-1} to 10^{-6} using peptone dilution blanks. A volume (0.1 ml) of each dilution is spread plated or pour plated (ISO method) on a suitable selective medium such as tryptose-sulphite-cycloserine (TSC) agar containing egg yolk emulsion (**Table 2**). TSC with or without egg yolk has been adopted as official first action for the presumptive enumeration of *C. perfringens* in foods by the Association of Official Analytical Chemists and also in the ISO standard method. After the inoculum has absorbed into the medium, the plates are overlaid with 10 ml TSC agar without added egg yolk. Plates are incubated for 24 h at 37°C in an anaerobic jar or hood. After incubation, sulphate reducing clostridial colonies are black due to the reduction of sulphite to H_2S and are further characterized by an opaque halo surrounding the colony. Opalescence or halo pro-

Table 2 Composition of tryptose–sulphite–cycloserine medium for presumptive identification and enumeration of *Clostridium perfringens*

	$g\ l^{-1}$
Tryptose (Difco)	15
Yeast extract	5.0
Soytone	5.0
Sodium metabisulphite	1.0
Ferric ammonium citrate	1.0
Agar	20
Cycloserine[a]	0.4

[a] Dissolved separately in 60 ml water at 50–60°C.
pH adjusted to 7.6 prior to autoclaving.
8 ml egg yolk emulsion (50% in saline) per 100 ml medium may be added, but is normally omitted from overlay.

duction is due to lecithinase (alpha toxin) activity during which the lecithin contained in egg yolk is broken down into phosphorylcholine and a diglyceride. This is termed the Nagler reaction.

The antibiotic cycloserine is added to TSC as a selective agent for *C. perfringens*. Many other clostridia are sensitive to this antibiotic and are therefore inhibited. Black colonies with haloes are counted in order to calculate the number of cells per gram of food. Additional tests for presumptive identification of *C. perfringens* include the Gram stain and examination of the 'stormy clot reaction' using iron-milk medium. In this test, modified iron milk medium is inoculated with 1 ml of an actively growing *C. perfringens* fluid thioglycollate culture and incubated at 46°C. After 2 h incubation, the sample is checked hourly for 'stormy fermentation' reaction (see above for details of this reaction).

Confirmatory procedures are required in order to exclude physiologically similar species of clostridia which are able to form black colonies on sulphite-containing media. Confirmation of *C. perfringens* involves inoculating a colony from TSC agar into buffered motility–nitrate and lactose–gelatin media and incubating for 24 h at 35°C. Gelatin liquefaction and lactose utilization is evaluated. Cultures are examined for gas production and a red to yellow colour change which is indicative of acid production. Since *C. perfringens* is non-motile, tubes of motility–nitrate medium should contain growth only in and along the stab line. *C. perfringens* is able to reduce nitrates to nitrites. If isolates are tested for sporulation, Duncan–Strong sporulation medium is inoculated with an actively growing culture and incubated for 18–24 h. Duncan–Strong medium is the most widely used sporulation medium for *C. perfringens* (**Table 3**). The sporulation broth is subsequently examined for the presence of spores by using a phase-contrast microscope. Additional biochemical reactions may be required in those cases where the isolates do not meet all the criteria for *C. perfringens*. Biochemical test strips are available from a number of suppliers for identification of *C. perfringens* strains.

Table 3 Composition of modified Duncan–Strong sporulation medium for *C. perfringens*

	$g\ l^{-1}$
Protease peptone	15
Yeast extract	4
Sodium thioglycollate	1
$Na_2HPO_4.7H_2O$	10
Raffinose	4

Add above components to 1 litre distilled water. Sterilize by autoclaving and adjust pH to 7.8 using filter-sterilized 0.66 M sodium carbonate.

The isolation of *C. perfringens* from the faeces of individuals with suspected *C. perfringens* food poisoning involves heating stool samples at 100°C for 60 min to select for heat-resistant spores. Strains of *C. perfringens* isolated from several persons in an outbreak and those recovered from a suspect food should be compared serologically for toxin production. Several molecular epidemiological techniques have been used in order to compare the identities of patient isolates relative to food sources. These include phage typing, bacteriocin typing, plasmid profiles and the use of DNA probes.

Enterotoxin Detection

Serological methods have been shown to be considerably more sensitive than biological methods in terms of ability to detect low levels of *C. perfringens* enterotoxin. Examples of biological methods for detection of *C. perfringens* enterotoxin include the rabbit ileal loop test, the guinea-pig test and the mouse test. These tests have been shown to detect enterotoxin in the microgram range. The rabbit ileal loop test is an example of a classical biological method for assaying *C. perfringens* enterotoxin. This test is based on the observation that enterotoxins elicit fluid accumulation in the small intestine of some animals which can be quantified. *C. perfringens* enterotoxin results in increased capillary permeability, vasodilation and intestinal motility. Permeability continues to increase and the injured cells eventually lyse. In the intestines, cell death leads to loss of fluid and diarrhoea.

The serological methods which have been used to detect *C. perfringens* enterotoxin include the microslide diffusion, single or double gel diffusion, electroimmunodiffusion, and the enzyme-linked immunosorbent assay (ELISA) technique. These methods involve the use of specific polyclonal or monoclonal antibodies and are able to detect enterotoxin in the nanogram range. Monoclonal antibodies to *C. perfringens* enterotoxin have recently been developed. The microslide diffusion, single or double gel diffusion and electro-immunodiffusion techniques are dependent on observing a precipitation 'line of identity' that occurs between the enterotoxin and the corresponding antiserum. The sensitivity of the ELISA for detecting and quantifying *C. perfringens* enterotoxin is quite good and a number of these assays have been developed. The assay uses rabbit anti-*C. perfringens* enterotoxin IgG to trap the enterotoxin on microtitre wells. The wells are then treated with antienterotoxin conjugated with horseradish peroxidase followed by a suitable chromogenic substrate. The absorbance for the unknowns can easily be compared to the absorbance for enterotoxin standards in order to quantify the enterotoxin. Detection of as little as 1 ng enterotoxin per gram of faecal material has been demonstrated.

To test *C. perfringens* isolates for enterotoxin production, cultures are inoculated into Duncan–Strong sporulation medium for 18–24 h at 35°C under anaerobic conditions. Rapidly metabolizable carbohydrates such as glucose should be avoided in sporulation media because they repress the sporulation process and are vigorously fermented to acid. Addition of starch, raffinose, methylxanthines, caffeine, theophylline and guanosine have been shown to increase sporulation and enterotoxin production by some *C. perfringens* strains. The sporulated culture is centrifuged for 15 min at 10 000 *g* and the cell-free culture supernatant is tested for the presence of enterotoxin by using reversed passive latex agglutination (RPLA). RPLA is a serological assay for *C. perfringens* enterotoxin which appears to be comparable to ELISA in terms of sensitivity. The RPLA technique involves the use of sensitized (antiserum to enterotoxin treated) latex beads that are exposed to serial dilutions of enterotoxin. The agglutination titre is determined after overnight incubation.

In addition, a number of DNA probe-based techniques for detection and identification of *C. perfringens* have recently been described. DNA probes are sensitive and avoid some of the difficulties associated with enterotoxin detection in biological materials. Specific assays include gene detection assays for identifying *C. perfringens* isolates that produce enterotoxin, 16S and 23S rRNA probes to identify DNA isolated from pure cultures of *C. perfringens*, and a method based on 16S rDNA probes for specific detection of *C. perfringens* in foods. The polymerase chain reaction (PCR) has been used to detect the presence of *C. perfringens* which contain the enterotoxin or other toxin genes. Data obtained using DNA probes for enterotoxin have suggested that the enterotoxin may have importance only in human disease. Isolates from sheep or piglets with *C. perfringens*-induced diarrhoea did not have the enterotoxin gene.

Alpha Toxin Detection

Recently, a monoclonal antibody-enzyme linked immunosorbent assay (ELISA) using IgG_1 monoclonal antibody to detect *C. perfringens* alpha-toxin in bacterial cell lysates and cultural supernatants was developed. Alpha toxin was detectable during the early vegetative growth stage and the amount of toxin production correlated with the bacterial cell concentration during exponential growth phase. Since all five *C. perfringens* biotypes (A–E) produce alpha toxin, the detection of this toxin by ELISA may serve as a more direct indicator for rapid identification of *C. perfringens* in foods. Because of the potential for low numbers of *C. perfringens* to be recovered from foods stored at cold temperatures, alpha toxin (phospholipase C) has been proposed as an indicator for the prior presence of high levels of *C. perfringens*. The presence of detectable activity would suggest that the *C. perfringens* population was greater than or equal to 10^5 cfu g^{-1} food at one time. Interestingly, this population level is used by the Centers for Disease Control for confirmation of a food-borne disease outbreak. The ELISA technique has been widely applied to detect, screen and quantitate various microorganisms and toxins in food products.

Federal Regulations

Recently, the US government issued new rules for meat and poultry inspection in order to set performance standards concerning cooked beef products, uncured meat patties, and certain poultry products. In order to ensure that vegetative spore-forming microorganisms do not have an opportunity to grow and produce toxin, the new Federal regulation performance standards allow for no more than a one order of magnitude increase of *Clostridium perfringens* in meat products. Limiting the growth of *C. perfringens* to a one order of magnitude increase would limit the multiplication of other slower growing spore-forming bacteria. The regulation thereby suggests that the monitoring of *C. perfringens* can be a useful indicator for limiting other harmful bacteria such as *Bacillus cereus*, *Clostridium botulinum* and *Staphylococcus aureus*.

It is anticipated that additional rapid and quantitative assays for the detection of *C. perfringens* will be added to the list of Federal Drug Administration/Association of Official Analytical Chemists approved standard detection methodologies. Such technological development will allow the meat industry to monitor the presence of *C. perfringens* in meats and thereby meet the new Federal performance standards.

See also: **Biochemical and Modern Identification Techniques**: Food-poisoning Organisms. **Clostridium**: Detection of Enterotoxins of *C. perfringens*. **Food Poisoning Outbreaks**. **Meat and Poultry**: Spoilage of Cooked Meats and Meat Products. **Nucleic Acid-based Assays**: Overview.

Further Reading

Banwart GJ (1981) *Basic Food Microbiology*. P. 271. Westport, Connecticut: AVI Publishing.

Blaschek HP and White BA (1995) Genetic systems development in the clostridia. *FEMS Microbiol. Rev.* 17: 349–356.

Foegeding PM (1986) In: Pierson MD and Stern NJ (1986) *Foodborne Microorganisms and their Toxins: Developing Methodology.* P. 393. New York: Marcel Dekker.

Food Safety and Inspection Service, USDA (1996) Performance standards for the production of certain meat and poultry products. *Federal Register* 61: 19564–19578.

Harmon SM, Kautter DA, Golden DA and Rhodehamel JE (1992) *Bacteriological Analytical Manual*, 7th edn. P. 209. Arlington, VA: AOAC International.

Jay JM (1996) *Modern Food Microbiology.* P. 451. New York: Chapman & Hall.

Labbe R (1989) In: Doyle MP (ed.) *Foodborne Bacterial Pathogens.* P. 192. New York: Marcel Dekker.

Liu ZL and Blaschek HP (1996) A monoclonal antibody-based ELISA for detection of *Clostridium perfringens* phospholipase C. *J. Food Protection* 59: 1–6.

Lorber B (1995) In: Mandell GL, Bennett JE and Dolin R (eds) *Principles and Practice of Infectious Diseases.* P. 2182. New York: Churchill Livingstone.

Rood JI and Cole ST (1991) Molecular genetics and pathogenesis of *Clostridium perfringens. Microbiol. Rev.* 55: 621–648.

Shone CC and Hambleton P (1989) In: Minton NP and Clarke DJ (eds) *Clostridia*, p. 274. New York: Plenum Press.

Steinhart CE, Doyle ME and Cochrane BA (1996) *Food Safety 1996*, p. 408. New York: Marcel Dekker.

Wrigley DM (1994) In: Hui YH, Gorham RJ, Murrell KD and Cliver DO (eds) *Foodborne Disease Handbook: Diseases Caused by Bacteria.* Vol. 1, p. 133. New York: Marcel Dekker.

Detection of Enterotoxin of *Clostridium perfringens*

L Petit, **M Gibert** and **M R Popoff**, Institut Pasteur, Unité Interactions Bactéries Cellules, Paris, France

Clostridium perfringens Enterotoxin and *C. perfringens* Food Poisoning

Clostridium perfringens is a spore-forming anaerobic bacterium that is widespread in the environment and is pathogenic to humans and animals. Strains are classified into five toxinotypes (**Table 1**) based on the production of four toxins (alpha, beta, epsilon and iota) which are produced during the exponential growth phase. In addition, some strains of types A to E synthesize an enterotoxin (CPE) which is only formed during sporulation.

Human food poisoning is caused by ingestion of food containing a large number of vegetative *C. perfringens* cells, and it is rarely due to preformed CPE in food. *C. perfringens* multiply and sporulate in the gastrointestinal tract and synthesize CPE which is released with the bacterial lysis. In humans, the illness is caused by enterotoxigenic *C. perfringens* type A strains which represent a small proportion (5%) of the global *C. perfringens* population. The *cpe* gene is located in a variable region of the chromosome in most of the strains involved in human food poisoning, whereas it is generally located in a plasmid in non-food-borne human gastrointestinal disease and veterinary isolates. In all the strains, the *cpe* gene is tightly expressed in a sporulation-associated manner. CPE is responsible for the symptoms (diarrhoea, abdominal pain, rarely nausea) which usually occur 8–24 h after the ingestion of contaminated food. Death is uncommon, but can occur in debilitated individuals, elderly people and young infants.

When orally administered to animals, CPE induces rapid fluid and electrolyte losses within 15–30 min. The ileum appears to be the segment of the intestine that is the most sensitive to CPE. In addition, CPE causes necrosis and desquamation of the tips of the intestinal villi.

The primary cytotoxic action of CPE involves a unique series of four early events. CPE binds to a proteinaceous receptor that is present on the surface of many mammalian cells. A 50 kDa membrane protein has been demonstrated in all CPE-sensitive cells and is absent from CPE-insensitive cells. Recently, a 22 kDa protein corresponding to the CPE receptor has been cloned. The biochemically identified 50 kDa protein may be a dimer of the 22 kDa protein. Binding of CPE to its receptor results in the formation of a 90 kDa complex which subsequently interacts with a 70 kDa membrane protein to form a large complex (160 kDa). CPE remains associated with the plasma membrane throughout its action, and inserts into the lipid bilayer of plasma membrane and/or undergoes a conformational change. It results in the breakdown of normal membrane permeability leading to the leakage of small molecules and subsequently to the inhibition of protein synthesis.

Enterotoxigenic *C. perfringens* strains are also associated with non-food-borne digestive diseases such as antibiotic-associated diarrhoea, chronic non-food-borne diarrhoea, and some cases of sudden infant death syndrome. Immunological immaturity of some infants could lead to a nonselective absorption of molecules including CPE from the intestine and to a rapid transport to the circulation responsible for the systemic effects of CPE.

A less common digestive disease is termed pig bel which occurs in young people of New Guinea. This

Table 1 *C. perfringens* toxins, typing and associated diseases

Toxin	Typing[a]	Disease
Alpha	A	Humans: gangrene
Enterotoxin (CPE)[b], Alpha	Enterotoxigenic A	Humans and animals: digestive diseases
Beta1, Alpha	C	Humans and animals: necrotic enteritis
Beta1, Beta$_2$, Alpha	C	Animals: necrotic enteritis
Beta 2, Alpha	A[c] or C	Animals: necrotic enteritis
Beta1, Epsilon, Alpha	B	Animals: diarrhoea
Epsilon, Alpha	D	Animals: enterotoxaemia
Iota, Alpha	E	Animals: enterotoxaemia

[a] Classical typing by mouse test.
[b] CPE is produced by a small proportion of strains from types A to E.
[c] Beta 2 Toxin is not detected by the mouse assay in some strains which are classified as type A.

is a necrotizing, haemorrhagic enteritis which is caused by *C. perfringens* C. The beta 1 toxin, which is responsible for the lesions, is very sensitive to protease digestion. People of New Guinea are usually vegetarian and consume sweet potatoes which contain trypsin inhibitors. In some circumstances, they have traditional pig feasting. *C. perfringens* C ingested from contaminated meat multiply in the intestine and produce beta 1 toxin which is a necrotizing and cytotoxic toxin. After World War II, *Darmbrand* which resembles pig bel, was associated with a *C. perfringens* infection precipitated by malnutrition in Northern Germany.

A beta 2 toxin has been found in *C. perfringens* strains involved in necrotizing enteritis in piglets and in typhlocolitis in horses. The prevalence in humans is unknown.

Importance of *C. perfringens* Food Poisoning

Clostridium perfringens is ubiquitous and can contaminate a wide variety of foods. Most of the *C. perfringens* outbreaks occur in collective restaurants (school canteens, hospitals, prisons, special gatherings). Meat and poultry products, particularly cooked with sauce, are foods at highest risk. In France, during the period 1990–1992, 56.1% of *C. perfringens* outbreaks were associated with the consumption of meat and poultry products. Contamination of meat by *C. perfringens* is common, but usually at a low level. This can be due to transfer of *C. perfringens* from the intestine to the muscles during the preparation of animals or to surface contamination of meat by dust at the slaughterhouse. However, food responsible for *C. perfringens* poisoning contains a large number of *C. perfringens* (at least 10^5 per gram), since most of the bacteria are killed by the acidic pH of the stomach and their multiplication in the intestine is hampered by the resident digestive microflora. The *C. perfringens* multiplication in food depends on the preparation and

storage conditions of meals. Since this microorganism sporulates, it can survive heating procedures. The multiplication rate is very rapid, and growth temperature ranges from 15 to 50°C, with an optimum temperature of 40–45°C. The generation time (7 min at 41°C in optimum conditions) is one of the shortest reported for any bacterium. Meat in sauce constitutes an excellent culture medium for *C. perfringens* which has fastidious growth requirements. The contributing factors involved in the proliferation of this bacterium in food include preparation in large amounts too far in advance of eating, inadequate cooling, cooked food being stored without adequate refrigeration and served again later.

The incidence of *C. perfringens* outbreaks varies according to countries and cooking practices. A recent report from WHO shows that *C. perfringens* is the second or third cause of reported food-borne disease outbreaks and represents 4–16% of the outbreaks (**Table 2**). In each country, the number of outbreaks has changed with time. This could be due to changes in cooking practices and/or in methods of preparation and storage of food. For example, there has been an important decrease of *C. perfringens* incidence in Canada in recent years. It is noteworthy that the number of cases in each *C. perfringens* outbreak is higher than in the other food-borne diseases. In the US during the period 1988–1992, 40 *C. perfringens* outbreaks were reported which concerned 3801 cases (an average of 95 cases in each outbreak). In France, 102 *C. perfringens* outbreaks included 6021 cases (an average of 59 cases in each outbreak) between 1989 and 1992, and 13 outbreaks concerning 518 cases (an average of 39.8 cases per outbreak) in 1997 (Table 2).

Enterotoxigenic *C. perfringens* is involved in sporadic diarrhoea cases: 7–31.1%. Risk factors associated with these infections are not known, but all intestinal disorders leading to perturbation of the digestive microflora can possibly induce a proliferation of *C. perfringens* and production of the

Table 2 Reported food-borne disease outbreaks in different countries according to Todd (1997) and Tremolieres (1996)

Bacteria	USA (1988–1992)			Australia (1980–1995)		
	Outbreaks (%)	Cases (%)	Cases/ outbreaks	Outbreaks (%)	Cases (%)	Cases/ outbreaks
Salmonella sp.	54.8	57.4	38.6	31.4	29.8	49
Staphylococcus aureus	5	4.6	33.6	10.5	2.2	10
Clostridium	4	10.3	95	16.3	6.3	20
Escherichia coli	1.1	0.7	22.2	1.2	0.5	23
Bacillus cereus	2.1	1.2	20.7	5.8	0.6	5.4
Other bacteria	12.07	15.9	–	13.8	6.8	–

Table 2 – continued

Bacteria	Europe (1990–1992)	France					
		(1989–1992)			(1997)		
	Outbreaks (%)	Outbreaks (%)	Cases (%)	Cases/ outbreaks	Outbreaks (%)	Cases (%)	Cases/ outbreaks
Salmonella sp.	76.8	72.6	50.3	11.1	75.6	62.9	12.3
Staphylococcus aureus	6.5	5.9	9.5	25.7	12	14.4	17.7
Clostridium perfringens	4.7	4	14.6	59	4.9	13.1	39.8
Escherichia coli	1.7	–	–	–	–	–	–
Bacillus cereus	1.4	0.1	0.5	62	0.4	0.6	25
Other bacteria	3.1	–	–	–	5.6	8.4	22

enterotoxin. *C. perfringens* counts in faeces of patients with sporadic diarrhoea are generally lower ($< 10^5$ per gram) than found in patients with food poisoning ($> 10^6$ per gram).

Identification of *C. perfringens* Food Poisoning

The identification of *Clostridium perfringens* food poisoning is based on the determination of CPE in stools of patients and bacteriological investigations of stools and incriminated food.

The bacteriological criteria are as follows:

- Food containing a large number ($> 10^5$ bacteria per gram) of vegetative *C. perfringens* cells.
- Isolation of large numbers ($> 10^6$ per gram) of the organism from faecal specimens. Faecal count in the normal human population is $< 10^3$ per gram. However, several reports indicate that *C. perfringens* spore counts $> 10^6$ per gram can also be found in debilitated, institutionalized patients who are neither acutely ill nor involved in a food poisoning outbreak.

Additional investigations to associate human illness and incriminated food are:

- the demonstration of a common *C. perfringens* serological type or ribotype in faecal specimens and in the incriminated food

- a common serological type or ribotype in faecal specimens from several people.

C. perfringens Enterotoxin Assays

Since CPE is only synthesized during sporulation, culture in special sporulation medium and control of the presence of sporulating cells are required for CPE detection in culture supernatant. Several sporulation media have been proposed with variable results according to the strains. A typical protocol of *C. perfringens* sporulation is as follows.

A 1 ml sample of *C. perfringens* growing culture in cooked meat medium is transferred to 10 ml fluid thioglycollate medium. The inoculated fluid thioglycollate medium is heat shocked for 20 min at 70°C, and then incubated overnight at 37°C. The fluid thioglycollate culture is transferred to 100 ml of Duncan–Strong sporulation medium and incubated overnight at 37°C. The culture is checked for the presence of spores by observation under phase-contrast microscopy and culture supernatant obtained by centrifugation is subjected to CPE detection.

The presence of CPE may also be detected directly in faecal samples prepared as follows. One volume of faecal specimen (approximately 1 g) is mixed in one volume (1 ml) of 0.01 M phosphate buffer, pH 7.2, containing 0.15 M sodium chloride (PBS) in a vortex mixer. The suspension is either centrifuged at 12 000 *g*

Table 3 Sensitivity of assay methods for *C. perfringens* enterotoxin

Method	Detection limit of:	
	Purified CPE (ng ml^{-1})	CPE in faeces (ng g^{-1})
Double gel diffusion	500–2000	
Counterimmunoelectrophoresis	200–2000	
Vero cell cytotoxicity	25–50	
Plating inhibition of Vero cells	1	40
RPLA	3	5–50
SLAT	0.1–3	
ELISA	0.1–3	5–10

RPLA, reverse passive latex agglutination; SLAT, slide latex agglutination; ELISA, enzyme-linked immunosorbent assay.

for 20 min at 4°C or passed through 0.45 or 0.22 μm membrane filters and the resulting supernatant or filtrate is tested.

Initially, biological techniques have been used for CPE detection including mouse lethality, Vero cell cytotoxicity and plating inhibition of Vero cells.

Specific polyclonal and monoclonal anti-CPE antibodies have been obtained, and a large variety of immunological tests have been proposed for the detection and titration of CPE. The first immunological tests were based on immunoprecipitation of CPE in agarose gel in the presence of specific antibodies: single gel diffusion, and double gel diffusion or Ouchterlony test.

Counterimmunoelectrophoresis

The sensitivity of the precipitation reactions is improved by using an electrical field (**Table 3**) and counterimmunoelectrophoresis (CIEP) is the most used of these techniques.

Two rows of wells separated by about 5 mm are cut in agarose gel. Serial dilutions of CPE and samples are dispensed into the wells of one row, and anti-CPE antibodies are distributed in the wells of the other row. An electrical field (10 V cm^{-1}) is applied (+ near the wells containing the antigen) for 30–60 min. A precipitation line is observed in the presence of CPE. The sensitivity is shown in Table 3.

Latex Agglutination Tests

Two latex agglutination tests have been described, a reverse passive latex agglutination (RPLA) which is achieved after overnight incubation, and a slide latex agglutination (SLAT) which requires only a few minutes.

RPLA   RPLA is commercially available (PET-RPLA, TD930, Oxoid, Basingstoke, UK). The sensitivity is about 3 ng ml^{-1} (Table 3).

The procedure is as follows:

1. For each sample, two rows of a 96 well V-type microtitre plate are used.
2. Place 25 μl of PBS containing 0.5% bovine serum albumin (BSA) in each well, except in the first well of each row. The last wells only contain PBS–BSA.
3. Add a 25 μl sample to the first and second wells of each row.
4. Serial twofold dilutions are done in each row from the second to the seventh well.
5. Add 25 μl of beads sensitized with immunopurified anti-CPE antibodies to each well of the first row.
6. Add 25 μl of control beads sensitized with non-immune rabbit immunoglobulins to each well of the second row.
7. Mix well by hand rotation of the plate or by using a plate shaker.
8. Cover the microplate with a lid or put the microplate in a humidified chamber.
9. Incubate the plate at room temperature for 20–24 h.

The results are interpreted as follows:

1. Agglutination is determined by visual inspection. This is easier with a black sheet under the microplate or with a test reading mirror.
2. The results are scored as +++ (complete agglutination), ++, +, +/– or – (absence of agglutination) (**Fig. 1**).
3. The row containing control latex must be negative. A nonspecific agglutination can be observed in some samples. A sample is considered to contain CPE when the positive agglutination in the sensitized row exceeds that in the control by two wells or more.

SLAT   The SLAT technique consists of latex bead agglutination in the presence of CPE on a glass slide.

Reagent preparation is as follows:

1. Dilute latex beads (0.8 μm) 1 : 3 in glycine buffer (0.1 M glycine, 0.15 M NaCl, pH 8.2).
2. Add anti-CPE immunoglobulins that have been purified by immunoaffinity on a Sepharose column containing immobilized CPE, (13 μg ml^{-1}, final concentration).
3. Agitate the mixture for 1 min at room temperature, then add an equal volume of PBS–0.1% BSA, and vortex vigorously to mix the suspension.

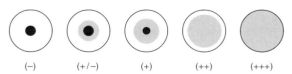

Figure 1 Interpretation of the agglutination results in RPLA.
+ corresponds to the agglutination of latex
– corresponds to the sedimentation of the particles

4. Use non-immune rabbit immunoglobulins G (Sigma) for the control latex.
5. Store the latex suspensions at 4°C.

The test procedure is as follows:

1. Mix 25 μl of samples and serial tenfold dilutions in PBS containing 0.5% BSA with 25 μl sensitized or control latex beads on a glass slide. Gently rotate each mixture and record the results after 3–5 min by visual inspection.
2. Score the results in a similar way to those for RPLA: ++ (complete agglutination), +, +/– and – (absence of agglutination).
3. Samples containing CPE do not agglutinate control latex beads. Note that samples containing a high concentration of CPE give negative or weakly positive results, and complete agglutination is observed with diluted samples.

The sensitivity depends on the purification of the immunoglobulins used for the latex bead preparation. When the immunoglobulin G fraction purified from rabbit anti-CPE serum is used for the sensitization of latex beads, the SLAT sensitivity with purified CPE is 100 ng ml^{-1}. However, by using specific anti-CPE immunoglobulins purified by immunoaffinity, a lower limit of detection of 0.1 ng ml^{-1} is attained.

Enzyme-linked Immunosorbent Assays

Several ELISA techniques have been proposed for the CPE titration in different samples including stool of patients.

A typical protocol is as follows:

1. Coat a microtitre plate with rabbit anti-CPE immunoglobulins (100 μl of a 5 μg ml^{-1} solution in PBS). Seal the plate, incubate it overnight at 22°C, and wash it four times with PBS containing 0.05% Tween20 (PBST).
2. Add CPE standard and test samples (100 μl diluted in PBST) to the antibody-coated plate, then seal it and incubate at 37°C for 90 min. Wash the plate as previously described and incubate for a further 90 min at 37°C in the presence of anti-CPE immunoglobulin G (IgG) horseradish peroxidase conjugate (100 μl diluted in PBST containing 1% normal rabbit serum).
3. After the washing procedure, add 100 μl of ABTS–H$_2$O$_2$ solution, containing 0.4 mM 2,2′-azino-di(3-ethylbenzthiazoline-6-sulphonate) (ABTS) and 1.3 mM H$_2$O$_2$ in 0.1 M citrate phosphate buffer, pH 4, to each well. Incubate the plate for 30 min at room temperature.
4. Read the absorbance at 405 nm. The sample is considered to contain CPE when the absorbance

is ⩾0.2 after correction for background which corresponds to the absorbance in control non-coated well. Estimate the CPE concentrations from a standard curve using purified CPE (0–50 ng ml^{-1}).

A variant procedure is the four layer sandwich ELISA procedure:

1. Coat each well of an immulon II enzyme immunoassay plate with 200 μl of goat anti-CPE serum (1 : 100 dilution in carbonate buffer 0.015 M Na$_2$CO$_3$–0.035 M NaHCO$_3$, pH 9.6), and incubate the plate overnight at 4°C in a humid chamber. Then after washing the plate with 100 μl of warm washing solution containing 0.85% NaCl, 0.05% Tween20 and 0.5% BSA per well, gently shake the plate on a rotary shaker for 2 min. Repeat this washing procedure three times.
2. To block the excess binding sites on the microtitre plate incubate at 37°C for 30 min with 100 μl of 3% BSA–1% normal goat serum diluted in PBS per well. Then wash the plate twice as described above.
3. Add samples (100 μl per well) containing CPE diluted in 0.05% Tween20 in PBS to each well, and incubate the plates at 37°C for 2 h. Wash each well once prior to repetition of the blocking procedure as described above for 30 min at 37°C.
4. Then wash the plate twice, and add 200 μl of rabbit antitoxin diluted 1 : 200 with 0.85% NaCl, 0.05% Tween20 and 1% BSA to each well and incubate for 2 h at 37°C. Wash three times, then add 200 μl of conjugate (goat anti-rabbit immunoglobulin G conjugated with alkaline phosphatase) of a 1 : 800 dilution in PBS–0.05% Tween20 for 2 h at 37°C. Wash a further three times, and add 200 μl of warm substrate (0.1% *p*-nitrophenol phosphate–10% diethanolamine–0.01% MgCl$_2$, pH 9.6).
5. Allow the reaction to progress at 37°C for 30 min and then terminate it by adding 50 μl of 2 M NaOH.
6. Read the results spectrophotometrically at 405 nm. For each sample, perform the test in duplicate. Determine the absorbances by subtracting the absorbances (< 0.02) in negative controls which do not receive CPE or sample. Values above 0.1 are considered as positive.

The sensitivity of ELISA is 1–25 ng ml^{-1} using purified CPE in aqueous solution and 5–500 ng g^{-1} CPE in faeces samples (Table 3). Protease activity of some samples is responsible for the decrease in sensitivity as a consequences of digestion of the IgG used for coating the polystyrene surface. This can be prevented by addition of serum albumin (1%) to the samples.

Detection of Enterotoxigenic *C. perfringens*

DNA-based methods including polymerase chain reaction (PCR) and DNA–DNA hybridization, have been developed for the identification of enterotoxigenic strains and *C. perfringens* typing. Multiplex PCR permits the simultaneous detection of several toxin genes. A duplex PCR has been designed to identify the alpha toxin gene, which corresponds to a marker of the *C. perfringens* species, since this gene is present in all strains except in very few rare strains, and the *cpe* gene which is characteristic of the strains involved in food poisoning. The detection level is approximately 10^5 *C. perfringens* cells per gram of stool or food sample, and 10 cells per gram when enrichment culture is used. The advantage of PCR is that *C. perfringens* sporulation is not required and reliable results are obtained with culture in usual growth medium.

Among the protocols of stool preparation, a rapid method is as follows: a 1 g stool sample is homogenized with 9 ml of distilled water; 1 ml is then centrifuged and the supernatant discarded. The pellet is resuspended in 0.2 ml of Instagen (BioRad). The mixture is incubated for 30 min at 55°C, vortexed vigorously for 10 s, and incubated for 10 min at 100°C. The mixture is vortexed again for 10 s and centrifuged (10 min at 10 000 r.p.m.). Supernatant (3 µl) both undiluted and diluted 1:10 in distilled water containing 3% BSA, is used for PCR amplification.

Advantages and Limitations of the CPE Detection Methods

Two situations have to be considered: *C. perfringens* food poisoning outbreaks and routine food control (**Fig. 2**).

In a *C. perfringens* food poisoning outbreak, the contaminated food contains at least 10^5 *C. perfringens* cells per gram and CPE is not detectable. The patients, during the two days after the onset of symptoms have $\geqslant 10^6$ enterotoxigenic *C. perfringens* cells per gram and CPE of $0.012–140 \, \mu g \, g^{-1}$ of stool. When faecal samples were collected on the first two days of an outbreak, 77% were enterotoxin positive, and among the specimens collected later than the second day, only 33% had detectable CPE. Enumeration and identification of *C. perfringens* from contaminated food and stool by the standard method is achieved in at least 24 h. The identification of enterotoxigenic strains requires sporulation by the strain and subsequently detection of CPE by an immunological or biological test (**Table 4**). Rapid methods can be used to identify *C. perfringens* outbreaks and set preventive measures in place. The presence of enterotoxigenic *C. perfringens* in food and stool is rapidly detected (about 6–8 h) by PCR without enrichment culture. The confirmation of *C. perfringens* food poisoning can be achieved by CPE detection in stool of patients in a few minutes by SLAT or several hours by ELISA or RPLA. The detection limits of these methods are in the range of the CPE levels found in patients, and CPE is not detectable in healthy individuals. ELISA

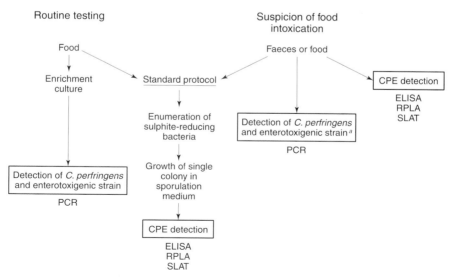

Figure 2 Schematic representation of the main methods for *C. perfringens* identification from food and faeces. [a]Direct PCR detection of *C. perfringens* and enterotoxigenic strains from food containing a large number of vegetative *C. perfringens* cells (> 10^5 bacteria per gram). PCR, polymerase chain reaction; ELISA, enzyme-linked immunosorbent assay; RPLA, reverse passive latex agglutination; SLAT, slide latex agglutination.

Table 4 Comparison of usual methods of CPE detection

	ELISA	RPLA	SLAT	Vero cell assay (plating inhibition of Vero cells)
Time required for complete test	< 8 h	24 h	30 min	24 h
Time spent on test	2.5 h	0.5 h	3–10 min	0.75 h
Specificity	Excellent	Good	Good	Good
Reproducibility	Yes	Yes	Yes	Yes/No (a fourfold change in cytotoxicity titres was observed)

ELISA, enzyme-linked immunosorbent assay; RPLA, reverse passive latex agglutination; SLAT, slide latex agglutination.

requires a longer time than SLAT but it can be automated.

The standard method for routine testing of food enumerates sulphite-reducing bacteria which includes *C. perfringens* and also other *Clostridium* spp. PCR with enrichment culture is a reliable and sensitive method (10 *C. perfringens* cells per gram). Within 24 h, *C. perfringens* can be detected and the enterotoxigenic strains discriminated. Moreover, PCR can be automated. The limitation is that the results are not quantitative. A quantitative method based on the most probable number method consists of inoculating serial dilutions of food samples into enrichment medium and performing PCR with each dilution culture. Quantitative PCR and standardization with the reference method of *C. perfringens* enumeration are required to use this method in food bacteriology, since certain levels of *C. perfringens* (50–200 per gram) are tolerated in some food products. Official regulations concern the anaerobic sulphite-reducing bacteria without distinction of *C. perfringens* from other bacteria, and without distinction of enterotoxigenic and non-enterotoxigenic *C. perfringens*. Since only enterotoxigenic *C. perfringens* strains have been recognized as being responsible for food poisoning and new methods, able to specifically identify these toxigenic bacteria, are available in routine testing, an adjustment of the regulation should be considered.

See also: **Clostridium**: *Clostridium perfringens*. **Enzyme Immunoassays**: Overview. **Food Poisoning Outbreaks**. **Immunomagnetic Particle-based Techniques**: Overview. **Nucleic Acid-based Assays**: Overview. **PCR-based Commercial Tests for Pathogens**.

Further Reading

Bartholomew BA, Stringer MF, Watson GN and Gilbert RJ (1985) Development and application of an enzyme linked immunosorbent assay for *Clostridium perfringens* type A enterotoxin. *Journal of Clinical Pathology* 38: 222–228.

Berry PR, Rodhouse JC, Hughes S, Bartholomew BA and Gilbert RJ (1988) Evaluation of ELISA, RPLA, and Vero cell assays for detecting *Clostridium perfringens* enterotoxin in faecal specimens. *Journal of Clinical Pathology* 41: 458–461.

Fach P and Popoff MR (1997) Detection of enterotoxigenic *Clostridium perfringens* in food and fecal samples with a duplex PCR and the slide agglutination test. *Applied and Environmental Microbiology* 63: 4232–4236.

Gibert M, Jolivet-Reynaud C and Popoff MR (1997) Beta2 toxin, a novel toxin produced by *Clostridium perfringens*. *Gene* 203: 65–73.

Haeghebaert S, Le Querrec F, Vaillant V, Delarocque-Astagneau E and Bouvet P (1998) Les toxi-infections alimentaires collectives en France en 1997. *Bulletin Epidemiologique Hebdomadaire* 41: 177–181.

Harmon SM and Kautter DA (1986) Evaluation of a reversed passive latex agglutination test kit for *Clostridium perfringens* enterotoxin. *Journal of Food Protection* 49: 523–525.

Katahira J, Inoue N, Horiguchi Y, Matsuda M and Sugimoto N (1997) Molecular cloning and functional characterization of the receptor for *Clostridium perfringens* enterotoxin. *Journal of Cell Biology* 136: 1239–1247.

Labbe RG (1989) *Clostridium perfringens*. In: Doyle MP (ed.) *Foodborne Bacterial Pathogens*. Pp. 191–234. New York: Marcel Dekker.

Lindsay JA, Mach AS, Wilkinson MA et al (1993) *Clostridium perfringens* type A cytotoxic-enterotoxin(s) as triggers for death in the sudden infant death syndrome: development of a toxico-infection hypothesis. *Current Microbiology* 27: 51–59.

Mahony DE, Gilliatt E, Dawson S, Stockdale E and Lee SHS (1989) Vero cell assay for rapid detection of *Clostridium perfringens* enterotoxin. *Applied and Environmental Microbiology* 55: 2141–2143.

McClane BA (1996) An overview of *Clostridium perfringens* enterotoxin. *Toxicon* 34: 1335–1343.

McClane BA and Snyder JT (1987) Development and preliminary evaluation of a slide latex agglutination assay for detection of *Clostridium perfringens* type A enterotoxin. *Journal of Immunological Methods* 100: 131–136.

McClane BA and Strouse RJ (1984) Rapid detection of *Clostridium perfringens* type A enterotoxin by enzyme-linked immunosorbent assay. *Journal of Clinical Microbiology* 19: 112–115.

Rood JI and Cole ST (1991) Molecular genetics and pathogenesis of *Clostridium perfringens*. *Microbiology Review* 55: 621–648.

Todd ECD (1997) Epidemiology of foodborne diseases: a worldwide review. *World Health Statistics Quarterly* 50: 30–50.

Tremolieres F (1996) Toxi-infections alimentaires en France métropolitaine. *Revue du Practicien* 46: 158–165.

Clostridium acetobutylicum

Hanno Biebl, GBF, National Research Centre for Biotechnology, Braunschweig, Germany

Introduction

The fermentation of carbohydrates to ethyl alcohol and lactic acid has been used since prehistorical times for beverages and food processing. Fermentation to butanol and acetone which is catalysed by *Clostridium acetobutylicum* was only discovered at the end of the nineteenth century and was exploited on an industrial scale in the first half of the twentieth century. The main products, butanol and acetone, do not have nutritional significance but are used as solvents for technical applications. Due to competition with more favourable petrochemical production lines and increasing prices for the necessary agricultural feedstocks, the butanol–acetone fermentation industry declined after World War II and was abandoned around 1960 in almost all the Western countries. As a consequence of the oil supply limitations at the end of the 1970s a revival of the process was contemplated, and major research activities were initiated in Europe, North America and elsewhere which still continue.

Description of the Species

Vegetative cells of *Clostridium acetobutylicum* are straight rods of $0.5–0.9 \times 1.5–6\ \mu m$ and are motile by peritrichous flagella (**Fig. 1**). They stain Gram-positive in growing cultures but Gram-negative in older cultures. During sporulation, cells swell markedly and form granulose, a polysaccharide reserve material. Spores are oval and subterminal. The optimum growth temperature is 37°C, and biotin and 4-aminobenzoate are required as growth factors. *C. acetobutylicum* cannot be identified by their metabolic products alone, as solvent may be absent and several related species are also able to form butanol. **Table 1** shows the biochemical tests presently in use to differentiate *C. acetobutylicum* from these species. *C. beijerinckii* was also used for industrial fermentations. It was formerly known as *C. butylicum* and included

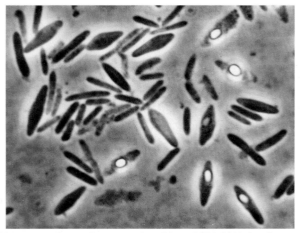

Figure 1 *Clostridium acetobutylicum* DSM 2152 (= 'C. saccharoberbutylacetonicum' N1) showing different stages of spore formation among vegetative cells.

strains that produced isopropanol instead of acetone *C. acetobutylicum* has recently been reclassified on the basis of phage biotyping, DNA fingerprint and 16S rRNA base sequencing. The existing strains were assigned to four groups of species rank, one of them being *C. beijerinckii*, another one *C. acetobutylicum* in the narrow sense; the remaining two are as yet unnamed.

Enrichment and Isolation

There is no selective enrichment procedure for butanol–acetone-forming clostridia. Nevertheless, they are easily obtained from soil, roots (especially of leguminous plants), cracked cereals and comparable sources using starchy mashes (4%) or media containing sugar. The samples are pasteurized for 2 min at 80°C to exclude non-spore-formers. Positive cultures are recognized by a characteristic sweet butylic odour or by gas-chromatographic analysis. Isolation is easiest on agar plates made of glucose (20–40 g l^{-1}) mineral medium with yeast extract (2–5 g l^{-1}) incubated in anaerobic jars (e.g. GasPak). Media for pure cultures have to be prepared under strictly anaerobic conditions preferably using the Hungate technique.

History of the Butanol–Acetone Fermentation Industry

The production of butanol and acetone is closely linked to the name of Chaim Weizman, the first president of Israel. Although the idea to exploit this fermentation economically was first realized by others, he isolated the first efficient strains, organized a research group and was involved in founding the first successful solvent factories in southern England in

Table 1 Biochemical characteristics of butanol- and acetone-producing *Clostridium* species

	C. acetobutylicum	C. beijerinckii	C. aurantibutyricum	C. puniceum
Growth in minimal medium	+	−	nk	nk
Growth factors required	B, pABA	a	nk	nk
Growth in peptone without carbohydrates	−	+	nk	−
Gelatin hydrolysed	−	−	+	+
Nitrate reduced	−	−	+	−
Lipase produced	−	−	+	−
Indol produced	−	−	−	+
Acetoin produced	+	−	−	nk

nk, not known; B, biotin; pABS, 4-aminobenzoic acid.
[a] Requires numerous vitamins in addition to amino acids.

1916. One year earlier a patent was issued, which was the very first that covered a biological process. Originally conceived for the production of butadiene, the monomer for synthetical rubber, from butanol the main interest shifted to acetone during World War I. Acetone was required in large amounts as a colloidal solvent for smokeless powder. The feedstock for the fermentation was maize meal, but other grain products were also used. After the war, the process was temporarily abandoned, but very soon a new application for butanol was found. Butanol and its ester butyl acetate are ideal solvents for the nitrocellulose lacquers that were required by the expanding automobile industry. Thus the stored butanol was salvaged, process facilities which had been erected in England, USA and Canada at the end of the war were reinstalled, and new factories were built. At the peak of the development, in 1927, a total of 148 fermenters, each with a capacity of 190 m³, were operating in two US plants, producing about 100 tons of solvents per day.

At the beginning of the 1930s there was a glut of molasses, and strains of *C. acetobutylicum* were isolated and developed able to convert higher amounts of carbohydrate and producing higher concentration of solvents than from maize, i.e. 6.5% sugar to 1.8–2.2% of solvents in contrast to 1.2–1.8 with starchy materials. During World War II, the butanol–acetone fermentation capacities in USA and England expanded again to fulfil the increased demand for acetone used for the manufacture of munitions, partly by commandeering alcohol distilleries. After 1945 the fraction of butanol and in particular acetone that was produced by fermentation declined progressively, but a few small facilities survived. The last factory in the western hemisphere, South Africa, closed in 1983 but there is reliable information acetone/butanol is currently being produced in China.

The Industrial Fermentation Process

Proper performance of the butanol-acetone fermentation requires expertise in a variety of fields including anaerobic culture technique, sterilization, distillation and waste disposal. Starting with a spore–sand mixture, the inoculum for the fermentation tank is scaled up through five stages of increasing size. To avoid degeneration of the culture (see below) the spores were always 'activated' by heat shock (e.g. 2 min at 100°C or 10 min at 80°C) after suspension in liquid medium, which was usually potato mash. For the final fermentation, maize meal and other starchy materials were used at a concentration of 8–10% without any supplements. Molasses media contained up to 6.5% sugar and had to be supplemented by a nitrogen source. Yeast water, corn steep liquor or distillation slop were used in combination with ammonium salts or gaseous ammonia which also served as pH control. A phosphorus source was necessary with beet and invert (=high test) molasses but not with backstrap molasses. The medium was sterilized by steam injection in continuous cookers, cooled to the fermentation temperature (37°C for maize mash, 30–33°C for molasses medium) through heat exchangers and pumped into the final fermenter. The fermenter, 90–750 m³, was steam sterilized, as were all other parts that come into contact with medium or inoculum, and gased with CO_2 before, during and after filling and inoculation. There was no mechanical agitation.

Depending on the strain and the inoculum size, the fermentation was complete after 30–60 h, and the beer was subjected to distillation. In a continuous process a concentrated solvent mixture was obtained which was separated and purified in fractionating columns. Usually a butanol : acetate : ethanol weight ratio of 6 : 3 : 1 was obtained with slight variations. Frequently the fermentation gases which consisted of about 60% CO_2 and 40% H_2 were also collected. CO_2 which accounts for 50% of the carbohydrate

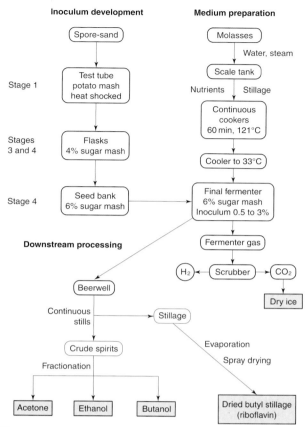

Figure 2 Flow diagram of the traditional acetone–butanol process using molasses.

The first phase is characterized by exponential growth, production of butyric and acetic acids and a concomitant decrease in pH. Hydrogen production is high. At a certain time, which varies between strains, growth slows down, and product formation switches from acidic to neutral products, butanol, acetone and ethanol; the hydrogen production reduces to one half of the former yield. The acids produced before are gradually converted, butyric acid faster than acetic acid. As a rule this second phase is also associated with marked changes in cell morphology. Cells begin to swell, accumulate reserve material in the form of granulose and transit through all the stages of spore formation (Fig. 1).

Several factors have been found necessary for the shift from acid to solvent production, a minimum concentration of the carbon source, a low pH and a minimum amount of butyric (and acetic) acid. pH and total acid concentration account for the deleterious undissociated acid fraction which explains why solvent can be formed not only at low pH and low acid concentration, but also at neutral pH, if high amounts of butyric acid are externally added. The sequence of physiological events is shown in **Figure 4**. The first step to relieve acid inhibition is to convert the butyric and acetic acids. The uptake requires co-enzyme A which is provided by the intermediate, acetoacetyl-CoA, and from the product of this reaction, aceto-acetate, acetone is released. Acetone formation, however, results in an excess of reducing

fermented was converted into dry ice, and the hydrogen was used for chemical synthesis, fat hardening or as fuel. The stillage that contains relatively high amounts of riboflavin and B vitamins was dried and sold as an additive to animal feeds. A flow sheet of the entire process is given in **Figure 2**.

A serious problem in the butanol–acetone fermentation was infection with bacteriophages. This is manifested as an unexpectedly early slowing of growth and gas production. As the phages have a narrow host range it was a common strategy to keep spores of a large number of strains and switch to a different strain if/when an infection was observed during inoculum preparation. Also phage-resistant mutants were isolated long before the infectious particles had become visible in the electron microscope.

Physiology of the Butanol–Acetone Fermentation

The fermentation of carbohydrates by *C. aceto-butylicum* typically proceeds in two phases (**Fig. 3**).

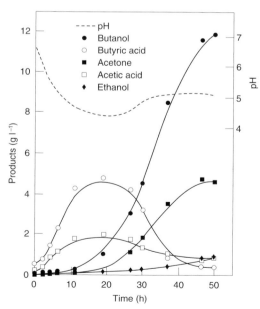

Figure 3 Typical course of a butanol–acetone batch fermentation. Adapted with permission from Ball H, Andersch W and Gottschalk G. (1982) Continuous production of acetone and butanol by *Clostridial acetobutylicum* in a two stage phosphate limited chemostat. *Appl. Microbiol. Biotechnol.* 15.

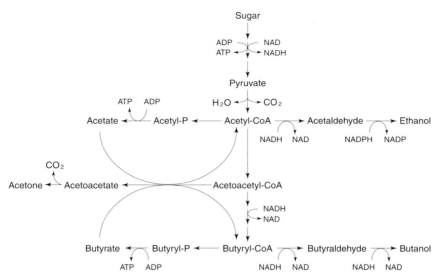

Figure 4 Metabolic pathways in *Clostridium acetobutylicum*.

equivalents, and consequently butanol is produced instead of butyric acid and ethanol instead of acetic acid. The shift from acids to solvents can also be described at a biochemical level in terms of fluctuations in the ATP and NAD(P)H pools and signal transduction to initiate synthesis of the relevant enzymes (which is, however, beyond the scope of this article).

The stationary solvent production phase may not occur. The cultures may miss the pH which is favourable for a shift and further acidify the medium until the cells are inactivated and lyse. This phenomenon can be observed in fast-growing cultures at near-optimum temperature or in rich medium. It cannot be confidently predicted and has, therefore, been aptly characterized as 'teetering on the edge of acid death'. However, under automatic pH control the pH can be held above the shift point and solvent formation can be secured. The shift pH which varies from strain to strain and with the culture conditions ranges between pH 4.3 and 5.5.

If cultures of *C. acetobutylicum* are regularly transferred as vegetative cells the ability to form butanol and acetone may be permanently lost. This unusual property is known as degeneration and has been circumvented by inoculating only from dry spores which were heat-shocked before incubation to eliminate the 'weak' spores. Nevertheless, solvent production can be retained in continuous cultures under conditions of organic substrate excess, but not under substrate limitation. This is particularly true for the type strain of *C. acetolbutyliaum* which was maintained for more than one year under phosphate limitation without changes in solvent productivity. Other strains, however, regularly shifted to acid formation after 20–

25 residence times independent of the limiting factor. Recently the molecular basis of degeneration has been elucidated. The genes that encode for the enzymes of solvent formation are located on one large plasmid which may be lost under an appropriate selective pressure.

The ratio of butanol to acetone is usually 2:1 and varies very little. However, under special conditions which include iron limitation and fermentation of whey (see below) the acetone fraction is reduced. Almost complete acetone repression is achieved, if hydrogen evolution is blocked by gassing with carbon monoxide. Attempts to generate hydrogenase-negative mutants that produce butanol only have not been successful, but strains with low acetone production have recently been isolated from nature.

As mentioned above, spore formation is linked to the solvent formation phase but solvent formation does not necessarily require sporulation. Asporogenous mutants have been isolated that still produce butanol and acetone, as do continuous cultures with vegetative cells under phosphate or product limitation.

In comparison to other fermentations, the maximum product concentration of 2% is relatively low. Growth experiments in the presence of individual end products have shown that cessation of the process is caused almost exclusively by butanol, whereas acetone and ethanol are not inhibitory at their physiological concentrations. The toxicity of butanol has been linked to an observed increase in the fluidity of the cell membrane impeding nutrient and product exchange. The final product concentration is also affected by an exoenzyme called autolysin which is produced during spore formation and may lead to premature cell lysis. Mutants that are deficient in autolysis formation exhibit an increased tolerance against butanol.

Recent Progress in Acetone–Butanol Research

Since 1980, the number of publications on the butanol–acetone fermentation has been substantial. Only three fields are considered here: usage of alternative fermentation substrates; development of better fermentation and recovery techniques; and genetical improvement of strains.

New Substrates

As the feedstock for the fermentation amounts to more than 60% of the production costs, efforts have been made to replace corn starch and molasses by cheaper substrates, preferably waste carbohydrates such as apple pomace, Jerusalem artichokes, lignocellulose and whey. The Jerusalem artichoke, a potato-like tuber, contains fructosans which, if hydrolysed enzymically and supplemented with ammonia, give excellent solvent yields. Lignocellulose, the most abundant carbohydrate source, also requires hydrolysis by acid or cellulases in addition to steam explosion to dissolve the hemicelluloses. Less pretreatment is necessary for sulphite waste liquor, a by-product of the paper industry. The hexoses and pentoses contained in this waste water are slowly but quantitatively fermented. Sweet whey from cheese production is one of the most promising substrates. It contains lactose in a concentration low enough to avoid inhibiting product concentrations (up to $5 \, g \, l^{-1}$). Although lactose is more slowly fermented than glucose the process including product recovery has been developed sufficiently that application in the near future seems possible.

Development of Fermentation and Product Recovery

Continuous fermentation in a chemostat mode has proved to be an effective tool to increase productivity in the butanol–acetone fermentation. Phosphate is an appropriate limiting factor but cultivation without nutrient limitation is also possible as the accumulating products limit growth and give rise to steady states. Usually lower product concentrations are obtained than in batch culture but by application of two stages, an acid-forming growth stage at high dilution rate and a solvent-forming fermentation stage at low dilution rate, a solvent concentration was achieved approaching the usual batch concentration of $20 \, g \, l^{-1}$ solvents.

To increase the relatively low productivity of chemostat cultures ($0.5–2 \, kg$ solvents/(h m^3)) two techniques, both designed to operate at elevated cell densities, were studied. With cell immobilization, spores are entrapped in gel beads or attached to solid particles using a low-growth medium, which is preferably nitrogen limited. Calcium-alginate beads and beechwood shavings have been successfully tested. Cell recycling involves permanent withdrawal of cell-free culture liquid into an external filtration unit and returning of the more concentrated culture to the fermenter. With both methods a productivity increase of about fourfold was achieved in comparison to the free-cell continuous culture. The rates obtained vary according to the amount of added complex substances such as yeast extract and peptone, the maximum being at $3 \, kg/(h \, m^3)$.

The low final solvent concentration attained in the butanol–acetone fermentation and the high energy requirement for distillation of butanol, the boiling point of which is greater than that of water, has initiated a search for alternative solvent recovery processes. The main emphasis was put on product removal procedures that are integrated in the fermentation and thus increase productivity by reducing the concentration of toxic products in the culture.

As suggested above in relation to the industrial production, extraction by a water-immiscible liquid in direct contact with the culture has the advantage of being simple to realize. Good results have been obtained with oleyl alcohol, diluted with decane to reduce viscosity. Octanol has also proved to be a useful extractant, but as this compound is slightly toxic to the clostridia, it was necessary to separate the cells from the culture liquid by microfiltration. The solvents are extracted selectively and can be recovered by distillation at a relatively low energy input. A modification of the liquid–liquid extraction is known as perstraction, the culture being separated from the extractant by a solvent-permeable membrane. It avoids formation of emulsions between the phases, and the extractant need not be sterilized and cannot affect the culture.

Inert gas is used to remove the solvents in variants with and without membranes. Gas-stripping, i.e. direct sparging of gas through the fermenter, is likewise attractive because of its simplicity. The microorganisms are not affected, and the products are easily recovered by condensation, with less energy consumption than with distillation of liquid extractants. It has been suggested that the self-produced fermentation gases, carbon dioxide and hydrogen, are used instead of expensive nitrogen. The membrane modification of gas-stripping, pervaporation, requires an extended tubing system which is immersed in the fermentation vessel. The solvents evaporate through the membrane and are drawn off by vacuum or sweep-gas. As the available membranes only allow passage of the solvents the acids accumulate in the culture and may stop the fermentation. This problem was solved by low-acid mutants that were able to reutilize all of the acids.

Adsorption to solid materials such as silicalite or polyvinylpyridine has also been tested. Relatively low loading capacity, high estimated costs for the adsorbants and the heat for desorption of the solvents presently diminish the chances for this method. For external application reversed osmosis has been evaluated and found to be more favourable than distillation.

Generally speaking the *in situ* recovery methods are interesting, but require high capital expenditure and permanent monitoring by the operator, and although their technical feasibility is established they need further development at the engineering level.

Genetic Strain Improvement

Greater expectations than can be achieved by biochemical engineering are envisioned in directed alteration of the bacterial metabolism, in particular by application of genetic engineering. Goals are to enhance butanol tolerance and thus increase the final solvent concentration, to influence the product ratio at will by manipulation of enzyme production and regulations, and to broaden the substrate range. Phage resistance is another useful objective.

Classical mutations exhibiting higher butanol tolerance have been generated by exposure to chemical mutagens and selected on medium with high butanol concentrations or were obtained by spontaneous alteration. Up until 1985 an increase of butanol tolerance was observed in a number of cases, and the mutant strains might have shown an increase in solvent concentration of up to 30% in comparison to the parent strain. However, the amounts obtained with these mutants did not appreciably exceed the previously obtained maximum values with non-mutated strains (**Table 2**, examples 1 and 2). In recent years mutant strains have been isolated that produced up to 2.6% of solvents of which up to 1.9% was butanol. In addition a strain was obtained that produced 2% of butanol and no acetone (Table 2, examples 3–5).

The recombinant DNA strains published so far have not quite reached the performance of the conventional mutants (Table 2, example 6), but undoubtedly the possibilities of inserting, deleting and overexpressing genes are far from being exhausted. Most of the prerequisites for successful genetic engineering have been created. Almost all the genes that encode for the enzymes involved in solvent and acid formation have been cloned and sequenced for their

nucleotides with a complete genome sequence almost established. Shuttle vectors for the cloned genes have been constructed that allow expression of genes in *Escherichia coli* or *Bacillus subtilis* and in *C. acetobutylicum*. The best method for DNA transformation appeared to be electroporation, conjugation was also successful in some cases but at a very low transformation frequency. A major problem was the high nuclease activity of the clostridia which could be overcome by methylating the DNA to be transformed. At present a complete genome analysis of the *C. acetobutylicum* type strain is in progress.

In the following paragraphs a few examples of genetically engineered strains are given which may mark the way for future development. A more detailed account of the present state of *C. acetobutylicum* genetics and the regulation of solventogenesis and associated phenomena is given by Woods (1995) (see Further Reading). The first successful attempt was started to increase acetone production. A synthetic operon that contained the genes for the two enzymes necessary for acetone formation, coenzyme A transferase and acetoacetate decarboxylase (see **Fig. 4**), was constructed using plasmid pFNK6 as vector. This plasmid was transformed into the wild-type strain of *C. acetobutylicum* ATCC 824 and caused an increase in the final acetone concentration, but, interestingly, the butanol and ethanol amounts were also increased (Table 2, example 6). Unlike the original strain, the final acid content of the mutant culture was almost zero. This is easily explained by the second function of the CoA transferase which releases CoA from its substrate to recycle produced carboxylic acids (see above).

Butanol formation alone was achieved in a solvent-negative mutant that was transformed with a plasmid containing a gene that encodes for the combined aldehyde/alcohol dehydrogenase. Although the butanol concentration was not very high, the complete absence of acetone formation is remarkable. A general increase of solvent production was also obtained in mutants which were inactivated in butyrate or acetate formation.

Strains that produce butanol only could be theoretically obtained by blocking the production of hydrogen, as suggested from experiments in which the hydrogenase was physiologically poisoned. Indeed inactivation of the hydrogenase(s) has been attempted, a respective gene has been cloned. Also available are genes that are responsible for autolysin production which allow the problem of premature cell lysis to be tackled on a genetic level in order to achieve higher cell densities and better butanol tolerance. Latest developments show that even better results can be expected from recombinant strains that

Table 2 Mutants of *Clostridium acetobutylicum* with enhanced final solvent concentration generated between 1983 and 1997

Example	Parent strain Mutant strain	Mutation method	Maximum concentration achieved (g l⁻¹)			
			Butanol	Acetone	Ethanol	Solvents
1	ATCC 824	Spontaneous	12.5	7.4	1.4	21.3
	ATCC 824 SA-1		13.6	7.3	2.0	22.9
2	903	Chemical mutagen	11	3.7	0	14.7
	904		14.1	6.6	0	20.7
3	B643	Chemical mutagen	12.4	7.5	1.8	21.7
	B643 B18		16.3	5.7	4.6	26.2
4	3	Spontaneous	10.8	4.1	0.3	15.2
	10		19.0	0	0.5	19.5
5	ATCC 4259	Chemical mutagen	16.8			22.6
	ATCC 4259 E-604		19.6			26.1
6	ATCC 824	Gene manipulation	9.8	4.2	0.9	14.9
	ATCC 824 pFNK6		13.0	8.6	1.8	23.4

are altered in solvent regulation. As a component of the *sol* operon, which involves important solvent formation genes, a regulator gene has been identified that acts as a DNA-binding repressor which negatively controls the transcription of the solvent genes. Inactivation of this gene led to solvent concentration which markedly exceeded all previously reported values.

Broadening of the substrate spectrum, in particular utilization of cellulosic material, is another field of genetic reconstruction of *C. acetobutylicum*. Cellulose is efficiently fermented by thermophilic clostridia like *C. thermocellum* with ethanol and acetate as fermentation products, and attempts have been made to express their cellulase system in *C. acetobutylicum*. Transformed cells were able to grow on lichenan, a β-glucan similar to cellulose, as sole carbon source.

It can be predicted that the progress achieved during the last 15 years of research in understanding physiology, genetics and regulation of *C. acetobutylicum* will bear fruit. Genetically engineered strains with essentially enhanced butanol production and broader substrate utilization in combination with novel developments in fermentation technology and product recovery will increase the chances for a revival of butanol–acetone fermentation as an industrial process.

See also: **Fermentation (Industrial)**: Basic Considerations; Media for Industrial Fermentations; Control of Fermentation Conditions; Production of Organic Acids. **Genetic Engineering**: Modification of Bacteria.

Further Reading

Bahl H and Gottschalk G (1988) Microbial production of butanol/acetone. In: Rehm H-J and Reed G (eds) *Biotechnology*. Vol. 6b, p. 1–30. Weinheim, Germany: VCH Verlagsgesellschaft.

Beesch SC (1952) Acetone–Butanol fermentation of sugars. *Ind. Eng. Chem.* 44: 1677–1682.

Jones TJ and Woods DR (1986) Acetone–butanol fermentation revisited. *Microbiological Reviews* 50: 484–524.

Prescott SC and Dunn CG (1959) *Industrial Microbiology*, 3rd edn. Ch. 13: The acetone–butanol fermentation. New York: McGraw-Hill.

Santangelo JD and Dürre P (1996) Microbial production of acetone and butanol: can history be repeated. *Chimica oggi/Chemistry Today* 14: 29–35.

Walton MT and Martin JL (1979) Production of butanol–acetone by fermentation. In: Peppler HJ and Perlmann (eds) *Microbial Technology*. Vol. 1, p. 187. New York: Academic Press.

Woods DR (ed.) (1993) *The Clostridia and Biotechnology*. Boston: Butterworth-Heinemann.

Woods DR (1995) The genetic engineering of microbial solvent production. *Trends in Biotechnology* 13: 259–263.

Clostridium tyrobutyricum

Martin Wiedmann and **Kathryn J Boor**, Department of Food Science, Cornell University, USA

Hartmut Eisgruber and **Klaus-Jürgen Zaadhof**, Institute for Hygiene and Technology of Foods of Animal Origin, Ludwig-Maximilians University, Munich, Germany

Clostridium tyrobutyricum is a predominant causative agent of structural and sensory defects in cheese, and, specifically, the late blowing defect in brine salted, hard and semi-hard cheeses (e.g. Emmental, Gouda, Edam). Therefore, this organism is of considerable economic concern to the food industry. *C. tyrobutyricum* is commonly found in milk and cheeses; other common *Clostridium* spp. in milk include *C. butyricum*, *C. sporogenes*, *C. beijerinckii*, *C. bifermentans* and *C. perfringens*.

The name of this species reflects the fact that this organism was first isolated from cheese (the Greek word *tyros* means cheese) and that it produces butyric acid. In comparison with many of the pathogenic and toxinogenic *Clostridium* spp., relatively little research has been directed toward *C. tyrobutyricum* and the development of detection methods for this organism. This article reflects the current state of knowledge regarding this organism and identifies key issues of primary concern to the food and dairy industries.

Characteristics of the Species

Clostridium tyrobutyricum is a Gram-positive, strictly anaerobic rod, occurring singly or in pairs, which is usually peritrichously flagellated and motile. Its size is in the range $1.9–13.3 \times 1.1–1.6 \, \mu m$. Spores are oval, subterminal and swell the cell. The optimum growth temperature is in the range 30–37°C, with moderate growth at 25°C and poor or no growth at 45°C. Surface colonies on anaerobically incubated blood agar plates are frequently β-haemolytic, circular, glossy and grey, with a diameter of 0.5 mm. This organism has been isolated from raw milk and dairy products, chicken, chicken salad, silage, gley soil and the faecal material of cattle, beagle dogs and human adults and infants. *C. tyrobutyricum* is non-pathogenic to humans and animals.

C. tyrobutyricum is a saccharolytic *Clostridium* spp. In peptone–yeast extract–glucose (PYG) broth, *C. tyrobutyricum* fermentation produces large amounts of butyric and acetic acids. Large volumes of gas are produced in PYG deep agar cultures. Pyruvate is converted into butyrate and acetate and lactate is fermented to butyrate, CO_2 and H_2. *C. tyrobutyricum* ferments lactate if acetate is also present in the growth medium, therefore this carbon source is commonly added to media to enhance detection of this organism. *Clostridium* spp. producing butyric acid are often referred to as 'butyric anaerobes'; these include *C. butyricum*, *C. sporogenes*, *C. beijerinckii* and *C. pasteuranium*. *C. tyrobutyricum* produces only traces of alcohol in comparison to *C. butyricum*, which can produce significant amounts of butanol in the late stages of fermentative growth.

C. tyrobutyricum is distinguished from *C. butyricum* and *C. beijerinckii* by its inability to ferment lactose, maltose and salicin. A limited number of *C. tyrobutyricum* strains have been shown to ferment lactose, however. Some *C. tyrobutyricum* strains ferment mannitol, xylose and mannose, and produce nitrite from nitrate. Characteristics of *C. tyro-*

Table 1 Characteristics of *C. tyrobutyricum*[a]

Characteristic	C. tyrobutyricum[a]
Spores	oval, subterminal
Motility	+
Haemolysis	–/weakly β-haemolytic
H_2S formation	–
Aesculin hydrolysis	–
Gelatin hydrolysis	–
Indole production	–
Nitrate reduction	–/(+)
Products from PYG	Butyric acid, in some cases acetic acid and/or small amounts of succinic, formic, lactic and propionic acid
Enzyme activities	
Lecithinase	–
Lipase	–
Urease	–
Acid production from:	
Arabinose	–
Cellobiose	–
Dulcitol	–
Fructose	+
Galactose	–
Glucose	+
Glycerol	–
Glycogen	–
Inositol	–
Lactose	–/(+)
Maltose	–
Mannitol	–/(+)
Mannose	+/(+)
Melezitose	–
Melibiose	–
Raffinose	–
Rhamnose	–
Ribose	–
Saccharose	–/(+)
Salicin	–
Sorbitol	–
Starch	–
Sucrose	–
Trehalose	–
Xylose	–/(+)

[a] +, positive reaction in > 95% of isolates; –, negative reaction in > 95% isolates; –/(+), negative or weakly positive reaction in > 95%; +/(+), positive or weakly positive reaction in > 95%. PYG, peptone–yeast extract–glucose broth.

butyricum are summarized in **Table 1**. Biochemical reaction patterns, as determined by the API 20 A system, are shown in **Table 2**. As *C. tyrobutyricum* shows very distinctive patterns in this test, this system provides differentiation from other *Clostridium* spp. commonly found in milk.

Table 2 Phenotypic characteristics of *C. tyrobutyricum* and selected other *Clostridium* spp. commonly found in milk

	ind	ure	gel	esc	glu	man	lac	sac	mal	sal	xyl	ara	gly	cel	mne	mlz	raf	sor	rha	tre
C. beijerinckii	−	−	v	v	+	v	+	+	+	+	+	v	v	+	+	−	+	v	v	v
C. bifermentans	v	−	+	v	+	−	−	−	v	−	−	−	−	−	v	−	+	−	−	−
C. butyricum	−	−	v	v	+	v	+	+	+	+	+	v	v	+	+	−	+	v	v	+
C. sporogenes	−	−	+	v	+	−	−	−	+	−	−	−	−	−	−	−	−	−	−	v
C. tyrobutyricum	−	−	−	−	+	+	−	−	−	−	v	−	−	−	+	−	−	−	−	−

ind: indole formation; *ure*: urease activity; *gel*: gelatin hydrolysis; *esc*: aesculin hydrolysis; *glu, man, lac, sac, mal, sal, xyl, ara, gly, cel, mne, miz, raf, sor, rha, tre*: acid formation from glucose, mannitol, lactose, saccharose, maltose, salicin, xylose, arabinose, glycerol, cellobiose, mannose, melezitose, raffinose, sorbitol, rhamnose, and trehalose.
+, Reaction positive for 90–100% of strains; v, reaction positive for 10–90% of strains; −, reaction negative for 90–100% of strains.

Detection Methods

Specific, sensitive and quantitative detection of *C. tyrobutyricum* in milk represents a major challenge for the food microbiologist. Detection and enumeration of *C. tyrobutyricum* spores in raw milk is necessary for monitoring and screening milk supplies used for cheese-making. Since spore numbers of *C. tyrobutyricum* should be less than one or two spores per 10 ml of raw milk to avoid the late blowing defect, very sensitive detection methods are crucial. Despite this need, no selective or differential media specific for *C. tyrobutyricum* are currently available. A detailed review of the various enumeration methods for *C. tyrobutyricum* used in different countries has been published by the International Dairy Federation and is included in this bibliography.

The presence of *C. tyrobutyricum* spores is currently monitored by applying methods for either the detection of lactate-fermenting anaerobic spore formers or for the detection of total numbers of anaerobic spore formers. Media appropriate for isolation of anaerobic spore formers can be made more selective for *C. tyrobutyricum* by (1) substituting lactate for glucose as the fermentable carbohydrate source; and by (2) adjusting the growth media to a pH of 5.3–5.5, which is similar to that of many cheeses. *C. butyricum*, which also ferments lactate in the presence of acetate, but is not responsible for the late-blowing defect, grows significantly more slowly than *C. tyrobutyricum* at pH 5.3–5.5 and is inhibited at pH 5.3 or below. A low pH in growth media also helps to avoid false positive results due to the growth of facultative anaerobic *Bacillus* spp. Therefore, it is advisable to adjust media for the specific detection of *C. tyrobutyricum* to a maximum pH of 5.4, as in the medium used in the NIZO Ede (Netherlands Institute for Dairy Research at Ede) method (see below).

Although detection of *C. tyrobutyricum* in cheese has been valuable in establishing this organism as the causative agent of the late-blowing defect, quantitative determination of *C. tyrobutyricum* spore numbers in cheeses (as achieved by the most probable number (MPN) methods described below) is of limited value, since vegetative cells are destroyed in the procedure. As a consequence, the *C. tyrobutyricum* numbers estimated by MPN do not reflect total numbers of vegetative cells and spores present in the cheese. Predictive capabilities for estimating relative numbers of *C. tyrobutyricum* spores to vegetative cells in cheeses have not been established. Spore numbers of 10^1–10^7 per gram have been found in cheese evolving butyric acid. Butyric acid production, which is an indicator of *C. tyrobutyricum* contamination in cheeses, can be evaluated by head-space gas chromatography or by high performance liquid chromatography (HPLC) techniques. These analytical techniques offer an additional approach for quantitatively screening for the presence and metabolic activity of *C. tyrobutyricum* in cheese. However, since fat degradation in cheese can also produce small amounts of butyric acid, determination of butyric acid values, alone, in cheeses with significant fat degradation might not be diagnostic for the presence of *C. tyrobutyricum*. To overcome this potential complication, quantification of capronic acid in addition to butyric acid (i.e. determination of an increase in butyric acid content, but no increase in the capronic acid content) will indicate fermentation of lactate to butyric acid and the absence of lipid degradation. Butyric acid values greater than 100 mg butyric acid per kilogram in Gouda cheese are indicative of fermentation of lactate to butyric acid.

A variety of media, including the Bacto-AC-medium, reinforced clostridial medium (RCM) and cooked-meat medium, are suitable for the cultivation and maintenance of *C. tyrobutyricum*. Chopped-meat agar slants or old PYG cultures are recommended for culture sporulation. Agar media containing sulphite (e.g. differential reinforced Clostridial medium (DRCM), sodium ferricitrate agar) are generally used for the detection of sulphite-reducing mesophilic *Clostridium* spp., but also permit growth of *C. tyrobutyricum*. It is important to note that, although many *Clostridium* spp. (e.g. *C. botulinum*, *C. sporogenes*,

C. bifermentans, *C. perfringens*) have the ability to reduce sulphite, *C. tyrobutyricum* is reported to be non-sulphite reducing.

The presence of some selective components commonly used in formulations for the detection of *Clostridium* spp. in agar media (e.g. crystal violet, neomycin, polymyxin B) is problematic for the detection and growth of *C. tyrobutyricum*. This species, or some strains within the species, is somewhat sensitive to many selective components. DRCM medium contains no selective components and can therefore be used for the detection and isolation of *Clostridium* spp., including *C. tyrobutyricum*, from milk. The absence of selective components mandates that the sample undergoes a heating step to inactivate vegetative cells that may be present. None of these media provide adequate selectivity or differentiation to allow direct quantitative detection of *C. tyrobutyricum* in milk. The utility of these media for the detection of *C. tyrobutyricum* from food products could be enhanced by combination with subsequent tests specific for this organism (e.g. colony hybridization, immunoblots).

MPN Procedures

Overview The MPN procedure using three or five tubes is currently the most common method for the estimation of *C. tyrobutyricum* numbers in milk. Generally, for MPN estimation of *C. tyrobutyricum*, milk sample volumes of 0.1 ml, 1.0 ml or 10 ml are added to an appropriate medium. Determination of the sample volumes (i.e. dilutions) used and the number of tubes per dilution depends on the specific application and purpose of each test. Since as few as one or two *C. tyrobutyricum* spores per 10 ml of milk can cause the late-blowing defect, a sensitivity of 2 spores per 10 ml is necessary for a raw milk screening assay. This level of contamination is indicated by a maximum of one positive tube out of three tubes containing 10 ml of milk per tube in an MPN test. Two dilutions (10 ml and 1 ml of milk) with three tubes per dilution are typically used in routine testing.

Sample Preparation and Incubation Conditions All MPN methods used for *C. tyrobutyricum* quantification detect the presence of spores, but not vegetative cells, since samples are heated to inactivate vegetative cells either before inoculation or immediately after addition to the medium. Although temperatures used for heat treatments vary widely between different protocols, recommended heat treatments are in the range 5–10 min at 75–80°C. Since *C. tyrobutyricum* spores are reportedly more heat sensitive than those of many other *Clostridium* spp.,

higher temperatures or longer heat treatments should be avoided.

Before incubation, inoculated MPN tubes are sealed, e.g. with paraffin, to exclude oxygen. Tubes are usually incubated at 37°C for 7 days. Tubes are positive if they show visible gas formation at the end of the incubation period. For the detection of *C. tyrobutyricum*, tubes are only designated positive if large volumes of gas have been produced, as indicated by obvious vertical displacement of the paraffin plug above the culture medium. *C. tyrobutyricum* spores generally produce positive results after incubation at 37°C for 4 days. In fact, a 4 day incubation is used for the NIZO Ede MPN method to optimize the likelihood that the growth of *C. tyrobutyricum* is predominantly responsible for positive results.

Media Commonly Used for MPN Estimations Media most suitable for quantitative detection of *C. tyrobutyricum* by MPN procedures include RCM with the substitution of lactate for glucose (also known as Fryer/Halligan Method), Bergère's modification of the lactate medium of Bryant and Burkey (BBMB-lactate), and the NIZO Ede media (**Table 3**). These media contain lactate as a carbon source to allow selective growth of lactate-fermenting spore formers. RCM-lactate and BBMB-lactate also contain acetate which facilitates lactate fermentation by *C. tyrobutyricum*. The pH of RCM-lactate can be adjusted to 5.4 to improve its selectivity for *C. tyrobutyricum*. The Weinzirl method is a classical MPN test for the detection of anaerobic spore formers. The Weinzirl approach uses milk, milk supplemented with glucose, milk supplemented with yeast extract, lactate and cysteine, or milk supplemented with glucose and lactate as growth media. However, determination of the presence of anaerobic spore formers by the original Weinzirl method generally does not correlate with the potential of the milk to cause the late-blowing defect in cheese. Because the original Weinzirl method uses milk as the primary growth medium, *C. tyrobutyricum* spores are not usually detected, since most strains are unable to ferment lactose. The NIZO-Ede method is a modification of the Weinzirl method which uses a lactic acid–glucose solution or a skim milk–lactic acid glucose solution adjusted to pH 5.45 to add lactate as a carbon source.

It is important to note that MPN techniques using RCM-lactate and BBMB-lactate do not allow specific detection of *C. tyrobutyricum*, but rather detect the presence of any spore formers which have the ability to ferment lactate in the presence of acetate. The modified RCM-lactate (pH 5.4) and NIZO-Ede utilize low pH (5.3–5.5) to improve selectivity for *C. tyrobutyricum*. The NIZO Ede method is reportedly

Table 3 Media used for estimation of *C. tyrobutyricum* by most probable number method

Media	Ingredients	Amount
Modified reinforced clostridial media (RCM-lactate) (adjust pH to 6.1[a])	Beef extract	10 g
	Tryptone	10 g
	Yeast extract	3 g
	60% Sodium lactate solution	23.3 ml
	Sodium acetate	8 g
	Starch	1 g
	L-Cysteine-HCl	0.5 g
	NaCl	5 g
	Agar-agar	2 g
	dist. H_2O	1000 ml
BBMB-lactate (adjust pH to 6.0)	Peptone	15 g
	Beef extract	10 g
	Yeast extract	5 g
	Sodium acetate	5 g
	60% Sodium lactate solution	8.4 ml
	L-Cysteine-HCl	0.5 g
	dist. H_2O	1000 ml
NIZO Ede[b]		
Solution 1	Glucose	5 g
	Lactic acid (1M)	20 ml
	dist. H_2O	up to 100 ml
Solution 2 (adjust pH to 5.45)	Skim milk	900 ml
	Solution 1	100 ml
Lactate–acetate–thioglycollate–ammonium sulphate (LATA) medium (adjust pH to 6.1)	Calcium lactate	20 g
	Sodium acetate	8 g
	Sodium thioglycollate	0.5 g
	Ammonium sulphate	1 g
	Agar	2 g
	Mineral supplement ($MgSO_4.7 H_2O$), 2.0%; $MnSO_4.4 H_2O$, 0.5%; $FeSO_4.7 H_2O$, 0.4%)	10 ml
	dist. H_2O	990 ml

[a] For increased selectivity for *C. tyrobutyricum* the pH can be adjusted to 5.4. For this medium the pH must be adjusted when using 10 ml of the sample to compensate for the pH increase due to sample addition.
[b] For 10 ml of the sample use 1 ml of solution 1; for 1 ml of the sample or dilution of it, use 10 ml of solution 2.

somewhat less sensitive for the detection of *C. tyrobutyricum* than the Fryer/Halligan method using modified RCM-lactate (pH 5.4).

Before inoculation, tubes containing the appropriate amount of media are either freshly sterilized or otherwise treated (i.e. by heating in a boiling-water bath or steaming for 10–20 min) to drive off dissolved oxygen which might inhibit growth of *Clostridium* spp. Although indicators such as resazurin can be used to indicate the redox status of the media (resazurin is colourless when reduced and pink when oxidized), these are generally omitted in routine MPN applications. The inoculation of 1 ml of milk, or more, to media containing lactate as a sole carbon source compromises the selectivity of the media due to the incorporation of lactose as an additional fermentable carbohydrate. Although confirmation tests on positive MPN tubes are not frequently performed on a routine basis, subculturing positive tubes in lactate–acetate–thioglycollate–ammonium sulphate medium (LATA)

is advisable. Plating on RCM plates containing 200 μg cycloserin per millilitre, followed by anaerobic incubation for 24–48 h and testing of selected colonies for lactate dehydrogenase activity using a colorimetric enzyme assay has also been proposed as a confirmation method. Currently, the most commonly used confirmation procedure is the inoculation of 1 ml of a 1 : 10 dilution prepared from a positive MPN tube into 10 ml LATA, followed by incubation under anaerobic conditions for up to 5 days. Enzyme-linked immunoassay (ELISA) tests for *C. tyrobutyricum* and gas chromatography for butyric and acetic acid also provide specific confirmation.

Antibody and DNA-based Detection Methods

Due to difficulties in identifying and differentiating *Clostridium* spp. and *C. tyrobutyricum* to species by classical approaches, novel methods for improving our abilities to quantitatively and rapidly identify and enumerate *C. tyrobutyricum* are under investigation.

Current classical methods require 4–7 days for quantitative estimation and are not specific for *C. tyrobutyricum*. Alternative antibody or DNA-based methods show significant promise. Although these approaches cannot currently replace standard MPN methods, some are well suited for reliable confirmation of the presence of *C. tyrobutyricum* spores in conjunction with the classical MPN methods. One particularly promising strategy for detection and quantification of *C. tyrobutyricum* in fluid milk samples could combine a membrane filtration step with subsequent incubation of the membrane on the surface of a suitable agar medium, followed by application of a *C. tyrobutyricum*-specific detection step using antibodies or DNA probes.

Antibody-based Methods Antibody-based tests, specifically ELISA tests, for detection of *C. tyrobutyricum* have been described. These tests are particularly useful for confirmation of the presence of this organism from positive MPN tubes. *C. tyrobutyricum* isolation using membrane filtration followed by direct detection of the organism on the membrane by a monoclonal antibody has also been reported. This strategy offers a promising approach for rapid, quantitative detection of this organism from fluid samples.

DNA-based Methods DNA probes based on specific 16S rRNA sequences have been shown to provide reliable identification of this species. Furthermore, polymerase chain reaction (PCR) primers based on unique 16S rRNA sequences have been used successfully to design a PCR assay for the specific detection of this species. The combination of these tools with the development of efficient methods for extraction of bacterial DNA from milk matrices could allow application of this strategy for rapid screening for the presence of *C. tyrobutyricum* in raw milk.

Importance in the Food Industry

Clostridium tyrobutyricum is an economic concern for the dairy industry because it causes structural and sensory defects in cheeses (the 'late-blowing' defect, **Fig. 1**) through production of large quantities of gas and butyric acid. The late-blowing defect, which is a consequence of the outgrowth of *C. tyrobutyricum* spores, occurs most frequently in brine-salted, hard and semi-hard cheeses (e.g. Gouda, Edam, Emmental, Gruyère). Butyric acid levels above $200 \, \mu g \, l^{-1}$ produce detectable off-flavours which result in the downgrading of cheese. In some cases, gas production is sufficient to rupture the entire cheese structure. Although other *Clostridium* spp., including *C. beijerinckii*, *C. butyricum* and *C. sporogenes*, have been

Figure 1 Late gas blowing in Gouda cheese. (Reproduced with permission from Kosikowski and Mistry 1997, F.V. Kosikowski, L.L.C., Westport, Connecticut.)

found in raw milk and cheeses, *C. tyrobutyricum* is considered the only *Clostridium* spp. responsible for the late-blowing defect in cheese. Not only has this species been isolated from cheeses exhibiting this defect, but also inoculation of *C. tyrobutyricum*, but not other species, into experimentally made cheeses can result in reproduction of the late-blowing defect. It is noteworthy, however, that not all cheeses artificially contaminated with *C. tyrobutyricum* developed the defect.

C. tyrobutyricum is thought to enter cheese in raw milk contaminated with silage or bovine faecal material. Spores of lactate-fermenting *Clostridium* spp. (including *C. tyrobutyricum*) are often found in high numbers (> 100 000 spores per gram) in improperly fermented silages. As a secondary indicator, a butyric acid content $> 1 \, g \, kg^{-1}$ silage suggests the likelihood of the presence of high numbers of clostridial spores, including *C. tyrobutyricum*. Grass silage has been more frequently associated with high spore counts than corn silage. This may be explained by the fact that a higher level of contamination with soil (containing clostridial spores) occurs when cutting grass as compared to harvesting corn. Improvement in the quality of grass silage, for example by using silage starters such as propionic or formic acid, can significantly improve feed quality and reduce the risk of transferring clostridial spores into raw milk. As there is a clear positive correlation between the feeding of poor quality silage and the presence of high spore numbers in the faecal matter of dairy cows, faecal material is likely the primary source of *C. tyro-*

butyricum contamination in milk. Attention to milking hygiene, including proper cleaning and disinfection of udders and teats, represents another critical point for reducing spore numbers in the raw milk.

Raw milk from cows fed silage is considered undesirable for the production of certain gourmet cheeses, such as Emmental. In some European countries, e.g. Switzerland, Austria and regions of Germany, cheese manufacturers will not purchase raw milk produced by silage-fed dairy cows. The ability to test quantitatively for the presence of *C. tyrobutyricum* is essential for screening milk for quality and compliance with the requirement for avoidance of silage feeding. High numbers of lactate-fermenting clostridial spores in raw milk are generally considered indicative of the presence of at least some amount of raw milk from cows fed silage.

Bactofugation (centrifugation > 5000 g) of raw milk can reduce spore numbers by about 98%, but cannot completely eliminate them. Therefore, this technology is only effective in preventing the late-blowing defect if the raw milk is of good microbial quality (< 5–10 spores per millilitre milk). In addition to improved sanitation and restriction of silage feeding, control measures for the late-blowing defect include the addition of sodium nitrate (which is not permitted in the US or in some other countries) and/or enzymes, particularly lysozyme, during the cheese-making process. The addition of lysozyme, which also negatively affects starter cultures, is effective in preventing late blowing only when low levels of *C. tyrobutyricum* spores are present. Pasteurization of the raw milk does not prevent the late-blowing defect since *C. tyrobutyricum* spores survive pasteurization. Even very low numbers of *C. tyrobutyricum* spores (1–2 in 10 ml) are sufficient to cause the late-blowing defect. Quality control for prevention of the late-blowing defect in the cheese-making process should therefore include two steps. (1) Clostridial spore numbers in raw milk should be monitored and these numbers should be included in the determination of quality premiums paid to milk producers. Rapid methods for screening raw milk for clostridial spores could also be used to exclude raw milk batches above a certain cutoff point for the production of specific types of cheeses. (2) At the manufacturing level, testing of the milk before fermentation could be used to monitor the effectiveness of bactofugation in the reduction of spore numbers.

In summary, *C. tyrobutyricum* is a significant economic problem for cheese-making industry. Since raw milk is the primary source of this organism and since the spores are able to survive pasteurization, beyond physical removal through bactofugation or through other means, in-plant quality assurance programmes

have minimal effect on reducing product contamination with this organism. Silage quality and milking hygiene are the most important factors with regard to the contamination of raw milk and therefore present potential critical control points for improvement of raw milk quality with regard to reducing levels of *C. tyrobutyricum*. Therefore, milk producers supplying manufacturers of gourmet cheeses will be increasingly called on to produce raw milk with low spore levels. As an incentive for producers, quality premiums for raw milk designated for the production of certain cheeses could be based on maintaining *C. tyrobutyricum* spore numbers below specific levels. Currently, clostridial spore numbers in raw milk are included in the calculation of milk quality premiums in certain areas of Germany, Italy and the Netherlands. Development and improvement of rapid detection methods will aid in monitoring, and ultimately, reducing the presence of *C. tyrobutyricum* in raw milk, and will, therefore help to minimize economic losses due to the late-blowing defect in high value cheeses.

See *also:* **Cheese**: In the Market-place. **Clostridium**: Introduction. **Enzyme Immunoassays**: Overview. **Milk and Milk Products**: Microbiology of Liquid Milk. **Nucleic Acid-based Assays**: Overview. **Sampling Regimes & Statistical Evaluation of Microbiological Results**. **Spoilage Problems**: Problems caused by Bacteria.

Further Reading

Bergère JL and Sivelä S (1990) Detection and enumeration of clostridial spores related to cheese quality – classical and new methods. In: *Methods of Detection and Prevention of Anaerobic Spore Formers in Relation to the Quality of Cheese*. Bulletin of the International Dairy Federation (IDF), no. 251, Brussels, Belgium.

Cato EP, George WL and Finegold SM (1986) Genus *Clostridium*. In: Sneath PNA, Mair NS, Sharpe ME and Holt JG (eds) *Bergey's Manual of Systematic Bacteriology*. Vol. 2, p. 1141. Baltimore: Williams & Wilkins.

Guricke S (1993) Laktatvergärende Clostridien bei der Käseherstellung. *Deutsche Milchwirtschaft* 15: 735–739.

Goudkov AV and Sharpe ME (1965) Clostridia in dairying. *J. Appl. Bacteriol.* 28: 63–73.

Halligan AC and Fryer TF (1976) The development of a method for detecting spores of *Clostridium tyrobutyricum* in milk. *NZ J. Dairy Sci. Technol.* 11: 100–106.

Heilmeier J (1985) Der Nachweis von käsereischädlichen Clostridien. *Deutsche Molkerei-Zeitung* 106: 196–199.

Herman LMF, de Block JHGE and Waes GMAVJ (1995) A direct PCR detection method for *Clostridium tyrobutyricum* spores in up to 100 milliliters of raw milk. *Appl. Environ. Microbiol.* 61: 4141–4146.

Hüfner J (1987) Neuere Erkenntnisse aus Forschung und Praxis zur Untersuchung der Käsereimilch auf anaerobe Sporenbildner. *Deutsche Molkerei-Zeitung* 108: 1230–1237.

Kammerlehner J (1995) Buttersäuregärung im Käse ganzjährig? *Deutsche Milchwirtschaft* 46: 903–908.

Klijn N, Bovie C, Dommess J et al (1994) Identification of *Clostridium tyrobutyricum* and related species using sugar fermentation, organic acid formation and DNA probes based on specific 16S rRNA sequences. *System. Appl. Bacteriol.* 17: 249–256.

Klijn N, Nieuwenhof FFJ, Hoolwerf JD, van der Waals CB and Weerkamp AH (1995) Identification of *Clostridium tyrobutyricum* as the causative agent of late blowing in cheese by species-specific PCR amplification. *Appl. Environ. Microbiol.* 61: 2919–2924.

Kosikowski FV and Mistry VV (1997) *Cheese and Fermented Milk Foods*, 3rd edn. F.V. Kosikowski L.L.C., Westport, Connecticut.

Kutzner HJ (1963) Untersuchungen an Clostridien mit besonderer Berücksichtigung der für die Milchwirtschaft wichtigen Arten. *Zbl. Bakteriol. Orig.* 191: 441–450.

Nedellec M, Cleret JJ, Robreau G, Talbot F, and Malcoste R (1992) Optimization of an amplified system for the detection of *Clostridium tyrobutyricum* on nitrocellulose filters by use of monoclonal antibody in a gelified medium. *J. Appl. Bacteriol.* 72: 39–43.

Rapp M (1987) Erfassung von Clostridien und Nachweis von *Clostridium tyrobutyricum* unter Praxisverhältnissen. *Deutsche Molkerei-Zeitung* 108: 76–82.

Zangerl P (1993) Nachweis käsereischädlicher Clostridien in Rohmilch und Käse. *Deutsche Milchwirtschaft* 44: 936–940.

Clostridium botulinum

Eric A Johnson, Department of Food Microbiology, Food Research Institute, University of Wisconsin, Madison, WI, USA

Characteristics of *Clostridium botulinum*

The genus *Clostridium* is a large and diverse group with more than 100 species. It includes anaerobic or aerotolerant rod-shaped bacteria that produce endospores and obtain their energy for growth by fermentation. Clostridia are classified on the basis of morphology, disease association, physiology, serologic properties, DNA relatedness and ribosomal RNA gene sequence homologies. Many species of clostridia produce protein toxins that are lethal to animals and are responsible for their pathogenicity. *C. botulinum* produces the most poisonous substance known; it is estimated that 0.1–1 µg of its characteristic botulinum neurotoxin (BoNT) is sufficient to kill a human and the lethal dose for most animals

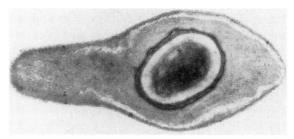

Figure 1 Characteristic spindle morphology of *C. botulinum*. The photograph shows a transmission electron micrograph (×50 000) of a longitudinal section through a spore and sporangium of *C. botulinum* type A.

is ca. 1 ng kg^{-1} body weight. The neurotoxins produced by *C. botulinum* cause the rare but severe neuroparalytic disease, botulism.

C. botulinum is a very diverse species comprising organisms differing widely in physiological properties and genetic relatedness. They all share the ability to produce BoNT and cause botulism in humans and animals. The neurotoxins are distinguished serologically and designated as types A to G. *C. botulinum* types A, B and E most commonly cause botulism in humans, whereas types B, C and D cause the disease in various animal species. *C. botulinum* and other clostridia are widely dispersed in nature by virtue of their ability to produce resistant endospores. The incidence of spores of *C. botulinum* varies according to geographic region. In the United States, type A is found most commonly west of the Rocky Mountains, and type B is found in certain regions of eastern USA. Type B from non-proteolytic strains of *C. botulinum* are also frequently found in Europe. Type A is infrequently found in the soils of England. Type A spores have also been detected in soils of China and South America. The principal habitat of type E spores appears to be freshwater and brackish marine habitats. It has commonly been found in the Great Lakes of the USA and in the western sea coasts of Washington State and Alaska. Type C strains occur worldwide, whereas the distribution of type D is more limited and it is especially common in certain regions of Africa.

C. botulinum consists of four physiological groups (I–IV) with diverse physiological and genetic characteristics. Group IV *C. botulinum* is the only group that has not been demonstrated to cause botulism in humans or animals and has been assigned to the species *C. argentinense*. The organisms in the other three groups are motile, produce lipase on egg yolk agar, liquefy gelatin and ferment glucose. The organisms are morphologically large rods, typically 1 × 4–6 µm with oval, subterminal spores that swell the rod giving the characteristic 'tennis-racket' or spindle-shaped cells (**Fig. 1**). Spores of most pathogenic

Table 1 Factors controlling growth and inactivation of *C. botulinum* in foods

Factor	C. botulinum group	
	I	*II*
Minimal pH	4.6	5.0
Minimal a_w	0.94	0.97
Required brine concn. (%)	10	5
Minimum temperature (°C)	10	3.3
Maximum temperature (°C)	50	45
D_{100} of spores (min)	30	< 0.2
D_{121} of spores (min)	0.2	–

species of clostridia can be produced in complex laboratory media such as chopped meat broth or tryptose–peptone–glucose broth.

Groups I and II are the cause of human botulism, whereas Group III causes botulism in various taxa of animals. The primary properties and limiting growth parameters of *C. botulinum* groups I and II pertaining to foods are presented in **Table 1**. Organisms in group I are proteolytic, and may produce type A, B or F BoNT. They may form highly heat-resistant spores, have an optimum growth temperature of 30–40°C, and are inhibited by 10% NaCl. Organisms in group II are commonly referred to as non-proteolytic, require sugars for growth, and may produce either type B, E, or F BoNT. They have a lower optimum temperature for growth (20–30°C), and some strains of types B and E can grow slowly in foods at temperatures as low as 3.3°C. Consequently, there has been considerable concern raised that group II organisms can grow and produce toxin in refrigerated foods that receive minimal processing and have extended shelf life. Strains that produce type E toxin are commonly associated with food-borne botulism transmitted in contaminated fish or marine products. Group II strains that produce type B toxin are commonly found in Europe and are associated with botulism from salt-cured meats.

Control of *C. botulinum* in Foods

The primary factors controlling growth of *C. botulinum* in foods are the temperature, pH, water activity, redox potential and oxygen level, the presence of preservatives and competing microflora. In the commercial setting, botulism can occur when a food is inadvertently exposed to temperatures that allow growth and toxin formation. Since BoNT is extremely potent, quantities sufficient to cause botulism can be formed without obvious spoilage of foods. In most foods *C. botulinum* is a poor competitor and other microorganisms such as lactic acid bacteria often grow more rapidly, commonly lowering the pH and preventing growth. However, the spores of *C. bot-*

ulinum are more resistant to heat, irradiation, and other processing methods than are vegetative cells of competing organisms. Therefore, minimal processing of foods can eliminate or reduce the numbers of competing microflora and increase the probability of *C. botulinum* growing and producing toxin. The critical level of oxygen that will permit growth of group I *C. botulinum* is 1–2% but this depends on other conditions such as a_w and pH.

Spores of group I *C. botulinum* have heat resistances ranging from $D_{121} = 0.03$–0.23 min and D_{100} approx. 30 min. Certain strains of *C. sporogenes*, which is genetically related to group I *C. botulinum*, produce spores with much higher heat resistance (maximum D_{121} approx. 1.0 min) than *C. botulinum*, and these strains can be used to determine the heat treatment required for obtaining a 12D inactivation or total lethality (F_0) as is recommended for shelf-stable low acid foods in cans, glass jars or pouches. The required treatment for achieving F_0 of a food from *C. botulinum* spores is ca. 3 min at 121°C or higher. The commercial processing of certain foods is less than an F_0 of 3 min since other factors control their safety from *C. botulinum*.

In preserved food products, *C. botulinum* growth can be prevented by a single factor, such as extensive thermal processing (a 'bot cook'). Often a combination of factors is used to prevent *C. botulinum* growth in low acid foods (pH $\geq$ 4.6). For example, in cured meats the combination of a mild heat treatment, and the presence of nitrite and salt prevents growth. Challenging foods with spores of *C. botulinum* and determining whether toxin is produced in optimal conditions or on temperature abuse is often a desired procedure to evaluate the botulinogenic safety of a food, particularly in new products or new formulations.

Due to the severity of botulinum poisoning, the food industry has devoted considerable research and resources to prevent botulism outbreaks in foods. The control of this organism is of such paramount importance to the safety of foods, that certain food laws and definitions such as thermal processing of low-acid foods in hermetically sealed containers were specifically designed to control *C. botulinum*. The organism has served as a 'barometer' by which to gauge certain advances in food formulation and processing. Thus, newly developed foods and food-processes may need to be evaluated for their impact on *C. botulinum* growth and toxin formation. These efforts and vigilance by the food industry have contributed to a safe food supply.

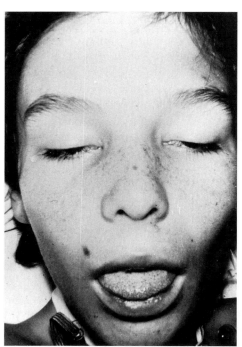

Figure 2 Patient with botulism. Photograph courtesy of Charles L. Hatheway (deceased), Centers for Disease Control and Prevention, Atlanta, Georgia, USA.

Clinical Features of Botulism

Botulism is categorized according to the route by which BoNT enters the human circulation. Classical food-borne botulism results from the ingestion of neurotoxin preformed in foods. Botulism caused by food poisoning generally has an incubation period of 12–36 h after consumption of a toxic food. Wound botulism is analogous to tetanus and occurs when C. *botulinum* grows and produces toxin in the infected tissue. Intestinal botulism results from the growth and toxin production by C. *botulinum* in the intestine (infant botulism and adult intestinal botulism). Since BoNT is entirely responsible for the clinical symptoms, the three types of botulism exhibit similar clinical symptoms. The characteristic symptomatology of botulism poisoning is a progressive descending symmetrical flaccid paralysis initially affecting musculature innervated by cranial nerves. The first signs are typically disturbances in ocula function including blurred and double vision, and the pupils become enlarged and unresponsive to light. As intoxication proceeds, a flaccid paralysis occurs in the facial and head region, characterized by weakness and drooping of the eyelids and facial muscles (**Fig. 2**). Speech becomes slurred, and swallowing and breathing become difficult. In severe cases, extreme muscular weakness causes the patient to become weak, fatigued and unable to lift their head and limbs. Death can occur, usually by respiratory failure or possibly by

cardiac arrest. Since BoNT affects alpha motor nerves and does not enter the central nervous system, sensory responses, mental function and consciousness are maintained. The inability of the patient to communicate the symptoms, and the awareness of the progression of the disease can cause mental depression and anxiety. In severe cases, intubation and respiratory assistance are required. If diagnosed early, administration of antibodies can scavenge the free toxin in the blood and prevent the disease from progressing to more severe symptoms. Equine antibodies are available from the Centers for Disease Control and Prevention in the United States and in various other public health laboratories throughout the world. Recovery from botulism is generally prolonged, requiring weeks to months for muscle activity to return to a normal level, but complete recovery is usually attained.

Food-borne botulism is quite rare in most areas of the world, although the actual incidence is probably higher than reported since mild cases are probably not diagnosed and botulism may be misdiagnosed as another neurological disorder. The prevalence of food-borne botulism throughout the world is probably associated with the prevalence of spores in the environment. The primary geographical regions of the world that have reported food-borne botulism are East Asia (China, Japan), North America, certain countries in Europe (Poland, Germany, France, Italy, Spain, Denmark, Norway), the Middle East (Iran), Latin America, Russia, and South Africa. Food-borne botulism is very rare in the UK, although certain outbreaks such as the Loch Maree incident, the Birmingham outbreak, and the hazelnut yogurt incident have attracted much attention and publicity. Recent examples of outbreaks of botulism in commercial or restaurant-prepared foods are presented in **Table 2**.

Infant botulism differs from food-borne and wound botulism in the ages of the affected persons, and usually the first symptom is constipation, indicated by not passing a stool in 3 days or longer. As the neurotoxin binds to motor nerves, the characteristic flaccid paralysis affects the baby's musculature in the head and neck regions. The baby has a weak cry and suck and the paralysis may render the baby unable to hold its head and limbs erect. Infants should receive intubation and respiratory assistance to prevent respiratory arrest. A recent study showed that administration of human antibotulinal antibodies shortened the hospital stay of infants with botulism. Botulism may be difficult to recognize in infants because of the baby's inability to communicate its symptoms, the rarity of the disease and inexperience of many doctors, and misdiagnosis of other neurological diseases such as Guillain–Barré syndrome, tick paralysis, drug reac-

Table 2 Examples of outbreaks of food-borne botulism from commercial foods or restaurant-prepared foods

Food product	Year	Location	Toxin type	No. of cases	No. of deaths
Canned peppers	1977	USA	B	58	0
Canned Alaskan Salmon	1978	UK	E	4	2
Kapchunka (salt-cured, uneviscerated whitefish)	1981	USA	B	1	0
Beef pot pie	1982	USA	A	1	0
Sauteed onions	1983	USA	A	28	1
Karahi-renkon (vacuum-packed, deep-fried, mustard-stuffed, lotus root)	1984	Japan	A	36	11
Chopped garlic-in-oil	1985	Canada	B	36	0
Kapchunka	1987	USA/Israel	E	8	2
Chopped garlic-in-oil	1989	USA	A	3	0
Hazelnut yoghurt	1989	UK	B	27	1
Faseikh (salted fish)	1991	Egypt	E	92	20
Cheese sauce	1993	USA	A	5	1
Skordalia (Greek salad with baked potato)	1994	USA	A	19	0
Marscapone cheese	1997	Italy	A	8	1

tions, or viral and bacterial infections of the nervous system.

Infant botulism has been reported from various countries throughout the world including all continents except Africa. Botulism is rare in most countries and the majority of cases have been detected in the USA. Within the USA, infant botulism occurs in clustered geographical regions with about half of the diagnosed cases occurring in California. The clustered geographic distribution of infant botulism may be related to the prevalence and type of spores in the environment. Nearly all cases of infant botulism have been caused by proteolytic strains of *C. botulinum* types A and B. BoNT-producing strains of *C. butyricum* and *C. baratii* have also successfully colonized the intestine of babies and caused botulism. The only food definitively shown to be associated with infant botulism is honey, and babies under one year of age should not be given this food.

Botulism is rare compared to many other food-borne microbial diseases but has a relatively high fatality rate in humans and animals. Human botulism outbreaks can have a dramatic impact on communities in which they occur and can lead to the demise of food companies, and outbreaks of animal botulism periodically devastate populations of domestic and wild animals. To prevent outbreaks, it is necessary for the food industry to properly formulate and process foods to prevent growth and toxin formation.

Properties and Detection of Botulinum Neurotoxin

The clostridia produce more types of protein toxins than any other genus of microorganisms. The outstanding feature of *Clostridium botulinum* is its ability to synthesize a neurotoxin of extraordinary potency. BoNT's comprise a family of pharma-

cologically similar toxins that bind to peripheral cholinergic synapses and block acetylcholine exocytosis at the neuromuscular junctions. BoNTs are produced in foods, in the intestine and in culture as progenitor toxin complexes that consist of botulinal neurotoxin associated with nontoxic proteins. The nontoxic components of the complexes have been demonstrated to impart stability to the neurotoxin and to prevent inactivation by digestive enzymes in the gut.

The diagnosis of botulism is generally accomplished by assessment of clinical symptoms in patients, and for food-borne outbreaks, on the clustering of cases involving a group of people who have eaten a common food. In most investigations of botulism, the primary goal is to detect the presence of BoNT since spores of *C. botulinum* are widespread in the environment and contaminate many foods. The detection of BoNT in the blood, gastric contents and food provides confirmation of botulism. Isolation of *C. botulinum* from a suspect food, from faeces of infants with botulism symptoms, or from wounds provides supporting evidence for the diagnosis of botulism. However, it does not provide a confirmation in most instances since spores are found in foods and occasionally in the faeces of healthy adults.

BoNT is preferably detected using a bioassay of the toxin extracted from a food or clinical sample. The extract is injected intraperitoneally into mice and the animals are observed periodically for typical signs of botulism for up to four days. Depending on the quantity of BoNT present, symptoms of botulism are generally observed within 4–24 h. Characteristic symptoms include decreased mobility of the animals, ruffling of the fur, difficulty in breathing, contraction of abdominal muscles giving the 'wasp' morphology, followed by convulsions and death. Animals demonstrating these signs usually die within 24–48 h. Animals that die sooner than 2 h or after 48 h should be considered as succumbing to substances other than BoNT. Death due to BoNT is confirmed by neutralization with serotype-specific antitoxins.

Complications are often encountered in the mouse bioassay of BoNT from clinical specimens and from certain foods. In particular, deaths caused by nonbotulinum interfering substances are common. These nonspecific fatalities can generally be avoided by diluting out the interfering lethal substance to an end point where death is caused by the more potent BoNT. Occasionally, more than one serotype of BoNT may be present in a sample being analysed, and confirmation would require neutralization by a mixture of antitoxins. With foods or clinical specimens, nonbotulinum deaths can occur by infection or by the presence of endotoxins. Infectious agents can be removed by membrane filtration or by addition of antibiotics to the extract being tested. Extracts containing endotoxins can generally be diluted to a proper end point, or the endotoxins can be removed by adsorption. With extracts from non-proteolytic strains of *C. botulinum* (group II), toxicity is increased by activation by a protease such as trypsin. In some foods, trypsin can generate toxic peptides and therefore the reaction should be terminated by addition of soybean trypsin inhibitor after 30–60 min.

Viable *C. botulinum* can be isolated from foods by enrichment in a suitable growth medium such as cooked meat–glucose broth or media containing peptones, yeast extract and glucose. *C. botulinum* has complex nutrient requirements and requires a rich medium for growth. For isolation, it is often useful to heat a portion of the food or clinical specimen at 80° or 60°C to select for spores of group I and II *C. botulinum*, respectively. Occasionally, 50% ethanol is used to inactivate vegetative cells in food samples analysed for group II *C. botulinum*. Following enrichment, the presence of BoNT is assayed by mouse bioassay as described previously. Selective isolation agars containing antibiotics including cycloserine, sulphamethoxazole and trimethoprim, have been used for the isolation of group I *C. botulinum* from clinical samples.

A variety of immunological methods have been developed for the detection of BoNT but most are not as sensitive as the mouse bioassay and they also have the potential drawback of detecting biologically inactive BoNT. Several advances in enzyme-linked immunosorbent assays (ELISA) have been made to alleviate these drawbacks and it is likely that ELISA will be used to complement but not replace the mouse bioassay.

Use of Botulinum Toxin as a Pharmaceutical

One of the most remarkable recent developments in medicine is the use of botulinum toxin to treat humans who suffer from dystonias and hyperactive muscle disorders. Botulinum toxin is increasingly being used to treat humans suffering from neurological diseases characterized by hyperactive muscle activity. In December 1989 the USA Food and Drug Administration licensed botulinum toxin as an orphan drug for the treatment of the human muscle disorders strabismus, hemifacial spasm and blepharospasm in patients 12 years of age and older, by direct injection of the toxin into the hyperactive muscle. Botulinum toxin is now being used worldwide for medical purposes. Botulinum toxin is being increasingly used for the treatment of a number of dystonias, movement disorders, cosmetic problems and pain disorders, all of which have been difficult to treat by traditional therapies. The important properties of botulinum toxin as a therapeutic are its high specificity for motor neurons, its very high toxicity which enables the injection of extremely low quantities thereby avoiding side effects and an immunological response, and its long (several months) duration of action. The treatment of neurological disorders with botulinum toxin stemmed from its properties as a food poison and its study as a potential biological terrorism agent. The use of the toxin as a drug has enabled thousands of humans to lead an enjoyable and productive life and has also opened a new field of investigation in the application of the toxin to nerve and muscle tissue in the human body.

See also: **Biochemical and Modern Identification Techniques**: Food-poisoning Organisms. **Clostridium**: Detection of Neurotoxins of *C. botulinum*. **Food Poisoning Outbreaks**.

Further Reading

Dickson EC (1918) Botulism. A clinical and experimental study. *Rockefeller Inst. Med. Res. Monog.* 8: 1–117.

Hatheway CL and Johnson EA (1998) *Clostridium*: the spore-bearing anaerobes. In: Collier L, Balows A and Sussman M (eds) *Topley and Wilson's Microbiology and Microbial Infections*, 9th edn. Vol. 2. *Systematic Bacteriology*. P. 731. London: Arnold.

Hauschild AHW (1989) *Clostridium botulinum*. In: Doyle MP (ed.) *Foodborne Bacterial Pathogens*. P. 111. New York: Marcel Dekker.

Hauschild AHW and Dodds KL (1993) *Clostridium botulinum. Ecology and Control in Foods*. New York: Marcel Dekker.

Johnson EA and Goodnough MC (1998) Botulism. In:

Collier L, Balows A and Sussman M (eds) *Topley and Wilson's Microbiology and Microbial Infections*, 9th edn. Vol. 3, *Bacterial Infections*. P. 723. London: Arnold.

Meyer KF and Eddie G (1951) Perspectives concerning botulism. *Z. Hyg. Infektionschr.* 133: 255–263.

Peck MW (1997) *Clostridium botulinum* and the safety of refrigerated processed foods of extended durability. *Trends Food. Sci. Technol.* 8: 186–192.

Schantz EJ and Johnson EA (1992) Properties and use of botulinum toxin and other microbial neurotoxins in medicine. *Microbiol. Rev.* 56: 80–99.

Smith LDS and Sugiyama H (1988) *Botulism*, 2nd edn, Springfield, Illinois: Charles C. Thomas.

van Ermengem E (1897) Ueber einen neuenn anaeroben *Bacillus* and seine Beziehungen zum Botulisms. *Z. Hyg. Infekt.* 26: 1–56. English reprinting: van Ermengem E (1979) A new anaerobic bacillus and its relation to botulism. *J. Infect. Dis.* 1: 701–719.

Table 1 Toxins produced by *Clostridium botulinum* organisms

Type	Subtype	Toxins	
		Major	Minor
A		A	–
	AB	A	B
	AF	A	F
B		B	–
	BA	B	A
	BF	B	F
C	(C$_\alpha$)	C$_1$	C$_2$, D
	(C$_\beta$)	–	C$_2$
D		D	C$_1$,C$_2$
E		E	–
F		F	–
G		G	–

Detection of Neurotoxins of *Clostridium botulinum*

S Notermans, TNO Nutrition and Food Research Institute, AJ Zeist, The Netherlands

Introduction

Botulism is a paralytic disease caused by one of the several potent protein exotoxins produced by the bacterium *Clostridium botulinum*. The illness usually occurs in one of the three clinical–epidemiological forms: (1) food-borne botulism, (2) infant botulism and (3) wound botulism. A small number of cases are of undetermined aetiology. The exotoxin produced by *C. botulinum* may be one of the seven different immunotypes, designated A–G (**Table 1**). All types share a common final pathogenesis: haematogenous circulation of toxin to peripheral cholinergic synapses where release of acetylcholine is blocked, impairing autonomic and neuromuscular transmission. Food-borne botulism is almost completely limited to botulinum toxins types A, B and E.

Assays for botulinum toxins have been developed primarily for diagnostic purposes as well as to increase knowledge of the aetiology of botulism. Assays are also used for developing rules for good manufacturing practices in the food industry.

Confirmation of food-borne botulism is based on the detection and identification of the toxin in the blood serum of patients as well as in the incriminated food. The quantities of toxin in blood serum are typically very low, whereas those in the incriminated food may be substantially higher. For diagnosis of infant botulism, detection and identification of the toxin in faecal material is necessary. However, detection of large numbers of toxin-producing organisms is also useful. Confirmation of wound botulism depends on the demonstration of *C. botulinum* organisms in wound exudate.

The detection of *C. botulinum* and the discrimination of these organisms from other clostridia are based on assays for toxin. This can sometimes be difficult because isolation of pure cultures is rather cumbersome. In the usual procedure, samples are enriched in suitable media and, after proper incubation, culture supernatants are tested for the presence of toxin. The quantity of toxin produced depends on several factors such as the type of sample, the presence of competitive microorganisms, and incubation temperatures. Generally, only small quantities of toxin are produced. Furthermore, production of toxic components by microorganisms other than *C. botulinum* has to be taken into account. For this reason, neutralization by specific antisera has to be tested, which then allows the identification of the infecting strain.

Botulinum toxins are extremely potent neurotoxins and generally occur at low concentrations in implicated foods. Therefore, only ultrasensitive assays are of interest. These include the bioassay in mice and the highly sensitive immunoassays like the enzyme-linked immunosorbent assay (ELISA).

Bioassay for Botulinum Toxin

The most sensitive and widely used biological assay of botulinum toxin is the intraperitoneal (i.p.) injection of material into mice that weigh 18–22 g. However, the test is unsuitable for examination of samples containing other substances that may cause interference or nonspecific death in mice. Furthermore, for quantitative determination of toxicity

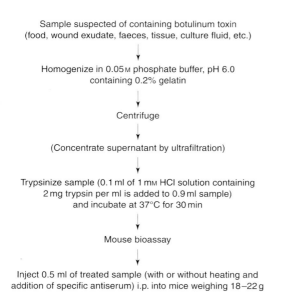

Sample suspected of containing botulinum toxin
(food, wound exudate, faeces, tissue, culture fluid, etc.)

↓

Homogenize in 0.05 M phosphate buffer, pH 6.0
containing 0.2% gelatin

↓

Centrifuge

↓

(Concentrate supernatant by ultrafiltration)

↓

Trypsinize sample (0.1 ml of 1 mM HCl solution containing
2 mg trypsin per ml is added to 0.9 ml sample)
and incubate at 37°C for 30 min

↓

Mouse bioassay

↓

Inject 0.5 ml of treated sample (with or without heating and
addition of specific antiserum) i.p. into mice weighing 18–22 g

Figure 1 Schedule for testing samples for the presence of botulinum toxin by the mouse bioassay. Adapted from Notermans and Nagel (1989).

by i.p. injection, relatively large numbers of animals and a period of 4 days are required.

Figure 1 presents a scheme for the mouse bioassay for botulinum toxin. To stabilize the toxin, samples to be tested are diluted in 0.05 M phosphate buffer, pH 6.0, containing 0.2% gelatin. The addition of bovine serum prevents nonspecific death in mice used to test toxicity of fish samples. After centrifugation of the homogenized samples, the supernatant can be concentrated by ultrafiltration. It has been shown that after centrifugation, a homogenate of canned beans could be concentrated at least 15-fold.

Botulinum toxin present in the (concentrated) supernatants can be potentiated considerably by addition of trypsin, which causes limited proteolysis (nicking) of the toxic molecule. This is especially true for type E toxin and for type B toxin produced by non-proteolytic strains of *Clostridium botulinum*. Other toxins originating from proteolytic strains are activated endogenously, but they are often partially nicked and additional trypsinization results in an increased toxicity. Trypsinization is usually omitted when stool samples are tested. However, it is not clear whether the activation of the toxin of non-proteolytic type B and E strains in the gut is maximal.

The test samples are injected i.p. into mice weighing 18–22 g. The symptoms in the mice often develop within 4 h after injection and include characteristic vibration of the abdominal wall, followed by the wasp-shaped abdomen and laboured breathing with or without paralysis of the limbs. Heating of the sample (80°C) for 5 min) or neutralization by specific antitoxin results in negative mouse bioassays. Samples

with antisera are incubated at 37°C for 30 min before i.p. injection into mice. When the toxicity of a sample is too high, it is diluted appropriately in the gelatin–phosphate diluent and neutralization tests are prepared. For identification of the toxin by neutralization reactions, account has to be taken of the seven immunologically different types of botulinum toxin (A–G). As a consequence, a sample (for instance an enrichment culture) may contain more than one type of toxin, and a large number of mice are needed to test all toxin–antiserum combinations.

For the quantitative determination of toxin by mouse bioassay, usually 0.5 ml volumes of serial twofold dilutions in the gelatin–phosphate medium are injected into four animals. After 4 days, the 50% lethal dose (LD_{50}) is calculated, often by using the method developed by Reed and Münch (1938). When mice are injected intravenously with 0.1 ml toxin solutions (about 10^3–10^5 i.p. LD_{50}), they are killed, within minutes, according to a definite and reasonably reproducible dose–survival time relationship. Standard curves have been prepared for different types of toxin. However, the intravenous injection method should only be applied to fully activated toxin, because activated and non-activated type E toxin give parallel, but distinct, curves.

Immunoassays for Botulinum Toxin

In both the serum of patients and in enrichment cultures, small quantities of the highly potent botulinum neurotoxin may be present. Therefore, only the most sensitive immunoassays are of value, such as the ELISA and the amplified ELISA. These techniques are based on a quantitative reaction of the toxin (antigen) with its antibody (antitoxin). The most widely applied technique is based on binding of the toxin present in a test sample to antibodies coated to a solid surface. The adsorbed toxin is then captured by a second antibody which is labelled with an enzyme. The enzyme activity is a quantitative indication of the amount of toxin present. Amplification of the ELISA reaction can be accomplished by among others the use of biotin–avidin reaction kinetics. In this case the capturing antibody is conjugated with biotin. Avidin, which is labelled with enzymes, reacts with the biotin. It has been demonstrated that the sensitivity is increased at least 10-fold. However, non-specific reactions will also be amplified. Therefore, well-selected antibodies, such as monoclonal antibodies, are necessary for success.

A general disadvantage of immunoassays, such as the ELISA, is that only the antigenicity is determined, and this may differ from the actual toxicity (**Table 2**). Specificity and sensitivity of the assays are mainly determined by the quality of the antiserum used. A number of systems for antibody production and selection have been developed to improve the quality of antiserum.

Production of Antiserum

Impure botulinum toxin is composed of nontoxic and toxic parts. The size of the progenitor toxin may be 19S, 16S or 12S whereas the homogeneous neurotoxin has a sedimentation constant of 7S. The nontoxic parts of some progenitor toxins are immunologically identical. Traditionally, antisera against botulinum toxin are produced by immunization with crude preparations of detoxified materials. Such antisera are suited for neutralization reactions but not for immunoassays. Type-specific antisera are prepared by immunization with homogeneous neurotoxins. Antibodies, produced as described above with the immunological sites (epitopes) in the whole molecule may still react with the toxin even if it is detoxified. In the experiments described in Table 1, botulinum toxin type B was added to surface water, at pH 8.1, and stored for several days at 20°C. There was a decrease in the mouse toxicity over time, but there was no associated decrease in immunogenicity. The same results were obtained with other types of botulinum toxins that were added to surface water. These results show that preferably antibodies which react with the toxic site(s) of the molecule should be used. This can be accomplished by using well-selected monoclonal antibodies.

Nonspecific Reactions

Each immunoassay is potentially sensitive to nonspecific reactions. These reactions occur if substances like staphylococcal protein A bind to the antibodies that are used in the assay. This protein A binds to the Fc fragments of immunoglobulin G (IgG) present in the coat of the solid surface as well as in the enzyme–antibody conjugate, giving rise to false positive reactions. These reactions can be avoided by adding neutral IgG to the test sample. False results may also be caused by lysozyme, which strongly associates with proteins with low isoelectric points, like immunoglobulin, and form bridges between the IgG in the coat and the enzyme-labelled antibodies.

Besides protein A and lysozyme, other unknown cross-reacting substances might be present in test samples. Consequently, the immunological detection of botulinum toxin may not be reliable and it is necessary to check for both false positive and false negative results. The addition of a known quantity of toxin to a negative sample can easily indicate false negative results. However, false positive results are more difficult to recognize.

Sensitivity of Immunoassays

To date, the sensitivities of all in vitro immunological methods are less than that of the mouse i.p. injection method, although some investigators have claimed techniques with a comparable sensitivity. With the mouse bioassay the minimum detectable quantity of toxin is approximately 20 pg. Using the ELISA the minimum detectable quantity is 1–10 ng. Using an amplification method the minimal detectable quantity is 0.1–0.8 ng.

Concluding Remarks

The mouse bioassay is the most sensitive and widely used method for assaying botulinum toxins. Other methods have been developed, but the sensitivity of all these in vitro methods is lower than the mouse i.p. injection method. Therefore, for diagnostic purposes, especially for botulism, the mouse bioassay is still the method of choice.

All in vitro methods so far described for botulinum toxin have an immunological basis, and thus their sensitivity is determined primarily by the kinetics of the antigen–antibody reaction. The sensitivity and reliability of these immunoassays may approach that of the mouse bioassay if high quality immunoglobulins (high specificity and high affinity) are used. Currently, active research is being done in this area by a number of groups which will undoubtedly lead to commercial products.

Table 2 Relation between ELISA and mouse bioassay for detection of type B botulinum toxin added to surface water. [a] Reproduced from Notermans and Nagel (1998)

| Incubation time (h) at 20°C | Quantity of toxin detected[b] | | |
| | | ELISA with coating antibodies | |
	Mouse bioassay	Polyclonal antibodies	Monoclonal antibody B-6-2
0	7100	7000	7100
24	2200	7200	2300
72	400	7100	500

[a] Sterile culture fluid of *C. botulinum* strain Okra was added to surface water (pH 8.1).
[b] Data expressed in i.p. mouse LD_{50} per ml.

See also: **Clostridium**: Detection of Enterotoxins of *C. perfringens*. **Enzyme Immunoassays**: Overview. **Food Poisoning Outbreaks**. **National Legislation, Guidelines & Standards Governing Microbiology**: European Union; Japan. **Nucleic Acid-based Assays**: Overview.

Further Reading

Boroff DA and Fleck U (1966) Statistical analysis of a rapid in vivo method for the titration of the toxin of *Clostridium botulinum*. *J. Bacteriol.* 97: 1580–1581.

DasGupta BR and Sugiyama H (1972) A common subunit structure in *Clostridium botulinum* types A, B and E toxins. *Biochem. Biophys. Res. Commun.* 48: 108–112.

Miyazaki S, Kozaki S, Sakaguchi S and Sakaguchi G (1976) Comparison of progenitor toxins of non-proteolytic with those of proteolytic *Clostridium botulinum* type B. *Infect. Immun.* 13: 987–989.

Notermans S and Nagel J (1989) Assays for botulinum and tetanus toxins. In: Simpson LL (ed.) *Botulinum Neurotoxin and Tetanus Toxin.* P. 319. San Diego: Academic Press.

Notermans S, Dufrenne J and van Schothorst M (1978) Enzyme-linked immunosorbent assay for detection of *Clostridium botulinum* toxin type A. *Jpn J. Med. Sci. Biol.* 31: 81–85.

Notermans S, Dufrenne J and van Schothorst M (1979) Recovery of *Clostridium botulinum* from mud samples incubated at different temperatures. *Eur. J. Appl. Microbiol. Biotechnol.* 6: 403–407.

Notermans S, Timmermans D and Nagel J (1982) Interaction of staphylococcal protein A in ELISA for detecting staphylococcal antigens. *J. Immunol. Methods* 55: 35–41.

Reed LJ and Münch H (1938) A simple method for estimating fifty percent end points. *Am. J. Hyg.* 24: 493–497.

Sakaguchi G, Sakaguchi S and Kondo H (1968) Rapid bioassay for *Clostridium botulinum* type E toxins by intravenous injection into mice. *Jpn J. Med. Sci. Biol.* 21: 369–378.

Shone C, Appleton N, Wilton-Smith P et al (1986) In vitro assays for botulinum toxin and antitoxins. *Dev. Biol. Standard* 64: 141–145.

Solberg M, Post LS, Furgang D and Graham C (1985) Bovine serum eliminates rapid nonspecific toxic reactions during bio-assay for stored fish for *Clostridium botulinum* toxin. *Appl. Environ. Microbiol.* 49: 644–649.

Sonnenschein B and Bisping W (1976) Extraction and concentration of *Clostridium botulinum* toxins from specimens. *Zentralbl. Bakteriol. Hyg.* Abt. I Orig. A 234: 247–259.

Suggi S and Sakaguchi G (1977) Botulogenic properties of vegetables with special reference to the molecular size of the toxin in them. *J. Food Safety* 1: 53–65.

Tacket CO (1989) Botulism. In: Simpson LL (ed.) *Botulinum Neurotoxin and Tetanus Toxin.* P. 351. San Diego: Academic Press.

COCOA AND COFFEE FERMENTATIONS

Poonam Nigam, School of Applied Biological and Chemical Sciences, University of Ulster, Coleraine, UK

Introduction

The main objectives of cocoa and coffee fermentations are the removal of mucilage from coffee and cocoa beans and the development of a number of flavour precursors in cocoa. Cocoa is made from the seeds (beans) of the cacao plant, the fruit of which is a pod containing up to 50 beans covered in a white mucilage. The mucilage is fermented by yeasts, and then beans – which darken during the week-long fermentation – are dried and roasted. The manufacture of coffee from ripe coffee fruits requires the initial removal of a sticky mucilaginous mesocarp from around the two beans in each fruit. The outer skin of the fruit is mechanically disrupted, and the whole is left to ferment. The mucilage is degraded by the fruit's own enzymes and by microbial extracellular enzymes. After fermentation, the beans are washed, dried, blended and roasted.

The popularity of cocoa and coffee is derived from their unique and complex flavours and possibly also from the presence of caffeine and similar compounds that may have a mild stimulatory effect. The flavours are initially developed during processing immediately after harvesting. This flavour development involves the action of various enzymes on the polyphenols, proteins and carbohydrates. Unlike many other fermented products it is the endogenous enzymes that are mainly responsible. In cocoa, the role of microorganisms is limited to removal of the pulp that surrounds the fresh seeds or 'beans'. The microbial activities result in the death of the bean and creation of the environment for development of flavour precursors. In coffee, their role is limited to removal of the pulp in some of the processing methods. During this initial processing a number of flavour precursors are formed, which in cocoa and coffee are further modified in Maillard reactions during roasting. In cocoa, there is also a reduction in bitterness and astringency caused by the oxidation of polyphenolic compounds.

Cocoa

Cocoa is used in a variety of products including:

- confectionery – milk chocolate tablets or bars, dark chocolate, white chocolate based on cocoa butter, milk and sugar, chocolate-coated products with various centres
- beverages – malted milk cocoa drinks, sweetened cocoa powder-based drinks
- ice cream and desserts.

Nature of the Crop

The cocoa tree (*Theobroma cacao*, family Sterculiaceae) is a small tree that grows naturally in the lower storey of the evergreen rainforest in the Amazon basin. Cocoa is commercially grown within the latitudes 20° north or south of the equator. There are several types or varieties of cocoa, usually classed in three main groups: Forastero, Trinitario and Criollo. Trinitario is considered to be derived from hybrids between various Forastero and Criollo varieties. The flavour potential of cocoa is determined genetically and depends mainly on the variety. Forastero types (e.g. Amelonado, Amazon varieties) are bulk cocoas used for milk chocolates and for cocoa butter and powder production. They comprise 95% of the crop. Criollo (light brown in colour) and Trinitario are fine cocoas; they are used for speciality dark chocolates because of their particular flavour or colour characteristics.

Cacao Fruit   *Theobroma cacao* bears small flowers in small groups on the trunks and lower main branches of the trees. Pollinated flowers develop into berries ('pods'), maturing over a 5–6 month period. The berry is a drupe 2.5–4.0 cm by 1.25–1.75 cm in size, containing 20–40 seeds (beans) and surrounded by a mucilaginous pulp (**Fig. 1**).

Harvesting   *Theobroma cacao* normally begins to bear berries after 3 years and the yield reaches a maximum after 8–9 years. Trees simultaneously bear flowers, developing berries and mature fruits. The pods develop on the trunk and branches. After about 5–6 months development the pods ripen and turn yellow or orange. Harvesting is carried out at varying frequencies (1–4 weeks). The pods are then opened either on the same day or after a few days in order to allow a sufficient quantity to accumulate for the fermentation stage. Beans are removed and separated from the placenta. At this stage they are covered in a sweet mucilaginous pulp.

Fermentation

The fermentation stage is of major importance in determining the quality of cocoa powder and chocolate confectionery. It has three purposes:

- liquefaction and removal of the mucilaginous pulp
- killing of the bean
- initiation of the development of aroma, flavour and colour.

Procedures of Fermentation   There are three main methods of fermentation used in various parts of the world.

1. The simplest method, used in West Africa, requires no special apparatus and is therefore cheap. In this method, beans are piled up underneath plantain leaves, covering the surface and bottom of the pile. To assist the sweatings to run away, the pile is built up over radially arranged pieces of wood. The pile is kept together for 6 days and turned on the second and fourth days. This has the effect of making the aerobic parts anaerobic and vice versa. Piles vary in size and can be 60–120 cm in diameter.
2. The method used extensively in South America involves fermentation of beans in large hardwood boxes holding up to 1.5 tonnes. These boxes have slatted bases or holes in the sides and base, which have a twofold function. They allow the sweatings to drain away and permit the access of air. Often these boxes are stacked stepwise and have removable sides, allowing easy transfer of beans to the box below. In this system, the first box often has twice the surface area of the other boxes and is half the depth. A covering of sacking or plantain leaves is placed over the surface of the beans. The

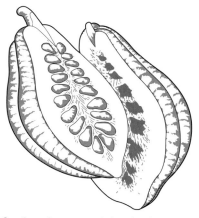

Figure 1   Section of cacao pod showing beans.

number of changes in this system is usually six, taking place once in 24 h.

3. In other methods beans may be placed in a plantain leaf-lined basket and left to ferment, or they may be placed in a hole in the ground. These methods have the disadvantage of low initial aeration and lack of drainage for the sweatings.

Changes Resulting from Fermentation During the course of fermentation the external appearance of the beans changes. Initially they are pinkish with a covering of white mucilage, but gradually they darken and the mucilage disappears. This colour change is oxidative; when a heap is disarranged, the beans on the outside are darker than those on the inside. As the beans are mixed, their colour becomes a more uniform orange-brown and, towards the end of the fermentation, nearly all the mucilage has disappeared leaving the beans slightly sticky; at this stage they are ready for drying.

Microflora Active in Cocoa Fermentation When the beans are removed from the pods, the pulp is inoculated with a variety of microorganisms from the environment. The pulp is an excellent medium for the growth of microorganisms since it contains plenty of sugars (**Table 1**). The following types of microorganisms have been found in the pulp fermentation (although only few are actively involved): *Acetobacter, Aerobacter, Arthrobacter, Azotomonas, Bacillus, Cellulomonas, Corynebacterium, Erwinia, Escherichia, Lactobacillus, Microbacterium, Micrococcus, Pediococcus, Propionibacterium, Pseudomonas, Sarcina, Serratia, Staphylococcus, Streptococcus, Zymomonas* and yeasts.

The fermentation consists of three overlapping phases. The total count of microorganisms increases in the first 24–36 h (10^5–10^6 organisms per gram) and then stabilizes or gradually reduces.

Phase 1: Anaerobic Yeasts In the first 24–36 h sugars are converted into alcohol in conditions of low oxygen and a pH of below 4.

$$C_{12}H_{22}O_{11} + H_2O \rightarrow 2C_6H_{12}O_6 + 18.8 \text{ kJ}$$
sucrose   glucose or fructose

$$C_6H_{12}O_6 \rightarrow 2C_2H_5OH + 2CO_2 + 93.3 \text{ kJ}$$
glucose or fructose   ethanol

Yeasts isolated from cocoa fermentations (**Table 2**) produce pectinolytic enzymes that break down the pulp cell walls. This causes the pulp to drain off the beans as 'sweatings'. The spaces formed between the beans allow air to enter. Bean death, usually on the second day, is caused by acetic acid and ethanol; the rise in temperature does not play any part.

Phase 2: Lactic Acid Bacteria Lactic acid bacteria are present at the start of fermentation (**Table 3**), although yeasts are dominant. The yeast activity becomes inhibited by alcohol concentration, increasing pH and greater aeration. After 48–96 h conditions become more favourable to lactic acid bacteria, which then dominate. Lactic acid bacteria convert a wide range of sugars and some organic acids (e.g. citric and malic acids) to lactic acid and – depending on the type of *Lactobacillus* – to acetic acid, ethanol and carbon dioxide.

Phase 3: Acetic Acid Bacteria Acetic acid bacteria occur very early in the fermentation (**Table 4**) and persist until the end. As aeration increases, acetic acid bacteria become more important. The main reaction is the conversion of ethanol to acetic acid.

$$C_2H_5OH + O_2 \rightarrow CH_3COOH + H_2O + 496 \text{ kJ}$$

This strongly exothermic reaction is mainly responsible for the rise in temperature up to 50°C.

Other Microorganisms Present During Fermentation and Drying Towards the end of fermentation the numbers of spore-forming bacteria increase, especially *Bacillus subtilis, B. circulans* and *B. lichen-*

Table 1 Composition of fresh pulp from cocoa. Adapted from Fowler et al (1998), with due acknowledgement to Professor B J B Wood.

Component	Fresh weight of pulp (%)
Water	82–86
Mono- and disaccharides	11–13
Plant cell-wall polymers	1.5–2.8
Proteins, peptides and amino acids	0.64–0.74
Fat	0.35–0.75
Citrate	0.29–1.3
Trace metals, vitamins, ethanol, etc.	Trace

Table 2 Various yeasts isolated from cocoas. Adapted from Fowler et al (1998), with due acknowledgement to Professor B J B Wood.

Yeasts	Ability to ferment	African cocoa	Malaysian cocoa
Hansenula spp.	+	Present	Present
Kloeckera spp.	+	Present	Present
Saccharomyces spp.	+	Present	Present
Candida spp.	±	Present	Present
Pichia spp.	Weak	Present	Absent
Schizosaccharomyces spp.		Present	Absent
Saccharomycopsis spp.	±	Present	Absent
Rhodotorula spp.	–	Absent	Present
Debaryomyces spp.	Weak	Absent	Present
Hanseniaspora spp.	+	Absent	Present

Table 3 Lactic acid bacteria of cocoa fermentation. Adapted from Fowler et al (1998), with due acknowledgement to Professor B J B Wood.

African cocoa	Malaysian cocoa	Trinidadian cocoa
Lactobacillus plantarum (homofermentative)	Lactobacillus plantarum	Lactobacillus acidophilus
Lactobacillus mali (homofermentative)		Lactobacillus bulgaricus
Lactobacillus collinoides (heterofermentative)	Lactobacillus collinoides	Lactobacillus casei
Lactobacillus fermentum (heterofermentative)		Lactobacillus fermentum
Unidentified strains (heterofermentative)	Unidentified strains	Lactobacillus lactis
		Lactobacillus plantarum (also Leuconostoc, Pediococcus and Streptococcus)

Table 4 Acetic acid bacteria of cocoa fermentation. Adapted from Fowler et al (1998), with due acknowledgement to Professor B J B Wood.

African cocoa	Malaysian cocoa	Trinidadian cocoa
Acetobacter rancens	Acetobater rancens	Acetobater acetie
Acetobacter xylinum	Acetobacter xylinum	Acetobacter roseus
Acetobacter ascendens	Acetobacter lavaniensis	Gluconobacter oxydans
Acetobacter lavaniensis	Gluconobacter oxydans	
Gluconobacter oxydans		

iformis. In Trinidad, *Streptococcus thermophilus* and *Bacillus stearothermophilus* accounted for more than half the isolates after 120 h. The most commonly present fungi, *Aspergillus*, *Mucor*, *Penicillium* and *Rhizopus*, are largely restricted to the outer surface of the fermenting and drying beans because they are strongly aerobic, tolerant of low water activity and can continue growth until the beans are nearly dry.

Effect of Fermentation on Product Quality

Development of Cocoa Flavour Precursors Flavour development occurs within the cotyledons in the bean. The compounds involved in flavour development are split between two types of cells: storage cells containing fats and proteins, and pigment cells containing the phenolic compounds and xanthines.

In the fresh, live cocoa seeds the cells and their contents are separated by membranes. During fermentation, there is first the initiation of germination, which causes water uptake by the protein vacuoles within the cells. Later, after bean death, the membranes break down. Various enzymes and substrates are then free to mix and the subsequent reactions produce the flavour precursors. The pH, determined mainly by diffusion of acetic acid, is important and the reaction rates are also increased by the warm temperatures during fermentation and drying.

Flavour-developing Compounds The following compounds are responsible for the main flavour attributes and precursors in cocoa:

1. Methylxanthines (caffeine and theobromine) impart bitterness. During fermentation the levels of methylxanthines fall by around 30% probably by diffusion from the cotyledons.
2. Polyphenolic compounds impart astringency. The

levels drop significantly during the fermentation and drying. Anthocyanins are rapidly hydrolysed to cyanidins and sugars (catalysed by glycosidases). This accounts for bleaching of the purple colour of the cotyledons. Polyphenol oxidases convert the polyphenols (mainly catechin) to quinones. Proteins and peptides complex with polyphenols give rise to the brown coloration typical of fermented cocoa beans.

3. Maillard reaction precursors are formed from sucrose and storage proteins. Sucrose is converted by invertase into reducing sugars. Fructose is found in fermented dried cocoa beans while glucose is utilized in further reactions. The storage proteins are initially hydrolysed by an aspartic endopeptidase (pH optimum 3.5) into hydrophobic oligopeptides. A carboxypeptidase (pH optimum 5.4–5.8) then converts these oligopeptides into hydrophilic oligopeptides and hydrophobic amino acids. These are cocoa flavour precursors involved in Maillard reactions during roasting to form cocoa flavour compounds.

Coffee

Coffee is not consumed for nutrition. Coffee gives the consumer pleasure and satisfaction through flavour, aroma and desirable physiological and psychological effects.

Nature of the Crop

The genus *Coffea* is a member of the family Rubiaceae and comprises evergreen trees and shrubs. Funnel-shaped flowers are followed by a pulpy fruit, the 'cherry', which contains two seeds, the coffee beans. *Coffea* grows wild in Africa and Madagascar and the genus includes a large number of species. Only three, *C. arabica*, *C. canephora* (Robusta) and *C. liberica* have been successfully used in commercial cultivation. *Coffea liberica*, however, was devastated during the 1940s by epidemics of tracheomycosis, due to infection by *Fusarium xylaroides*, and commercial growth of this species has effectively ceased.

Both *C. arabica* and *C. canephora* are available in a large number of varieties and cultivars. A number of both intra- and interspecific hybrids have been developed, of which the Arabica-Robusta hybrid, Arabusta, is intended to produce a coffee of better quality than Robusta, while being more vigorous and disease resistant than Arabica. The beans are also of low caffeine content. Although only *C. arabica* and *C. canephora* are grown commercially, the gene pool of *Coffea* consists of all species. Species such as *C. stenophylla* and *C. congenis* are thus of importance as sources of novel genetic material in breeding improved strains of *C. arabica* and *C. canephora*.

Coffee Fruit   A mature coffee fruit is a fleshy, spheroidal berry, a drupe about 15–20 mm in diameter. It changes colour from green to cherry-red while ripening. Fruits reach their maturity within an average of 9 months, depending on the variety. Arabica coffee fruits are oval and long, while robusta fruits are smaller, of round to irregular shape. They are covered by a skin-like, smooth red film (the epicarp) which covers the mesocarp. Depending on the variety, the mesocarp represents 40–65% of the weight and is composed of water (70–85%), sugars and pectin. The bean is rich in polysaccharides, lipids, reducing sugars, sucrose, polyphenols and caffeine.

The fruit normally contains two beans (endosperm) surrounded by a thin membrane known as the 'silver skin' (spermoderm). The beans and the silver skin are protected by a hard, horny endocarp generally referred to as the 'parchment'. Adhering firmly to the outside of the parchment is a pulpy, mucilaginous mesocarp which is covered by the fruit skin or pulp (exocarp) (**Fig. 2**).

Harvesting   To obtain the highest quality end product, coffee is harvested when the berries are fully red ripe. Underripe and overripe berries are difficult to process and result in a poor-quality product. Coffee berries come to full ripeness over an extended period and it is usual to pick red berries individually and to repeat picking at intervals of 7–14 days. A maximal yield is normally obtained from 7-year-old trees. Coffee trees produce an average of 2.5 kg of berries per year, yielding around 0.5 kg of green coffee or the equivalent of 0.4 kg of roasted coffee, which corresponds to about 40 cups of beverage.

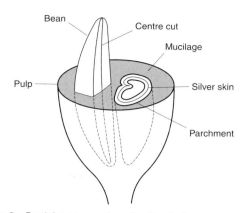

Figure 2   Partial cross-section of coffee fruit.

Fermentation

Pulping involves removal of the coffee skin by suitable mechanical methods such as using a machine aquapulper. After pulping, coffee is fermented. Fermentation of coffee is the process by which the mucilaginous mesocarp adhering to the coffee parchment is degraded by enzymes. The mucilage is subsequently washed off to leave parchment coffee which is subjected to a drying regime to obtain a moisture content of 10–11%. Coffee fermentation accomplishes two important objectives. It removes the sticky mucilage layer allowing for quick drying of the parchment coffee, and improves the appearance of the raw beans.

Procedure of Coffee Mucilage Removal

By Natural Fermentation   Coffee fermentation is required for the removal of mucilage from parchment coffee. Natural fermentation refers to the process of mucilage removal by enzymes naturally occurring in the coffee fruit and/or elaborated by the natural microflora acquired from the environment. Pulped coffee is placed in concrete or wooden tanks and left to ferment, either under water or with constant drainage of water and mucilage liquors. The latter process, known as 'dry fermentation', is preferred; underwater fermentation is slower and results in a greater production of volatile acids, which may taint the final coffee beverage. Natural fermentation takes 20–100 h; its duration varies with the stage of ripeness, temperature, pH value, concentration of ions, coffee variety, microflora population and aeration. It has been demonstrated that lowering the temperature and pH value retards the rate of fermentation and that aerobic fermentations are faster than anaerobic fermentations. It would be expected that the availability of oxygen under water is restricted by the amount that can dissolve in the water at any given time.

By Commercial Enzymes   Several commercial enzymes are available for coffee fermentation. The earliest one was marketed under the trade name Benefax. Later brands have included Pectozyme, Cofepec and Ultrazym. These are mould–enzyme preparations with appropriate inert fillers. The commercial enzymes are generally mixtures of pectic enzymes but may contain hemicellulases and cellulases. Because of financial constraints, these enzymes have not been widely used. Most factories restrict the use of commercial enzymes to peak production periods or when natural fermentations are slow. Conditions created by overproduction and slow fermentations usually upset the smooth running of a factory. Congestion can occur either in fermentation and soaking tanks or on drying tables. These conditions affect the coffee quality adversely because of the concomitant physiological activities by wild microorganisms and those in the bean. Since commercial enzymes are applied by mixing with coffee in fermentation tanks, only a little saving of space is afforded by their use in the normal factory routine.

Stages of the Coffee Fermentation Various factory practices increase the rate of fermentation. These include dry feeding of pulped coffee into fermentation tanks, and using recirculated water which is rich in enzymes. Sophisticated factories aerate or use other additives that enhance enzyme activity. The addition of lime provides calcium ions which activate specific enzymes. After the mucilage has been degraded, parchment coffee is washed and graded by water in concrete canals.

In East Africa, a two-stage fermentation procedure includes a quick dry fermentation stage, washing off the mucilage, followed by a 24 h underwater soak. The advantages of this procedure were shown to be improvement of the raw bean appearance through outward diffusion of undesirable browning compounds from the beans, specifically from the centre cut and the silver skin. Coffee fermented under water or processed by the two-stage fermentation procedure tends to deteriorate in quality during drying because of the preponderance of cracked parchment. This may be avoided by subsequent carefully controlled drying.

Natural fermentation of coffee is carefully controlled, otherwise off flavours can develop and be reflected in the final liquor quality. Onion flavour develops in coffee as a result of the production of propionic acid. Production of propionic and butyric acids during the final stages of fermentation is greater during underwater fermentation, and is also dependent on a heavy initial washing before fermentation. The incidence of an off flavour referred to as 'stinkers' may be associated with high temperatures obtainable during fermentation. 'Sourness' and 'stinkers' in

Table 5 Chemical composition, on a wet and dry basis, of coffee mucilage

Mucilage components	Chemical composition (%)
Wet basis	
Moisture	85.0
Total carbohydrates	7.0
Nitrogen	0.15
Acidity (as citric acid)	0.08
Alcohol-insoluble compounds	5.0
Pectin (as galacturonic acid)	2.6
Dry basis	
Pectic substances	33
Reducing sugars	30
Non-reducing sugars	20
Cellulose and ash	17

coffee are caused during fermentation under anaerobic conditions created by high proportions of reducing agents in the fermentation waters. Off flavours in coffee are caused by various factors which need proper investigation based on a correct understanding of the biochemistry involved in the fermentation process. This has led to the introduction of various methods of coffee processing which are not dependent on natural fermentation and are therefore easier to control. However, the delicate nature of the coffee-bean tissue defies any attempts to rid it completely of off flavours as detected by a subjective human palate.

Biochemistry of Coffee Fermentation

Composition of Mucilage The chemical and physical characteristics of coffee mucilage are basic to an understanding of coffee fermentation. Mucilage forms 20–25%, wet basis layer of 0.5–2.0 mm thickness. Chemically, coffee mucilage consists of all the higher plant cell materials including water, sugars, pectic substances, holocellulose, lipids and proteins (**Table 5**). The most important chemical components of mucilage are pectic substances together with carbohydrates and their breakdown products.

The important component in the coffee fermentation is mainly the cell wall and intercellular material characteristic of the parenchymatous cells of fruits. The middle lamella of coffee mucilage cells is primarily pectinic, and the cell contains pectin and cellulose materials. The insoluble fraction of coffee mucilage is expected to consist mainly of pectic substances in close association with other cell wall and intercellular materials, including hemicelluloses and phospho- and galactolipids. Breakdown of this cellular material and its detachment from coffee parchment are the important biochemical processes in coffee fermentation.

Changes Resulting from Fermentation

1. When coffee is pulped and left in a dry heap or under water, fermentation occurs. After a period of 20–100 h, depending mainly on the environmental temperature, the mucilage detaches from the parchment and can be readily washed with water.

2. On completion of fermentation, a few beans when rubbed in the hand feel gritty. Various chemical changes occur during the process of fermentation (**Table 6**).

3. Production of carboxylic acids changes the pH value of the fermentation liquor from 5.9 to 4.0 Acetic and lactic acids (also sometimes propionic acid) are produced early in coffee fermentation, and propionic and butyric acids are elaborated later.

4. There is a close positive correlation between the appearance of propionic acid in the fermentation stage and the incidence of 'onion' flavour in coffee beverage.

5. The carboxylic acids are produced through degradation of sugars by microorganisms.

6. Ethanol is one of the products of coffee fermentation (**Fig. 3**). The evolution of hydrogen and carbon dioxide occurs during both dry and underwater fermentations. Hydrogen is produced through breakdown of sugars by bacteria of the coliform group. *Escherichia coli* metabolizes glucose by a mixed acid fermentation at pH 7.8.

7. *Aerobacter aerogenes* gives a lower yield of mixed acids, particularly of lactic acid, because some pyruvic acid is converted into acetylmethyl-carbinol and butanediol.

8. The presence of reducing and non-reducing sugars in soluble mucilage fractions is observed after complete fermentation. Some of the sugars forming part of the structure of mucilage are arabinose, xylose, galactose, fructose and glucose. Arabinose, xylose and galactose are part of the insoluble structure of mucilage. The soluble sugars form an excellent medium for growth of microorganisms.

9. A lipid fraction isolated from fermented mucilage indicated the presence of an esterified sterol glycoside. Since pectic acids with four or fewer galacturonic acid units are not found in natural fermentation liquors, mucilage degradation involves breakages in cross-linkages which may implicate lipids and hemicellulose materials.

10. Changes in the quality of the coffee bean are fundamental to the continued practice of naturally fermenting coffee. In the two-stage fermentation (in East Africa), the raw bean quality improves and this improvement is reflected in the roast and final beverage quality. The improvement in raw appearance is dependent on diffusion of various compounds from the bean which also result in weight losses of 3–12%.

11. The higher weight losses are observed in underwater fermentations. This magnitude of loss would make fermentation an expensive exercise, thus nullifying the gains in raw bean quality. Despite these observations, natural fermentation of *Coffea arabica* is the preferred dimucilaging method.

Microflora Active in Coffee Fermentation The major factors in natural fermentations are the extracellular enzymes elaborated by microorganisms. Since mucilage contains simple sugars, polysaccharides, minerals, protein and lipids, it forms a good medium for microbial growth. Bacteria observed in fermenting coffee include lactic acid-producing bacteria of the genera *Leuconostoc* and *Lactobacillus*; coliform bacteria resembling species of the genera *Aerobacter* and *Escherichia* (in Brazilian coffee), and pectinolytic species of the genus *Bacillus*.

A microbial succession, involving members of the Enterobacteriaceae, species of *Enterococcus* and lactic acid bacteria, plays some part in the lowering of the pH value to about 4.3, which tends to inhibit the activity of pectinolytic enzymes. This prevents the growth of many spoilage microorganisms. Their extensive growth is likely to lead to the development of undesirable flavours.

Bacteria belonging to the family Enterobacteriaceae

Table 6 The composition of coffee mucilage before and after complete fermentation

Component	Percentage on dry basis	
	Before fermentation (%)	After fermentation (%)
Water-soluble	35.3	50.7
Lipid	6.0	4.0
Pectin	47.0	36.2
Holocellulose	9.4	8.0
Unaccounted	2.3	1.1

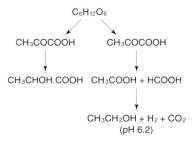

Figure 3 Products of coffee fermentation.

found in Congo coffee are very similar to those isolated from fermenting Brazilian coffee. They resemble closely *Erwinia dissolvens* and *Erwinia atroseptica*.

Pectinolytic microorganisms isolated from coffee fermentations belong to the genera *Bacillus*, *Erwinia*, *Aspergillus*, *Penicillium* and *Fusarium*. Bacterial isolates from coffee closely correspond to *E. dissolvens*.

Yeasts in fermenting coffee have no ability to degrade pectin; however, some mucilage-degrading yeasts are found on the surface of *Coffea canephora*.

Mould enzymes are known to speed up mucilage breakdown. Fungi of the genera *Aspergillus*, *Fusarium* and *Penicillium* were isolated from depulped coffee.

Effect of Fermentation on Product Quality The aim of the fermentation is the degradation of the residual mucilage layer which contains up to 30% pectin. The positive aspects linked with the development of flavours, tastes, change in texture, etc. normally associated with fermentation processes are not considered important for coffee. However, certain organoleptic and visual deviations are due to the formation of aliphatic acids, which is increased by underwater fermentation. This is in contrast to dry fermentation, where water is drained away immediately. Washing or soaking to eliminate undesirable components is thus recommended, although some losses in caffeine and chlorogenic acids may be observed. Apart from aspects related to fermentation, the growth of microorganisms in beans has been linked with the development of off flavours and off tastes and the presence of mycotoxins. Beans causing 'rio taste' showed the presence of bacteria and moulds. The presence of 2,4,6-trichloroanisole, which can be produced by moulds, has been detected in beans showing organoleptic deviations.

See also: **Lactobacillus**: Introduction. **Leuconostoc**.

Further Reading

Arunga RO (1982) Coffee. In: Rose AH (ed.) *Fermented Foods, Economic Microbiology*. Vol. 7, p. 259. London: Academic Press.

Carr JG (1982) Cocoa. In: Rose AH (ed.) *Fermented Foods, Economic Microbiology*. Vol. 7, p. 275. London: Academic Press.

Carr JG (1985) Tea, coffee and cocoa. In: Wood BJB (ed.) *Microbiology of Fermented Foods*. Vol. 1, p. 133. London: Elsevier.

Castelein J and Verachtert H (1983) Coffee fermentation. In: Rehm HJ and Reed G (eds) *Biotechnology*. Vol. 5. p. 588. Weinheim: VCH.

Fowler MS, Leheup P and Cordier JL (1998) Cocoa, coffee and tea. In: Wood BJB (ed.) *Microbiology of Fermented Foods*, 2nd edn. Vol. 1, p. 128. London: Blackie.

Haarer AE (1962) *Modern Coffee Production*, 2nd edn. P. 492. London: Leonard Hill.

Lopez AS and Dimick PS (1995) Cocoa fermentation, In: Reed G and Nagodawithana TW (eds) *Biotechnology*, 2nd edn. Vol. 9, *Enzymes, Biomass and Feed*. P. 561. Weinheim: VCH.

Varnam AH and Sutherland JP (1994) Cocoa, drinking chocolate and related beverages. In: *Beverages: Technology, Chemistry and Microbiology*. P. 256. London: Chapman and Hall.

Wrigley G (1988) *Coffee*. Harlow: Longman.

Coffee *see* **Cocoa and Coffee Fermentations**.

Colorimetric DNA Hybridization *see* **Listeria**: Detection by Colorimetric DNA Hybridization; **Salmonella**: Detection by Colorimetric DNA Hybridization.

Colours *see* **Fermentation (Industrial)**: Colours/Flavours Derived by Fermentation.

CONFECTIONERY PRODUCTS – CAKES AND PASTRIES

Philip A Voysey, Microbiology Department, Campden & Chorleywood Food Research Association, Chipping Campden, Gloucestershire, UK
J David Legan, Microbiology Department, Nabisco Research, East Hanover, USA

Cakes and pastries provide a nutritious environment for microbial growth but probably show a greater diversity of moisture content, water activity (a_w) and pH than most other food groups. Hence, cakes and pastries offer a wide range of different habitats for microbial growth. Nevertheless, they have an excellent public health record. In part this is because factors intrinsic to the products, such as a_w, pH or preservative content, prevent the growth of bacterial pathogens. In part it is because the baking process inactivates most organisms that would be present in the raw materials. A disproportionate number of the microbiological problems affecting these products are associated with perishable, unbaked fillings such as dairy cream or certain types of custard. This article discusses the factors affecting the spoilage of cakes and pastries, including a_w, pH, use of preservatives and use of atmosphere modification, with reference to their effects on both the rate and type of spoilage. It also examines outbreaks of food poisoning that have been associated with cakes and pastries and some measures for maximizing the safety of these products.

What are Cakes and Pastries?

Cakes and pastries are sweet baked goods (of a class often called flour confectionery). Cakes are made by baking a batter of flour, sugar, fat and water (possibly with eggs, milk, fruit or other flavourings). Pastries are baked from a dough or paste of flour and fat that may be enriched with other ingredients. Both cakes and pastries may be filled or coated with a variety of materials.

Products include rich fruit cakes which may be stable for many months or even years as a result of a combination of reduced a_w, low pH and antimicrobial effects of the fruit that are probably linked to caramelization products formed on baking. Less stable are plain sponge cakes like Madeira cake or pound cake, which have a shelf life of a few days to several weeks. Least stable of all are cakes or pastries filled with cream, custard or fresh fruit that are highly perishable (high a_w) and this restricts the life of the products to only a day or so at ambient temperatures.

These perishable fillings support bacterial growth and have occasionally given rise to spoilage and food-poisoning incidents. Fondant, fudge, sugar paste and chocolate coatings may also be susceptible to microbial spoilage. The microbiology of chocolate is covered elsewhere in this book.

Effects of Baking

Cakes are made in a variety of formats and the bake time and temperature vary widely. In each case, baking is sufficient to kill any vegetative microbes which are present prior to baking. There are a number of bacterial spores (produced, for example, by species of *Bacillus*) which are able to survive baking. The outgrowth of bacterial spores is inhibited by a_w below 0.97–0.93. Some fungal ascospores such as those of *Xeromyces bisporus* and *Byssochlamys fulva* may survive some baking processes if present. These are potentially significant spoilage organisms, but are not frequently encountered.

Effects of Post-bake Operation

The commonest sources of microbial contamination of cakes and pastries are in the handling and processing that occur after baking but before packaging. These include cooling, slicing, filling and decorating. Pastries are produced in three basic ways:

1. Fillings are dispensed into pre-baked pastry tubes or shells, then icing is added, e.g. chocolate éclairs.
2. A preformed pastry shell is filled with uncooked filling; the entire pastry is then baked, e.g. custard tarts.
3. A pre-baked pastry is filled with ingredients. Other ingredients are then added without a final bake, e.g. iced cakes.

Cooking fillings to 76–86°C (170–187°F) kills most microorganisms except bacterial spores, assuming that the minimum temperature in the entire batch reaches this temperature. In type 1 pastries, there is opportunity for recontamination during cooling and dispensing. There is more risk associated with type 3 pastries, since some ingredients are not cooked at all.

Meringue is an important exception to these rules. It can be made by heating at 230°C (446°F) for 6 min or temperatures as low as 60°C (140°F) for several hours. The high sucrose concentration significantly increases the heat resistance of many strains of *Salmonella*. This, coupled with a process at the lower end of the temperature range, has allowed *Salmonella* to survive in laboratory challenge studies. Of course, meringue is also an excellent insulator (it consists of a foam of air bubbles) and this property has the potential to allow the survival of bacterial pathogens; the insulation protects them from high temperatures.

Factors Affecting Microbial Growth

From a microorganisms's point of view, cakes and pastries offer a range of tempting environments for growth. Several factors influence the type and rate of microbial growth in cakes and pastries and their coatings, fillings or raw ingredients, including a_w, temperature, pH, concentration of preservatives and gaseous environment around the product. Of these, the most important is a_w. In simple terms, this is a measure of the amount of free water available in a foodstuff. As the a_w value increases, the ease with which microorganisms can extract water from the product increases. Water activity is normally derived from a measurement of the relative humidity (RH) that develops in the head space around the product in a sealed chamber ($a_w = RH/100$). This value is easy to obtain and useful for predicting microbial growth. It is often not the true a_w however, because a_w is defined under equilibrium conditions, whereas the products that we are interested in are never truly in equilibrium. The a_w range of a number of cake and pastry items is given in **Figure 1**, together with the minimum a_w which permits the growth of various groups of microorganisms.

Below a_w 0.6, no microbial growth occurs, thus dry ingredients such as flour, cocoa powder, coconut, sugar, and low-moisture products such as biscuits (cookies), crackers, meringues and shortbread are not subject to microbiological spoilage as long as they are packaged and stored to prevent moisture migration from the environment; however, pathogens, if present, may survive for considerable periods.

At a_w levels below 0.7 the range of microbes capable of growth is restricted and flour confectionery items can be considered to be safe from microbial growth for most practical purposes, providing that condensation is avoided, as this can lead to localized regions of higher a_w. Nevertheless, a few organisms can cause spoilage if present, e.g. fermentation of jam fillings caused by growth of the yeast *Zygo-*

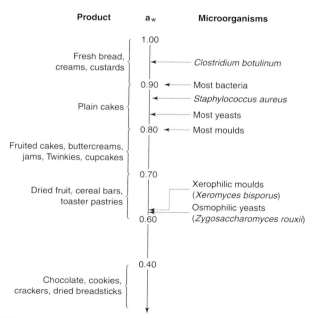

Figure 1 Water activity (a_w) ranges for various types of food, and the a_w ranges at which microorganisms can grow.

saccharomyes rouxii (minimum a_w 0.65), and growth of the mould *Xeromyces bisporus* (minimum a_w 0.61) on fruit cakes, dried fruit and chocolate-enrobed products.

As the a_w of products and ingredients increases, so does the range of microorganisms that are able to grow, until at an a_w of 0.95–0.99 almost all bacteria, yeasts and moulds are able to proliferate.

Temperature also influences the rate of growth of microorganisms found in association with flour confectionery items. Chill temperatures of 0–5°C (32–41°F) are needed to restrict microbial growth in perishable items such as cakes and pastries with dairy cream fillings.

Bacteria tend to be more sensitive to low pH than are yeasts and moulds. Consequently certain acidic fresh fruit fillings are not subject to bacterial spoilage despite the fact that they have an a_w high enough to support growth. These fillings may still be spoiled by yeast or mould growth. The pH of flour confectionery items is also important when considering the use of preservatives, which will be discussed later. Environmental conditions such as the make-up of the gaseous atmosphere around the product or item may also be important.

Food Poisoning

Cakes and pastries with perishable fillings such as cream (dairy and imitation) or custard have been implicated in cases of food poisoning. Out of 2226 outbreaks of food poisoning in the US between 1973 and 1987, 51 (2%) were attributed to bakery products – a low proportion, though no justification for complacency. Even in large industrial bakeries there can be a high proportion of hand operations involved in filling and finishing these products. Microbiological surveys of product purchased in retail stores in Europe and the US have found coliforms and *Escherichia coli* in up to 30% of cakes and pastries, especially those containing cream fillings. *Staphylococcus aureus* and *Bacillus cereus* have been found in between 4 and 25% of cream-filled pastries. This incidence of potential pathogens and indicator organisms is not reflected in the incidence of food poisoning associated with the products in the UK or the US. The discrepancy may reflect the selection of relatively perishable products for the surveys. Or it may reflect the fact that food poisoning by both *S. aureus* and *B. cereus* is mediated by toxin formed in the food when the bacteria grow to large populations, typically > 1×10^5 per gram. Both dairy and imitation creams have been implicated in food-poisoning cases (with imitation creams, however, it is not always as easy to tell if microbial growth is occurring since these creams do not go sour).

Custard-filled flour confectionery items have also been implicated in incidents of food poisoning. In one survey of 133 samples of vanilla slices (carried out in the UK in 1977) containing custard, 41% were found to contain *B. cereus*. A large outbreak of *B. cereus* food poisoning associated with vanilla slices was reported in 1984. *S. aureus* has also been found in a high proportion of custard-containing pastries. In the investigation of one food-poisoning incident, 20% of the products sampled from small and plant bakeries contained coagulase-positive *S. aureus*. In another outbreak, 17 people contracted *Salmonella enteritidis* phage-type 4 food poisoning from custard slices from a small bakery. The custard had been made with fresh shell-eggs and had not been properly cooked.

With the exception of unbaked fillings, bacterial pathogens mainly gain access to cakes and pastries through post-baking contamination. The incidence in foods of pathogens with very low infectious doses, such as *E. coli* O157:H7 and *Salmonella typhimurium* DT104, has increased during the 1990s. This requires even more stringent attention to plant and personal hygiene to maintain the safety record of these products. It is also necessary to take steps to control the growth of pathogenic microorganisms between product manufacture and consumption. The use of chilled or frozen storage and display is possibly the easiest means for this, although it may adversely affect product eating quality. Chilled distribution of short-life products is more readily achieved in geographically small markets such as the UK, European national markets and around US cities. The logistics can become prohibitive for geographically large markets, including pan-European or US national distribution.

Spoilage

Many flour confectionery products are designed to be distributed, sold, stored and consumed at ambient temperatures.

These are expected to have shelf lives of anything from a few days to several months and are generally very safe because their a_w is too low to support bacterial growth. In most flour confectionery products, the primary factor limiting shelf life is mould or yeast growth. However non-microbial rancidity, staling, drying out or softening due to moisture gain are all factors capable of limiting the life of these products and should not be forgotten.

The rate at which moulds and yeasts spoil flour confectionery is defined by the product a_w. Typically, mould or yeast spoilage of flour confectionery can manifest itself in several ways:

1. As typical visible mould colonies, for example of the moulds *Penicillium* or *Aspergillus* spp. or, at lower a_w of more xerotolerant moulds, including *Eurotium* spp. and *Wallemia sebi*. *Xeromyces bisporus* is rarely seen but its extreme xerotolerance (minimum a_w for growth 0.61) means that it occasionally causes severe spoilage in products generally considered stable.
2. As bubbling in jams, fondants or fruit fillings or as pitting or cracking of icings as a result of the pressure of carbon dioxide gas formed by yeast fermentation. Yeast fermentation also produces alcohol and may produce other compounds with strong odours. For example, *Pichia anomala* can produce ethyl acetate which may give the impression of a product suffering from a chemical adulteration. *P. burtonii* has been known to produce styrene from cinnamaldehyde when fermenting syrup spiced with cinnamon was used for glazing hot cross buns.
3. As low white or off-white 'dusty' growth of one of a number of 'pseudomycelial' yeasts such as *Candida guilliermondii*, *C. parapsilosis*, *Debaromyces hansenii*, *P. anomala*, *P. burtonii*, *Saccharomycopsis fibuligera* and even bakers' yeast *S. cerevisiae* on the product surface. This is especially visible on the surface of dark products and is known as chalk mould because of its resemblance to a sprinkling of chalk dust. Since it is white in colour, it is often missed on white-coloured products. It is more frequently seen on breads than on flour confectionery.

Of all the microbiological spoilage problems encountered by the cake and pastry manufacturer, mould growth is most frequently encountered and is often the major factor governing shelf life. The work of Seiler and colleagues in the 1960s identified a logarithmic relationship linking a_w and the mould-free shelf life of preservative-free cakes when incubated at different temperatures. The relationship is represented in simple form in **Figure 2**, and is widely used to estimate the mould-free shelf life of existing and new products, without the need for expensive and long storage trials. It is also used during new product development to identify the a_w needed to achieve the desired mould-free shelf life. This work forms the basis of the software package ERH-Calc, marketed by CCFRA (Gloucestershire, UK).

Water activity is also very influential in determining the rate at which yeasts spoil flour confectionery. Fermentative spoilage problems are less common than mould spoilage but, at a given a_w, fermentation tends to occur more quickly than mould spoilage. Since the materials which are most susceptible to fermentative spoilage, such as jams and icings, are used as fillings and coatings, this is very important because moisture migration from the product crumb to the filling can increase its a_w and reduce its expected fermentation-free life.

The number of yeasts initially present in a product or filling is important in determining the spoilage potential of that filling. **Figure 3** shows the effect of jams at different water activities on the growth of an osmophilic yeast over time. It also illustrates the effect of a_w and inoculation level in the rate of fermentation of jam. The yeast strain used was *Zygosaccharomyces rouxii*.

Preservation Methods

The easiest and cheapest way of preventing microbial growth on cakes and pastries is through use of permitted preservatives. The more commonly used pre-

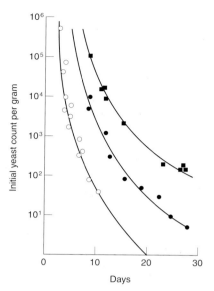

Figure 3   Effect of water activity (a_w) and initial yeast count on the time needed for fermentative spoilage of jam at 25°C. Filled squares, a_w 0.73–0.74; filled circles, a_w 0.76–0.77; open circles, a_w 0.82–0.83.

servatives worldwide for flour confectionery products are propionic acid and sorbic acid and certain of their salts. Their regulatory status varies from market to market both for concentration permitted and product types in which they are allowed. Since both are organic acids (or their salts), their antimicrobial action is heavily influenced by the concentration of undissociated acid (or salt) present rather than the total concentration. The percentage of undissociated acid (the effective species) increases as the pH decreases (**Figure 4**). Thus, a manufacturer seeking to increase product shelf life by using a preservative will consider pH when deciding how much preservative to add.

Figure 5 shows the effect of pH on the increase in mould-free shelf life in cake containing 1000 mg kg^{-1} of sorbic acid. Dramatic increases in mould-free life

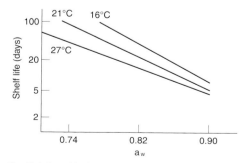

Figure 2   Relationship between water activity (a_w) and shelf life of cake at 16, 21 and 27°C (60, 70 and 80°F respectively). The cake contained no mould inhibitor and was protected from moisture loss during storage.

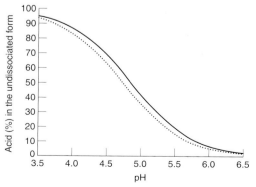

Figure 4   Dissociation curves for sorbic (dashed line) and propionic acid (continuous line).

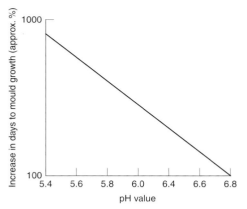

Figure 5 Effect on the approximate per cent increase in the mould-free shelf life of cake at a water activity (a_w) of 0.85 treated with sorbic acid at 1000 mg kg^{-1}.

are theoretically possible, particularly in products with low a_w. However, high concentrations of preservative can cause 'off' odours and flavours within the product. A level calculated to give a 50% increase in mould-free shelf life rarely causes such problems, but sensory evaluation of a test batch of product is always recommended.

Reformulation of recipes is sometimes useful for extending the shelf life of flour confectionery items. Water activity (cakes) and/or pH (fillings) are commonly manipulated to restrict microbial growth. However, care must be taken not to interfere to any great extent with the sensory properties of the product being developed. *S. aureus* (a toxin-producing bacterium) can be a particular problem with this approach, since it can grow at an a_w as low as 0.86 and a pH of 4.3–4.8 (although not both together; **Table 1**).

Gas packaging is a technique which is now widely used for products in the UK and Europe. Typically, carbon dioxide and an inert gas such as nitrogen are used in differing percentages for different product types. These gases are flushed into a film sealed around a product such that they replace the air surrounding the product. Carbon dioxide is used as it has an inherent antimicrobial effect, and nitrogen as it helps to prevent organoleptic deterioration of the products. Since moulds require oxygen to grow, and this is limited in a modified atmosphere pack, very significant increases in the mould-free shelf life of flour confectionery items can be achieved using this technique.

The use of carbon dioxide to replace the air around products with a_w below 0.90 has increased a given mould-free life up to five times that in air packs, provided that seal integrity is maintained (**Fig. 6**).

Another approach to mould control by restricting the oxygen content of the package is to include an oxygen scavenger. Currently this consists of a small sachet of iron-based material that is added to the package. As the iron rusts, it removes oxygen from the package, creating an atmosphere with oxygen < 0.1%. The sachet then acts as a sink to remove any oxygen that diffuses through the film during storage. Very long extensions in life are possible using this technology, which is well accepted in Japan and gaining acceptance in other markets.

The use of alcohol is effective at preventing mould growth, especially on a number of bakery items. The

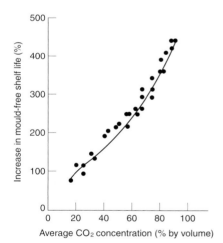

Figure 6 The effect of packaging in carbon dioxide on the mould-free shelf life of cake with a water activity (a_w) of 0.9.

Table 1 Limiting water activity (a_w) for *Staphylococcus aureus* grown aerobically in brain heart infusion at different pH and temperatures

Limiting a_w	a_w controlled by	Temperature (°C)	pH	Duration of experiment (days)
0.85	NaCl	45	7.5	Not stated
0.86	NaCl	30	7.0	28
0.85–0.87	Sucrose or NaCl	30	7.0	10
0.89	Glycerol	30	7.0	28
0.90–0.93	Sucrose or NaCl	30	4.9	10
0.93–0.96	Sucrose or NaCl	12	5.5	25
0.94	NaCl	22	5.0	30
0.96–0.99	Sucrose or NaCl	12	4.9	25

Table 2 International Commission on Microbiological Specifications for Foods recommendations for production of microbiologically acceptable pastries

1. Use only pasteurized eggs and dairy products
2. Cook the pastry filling thoroughly, with appropriate mixing to ensure uniformity of temperature
3. Keep raw ingredients and processes separate from cooked products
4. Control dusts and aerosols by establishing air movement away from the cooked product area
5. Clean and sanitize equipment that contacts fillings on a frequent basis
6. Cool cooked fillings rapidly to 5°C (40°F) or below by refrigeration, while mixing; or fill pastry shells with hot filling and refrigerate immediately
7. Maintain refrigeration of fillings and filled pastries that are capable of supporting growth of *Staphylococcus aureus* until they are consumed
8. As an alternative to refrigeration, alter the formulation by reducing the pH, reducing the a_w, or using preservatives to control the growth of pathogens
9. Wash and sanitize hands before handling cooked product
10. Minimize hand contact with cooked products, and keep persons with respiratory or skin infections away from the cooked product area

Table 3 Institute of Food Science and Technology microbial specifications guidelines for cakes and pastries

	GMP	Maximum
Pathogens		
Salmonella spp.	ND in 25 g	ND in 25 g
Staphylococcus aureus	1×10^2 per g	1×10^4 per g
Indicators and spoilage organisms		
TVC	1×10^3 per g	a
Enterobacteriaceae	10 per g	1×10^3 per g
Yeast (fondants, etc.)	1×10^2 per g	1×10^5 per g
Moulds	1×10^2 per g	1×10^4 per g

GMP = Good manufacturing practice; ND = not detected.
a For TVCs, monitoring levels over time is a useful means of building up trend analysis, which can be a powerful tool in picking up changes in levels of microorganisms throughout production.

alcohol acts as a vapour phase inhibitor rather than a surface sterilant and can be sprayed on to the product or applied indirectly, e.g. on a saturated pad. Some popular cake products in Argentina use this technology.

Good hygienic practice has a major part to play in achieving and even extending the shelf life of flour confectionery products. Hand-finished cakes and pastries are especially susceptible to contamination from pathogenic bacteria such as *S. aureus* (which is carried by up to 50% of the population) and yeasts.

The International Commission on Microbiological Specifications for Foods made a number of recommendations for controlling the quality and safety of cream- and custard-filled pastries (**Table 2**). Many of these also apply to cakes and other baked foods.

One of these recommendations which needs special emphasis is adequate baking of the product. This will kill many organisms in the dough used to formulate the products. Good hygiene is still needed post-baking to ensure that the level of problems associated are restricted. Post-baking techniques, such as passing product under ultraviolet or high-intensity light to kill off surface contamination, especially from moulds, have been reported to have some success at restricting microbial growth on some products.

Microbial Specifications

In the UK, the Institute of Food Science and Technology (IFST) has drawn up generalized microbiological specifications for cakes and pastries (**Table 3**). The specifications include a level GMP, which indicates the level expected immediately following production of the food under Good Manufacturing Practices and a level Maximum which specifies the maximum acceptable levels at any point in the shelf life of the product. The specifications give a useful benchmark but it is important to recognize that no amount of end product testing will ensure product safety. The excellent safety record of cakes and pastries is a testament to the inherent properties of the products and the traditional skills of bakers in the days before formal safety management systems such as hazard analysis of critical and control points (HACCP).

Acknowledgements

We gratefully acknowledge helpful comments on the manuscript from Dr Martin Cole and Dr Steven Walker. Karen Tudor cheerfully converted several faxed drafts into readable typed copy.

See also: **Bacillus**: Bacillus cereus. **Food Poisoning Outbreaks**. **Salmonella**: *Salmonella enteritidis*. **Spoilage Problems**: Problems caused by Fungi. **Staphylococcus**: *Staphylococcus aureus*.

Further Reading

Bennion EB, Bamford GST and Bent AJ (1997) *The Technology of Cake Making*, 6th ed. London: Blackie.

International Commission on Microbiological Specifications for Foods (1996) Characteristics of microbial pathogens. In: *Microorganisms in Foods 5*. London: Blackie Academic and Professional.

International Commission on Microbiological Specifications for Foods (1998) Cereals and cereal products. In: *Microorganisms in Foods 6*. Microbial Ecology of Food Commodities. London: Blackie Academic and Professional.

Legan JD (1999) Cereals and cereal products. In: Lund BM, Baird-Parker AC and Gould GW (eds) *The Microbiological Safety and Quality of Foods*. Gaithersburg, MD: Aspen Publishers Inc.

Seiler DAL (1976) The stability of intermediate moisture foods with respect to mould growth. In: Davies R, Birch GG and Parker KJ (eds) *Intermediate Moisture Foods*. Applied Science.

Seiler DAL (1988) Microbiological problems associated with cereal-based food. *Food Science and Technology Today* 2 (1): 37–43.

Shapton DA and Shapton NF (1991) *Principles and Practices for the Safe Processing of Foods*. Oxford: Butterworth-Heinemann.

Street CA (1991) *Flour Confectionery Manufacture*. Glasgow: Blackie Publishing.

Confocal Laser Microscopy *see* **Microscopy**: Confocal Laser Scanning Microscopy.

COSTS/BENEFITS OF MICROBIAL ORIGIN

Jill E Hobbs, George Morris Centre, Calgary, Alberta, Canada
William A Kerr, Department of Economics, University of Calgary, Alberta, Canada

A number of major food industries would not exist without organisms of microbiological origin. These include industries producing dairy products such as cheese and yoghurt, alcoholic beverages such as wine and beer, and leaven breads, among others. On the other hand, microbiological organisms are a major cause of food spoilage and food safety concerns. Considerable resources are devoted to reducing the potential losses arising from food spoilage and ensuring the safety of food. Hence, microbiological organisms provide benefits to the food industry as well as imposing considerable costs. In this article, economic approaches to issues of spoilage and food safety are explained. The losses imposed on the food industry due to spoilage in both production and storage are examined. The costs of ensuring high-quality and safe foods are then explored. The losses suffered by society as a result of food-borne diseases are then discussed. The added value, which arises from beneficial microbial agents, begins a section on the benefits arising from microbial agents. The size and importance of the major industries, which have resulted from the beneficial properties of microbiological organisms, are laid out. Finally, a discussion of the major international issues facing the food industry pertaining to the use of microbiological agents is presented.

Costs

The major costs arising from microbiological agents are associated with food spoilage and food safety. The former tend to be borne privately by food industry firms and consumers. Food safety has both private and public costs when food-borne diseases are considered. Mitigation of food spoilage and reducing the incidence of food-borne diseases is a major food industry cost.

Economic Approaches to Food Spoilage and Food Safety

While calculating the costs associated with food spoilage and food-borne diseases, as well as the costs of strategies used to reduce or mitigate their effects can be very complex, the conceptual model used by economists to depict the choices available to food industry firms is quite simple. The two types of costs associated with food spoilage are depicted in **Figure 1**. Food spoilage is taken to mean food whose condition has deteriorated to the point where it cannot be sold, either because it is unpalatable or because it is no longer safe. In the case of food spoilage it is assumed that the firm can easily determine if the food is safe either by visual inspection (or perhaps odour detection) or simple testing. This differs from the food safety case where determining whether food is safe and will remain safe once it leaves the control of the firm is much more costly. In Figure 1, monetary cost

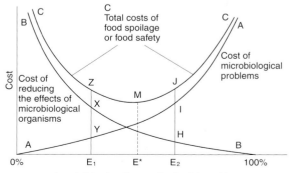

Figure 1 Economics of food spoilage and food safety.

is depicted on the vertical axis. The effect of micro-biological problems in terms of the extent of food spoilage (or food safety problems) is depicted on the horizontal axis. The right-hand end of the horizontal axis indicates that no effort has been expended to reduce the spoilage arising from microbiological organisms, resulting in complete food spoilage (100%). The point where the horizontal axis intersects the vertical axis represents total elimination of spoilage (0% spoilage).

The cost of microbiological problems (in this case spoilage) associated with each level of spoilage is the curve labelled AA in Figure 1. If no effort is made to control harmful microbial agents, so that a high level of spoilage occurs, then the costs in terms of spoiled food will be very high. The cost will be the revenue lost from not being able to sell spoiled food and the costs of disposing of products that cannot be sold. On the other hand, if considerable effort has been undertaken to reduce spoilage, then spoilage will be reduced and the losses small (moving us closer to 0% spoilage on the graph). The curve is normally expected to have the shape depicted because (reading from right to left), the initial low levels of effort to reduce food spoilage are likely to produce large reductions in spoilage, giving large reductions in the cost of microbiological problems. Additional efforts (as we move closer to 0% spoilage) are unlikely to bring equally large returns. This is because firms will take the easiest steps first to reduce food spoilage – those which have the biggest impact. The shape of the curve will vary depending upon the individual food product.

The costs associated with mitigating the effects of microbial agents for each level of reduction in spoilage are labelled BB in Figure 1. These costs might be those associated with installing refrigeration, pasteurizing, chemical treatments, etc. These costs typically will encompass a number of activities undertaken to reduce spoilage in a particular food. Again, the curve is expected to have the shape depicted in Figure 1 because (reading from right to left) the costs associated with achieving modest gains in spoilage reduction are low, while the costs of achieving reductions to, say, only 5% or 0% spoilage are extremely high. Of course, the curves may have other configurations depending upon the particular technology employed.

The costs associated with any level of food spoilage, such as point E_1, can be read off the graph. The cost of the remaining spoilage is E_1–Y while the cost associated with achieving that level of spoilage is higher: E_1–X. The total cost of the E_1 level of spoilage would be the sum of E_1–X and E_1–Y or E_1–Z. Point E_2, in comparison, represents a lower level of reduction cost, E_2–H, but a higher level of spoilage cost,

E_2–I. The total cost associated with the level of food spoilage E_2 is E_2–J.

The objective of food industry firms' strategies regarding spoilage is to minimize total cost – to find the minimum value of curve CC. This is depicted at point E^* with total cost E^*–M. Figure 1 clearly illustrates that it is inappropriate to focus exclusively on either the cost of reducing spoilage or the losses associated with spoilage when making business decisions. Further, it shows that some positive level of food spoilage is likely to be optimal. The importance of this point is that, as a business strategy, attempting to eliminate spoilage totally will be suboptimal. Technological changes which improve spoilage reduction should lower costs, shifting the BB curve down and leading to a leftward movement in E^*, meaning less spoilage.

The economics of food safety can be approached in a similar manner and Figure 1 can also be used to depict food safety (as opposed to spoilage problems). In this case, however, the cost of microbiological problems are those associated with food-borne diseases. These costs tend to be more complex because they can involve not only health-care costs such as visits to the doctor and medicines but also the costs associated with the lost earnings of those infected, lost productivity and death. Often these costs are difficult to measure. Governments may become involved in food safety because it may not be possible to attribute the cause of a disease directly to the perpetrator (meaning the legal system will not be effective in awarding damages) and firms will then underinvest in food safety. This market failure means that governments will become involved in food safety issues in the name of increasing society's welfare through regulation or through activities such as direct food inspections. Thus, the cost of microbiological problems may only be partially reflected in the costs faced by firms and the costs of reducing the effect of microbiological organisms will be shared by firms and the government. Again, however, the objective of food safety strategies (of food industry firms, consumers and governments combined) should be to minimize the sum of the two costs for society.

The model suggests that the objective of a food safety strategy should not normally be the total removal of risks of food-borne diseases. Probably the most obvious example of this trade-off is with *Salmonella*. Total eradication is not considered a commercially viable option by either industry or government regulators, yet *Salmonella* represents the main cause of documented food-borne illness in developed countries. Eradication would only be the correct strategy if the cost of a food-borne disease was very high at any level other than reduction to 0%. This would mean that, at any level greater than 0% food safety problems, the AA curve would be above where the BB curve cuts the vertical axis.

Losses in the Production and Storage of Food

In developed countries it may appear that the main threat to foodstuffs arises as a result of the activities of microbial agents. However, this may not be the case in developing countries where considerable losses and damage to food during production and storage are the result of rodents, insects and other non-microbial organisms such as maggots. For example, it has been estimated that up to one-third of the Indian grain harvest is lost to rats, primarily due to poor storage facilities. In developed countries these types of losses have been reduced to the point that they are almost ignored. While their control has simply become routine, one should not ignore the fact that considerable resources are spent on their control.

Food spoilage arising from microbial origins, however, is the primary focus of modern food industries. As with the control of non-microbial pests, many of the food-industry activities to control spoilage arising from microbial origins have become so routine as to be considered normal production practices – cooking of canned meat, pasteurization of milk, the addition of chemical preservatives, etc. As older processes become routine, new challenges for the food industry arise and become the focus of attention. Changes in the focus of food spoilage mitigation activities arise for a number of reasons. Technological change in the commercial food and related industries (e.g. transportation, home food storage and preparation) is a major initiator of change. For example, the advent of relatively low-cost refrigeration spawned the need to develop 'cold chain' capabilities for the meat industry. The widespread availability of microwaves led to the need to think about an entirely new range of spoilage situations.

Changes in consumer tastes can also alter the way spoilage is viewed. If consumers become resistant to the use of chemical additives in food, then firms are faced either with higher incidents of spoilage or with finding alternative control mechanisms. The activities of the food industry itself may be responsible for raising the expectations of consumers regarding what is an acceptable product, hence inadvertently increasing the proportion of food which is considered spoiled. The marketing of the cosmetic aspects of fruits and vegetables is the most obvious example.

Supply chains in the food industry have also become longer over time. The lengthening of supply chains has led to increased opportunities for food to spoil. This lengthening has two aspects – longer geographic distances (hence, often a lengthening of shelf-life time

requirements) and a greater number of participants along the supply chain handling food. One example of the latter is the increase in the proportion of meals eaten outside the home. The problems faced by restaurants, both in matching quickly available food quantities to unpredictable customer numbers and preferences (how many steaks to unthaw on any given day) and in the training of low-paid and often transient staff, lead to increases in spoilage relative to within-home food preparation.

It should be clear that the degree of food spoilage cannot be separated from the efforts expended to reduce it. Further, increased efforts to reduce spoilage at one point in the supply chain (food processing) may be offset by a different set of trade-offs between the quantity of food allowed to spoil and the costs of reducing spoilage at another point in the supply chain (restaurants). Arriving at a reasonable estimate of the proportion of foods spoiled or their value is almost impossible. Food spoilage within the commercial food chain arises from two sources: first, from firms making routine calculations which implicitly recognize the total cost-minimizing trade-offs illustrated in Figure 1. This means that a certain quantity of food is routinely expected to spoil because the cost of reducing spoilage further exceeds the benefits of reducing spoilage. This is represented by points to the left of E* in Figure 1. Spoilage also arises from unforeseen breakdowns in the systems put in place to reduce the problem. These can lead to a high degree of spoilage for a particular crop or the withdrawal of entire production runs when, for example, samples of meat products are found in a deteriorated condition and all of the product must be destroyed to restore consumer confidence. While much of this product is not technically spoiled, it still represents a loss arising from a breakdown in spoilage prevention.

The Costs of Quality Control

Food quality control problems arising from microbial agents encompass two main cost issues. The first is the absolute cost of food quality control relative to the total cost of food. The second relates to who bears those costs. In the case of the latter, three broad groups can be identified: the food industry, government and consumers (broadly defined to include their significant role in the home preservation and preparation of food). The food industry makes substantial expenditures to ensure the quality of food. Governments take considerable responsibility for establishing food safety standards and for inspection (everything from meat inspectors in slaughter plants to visits to restaurant kitchens). Despite all of these efforts, the main line of defence, as it were, against many food-borne diseases is the consumer preparing food in the home.

The simple precautions taken to cook meat adequately, for example, form the major defence against the health hazard posed by *Salmonella*. In a similar fashion, home food preservation methods such as canning, pickling, and latterly, home freezing, can constitute a significant proportion of the resources expended by a society to maintain the quality of food. Of course, in the past when food supply chains were shorter and simpler – when home-produced food on farms constituted the major portion of the society's diet and most food transactions took place at local markets – food quality maintenance activities were almost entirely the responsibility of the consumer. This is still the case in many developing countries. As technology changed and food supply chains lengthened when countries developed, more and more of the responsibility for food quality maintenance fell on firms. As it is often very difficult to isolate the source of food-borne disease, and therefore difficult to prove liability, governments increasingly became involved in ensuring the maintenance of food quality in order to protect consumers better. Cost efficiencies relating to prevention in processing plants and the strategic use of individuals with scientific knowledge (relative to the cost of broad-based scientific education of consumers) moved food safety firmly into the public realm.

The proportion of the total cost of quality maintenance borne by the actors along the food chain is constantly changing. Recently, for example, government-funded meat inspection has been moved into the private sector in some countries, while the burden of government-run inspection services has been transferred to industry through cost recovery programmes in others. The move to hazard analysis critical control points (HACCP) systems in some sectors in the US, while probably improving the efficacy of food safety systems, was also motivated by the government's desire to move more of the cost of maintaining food quality to the private sector in times of government budget difficulties.

In some cases, consumer preferences (or prejudices) keep the total costs of maintaining food quality higher than necessary. The poor image of irradiated foods among consumers has effectively restricted the use of this technology in controlling microbial agents. In general, however, technological changes tend to shift the BB curve in Figure 1 downward, moving E* to the left and increasing food quality while lowering its cost. The US government budget for food safety regulatory activities was US$1.2 billion in 1994, and the food industry will expend many times that amount to comply with food safety regulations and in private initiatives to reduce the spoilage of foods. Progress continues to be made in ensuring the maintenance of food quality. For example, it has been estimated that listeriosis cases fell 44% in a decade due to industry, regulatory and educational efforts to reduce *Listeria* contamination in foods.

Societal Losses from Food-borne Disease

As suggested above, there are three costs which make up the BB curve in Figure 1: costs incurred by individuals/households, the private food industry and the public regulatory system. When food-borne disease is considered, the costs which make up the AA curve in Figure 1 – the costs associated with problems caused by microbial agents – extend far beyond those associated with spoiled food. These costs include the direct cost of disease (medical visits, drugs, hospitalization), the costs associated with death (including the loss of a breadwinner's income and trauma for family members and friends), unmitigated pain and suffering of those who become ill prior to treatment or for whom no treatment is available, losses in worker productivity and the anxiety which exists about food-borne health risk. Looked at another way, these could all be reduced from increased activities to maintain the quality of food and, hence, represent the true costs associated with food-borne disease.

Research dealing with the costs of food-borne disease often uses cost of illness (COI) estimates. These estimates, however, are often incomplete because they only include medical costs and the cost of lost productivity. The other costs are usually omitted due to a lack of suitable measures. Estimating the costs associated with death are particularly difficult. As a result, COI estimates usually underestimate the true costs of food-borne disease caused by microbial agents. An alternative method sometimes used to determine the value (or costs) associated with food-borne disease is willingness to pay (WTP) studies which attempt to estimate the value that individuals place on reductions in the risks associated with food-borne disease. Again, these studies provide only ballpark estimates due to the difficulties associated with individuals' understanding of the actual (as opposed to the perceived) risks involved and the costs which would be imposed on them. In other words, the estimates attained in WTP studies represent only the individual's perception of the costs associated with food-borne disease and may either over- or underestimate the true cost.

In COI studies, food-borne illness costs from six major bacterial pathogens have been estimated at US$2.9–6.7 billion per year in the US alone. While estimates vary considerably due to under-reporting and failure to identify the true cause of illness, microbial pathogens in food cause an estimated 6.5–33 million cases of human illness and up to 9000 deaths annually in a US population of approximately 260 million. It should be remembered that these are the costs that remain after the extensive expenditure of resources by individuals, the food industry and the public sector to maintain the quality of food. The foods most likely to cause human illness are animal products such as red meat, poultry and eggs, seafood and dairy products. Meat and poultry are estimated as the source of approximately 80% of the annual costs of human illness from food-borne pathogens.

The major microorganisms that cause disease are bacteria, fungi, parasites and viruses. More than 40 different food-borne pathogens are believed to cause human illness. Over 90% of confirmed food-borne illnesses have been attributed to bacteria. Six bacterial pathogens are considered of most importance – *Salmonella*, *Campylobacter jejuni*, *Escherichia coli* O157:H7, *Listeria monocytogenes*, *Staphylococcus aureus*, and *Clostridium perfringens*. The potential pathways of human exposure to pathogens found in animals, for example, include direct contact with live food animals (putting at risk farmers, livestock transporters, etc.), indirect contact with live food animals such as touching animal waste (farmers, processing plant workers), direct contamination by the carcass (processing plant workers, government inspectors), indirect contamination by the carcass through contact with knives or contaminated clothing (processing plant workers, laundry employees), cross contamination of food products in slaughterhouses, food preparation establishments and in the home (food industry workers, consumers), consumption of meat, poultry and dairy products (consumers) and person-to-person transmission (restaurant staff, consumers). Hence, the costs of food-borne disease can be found all along the food supply chain and prevention measures cannot be exclusively centred on the final consumer. Measures taken near the end of the food supply chain which could eliminate the risk to consumers at a low cost may not be appropriate when the risks along the entire food chain are considered.

There are three major classifications of food-borne diseases. Food-borne intoxications are caused by consuming food that contains toxins released during the growth stages of specific bacteria or microtoxins produced by moulds. Food-borne toxic infections arise when the pathogens produce harmful or deadly toxins while multiplying in the human intestinal tract. Food-borne infections result when pathogens are eaten, become established and multiply in the body. Most cases of food-borne illness are classified as acute because they occur quickly after ingestion and are self-limiting. Symptoms are gastrointestinal problems and/or vomiting. Unless the acute food-borne illness results in death, most costs relate to health-care expenditures and short-run losses in worker productivity.

Approximately 2–3% of acute cases develop long-term illness or chronic problems of the rheumatoid, cardiac and neurological systems. These chronic illnesses may afflict individuals for the rest of their life or cause premature death. In these cases the costs are ongoing and include the reduced productivity and incomes associated with long-term disability. The costs may even be borne intergenerationally with reduced incomes eliminating the possibility of university education for children and their subsequent ability to earn income. Quality of life for the individual and family members may be considerably reduced. Clearly, an individual's WTP to avoid the chronic effects of food-borne illness might considerably exceed COI estimates based on medical costs and lost productivity.

Benefits

While attention is often focused on the costs associated with food spoilage and food safety that result from the activities of microbiological agents, it is clear that considerable benefits also accrue from the existence of other microbial organisms. Beyond the basic fact that life as we know it could not exist without microorganisms – cows would not be able to digest grass, there would be no oil to fuel industrial processing and food distribution, no compost recycling of nutrients would take place – there are specific industries which are based on the activities of microorganisms. Fermented milk products such as cheese and yoghurt are based on species of *Lactobacillus* and *Streptococcus*. Further, some cheeses such as Stilton, Camembert, Brie and Limburger have specific bacterial ripeners added. All three of the major alcoholic beverage industries – wine making, brewing and spirit production – depend upon the actions of microbial agents. Vinegar production is directly dependent on *Acetobacter*. The quality of bread is considerably enhanced by leavening, which is based on having yeast act on sugars in the dough. The resulting carbon dioxide forms tiny bubbles in the dough, which lightens it and gives the bread a more open texture. Citric acid used in the manufacture of, for example, lemonade is produced by fermentation of glucose by the mould *Aspergillus*. Glutamic acid is a flavour enhancer. Yeast extracts are marketed directly as food products. Many other food production processes benefit directly from the actions of microbiological organisms. Clearly, food consumers' choices are considerably increased and their quality of life enhanced by the existence of microbial agents.

Added Value from Microbiological Agents

In assessing the benefits accruing from industries based on microbiological organisms it is important to keep in mind that it is the added value which can be attributed to the industry which is important and not the total value of the industry. Added value is the difference between an industry's sales and the costs of its raw materials. In other words, the cheese industry can only be assessed for its additional value above the raw milk which is used to produce it. The wine industry adds value to the grapes which are produced. Of course, if no wine were produced, fewer grapes would probably also be produced due to saturation in the table grape or grape juice markets. This does not mean, however, that the additional grapes plus the wine produced represent the net gain attributable to the wine industry because the resources used to produce the grapes would probably be used to produce other products. Those products are forgone as a result of wine production. The forgone products represent what economists call opportunity costs. The added value of an industry arises in relation to its opportunity cost. Despite this, the added value in, for example, the cheese industry is still considerable when the wholesale value of cheese is compared to its raw milk inputs. The value of brewery products far outstrips its input costs. In the case of industries such as bakeries the added value seems much more modest when one compares, for example, the prices of leaven and unleaven bread.

Size of the Industry

The size of industries based on microbiological organisms is very large. For example, in 1997 cheese production was reported in almost 100 countries and exceeded 15 000 000 metric tonnes. Bread and related products reached almost 50 000 000 metric tonnes, much of it leaven. Vinegars exceeded 20 000 000 metric tonnes. Distilled alcoholic beverage production reached 50 000 000 hl, wine 225 000 000 hl and beer 1 150 000 000 hl. In the US annual cheese consumption is approximately 12.15 kg per capita. Per capita, cottage cheese consumption is 1.3 kg and yoghurt 2.1 kg.

In 1993, 11 of the 50 largest food-processing firms in the world had as their primary business microbiological based products. The 11 firms had combined processed food sales of over US$66 billion and included such industry giants as Kirin, Anheuser-Busch and Seagrams. Microbiological foods and beverages are extensively traded in international markets. Global cheese exports were valued at US$11 500 million in 1996, wine exports US$12 000 million and beer exports US$5000 million. The heavy degree of international trade and investment in microbiological based food products raises significant international issues.

National and International Issues

The major national and international issues pertaining to microbiological organisms relate to rising consumer concerns regarding food safety and the technological changes embodied in agri-food biotechnology. Food safety awareness has been rising among consumers in developed countries. Awareness is not synonymous with being informed, and those charged with ensuring food safety both at the political and technical level are often faced with what appears to be irrational consumer concerns. The food industry has long been a beneficiary of a virtual consensus among food scientists, consumers and policy makers regarding what was considered *appropriate science* as it related to food safety. In practice, this often meant that the latter two groups deferred to the former for the establishment of food safety standards and protocols. In recent years, however, the consensus on what constitutes appropriate science has declined. While the degree to which this consensus has been diluted varies among developed countries – less trust in western Europe than in the US, for example – there is little doubt that sufficient numbers of consumers (where sufficient numbers means that they cannot easily be ignored by politicians) are no longer willing to defer to the judgments of scientific experts regarding what constitutes appropriate science. The decline in consumer confidence is related to well-publicised breakdowns in the food safety system – tainted fast-food hamburgers, *Salmonella* in eggs – and, in the UK, the industry and regulatory system's apparent inability to deal with the 'mad cow disease' crisis. This deterioration in confidence in food safety regimes has left policy makers with the unenviable job of attempting to restore confidence through tightening and redesigning food safety regimes. Often this has been accomplished in times of severe budgetary restraints. Food processors have been faced with rising regulatory costs and the spectre of large liability awards arising from the legal system (particularly juries) which are more willing to assign blame directly than in the past. A period of rapid change in food safety initiatives, both public and private, has begun. Individual countries have been unilaterally altering standards and protocols leading to differences among countries – differences which can inhibit trade.

The international harmonization of food safety standards is a long and resource-intensive process. Differences in food safety standards and protocols increase costs for firms wishing to export because they must undertake a separate set of food safety procedures for each foreign market they wish to enter. This may be the case even if food safety standards are more stringent in their domestic market because foreign regulations specify different procedures. In some cases, the extra cost may not be justified by the level of foreign sales, effectively shutting the firm out of foreign markets. In other cases, it may not be technically feasible to satisfy foreign requirements. Food processors in developing countries may be particularly disadvantaged. Food safety regulations can also be used strategically as non-tariff barriers to international trade.

One of the major accomplishments of the recent Uruguay round of GATT/WTO negotiations was the Agreement on the Application of Sanitary and Phytosanitary Measures (SPS) and the Agreement on Technical Barriers to Trade (TBT). In the SPS Agreement, any trade restrictions based on food safety issues should be based solely on scientific principles. In the TBT, countries have agreed that the costs associated with regulations, for example relating to the labelling of food, and in particular the verification of labels, should not exceed the benefits consumers receive from food labelling.

Biotechnology holds great promise for microbiological foods. This relates both to the creation of specifically tailored microorganisms to enhance food production and the building-in of genetic resistance to existing harmful microbial agents. Consumer acceptance of genetically modified foods, particularly transgenic products, however, is far from universal and regulators are struggling with how to deal with the potential for unforeseen toxic build-ups over the long run and possible environmental risks. Nowhere is the lack of consumer confidence in the food safety system more evident than on the topic of transgenic microbiological organisms. Again, levels of consumer concern vary internationally, bringing the issue to the top of the trade agenda.

See also: Acetobacter. **Bovine Spongiform Encephalopathy (BSE)**. **Bread**: Bread from Wheat Flour. **Cheese**: Microbiology of Cheese-making and Maturation. **Chilled Storage of Foods**: Principles. *Escherichia coli* **O157**: *Escherichia coli 0157:H7*. **Fermentation (Industrial)**: Basic Considerations. **Fer-**mented **Foods**: Origins and Applications. **Fermented Milks**: Range of Products. **Food Poisoning Outbreaks**. **Genetic Engineering**: Modification of Yeast and Moulds; Modification of Bacteria. **Hazard Appraisal (HACCP)**: The Overall Concept. **Heat Treatment of Foods**: Thermal Processing Required for Canning; Ultrahigh Temperature (UHT) Treatments; Principles of Pasteurization; Action of Microwaves; Synergy Between Treatments. **International Control of Microbiology**. *Listeria*: Introduction. **National Legislation, Guidelines & Standards Governing Microbiology**: Canada; European Union; Japan. **Preservatives**: Classification and Properties. **Process Hygiene**: Designing for Hygienic Operation. *Salmonella*: Introduction. **Spoilage Problems**: Problems caused by Bacteria; Problems caused by Fungi. **Vinegar**. **Wines**: Microbiology of Wine-making. **Yeasts**: Production and Commercial Uses.

Further Reading

Benenson AS (ed.) (1990) *Control of Communicable Diseases in Man*, 15th edn. Washington, DC: American Public Health Association.

Buzby JC, Roberts T, Lin CJ and MacDonald JM (1996) *Bacterial Foodborne Disease: Medical Costs and Productivity Losses*. Agricultural economic report no. 741. Washington, DC: United States Department of Agriculture.

Caswell JA (ed.) (1995) *Valuing Food Safety and Nutrition*. Boulder, CO: Westview Press.

Haddix AC, Teutsch PA, Shaffer PA and Dunet DO (eds) (1996) *Prevention Effectiveness: A Guide to Decision Analysis and Economic Evaluation*. New York: Oxford University Press.

Henderson DR, Handy CR and Neff SA (eds) (1996) *Globalization of the Processed Food Market*. Agricultural economic report no. 752. Washington, DC: United States Department of Agriculture.

Jones JM (1992) *Food Safety*. St Paul, MN: Egan Press.

Just RE, Heuth DL and Schmitz A (1982) *Applied Welfare Economics and Public Policy*. Englewood Cliffs, NJ: Prentice-Hall.

Kerr WA and Perdikis N (1995) *The Economics of International Business*. London: Chapman & Hall.

Noble WC (1979) *Microorganisms and Man*. Studies in biology no. 111, London: Institute of Biology.

Organisation of Economic Cooperation and Development (1998) *The Future of Food*. Paris: OECD.

Cream *see* **Milk and Milk Products**: Microbiology of Cream and Butter.

Critical Control Points *see* **Hazard Appraisal (HACCP)**: Critical Control Points.

Crustacea *see* Shellfish (Molluscs and Crustacea): Characteristics of the Groups; Contamination and Spoilage.

CRYPTOSPORIDIUM

R W A Girdwood and **H V Smith**, Scottish Parasite Diagnostic Laboratory, Stobhill Hospital, Glasgow, UK

Characteristics of the Genus

Organisms found in the gastric crypts of an experimental laboratory mouse were first described and named *Cryptosporidium* by Tyzzer in 1907. Recognized as a cause of morbidity and mortality in young turkeys in 1955 (*C. meleagridis*) and as a cause of scouring in calves in 1971 it was not until 1976 that the first two (separate) human cases of cryptosporidiosis were described.

More than 20 'species' have been described on the basis of the animal hosts from which they were isolated. This proliferation of speciation was deemed to be ill-founded in that many morphologically identical 'species' were shown by experiment to lack host specificity. Currently, there are eight valid species, namely *C. parvum* and *C. muris* which infect mammals, *C. baileyi* and *C. meleagridis* which infect birds and *C. serpentis* and *C. nasorum* which infect reptiles and fish, respectively. *C. wrairi* has been described in guinea pigs and *C. felis* in cats. Currently, there is pressure to reclassify the species of *Cryptosporidium* which infects fish into a new genus *Piscicryptosporidium*, on the basis of ultrastructural and development differences in the parasites. The morphometry of the oocysts of some species of *Cryptosporidium* are outlined in **Table 1**.

Human isolates are currently described as *C. parvum* or *Cryptosporidium* sp. Based on increasing molecular evidence it has been suggested that within *C. parvum* there are two genotypes with different mammalian host ranges. Thus some mammalian strains (genotype C) are capable of infecting humans but some human strains (genotype H) cannot infect non-human hosts. Genotyping of oocysts has been performed using the following genes: *Cryptosporidium* oocyst wall protein, dihydrofolate reductase, *Cryptosporidium* thrombospondin related adhesive protein-1 and -2, ribonuclease reductase and the internal transcribed spacer 1 of the 18s rRNA gene. Whether strains can transform or segregate within a given host is currently the subject of research. *Cryptosporidium* spp. are classified as coccidians and they make up the order Emeriida with four other genera, *Isospora*, *Sarcocystis*, *Cyclospora* and *Toxoplasma* (**Table 2**).

Life Cycle (Fig. 1)

C. parvum is monoxenous. Unlike many other coccidians the life cycle is completed in a single host. All stages of the cycle, which include both asexual and sexual forms, occur in the intestinal or, less frequently, respiratory tract either at the most superficial level of the epithelial lining, within parasitophorous vacuoles or in the lumen. The vacuoles are at the microvillous surface and are intracellular but extracytoplasmic. Infection is initiated by the ingestion of the environmentally resistant transmissive stage – the oocyst. The oocyst of *C. parvum* is spherical or subspherical,

Table 2 Classification of the genus *Cryptosporidium*

Kingdom	: Protozoa
Phylum	: Apicomplexa (Sporozoa)
Class	: Coccidea
Order	: Eimeriida
Genus	: *Cryptosporidium*

Table 1 Some differences between species within the genus *Cryptosporidium*

Species of Cryptosporidium	Dimensions of oocysts (μm)	Site of infection	Type of vertebrate infected
C. parvum	4.5 × 5.5	Small intestine	Mammals
C. muris	5.6 × 7.4	Stomach	Mammals
C. baileyi	4.6 × 6.2	Bursa of Fabricius, cloaca	Birds
C. meleagridis	4.0 × 5.2	Intestine	Birds
C. serpentis	5.4 × 6.0	Stomach	Reptiles

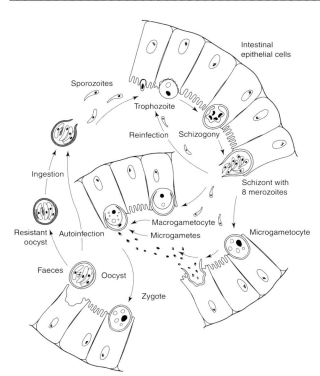

Figure 1 Life cycle of *Cryptosporidium parvum* (adapted from Smith and Rose, 1998).

smooth walled, measures 4.5–5.5 μm in diameter and contains four banana-shaped naked sporozoites which, on release, penetrate the microvillus surface of the epithelium to become trophozoites (meronts) within the created parasitophorous vacuole. Asexual multiplication (merogony or schizogony) results in the release of merozoites which invade other epithelial cells to repeat this process. This repetitive asexual multiplication is derived from from type I meronts but merozoites can produce type II meronts which in turn produce microgamonts and macrogamonts. The microgamonts rupture to release microgametes which fertilize the macrogamonts to produce zygotes. The majority of zygotes mature into thick-walled oocysts which on being released into the lumen are voided in the faeces as the environmentally robust transmissive stage. It is thought that a small proportion of zygotes produce thin-walled oocysts which rupture within the lumen and perpetuate epithelial invasion and infection.

Detection in Faeces

Oocysts are excreted in the faeces of human and animal hosts and they can be preserved in 10% formalin or SAF (sodium acetate–acetic acid–formalin) fixatives. Oocyst viability can be retained in 2.5% potassium dichromate. Fresh or preserved stools can be concentrated to increase the yield of oocysts by sedimentation using the formol–ether or formol–ethyl acetate techniques or by any of the conventional faecal parasite flotation methods such as zinc sulphate, saturated sodium chloride or sucrose solutions. Because the oocysts are small and lack distinctive visible internal or external structures, staining techniques are required to differentiate these organisms from similarly sized objects present in faeces or faecal concentrates. A modified Ziehl–Neelsen acid fast (mZN) stain and the auramine phenol fluorescent stain are the most widely used, although safranin–methylene blue, Wright–Giemsa, Kinyoun modification of mZN (KmZN) and quinacrine stains have also been recommended. Fluorescence using fluorescein-labelled polyclonal or monoclonal antibodies have the advantage of genus specificity and any of the techniques which utilize fluorescence have the advantage of requiring lower total magnifications for detection thus enabling preparations to be scanned more rapidly. Although oocysts can be excreted in very large numbers in both humans and animals (up to 10^7 per gram) during the acute phase of infection, it should be appreciated that the threshold for 100% detection efficiency using conventional methods can be as high as 5×10^4 oocysts per gram. These limitations result in the underdiagnosis of infection and consequently miss sources contributing to environmental and thereby water and food contamination. Although complete development in vitro has been described (see Table 4), in vitro culture is not used for diagnostic purposes.

Diagnosis can be augmented by light microscopy of haematoxylin and eosin stained intestinal and (in the immunocompromised) respiratory mucosal biopsies. Transmission electron microscopy of similar biopsies can reveal the parasitophorous vacuoles containing the replicating stages but is not used routinely for diagnostic purposes. The demonstration of circulating specific antibodies is seldom of diagnostic use because, in the absence of detecting seroconversion from negative to positive, positive reactions can only indicate current or past infections. Accordingly, serological methods tend to be used for epidemiological surveys.

Sensitivity of Detection

In unconcentrated faecal smears a detection limit of 10^6 oocysts per millilitre of faeces was reported using KmZN. After stool concentration, 1×10^4–5×10^4 oocysts per gram of unconcentrated stool are necessary to obtain a 100% detection efficiency using either KmZN or immunofluorescence using a commercially available fluorescein isothiocyanate-conjugated anti-*Cryptosporidium* monoclonal antibody (FITC-C-mAb) method. For bovine stools, the threshold for

100% detection efficiency using auramine phenol or FITC-C-mAb is 1×10^3 oocysts per gram. Variations in faecal consistency also influence the ease of detection, with oocysts being more easily detected in concentrates made from watery, diarrhoeal specimens than from formed stool specimens. A number of immunological methods have also been developed for detecting Cryptosporidium oocysts. The use of a FITC-C-mAb offers no increase in sensitivity over conventional stains. In addition to microscopic techniques, antigen capture enzyme-linked immunosorbent assays (ELISAs) have been reported with detection limits in the region of $3 \times 10^5 – 10^6$ oocysts per millilitre, which is similar in sensitivity to microscopy. Antigenic variability between clinical isolates of Cryptosporidium could further compromise immunodiagnostic tests. Although effective for diagnosing symptomatic infection, these methods remain insensitive and unable to detect small numbers of oocysts.

Alternative Detection Procedures

Flow cytometry coupled with cell sorting (FCCS) has recently been used to detect oocysts in human stool samples with a fourfold increase in sensitivity over direct fluorescence antibody tests, and offering, in some instances, a detection limit of 10^3 oocysts per millilitre of sample. However, FCCS is a costly procedure, and being based on the use of FITC-C-mAb might be subject to antigenic variability in oocyst epitopes.

Polymerase chain reaction (PCR) is more sensitive than conventional and immunological assays for the detection of oocysts in faeces. Approaches to developing diagnostic PCR primers include (1) design from sequence information for the 18S rDNA gene from the rRNA WWW server; (2) construction and screening of genomic DNA libraries and (3) use of the random amplified polymorphic DNA (RAPD) technique, which detects nucleotide sequence polymorphisms without requiring previously determined nucleic acid sequence information.

In one study, microscopic detection of Cryptosporidium oocysts in human faecal samples was 83.7% sensitive and 98.9% specific compared with 100% sensitivity and specificity for PCR. A further advantage of PCR is the identification of asymptomatic carriers; as well as fomites, food and water contaminated with small numbers of oocysts. Rapid detection of carriers could limit clinical sequelae in 'at risk' groups and reduce environmental spread. At present, in the clinical diagnostic laboratory, PCR is more likely to become a tool to assist epidemiological investigations than a routine method. The sensitivity of published methods range from 1–10 oocysts by

RAPD to approximately 5×10^7 oocysts by nested PCR.

Before adoption in clinical laboratories, both the variability between methods and the recognized difficulties in amplifying nucleic acids from faecal specimens by PCR will have to be overcome. Faecal samples can contain many PCR inhibitors. In addition to bilirubin and bile salts, complex polysaccharides are also significant inhibitors. For Cryptosporidium, boiling faecal samples in 10% polyvinyl-polypyrrilidone (PVPP) before glassmilk extraction can reduce inhibition. The use of Catrimox-14 (TM) (Iowa Biotechnology, Iowa), a cationic surfactant, during extraction of nucleic acids can also eliminate inhibitory substances from faecal samples.

Fluorescence in situ hybridization (FISH) technology has been used to identify C. parvum oocysts in a laboratory-based approach. Probes to unique regions of C. parvum rRNA were synthesized and used to detect oocysts on glass microscope slides. Two probes, with similar hybridization characteristics, were labelled with a novel fluorescent reporter, 6-carboxyfluorescein phosphoramadite (excitation 488 nm, emission 522 nm) and viewed by epifluorescence microscopy.

Infectivity and In Vitro Culture

The infectious dose is thought to be small. A human volunteer study indicated that the ID_{50} for C. parvum is 132 oocysts, although infection occurred in 18 of 29 (62%) of volunteers ingesting 30 oocysts or more of a bovine (Iowa) strain of C. parvum. Ten oocysts produced infection in two out of two infant non-human primates, whereas five oocysts produced disease in gnotobiotic lambs. The ID_{100} for outbred neonatal (CD-1 strain) mice is approximately 300 oocysts and the ID_{50} ranges between 60 and 87 oocysts of either an Iowa (USA, bovine, genotype C) or Moredun (UK, cervine–ovine, genotype C) strain of C. parvum.

Complete development (from sporozoite to oocyst) of C. parvum has proven difficult in in vitro culture. Although incomplete development in vitro (primarily asexual developmental stages) has been documented in a variety of human and animal derived cell lines, reports of oocyst production in vitro are scarce (**Table 3**). Currently, in vivo infectivity of isolates can be determined in neonatal or infant mice; 4–7-day-old neonatal mice of the CD-1 strain are commonly used because the strain is outbred and has large litter sizes. Either conventional haematoxylin and eosin histology of the small intestine, to demonstrate endogenous stages and oocysts or homogenization of the small intestine, followed by concentration and clarification

Table 3 Development of *Cryptosporidium* in cell culture

Complete development (asexual and sexual)	Partial development (asexual)
Human fetal lung (HFL)	Human ileocaecal adenocarcinoma (HCT-8)
Human endometrial carcinoma (RL95-2)	Madin–Darby bovine kidney (MDBK)
Bovine Fallopian tube epithelial cells	Madin–Darby canine kidney (MDCK)
Primary chicken kidney (PCK)	Human colonic adenocarcinoma (Caco-2)
Porcine kidney (PK-10)	Human pelvic adenocarcinoma (Hs-700T)
Human monocyte cell line (THP-1)	Human breast infiltrating ductal carcinoma (BT-549)
Human colonic adenocarcinoma (Caco-2) (asexual and sexual stages)	Human colonic adenocarcinoma (Ls-174T)
	Human colonic adenocarcinoma (HT-29)
	BALB/c mouse embryo (BALB/3T3)
	Human endometrial carcinoma (RL95–2)
	Human fibrosarcoma (HT-1080)

of oocysts and visualization by epifluorescence with a commercially available FITC-C-mAb reactive with oocyst epitopes can be used to demonstrate infection.

Determination of Viability

The conventional techniques of animal infectivity and excystation in vitro are not applicable to the small numbers of organisms found in water and food concentrates. Examination by phase contract or Nomarski differential interference contrast (DIC) microscopy, which can determine gross differences between viable and non-viable organisms, such as the morphological integrity of the parasites within the oocyst, is subjective and often compromised by the presence of occluding particulates and other debris. Recent attempts to develop rapid, objective estimates of organism viability have revolved around the microscopical observation of fluorescence inclusion or exclusion of specific fluorogens as a measure of viability. A fluorogenic vital dye assay for the determination of viability of *C. parvum* oocysts, based on the inclusion/exclusion of two fluorogenic dyes, 4,6-diamidino-2-phenylindole (DAPI) and propidium iodide (PI) has been developed. Discrimination between non-viable and viable oocysts is based on the former including DAPI (see Fig. 3) into the nuclei of the four sporozoites, but excluding PI, whereas the latter include PI either into the nuclei or cytoplasm as well as DAPI. A UV filter block is used to visualize DAPI (350 nm, excitation; 450 nm, emission) and a green filter block for PI (535 nm, excitation; >610 nm,

emission). Results correlated closely with optimized in vitro excystation.

Two further fluorogens (SYTO® 9, SYTO® 59; Molecular Probes, Eugene, Oregon, USA) have also been developed as surrogates of oocyst viability. SYTO® 9 fluoresces light green–yellow under the blue filter (480 nm, excitation; 520 nm, emission) and SYTO® 59 fluoresces red under the green filter block (535 nm, excitation; >610 nm, emission) of an epifluorescence microscope. Both are nucleic acid intercalators and have been shown to correlate closely with animal infectivity in neonatal CD-1 mice, but not with in vitro excystation. SYTO® 9 and 59 are permeable to oocysts with damaged membranes and, therefore, stain non-viable oocysts.

The detection of sporozoites following in vitro excystation (where sporozoites are released from the oocyst into a suspending medium) has been used as the first stage in developing a PCR-based viability assay. Two approaches have been described for *C. parvum* oocysts. In the first of two approaches, sporozoites are excysted in vitro according to a standardized protocol, lysed and the DNA is amplified using previously published primers for a repetitive oocyst protein gene sequence. In the second approach, immunomagnetizable separation (IMS) is used to concentrate oocysts from the inhibitory matrix of faeces, the sporozoites excysted, their DNA released and amplified by PCR using the 'Laxer et al' primers in a nested PCR (IC-PCR). The sensitivity of the first approach was approximately 25 oocysts (100 sporozoites) in an experimental system, but no information was available on its sensitivity in environmental samples. The sensitivity of the second approach was 1–10 oocysts in purified samples and 30–100 oocysts inoculated into stool samples.

A reverse transcription PCR (RT-PCR), which amplifies messenger RNA (mRNA) of the *C. parvum* heat-shock protein 70 (*hsp 70*) has also been used to determine the viability of oocysts in four different water types. Synthesis of *hsp 70* mRNA was induced by a 20 min incubation at 45°C followed by five freeze–thaw cycles (1 min liquid N_2, 1 min 65°C) to rupture the oocysts and mRNA was hybridized on oligo $(dT)_{25}$-linked magnetizable beads. RT-PCR was undertaken according to the manufacturer's instructions using a primer specifically designed to prime *C. parvum hsp 70*. The RT reaction mixture was amplified by PCR and the amplicons detected visually on gels and by chemiluminescent detection of Southern blot hybridizations. An RNA internal positive control was also developed and included in each assay in order to safeguard against false negative results caused by inhibitory substances.

Fluorescence *in situ* hybridization (FISH) technology has been used to determine the viability of *C. parvum* oocysts. The fluorescently labelled oligonucleotide probe targets a specific sequence in the 18S rRNA of *C. parvum* and causes viable sporozoites (capable of in vitro excystation) to fluoresce. Neither dead oocysts, nor organisms other than *C. parvum* fluoresced following *in situ* hybridization. FISH-stained oocysts did not fluoresce sufficiently brightly to allow their detection in environmental water samples; however, simultaneous detection and viability could be undertaken when FISH was used in combination with a commercially available FITC-C-mAb.

Biophysical methods have also been applied to the determination of oocyst viability. Both dielectrophoresis and electrorotation have been used to demonstrate differences between viable and non-viable oocysts. For both techniques oocysts have to be partially purified and suspended in a low conductivity medium. A spiral oocyst concentrator, based on the technique of travelling wave dielectrophoresis, can concentrate oocysts and determine their viability simultaneously. Further developments in the application of dielectrophoresis and electrorotation to oocyst concentration, detection and determination of viability are awaited.

Food-borne Transmission (see Fig. 2)

Three outbreaks of food-borne transmission have been documented. Two occurred in the USA following the consumption of non-alcoholic, pressed, apple cider, in 1993 and 1996 affecting a total of 185 individuals. In the first outbreak, apples were collected from an orchard in which an infected calf grazed. Some apples had fallen onto the ground (windfalls) and had probably been contaminated with infectious oocysts then. The source of oocysts in the second outbreak is less clear as windfalls were not used and waterborne, as well as, other routes of contamination were suggested. A food-borne outbreak, which affected 15 individuals, occurred in 1995 with chicken salad, contaminated by a food handler, being the probable vehicle of transmission. Food and water may also serve to prevent oocysts from becoming desiccated.

Methods of Detection in Food

As would be expected from the life cycle of *Cryptosporidium*, there are no records of intrinsically contaminated food being implicated as a source of human infection. Food becomes a potential source of human infection by contamination during production, collection, transport and preparation (e.g. milk, fruit, vegetables etc.) or during processing. The sources of such contamination are usually faeces, faecally contaminated soil or water or infected food handlers. Current methods of detection use either examination of washings from the surface of foods or of foods in a liquid phase. Thus the methods are those, or modifications of those, which apply to water. Both conventional (filtration, centrifugation, clarification) and developing (IMS, PCR) techniques have been used to demonstrate the presence of oocysts in liquid foods.

Conventional Methods

Conventional methods including small volume filtration and centrifugation have been used to concentrate *C. parvum* oocysts seeded into a variety of liquid foods, including whole, semi-skimmed and skimmed milk, orange juice, carbonated beverages, wine and natural mineral waters ($\leqslant 1$ l). Oocysts were detected by epifluorescence microscopy following the application of a commercially available FITC-C-mAb according to the manufacturer's instructions. With the exception of natural mineral waters, recoveries were low (4–15%), and often the matrix required some form of pre-treatment to render the sample filterable, to remove particulates and/or to facilitate sampling of a larger volume. Pre-filtration, trypsin digestion, and surfactant (sodium dodecyl sulphate, SDS) treatment increased the volume which could be analysed. None of the pre-treatments affected the ability of the FITC-C-mAb to react with the oocysts.

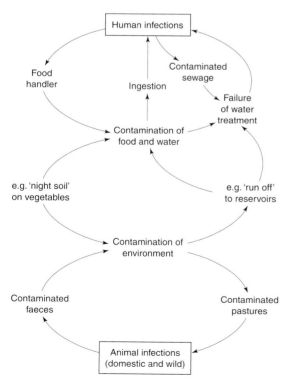

Figure 2 Transmission cycles of *Cryptosporidium parvum*.

Between 56 and 93% of oocysts can be recovered from 1 l volumes of four natural mineral waters (with total dissolved solids (TDS) of 91–430 mg l^{-1}), seeded with ca. 50 oocysts, by filtration through 13 mm polycarbonate membranes and 25% from 10 l samples (seeded with ca. 50 oocysts) filtered through 47 mm cellulose nitrate membranes. Oocysts were enumerated by epifluorescence microscopy using a commercially available FITC-C-mAb according to the manufacturer's instructions.

Cryptosporidium oocysts have been detected on a variety of vegetables purchased from 13 markets in a Peruvian periurban slum. Vegetables were washed twice with 50 ml distilled water, the washings centrifuged at 1200 g and the pellet examined for the presence of oocysts using a commercially available FITC-C-mAb. Of 110 samples, 16 (14.5%) were contaminated with oocysts and consisted of cabbage, lettuce, parsley, green onions and leeks. Recovery efficiencies, determined on lettuce leaves seeded with either 50 or 100 oocysts were 36.4% and 25.2%, respectively.

Developing Techniques

IMS and PCR techniques have both applied to detecting oocysts in milk and natural mineral waters.

Milk An 'in house' IMS, developed in the authors' laboratory, recovered 38.4 ± 14.7% of oocysts seeded into full-fat cows' milk. Oocysts seeded into raw milk concentrate (prepared by centrifuging 1 litre of pasteurized cows' milk and seeding 10^3 oocysts into 500 µl of the pellet) were labelled with FITC-C-mAb according to the manufacturer's instructions and excess FITC-C-mAb was removed by sequential washing. Magnetizable anti-FITC mAb-coated beads (10^6 beads) were added to FITC-C-mAb-labelled oocysts, and incubated in an end-over-end rotator for 30 min at room temperature. Bead-bound oocysts were concentrated by magnet on the inside of the specimen tube and the residue was removed by aspiration. The bead–oocyst complexes were washed three times with Hank's balanced salt solution concentrated to a minimum volume and the oocysts were enumerated by epifluorescence microscopy. At least two commercially available IMS kits are available currently and could be used to concentrate oocysts from milk and other turbid concentrates.

A PCR method which can detect one oocyst in 20 ml of raw milk has been published. Oocysts (100, 50, 10, 5 and 1) were seeded into 20 ml raw milk which was pre-treated with 1 ml Bacto-Trypsin and 5 ml Triton X-100 for 30 min at 50°C. Samples were centrifuged (500 g; 10 min), the supernatant discarded and the pellet resuspended in 100 µl lysis buffer

(120 mM NaCl, 10 mM EDTA, 25 mM Tris, pH 7.5, 1% sarcosyl) containing 5 mg proteinase K per ml, then exposed to 10 cycles of freezing and thawing (5 min liquid N$_2$, 5 min 65°C) to release sporozoite DNA. The freeze–thaw extract was further digested in proteinase K (5 mg ml^{-1}) at 55°C for 1 h, the DNA purified (Isogene kit, Perkin-Elmer) then diluted in water to 60 µl. Primers developed from a published sequence of CpR1 (coding for an oocyst wall protein) amplified a 358 bp sequence. After 40 cycles (denaturation, 94°C for 1 min; annealing, 45°C for 1 min; extension, 72°C for 1 min) and final elongation at 72°C for 2 min, amplicons were identified in ethidium bromide-stained agarose gels. PCR products were also hybridized with a 25 bp oligonucleotide probe homologous to an internal portion of the amplicon. PCR products were denatured (5 min boiling water), chilled on ice, diluted in an equal volume of 20 × SSC and cross-linked (UV light, 5 min) onto nylon, prehybridized (30 min, 50°C) in 30 ml prehybridization solution (5 × SSC, 2% (w/v) blocking reagent (1096 176, Boehringer), 0.02% (w/v) SDS, 0.1% (w/v) lauryl sarcosine) then hybridized in 2 ml hybridizing solution containing 50 pmol digoxygenin-labelled probe at 50°C for 4 h. AMPPD chemiluminescent signals were detected on an image quantifier. A previously published set of *Cryptosporidium* primers (Laxer et al) amplified genomic DNA from *Elmeira acervulina*, whereas the published probe hybridized with *Eimeria acervulina* amplicons and *Giardia intestinalis* DNA. Thus, the 'Laxer et al' primers should not be used in matrices where these organisms are likely to be contaminants.

Natural Mineral Waters The effectiveness of two commercially available IMS kits to concentrate C. *parvum* oocysts from natural mineral waters has been assessed in the authors' laboratory. Oocysts (ca. 50) were seeded into 1 l volumes of four natural mineral waters (TDS = 91–430 mg l^{-1}) and the samples were filtered through 13 mm polycarbonate filters. Oocysts were eluted from the filter by agitating the filter in 10 ml phosphate-buffered saline, pH 7.2, concentrated by IMS and enumerated by epifluorescence microscopy. Kit A, which utilizes an IgM isotype mAb to capture oocysts, had a percentage recovery range (±SD) of 56.3 ± 13.2–68.7 ± 25.0%, whereas kit B, which utilizes an IgG isotype mAb to capture oocysts, had a percentage recovery range (±SD) of 46.7 ± 8.4–72.5 ± 9.2%. Some divalent cations in TDS can affect the performance of IMS, and kits containing mAb paratopes with higher affinities for their epitopes can outperform lower affinity mAbs.

A PCR method for detecting small numbers (< 20) of oocysts in 1000× concentrates of natural mineral waters has been developed in the authors' laboratory. DNA is released from *Cryptosporidium* oocysts by multiple (10–15) cycles of freezing and thawing in lysis buffer, causing the release of up to 90% of nuclei from oocysts. A competitor molecule, which serves as an internal control and co-amplifies with *Cryptosporidium* primers has been developed. The primers co-amplify a sequence of 435 bp from *Cryptosporidium* genomic DNA and a sequence of 503 bp for the internal control. The amplicons are readily resolved by agarose gel electrophoresis and are identified as two distinct bands after staining with ethidium bromide. By titrating serial dilutions of competitor with a fixed amount of *Cryptosporidium* DNA and amplifying by PCR, a dilution of competitor which fluoresces at a similar intensity to the (test) *Cryptosporidium* DNA can be identified. The products can be measured by eye, on an agarose gel, following UV excitation of the ethidium bromide-stained amplicons. The dilution of competitor amplicon which fluoresces at a similar intensity as the test DNA contains an equimolar concentration of DNA. A dilution of 10^{-9} of the cloned competitor contains 0.13 attomoles of DNA, and can be used for the detection of *Cryptosporidium* DNA extracted from between 1 and 19 oocysts. The subsequent analysis of amplicons, by restriction fragment length polymorphism (RFLP), provides a means of distinguishing between *C. parvum*, *C. muris* and *C. baileyi* DNA. The advantages of this assay, when compared with other previously published assays include: (1) the high sensitivity of detection offered by a multi-copy gene such as the 18S rRNA (the *Cryptosporidium* genome contains between 100 and 200 copies of this gene); (2) identification to species level by RFLP, with the simultaneous detection of (at least) three species of *Cryptosporidium*; and (3) the ability to use the competitor molecule as a positive control when testing for inhibitory effects produced by naturally occurring components in water, environmental and food samples, thus increasing the level of confidence from negative results.

Methods of Detection in Water and Other Liquids

Only the general considerations will be discussed below. As *Cryptosporidium* oocysts occur in low numbers in the aquatic environment and no practicable in vitro culture enrichment techniques are available, large volumes are sampled. All methods contain separate sampling, elution, clarification and concentration and identification elements. 'Standardized' methods, which continually evolve, are available in the UK and the USA. Large volume (depth filter cartridges) and small volume (flat bed membranes, pleated membrane capsule, flocculation) sampling strategies can be used. Sampling can be performed for various reasons including monitoring, increased surveillance, outbreak situations, etc. and different strategies can apply for different rationales. The eluted retentate is concentrated and analysed for the presence of the organism. In addition to oocysts, size-based isolation methods concentrate large amounts of extraneous particulate material which interfere with organism detection and identification. Detergents and surfactants (0.01% Tween 20, 0.01% Tween 80, 1% SDS) are included to prevent oocysts and particulates from clumping. Flotation media such as sucrose (1.18 sp. gr.) or Percoll–sucrose (1.1 sp. gr.), on which oocysts float, are used to concentrate organisms and separate them from extraneous debris. Flow cytometry with cell sorting and IMS can also be used to concentrate cysts.

Identification is performed by epifluorescence microscopy and differential interference contrast microscopy, where possible, and putative objects are identified as oocysts using a defined series of criteria (**Table 4**) based on morphometry (the accurate measurement of size and shape) and morphology. Specificity and sensitivity are of paramount importance when attempting to detect small numbers of oocysts in concentrates. At present, all commercially available monoclonal antibodies are genus specific and no species specific monoclonal antibody is available. As the antibody paratopes bind surface-exposed oocyst epitopes, the fluorescence visualized defines the maximum dimensions of the organism, enabling morphometric analyses to be undertaken. The nuclear fluorochrome DAPI, which binds to sporozoite DNA, is an effective adjunct to FITC-mAbs for highlighting the four nuclei of the sporozoites in sporulated oocysts (Fig. 3).

Physical changes including distortion, contraction, collapse and rupture of the oocyst reduce the number of organisms that conform to accepted criteria (Table 4), resulting in the under-reporting of positives. Currently, the ability to determine species, viability and infectivity of oocysts, necessary for the interpretation of results obtained through microscopic examination, is lacking. We must assume that each intact oocyst

Table 4 Characteristic morphological features of *C. parvum* oocysts by Nomarski differential interference contrast (DIC) microscopy

- Spherical or slightly ovoid, smooth, thick walled, colourless and refractile
- 4.5–5.5 μm
- Four elongated, naked (i.e. not within a sporocyst(s)) sporozoites and a cytoplasmic residual body within the oocyst. Intact sporulated oocysts contain four nuclei

Figure 3 An oocyst of *Cryptosporidium parvum*. Bar = 5 microns (see also color **Plate 13**.)

detected in the sample is potentially infectious to humans. In order to overcome such difficulties and to present regulators with more definite information regarding the biological status of oocysts, the use of more discriminating techniques is necessary.

Nucleic Acid-based Methods

The detection of small numbers of *Cryptosporidium* oocysts, by amplifying, by PCR, specific regions of DNA has been reported, and the potential for enhanced sensitivity and specificity (see above) has stimulated interest in its application for the detection of environmental pathogens. More than 20 PCR assays have been published, using various targets and for various matrices (primarily faeces and water). In a comparison of four pairs of published PCR primers, the sensitivity of optimized reactions ranged between 1 and 10 oocysts in purified preparations and 5 to 50 oocysts in seeded environmental water samples, with the maximum sensitivity achieved with two successive rounds of amplification followed by hybridization of an oligonucleotide probe to the PCR amplicon and chemiluminescent detection of the hybridized probe. Some primers amplified genus specific amplicons whereas others amplified species specific amplicons. Of the primers tested, the 'Laxer et al', primers provided the best combination of sensitivity and specificity. Primers which amplify a sequence of the 18S rDNA gene have been used, with or without IMS, to amplify DNA from oocysts seeded into water, and possess a sensitivity similar to that described above. IMS helped reduce the inhibitors of PCR (humic and fulvic acids, organic compounds, salts and heavy metals, etc.) frequently found in water concentrates.

FISH has also been used to detect oocysts in water concentrates. The simultaneous use of FISH and FITC-C-mAb for oocyst detection and determination of viability has been described previously.

Importance to the Food and Water Industries

Both surface contamination of fresh produce as well as contaminated water used in food preparation are transmission routes that are significant to the food industry. Surface contamination can occur following contamination by the infected host or transport host (birds, flies), the use of manure or night soil, water for irrigation, fumigation and pesticide application, etc. Freshwater bivalves can concentrate viable oocysts from their environment and should be added to the list of transport hosts. Contaminated water is an important source of human infection either by direct consumption or by the use of contaminated water in food processing or preparation. Features of *C. parvum* oocysts which predispose to contamination of water sources are listed in **Table 5** and the importance of the waterborne route of infection is exemplified in **Table 6**. Contamination by infected food handlers is minimized by adherence to effective hygienic practices. It is not known whether there is seasonal variation in surface contamination of foods, but spring and/or autumn peaks in infection of humans and livestock might influence when foods become surface contaminated. Increased humidity can enhance cyst survival. Food and water can enhance

Table 5 Potential for waterborne transmission (adapted from Smith et al, 1995)

Characteristics of Cryptosporidium parvum

- Approximately 10^{10} oocysts are excreted during symptomatic infection. Infected calves excrete up to 10^7 oocysts per gram of faeces
- *C. parvum* infections have been reported from a variety of mammals including human beings, domestic livestock, pets and feral animals
- Significant densities of oocysts can be discharged in sewage effluent. At an average density of 10 oocysts per litre of sewage effluent, up to 65×10^6 oocysts per day can be discharged from a sewage treatment works servicing a population of 50 000 in the UK
- Oocyst survival is enhanced in moist cold environments (such as those found in temperate regions), e.g. a small proportion of viable oocysts can survive for 6 months suspended in water
- Waterborne outbreaks indicate that oocysts can survive physical water-treatment processes and disinfection. Oocysts are insensitive to disinfectants commonly used in water treatment.
- *C. parvum* oocysts are 4.5–5.5 µm and are not strained effectively by sand filters. Oocysts pass through the pores between sand grains (60–500 µm) in a sand bed
- 30 oocysts can cause infection in humans (ID_{50} = 132 oocysts); 10 oocysts can cause infection in juvenile non-human primates; five oocysts can cause infection in gnotobiotic lambs
- Viable oocysts excreted by transport hosts such as seagulls

Table 6 Some documented outbreaks of waterborne cryptosporidiosis (adapted from Smith et al, 1995)

Year	Country	No. of people affected	Oocysts detected in implicated water supply	Postulated reasons for outbreak occurring
1984	USA, (Bexar County, Texas)	79	None detected	Sewage contamination of well water
1987	USA (Carroll County, Georgia)	13 000	Oocysts detected in raw and treated water at densities of up to 2.2 per litre	Faults in operational procedures
1988	UK (Ayrshire, Scotland)	27	0.04–4.8 oocysts per litre of treated water	Post-treatment contamination: slurry contamination of water pipe line
1989	UK (Swindon/ Oxfordshire)	>515	Oocysts detected in treated water at densities of 0.002–77 per litre	Possible contamination of raw water by cattle slurry/muck
1990	UK (N. Humberside)	447	Oocysts detected in raw and treated water	Bypassing of slow sand filters
1990	UK (Isle of Thanet)	>47	None detected	Unknown
1991	USA (Berks County, Pennsylvania)	551	Oocysts detected in raw but not treated water	Chlorination as sole disinfectant for 'on site' well at picnic site
1992	USA (Jackson Co., Oregon)	15 000	Oocysts detected in raw spring water initially, but not in chlorinated water	Chlorination as sole disinfectant for spring water source
1992	UK (Warrington)	47	None detected	Surface contamination of borehole from agricultural land during heavy rainfall; deficiencies in monitoring of water supply
1992	UK (Bradford, Yorkshire)	125	Oocysts detected at 0.28 per litre in raw, 0.01–0.18 per litre in treated and 0.03 per litre in distribution waters	Heavy rainfall in catchment area of the raw water reservoir immediately prior to probable time of infection
1993	UK (Poole, Dorset)	40	Sporadic low levels of oocysts detected in service reservoir but not in settled silt	No likely mechanism for contamination of borehole water supply identified
1993	USA (Milwaukee, Wisconsin)	403 000	Oocysts detected in ice made at the time of the outbreak at densities of up to 0.13 per litre	Possible contamination from either sewage, agricultural waste or slaughterhouse effluent
1993	USA (Yakima County, Washington)	3	Oocysts detected in well water	Melting snow and spring rains containing faeces (cattle, sheep, elk) contaminated well water
1994	USA (Clark County, Nevada)	78	Not determined	No identifiable treatment deficiency
1994	USA (Walla Walla County, Washington)	86	Not determined	Seepage of treated wastewater into untreated/chlorinated well water
1994	Japan, (Kanagawa Prefecture)	461	Not stated	Post-treatment contamination of municipal drinking water due to cross-connection following malfunction of wastewater pump in private building
1995	USA (Alachua County, Florida)	72	Oocysts detected at the tap	Inadequate backflow prevention allowing wastewater from garbage can washer to enter the camp kitchen's potable water distribution system
1995	UK (Torbay area, Devon)	575	Oocysts detected in raw and treated water	Probable oocyst contamination of raw water from sewage effluent
1996	Japan (Saitama Prefecture)	8705	Oocysts detected in raw and treated water	Sewage contamination
1997	UK (West Hertfordshire and north London)	345	Oocysts detected in raw (max. density 0.2 per litre), backwash (max. density 0.2 per litre) and treated (max. density 0.3 per litre) waters	Raw water from deep chalk borehole under influence of surface water

cyst survival by preventing cysts from becoming desiccated. In addition, acquisition of foods from global markets, consumer vogues such as consumption of raw vegetables and undercooking to retain natural taste and preserve heat-labile nutrients can increase the risk of food-borne transmission.

Exposure to a temperature of 65°C for 2 min or 73°C for 1 min kills oocysts. A proportion of oocysts can survive at –20°C for 12 h, but storage at –70°C for 1 h kills oocysts. Air drying (room temperature) for 4 h kills oocysts. Oocysts are resistant to the concentrations of chlorine disinfection used in water treatment, partially susceptible to chlorine dioxide and more susceptible to ozone disinfection. Whereas conventional ultraviolet (UV) systems appear ineffective in killing oocysts, commercially available, higher energy systems have been shown to kill oocysts. Oocysts remain unaffected by storage in water (including chemical water treatment) at pH 3–10. Exposure to hydrogen peroxide (10 vol. for 10 min) and ammonia (5% for 18 h) kills oocysts.

Some preliminary data on oocyst survival in beverages and foods are available. About 8–20% of oocysts remained viable after 24 h storage in beer at 4°C, whereas 7–12% of oocysts remained viable after 24 h storage in beer at 22°C. An average of 14% of oocysts remained viable when stored at 4°C in carbonated beverages compared with 9% viable when stored in carbonated beverages at 22°C. After 24 h storage in orange juice, 65% of oocysts remained viable, whereas 89% remained viable in an infant formula.

Importance to the Consumer

Human infection is contracted following the ingestion of viable *Cryptosporidium parvum* oocysts (**Fig. 2**). Oocysts are infective when voided and transmission is either directly by the faecal (human or animal) oral route or indirectly via faecally contaminated water of food. Person-to-person transmission, is a major route and has been documented between family/household members, sexual partners, health workers and their patients, and children in day-care centres and other institutions. Transmission in day-care centres appears particularly common (more than 20 documented cases) probably due to the lower standards of personal hygiene exhibited by pre-school children.

There is an important zoonotic component in the epidemiology of the disease. Animals can be infected, experimentally, with oocysts of human origin (genotype C). Cryptosporidiosis has been reported in a variety of domesticated animals, livestock and wildlife, including companion animals which may be reservoirs of human infection. Zoonotic transmission has been documented in children on farm visits, and also from laboratory animals and household pets.

The most important route of environmental transmission is through the contamination of water by human (sewage) or animal faecal material (slurry, farm drains, run-off from fields, etc.) from infected individuals (Table 5). The importance of the waterborne route is amplified by the failure of conventional water treatment processes to kill or remove all oocysts and also by the fact that less than 100 oocysts can cause disease. The seasonal variation in incidence in some temperate countries, such as the UK, where spring and autumn 'peaks' occur have been attributed to animal husbandry practices (e.g. the release of cattle onto pastures from indoor winter quarters and the high incidence of calving in the spring), which result in contamination of farm drains, pasture and, thereby, waterborne contamination.

Although there are few documented food-borne outbreaks of cryptosporidiosis, the potential for food-borne transmission of *Cryptosporidium* should not be underestimated. Waterborne outbreaks can affect many thousands of individuals, the largest documented to date being reported from Milwaukee, USA in 1993 when more than 400 000 people were estimated to have been infected. *Cryptosporidium* has been implicated in at least 20 waterborne outbreaks, affecting over an estimated 427 100 individuals (Table 6).

Because of the mode of transmission of the parasite, cryptosporidiosis is endemic in children in countries where poor sanitation prevails. In such circumstances, *C. parvum* has a major impact in contributing to malnutrition and morbidity and mortality. Cryptosporidiosis has a worldwide distribution and is recognized as the most important (numerically) protozoan human intestinal pathogen. Prevalence figures from surveys of individuals with diarrhoea range from approximately 2.5% in North America and Europe to more than 10% in Africa.

In immunocompetent individuals the disease is self-limiting with symptoms ceasing within two weeks. The duration of the illness can be as short as three days. The incubation period is about seven days. Diarrhoea, present in more than 90% of infected individuals, can be accompanied by nausea, vomiting, headache, low-grade fever and abdominal pain. The diarrhoea with a frequency of about five to ten motions per day is watery, and lacking in faecal content. In young children or the elderly the diarrhoea and vomiting (which tends to be more prominent) can lead to dehydration. In the immunocompromised, cryptosporidiosis is a severe and life-threatening disease. The diarrhoea becomes cholera-like with often in excess of 20 motions per day. Severe nausea is a prominent and troublesome symptom. In immunodepressed individuals (e.g. AIDS patients, transplant patients, leukaemia patients) the infection can spread to the gall bladder and respiratory epithelium resulting in cholecystitis in the former and wheeze, cough and shortness of breath in the latter. Occasionally respiratory infection occurs in the absence of intestinal disease.

Treatment

As yet, there is no effective anti-*Cryptosporidium* agent, therefore treatment revolves around supportive therapy such as a rehydration and symptomatic treatment and, where possible, reversal of the underlying immunodepression.

Regulations

The third Report of the Group of Experts on 'Cryptosporidium in water supplies' recommended that human cryptosporidiosis be made a laboratory reportable disease in England and Wales (human cryptosporidiosis was made a laboratory reportable disease in Scotland in 1989).

The UK Food Safety Act (1990) requires that food, not only for resale but throughout the food chain must not have been rendered injurious to health, be unfit, or be so contaminated – whether by extraneous matter or otherwise – that it would be unreasonable for it to be eaten. A set of horizontal and vertical regulations (which are provisions in food law) covering both foods in general and specific foodstuffs also ensure that that food has not been rendered injurious to health. Other than being a potential microbiological contaminant, *Cryptosporidium* is not identified in these regulations.

The UK Government intends to propose regulation for *Cryptosporidium* in drinking water. The regulation will set a treatment standard at water treatment sites determined to be of significant risk following risk assessment. Daily continuous sampling of at least 1000 l over at least a 22 h period from each point at which water leaves the water treatment works is required and the goal is to achieve an average density of less than one oocyst per 10 l of water.

In the USA, the Food and Drug Administration has responsibility for enforcing regulations as detailed by the Federal Food, Drug and Cosmetic Act. As for *Cyclospora*, no regulations specifically address *Cryptosporidium*, but products contaminated with the organism are covered by sections of the Act depending on whether the foods are domestic or imported. The reader is referred to the section on *Cyclospora* for a detailed account of regulations in the USA.

Apart from regulations governing agricultural wastes, current regulations in the UK and USA do not require risk-based standards or guidance based on protection from microbial contaminants such as *Cryptosporidium*. Good management practices (GMPs) are also suggested for farms and agricultural wastes. The EC Agri-Environment Regulation, under the Common Agricultural Policy, promotes schemes which encourage farmers to undertake positive measures to conserve and enhance the rural environment in Europe.

See also: **Milk and Milk Products**: Microbiology of Liquid Milk. **Nucleic Acid-based Assays**: Overview. **PCR-based Commercial Tests for Pathogens**. **Waterborne Parasites**: Detection by Classic and Modern Techniques.

Further Reading

Anonymous (1998) Cryptosporidium in water supplies. Third Report of the Group of Experts; Chairman, Professor Ian Bouchier. Department of the Environment, Transport and the Regions, Department of Health. London: HMSO. 171 pp.

Fayer R (ed.) (1997) Cryptosporidium *and Cryptosporidiosis*. Boca Raton, Florida: CRC Press.

Smith HV, Robertson LJ and Ongerth JE (1995) Cryptosporidiosis and giardiasis, the impact of waterborne transmission. Journal of Water Supply Research and Technology – *Aqua* 44: 258–274.

Smith HV and Rose JB (1998) Waterborne Cryptosporidiosis, current status. *Parasitology Today* 14: 14–22.

Tzipori S (ed.) (1998) Opportunistic protozoa in humans. *Advances in Parasitology*, Part 1. Cryptosporidium parvum *and Related Genera*. P. 1. London: Academic Press.

Cultural Techniques see **Aeromonas**: Detection by Cultural and Modern Techniques; **Bacillus**: Detection by Classical Cultural Techniques; **Campylobacter**: Detection by Cultural and Modern Techniques; **Enrichment Serology**: An Enhanced Cultural Technique for Detection of Food-borne Pathogens; **Fungi**: Food-borne Fungi – Estimation by Classical Cultural Techniques; **Listeria**: Detection by Classical Cultural Techniques; **Salmonella**: Detection by Classical Cultural Techniques; **Shigella**: Introduction and Detection by Classical Cultural Techniques; **Staphylococcus**: Detection by Cultural and Modern Techniques; **Verotoxigenic E. coli and Shigella spp.**: Detection by Cultural Methods; **Vibrio**: Detection by Cultural and Modern Techniques.

CULTURE COLLECTIONS

F M Dugan, USDA-ARS Western Regional Plant Introduction Station, Washington State University, USA
J S Tang, American Type Culture Collection, Manassas, USA

Introduction

Since microbial cultures play a vital role in food microbiology, it is important that cultures are properly preserved and stored to ensure that viability is retained and useful properties not altered. Many food and beverage companies have their own culture collections. Such collections usually contain starter cultures, including genetically manipulated strains that yield better performance and confer special features, such as unique flavour or resistance to bacteriophage infection. However, the science of conservation of microbial germ plasm has become increasingly complex. Legal and regulatory aspects, public and workplace safety, *ex situ* biodiversity conservation, plus data management and bioinformatics, have all driven the need for organizations specifically dedicated to the conservation of microbial germ plasm.

The Mission of Culture Collections

Culture collections are heavily utilized as sources of biological materials and standards by academia, industry, agriculture, medicine and research. Such collections specialize in the long-term preservation and storage of microorganisms, cell lines and related biological materials, as well as the documentation associated with these materials. Depending on the collection, users may acquire a given microorganism, instructions for its cultivation, and ancillary documentation concerning its history, role in research and practical applications. Collections specialize to varying degrees regarding the taxa under their custody, the markets and applications for which the microorganisms are provided, and areas of research pursued by collection scientists. Common areas of emphasis include *ex situ* conservation (biodiversity), provision of organisms functioning as experimental models, reference strains for quality control, and tools for biotechnology. Collections may be private and profit-oriented, private and non-profit, or public (academic or governmental); cultures may be distributed for a fee or, more rarely, free of cost.

Types and Ex-types and Authentic, Representative and Reference Cultures

The stability and utility of species names are essential to biological science. For example, the name *Saccharomyces kluyveri* Phaff, M.W. Miller & Shifrine

refers to a specific microorganism, a yeast closely related to the common brewing and baking yeast. The yeast was given the species name *klyuveri* by its authors Phaff, Miller and Shifrine, who first published the name, which they placed in the genus *Saccharomyces*. When a species name is published, authors are required to designate a given strain as the *type* strain for that species. The type strain should exhibit characters in agreement with those published in the species description, and, ideally, should be representative of the majority of strains belonging to the species. Although it is common to speak of 'type cultures', strictly speaking only the original specimen is the type and the cultures descended from it are *ex-types*, meaning 'derived from the type'. Although the original specimen is often (nearly always for fungi) preserved as a non-viable herbarium specimen, slide or illustration, ex-type cultures can be used for experiments requiring a living organism. Living ex-type cultures of *Saccharomyces kluyveri* are preserved in various culture collections throughout the world and are available to researchers.

Types may be further designated with a prefix according to the nature of the type material. For example, a *neotype* is a strain accepted by international agreement to replace a type strain that is no longer in existence. The neotype should possess the characteristics as given in the original description; any deviation should be explained.

Authentic strains or cultures are identified by the author(s) of the names to which the cultures are assigned. *Representative* or *reference* strains or cultures are those used in published taxonomic or physiological studies and are putatively representative of the species of which they are a member. Rules governing the naming of bacteria and fungi (including yeasts, which are, in fact, fungi) and the designation of types are given in the International Code of Nomenclature of Bacteria and the International Code of Botanical Nomenclature. Major culture collections house and distribute thousands of such type and reference strains.

Types, ex-types, and authentic cultures are commonly used in research in systematics and taxonomy, and for quality control of commercial systems for identification of microorganisms. Representative or reference cultures are commonly used for materials testing or as positive controls in the clinical market.

Materials testing involves measuring the degree to which a product is resistant to microbial degradation; a positive control is used to measure the capacity of a product to grow or detect a microorganism. Reference cultures also provide a standard for testing antimicrobial substances, or serve as quality control for diagnostic kits and panels. Many organizations, such as the British Pharmacopoeia, the Deutsches Institut für Normung, the American Society for Testing and Materials, National Committee for Clinical Laboratory Standards, and many others, publish protocols mandating the use of specific reference strains. Given strains are also used as bioassays for contaminants or additives in foodstuffs and beverages, for the microbiological analysis of nutrients and vitamins, or to test disinfectants in food hygiene.

Preservation Methods

The standard methods of long-term preservation are storage in liquid nitrogen vapour, freeze-drying (lyophilization) and storage in a −80°C freezer. Ancillary methods are storage in distilled water, on silica gel, on agar slants under mineral oil, on dried filter paper or in dried soil or dried plant material, or in a −20°C freezer. Depending on the method, microorganisms may be protected by a cryoprotectant so that cells better survive the preservation process.

Storage in liquid nitrogen vapour (−130°C or lower) is the most satisfactory method for most microorganisms. Sterile, plastic, screw-capped vials are typically used to contain cells or spores suspended in cryprotectant. The cryoprotectant is usually 10% glycerol, or occasionally 5% dimethyl sulphoxide (DMSO), depending on the requirements of the individual organism. Materials are taken to −130°C in a systematic, programmed manner: from approximately 5°C to −40°C at 1°C min^{-1}, then at 10°C min^{-1} until −90°C, after which the materials are transferred to liquid nitrogen vapour (−130°C to −196°C, the latter being the temperature of liquid nitrogen). Stainless-steel liquid nitrogen containers are commercially available in a large range of sizes. The vials themselves may be stored within the containers in a variety of racks or canes. It is essential that labels are able to withstand long periods in liquid nitrogen and/or liquid nitrogen vapour.

Lyophilization utilizes the same principles of controlled decline in temperature and provision of a cryoprotectant, but adds the component of removal of liquid via application of a vacuum. The most common types of cryoprotectant are 20% double-strength skim milk or solutions of sucrose and bovine serum. Some environmental bacterial isolates and fastidious organisms prefer 5% dextran plus 5% trehalose as cryo-

protectant for better survival during the lyophilization process. There are many types of commercially available freeze-dryers. One commonly employed system involves placing materials at −70°C in a mechanical freezer for 1 h, then transfer to liquid nitrogen vapour (−130°C) for a minimum of 15 min. Material is transferred to the freeze-dryer, where it is kept at approximately −30°C to −40°C for several hours, then at approximately 0°C for several more hours. Vacuum is applied during the freezing process, and the vials sealed after completion. Glass vials may be hermetically sealed or sealed with a rubber plug and metal crimp cap, but only the former permanently excludes oxygen, an essential condition for truly long-term preservation. Lyophilized materials should be stored in dry conditions at 5°C.

Many smaller collections utilize low temperature mechanical freezers, which maintain organisms at −70°C to −80°C. Materials, provided with a cryoprotectant (usually 10% glycerol) are often placed directly into the freezer, but some materials may require programmed freezing. Materials may be stored in plastic screw-capped vials, multi-well plastic plates, or other, sturdy, shatter-resistant materials. Some organisms, such as many yeasts, can be stored on filter paper saturated with 10% glycerol, and a few organisms can be stored on dried filter paper at room temperature. Some microorganisms can be safely stored in a conventional (−10°C to −20°C) freezer, especially with proper prior treatment; however this temperature is highly detrimental to many organisms and the ability to survive even short storage in a conventional freezer should not be assumed.

Storage on silica gel is convenient for bacteria or fungi that possess highly resistant spores or cells. Sterile, granular silica gel is thoroughly dehydrated in a drying oven. A few drops of dense suspension of spores in 5% skim milk are placed onto the gel and rapidly cooled by placement into an ice bath (the reaction is exothermic). Vials are capped and may be stored in a conventional freezer. An analogous method substitutes sterile soil for silica gel, but extreme care must be taken to ensure soil sterility. These methods are best regarded as medium-term rather than long-term preservation. Storage of dilute spore or cell suspensions in sterile water at 5°C also provides good medium-term preservation for many microorganisms. Storage on agar slants under mineral oil is a well-established technique for medium-term preservation, although it is not as space efficient and is more labour intensive than some alternatives. Storage on agar slants without oil is not recommended, because shelf life is short and because the cultures are susceptible to invasion by mites.

Some useful generalites may be formulated with regard to cryopreservation. In general, freeze-drying or silica-gel (or sterile soil), and –70°C are suitable for bacteria, or for fungi producing conidia or ascospores. Fungi producing zoospores are not well preserved by these methods, and require liquid nitrogen vapour or storage on slants under mineral oil, as do many basidiomycetes. Bacterial spores are more resistant to the rigours of freeze-drying than vegetative cells. Gram-positive bacteria appear to be more resistant than Gram-negative bacteria. Most of the filamentous bacteria, like cyanobacteria, and long spiral rods of *Spirillum* and *Borrelia*, are preserved in liquid nitrogen. They do not survive freeze-drying. Similarly, fungal cultures consisting mostly of hyphae also preserve poorly in a lyophilized state. Larger cells are more difficult to preserve than smaller cells. Bacterial and yeast cells are usually harvested for preservation at the late exponential or beginning stationary phase of the growth curve. These cells seem to best withstand the rigours of freezing or freeze-drying. Anaerobic bacteria that are grown in an atmosphere of 80% nitrogen, 10% carbon dioxide and 10% hydrogen, should be processed as rapidly as possible and immediately frozen so that oxygen does not become a critical, damaging factor.

In lyophilization, a greater initial concentration of cells yields a higher concentration of cells surviving freeze-drying. A high concentration of cells also reduces the amount of unnecessary extracellular water that needs to be removed. For bacteria it is best to start with a population of 10^6–10^8 cells per millilitre; the same range is achievable with many of the more prolifically sporulating fungi and yeasts.

As a rough guide, one may inspect the catalogue of any major culture collection; if the organism is distributed as a freeze-dried culture one can attempt lyophilization, but if it is distributed as an agar slant then long-term storage in liquid nitrogen vapour is a better option.

The Recovery of Preserved Cultures

The factors that play a role in cell survival during preservation and long-term storage will also affect the recovery of preserved cultures to a metabolically active state. These include cell size, physiological age of cells, cell concentration and the type of cryoprotectant.

There are a number of chemical compounds that have been used to confer some degree of protection on cells during the preservation process. Good cryoprotectants reduce the size and quantity of ice crystals, dehydrate cells before freezing, lower the freezing point of cells, stabilize cell membranes, and buffer the effects of pH changes. Cryoprotectants for freeze-dried cultures should result in formation of a compact, solid pellet and promote easy re-hydration of the culture.

The method of recovery of a metabolically inactive culture back to a metabolically active state depends on the method under which it was stored. Cultures of fungi and yeasts stored in liquid nitrogen vapour or in a low temperature mechanical freezer should be revived by immersion of the vial in a water bath at the appropriate temperature (usually 35–55°C) until the culture has thawed, followed by immediate transfer to the appropriate culture medium. It is critical not to leave the vial at elevated temperature after thawing, especially at higher temperatures. Lyophilized cultures of fungi and yeasts should be re-hydrated with sterile, distilled water, left for 2 or more hours (overnight is not too long) at room temperature or 5°C, and transferred to the appropriate cultivation medium. Fungal cultures on dried filter paper may require re-hydration with sterile, distilled water prior to inoculation of cultivation medium. Crystals of silica gel containing fungi, yeasts or bacteria may be placed directly onto the cultivation medium. Fungi stored on agar slants or in distilled water can be transferred directly to fresh cultivation media.

In contrast to fungi and yeasts, lyophilized cultures of bacteria do not require re-hydration in water prior to transfer to cultivation medium. Freeze-dried bacterial cultures can be recovered by re-suspending the pellet directly in their maintenance or propagation media. It is not uncommon to require longer incubation time for the initial recovery of bacterial cells in the freeze-dried form; however, the lag phase of subsequent transfers is significantly diminished.

Sources of Deterioration During Storage

Stability is never completely assured between the time of cell preparation for preservation and the use of the final product. It is important to understand the factors that erode the stability of preserved cultures and make every attempt to prevent or minimize them.

Excessive water can be deleterious because it results in a higher sensitivity to environmental conditions, and hence in a shorter shelf life. To insure optimum conservation conditions, it is important to find adequate methods for the control and determination of the residual moisture of a product; 1–3% of residual moisture is generally acceptable. Another factor that affects the stability of the preserved culture is the presence of oxygen. Free radical production is detrimental to freeze-dried cultures, so every effort should be made to eliminate oxygen by processing

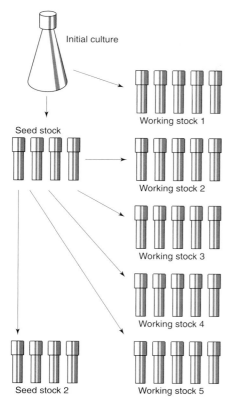

Figure 1 Origin, production and utilization of seed stock. Some collections produce an initial batch of working stock straight from the initial culture, as illustrated here. Others produce all batches of working stock from seed stock.

and preserving the products either in a vacuum or in an atmosphere of inert gas.

Decline in viability is not uncommon for freeze-dried cultures. Organisms which do not survive freeze-drying or which lose a significant percentage of viable cells each year should be preserved in the vapour phase of liquid nitrogen.

Minimizing Genotypic and Phenotypic Instability

There is the potential for genetic change during pre-servation, storage and recovery. Two possible mech-anisms account for this genetic change: new damage to DNA and selection for existing mutations. Essential to the maintenance of phenotypic and genotypic traits are optimization of the preservation method, the selection of cultivation media, and the strategy for conservation of seed stock (that stock closest to the original isolate). In general, liquid nitrogen vapour is the least deleterious to nucleic acids in the meta-bolically inactive cell and therefore recommended for preservation of genotypic properties. Preservation of phenotypic characters is contingent on cultivation media, and many microorganisms, especially patho-

gens, may require periodic passage through the natural host in order to maintain the virulence of newly produced batches. Sometimes cultivation on natural substrate killed and sterilized by autoclaving, radiation or other means is suitable for preservation of phenotypic characters. The maintenance and sta-bility of mutations in organisms used for genetic studies is especially contingent on correct media for-mulation, and optimally utilizes periodic application of replica plating or other methods to select from the population those cells still possessing the desired mutation(s). If economy does not allow consistent application of a selection system, a rich cultivation medium should be used so that cultivation does not select for revertants. Finally, conservation of seed stock will minimize the number of cultivation pas-sages between the original seed stock and distributed material (**Fig. 1**). Frequent subculturing increases the risk of contamination and mutation. Seed stock should be conserved and used only to replenish dis-tribution or working stocks. The seed stocks will be replenished when all seed vials are depleted. It is imperative to characterize new seed stocks and assure they have identical traits to the original culture. A properly designed seed stock system will result in material that, although separated by years, may only be one or two passages from the original culture. The size of a seed batch is usually a compromise between this principle and the space and equipment available for long-term storage.

Table 1 Major culture collections worldwide

Agricultural Research Service Culture Collection, Peoria, Illinois, USA
American Type Culture Collection, Manassa, Virginia, USA
Belgian Coordinated Collection of Microorganisms:
   IHEM Biomedial Fungi and Yeasts Collection, Brussels
   LMG Bacteria Collection, Ghent
   MUCL Agroindustrial Fungi and Yeasts Collection, Université Catholique de Louvain
Canadian Collection of Fungal Cultures, Ottawa, Canada
Centraalbureau voor Schimmelcultures, Baarn, The Netherlands
CSIRO Culture Collection, North Ryde, New South Wales, Australia
Deutsche Sammlung von Mikroorganismen und Zellkulturen, Braunschweig, Germany
Institute of Applied Microbiology, Tokyo, Japan
Institute for Fermentation, Osaka, Japan
International Mycological Institute, Surrey, UK
Institut Pasteur, Paris, France
Japan Collection of Microorganisms, Wako, Saitama, Japan
The National Collections of Industrial and Marine Bacteria Ltd, Aberdeen, Scotland, UK
The National Collection of Type Cultures, London, UK

Principal Collections Worldwide

There are numerous collections worldwide. Most are located at universities, museums, herbaria or governmental laboratories. Many collections reflect the activities of individual researchers. Names and addresses of collections, together with names and contact information for curators, can be located in directories of collections and herbaria. Smaller in number are collections dedicated to large-scale production and provision of services on a global scale. These latter are listed in **Table 1**. The most progressive collections produce cultures under standard operating procedures, either conforming to Good Laboratory Practices (GLP), Good Manufacturing Practices (GMP) or ISO 9000 practices, or adopting elements of one or more such practices, depending on conditions.

See also: **Good Manufacturing Practice**.

Further Reading

Ashwood-Smith MJ and Grant E (1976) Mutation introduction in bacteria by freeze-drying. *Cryobiology* 13: 206–213.

Greuter W, Barrie FR, Burdet HM et al (1994) International code of botanical nomenclature. *Regnum Vegetabile* 131: xviii + 389 pp.

Hall GS and Minter DW (1994) *International Mycological Directory*, 3rd edn. Wallingford, Oxon, UK: CAB International.

Hawksworth DL, Kirk PM, Sutton BC and Pegler DN (1995) *Ainsworth & Bisby's Dictionary of the Fungi*, 8th edn. Wallingford, Oxon, UK: CAB International.

Holmgren PK, Holmgren NH and Barnett LC (1990) Index herbariorum. Part I: The herbaria of the world, 8th edn. *Regnum Vegetabile* 120: x + 693 pp.

Hunter-Cevera JC and Belt A (1996) *Maintaining Cultures for Biotechnology and Industry*. New York: Academic Press.

Jong SC, Dugan FM, Birmingham JM and Cypress RH (1998) Internal quality control audits for microbiology laboratories in culture collections. *SIM News* 48(2): 66–69.

Lapage SP, Sneath PHA, Lessel EF, Skerman VBD, Seeliger HPR and Clark WA (1992) *International Code of Nomenclature of Bacteria*. Washington, DC: American Society for Microbiology.

MacKenzie AP (1976) Comparative studies on the freeze-drying survival of various bacteria: Gram type, suspending medium and freezing rate. *Developments in Biological Standardization* 36: 263–277.

Nei T and Osiro R (1979) Effects of freezing and drying on genetic characteristics of yeast cells. *Cryoletters* 1: 24–27.

Ohnishi T, Tanaka Y, Yog M, Takeda Y and Miwatani T (1977) Deoxyribonucleic acid strand breaks during freeze-drying and their repair in *Escherichia coli*. *Journal of Bacteriology* 130: 1393–1396.

Simione F and Brown EM (1991) *ATCC Preservation Methods: Freezing and Freeze-drying*, 2nd edn. Rockville, MD: ATCC.

Smith D and Onions AHS (1994) *The Preservation and Maintenance of Living Fungi*, 2nd edn. Wallingford, Oxon. CAB International.

Sugawara H, Ma J, Miyazaki S, Shimura J and Takishima Y (1993) *World Directory of Cultures of Microorganisms – Bacteria, Fungi and Yeasts*, 4th edn. Hirosawa, Japan: World Federation of Culture Collections.

Curing *see* **Meat and Poultry**: Curing of Meat.

CYCLOSPORA

Ann M Adams and **Karen C Jinneman**, Seafood Products Research Center, USDA, Bothell, USA
Ynes R Ortega, University of Arizona, Department of Veterinary Science and Microbiology, Tucson, USA

Cyclospora cayetanensis is a recently recognized pathogen of humans. Although the protozoan is probably closely associated with contaminated water, the majority of outbreaks in North America has been traced to the consumption of fresh fruits and vegetables. The characteristics of the organism pose distinct challenges to the study, detection and regulation of the pathogen. This article provides background on the biology, epidemiology and isolation of the protozoan. The current methods and the available variations are described for the detection and iden-

tification of oocysts in food products. The current regulatory status in the US of products contaminated with *Cyclospora* is considered. Lastly the impacts on the consumer and the food industry are discussed, including suggestions for the avoidance or alleviation of contamination.

Characteristics of the Genus and Relevant Species

The genus *Cyclospora* was erected by Schneider in

1881 when he described a new species of coccidian parasite from a myriapode, *Glomeris* sp. *Cyclospora* belongs in the family Eimeriidae, subphylum Apicomplexa. The family Eimeriidae is composed of about 16 genera which can be distinguished by the number of sporocysts and sporozoites within the oocysts. Of those genera, *Cyclospora* has been shown by phylogenetic analysis to be most closely related to the genus *Eimeria*, particularly to those species infecting chickens. Oocysts of *Cyclospora* have two sporocysts; oocysts of *Eimeria* have four sporocysts. Both genera have two sporozoites per sporocyst, resulting in a total four sporozoites in an oocyst of *Cyclospora* (**Fig. 1**) and eight within an oocyst of *Eimeria* sp. In addition, sequences for ribosomal RNA from *C. cayetanensis* and *Eimeria* spp. demonstrate considerable similarities. Some researchers propose that, regardless of the morphological differences, *Cyclospora* should be considered a member of the genus *Eimeria*. Further analyses are necessary to reconcile these results with the morphological differences.

Eleven species of *Cyclospora* have been described and have been reported from moles, rodents, insectivores, snakes and humans. Another coccidian belonging to this genus has been reported from nonhuman primates, but has not been fully described. Distinction between species of *Cyclospora* is based on the size and morphology of the oocysts (**Table 1**). The recognition of *Cyclospora* as a protozoan pathogenic to humans is relatively recent. In 1979, Ashford reported an *Isospora*-like coccidian infecting humans in Papua, New Guinea. Throughout the 1980s, investigators found similar structures in faecal samples from patients with diarrhoea and they soon determined that the organism was the causal agent. Because of the appearance and staining characteristics of the unsporulated oocysts, these infections were initially attributed to cyanobacterium-like bodies or coccidian-like bodies (CLBs). In 1993, these CLBs were characterized as oocysts belonging to a species of *Cyclospora*. The following year, the new species was designated as *C. cayetanensis* after the location of the author's principal studies, Cayetano Heredia University in Lima, Peru.

C. cayetanensis appears to be endemic in subtropical countries, although it has also been reported from temperate countries. Cyclosporiasis has been diagnosed in Nepal, Indonesia, Bangladesh, China, Viet Nam, Peru, Guatemala, Haiti, Honduras, Brazil, Mexico, England, Australia, the United States and Canada. Foreign tourists and expatriots from Europe and North America were found to be infected after returning from endemic countries. Recent infections in the USA and Canada were epidemiologically traced to imported produce. Although *C. cayetanensis* is not considered to be endemic in the USA, some cases cannot be traced to a foreign source. For example, a

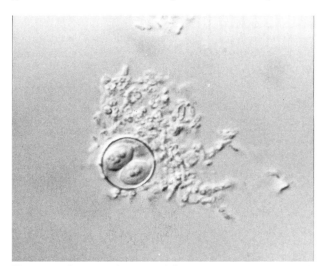

Figure 1 Sporulated oocyst of *Cyclospora cayetanensis*, with two sporocysts. Diameter of oocyst = 10 μm.

Table 1 Species of *Cyclospora* (adapted with permission, from Ortega et al, 1993)

Species	Host species	Common name	Oocyst size (μm)	Authorities
C. glomericola	Glomeris sp.	Millipede	25–36 × 9–10	Schneider, 1881
C. caryolytica	Talpa europaea	European mole	16–19 × 13–16	Schaudinn, 1902
C. viperae	Vipera aspis	European asp	16.8 × 12.6	Phisalix, 1923
C. babaulti	Vipera berus	European adder	16.8 × 10.5	Phisalix, 1924
C. tropidonoti	Tropidonotus natrix (= Natrix natrix)	Grass snake	16.8 × 10.5	Phisalix, 1924
C. talpae	Talpa europaea	European mole	12–19 × 6–13	(Pellérdy and Tanyi, 1968) Duszynski and Wattam, 1988
C. megacephali	Scalopus aquaticus	Eastern mole	14–21 × 12–18	Ford and Duszynski, 1988
C. ashtabulensis	Parascalops breweri	Hairy-tailed mole	14–23 × 11–19	Ford and Duszynski, 1989
C. parascalopi	Parascalops breweri	Hairy-tailed mole	13–20 × 11–20	Ford and Duszynski, 1989
C. angimurinensis	Chaetodipus hispidus	Hispid pocket mouse	19–24 × 16–22	Ford, Duszynski and McAllister, 1990
C. cayetanensis	Homo sapiens	Human	8–10	Ortega, Gillman and Sterling, 1994

cluster of cases in Chicago in 1990 was traced to a contaminated water tank but the original source of the organism was not determined.

Research is underway to determine possible reservoir hosts for *C. cayetanensis*. This work focuses on experimental infections and on surveys of mammals and birds, both domestic and wild, in endemic areas. The former includes pigs, rabbits, ducklings, chickens, rats, gerbils and mice (nude, neonatal, and adult) with no successful infections to date. Wild birds, rodents, cats, chickens, ducks and dogs have been surveyed. Oocysts resembling those of *C. cayetanensis* have been recovered from chickens in Mexico, a duck in Peru, and two dogs in Brazil. No evidence of intestinal involvement is available although experimental infections with some of these animals were unsuccessful. Research is continuing, but given the habits of these animals, oocysts could have been ingested from the environment and passed through the gastrointestinal system. Although further work may determine that these animals were not infected with *C. cayetanensis*, they might act as important vectors in the dissemination of oocysts.

C. cayetanensis is most likely host specific to humans, although further investigation is needed. In addition to the inability to confirm infections in other hosts, consideration of the high degree of host specificity demonstrated by other species of *Cyclospora* and *Eimeria* supports this hypothesis.

Life Cycle

The life cycle for cyclosporans has not been fully described (**Fig. 2**). In general, sporulated oocysts are ingested by the host. The oocysts excyst within the intestine and release the sporocysts, and subsequently, the sporozoites (**Fig. 3**). The sporozoites enter intestinal or hepatic cells, depending on species, and undergo merogony which is a form of asexual reproduction. Merozoites break out of the host cell and enter new cells. Numerous cycles of asexual reproduction may occur, depending on the species of coccidian and the immune status of the host. Eventually, gametogony transpires in which sexual reproduction occurs and oocysts are formed.

Few reports on the intracellular stages of the parasites have been reported. *C. cayetanensis* infects epithelial cells of the duodenum and jejunum of humans. Merogony and gametogony occur intracyto-

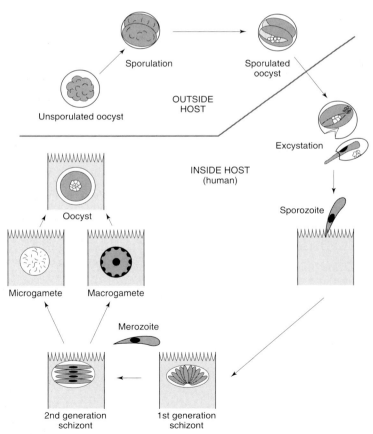

Figure 2 Proposed life cycle of *Cyclospora cayetanensis* in humans. Excystation occurs in the intestine, as do the intracellular stages.

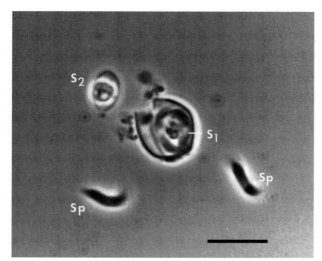

Figure 3 Excystation of *Cyclospora cayetanensis*, bar = 10 μm. One sporocyst (S₁) has two sporozoites inside the ruptured oocyst. The second sporocyst (S₂) is ruptured outside of the oocyst. The residual body remains inside the second sporocyst; the two sporozoites (Sp) are free.

plasmically in intestinal cells. *C. vipera* and *C. glomericola* infect host intestinal epithelium. The parasitic vacuoles of *C. caryolitica* and *C. talpae* are localized intranuclearly; the first invades the small intestine whereas merogony for *C. talpae* occurs in mononuclear cells in the capillary sinusoids of the liver and gametogony is localized in the epithelial cells of the bile ducts.

The oocysts are unsporulated and noninfectious when shed by the host in faeces (**Fig. 4**). Sporulation times for viable oocysts varies with the species. *C. caryolitica* sporulates at room temperature at 4–5 days, whereas *C. talpae* and *C. cayetanensis* require 2 weeks. This period required for the oocyst to become infectious suggests that the contamination of produce

usually occurs with fully sporulated oocysts. Because unsporulated oocysts require optimal time and conditions in the environment for sporulation, the shelf-life for fresh produce would be exceeded before unsporulated oocysts became infectious.

Isolation and Culturing of Oocysts

Faecal samples from suspected infections can be preserved in 10% formalin; polyvinyl alcohol; or sodium cacetate, acetic acid and formalin solution (SAF). However, the oocysts will no longer be viable, sporulation will not occur, and diagnosis will be restricted to staining and autofluorescence. Produce suspected of being contaminated with *Cyclospora* oocysts can also be fixed and preserved as described for faecal samples. If viable oocysts are desired, faecal samples containing *Cyclospora* oocysts can be maintained under refrigeration in 2.5% potassium dichromate or 1% sulphuric acid.

To isolate oocysts of *Cyclospora*, faecal samples are strained through sterile gauze or screen mesh to eliminate the debris. Following this initial separation, oocysts can be concentrated by the Ritchie procedure (chloroform: ethyl acetate), standard Sheather's sucrose flotation method, and discontinuous sucrose gradients. For final purification, a caesium chloride gradient is recommended.

Sporulation can be accomplished with oocysts stored in the potassium dichromate or sulphuric acid solutions and maintained at room temperature for 7–15 days. Sporulation can also occur in water; however, the growth of faecal bacteria or fungi will not be inhibited. A sterile sample of purified oocysts can be achieved by exposing the oocysts to a straight bleach solution for 15 min prior to washing and storing the oocysts in potassium dichromate. Excystation is

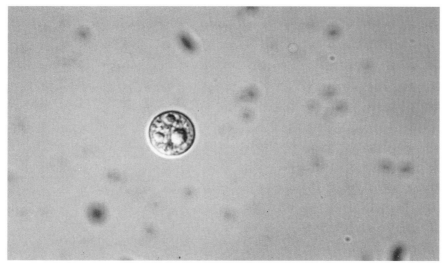

Figure 4 Unsporulated oocyst of *Cyclospora cayetanensis*. Diameter of oocyst = 10 μm.

achieved by using fully sporulated oocysts in a buffer containing sodium taurocholate and trypsin.

Methods of Detection in Foods

The analysis of food samples for the presence of *Cyclospora* is a relatively new procedure, posing a range of problems for the scientist. In clinical samples, numerous oocysts may be detected in a faecal smear. In contrast, the number of oocysts within a food sample is likely to be considerably lower, such that a slide may have few, if any, oocysts on it. The detection and/or enumeration of bacteria often includes an enrichment protocol. However, *Cyclospora* is an obligate intracellular parasite and no replication or reproduction occurs outside of the host. Therefore, no enrichment methods currently exist for this protozoan and the possibility of detection in a food sample is lower in comparison to analyses for bacterial contamination.

Sample size may also affect the possibility of detecting the parasite. *Cyclospora* is considered to have a relatively low infectious dose – approximately 100 oocysts and possibly as few as 10. With the present methods, it is unlikely that oocysts of *Cyclospora* would be detected at such low levels within a sample. In addition, protocols generally require results from an analysis to be confirmed by a separate method. *Cyclospora* has been recovered from produce samples in Peru, but no laboratories have reported positive results from commercial samples. Analytical procedures for the detection of *Cyclospora* are continually being tested and refined to improve sensitivity. The following protocols are those presently in use. Information concerning variations within these protocols is also given.

Wash Procedure

Two approaches have been developed to detect and identify oocysts of *Cyclospora* using microscopy or a polymerase chain reaction (PCR) test. The efficacies of both microscopy and PCR are dependent on the recovery of oocysts from produce, commonly through a wash procedure. This entails placing produce in a bag with the wash solution, agitating the sample for a period of time, and pouring the solution off. After centrifugation of the wash solution, the liquid is decanted and aliquots of the resulting sediment are then analysed by microscopy or PCR. The amount of sample for each wash can vary according to the amount of sample available and protocol being used, but 250 g of produce is commonly used. Some spiking protocols may require as little as 25 g of produce. The produce should be adequately covered by the wash solution; this generally requires an equal volume of liquid to sample weight.

Samples are washed by agitation for 20 min on an orbital (platform) shaker at 100 cycles per minute. The bag is inverted after 10 min. Care should be taken to minimize the fragmentation or destruction of the sample. For produce with greater structural integrity, such as lettuce, the number of cycles per minute can be increased. The use of a 'stomacher' or 'pulsifier' may also be considered, although the treatment period should be decreased to prevent destruction of the sample. Homogenizing the sample completely with a stomacher will actually result in the dilution of the sample and lower the chance of detection. Sonication has been suggested as a means of dislodging oocysts and other material from the surfaces of produce; however, care must be taken to keep the energy level low to prevent disruption of oocysts at this point in the protocol.

After completion of the agitation step, the liquid is decanted and centrifuged at $1500 \times g$ for 20 min. The supernatant is aspirated and discarded; the sediment is measured and can be stored at refrigeration temperature (4°C) until further analysis. If storage is for an extended time, the addition of a 2.5% solution of potassium dichromate will retard the growth of bacteria and yeast. Pellets can be fixed and preserved in 10% formalin, but PCR analysis of the material is then no longer possible.

The wash procedure has been modified by several laboratories to increase the recovery of oocysts. One of the most common approaches is the use of different solutions during the wash procedure. These different solutions are evaluated by the percentage recovery of oocysts from spiking environments. In these, a known number of oocysts is placed on produce and the wash procedure outlined above is then followed. Recoveries from spiking experiments for various parasites (e.g. *Ascaris*, *Cryptosporidium*, *Giardia*, etc.) generally range from approximately 1 to 40%. In spiking experiments using raspberries, approximately 30% of oocysts of *Cyclospora* are recovered.

Surfactants are generally used as a means to dislodge eggs and cysts of parasites from the surfaces of produce. However, in spiking experiments comparing distilled water and detergent solutions ($1 \times$ Tris-EDTA, pH 7.4 with 0.1% Laureth 12 (PPG Industries), or sodium dodecyl sulfate with 1% Tween), no appreciable differences were detected in the recoveries of oocysts of *Cyclospora* from raspberries. At the present time, distilled water is the solution of choice.

Another approach still under development, is the use of enzyme solutions to treat the surfaces of produce, particularly those surfaces covered by hairs or heavily textured. For example, raspberries are not only variously covered by fine hairs, but the product also has an uneven surface formed by individual drupelets (**Fig. 5**). The hairs are very resistant to breakage and contribute to the cohesion of the drupelets forming the raspberry. Oocysts may be caught under or attached to the hairs, or may be within the crevices between the drupelets. The purpose of the enzyme solutions is not to completely digest the produce to form a homogenous sample. Rather, the raspberries would be exposed to the enzymes long enough to detach or dissolve the hairs and outer surface while leaving the drupelets and berries essentially intact. Exposure times (agitation period) and concentrations of the enzymes can be manipulated to optimize the treatment of the surfaces. Enzymes presently being tested include a 1% pepsin solution, pH 1; a 1% pectinase (Sigma) solution; and a commercially available 'cocktail' of enzymes which includes pectinase and pepsin. The agitation period is increased for an exposure time of 30 min. In early experiments, approximately 20% of the oocysts of *Cyclospora* were recovered.

Lastly, the use of sucrose or Percol gradients to further separate the oocysts from other components of the wash sediment have been attempted. The goal is to increase the possibility of detection of oocysts by microscopy or PCR. Thus far, subjecting the wash sediments to procedures involving gradients have resulted in lower recoveries than if the sediments were tested by microscopy or PCR directly.

Microscopy

Oocysts of *Cyclospora* are acid-fast variable, ranging from a clear to a reddish colour after the use of stains such as the modified Ziehl–Neelsen or the Kinyoun acid-fast. Preparation of permanent slides for analysis is attractive because deterioration of the sample

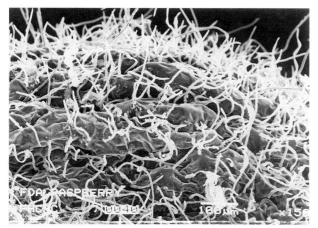

Figure 5 Scanning electron micrograph of surface of raspberry, illustrating the uneven surface and numerous hairs.

would no longer be a factor, the slides could easily be sent to other laboratories, and the material would no longer be infectious after the fixation step. However, through the experiences of several laboratories, permanent slides such as those made with acid-fast stains were found to be generally unacceptable for food substrate samples. Other components of the sediment from the produce, such as pollens and yeasts, may also take up stain and interfere with the detection of oocysts. In addition, the fixation and staining process can cause the oocysts to shrink and/or collapse. Internal structures of the oocysts are no longer visible and the characteristic shape and size of the oocysts are altered. In clinical samples, the scientist or technician can observe numerous oocysts prior to making a diagnosis. In contrast, the slide made from the wash sediment of a food sample may contain few, if any, oocysts. Determination of oocysts on such a slide is very difficult and false results are commonplace.

For detection of *Cyclospora*, fluorescent microscopy of wet mounts using ultraviolet (UV) epifluorescence is more sensitive than scanning permanently stained preparations. The microscope should be equipped with a mercury lamp; a tungsten bulb will not provide the appropriate wavelengths and the oocysts will be difficult to observe. The excitation filter should be a 365/10, although a 330–380 nm filter will also be adequate; the dichroic mirror should be 400 nm; and the barrier filter should be 420 nm.

Unlike a clinical sample, microscopical analysis of a food sample requires that the entire slide be scanned. To prevent the wet mount from drying out during analysis, the cover slip should be ringed with silicone grease. Generally, a 10 μl of sediment is analysed per slide. The wet mount is examined at 400×. Oocysts of *C. cayetanensis* are characteristically spherical, 8–10 μm in diameter, and autofluoresce cobalt blue. No fluorescent stains are necessary. The interior of the cyst does not fluoresce, or fluoresces very little. If a suspected oocyst of *Cyclospora* is detected, confirmation should be made with bright field illumination at 1000× (tungsten illumination is used at this time). Internal structures are more clearly elucidated with differential interference contrast. The oocyst may or may not be sporulated; the analyst should consider the morphology of the oocyst accordingly.

Microscopical examination of wash sediment from produce is strikingly different from that of clinical samples. Sediment from produce lacks the homogeneity encountered in other samples. In addition to soil, the wash sediment has many components, including pollens, yeasts, fungi, moulds and other organisms. The pollens may vary in size and shape, but they generally fluoresce a much brighter blue than

Cyclospora. Yeasts may be almost perfectly spherical and vary considerably in size. Yeasts in the size range for oocysts of *Cyclospora* are not uncommon, but they do not fluoresce in a similar fashion. Other organisms, such as free-living nematodes, mites from pollinating bees, other insects, eggs (often nematode eggs), cysts, and other oocysts, may also be observed. The cysts and oocysts may fluoresce blue or red. Coccidian parasites of other animals may be naturally present at the agricultural site and in the resulting wash sediment of the produce. Oocysts of species of *Eimeria*, a coccidian parasite closely related to *Cyclospora*, have been isolated from raspberries. Microscopically, the oocysts of the two genera can be distinguished by size and shape, and when sporulated, by the number of sporocysts (two for *Cyclospora* and four for *Eimeria*). Members of these genera have two sporozoites per sporocyst, resulting in four sporozoites in an oocyst of *Cyclospora* and eight within an oocyst of *Eimeria*. Species of *Eimeria* are generally oval in shape, although some may be imperfectly round after sporulation (e.g. *E. mitis*), and measure 11–35 µm in greater diameter. Oocysts of *Eimeria* also autofluoresce a blue, but the oocysts are not as distinctive as those of *Cyclospora* and may be missed while scanning with UV.

Polymerase Chain Reaction

In addition to microscopy, a molecular approach is needed for screening of samples or confirmation of microscopical results. There is a PCR test which detects *Cyclospora* by amplifying a region of the *Cyclospora* 18S ribosomal RNA gene. The procedure does not produce an amplified fragment from other closely related coccidian species such as *Cryptosporidium parvum*, *Toxoplasma gondii* or *Isospora felis*. However, the PCR method does yield an amplicon of the same molecular size as *Cyclospora* and *Eimeria*, rendering the two genera indistinguishable when using these primers for the PCR. The *Eimeria* genus is composed of a number of species, each of which infects particular mammalian and avian hosts. Therefore, the application of this PCR method to the analysis of food and environmental samples could produce an amplified product derived from either *Cyclospora* or *Eimeria* spp. The ability to distinguish between the human pathogen *C. cayetanensis* and possible contamination by members of the genus *Eimeria*, which are not infectious to humans, will be essential for any analysis of food products. The nucleotide sequences of *Cyclospora* and *Eimeria* spp. are 94–96% similar, but specific nucleotide differences in the amplified segment do exist. A restriction fragment length polymorphism (RFLP) analysis can differentiate between the PCR amplicons produced by *Cyclospora* and *Eimeria* spp. In addition, PCR is hindered by the low amounts of the target oocysts and the presence of inhibitory substances in the sample matrix. The following protocols include information on template preparation, the PCR test and the RFLP analysis for positive PCR results.

DNA Template Preparation for PCR   Template preparation for PCR of food samples entails concentrating oocysts from the wash sediment, disrupting the oocysts to expose the DNA, and overcoming the effects of PCR inhibitors which may be in the sample. An aliquot of the sediment, approximately 100 µl, is commonly washed and resuspended in a buffer solution, such as 1 × PCR buffer I (Perkin-Elmer, Branchburg, NJ) or Tris EDTA, to assist the concentration of oocysts and to remove soluble components that may cause inhibition. The use of Sheather's sugar flotation technique to concentrate oocysts in the wash sediment before template preparation is of limited benefit with produce samples and is generally not used. Concentration of oocysts of another coccidian parasite, *Cryptosporidium parvum*, has been accomplished through the use of magnetic antibody techniques. Although this is an attractive method, antibodies to *C. cayetanensis* are not available at this time.

The DNA is released from the oocysts by mechanically breaking the oocysts open. A common method, adapted from PCR analysis for *Cryptosporidium* oocysts, involves six cycles of freeze/thaw procedure in which the aliquot of sediment is subjected to liquid nitrogen or a dry ice/ethanol bath for 2 min followed by a 2 min exposure in a water bath at 98°C. The mixture is vortexed and then centrifuged at 14 000 r.p.m. for 3 min. The supernatant is retained for PCR analysis and may be stored at –20°C until needed. Mechanical disruption may also be accomplished with the addition of siliconized glass beads and vigorous vortexing. To break open the oocysts, some protocols consist of three freeze/thaw cycles followed by the addition of glass beads and vortexing. This procedure appears to be as efficient as the six cycles of freeze/thaw. However, the yield of supernatant is generally less than with the latter because the sediment (including the beads) does not pellet from centrifugation as well as in the latter. Sonication (2 min at 120 W) may be used to disrupt the oocysts, but there are concerns over the ability to control the process so as to avoid fragmenting the DNA.

The amount and type of PCR inhibitors vary from sample to sample. A number of strategies have been tried to reduce the inhibitory effects. Dilution of the template is effective, but the concentration of target oocysts decreases and the sensitivity of the PCR is reduced. For example, for raspberry samples, a dilution of 1:1000 of the template may be required to overcome inhibition of the PCR. The addition of a 6% Chelex resin matrix (Instagene, BioRad, Hercules, CA) to the template preparation before the disruption of the oocysts is able to reduce some of the inhibitory effects of the sample. An effective means of minimizing inhibition from the wash sediments is through the addition of 2 μl of non-fat dried milk (50 mg ml^{-1}) to a maximum of 20 μl of the supernatant before the amplification reaction. Non-fat dried milk solutions overcome the inhibitory effects of PCR by plant and soil material template extracts, although the mechanism by which the inhibitory effects are reduced is unknown. For raspberry samples, the addition of the non-fat dried milk solution results in a 400-fold increase in the amount of template which can be analysed per reaction.

The inclusion of controls is important to determine if compensation has been made for any inhibition present. The use of intact oocysts for spiking control sediments provides a useful measure of the efficacy of the oocyst-disruption protocol, as well as an indicator of the degree of inhibition associated with the sample. Oocysts of *Eimeria* spp. are generally more available and can be used for these controls. Purified DNA from oocysts of *Cyclospora* or *Eimeria* is also used in controls for the PCR and RFLP. DNA from as few as ten oocysts of *Eimeria* spp. and as few as 19 oocysts of *Cyclospora* sp. per reaction can be detected by the PCR analysis. The effect of inhibitors present in the food sample may reduce this level of PCR sensitivity.

PCR Amplification Amplification of the region of the 18S rRNA gene of *Cyclospora* or *Eimeria* is accomplished with a nested PCR. The original method used primers with restriction site 'leaders'; allowing the amplified region to be sequenced and resulting in a predicted PCR product of 308 base pairs. When sequencing is unnecessary, the use of primers synthesized without these 'leaders' appears to increase the efficiency of the amplification, and the necessary number of reaction cycles for analysis of food samples is reduced from 45 to 35. The predicted product from the use of these modified primers is 294 base pairs. The first round of PCR requires: 50 mM KCl; 10 mM Tris-HCl, pH 8.3; 0.001% gelatin; 1.5 mM MgCl$_2$; 200 μm each dATP, dCTP, dGTP and dTTP; 1.5 units AmpliTaq® polymerase; up to 20 μl of prepared template; and 0.2 μm each of the forward and reverse primers. The sequence of the modified forward primer (mCYCF1E) is 5′-TACCCAATGAAAACAGTTT-3′ and that of the reverse primer (mCYCR2B) is 5′-

CAGGAGAAGCCAAGGTAGG-3′. Conditions for the reaction cycling consist of an initial inactivation/denaturation step of 5 min at 95°C followed by 35 cycles of denaturation at 94°C for 30 s, annealing at 53°C for 30 s, and extension at 72°C for 90 s. A final extension of one cycle at 72°C for 10 min is completed before the reactions are held at 4°C until used in the second round of PCR. The amplification components and cycling conditions are the same in the second round except that the template consists of 5 μl of the product from the first round, the annealing temperature is increased to 60°C, and different primers are used. The forward primer (mCYCF3E) for the second round has the sequence 5′-CCTTCCGCGCTTCGCTGCGT-3′ and the reverse primer (mCYCR4B) has 5′-CGTCTTCAAACCCCCTACTG-3′. After the nested PCR is completed, the amplification products are electrophoresed in an agarose gel, stained with ethidium bromide and visualized on a UV transilluminator (**Fig. 6**).

RFLP Analysis As shown in Figure 6, the amplification products of *Cyclospora* and *Eimeria* cannot be distinguished by PCR. For positive results in the PCR, differentiation between *Cyclospora* and non-human pathogens of the genus *Eimeria* by a method such as RFLP is necessary. The restriction endonuclease *Mnl*I (Amersham Life Sciences Inc., Arlington Heights, IL) is used with this PCR amplicon because it provides one unique cut site for *Cyclospora* sp., a different unique cut site for most species of *Eimeria* and one cut site in common for both *Cyclospora* and *Eimeria* spp. sequences (**Fig. 7**).

The restriction digest is prepared using 10 μl of PCR-amplified DNA from the second PCR amplification, one unit of *Mnl* I, 5 μl 10 × buffer M supplied with the restriction enzyme, and sterile deionized water to a final volume of 50 μl. The digest is incubated for 1 h at 37°C. A 10 μl sample of the digest is electrophoresed in an agarose or polyacrylamide gel to provide good separation of fragments in the 200–50 base pair size range. Molecular-size standards should be included in multiple lanes for interpolation of fragment sizes. The gel is electrophoresed until the dye front is approximately 1 cm from the end of the gel and stained with ethidium bromide. The resulting bands are visible with a UV transilluminator and can be photographed or subjected to image analysis. Restriction digest of PCR amplified products from *Cyclospora* sp. and *Eimeria* spp. reveals distinct RFLP banding patterns discernible by visual examination of the gel photograph (**Fig. 8**). The molecular base pair sizes of the restriction fragments generated by restriction endonuclease digestion of the PCR products

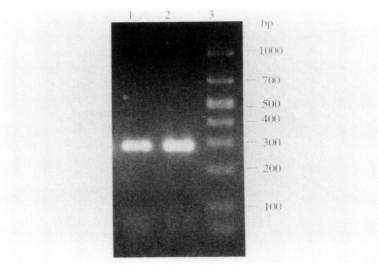

Figure 6 A 294 bp amplification product is produced from the 18S rRNA gene of *Cyclospora* and *Eimeria* spp. Lane 1, *Cyclospora* sp; lane 2, *E. tenella*; lane 3, molecular size standards.

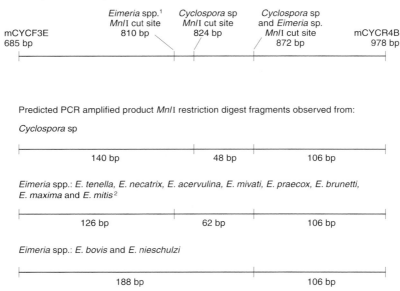

Figure 7 Molecular sizes predicted for restriction fragments from *Cyclospora* and *Eimeria* spp. PCR amplified products digested with *Mnl*I. [1] Based on GenBank® DNA sequence information this *Mnl*I cut site would exist for *E. tenella*, *E. necatrix*, *E. acervulina*, *E. mivati*, *E. praecox*, *E. brunetti*, *E. maxima* and *E. mitis*, but not for *E. bovis* or *E. nieschulzi*. [2] *E. mitis* would have a predicted fragment size of 125 bp based on GenBank® sequences U67118 and U40262.

should be calculated and compared to the sizes predicted from GenBank® sequence data. For the sample to be designated as containing oocysts of *Cyclospora* or *Eimeria* spp., the calculated band sizes should be within ±5% of those predicted.

Regulations

Although *Cyclospora cayetanensis* has only been recently recognized as a human pathogen, the organism is covered by several rules and regulations within the US. With the numerous outbreaks in 1996 and 1997 in the eastern US (and Canada), the Centers for

Disease Control and Prevention established cyclosporiasis as a reportable disease. *C. cayetanensis* is also included as an emerging pathogen in the Food Safety Initiative. This programme focuses on the monitoring of outbreaks, on research of the selected pathogens, and on the enforcement of regulations.

In the USA, the Food and Drug Administration is responsible for the enforcement of regulations as detailed by the Federal Food, Drug and Cosmetic Act. No regulations specifically address *C. cayetanensis*, but products contaminated with the organism are covered by sections of the Act depending on whether the foods are domestic (either produced within the

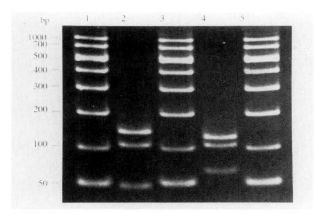

Figure 8 Restriction fragments from *Mnl*I digestion of PCR amplified products from *Cyclospora* and *Eimeria* spp. Lane 2, *Cyclospora*; lane 4, *E. tenella*; lanes 1, 3, and 5, molecular size standards.

US, or already imported and in the domestic market) or imported (at the port of entry). If domestic, produce contaminated with oocysts of *C. cayetanensis* would be in violation of section 402(a)(1): 'bears or contains any poisonous or deleterious substance which may render it injurious to health'. Imports found to be contaminated with the protozoan can be detained (prevented from entering the country) under section 801(a)(1): 'has been manufactured, processed, or packed under insanitary conditions'. Analysis of regulatory samples by FDA follows the procedures contained within the *Bacteriological Analytical Manual*, 8th edition, revision A (1998).

As part of the Food Safety Initiative, efforts are being undertaken to assure the safety of produce consumed within the US. As a result, a draft guidance on good agricultural practices and good manufacturing practices for fruits and vegetables has been issued. The guide does not contain additional regulations, rather, it consists of recommendations to growers, packers, transporters, and distributors of produce to minimize the risks of food-borne diseases. The purpose of the guide is the prevention of microbial contamination, including that of *Cyclospora*, by applying basic principles to the use of water and organic fertilizers, employee hygiene, field and facility sanitation, and transportation. Advice is given on the establishment of a system for accountability to monitor personnel and procedures from the producer to the distributor.

Importance to the Food Industry

The presence of *Cyclospora* and other food-borne pathogens can have serious impacts on businesses within the food industry. Because the majority of cases and outbreaks have implicated fresh produce (raspberries, lettuce and basil), those businesses

involved in the production and shipment of such produce should consider possible routes of contamination. Sources of water used for irrigation, fumigation and pesticide application should be inspected. If necessary, treatment of water by filtration, heating, or by ozone exposure should be pursued. Chlorination, although effective against many bacteria, is not an appropriate treatment for *Cyclospora* because the oocysts are highly resistant to the chemicals. Similarly, the use and application of fertilizers should be monitored. Raw manure or night soil should be adequately processed or composted to eliminate possible contamination of crops. Contamination by infected personnel can be avoided by proper hygiene and timely treatment of symptoms. Exposure of the produce to animals, both domestic and wild, should be avoided as much as is reasonable. Although no reservoir host for *C. cayetanensis* has been found, evidence indicates that domestic animals can distribute oocysts with their faeces.

The choice of produce grown in endemic areas should also be carefully investigated. Although all fresh produce grown in endemic regions can theoretically be contaminated with oocysts of *Cyclospora*, some products by virtue of their surface structures or growth requirements appear to have a greater probability of transmitting the organism. For example, although raspberries and blackberries are grown in similar areas, raspberries have primarily been implicated in outbreaks.

Infections by *C. cayetanensis* show a marked seasonality, with most cases in late spring or early summer and corresponding to the rainy season. Similar crops of produce shipped from endemic regions in the drier season (autumn) have not been implicated with any outbreaks of cyclosporiasis. Agricultural companies may consider timing crops for shipment during autumn when possible. Importers and distributors may consider acquiring some produce from sources in non-endemic regions during the rainy season. Although fresh produce often brings a better price for growers and importers, the use of spring crops for frozen or cooked products may be a viable option to alleviate the transmission of *C. cayetanensis*. No cooked or frozen product has been implicated in any cases of cyclosporiasis.

Importance to the Consumer

Cyclosporiasis is characterized by mild to severe nausea, anorexia, abdominal cramping, mild fever and watery diarrhoea. Diarrhoea alternating with constipation has been commonly reported. Some patients present with flatulent dyspepsia and less frequently joint pain and night sweats. Onset of illness is usually abrupt in patients 7 to 14 days after ingestion of oocysts and symptoms may persist an average of 7 weeks.

Duodenal and jejunal biopsies of patients with cyclosporiasis have shown varying degrees of jejunal villous blunting, atrophy and crypt hyperplasia. The widening is due to a diffuse oedema and infiltration of the villous mucosa by a mixed inflammatory infiltrate. Numerous plasma cells, lymphocytes and eosinophils are frequently observed. Extensive lymphocytic infiltration into the surface epithelium is present, particularly at the tip of the shortened villi. Reactive hyperaemia with dilation and congestion of the villar capillaries are also observed. In Nepalese patients, but not in Peruvians, the surface epithelium shows focal vacuolation, loss of brush border and an alteration of epithelial cells from a columnar to cuboidal shape. The reactive response of the host is not associated with the number of intracellular parasites present in the tissues.

To date the only successful antimicrobial treatment for *Cyclospora* is trimethoprim–sulfamethoxazole (TMP–SMX). This therapy has been tested in children, immunocompetent and immunocompromised adults. In the initial report from Peru, TMP–SMX brought about the cessation of symptoms and oocyst-excretion in five patients within 4 days after initiation of treatment. In Nepal, expatriates infected with *Cyclospora* had cessation of symptoms and excretion of oocysts after 7 days of treatment. Antimicrobials that have failed in the treatment of *Cyclospora* include metronidazole, tetracycline, pyrimethamine, and quinolone derivatives such as ciprofloxin and norfloxin.

Patients with acquired immune deficiency syndrome (AIDS) appear to have a higher density of parasites than immunocompetent individuals infected with *Cyclospora*. However, the prevalence of *Cyclospora* in patients positive for human immunodeficiency virus is not higher than in immunocompetent populations; probably due to the frequent use of TMP-SMX for *Pneumocystis carinii* prophylaxis. This is further supported by the high prevalence of *C. cayetanensis* in adult AIDS patients in Haiti where TMP-SMX prophylaxis is infrequent.

Routes of transmission for *Cyclospora* are still undocumented, although the faecal–oral route, either directly or via water, is probably the major one. In the US, epidemiological evidence suggested that water was responsible for sporadic cases of cyclosporiasis. After heavy rains in Utah, a man became infected after cleaning his basement flooded with runoff from a nearby farm. In another incident, an 8-year-old Chicago boy became ill and passed oocysts in his faeces one week after swimming in Lake Michigan.

An outbreak involving 20 individuals, most of whom were physician residents, occurred in a Chicago hospital in 1990. Despite the implications of water in transmission, organisms confirmed as *Cyclospora* have rarely been identified from water samples.

The prolonged sporulation time, 1–2 weeks, would further support the hypothesis that *Cyclospora* can be acquired by consumption of contaminated water or produce that has been in contact with contaminated water. Oocysts are excreted unsporulated and are noninfectious at that time. The rate at which sporulation occurs depends on a variety of environmental factors, including temperature and humidity. Because sporulated oocysts are needed for infection, person-to-person transmission is unlikely. The minimum number of oocysts required to infect an individual is unknown, as is how long the organism can survive under different environmental conditions.

Food-borne outbreaks are more common than those traced to contaminated water. In 1996, *Cyclospora* outbreaks occurred in the US and Canada and affected more than 1400 individuals. Many of the outbreaks were clustered, but isolated cases were also observed. Initially, these outbreaks were thought to be associated with the consumption of strawberries, but later epidemiological data implicated imported raspberries. In 1997, outbreaks in the US were associated with imported raspberries, and later that year, to contaminated basil and lettuce.

In Nepal, the prevalence of *Cyclospora* was highest in adult expatriates, whereas in endemic areas in Peru children under 10 years old were most susceptible to infection, but most of them were asymptomatic. Adults from endemic areas did not present the infection, but adults from medium to high socioeconomic status as well as travellers would be symptomatic. In Peru, few vegetables from markets of endemic areas were contaminated with oocysts of *C. cayetanensis*.

Seasonality of infection is extremely strong. In over 6 years of charting cyclosporiasis in Peru, nearly all infections occur from December to July. It is extremely rare to document infection at any other time of the year. In the US, the majority of outbreaks occur from May to July. In Nepal, infections and illness occur most frequently from May to August. The reasons for this marked seasonality have not been defined, although the relationship with the rainy season has been observed.

Consumers can take some measures towards avoiding infection by *C. cayetanensis*. Produce that is properly cooked or frozen has not been implicated in any cases of cyclosporiasis. Few cases in North America or Europe have indicated a domestic source of contamination, so produce from these areas is unlikely to transmit the protozoan. Although still

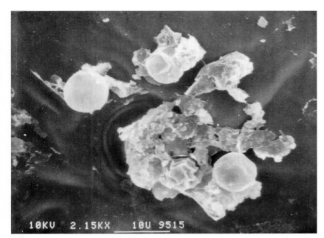

Figure 9 Scanning electron micrograph of oocysts of *Cyclospora cayetanensis* remaining on surface of lettuce after washing. Reproduced with permission, from Ortega et al (1997).

under study, irradiation of produce may provide some protection against *Cyclospora*. Consumers should always wash fresh vegetables and fruit, but this may not be effective in the prevention of cyclosporiasis. Numerous people affected by cyclosporiasis in 1996 stated that they had washed raspberries prior to consumption. Not only does *Cyclospora* probably have a low infectious dose, but washing of vegetables experimentally contaminated with *C. cayetanensis* oocysts does not remove all the oocysts (**Fig. 9**). Lastly, when travelling in endemic regions, consumers should take care to consume only fully cooked foods, or properly washed and peeled vegetables and fruit. The purity and source of all liquids should be considered.

See also: **Cryptosporidium. Direct Epifluorescent Filter Techniques (DEFT). Good Manufacturing Practice. International Control of Microbiology. Microscopy**: Light Microscopy. **National Legislation, Guidelines & Standards Governing Microbiology**: Canada. **PCR-based Commercial Tests for Pathogens**.

Further Reading

Bier JW, LeClerc E, Jinneman KG et al (1998) Concentration and Preparation of *Cyclospora* from Berries for the Polymerase Chain Reaction (PCR) and Microscopy. In: Jackson G (ed.) *FDA Bacteriological Analytical Manual*, 8th edn, revision A. Ch. 19A. Gaithersburg, MD: AOAC International.

DeBoer SH, Ward LJ, Li X and Chittaranjan S (1995) Attenuation of PCR inhibition in the presence of plant compounds by addition of BLOTTO. *Nucleic Acids Research* 12: 387–395.

Herwaldt BL, Ackers M-L and the *Cyclospora* Working Group (1997) An outbreak in 1996 of cyclosporiasis associated with imported raspberries. *New England Journal of Medicine* 336: 1548–1558.

Jinneman KC, Wetherington JH, Hill WE et al (1998) Template preparation for PCR and RFLP of amplification products for the detection and identification of *Cyclospora* sp. and *Eimeria* spp. oocysts directly from raspberries. *Journal of Food Protection* 61: 1497–1503.

Johnson DW, Pieniazek NJ, Griffin DW, Misener L and Rose JB (1995) Development of a PCR protocol for sensitive detection of *Cryptosporidium* oocysts in water samples. *Applied Environmental Microbiology* 61: 3849–3855.

Ortega Y, Sterling CR, Gilman RH, Cama VA and Diaz F (1993) *Cyclospora* species – a new protozoan pathogen of humans. *New England Journal of Medicine* 328: 1308–1312.

Ortega Y, Sterling CR and Gilman RH (1994) A new coccidian parasite (Apicomplexa: Eimeriidae) from humans. *Journal of Parasitology* 80: 625–629.

Ortega YR, Nagle R, Gilman RH et al (1997) Pathologic and clinical findings in patients with cyclosporiasis and a description of intracellular parasite life-cycle stages. *Journal of Infectious Diseases* 176: 1584–1589.

Ortega YR, Roxas CR, Gilman RH et al (1997) Isolation of *Cryptosporidium parvum* and *Cyclospora cayetanensis* from vegetables collected from markets of an endemic region in Peru. *American Journal of Tropical Medicine and Hygiene* 57: 683–686.

Relman DA, Schmidt TM, Gajadhar A et al (1996) Molecular phylogenetic analysis of *Cyclospora*, the human intestinal pathogen, suggests that it is closely related to *Eimeria* species. *Journal of Infectious Diseases* 173: 440–445.

Soave R (1996) Cyclospora: an overview (review). *Clinical Infectious Diseases* 23: 429–435.

US Food and Drug Administration (1998) *Guidance for industry: Guide to minimize microbial food safety hazards for fresh fruits and vegetables*. US Department of Health and Human Services, Washington DC, 39 pp.

Yoder KE, Sethabutr O and Relman DA (1996) PCR-based Detection of the Intestinal Pathogen *Cyclospora*. In: Persing DH (ed.) *PCR Protocols for Emerging Infectious Diseases*, a supplement to *Diagnostic Molecular Microbiology: Principles and Applications*. Pp. 169–176. Washington, DC: ASM Press.

Cytometry *see* **Flow Cytometry**.

D

Dairy Products *see* **Brucella**: Problems with Dairy Products; **Cheese**: In the Market Place; Microbiology of Cheese-making and Maturation; Mould-ripened Varieties; Role of Specific Groups of Bacteria; Microflora of White-brined Cheeses; **Fermented Milks**: Yoghurt; Products from Northern Europe; Products of Eastern Europe and Asia; **Probiotic Bacteria**: Detection and estimation in fermented and non-fermented dairy products.

DEBARYOMYCES

W Praphailong, National Center for Genetic Engineering and Biotechnology, Bangkok, Thailand
G H Fleet, Department of Food Science and Technology, The University of New South Wales, Sydney, Australia

Copyright © 1999 Academic Press

Characteristics of the Genus and Relevant Species

Species of *Debaryomyces* are commonly found in soils, waters, plants, foods and clinical specimens. Present taxonomic classification accepts 15 species within the genus (**Table 1**), although an additional species *Debaryomyces prosopidis* has been proposed. *Debaryomyces hansenii* (imperfect form *Candida famata*) is, by far, the most significant species found in foods. Of the other species, there are only occasional reports on the isolation of *D. polymorphus*, *D. etchellsii*, *D. maramus* and *D. carsonii* from foods. Species of the genus undergo asexual reproduction by multilateral budding, with cells occurring singly, in pairs, short chains or small clusters. Pseudomycelium is usually lacking, but primitive or even well-developed pseudohyphae may be produced in some species. Sexual reproduction characteristically occurs by conjugation between a mother cell and its bud, but occasionally conjugation between separated cells is observed. Variation in the morphology and number of ascospores per ascus provides a good criterion for differentiation between the species. Ascospores are usually spheroidal to ovoidal in shape and are often distinguished by a warty or roughened surface. The ascospores of some species (e.g. *D. occidentalis*) have a distinct equatorial ledge. The number of ascospores per ascus varies from one to four depending on the species and, with the exception of three species, they are not usually liberated from the ascus.

The ability to ferment sugars varies from an absent, weak to vigorous reaction, nitrate is not assimilated but strains of some species assimilate nitrite. Ubiquinone Q-9 is present and the diazonium blue B reaction is negative. Lipid composition is characterized by the presence of linoleic (C18 : 2) and linolenic (C18 : 3) fatty acids. The mol% G+C content is in the range of 33–43%. Karyotyping of the genus is not complete, but three to seven chromosomes are generally present within the species.

Inclusion of species within the genus has undergone significant revision over the years. The first major description of the genus by Lodder and Kreger-van Rij in 1952 included only five species, namely *D. hansenii*, *D. kloeckeri*, *D. subglobosus*, *D. nicotianae* and *D. vini*. Lodder's 1970 classification, based largely on fermentation and assimilation tests, combined the first four of these species into one species *D. hansenii*, and introduced seven new species, *D. castellii*, *D. coudertii*, *D. marama*, *D. phaffii*, *D. cantarellii*, *D. tamarii* and *D. vanriji*. Price and co-workers in 1978 proposed a major revision of species within the genus after a detailed study of DNA sequence similarity by reassociation/hybridization kinetics. In particular, they showed that several species of *Pichia* were related to some *Debaryomyces* species. Consequently, *D. cantarellii* and *D. phaffii* were merged with *Pichia polymorpha* to become *D. polymorphus*. *Pichia pseudopolymorpha* became *D. pseudopolymorphus*. Using data from partial sequencing of ribosomal RNA subunits, Kurtzman and co-workers and Yamada and co-workers have further refined the description of the genus, and this has given the current recognition of 15 species (Table 1). A notable outcome from these studies was the close similarity of *Schwanniomyces occidentalis* with *Debaryomyces* species and its redefinition as *D. occidentalis* with two varieties. However, some authors do not agree with classification of *Schwanniomyces*

Table 1 Key properties[a] of species within the genus *Debaryomyces*

Species	Mol% G+C	37°C	NaCl (10%)	Vit	Fermentation G	Su	Tr	Assimilation Su	La	Me	Ra	Xy	Gl	Er	Ascospores Shape	Number[b]	Lib[c]
D. carsonii	36.8–39.7	+	+	–	–	–	–	+	–	v	v	+	–	–	Spheroidal	1–4	+
D. castellii	37.1	–	+	–	+	+	–	+	+	+	+	+	+	–	Spheroidal	1–3(1)	–
D. coudertii	37.4	–	+	–	+	–	–	–	–	–	–	+	+	+	Spheroidal	1	–
D. etchellsii	38.5–40.6	+	+	–	x	w/–	–	+	–	–	–	+	–	–	Spheroidal	1–4	+
D. hansenii																	
var. *hansenii*	37.3–38.6	–	+	–	w/–	w/–	w/–	+	v	v	+	+	x	v	Spheroidal	1–2(1)	–
var. *fabryi*	36.4–36.8	+	+	–	w/–	w/–	w/–	+	v	v	+	+	x	v	Spheroidal	1–2(1)	–
D. maramus	39.1	–	+	–	w/–	–	–	+	v	v	s	+	+	+	Ovoidal	1–4(2)	–
D. melissophilus	39.8	–	+	–	–	–	–	+	–	–	–	–	+	+	Spheroidal	1–4	–
D. nepalensis	37.6–38.0	–	+	–	w/–	w/–	w/–	+	v	+	+	+	+	+	Spheroidal	1	–
D. occidentalis																	
var. *occidentalis*	35.2	+	–	–	+	+	–	+	v	v	+	+	+	–	Spheroidal (L)	1–2(1)	–
var. *persoonii*	35.4	+	–	+	+	+	–	+	+	v	+	–	+	–	Spheroidal (L)	1–2(1)	–
D. polymorphus	35.7–35.9	+	+	+	s	s	–	+	v	v	+	+	+	+	Spheroidal	1–2(1)	–
D. pseudopolymorphus	35.7	–	+	–	+	s	–	+	+	+	+	+	+	+	Spheroidal	1–4	–
D. robertsiae	42.7	–	+	+	+	+	+	+	–	–	+	+	+	+	Spheroidal	1–4	–
D. udenii	35.8	–	+	–	ws	w/–	w/–	+	–	+	+	+	+	+	Spheroidal	1–4	+
D. vanrijiae																	
var. *vanrijiae*	33.2–33.3	+	+	+	w/–	w/–	–	+	–	+	+	+	+	+	Spheroidal	1–4	–
var. *yarrowii*	33.0	–	+	+	w/–	w/–	–	+	–	–	+	+	+	v	Spheroidal	1–4	–
D. yamadae	34.5	–	+	–	s	–	–	+	+	–	+	+	w	–	Spheroidal	1–4	–
D. prosopidis	37–38	+	+	–	–	–	–	+	–	–	+	+	+	+	Globose	1–2	–

Table adapted from Nakase et al (1998) with permission from Elsevier Science.

[a]Abbreviations: 37°C, growth at 37°C; NaCl (10%), growth in 10% NaCl + 5% glucose; Vit, growth in vitamin-free medium; G, glucose; Su, sucrose; Tr, trehalose; La, lactose; Me, melibiose; Ra, raffinose; Xy, D-xylose; Gl, gluconate; Er, erythritol; +, positive; s, positive but slow; x, positive or weak; w, weak; ws, weak and slow; w/–, weak or negative; v, variable; –, negative.

[b]Numbers of ascospores per ascus; numbers in parentheses refer to the number of ascospores most frequently observed. (L) indicates ascospores with equatorial ledge.

[c]Ascospores liberated by lysis of asci.

occidentalis within the genus *Debaryomyces* because it has some quite distinct phenotypic properties. Also *Wingea robertsii* was described as *D. robertsiae*. Other changes were the description of two former species of *Pichia* as *D. carsonii* and *D. etchellsii* and the removal of *D. tamarii*. The key properties that differentiate species within the genus are shown in Table 1.

Physiological and Biochemical Properties

Of the 15 species in the genus, only *Debaryomyces hansenii* and *D. occidentalis* have attracted significant study of their physiological, biochemical and molecular properties. These studies reflect the substantial diversity in growth and metabolic behaviour of yeasts within the genus.

Debaryomyces hansenii is considered to be non-fermentative. It metabolizes sugars to pyruvate by the Embden–Meyerhof–Parnas (EMP) pathway and then oxidizes pyruvate through the tricarboxylic acid (TCA) cycle. Organic acids such as citric, lactic and succinic are assimilated through the TCA cycle. The pentose phosphate pathway also operates in this yeast. Contrary to the general view, there are reports of some strains of *D. hansenii* and *C. famata* that ferment glucose and other hexoses. Extracellular protease and lipase production have been reported in some but not all strains. These enzymes have not been isolated and characterized. Amylolytic and pectinolytic activities are absent. The most distinguishing feature of *D. hansenii* is its ability to grow in the presence of extremely high concentrations of salt (NaCl). Although the growth response to NaCl varies with the strain, most grow in the presence of 15% (w/v) NaCl and there are some strains that grow at 20–24% (w/v) NaCl. High salt tolerance has also been reported for *D. etchellsii*. Salt tolerance of *D. hansenii* is greatest at pH values near 5.0 and decreases at pH 3.0 and pH 7.0. The molecular basis of salt tolerance in *D. hansenii* has been extensively studied and is related to the ability of this yeast to accumulate high intracellular concentrations of glycerol as an osmo-protectant or compatible solute. Substantial amounts of this glycerol are excreted into the extracellular medium, especially during the stationary phase, but it is re-utilized when glucose substrate is exhausted. The pathway of glycerol production has been studied and it originates from glucose by the EMP pathway. Intracellular arabitol is also accumulated and excreted, but its production (via the pentose phosphate pathway) appears constitutive and occurs in the absence of salt stress. It has been suggested that *D. hansenii* also has an appropriate

membrane bound ATPase that accomplishes an effective extrusion of Na^+ ions.

With the exception of *D. hansenii*, little is known about the environmental factors which limit the growth of species listed in Table 1. Other than *D. occidentalis*, all grow in the presence of 10% NaCl. *D. hansenii* and *D. etchellsii*, at least, are also tolerant of very high concentrations of sugars and grow in the presence of 60% (w/v) sucrose. Growth of *D. hansenii* is very weak at pH 2.5 but strong in the pH range 3.0–8.0. Many authors have made the qualitative observation that *D. hansenii* exhibits faster growth at 1–5°C compared with other yeast species, and there is a report of growth at –12.5°C. *D. hansenii* is not particularly tolerant of preservatives or heat treatment. It is inhibited at pH 5.5 by 250–500 mg l⁻¹ of benzoic or sorbic acids and has a D value of 12 min at 48°C. Some strains have a strong tendency to flocculate and this could be a potential survival mechanism in hostile environments.

In contrast to *D. hansenii*, *D. occidentalis* is a vigorous fermenter of sugars under non-aerated conditions. It is a Crabtree negative yeast and, under aerated conditions, it channels the sugars into the TCA cycle. Unlike *D. hansenii*, this species is not particularly tolerant of high salt or high sugar environments. The most distinctive property of *D. occidentalis* is its efficient degradation of starch by the production of extracellular α-amylases and a glucoamylase that can by-pass the α-(1→6)-linked branch points in amylopectin. Because of this property, there has been substantial scientific and industrial interest in this yeast. The kinetics of production and properties of these amylolytic enzymes have been well characterized and their genes have been cloned and sequenced. Techniques for manipulating the expression of these genes and for transferring them to other yeast species have been developed.

Significance in Foods

Literature on the occurrence of *Debaryomyces* species in foods is largely unfocused and scattered over many years. It is difficult to track because of the numerous changes of name of the species. Most studies concern *D. hansenii* and there are only occasional reports on the occurrence and significance of other species, such as *D. etchellsii*, *D. polymorphus*, *D. maramus* and *D. carsonii* in foods (**Table 2**). There is no reason to explain why the other species listed in Table 1 (e.g. *D. occidentalis*) are not found in food ecosystems, but more systematic and focused study will probably reveal their presence.

The early literature reveals the frequent isolation of *D. hansenii* from meat products, especially processed

Table 2 Significance of *Debaryomyces* species in the food and beverage industries

Species	Significance
Debaryomyces hansenii (Candida famata)	*Occurrence/spoilage*: delicatessen, cured, fermented, minced meats; seafoods, fish sauces; yoghurts, cheeses; brined vegetables; mayonnaise-based salads; silage *Biotechnological*: starter cultures for meat sausage fermentation; starter cultures for maturation of cheeses; biocontrol agent of bacterial and fungal spoilage; xylitol production
Debaryomyces etschellsii	*Occurrence/spoilage*: carbonated soft drinks; sugar syrups; brined vegetables; mayonnaise-based salads, soy sauce; fermented meat products
Debaryomyces polymorphus	*Occurrence/spoilage*: carbonated soft drinks; fruit products delicatessen, cured and fermented meats
Debaryomyces maramus	*Occurrence/spoilage*: meat products; cheese
Debaryomyces carsonii	*Spoilage*: salted fish paste
Debaryomyces occidentalis	*Biotechnological*: amylase production; waste utilization; single-cell protein

products, such as frankfurters, bacon, hams and fermented and unfermented sausages. In some cases, presence of the yeast was associated with the development of a slimy surface layer on the product. Recent, more extensive studies have confirmed the predominance of *D. hansenii* in meat products compared with other yeasts, and these conclusions have been extended to include seafoods such as fresh fish. Populations in the range 10^4–10^6 cfu g^{-1} (or even higher in fermented salami) are often reported. *D. etschellsii*, *D. polymorphus* and *D. maramus* are also found in these products, but less frequently. The impact of this yeast growth on the flavour of meat products is not clear, but cannot be assumed to be negative. Indeed, there is a positive correlation between the desired flavour of some Italian salami sausages and the presence of *D. hansenii*. Some, but not all, of the strains of *D. hansenii* isolated from meat products produce extracellular proteases and lipases that could contribute to flavour development by the breakdown of meat proteins and fats. The ability of these enzymes to operate well at low pH may be an appealing property. Consequently, consideration has been given to the use of selected strains of *D. hansenii* as starter cultures in the production of fermented sausages. Factors thought to select for the growth of *D. hansenii* in meat products include its tolerance of salt, utilization of organic acids (e.g. lactic), protease and lipase production, good growth at low temperatures, and the ability of some strains to utilize sodium nitrite which is added as a curing agent in some products.

D. *hansenii*/*C. famata* have now emerged as the most important yeasts in the dairy industry. Weakly fermenting species have been linked to the spoilage of yoghurts, but their greatest significance is in cheese production, especially with the mould-ripened soft cheeses such as Camembert, Brie and blue-veined varieties. Many surveys of these and other types of cheeses have revealed a consistently high incidence of *D. hansenii*, often at populations of 10^6–10^7 cfu g^{-1} or higher. The yeast originates as a natural contaminant of the cheese brine and grows at both the outer and inner parts of the cheese curd during the maturation stage. Again, the ability of the yeast to tolerate the high salt environment of the cheese, utilization of lactic acid, protease and lipase production, growth at low temperature and, possibly, production of polyols such as glycerol are key factors that favour its growth and contribution to the biochemistry of cheese maturation. A clear link between such activity and a sensory outcome remains to be established, but the relationship is assumed to be positive since commercial starter cultures of *D. hansenii* are available for encouraging the maturation process. An important property of these strains might be the ability of their proteases and lipases to operate at high salt concentrations and low temperatures.

Debaryomyces species, especially *D. etschellsii*, are frequently isolated from brines used to ferment products such as olives and cucumbers, and they are also associated with traditional Japanese fermented products such as soy sauce and miso. Curiously, a high proportion of killer strains of *D. hansenii* with broad-spectrum killer activity has been isolated from the latter ecosystems. Presumably, these yeasts grow on the surface of brine solutions, utilizing lactic acid produced by the lactic acid bacteria involved in these fermentations. However, some strains of *D. etschellsii* also ferment sugars. This latter property may also explain the occasional association of *D. etschellsii* with the spoilage of high sugar syrups. There are occasional reports of the isolation of *Debaryomyces* species from soft drinks, beer, wine and vegetable salads but, generally they are not significant spoilers of these products. An unusual but significant form of spoilage has been reported for *D. carsonii*, which grew in a Japanese chickuwa fish paste, transforming *trans*-cinnamic acid to styrene which gave the product an unacceptable petroleum-like aroma.

As noted already, the association of *D. hansenii* with foods does not necessarily have negative implications and there is significant interest in exploiting this species as a starter culture in meat and cheese production. In the case of cheese production, it has been reported to have good biocontrol over spoilage species of *Clostridium*. Also in the context of biocontrol, several papers in the late 1980s suggested that *D. hansenii* was an effective natural antagonist for controlling the fungal spoilage of various fruits by species of *Penicillium*, *Botrytis* and *Rhizopus*. However, later study revealed that the yeast is an unusual strain of *Candida guilliermondii* and not *D. hansenii*. Nevertheless, this work has stimulated interest in the yeast as a potential novel biocontrol agent. In another industrial application, *D. hansenii* has potential value in bioconversion of xylose into the sweetener, xylitol. The enzymes, xylose reductase and xylitol dehydrogenase, associated with this process have been examined.

The amylases of *D. occidentalis* have potential application in the production of sugar syrups for food and beverage processing. The genes for these amylases have been incorporated into brewing strains of *S. cerevisiae* for the purpose of using these strains in the production of low calorie or dextrin-free beers. *D. occidentalis* could be used to process starchy waste material into single-cell protein.

Debaryomyces spp. are not generally regarded as pathogenic to humans and no food-borne disease outbreaks have been attributed to these organisms. However, *D. hansenii*/*C. famata* have been implicated in isolated cases of septicaemia and skin and mucosal surface infections where they are considered as weak opportunistic pathogens, especially for immunocompromised patients.

Enumeration and Identification

Food sample (10 g) is suspended in 90 ml of 0.1% peptone water, homogenized for approximately 1 min and then diluted 10-fold, as necessary, in 0.1% peptone water. Aliquots (0.1 ml) of the dilution are then spread inoculated over the surface of plates of media such as malt extract agar, glucose–yeast extract agar or tryptone glucose yeast extract agar. Bacterial antibiotics, such as chloramphenicol, oxytetracycline, chlorotetracycline, gentamicin and streptomycin can be added to these media at concentrations up to 100 µg ml^{-1} to suppress bacterial growth. For the isolation of *Debaryomyces* from products like cheese, overgrowth of moulds (*Penicillium* spp.) on the plates can occur. Incorporation of the mould inhibitor, biphenyl (50 mg l^{-1}) into the medium can overcome this problem. Plates are incubated at 25°C for 4–7 days and yeast colonies counted. Virtually all yeast species will grow on the media just described. There-

fore, it will be necessary to isolate and identify individual colonies. The identification of *Debaryomyces* spp. follows standard morphological biochemical and physiological tests and keys as outlined in *The Yeasts, a Taxonomic Study*, 4th edition, edited by CP Kurtzman and JW Fell, Elsevier Science (1998) (Table 1). *D. hansenii*/*C. famata*, at least, identifies very well in the rapid computer-based Biolog (Biolog Inc California) and ATB 32C (bioMérieux) systems that incorporate a range of these tests in kit form.

To avoid potential osmotic shock and stress, it has been suggested that 5–10% NaCl be included in the diluent and plating medium when isolating these yeasts from high salt foods. However, we and others have not found these steps to offer any benefit.

No selective or differential media have been reported for these yeasts. However, inclusion of 10–15% (w/v) NaCl into the medium formulation could assist in selecting for the growth of these species, except *D. occidentalis*. A differential medium based on the hydrolysis of starch could be developed for the isolation of *D. occidentalis*.

A PCR method that differentiates *D. hansenii*/*C. famata* from *Candida guilliermondii* has been reported and is based on amplification of the large subunit rDNA between base positions 402 and 669. A *D. hansenii* nucleic acid probe based on sequences in the 18S rRNA has been reported. As yet, neither of those molecular methods has been developed for routine use.

See also: **Fermented Foods**: Fermentations of the Far East. **Fermented Milks**: Yoghurt. **Meat and Poultry**: Spoilage of Cooked Meats and Meat Products.

Further Reading

Andrews S, de Graaf H and Stamation H (1997) Optimisation of methodology for enumeration of xerophilic yeasts from foods. *International Journal of Food Microbiology* 35: 109–116.

Dillon VM and Board RG (1991) Yeasts associated with red meats. *Journal of Applied Bacteriology* 71: 93–108.

Dohmen RJ and Hollenberg CP (1996) *Schwanniomyces occidentalis*. In: Wolf K (ed.) *Nonconventional Yeasts in Biotechnology. A Handbook*. P. 117. Berlin: Springer-Verlag.

Girio FM, Pelica F and Amaral-Collae MT (1996) Characterisation of xylitol dehydrogenase from *Debaryomyces hansenii*. *Applied Biochemistry and Biotechnology* 56: 79–87.

Kosse D, Ostenrieder I, Seiler H and Scherer S (1998) Rapid detection and identification of yeasts in yogurt using fluorescently labelled oligonucleotide probes In: Jakobsen M, Narvhus J and Viljoen BC (eds) *Yeasts in the*

Dairy Industry: Positive and Negative Aspects. P. 132. Brussels: International Dairy Federation.

Kurtzman CP and Robnett CJ (1991) Phylogenetic relationships among species of *Saccharomyces*, *Schizosaccharomyces*, *Debaryomyces* and *Schwanniomyces* determined from partial ribosomal RNA sequences. *Yeast* 7: 61–72.

Kurtzman CP and Robnett CJ (1994) Synonymy of the yeast genera *Wingea* and *Debaryomyces*. *Antonie van Leeuwenhoek* 66: 337–342.

Martinez XC, Narbad A, Carter AT and Stratford M (1996) Flocculation of the yeast *Candida famata* (*Debaryomyces hansenii*): an essential role for peptone. *Yeast* 12: 415–423.

Nakase T, Suzuki M, Phaff HJ and Kurtzman CP (1998) *Debaryomyces* In: Kurtzman CP and Fell JW (eds) *The Yeasts, a Taxonomic Study*, 4th edition. P. 157. Amsterdam: Elsevier Science.

Nishikawa A, Sugita T and Shinoda T (1997) Differentiation between *Debaryomyces hansenii/Candida famata* complex and *Candida guilliermondii* by polymerase chain reaction. *FEMS Immunology and Medical Microbiology* 19: 125–129.

Phaff HJ, Vaughan-Martini A and Starmer WT (1998) *Debaryomyces prosopidis* sp. nov., a yeast from exudates of mesquite trees. *International Journal of Systematic Bacteriology* 48: 1419–1424.

Praphailong W and Fleet GH (1997) The effect of pH, sodium chloride, sucrose, sorbate and benzoate on the growth of food spoilage yeasts. *Food Microbiology* 14: 459–468.

Praphailong W, van Gestel M, Fleet GH and Heard GM (1997) Evaluation of the Biolog system for identification of food and beverage yeasts. *Letters in Applied Microbiology* 24: 455–459.

Price CW, Fuson GB and Phaff HJ (1978) Genome comparison in yeast systematics: delimitation of species within the genus *Schwanniomyces*, *Saccharomyces*, *Debaryomyces* and *Pichia*. *Microbiological Reviews* 42: 161–193.

Roostita R and Fleet GH (1996) The occurrence and growth of yeasts in Camembert and Blue-veined cheeses. *International Journal of Food Microbiology* 28: 393–404.

Shimodza K, Kimara E, Yasui Y (1992) Styrene formation by the decomposition by *Pichia carsonii* of *trans*-cinnamic acid added to a ground fish product. *Applied and Enviroonmental Microbiology* 58: 1577–1582.

Sorensen BB (1997) Lipolysis of pork fat by the meat starter culture *Debaryomyces hansenii* at various environmental conditions. *International Journal of Food Microbiology* 34: 187–193.

Tudor A and Board RG (1993) Food spoilage yeasts. In: Rose AH and Harrison JS (eds) *The Yeasts*, 2nd edn. Vol. 5, p. 435. London: Academic Press.

van den Tempel T and Jakobsen M (1998) The technological characteristics of *Candida famata* isolated from Danablu. In: Jakobsen M, Narvhus J and Viljoen BC (eds) *Yeasts in the Dairy Industry: Positive and Negative Aspects.* P. 59. Brussels: International Dairy Federation.

DESULFOVIBRIO

M D Alur, Food Technology Division, Bhabha Atomic Research Centre, Mumbai, India

This article deals with morphological, cultural and physiological traits of the genus *Desulfovibrio* and its related species. Methods of enrichment, isolation and enumeration are briefly described. The influence of overproduction of these bacteria on agricultural and animal production is briefly discussed.

Dissimilatory sulphate-reducing bacteria are the oldest forms of bacterial life on earth. In contrast to other bacteria, which require sulphate only in amounts necessary for the formation of their sulphur-containing cell constituents (assimilatory sulphate reduction), dissimilatory sulphate-reducing bacteria utilize sulphate mainly as the terminal electron acceptor in the anaerobic oxidation of organic substrates. As a consequence, they accumulate large amounts of sulphide in their natural habitats. In contrast to the facultative anaerobic dissimilatory nitrite-reducing bacteria, sulphate-reducing bacteria are strict anaerobes. Pure cultures not only require the absence of oxygen for growth, but also low redox potential of 0–100 mV in the medium.

Characteristics of the Genus

On the basis of their oxidative and metabolic capabilities, the strains and species that metabolize even-numbered higher and lower fatty acids, propionate, lactate and pyruvate to acetate are assigned to the genus *Desulfovibrio*.

Various cell types have been described in the *Desulfovibrio* group including cocci, oval or long straight rods, curbed rods or spirillum cells with gas vesicles and gliding multicellular filaments. However, they are normally curved rods of variable length, measuring $0.5–1.5 \times 2.5–10 \, \mu m$, usually occurring singly but sometimes in chains which have the appearance of spirilla. Swollen pleomorphic forms are common, actively motile by means of a single polar flagellum, strict anaerobes which reduce sulphates to hydrogen sulphide.

Physiological Properties

The bacterium does not liquefy gelatin; it grows in media prepared with sea water or 3% mineral salt solution, enriched with sulphate and peptone. *Desulfovibrio aestuarii* can tolerate temperatures above 45°C. Agar colonies are small, circular and raised with dark centres.

In the presence of sulphates, desulfovibrios utilize peptones, asparagine, glycine, alanine, glucose, fructose, ethanol, butanol, glycerol, acetate, lactate and malate. Some strains utilize molecular hydrogen as the sole source of energy to reduce sulphate to hydrogen sulphide.

These bacteria produce 950 mg H_2S per litre of the medium. They do not reduce nitrates to nitrites. *D. desulfuricans* reduces sulphate to hydrogen sulphide as it oxidizes carbohydrates to acetic acid. Temperature optima for growth is in the range 25–40°C.

These bacteria utilize either sulphate and other oxidized sulphur compounds or elemental sulphur as electron acceptor. Electron donors used for sulphate reduction include H_2, alcohols, fatty acids, other monocarboxylic acids, dicarboxylic acids, some amino acids, a few sugars and some other aromatic compounds. Long-chain alkanes can be utilized anaerobically by a particular type of sulphate-reducing bacterium. Carbohydrates are rarely degraded. Cells may contain *c*-type or *b*-type cytochromes. Many species of the sulphate-reducing bacteria contain sulphide reductase, desulfoviridin, desulforubidin or P582. Hydrogenase is present in many species. Nitrate reduction to ammonia is rare. Some species have the ability to fix nitrogen. Genera and species differ with respect to the utilization of organic compounds. Thus, incomplete oxidation of substrates such as lactate to carbon dioxide and acetate occurs, the latter cannot be oxidized further.

The mol% G+C of the DNA is in the range of 37–67%.

Most species are positive when tested for desulfoviridin. Desulfoviridin (green) sulphite reductase occurs in the cytoplasm of *Desulfovibrio* and is involved in the dissimilatory sulphate reduction.

The type species are *D. sulfuricans*, *D. vulgaris*, *D. africans*, *D. gigas* and *D. thermophilus*. Comparative physiological properties of some species of the genus *Desulfovibrio* are given in **Table 1**.

Methods of Isolation and Enumeration

Enrichment and Isolation Procedures

A bacteriological salt mixture containing a suitable carbon source and sulphates will yield *Desulfovibrio* spp., if it is incubated anaerobically with inoculum at pH 7.5. E_h should be brought below −150 mV by adding chemical-reducing agent. Ferrous salts are present in excess to indicate sulphate reduction by blackening. The composition of media for the cultivation of *Desulfovibrio* spp. is given in **Table 2**.

Media containing a precipitate such as medium B are more satisfactory than those without a precipitate and growth is often seen to begin as a zone of blackening in the precipitate.

Glass-stoppered bottles (30–60 ml capacity) are satisfactory containers for enrichment culture. The stopper should be coated with silicone grease which resists autoclaving. The bottles should be filled to the brim with medium B plus inoculum and the stopper pressed to eliminate all air. Unequivocal blackening after incubation indicates enrichment of sulphate-reducing bacteria.

The oxygen sensitivity of slow growth of *Desulfovibrio* makes isolation inconvenient. Hence, sequential dilution of dispersed colonies from deep agar or sediments from a liquid enrichment culture is often preferred.

Isolation of *Desulfovibrio* spp.

From a stock solution of medium B of appropriate salinity take 25 ml and supplement with 1% agar, 0.01% sodium ascorbate and 0.01% sodium thioglycollate, and autoclave for 5 min at 121°C. This medium should be used at once. Place 4 ml portions into long sterile tubes (10 × 150 mm or vanilla tubes) and keep molten at 40°C. Dip a flamed Pasteur pipette into the enrichment culture and then successively into tubes 1 to 6. Incubate at 30°C. The anerobes will grow in the depths of agar. After 4–5 days, distinct black colonies are visible in the fourth or fifth tube. Break the tube at a convenient point and withdraw two or three colonies with Pasteur pipette.

Culture Media

Many different types of media have been used for the cultivation of sulphate-reducing bacteria. However, no one medium is well suited for all species. Two media which are frequently employed are described here.

Lactate Medium for *Desulfovibrio* A lactate medium that can be prepared simply has long been used successfully for cultivation of the classical sul-

Table 1 Comparative morphological and physiological properties of some *Desulfovibrio* species

Characteristics	D. africans	D. vulgaris	D. desulfuricans
Morphology	Vibrio	Vibrio	Vibrio
Width × length (µm)	0.5–0.6 × 2–3	0.5–0.8 × 1.5–4	0.5–0.8 × 1.5–4
Motility	+	+	+
Hydrogen	+	+	+
Optimum temp. °C	30–36	30–36	30–36
Sulphate reductase	+	+	+
G+C content (mol.%)	65	65	50
NaCl requirement	–	–	–
Electron donors for sulphate reduction			
Formate	+	+	+
Acetate	–	–	–
Fatty acids	–	–	–
Ethanol	+	+	+
Lactate	+	+	+
Fumarate	–	+	+
Malate	+	+	+

Table 2 Composition of media for the cultivation of *Desulfovibrio* spp.

Component	Concentration in distilled water (g l^{-1})		
	Medium B	Medium C	Medium D
KH_2PO_4	0.5	0.5	0.5
NH_4Cl	1.0	1.0	1.0
$CaSO_4$	1.0	–	–
Na_2SO_4	–	4.5	–
$MgSO_4.7H_2O$	2.0	2.0	–
$CaCl_2.2H_2O$	–	0.06	0.1
$MgCl_2$	–	–	1.6
$FeSO_4.7H_2O$	0.5	0.004	0.004
Sodium lactate	3.5	3.5	–
Sodium pyruvate	–	–	3.5
Yeast extract	1.0	1.0	1.0
Sodium citrate	–	0.3	–

Table 3 Preparation of lactate medium

Component	Concentration in distilled water (g l^{-1})	
	Medium B	Medium C
KH_2PO_4	0.5	0.5
NH_4Cl	1.0	1.0
$CaSO_4$	1.0	–
Na_2SO_4	–	4.5
$MgSO_4.7H_2O$	2.0	2.0
$CaCl_2.2H_2O$	–	0.06
NaCl (for marine strains)	2.5	2.5
Sodium lactate (50%)	7.0 (5.5 ml)	12.0 (9.5 ml)
Yeast extract	1.0	1.0
Sodium citrate	–	0.3

phate reducer *Desulfovibrio*. Lactate or pyruvate serves as both electron donor and carbon source. The medium is buffered with phosphate and contains iron. Further trace elements or vitamins are not added. They may be present in the yeast extract or as impurities in chemicals or water.

Stock Solutions Ferrous sulphate solution is prepared by dissolving 0.5 g $FeSO_4.7H_2O$ in 10 ml of distilled water. To prevent oxidation the solution should be autoclaved under a headspace of nitrogen in tubes with a fixed stopper. Also, acidification with 1 ml of 1 M H_2SO_4 counteracts oxidation.

To prepare the reductant solution dissolve sodium thioglycollate (0.1 g) and ascorbic acid (0.1 g) in 10 ml distilled water. Autoclave under nitrogen as described for ferrous sulphate solution. Stocks of sodium thio-glycollate should be freshly prepared. The components of the culture are as listed in **Table 3**.

To avoid formation of precipitates, add dry salts successively to the water during stirring. Adjust the pH to 7.2 and then autoclave the solution. Further components are added from the sterile stock solutions after autoclaving: $FeSO_4$ solution, 10 ml (for medium B) and 0.08 ml (for medium C); and reductant solutions 10 ml (medium B) and 1 ml (medium C). The completed medium is aseptically distributed into bottles or tubes.

For plates, media B and C are prepared with 10 g agar per litre. Before autoclaving, the agar should be dissolved by stirring in a boiling-water bath. Medium B turns black upon bacterial sulphide production (precipitation of FeS) and can thus be used for diagnostic purpose, e.g. for detection of sulphate reducer in counting series and also for maintenance of stock culture. Medium C is clear and hence is suitable for mass culture of *Desulfovibrio*.

For cultivation of *Desulfovibrio* with hydrogen, media B and C are supplemented with 0.2–0.4 g sodium acetate per litre. The hydrogen gas space should contain some CO_2. Acetate and CO_2 are the main carbon sources required for lithotrophic growth.

Desulfoviridin Test   The test is used to detect desulfoviridin in certain sulphate-reducing bacteria. The pellet from a centrifuged culture (15 ml) is resuspended in growth medium, a drop of 2 M NaOH is added and it is examined at 365 nm. If the test is positive, the suspension fluoresces red due to the presence of the free sinohydrochloride chromophore of desulfoviridin.

Importance to Food Production

A wide variety of sulphur reducers are present in sulphate-containing sediments, especially marine and estuarine habitats because these sediments are constantly perfused with sulphate-containing sea water. The greatest variety of species has been isolated from marine sediments, where, due to its high concentration in sea water (28 mM) sulphate is seldom a growth-limiting factor. In addition, sulphate reducers often occur in the rumen and intestinal tract of non-ruminant animals. They are important because they are the terminal members of the anaerobic food chain. The sulphur-reducing bacteria are widely distributed in nature and produce large quantities of sulphides. Sulphate-reducers are also found in rice paddies and in anaerobic digesters or sewage plants. Significant activities and cell densities of sulphate-reducing bacteria have even been observed in the upper oxic zones of freshwater and marine sediments. Sulphide is quite toxic and consequently these bacteria are responsible for massive mortalities of fish and estuarine water fowl. In waterlogged soils, sufficient sulphide may be produced to damage plants. This is a problem in the cultivation of rice, which is normally grown in flooded paddies to control weeds.

On the other hand, sulphide production by the sulphur-reducing bacteria can provide a source of reductant to support the growth of aerobic chemoautotrophic or anaerobic green or purple bacteria in illuminated habitats.

The anaerobes of the genus *Desulfovibrio* are capable of reducing S° to H_2S. Hydrogen sulphide is thus likely to accumulate in waterlogged soils. Since hydrogen sulphide is toxic to higher organisms, sulphate reduction is undesirable from an agricultural point of view. Hydrogen sulphide reacts with lead, iron and zinc to form insoluble precipitates. Thus, in areas where H_2S is prevalent, iron metal will not be available in the soluble form which is required for all organisms. In anaerobic marine sediments, due to higher concentrations of H_2S, soluble metals are in short supply. *Desulfovibrio* species are important hydrogen scavengers in the anaerobic degradation of organic matter in sediments. High densities of *Desulfovibrio* have been found in methane-producing whey fermenters without added sulphate. This suggests that even in the absence of sulphate, *Desulfovibrio* is involved in anaerobic degradation by channelling lactate into methanogenesis.

See also: **Methanogens**.

Further Reading

Balows A, Truper HG, Dworkin M, Harder W and Schleifer KH (eds) (1992) *The Prokaryotes*. Vol. IV. London: Springer-Verlag.

Hurst CJ, Knudsen GR, McInerney MJ, Stetzenbach LD and Walter MV (eds) (1997) *Manual of Environmental Microbiology*. Washington, DC: ASM Press.

Nester EW, Roberts CE, Lidstrom ME, Pearsal NN and Nester MT (eds) (1983) *Microbiology*. Japan: Saunders College Publishing, Holt-Saunders.

Stanier RY, Ingraham JL, Wheelis ML and Painter PR (eds) (1993) *General Microbiology*. London: MacMillan.

Starr MP, Stolp H, Truper HG, Balow A and Schlegel HG (eds) (1986) *The Prokaryotes*. Vol. I. New York: Springer-Verlag.

Deuteromycetes *see* **Fungi**: Classification of the Deuteromycetes.

DIRECT (AND INDIRECT) CONDUCTIMETRIC/ IMPEDIMETRIC TECHNIQUES

Food-borne Pathogens

D Blivet, AFSSA, Ploufragan, France

An uninoculated cultural medium is composed of uncharged or weakly charged substrates. A bacterial population metabolizes these substrates, to yield charged end products; impedimetry measures the resulting conductance changes of the cultural medium. The conductance curve as a function of time is similar to a growth curve, and the start of the exponential phase of the curve is called the detection time (DT). The DT is inversely proportional to the initial bacterial population present in the cultural medium, thus enabling a quantitative evaluation of the inoculum.

Impedimetric methods can be applied to various analyses, in microbiology as well as in biochemistry. The most significant developments have been observed in food analysis, which requires the detection and/or the enumeration of microorganisms. The most relevant uses are the replacement of standard enumeration methods (agar plating or most probable number techniques) in the analysis of raw materials or finished products, the estimation of food spoilage microorganisms, and the detection of groups of microorganisms indicative of sanitary status of both equipment and products. Interest in the detection of food pathogens such as *Salmonella* is also increasing.

Detection and Enumeration of Food Pathogens

Escherichia coli

Both direct and indirect impedimetric techniques have been evaluated for the detection of *Escherichia coli* in drinking water. The direct technique employs a medium containing trimethylamine N-oxide and glucuronic acid; it yields data correlating well with the traditional technique (membrane filtration and isolation on selective agar), although the presence of *Salmonella* spp. may give false positive results. *Salmonella* can also give false positive reactions in the indirect impedimetric method, but these problems of false positive results are not of great concern in practice, because *Salmonella* is also a major pathogen in water.

Listeria

Two growth media, modified by the addition of antibiotics, have been tested for the detection of *Listeria* spp. Although they are not absolutely specific, the differentiation between *Listeria* spp. and the other bacterial species can be achieved by the analysis of conductance curves; however, it is not possible to distinguish *L. monocytogenes* from other *Listeria* species.

In 1991, Bolton described a new detection medium able to be used by direct conductivity. Although developed for the specific detection of *L. monocytogenes*, the method also detected other *Listeria* species (*innocua*, *ivanovii* and *seeligeri*). Initial comparisons of Bolton's method against the traditional method on a variety of food products were encouraging.

The second medium was developed for the detection of *Listeria* spp. in cheese, using capacitance measurements; results were considered to be positive when the capacitance changes during the 30 h incubation exceeded 30%. The efficiencies of direct and indirect impedimetric standard methods using this broth are roughly equivalent.

Campylobacter

A protocol based on conductance measurements has been developed for the detection of *Campylobacter*. It includes a pre-enrichment in a selective broth, followed by incubation at 42°C in a specially formulated selective medium. The results of analysis of naturally contaminated samples were satisfactory.

Staphylococcus aureus

Media have been developed for the detection of *Staphylococcus aureus* in various food products by conductance or capacitance measurements. Many broths for *Staphylococcus* detection have high salt concentrations, but Bolton in 1990 first showed that the detection of pure *Staphylococcus aureus* strains was possible in Whitley Impedance Broth (WIB) supplemented with NaCl $100 \, g \, l^{-1}$ or LiCl $5 \, g \, l^{-1}$. However, these formulations have not been tested for their selectivity against other microorganisms. Mueller–Hinton medium supplemented with oxacilline was used in the direct technique to differentiate oxacillin-resistant strains from sensitive ones. From positive blood cultures showing a growth of Gram-positive cocci and a thermonucleasic activity, oxacillin-resistant *Staphylococcus aureus* (ORSA) strains were detected within 24 h, while oxacillin-sensitive

Staphylococcus aureus (OSSA) strains did not give any detection time, and consequently were repressed.

Clostridium perfringens

Detection methods for *Clostridium perfringens* are at an early stage of development. Two media derived from the tryptose–sulphite–cycloserine (TSC) medium for the detection of pure *Clostridium perfringens* strains have been tested: TSC-LP (TSC with lactose, phenolphthalein diphosphate and trimethyl aminoxide) and TSC-MUP (TSC with lactose and methyl-umbelliferyl phosphate). These two media inoculated with pure strains of different clostridial species were incubated in microaerophilic conditions at 46°C in impedimetric cells. To confirm positive impedance cells the authors used a phosphatase test, followed by a catalase test for the cultures in TSC-LP, and examination under ultraviolet light for the cultures in TSC-MUP. The correlation coefficient between the enumerations on TSC agar and the detection times by impedimetry was 0.90, but no naturally contaminated samples were tested. The growth of pure cultures of *Clostridium perfringens* and *C. sporogenes* in five non-selective media was followed by direct impedimetry; two media gave satisfactory results – thioglycollate broth and Columbia Broth Malthus (CBM). However, for these media to be useful for the detection of *C. perfringens* in different samples, means of repressing competitive flora must be found.

Salmonella

Two media for *Salmonella* detection are approved by the Association of Official Analytical Chemists (AOAC): the Easter and Gibson medium and the modified medium of Ogden.

Easter and Gibson Medium   The Easter and Gibson medium exploits the ability of *Salmonella* to reduce trimethylamine *N*-oxide (TMAO) to trimethylamine (TMA), a property shared by many Enterobacteriaceae. Three enrichment broths, selenite–cystine (SC), Rappaport–Vassiliadis (RV) and Müller–Kauffmann tetrathionate (MKT), were supplemented with TMAO.HCl (0.5%) and tested on *Salmonella* strains and other genera. For *Salmonella*, the SC medium TMAO gave the best response. This formulation was then tested with the addition of either lactose or dulcitol. The final method involves pre-enrichment in buffered peptone water supplemented with TMAO 0.1% and dulcitol 0.5% (90% of *Salmonella* strains can use this sugar). The enrichment broth (SC, TMAO and dulcitol, SC/T/D) contains bacteriological peptone $5 \, g \, l^{-1}$, $Na_2HPO_4.2H_2O$ $10 \, g \, l^{-1}$, dulcitol $5 \, g \, l^{-1}$, TMAO.HCl $5.6 \, g \, l^{-1}$, and

sodium biselenite $4 \, g \, l^{-1}$. Just before use, an L-cystine solution is added to the broth. After incubation, 200–250 µl are inoculated in 10 ml of SC/T/D in impedimetric cells at 37°C. A response greater than 600 µs and a conductance change rate greater than or equal to 100 µs h^{-1} indicate a *Salmonella*-positive result. Positive and negative results are confirmed by streaking on selective agar. This protocol was tested in the analysis of naturally or artificially contaminated samples, and it appeared that some *Citrobacter freundii* strains produced false positive detections.

Reasoning that some *Salmonella* strains were not detected in the Easter and Gibson medium because of their inability either to reduce TMAO or to ferment dulcitol, and the total or partial inhibitory role of selenite, Ogden and Cann in 1987 replaced dulcitol by mannitol, and obtained good conductance curves. However, despite success in detecting those *Salmonella* strains not fermenting dulcitol, *Citrobacter freundii* and *Escherichia coli* strains continued to give false positive results. Other researchers tried to replace dulcitol: deoxyribose allows the detection of more *Salmonella* strains than does dulcitol, reducing the number of false positive results. A number of studies using the formula of Easter and Gibson have been reported. Some recorded false positive reactions due to *Citrobacter freundii*, in the analysis of animal feed samples. Other studies reported rates of detection of 95% for raw meat samples but only 59% for animal feed samples contaminated by *Salmonella* (after confirmation of positive impedimetric cells by streaking on brilliant green agar and confirmation of negative cells by enzyme-linked immunosorbent assay, ELISA). False positive reactions ranged from 5.8% to 13.5%.

More recently the use of immunomagnetic separation to improve *Salmonella* detection using the Easter and Gibson medium has been described. Studies indicate that this separation traps *Salmonella*, but also other species like *Citrobacter*.

Ogden Medium   In 1988, Ogden described a medium that could be used to distinguish between *Salmonella* and *Citrobacter freundii*. One of the biochemical properties that differentiate *Salmonella* from *Citrobacter* is the ability to decarboxylate lysine: 90–100% of *Salmonella* strains and only 0–10% of *Citrobacter* strains exhibit this decarboxylation.

Using Moeller decarboxylase medium as a growth medium, Ogden tested the effects of selenite, phosphate, pH indicators, pyridoxal and beef extract, finally developing a medium containing lactalbumin hydrolysate $5 \, g \, l^{-1}$, glucose $10 \, g \, l^{-1}$, L-lysine $10 \, g \, l^{-1}$ and sodium biselenite $4 \, g \, l^{-1}$. Before use, a sterilized solution of L-cystine was added. The cultures were pre-enriched in buffered peptone water supplemented with glucose and lysine. By this method, problems with *Citrobacter freundii* strains were eliminated, but some *Hafnia alvei* strains were detected. In 1990, Ogden modified this lysine medium to improve, not the specificity, but the efficiency by using an appropriate statistical technique. The following composition was proposed: lactalbumin hydrolysate $5 \, g \, l^{-1}$, glucose $1 \, g \, l^{-1}$, L-lysine $30 \, g \, l^{-1}$ and sodium biselenite $4 \, g \, l^{-1}$. This new medium reduces the rate of false negative reactions. Arnott and co-workers also developed a medium based on lysine decarboxylation, containing yeast extract $3 \, g \, l^{-1}$, glucose $1 \, g \, l^{-1}$, L-lysine $5 \, g \, l^{-1}$, sodium biselenite $0.8 \, g \, l^{-1}$, ferric ammonium sulphate $0.075 \, g \, l^{-1}$ and sodium thiosulphate $0.075 \, g \, l^{-1}$. This broth was first developed to be used in combination with the Easter and Gibson medium, in order to optimize detection of samples contaminated by *Salmonella*. In a study of confectionery products, results confirmed the false positive reactions due to the growth of *Citrobacter freundii* and *Escherichia coli* strains in the Easter and Gibson medium, and to the growth of *Serratia marcescens* and *Klebsiella pneumoniae* strains in the Arnott medium.

A comparative study between the conductance method (with the Easter and Gibson and the Ogden media described above) and the conventional method did not show any significant difference, but we must bear in mind that it dealt with artificially contaminated samples. In an analysis of 210 milk powder samples, the method using these two conductance media indicated 135 positive samples, but only 66 of these were confirmed by streaking on a selective agar, showing a high rate of false positive results. Another study showed a lack of sensitivity (69.7%) and specificity (80%) of this protocol in the analysis of 303 animal feed samples.

The combination of both media with a test based on the use of bacteriophages, was compared with a conventional method from artificially contaminated confectionery product samples; this 'bacteriophage test' consisted of the addition of a preparation of *Salmonella*-specific bacteriophages in a medium containing dulcitol. A delay in the detection or in gas production (visible on the curve) and the decrease of the amplitude of the conductance curve, was considered to be positive. This modification permits a negative screen in 2 days. The use of this bacteriophage test and a medium enriched in inhibitors allowed the detection of *Salmonella senftenberg* and *S. london* but with a rate of false positive reactions of 13.5%, probably due to the presence of *Citrobacter freundii* and *Hafnia alvei*.

A further modification of the protocol considers the results to be positive in Easter and Gibson medium if the acceleration rate exceeds 25 units h^{-1} and if the conductance changes are greater than 250 units, and in Ogden medium if the acceleration rate exceeds 100 units h^{-1} and the conductance changes are greater than 500 units. Unfortunately, this protocol does not eliminate false negative and false positive reactions due to the presence of *Pseudomonas* spp. and *Citrobacter amalonaticus*.

Other Media The lysine–iron–cystine–neutral red (LICNR) medium has been evaluated as a conductance medium for *Salmonella* detection; many bacterial strains gave false positive reactions. It was later modified by the addition of inhibitors (novobiocin and mandelic acid) which eliminated some false reactions, but not those due to *Citrobacter freundii*. A comparison with the Easter and Gibson medium and the Ogden medium in tests of confectionery product samples indicated that the most specific medium was the Easter and Gibson medium and the least specific the LICNR, with respective false positive rates of 1.0% and 3.8%. The authors concluded that their protocol was very sensitive and specific owing to the combination of the results of three different broths (Easter and Gibson, Ogden and LICNR).

Bolton initiated the research on *Salmonella* detection by the indirect conductance technique, showing that Whitley Impedance Broth (Don Whitley Scientific Ltd, Shirley, UK) allowed good negative conductance changes, and the detection of *S. typhimurium* and *S. virchow* pure strains.

The specificity and the sensitivity of four different formulations of RV broth were evaluated by the indirect technique on animal feed and raw meat samples: the analysis used both impedimetry (5 ml at 42°C for 36 h) and conventional methods (RV broth and isolation on brilliant green agar). Some commercial formulations yielded longer detection times, decreasing the growth of pure *Salmonella* strains probably as a result of inhibition by high $MgCl_2$ concentrations. It was concluded that this indirect system does not increase the rate of false negative detections, although potential false positive detections were not discussed. The direct method (with the Easter and Gibson and Ogden media) was compared with the indirect technique using RV broth in a study of 400 animal feed samples. The direct method showed lower selectivity and specificity, detecting 70% of samples indicated to be positive by the indirect technique. The rate of false positive results is 10% higher with the conductance technique.

A system developed for the detection of *Salmonella* by indirect conductance, using the specially formulated KIMAN broth (WIB supplemented with novobiocin $20 \, mg \, l^{-1}$, malachite green $10 \, mg \, l^{-1}$ and potassium iodide $40 \, g \, l^{-1}$) and followed by a confirmation on selective agar, yielded 100% sensitivity and a specificity up to 98.8% depending on the type of food analysed.

Conclusion

Impedance techniques offer the advantages of rapid detection and enumeration of microorganisms. Many researchers have developed specific media for the detection of different food pathogens, as described above. However, the specificity of the method is limited by the selectivity of the media so far developed, and for this reason most impedance protocols have shown inadequate specificity. Currently, impedance techniques are largely restricted to the enumeration of general flora such as total coliforms, faecal coliforms, Enterobacteriaceae or yeasts and moulds.

See also: **Campylobacter**: Introduction. **Clostridium**: Clostridium perfringens. **Escherichia coli**: Escherichia coli. **Listeria**: Introduction. **Salmonella**: Introduction; Detection by Enzyme Immunoassays. **Staphylococcus**: Staphylococcus aureus.

Further Reading

Bolton FJ (1990) An investigation of indirect conductimetry for detection of some food-borne bacteria. *Journal of Applied Bacteriology* 69: 655–661.

Bolton FJ (1991) Conductance and impedance methods for detecting pathogens. In: Vaheri A, Tilton RC and Barlows A (eds) *Rapid Methods and Automation in Microbiology and Immunology*. Finland: P. 176.

Easter MC and Gibson DM (1985) Rapid and automated detection of *Salmonella* by electrical measurements. *Journal of Hygiene* (Cambridge) 94: 245–262.

Firstenberg-Eden R and Eden G (1984) *Impedance Microbiology*. Letchworth: Research Studies Press.

Gibson DM, Coombs P and Pimbley DW (1992) Automated conductance method for the detection of *Salmonella* in foods: collaborative study. *Journal of AOAC International* 75: 293–302.

Ogden ID (1988) A conductance medium to distinguish between *Salmonella* and *Citrobacter* spp. *International Journal of Food Microbiology* 7: 287–297.

Ogden ID (1990) *Salmonella* detection by a modified lysine conductance medium. *Letters in Applied Microbiology* 11: 69–72.

Ogden ID and Cann DC (1987) A modified conductance medium for the detection of *Salmonella* spp. *Journal of Applied Bacteriology* 63: 459–464.

DIRECT EPIFLUORESCENT FILTER TECHNIQUES (DEFT)

Barry H Pyle, Montana State University, Bozeman, USA

Copyright © 1999 Academic Press

The direct epifluorescent filter technique (DEFT) was introduced in the early 1980s for the enumeration of bacteria in milk. Since then, the method has been adapted for counting bacteria in a variety of foods, including meat, fruit, vegetables and beverages. In addition to bacteria, it is possible to enumerate yeasts and moulds. These techniques are rapid, and facilitate enumeration of low cell numbers, especially in filterable samples such as beverages. A similar technique is referred to as the acridine orange direct count (AODC).

Direct microscopic counts of microorganisms in foods avoid some of the inherent deficiencies of traditional culture methods. More than 90% of viable microbes may be missed by current culture techniques, so direct counts are typically 10 times or more greater than total viable counts. The differences tend to be larger when bacteria have been injured by stressors such as heat, dehydration, disinfection and osmotic conditions. Some cells may become viable but non-culturable (VNC), in which case they fail to grow in routine culture but can be detected following special pre-incubation treatments or direct activity measurements.

Principles of the Test

For food samples, the procedure involves sample pretreatment, usually with buffer containing detergents and enzymes, filtration through a microporous membrane filter, staining with a fluorochrome, and epifluorescent microscopy for examination and enumeration. Fluorescence microscopy is mainly used for counting single cells or clumps. In addition, filter membranes can be transferred to solid media and incubated for a few hours for microcolony formation by viable cells. DEFT has also been used after enrichment to detect low numbers of bacteria in foods.

Equipment

A compound microscope with an epifluorescent illuminator, appropriate light filters, stage micrometer and eyepiece counting graticule (10×10 square) is required to perform DEFT. Filter assemblies and vacuum systems (100 kPa or less) are also needed.

Procedures

Some procedural steps vary depending on the type of food, microbes to be enumerated and whether stained cells or microcolonies are to be counted. The following procedure is recommended by the American Public Health Association for milk samples.

Sample Pre-treatment

Prefiltration or hydrolytic enzyme digestion may be required to facilitate membrane filtration. For milk, somatic cells are lysed by adding 0.5 ml rehydrated trypsin and 2 ml 0.5% Triton X-100 to 2 ml of sample, and incubating for 10 min at 50°C.

Filtration

A filter assembly is warmed with 5 ml of 50°C Triton X-100 before sample filtration through a 25 mm diameter black polycarbonate membrane (0.6 μm pore size). The filter assembly is then rinsed with a second 5 ml of 50°C 0.1% Triton X-100.

Staining

The membrane filter is overlaid for 2 min with 2 ml of stain (0.025% acridine orange (AO) and 0.025% Tinpal AN in $0.1 \, \text{mol} \, \text{l}^{-1}$ citrate-NaOH buffer, pH 6.6). This is followed by rinsing with 2.5 ml $0.1 \, \text{mol} \, \text{l}^{-1}$ pH 3 citrate-NaOH buffer, and 2.5 ml 95% ethanol. The filter is air-dried and mounted on a slide with non-fluorescent immersion oil.

Microscopy

The slide is examined either with a dry 60× fluorescence objective, or an oil immersion 100× objective, through a fluorescence microscope with light filters for AO, and an eyepiece counting graticule which has been calibrated with a stage micrometer. While some standard methods recommend counting only orange fluorescent cell clumps and single cells, it is advisable to count both orange and green cells to obtain the total direct microscopic count. A clump is a group of cells separated by at least twice the distance of the two cells nearest each other. Typically, at least 300 cells and at least three microscopic fields should be counted. The sensitivity of direct microscopy is ca. 10^4–10^5 cells per millilitre of sample.

Calculation

The number of cells in the original sample is obtained by multiplying the average count per field by the number of fields on the filtrable area of the filter (ca. 18 mm diameter, depending on the filtration assembly), and dividing by the volume of sample filtered.

Microcolony Count

Either selective or nonselective media may be used for microcolony formation. A 10 g sample of food is homogenized in 90 ml 0.1% peptone water, prefiltered through 5.0 μm pore size nylon mesh, then filtered through a 0.4 μm pore size black polycarbonate membrane. The filter is incubated on an agar medium or lipid medium support pad for 3–6 h at 30°C, depending on the medium and target organism. After AO staining, microcolonies that have $\geqslant 4$ bright orange cells are enumerated microscopically.

Direct Viable Count (DVC)

The sample is incubated with dilute nutrients and nalidixic acid or another similar antibiotic that inhibits DNA gyrase, preventing completion of the cell division cycle. Substrate responsive cells elongate or enlarge because of failure to septate. Following staining, cells that are more than 1.5× the typical size are enumerated by microscopy.

Alternative DEFT Stains

Differences in the numbers of bacteria detected may depend on the staining method and the sample characteristics. While AO has been used widely in DEFT procedures, alternatives such as 4′,6-diamidino-2-phenylindole (DAPI) are replacing AO in many applications. A variety of other stains including acriflavine, bisbenzimide dyes, erythrosine and fluorescein isothiocyanate have also been used.

Viability Stains

A number of fluorescent stains are available that can indicate bacterial cell viability or metabolic activity. These include the use of dual staining such as the Live/Dead *Bac*Light viability kit (Molecular Probes, Eugene, OR), which is used to distinguish live bacteria with intact plasma membranes from dead bacteria with compromised membranes. Stains such as rhodamine 123 can be used to detect cells with a membrane potential, while $DiBAC_4(3)$ (Molecular Probes) permeates cells that lack a membrane potential. Cyanoditolyl tetrazolium chloride (CTC) is taken up and converted to intracellular fluorescent CTC-formazan crystals by dehydrogenase activity in respiring cells. Esterase activity can be detected by uptake and cleavage of fluorescein diacetate, which forms free fluorescein in active cells. Although the colour of AO staining was proposed as a means of determining viability or physiological activity, results should be interpreted with caution because of the effects of staining methods.

Ab-DEFT and Immunomagnetic Separation

Use of fluorescent antibodies permits rapid enumeration and identification of specific bacteria such as *Escherichia coli* O157:H7 in some foods, including milk, juice and beef. *Listeria* in fresh vegetables have also been quantified by Ab-DEFT. Immunomagnetic separation (IMS) methods have been combined with AB-DEFT to improve sensitivity. It may be possible to detect as few as 10^1–10^2 cells per millilitre or per gram of sample using IMS methods.

Automated Methods

At least two automated systems are available for DEFT. The BactoScan (Foss Electric) performs a total count of bacteria in raw milk by pre-treatment, staining and detection on the outer edge of a rotating disc. Up to 80 raw milk samples may be analysed per hour. COBRA (Biocom) automates the filtration, staining, rinsing, drying and counting procedures using automated microscopy and image analysis. Over 100 samples per hour may be processed. Results obtained with these systems correlate well with colony counts. Image analysis has also been used to automate the DVC procedure. The MicroStar (Millipore) is an instrument for enumerating bacterial microcolonies and individual yeasts using ATP luminescence. Flow cytometry techniques have been used to enumerate fluorochrome-stained cells, in addition to solid-phase laser scanning cytometry (ChemScan or ScanRDI, Chemunex). ChemChrome V3 (Chemunex) which indicates esterase activity may be used to detect the total number of metabolically active cells with this system. A hybrid method that includes IMS with CTC incubation and fluorescent antibodies has been used with the solid-phase cytometer to detect low numbers ($< 10\,g^{-1}$) of *E. coli* O157:H7 with an indication of their respiratory activity in ground beef within 5–7 h.

See also: **ATP Bioluminescence**: Application in Meat Industry; Application in Hygiene Monitoring; Application in Beverage Microbiology. **Electrical Techniques**: Food

Spoilage Flora and Total Viable Count (TVC). *Escherichia coli* **O157**: Detection by Commercial Immunomagnetic Particle-based Assays. **Flow Cytometry. Hydrophobic Grid Membrane Filtrate Techniques (HGMF). Immunomagnetic Particle-based Techniques**: Overview. **Total Viable Counts**: Pour Plate Technique; Spread Plate Technique; Specific Techniques; MPN; Metabolic Activity Tests; Microscopy; **Verotoxigenic *E. coli***: Detection by Commercial Enzyme Immunoassays.

Further Reading

Boisen F, Skovgaard N, Ewald S, Olsson G and Wirtanen G (1992) Quantitation of microorganisms in raw minced meat using the direct epifluorescent filter technique: NMKL collaborative study. *J. AOAC Int.* 75: 465–473.

Duffy G and Sheridan JJ (1998) Viability staining in a direct count rapid method for the determination of total viable counts on processed meats. *J. Microbiol. Methods* 31: 167–174.

Grigorova R and Norris JR (eds) (1990) *Methods in Microbiology*. Volume 22: *Techniques in Microbial Ecology*. P. 1. London: Academic Press.

Kepner RL Jr and Pratt JR (1994) Use of fluorochromes for direct enumeration of total bacteria in environmental samples: past and present. *Microbiol. Rev.* 58: 603–615.

Kogure K, Simidu U and Taga N (1979) A tentative direct method for counting living marine bacteria. *Can. J. Microbiol.* 25: 415–420.

Lisle JT, Broadaway SC, Prescott AM, Pyle BH, Fricker C and McFeters GA (1998) Effects of starvation on physiological activity and chlorine disinfection resistance in *Escherichia coli* O157:H7. *Appl. Environ. Microbiol.* 64: 4658–4662.

McFeters GA, Singh A, Byun S, Callis PR and Williams S (1991) Acridine orange staining reaction as an index of physiological activity in *Escherichia coli*. *J. Microbiol. Methods* 13: 87–97.

Mossel DAA, Corry JEL, Struijk CB and Baird RM (1995) *Essentials of the Microbiology of Foods*. P. 302, 339. Chichester: Wiley.

Pettipher GL (1983) *The Direct Epifluorescent Filter Technique for the Rapid Enumeration of Micro-organisms*. Hertfordshire: Research Studies Press/New York: Wiley.

Pettipher GL and Roderigues UM (1982) Rapid enumeration of microorganisms in foods by the direct epifluorescent filter technique. *Appl. Environ. Microbiol.* 44: 809–813.

Pettipher GL, Fulford RJ and Mabbitt LA (1983) Collaborative trial of the direct epifluorescent filter technique (DEFT), a rapid method for counting bacteria in milk. *J. Appl. Bacteriol.* 54: 177–182.

Pettipher GL, Watts YB, Langford SA and Kroll RG (1992) Preliminary evaluation of COBRA, an automated DEFT instrument, for the rapid enumeration of micro-organisms in cultures, raw milk, meat and fish. *Lett. Appl. Microbiol.* 14: 206–209.

Pyle BH, Broadaway SC and McFeters GA (1999) Sensitive

detection of *Escherichia coli* O157:H7 in food and water using immunomagnetic separation and solid-phase laser cytometry. *Appl. Environ. Microbiol.* 65: 1966–1972.

Restaino L, Castillo HJ, Stewart D and Tortorello ML (1996) Antibody-direct epifluorescent technique and immunomagnetic separation for 10-h screening and 24-h confirmation of *Escherichia coli* O157:H7 in beef. *J. Food Prot.* 59: 1072–1075.

Roderiguez-Otero JL, Hermida M, Cepeda A and Franco C (1993) Total bacterial count in raw milk using the BactoScan 8000. *J. AOAC Int.* 76: 838–841.

Singh A, Pyle BH and McFeters GA (1989) Rapid enumeration of viable bacteria by image analysis. *J. Microbiol. Methods* 10: 91–101.

Tortorello ML and Gendel SM (1993) Fluorescent antibodies applied to direct epifluorescent filter technique for microscopic enumeration of *Escherichia coli* O157:H7 in milk and juice. *J. Food Prot.* 58: 672–677.

Tortorello ML, Reineke KF and Stewart DS (1997) Comparison of antibody-direct epifluorescent filter technique with the most probable number procedure for rapid enumeration of *Listeria* in fresh vegetables. *J. AOAC Int.* 80: 1208–1214.

Vanderzant C and Splittstoesser DF (1992) *Compendium of Methods for the Microbiological Examination of Foods.* P. 102. Washington, DC: American Public Health Association.

Disinfectants *see* **Process Hygiene**: Testing of Disinfectants.

DRIED FOODS

M D Alur and **V Venugopal**, Food Technology Division, Bhabha Atomic Research Centre, Mumbai, India

This article describes different methods of dehydration of foods, the influence of water activity on growth and toxin formation of pathogenic bacteria as well as physical, chemical and microbiological changes occurring in food as a result of drying. The importance of the temperature of rehydration to prevent health hazard and the measures to be taken to gain advantages of dehydration of foods are also discussed briefly.

Advantages of Ambient Temperature Storage

Dehydration is a method of food preservation that arrests the activity of food-borne microorganisms and retards chemical changes in the food. The preservation of food by dehydration is of ancient origin. The method is important since practically every type of food can be prepared in dehydrated form, including nuts, vegetables, fruits, meats, fish, eggs, milk and soups.

The most economical method of drying is sun-drying, which is commonly used in hot, dry areas. The product is spread out in a thin layer on a cloth or plastic sheet in the sun. The process is slow and uncertain and may be lengthened by unexpected rains. It usually takes several days for the moisture content to be reduced to a level low enough to prevent spoilage.

Drum drying is the most common commercial method of drying foods. The product is tumbled in a large drum and warmed, dry air is passed over the product and carries the moisture away. Drum drying is more expensive than sun drying, but can be done anywhere and the moisture content can be reduced to a lower level.

Dried foods find their greatest use in times of disaster, either natural or man made. Though it is true that drying is one of the principal methods of food preservation, and a natural force used in preservation of seeds and fruits, dried vegetables have not been popular unless war time or national disaster enforces their use. Current drying techniques yield highly acceptable food product such as dried skim milk, potato flakes, soup mixes, soluble coffee, prepared baby foods and freeze-dried prepared meals. The obvious economic benefit in the distribution of acceptable dried foods should be emphasized in this method of food preservation. The caramelization, loss of texture and physical form, loss of flavour characteristics and poor rehydration ability of dried foods have left an imprint on the minds of consumers. However, improved technology can be used to produce high quality dried foods.

Drying is a complex process involving the application of thermal energy which causes evaporation of water into the vapour phase, is therefore, associated with reduction in volume or weight and results in changes in shape of the foods (powder, flakes, granules, texture), physical changes (phase transition, crystallization) and chemical or biochemical changes (resulting in changes in flavour, colour and texture). The dehydrated foods may contain as little as 2–3%

moisture. The actual bulk of certain dehydrated foods may be further reduced by compression; for example the ratios of raw cabbage to packed dehydrated product can be 18 : 1 (by weight) or 38 : 1 (by volume) when compressed, compared to 12 : 1 and 7 : 1, respectively, when compression techniques are omitted. Similarly, one tonne of chilled carcass beef occupies 3.07 m³ of actual whole space, one tonne of frozen carcass beef occupies 2.5 m³ and one tonne of boneless frozen beef, 2.23 m³. The dehydrated equivalent in the form of mince weighs 242 kg, compressed and packed into wooden cases. The total weight is 280 kg and the volume 0.35 m³. The advantages and disadvantages of dehydration of foods are shown in **Figure 1**.

Dryers used for drying food items can be classified based on the mode of heat transfer, namely, convection (direct), conduction (contact) or radiative, which supply heat at the surface of the material. Dielectric heating (microwave or radio frequency) induces heating throughout the volume of the food. The methods of classification of dryers based on various criteria are given in **Table 1**.

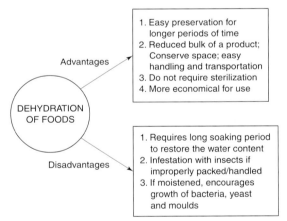

Figure 1 Advantages and disadvantages of dehydration of foods.

Table 1 Criteria used for classification of dryers

Criterion	Type
Mode of operation	Batch, continuous
Heat in-put	Convection, conduction, radiation, electro-magnetic field, combination of one or more of the above
Operating pressure	Vacuum, atmospheric, high pressure
Drying medium	Hot air, super-heated steam, flue gases
Drying temperature	Below boiling point of water, above boiling point, below triple point (freeze-drying)
Residence time	Short (< 1 min), medium (1–30 min), high (> 30 min up to several hours)
State of material in the dryer	Stationary, moving, agitated, vibrated

State of Water in Foods

Food constituents interact in various ways with water depending on the functional groups available in the food constituents. The interactions may be broadly ionic or non-ionic (polar or non-polar) in nature. An ion immobilizes water in oriented layers surrounding the charge centre. For example, the higher the amount of available polar amino acids in a food system, the larger the volume of bound water. Non-polar side chains of food macromolecules interact with water through hydrophobic interactions, the cages of water molecules being held by a multitude of non-covalent interactions such as van der Waals' forces, hydrogen bonds and other weakly polar bonds. The hydrophobic interaction of non-polar groups in aqueous solution has a major influence on the stability of macromolecules in the food system. These interactions help the food systems to hold significant amounts of water in their matrices. For example, muscle proteins can hold three to four times their weight of water; gelatin and pectin gels can hold as much as 20 times their weights of water. Heating (drying at high temperature) or incorporation of electrolytes such as salts affect the non-covalent interactions of food macromolecules and water resulting in phase separation, shrinkage and release of water. Moisture contents of various foods before and after drying are given in **Table 2**.

Water Activity

Water activity is defined as the ratio of the vapour pressure exerted by water contained in the food (P) to the vapour pressure of pure water (P_o) at the same temperature. Water activity is also defined as the equilibrium relative humidity in which a food could neither gain nor lose moisture. The water sorption phenomenon of the food is a true reflection of such interactions. The sorption property of the food is best described by the sorption isotherm, which is a plot of

Table 2 Moisture content of various foods before and after drying

Product	Moisture (%)	
	Before drying	After drying
Yolk	51.0	1.1
Roast beef	60.0	1.5
Broiled chicken	61.0	1.6
Whole egg	74.0	2.9
Cooked corn	76.5	3.2
Raw figs	78.0	3.6
Boiled potato	80.0	4.0
Apple juice	86.0	6.2
Whole milk	87.0	5.0
Cooked beans	92.0	11.5

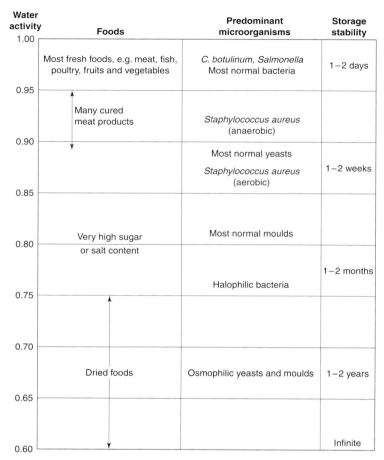

Figure 2 Relationship between water activity and growth of organisms.

the amount of water sorbed as a function of water activity.

Water activity, $a_w = P/P_0 = ERH/100$

The a_w of any food is in the range 0–1.0. The relationship between water activity and growth of organisms is shown in **Figure 2**.

Water Sorption Phenomenon

The sorption property of a food is best described by the sorption isotherm which is a plot of the amount of water sorbed as a function of water activity. The water activity of a food can be controlled by removal or addition of water to reach the desired point in the water sorption isotherm. The isotherm can be divided into three regions describing the state of the water present. Region A at low moisture content, corresponds to the adsorption up to a mono-molecular layer of water. Region B above this moisture content, is the adsorption of additional water over the mono layer giving a multilayer or intermediate layer. Region C, at high moisture content, refers to the condensation of water in the pores of the material followed by

dissolution of the soluble material. It is also described as mobile, free or capillary water.

The differences between adsorption and desorption curves are called sorption hysteresis. The desorption curve always has a higher moisture content than that of adsorption curve at the same water activity. The phenomenon may be explained by the fact that drying causes shrinkage of the food leaving behind less polar sites for water binding in a subsequent adsorption cycle.

Microbiology of Dried Foods

Microorganisms require water in order to grow; however the amount of moisture that permits growth varies with different kinds of microbes. Food microbiologists use the term water activity to designate the amount of moisture available in foods. Salted or dried foods have lower water activity or lower amounts of water available for microbial growth, compared with most fresh foods. Consequently, many bacteria are inhibited in salted or dried foods. Drying decreases the water activity of a food below the limits required for microbial growth. Drying by natural means (sun

Table 3 Effect of water activity (a_w) on growth and toxin production by some food-borne pathogens

Pathogenic microorganisms	Minimum a_w	
	Growth	Toxin production
Aspergillus flavus	0.82	0.78
Staphylococcus aureus	0.84	0.86
Clostridium botulinum types A and B	0.92	0.82
Listeria monocytogenes	0.92	–
Bacillus cereus	0.93	–
Vibrio parahaemolyticus	0.95	–
Salmonella sp.	0.96	–
Clostridium perfringens	0.96	–

drying) or by artificial means is often supplemented by other methods, such as salting or adding high concentrations of sugar.

During drying, the rate of removal of water, which is held by the food constituents, depends not only on the drying conditions, but also the chemical nature of the food. In addition to drying, the amount of moisture available for microbial growth may also be reduced by incorporating certain additives that can bind the water and thus reduce a_w. Such additives include glycerol, sugar or salt, which may not affect flavour of the foods. In addition to water activity the nature of the solutes in the food system also influence the growth of surviving flora. For example, at the same water activity, glycerol may permit the growth of organisms such as *Clostridium* spp., whereas sodium chloride may prevent microbial growth.

Dehydrated foods, unlike canned foods, are not sterile. Therefore, it is of utmost importance to prevent the entrance of organisms which are capable of producing toxins. Microorganisms respond to reduced water activity in a number of ways. Microbial growth fails to occur at reduced water activity levels. The effect of reduced water activity is characterized by an extension of the lag phase, a suppression of the log phase and a reduction in the total number of viable microorganisms. The minimal water activity levels for growth and toxin production of a number of food-borne pathogens are shown in **Table 3**.

Heat applied during a drying process causes a reduction in the total number of microorganisms, but the effectiveness varies with the kinds and numbers of organisms originally present. Usually, yeasts and most bacteria are destroyed but spores of bacteria and moulds survive together with a few vegetative cells of heat-resistant bacteria. *Salmonella* spp. can survive in dried eggs and these organisms can cause diseases in susceptible individuals who eat inadequately cooked egg products. Some countries, therefore, have laws stating that only pasteurized dried eggs can be sold.

More organisms are killed by freezing than by dehy-dration during the freeze drying process. If the drying process and storage conditions are adequate, there will be no growth of microorganisms in the dried food. During storage, there is a slow decrease in numbers of organisms. The microorganisms that are resistant to drying will survive. In particular, resistant spores of bacteria, moulds and some of the micrococci survive (**Fig. 3**).

In conventional drying used for vegetables, the microbial flora will be modified by drying, and if the equipment is not clean, increases in counts may be anticipated. Temperatures up to 90°C are used during the first stage of drying; the rapid moisture loss from foods during this period induces a cooling effect, and thus the temperature is maintained at 40–50°C. Hence, a small reduction in microbial counts occurs. In the second stage of drying the food temperatures are higher (60–70°C) and yeasts and many bacteria are killed. The residual flora after drying consists of spore-formers (*Bacillus* and *Clostridium* spp.), enterococci and moulds (e.g. *Aspergillus, Penicillium, Alternaria* and *Cladosporium* spp.). With dried spray-dried milk and egg, the temperatures reached are not so high, and hence a much more varied flora, including *Salmonella* spp., may persist. To overcome this danger, milk is now pasteurized before drying. On the other hand, microbial counts in freeze-dried foods are often high.

To prevent the formation of bacterial toxins or the development of organisms pathogenic to humans, the food product should be dried at temperatures where growth is unlikely to occur. At temperatures of 50°C,

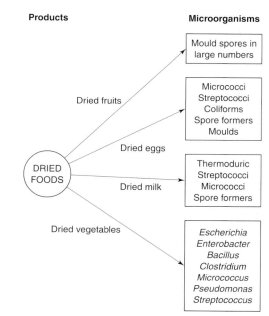

Figure 3 Microbiological status of specific dried foods.

it is probably the minimum. Where some heating below this temperature is unavoidable because of loss of quality in the product, the period of dehydration should not exceed 4 h. Bacterial growth generally does not occur in foods containing less than 15% water. Foods at a_w levels below 0.78–0.80 should not be a safety hazard. They also indicate that mould species are generally more osmotolerant than bacterial species both in their ability to grow and produce toxins at low a_w levels. *Staphylococcus aureus*, the bacterial species capable of growth at the lowest a_w has been exploited in some food regulations, e.g. the requirement of $a_w < 0.85$ for the transportation of dried foods in Germany and the USA.

Foods can be protected against microbial deterioration by keeping their a_w below a certain value. The approximate lower limit of a_w at which bacteria, yeast and moulds can grow are 0.91, 0.88 and 0.80, respectively. However, apart from water activity, the nature of the solute also determines the survival of residual organisms.

Alarm Water Contents

As a guide to the storage stability of dried foods, the 'alarm water content' has been suggested. This is the water content which should not be exceeded if mould growth is to be avoided. Although drying stops growth, it does not kill all bacteria in the foods. For example, a number of cases of salmonellosis have been traced to dried eggs. This occurs because eggshells may be heavily contaminated with species of pathogenic *Salmonella* from the gastrointestinal tracts of the hen. When the eggs are cracked, many of the bacteria enter and subsequently survive drying. These bacteria can cause disease in susceptible individuals who eat inadequately cooked egg products.

Dehydrated foods are never sterile. In this respect, they differ from canned foods. Therefore, it is of utmost importance to prevent the entrance of organisms which are capable of producing food poisoning in such foods. This applies to the toxigenic forms of *Clostridium botulinum* and certain strains of *Staphylococcus aureus*. Dehydrated foods should also be free from certain intestinal microorganisms that are likely to be pathogenic to humans when taken by mouth, e.g. members of the genera, *Salmonella* and *Shigella*. The bacterial counts of dried foods should be low to avoid development of undesirable flavours during the period of reconstitution.

Most food poisoning bacteria, including *C. botulinum*, *Salmonella* and *Shigella*, are unable to grow or produce toxins below an a_w of 0.95 in foods. Some aerobic spore-forming bacteria such as *Bacillus cereus* can tolerate much lower water activity (0.93) and the

most xerotolerant species like *Bacillus subtilis* are able to develop at a_w 0.90.

Staphylococcus aureus is the most xerotolerant and toxin-producing bacterium. The limit of growth in laboratory media is 0.86, but enterotoxin A which is frequently encountered in foods is not produced below an a_w of 0.88 at 25°C. Such a difference between a_w levels permitting growth and those allowing the production of toxins is also frequent for mycotoxins produced by toxigenic moulds.

Mycotoxins are non-antigenic toxic substances for humans and animals which can be elaborated by moulds in foodstuffs. The lowest a_w levels for mycotoxin biosynthesis are 0.80 and concern aflatoxin B_1. It is well established that contamination by mycotoxins never occurs without an evident growth of moulds which lead to the rejection of the product. Nevertheless, mycelial growth and toxinogenesis of *Aspergillus flavus* without any evident sporulation have been reported. This forms an important problem of public health.

Mycotoxins also represent a potential problem in the cheese industry, especially, blue cheeses which are normally processed with the help of moulds. Most *Pencillium* strains produce numerous mycotoxins under laboratory conditions. However, it seems that mycotoxins in cheese have not represented a problem since it has been calculated that an adult of 70 kg must eat daily 7 kg of blue cheese contaminated by 5 p.p.m. of mycophenolic acid in order to show early symptoms.

S. aureus forms five enterotoxins, A, B, C, D and E of which enterotoxin A is the most encountered in foods. It has been shown that an increase in NaCl concentration results in a decrease in growth rate of *S. aureus* and even greater suppression of toxin production. For example when growth was suppressed at 16% NaCl, enterotoxin B was suppressed more, even at 10% NaCl.

The minimum a_w for toxin produced by *Clostridium botulinum* strains is complicated by interacting factors such as NaCl concentration, pH, oxidation–reduction potential and proteolytic activity. However, formation of toxin by type E is more sensitive to reduced water activity than that of either type A or B, and neither growth nor toxin production has been reported at a_w less than 0.93.

Similarly, mycotoxin formation is more sensitive to a reduction in water activity than is the growth rate of the mould forming it. To prevent the formation of bacterial toxins or the development of organisms pathogenic to humans, the food product should be dried at a temperature where growth is unlikely to occur. The 'alarm water' contents for miscellaneous foods are given in **Table 4**. These values are equivalent

Table 4 The 'alarm water' content of miscellaneous foods

Food product	Water (%)
Dehydrated fruits	18–25
Starch	18
Dehydrated vegetables	14–20
Pulses, milk powder, dehydrated meat	15
Rice, wheat flour	13–15
Dehydrated whole eggs	10–11
Whole milk powder	8

Table 5 Quality changes that occur in foods during drying

Physical changes	Chemical/biochemical changes
Shrinkage	Lipid oxidation
Density	Denaturation of proteins
Porosity, shape, size	Enzymatic browning
Reduced rehydration	Maillard reaction
Destruction of all cell structure and hence textural changes	Decomposition of vitamins and hence loss of nutritional quality

to an a_w of 0.7, at which point microbiological stability can be expected.

Changes in Quality of Foods during Drying and Storage

Dehydrated meats when exposed to air, develop a rancid flavour. Oxidation also leads to bleaching of meat pigments. Dehydrated fish, when exposed to air, gradually develop a strong fish odour, to ultimately become ammoniacal.

The quality changes that occur in foods during drying are given in **Table 5**.

In dried foods, the microbial activity is arrested by lowering water activity below the values required for their growth. In freeze-dried foods, although microorganisms may survive the operation, a continual 'drying off' may occur during storage. Under temperate conditions, most dehydrated foods, if properly packed, have a shelf life of over two years. However, at storage temperatures of 37°C, the shelf life may be only 3–6 months. The lower the moisture content, the longer the shelf life at high temperatures. Dehydrated foods are susceptible to changes, and if left in the open, will tend to stale in the first instance. This is because oxidative rancidity and enzyme action can occur at water activities lower than that required for the growth of moulds, yeasts or bacteria. In tropical climates, exposure to high temperature gives rise to brown discoloration. The rate and the extent of browning depends on the reducing sugar content, e.g. dried eggs, dehydrated potatoes. In addition to browning, other types of deterioration such as development of off flavours and toughening of the products due to some loss of reconstitution may occur. Simi-

larly, deterioration of foods and control strategies are depicted in **Figure 4**.

Packaging

Ideal packaging materials used to control quality changes are those which can prevent penetration of microorganisms as well as transport of water vapour, aroma and oxygen. Susceptibility of polymer films to bacterial penetration decreases with increasing density. For example, polyethylene film with a weight of $11.5 \, g \, m^{-2}$ has a penetration value of 40; this decreases to 5 when the weight is increased to $22 \, g \, m^{-2}$. Similarly, bacterial penetrability of polyvinylidene chloride (PVDC) decreases from 95–100 to zero as the weight of the film increases from 7–21 to $45 \, g \, m^{-2}$.

If dehydrated vegetables are kept in packs for too long, oxidation can be detected by the odour of the contents; e.g. a hay-like smell from cabbage, a rancid odour from potatoes, a perfumed odour from carrot. Nonetheless, since the general purpose of food storage is to extend the product of a glut season to avoid possible shortages during the next season, minor changes in quality may be acceptable.

Rehydration

Rehydration is the part of food processing that normally takes place in the home or in an institution where foods are prepared for serving. In dried foods, the normal microflora is at least partially reduced by the drying conditions. Therefore, rehydration may pose microbial hazards, since such foods may favour rapid growth of pathogens in the absence of competition from normal microflora. For example, it has been reported that in rehydrated chicken, *Staphylococcus aureus* and faecal streptococci can compete with a total aerobic flora at temperatures above 20°C, even though the pathogens may be present initially as low as 0.01% of total flora.

A bacterial cell in a completely dried state has an infinite concentration of solutes. As the solvent water is reintroduced, the solute concentration will decrease rapidly, but not uniformly. This may result in a loss in the loss in the control of permeability by the membrane. Permeability alterations have been detected in freeze-dried bacterial cells. Freeze-dried *Escherichia coli* become sensitive to chloramphenicol, streptomycin and actinomycin D as well as to sodium deoxycholate. Undried cells of this species are not affected by these compounds upon rehydration. Freeze-dried *Salmonella typhimurium* also showed altered membrane permeability as measured by increased sensitivity towards antibiotics and ribonuclease.

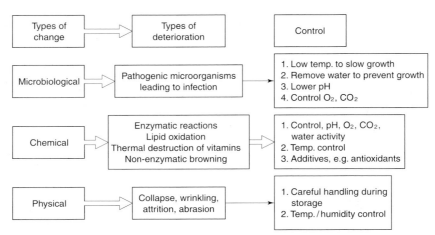

Figure 4 Deterioration of foods and control strategies.

The microbial content and temperature of water used for rehydration of dried foods will affect the keeping quality of the rehydrated product. Bacteria in freeze-dried chicken meat are further reduced by rehydration with water at 50°C and almost eliminated when the water is at 85–100°C. Growth of bacteria in the rehydrated meat will occur at favourable temperature, but there is a good shelf life at 4°C. *Staphylococcus aureus* has been found to survive freeze-drying and rehydration at 60°C. Therefore, rehydration at 100°C is recommended.

When dried foods are rehydrated, similar responses are shown by the contained microorganisms, as with thawed frozen foods. There is a lag phase of growth and many organisms exhibit metabolic injury. However, the temperature of the water used for rehydration has an important effect on the flora and on the subsequent rate of spoilage of the food. If boiling water is used, *Bacillus* spp. will predominate and cause spoilage and at lower rehydration temperatures, the flora will become most varied and contain heat-labile organisms. When refrigerated storage is used, the storage life of most rehydrated foods is restricted to 1–2 days, but storage at room temperature is limited to a few hours.

The preventive measures that need to be taken to ensure safety of dried foods include the following:

- Keep the moisture content as low as possible, e.g. the moisture content of cabbage should be 3% which will double its storage life.
- Keep the level of reducing sugar as low as possible. These compounds are involved in the non-enzymatic browning and, if they are reduced, the storage stability of the product is increased.
- While blanching, keep leached soluble solids low.
- Before dehydration treat vegetables with SO_2 to protect vitamin C and retard the browning reaction.

One of the most important considerations in preventing fungal spoilage of dried foods is relative humidity (RH) of the storage environment. If improperly packaged and stored, under conditions of high RH dried foods will pick up moisture from the atmosphere and spoilage will be inevitable since surface growth is the characteristic of moulds.

See also: **Bacillus**: *Bacillus cereus*; *Bacillus subtilis*. **Clostridium**: *Clostridium botulinum*. **Eggs**: Microbiology of Egg Products. **Mycotoxins**: Occurrence. **Preservatives**: Traditional Preservatives – Sodium Chloride. **Salmonella**: Introduction. **Staphylococcus**: *Staphylococcus aureus*.

Further Reading

Baker CGJ (ed.) *Industrial Drying of Foods*. New York: Chapman and Hall.

Dalgleish JM (1984) Dehydration and Dried Products. In: Rankin MD (ed.) *Food Industries Manual*, 21st edn. Glasgow and London: Leonard Hill.

Duckworth RB (ed.) (1975) *Water Relations of Foods*. London: Academic Press.

Hardman TM (ed.) (1989) *Water and Food Quality* London: Elsevier Applied Science.

Hayes PR (1985) *Food Microbiology and Hygiene*. London: Elsevier Applied Science.

Jay JM (1992) *Modern Food Microbiology*. New York: Van Nostrand.

Karel M (1991) Physical Structure and Quality of Dehydrated Foods. In: Majumdar AS (ed.) *Drying '91*. Amsterdam: Elsevier.

Majumdar AS (1997) Drying Fundamentals. In: Baker CJB (ed.) *Industrial Drying of Foods*. New York: Chapman and Hall.

Mathlouthi M (ed.) (1986) *Food Packaging and Preservation. Theory and Practice*. New York: Elsevier Applied Science.

Montville JJ (ed.) (1987) Concepts in Physiology and Metabolism. In: *Food Microbiology*. Vol. 1. Boca Raton, Florida: CRC Press.

Nester EW, Roberts CE, Lidstrom ME, Pearsall NN and Nester MT (eds) (1983) *Microbiology*. Japan: Saunders College Publishing, Holt-Saunders.

Nickerson JTR and Sinskey AJ (eds) (1977) *Microbiology of Foods*. New York: Elsevier.

Salame M (1986) The Use of Barrier Polymers in Food and Beverage Packaging. In: Finlayson KM (ed.) *Plastic Film Technology*. Vol. 1, p. 132. New York: Technomic.

Simatos D and Multon JL (eds) (1985) *Properties of Water in Foods in Relation to Quality and Stability*. Boston: Martinus Nijhoff.

ECOLOGY OF BACTERIA AND FUNGI IN FOODS

Contents

Influence of Available Water

K Krist, Meat and Livestock Australia, Sydney,
Australia

D S Nichols and **T Ross**, School of Agricultural
Science, University of Tasmania, Hobart, Australia

Introduction

Although there is a variety of resting or survival stages of microorganisms that are resistant to drying, all organisms need water to remain metabolically active. The availability of water to an organism in an environment is not simply a function of how much water is present, but the degree to which it is adsorbed to the insoluble components of the environment or chemically associated with solutes in that environment. For this reason, the concept of water activity (a_w), a measure of the availability of water to participate in chemical reactions, was devised. Though a_w is not a perfect predictor of the behaviour of microorganisms in a specified environment (knowledge of the solutes and factors that contribute to the a_w is also required), it is widely used to describe the relationship between the water in an environment and its microbial ecology.

The reduction of a_w to increase the microbiological stability of foods has probably been used since antiquity. The drying effect of the air and the sun required no special technology and is still used today. Similarly, the addition of salt or sugars requires no special technology and has been used for centuries to preserve food. Those techniques are still in use in many parts of the world, using free energy and providing safe products. More recently, technology (e.g. hurdle technology) has sought to maximize the potential of drying techniques while minimizing the severity of treatments to develop shelf-stable products that are less altered from the fresh state.

This article considers the microbial ecology of bacteria and fungi in relation to a_w. a_w and related terms, are defined. Methods for manipulating a_w in foods are discussed, and the effects of a_w on growth rate, lag-phase duration, yield and death rate of microorganisms described. The physiology of the response of microbial cells to a_w stress is also discussed.

Concept of Water Activity/Available Water

Water activity can be affected by both solutes and adsorption. The solutes effect is called osmotic potential. The adsorption effect is called matric water potential but it is not widely considered in food microbiology. None the less, insoluble materials such as wood, paper, metal and glass, and including foods, adsorb water. The strength of the attachment is a function of the physical and chemical properties of the material. Those materials will tend to take water up from, or release water to, the atmosphere until an equilibrium is reached between the atmosphere and the material. Foods will tend to equilibrate with the relative humidity of the container or environment they are stored in. Thus, dry foods can take up water from humid environments, or moist foods will tend to dry out in dry environments. If a food is allowed to equilibrate with the humidity of the storage atmosphere, the matric a_w will affect the organism just as if the osmotic a_w had been altered to the same relative humidity.

The terms water activity, water potential, osmotic pressure and solute concentration are often used interchangeably by microbiologists to refer to the availability of water to microorganisms. Although each of these concepts is related, they are different. Solute concentration is self-explanatory, although it may be expressed in different ways (e.g. w/w, w/v, molarity, molality, etc.). High solute concentrations result in decreased a_w, and less water available to microorganisms for metabolism. Solutes that alter a_w are termed humectants.

Water Activity

A_w is a fundamental property of aqueous solutions. It is defined as:

$$a_w = \frac{\rho}{\rho_0} \qquad \text{(Equation 1)}$$

where ρ = vapour pressure of the solution; ρ_0 = vapour pressure of the pure water under the same conditions of temperature, etc. And where:

$$\frac{\rho}{\rho_0} = \text{relative humidity}$$

The a_w of most solutions is temperature-dependent. Equilibrium relative humidity, a measure widely used in meteorology and building environmental control, is related to a_w by the simple expression:

$$a_w = \frac{\text{Equilibrium relative humidity (\%)}}{100} \qquad \text{(Equation 2)}$$

When solutes are dissolved in water, some of the water molecules become more ordered as they become oriented on the surfaces of the solute molecule. This reduces the vapour pressure of the solution, since on average the water molecules then have less entropy. In turn, a_w is reduced. The a_w of a solution decreases with increasing solute concentration. The effect of solute concentration on a_w is expressed mathematically:

$$a_w = \frac{-vm\Phi}{e^{55.51}} \qquad \text{(Equation 3)}$$

where v = the number of ions generated by each molecule of solute (e.g. for non-electrolytes, $v = 1$; for NaCl, $v = 2$; for H_2SO_4, $v = 3$); m = molal concentration of the solute; φ = molal osmotic coefficient.

Equation 3 reveals that the a_w at a given solute concentration is dependent on the specific solute, because each solute has its own osmotic coefficient and will dissociate into a different number of ions.

Osmotic Pressure

The osmotic pressure of a solution is related to its a_w and includes this term in its definition:

$$\text{Osmotic pressure} = \frac{-RT \ln a_w}{\overline{V}} \qquad \text{(Equation 4)}$$

where R = the universal gas constant ($8.314\,\text{J}\,\text{k}^{-1}\,\text{mol}^{-1}$); T = absolute temperature (K); $\overline{V}$ = partial molar volume of water and all other terms are as previously defined.

Increased osmotic pressure literally means that the cell is subjected to an increased external pressure, or alternatively, a decreased internal pressure. Increased extracellular osmotic pressure refers to a situation where the availability of water to bacteria is decreased.

The term water potential, widely used by soil microbiologists, also expresses the availability of water, but is defined as the difference in free energy of the environment being considered, and a pool of pure water at the same temperature: the terms water activity and water potential are measures of the energy of water. Water potential may be expressed in a number of units, of which the most widely used is the bar ($10^6\,\text{dyn}\,\text{cm}^{-2}$). Water potential is always a negative value or zero.

As shown in **Table 1**, a_w and water potential are not directly proportional, however, a 0.01 decrease in a_w corresponds to a decrease of approximately 15 bar water potential in the range of a_w typical of foods.

Tables of a_w for various solutes and solute mixtures are available in the literature. The effect on a_w of solutions containing several solutes can be estimated from the concentration of each solute, using the following formula:

$$a_{w\,\text{total}} = a_{w1} \times a_{w2} \times a_{w3} \times \ldots\ldots\ldots\ldots \times a_{wn} \qquad \text{(Equation 5)}$$

where a_{w1}, a_{w2}, a_{w3}, a_{wn} are the a_w calculated from the concentration of each solute independently.

This equation can readily be applied to liquid foods, e.g. broths, juices and syrups and can also be used for solid foods by determining the concentration of solutes in the aqueous phase.

Water potential, ψ, is related to water activity by the equation:

$$\Psi = (RT/M) \ln \left(\frac{\rho}{\rho_0} \right) \qquad \text{(Equation 6)}$$

where M = the molecular weight of water ($0.018\,\text{kg}\,\text{mol}^{-1}$) and all other terms are as previously defined.

Factors Affecting Water Activity

Addition of water or removal of solutes can increase a_w. In food microbiology, however, one is usually interested in reducing a_w to improve the microbiological stability of the product. The a_w of an environment can be reduced by the addition of solutes, or water binding substances that decrease matric water potential, or by the removal of liquid water.

Table 1 Comparison of water activity (a_w) and water potential (ψ) values and concentration of solutes required to achieve them

a_w	Water potential (bar)[a]	NaCl concentration (g l^{-1})	Sucrose concentration (g l^{-1})	Other solutes (a_w at 25°C) (g l^{-1})
0.995	−7	8.7	92	
0.980	−28	35	342	
0.850	−224	190	2050 (saturated)	
0.843	−235			KCl (saturated, 357)
0.753	−390	260 (saturated)		
0.577	−757			NaBr (saturated, 909)
0.328	−1534			MgCl$_2$ (saturated, 1667)
0.113	−3000			LiCl (saturated, 769)
0.100	−3168			

[a] 1 bar = −100 J kg^{-1}.

Freezing

Liquid water can be removed, in effect, by freezing. The preservative effects of freezing are due not only to temperature depression, but also to the effect of decreasing a_w in the remaining liquid water. As the water in the food freezes it increases the effective concentration of solutes in the remaining liquid water. Those organisms remaining in the liquid phase are exposed to increasingly severe osmotic challenge as freezing proceeds. The same ecological challenges apply to bacteria naturally present in Arctic and Antarctic environments. The physiology of the organisms naturally present in those extreme environments is instructive for understanding the effects on microorganisms present in frozen foods and is discussed briefly later.

Drying

The removal of water by evaporation also increases the concentration of the solutes in the remaining water. As described below, the effect on a_w of the remaining free water will depend on the level and type of solutes initially present.

Specific Solutes

The a_w-modifying effects of several different solutes are shown in Table 1. Addition of solutes increases the osmotic potential of the water. As suggested by Equation 3, the effect of specific ionic solutes is related to their concentration, the number of ions that the molecule dissociates into, its dissociation constant, and also specific properties of the solute. Non-ionic solutes also reduce water activity.

Generally, NaCl, KCl, glucose and sucrose show similar patterns of effect on microbial responses while glycerol usually permits growth at lower a_w, although there are specific exceptions, e.g. *Staphylococcus aureus* is more inhibited by glycerol than NaCl. Glycerol differs from other solutes in that it is able to permeate the cell freely.

NaCl is somewhat unique in terms of humectants

in that the ionic species Na$^+$ is also a primary ion in cell function. Symporters are proteins that transport selected substances across the cell membrane, in a manner dependent on the co-transport of a second substrate in the same direction. A number of symporter systems are Na$^+$-driven. Cytoplasmic levels of Na$^+$ are also tightly regulated in most species, and fluctuating external Na$^+$ levels challenge microbial cells beyond the osmotic effect of a_w. Much of the research in this area has been conducted using bacteria; however, the general principles also hold for fungi.

Within *Escherichia coli*, an active extrusion mechanism is responsible for the regulation of intracellular Na$^+$ concentration which enters via symporter systems. The primary mechanism consists of a series of membrane-associated transport proteins known as antiporters. As protons flow into the cell (through the antiporter channel) along the concentration gradient established by respiratory chains, Na$^+$ is extruded from the cytoplasm. Many marine and anaerobic bacteria rely heavily on Na$^+$ cycling, with additional Na$^+$-translocating respiratory chains and ATPases responsible for Na$^+$ removal from the cell interior. Most, if not all, symporters in these bacteria are coupled to Na$^+$ influx.

The linkage between Na$^+$/H$^+$ antiporters results in an increased interaction between pH and NaCl in marine and anaerobic bacteria, so that their growth tends to be increasingly inhibited by NaCl as the pH of the medium increases. This is an example of specific effects of the humectant itself other than its direct effect on a_w.

Levels in Typical Foods

Representative a_w of foods are shown in **Table 2**. Foods range from those with very little free water (freeze-dried products, cereals, powdered products) to almost completely free water (e.g. fresh meat and produce, bottled water products). Most fresh produce has a_w close to 1.00 if the tissues are cut but may have

Table 2 Representative water activity of foods

Food	Typical water activity
Milk, fruit, vegetables	0.995–0.998
Fresh meat, fish	0.990–0.995
Cooked meat, cold smoked salmon	0.965–0.980
Liverwurst	0.96
Cheese spread	0.95
Caviar	0.92
Bread	0.90–0.95
Salami (dry)	0.85–0.90
Soft, moist pet food; chocolate syrup	0.83
Fruit cakes, preserves, soy sauce	0.80
Salted fish, honey	0.75
Dried fruit	0.60–0.75
Dried milk (8% moisture)	0.70
Cereals, confectionery, dried fruit, peanut butter	0.70–0.80
Ice at −40°C	0.68
Dried pasta, spices, milk powder	0.20–0.60
Freeze-dried foods	0.10–0.25

significantly lower *surface* water activity, e.g. on intact fruits and vegetables due to the presence of the cuticle. Meat carcass surfaces can also dry during processing, lowering the a_w sufficiently to inhibit microbial activity greatly. Thus, it is important to know not only the type of food but also the form and packaging that it is in to understand the microbial ecology.

General Reactions of Bacteria, Yeasts and Mycelial Fungi

Most microorganisms are active over only a relatively narrow range of a_w and a_w differences in the order of 0.001–2 are significant on the microbial ecology of an environment. Thus, a_w values in food microbiology are normally quoted to three significant figures.

Gram-negative bacteria, typically, are only able to grow in environments of a_w greater than about 0.95. Many Gram-positive bacteria can withstand a_w as low as about 0.9, but few can grow at a_w lower than 0.8. Some, specifically adapted to life in hyper-saline environments, are active at a_w as low as 0.75 and might be found, e.g. in dried salted fish. Fungi are generally more tolerant of reduced a_w than are bacteria. Some yeasts and moulds are able to withstand a_w as low as 0.60. Growth rates of bacteria are typically faster than those of eukaryotes. Thus, despite the fact that many yeasts and moulds are able to grow on foods of high a_w, such foods are usually rapidly dominated and spoiled by bacterial contaminants. Fungi have a selective advantage at lower a_w, and are more usually associated with the spoilage of reduced a_w products, e.g. bread, cheese, jams, syrups, fruit juice concentrates, grains, etc. As indicated above, the effect of a_w depends on the major solutes responsible for the reduced a_w. Ionic solutes (salts) have a greater inhibitory effect on microbial metabolism than non-ionic solutes (e.g. sugars).

Range of Growth

Each microorganism has a minimum and maximum a_w for growth. For many species, the maximum a_w for growth is effectively 1.000. Although growth could not occur in pure water, some organisms are able to grow in the presence of very low levels of nutrients. Pseudomonads, and even algae, are able to grow in some types of bottled water, indicating the need for techniques to eliminate viable organisms from these products during production. A range of terms used to describe the response and tolerance of microorganisms to a_w and specific solutes is shown in **Table 3**.

Table 3 Classification of microorganisms according to their preferred water activity range for growth

Nomenclature	Water activity range for growth
Haloduric	Able to withstand, but not grow at, high concentrations of salt
Halophile	Requiring salt for growth
Extreme halophile	Requiring 15–20% salt for growth
Osmotolerant	Able to withstand, but not grow at, high concentrations of sugar
Osmophile	Organisms that grow best, or only, under high osmotic pressure, due to sugars
Xerophilic	Requiring reduced water activity (as distinct from requiring high osmotic pressure)

The a_w range of growth is solute-dependent. Many bacteria, for example, are more tolerant of reduced a_w if the solute is glycerol. This characteristic is not, however, universal. Tolerance to a_w is greatest when all other factors in the environment are optimal for growth. As other environmental factors become less optimal the range of a_w that supports growth is reduced. Examples are presented in Figures 3 and 5 of the related entry 'Predictive microbiology'. The effects, however, are not always intuitive.

Representative tolerance ranges under otherwise optimal conditions for various microbial groups are shown in **Table 4**.

Combinations of Factors

It is common in some foods for a variety of factors to be used to control microbial growth. This approach exploits the interaction of a_w and other physico-chemical parameters such as temperature and pH in food environments. Such interactions form the basis of the hurdle concept.

The physico-chemical factors of NaCl and temperature have a close interaction, with the temperature range for growth of most organisms displaying a dependence on salinity. In general,

Table 4   Representative tolerance ranges for various microbial groups and species

Organism or group	Lower a_w limit (Solute)
(Most) Gram-negative rods	0.95–0.96 (NaCl)
Escherichia coli	0.95–0.955 (NaCl)
Pseudomonas fluorescens	0.97 (Sucrose)
Pseudomonas fluorescens	0.96 (NaCl)
Vibrio parahaemolyticus	0.96 (Glucose)
Vibrio parahaemolyticus	0.93 (NaCl)
(Most) Gram-positive bacteria	0.90–0.94 (NaCl)
Listeria monocytogenes	0.92–0.93 (NaCl)
Staphylococcus aureus	0.89 (Glycerol)
Staphylococcus aureus	0.87 (Sucrose)
Staphylococcus aureus	0.86 (NaCl)
Bacillus cereus	0.95 (Glucose)
Bacillus cereus	0.94 (NaCl)
Bacillus cereus	0.92 (Glycerol)
Yeasts	0.65–0.92 (NaCl)
Zygosaccharomyces rouxii	0.65 (Sucrose)
Saccharomyces cerevisiae	0.90 (Sucrose)
Moulds	0.65–0.90 (NaCl)
Penicillium chrysogenum	0.80 (KCl, glucose)
Wallemia sebi	0.75 (Glycerol)
Eurotium spp.	0.66 (Glucose and fructose)
Algae	0.75–0.90
Most groups	0.90–0.95 (NaCl)
Dunaliella	0.75 (NaCl)

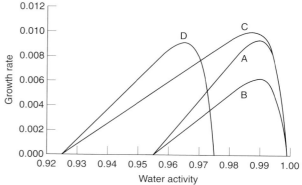

Figure 1   Effect of water activity on the growth rate of bacteria. Curves A and D represent two organisms, each adapted to a different water activity range for growth. Curve B represents the effect of a second suboptimal environmental factor on the growth rate of organism A. The water activity range is unaltered, the relative response remains the same, but the absolute growth rate is reduced at all water activities. Curve C represents the effect of a different, non-ionic solute (or humectant) on the growth response of organism A. That humectant permits A to grow over a wider range of water activities. After Ross T (1999) *Predictive Food Microbiology Models in the Meat Industry*. Sydney: Meat and Livestock Australia.

reduced a_w confers enhanced heat resistance on microbial cells. The basis for this behaviour is perhaps due to the cross-protection that osmotic stress affords against temperature stress, believed to be mediated by a general stress response under the control of the *Rpos* gene. (Interestingly, if grown at suboptimal salinities, a number of marine bacteria exhibit a *lowered* maximal temperature for growth compared to growth at the optimal salinity.) The minimum temperature for growth for many food-borne organisms is, however, *increased* by decreasing a_w. This raises the intriguing possibility that the basis of these effects lies in the energy of the water itself, i.e. if the kinetic energy of water molecules mediates the lethal effect of temperature, then the reduction of water energy by solutes may have the same effect as reducing temperature.

The growth rate response of microorganisms to water activity is illustrated in **Figure 1**. Growth rate increases in proportion, approximately, with increasing a_w above the minimum a_w for growth, and up to an optimum growth rate value. Beyond this value the growth rate declines, usually rapidly as a function of increasing a_w, until the maximum a_w is reached. Growth rate is a characteristic of the environment, and is not affected by the previous history of the cell, unlike lag time. The effect of a_w on growth rate is affected by the specific humectant.

There is no specific correlation between a_w tolerance and tolerance to other environmental factors. Thus, the manipulation of a_w in a product could have different consequences for the microbial ecology of the foods at different temperatures. An illustration of the selective effect of temperature and a_w on different organisms is presented in **Figure 2**.

Lag, Germination and Sporulation, Toxin Production

The lag time is generally considered to be a period of adjustment to a new environment, requiring the synthesis of new enzymes and cell components to enable the maximum rate of growth possible in that environment. As indicated above, the growth rate and, by inference the metabolic rate, is a function of the environment. As such, the lag time observed upon transfer of a cell to a new environment could be expected to result from both the amount of adjustment to that new environment and the rate at which those adjustments could be made. In general, lag times are longer at a_w that are less optimal for growth and where the difference between the old and new growth environment is larger, especially when the new environment is less favourable for growth than the old.

Generally the limits for microbial sporulation are the same as the limits for growth, although sporulation may occur at a_w slightly lower than that required for growth. Spores can sometimes also *germinate* at a_w below those which permit growth. Toxin production does not occur at a_w below those which permit growth, and is often prevented at a_w con-

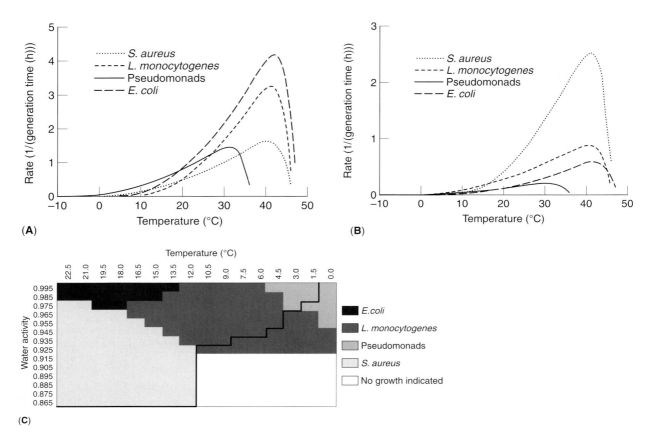

Figure 2 Comparison of the combined effect of environmental factors on growth rate of psychrotrophic spoilage pseudomonads, *Listeria monocytogenes*, *Escherichia coli* and *Staphylococcus aureus*. (**A**) The predicted effect of temperature on rates of aerobic growth at $a_w = 0.990$. (**B**) The predicted effect of temperature on rates of growth at $a_w = 0.960$. (**C**) Interactive effects of temperature and water activity on the microbial ecology of foods. Dominance domains for selected microorganisms potentially present on raw foods were estimated from predictive models for the aerobic growth of psychrotrophic spoilage pseudomonads, *L. monocytogenes*, *E. coli* and *S. aureus* at many combinations of water activity and temperature. The shaded areas represent that combination of factors in which the indicated organism would be expected to limit the acceptability of the product. The limits imposed for acceptability were that the predicted increase in the pathogen should not exceed a factor of 10 after 7 days storage. The limits for pseudomonads were that the increase in 7 days should be not more than 1000-fold, assuming an initial level of 1000 cfu cm^{-2}. All organisms were assumed to experience a lag time equivalent to one generation time at the nominated conditions. The part of the plot to the left of the bold line shows those sets of conditions under which the required bacterial growth limits are exceeded. For all conditions the organism closest to attaining the tolerance limit, and hence posing the greatest risk, is indicated. Note: The growth rate of pseudomonads was scaled to reflect the greater tolerance of this organism on the product, i.e. approx. 10 doublings of pseudomonads but only approx. 3 doublings of pathogens are tolerable by the criteria described. After McMeekin and Ross (1996) with permission from Elsevier Science.

siderably higher than those required to prevent growth.

Yield

At a_w less than the optimum, cell yield declines. The decline is not always a direct function of the a_w stress applied and it appears that some bacteria, at least, can tolerate a range of a_w without a change in yield. In *E. coli*, for example, in the a_w range from approximately 0.970 to 0.997 (using NaCl as the humectant), yield declined slightly ($\leqslant 20\%$) with decreasing a_w compared to the optimum a_w (ca. 0.995). At a_w lower than approximately 0.970, yield declines dramatically

as a function of a_w, until the lower a_w limit for growth, approximately 0.955, is reached.

Inactivation

At a_w lower than the minimum for growth, the cell either remains dormant or dies. Compounding this action, however, is the effect of a_w on the cell and the environment itself. Reduced a_w usually correlates with decreased chemical activity, with the result that the preservative effect of low a_w on foods may also preserve microorganisms present in the foods. This is particularly true for low a_w (i.e. < 0.7) products, in which microbial survival may be enhanced in comparison to that at higher a_w.

Mechanisms

While the changes in cell physiology that accompany osmotic stress are known in some detail, the physico-chemical mechanisms that underlie the effects of those responses are not well understood. One interpretation of the effects of a_w on the ecology of microorganisms considers that a_w creates a homeostatic burden. To maintain homeostasis, the cell must expend energy, whether to import or synthesize compatible solutes, modify membrane components, etc. This energy is unavailable for synthesis of new biomass and leads to reduced yield. This hypothesis further proposes that the cells' homeostatic demands ultimately consume all the available energy and the cell is able only to survive. Extending this paradigm, cell death could be interpreted to result when the homeostatic demands are unable to be met and the cell is unable to maintain the functional integrity of those enzymes and pathways necessary for continued viability.

Effect of Water Activity on Intracellular Structures and Chemical Composition of Cells

To remain viable, microorganisms, like plant cells, need to maintain a positive turgor pressure, possibly to provide a stimulus for cell elongation and growth. When a cell experiences an osmotic 'upshock' (i.e. transfer to lower a_w), the cell loses water due to osmosis because the microbial cell membrane is permeable to water and relatively impermeable to solutes. Water moves out of the cell to restore osmotic equilibrium, resulting in shrinkage of the cells. In extreme cases the cell membrane shrinks away from the cell wall, a process termed plasmolysis. Microbial cells must counteract the osmotic stress to restore the turgid, pre-stress state and have evolved a number of physiological responses to reduced a_w including changes in:

- cell membrane composition
- protein synthesis
- adjustment of cytoplasmic a_w.

The cell membrane is the main barrier to water and solute exchange between the cytoplasm and the external environment. It plays an important role in the physiological response to osmotic stress, responding with changes to both its lipid and protein components.

The synthesis of some proteins is induced by osmotic stress. Increased levels of solute transport proteins (porins) are likely during the osmoregulatory response. Like porins, many other osmotically induced proteins form the cellular machinery to facilitate a change in cytoplasmic a_w.

Macromolecular conformation, and therefore function and activity, is affected by intracellular a_w due, in part, to the effects of humectants on the physical structure of water. Some changes to membrane structure in response to a_w stress appear to enable membrane-bound enzymes to retain the conformation required for catalytic activity. The role of compatible solutes in optimizing molecular conformation is discussed below.

Cell Membrane Composition

The chemical composition of microbial cell membranes is described elsewhere in this volume. In response to high salinity there is an increase in the proportion of negatively charged phospholipids, often phosphatidylglycerol and/or glycolipids. This alteration is needed to maintain the membrane in the proper lipid bilayer phase for normal function.

Apart from the extreme halophiles of the Archaea there does not appear to be a correlation between microbial membrane composition and intrinsic a_w tolerance. However, the effect of a_w on a given membrane composition does depend to a large extent on the type of membrane (correlated with chemotaxonomic grouping, e.g. Bacteria, Archaea, yeast, fungi) and to a lesser extent, the nature of the humectant.

There are several elements common to cell membrane responses to changing a_w. The first of these is membrane surface charge. The head groups of the major microbial membrane lipids (phospholipids and phosphoglycolipids) are negatively charged from the associated phosphate residue. Certain phospholipid classes also contain positively charged head-group moieties, resulting in all polar lipid classes being either anionic or zwitterionic. The membrane surface of all microbes therefore possesses a net surface charge dependent on the phospholipid classes present. Ionic humectants may disrupt the membrane surface charge by interaction with phospholipid groups, requiring an alteration in membrane composition. Many halotolerant and moderately halophilic bacteria respond to reduced a_w by increasing the proportion of anionic phospholipids in the membrane at the expense of zwitterionic components, believed to aid the membrane in maintaining a functional bilayer phase.

The fatty acid composition of the cellular membrane also influences functionality and is actively modified in response to changing environmental factors. In general, in response to decreasing a_w most bacteria increase fatty acid chain length and/or decrease fatty acid unsaturation. Again, this is thought to maintain the membrane in a functional bilayer phase. In certain cases, the mechanism may involve direct inhibition of fatty acid biosynthetic enzymes by increased levels of NaCl.

Archaeal membranes possess phosphorus-containing lipid species as in other microorganisms but consisting of a glycerol backbone with two ether-linked C_{20} prenyl chains. This Domain contains all the extremely halophilic bacteria, with their membranes characterized by diphytanylglycerol diethers. Dephosphorylated derivatives may be present with a significant proportion of glycolipids. Extreme halophiles are characterized (but not exclusively) by the presence of neutral lipid components, mostly isoprenoid hydrocarbons, such as squalene. The resulting membrane bilayer is more ordered and less flexible than those formed from other lipid types. The C_{20} phytanyl residues may be present as branched or ring-containing structures which act as similar adaptive responses to fatty acid structure within other microorganisms. It is believed that the close packing exhibited by phytanyl residues in Archaeal membranes is the basis for their resistance to extreme environmental conditions.

While yeasts and fungi, as eukaryotes, contain many additional lipid types as storage and intracellular membrane components, their cellular membrane is dominated by phospholipid species as for the Bacteria. Thus, the common changes in fungal cell membrane composition to changing a_w are similar to those of the Bacteria, both in terms of polar lipid class manipulation and adaptation of fatty acid composition.

Cytoplasmic Water Activity

Moulds and yeasts accomplish the restoration and maintenance of turgor pressure by accumulation from the environment, or by *de novo* synthesis, of intracellular polyols to establish equivalent osmotic pressure intracellularly as exists extracellularly. Bacteria also accumulate or synthesize a range of compounds for the same purpose. Compounds used in this way share the property that they do not interfere with metabolic processes. As such, they have been termed compatible solutes.

Microorganisms adjust their cytoplasmic a_w using one of two strategies: the salt-in-cytoplasm type and the organic-osmolyte-in-cytoplasm type. Most, like the yeasts and moulds, use the organic-osmolyte-in-cytoplasm strategy for osmoadaptation. In this strategy salts are excluded, while organic solutes are synthesized or accumulated from the environment. Some bacteria can also adjust their cytoplasmic water by accumulating KCl to high intracellular concentration. This is considered a primitive strategy because it does not provide a normal cytoplasmic environment. This salt-in-cytoplasm strategy requires that the cell should

make additional physiological adjustments, especially in regard to enzyme function. The enzymes of prokaryotes that use the salt-in-cytoplasm strategy have additional negative charge that makes them stable at high solute concentrations but unstable at low concentrations.

Compatible Solutes

The activity of water is significantly influenced by the molecular structure of the solution. Water as a liquid is characterized by a (relatively) high degree of molecular motion resulting in a dynamic random distribution of molecular orientation. The potential degree of hydrogen bonding between water molecules is therefore not fully realized, allowing water molecules to pack together in a relatively tight manner or higher density. As the degree of molecular motion decreases (e.g. with lower temperature), a higher degree of hydrogen bonding between water molecules becomes possible and molecules adopt a more ordered array with a decreased density. With decreased temperature this process continues until the ordered molecular array of ice is achieved. Solute molecules decrease the activity of water by the same process.

The organic compounds synthesized or accumulated by microorganisms to balance their intracellular osmotic potential to that of their environment share the property that they do not affect the function of normal salt-sensitive enzymes. The use of compatible solutes to counter osmotic stress is not limited to microorganisms. Plants and animals also use the organic-solute-in-cytoplasm strategy and employ the same compounds as compatible solutes, suggesting that these compounds share fundamental properties that make them suitable for this role.

The compatible solutes have low molecular weights and polar functional groups, properties which make them highly soluble and facilitate their accumulation to high intracellular concentration. They are uncharged at normal cytoplasmic pH – an important property because high cytoplasmic ionic strength would be detrimental to enzyme function. These requirements limit the range of compounds that can be utilized as compatible solutes. Classes of compounds that are known to perform this function and specific examples are presented in **Table 5**.

Compatible solutes do not hinder the function of normal (salt-sensitive) enzymes and, in fact, protect proteins from the denaturation that would otherwise occur in solutions of high ionic strength. That protection also extends to the denaturing effects of freezing, heating and drying.

The mechanism of this protective effect is unknown. One observation, fundamental to attempts to resolve

Table 5 Classes of compounds, and examples, which can act as compatible solutes

Compound class	Example
Cations	K+
Sugars	Trehalose, sucrose, sorbitol
Sugar polyol derivatives	Glucosyl glycerol
Zwitterionic trimethylammonium and demethylsulphonium compounds	Betaines, thetaines
Natural amino acids	Proline, glutamine
Glutamine amide derivatives	Nα-carbamoylglutamine amide
N-acetylated diamino acids	Nδ-acetylornithine
Ectoines	Ectoine, β-hydroxyectoine

that mechanism, is that compatible solutes are excluded from the layer of water immediately surrounding macromolecules. Several hypotheses exist. Common to the hypotheses is the effect of compatible solutes on the physical structure of liquid water, whether through the surface tension-modifying effects of compatible solutes, or through alteration in the ratios of high- and low-density regions of water, or free and bound regions within the cell. Thus, it is widely considered that compatible solutes alter the physical environment within the cell at the molecular level, rather than the physiology of the cell itself.

Conclusions

Extracellular a_w has a profound influence on the microbial ecology of foods, a fact that has been exploited empirically since antiquity. Superimposed on the relatively simple ecological responses of individual microorganisms are complex interactions due to specific solutes, other environmental conditions and other microorganisms that may be present in the food. Equally, though superficially simple, those responses belie complex physiological responses, and the changes that occur in the physical structure of water itself. Food scientists are increasingly seeking ways to exploit the microbial ecology of foods to satisfy consumer preferences and the need for safety and stability of products. Accurate manipulation of the microbial ecology of foods will, however, require a detailed understanding of the mechanisms controlling microbial responses, with which we are only beginning to come to terms.

See also: **Ecology of Bacteria and Fungi in Foods**: Influence of Temperature. **Freezing of Foods**: Damage to Microbial Cells; Growth and Survival of Microorganisms. **Hurdle Technology. Predictive Microbiology & Food Safety**.

Further Reading

Board RG, Jones D, Kroll RG and Pettipher GL (eds) (1992) *Ecosystems: Microbes: Food*. Oxford: Blackwell Scientific Publications.

Brock TD *Biology of Microorganisms*. London: Prentice/Hall International.

Csonka LN and Hanson AD (1991) Prokaryotic osmoregulation: genetics and physiology. *Annual Review of Microbiology* 45: 569–606.

Gould GW (1989) *Mechanisms of Action of Food Preservation Procedures*. London: Elsevier Applied Science.

International Commission for the Microbiological Specifications for Foods (1996) *Microorganisms in Foods 5. Microbiological Specifications of Food Pathogens*. London: Blackie Academic and Professional.

International Commission for the Microbiological Specifications for Foods (1978) *Microorganisms in Foods 1. Their Significance and Methods of Enumeration*. London: Blackie Academic and Professional.

McMeekin TA and Ross T (1996) Shelf life prediction – status and future possibilities. *International Journal of Food Microbiology* 33: 65–83.

Mossel DAA, Corry JEL, Struijik CB and Baird RM (1995) *Essentials of the Microbiology of Foods. A Textbook for Advanced Studies*. Chichester: John Wiley.

Rockland LB and Beuchat LR (1986) *Water Activity: Theory and Applications to Food*. New York: Marcel Dekker.

Russell NJ, Evans RI, ter Steeg PF, Hellemons J, Verheul A and Abee T (1995) Membranes as a target for stress adaptation. *International Journal of Food Microbiology* 28: 255–261.

Troller JA (1985) Water relations of food-borne bacterial pathogens: an update review. *Journal of Food Protection* 49: 656–670.

Troller JA and Christian JHB (1978) *Water Activity and Foods*. New York: Academic Press.

Influence of Temperature

T Ross and **D S Nichols**, School of Agricultural Science, University of Tasmania, Australia

Introduction

Temperature control is perhaps the most widely used method of manipulating the microbial ecology of foods. It can be used to inhibit growth of spoilage or pathogenic organisms, to inactivate or kill unwanted microorganisms or to optimize growth or metabolism of microorganisms in fermentations. The patterns of the effect of temperature on the ecology of bacteria, yeasts and filamentous fungi are remarkably uniform. In general:

- Although microorganisms have evolved to grow within different temperature ranges, those preferred ranges typically span only ~ 35–40°C for bacteria and 25–30°C for fungi

- Within most of that range, an increase in temperature increases the rate of the microbial response, whether growth/metabolism or inactivation
- Although microorganisms have some capacity to alter their structure and biochemistry to moderate the effects of temperature on their activity and metabolism, they are unable to achieve temperature homeostasis

This entry considers the temperature limits for microbial growth, and the effect of temperature on microbial growth rate, metabolic rate and composition. The physiological basis of those responses and their consequences for the microbial ecology of foods are also discussed. Separate entries consider the effects of heating and freezing on microbial populations and physiology.

Microbial Ecology of Foods: Evolution of Specific Microbiota

Environmental microbial ecology tends to be concerned with open systems through which energy and chemicals flow. Food microbiology is more often concerned with batch processes, whether daily production runs, or the resulting individual units of foods for retail sale. Those batches are closed systems having finite resources of carbon and energy, and negligible capacity for removing the waste products of microbial metabolism. The features of populations growing in a batch culture are shown in **Figure 1**. Under constant environmental conditions, the pattern of population growth is S-shaped, and can be described math-

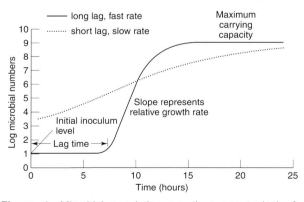

Figure 1 Microbial population growth curves typical of growth in foods. Microorganisms may exhibit a lag phase before the full growth rate potential is reached. Each species of microorganism will have a characteristic maximum growth rate, governed by the conditions in the food. When the total population in the food reaches 10^9–10^{10} cfu g^{-1}, then the growth of all components of the population will slow markedly or cease. After Ross T (1999) *Predictive Food Microbiology Models in the Meat Industry*. Sydney: Meat and Livestock Australia.

ematically in terms of four properties, shown in Figure 1 – initial inoculum level, lag time, growth rate and maximum carrying capacity.

The values of those properties are variable. The maximum carrying capacity of the system is usually a property of the food. When this level is reached, the growth of all groups of organisms in the product slows greatly or ceases. Thus, a slow-growing organism or one initially present in very low numbers may never reach that level. Conversely, if the growth rate of organisms initially present in very low numbers is much faster than that of all the other elements of the microbiota initially present, then it may still achieve numerical dominance. The steeper curve in Figure 1 is an example.

That the microbial ecology of foods is often concerned with batch processes tends to simplify understanding of the ecology of the system. In most foods there are relatively few microorganisms present initially and there is little competition for resources. Thus, bacteria and fungi will grow at the fastest rate possible in that environment, until the environment is either depleted of essential nutrients or until it is contaminated by the toxic metabolites of microbial growth so that growth is no longer possible. In many cases, this level is reached when the total microbial population is of the order of 10^9–10^{10} cfu g^{-1} or cfu ml^{-1} of the food product. Attainment of maximal population densities of desirable organisms, at the expense of other organisms potentially present, is the aim of fermented food production and is an example of manipulation of the microbial ecology of a food to select for the desired fermentative microbiota. That selection is achieved by optimization of the growth rate of the desired organisms in comparison to those of other organisms. It can also be achieved by using a high level of inoculum, or a combination of both.

For an organism to contribute to the ecology of an environment it must be metabolically active in that environment. That, in turn, requires that the physicochemical conditions of the environment remain within the tolerance range of that organism. In the current context, if temperatures exceed the minimum or maximum tolerance of the organism, the organism will fail to thrive and will eventually be eliminated. Temperature control can be used to slow the growth of a target organism relative to others present and suppress its potential effect on the environment.

Thus, knowledge of the environmental tolerance ranges of microorganisms, and the effects of environmental factors on the growth rate within that tolerance range, can be used to manipulate the microbial ecology of foods, e.g. to extend shelf life. Shelf life can be extended by promoting high levels of desirable organisms that stabilize the microbiology of the product, as with fermented foods, or by delaying the attainment of high levels of undesirable organisms that would cause spoilage of the product. The microbial ecology of foods, however, is concerned not only with organisms that become numerically or metabolically dominant but equally involves manipulating the food environment to suppress the growth of, or even eliminate, pathogenic microorganisms whose significance bears no relationship to their contribution to the microbial ecology of the product.

Thus, to determine the microbial ecology of a food requires information about the types and numbers of organisms initially present, their tolerance ranges and growth rates, the properties of the food, and the environmental conditions to which the food was exposed between production and consumption, and the *time* involved. Microbial tolerance limits to fluctuations of chemical and physical factors in an ecosystem do not determine which microorganisms are present at any given moment, but rather which microorganisms can be present on a sustained basis in that ecosystem. The interactions, over time, of the environment and the microorganisms present lead to the evolution of a specific microbiota.

Frequently, temperature will be the most variable feature of the environment of a microorganism in food. Thus, to understand the effects of temperature on the microbial ecology of foods, and to manipulate that ecology, it is necessary to consider the effects of temperature on the rates and limits of microbial growth.

Effect of Temperature on Microbial Growth

Growth Rate

Temperature affects the *potential* for microbial growth, the *rate* of growth or death and the production of metabolites. Bacterial and fungal growth rates respond to temperature as shown in **Figure 2**. There are upper and lower limits to growth, at which the growth rate becomes zero, and an optimum temperature at which the growth rate is maximal. The minimum, maximum and optimum temperatures for growth are known as the cardinal temperatures for growth. As will be discussed later, the temperature at which growth rate is maximal is not, of necessity, the optimum temperature for growth.

Between the minimum and optimal temperatures, the growth rate increases with increasing temperature. The growth rate increase is not proportional to the temperature, but increases more rapidly as the temperature is increased. As the temperature increases above the optimum, the growth rate decreases rapidly, due to thermal inactivation of cellular macro-

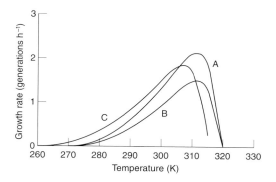

Figure 2 Interaction of environmental factors in determining the growth rate of bacteria and fungi. Curves A and C represent two organisms, each of which has adapted to a different temperature range for growth. Curve B represents the effect of a second suboptimal environmental factor on the growth rate of organism A. The relative response remains the same, and the absolute growth rate is reduced at all temperatures.

Table 1 Classification of microorganisms according to their preferred temperature region for growth

Classification	Temperature at which growth rate is maximal
Psychrophile	15°C or less
Psychrotroph	25–30°C
Mesophile	35–45°C
Thermophile	>45°C

molecules needed for growth. At low temperatures the growth rate does not necessarily decrease indefinitely to zero and there may be a critical threshold temperature below which growth suddenly is not possible. The pattern of response depicted in Fig. 2 is generally true for poikilothermic organisms.

Each organism has its own preferred temperature range for growth, related to its usual growth habitat. For bacteria, the range of growth usually spans 35–40°C, irrespective of the preferred temperature region for growth. Fungi typically grow over a range of 25–30°C. According to the preferred temperature region for growth, organisms are classified as psychrophiles, psychrotrophs, mesophiles or thermophiles, as shown in **Table 1**.

Despite the fact that organisms have adapted to different temperature ranges for growth, bacteria do not exhibit complete growth rate compensation for that preferred temperature range. Among the fastest-growing organisms known are those that are selected for and cause problems in moist, proteinaceous foods. Those foods are very nutritious and temperature is the only constraint to microbial growth. Among those organisms, strains that grow fastest at low temperatures nonetheless grow more slowly at their optimum than those species best adapted to growth at higher temperature. For example, the fastest known

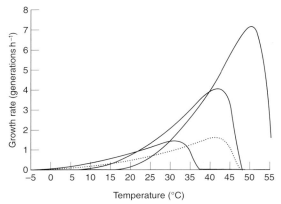

Figure 3 Comparison of the growth rates of selected food-borne bacteria of different thermal adaptation in nutrient-rich environments, showing the lack of temperature compensation. The organisms depicted are among the fastest-growing in their respective preferred temperature regions. The solid curve in the lower region (–5 to 37°C) represents the growth rate of psychrotrophic spoilage pseudomonads; the middle solid line (7–48°C) is for growth of *Escherichia coli*, a mesophile, and the upper curve is representative of the nearly thermophilic *Clostridium perfringens*. The dashed line represents the growth of *Listeria monocytogenes* and is included to show that *L. monocytogenes* is not fast-growing relative to other food-borne organisms. None the less, under appropriate conditions, it can multiply sufficiently to cause problems, particularly in foods of reduced water activity (e.g. <0.96). Under such conditions *L. monocytogenes* does have a growth rate advantage over other food-borne organisms, particularly in the chill temperature range.

bacterial growth rates recorded are for *Clostridium perfringens*, which has a generation time of ~7 min in the temperature range 40–45°C. Conversely, psychrotrophic pseudomonads which are the dominant species and cause of spoilage of aerobically stored, chilled fresh foods have generation times at 5°C in the range 4–5 h. **Figure 3** illustrates this behaviour.

Lag Times

For a given population, the lag time responds to temperature in the same way as growth rate, and it has often been observed that there is a direct proportionality between the lag time of a culture and its generation time at any temperature. However, the previous history of the cell or population can also affect the lag time. The lag time may be considered to be determined both by the amount of work to be done to equip the cell to adjust a new environment, and the rate at which that work can be done. The former component is related to the magnitude of the change in environment, the latter by the environment itself. In an otherwise constant environment, an abrupt temperature shift can induce a lag time in an exponentially growing microbial population (see later).

Interactions with Other Factors

The patterns of microbial responses to water activity and pH are described separately in this volume. In terms of growth rate, these responses are superimposed on the effect of temperature. The combined effects can be understood in terms of the gamma concept, sometimes used in predictive microbiology. That concept proposes that the effects of multiple inhibitory factors are additive (in terms of their relative effects) so that microbial growth rate can be described by an expression based on the growth rate at the optimum conditions (i.e. optimum temperature, water activity, pH, etc.) multiplied by a sequence of terms (each of which can take values between 1 and 0) which model the degree of 'non-optimality' of each of the other environmental conditions (i.e. the distance from the respective optima).

Thus the model has the general form:

Growth rate = optimum rate
 × relative inhibition due to suboptimal temperature
(1 – 0)
 × relative inhibition due to suboptimal water activity
(1 – 0)
 × relative inhibition due to suboptimal pH
(1 – 0)

This concept can also be extended to include the inhibitory effects of conditions beyond the optimum.

There is synergism, however, when multiple inhibitory factors are present in the environment. That synergism is manifest in modification on the environmental *limits* for growth of microbes.

Growth Limits

Each organism has reasonably well-defined limits for growth in response to individual environmental factors, when all other factors are optimal for growth or survival. These limits are altered by the other environmental conditions. For example, the potential temperature limits for growth are reduced by a suboptimal level of a second environmental factor. An example of these interactions is shown in **Figure 4**.

The hurdle concept and more specifically its application in hurdle technology, seeks to exploit this phenomenon by using combinations of levels of environmental factors, each of which on its own is insufficient to prevent growth, but which acting in combination do prevent the growth of target microorganisms. Quantitative knowledge of the combination of levels of environmental factors required to prevent growth is scarce. A limited amount of data exists for a few food-borne bacteria which are pathogenic to humans. It is believed, however, that as conditions become less optimal, i.e. as the environmental conditions move further from the interface

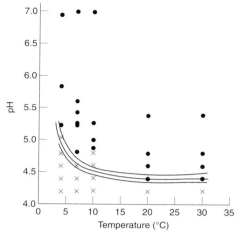

Figure 4 Interaction of environmental factors determining the boundary between growth and death of bacteria. Based on the effects of temperature and pH on the growth range of *Listeria monocytogenes*, the minimum temperature for growth depends on the pH of the environment. Closed circles are conditions under which growth was possible; crosses are conditions under which growth could not be demonstrated. The lines on the figure are the predicted boundary between growth-permissive and growth-preventing conditions at three levels of probability. Upper line: 90% probability of growth; middle line: 50% probability of growth; lower line: 10% probability of growth.

and into the no-growth region, the rate of microbial death increases. There are, however, minor exceptions to this pattern. For example, if a factor other than temperature prevents microbial growth, reduction of the temperature (i.e. moving further into the 'no-growth' region) will reduce the rate of death.

An interesting observation from studies on the interaction of environmental factors is that the optimum temperature for tolerance to a second inhibitory factor is often at a lower temperature than for optimum growth rate. Evidence of this behaviour can be seen in Figure 4. Suboptimal water activity due to the presence of salt may increase the maximum temperature at which growth is possible. This effect may be due to the synthesis of stress proteins which provide cross-protection to temperature stress or, alternatively, to the alteration of the physicochemical structure of water due to the presence of a humectant, or due to membrane rigidification.

Interpretation of the Effect of Temperature on Microbial Growth

Effect of Temperature on Reaction Rate

Microbial growth can be considered as a complex sequence of chemical reactions. The chemical reac-

tions that occur within bacterial or fungal cells are geared towards either:

- the provision of energy and reducing power from the environment to the cell (catabolism) or
- the synthesis of structural and other macro-molecules required for growth (anabolism)

The rates of those reactions, and hence of microbial growth, are dependent on temperature, as may be described by Eyring's absolute reaction rate equation:

$$V = \frac{kT}{h} \times [r] \times e^{-\Delta G^{\dagger}/RT} \qquad \text{(Equation 1)}$$

where: V = rate of reaction, k = Boltzmann's constant, T = temperature, h = Plank's constant, $[r]$ = concentration of reactant, $\Delta G^{\dagger}$ = Gibbs free energy of activation, or activation energy and R = gas constant. This relationship is based on the Arrhenius-van't Hoff equation.

Most metabolic reactions within cells do not occur at measurable rates without the catalytic assistance of enzymes. Enzymes are proteins and are fundamental to all metabolic functions. They mediate the transformation of different forms of chemical energy.

A biochemical reaction proceeds from reactants to products via one or more transition states that possess a higher free energy than that of the reactants (**Fig. 5**). Enough energy must be supplied to the system to overcome this barrier and allow the formation of the transition state. Thermodynamically, this is quantified by the free energy of activation, $\Delta G^{\dagger}$. The Gibbs free energy function is derived from a combination of the first and second laws of thermodynamics:

$$\Delta G = \Delta H - T\Delta S \qquad \text{(Equation 2)}$$

where: ΔG = change in free energy of the system, ΔH = change in enthalpy of the system, ΔS = change in entropy of the system and T = temperature (K).

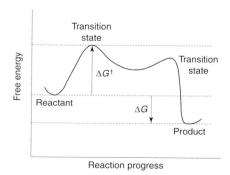

Figure 5 Metabolic reaction in terms of Gibbs free energy (*G*). The change in free energy of the system (ΔG) determines whether the reaction is possible. The magnitude of the free energy of activation ($\Delta G^{\dagger}$) influences the likelihood of the reaction and the rate. ΔG and $\Delta G^{\dagger}$ are both functions of temperature.

For a chemical reaction to occur spontaneously, the change in free energy, ΔG (free energy of the products minus the free energy of the reactants), must be negative. This requirement is independent of the path of the reaction (Fig. 5).

While ΔG indicates whether a given reaction is possible, $\Delta G^{\dagger}$ describes the amount of energy needed to drive the reaction. The kinetic energy of the reactants determines whether they have sufficient energy to overcome the Gibbs free energy, which is also often termed the activation energy. The kinetic energy is related to the temperature of the system, but not all the reactant molecules have the same kinetic energy at a given temperature. Rather, the energies of the reactant molecules form a distribution of kinetic energies, the average of which increases with temperature. A higher temperature increases the probability that the reactants will have sufficient energy to overcome $\Delta G^{\dagger}$ so that the reaction can proceed to completion. Thus, the probability of reaction, and therefore the rate, is also dependent on temperature.

Enzymes accelerate biochemical reactions by decreasing $\Delta G^{\dagger}$. Decreasing $\Delta G^{\dagger}$ effectively increases the number of substrate molecules with sufficient energy to complete the reaction. Consequently, the reaction is perceived to occur at an increased rate.

From Equation 1, the logarithm of rate is expected to be linearly related to the reciprocal of temperature, with the slope of that line being equal to the activation energy of the response. A plot of 1n(rate) vs. 1/temperature is known as an Arrhenius plot. **Figure 6** is an Arrhenius plot of the growth rate of *Escherichia coli*, and is typical of Arrhenius plots of microbial growth rate. That plot, however, shows a deviation from the Arrhenius relationship at high and low temperatures. This deviation has often been attributed to the denaturation of one or several key macromolecules required by the organism for growth, as shown in **Table 2**. An alternative hypothesis is that the co-ordination of catabolic and anabolic reactions within the cell breaks down at high and low temperatures, leading to a reduction in the efficiency of metabolism, and eventually to complete breakdown of balanced growth.

Whatever the reason, the Arrhenius plot of microbial growth rate can be considered in terms of three regions related to temperature. The normal physiological range is that region where the growth rate responds to temperature as predicted by Equation 1, i.e. the central straight-line portion. At any temperature within the normal physiological range the chemical composition of the cell is essentially constant. Beyond this range are the high- and low-temperature regions. Cells grown in the high- and low-temperature region not only have growth rates that deviate from that predicted by Equation 1 but also are increasingly different in composition to those grown in the normal physiological range. Transitions to the high- and low-temperature regions have been shown to result in synthesis of proteins not expressed in the normal temperature region. As discussed in greater detail below, membrane lipid composition is also altered by the synthesis and incorporation of different fatty acids into the membrane which have the effect of maintaining membrane fluidity.

Unification of the Microbial Response to Temperature

The above observations and discussion offer a consistent interpretation of the effects of temperature on microbial growth rates and limits. The temperature *limits* for growth are governed by the high- and low-temperature stability of one or several key macromolecules without which growth cannot proceed.

Growth *rate* increases with increased temperature in accordance with Equation 1 until the increase in temperature disrupts the conformation of enzymes, and/or the integration of anabolic and catabolic rates. Thus, metabolic efficiency decreases, leading to the observed high- and low-temperature deviations. In this interpretation, the optimum growth temperature is viewed as the point of equilibrium between increasing reaction rates due to temperature, and the deleterious effects of temperature on macromolecular stability and/or integration of metabolism. This interpretation also leads to an explanation of why the

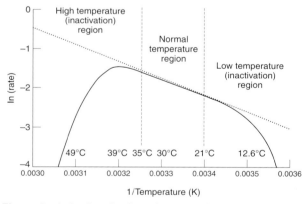

Figure 6 Arrhenius plot, based on a predictive model fitted to *Escherichia coli* growth rate data, and typical of microbial growth rate data. The dashed line is the predicted growth rate based on data in the normal temperature region for growth. If growth rate data are collected over the full biokinetic region, however, deviations from this prediction are observed at high and low temperatures. The normal temperature region depicted was judged subjectively, based on the deviation of the observed data from that predicted by extrapolation of equation 1, but does correspond to temperatures beyond which measurable changes in the biochemical composition of *E. coli* occur.

Table 2 Hypothetical physiological basis for the effects of temperature on microbial growth rate

Effect of temperature on enzyme-structure and efficiency

The rate of enzyme-catalysed reactions is also dependent on the concentration of active enzyme, itself a function of temperature. Enzymes are proteins. The functional activity of enzymes is dependent upon their shape or conformation, but they are flexible – the flexibility is required to achieve their catalytic function. Temperature affects the bonds in the molecule and, if the temperature changes too much, the conformation becomes so distorted that the enzyme is no longer catalytically active. This process is called denaturation. Denaturation can be visualized as unfolding of the protein, and can occur both when temperature becomes too high and also when it becomes too low. That denaturation is reversible and the protein can refold spontaneously if the temperature returns to within the range for stability. If the temperature becomes too high, however, irreversible denaturation takes place

Hypothetical physiological basis of temperature on microbial metabolism

A number of theoretical models have been advanced since the 1930s to explain the effect of temperature on bacterial growth rate. Most have as their basis the idea of a rate-limiting, enzyme-catalysed master reaction for growth. The concept of the models for the temperature dependence of poikilothermic growth mentioned earlier is that there is a single enzyme-catalysed reaction that limits microbial growth rate under all conditions. This putative reaction and the enzyme that catalyses it have been termed the master reaction and the master enzyme respectively. The activation energy of the master reaction is considered to be the ultimate limit to growth rate at all temperatures.

The hypothesis continues that the master enzyme itself is subject to the effects of temperature, so that as temperature increases above the optimum for conformational stability or decreases below it, the enzyme progressively becomes denatured. The transition of the master enzyme between conformationally active and inactive states is a function of temperature. The effect of this is a reduced level of sites available for catalysis, perceived as a reduction in the rate of reaction as seen at high and low temperatures beyond the normal physiological range

temperature for maximum growth rate does not necessarily correspond to the temperature of *maximum tolerance* to a second, suboptimal environmental constraint, i.e. the temperature of maximum tolerance is in the mid-range of the normal temperature region, where one would expect the greatest metabolic efficiency, and greatest capacity to overcome an environmental hurdle by homeostatic mechanisms.

If the lag time is a period of metabolic adjustment, requiring synthesis of new protein, it follows that the effect of temperature on those processes will be similar to the effect of temperature on growth rate. The induction of lag times due to abrupt temperature shift corresponds to whether the temperature shift involves a transition from one temperature region to another, particularly from the normal to low region.

Effects of Temperature on Metabolism

Enzymes energetically stabilize transition states of reaction intermediates. By catalysing specific reactions between selected substrates, enzymes can act as molecular switches in determining both the rate and direction of metabolic pathways.

The preceding discussion has described how temperature may act at a fundamental level of metabolism by directly influencing the type and rate of biochemical reactions. Temperature is an extrinsic ecological factor influencing microbial growth and metabolism. While most bacteria and fungi can actively regulate intrinsic physicochemical parameters within the cell (e.g. water activity, pH, redox), the microbial cell can only react to changes in environmental temperature in an effort to maintain functional metabolism. The regulation of cellular

metabolism in response to changes in environmental temperature is of primary importance to survival and growth. There are two broad areas in which metabolic regulation in response to temperature variation occurs:

● The maintenance of functional cell integrity to provide a suitable physicochemical environment for metabolic function
● The regulation of enzyme activity to maintain co-ordination between catabolic and anabolic processes

Maintenance of a physicochemical environment compatible with enzyme function necessitates a functional cell membrane assembly for both bacteria and fungi. Under the fluid mosaic membrane model this requires:

● that the cell regulate its lipid composition to ensure that a stable bilayer is formed with membrane proteins
● that this bilayer remains in a sufficiently fluid state

A great deal of research has been aimed at the latter requirement. The former point has been largely ignored due to the general assumption that natural lipid mixtures spontaneously form a bilayer arrangement. Generally, during normal cell growth this is true.

Lipid Composition

In the fluid state, the lipid components of a membrane bilayer remain miscible in all proportions. However, as the fluid–crystalline phase transition occurs during

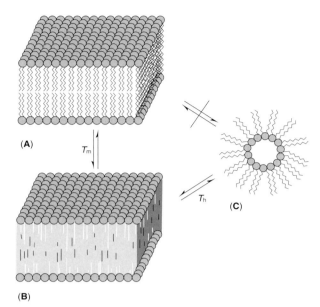

Figure 7 Illustration of the lipid bilayer structure in (A) the crystalline (gel) phase where acyl motion is low; (B) the fluid phase where acyl residue motion is high; (C) a non-bilayer (inverted hexagonal) lipid phase. T_m refers to the liquid-crystalline-phase transition which may occur to the membrane due to a temperature downshift (B → A). T_h refers to the bilayer–non-bilayer-phase transition which may occur due to a temperature upshift (B → C).

a temperature downshift (**Fig. 7**), individual lipid components begin to separate and crystallize, depending on their individual thermodynamic properties. As the bilayer freezes, crystalline regions grow at the expense of fluid ones, with lipids of progressively lower melting point moving out of fluid regions into crystalline ones.

The crystallization, or freezing, of the membrane causes large changes in the viscoelastic properties of the cell. In the predominantly crystalline state, cellular membranes can become osmotically fragile with leakage of intracellular components as the bilayer loses its ability to act as an efficient permeable barrier. Such leakage is believed to occur from two main sources; the formation of microscopic fissures in crystalline regions and the formation of grain boundary effects. Grain boundary effects represent areas of disorder occurring at interfaces between differently oriented crystalline regions formed during the fluid–crystalline-phase transition. These areas may provide leakage sites for small molecules and ions. In addition the crystalline regions formed are, in effect, semi-solid regions within the membrane (Fig. 7). As the membrane loses the flexibility previously afforded by the more viscous fluid state, mechanical deformation and shrinkage can result in microscopic fissures forming through which additional leakage may occur.

Second, the physical state of the membrane lipids

also has the potential to exert a large effect on many essential physiological cell processes such as sugar, amino acid and ion transport, chemotaxis and membrane-associated oxidation and reduction enzymes. The sensitivity of these processes to the physical state of the membrane derives from the fact that major protein components and/or assemblies of these systems are located within, or in close association with, the lipid bilayer.

Liquid–crystalline-phase transitions caused by temperature downshifts are therefore detrimental for the cell. Indeed, for almost all microbes the membrane bilayer is present in a fluid or predominantly fluid state at growth temperature. For example, in *E. coli* the bulk crystalline–liquid-phase transition is completed 7–15°C below the ambient temperature, ensuring that the membrane is completely fluid at the temperature of growth. Upshifts in temperature may also be detrimental to metabolic function, as the membrane may also lose proper function from excess fluidity. If the temperature upshift is of sufficient severity, the increased fatty acid chain motion within the bilayer may also result in the formation of non-bilayer lipid phases and the loss of membrane function (Fig. 7). The transition temperature is mainly controlled by the melting point of the constituent membrane phospholipid fatty acids, as the phase transition is essentially a hydrocarbon-mediated event. By manipulation of phospholipid fatty acid composition, bacteria and fungi may alter their physical membrane characteristics to maintain metabolic functions during and after changes in environmental temperature.

Regulation of Enzymes in Response to Temperature

The manipulation by bacteria and fungi of one aspect of their cellular composition, to retain functionality in response to changes in temperature, has been described in general terms above. The innate effect of temperature on enzyme function and the critical role of enzymes in metabolic processes have also been discussed. There are multiple mechanisms by which metabolic regulation occurs. Enzymes also play a fundamental role in the regulation of metabolism in response to temperature. Two major modes of enzyme regulation occur:

- a change in the cellular concentration of a given enzyme (usually achieved by a change in the rate of enzyme synthesis), thereby altering the overall level of cellular activity
- modulation of existing enzyme activity primarily via covalent modification and/or allosteric interactions

An example of such an enzymatic regulatory mech-

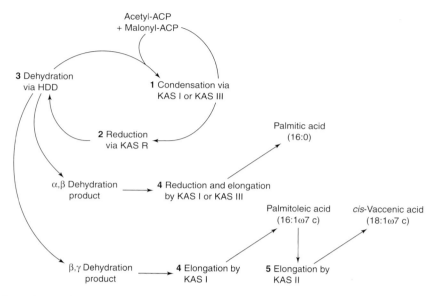

Figure 8 Schematic pathway of fatty acid biosynthesis in *Escherichia coli* denoting the major products involved in thermal regulation. The synthesis of *cis*-vaccenic acid is a critical metabolic function in response to decreasing environmental temperature. This process is regulated by the activity of KAS II. KAS I, β-ketoacyl-ACP synthase I; KAS II, β-ketoacyl-ACP synthase II; KAS III, β-ketoacyl-ACP synthase III; KAS R, β-ketoacyl-ACP reductase; HDD, β-hydroxydecanoyl-ACP dehydrase.

anism is found in the adaptation of membrane fatty acid composition to temperature within *E. coli*. **Figure 8** summarizes the major steps in the fatty acid biosynthetic pathway of *E. coli*. The major fatty acids produced by *E. coli* are palmitic acid (16:0), palmitoleic acid (16:1ω7c) or *cis*-vaccenic acid (18:1ω7c): the ratio of these major components varies with temperature. In response to a decrease in growth temperature, the amount of *cis*-vaccenic acid in the membrane increases while the level of palmitic acid declines. That is, there is an increase in unsaturated fatty acids at the expense of saturated components.

However, studies with temperature-sensitive mutants of *E. coli* revealed that the specific ability to produce *cis*-vaccenic acid (rather than simply palmitoleic acid) was essential for thermal regulation. Further investigations revealed that the presence of a single enzyme, β-ketoacyl-ACP synthase II (KAS II) was responsible for the critical step of converting palmitoleic acid to *cis*-vaccenic acid during low-temperature thermal regulation (Fig. 8). Inhibitor studies demonstrated that neither mRNA nor protein synthesis was required to achieve an increased rate of *cis*-vaccenic acid synthesis within 30 s of a temperature downshift. That is, KAS II is present within *E. coli* at all growth temperatures but is regulated so that it only becomes active under low-temperature conditions.

Similar studies of fatty acid modification in fungi have highlighted a further important aspect of meta-

bolic response to temperature. Experiments using the mycelial fungi *Tetrahymena pyriformis* and *Cunninghamella japonica* indicate that the production of unsaturated fatty acids by enzyme-linked desaturation relies upon temperature-induced changes in membrane fluidity. A decrease in membrane fluidity (with decreasing temperature) results in the activation of membrane-associated desaturase enzymes, which undergo a change in conformation allowing them to act upon surrounding fatty acid substrates. The increase in unsaturated fatty acids consequently restores a functional level of fluidity to the membrane. Importantly, in this case, the activation of desaturase enzymes is directly modulated by the degree of membrane fluidity which is influenced by temperature.

Summary

Bacteria and fungi are unable to achieve temperature homeostasis and, within a narrow range of temperature, their metabolic rates respond to temperature in the same manner as simple chemical reactions, increasing in rate with increasing temperature. Beyond this range the effects of temperature become more pronounced, requiring microorganisms to manipulate their composition to minimize the effects of temperature on their metabolism. Different species have adapted to growth in different temperature regions, but even with the capacity for manipulation, most bacteria can grow within a range of temperature that spans 35–40°C only, and fungi a range of 25–30°C only. Those temperature tolerance limits may be further reduced by other environmental constraints. Thus, there is a complex interaction of growth rates and tolerance ranges to temperature and other factors that affect the microbial ecology of a food. None the less, the known patterns of microbial response to temperature enable the microbial ecology of foods to be reasonably well understood, and to enable that ecology to be manipulated by temperature control.

See also: **Clostridium**: *Clostridium perfringens*. **Ecology of Bacteria and Fungi in Foods**: Influence of Available Water; Influence of Redox Potential and pH. **Fermented Foods**: Origins and Applications. **Freezing of Foods**: Damage to Microbial Cells; Growth and Survival of Microorganisms. **Heat Treatment of Foods**: Ultra-high Temperature (UHT) Treatments; Principles of Pasteurization; Action of Microwaves; Synergy Between Treatments. **Hurdle Technology. Predictive Microbiology & Food Safety**.

Further Reading

Berry ED and Foegeding PM (1997) Cold adaptation and growth of microorganisms. *Journal of Food Protection* 60: 1583–1594.

Board RG, Jones D, Kroll RG and Pettipher GL (eds) (1992) *Ecosystems: Microbes: Food*. Oxford: Blackwell Scientific Publications.

McMeekin TA and Ross T (1996) Shelf life prediction – status and future possibilities. *International Journal of Food Microbiology* 33: 65–83.

Mossel DAA, Corry JEL, Struijik CB and Baird RM (1995) *Essentials of the Microbiology of Foods. A Textbook for Advanced Studies*. Chichester: John Wiley.

Neidhart FC, Ingraham JL and Schaechter M (1990) *Physiology of the Bacterial Cell. A Molecular Approach*. Sunderland, Massachusetts: Sinauer.

Precht H, Christopherson and Hensel H (1973). *Temperature and Life*. Berlin: Springer-Verlag.

Influence of Redox Potential and pH

Alexandra Rompf and **Dieter Jahn**, Institute for Organic Chemistry and Biochemistry, Albert Ludwigs University Freiburg, Germany

Concepts of Redox Potential and pH

Redox potential and pH are inherent or intrinsic factors of food habitat. Reduction–oxidation or redox potential (E'_0) describes the differences in electrical units measured in millivolts (mV) or volts (V) generated by a system in which one substance is oxidized and a second substance is reduced. During this process a reduced substance loses electrons (thus it is oxidized) to an oxidized substance (thus it is reduced). In biological systems, reduction–oxidation of substances is the basic principle of energy generation. Energy-rich compounds are oxidized stepwise in cellular catabolism. In the presence of sufficient oxygen most microorganisms perform classical oxidative phosphorylation. Electrons from various oxidized electron donors are transported via an electron transport chain to the electron acceptor oxygen, which is reduced to water (**Fig. 1**). The process is coupled to proton translocation via the cytoplasmic membrane in bacteria or the inner mitochondrial membrane in fungi. The electrochemical pH gradient formed is used to drive ATPase to form ATP from ADP and P_i. Therefore, changes in the environment oxygen tension have a drastic impact on the catabolic functions described.

In the absence of oxygen, it is possible for bacterial and some fungal cells to achieve anaerobic ATP formation in two ways. Oxygen can be replaced by alternative electron acceptors such as nitrate, fumarate, dimethyl sulphoxide (DMSO), trimethylamine N-oxide (TMAO) or thiosulphate to sustain further electron transport and membrane potential-driven ATP synthesis (Fig. 1). In biological systems electrons are carried from the electron donor with more negative/less positive redox potential stepwise to acceptors which have less negative/more positive redox potential. The covered redox potential differences combined with proton gradient formation determine the rate of energy conversion. The larger the differences in redox potential between the electron donor and acceptor, the better the chances of efficient energy conservation. This explains why oxygen is the most valuable electron acceptor in nature (**Table 1**).

The second possibility is characterized by ATP formation at the level of substrate degradation and is called fermentation. In the absence of electron transport chains for energy generation and NAD$^+$ regeneration, fermentation processes are only partially oxidizing their substrate, leading to the excretion of energy-rich metabolic intermediates and overall poor energy conservation (Fig. 1). On the basis of their growth capabilities in the presence and absence of free oxygen, microorganisms are grouped as aerobes, anaerobes, facultative anaerobes and microaerophiles. Strict aerobes require oxygen as an electron acceptor for energy generation. While strict or obligate anaerobes cannot grow in the presence of small amounts of oxygen due to the lack of superoxide dismutase activity which is necessary for oxygen radical scavenging, strict anaerobic organisms mostly grow on the basis of fermentation processes. Fac-

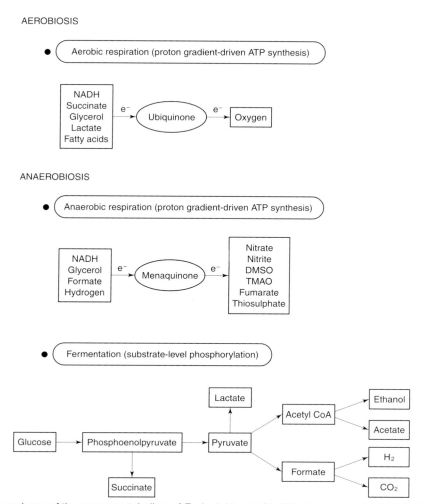

AEROBIOSIS

- Aerobic respiration (proton gradient-driven ATP synthesis)

NADH / Succinate / Glycerol / Lactate / Fatty acids → e^- → Ubiquinone → e^- → Oxygen

ANAEROBIOSIS

- Anaerobic respiration (proton gradient-driven ATP synthesis)

NADH / Glycerol / Formate / Hydrogen → e^- → Menaquinone → e^- → Nitrate / Nitrite / DMSO / TMAO / Fumarate / Thiosulphate

- Fermentation (substrate-level phosphorylation)

Glucose → Phosphoenolpyruvate → Pyruvate
Phosphoenolpyruvate → Succinate
Pyruvate → Lactate
Pyruvate → Acetyl CoA → Ethanol
Acetyl CoA → Acetate
Pyruvate → Formate → H_2
Formate → CO_2

Figure 1   Redox dependence of the energy metabolism of *Escherichia coli*. DMSO, dimethyl sulphoxide; TMAO, trimethylamine *N*-oxide.

Table 1   Reduction potentials (E'_0) of redox pairs important in microbial catabolism

Redox pair	E'_0 (V)
CO_2/Formate$^-$	−0.43
$2\,H^+/H_2$	−0.41
$S_2O_3^-$/HS^- + HSO_3^-	−0.40
NAD^+/NADH	−0.32
Pyruvate$^-$/Lactate$^-$	−0.19
Menaquinone ox/red	−0.075
Fumarate$^-$/Succinate 2^-	+0.033
Ubiquinone ox/red	+0.113
DMSO/DMS	+0.16
NO_2^-/NO	+0.36
NO_3^-/NO_2^-	+0.43
O_2/H_2O	+0.82

ultative anaerobes use oxygen if available and grow anaerobically using alternative electron acceptors and/or fermentation. Microaerophiles grow best at low oxygen tension. The redox potential of the environment at which different groups of microorganisms

usually grow is between +500 and +300 mV for aerobes, between +300 and −100 mV for facultative anaerobes and between +100 and −250 mV or more for anaerobes. Clearly, the presence and absence of oxygen and, as a consequence, the changing redox potential determine the mode of energy generation in different environments, including food.

The pH indicates the concentration of hydrogen ions in a system. It is expressed as a negative logarithm of hydrogen ion or proton concentration in a range from 0 to 14: 7 is neutral pH. The proton concentration of a habitat depends on the concentration and nature of the acids present. Strong acids like hydrochloric acid or phosphoric acid dissociate completely, while weak acids such as lactic acid or acetic acid remain in equilibrium of dissociated and undissociated form.

The pH of the environment has a profound effect on the growth and survival of all microorganisms. The pH range and optimum for growth is species-

Table 2 Minimum and maximum pH values for growth of microorganisms

Microorganism (examples)	Minimum pH	Maximum pH	Acid tolerance
Micrococcus sp.	5.6	8.1	Low
Pseudomonas aeruginosa	5.6	8.0	Low
Bacillus stearothermophilus	5.2	9.2	Low
Clostridium sporogenes	5.0	9.0	Medium
Clostridium perfringens	5.0	8.3	Medium
Campylobacter jejuni	4.9	8.0	Medium
Listeria monocytogenes	4.5	9.0	Medium
Yersinia enterocolitica	4.5	9.0	Medium
Vibrio parahaemolyticus	4.8	11.0	Medium
Clostridium botulinum type A, B	4.5	8.5	Medium
Staphylococcus aureus	4.0	9.8	Medium
Salmonella spp.	4.0–4.5	8.0–9.6	Medium
Escherichia coli	4.4	9.0	Medium
Proteus vulgaris	4.4	9.2	Medium
Streptococcus lactis	4.3–4.8	9.2	Medium
Bacillus cereus	4.3–4.9	9.3	Medium
Lactic acid bacteria			
Lactobacillus spp.	3.8–4.4	7.2	High
Acetic acid bacteria			
Acetobacter acidophilus	2.6	4.3	High
Alicyclobacillus acidoterrestris	2.2	6.0	High
A. acidocaldarius	2.2	6.0	High
Yeasts			
Saccharomyces cerevisiae	2.3	8.6	High
Fungi			
Penicillium italicum	1.9	9.3	High
Aspergillus oryzae	1.6	9.3	High

specific. Moulds grow between values of 1.5 and 9, yeasts between 2.0 and 8.5, Gram-positive bacteria between 4.0 and 8.5 and Gram-negative bacteria between 4.5 and 9.0 (**Table 2**). Changes in the external pH directly influence the status of the proton gradient at the cytoplasmic membrane of bacteria, and this is responsible for ATP synthesis, transport processes and motility. Since many microbial species require homeostasis of their internal pH for growth, changes in the external pH induce complex adaptation processes, which are outlined below.

In order to adapt to the prevalent living conditions efficiently, the cells must first detect the environmental stimuli and, in turn, respond with changes in their metabolism. This control is mostly due to transcriptional regulation of the genes that encode the individual proteins involved in biochemical functions specific to certain environmental conditions. Very sophisticated control mechanisms ensure that enzymes which are no longer required are not synthesized, whereas others that are needed under the present conditions are formed in sufficient amounts (see below).

Different Redox Potentials and pH Values in Different Foods

The redox potential of food depends on its chemical composition, the treatment given and the storage conditions in relation to oxygen tension. For example, in fresh muscle foods, respiration in the tissue utilizes most of the remaining oxygen. The aerobic or facultative components of the bacterial microflora are also utilizing oxygen. However, their impact on the oxygen content of meat is minimal compared with the respiratory activity of the surrounding muscle tissue due to the relatively low population which is initially present. Under aerobic storage conditions oxygen will diffuse into the tissue and change the redox potential again. The aerobic spoilage microflora is dominated by species of *Pseudomonas, Alcaligenes, Acinetobacter, Moraxella* and *Aeromonas*. In vacuum-packed products, the initial oxygen present in the meat is breathed by the tissue. The spoilage microflora under anaerobic conditions is dominated by lactic acid bacteria such as *Lactobacillus coryneformis, L. plantarum* and *L. cellobiosis*. If the pH of the muscle tissue is higher, as in pork or lamb meat,

Table 3 pH Values of different foods

pH area	Food	pH value
Alkaline pH > 7.0	Egg white	Up to 9.6
Neutral pH 7.0–6.5	Fresh meat	7.2
	Milk	7.0–6.8
	Shrimps	7.0–6.8
	Oysters	6.7–6.3
	Poultry	6.7–6.3
Slightly acidic pH 6.5–5.3	Fish	6.6–5.7
	Boiled ham, sausages	6.4–6.0
	Pork meat	6.0–5.6
	Beef	5.8–5.4
	Gammon	5.8–5.3
	Meat paste	5.8–5.0
	White bread	6.0–5.0
	Vegetables	6.5–5.0
Medium acidic pH 5.3–4.5	Canned food, smoked sausages	5.2–4.9
Acidic pH 4.5–3.7	Mixed pickles	4.5–3.5
	Tomatoes	4.4–4.0
	Yoghurt	4.2–3.8
	Mayonnaise	4.1–3.0
	Fruits	4.5–3.0
Strongly acidic pH < 3.7	Sauerkraut	3.7–3.1
	Apples, apple juice	3.5–3.3
	Plums	3.0–2.8
	Lemons	2.4–2.2

or residual amounts of oxygen are present, other microorganisms such as *Brochothrix thermosphacta* and *Shewanella putrefaciens* may make substantial contributions to product spoilage. However, the growth rate of bacteria under anaerobic conditions is considerably reduced compared to aerobic conditions. Therefore, the shelf life of muscle foods can be extended by storage under vacuum or modified atmospheres. Typically, this atmosphere contains elevated amounts of carbon dioxide, resulting in a drastically reduced growth rate of microorganisms. The specific mechanism of this bacteriostatic activity of carbon dioxide is poorly understood. Moreover, the use of carbon monoxide, nitrous oxide and sulphur dioxide in the gaseous atmosphere of packed muscle foods has been suggested.

Fresh vegetables contain microorganisms which originate in soil, water, air or other environmental sources. They have a high pH, with the exception of tomatoes. In the presence of air the most common spoilage is caused by different types of moulds from the genera *Penicillium*, *Phytophthora*, *Alternaria*, *Botrytis* and *Aspergillus* and bacteria from the genera *Pseudomonas*, *Erwinia*, *Bacillus* and *Clostridium*. Fruits have a pH of 4.5 or below. This microbial spoilage of fruits and fruit products is confined to

various moulds and yeasts and acidic bacteria (lactic acid bacteria, *Acetobacter*, *Gluconobacter*). The pH values for foods are given in **Table 3**. In general, the pH of food can vary greatly. Most fruits, fruit juices, fermented fruits, meat and dairy products are highly acidic foods (pH below 4.6) while most vegetables, fresh meat, fish, milk and canned foods are weakly acidic foods (pH 4.6 and above).

Molecular Adaptation of Microorganisms to Changes in Redox Potential and pH

As described above, oxygen tension and pH influence the growth and survival of microorganisms in food. The success of various microbial species under a variety of different environmental conditions is mostly correlated to quick and efficient adaptation mechanisms. In general, external signals such as change in oxygen tension or pH are detected by sensor proteins which are located at the cell surface or intracellularly. Subsequently, a process called signal transduction is initiated. Many of the regulatory systems (**Fig. 2**) by which cells sense and then respond to environmental signals are called two-component regulatory systems. Such systems are characterized by having a signal-specific sensor kinase located in the cell membrane, and an intracellular response regulator protein. The sensor kinase detects an external signal from the environment and in response phosphorylates itself at a specific histidine residue on its cytoplasmic surface. This phosphoryl group is then transmitted to the

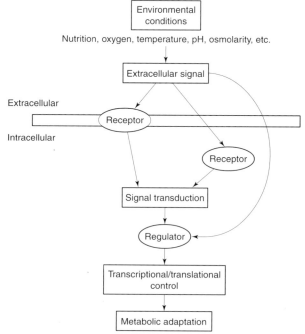

Figure 2 Molecular strategies of microorganisms to adapt to changing environmental conditions.

response regulator. The response regulator is typically a DNA-binding protein that regulates the transcription of target genes. Termination of the response requires a specific phosphatase, which removes the phosphoryl group from the response regulator protein. In some cases this reaction is carried out by the sensor kinase itself, while in other systems an additional third protein carries out the reaction. Two-component regulatory systems are now known to be involved in the environmental control of gene expression in many bacteria and yeasts. Additionally, there are intracellular single sensor proteins which directly regulate gene expression as a consequence of environmental activation. All systems have in common excitation by an extracellular signal which is transduced and leads to a transcriptional or translational control of gene expression and, in consequence, to a metabolic adaptation.

The best-investigated bacterial organism *Escherichia coli* contains several regulatory systems which respond to changes in oxygen tension. After a change from aerobic to anaerobic growth the majority of anaerobic systems for the utilization of alternative electron acceptors in the form of primary dehydrogenases (hydrogenases, anaerobic formate dehydrogenases and glycerol 3-P-dehydrogenases) and terminal oxidases (various nitrate and nitrite reductases, TMAO reductase, DMSO reductase, fumarate reductase) are induced by the one-component regulator Fnr (which stands for fumarate nitrate reduction). Perception of oxygen tension changes is mediated by an Fnr-bound iron sulphur cluster. Under reducing conditions the iron sulphur cluster is intact and the regulator binds as a dimer to a highly conserved palindromic DNA sequence. Depending on the location of these recognition sequences with regard to promoter Fnr is activating or inactivating gene expression. The two-component regulatory system ArcA/B (which stands for aerobic respiration control) is mainly responsible for the repression of genes of the aerobic metabolism (cytochrome *o* oxidase, citric acid cycle enzymes, NADH : quinone oxidoreductase) and the induction of enzymes of mixed acid fermentation (pyruvate formate lyase, etc.). ArcA is located in the membrane and is the receptor responsible for transferring the signal via transphosphorylation to the transcriptional regulator ArcB. It was proposed that changes in the electron flow in the membrane constitute the signal for this system. The broad answer to anaerobic conditions induced by Fnr and ArcA/B is subsequently differentiated by the presence of the energetically most efficient alternative electron acceptor nitrate.

As one can deduce from Table 1, the yield of various modes of anaerobic growth varies drastically. If possible, *E. coli* uses only the most efficient mode of nitrate respiration. As a consequence, proteins involved in this process (nitrate reductase, nitrate transporter, nitrite reductase) are further induced while most other functions (fumarate reductase, DMSO reductase, TMAO reductase) are repressed. Again, two-component regulatory systems (NarX/L, NarQ/P, which stands for nitrate reduction) are responsible for the economic differentiation of this anaerobic answer on the molecular level. NarX and NarQ are the membrane-localized sensor proteins for nitrate and nitrite which transfer their signal via phosphorylation to the transcriptional regulators NarL and NarP, which in turn are responsible for appropriate gene activation or repression (**Fig. 3**). This already complex molecular response is further coordinated with oxygen stress and other regulators.

The mechanism employed by a microbial cell to sense a change in environmental pH can be divided into three broad categories. First, pH influences the structure and activity of cellular components (**Fig. 4**). For example, the protonation or deprotonation of amino acid residues in proteins has profound effects on the secondary and tertiary structure and as a consequence on the activity of such proteins. In that case, the protein itself serves as a direct sensor of changes in pH. The second mechanism involves the ability of cells to respond to small molecules in one particular ionization state. An example of this is the ability of certain weak acids to cross the membrane in their protonated form. After entry in the cell these acids often dissociate and change the intracellular pH. Therefore, an increased cellular concentration of such compounds indicates increased environmental acidity (Fig. 4). Finally, membrane proteins are well suited to sense pH-related signals of all three classes. In particular, proteins spanning the cytoplasmic membrane can measure ΔpH and the concentration of molecules in the environment which are important for the momentary pH. They react directly to changes in pH via changes of their protonation state (Fig. 4).

Salmonella typhimurium uses three progressively more stringent mechanisms to maintain a pH value consistent with viability. First, at pH > 6.0, the pH is adjusted through the homeostatic response, which allosterically modulates the activity of proton pumps to increase the rate of proton extrusion from the cytoplasm. The following response is induced by pH values of 5.5–6.0 and includes the synthesis of at least 12 new proteins and the repression of six proteins. Amino acid decarboxylases (such as the lysine or arginine decarboxylase CadA and SpeA) are induced, which neutralize the medium under acidic conditions. The positive regulator involved is called CadC. Fermentation genes such as for lactate dehydrogenase

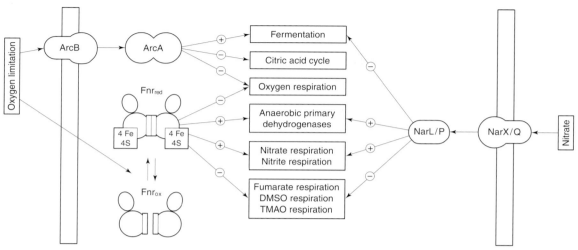

Figure 3 Simplified model of oxygen regulation in *Escherichia coli*. ArcA/B, Aerobic respiration control; Fnr, fumarate nitrate reduction; Nar, nitrate reduction; DMSO, Dimethyl sulphoxide; TMAO, trimethylamine *N*-oxide.

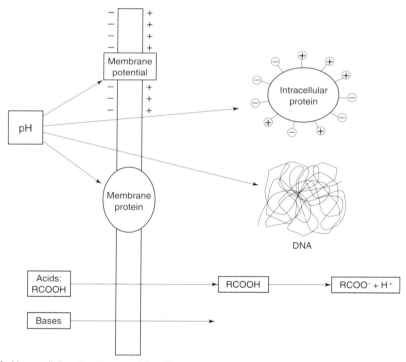

Figure 4 Influence of pH on cellular structures and functions.

(LDH) are acid-dependent and repressed to generate fewer acidic end products. Finally, the synthesis of acid shock proteins is triggered at pH values from 3.0 to 5.0. Mainly various transcriptional activators are induced and subsequently regulate their target genes. This final response was found to be similar to a cold shock reaction.

The best-characterized base-inducible system in *E. coli* is that of sodium proton antiporter NhaA, which regulates the intracellular sodium concentration. Expression of *NhaA* is increased over an external pH

range of 7.0–9.0. Also induced by alkaline pH are the aerobic electron transport systems. Terminal oxidases (like cytochrome *o* or cytochrome *d*) respond to basic environmental conditions in co-ordinated regulation via Fnr and ArcA. The combined regulation of oxygen and pH has frequently been observed.

Control of Microbial Growth by Low pH and Organic Acids

In the past our ancestors recognized that some foods

from plant sources like fruits or fermented foods and beverages are more resistant to food spoilage compared to raw material. Later it was recognized that most microorganisms present in food grow over a rather restricted pH range. Especially at lower pH, many do not survive (Table 2). This knowledge led to the use of many organic acids as food additives. Some are present naturally, such as citric acid in citrus fruits, sorbic acid in rowanberries and benzoic acid in cranberries. Others are produced during fermentation processes by starter culture microorganisms, such as propionic, acetic or lactic acids. Finally, naturally occurring acids are added as preservatives to prepared foods and beverages to reduce the pH. However, the antimicrobial activity is due to the combined action of undissociated and dissociated parts of the acid used (Fig. 4). Important parameters for antimicrobial effectiveness are the pK of the acid, determining the dissociation state at various pHs, the water solubility and the lipophilic properties which are responsible for effectiveness in entering a target cell. Acid with the higher pK and a higher amount of undissociated molecules at food pH showed the best antimicrobial properties. Undissociated molecules, being lipophilic, freely enter through the membrane simply as a function of the concentration gradient. Inside the cytoplasm pH values around 6.5–7.0 in acidophiles and 7.5–8.0 in neutrophiles lead to dissociation of the acid and release of protons and anions (Fig. 4). Anions are either removed from the interior of the cell or used as carbon sources, such as acetate or lactate. More importantly, protons will reduce the internal pH and adversely affect the proton gradient, interfering with energy generation, nutrient transport and mobility. Removal of protons requires high amounts of energy which the targeted cells may not be able to generate. Moreover, the low intracellular pH acts on the structure of cellular components by interfering with ionic bonds, damaging macromolecules and inflicting lethal injuries to the cell.

The antimicrobial effectiveness of four acids (**Table 4**) follows the order acetic > propionic > lactic > citric acid. All four are very soluble in water, while benzoate

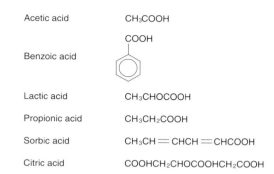

Figure 5   Structures of organic acids.

and sorbate are poorly soluble (**Fig. 5**). Acetic acid is usually employed as vinegar (5–10% acetic acid) in pickles, dressings and sauces. Inhibitory concentrations of the undissociated acid are between 0.01% against *Staphylococcus aureus* and 0.5% against *Saccharomyces* spp. Propionic acid is used at 0.1–0.2% in bakery products, cheeses, jam and tomato purée. Inhibitory concentration of the undissociated acid is 0.05% against moulds and bacteria. Lactic acid is used up to 2% in carbonated drinks, salad dressings, pickles, low-heat-processed meat products and sauces. However, it is quite ineffective against yeast and moulds. Citric acid is used up to 1% in non-alcoholic drinks, jams, jellies, bakery products, cheeses, canned vegetables and sauces. The ability of citric acid to chelate divalent ions is partially responsible for its antimicrobial activity. Sorbic acid is an unsaturated acid and is used between 0.05 and 0.2% in non-alcoholic drinks, some alcoholic drinks, processed fruits and vegetables, dairy desserts, confectioneries, mayonnaise, salad dressing, spreads and mustard. It is more effective against moulds and yeasts than against bacteria. The antimicrobial effect of sorbate is due to the inhibition of various central enzymatic functions of the cell, such as cell wall, protein, RNA and DNA synthesis. Finally, benzoic acid inhibits the function of membrane proteins, especially of the respiratory electron transport chains. It is employed at concentrations of 0.05–0.2% for the preservation of beverages, pickles, mayonnaise, confectioneries, mustard, cottage cheese and salad dressings.

See also: **Acinetobacter**. **Aeromonas**: Introduction. **Alcaligenes**. **Escherichia coli**: *Escherichia coli*. **Lactobacillus**: Introduction. **Meat and Poultry**: Spoilage of Meat. **Moraxella**. **Preservatives**: Permitted Preservatives – Benzoic Acid. **Pseudomonas**: Introduction. **Salmonella**: Introduction. **Spoilage Problems**: Problems caused by Bacteria.

Table 4   Influence of pH on the amount (%) of dissociated ions of weak organic acids

Acid	pK	% dissociated at pH		
		4	5	6
Acetic	4.8	15.5	65.1	94.9
Propionic	4.9	12.4	58.3	93.3
Lactic	3.8	60.8	93.9	99.3
Citric	3.1	81.1	99.6	>99.1
Sorbic	4.8	18.0	70.0	95.9
Benzoic	4.2	40.7	87.2	98.6

Further Reading

Böck A and Sawers G (1996) Fermentation. In: Neidhardt FC, Curtiss III R, Ingraham JL et al (eds) *Escherichia coli and Salmonella Cellular and Molecular Biology.* Washington, DC: American Society of Microbiology.

Brown MH and Emberger O (1980) Oxidation reduction potential. In: Silliker JH (ed.) *Microbial Ecology of Foods.* Vol. 1. New York: Academic Press.

Corlett DA Jr and Brown MH (1980) pH and acidity. In: Silliker JH (ed.) *Microbial Ecology of Foods.* Vol. 1. New York: Academic Press.

Doores S (1993) Organic acids. In: Davidson PM and Branen AL (eds) *Antimicrobials in Foods*, 2nd edn. P. 95. New York: Marcel Dekker.

Eklund T (1989) Organic acids and esters. In: Gould GW (ed.) *Mechanisms of Action of Food Preservation Procedures.* P. 181. New York: Elsevier Applied Science.

Farber JM (1991) Microbiological aspects of modified atmosphere packaging technology – a review. *Journal of Food Protection* 54: 58.

Foster JW, Park YK, Bang LS et al (1994) Regulatory circuits involved with pH-regulated gene expression in *Salmonella typhimurium. Microbiology* 140: 341–352.

Gennis R and Stewart V (1996) Respiration. In: Neidhardt FC, Curtiss III R, Ingraham JL et al (eds) *Escherichia coli and Salmonella Cellular and Molecular Biology.* Washington, DC: American Society of Microbiology.

Leyer GJ, Wang L-L and Johnson EA (1995) Acid adaptation of *Escherichia coli* O157:H7 increases survival in acid foods. *Applied and Environmental Microbiology* 61: 3752–3755.

Lynch AS and Lin ECC (1996) Responses to molecular oxygen. In: Neidhardt FC, Curtiss III R, Ingraham JL et al (eds) *Escherichia coli and Salmonella Cellular and Molecular Biology.* Washington, DC: American Society of Microbiology.

Ooraikul B and Stiles ME (1991) *Modified Atmosphere Packaging of Food.* New York: Ellis Harwood.

Sloncewski JL and Foster JW (1996) pH-Regulated genes and survival at extreme pH. In: Neidhardt FC, Curtiss III R, Ingraham JL et al (eds) *Escherichia coli and Salmonella Cellular and Molecular Biology.* Washington, DC: American Society of Microbiology.

Sofos JN (1994) Microbial growth and its control in meat, poultry, and fish. In: Pearson AM and Dutson TR (eds) *Advanced in Meat Research.* Vol. 9, p. 359. New York: Chapman & Hall.

Unden G, Becker S, Bongaerts J, Holighhaus G, Schirawski J and Six S (1995) O$_2$-Sensing and O$_2$-dependent gene regulation in facultatively anaerobic bacteria. *Archives of Microbiology* 154: 81–90.

EGGS

Contents
Microbiology of Fresh Eggs
Microbiology of Egg Products

Microbiology of Fresh Eggs

N H C Sparks, Department of Biochemistry and Nutrition, Scottish Agricultural College, Auchincruive, Ayr, Scotland

Eggs are one of the few foods that, in the UK, pass from the farm to the consumer with minimum treatment. Even the washing or sanitizing of the shell is prohibited if the egg is to be sold as a Grade A product. The ability of the chicken to produce a food that can be stored for up to 3 weeks without adversely affecting its eating or bacteriological safety is an indication of the complex antimicrobial systems that have evolved to protect the egg, and in particular the yolk, from both pathogens and spoilage organisms. The ability of microorganisms to adapt, however, as evidenced by the increasing impact of *Salmonella enteritidis* on egg production, poses a constant challenge to the safety of fresh eggs as a food product.

Structure and Composition of Fresh Eggs

The hen's egg is formed in the ovaries and oviduct (**Fig. 1**). The oviduct is some 60 cm in length and for functional purposes is divided into the infundibulum or neck, the magnum, isthmus, uterus or shell gland and the vagina. The ova or yolk is formed over a period of approximately 9 days. The constituents (approximately 49% water, 16% protein, 34% lipid, 0.1% carbohydrate and 1% ash) are transported via the blood stream to the ovaries, where the material is taken up and contained within the perivitelline membrane. Following its release, the yolk should be gathered into the neck of the oviduct by the infundibulum and there begins its passage down the oviduct. Immediately after its entry into the oviduct, the second vitelline membrane is deposited and then, on entry to the magnum, albumen deposition begins. The albumen is laid down in three distinct layers; the inner and outer thin albumen differ from the more viscous middle thick albumen in the percentage of the protein ovomucin (four times greater). With this

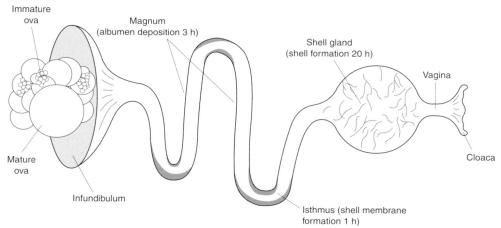

Figure 1 Schematic diagram of the reproductive organs of the hen.

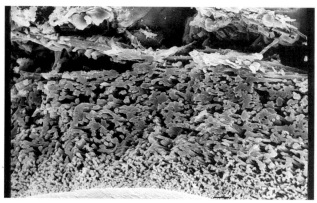

Figure 2 Radial section of fibrous egg shell membranes showing the relatively thin fibres of the inner membrane and the thicker fibres of the outer membrane.

exception, the different albumen layers are relatively similar in composition, consisting of approximately 88% water, 10% protein, 0.03% lipid, 0.6% carbohydrate and 0.5% ash.

The egg spends approximately 3 h in the magnum before moving on to the isthmus, where the sheet-like limiting membrane is deposited, followed by two fibrous shell membranes. The fibres lie parallel to each other and are arranged in a random manner within the tangential plain (**Fig. 2**). The fibres of the inner membrane, which oppose the limiting membrane, differ from those in the outer shell membrane only in that they are thinner ($< 2 \,\mu$m cf. $< 3.6 \,\mu$m) and more tightly packed. The inner membrane is approximately $20 \,\mu$m thick, while the outer membrane is thicker – approximately $50 \,\mu$m. Towards the end of the hour that the egg spends in the isthmus, water is moved into the albumen. This process of plumping continues throughout the early stage of the shell formation in the shell gland (or uterus).

The egg spends some 18 h in the uterus – approximately 75% of the total time it spends in the oviduct.

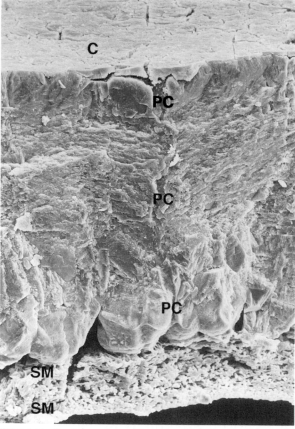

Figure 3 Radial section of hen egg shell showing the cuticle (C), pore canal (PC) and shell membranes (SM).

Once the initial layer of shell has formed across the surface of the outer shell membrane, the plumping process ceases. The shell consists of 98% calcium carbonate in the calcite form and 2% organic matrix (**Fig. 3**). Permeating the shell are between 7000 and 17 000 trumpet-shaped pores that are approximately $10 \,\mu$m in diameter and which are essential in the fertile egg for the exchange of respiratory gases. Immediately prior to oviposition, the organic cuticle is deposited, forming a relatively thin (0.5–13 µm) layer over the shell, which is normally 300–400 µm thick. Where the cuticle bridges the mouth of the pore canal, it forms a loose plug, rather like a loose cork in the neck of a bottle.

Antimicrobial Defence Systems

Physical Defence

The egg's antimicrobial defence mechanisms are both physical and chemical in nature. If we consider bacteria located on the surface of the shell, the cuticle presents the first line of defence. Immediately following oviposition, the cuticle has a fragile, sponge-like moist structure. While it is in this condition any bacteria that come into contact with the surface of the shell will be rapidly translocated through the shell to the shell membranes (**Fig. 4**). This occurs as a result of the cuticle's open structure and the presence of water in the cuticle and pore canals. Normally, however, the cuticle would have dried before bacteria came into contact with it. Under these circumstances, the bacteria would tend to be restricted to the relatively dry and hostile environment of the shell surface. This environment is reflected in the predominantly Gram-positive bacteria commonly isolated from the shells of eggs: *Micrococcus, Achromobacter, Aerobacter, Alcaligenes, Arthrobacter, Bacillus, Cytophaga, Escherichia, Flavobacterium, Pseudomonas* and *Staphylococcus*.

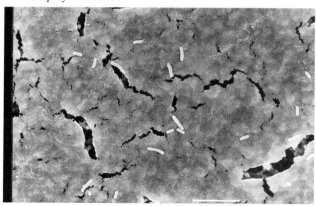

Figure 4 *Salmonella* on the inner surface of the shell's limiting membrane.

Once the cuticle has dried, probably the most common cause of bacteria being drawn through the shell is the presence of water on the shell and in the pore canals. This can come about either through condensation forming on the shell (so-called sweating) or through washing or sanitizing procedures.

Fungal growth on the shell tends to occur only when the eggs are held at relatively high levels of humidity (> 80%RH), and the following species have been reported to be associated with shell eggs: *Aspergillus, Penicillium, Cladosporium, Rhizopus* and *Mucor*. Once fungi have colonized the shell's surface the pores can be penetrated relatively rapidly by the hyphae. However, whereas while the shell remains intact microorganisms are forced to traverse the pore canals, once the shell is cracked the shell may offer very little resistance and the gross contamination of the egg contents can be extremely rapid.

The fibrous shell membranes that form the foundation for, and hence are crucial to, the correct formation of the over-lying shell offer very little defence against microorganisms. Studies have shown that bacteria can grow within the membranes, albeit that the nutritionally poor environment within the membranes appears to favour the Gram-negative rather than the Gram-positive organisms (which dominate the organisms recovered from the surface of shells). Ultimately the growth of the contaminants is constrained by either the physical presence of the limiting membrane or the chemostatic nature of the albumen or a combination of both. Irrespective of the mechanism, if contamination of the albumen is to occur, bacteria must eventually pass through the limiting membrane. Whether this is achieved by the contaminants degrading the limiting membrane or passing through naturally occurring holes in the membrane is uncertain. Once the organisms have passed through the limiting membrane, they are presented with the hostile environment of the albumen which spatially separates the yolk (which is rich in nutrients and has little, if any, inherent antimicrobial properties) from the shell.

The albumen's viscosity, or physical defence, is the result of the interaction between the proteins ovomucin and lysozyme at neutral pH. However, during the days that follow oviposition, the loss of carbon dioxide from the albumen by diffusion brings about an increase in pH. As the pH rises, so the interaction between ovomucin and lysozyme decreases and the viscosity is lost. This is important in the fertile egg for the successful development of the embryo and, although the antimicrobial benefits of the high viscosity are lost, the increased pH increases the efficacy of the chemical defence provided by the numerous antimicrobial proteins (**Fig. 5**), which are discussed below.

Chemical Defence

Although the lysozymes are arguably the best known of the albumen's antimicrobial proteins, they are probably less efficacious than the protein ovotransferrin. Lysozyme acts on the β(1–4) glycosidic bond between *N*-acetylglucosamine and *N*-acetylmuramic acid in the water-insoluble peptidoglycan

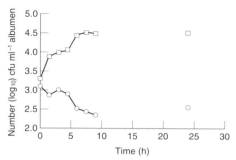

Figure 5   The effect of albumen pH on the growth of *Salmonella typhimurium* when incubated in vitro at 37.5°C. Squares, pH 8.54; circles, pH 9.61.

of eubacterial cell walls while ovotransferrin, as the name suggests, chelates a number of metal ions, including iron. By making iron unavailable to bacteria, the ability of microorganisms to replicate and grow is restricted. The importance of ovotransferrin's ability to bind the available iron within the albumen is exemplified by experiences in the US. There, eggs which had been washed or so-called sanitized, were rotting in relatively large numbers held in store. Upon investigation, it was shown that the increased incidence of rots resulted from the eggs being washed in water containing relatively high levels (> 4 p.p.m.) of iron. The wash water was penetrating the shell and providing sufficient iron to negate the ovotransferrin.

Other proteins, such as ovomucoid, ovoinhibitor, ovoflavoprotein and avidin, will inhibit trypsin, inhibit proteases, chelate riboflavin and chelate biotin, respectively.

Under normal production and storage conditions the physical and chemical defence systems combine to delay the growth of contaminants for about 21 days. Even when abused, for example by inoculating bacteria on to the shell membranes and incubating the egg at 37°C, the egg's antimicrobial systems prevented gross contamination for over 12 days (**Fig. 6**). The mechanisms that result in this delay are, however, the

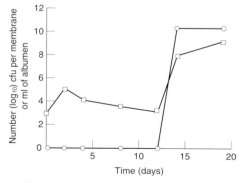

Figure 6   The growth of a mixed culture in the shell membranes and subsequent growth within the albumen (egg incubated at 37.5°C). Squares, membranes; circles, albumen.

subject of debate. Generally, it is agreed that the quiescent period is terminated when the contaminants make contact with the nutrients originating from the yolk. But while some researchers contended that growth occurred following penetration of the vitelline membrane by contaminants, others postulated that it was the leaching of iron and nutrients from a deteriorating membrane that resulted in the onset of a rapid growth of the contaminants.

Contamination Problems with *Salmonella*

Pathogens recovered from eggs in the past include species of *Aeromonas*, *Campylobacter*, *Listeria* and *Salmonella*. However, in terms of reported outbreaks of food poisoning attributed to eggs it is only *Salmonella* that is of significance. Shells of eggs have been reported to be contaminated with a range of *Salmonella* spp. including *S. anatum*, *S. bareilly*, *S. enteritidis*, *S. derby*, *S. essen*, *S. heidelberg*, *S. montevideo*, *S. oranienburg*, *S. thompson*, *S. typhimurium* and *S. worthington*. The relatively low water activity normally associated with the surface of the 'nest-clean' (free of visible contaminants) shell means that in practice most *Salmonella* spp. die relatively soon after contamination.

Up until the 1980s salmonellosis associated with hens' eggs was relatively infrequent and it was duck eggs that had long been implicated in salmonellosis outbreaks, presumably due to the environment under which they were produced. There was concern therefore in the UK when in the late 1980s an increase in the incidence of salmonellosis was attributed to hens' eggs, specifically an increase in outbreaks due to *S. enteritidis*. In the last 3 weeks of November 1988, for example, this organism accounted for 1167 or 57.2% of the *Salmonella* reports (although eggs were not implicated in all outbreaks of poisoning due to *S. enteritidis*). Of these, 890 reports were associated with *S. enteritidis* type 4 (PT4). In 1997 over 32 000 *Salmonella* infections were reported in England and Wales – an increase of 11% on the figure for the previous year. During the same period *S. enteritidis* PT4 infections rose by over 2000 (16%) while infections associated with other *S. enteritidis* phage types rose by 2500 (48%); infections associated with *S. typhimurium* and other *Salmonella* spp. fell by 16% and 9% respectively.

However, while it has been suggested that by adapting to the conditions found in the oviduct or ovaries, *S. enteritidis* PT4 has managed to circumvent many of the egg's natural antimicrobial systems (unlike *S. enteritidis* PT13A, for example, which is more commonly associated with faecal contaminant), the role of eggs as a food-poisoning vector is disputed by many in the egg production industry. Studies have shown, for example, that the correlation between the number of hens infected with *S. enteritidis* and the number of infected eggs laid is very variable. For infected flocks the number of eggs whose contents test positive ranged in one study from 0.1 to 1.0%. Furthermore, because *Salmonella* spp. are commonly associated with meat (both red and white) and dairy products as well as eggs, the potential for cross contamination to occur post-production, and in particular in the home, exists and needs to be considered. Poor hygiene can also result in secondary outbreaks.

Implications for Human Health

As far as human health is concerned, the main pathogen associated with shell eggs is *Salmonella*. While the common clinical features of salmonellosis are diarrhoea, vomiting and fever, infection may result in forms ranging from mild gastroenteritis to septicaemia or death (of the approximately 30 000 *Salmonella* recorded in humans in England and Wales in 1990, 68 people died from *Salmonella*-related illness).

Although transmitted predominantly in foodstuffs, the cooking process normally kills the organism. The rise in the UK in the late 1980s in the reported numbers of food poisoning attributed to *Salmonella* in eggs or egg products culminated in the Chief Medical Officer advising consumers to stop eating raw eggs and food containing uncooked eggs. Furthermore, it was recommended that those who could be considered vulnerable, such as the sick, elderly, pregnant women and babies, should only eat eggs when they had been cooked sufficiently to solidify the yolk and albumen. At that time the dominant causal organism was *S. enteritidis* and data for the period July to September 1998 show that *S. enteritidis* still dominates, accounting for 284 out of a total of 353 food-poisoning outbreaks attributed to *Salmonella*. Of the 284 outbreaks caused by *S. enteritidis*, 173 were due to *S. enteritidis* PT4. *S. typhimurium* accounted for 44 of the 69 outbreaks caused by *Salmonella* other than *S. enteritidis*. While not all of the outbreaks caused by *S. enteritidis* identified eggs as the suspect vehicle of infection, foods incorporating eggs (e.g. mayonnaise, egg sandwiches, egg fried rice, mousse) are frequently listed.

Reducing Infection of Flocks and the Risks in Storage

Eggs may become contaminated with microorganisms through either the vertical (e.g. infection of the ovaries or oviduct) or horizontal routes (e.g. cross contamination due to dust, faecal material, etc.).

While the vertical transmission of viruses (e.g. Oncornavirinae, paramyoviruses, picornavirus) and mycoplasmas (e.g. *Mycoplasma meleagridis*, *M. gallisepticum*, *M. synoviae*, *M. iowae*) can have a major impact on poultry production, these organisms do not affect the human population. The vertical transmission of food-poisoning bacteria has tended to be restricted to organisms such as *S. enteritidis*, although there have been some recent accounts of *S. typhimurium* serotypes infecting the ovaries.

Organisms associated with the contamination of eggs by horizontal transmission are far more numerous. Spoilage organisms that have been recovered from eggs include *Pseudomonas*, *Aeromonas*, *Acinetobacter*, *Alcaligenes*, *Citrobacter*, *Cloaca*, *Escherichia*, *Hafnia*, *Proteus* and *Serratia*. Pathogens recovered from shells include a wide range of *Salmonella* spp. (e.g. *S. heidelberg*, *S. montevideo*, *S. typhimurium*, *S. enteritidis*, *S. bareilly*), *Campylobacter* spp., *Listeria* spp. and *Aeromonas* spp. However, in recent years it is the upsurge in food poisoning due to *S. enteritidis* that has been notable, and when considering horizontal transmission it is notable that *S. enteritidis* PT13A is more often associated with faecal contamination of the shell than the ovaries (cf. *S. enteritidis* PT4).

Control measures designed to reduce the risk of product contamination must therefore encompass two approaches: first, they should ensure that replacement laying hens are free of the microorganisms which would be of concern and second, once housed, the risk of infection of the stock or cross contamination of the product must be minimized.

To appreciate the control measures it is necessary to understand the commercial production process. In brief, day-old chicks, hatched in a dedicated layer hatchery, will normally be reared in one of two systems: either on the floor (the majority of UK birds are reared in this way) or in cages. Birds will be reared at dedicated rearing sites from day-old to some 16 weeks of age, the point at which they are approaching sexual maturity. The 'point-of-lay' bird will then be transferred to a laying farm. In the UK eggs are produced using one of a number of systems: intensive (e.g. cage: 80%), perchery and other (15.5%) and free-range (4.4%). The pullet will be stimulated into lay by increasing the number of hours of light that the bird is exposed to in a 24-h period. Once in lay, hens will typically be kept until 70–76 weeks of age, at which time the house will be depopulated, cleaned and then restocked.

Measures aimed at reducing the risk of contamination of the product are often targeted at the *Salmonellae*, since these are particularly high-profile organisms as far as egg products are concerned. The hatching eggs which are used to produce the day-old layer bird should come from parent flocks which are managed according to Codes of Practice and Regulations, such as Ministry of Agriculture, Fisheries and Food Codes of Practice for the Prevention and Control of *Salmonella* in Breeding Flocks and Hatcheries (PB 1564), the Prevention of Rodent Infestations in Poultry Flocks (PB 2630) and the Poultry Breeding and Hatcheries Order 1993. These cover such points as microbiological monitoring of the parent flock for *S. enteritidis*, the testing of day-old chicks at the hatchery for *S. enteritidis* and action (normally slaughter of the flock) to be taken in the event of a positive result.

It is advisable that replacement pullets should be vaccinated against *S. enteritidis* PT4 (at day-old, 4 and 18 weeks of age). Before day-olds are placed in the rearing house the house should be checked microbiologically for *S. enteritidis* (the area should be resanitized if a positive result occurs). Standard biosecurity measures should be adopted, including the use of footbaths, operation of an effective rodent control programme and all employees should wear protective clothing. The risks associated with people, vehicles or materials coming on site and acting as vectors for microorganisms such as *S. enteritidis* are substantial. Therefore the number of visitors to the site should be minimized and those who do visit should wear protective clothing. Any vehicle coming on to the site should be cleaned externally and the wheels and lower portion of the vehicle sprayed with a sanitizer. The feed should be treated to minimize the risk of *S. enteritidis* contamination. This can be achieved in a number of ways; two of the more common techniques are heat treatment of the feed (mash) and the addition of an organic acid. The use of heat (e.g. 85°C for 3 min or 75°C for 6 min) can achieve a total kill of Enterobacteriaceae and moulds (both from initial levels of 10^6).

Before the point-of-lay pullets can be transferred from the rearing to the laying farm a statistically valid number of birds should be tested for *S. enteritidis* by taking cloacal swabs. Once the laying house has been tested and shown to be negative for *S. enteritidis*, the point-of-lay pullets can be housed. Strict biosecurity is imperative as most commercial sites are multi-age, that is to say, the poultry houses on a site will contain flocks that range in age (between houses) from approximately 16 to 76 weeks of age. Measures such as those outlined above for rearing sites should be adopted so as to minimize the risk of cross contamination.

Depending on the system, eggs may be laid in a range of environments. Cage eggs can be laid on to an inclined wire floor that causes the egg to roll away to the front or rear of the cage, away from the bird. The eggs would normally roll on to a belt, which conveys the egg via a series of lifts and belts to the egg store or packing station. Eggs which are produced on barn/perchery or free-range systems are normally laid into a nest box which can either be individual (one box per five or fewer hens) or communal (1 m² of nesting space per 120 hens). Depending on the design of the nest box, eggs may be collected manually or, as in the cage systems described above, automatically. If collected manually the eggs are normally laid on to either white wood shavings or similar material or plastic Astroturf-like matting. Automated systems can also use plastic matting or a similar material; the main criteria are that the eggs will roll over the surface and that the substrate is not lost from the nest box as the egg moves on to the belt.

At the moment the egg emerges from the bird it is warm (ca. 41°C) and moist. The moisture arises from the cuticle which at this stage has been shown to have an immature structure (see the above section on antimicrobial defence). In essence, the presence of water within the cuticle, combined with the open structure, allows bacteria to penetrate the egg to the level of the membranes in relatively large numbers. It is therefore essential that the environment into which the egg is laid contains as few pathogens and spoilage organisms as possible.

Once laid, the risk of cross contamination (e.g. from other hens and/or other eggs) should be minimized by removing the egg as soon as is feasible from the bird and moving the egg to either a store or the packing station. The main contaminants isolated from the shell of hens' eggs have been shown to be *Micrococcus*, *Achromobacter*, *Aerobacter*, *Alcaligenes*, *Arthrobacter*, *Bacillus*, *Cytophaga*, *Escherichia*, *Flavobacterium*, *Pseudomonas* and *Staphylococcus*. These bacteria are associated with dust, faeces and soil and reflect the relatively dry environment of the shell. To minimize the incidence of these contaminants, staff handling eggs need to wash their hands before and after collecting eggs, segregate (and handle separately) 'nest-clean' and dirty, cracked or broken eggs and collected eggs on to visibly clean trays.

Whether eggs are collected from egg stores and taken by road to the packing centre or conveyed directly from the poultry house to the packing centre, care must be taken to ensure that the temperature to which the eggs are exposed remains constant and above 5°C but below 20°C. This is important as it is a means of controlling the growth of organisms within the egg and because it reduces the risk of condensation forming on the shell. In the egg industry condensation, or 'sweating', as it is referred to colloquially, can occur on the shell when eggs are moved out of cool stores into a warmer environment. Water on the shell is of particular concern because of the ease with which bacteria, in the presence of water, can move along the pores of the shell. If condensation is allowed to remain on the shell for prolonged periods the risk of fungal (e.g. *Cladosporium* spp.) growth on the shell becomes significant.

While the washing of Grade A eggs is forbidden in the UK, egg washing or sanitizing is common practice in other parts of the world. If eggs are to be sanitized it is important that certain criteria are met, because water can facilitate the movement of bacteria through the shell. These criteria include ensuring that the wash water is maintained at a constant temperature of ca. 42°C; that the temperature of the eggs is less than that of the wash water but the differential should not be greater than ca. 35°C (as this will increase the incidence of shell cracks), and that the sanitizing solution always contains sufficient active sanitizer.

Recommended conditions for the storage of table eggs on the farm or at the packing station are < 20°C (say, 15°C) and approximately 75% RH. If they are to be stored on the farm, it is good practice to ensure that eggs are transported to the packing station as soon as possible after lay and within a maximum of 3 days.

Once in the packing station the eggs are graded. At the time of publication (1999) eggs sold in the UK are classified as Grade A (fresh eggs), Grade B (second-quality or preserved eggs) or Grade C (non-graded manufacturing eggs: eggs which are dirty or have a cracked shell but intact membrane can be sold to egg-breaking or processing plants). Eggs that are broken cannot enter the human food chain.

Grading consists of removing those with visible signs of contamination on the shell; 'candling' (shining a bright light through the egg, allowing an operator to detect and remove eggs with inclusions, cracked shells and other imperfections) and then sorting according to weight and packing. Following packing and boxing or film-wrapping on a pallet, the eggs are held in store (as described above) and dispatched as rapidly as possible. Among other information, such as the packing station details, the packaging should show the 'best before date'. This date is a maximum of 4 weeks from the point of lay and in some instances can be 3 weeks from the point of lay. It is increasingly common for eggs also to carry 'best before dates' printed on the shell using an ink jet printer.

See also: **Eggs**: Microbiology of Egg Products. **Food Poisoning Outbreaks**. *Salmonella*: Introduction; *Salmonella enteritidis*. **Viruses**: Introduction.

Further Reading

Anonymous (1989) *Salmonella in Eggs – A Progress Report*. London: HMSO.

Anonymous (1993) *Report on Salmonella in Eggs*. London: HMSO.

Anonymous (1998) *Code of Practice for Lion Eggs*, 2nd edn. London: British Egg Industry Council.

Board RG (1966) Review article. The course of microbial infection of the hen's egg. *Journal of Bacteriology* 29: 319–341.

Board RG and Fuller R (eds) (1994) *Microbiology of the Avian Egg*. London: Chapman & Hall.

Board RG, Sparks NHC and Tranter HS (1986) Antimicrobial defence of the avian egg. In: Gould G, Rhodes-Roberts ME, Charnley AK, Cooper RM and Board RG (eds) *Natural Antimicrobial Systems*. P. 82. Bath: Bath University Press.

Duguid JP and North RAE (1991) Eggs and *Salmonella* food-poisoning: an evaluation. *Journal of Medical Microbiology* 34: 65–72.

Humphrey TJ (1994) Contamination of egg shell and contents with *Salmonella enteritidis*: a review. *International Journal of Food Microbiology* 21: 31–40.

Sparks NHC (1996) Eggs. In: Milner J (ed.) *Meat Products*. P. E1. *LFRA Microbiology Handbook*. Surrey: Leatherhead Food Research Association.

Wells RG and Belyavin CG (eds) (1987) *Egg Quality – Current Problems and Recent Advances*. London: Butterworths.

Microbiology of Egg Products

J Delves-Broughton, Aplin & Barrett, Cultor Food Science, Devon, UK

R G Board, Bradford-on-Avon, Wilts, UK

The albumen and yolk separated from the egg shells are used to produce a variety of liquid and dried products (**Table 1**). With liquid products, especially whole yolk, destined for further processing, salt, sugar (both prevent gelling) or acidulants can be added. When shell eggs are converted into egg products both the health and spoilage risks can increase enormously.

Effects of Processing on Microorganisms

The initial flora of raw liquid egg will consist of a diverse flora of Gram-negative and Gram-positive bacteria that originate from the shell, the occasional infected egg, processing equipment (egg breakers, pipes, shell filters, etc.), and from those who handle eggs. All necessary practical attempts to reduce the bacterial load originating from the above sources by operation of Good Manufacturing Practice (GMP)

Table 1 Egg products and their uses

Product	Examples of use
Whole egg	
Frozen	Baked goods, institutional cooking, mayonnaise
Dried[a]	Above, plus ice cream manufacture, preparation of dry mixes for cake
Liquid – extended-shelf-life products	250–1000 ml cartons for use by bakers, caterers, home bakers, home use, etc. Shelf life of 28 days plus at 4°C
Value-added liquid extended-shelf-life products	As above, including low-cholesterol products, ready-prepared liquid scrambled egg, omelette and crêpe mixes, peeled boiled eggs
Albumen[a]	
Chilled (storage life approx. 4 days at 4°C)	Baked goods, icings, chocolate
Yolk	
Frozen – plain[a]	Baked goods, noodles, ice cream
Chilled or frozen – salted	Salad dressings, soups, mayonnaises
Chilled or frozen – sugared	Baked goods, egg nog, ice cream
Chilled – liquid with extended shelf life	1 kg cartons or larger amounts in 'bag-in-box' for use by bakers, caterers, home bakers, etc.

[a] Glucose must be removed before freezing or drying (see Table 5). Based on Board (1999).

and other means should be undertaken so that the bacteriological quality of the raw egg prior to further processing is as good as possible. Thus, in the selection of eggs for breaking, spoiled eggs should, if possible, be discarded, as a single rotten egg can add millions of bacteria to egg products and contaminate equipment. In most cases, candling – an unbroken egg examined with transmitted light – can identify grossly contaminated eggs. Certain types of rots, e.g. fluorescent rots caused by *Pseudomonas* spp., and especially those caused by the *Acinetobacter-Moraxella* group, are difficult or impossible to detect by candling alone.

Some egg processors operate an inspection system. A person sitting alongside an egg-cracking machine can stop the process and remove a contaminated egg if one is detected by appearance or smell. Some apparatus permits dumping of spoiled eggs manually without stopping the process. Even so, a colourless rot produced by *Acinetobacter/Moraxella* is unlikely to be detected.

Eggs with dirty or cracked shells, or those that have been incubated, or stored for a long time, will considerably increase the initial level of bacterial contamination of raw egg. Newly laid eggs contain significantly fewer bacteria than older ones. Indeed, it is recognized that in order to produce ultrapasteurized egg products of good bacteriological quality it is essential to use eggs within a few hours of being laid by dedicated flocks of hens and to pay critical attention to the cleanliness and hygiene of the processing equipment.

Eggs washed in a correct manner can significantly reduce the levels of bacterial contamination of liquid egg. Egg washing on a large scale began in the US in the 1940s. It became evident that the practice could be counterproductive if the temperature of the wash water was lower than that of the egg contents. This resulted in water, bacteria and iron in water from bore holes being pulled into the egg and triggered gross contamination of the white and yolk.

Appropriate codes of practice on how to wash eggs have been introduced. Washing is mandatory practice in the US and Canada. It is prohibited in many countries because of the fear that, if improperly applied, washing can increase contamination within the egg and lead to unacceptable levels of rots in stored eggs. Bactericidal agents such as chlorine, iodine or quaternary ammonium compounds can be added to the wash water.

Contemporary egg-breaking machines break large numbers of eggs quickly, producing whole egg or separating the yolk from the albumen. Whole egg may also be produced by crushing eggs and separating the egg contents from the shell debris by centrifugation. Mixing of contents with the broken and relatively clean shells can lead to significant contamination of liquid egg. Indeed, this practice is prohibited in many countries. Filters are often used to remove shell debris from the liquid egg. It is important that these are back flushed, cleaned and sanitized regularly. If they are not, accumulated debris will support the growth of bacteria that continuously inoculate the product. Homogenization of the liquid whole eggs, albumen or yolks will ensure that microbial contaminants are distributed uniformly throughout a batch. Liquid eggs should be processed with minimum delay or, if not, be stored at temperatures of not more than 4°C.

Pasteurization of Liquid Egg

Pasteurization equipment (heat plate and tubular heat exchangers) used by the egg-processing industry is basically similar to that used in the dairy industry to prepare pathogen-free milk with an extended shelf life at chill temperatures. Heat-processing regimes are designed to ensure the destruction of the bacterial pathogen, *Salmonella*, that may be derived from the

Table 2 Heat resistance characteristics of *Salmonella* and *Listeria monocytogenes* in liquid egg products

Organism	Medium	D°C value (min)	Z value (°C)
Salmonella enteritidis (17 strains)	Whole egg	$D_{57.2}$ 1.21–2.81	
		D_{60} 0.20–0.52	
Salmonella typhimurium	Whole egg	D_{60} 0.27	
	Yolk	D_{60} 0.40	
Salmonella enteritidis	Yolk	D_{60} 0.55–0.75	4.6–6.6
Salmonella seftenberg	Yolk	D_{60} 0.73	4.1
Salmonella typhimurium	Yolk	D_{61} 0.67	3.2
Salmonella typhimurium	Albumen	$D_{54.8}$ 0.64	
		$D_{56.7}$ 0.25	
Eight isolates	Albumen	$D_{56.6}$ 1.44	4.0
Listeria monocytogenes	Whole egg	D_{51} 14.3–22.6	5.9–7.2
		$D_{55.5}$ 5.3–8.0	
		D_{60} 1.3–1.7	
		D_{66} 0.06–0.20	
Listeria monocytogenes	Yolk	$D_{61.1}$ 0.7–2.3	5.1–11.5
		$D_{63.3}$ 0.35–1.28	
		$D_{64.4}$ 0.19–0.82	
Listeria monocytogenes	White	$D_{55.5}$ 13.0	11.3
		$D_{56.6}$ 12.0	
		$D_{57.7}$ 8.3	

Based on ICMSF (1998).
D value is the time in minutes at a stated temperature to kill 1 log (90%) of the number of bacteria.
Z value is the temperature increase required to reduce the D value by a factor of 10.

surface of the shell, or in the case of some *Salmonella* spp., including the virulent pathogen *S. enteritidis* PT4, from egg contents infected in the oviduct. A summary of studies on the heat resistance of *Salmonella* is presented in **Table 2**. Many studies have shown that the heat resistance of *Salmonella* will vary between species and strains, depending upon the physiological state of the cells used, and the physical and chemical characteristics of the individual egg product. An atypical strain that is not destroyed by current heat processes is *S. seftenberg*, but fortunately this organism has been found to be so rare that a decision has been made that the functional quality of the pasteurized product need not be sacrificed to protect against a strain that occurs so infrequently. It is notable that the pasteurization regimes developed to control *Salmonella* were found to be effective during the pandemic caused by *S. enteritidis* PT4. In the US, eggs from *Salmonella*-suspected flocks are often sent to processing for safety's sake.

The psychrotrophic *Listeria monocytogenes* is another pathogen of concern. Investigations into the heat resistance of *L. monocytogenes* (Table 2) indicate it will be controlled by the current heat processes used in the egg industry, even though it is more heat-resistant than *Salmonella*. Investigators have concluded that the heat processes used currently will ensure the eradication of *L. monocytogenes* in the processing of liquid egg, providing the initial levels of contamination with the pathogen are low (**Table 3**). These processes ensure a good kill of *Salmonella*

without adversely affecting the functional properties of the egg (whipping, emulsifying, binding, coagulation, flavour, texture, colour and nutrition).

In some countries a test for α-amylase present in the yolk is used to verify the efficacy of pasteurization. Pasteurization processes used in the UK (64.4°C for 2–5 min) destroy this enzyme, but those used in the US (60°C for 3.5 min) do not. The α-amylase test cannot be used with salted or sugared eggs. Pasteurization reduces the bacterial count in liquid eggs by 100- to 1000-fold, usually to a level of 100 cfu g^{-1}. Survivors are mostly *Micrococcus*, *Staphylococcus*, *Bacillus* and a few Gram-negative rods. Most survivors are incapable of growth at temperatures below 5°C. Psychrotrophic strains of *Bacillus cereus* can be of concern, and the addition of the bacteriocin nisin has been used to control the growth of this spore-forming food-poisoning bacterium as well as to extend the shelf life of the refrigerated pasteurized liquid egg products. Ultrapasteurization has permitted the production of long-shelf-life, refrigerated liquid egg products for use in the institution or home. The process uses novel temperature/time combinations that result in greater destruction of bacteria without impairment of the functional properties of the egg. Examples of ultrapasteurization processes are 70°C/90 s (liquid whole egg), 65.5°C/300 s (liquid yolk) and 57°C/300 s (albumen). Such products have an extended shelf life of 3–6 months at refrigerated temperatures.

It is evident from Table 1 that products other than

whole liquid egg are pasteurized. Generally, the pasteurization processes have to be more severe for modified egg products as the heat resistance of bacteria is increased by solutes such as sugar and salt. Of course, the growth of surviving bacteria, particularly if heat-damaged, will be impeded by the solute-rich products, even at temperatures conducive to the growth of the survivors.

Table 3 Minimum pasteurization temperatures and times for whole-egg products required by regulations in various countries

Country	Time (s)	Temperature (°C)
Whole egg		
Australia	150	62
China	150	63
Denmark	90–180	65–69
Poland	180	68
UK	150	64
US	210	60
Albumen		
UK	150	57.2
US	372	56
US	210	57
Yolk		
US	210	61
Yolk with 2% or more added sugar		
US	210	63
Yolk with 2–12% added salt		
US	210	63

Based on ICMSF (1998) and Board (1999).

Care must be taken to avoid impairment of the functional properties of egg white by pasteurization. Pasteurized salted and sugared egg products, because of their low water activity, have significantly increased shelf life, even at ambient temperatures. The addition of aluminium sulphate in solution protects the egg white from damage by the effect of heat; the addition of hydrogen peroxide allows the use of a lower heat process (52–53°C for 1.5 min) with a similar bactericidal effect, as does a vacuum process combined with a temperature of 57°C for 3.5 min. Novel but as yet unexploited methods of pasteurization of liquid egg include electroporation and nanothermosonication.

Unpasteurized liquid egg can be used as an ingredient in acid salad dressings and mayonnaises. Salmonella and staphylococci derived from eggs will die in a few days, providing the pH is below 4.0. The death rate of the bacteria – in other words, the autosterilization of a product – will be faster if products are held at ambient rather than refrigerated temperatures. Salad dressings and mayonnaises prepared from unpasteurized liquid egg for domestic purposes should not be consumed freshly prepared but stored for 3–4 days at ambient temperature before consumption.

Dried Eggs

Spray drying is the most common method of drying egg products. In this process the liquid is finely atomized into a stream of hot air. The enormous surface area created by atomization allows water evaporation to take place rapidly. Other, less commonly used methods include pan and belt drying. Freeze drying is a new method of drying egg products. Water is removed from a product while it is still in the frozen state. The product is frozen and then subjected to a high vacuum. Heat is supplied to the product while it is drying. This product is more popular commercially in the US than in Europe.

The microbiology of all these methods is similar. Drying kills many of the bacteria present initially in the liquid egg. Once the product is dry, stabilization of the microbiological population occurs and further decline occurs only slowly, even at ambient temperature. The predominant bacteria in the dried product are enterococci and *Bacillus* spp. The number of *Salmonella* can be reduced by 10 000-fold during drying. *Salmonella* can be a problem in dried eggs. The problem can be exacerbated if salmonellas grow during fermentation for glucose removal (see below). Despite the fact that *Salmonella* should be absent in pasteurized liquid egg prior to the drying process, salmonellas can often contaminate a finished dry-packaged product. After the product has been dried, the salmonellas can be destroyed by hot storage (hot-room treatment). Examples of times and temperatures for pasteurization of dried egg white are given in **Table 4**. These combinations have no demonstrable effect on functional qualities. *Salmonella* in egg powders can also be inactivated by irradiation. During and immediately following World War II when

Table 4 Time and temperatures of hot-room storage to destroy salmonellas in dried egg albumen

Pre-treatment	Temperature (°C)	Time (days)
Fermented, pan-dried	48.9	20
	54.4	8
	57.2	4
3% moisture	50	9
6% moisture	50	6
Spray-dried	54.4	7
Pan-dried	51.7	5
Adjusted to pH 9.8, with ammonia, pan-dried	49	14
Treated with citric acid	55	14
Spray-dried	49	14

Based on ICSMF (1998).

dried egg was in widespread use, especially in the home, salmonellosis arising from ingestion of contaminated products was extremely common. Since that time the control measures of pasteurization and hot-room storage enforced by legislation have made it a negligible problem in developed countries.

Glucose Removal

Dried egg whites contain about 0.6% free carbohydrate, primarily as glucose. On warm storage the carbohydrate can react with proteins by the Maillard reaction, which causes off flavours, insolubility and a brown discoloration. These problems are prevented by the removal of glucose from the liquid egg white before the product is dried. An early method of glucose removal allowed natural fermentation to take place. This caused problems if salmonellas grew. More recent methods (**Table 5**) exploit fermentation by yeast or bacteria, or the use of the enzyme glucose oxidase. Of these, bacterial fermentation appears to be the most commonly used method.

Table 5 Methods for removal of glucose from liquid eggs

Microorganism or agent	Comment
Natural flora	Traditional method used in China until 1940s. Principal bacteria Enterobacter, enterococci and other bacteria
Controlled bacterial fermentation, e.g. Klebsiella pneumoniae	Most commonly used method. Lactobacillus spp. have also been used
Yeast fermentation, e.g. Saccharomyces cerevisiae	3 h fermentation at 37°C. Can be followed by centrifugation to remove the yeast
Enzyme fermentation: glucose oxidase and catalase	Glucose oxidized to gluconic acid. Catalase destroys the hydrogen peroxidase that is formed

Based on ICSMF (1998) and Board (1999).

Product innovation includes the production of added-value products such as liquid scrambled egg mixes, omelette mixes and pancake mixes. Ready-prepared egg products include hard-cooked eggs, diced egg, scrambled egg and omelettes. Hard-boiled eggs with shell removed followed by immersion in boiling water or exposure to steam can have their shelf lives extended by packing in a solution of citric acid and benzoates. Other products mentioned above need to be stored either chilled or frozen, and treated basically in the same way as cooked meat products. Final preparation of egg products in the home using shell egg should entail thorough cooking procedures to ensure destruction of *S. enteritidis* PT4, and eggs that are no more than 3 weeks old and have been stored at chill temperatures.

*See also: **Acinetobacter**. **Chilled Storage of Foods**: Principles. **Dried Foods**. **Eggs**: Microbiology of Fresh Eggs. **Heat Treatment of Foods**: Principles of Pasteurization. **Listeria**: Listeria monocytogenes. **Natural Antimicrobial Systems**: Lysozome and other Proteins in Eggs. **Salmonella**: Introduction; Salmonella enteritidis.*

Further Reading

Board RG (2000) Microbiology of eggs. In: Lund BM, Baird-Parker AC and Gould, GW (eds) *The Microbiological Safety and Quality of Foods*. Gaithersburg, MD: Aspen. (In press)

Board RG and Fuller R (eds) (1994) *Microbiology of the Avian Egg*. London: Chapman & Hall.

International Commission on Microbiological Specification for Foods (ICMSF) (1998) Eggs and egg products. In: Roberts TA, Pitt JI, Farkas T and Grau FH *Microorganisms in Foods 6. Microbial Ecology of Food Commodities*. Ch. 15, p. 475. London: Blackie Academic.

ELECTRICAL TECHNIQUES

Contents

Introduction

D Blivet, AFSSA, Ploufragan, France

Copyright © 1999 Academic Press

Impedimetry, now undergoing important development, is not in fact a new technique. Impedimetry associated with microbiology was first mentioned during a congress of the British Medical Association in Edinburgh in July 1898, at which Stewart presented a paper (later published in the *Journal of Experimental Medicine*) entitled 'The changes produced by the growth of bacteria in the molecular concentration and electrical conductivity of culture media'. The curves presented followed the putrefaction of blood and serum, and were very similar to those obtained with the current apparatus. Today, impedimetry is considered to be a rapid method, whereas Stewart's impedance changes were recorded over 30 days. Serious development of the technique began in the 1970s. Today, four systems are commercially available in (**Table 1**).

Impedance, Conductance and Capacitance

Definitions

The *impedance* (Z) can be simply defined as the resistance to flow of an alternating current as it passes through a conducting material. When two metal electrodes are immersed in a conductive medium, the system behaves either as a resistor and capacitor in series, or as a conductor and capacitor in parallel. Three components determine the flow of current: the real resistance of the solution, a capacitance in series with a resistance arising from an oxidization at the surface of the electrodes, and a resistance due to the accumulation of charge dipoles just close to the electrodes. The impedance of such a circuit, and therefore, of a measurement cell, is given by Equation 1.

$$Z = \sqrt{\left\{R^2 + (1/2\pi fC)^2\right\}}$$
$$= \sqrt{\left\{(1/G)^2 + (1/2\pi fC)^2\right\}} \qquad \text{(Equation 1)}$$

where Z is the impedance expressed in ohms (Ω), R is the resistance expressed in ohms, C is the capacitance expressed in farads (F), G is the conductance

expressed in reciprocal ohms (Ω^{-1}) or siemens (S), and f is the frequency expressed in hertz (Hz).

The *conductance* (G) is defined as the inverse of resistance: G = (1/R). When an electric field is imposed on a electrolytic solution, the ions tend to migrate – the cations towards the cathode and the anions towards the anode. This migration of ions constitutes the flow of current in the solution, and each ion carries a fraction of the current proportional to its motility and concentration.

The *capacitance* (C) is a property of an element that stores electrical energy without dissipating it. It is linked to the accumulation of electric charges around the electrodes (**Fig. 1**). An increase in conductance or capacitance results in a decrease in impedance.

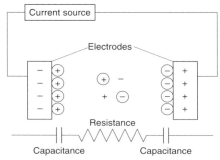

Figure 1 Schematic definition of capacitance.

Factors of Variation

The capacitance and the conductance vary with the current frequency (see Equation 1 and Figs 1 and 2). At low frequency, the impedance is principally affected by the capacitance, while at high frequency, it depends mainly on the conductance (**Fig. 2**).

The electrochemical perturbation required to create a detectable impedance change depends on the type

Table 1 Commercially available impedimetry systems for microbiology

	RABIT	Bactometer	Malthus 2000	BACTRAC
Measuring cells:				
Material	Propylene	Polypropylene	Glass	Propylene
Volume	10 ml	Modules of 2×8 ml	8 ml or 100 ml	10 ml or 100 ml
Use	Reusable	Single use	Reusable	Reusable
Electrodes	Stainless steel	Stainless steel	Platinum	Stainless steel
Incubator	Aluminium block with stabilized temperature	Air pulse convection incubator	Water bath	Aluminium block with stabilized temperature
Temperature accuracy	$\pm 0.005°C$	$\pm 0.1°C$	$\pm 0.006°C$	$\pm 0.1°C$
Configuration	16×32 cells	4×128 cells	5×240 cells	4×40 cells
Current frequency	21 kHz	1.2 kHz	10 kHz	1.0 kHz
Measures	Conductance	Impedance Conductance Capacitance	Conductance	Conductance
Temperature range	Room temperature to 55°C	8°C to 55°C	4°C to 56°C	Room temperature to 65°C

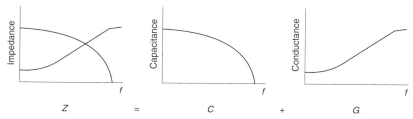

Figure 2 Impedance (Z), capacitance (C) and conductance (G) change as a function of frequency (f).

and position of the electrodes. It has been shown that electrodes located at the bottom of a measurement cell allow detection thresholds log 10 lower compared with the same electrodes located at the top of the cell, near the surface of the bacterial solution.

Bacterial generation times are affected by the temperature. As a result, changes in temperature affect the detection time, owing to the increasing motility of ions (molecular agitation). A temperature increase of 1°C causes a mean increase of 0.9% in the capacitance and of 1.8% in the conductance. If the temperature is not constant, the impedance curves may reflect temperature changes rather than bacterial metabolism.

For a given initial bacterial concentration, conductance change frequently can be correlated with bacterial metabolism. The time required before an acceleration in conductance is observable (the 'detection time') is shorter for samples of high initial bacterial density than for samples of low bacterial density.

Differences Between the Measurements

The impedance is a function of the capacitance, the conductance and the frequency. It has been established that the capacitance is relatively insensitive to the changes due to the bacterial growth, and that it is subject to random fluctuations of the same amplitude as those due to bacterial growth. However, more recently, it has been shown that under certain conditions (such as the use of stainless steel or low-frequency electrodes) the capacitance effect can be useful in the study of microbial growth.

The impedimetric systems available record either the impedance, conductance and capacitance signals, or only the conductance. The choice may be guided by the following observations: when employing a medium of low ionic strength, the bacterial metabolism results in easily detectable conductance changes, associated with an accumulation of ionized end products in the medium. In such a case, measurement of the conductance signal alone is usually sufficient to detect the bacterial growth. The situation is different for detection of yeasts, which produce significant capacitance changes, but only minor conductance changes – a capacitance change of 20% due to yeast metabolism, with a low conductance change (2%),

has been described. Moreover, conductance variations (decrease or increase) vary between yeast species. The low conductance changes obtained with yeasts may be due to the fact either that they do not produce highly ionized metabolites, or that they absorb ions from the medium. Except for yeasts and moulds, conductance is the more frequently used today.

Relationship Between Conductance and Conductivity

For many years the application of impedimetric techniques for the detection of impedimetric techniques for the detection of microorganisms was completely dependent on the empirical development of adequate culture media. There was insufficient theoretical knowledge to foresee that a given combination of microorganism and medium would increase or decrease the medium conductivity. In 1985 Owens described a theory of solution conductivity which permits the rational formation of culture media destined to measure conductance changes. Although the theory is only suitable for dilute solutions containing few ionic species (unusual in a specific culture medium), it can nevertheless direct the researcher to a useful choice of ingredients when developing a medium.

Electrolytic solutions are characterized by their capacity to conduct current when they are introduced in a circuit. The ability of an electrolyte to conduct current determines the resistance R of the solution. The resistance of any conductor is given by Equation 2:

$$R = \rho \ (l/A) \qquad \text{(Equation 2)}$$

where R is the resistance expressed in Ω, ρ is the specific resistance or resistivity of the material expressed in $\Omega \, cm$, l is the distance between two electrodes expressed in cm, and A is the surface area of the electrodes expressed in cm^2.

The conductivity (k) is the reciprocal of the resistivity of the material (Equation 3):

$$k = 1/\rho = (1/R)(l/A) = G(l/A) \qquad \text{(Equation 3)}$$

where k is the conductivity expressed in $\Omega^{-1}\,cm^{-1}$ or $S\,cm^{-1}$. Consequently, the electrolytic conductivity of a solution is equal to the conductance of a length of $1\,cm$ and a surface of $1\,cm^2$.

Conductivity of Electrolytic Solutions

An empirical relationship exists between the molar conductivity and the electrolyte concentration (Equation 4):

$$\Lambda = \Lambda_0 - K\sqrt{c} \qquad \text{(Equation 4)}$$

where Λ is the molar conductivity expressed in $S\,cm^2\,mol^{-1}$, Λ_0 is the molar conductivity at infinite dilution expressed in $S\,cm^2\,mol^{-1}$, and K is a constant mainly controlled by the valency of the ions; this is the concentration of the solution expressed in $mol\,l^{-1}$.

In theory, the conductivity of electrolyte mixtures can be calculated by adding the respective conductivities of the different ions. For example, for a mixture of KCl $0.001\,mol\,l^{-1}$ and NaCl $0.001\,mol\,l^{-1}$.

$$1000k = (\lambda_{Na} \times 0.001) + (\lambda_k \times 0.001)$$
$$+ (\lambda_{Cl} \times 0.002) \qquad \text{(Equation 5)}$$

where k is the conductivity expressed in $\Omega^{-1}\,cm^{-1}$ or $S\,cm^{-1}$, and λ is the molar conductivity of respective ions expressed in $S\,cm^2\,mol^{-1}$.

However, while it is possible to calculate the molar conductivities of ions in diluted solutions containing three or four kinds of ions, it is not possible to calculate such values accurately in more complex solutions, especially at the concentrations encountered in culture media.

Evaluation of Conductimetric Data

Variation of Conductivity Linked to Metabolic Activity Culture media contain various ionic species, and as a consequence calculation of their absolute conductivities is difficult. However, because the conductivity changes resulting from the metabolic activities of microorganisms are the main interest, it is not necessary for the calculated conductivities to be accurate values.

A microorganism can be represented as a compartment engaged in exchanges with the external environment (**Fig. 3**). The compounds of the external environment may be classed as electron donors, electron acceptors, carbon or nitrogen sources, other inorganic nutrients or metabolites, all of which may be charged or not. However, the conductivity of the microbial cell is negligible compared with the conductivity changes of ions associated with the growth.

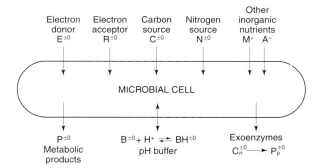

Figure 3 Interactions of a microbial cell with its external environment. $E^{\pm 0}$, $R^{\pm 0}$, etc. represent compounds with net positive, negative or zero change. From Owens (1985).

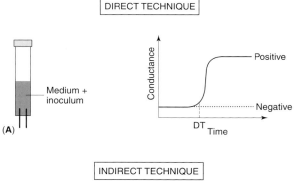

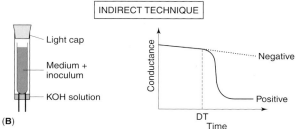

Figure 4 Characteristics of (A) direct and (B) indirect conductance techniques. DT, detection time.

Variation of Conductance Linked to Metabolic Activity The conductance changes are linked to the changes occurring in the culture medium. Bacterial metabolism gives rise to new compounds in the medium: the weakly charged or neutral substrate molecules are transformed into charged end products; this phenomenon is observed during the transformation of proteins into amino acids, carbohydrates into lactate, and lipids into acetate. If conductance changes are plotted as a function of time, the resulting curve is similar to a bacterial growth curve (**Fig. 4**).

The point where the conductance change rate exceeds a predetermined value is referred to as the detection time (DT). Because DT is a function of both the type of growth medium and the initial bacterial population, it is generally shorter than the time required for the visual detection of bacterial development on agar media. Samples likely to contain low

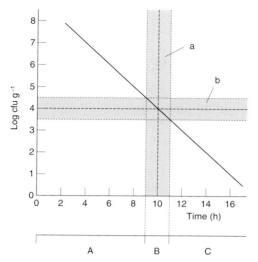

Figure 5 Calibration curve for the hypothetical product X. The permissible level of 10^4 cfu g^{-1} and the shaded zones (a for the impedance method, b for the standard plating method) allow the samples to be classified into three groups: A, acceptable samples; B, doubtful samples; C, unacceptable samples.

microbial populations are detected later than those that are highly contaminated. The detection time correlates well with the number of bacterial cells present initially. If a good correlation between the DT and colony count on agar medium is obtained, it will be possible, for a given product, to classify samples by choosing a contamination level – for example 10^4 cfu g^{-1} – and determining the corresponding detection time ('acceptable' detection time). Samples with DTs exceeding this value will be recorded as 'acceptable' samples. Because all microbial count estimations are subject to some uncertainties, impedimetric results are graded into three levels: acceptable samples, doubtful samples (in the intermediate zone) and unacceptable samples. The concept of the intermediate zone allows the operator to take into account the uncertainty of some detection times and to subject these samples to further tests, eventually by traditional methods (**Fig. 5**).

The growth of some microorganisms, such as yeasts, does not result in high conductance changes; these microorganisms produce non-ionized metabolites such as ethanol rather than highly ionized products. This highlights the fact that the range of media for impedimetry is limited by the requirement for a low initial conductivity. Culture media of high inherent conductivity do not permit the visualization of a conductance curve due to bacterial metabolism, even if growth has occurred. An alternative, 'indirect', technique was therefore developed.

Indirect Impedimetry

The indirect impedance technique measures conductance changes not in the culture medium but rather in a solution of potassium or sodium carbonate or hydroxide. Such solutions are carbon dioxide released by the microorganisms growing in the culture medium positioned just above the alkaline solution. This ingenious technique eliminates the problem of saturation of the impedance apparatus by highly conductive media (see Fig. 2).

Principle

The CO_2 produced by the microorganisms is absorbed by the alkaline solution. The hydroxide (OH^-) ions react with CO_2 (Equation 6):

$$CO_2 + 2OH^- \rightarrow CO_3^{2-} + H_2O \qquad \text{(Equation 6)}$$

and the negative conductance variations recorded can be explained by the molar ion conductivity:

$$\Delta\Lambda_0 = \Lambda_0 CO_3^{2-} - 2\Lambda_0 OH^-$$
$$\Delta\Lambda_0 = 138.6 - 2(198.6) = -258.6 \text{ S cm}^2 \qquad \text{(Equation 7)}$$

per mole CO_2 absorbed, and the observed variations are mainly due to the conductivity of OH^- ions. These ions disappear from the solution to give CO_3^{2-} ions which contribute less to the overall conductivity.

Measurement Systems

The stripping solution is an alkaline solution of potassium or sodium salts. The pH must be 11 or more. The conductance change is proportional to the amount of CO_2 produced, the volume and concentration of the absorbing solution, and the cell constant of the electrode. For optimal CO_2 transfer from the culture to the absorbing solution, the culture medium pH must be around 5, the head space volume must be low and the exchange surface large.

The company Don Whitley Scientific Ltd (Shirley, UK) was the first to adapt its RABIT apparatus to the indirect technique. The culture medium is held in a test tube above the electrodes, which are in contact with a KOH solution. The acid–base reaction between CO_2 and KOH results in a negative conductance change. This technique can be used for all microorganisms that produce CO_2, regardless of the type of metabolism, and the nature of the culture medium. The best results are obtained with a KOH volume sufficient to cover the electrodes (0.7–1.2 ml), and with concentrations up to $7 g l^{-1}$, although $5–6 g l^{-1}$ is preferred. This technique is applicable to the detection of numerous microorganisms, including *Staphylococcus aureus*, *Listeria monocytogenes*, *Enterococcus faecalis*, *Bacillus subtilis*, *Escherichia coli*, *Pseudomonas aeruginosa*, *Aeromonas hydrophila* and *Salmonella* spp.

Conclusion

Impedimetry is a useful way to estimate the amount of bacteria in a product in a short time (less than 24 h). The detection time is shorter when bacterial levels are high. This technique is therefore of prime interest for industrial monitoring of quality assurance or hazard analysis critical control point systems, as it is able to predict shelf life or contamination with pathogens before the sale of the product.

See also: **Electrical Techniques**: Food Spoilage Flora and Total Viable Count (TVC); Lactics and other Bacteria. **Hazard Appraisal (HACCP)**: The Overall Concept.

Further Reading

Bolton FJ (1990) An investigation of indirect conductimetry for detection of some food-borne bacteria. *Journal of Applied Bacteriology* 69: 655–661.

Firstenberg-Eden G, Eden R and Eden G (1984) *Impedance Microbiology*. Letchworth: Research Studies Press.

Owens JD (1985) Formulation of culture media for conductimetric assays: theoretical considerations. *Journal of General Microbiology* 131: 3055–3076.

Owens JD, Thomas DS, Thompson PS and Timmerman JW (1989) Indirect conductimetry: a novel approach to the conductimetric enumeration of microbial populations. *Letters in Applied Microbiology* 9: 245–249.

Richards JCS, Jason AC, Hobbs G, Gibson DM and Christie RH (1978) Electronic measurement of bacterial growth. *Journal of Physics E: Scientific Instruments* 11: 560–568.

Food Spoilage Flora and Total Viable Count (TVC)

G Salvat and **D Blivet**, AFSSA, Ploufragan, France

Impediometry was first used in 1898 by Stewart to monitor the changes of electrical conductivity during the putrefaction of blood and serum over a period of 30 days.

The first commercial applications concerned the prediction of shelf life for food. Methods used to appreciate food spoilage nowadays were nonspecific (TVC) or concerned the detection or counting of specific identified food spoilage agents such as pseudomonads, yeasts, *Lactobacillus* or *Brochothrix thermosphacta*. This article will present both specific and nonspecific techniques for use in shelf-life prediction.

Nonspecific Impedance Technique: Total Viable Count (TVC)

Each commercial firm developing impedance apparatus has formulated its own media for TVC application. This method uses simple culture media able to yield charged metabolites through the growth of the majority of bacterial species. The first step in applying the method is to develop the correlation curve between classical TVC and impedance measurements. This requires that at least 100 samples are treated with both methods to obtain a reliable correlation curve. A correlation curve must be plotted for each specific application, and attention must be paid to the identity of the spoilage flora encountered. The potential level of Enterobacteriaceae is an important parameter to consider if this organism is present in great quantity, it may grow more rapidly than the real spoilage flora and interfere with the shelf-life prediction as the incubation temperature mostly used favoured Enterobacteriaceae (30°C).

The user of impedance technique has to bear in mind that this method measures the metabolic activity of bacteria that may not precisely correlate with the amount of bacteria. There are two ways of obtaining well-fitted correlation curves. The first is to build a curve for each family of sample analysed – one for poultry, one for meat, one for cooked meals. The second way to improve the reliability of the curves is to lower the incubation temperature (ca. 20°C to 25°C) in order to favour the growth of spoilage microorganisms rather than those of Enterobacteriaceae. This approach may significantly increase time to detection.

Other parameters, such as sample preparation or storage conditions, may interfere with the time to detection. Diluting samples with the media used for impedance measurement shortens the detection time and avoids another dilution of the sample. This parameter may be important when high volumes of diluted samples are added to the impedance growth medium. Diluents of low conductivity must be used for impedance or it may be impossible to detect impedance changes.

Refrigeration or other stresses applied to bacteria present in the diluted samples before analysis may increase the detection time. Nevertheless, the calibration curve has to be built with a wide range of data covering a range of levels of confirmation as great or greater than expected data.

Refrigeration of the samples or of the bacterial suspension before impedance analysis may extend the detection time and result in an underestimate of the TVC.

The presence of inhibitory substances may interfere with the detection of microorganisms: appropriate dilution and neutralizing solutions must be used in such cases, in particular when the samples are obtained from cleaned and desinfected surfaces.

The last point to consider when inspecting a calibration curve obtained from a wide range of contamination levels (from 10^8 to 10^1 cfu ml^{-1}) is the mathematical function describing the curve. Impedance curves are usually better described by a decreasing exponential function than a straight line, especially in the range of lower counts (high TTD). When a linear regression of the $\log_{10}$ versus the $\log_{10}$ cfu ml^{-1} was calculated, an increase in the linear regression coefficient was generally noticed. This phenomenon of non-linear conductance curve response has been described for high and low contaminated samples. It was concluded that accurate linear results could not be guaranteed for bacterial counts of $> 10^7$ or $< 10^2$ cfu g^{-1}. For high contamination levels, the threshold level for impedance change (TTD) may be reached before establishment of a good baseline, making accurate determination of short TTD more difficult; for low levels of contamination, a 'tail' may be produced in the scattergram, due to the imprecision of both standard plate counts and distribution of microorganisms in the impedance-measuring wells.

Despite this, correlation coefficients obtained for the TVC by linear regression with the impedance technique are frequently good enough ($r^2 > 0.90$) for the impedance technique to be used to evaluate TVC in many kind of food products.

Yeasts

The growth of some microorganisms, such as yeasts, does not result in large conductance changes; these microorganisms produce non-ionized metabolites such as ethanol rather than highly ionized products. Moreover, the growth media for these organisms tend to have high conductivities, which prevents visualization of a conductance curve due to bacterial metabolism, even if growth occurs. That is why alternative techniques, the indirect impedance technique (conductance) or capacitance measurements are becoming popular for the determination of yeast counts.

Lactic Acid Bacteria (LAB)

Impedance techniques are commonly used to count LAB in products such as milk, fruit juices, wheat sourdough or starter cultures. When applied to milk or milk products the impedance medium is usually milk itself, milk acidified at pH 5.0 for specific *Lactobacillus* count from yoghurt, or milk added with 15% saccharose for a specific streptococci count from yoghurt. Generally speaking, for LAB impedance counting applications, a general-purpose impedance medium could be enriched with the sugar specific to the fermentative metabolism of the LAB, which represent the main flora of the test sample.

We are unaware of the development of a specific medium for estimating LAB in meat. This is probably due to the possibility of cross-reactions with other spoilage flora or Enterobacteriaceae that may be encountered in meat and poultry products due to the difficulty in formulating a specific medium for LAB.

Pseudomonads

Some authors have tested, with some success, an impedance technique which is able to detect *Pseudomonas* on meat products within less than 24 h. A conductance broth, supplemented with cephaloridine, fusidin and cetrimide (CFC) is used, but with higher concentrations of fusidin and cetrimide in order to obtain a better inhibition of *Pseudomonas* competitors (Table 1). In our experience, CFC agar is not able to inhibit all Enterobacteriaceae encountered on refrigerated poultry carcasses, and the oxidase test is required to provide a presumptive identification of the colonies. Such a test cannot be used, however, in an impedance liquid medium. For these reasons, a new medium specifically designed for the detection of *Pseudomonas* in poultry product has been developed. The final medium, designated MCCCD was first validated for its ability to promote the growth of 16 *Pseudomonas* strains. The TTD of 0.2 ml of a 10^{-6} dilution of a 24 h culture of the pure strains was approximately 10 h, except for one strain of *P. maltophila*, whose TTD was 15 h. MCCCD medium was then compared to the standard plating procedure in 106 samples of poultry neck skin originating from two different processing plants. The linear regression (**Fig. 1**) was established between the CFC agar count of confirmed *Pseudomonas* colonies ($\log_{10}$) and the time to detection obtained with the impedance technique using the MCCD medium.

In order to ensure the specificity of the MCCCD medium, 67 bacterial strains isolated from analysis of poultry neck skins with the MCCCFD medium, were identified. All were *Pseudomonas* strains.

Samples contaminated with approximately 10^3 *Pseudomonas* were detected within 18 h 45 min. This impedance technique, using MCCCD medium, could be used to count *Pseudomonas* from poultry neck skins sampled in processing plants. The technique

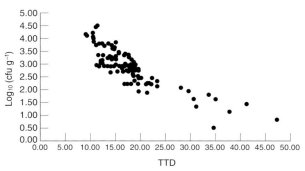

Figure 1 Correlation between impedance technique and ISO/DIS 13720 standard plating method evaluated on poultry neck skin samples. Characteristics of the figure: $\log_{10} cfu\,g^{-1} = 4.69 - 0.09 \times TTD$. $r = -0.848$; standard error $= 0.39$ ($\log_{10} cfu\,g^{-1}$). Number of samples $= 106$. TTD, time to detection.

enables the shelf life of poultry products to be predicted within 19 h and could be of value for Hazard Analysis and Critical Control Point monitoring and verification purposes.

See also: **Adenylate Kinase**. **ATP Bioluminescence**: Application in Meat Industry; Application in Dairy Industry; Application in Hygiene Monitoring; Application in Beverage Microbiology. **Electrical Techniques**: Introduction. **National Legislation, Guidelines & Standards Governing Microbiology**: European Union; Japan. **Rapid Methods for Food Hygiene Inspection**. **Sampling Regimes & Statistical Evaluation of Microbiological Results**. **Total Viable Counts**: Pour Plate Technique; Spread Plate Technique; Specific Techniques; MPN; Metabolic Activity Tests; Microscopy. **Ultrasonic Imaging**: A Non-destructive Method for Assessing Microbial Spoilage of Dairy Products. **Ultrasonic Standing Waves**.

Further Reading

Blivet D (1997) Intérêt de l'impédancemétrie pour la détection et le dénombrement de micro-organismes dans les denrées alimentaires. PhD. Compiègne, France: University of Technology of Compiègne.

Firstenberg-Eden G, Eden R and Eden G (1984) *Impedance Microbiology*. Letchworth, UK: Research Studies Press.

Richards JCS, Jason AC, Hobbs G, Gibson DM and Christie RH (1978) Electronic measurement of bacterial growth. *Journal of Physics E: Scientific Instruments* 11: 560–568.

Salvat G, Rudelle S, Humbert F, Colin P and Lahellec C (1997) A selective medium for the rapid detection by an impedance technique of *Pseudomonas* spp. associated with poultry meat. *Journal of Applied Microbiology* 83: 456–463.

Lactics and Other Bacteria

Ladislav Čurda, Department of Dairy and Fat Technology, Prague Institute of Chemical Technology, Czech Republic

Bacteria are unicellular prokaryotic organisms that usually multiply by transverse divisions. The daughter cells either separate from the parental cell or they remain attached to each other forming pairs, chains or irregular aggregates. Bacteria of importance in foods are heterotrophic, i.e. they assimilate small organic molecules and utilize macromolecules such as starch, cellulose, proteins or lipids by means of extracellular hydrolytic enzymes. The occurrence and activity of bacteria in foods may be positive (e.g. starter cultures in the dairy industry) as well as negative (e.g. food-borne infections and intoxications, food spoilage).

Lactic acid bacteria (LAB) play an important role in the food industry. LAB produce lactic acid in L(+), D(−) or DL form from pyruvate according to different metabolic pathways. Homofermentative LAB ferment glucose via fructose diphosphate pathway to lactic acid that is the only end metabolite (e.g. the genera *Lactococcus*, *Streptococcus*, *Pediococcus* and homofermentative lactobacilli). The 6-phosphogluconate pathway (*Leuconostoc*) and bifidus pathway (*Bifidobacterium*) are used by heterofermentative bacteria that produce, in addition to lactic acid, a number of products such as acetic acid, ethanol, CO_2 or acetoin.

LAB and other bacteria are particularly suited to electrical techniques (ET), because, under proper conditions, they usually produce strongly ionized metabolites which cause changes in the electrical properties (e.g. impedance) of the culture medium. The measurement of the electrical properties is a function of the bacterial viable biomass. On the other hand, microorganisms such as yeasts and moulds give a weak impedance signal only and are usually measured by indirect methods (indirect impedimetry).

Range of Food Applications

The versatility of ET enables their broad application in food microbiology including assessment of the quality of incoming raw materials, evaluation and control of the production process and assessment of the quality of finished products and their shelf-life.

Microorganisms Analysed by Electrical Techniques

Most applications relate to the estimation of the contamination level of a food sample. In addition to the total viable count different groups of bacteria may be studied by ET using selective media. Applications for detection or estimation of coliforms, Enterobacteriaceae, *Escherichia coli*, *Salmonella*, *Listeria*, *Staphylococcus aureus*, *Clostridium*, LAB including dairy starter cultures, aerobic sporeformers and psychrotrophs are known. For quantitative evaluation, it is essential to perform a calibration step. This is described below.

Types of Samples

A broad spectrum of food samples have been analysed by ET including milk and milk products (raw, pasteurized, UHT or dried milk, whey powder, butter, yoghurt, cheese, ice-cream), meat and meat products (minced meat, sausage), fish and fish products, margarine, eggs, confectioneries, chocolate, beverages (beer, fruit juices, mineral water), tomato products, spices, cereals and bakery products.

Shelf-life Prediction and Sterility Test of UHT Products

Since the ET are based on the end products of microbial metabolic activity, they are suitable for the estimation of shelf-life and sterility of various products. Shelf-life of pasteurized milk and other dairy products is estimated by the impedance method, usually after pre-incubation. The test needs a total time of 13–48 h. The results are correlated with a shelf-life similar or better than traditional methods, e.g. standard plate count method or Mosely test, that however take up to 10 days.

The quality of products fermented by LAB is often deteriorated by the Gram-negative bacteria that can be estimated using a selective medium. Shelf-life of these products can be predicted from calibration and relationship between shelf-life and DT.

UHT products have been widely available during the last few years. These products do not contain any viable microorganisms, and the impedance measurement is a suitable tool for the sterility check. The analysis involves a pre-incubation step (24 h) of the UHT product which ensures that all microorganisms (including those sublethally damaged) will be detected. The whole test is performed within 48 h. In UHT milk some heat-stable microbial proteases may not be completely destroyed by the UHT process and hence, may cause sensory defects during storage. Thus, the increase of impedance sometimes observed in a sterile product (baseline drift) may indicate significant enzymatic or chemical changes of the product. The method is also used for the estimation of LAB in UHT-treated fruit juices.

Activity of Starter Cultures

Starter cultures for the dairy, meat and wine industries are, from a technological point of view, better characterized by their metabolic activity than by viable cell counts. The changes in electrochemical properties of culture media can be used for the evaluation of metabolic activity and the stability of the starter culture. Reconstituted skim milk (10% wt) can be used as a culture medium for determining activity of dairy starter cultures. Significant reduction of detection time (DT) is achieved by addition of yeast extract (0.1–0.2%). Impedance or conductance measurements are less sensitive to buffering properties than pH. The main parameters responsible for the activity of starter cultures are DT, generation time (GT), inflectiontime (IT) and intercept on the log efn-axis in calibration equation (q). The stability could be judged besides these parameters, from the general shape of the impedance curve.

The activity of starter cultures is closely related to the metabolic activity of LAB used in fermentation processes, which are controlled on the basis of results of the impedance measurement. The activity of the starter culture, besides cultivation conditions, is influenced mainly by the presence of antibacterial substances or by phage infection.

Antibacterial Substances

Many antibacterial substances occur in food samples, e.g. preservatives, antibiotics or bacteriocins. Because the growth and metabolic activity of microorganisms are suppressed in the presence of these substances, their content in the food can be estimated by monitoring the growth kinetics of a selected test microorganism, and ET are a suitable tool for it. The inhibitory activity of the substance is shown by an increased DT. Other parameters also influence DT, e.g. the microbial generation time. The sensitivity of the method increases if a lower initial count of a test microorganism is used. Longer cultivation time is needed in this case for sufficient growth. Therefore a suitable compromise should be selected (usually between 10^3 and 10^5 cfu ml^{-1}), allowing sufficiently accurate determination in a short time. From a technological point of view it is important to pay attention to specific inhibitory action exhibited by some antibacterial substances, e.g. bacteriocins with a narrow spectrum of activity may induce some imbalance in mixed LAB cultures.

ET are also suitable for estimating antibiotics and preservatives (e.g. benzoic or sorbic acids, nisin) quickly and cost-effectively. Direct effects of the preservatives on bacterial species responsible for spoilage of preserved food can also be investigated. For this purpose the conditions of measurement should be as near as possible to those in the real sample (e.g. milk can be used as a culture medium for psychrotrophic spoilage flora). It should be noted that pH has a significant effect on the efficacy of many preservatives and on bacterial growth.

As test microorganisms for detection of antibacterial substances, the commonly used strains include *Lactobacillus delbrueckii* subsp. *bulgaricus*, *L. acidophilus*, *Bacillus stearothermophilus*, *B. subtilis* etc.

The efficacy of disinfectants for the removal of microbial biofilms can be determined by ET inserting a test disc directly into a sample cell. Biofilms develop on the metal or rubber surface and are potential sources of food contamination.

Phage Infections

Phage infections of lactic starters and cheese milk account for severe problems and product defects in the dairy industry. As these defects are caused especially by a serious failure of acid development or proteolytic activity, ET are useful for detection of phage infection. A suitable medium for this purpose seems to be reconstituted skim milk in which LAB are cultured at optimal temperature. Phage activity on a sensitive strain of LAB is shown by a delay in the *DT* and a decrease in final conductance response.

When phages stop microbial growth at less than 10^7 cfu ml^{-1}, *DT* is not observed. For a quantitative result it is important to correlate the number of LAB cells and the initial phage number. The quantitative evaluation of phages is based on an inversely proportional relationship between phage number and impedance or conductance after a defined time interval. Final conductance or impedance is very sensitive to the presence of phages, with 10 phages per millilitre being detectable. The traditional methods may be efficiently replaced by this method.

ET are also useful for the detection and selection of phage-resistant strains in the culture. Different values of delay in *DT* are obtained in this case, *DT* depends on proportion of resistant cells in the inoculum.

Further research in field of phage infections is needed. The method using specific bacteriophage as selective agent is also remarkable.

Optimization of Cultivation Conditions

The type and composition of culture media are important requirements of the ET. For a given analysis the media usually require optimization to give the desired conductance response and selectivity. Estimation of growth factors, e.g. vitamins, can also be

carried out using microbiological media combined with impedance measurements. This technique is further optimized by varying conditions such as temperature, stirring rate and extent of aeration.

Identification of Bacteria

ET have recently shown potential in the field of microbial identification. In this case bacterial growth is measured in a number of media with different nutrient status and under different growth conditions, the results of which can be used for identification purposes.

Other Applications

ET have been used to detect other microorganisms, including *Salmonella*, *Listeria*, *Staphylococcus aureus*, *Clostridium perfringens*, Enterobacteriaceae, coliforms and *Escherichia coli*. Many of these applications are described in more depth elsewhere.

Electrical Media

The composition of the culture medium is a fundamental requirement for ET. In formulating the electrical medium we need to consider that any uptake or excretion of a charged ion by the bacterial cell must be balanced by the outward flux of oppositely charged ions, or by excretion of similarly charged ones. In the case of LAB the major outward H$^+$ flux is associated with the excretion of lactic acid. This flux may be enhanced by using a positively charged donor of electrons or nitrogen source (some amino acids and NH$_4$). The electron donor has a major influence on conductivity changes, because it is metabolized in a large quantity. The selected buffer system should amplify H$^+$ flux resulting from the metabolic activity which in turn produces a large change of its conductivity. The direction of this change should support other conductivity changes. Tris or histidine buffers are suitable for LAB because their conductivity increases with decrease in pH. A phosphate buffer is not recommended as its conductivity change counteracts the increase associated with acid production. Conductivity increases in the presence of small or multicharged ions, but decreases with ion-pair formation. The hydrogen ion is a more effective conductor than other ions.

In general, the medium needs to be suitable for providing maximal metabolic activity for the test bacteria; the medium should produce a strong electrical signal, and it should be selective if microorganisms other than that being studied are present in the sample. Media and cultivation conditions used in standard classical methods are generally optimized for the development of a maximal amount of biomass. However, ET measures metabolic activity and hence the composition of the electrical medium, pH or temperature may be different from the classical techniques. The standard classical media often have high salt content and high initial conductivity, and for that reason they are used in conjunction with indirect electrical methods or with electrode impedance measurement only. For the direct electrical method a low ionic strength medium is recommended.

The simplest medium suitable for ET is reconstituted skim milk (10% wt). The electrical change associated with growth of LAB in this medium is satisfactory. Milk is also optimal for the growth of LAB. The results for LAB growth using ET are closely related to that associated in a number of dairy products. Shorter *DT* and, therefore, quicker results are obtained with addition of yeast extract (0.1–0.2%). Yeast extract is prepared as a stock solution (10–20%), which is autoclaved separately (121°C for 15 min). *Leuconostoc mesenteroides* subsp. *cremoris* is stimulated by addition of $0.14 \, g \, l^{-1}$ of $MnSO_4$. Production of CO_2 can be measured by indirect methods in milk with yeast extract, $MnSO_4$, and 0.5% sodium citrate. Gram-negative bacteria are cultivated selectively after addition of 0.1% of benzalkoniumchloride in 10% (v/v) sterile solution to milk.

Short DT was found for *Lactococcus lactis* subsp. *lactis* in a sterile medium composed of 3% special peptone and 0.25% yeast extract (pH 7), 12.5 ml of 5% urea, 25 ml of 5% arginine made up to a final volume of 1000 ml. Conductance change in a carbohydrate-deficient medium is increased by ammonia production. Addition of easily metabolizable nitrogen sources are therefore advantageous.

Producers of instruments for ET supply a range of dehydrated culture media which has been specifically developed to obtain optimal results, e.g. BiMedia 620 (SY-LAB GmbH, Austria) for lactobacilli and Bimedia 630 for beer spoiling bacteria. Pre-prepared conductance or impedance cells containing microbiological media are also available for some systems.

Techniques and Protocols

Instruments for Electrical Techniques

Commercial availability of instrumentation and media has enabled the use of ET in food microbiology. In general the instruments consist of an incubator unit and personal computer. The incubator unit is equipped for the measurement of the electrical quantity and ensures the temperature control. Precise temperature control is a critical requirement for this technique, because for common culture media used

in classical techniques the rate of change of conductivity is about $1.016°C^{-1}$. The software automatically acquires data from incubators and stores them on hard disk. This software allows the user to view impedance curves, print them, create reports and evaluate calibration curves. Some instruments use disposable sample cells which reduce operator exposure to pathogens and microbial contaminants. An overview of instruments designated for ET is shown in **Table 1**. Manufacturers offer a range of application notes or develop specific application according to the customers' requirements.

Standardization of Electrical Techniques

Only a few methods based on ET are currently accepted as standard validated methods. Standard methods are available mainly for total viable count and estimation of certain pathogens (e.g. *Salmonella*). First steps to worldwide acceptance of ET have been made in recent years. The first condition is to find a common validation procedure for an alternative microbiological method. This procedure has been prepared by MicroVal, which will work as the European validation authority after final adoption by European Committee for Standardisation (CEN). Within the MicroVal project a successful trial validation has been performed for the enumeration of the aerobic mesophilic counts in minced meat using the BacTrac instrument (10 participating laboratories in five European countries). ET for detection of *Salmonella* have been validated by AOAC as a first action method, the British Standard Institute (BS4285), the United States Department of Agriculture (USDA) and the German Standards Institute (DIN10115). The American Public Health Association (APHA) has incorporated conductance and impedance method as Class B in its Standard Methods for The Examination of Dairy Products. There are assays for bacterial count in milk, shelf-life of pasteurized milk, sterility test of UHT milk and for the analyses of Gram-negative bacteria, coliforms and *Salmonella*. Still, there is a growing demand on rapid microbial test. This is supported by expectation of the worldwide acceptance of the ET as a standard method for detection of bacteria in foods.

Sample Preparation

Sample preparation for ET is usually very simple. No dilution is required for liquid samples (milk, beer, juice), as they are analysed directly after shaking. Pulpy or solid samples are diluted with Ringer solution or peptone water and homogenized in a Stomacher or Ultra-Turax. If the dilution step is omitted, then a pH adjustment of the culture media might be necessary especially for samples with low pH (yoghurt). Butter (5 g + 9 ml Ringer solution) is melted

Table 1 Instruments for electrical techniques and their parameters

	Bactometer	BacTrac	Malthus	Rabit
Producer	bioMérieux Vitek, Inc., Hazelwood, USA	SY-LAB VGmbH, Purkersdorf, Austria	Malthus Instruments Ltd, Bury, UK	Don Whitley Scientific Ltd, Shipley, UK
Measured signal	Impedance or conductance or capacitance	Medium impedance (conductance) and electrode impedance (capacitance)	Conductance	Conductance
Measured value (units)	Relative change (%)	Relative change (%)	Absolute values (μS)	Absolute values (μS)
Direct/indirect method	+/−	+/+	+/+	+/+
Growth recognition	Rate change of signal basis	Threshold of change (set by operator)	Rate change of signal basis	Rate change of signal basis
Sample capacity	64–512	40–240	60–1200	32–512
Incubators per one PC	4	6	10	16
Incubator dimensions		400 × 540 × 235 mm	730 × 340 × 510 mm	400 × 600 × 400 mm
Temperature range	10°C below ambient to 55°C[a]	0–65°C	5–56°C	25–45°C
Thermostat type	Air cabinet	Dry aluminium block	Water bath	Dry block
Cooling system	Peltier element	Tap water or any external cooling system	Special cooling system	Not possible
Sample cell	Disposable modules with 16 wells	Re-useable and disposable	Re-useable and disposable	Re-useable
Cell volume	1.5–2 ml	1–10 ml and 5–100 ml	2–10 ml and 20–200 ml	2–10 ml
Electrodes	2, stainless steel	4, stainless steel	2, Pt on ceramic carrier (re-useable), stainless steel (disposable)	Stainless steel
Operating system	DOS	DOS, Windows NT in preparation	DOS, Windows	Windows

[a]Two separate compartments with different temperatures in one incubator.

at 45°C in a water bath and the aqueous phase (2 ml corresponds to 1 g of butter) is used for inoculation. Resuscitation (4 h at 37°C in buffered peptone-water) of bacteria stressed by high temperature and by low water activity is recommended for powdered samples, e.g. dried milk. Raw meat is analysed after dispersion in a peptone–water with a stomacher.

An Example of Protocol of ET Assay

An example of the protocol for the detection of aerobic spore-formers in a food sample is shown in **Figure 1**. The basic steps of the procedure are universal, excluding selective heating. Other assays differ, namely in the instrument setting and culture medium used. Samples with low numbers of bacteria need pre-enrichment before inoculation. The presumptive positive results require further confirmation steps as described elsewhere.

Interpretation and Presentation of Results

Data Acquisition

As mentioned above ET monitor microbial metabolic activity through specific changes in the electrical properties (conductance, capacitance or impedance) of the growth media. Most systems use impedance, which is a measure of the total opposition to the flow of a sinusoidal alternating current in a circuit. Impedance comprises the vectorial combination of a conductive and capacitive element. Their combination depends on the frequency used. This varies between 400 and 25 kHz for conductance signal. The conductance is associated with changes in the bulk ionic medium (so-called media impedance), and the capacitance with changes near the electrode surface (capacitance, electrode impedance). The units of impedance are S^{-1}. Conductance is recommended for monitoring bacterial growth in media with low conductivity.

The capacitance is directly proportional to the area of the double layer near the electrode surface and inversely proportional to the thickness of the double layer. Both these factors are strongly influenced by pH, because hydrogen ions increase the effective area and decrease distance between inner and outer layers on the electrode surface. Electrode impedance is useful only if a more conductive medium is available or if the inoculated sample contains many ions. Greater sensitivity of electrode impedance results in quicker response to microbial growth, but it is more prone to scattering which results in high noise to signal ratio.

Changes in electrical properties of inoculated culture medium are measured in cells equipped with one or two pairs of stainless-steel electrodes. Data are

Instrument setting

Temperature for mesophilic bacilli: 30–37°C (preferably 37°C)
Temperature for thermophilic bacilli: 55°C
Threshold of medium impedance change: 5%
Threshold of electrode impedance change: 10%
Maximum time limit: 24 h

↓

Sample preparation

Add 10 g sample to 90 ml of sterile 0.9% NaCl solution or Ringer solution
Solid samples – homogenize in a Stomacher or Ultra-Turax

↓

Selective heating procedure

Heat homogenate to 70°C for 10 min in a water bath for 10 min
Cool down to ambient temperature

↓

Inoculation

Fill the BacTrac measuring cell with 9 ml of BiMedia 001A or 002A
(general impedance broth)
Add 1 ml of heat treated sample to measuring cell,
include negative control (blank)
Mix carefully (twisting the cell, not inverting them)
Insert the measuring cells into BacTrac

↓

Measurement

One hour after a cell has been placed in the BacTrac,
that individual measurement will start automatically

↓

Evaluation of results

After calibration cfu will be calculated automatically as soon
as the impedance signal exceeds the pre-selected threshold

Figure 1 Detection of aerobic spore-formers by impedance method using BacTrac 4000 Series (according to Application Note of SY-LAB GmbH, Purkersdorf, Austria, with kind permission).

collected at a present interval (e.g. every 10 min) and stored in a computer. Some systems convert the impedance data into relative change of impedance. The shape of the resulting impedance curve resembles a growth curve in a normal culture medium. An example of impedance curve is shown in **Figure 2**. Relative changes are better comparable. Similarities in appearance show conductance or conductivity.

The first part of the impedance curve is stabilizing time, which is required for temperature equilibration between sample and incubator. It depends on the incubator type and sample volume. Some systems do not register this phase. The growth curve of bacteria starts with a lag phase. The number of cells in this phase remains practically the same. Impedance may also be constant or a drift is observed owing to weak metabolic activity of bacteria. This drift can occur even in a sterile medium without microorganisms. The impedance change can take on negative values, e.g. in case of the uptake of some ions by bacteria. A decrease of impedance may denote synthetic activity of bacteria.

As soon as the bacterial count reaches a level of approximately 10^5–10^6 cells ml^{-1}, the impedance curve accelerates and DT is registered. This acceleration is related to the production of low-molecular-weight metabolites above a certain threshold level. Variability of microbial count estimated as DT for different LAB strains can be explained by dissimilarity of metabolic activity – some strains need fewer cells to achieve a threshold concentration of ionized metabolites detected at DT. Cell multiplication occurs at DT which corresponds to a log phase. DT is mainly dependent on the initial bacterial count, but is also affected by physiological state of bacteria. An inverse linear calibration curve is obtained between logarithm of initial bacterial count per millilitre and the DT.

The impedance curve takes on an approximately linear shape after an acceleration phase and is characterized by slope K (**Fig. 3**). The time IT span to the turning point of impedance curve, i.e. to the inflection point of this more or less sigmoidal curve, provides information about the maximal metabolic activity of the bacterial culture. The IT is the maximum point on the curve obtained by plotting differences of impedance measurement against time. The parameter IT is useful, e.g. for the estimation of preservative concentration, particularly if the preservative counteracts the metabolic activity of bacteria and has no killing activity. The correlation between IT and the preservative concentration is better than comparison with DT.

When the nutrients in the sample are exhausted or the end-product metabolites inhibit multiplication of the bacterial population in the stationary phase, the slope of the impedance curve decreases, but it is still positive. In the death phase, the number of viable bacterial cells decline. Despite this, metabolic products increase. Impedance can be increased in this phase by lysis of the cells that releases ions. Metabolic pathways may be changed in the stationary and death phases by the lack of some nutrients, and the course of the impedance curve is often unpredictable at this stage.

The ideal impedance curve possesses no noise, the baseline is without drift, and there is a short and acute acceleration phase. These properties enable an accurate determination of DT. Evaluation of the impedance curve can be complicated by the presence of two or more accelerations caused by a change of the metabolic pathway or by the presence of miscellaneous types of bacteria with different generation times. The formation of gas by some bacteria may cause noise in the impedance signal.

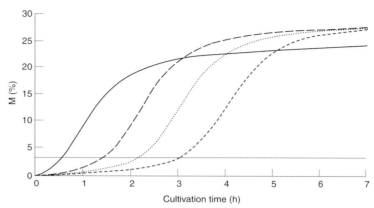

Figure 2 Examples of the impedance curves for yoghurt culture in reconstituted skim milk (10% wt). M, impedance change (%); inoculum 1%, $DT = 0.53$ h (—); 0.1%, $DT = 1.39$ h (– – –); 0.01%, $DT = 2.14$ h (. . . . .); 0.001%, $DT = 2.98$ h (- - - -). DT values for impedance change limit 3%. Average values from three measurements at 42°C (BacTrac 4100).

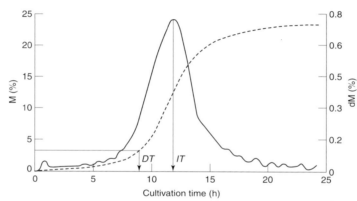

Figure 3 Evaluation of impedance curve. M, impedance change (%) (- - - -); dM, differences of two subsequent M values within 10 min measurement interval (%) (—). DT, detection time for impedance change limit 3% (8.91 h), IT, time of inflection point (12.00 h). Inoculum: *Lactobacillus acidophilus* (1%) in reconstituted skim milk (10% wt). Average values from three measurements at 37°C (BacTrac 4100).

Calibration

The calibration procedure consists of estimating the bacterial count by the standard cultural method and the determination of DT on a set of samples contaminated by bacteria of interest. As mentioned above, the relationship between a logarithm of colony forming units (*log cfu*) and DT is calculated. A linear calibration equation is predominantly applied:

$$log\ cfu = q + \delta \cdot DT$$

The reliability of a calibration curve described by the correlation coefficient r depends on the accuracy of standard and electrical methods, number of samples analysed, range of bacterial counts, etc.

Inoculation of nutrient broth by food sample or by food extract may influence the initial impedance value and microbial growth kinetics, and must be accounted for during the calibration procedure. In particular, the presence of uneven concentrations of antibacterial substances might lead to impairment of the correlation between DT and the standard plate count, where the influence of inhibitors is lowered by the dilution step. DT and consecutively the calibration or results might be negatively influenced, if time between mixing the food sample with culture medium and the insertion of the measuring cell into the instrument is not constant.

The correlation between the plate count and DT is improved if differences in the mean generation times (GT) of all bacteria in the sample are minimized. It can be reached by a proper choice of cultivation conditions and by a modification of the culture medium. The correlation coefficient r may achieve value > 0.97 for a single strain culture, but for multiple strain culture of one species r may be about 0.9, and for samples containing bacteria belonging to different species (e.g. raw milk) the expected r value could be roughly 0.8.

The required number of samples depends on the desired reliability of the calibration. The samples should evenly cover the calibration range. The recommended calibration range is 4 or 5 log cycles.

The calibration curve is used for the rapid assessment of cfu and for a rough classification of food samples into three groups: (1) samples having a contamination level above the permissible level; (2) suspect samples; and (3) samples having a contamination level below the permissible level. The limit detection times are estimated from the calibration curve and from the permissible contamination level, to this value of DT is added and subtracted one standard deviation. By means of calibration curves the automatic determination of initial bacterial concentration is also enabled.

The calibration line may be further utilized for the estimation of the generation time and metabolic activity of test bacteria in test samples. The generation time (GT) of the studied bacterial strain can be estimated from the calibration equation:

$$GT = \frac{\log 2}{|a|}$$

where a is slope in the calibration curve. A rapid procedure for the estimation of GT is based on inoculation of the culture into a suitable medium, with a simultaneous inoculation of a 100-fold dilution of the culture. DT is determined from two or more replicates and the average difference in δDT is calculated. GT is estimated as follows:

$$GT = \frac{\delta DT \cdot \log 2}{2}$$

It should be noted that GT estimated by this method is not influenced by metabolic activity because it depends only on the difference of DT and on the slope of calibration line.

The intercept on the log cfu-axis q represents the number of bacteria that may determine a DT at time zero. Although it is a theoretical extrapolation, this value can serve for characterization of the strain, because it is related to the rate of production of ionized compounds.

Advantages and Limitations of Electrical Techniques

Versatility   A wide range of bacteria in food samples can be determined by ET, as described previously. Many advantages accompany the combination of ET with other methods for confirmation of results. For instance, time is saved, because ET serves as an enrichment step (min. 10^6 cfu ml^{-1}) and costs are kept at a low level, since expensive traditional or alternative rapid methods (e.g. immunological methods) need only be used on presumptive positive samples identified by ET. Low contaminated samples can be analysed by the impedance method in conjunction with filter techniques where a filter containing the test sample is inserted directly into the impedance cell with culture medium. It is possible to analyse turbid or opaque samples and samples with small particles by ET. However, other methods are more suitable for some applications, e.g. the bioluminescence method for hygiene monitoring or the Bactoscan method for total viable flora in raw milk samples.

Rapidity DT of most assays takes a few hours whereas traditional methods usually require several days. Potential risk from heavily contaminated food samples can be reduced, because these samples have a shorter DT than less contaminated ones. A shorter time of analysis of raw materials and products reduces storage space requirements and allows the products to be moved to the market more rapidly. Rapidity of the impedance method depends on the sensitivity of the instrument and proper optimization of the culture medium. ET are not as fast as some non-cultural techniques because they involve a cultivation step, thus, several hours are required to obtain results. If differences exist among GT of bacteria in the samples, a low correlation with standard plate count could be observed.

Costs   ET require high capital expenditure. However, they bring cost savings in terms of reduction in labour and chemicals because they only require a simple preparation of the sample. Dilution of the sample prior to its insertion into the instrument is often omitted. The traditional methods are labour and material intensive, time-consuming and cumbersome.

Computer Control   This allows automatic measurement and evaluation procedures. Data on previously analysed samples are easily available for further evaluation. Computer control also reduces risk of operator error.

Precision, Accuracy, Reproducibility   Traditional methods tend to have relatively low reproducibility, their precision and accuracy are highly operator dependent. The sensitivity of ET is, in some cases, greater than for traditional methods, e.g. impedance assay is capable of detecting lower concentration of antibacterial substance. The analytical parameters of ET could be impaired by some factors. The food sample or its extract used as inoculum may influence microbial growth kinetics and should be taken into consideration during the calibration procedure. Frozen samples can show longer DT for a similar plate count, because stressed bacteria have a lower initial growth rate. Changes in DT of frozen samples can be caused by the presence of different kinds of bacteria.

Capacity   ET provide simultaneous analysis of large numbers of food samples. The sample capacity is flexible unlike techniques such as direct epifluorescent filter techniques (DEFT) and ATP bioluminescence.

Growth Analysis   Impedance assay and other ET are dynamic methods with nearly continuous measurement that provide information about microbial activity and growth kinetics as a function of time. Information concerning metabolic activity may have greater importance than information about cfu from standard plate count method.

See also: **ATP Bioluminescence**: Application in Dairy Industry; Application in Hygiene Monitoring; Application in Beverage Microbiology. **Bacteriophage-based Techniques for Detection of Food-borne Pathogens**. **Biochemical and Modern Identification Techniques**: Food-poisoning Organisms. **Direct (and Indirect) Conductimetric/Impedimetric Techniques**: Food-borne Pathogens. **Electrical Techniques**: Introduction. **Hydrophobic Grid Membrane Filter Techniques (HGMF)**. **Immunomagnetic Particle-based Techniques**: Overview. **Petrifilm – An Enhanced Cultural Technique**. **Reference Materials**. *Salmonella*: Detection by Classical Cultural Techniques; Detection by Enzyme Immunoassays; Detection by Colorimetric DNA Hybridization; Detection by Immunomagnetic Particle-based Assays. **Sampling Regimes and Statistical Evaluation of Microbiological Results**. *Staphylococcus*: Detection by Cultural and Modern Techniques. **Total Viable Counts**: Pour Plate Technique; Spread Plate Technique; Specific Techniques; MPN; Metabolic Activity Tests; Microscopy. **Ultrasonic Standing Waves**.

Further Reading

Carmiati D and Neviani E (1991) Application of the conductance measurement technique for detection of *Streptococcus salivarius* ssp. *thermophilus* phages. *Journal of Dairy Science* 74: 1472–1476.

Čurda L, Plocková M and Šviráková E (1995) Growth of *Lactococcus lactis* in the presence of nisin evaluated by impedance method. *Chem. Microbiol. Technol. Lebensm.* 17: 53–57.

Lanzanova M, Mucchetti G and Neviani E (1993) Analysis of conductance changes as a growth index of lactic acid bacteria. *Journal of Dairy Science* 76: 20–28.

Marshall RT (ed.) (1992) *Standard Methods for The Examination of Dairy Products*, 16th edn. Washington: American Public Health Association.

Okigbo LM, Oberg CJ and Richardson GH (1985) Lactic culture activity tests using pH and impedance instrumentation. *Journal of Dairy Science* 68: 2521–2526.

Owens JD (1985) Formulation of culture media for conductimetric assays: theoretical considerations. *Journal of General Microbiology* 131: 3055–3076.

Silley P and Forsythe S (1996) Impedance microbiology – a rapid change for microbiologists. *Journal of Applied Bacteriology* 80: 233–243.

Suhren G and Heeschen W (1987) Impedance assays and the bacteriological testing of milk and milk products. *Milchwissenschaft* 42: 619–627.

Svensson U (1994) Starter culture characterization by conductance methods. *Journal of Dairy Science* 77: 3516–3523.

Electron Microscopy *see* **Microscopy**: Scanning Electron Microscopy; Transmission Electron Microscopy.

Electroporation *see* **Minimal Methods of Processing**: Electroporation – Pulsed Electric Fields.

Endospores *see* **Bacteria**: Bacterial Endospores.

ENRICHMENT SEROLOGY
An Enhanced Cultural Technique for Detection of Food-borne Pathogens

Clive de W Blackburn, Microbiology Unit, Unilever Research Colworth, Bedford, UK

Introduction

Conventional methods for the detection of food-borne bacterial pathogens in food rely on a series of cultural enrichment steps. In the case of *Salmonella*, the conventional cultural method (CCM) consists of:

- pre-enrichment (16–26 h) to allow the resuscitation and multiplication of sublethally injured *Salmonella* cells
- selective enrichment (22–52 h) to increase the ratio of salmonellae to competitor organisms
- plating on selective/differential agar media (22–48 h) to enable the recognition of *Salmonella* colonies while suppressing the growth of the background microflora
- biochemical and serological confirmation (4–48 h) of presumptive-positive *Salmonella* colonies.

The definitive identification of salmonellae is for the most part serological, and one of the strategies for rapid detection has been to apply this stage directly to liquid cultures, thereby omitting the selective/differential agar plating stage.

Original Enrichment Serology Method

In 1969 an accelerated *Salmonella* detection procedure was reported by Sperber and Deibel that involved standard pre-enrichment and selective enrichment followed by application of direct serological testing. The standard tube agglutination test, which has been shown to require 2×10^8 salmonellae per millilitre for a positive result, was modified to give a fourfold increase in sensitivity and was applied initially to *Salmonella*-selective enrichment cultures. However, the test was found to be unreliable owing to carry-over of precipitates and insufficient cell numbers and/or poor antigen development because of the toxicity of the media. The inclusion of a 6 h elective enrichment step in brain–heart infusion broth provided a nonselective environment in which flagella production was not inhibited, but there were problems with autoagglutination of some bacteria. This was overcome by the use of a broth containing 0.2% D-mannose (M broth), which had been used pre-

viously to prevent fimbrial agglutination of *Salmonella* cultures, and the inclusion of nonspecific agglutination controls (physiological saline instead of antiserum). Using this procedure, termed enrichment serology (ES) by its originators, results could be obtained within 50 h compared with 96–120 h for the CCM (**Fig. 1**).

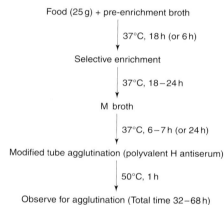

Food (25 g) + pre-enrichment broth

↓ 37°C, 18 h (or 6 h)

Selective enrichment

↓ 37°C, 18–24 h

M broth

↓ 37°C, 6–7 h (or 24 h)

Modified tube agglutination (polyvalent H antiserum)

↓ 50°C, 1 h

Observe for agglutination (Total time 32–68 h)

Figure 1 Enrichment serology method for the detection of *Salmonella* in foods.

Initial application of the ES procedure to the detection of *Salmonella* in foods and animal feeds showed good correlations with the CCM, but in some subsequent evaluations ES yielded large numbers of false negative results (**Table 1**). Increasing the M broth enrichment to 24 h, or using the ES procedure in combination with the fluorescent antibody technique, was found to improve detection rates. As with the CCM, the sensitivity of the ES procedure was dependent on the selective enrichment broth, but a combination of selenite–cystine broth and tetrathionate broth gave the most positive results. A further decrease in analysis time has been achieved using a modified ES procedure (6 h pre-enrichment, 18 h selective enrichment in tetrathionate broth, 6 h M broth enrichment), which when applied to the detection of *Salmonella* in soy products yielded fewer false negative results than the CCM.

The ES method is rapid and less labour-intensive than the CCM because the presumptive positive colony stage is avoided, although a pure culture of

Table 1   Examples of evaluations of Sperber and Deibel's enrichment serology (ES) method

Authors	Foods	No. of samples	Modifications	ES positive results	CCM positive results	Agreement (%)
Sperber & Deibel (1969)	Dried foods and feeds (nc)	105	None	37	37	100
Fantasia et al (1969)	Foods, feed and pharmaceutical products (nc)	689	None	132 (1 f–)	132 (1 f–)	99.7
Boothroyd & Baird-Parker (1973)	Raw materials and products (nc)	2005	None	209 (93 f–)	302	95.4
		769	24 h M broth	184 (11 f–)	195	98.5
Hilker & Solberg (1973)	Condiments, food products, animal feeds (nc)	126	None	64 (2 f–, 1 f+)	66	97.6
Surdy & Haas (1981)	Soy products (nc)	3486	6 h PE	3475 (11 f–)	3382 (104 f–)	96.7
Humbert et al (1990)	Poultry meat products (nc)	72	None	29 (13 f–)	41 (1 f–)	80.5

Key: nc, naturally contaminated; f–, false negatives; f +, false positives; PE, pre-enrichment.

Salmonella cells can be obtained by streaking from the M broth culture. No specialized equipment or training is required for the method and there is no increased expense; in fact a 37% cost reduction has been claimed. The method requires a minimum of about 10^7 colony forming units (cfu) per millilitre in the enrichment broth and failure to reach these levels may account for the high false negative rate for some products. Non-motile strains will not be detected, although this can be overcome by the use of a polyvalent O antiserum, but at the expense of a likely increase in the false positive rate. Although the technique is not widely used by the food industry, its principle has led to the development of several commercially available methods and 'enrichment serology' has become the generic term for these methods.

Commercial ES Methods

Latex Agglutination

To improve the sensitivity and visualization of serological agglutination reactions, specific somatic or flagella antibodies have been coupled to latex particles and there are now many commercially available latex agglutination tests covering a range of microorganisms. These kits are intended for use with dense cell suspensions prepared from isolated colonies as a means of confirming a presumptive pathogen identification (**Fig. 2**). The tests are quick and easy to perform and the agglutination reaction typically takes place within 2–10 min. Most kits consist of a single colour latex preparation, although a coloured latex test for the detection of *Salmonella* (Spectate Salmonella Coloured Latex Test, Rhône-Poulenc Diagnostics Ltd., Glasgow, UK) has been developed. The test consists of a mixture of red, blue and green latex

particles; each colour of latex is sensitized with specific antibodies to different groups of *Salmonella* which agglutinate to produce a crescent of colour depending on the serogroup present.

Latex agglutination has the advantages of being very simple and rapid, but the minimum detection limit (about 10^7 cfu ml^{-1}) in the final broth means that it is limited in its point of application during cultural enrichment. A number of *Salmonella* latex kits have been evaluated for use at various stages of cultural enrichment (**Table 2**). When applied to *Salmonella*-selective enrichment broths, the latex tests often gave a high false negative rate owing to the inability of the broths to produce detectable numbers. Application at progressively earlier stages of cultural enrichment (6 h selective enrichment, 18 h pre-enrichment) only compounded the problem. The kits also suffered from the presence of suspended particulate matter in the food enrichment cultures and the colour of the selective enrichment broth occasionally hampered interpretation of reactions. Application of the latex kits after 6 h post-enrichment in either M broth or nutrient broth generally gave the best agreement with the CCM. Cross reactions of the *Salmonella* antibodies with certain strains of *Citrobacter freundii*, *Escherichia coli* and *Proteus mirabilis* accounted for some false positive reactions, which mainly occurred with environmental samples.

The limitation of sensitivity has led to the use of latex agglutination tests for the confirmation of presumptive positive biochemical tests in which large numbers of the target organism are required to give a reaction. For example, a latex agglutination test is used as part of the Oxoid Salmonella Rapid Test (see below). Latex tests have also been used to confirm

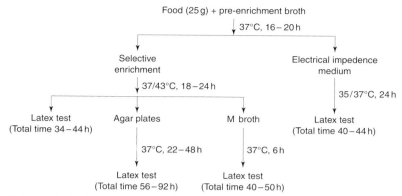

Figure 2 Application of latex agglutination kits for the detection of *Salmonella* in foods.

Table 2 Examples of evaluations of methods using latex agglutination kits after liquid enrichment

Authors	Foods	Latex kit	Preceding enrichment	No. of samples	Latex positive results	CCM positive results	Agreement (%)
Blackburn & Patel (1988)	Milk powder, turkey, prawns (ac)	Microscreen Salmonella	SE broths M broth	9 9	2 (2 f–) 4	4 4	77.7 100
Clark et al (1989)	Raw/cooked products (nc)	Spectate	SE broths	40	16	16	100
Bird et al (1989)	Environmental, powders (ac)	Spectate	SE broths	501	203[a] (2 f–, 10 f+)	205	97.6
Reid (1991)	Beef, beef by-products (nc)	Serobact Salmonella	TBG/M broth RV/M broth	81 81	11 (3 f+) 16 (1 f+, 1 f–)	11 17	96.3 97.5
D'Aoust et al (1991)	Meat, animal feed, egg, dried products (nc)	Bactigen Spectate Microscreen Salmonella	M broth Nutrient broth M broth	55 55 55	21 (3 f–) 19 (5 f–) 18 (6 f–)	24 24 24	94.5 90.9 89.1
Davda & Pugh (1991)	Confectionery products (nc/ac)	Microscreen Salmonella	Bactometer conductance positives	90	44	44	100
Sutcliffe et al (1991)	Water (nc)	Microscreen Campylobacter	Filtration, centrifugation	76	12 (11 f–)	17 (6 f–)	77.6
Baggerman & Koster (1992)	Fresh/frozen meat (nc)	Microscreen Campylobacter	Filtration, CCD broth	75	62	46 (16 f–)	78.7

[a] Eight negative samples caused autoagglutination.
Key: ac, artificially contaminated; nc, naturally contaminated; f–, false negatives; f+, false positives; SE, selective enrichment; TBG, tetrathionate–brilliant green broth; RV, Rappaport–Vassiliadis broth; CCD, charcoal cefoperazone desoxycholate.

presumptive positive *Salmonella* samples using methods based on impedance or conductive measurement.

Although it is primarily *Salmonella* latex tests that have been evaluated as enrichment serology methods, the Microscreen Campylobacter test (Microgen Bioproducts Ltd., Camberley, UK) has been used for the detection of *Campylobacter* in fresh and frozen raw meat. The method involved incubation in CCD broth (42°C, 8 h), filtration (0.45 μm) and incubation in blood-free modified charcoal cefoperazone desoxycholate (CCD) broth (42°C, 16–40 h) prior to application of the latex test. The *Campylobacter* enrichment serology method was more rapid and sensitive than the CCM. The Microscreen Campylo-

bacter latex kit has also been used for the testing of water samples following physical enrichment (filtration and centrifugation) rather than cultural enrichment. The latex test was found to be 1000 times more sensitive for *Campylobacter jejuni* than for *Campylobacter coli*, but the prevalence of this latter species in a number of the samples accounted for the high rate of false negative results.

The ease of use and specificity of latex agglutination kits has resulted in their widespread use in the food industry, although their application is primarily for rapid confirmation of presumptive positive colonies from agar plates. The agglutination technique itself is rapid and requires no additional skills or equipment.

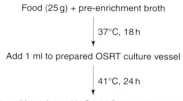

Food (25 g) + pre-enrichment broth

↓ 37°C, 18 h

Add 1 ml to prepared OSRT culture vessel

↓ 41°C, 24 h

Test positive tubes with Oxoid Salmonella Latex Test
(Total time 42 h)

Figure 3 Oxoid Salmonella Rapid Test method for the detection of *Salmonella* in foods.

Oxoid Salmonella Rapid Test

Several enrichment serology techniques have utilized the fact that most *Salmonella* serotypes are motile. In 1969 a glass apparatus was developed that relied upon the migration of salmonellae through selective and/or differential semisolid agars and the serological testing of the resulting presumptive positive broth cultures. A commercially available method, based on similar principles, has since been developed. The Oxoid Salmonella Rapid Test (OSRT; Oxoid Ltd., Basingstoke, UK) consists of a disposable culture vessel containing two tubes, each of which contains dehydrated selective media in the lower compartment and dehydrated selective/differential media in the upper compartment, separated by a porous partition. The media are hydrated with sterile distilled water and a *Salmonella*-elective medium is added to the culture vessel along with a novobiocin disc. The unit is inoculated with food pre-enrichment broth culture and during incubation at 41°C for 24 h any salmonellae present migrate into the tubes containing selective and diagnostic media. Cultures in the tubes in which the biochemical tests are positive are tested by serological agglutination with a 2 min antibody-coated latex test (**Fig. 3**). Confirmation of OSRT-positive results can be obtained by conventional streaking from the positive tubes.

The OSRT has been evaluated using a wide range of foods (**Table 3**). The incidence of false positive results was generally low and in many studies there was a sensitivity equivalent to, or greater than, the CCM. In one evaluation a high level of false negative results was obtained from minced meat and poultry samples and it was suggested that overgrowth of *Salmonella* by competing flora had occurred. In a separate study it was noted that although discrimination between positive and negative results was generally obvious, occasionally with raw foods the colour change was less distinct and these samples might otherwise be reported as negative. With raw foods it has also been reported that the percentage of OSRT colour-positive samples that were subsequently latex-negative (11–53%) was greater than that for processed foods.

Although quite manipulative, the test is quick (3–5 min) and easy to set up; it provides results after 42 h and therefore a time saving of 1–3 days for *Salmonella*-negative results compared with the CCM. Confirmation of presumptive positive samples requires a further 1–2 days. It has been estimated that the 'hands on' time for the test is 6 min per sample compared with 20 min for the CCM. As with all tests based on motility enrichment, non-motile strains of *Salmonella* will not be detected, but their incidence comprises less than 0.1% of clinical isolates.

Modified Semisolid Rappaport–Vassiliadis Medium

Rappaport–Vassiliadis (RV) broth was developed for the selective enrichment of *Salmonella* and in 1986 de Smedt developed a modified semisolid RV (MSRV) medium by adding agar. The detection principle was based on the ability of *Salmonella* to migrate through the MSRV, forming halos of growth, while the motility of other organisms was largely inhibited by selective agents (magnesium chloride, malachite green, novobiocin and a 42°C incubation temperature). Motility enrichment on MSRV medium was applied, after conventional pre-enrichment (direct motility enrichment) by placing drops of culture on the surface of an MSRV

Table 3 Examples of evaluations of the Oxoid Salmonella Rapid Test

Authors	Foods	No. of samples	OSRT positive results	CCM positive results	Agreement (%)
Holbrook et al (1989a)	Poultry, raw/cooked meat, offal, vegetables, dried products, ice cream, animal feed environmental (nc/ac)	820	216 (10 f+, 7 f–)	201 (22 f–)	95.2
Holbrook et al (1989b)	Meat, poultry, seafood, dairy products, dried foods (nc/ac)	96	46 (1 f–)	47	99.0
Hirata et al (1991)	Chicken (nc)	77	29 (1 f–)	30	98.7
Blackburn & Patel (1991)	Raw/cooked meat and seafood, powders, chocolate (nc/ac)	38	16	16	100
In't Veld & Notermans (1992)	Mayonnaise, milk powder, minced meat, poultry (nc/ac)	80	28 (13 f–)	40 (1 f–)	85.0

Key: ac, artificially contaminated; nc, naturally contaminated; f–, false negatives; f+, false positives.

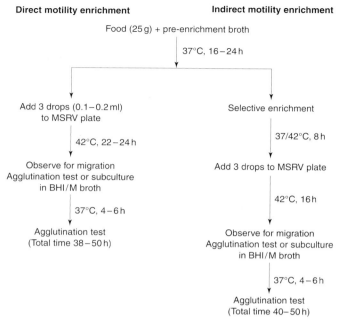

Figure 4 Direct and indirect methods using modified semisolid Rappaport–Vassiliadis medium for the detection of *Salmonella* in foods. BHI, brain–heart infusion.

plate and incubation at 42°C for 24 h (**Fig. 4**). If migration occurred the culture was tested by slide agglutination either directly, by cutting a well in the outer edge of the migration zone and allowing it to fill with liquid, or after inoculation and growth in brain–heart infusion broth for 4–6 h. This MSRV method gave 39% more *Salmonella*-positive samples than a CCM using tetrathionate–brilliant green (TBG) broth.

Subsequent to this initial study, however, direct motility enrichment was shown to be less productive than the CCM for some foods and environmental samples, possibly owing to overgrowth by a competing flora (**Table 4**). As a result, the application of MSRV after pre-enrichment and 8 h selective enrichment (indirect motility enrichment) was developed. Although no more productive than direct motility enrichment, when the two MSRV methods were used in combination they proved to be as effective as conventional procedures. Some of these studies have included collaborative trials and as a result MSRV has now been approved as an official method by the Association of Official Analytical Chemists (AOAC) for the detection of motile *Salmonella* in dried milk products, cocoa and chocolate.

The MSRV method has the advantages of being able to detect atypical salmonellae (lactose-fermenting and non-H_2S-producing) which might otherwise be missed on some *Salmonella*-selective agars. However, non-motile strains and some type cultures, e.g. *Sal-monella typhimurium* NCTC 74, are not detected. In one study a large number of strains (11%) from naturally contaminated samples failed to migrate on MSRV and it was suggested that the highly selective environment of the MSRV medium can affect the development of flagella. It has been noted that it is important to record motility soon after removal from incubation at 42°C because migration of some motile non-salmonellae can occur at lower temperatures.

A modification of MSRV has been developed with the inclusion of a differential system. Diagnostic Semi-solid Salmonella Agar (DIASSALM, Lab M, Bury, UK) has two indicator systems: saccharose combined with bromocresol purple and ferrous iron in combination with thiosulphate. After incubation, the plates are examined for a motility zone with a purple/black colour change (due to H_2S production). When the motility zone is absent, but the centre of the drop is blackened, non-motile salmonellae may be present. Confirmation is done by taking culture from the edge of the motility zone and streaking on to *Salmonella*-selective agar or applying a latex agglutination test directly. The addition of ferrioxamine E to buffered peptone water has been shown to increase the motility of *Salmonella* on both DIASSALM and MSRV and a modified direct motility enrichment method using a 6 h pre-enrichment has been proposed.

The MSRV technique is more rapid and less labour intensive than the CCM, although if a number of

Table 4 Examples of evaluations of the MSRV method

Authors	Method	Foods	No. of samples	MSRV positive results	CCM positive results	Agreement (%)
De Smedt et al (1986)	Direct	Minced meat, egg, cocoa, chocolate, milk powder (nc)	448	75 (1 f–)	54 (22 f–)	94.9
De Smedt & Bolderdijk (1990)	Direct and indirect	Cocoa, chocolate products (ac)	450 (15 labs)	347 (24 f–)	320 (51 f–)	83.3
De Zutter et al (1991)	Direct	Meat, poultry, cocoa, milk powder, environmental (nc)	430 (8 labs)	154 (7 f–)	145 (16 f–)	94.7
In't Veld & Notermans (1992)	Direct	Mayonnaise, milk powder, meat, poultry (nc/ac)	80	39 (2 f–)	40 (1 f–)	96.3
O'Donoghue & Winn (1993)	Direct and indirect	Meat, dried products, ready meals (nc/ac)	237	165 (1 f–)	166	99.6
Joosten et al (1994)	Direct	Environmental (nc/ac)	210	82 (18 f–)	100	91.4
De Smedt et al (1994)	Direct and indirect	Cocoa powder, chocolate (ac)	750 (13 labs)	407 (8 f–)	394 (21 f–)	96.1
Bolderdijk & Milas (1996)	Direct and indirect	Dried milk products (ac)	860 (19 labs)	828	820 (8 f–)	99.0
Fierens & Huyghebaert (1996)	Direct	Animal feeds (nc)	217	19 (2 f–)	17 (4 f–)	>97.2
Wiberg & Norberg (1996)	Direct	Meat, poultry, dried products, liquid egg, red pepper (nc)	419	134 (20 f–)	153 (1 f–)	95.2

Key: ac, artificially contaminated; nc, naturally contaminated; f–, false negatives.

samples show migration but are subsequently negative then the degree of time saving is reduced. In one study, for example, 70 out of 217 naturally contaminated feed samples showed migration, but only 19 of these were found to be contaminated with *Salmonella*. The cost of the MSRV method is similar to the CCM, and dehydrated MSRV medium is available from a number of manufacturers.

Salmonella 1-2 Test

The Salmonella 1-2 Test (BioControl Systems, Inc., Bothel, USA) is a two-chamber plastic vial for the detection of *Salmonella* and is based on selective and motility enrichment combined with immuno-precipitation. Pre-enrichment or selective enrichment culture is added to an inoculation chamber containing tetrathionate–brilliant green serine broth, and salmonellae migrate through a chamber containing a non-selective semisolid medium and are immobilized by polyvalent anti-*Salmonella* flagella antibodies giving a U-shaped precipitation band. The 1-2 Test is read after 14–30 h and presumptive positive results are confirmed using conventional procedures by streaking from the inoculation chamber (**Fig. 5**).

The Salmonella 1-2 Test protocol has been modified since it was first launched. Originally, the 1-2 Test vial was inoculated with direct selective enrichment cultures for raw flesh and highly contaminated foods, and pre-enrichment cultures for all other foods. In several evaluations high rates of false negative results for animal feeds, environmental samples and raw meats were obtained (**Table 5**). This occurred when either pre-enrichment or direct selective enrichment cultures were used to inoculate the 1-2 Test vial and it was attributed to the presence of large numbers of competitor organisms. The use of a two-step enrichment (pre-enrichment and selective enrichment in TBG broth for 18–24 h) increased the reliability of the 1-2 Test for these samples and the manufacturer modified the enrichment protocol for animal feeds and flour-based products accordingly. This prolonged the time required for testing by 24 h, but a further modification was made in order to obtain presumptive results within 48 h. After pre-enrichment and a 7 h incubation in TBG broth, 1.5 ml culture was added to the emptied inoculation chamber of the 1-2 Test vial. This modified 1-2 Test method has been found to be more reliable in subsequent studies and it has been adopted by the manufacturer for use with raw flesh and highly contaminated products. Although the 1-2 Test has been reported as being easy to read, in one evaluation using frozen shrimp a variation in interpretation of results between analysts was dem-

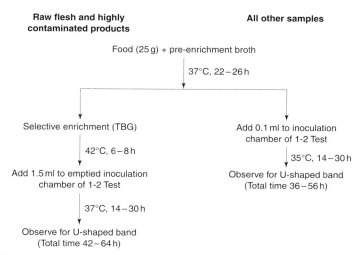

Figure 5 Salmonella 1-2 Test method for the detection of *Salmonella* in foods. TBG, tetrathionate–brilliant green broth.

Table 5 Examples of evaluations of the Salmonella 1-2 Test

Authors	Foods	Preceding enrichment	No. of samples	1-2 Test positive results	CCM positive results	Agreement (%)
D'Aoust & Sewell (1988)	Meat, chocolate, dried products, animal feeds (nc)	PE or DSE	186	25[a] (21 f–)	43 (3 f–)	87.0
Nath et al (1989)	Environmental, animal feeds, milk powder (nc)	PE	196	26 (8 f–)	34	95.9
		PE/SE (24 h)	314	82 (2 f–)	81 (3 f–)	98.4
Oggel et al (1990)	Animal feeds, environmental, egg products (nc)	PE/SE (7 h)	283	70 (3 f–)	73	98.9
St Clair & Klenk (1990)	Animal feed, chicken, nuts (nc)	PE/SE (24 h)	250	128 (5 f+, 3 f–)	120 (11 f–)	92.4
Humbert et al (1990)	Poultry products (ne)	PE	72	19 (2 f+, 23 f–)	41 (1 f–)	63.8
		DSE	24	11 (4 f–)	14 (1 f–)	79.2
		PE/SE (24 h)	24	14 (1 f–)	14 (1 f–)	91.6
Allen et al (1991)	Frozen shrimp (ac)	PE/SE (24 h)	200	115 (9–12 f+, 1–6 f–)[b]	110 (5 f–)	92.0–94.5
Feldsine et al (1995)	Animal feed, dried products, chocolate, cheese (nc/ac)	PE	1735 (3 labs)	1016 (15 f–)	1029 (2 f–)	99.0
		PE/SE (6–7 h)	1735 (3 labs)	1029 (3 f–)	1029 (3 f–)	99.7
Feldsine & Falbo-Nelson (1995)	Meat, fish, animal feed (nc/ac)	PE/SE (6–7 h)	320	213 (6 f–)	211 (4 f–)	96.9

[a] 20 positive after 8 h incubation of 1-2 Test vial.
[b] Variation due to analysts' interpretation.
Key: ac, artifically contaminated; nc, naturally contaminated; f–, false negatives; f+, false positives; PE, pre-enrichment; SE, selective enrichment; DSE, direct selective enrichment.

onstrated and this degree of subjectivity was thought to explain some of the false positive results that occurred.

In the original protocol, the 1-2 Test vial was read after both 8 h and 24 h incubation. A number of studies demonstrated that early (8 h) examination of the 1-2 Test vials led to false positive results and an increase in the false negative rate for both high and low moisture foods compared with examination after 24 h. Since then the manufacturer has modified the incubation step to 14–30 h.

The Salmonella 1-2 Test is easy to use and provides results more rapidly than the CCM. The 'hands on' time has been estimated to be 4 min per sample for processed foods and 9 min per sample for raw foods. Although a number of evaluations have shown the reliability of the method to be poor, subsequent protocol modifications have led to improvements in its performance and it is now an AOAC approved method.

Conclusions

Most enrichment serology methods have been developed for the detection of *Salmonella*, and by obviating the need for the isolation of colonies, they provide results more rapidly and are less labour intensive than conventional cultural methods. However, confirmation of enrichment serology positive results by streaking from liquid culture increases the labour and test time for presumptive positive samples. Subsequent inability to isolate the target organism may indicate a false positive reaction, but it can sometimes reflect the deficiencies of selective or differential agars in the presence of high numbers of competitive organisms. The reliability of enrichment serology methods has been shown to depend on a number of factors, including the length of cultural enrichment, the choice of enrichment media, food products, level of competitor organisms, injured cells and the presence of non-motile strains. Some of these factors may need to be considered before a choice of enrichment serology method is made.

See also: **Bacteriophage-based Techniques for Detection of Food-borne Pathogens. Biochemical and Modern Identification Techniques**: Introduction. ***Campylobacter***: Detection by Cultural and Modern Techniques; Detection by Latex Agglutination Techniques. ***Escherichia coli* O157**: Detection by Latex Agglutination Techniques. **Food Poisoning Outbreaks. Hydrophobic Grid Membrane Filter Techniques (HGMF). National Legislation, Guidelines & Standards Governing Microbiology**: Japan. **Nucleic Acid-based Assays**: Overview. **PCR-based Commercial Tests for Pathogens. Reference Materials. *Salmonella***: Detection by Classical Cultural Techniques; Detection by Latex Agglutination Techniques; Detection by Enzyme Immunoassays; Detection by Colorimetric DNA Hybridization; Detection by Immunomagnetic Particle-based Assays. **Sampling Regimes & Statistical Evaluation of Microbiological Results**.

Further Reading

Alen G, Bruce VR, Andrews WH, Satchell FB and Stephenson P (1991) Recovery of *Salmonella* from frozen shrimp: evaluation of short-term selective enrichment, selective media, postenrichment, and a rapid immunodiffusion method. *Journal of Food Protection* 54: 22–27.

Baggerman WI and Koster T (1992) A comparison of enrichment and membrane filtration methods for the isolation of *Campylobacter* from fresh and frozen foods. *Food Microbiology* 9: 87–94.

Bird JA, Easter MC, Hadfield SG, May E and Stringer MF (1989) Rapid *Salmonella* detection by a combination of conductance and immunological techniques. In: Stannard CJ, Petitt SB and Skinner FA (eds) *Rapid Microbiological Methods for Foods, Beverages and Pharmaceuticals*, SAB Tech. Series 25. P. 165. Oxford: Blackwell.

Blackburn C de W (1993) Rapid and alternative methods for the detection of salmonellas in foods – a review. *Journal of Applied Bacteriology* 75: 199–214.

Blackburn C de W and Patel PD (1989) Brief evaluation of the Microscreen *Salmonella* latex slide agglutination test for the detection of *Salmonella* in foods. Tech. Note 86. Leatherhead: Leatherhead Food RA.

Blackburn C de W and Patel PD (1991) Brief evaluation of the *Salmonella* Rapid Test (Oxoid) and hydrophobic grid membrane filter technique for the detection of salmonellae in foods. Tech. Note 91. Leatherhead: Leatherhead Food RA.

Bolderdijk RF and Milas JE (1996) *Salmonella* detection in dried milk products by motility enrichment on modified semisolid Rappaport–Vassiliadis medium: collaborative study. *Journal of AOAC International* 79: 441–450.

Boothroyd M and Baird-Parker AC (1973) The use of enrichment serology for Salmonella detection in human foods and animal feeds. *Journal of Applied Bacteriology* 36: 165–172.

Clark C, Candlish AAG and Steell W (1989) Detection of *Salmonella* in foods using a novel coloured latex test. *Food and Agricultural Immunology* 1: 3–9.

D'Aoust JY and Sewell AM (1988) Reliability of the immunodiffusion 1-2 Test™ system for detection of *Salmonella* in foods. *Journal of Food Protection* 51: 853–856.

D'Aoust JY, Sewell AM and Greco P (1991) Commercial latex agglutination kits for the detection of foodborne *Salmonella*. *Journal of Food Protection* 54: 725–730.

Davda C and Pugh SJ (1991) An improved protocol for the detection and rapid confirmation within 48 h of salmonellas in confectionery products. *Letters in Applied Microbiology* 13: 287–290.

De Smedt JM and Bolderdijk R (1990) Collaborative study of the international office of cocoa, chocolate, and sugar confectionery on the use of motility enrichment for *Salmonella* detection in cocoa and chocolate. *Journal of Food Protection* 53: 659–664.

De Smedt JM, Bolderdijk RF, Rappold H and Lautenschlaeger D (1986) Rapid *Salmonella* detection in foods by motility enrichment on a modified semi-solid Rappaport-Vassiliadis medium. *Journal of Food Protection* 49: 510–514.

De Smedt J, Bolderdijk R and Milas J (1994) *Salmonella* detection in cocoa and chocolate by motility enrichment on modified semi-solid Rappaport-Vassiliadis medium: collaborative study. *Journal of AOAC International* 77: 365–373.

De Zutter L, De Smedt JM, Abrams R et al (1991) Collaborative study on the use of motility enrichment on modified semisolid Rappaport-Vassiliadis medium for the detection of *Salmonella* from foods. *International Journal of Food Microbiology* 13: 11–20.

Fantasia LD, Sperber WH and Deibel RH (1969) Comparison of two procedures for detection of *Salmonella* in food, feed, and pharmaceutical products. *Applied Microbiology* 17: 540–541.

Feldsine PT, Falbo-Nelson MT, Hustead DL, Flowers RS and Flowers MJ (1995) Comparative and multi-laboratory studies of two immunodiffusion method enrichment protocols and the AOAC/*Bacteriological Analytical Manual* culture method for detection of *Salmonella* in all foods. *Journal of AOAC International* 78: 987–992.

Fierans H and Huyghebaert A (1996) Screening of *Salmonella* in naturally contaminated feeds with rapid methods. *International Journal of Food Microbiology* 31: 301–309.

Hilker JS and Solberg M (1973) Evaluation of a fluorescent antibody-enrichment serology combination procedure for the detection of salmonellae in condiments, food products, food by-products, and animal feeds. *Applied Microbiology* 26: 751–756.

Hirata I, Suzuki K, Ikejima N et al (1991) Rapid detection of *Salmonella* from meats by Oxoid Salmonella Rapid Test. *Japanese Journal of Food Microbiology* 8: 151–156.

Holbrook R, Anderson JM, Baird-Parker AC et al (1989a) Rapid detection of salmonella in foods – a convenient two-day procedure. *Letters in Applied Microbiology* 8: 139–142.

Holbrook R, Anderson JM, Baird-Parker AC and Stuchbury SH (1989b) Comparative evaluation of the Oxoid Salmonella Rapid Test with three other rapid *Salmonella* methods. *Letters in Applied Microbiology* 9: 161–164.

Humbert F, Salvat G, Lalande F, Colin P and Lahellec C (1990) Rapid detection of *Salmonella* from poultry meat products using the '1.2. Test®'. *Letters in Applied Microbiology* 10: 245–249.

In't Veld PH and Notermans S (1992) Use of reference materials (spray-dried milk artificially contaminated with *Salmonella typhimurium*) to validate detection methods for *Salmonella*. *Journal of Food Protection* 55: 855–858.

Joosten HMLJ, van Dijck WGFM and van der Velde F

(1994) Evaluation of motility enrichment on modified semisolid Rappaport-Vassiliadis medium (MSRV) and automated conductance in combination with Rambach agar for *Salmonella* detection in environmental samples of a milk powder factory. *International Journal of Food Microbiology* 22: 201–206.

Nath EJ, Neidert E and Randall CJ (1989) Evaluation of enrichment protocols for the 1-2 Test™ for Salmonella detection in naturally contaminated foods and feeds. *Journal of Food Protection* 52: 498–499.

O'Donoghue D and Winn E (1993) Comparison of the MSRV method with an in-house conventional method for the detection of *Salmonella* in various high and low moisture foods. *Letters in Applied Microbiology* 17: 174–177.

Oggel JJ, Nundy DC and Randall CJ (1990) Modified 1-2 Test™ system as a rapid screening method for detection of *Salmonella* in foods and feeds. *Journal of Food Protection* 53: 656–658.

Reid CM (1991) Evaluation of rapid methods for the detection of salmonellae from meat and meat products. Meat Industry Research Institute of New Zealand No. 864 Hamilton, New Zealand.

Sperber WH and Deibel RH (1969) Accelerated procedure for *Salmonella* detection in dried foods and feeds involving only broth cultures and serological reactions. *Applied Microbiology* 17: 533–539.

St Clair VJ and Klenk MM (1990) Performance of three methods for the rapid identification of *Salmonella* in naturally contaminated foods and feeds. *Journal of Food Protection* 53: 961–964.

Surdy TE and Haas SO (1981) Modified enrichment-serology procedure for detection of salmonellae in soy products. *Applied and Environmental Microbiology* 42: 704–707.

Sutcliffe EM, Jones DM and Pearson AD (1991) Latex agglutination for the detection of *Campylobacter* species in water. *Letters in Applied Microbiology* 12: 72–74.

Wiberg C and Norberg P (1996) Comparison between a cultural procedure using Rappaport-Vassiliadis broth and motility enrichments on modified semisolid Rappaport-Vassiliadis medium for *Salmonella* detection from food and feed. *International Journal of Food Microbiology* 29: 353–360.

Entamoeba *see* **Waterborne Parasites:** Entamoeba.

ENTEROBACTER

Thomas W Huber, Medical Microbiology and Immunology, Texas A&M College of Medicine, Temple, USA

Enterobacter are coliforms and members of the family, Enterobacteriaceae. They are not considered to be faecal coliforms because they are not usual constituents of human faeces. *Enterobacter* are probably normal intestinal flora of animals and especially birds. Faecal contamination by fauna may account for the presence of *Enterobacter* in forested and agricultural soils and run-off, or *Enterobacter* may occur naturally in soil. *Enterobacter* spp. are frequently isolated from fruits and vegetables which may be the source of organisms involved in nosocomial infections. The significance of *Enterobacter* spp. in foods ranges from benefactor to casual contaminant to potential pathogen. *Enterobacter* spp. may have a probiotic effect, have been tried as foodstuff, and are used as leavening agents. *Enterobacter* may cause loss of aesthetic appeal or even spoil foods. Food or water may serve as the source of nosocomial *Enterobacter* infections. Enterotoxigenic strains of *Enterobacter* have the potential to cause food-borne illness.

Taxonomy and History

In the 1920s when *Escherichia coli* was called *Bacillus coli*, *Enterobacter aerogenes* and *E. cloacae* were known as *Bacillus aerogenes* and *B. cloacae*, respectively. It was certain that the presence of *B. coli* in water represented human faecal contamination but the meaning of the occurrence of *B. aerogenes* or *B. cloacae* was less clear. Later *B. aerogenes* and *B. cloacae* were assigned to the genus *Aerobacter*. The differentiation of *A. aerogenes* from *Klebsiella* was difficult and *A. aerogenes* was also referred to as *Klebsiella aerogenes* in older literature. The term, *Paracolobactrum* was used for bacteria which seemed to be identical to *Aerobacter aerogenes* or *A. cloacae* except for delayed fermentation of lactose. The fifth edition of *Bergey's Manual of Determinative Bacteriology* published in 1939 clearly distinguishes *Aerobacter* from *Klebsiella* but misnomers of *E. aerogenes* as *K. aerogenes* and/or *A. aerogenes* persist today, especially in environmental literature.

Enterobacter aerogenes represents the type species of *Enterobacter*. The currently used genus name, *Enterobacter*, was proposed in 1960 and accepted in *Bergey's Manual of Determinative Bacteriology* in 1974. *E. aerogenes* and *E. cloacae* were the only species recognized. Many variants or newly discovered *Enterobacter* species have emerged with the advent of studies of DNA-relatedness in the 1970s. *E. aerogenes* and *E. cloacae* are by far the most common *Enterobacter* species encountered in clinical infections, food or the environment. *Pantoea agglomerans*, previously *Enterobacter agglomerans*, is frequently isolated from fresh fruits and vegetables and will be covered in the *Pantoea* chapter of this work. *Enterobacter sakazakii* has been found to contaminate infant formula and cause neonatal meningitis. *E. cancerogenus* has been placed in synonymy with *E. taylorae*, formerly CDC Enteric group 19. *E. amnigenus* 1 and 2 and *E. hormaechei*, *E. intermedium* and *E. gergoviae* have been only rarely reported as agents of human infection and their occurrence in nature or in foods has not been fully delineated. *E. dissolvens* and *E. nimipressuralis* have been shown to be distinct species that are less frequently isolated and their ecologic niche is unknown.

Enterobacter is fermentative, facultatively anaerobic, and has a propensity to utilize most sugars and a variety of other substrates important to the food industry. **Figure 1** shows a flow chart for differentiation of *Enterobacter* from other coliform Enterobacteriaceae. **Figure 2** is a flow chart for presumptive identification of the currently recognized species of *Enterobacter*. **Table 1** shows the reactions most useful in the identification of species *Enterobacter*.

Natural Occurrence

Enterobacter spp. can be isolated from water contaminated by sewage suggesting its presence in the human gastrointestinal tract and faeces. However, *Enterobacter* is not a major constituent of normal bowel flora. *Enterobacter* has been isolated from soil associated with the roots of legumes. It has also been isolated from water, lettuce, milk, fruits, vegetables, meat and poultry and can be recovered from the hands of about 25% of food handlers. The occurrence of *Enterobacter* on the hands of food handlers seems to be the result of working with foods containing the organisms rather than faecal contamination. *Enterobacter* was recovered less often from workers with a recent bathroom break than from workers who had remained on the job handling fruits and vegetables. Occasionally more organisms have been recovered from washed than from unwashed vegetables.

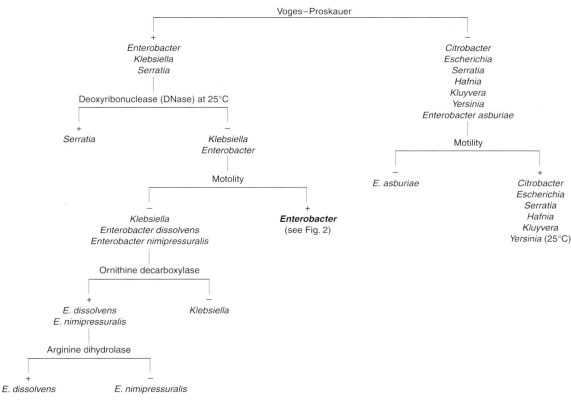

Figure 1 Flow chart for the presumptive identification of coliform organisms. Enterobacteriaceae which hydrolyse *o*-nitrophenyl-β-D-galactopyranoside (ONPG) are considered to be coliforms (lactose fermenters). The differentiation of coliform genera is based on reactions with at least 90% reliability.

E. aerogenes and *E. cloacae*

Though *E. aerogenes* and *E. cloacae* are distinct species, their distribution, presence in food and occurrence in nosocomial infection presents such a similar pattern that they are discussed together. *E. cloacae* has been isolated from the intestinal contents associated with 12 000-year-old mastodon remains. *Enterobacter* were recovered from 2000-year-old glaciers in Canada's High Arctic where no human has been known to visit, much less inhabit. The genetic material for nontransferable ampicillin resistance, common in today's clinical isolates, was found in these isolates from prehistoric and pristine sources. *E. cloacae* capable of degrading pentaerythritol tetranitrate was isolated from soil. *Enterobacter* was isolated from fish intestines and water in catfish ponds. *E. cloacae* has been shown to arise in coliform-free water by growing in the water in a distribution system with no breaks or outside source of enteric contamination. *E. aerogenes* and *E. cloacae* which were isolated from an operating drinking water system were found to be able to multiply under the conditions found in municipal water systems. *E. aerogenes* was isolated from the water of a pristine lake and was found to actively degrade propanil, a com-

monly used herbicide. *E. cloacae* has been found in natural fermentation of soya beans soaked in tap water for up to 36 h. *E. cloacae* and *E. amnigenus* were the most common organisms in bottled mineral, spring and table waters. Survival studies of *E. cloacae* in water show a slow death rate, so that the organism could be viable for long periods of time. Some *E. cloacae* isolated from environmental water has been shown to produce enterotoxin which is antigenically and genetically related to the heat labile (LT) enterotoxin produced by enterotoxigenic *E. coli*.

Water is often involved in the transmission of *Enterobacter* because it is easily contaminated and supports survivability. The source of *E. cloacae* bacteraemia has been traced to contamination by tap-water of pressure-monitoring equipment used in open-heart surgery. *E. cloacae* has also been isolated from Holy water used during patient visitation in hospitals and could present a risk for hospital-acquired infections. A drainpipe leaking water in a hospital resulted in 129 infections with *E. cloacae*. *E. cloacae* with an identical antibiogramme to patient isolates was isolated from the leaking water; repair of the leak resulted in a lower infection rate. *E. cloacae* contaminating distilled water that was used in ven-

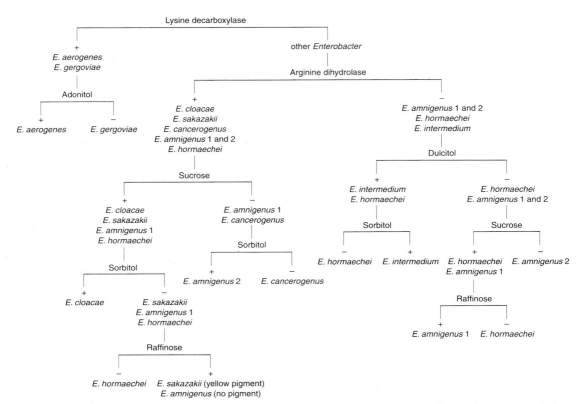

Figure 2 Flow chart for the presumptive identification of species of *Enterobacter*. Dichotomy is based on characteristics with at least 90% reliability. Confirmatory tests for complete identification are presented in Table 1.

Table 1 Biochemical reactions to differentiate species of *Enterobacter*

Test	aerog-enes	cloacae	gergoviae	saka-zakii	taylo-rae	amnig-enus 1	amnig-enus 2	asbur-iae	hormaechei	intermedium	cancero-genus	dissolvens	nimipres-suralis
Methyl red	−	−	−	−	−	−	V	+	V	+	−	−	+
Citrate	+	+	+	+	+	V	+	+	+	V	+	+	−
Urea	−	V	+	−	−	−	−	V	V	−	−	+	−
Lysine decarb-oxylase	+	−	+	−	−	−	−	−	−	−	−	−	−
Arginine dihydrolase	−	+	−	+	+	−	V	V	V	−	+	+	−
Ornithine decarboxylase	+	+	+	+	+	V	+	+	+	V	+	+	+
Motility	+	+	+	+	+	+	+	−	V	V	+	−	−
Lactose fer-mentation	+	+	V	+	−	V	V	V	−	+	−	−	−
Sucrose fer-mentation	+	+	+	+	−	+	−	+	+	V	−	+	−
Dulcitol fer-mentation	−	V	−	−	−	−	−	−	V	+	−	−	−
Adonitol fer-mentation	+	V	−	−	−	−	−	−	−	−	−	−	−
Sorbitol fer-mentation	+	+	−	−	−	−	+	+	−	+	−	+	+
Raffinose fer-mentation	+	+	+	+	−	+	−	V	−	+	−	+	−
Yellow pigment	−	−	−	+	−	−	−	−	−	−	−	−	−

− and +, minimum of 90% reactivity.
v, variable reactions, 11–89%.

tilators, was the source of a nosocomial outbreak in a paediatric intensive care unit. The probe and probe cover of a blood gas machine harboured *E. cloacae* that was the source of infection in a neonatal unit. *E. cloacae* has been isolated from floors, walls and pharmaceutical products in surgery, renal and neo-natal wards and the pharmacy of a hospital. *E. cloacae* contaminating defective flush valves of haemodialysis machines caused sepsis in haemodialysis patients in Canada, the USA and Israel. *Enterobacter* was found to be resident in saline waste drains and infected patients when defective valves allowed back flow from the waste drains to enter their bloodstream.

Enterobacter can be found in association with many animals. *E. aerogenes* infected the dermis and retarded the growth of wool in Merino sheep. *E. aerogenes* causes mastitis in cows but is not the most common agent. *E. cloacae* and *E. aerogenes* were the most frequent bacterial isolates from meat and hides. *Enterobacter* were isolated from frozen tail meat samples of Nile crocodiles. *Enterobacter* exhibiting antimicrobial resistance were isolated from carcasses of commercially slaughtered chickens. Twenty freezing and thawing cycles of plasmid containing *E. cloacae* reduced the number of viable cells in food samples but did not reduce plasmid carriage; nor did storage at 4°C. *E. cloacae* was found to constitute 10–25% of the intestinal flora of rabbits. About 40% of rabbits that were administered ampicillin died and *E. aerogenes* was the predominant organism isolated from their caecum. *E. cloacae* was the predominant enteric found in the faeces of domestic passeriform birds but only 17% of the grain-eating birds had enteric organisms in their flora.

E. cloacae and *E. aerogenes* were isolated from foods obtained from street vendors or restaurants, as well as home prepared meals in Mexico and in the USA. Some *Enterobacter* isolates produced enterotoxin, which perhaps explains why the experimental ingestion of *Enterobacter* as food caused diarrhoea.

E. cloacae was the most prevalent bacterial isolate from cockroaches collected in food-handling establishments, low income flats, hospital outpatient and nursing areas, a swimming pool and the toilet area of the swimming pool. The implication is that *Enterobacter* is widespread in urban environments and may be spread by cockroaches. *E. aerogenes* and *E. cloacae* were commonly encountered when surfaces and formites such as eating utensils, hospital food and food from caterers were cultured.

Enterobacter sakazakii

E. sakazakii attained unique species designation in 1980 and is no longer referred to as the yellow pigmented *E. cloacae*. Distinctive DNA-relatedness results confirmed that pigment production and biochemical differences were sufficient to justify a new species. *E. sakazakii* has caused bacteraemia in an elderly man and several outbreaks of neonatal meningitis. The source was not found for the elderly patient but the powdered milk used in infant formula was found to harbour the epidemic DNA type *E. sakazakii* when restriction endonuclease patterns were analysed. Dried infant formula was found to be the source in many outbreaks of neonatal meningitis due to *E. sakazakii*. It is believed that low numbers of *E. sakazakii* multiplied enough to present an infectious dose, especially when fed to premature or low-birth-weight infants. Recommendations to avoid *E. sakazakii* intestinal and/or disseminated infection were to prepare formula just before use, or to refrigerate formula and warm it just before feeding. How formula powders become contaminated with *E. sakazakii* is not known but milk, cows and the environment are likely sources. *E. sakazakii* is destroyed by pasteurization but is more thermotolerant than many other enteric species. *E. sakazakii* has been isolated from mesothermal mineral springs. *E. sakazakii* was isolated from two-thirds of vat-washed and about one-third of conveyor belt-washed beer mugs. The occurrence was blamed on unhygienic practice but could also be accounted for by contamination from the environment.

Enterobacter cancerogenus (E. taylorae)

E. cancerogenus (formerly *Erwinia cancerogenus*) was proposed as synonymous with *E. taylorae* in 1994. *E. taylorae* was previously designated as CDC Enteric Group 19. DNA-relatedness of all organisms with these designations placed them in a single group and the name *E. cancerogenus* was found to have precedence. *E. cancerogenus* has been reported to cause wound infections and septicaemia in traumatized humans. It is believed to have entered trauma or crush wounds from environmental sources at the time of the accident.

Enterobacter hormaechei

E. hormaechei was proposed as the name for CDC Enteric Group 75 because of DNA-relatedness grouping and because biochemical reactions distinguish it from other species of *Enterobacter*.

Enterobacter asburiae

E. asburiae is the name for CDC Enteric Group 17 and is a distinct species by DNA-relatedness grouping. *E. asburiae* is most closely related to *E. cloacae* but is distinguished biochemically by being methyl red positive and non-motile.

Enterobacter intermedium

E. intermedium has been isolated from surface water and soil and its biochemical characteristics were described in 1987. *E. intermedium* has been isolated from scombroid fish species purchased at retail markets. It was found to be capable of decarboxylating histidine to form histamine, the putative agent of scombroid poisoning, in market fish.

Use as a Foodstuff

Bacteria are easily grown in large numbers and contain all of the constituents required for complete nutrition. If bacteria could be used as food, world starvation problems could be easily alleviated. Heat-killed *E. aerogenes* cultures in 6–12 g amounts were eaten by human volunteers. Despite potential nutritional benefit, *E. aerogenes* were poorly tolerated and caused nausea and diarrhoea following ingestion. The time of onset of diarrhoea suggested that *Enterobacter* is toxic to the human gastrointestinal tract thus making it unsuitable foodstuff.

Use in Food Preparation

Chinese buns called 'mantou' are prepared by a natural fermentation method using water in which fruits or vegetables have been steeped as starter (the Lao-mien method). The Lao-mien method results in buns with good rising properties, little acid and good flavour. Furthermore, dough with Lao-mien leavening rises better after refrigerator or freezer storage than dough leavened with traditional yeast. The inflation characteristics of risen dough are more desirable with Lao-mien but more incubation time is required. The active principle in Lao-mien is *E. cloacae*. Leavening of mantou by *E. cloacae* isolated from water in which apples had been steeped was enhanced by the addition of casamino acids to the dough; L-asparagine was the amino acid which was most rapidly consumed during cell growth. Hydrogen is the by-product of metabolism responsible for inflation of the dough.

Food Interactions and Growth Control

Meat, seed and pulp of *Carica papaya* Linn has been shown to be bacteriostatic for *E. cloacae*. Extracts of the Chinese medicinal plants, Tin Men Chu, Siu Mao Heung and Sey Lau Pai, inhibited the growth of *E. aerogenes*. Sorbate inhibited growth of *E. cloacae* in 'in-use' enteral feeding bags and reduced the number of organisms by 75%. *E. aerogenes* and *E. cloacae* fermented carrageenan, resulting in loss of gel rigidity, an increase in available calories and loss of aesthetic appeal of foods. *E. cloacae* and *E. aerogenes* rapidly reduced the food colouring agents, metanil yellow and indigo carmine, thus altering the colour and the aesthetic appeal of foods in which those agents were used. Essential oils which occur naturally in foods can inhibit *Enterobacter*. Cinnamic aldehyde at $500 \,\mu g \,ml^{-1}$, citral at $1000 \,\mu g \,ml^{-1}$, geraniol at $1000 \,\mu g \,ml^{-1}$, and butylated hydroxy anisole at $200 \,\mu g \,ml^{-1}$ prevented the growth of *Enterobacter* in media incubated for 30 days.

Detection of Coliforms and *Enterobacter* in Water and Foods

There are many methods for the detection of coliforms in food and water. Preliminary screening methods usually capitalize on the ability of coliforms to resist bile and to ferment lactose. Growth and/or acid and/or gas from lactose in brilliant green lactose bile broth (BGL) or lauryl sulphate tryptose broth (LST) are preliminary evidence of coliforms. The ability of isolates to hydrolyse o-nitrophenyl-β-D-galacto-pyranoside (ONPG) provides a single test to identify coliforms. Probes to detect the ONPG gene, which codes for the synthesis of β-galactosidase, have also been used to detect coliforms. Conventional manual or automated biochemical tests are used when the complete identity of organisms is required.

Coliforms can be separated into faecal and non-faecal coliforms. The production of gas in *E. coli* (EC) broth incubated at 44.5°C has been used to distinguish faecal coliforms. Since *E. coli* is considered to be faecal and all other coliforms non-faecal in origin, it is important to confirm the presence of *E. coli*. *E. coli* is unique among coliforms in that it hydrolyses 4-methylumbelliferyl-β-D-glu-curonide (MUG). Subjecting isolates to the MUG test can be used to confirm the presence of *E. coli*. Probes to detect the gene responsible for MUG hydrolysis have also been used to detect *E. coli* in samples. Conventional biochemical testing may also be used for unequivocal identification of *E. coli*.

Enterobacter are detected by growth in BGL or LST broths. *Enterobacter* and *Klebsiella* can be differentiated from other enterics and non-enterics, such as *Pseudomonas*, *Aeromonas* and *Vibrio*, by their ability to produce β-D-xylosidase. The substrates for xylosidase detection, p-nitrophenyl-β-D-xylo-pyranoside or 4-trifluoromethylumbelliferyl-β-D-xyl-opyranoside, can be incorporated into plating media for rapid presumptive identification of *Klebsiella* and *Enterobacter*. Conventional biochemicals must be used for complete identification because *Enterobacter* and *Klebsiella* are not differentiated and about half of *Pantoea agglomerans* strains exhibit xylosidase activity.

Probiotics

It is possible that *Enterobacter aerogenes* and other *Enterobacter* species may exert a probiotic effect in the gastrointestinal tract of humans and animals. Probiotics are preparations of living organisms that benefit the host by improving intestinal microbial balance. Mechanisms include the production of substances which inhibit pathogens, compete for their substrates or compete for intestinal adherence sites.

The probiotic effect of normal enteric flora increases resistance to *Salmonella* infection. In a simulated human colonic ecosystem using six microbial competitors and five substrates, *E. aerogenes* reached a higher density than *Salmonella typhimurium*, but usually not as high as *E. coli* or *Bacteroides ovatus*. When *Enterobacter* is present in the human intestine, it probably adds to the probiotic effect exhibited by the normal flora. Whether the use of probiotics can reduce the carriage of *Salmonella* in animals used for food remains to be determined.

Antimicrobial resistance

Production of cephalosporinase is a chromosomally mediated characteristic of *E. cloacae*, *E. aerogenes*, *E. cancerogenus* and *E. asburiae*. Some but not all *E. sakazakii* isolates produce a chromosomally mediated cephalosporinase. Cephalosporinase has not been detected in *E. gergoviae*. *Enterobacter* species are common in foods, water and the environment, making it likely that humans ingest them daily. *Enterobacter* species usually fail to establish prominence in the intestine because normal flora suppress or outgrow them. Treatment of hospitalized patients with antibiotics increases the *Enterobacter* component of normal gut and skin flora. Treatment of hospitalized patients with beta-lactam antibiotics favours *Enterobacter* over normal flora which do not produce beta-lactamase. The incidence of nosocomial infections due to *Enterobacter* is increasing. *Enterobacter* is now the third most common Gram-negative nosocomial pathogen and the most frequent cause of nosocomial pneumonia. Colonization of the stomach usually precedes the development of Gram-negative pneumonia. *Enterobacter* are among the more antimicrobial-resistant nosocomial isolates. Typing studies show that sometimes resistant *Enterobacter* outbreaks are due to a point source, but just as often no relatedness of the isolates is found. In either circumstance, *Enterobacter* in the intestine may thrive during antimicrobial therapy and cause an endogenous nosocomial infection in the host which may or may not be spread to other patients causing exogenous nosocomial infection.

See also: **Bacillus**: Introduction. **Biochemical and modern identification techniques**: Enterobacteriaceae, Coliforms and *E. coli*. **Enterobacteriaceae, Coliforms and *E. coli***: Introduction; Classical and Modern Methods for Detection/Enumeration. **Fermented Foods**: Fermentations of the Far East. **Klebsiella**. **Microflora of the Intestine**: The Natural Microflora of Humans; Detection and Enumeration of Probiotic Cultures. **Pantoea**. **Probiotic Bacteria**: Detection and Estimation in Fermented and Non-fermented Dairy Products. **Waterborne Parasites**: Detection by Classic and Modern Techniques.

Further Reading

Abdelnoor AA, Batshoun R and Roumani BM (1983) The bacterial flora of fruits and vegetables in Lebanon and the effect of washing on the bacterial content. *Zbl. Bakt. Hyg. I. Abt. Orig.* B 177: 342–349.

Coleman ME, Dreesen DW and Weigert RG (1996) A simulation of microbial competition in the human colonic ecosystem. *Appl. Environ. Microbiol.* 62: 3632–3639.

DeWit JC and Kampelmacher EH (1981) Some aspects of microbial contamination of hands of workers in food industries. *Zbl. Bakt. Hyg., I. Abt. Orig.* B 172: 390–400.

Dumavibhat B, Tiengpitak B, Srinkapibulaya S, Nilakul C and Tuntimavanich S (1989) Hygienic status of food handlers. *J. Med. Assoc. Thai.* 72: 577–581.

Farmer III JJ (1995) Enterobacteriaceae: Introduction and Identification. In: Murray PR et al (eds) *Manual of Clinical Microbiology*. Ch. 32, p. 438. Washington, DC: American Society for Microbiology.

Manafi M and Rotter ML (1991) A new plate medium for rapid presumptive identification and differentiation of Enterobacteriaceae. *Int. J. Food Microbiol.* 14: 127–134.

Nazarowec-White M and Farber JM (1997) Thermal resistance of *Enterobacter sakazakii* in reconstituted dried-infant formula. *Lett. Appl. Microbiol.* 24: 9–13.

Rivault C, Cloarec A and LeGuyader A (1993) Bacterial load of cockroaches in relation to urban environment. *Epidemiol. Infect.* 110: 317–325.

Schlaes DM (1993) The clinical relevance of *Enterobacter* infections. *Clin. Ther.* 15 (Suppl. A): 21–28.

Tamura A, Nagano H, Omori M et al (1995) Improvement in hydrogen productivity by a leavening bacterium, *Enterobacter cloacae* GAO, and its application to *Mantou*. *Biosci. Biotech. Biochem.* 59: 2137–2139.

Venkateswaran K, Murakoshi A and Satake M (1996) Comparison of commercially available kits with standard methods for the detection of coliforms and *Escherichia coli* in foods. *Appl. Environ. Microbiol.* 62: 2236–2243.

Wood LV, Ferguson LE, Hogan P et al (1983) Incidence of bacterial enteropathogens in foods from Mexico. *Appl. Environ. Microbiol.* 46: 328–332.

ENTEROBACTERIACEAE, COLIFORMS AND *E. COLI*

Contents
Introduction
Classical and Modern Methods for Detection/Enumeration

Introduction

Ashok Pandey, Biotechnology Division, Regional Research Laboratory, Trivandrum, India

Vinod K Joshi, Department of Post-harvest Technology, Dr YSP University of Horticulture and Foresty, Nauni, India

Poonam Nigam, Biotechnology Research Group, University of Ulster, N. Ireland, UK

Carlos R Soccol, Laboratorio de Processos Biotecnologicos, Universidade Federal do Parana, Curitiba, Brazil

Enterobacteriaceae

The members of Enterobacteriaceae occupy a central position in current biology and, due to their diverse properties and historical role, are of great significance for medical researchers, food microbiologists and technologists, biochemists and molecular biologists. The members of this family, especially *Escherichia coli* and *Salmonella*, have most widely been used to study the fundamentals of biology, whether genetic exchange, biochemical pathway elucidation, genetic map sequencing, gene regulation or genetic engineering or molecular portrayal of viral morphogenesis. The family Enterobacteriaceae is the largest of three families in Section 5 of *Bergey's Manual*. Members of this family are sometimes referred to as enteric bacteria as many are inhabitants of the intestines of humans or animals. Because of this, they are also called coliforms. While some of these members are free-living organisms, others live in co-operation with or at the expense of their host, and yet others decompose dead organic matter.

The microflora, such as *E. coli*, *Fusobacterium* and *Bacteroides*, that colonize the inner surface and cavities of the gastrointestines of humans and other animals (referred to as normal microflora), provide their host with a certain amount of protection against invading pathogens like *Salmonella* and *Shigella*. This is partly achieved by competing for space and nutrients and partly by the antimicrobial substances, such as colicins, produced by them. Human faeces, with an estimated 400 different species and composition, obviously provide a good source of intestinal bacteria.

Table 1 Genera and species of Enterobacteriaceae

Genera	Type species	Species
Arsenophonus	A. nasoniae	Only species
Budvicia	B. aquatica	Only species
Buttiauxella	B. agrestis	Only species
Cedecea	C. davisae	5, including unnamed spp. 3 and 5
Citrobacter	C. freundii	4
Edwardsiella	E. tarda	3
Enterobacter	E. cloacae	14
Erwinia	E. amylovora	17
Escherichia	E. coli	6
Ewingella	E. americana	Only species
Hafnia	H. alvei	Only species
Klebsiella	K. pneumoniae	6
Kluyvera	K. ascorbata	2
Leclercia	L. adecarboxylata	Only species
Leminorella	L. grimontii	2
Moellerella	M. wisconsensis	Only species
Morganella	M. morganii	Only species
Obesumbacterium	O. proteus	Only species
Pantoea	P. agglomerans	2
Pragia	P. fontium	Only species
Proteus	P. vulgaris	4
Providencia	P. alcalifaciens	5
Rahnella	R. aquatilis	Only species
Salmonella	S. choleraesuis	7
Serratia	S. marcescens	11
Shigella	S. dysenteriae	4
Tatumella	T. ptyseos	Only species
Xenorhabdus	X. nematophilus	5
Yersinia	Y. pestis	11
Yokenella	Y. regensburgei	Only species

Source: *Bergey's Manual of Systematic Bacteriology* (1994).

Members of the Family Enterobacteriaceae

In *Bergey's Manual of Systematic Bacteriology* (1994), facultative anaerobic Gram-negative rods have subgroup 1 family Enterobacteriaceae, subgroup 2 family Vibrionaceae, subgroup 3 family Pasteurellaceae and subgroup 4 containing other genera. The family Enterobacteriaceae has 30 genera and many species included in it. These genera, with their type species, are listed in **Table 1**.

There are several genera which were not included in *Bergey's Manual of Determinative Bacteriology* but have presently been included in the Enterobacteriaceae family. These include *Buttiauxella* (1981), found in fresh water, *Ewingella* (1983) found in clinical sources like sputum, wounds and blood, *Budvicia* (1983), mostly found in fresh water, and also in human stool, *Moellerella* (1984), found in human stool and water, *Yokenella* (1984), found in human stool, urine and insect intestine, *Leminorella* (1985) from human stool and urine samples, *Leclercia* (1986), found in human clinical specimens like sputum and blood, *Pragia* (1988) in water (one from human stool also), *Pantoea* (1989) in plants, animals and humans, and *Arsenophonus* (1991), in female wasps.

General Characteristics of the Family Enterobacteriaceae

The bacteria belonging to this family are Gram-negative, petrichously flagellated (motile) or non-motile, facultative anaerobic straight rods. Members of this family convert glucose into acid or acid and gas. Nitrate is converted to nitrite. The indolephenol test is negative and most members of the family produce catalase. **Table 2** gives general characteristics of some genera of the family Enterobacteriaceae.

Nutritional Requirements

The minimal nutritional requirements of the members of the family Enterobacteriaceae are often very simple. However, *Enterobacter*, *Proteus* and *Shigella* need nicotinic acid frequently, *Salmonella* needs tryptophan and *Photobacterium* needs methionine. Growth under aerobic condition is easily achieved but, in anaerobic mode, growth is severely dependent on the availability of fermentable sugars. Biosynthesis of amino acids has a distinct regulatory pattern in these bacteria, which is not found outside the enteric group.

Fermentative Metabolism

The members of the family universally utilize carbohydrates. Fermentation of sugars takes place via the Embden–Meyerhof–Parnas pathway through mixed acid fermentation resulting in lactic acid, acetic acid, succinic acid, formic acid and ethanol. There may be a large variation in the end products formed quantitatively among different strains and even within strains under different fermentation conditions as end products are formed by independent pathways. One unique fermentative metabolic characteristic of *Enterobacter* and *Serratia* and some species of *Erwinia* is the formation of a neutral end product (butanediol). Metabolic properties of members of the Enterobacteriaceae family are useful in characterizing and distinguishing them (**Table 3**). During fer-

Table 2 Characteristics of important members of the family Enterobacteriaceae

Microorganism	Salient characteristics
Escherichia coli	Straight rods, inhabitant of gastrointestinal tract of mammals, may cause enteric disease, indicator organism for faecal contamination
Salmonella S. typhi S. enteritidis S. arizona S. paratyphi	Known pathogen, food-poisoning agent, causes typhoid and gastroenteritis
Shigella S. dysenteriae S. sonnei S. flexneri S. boydii	Cause of shigellosis or bacillary dysentery; some species produce exotoxins, inhabitant of gastrointestinal tract, transmitted through water and food
Citrobacter freundii	Found in water and food, associated with many infections
Klebsiella pneumoniae	Widely distributed in nature, commensal in the intestinal tract of mammals, can cause gastroenteritis, pneumonia and urinary tract infections
Enterobacter E. aerogenes E. cloacae	Inhabitant of gastrointestinal tract of mammals, can cause enteric and urinary tract infections
Serratia marcescens	Widely distributed, forms red-colour colonies, opportunistic pathogen
Proteus P. vulgaris P. mirabilis P. inconstans	Found in intestine of mammals; some species can cause urinary infections while others cause diarrhoea
Yersinia Y. pestis Y. enterocolitica	Found in nature infecting small feral rodents from where it is transmitted to humans by fleas, causing bubonic plague and enteric disease: enterocolitis
Erwinia	Plant pathogen

Modified from Cano RJ and Colome JS (eds) (1986).

Table 3 Biochemical tests and response of selected members of Enterobacteriaceae

Member	Indole (from tryptophan)	Methyl red (acid production to bring pH below 4.4)	Voges–Proskauer (acetoin-production)	Citrate utilization
Escherichia coli	+ve	+ve	–ve	–ve
Shigella	+ve	+ve	–ve	–ve
Salmonella typhimurium	–ve	+ve	–ve	+ve
Citrobacter freundii	–ve	+ve	+ve	+ve
Klebsiella pneumoniae	–ve	+ve	–ve	+ve
Enterobacter aerogenes	–ve	–ve	+ve	+ve

mentation, the production of gas (carbon dioxide) is a tool to differentiate between _Escherichia coli_ from pathogens like _Shigella_ and _Salmonella_, which do not produce gas. Similarly, because they possess formic hydrogenylase, members of the genus _Enterobacter_ are vigorous gas-producers but paradoxically _Serratia_ does not produce it (in fact, the gas is produced but remains solubilized in the medium). Lactose fermentation is a characteristic of _Escherichia_ and _Enterobacter_ but is absent in _Shigella_, _Salmonella_ and _Proteus_.

Genetic Relations amongst Enterobacteriaceae

The genetic relationship between enteric bacteria has been facilitated by the discovery of conjugational and transductional gene transfers. This group of bacteria can acquire plasmids by conjugation from the donor (_Escherichia coli_) and maintain it as an extra-chromosomal element. Both substituted (F-_lac_) and R-factors (drug resistance) can be widely disseminated amongst the enteric group. Production of chromosomal hybrids and genetic maps indicates a high degree of homology in _E. coli_ and other members of the genera _Salmonella_ and _Shigella_. Intergenic DNA–DNA hybridization obtained in vitro also confirms the close relationship of this group. However, as evidenced by rare chromosomal hybrid formation and DNA–DNA hybridization, members of the genera _Enterobacter_, _Proteus_ and _Serratia_ are different from _E. coli_, _Salmonella_ and _Shigella_.

Enterobacteriaceae and Nosocomial Infections

Many members of the family Enterobacteriaceae are considered to be opportunistic microorganisms responsible for nosocomial infections (hospital-acquired infections). Nosocomial infections are a serious problem for hospitals. Such infections are transmitted to patients in hospital through contaminated objects like surgical instruments, respirators, linens and dressing materials or the hospital staff, including doctors. Contaminated food and medication have sometimes been implicated. Typical examples of such infections and their causative agents are lower respiratory infection (_Klebsiella_

pneumoniae), gastroenteritis (_E. coli_, _Salmonella_, _Shigella_), septicaemia, burns and wounds (_E. coli_) and urinary tract infections (_Enterobacter aerogenes_ and _Escherichia coli_).

Coliforms

Coliforms are an important group of the family Enterobacteriaceae, which constitute about 10% of intestinal microflora. These bacteria are facultative anaerobes, Gram-negative, non-motile and rod-shaped, which ferment lactose with gas formation (35°C, 48 h). The test was designed to meet the definition, as detailed later in this section. For almost a century, coliforms, especially _E. coli_, were thought to be of intestinal origin in humans and other animals; however, there are coliforms which do not have any history associated with faeces and are found in fresh water. When coliforms are not detected in a specified volume of water, it is considered to be safe to drink. In view of this, it was necessary to revive the coliform concept to establish water quality, or do we need a superior alternative to these organisms? Nevertheless, it remains a widely accepted indicator of the microbial quality of water.

As a matter of fact, the coliform group remains an artificial group of convenience rather than a precise indicator of sanitary significance. Instead of challenging its usefulness, confusion has ensued.

Escherichia

The genus _Escherichia_ consists of both motile and non-motile bacteria, which conform to the definition of the family Enterobacteriaceae, and the tribe Eschericherieae. Both acid and gas are formed from fermentable carbohydrates. Salicin is fermented by many species but inositol is not utilized and adonitol is used by only one species. Lactose is rapidly fermented by most members, although there are also slow- or non-fermenting strains. Sodium acetate is frequently used as a sole carbon source. _E. coli_ (Migula) Castellani and Chalmers is the type strain of this genus.

Escherichia coli

E. coli are Gram-negative, facultative anaerobic, non-spore-forming bacteria. Based on serological properties or the presence of virulence factors, these have been grouped into many subdivisions, among which the following four deserve special attention:

1. Enteropathogenic (EPEC)
2. Enteroinvasive (EIEC)
3. Enterotoxigenic (ETEC)
4. Enterohaemorrhagic (EHEC).

While in the first two cases the pathogenic mechanisms are not fully understood and are still being studied, in the latter two cases it has been established that pathogenicity is related to toxin production.

E. coli is the most important member of the family Enterobacteriaceae and probably the best-understood organism. First isolated by the German bacteriologist, Theodar Escherich in 1885 from children's faeces, it shows remarkable power in colonizing its host, the intestine of mammals and birds. It does not survive long in water and soil. It is a universal inhabitant of the human gut (less than 1% of total microbial population) and is predominantly a facultative anaerobe. Not all strains of *E. coli* live peacefully in the gut of its host and it is responsible for many diseases. It is used as an indicator organism to determine the faecal contamination of water and the presence of enteric pathogens.

General Characteristics of *E. coli* *E. coli* cells are rod-shaped, non-motile, non-sporulating. They grow at mesophilic temperature and 37°C is the optimum. They show a positive result to the fermentation test and catalase reaction and negative to oxidase test. Their D-value at 60°C is 0.1 min. Although they can grow at pH 4.4, they grow well in media with near-neutral pH and water activity (a_w) 0.95.

E. coli can be differentiated from the other members of Enterobacteriaceae on the basis of different sugar fermentation and other biochemical reactions. The classical IMVIC group of tests is commonly employed for differentiation; some of these tests are available in modern miniaturized test systems. In the IMVIC (indole, methyl red, Voges-Proskauer, citrate) test, most strains of *E. coli* are methyl red-positive and VP (Voges-Proskauer) and citrate-negative.

The plasmids of *E. coli* have been studied in detail. The enterotoxigenic strains are known to carry five or more plasmids, including those for antibiotic resistance, enterotoxin production and adherence to antigens. Col. V is a specific plasmid which controls a sequestering mechanism, possibly by enterochelin or enterobactin-serum resistance. The serotyping scheme (lipopolysaccharide somatic O, flagellar H, poly-saccharide capsular K antigen) shows that in the currently applied O:H system, O antigen defines the principal group while H signifies the serovars. The strain tends to fall in this group and thus plays an important role in detecting pathogens in epidemiological investigations. The enteropathogenic serotypes of *E. coli* (018, 044, 055, 086, 0111, 0114, 0119, 0126, 0127, 0128ab, 0142, 0158) produce toxins, adhere to intestinal mucosa disturbing the function of microvilli and cause diarrhoea, while enteroinvasive serotypes (028ac, 029, 0124, 0136, 0143, 0144, 0152, 0164, 0167) invade and proliferate within epithelial cells, eventually causing death of the cells. Enterotoxigenic serotypes (06, 08, 020, 025, 027, 063, 078, 080, 085, 0115, 0128ac, 0139, 0148, 0153, 0159, 0167) and enterohaemorrhagic serotypes (01, 026, 091, 0111, 0113, 0121, 0128, 0145, 0157) of *E. coli* are associated with diarrhoeal disease.

Pathogenic *E. coli* capable of producing diarrhoea can be transmitted through the faecal–oral route. These strains possess virulence factors such as adherence factor, fimbriae and a variety of toxin products. Based on their phenotype characteristics and nucleotide sequence, these toxins could be grouped under two different categories, heat-labile (LT) and heat-stable (ST). The heat-stable toxins (ST) can be divided into ST I and ST II based on their solubility in methanol and activity in the infant mouse intestine. Further division of ST I is made into ST Ia (STP), found in exotic and farm animals and humans, and ST Ib (STH), found only in humans. Thus, the incidence of STH toxin could have potential as an indicator of human versus animal faecal sources of *E. coli*.

Survival of Coliforms/*E. coli*

Coliforms, especially *E. coli*, have the capability to survive during nutritional starvation and adverse conditions. These bacteria have evolved a sophisticated system of physiological and morphological changes, which they undergo when they pass through stationary stresses. Their modified cells have characteristics of the endospores of Gram-positive bacteria, such as resistance to a wide range of environmental stresses. Survival responses are directed to ensure survival of the stress and to ensure growth after removal of the stress. Some strategies adopted to ensure their survival are given here:

- Reproduction in large numbers.
- Growth in a variety of habitats, thus serving as environmental reservoirs from which animals can be infected.
- Protection by sheltering from unfavourable stresses, e.g. shade from the floating mat of *Lemna*

gibba L provides shelter to *E. coli* from high-intensity sunlight.

- Stresses such as osmotic and temperature shocks and nutrient limitations are dependent on position in the growth cycle, e.g. during log-phase, stress impact is greater.
- During stationary starvation, survival is controlled at a molecular level by bringing about physiological and morphological changes, e.g. starved *E. coli* cells are smaller than normal cells. Some starved cells also produce curly fibres, causing bacteria to clump together.
- After passing through a stationary phase, *E. coli* cells show more resistance to other stresses caused by nutrient limitation, osmotic and temperature shocks, acid and salinity, UV radiation, oxidative stress and uptake of antibiotics.
- When exposed to high temperatures, *E. coli* quickly produces heat-shock proteins (HsPs) to alleviate damage to proteins.
- Osmotic stress is overcome by osmoregulation, which is either controlled by moving away from unfavourable concentrations of osmolytes or by maintaining the constant cell volume.
- When grown in acid or alkaline media, the cells produce decarboxylase and deaminase to neutralize the acid or alkali, respectively.
- Formation of viable but non-culturable cells (VBNC) is an important strategy. These cells are metabolically active, incapable of division, have characteristics of stationary starving cells and do not grow unless ingested by a suitable host.

Coliforms/*E. coli* and Water Supplies

Water contains a large number of microorganisms, some of which are harmful to human health, while others are indicative of the level of contamination. It is not only impractical but also economically not feasible to monitor water for each and every type of microorganism. Thus, some selected representative microorganisms are monitored and these are termed indicator organisms. However, there is not a satisfactory performance standard for using coliforms to characterize the effectiveness of the water supply since coliforms are captured in the treatment process. Coliphages mimic many properties of viruses and these can be used in the evaluation process. Alternatively, *Clostridium* may be another promising candidate for this purpose.

The major human pathogens belonging to the Enterobacteriaceae family transmitted in water include *Salmonella* (causing salmonellosis – gastroenteritis and typhoid), *Shigella* (causing shigellosis – bacillary dysentery), *E. coli* (gastroenteritis) and *Yer-sinia enterocolitica*, which also causes gastroenteritis.

Coliform Biofilm

A surface exposed to water containing populations of microorganisms can result in the establishment of these microorganisms in an immobilized form. The immobilized microorganisms are capable of growth, reproduction and production of extracellular polymeric substances (EPS). EPS frequently extend from the cell, forming a mass of tangled fibrous structure to the entero assemblage, which is called biofilm. Biofilm proiling of coliforms is another problem. *Klebsiella*, *Enterobacter* or *Citrobacter* could prevent the detection of faecal coliforms or *E. coli*. Biofilm-derived *E. coli* is capable of surviving in large populations at free chlorine levels several times higher than that needed to kill a planktonic culture. These *E. coli* are protected either by a component of the biofilm, or by their own physiological characteristics.

Thus, biofilm causes problems in the water supply as it provides opportunities to the pathogens to adhere and reproduce, despite the high concentration of free chlorine. The sloughing of biofilm might release aggregation of cells with pathogens into the potable water supply system. Consumption of such water could lead to infection and health problems in humans.

Injured Coliforms and their Significance

When enteric bacteria are subjected to sublethal levels of acute antibacterial agents and conditions, phenotypes of these bacteria are altered. However, they may adopt their normal growth under favourable conditions. This leads to the formation of injured coliforms. For example, *E. coli*, when exposed to phenolic antiseptics, does not form colonies when grown on the commonly accepted media and conditions but shows revival when grown subsequently under favourable conditions. Estimation of enteric bacteria with sublethal stress conditions could provide additional safety in water supplies.

Detection of Coliforms and *E. coli* in Water

The routine approach to monitoring water supplies for the presence of faecal pollution is by testing the indicator microorganisms. The most probable number (MPN) test is most commonly applied.

Coliforms/*E. coli* and Foods

It is well established that many microbial diseases are of food-borne nature. Food-borne diseases may be of two types: in the first, infective microorganisms produce toxins which are eaten with the food, resulting in food poisoning. In the other type, infective microorganisms are transmitted to the host with the food.

Coliforms/*E. coli* and Food Quality

Coliforms and *E. coli* assume significance for their role in food and food quality. Their presence and population indicate the microbiological quality of the raw material as well as the efficacy of the processing techniques, such as pasteurization. Since all the operations in food processing involve water, the microbiological quality of water will have a great influence on the quality of the final product. There are a number of related factors, including contamination by the food handler, hygienic conditions prevailing in the processing plant and post-processing contamination, or inadequate processing. These are relevant where products like canned and packaged products such as vegetables, fruits, milk and milk products are concerned. If coliforms come into contact with such products, they result in spoilage or can transmit bacteria to humans, causing gastrointestinal problems or even food poisoning. *E. coli* has been found to be a causative agent in many such diseases. It is also an indicator of the bacteriological quality of milk.

In heat-treated food, although the coliform test is a useful means of assessing inadequate processing, poor sanitary practices or post-processing contamination, it is a common practice to use the total Enterobacteriaceae count instead of coliforms alone. Since some members of the family (e.g. *Erwinia*) are associated with soil or plants, their presence in some foods, especially vegetables, may be unavoidable. The presence of such coliforms, thus, in foods would not indicate faecal contamination. It is important to note that in respect to sublethal injury to some *E. coli* in food, the situation is similar to that arising in contamination of water, as it may not be detected by conventional tests. It should be kept in mind that *E. coli* may not be as resistant as other enteric pathogens. It is killed during pasteurization and dies in storage under conditions of drying and freezing. Even in the environment where most pathogens persist, *E. coli* disappears. A similar situation is encountered in the marine environment. Thus, it is a poor indicator of pathogens of marine origin.

Detection of Coliforms/*E. coli* in Foods

Coliforms/*E. coli* are normally detected by growing in different media and determining various biochemical tests. Serotyping provides a useful guide to identification.

Infective Bacterial Food Poisoning and Enterobacteriaceae

A group of bacteria, including some genera of Enterobacteriaceae, are responsible for infective bacterial food poisoning. This mode of pathogenesis is not mediated by toxins, although they may be produced. Commonly involved genera are *Salmonella*, *Yersinia* and *Escherichia*.

Salmonella is responsible for salmonellosis, which is the most frequently occurring bacterial infection and food-borne illness. In *Salmonella* infection, when there is a high increase in the number of *Salmonella*, the chances of an outbreak of this disease increase. However, it is dependent on many factors, such as consumer resistance, number of organisms ingested and their ineffectiveness. Apparently *Salmonella* attains quite a high number without causing an alteration in the sensory qualities of the food. *S. typhimurium*, which causes human gastroenteritis, is the species which is most frequently isolated. The organism originates from a host of sources, primarily including foods, poultry and their eggs and rodents, although there are also other sources such as cats, dogs, swine and cattle. Changes in the processing, packaging and compounding of food and feeds in recent years have apparently increased the incidence of salmonellosis. Large-scale handling of foods also tends to increase the spread of salmonellosis. Food-vending machines add to the risk, as does precooked food.

Yersinia enterocolitica is the species which is generally associated with meat and meat products, milk and milk products and vegetables. This organism has been isolated from almost 50% of samples of raw milk analysed in the UK. It has even been isolated from milk pasteurized at high temperature for a short time, causing serious concerns. It is also capable of multiplying at low temperatures (in cold storage), causing concern for the contamination of other foods stored together. The organism can come from contaminated pork or meat products containing pork, although human carriers have also been implicated. *Y. pseudotuberculosis* is another closely related species of concern. It is believed that *Y. enterocolitica* has probably been the organism responsible for sporadic cases of food poisoning in Europe and Japan, in which abdominal pain was the major symptom.

Escherichia is the third group of the family, which has been found associated with food contamination. *E. coli* is the species responsible; besides producing enterotoxin and causing related diseases, it is responsible for infective food poisoning. However, to cause infection the number must be quite high. Faecal contamination of foods, either by direct contact or indirectly through contaminated water, is the most common method of transmission. Although there may be a range of food products which are the source of infection (as described below), the most likely contaminated foods are meat, meat products and fresh vegetables.

E. coli and Food-borne Outbreaks

E. coli has been incriminated as the aetiological agent of food poisoning involving diverse foods such as raw milk, cream, cream puffs, creamed fish, pie, mashed potatoes, dates, vegetables, mould-ripened cheese, uncooked or poorly cooked meat and poultry. The main source of contamination of this organism is apparently beef.

Several strains of *E. coli* have emerged as the potent food-borne pathogens. One particular strain (O157:H7) has been identified as one of the most devastating for humans, causing several deaths each year. It generally causes bloody diarrhoea but it is also responsible for kidney failure in children.

This organism came into sharp focus in 1971 when an outbreak of gastroenteritis of food origin was traced to imported cheese in the US. This led to the development of specific and accurate methods of assessing toxic components as well as the virulence of the pathogenic strains of *E. coli*. In the nursery epidemics which occurred in the UK in 1945, the mortality rate was as high as 50%. Some *E. coli* strains were shown to produce responses with rabbit ileal and led to studies on *E. coli* as aetiological agent of a cholera-like disease in India. The disease symptom is usually diarrhoea, taking about 12–72 h for manifestation. The usual duration of the disease is 1–7 days. Traveller's diarrhoea is another common enteric infection among North Americans and Europeans travelling to less-developed countries. The infection rate is as high as 50%. The local population is immune to this infection.

Preventive Measures

While considering preventive measures, it should be kept in mind that generally these strains are widely distributed in the food environment, although in small numbers. Even when the number of *E. coli* in food is low, it does have a potential as a food-borne pathogen and will proliferate, if conditions permit. In general, thus, preventive measures include avoiding direct and indirect contamination of foods, strict personal hygienic practices, proper cooking of processed foods and reasonably good packaging and storage conditions.

See also: **Biochemical and Modern Identification Techniques**: Enterobacteriaceae, Coliforms and *E. coli*.

Biofilms. **Enterobacteriaceae, Coliforms and *E. coli***: Classical and Modern Methods for Detection/Enumeration. **Escherichia coli**: *Escherichia coli*. **Escherichia coli O157**: Detection by Latex Agglutination Techniques; Detection by Commercial Immunomagnetic Particle-based Assays. **Food Poisoning Outbreaks**. **Salmonella**: Introduction. **Water Quality Assessment**: Routine Techniques for Monitoring Bacterial and Viral Contaminants; Modern Microbiological Techniques. **Yersinia**: *Yersinia enterocolitica*.

Further Reading

Adams MR and Moss MO (eds) (1996) *Food Microbiology*. New Delhi: New Age International.

Balows A, Truper HG, Harder W and Schliefer KH (eds) (1992) *The Prokaryotes*. Vol. III. New York: Springer-Verlag.

Black JG (ed.) (1996) *Microbiology: Principles and Applications*. New Jersey: Prentice Hall.

Cano RJ and Colome JS (eds) (1986) *Microbiology*. New York: West Publishing.

Eley AR (ed.) (1996) *Microbial Food Poisoning*. London: Chapman & Hall.

Ewing WH (ed.) (1986) *Identification of Enterobacteriaceae*. Oxford: Elsevier.

Frozier WC and Westhoff DC (1996) *Food Microbiology*. New Delhi: Tata McGraw Hill.

Hort JG, Krieg NR, Sneath PHA, Staley JT and Williams ST (1994) *Bergey's Manual of Determinative Bacteriology*. Baltimore, MD: Williams & Wilkins.

Joshi VK and Pandey A (1999) *Biotechnology: Food Fermentation*. Vol. I. New Delhi: Educational Publishers.

Kay D and Fricker C (eds) (1997) *Coliforms and E. coli*. Cambridge: Royal Society of Chemistry.

Lim DV (ed.) (1989) *Microbiology*. New York: West Publishing.

Madigam MT, Martinko JM and Parker J (eds) (1977) *Biology of Micro-organisms*. Prentice Hall International.

Prescott LM, Harley JP and Klein DA (eds) (1993) *Microbiology*. Oxford: Wm. C. Brown.

Stanier RY, Adelberg EA and Ingraham JL (eds) (1976) *General Microbiology*. London: Prentice Hall.

Classical and Modern Methods for Detection/Enumeration

Enne de Boer, Inspectorate for Health Protection, Zutphen, The Netherlands

Marker organisms like Enterobacteriaceae, coliforms and *Escherichia coli* are used either as indicators for inadequate processing of food and water, or as an index for the possible presence of enteric pathogens. The entire Enterobacteriaceae family is used as indi-

Table 1 Present status of International Organization for Standardization (ISO) standards for Enterobacteriaceae, coliforms and *Escherichia coli*

ISO standard	Organisms	Product	Technique	Media
5552:1997	Enterobacteriaceae	Meat	p/a, MPN, colony count	EE, VRBG
7402:1993	Enterobacteriaceae	Foods	MPN + colony count	EE, VRBG
8523:1991	Enterobacteriaceae	Foods	p/a	EE, VRBG
4831:1991	Coliforms	Foods	MPN	LST, BGB
4832:1991	Coliforms	Foods	Colony count	VRB
5541-1:1986	Coliforms	Milk	Colony count	VRB
5541-2:1986	Coliforms	Milk	MPN	LST, BGB
6391:1997	*E. coli*	Meat	Membranes	MMG, TB
7251:1993	*E. coli*	Foods	MPN	LST, EC
11866-1:1997	*E. coli*	Milk	MPN	LST, EC
11866-2:1997	*E. coli*	Milk	MPN	LST + MUG
11866-3:1997	*E. coli*	Milk	Membranes	MMG, TB

p/a = Presence/absence test; MPN = most probable number technique; EE = Enterobacteriaceae enrichment broth; VRBG = violet red bile glucose agar; LST = lauryl sulphate tryptose broth; BGB = brilliant green bile broth; VRB = violet red bile lactose agar; MMG = mineral-modified glutamate agar; TB = tryptone bile agar; EC = EC broth; MUG = 4-methylumbelliferyl-β-D-glucuronide.

cator organisms in the evaluation of processed foods, such as cooked meals, meat products and egg products. The presence of these organisms indicates post-process contamination. Testing for coliform organisms, especially 'faecal' coliforms, is often used as a presumptive test for *E. coli*, and is applied in several products such as milk products, baby foods, ice cream and mineral waters. *E. coli* is considered primarily as an index organism, indicating the possible presence of ecologically similar pathogens, because this organism is always present in faeces. Testing for *E. coli* is done in products such as raw vegetables, raw milk, cheeses and shellfish.

Classical methods for the enumeration of Enterobacteriaceae, coliforms and *E. coli* rely on the use of specific microbiological media to isolate and enumerate viable cells in food and water. These methods are very sensitive and can give both qualitative and quantitative information. However, the major concern is that these methods require several days to give results and are labour-intensive. Modern rapid methods deal with the early detection and enumeration of these microorganisms by use of different techniques, like membrane filter methods, impedimetry, immunological methods and nucleic acid-based assays.

As the detection and enumeration of marker organisms like Enterobacteriaceae, coliforms and *E. coli* is widely applied in many laboratories for food and water microbiology there is a great interest in the development of both classical and modern techniques for the detection of these organisms.

Culture Media

Different microbiological media are used in the detection and enumeration of Enterobacteriaceae, coliforms and *E. coli* in foods and water. **Table 1** shows

Table 2 Selective components in media for Enterobacteriaceae, coliforms and *Escherichia coli*

Selective component	Media
Bile	EE, TB, BGB, VRB, VRBG
Lauryl sulphate	LST
Brilliant green	EE, BGB
Crystal violet	VRB, VRBG

EE = Enterobacteriaceae enrichment broth; TB = tryptone bile agar; BGB = brilliant green bile broth; VRB = violet red bile lactose agar; VRBG = violet red bile glucose agar; LST = lauryl sulphate tryptose broth.

the current applied International Organization for Standardization (ISO) standards for these organisms. It has been decided within ISO that the revision of these standards within the coming years will follow a horizontal approach. This means that the aim will be to have one standard for, say, *E. coli* in foods and water instead of different standards for different product groups.

Enrichment Media

For the detection of low numbers of Enterobacteriaceae and coliforms, selective enrichment is required. MacConkey was the first to formulate a medium for coliform bacteria containing lactose as sugar for fermentation and bile as selective component. MacConkey broth with either neutral red or bromocresol purple as indicators for acid production is still in use for the detection of coliforms and *E. coli*, especially in water and milk. Enterobacteriaceae enrichment (EE) broth for Enterobacteriaceae and brilliant green bile broth for coliforms are modifications of MacConkey's liquid medium. In these media the triphenylmethane dye brilliant green and bile are used as inhibitors, especially of lactose-fermenting Gram-positive organisms (**Table 2**). EE broth is used as an enrichment medium following pre-enrichment in tryptone soya broth or buffered peptone water.

Most of the culture methods used for the detection of *E. coli* are based on the principle of lactose fermentation, gas production and indole formation at 44 or 45°C. For liquid media an incubation temperature of 45°C is preferred, as lower temperatures result in less specificity and consequently more false-positives; higher temperatures may result in loss of *E. coli*, so false-negatives. The incubation time may be limited to 24 h for some products, like water and shellfish, but generally more positive tubes are found after 48 h of incubation.

These tests do not detect lactose-negative or non-aerogenic *E. coli*, which comprise 5–10% of *E. coli* strains, and indole-negative strains (3–5%). In general, formation of indole is a more reliable characteristic for *E. coli* than lactose fermentation. Standardized methods for the detection and enumeration of *E. coli* comprise a most probable number (MPN) technique using lauryl sulphate tryptose broth as a mildly selective enrichment medium at 37°C, followed by enrichment at 45°C in *E. coli* (EC) broth, containing 0.15% bile salts and rely on the fermentation of lactose with the production of gas, and the production of indole from tryptophan (ISO 7251: 1993; **Fig. 1**). As the MPN procedure requires several days and is very labour-intensive, alternative methods are often used.

For the recovering of sublethally injured organisms

Prepare a 10^{-1} food suspension and one
further decimal dilution (10^{-2})

↓ Inoculate

10 ml of the 10^{-1} suspension to three tubes of
double-strength lauryl sulphate tryptose (LST)
broth containing a Durham tube

1 ml of the 10^{-1} suspension to three tubes of
single-strength LST broth + Durham tube

1 ml of the 10^{-2} suspension to three tubes of
single-strength LST broth + Durham tube

↓ 45°C, 24 h

Tubes showing gas formation

↓ Inoculate (loopful)

Tubes with 5 ml tryptone water

↓ 45°C, 48 h

Add indole reagent, mix, examine for red colour
after 1 min

↓

Count positive tubes for each dilution;
use MPN tables to calculate MPN

Figure 1 Enumeration of *Escherichia coli* using the most probable number (MPN) technique.

Food suspension

↓ 1 ml

Spread on cellulose acetate membrane
on dried surface of plate with
minerals-modified glutamate agar

↓ 30 or 37°C, 4 h

Using forceps, transfer membrane to
plate with tryptone bile agar (TBA)
or TBA + BCIG

↓ 44°C, 18–24 h

Lift the membrane from the agar surface
and lower it into the upturned lid of the
plate containing 2 ml of indole reagent.
Indole-positive colonies develop a
pink colour within a few minutes.
Or count blue colonies on TBA + BCIG

Figure 2 Enumeration of *Escherichia coli* using membranes. BCIG, 5-bromo-4-chloro-3-indolyl-β-D-glucuronide.

on selective media, suitable (pre-) enrichment media must be used. Mineral-modified glutamate medium, a chemically defined medium based on glutamic acid, was shown to be an efficient enrichment medium for the detection of small numbers of *E. coli* in chlorinated waters. This medium is also used for resuscitation in the direct plate method with membranes (**Fig. 2**).

Plating Media

When large numbers of *E. coli* are expected, samples can be plated directly on isolation media for coliforms and *E. coli*.

The most commonly used plating medium for coliforms and *E. coli* is violet red bile (VRB) agar. The medium is modified from MacConkey's medium, mainly by the addition of crystal violet as an inhibitory component. Violet red bile glucose (VRBG) agar for the enumeration of Enterobacteriaceae has the same composition as VRB agar, except that lactose has been replaced by glucose. Some related Gram-negative bacteria may be suppressed by pouring an overlayer of the same medium on the inoculated medium. VRB and VRBG agar plates are usually incubated at 37°C. For enumeration of *E. coli* a temperature of 44°C is recommended and for the enumeration of the full range of coliforms or mesophilic Enterobacteriaceae the media are incubated at 30°C. Psychrotrophic Enterobacteriaceae can be detected by incubation of VRBG plates at 7°C for 10 days. Typical colonies of coliforms on VRB and Enterobacteriaceae on VRBG are round, purple-red, usually surrounded by purple-red haloes. Typical colonies on VRBG must be further confirmed as some other organisms, especially *Aeromonas* spp., may show specific growth on this medium. The suspect colonies must be tested for oxidase reaction (negative) and glucose fermentation (positive) for confirmation as Enterobacteriaceae. For identification to genus and species level modern biochemical identification techniques can be used.

Eosine methylene blue agar modified by Levine is used for the differentiation of *E. coli* and *Enterobacter aerogenes*. Colonies of *E. coli* show a typical greenish metallic sheen by reflected light and dark purple centres by transmitted light. However, some biotypes of *E. coli* do not produce typical colonies with green sheens and slow or non-lactose fermenters produce colourless colonies.

E. coli also has a characteristic colour (red with green metallic sheen) on Endo agar. However, this medium is not widely used because it contains the potentially carcinogenic compound fuchsin.

Generally there is no need to sterilize media for coliforms and Enterobacteriaceae. Pasteurization of these media results in sufficient decontamination, as surviving spore-forming bacteria will not be able to develop in media with bile salts and dyes. Moreover sterilization may inactivate some of the inhibiting substances, making the media less selective.

The quality control of these media can be done by an ecometric procedure or the modified Miles–Misra method and for liquid media by a dilution to extinction method, described by the International Committee on Food Microbiology and Hygiene in their book *Culture Media for Food Microbiology*. Test strains which can be used for quality control are listed in **Table 3**.

Table 3   Test strains for quality assessment of media for Enterobacteriaceae, coliforms and *Escherichia coli*

Media	Test strains				
	Productivity			Selectivity	
LST	*Escherichia coli* 9001	NCTC		*Enterococcus faecalis* NCTC 8213	
	Hafnia alvei NCTC 8105				
BGB	*Escherichia coli* 9001	NCTC		*Enterococcus faecalis* NCTC 8213	
	Citrobacter freundii 6272	NCTC		*Staphylococcus aureus* NCTC 7447	
EE	As for LST			As for BGB	
VRB	*Escherichia coli* 9001	NCTC		As for BGB	
VRBG	As for LST			As for BGB	
TB	*Escherichia coli* 9001	NCTC		*Klebsiella oxytoca* NCIMB 12819	
				Pseudomonas aeruginosa NCTC 10662	
				Staphylococcus aureus NCTC 7447	

LST = lauryl sulphate tryptose broth; BGB = brilliant green bile broth; EE = Enterobacteriaceae enrichment broth; VRB = violet red bile lactose agar; VRBG = violet red bile glucose agar; TB = tryptone bile agar.

Defined Substrate Technology

Recently a new generation of culture media has been described using fluorogenic and chromogenic substrates. These substrates yield brightly coloured or fluorescent products when acted on by bacterial enzymes and often make subculturing and further confirmation unnecessary.

The enzyme β-D-galactosidase which catalyses the fermentation of lactose by coliforms can be detected by different chromogenic substrates. In practice, 5-bromo-4-chloro-3-indolyl-β-D-galactopyranoside (X-GAL) is added to media for the detection and enumeration of coliforms. Splitting of X-GAL results in a colour change from blue-green to indigo and indicates the presence of coliforms.

Over 95% of *E. coli* strains, but also some *Salmonella* spp., *Shigella* spp. and *Yersinia* spp., possess the enzyme β-D-glucuronidase (GUD). *E. coli* O157:H7 strains do not possess GUD and this characteristic is used in the confirmation of these strains, especially to discriminate *E. coli* O157:H7 from other *E. coli* strains. GUD is an enzyme that catalyses the hydrolysis of β-D-glucopyranosiduronic acids into their corresponding aglycons and D-glucuronic acid. For the detection of GUD activity the fluorogenic substrate 4-methylumbelliferyl-β-D-glucuronide (MUG) and the chromogenic substrate 5-bromo-4-chloro-3-indolyl-β-D-glucuronide (X-GLUC or BCIG) are most frequently used. MUG is broken down by GUD to release 4-methylumbelliferone, which fluoresces blue under ultraviolet light. Indoxyl released from X-GLUC is rapidly oxidized to indigo, which is insoluble and builds up within the cells, resulting in blue *E. coli* colonies. These substrates have been incorporated into several selective media for rapid detection of *E. coli*, like VRB and tryptone blue (TB) agar (**Fig. 3**).

Adding MUG to lauryl tryptose broth, a medium for the detection of coliforms in water means it is now suitable for detection of presumptive *E. coli*. Application of this medium, called modified lauryl sulphate tryptose broth, in an MPN technique is recommended for samples of milk and milk products in which comparatively low numbers of presumptive *E. coli* and/or other coliforms are suspected (ISO 11866-2). Chromogenic or fluorogenic substrates are applied in several commercial systems for the detection and enumeration of coliforms and *E. coli* (**Table 4**).

Petrifilm *E. coli* count plates (3M, Minneapolis, MN) contain X-GLUC for GUD detection, the ingredients of VRB agar (without agar) and a cold-water-soluble gelling agent coated on to a plastic film. These reagents are hydrated when 1 ml of the test portion is added. Pressure applied to a plastic spreader placed

Preparation of the initial suspension
and further dilutions of the sample

↓ 1 ml

Mix with 15 ml TBX medium cooled at 50°C

TBX

Tryptone	20	g
Bile salts	1.5	g
X-glucuronide (BCIG)	0.075	g
Agar	12	g
Water	1000	ml

↓ 44°C, 18–24 h

Count blue/green colonies

Figure 3 Enumeration of presumptive *Escherichia coli* using the chromogenic medium TBX (tryptone, bile salts and X-glucuronide).

on an overlay film distributes test portions over 20 cm². After incubation for 24 h at 37°C, coliforms appear as red colonies with gas bubbles, and *E. coli* as blue colonies with gas bubbles. In an evaluation of the Petrifilm system for the enumeration of *E. coli* from shellfish, this system was found to be unsuitable because of inhibition of the blue coloration of *E. coli* colonies by undiluted mussel homogenate.

False-positive results with GUD-based *E. coli* detection methods are sometimes found because of the presence of GUD in some raw foods. In these cases the use of a pretreatment procedure to eliminate auto-fluorescent substances from the sample is recommended.

Although not as sensitive for GUD as MUG, the principal advantage of X-GLUC over MUG is that typical colonies can be recognized without using ultraviolet light illumination. Another disadvantage of the use of MUG is the possible diffusion of fluorescence over the entire agar surface, complicating colony differentiation. A disadvantage of both chromogenic and fluorogenic media is the high costs.

In GUD detection systems a number of factors may influence the assay substantially and may cause false-positive and false-negative identifications. These include: strain differences in response to particular substrates and substrate concentration; effects of carbohydrate content and selective agents in the medium; incubation time and temperature; pH changes; interference by large numbers of competing bacteria or substances in the sample itself.

Table 4 Commercially available media for the simultaneous detection of coliforms and *Escherichia coli* (data from Dr M Manafi, University of Vienna, Austria)

Medium	Substrate/colour coliforms	Substrate/colour E. coli	Manufacturer
Liquid media			
Fluorocult[R] LMX broth	XGAL/blue-green	MUG/blue fluorescence	Merck (Germany)
Readycult coliforms	XGAL/MUG	MUG/blue fluorescence	Merck (Germany)
ColiLert	ONPG/Yellow	MUG/blue fluorescence	Idexx (USA)
Coliquick	ONPG/Yellow	MUG/blue fluorescence	Hach (USA)
Colisure	CPRG/red	MUG/blue fluorescence	Millipore (USA)
Solid media			
EMX-agar	XGAL/blue	MUG/blue fluorescence	Biotest (Germany)
C-EC-MF agar	XGAL/blue	MUG/blue fluorescence	Biolife (Italy)
Chromocult	SalmonGal/red	XGLUC/blue-violet	Merck (Germany)
Coli ID	XGAL/blue	SalmonGlu/rose-violet	bioMérieux (France)
CHROMagar ECC	SalmonGal/red	XGLUC/purple	Chromagar (France)
Rapid *E. coli* 2	XGAL/blue	SalmonGlu/purple	Sanofi (France)
E. coli/coliforms	SalmonGal/red	XGLUC/purple	Oxoid (UK)
ColiScan	SalmonGal/red	XGLUC/purple	Micro.Lab. (USA)
MI agar	MUGal/blue fluores.	Indoxyl/blue	
Other systems			
ColiComplete	XGAL/blue	MUG/blue fluorescence	Biocontrol (USA)
ColiBag/Water check	XGAL/blue-green	MUG/blue fluorescence	Oceta (Canada)
Pathogel	XGAL/blue	MUG/blue fluorescence	(USA)
E. Colite and ColiGel	XGAL/blue	MUG/blue fluorescence	Charm Sci. (USA)
ColiChrome2Redigel	SalmonGal/red	XGLUC/purple	
m-Coliblue	TTC/red	XGLUC/blue	Hach (USA)

XGAL = 5-bromo-4-chloro-3-indolyl-β-D-galactopyranoside; MUG = 4-methylumbelliferyl-β-D-glucuronide; ONPG = o-nitrophenyl-β-D-galactopyranoside; CPRG = chlorophenol red β-galactopyranoside; SalmonGal = 6-bromo-3-indolyl-β-D-galactopyranoside; X-GLUC = 5-bromo-4-chloro-3-indolyl-β-D-glucuronide; TTC = triphenyl tetrazolium chloride.

Membrane Methods

The direct plating method for the rapid and relatively simple enumeration of *E. coli*, described by Anderson and Baird-Parker, utilizes the ability of *E. coli* to produce indole from tryptophan at 44°C when grown on a cellulose acetate membrane on plates of TB agar. For the enumeration of sublethally damaged cells of *E. coli* in dried, frozen or heat-processed foods the inoculum is first applied to a cellulose acetate membrane on minerals-modified glutamate agar and incubated for 4 h at 37°C. After this resuscitation step the membrane is transferred to a TB agar plate. After incubation *E. coli* colonies are confirmed by adding indole reagent to the colonies on the membrane, resulting in pink colonies for all the indole-positive strains (Fig. 2). Discrimination between indole-positive and negative strains may be difficult when high numbers of colonies are present. The growth of indole-positive coliforms other than *E. coli* is inhibited by the elevated incubation temperature. Incidentally, some indole-positive *Hafnia oxytoca* strains may appear, especially when the membranes are incubated below 44°C. The indole treatment is lethal for bacterial cells, making isolation of *E. coli* strains for further characterization impossible. The direct plating method is simple and sensitive, though 85 mm cellulose membranes are rather expensive and not always easily available.

A hydrophobic grid membrane filter (HGMF) technique was developed for the enumeration of *E. coli*. These filters are divided into 1600 or more grid cells. The sample is filtered through the membrane filter which traps target organisms within the grid cells. The inoculated HGMF is placed on an agar medium appropriate for the isolation of *E. coli* and colonies are counted and confirmed after incubation. The HGMF technique has the advantage of removing inhibitors or unwanted ingredients, concentrating organisms, and a 3-log counting range.

Impedimetry

Impedance technology like the Bactometer system is based on electrical responses in a culture medium caused by microorganisms. These electrical variations, measured as impedance, conductance or capacitance, depend on the number and metabolic activities of the microorganisms. Using specific media, this test system can also be used for the detection and enumeration of coliforms and Enterobacteriaceae. The protocol is as follows: coliform medium or entero medium (bioMérieux) is dispensed in modules made up of 16 disposable wells each with measurement electrodes fixed to their base. Then 1 ml of sample suspension is added to the module. The module is inserted into the Bactometer processing unit which simultaneously incubates (18 h at 35°C) and reads (every 6 min) the samples in the modules. Detection times, which correlate with counts of coliforms and Enterobacteriaceae, appear automatically on the screen of the computer. This system gives rapid results and minimal sample handling and using disposable modules avoids cleaning and sterilizing. Before the system can be applied the user must validate the procedure according to the products to be analysed and the microbiological criteria selected.

Immunological Methods

Polyclonal or monoclonal antibodies suitable for immunological detection of commensal *E. coli* strains can be easily developed. However, in practice immunoassays are usually not applied for the detection and enumeration of *E. coli* as an index organism. Several enzyme-linked immunosorbent assays (ELISAs) have been developed for the detection of toxins produced by pathogenic *E. coli*.

Molecular Methods

Recently molecular methods have been developed which offer considerable potential for obtaining a rapid result. These methods include the use of DNA hybridization and the polymerase chain reaction (PCR) technique.

DNA Hybridization

The induction of GUD synthesis in *E. coli* is the product of the *uidA* structural gene. With labelled DNA probes specific for the *uidA* region the colony-blotting technique has been applied in detecting gene sequences in both phenotypically GUD-positive and negative *E. coli* strains.

A commercial DNA hybridization test for the semi-quantitative detection of *E. coli* in foods has been developed by Gene-Trak Systems. The test employs probes which are specific for the 16S ribosomal RNA of *E. coli* and a colorimetric detection system. The label is fluorescein isothiocyanate, which can be detected with horseradish peroxidase-labelled antibodies. The Gene-Trak DNA hybridization assay protocol consists of the following steps:

1. Enrich the food sample in lauryl sulphate tryptose broth.
2. Add enriched sample to test tubes.
3. Add lysis solution to sample tubes and incubate to release rRNA target.
4. Add probe solution and incubate. The DNA probes hybridize to the rRNA target.
5. Place dipsticks into tubes and incubate to attach probe hybrids to dipsticks.

6. Wash dipsticks and place in tubes containing enzyme conjugate and incubate.
7. Wash dipsticks and place in tubes containing substrate-chromogen. Incubate for colour development. Discard dipsticks. Add stop solution to tubes.
8. Read absorbance at 450 nm on the Gene-Trak photometer.
9. An assay producing an absorbance value greater than 0.10 indicates the presence of *E. coli* in the test sample at a level greater than 3 cells per gram of original sample. A positive assay may be confirmed by standard culture procedures if desired. *Shigella* and *Escherichia fergusonii* may give false-positive results.

PCR Technique

Successful PCR amplifications were achieved when using *lacZ* primers for the detection of all coliform bacteria and *lamB* primers for *E. coli*.

A PCR assay using a primer set for *E. coli*, which was derived from the *uidA* gene, also correctly identified *E. coli* strains. The direct detection by PCR of *E. coli* in drinking water may cause problems. The DNA of non-viable *E. coli*, killed by chlorine or heat, is usually detectable by PCR, leading to false-positive results. This problem can be solved by enrichment for several hours before detection by PCR.

Detection of Pathogenic *E. coli*

The species *E. coli* contains both pathogenic and non-pathogenic strains and it is very important to have methods available which can differentiate between them. The detection and enumeration of pathogenic *E. coli* in foods are quite difficult. Procedures for food samples using selective enrichment broth at elevated temperatures (44–45.5°C) seem to favour the isolation of non-pathogenic environmental strains compared to pathogenic strains. Some pathogenic strains do not have the typical characteristics of *E. coli*, such as fermentation of lactose and gas production. Adequate methods have been developed for entero-haemorrhagic *E. coli*, especially of serotype O157:H7. There are still no simple sensitive procedures available for the direct cultivation of enteropathogenic (EPEC), enterotoxigenic (ETEC) and enteroinvasive (EIEC) *E. coli* in foods. For the isolation of these organisms resuscitation in brain–heart infusion broth for 3 h at 35°C, followed by enrichment in tryptone phosphate broth for 20 h at 44°C and plating on Levine's eosin-methylene blue agar and MacConkey agar is advised. Both typical (lactose-fermenting) and non-typical (non-lactose-fermenting) colonies have to be picked and characterized by biochemical, serological and

pathogenic properties. This means that from one sample numerous isolates must be assayed.

The majority of EPEC strains fall into certain well-recognized O:H serotypes. DNA probes and PCR primers have been developed for the three major characteristics of EPEC: production of attaching-and-effacing lesions, EPEC adherence factor and lack of Shiga toxin.

Detection of ETEC relies mainly on detection of the enterotoxins LT and/or ST or their encoding genes by the suckling mouse assay, cell assays, immunoassays, DNA probes and PCR assays.

Identification of EIEC entails demonstrating that the organism possesses the biochemical profile of *E. coli*, yet with the genotypic or phenotypic characteristics of *Shigella* spp. Phenotypic assays for testing the invasive potential of EIEC are the Sereny (guinea pig keratoconjunctivitis) test and a HeLa cell monolayer test. Polynucleotide probes, a PCR assay and an ELISA test for EIEC have been described.

See also: **Biochemical and Modern Identification Techniques**: Enterobacteriaceae, Coliforms and *E. coli*. ***Escherichia coli***: *Escherichia coli*; Detection of Enterotoxins of *E. coli*. ***Escherichia coli* O157**: Detection by Latex Agglutination Techniques; Detection by Commercial Immunomagnetic Particle-based Assays. **Petrifilm – An Enhanced Cultural Technique**. ***Salmonella***: Detection by Colorimetric DNA Hybridization. **Verotoxigenic *E. coli* and *Shigella* spp.**: Detection by Cultural Methods.

Further Reading

Bej AK, Mahbubani MH, Dicesare JL and Atlas RM (1991) Polymerase chain reaction-gene probe detection of microorganisms by using filter-concentrated samples. *Applied and Environmental Microbiology* 57: 3529–3534.

Corry JEL, Curtis GDW and Baird RM (eds) (1995) *Culture Media for Food Microbiology.* Pp. 473–478. Amsterdam: Elsevier.

Frampton EW and Restaino L (1993) Methods for *Escherichia coli* identification in food, water and clinical samples based on beta-glucuronidase detection. *Journal of Applied Bacteriology* 74: 223–233.

Hitchins AD, Hartman PA and Todd ECD (1992) Coliforms – *Escherichia coli* and its toxins. In: Vanderzant C and Splittstoesser DF (eds) *Compendium of Methods for the Microbiological Examination of Foods*. 3rd edn. Washington: American Public Health Association.

Hitchins AD, Feng P, Watkins WD, Rippey SR and Chandler LA (1995) *Escherichia coli* and the coliform bacteria. In: *Bacteriological Analytical Manual*, 8th edn. Gaithersburg: AOAC International.

Hofstra H and Huis in 't Veld JHJ (1988) Methods for the detection and isolation of *Escherichia coli* including pathogenic strains. *Journal of Applied Bacteriology Symposium Supplement* 65: 197S–212S.

Manafi M (1996) Fluorogenic and chromogenic enzyme substrates in culture media and identification tests. *International Journal of Food Microbiology* 31: 45–58.

Nataro JP and Kaper JB (1998) Diarrheagenic *Escherichia coli*. *Clinical Microbiology Reviews* 11: 142–201.

Roberts D, Hooper W and Greenwood M (eds) (1995)

Practical Food Microbiology, 2nd edn. London: Public Health Laboratory Service.

Weiss KF, Chopra N, Scotland P, Riedel GW and Malcolm S (1983) Recovery of fecal coliforms and of *Escherichia coli* at 44.5, 45.0 and 45.5°C. *Journal of Food Protection* 46: 172–177.

ENTEROCOCCUS

Giorgio Giraffa, Istituto Sperimentale Lattiero Caseario, Lodi, Italy

Bacteria of the genus *Enterococcus*, or enterococci (formerly the 'faecal' or Lancefield group D streptococci) are predominantly inhabitants of the gastrointestinal tract of humans and animals and also commonly occur in vegetables, plant material and foods, especially those of animal origin. However, there is no consensus on the significance of their presence in foodstuffs. The importance of the enterococci for food and public health microbiologists is related to their enteral habitat, their use as indicators for food safety and their possible involvement in food-borne illness.

In this article, the following topics are covered: the characteristics of the genus and relevant species of enterococci; methods of detection and enumeration in foods; details of procedures specified in national or international regulations; and the microbial ecology of enterococci in foods and importance of the genus and individual species in the food industry.

Characteristics of the Genus and Relevant Species

Taxonomy

Bacteria of the genus *Enterococcus* ('enterococci') have been recognized since Thiercelin (1899) described them as the 'entérocoque' to emphasize their intestinal origin. However, the taxonomy of this group of bacteria has always been vague. Development of genotypic methods for studying phylogenetic relationships of food-associated lactic acid bacteria has resulted in considerable changes in their taxonomy. On the basis of 16S rRNA cataloguing, the genus *Streptococcus* was separated during the 1980s into the three genera *Enterococcus*, *Lactococcus* and *Streptococcus*. Consequently, bacteria previously named as *Streptococcus faecalis*, *Streptococcus faecium*, *Streptococcus avium* and *Streptococcus gallinarum* were transferred to the revived genus *Enterococcus* as *Enterococcus faecalis*,

Enterococcus faecium, *Enterococcus avium* and *Enterococcus gallinarum*, respectively.

In the same years, other streptococci that possessed the characteristics of the *Enterococcus* division as defined by Sherman (see later) were transferred to the new genus. More recently, however, other species of enterococci have been described on the basis of chemotaxonomic studies and phylogenetic evidence provided by 16S rRNA sequencing. Since the transfer of *S. faecalis* and *S. faecium* to the revived genus *Enterococcus* in 1984, the total number of species presently included here is 20. Several of these species have been transferred from the genus *Streptococcus* and 11 were newly added (**Table 1**). This situation continues to fluctuate from time to time as individual species are moved into other genera or new taxa are discovered. In addition to providing a more definitive delineation of the genus *Enterococcus*, comparative 16S rRNA sequence analysis has also revealed the presence of several 'species groups' within the genus (**Table 2**). The *faecium* 'species group' contains *E. durans*, *E. faecium*, *E. hirae* and *E. mundtii*. The *avium* group contains *E. avium*, *E. raffinosus*, *E. malodoratus* and *E. pseudoavium*. The *gallinarum* group consists of the species *E. casseliflavus*, *E. gallinarum* and *E. flavescens*. Some species, however, do not cluster with these groups and form individual lines of descent, such as *E. faecalis*, the type species of the genus, *E. sulfureus*, *E. saccharolyticus* and *E. dispar*. From sequence homology determination of 16S rRNA, *E. cecorum* and *E. columbae* are also allocated to two phylogenetic distinct lineages but the high relatedness between the two species led to them being clustered into a common group, i.e. the *cecorum* group (Table 2).

Biochemical and Physiological Attributes

Since the transfer in recent years of new species that do not show the classical phenotypic traits of the genus *Enterococcus*, there are no phenotypic or physiological characteristics that clearly separate the

Table 1 Species included in the genus *Enterococcus*. (Reproduced with permission from Stiles and Holzapfel 1997)

Species	Source[a]	Principal habitats	Lyo-Group E[b]
E. faecalis		Human and other animal intestines	I
E. faecium		Human and other animal intestines, including poultry	II
E. avium	Transfer	Poultry (rare) and mammalian intestines	IV
E. casseliflavus[c,d]	Transfer	Grass, silage, plants, soil	III
E. cecorum[e]	Transfer	Clinical origin, animals	
E. columbae	New	Pigeon intestine	
E. dispar	New	Human origin	
E. durans	Transfer	Clinical isolate	II
E. fallox	New		
E. flavescens[c]	New	Clinical origin	
E. gallinarum[d]	Transfer	Poultry intestine	VI
E. hirae	New	Animal intestines	V
E. mundtii[c]	New	Grass, silage, plants, soil	VII
E. malodoratus	Transfer	Originally from Gouda cheese	I
E. pseudoavium	New	Bovine mastitis	IV
E. raffinosus	New	Clinical isolate, endocarditis	IV
E. saccharolyticus	Transfer	Bedding and skin of cattle	
E. seriolicida[f]	New		
E. solitarius[g]	New		IV
E. sulfureus	New	Plant material	

[a]New, new species; transfer, transfer from group D streptococci.
[b]Lyo-groups E I–VII marker reactions.
[c]Pigmented.
[d]Motile.
[e]Not group.
[f]Syn. *Lactococcus garvieae* (?).
[g]Phylogenetically related to *Tetragenococcus*.

genus from the other Gram-positive, catalase-negative cocci. Since the identification scheme described by Sherman in 1937, enterococci grow at 10°C and at 45°C, in the presence of 6.5% NaCl and at pH 9.6, they survive at 60°C for 30 min and present the Lancefield group D antigen. These characteristics were subsequently adopted in revising the genus by Schleifer and Killper-Balz in 1984. However, several species and strains of *Enterococcus* do not share all of these criteria. For example, the Lancefield group D antigen is not peculiar to *Enterococcus* because the faecal streptococci *Streptococcus bovis* and *S. equinus* also react with the group D antisera. For this reason, they are also called 'the group D non-enterococci'. Conversely, *E. cecorum*, *E. columbae*, *E. dispar*, *E. saccharolyticus*, many strains of the *avium* 'species group' and even *E. durans* fail to react with Lancefield group D antisera. *E. dispar* and *E. sulfureus* do not grow at 45°C, and *E. cecorum* and *E. columbae* fail to grow or grow poorly at 10°C or in 6.5% NaCl broth.

Pyrolidonylarylamidase (pyroglutamyl aminopeptidase), and aesculin hydrolysis in combination with resistance to 40% (v/v) bile, have found widespread acceptance as basic tests for valuable identification of most *Enterococcus* species. Other characteristics may be useful, especially those gen-

erally seen in all presently known enterococci, for a presumptive phenotypic delineation of the genus *Enterococcus* (**Table 3**). Unfortunately none of these criteria or their combinations are unique to the enterococci. As a rule of thumb, the following scheme could be used: Gram-positive, catalase-negative cocci, able to grow in the presence of 6.5% NaCl broth and on media containing 0.04% sodium azide, commonly used in the selective isolation of enterococci, and able to hydrolyse aesculin in the presence of bile, can be identified as presumptive *Enterococcus* spp. Under these conditions, only bacteria of the group D non-enterococci (*S. bovis*, *S. equinus*) usually show vigorous growth similar to the 'classical' enterococci. This procedure, however, excludes several new enterococcal species, such as those belonging to the *avium* and *cecorum* species groups, which grow poorly on the enterococcal selective media.

With regard to species identification, the different 'species groups' that usually comprise *Enterococcus* species do not have easily observable positive biochemical characteristics useful for their differentiation. Only the phylogenetically distinct species *E. faecalis*, *E. sulfureus*, *E. saccharolyticus* and *E. dispar* differ in several characters from each other and from the enterococcal 'species groups' (Table 2). Enterococci also produce bacteriolytic enzymes and

Table 2 Tests differentiating between the *Enterococcus faecium* (*E. faecium*, *E. durans*, *E. hirae*, and *E. mundtii*), the *Enterococcus avium* (*E. avium*, *E. pseudoavium*, *E. malodoratus* and *E. raffinosus*), the *Enterococcus gallinarum* (*E. gallinarum*, *E. casseliflavus* and *E. flavescens*) and the *Enterococcus cecorum* (*E. cecorum* and *E. columbae*) species groups and phylogenetically distinct enterococcal species. (Reproduced with permission from Devriese et al 1993)

Test	*Enterococcus* faecalis	faecium group	avium group	gallinarum group	cecorum group	sulfureus	saccharolyticus	dispar
Motility	–	–	–	+	–	–	–	–
Group D antigen	+	+	D+	+	–	–	–	–
APPA	–	–	+	–	–	?	?	?
PYRA	+	+	+	+	–	+	–	+
Alkaline phosphatase	–	–	–	–	D+	–	–	–
α-Galactosidase	–	D	D	+	+	+	+	+
Arginine dihydrolase	+	+	–	+	–	–	–	+
Acid production from								
Adonitol	–	–	+	–	–	–	–	–
L-Arabinose	–	D	D+	+	D	–	–	–
D-Arabitol	–	–	+	–	D	–	+	–
L-Arabitol	–	–	D	–	–	–	–	–
D-Cyclodextrin	+	D+	D–	D	+	?	?	?
Dulcitol	–	–	D	–	–	–	–	–
Gluconate	D+	D	D+	D	+	+	–	–
Glycerol	+	–[a]	D	D	–	–	–	+
Inulin	–	–(D–?)[b]	–	+(D?)[b]	+	–	+	–
2-Ketogluconate	D	–	+	–	D	+	+	+
D-Lyxose	–	–	D	–	–	–	–	–
Mannitol	+	D	+	+	D	–	+	–
Melezitose	D+	–	D	D–	D	+	+	–
Melibiose	–	D	D	+	D	?	+	–
α-Methyl-D-glucoside	–	–	+	+	D	+	+	+
D-Raffinose	–	D	D+	D	+	?	+	+
Sorbitol	D+	D–	+	D–	D	–	+	–
L-Sorbose	–	–	+	–	–	?	D–	–
Xylitol	–	–	D+	–	D–	–	–	–
D-Xylose	D–	D	D–	+	D	–	–	–
Pyruvate fermentation	+	–	D	–	?	?	?	?

[a]In anaerobic conditions (under paraffin cover).
[b]Unclear in one species description of *E. mundtii*, stated to be positive in the species description of *E. gallinarum* and *E. casseliflavus* but listed as negative in *E. gallinarum* and usually positive in *E. casseliflavus* in other identification schemes.
D, Different or variable; D+, usually positive; D–, usually negative; APPA, alanyl-phenyalanyl-prolinarylamidase (not yet reported for all species); PYRA, pyrolidonylarylamidase.

these bacteria can be placed into different so-called lyo-groups on the basis of such activities. Results of lyo-group analysis of *Enterococcus* bacteriolytic enzymes is consistent with current enterococcal taxonomy (Table 1).

Methods of Detection and Enumeration in Foods

Detection Media

Up to more than 100 modifications of selective media have been described for enterococci and other group D streptococci. It is therefore difficult to recommend a universal medium. The choice of a medium and a method of detection depend on whether the group D non-enterococci and total enterococci or only entero-

cocci have to be detected, the level of contamination of the habitat, and the type of product to be examined. Also, methods for recovering environmentally stressed cells should be carefully considered. Failure to detect stressed enterococci could underestimate their densities. **Table 4** lists the relevant characteristics of the most commonly used media for selective detection and enumeration of enterococci in foods.

Selective Cultivation Procedures

Selective agents commonly used are sodium azide, thallous acetate, kanamycin or gentamycin. Other limiting ingredients or factors are crystal (or ethyl) violet, Tween 80, sodium carbonate (or citrate), bile salts, acidic pH values (e.g. 6.0 or 6.2) or elevated incubation temperatures (42–45°C). Because all

Table 3 Characteristics generally or with few exceptions seen in all enterococci. (Reproduced with permission from Devriese et al 1993)

Characteristics	Result
Resistance to 40% (v/v) bile	+
β-Glucosidase	+
Urease	−
Voges–Proskauer reaction	+[a]
Leucinearylamidase	+
β-Glucuronidase	−[b]
Aesculin hydrolysis	+
Acid from:	
N-Acetylglucosamine	+
Amygdalin	+[c]
D-Arabinose	−[d]
Arbutin	+
Cellobiose	+
Erythritol	−
D-Fructose	+
Galactose	+
β-Gentiobiose	+
Glucose	+
Glycogen	−[e]
D-Fucose	−
L-Fucose	−
Lactose	+[f]
Maltose	+
D-Mannose	+
Methyl-β-D-glucopyranoside	+[g]
α-Methyl-D-xyloside	−
Pullulan	−[g]
Ribose	+[h]
Salicin	+
Trehalose	+[f]
L-Xylose	−

[a]Negative only in *Enterococcus saccharolyticus*.
[b]Positive only in most *E. cecorum* strains.
[c]Negative in some *E. columbae* strains.
[d]Delayed positivity in some *E. avium* group strains.
[e]Positive only in some *E. gallinarum*, *E. cecorum* and *E. columbae* strains.
[f]*Enterococcus faecalis* (saccharolytic variant) strains from human infections may be lactose- and trehalose-negative.
[g]Not reported in some of the newer species.
[h]Negative only in *E. flavescens*.

enterococcal species do not tolerate incubation at 45°C, a short pre-incubation at a lower temperature (e.g. 37°C) can be beneficial in some circumstances (see later). The time of incubation when the results are assessed may also have a differentiating effect. Usually, prolonged incubation (more than 24 h) increases the number of unwanted species capable of growth in enterococcal selective media. Increasing or decreasing concentrations of selective agents will also yield different degrees of selectivity.

Indicator substances added to the media are useful for the recognition of enterococci and for rapid identification of single species according to colonial appearance. Reduction of triphenyl-tetrazolium chloride (TTC), or potassium tellurite and the formation of a brownish black precipitate due to the formation of an aesculin (6,7-dihydroxycoumarin-6-glucoside)-ferrous complex after hydrolysis of aesculin to glucose and aesculetin, are most commonly used.

These different factors explain why there are so many varieties of selective media for enumeration of enterococci (Table 4). The main problems are insufficient selectivity and differential capacities. These *Enterococcus* selective media allow the growth of several other genera of Gram-positive cocci and bacilli. Growth of these bacteria is usually slower, colonies are smaller, and aesculin splitting or TTC reduction are less intense. On the other hand, several enterococcal species do not grow well on these media *E. cecorum* and *E. avium* do not grow on commonly used enterococcal selective media such as M-enterococcus (also called Slanetz and Bartley) agar, and colonies of these species on kanamycin–azide–aesculin agar are small without or with only slight browning.

In KF and Slanetz and Bartley media the growth of lactococci, some streptococci, and lactobacilli is not inhibited. These media are not satisfactory in food microbiology because the lack of selectivity is likely to overestimate the real number of enterococci. In kanamycin–azide–aesculin, the typical medium for isolation of enterococci from foods, selectivity mainly depends on the presence of sodium azide, which is inhibitory to Gram-negative bacteria and many strains of group D non-enterococci. However, some growth of mesophilic lactobacilli, some of which split aesculin, is possible. This can cause errors in all specimens, such as fermented food products, where a high level of lactobacilli and 'background microflora' is expected. Increasing the content of sodium azide and raising the incubation temperature to 42–45°C enhance the selectivity of the media but also reduce the recovery rate of enterococci. This is particularly true for several new enteroccocal species, such as *E. avium* or *E. cecorum*, or for isolates from foods, which are sublethally stressed from previous technological treatments (e.g. heat treatments, acidity changes, salting). In this respect, the recovery and detection rates of stressed enterococci from foods can be enhanced by 'resuscitation' techniques. Basically, they consist of cultivating food suspensions in non-selective nutrient media (i.e. brain heart infusion, or BHI) at 30–37°C for a short incubation period (4–6 h), followed by overlaying with a selective medium.

As stated above, growth on selective media can also be performed by shortening the incubation time to 18–24 h, or by changing the amount and/or type of the selective agent in the medium. For example, the

Table 4 Selective media widely used in routine monitoring of foods for quantitative isolation of enterococci

Media	In full	Selective agent(s) (%)	Differential agent(s)	Culture conditions	Recovery ability (sources)	Main contaminant(s)
KF	KF-Streptococcus	NaN$_3$[a] (0.04)	TTC[b]; bromocresol-purple	35–37°C; 48 h	Meat, fish, dairy, water	Mesophilic lactobacilli
KAA	Kanamycin–azide–aesculin	Kanamycin (0.002); NaN$_3$ (0.015)	Aesculin	42°C; 18 h or 37°C; 72 h	Meat, fish, dairy	Mesophilic lactobacilli
TITG	Thallous–acetate–tetrazolium–glucose	Thallous acetate (0.01); pH 6.2	TTC	42°C; 18 h or 37°C; 24 h	Meat, fish, dairy	Staphylococci; Enterobacteriaceae; Pseudomonadaceae
GTC	Gentamycin–thallous–acetate	Gentamycin (0.0025); thallous acetate (0.05)	Aesculin	42°C; 18 h or 37°C; 18–24 h	Meat, fish, dairy	Staphylococci; Enterobacteriaceae; Pseudomonadaceae
F-GTC	Fluorogenic–gentamycin–thallous–acetate	Gentamycin (0.0025); thallous acetate (0.05)	MUG[c]	37°C; 18–24 h	Meat, fish, dairy	Staphylococci; Enterobacteriaceae; Pseudomonadaceae
CA	Citrate–azide	NaN$_3$ (0.04); citrate (1.0)	TTC	37°C; 48–72 h	Dairy products	Other group D non-enterococci
CATC	Citrate–azide–Tween–carbonate	NaN$_3$ (0.04); sodium carbonate (0.2); Tween 80 (0.1); sodium citrate (1.5)	TTC	37°C; 24 h	Meat, fish, dairy	Other group D non-enterococci
ME	M-enterococcus or Slanetz and Bartley	NaN$_3$ (0.05)	TTC	35–37°C; 24–48 h	Meat, dairy	Staphylococci; lactic streptococci; some lactobacilli
ABA (BEA)	Aesculin–bile–azide	NaN$_3$ (0.025); bile salts (1.0)	Aesculin	37°C; 24–72 h	Meat, dairy	Staphylococci; Enterobacteriaceae; Pseudomonadaceae
KA	Crystalviolet–azide	NaN$_3$ (0.09); crystalviolet (0.0002)	Crystalviolet	37°C; 24–48 h	Water, dairy products and other foods	
EVA	Ethylviolet–azide	NaN$_3$ (0.04); ethylviolet (0.0083)	Ethylviolet	37°C; 24 h	Water and other specimens	

[a]Sodium azide.
[b]2,3,5-Triphenyltetrazolium chloride.
[c]4-Methyl-umbelliferyl-α-D-galactoside.

gentamycin–thallous–acetate medium (Table 4) is selective for the group D non-enterococci and enterococci in total. This medium also allows good recovery of stressed cells because the combined selective action of thallous acetate and gentamycin is better tolerated than that of sodium azide. Selectivity of gentamycin–thallous–acetate medium, or thallous–acetate–tetrazolium–glucose medium, can be further enhanced by incubating plates at 42°C for a maximum of 18 h and adjusting the pH of the medium at 6.2. Recently, gentamycin–thallous–acetate has been improved in its differential ability by introducing fluorogenic–gentamycin–thallous–acetate medium (Table 4), in which aesculin is substituted by a fluorogenic substrate (4-methyl-umbelliferyl-α-D-galactoside, MUG) for detection of glycosidic activity of enterococci. The use of chromogenic and fluorogenic substrates to differentiate enterococci has recently received considerable attention. This is because the most commonly used indicators are not always satisfactory as differential agents. For example, the citrate–azide or citrate–azide–Tween–carbonate media (Table 4), which are highly selective for enterococci, are more selective for *E. faecalis* than for *E. faecium* because this latter species shows a weaker ability to reduce TTC.

It is clear that a highly selective and differential medium for enterococci still does not exist. Most commonly used media either allow growth of non-group D streptococci or inhibit growth of some species and strains of enterococci, especially if new habitats or samples with unknown microflora are investigated. Sometimes, growth of non-streptococci also occurs, and group D non-enterococci are almost always recovered and detected together with the enterococci. It should, however, be noted that group D non-enterococci, such as *S. bovis*, are more sensitive to commercial food-processing treatments and do not occur as frequently and are not as numerous in foods as the enterococci.

Procedures Specified in National or International Regulations

Methods for the Enumeration of Enterococci in Water Samples

International standard procedures for detection and enumeration of enterococci for determination of water quality are available. The International Organization for Standardization (ISO) describes two methods which can be applied to all types of water: a method of enrichment in liquid medium (ISO-7899–1) and a method of detection and enumeration by membrane filtration (ISO-7899–2).

ISO 7899–1 consists of two steps. The first step comprises enrichment culture, which is performed by incubating the sample in the selective liquid medium azide–glucose broth at 35 or $37 \pm 1°C$ for 24–48 h. Composition of the azide–glucose broth is (g l^{-1}): pancreatic digest of casein, 15; glucose, 7.5; beef extract, 4.5; NaCl, 7.5; sodium azide, 0.2; bromocresol purple, 0.0032. Enterococci grow in this medium and ferment glucose with the formation of acid, which causes a change in the colour of the pH indicator bromocresol purple from purple to yellow. The second step confirms the identification. All enrichment tubes showing positive reactions after 24 or 48 h are subcultured on a confirmation medium to eliminate false positive reactions such as those by other Gram-positive cocci or rods. The confirmation medium, aesculin–bile–azide agar is then incubated at $44 \pm 0.5°C$ for 48 h. Black–brown colonies are then subjected to a catalase test, which consists of placing a drop of hydrogen peroxide solution on the colonies. Evolution of bubbles of oxygen indicates catalase-positive organisms. Only catalase-negative colonies may be regarded as enterococci.

ISO 7899–2 consists of a presumptive determination of enterococci based on filtration of a specified volume of water sample through a membrane filter (pore size 0.45 μm) sufficient to retain the bacteria. The filter is placed on a solid selective medium (KF–Streptococcus or Slanetz and Bartley agar) containing sodium azide and TTC (Table 4). After incubation at $37 \pm 1°C$ for 48 h all raised colonies, which show a red, maroon or pink colour, either in the centre of the colony or throughout, are counted as 'presumptive' enterococci. Confirmation is carried out by plating on aesculin–bile–azide agar and incubating at $44 \pm 0.5°C$ for 48 h. Typical black colonies which are also catalase negative may be regarded as enterococci. Italian guidelines indicate the absence of enterococci (here defined as 'group D faecal streptococci') in 100 ml of water.

In Standard Methods for the Examination of Water and Wastewater (1995), the approved method for enterococci enumeration in recreational as well as drinking waters is a two-step procedure similar to ISO 7899–2: (1) growth on a selective and differential agar medium (M-enterococcus agar) at $41 \pm 0.5°C$ for 48 h by membrane filter procedure; (2) an *in situ* confirmation test for aesculin hydrolysis. Confirmation consists of transferring filters on EIA agar medium containing 1 g l^{-1} of aesculin and 0.5 g l^{-1} of ferric citrate, followed by incubation at $41 \pm 0.5°C$ for 20 min. Pink to red enterococcal colonies develop a black or reddish-brown precipitate on the underside of the filter. Colonies are counted using a fluorescent lamp and a magnifying lens. A 24 h one-step procedure that eliminates the *in situ* aesculin test through addition of 0.75 g l^{-1} of indoxyl-β-D-glucoside to the primary Slanetz and Bartley agar medium has been proposed. Enterococci growing on this modified medium and producing the appropriate glycosidase will form blue colonies through release of an aglycone that is converted into indigo.

Enumeration of Enterococci in Foods

KF-enterococcus or Slanetz and Bartley agar have been accepted by most food industries and regulatory agencies for the quantitative estimation of enterococci in non-dairy foods. For example, detection of enterococci (here called 'faecal streptococci') in foodstuffs according to the Nordic Committee for Methods in Food Analysis (NMKL) is performed by using Slanetz and Bartley agar as selective medium. Briefly, 10 g of the food sample is weighed into a sterile stomacher filter bag and diluted with 90 ml of saline–peptone solution. After homogenization, 100 μl is transferred to Slanetz and Bartley agar plates by loop streaking and plates are incubated at $44 \pm 0.2°C$ for 48 h. After incubation, typical *Enterococcus*-like colonies (red or pink coloured colonies) are identified as presumptive enterococci.

More selective conditions are necessary in dairy or meat products to reduce background growth of other streptococci and lactobacilli. For example, Slanetz and Bartley agar is suggested for membrane filtration procedures and has been used to select for enterococci in dried foods, including non-fat dry milk, and in meat products. For the isolation of enterococci from meat and meat products, the ISO recommends the use of Slanetz and Bartley. In the ISO-version, the medium contains 0.05% sodium azide, 0.01% TTC, and the final pH is 7.2. Homogenized meat samples are inoculated by a loop-streaking technique and plates are incubated at $37 \pm 1°C$ for 24 h. However, this medium may not inhibit the growth of lactobacilli and streptococci, such as *Lactobacillus acidophilus* and *S. bovis*.

The following procedure has been suggested in Standard Methods for the Examination of Dairy Products (1992) for its ability to enumerate enterococci in dairy products. After Petri dishes have been inoculated with samples of product, 10–12 ml of sterile, tempered (45°C) citrate–azide agar per plate are added. This medium contains $(g\,l^{-1})$: pancreatic digest of casein, 20; sodium citrate, 10; yeast extract, 5; glucose, 5; NaCl, 5; K_2HPO_4, 4; KH_2PO_4, 1.5; sodium azide, 0.4; TTC, 0.01; agar, 15. Then, 3–4 ml of medium is added as an overlay to cover the surface of the solidified agar completely. After the overlay has solidified, plates are incubated at $37 \pm 1°C$ for 48–72 h. Colonies with a red, maroon, or pink colour are counted as presumptive enterococci. Results are reported as 'enterococcus count per gram' or 'per millilitre'. This method is useful for evaluating sanitation in butter manufacturing plants.

Importance of the Genus and Individual Species in the Food Industry

Enterococci may be considered an obligatory part of the intestinal cavity of humans and animals. Therefore, these bacteria have been useful as indicators of poor hygiene in water and in foods. However, the use of enterococci as 'hygienic indicators' requires better knowledge of their microbial ecology. For example, *E. faecium* is widely distributed in the intestines of humans and other animals, whereas *E. faecalis* is more generally associated with humans. However, both species are widely distributed in the environment and they are also associated with plants. In vegetable foods, interpretation of enterococcal findings is difficult because many strains remain undefined and because species of plant-associated enterococci, such as *E. mundtii* or *E. casseliflavus* may occur together with enterococci of faecal origin, such as *E. faecalis* or *E. faecium*. With this wide distribution it is not surprising that enterococci occur in different foods, especially those of animal origin.

The value of most *Enterococcus* spp. as indicators of faecal contamination of foods is also limited by their ability to survive in adverse environmental conditions such as extreme pH, temperatures and salinity. This means that these bacteria could withstand normal conditions of food production. Therefore, although enterococci remain good indicators of faecal pollution in drinking water and some dairy products, such as yoghurt, butter and milk powder, they can be considered as faecal indicator organisms in other foods only in a broad sense because the occurrence and significance of their presence may vary according to the species, the habitat, and the technology of food production.

In poultry, few enterococci (less than $5000\,cm^{-2}$ of breast skin) suggest good evisceration technique and good hygiene. In severely heat-treated dairy products, such as yoghurt, pasteurized creams, milk powders and butter, enterococci can be considered as indicators of recontamination. The enterococcal count is more reliable than the coliform count as an index of the sanitary quality of churned butter. This is because enterococci are better able than coliforms to survive in the unfavourable microenvironment of salted butter. In addition, the enterococcal count may be a more reliable indicator of the sanitary quality of yoghurt than the coliform count because coliforms are inactivated in the low pH environment; whereas enterococci are not.

In yoghurt, in which enterococci represent a consistent part of the thermoduric spoilage microflora, Italian guidelines indicate acceptable levels of 10–$100\,cfu\,g^{-1}$ of these bacteria. In butter, an enterococcal count of fewer than 10 colonies per gram of product is considered duly stringent for a well-managed butter manufacturing plant. The meaning of the presence of enterococci in cheeses depends on different cheese technologies. In cheeses, enterococci are found in soft or semi-hard cheeses made with both raw and pasteurized milk. In raw milk cheeses, enterococci present in raw milk and/or in natural lactic starter cultures may develop during cheese manufacture and ripening and, depending on the different technologies, may represent the predominant microflora found in cheese. No indications are available for acceptable levels of enterococci in these cheeses. In pasteurized milk cheeses, the presence of enterococci is undesirable because they can cause spoilage problems. In these cheeses, which are usually fresh or soft cheeses, the presence of more than 10–$100\,cfu\,g^{-1}$ of enterococci can be considered due to poor hygienic conditions during cheese production.

Several strains of *Enterococcus* species associated with food systems, such as *E. faecium* and *E. faecalis*, have been reported to produce antibacterial proteins (bacteriocins) inhibitory against food spoilage or pathogenic bacteria, such as *Listeria monocytogenes*, *Staphylococcus aureus*, *Vibrio cholerae*, *Clostridium* spp. and *Bacillus* spp. Enterococcal bacteriocins, which are shown to be produced during cheese manufacture or sausage production, could offer useful protection against unwanted bacteria in food technology.

Many food hygienists are reluctant to accept the presence of enterococci in foods, because there is not consensus on whether enterococci can be fully considered as GRAS (generally recognized as safe) microorganisms. The potential pathogenicity of these bacteria has recently become a matter of controversy despite the long history of enterococci in foods that have not caused health concerns.

Enterococci are normally of relatively low virulence but the emergence of many strains of *E. faecium* and *E. faecalis* resistant to all currently available antibiotics is posing serious problems in clinical therapy. The presence and the chronic usage of antibiotics as nutritive food additives or as therapeutic agents have selected microbial flora with multiple, often transferable, resistance determinants. Antibiotic-resistant enterococci, staphylococci, and even lactic acid bacteria, have entered the human food chain. The formation of biogenic amines, especially tyramine, by *E. faecalis* and other enterococcal species has been reported. Decarboxylating enterococci can find suitable conditions to produce biogenic amines in a range of fermented food products, including cheese, meat and fish products.

See also: **Cheese**: Microbiology of Cheese-making and Maturation. **Milk and Milk Products**: Microbiology of Liquid Milk. **Waterborne Parasites**: Detection by Classic and Modern Techniques.

Further Reading

Atlas RM (1997) In: Parks LC (ed.) *Handbook of Microbiological Media*, 2nd. edn. New York: CRC Press.

Collins CH, Lyne PM and Grange JM (1989) *Collins and Lyne's Microbiological Methods*, 6th edn, p. 307, London: Butterworth.

Devriese LA, Pot B and Collins MD (1993) Phenotypic identification of the genus *Enterococcus* and differentiation of phylogenetically distinct enterococcal species and species groups (review article). *Journal of Applied Bacteriology* 75: 399–408.

Devriese LA and Pot B (1995) The genus *Enterococcus*. In: Wood BJB and Holzapfel WH (eds) *The Lactic Acid Bacteria*, vol. 2, *The Genera of Lactic Acid Bacteria*, p. 327. New York: Blackie Academic & Professional.

Eaton AD (ed.) (1995) Standard Methods for the Examination of Water and Wastewater, 19th edn, p. 72. Washington: American Public Health Association.

Giraffa G, Carminati D and Neviani E (1997) Enterococci isolated from dairy products: a review of risks and potential technological use (review article). *Journal of Food Protection* 60: 732–737.

ISO 7899 (1984) *International Standard. Water Quality – Detection and Enumeration of Faecal Streptococci*. Part 1: Method by enrichment in a liquid medium, 3 pp. Part 2: Method by membrane filtration, 4 pp. Geneva: International Organisation for Standardisation.

Marshall RT (ed.) (1992) Standard Methods for the Examination of Dairy Products, 16th edn, p. 279. Washington: American Public Health Association.

Orvin Mundt J (1986) Enterococci. In: Sneath PH, Mair NS, Sharpe ME and Holt JH (eds) *Bergey's Manual of Systematic Bacteriology*, vol. 2, Baltimore: Williams & Wilkins.

Reuter G (1985) Selective media for group D streptococci (review article). *International Journal of Food Microbiology* 2: 103–114.

Reuter G (1992) Culture media for enterococci and group D streptococci (review article). *International Journal of Food Microbiology* 17: 101–111.

Stiles ME and Holzapfel WH (1997) Lactic acid bacteria of foods and their current taxonomy (review article). *International Journal of Food Microbiology* 36: 1–29.

Enterotoxins *see Bacillus*: Detection of Enterotoxins; *Staphylococcus*: Detection of Staphylococcal Enterotoxins.

Enteroviruses *see Viruses*.

ENZYME IMMUNOASSAYS: OVERVIEW

Arun Sharma, Food Technology Division, Bhabha Atomic Research Centre, Mumbai, India

An immunoassay is a technique used for the estimation of small quantities of a substance using immunological methods and principles. Immunology is the study of the mechanisms that provide immunity to the body against foreign substances that find entry into the body. The immune response that the body invokes can be divided into two main categories. The humoral response involves production of a protein, an immunoglobulin, known as antibody, by the lymphocytes. The antibody binds the foreign substance in response to which it is produced very specifically, but non-covalently. The other response is called the cell-mediated immunity in which the lymphocytes directly attack the foreign substance and is of little importance to immunoassays.

Immunoassay could therefore be broadly defined as an analytical technique based on the highly specific binding of a substance to be analysed (analyte) to its antibody. Therefore, antibody to the analyte is the main reagent used in an immunoassay.

Antigen and Antibody

Antigen

An antigen is any foreign substance which when introduced into the body of an animal elicits an immune response by producing specific antibody molecules against it. Antigen should therefore essentially be an immunogen. Depending on the innate ability of a molecule to generate immune responses antigens could be divided into two categories.

Immunogen

For a substance to be an immunogen it should be non-self or foreign to the body and it should be sufficiently intricate, that is it should have a relatively large molecular weight and structural complexity. An immunogen has the innate ability to generate an immune response. A macromolecule such as a protein, a nucleic acid, a polysaccharide, or a lipid when injected directly into the body of an animal is capable of invoking an immune response. Therefore, these antigens are true immunogens. Good immunogens have an $M_r > 10\,000$.

Hapten

A hapten is a substance that can combine with a specific antibody but lacks antigenicity of its own. Many small molecules of $M_r < 1000$ such as toxins, drugs and hormones are not capable of invoking immune response when injected directly into animals. They are thus not immunogenic by themselves, and are called haptens. These small-molecular-weight compounds need to be linked to a large molecule, such as a protein like bovine serum albumin or keyhole limpet haemocyanin to make them immunogenic before injecting into an animal to get the desired antibody. The reactive carboxylic or amino group in a hapten could be used to conjugate it to the protein molecule. Alternatively, a reactive group such as carboxylic acid may be added to a hapten for this purpose. Lysine, tyrosine or histidine residues in the carrier proteins may be used for linking. Most of the antibody will be generated against the carrier protein, but a good proportion will also be generated against the hapten molecule.

Adjuvant

Adjuvant is a preparation that enhances the immunogenicity of an antigen. Adjuvant is used in the production of both polyclonal and monoclonal antibodies. The adjuvant permits a prolonged and slow release of the antigen in the body. Adjuvant also protects the antigen through non-specific stimulation of immune response. It also allows smaller quantities of the antigen to be used. Some of the common adjuvants are Freund's complete adjuvant, a preparation containing inactivated *Mycobacterium tuberculosis* cells in oil, and alum salt containing inactivated *Bordetella pertussis* cells. Some non-ionic surfactants, and muramyl peptides are also used.

Antibody

Antibody is an immunoglobulin protein capable of specific combination with the antigen that caused its production in a susceptible animal. Antibody molecule contains specific binding site capable of binding tightly but non-covalently to the original antigen, and hence causing its precipitation, neutralization or death, via phagocytosis, complement-mediated cell lysis, depending on whether antigen is a macromolecule, toxin or microorganism.

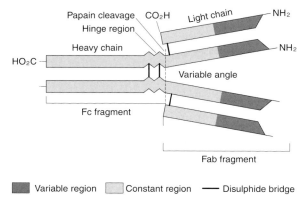

Figure 1 Structure of an antibody molecule.

Antibodies belong to a group of proteins known as immunoglobulins (Ig), which are further divided into five classes namely, IgG, IgM, IgA, IgD and IgE. IgG constitutes about 80% of the total immunoglobulins in the serum. All immunoglobulins contain four polypeptide chains (**Fig. 1**). Two of the polypeptide chains are heavy and two light. IgG is a glycoprotein of M_r 150 000 with a carbohydrate content of 3%. It shows complement fixation activity and has a half-life of 23 days. The proteolytic enzyme papain cleaves the IgG molecule into three fragments of equal sizes. Two identical antigen binding fragments (Fab), and one crystallizable fragment (Fc), which is capable of binding and activating the complement system.

Papain actually cleaves the antibody in the hinge region. The region also contains more than one interchain disulphide bridge. About a hundred residues in the N-terminal region of both heavy and light chains vary in different IgG molecules and are called variable regions. One IgG molecule contains two identical light chains and two identical heavy chains. The molecule has two identical antigen binding sites.

Production of Antibodies

Polyclonal Antibodies (Antisera)

Antibodies are raised by injection of a solution or suspension of an antigen into a suitable animal such as mouse, rabbit, sheep, goat or horse. This will generate a primary humoral response producing mainly IgM type antibodies. A further injection (booster) of the same antigen given after 25–30 days generates high titre of IgG type antibodies. More booster injections at fortnightly intervals can lead to hyperproduction of IgG. Such animals are known to be hyperimmunized. A suitable incision is made in the posterior marginal vein of the ear of the animal and blood is drawn. The blood is allowed to clot at 37°C for 1 h and further contract at 4°C. Nearly half the blood volume is recovered as serum by centrifugation.

The serum is incubated at 56°C for 1 h to inactivate the complement and proteases. The serum is stored at –20°C in small aliquots.

For weakly antigenic molecules, the period of exposure to the antigen could be extended either by repeated injections or by the use of an adjuvant. The serum contains more than 50 proteins. Serum albumin constitutes more than 60% of the total protein present. There may be more than 10 000 types of IgG molecules, some of which are specific for the antigen of interest. Normally an antigen can induce production of a number of different antibodies, each capable of reacting with a different epitope of the antigen. An epitope is a small site on an antigen, to which a complementary antibody may bind specifically. There may be a number of such sites present on an antigen molecule. This can lead to slight cross-reactivity of the antiserum to substances with structural features similar to the antigen (analogues). Efforts to produce specific, homogeneous, single molecular species of an antibody, a monoclonal antibody, did not succeed until the discovery of hybridoma technology in 1975.

Monoclonal Antibody

Monoclonal antibody is an immunoglobulin derived from a single clone of a hybrid cell line and is therefore homogeneous. Monoclonal antibody is produced by a cell line created by fusion of B lymphocytes with myeloma cells. Myeloma cells are neoplastic B lymphocytes which are virtually immortal. They continue to divide and secrete a single species of immunoglobulin in cell culture. Aminopterin, a metabolic poison can block *de novo* synthesis of DNA in the cell. However, the aminopterin-poisoned cells can still survive with the help of the enzyme hypoxanthine guanine phosphoribosyl transferase (HGPRT) through a pathway called the salvage pathway, if the medium is fortified with hypoxanthine and thymidine. For the creation of a hybrid cell line (hybridoma), a myeloma cell is fused with an antibody-producing lymphocyte. For this purpose special mouse myeloma (HGPRT⁻) cells that lack the enzyme HGPRT and produce little or no unwanted immunoglobulins are used. Such cells cannot grow in a medium that contains hypoxanthine, aminopterin and thymidine (HAT medium), as they are incapable of using the salvage pathway. The spleen from a mouse immunized with the desired antigen is used as a source of B lymphocytes. These cells are fused with the HGPRT⁻ myeloma cells in the presence of polyethylene glycol. The fused cells are allowed to grow in the HAT medium. In this medium only hybrid cells are able to grow. The spleen cells cannot grow. Lymphocytes and fused lymphocytes die because they cannot grow in vitro. Myeloma cells (HGPRT) and fused myeloma cells are poisoned by aminopterin and hence cannot grow. The clones producing the desired antibody are subcultured and frozen to prevent contamination. Single clones could be obtained by carrying out limiting dilutions of the cultures in a suitable medium in microtitre plates with feeder cells (normal macrophages) or nutrient agar. The desirable clones could be used to produce large quantities of monoclonal antibody using cell culture techniques or by growing these cells as ascites (single cell) tumour in the peritoneal cavity of a suitable mouse.

Characterization of the Antibody

The recovered antibody should be characterized for specificity to the antigen (analyte) or in other words for its cross-reactivity toward the structural analogues of the analyte. In a polyclonal serum removal of the non-specific antibodies can improve specificity. The main differences between monoclonal and polyclonal antibody are given in **Table 1**.

Analytical Use of the Antigen–Antibody Reaction

The binding of an antigen to its antibody is accomplished by weak non-covalent forces. These include, hydrogen bonds, Van der Waals forces, and hydrophobic interactions. The binding of an antibody with its antigen, say a hapten with one antigenic determinant, can be represented by the following equation.

$$\text{Antibody} + \text{Antigen} \underset{k_d}{\overset{k_a}{\Leftrightarrow}} \text{Antibody–Antigen}$$

The k values are normally in the range 10^8–10^{10} l mol^{-1}. When an antigen is mixed with its antibody, at a suitable concentration of the antigen the antibody may form a precipitate. The precipitate may be isolated by centrifugation or could be visualized in a solid medium such as agar. In free solution this reaction between antigen and antibody is known as the precipitin reaction. The amount of antigen–antibody complex increases with increasing amount of antigen added. Further increase in antigen concentration may lead to solubilizing the precipitate. Therefore, only in the zone of equivalence is the maximum precipitate formed. It is possible to carry out both qualitative and quantitative analysis of an antigen using this principle. However, precipitin reaction requires standardization of the zone of equivalence, thus consuming large amounts of antigen and antisera. Use of a solid medium to visualize the precipitin reaction with the development of immunodiffusion techniques was later found to be simpler and reliable. Simple immunodiffusion usually involves the diffusion of the antigen from a solution into a gel containing antibody. In double immunodiffusion also known as the Ouchterlony technique, both antigen and antibody placed in opposite wells move toward each other.

When the antigen is particulate, as in the case of a bacterium, the reaction is known as agglutination, which can be easily visualized.

It was realized in 1960 that an antigen or antibody could be labelled with a radioisotopic tracer allowing detection and quantification of substances at very low levels. The technique, known as radioimmunoassay, became a powerful analytical tool. Later, in 1971, the use of enzyme labelling of an antigen or antibody led to the development of enzyme immunoassays which are becoming increasingly popular.

With an adequate method to separate the antibody–antigen complex from the free antibody or antigen and a suitable detection method for the marker present on either the antigen or antibody, it is possible to determine with ease very low concentrations of the

Table 1 Comparative features of polyclonal and monoclonal antibody

Characteristics	Polyclonal	Monoclonal
Animals used	Rabbit, sheep, goat, horse	Mouse
Antigen purity	To make a specific antibody highly pure antigen is required	Not important
Time required for production	1–2 months	2–3 months
Nature of antibody produced	Various classes (IgG, IgM and so on) of antibody are produced	Single class of antibody is produced
Purity of antibody	Contains many antibodies recognizing many determinants on an antigen	Contains single antibody recognizing only a single determinant
Concentration produced	Up to mg ml^{-1}	Up to a few μg ml^{-1} in culture, and mg ml^{-1} in ascites fluid
Cross-reactivity	Positive	Absent
Affinity	Heterogeneous	Homogenous
Cost of production	Low	High
Reproducibility and standardization	Difficult	High

analyte (antigen) in a given sample. A variety of solid supports have been developed for carrying out the immunoassays. These include polystyrene, polyvinyl chloride, polycarbonate and copolymer beads as well as microtitre plates.

Radioimmunoassay (RIA)

The RIA technique was first described in 1960. The Nobel Prize for Medicine in 1977 was awarded to Rosalyn Yalow for her invention of the RIA technique. Radioimmunoassays were the earliest of the immunoassays developed for the estimation of small quantities of biological substances.

RIA is a limited reagent heterogeneous immunoassay requiring separation of the antibody-antigen complex. The assay is basically carried out in antigen capture format. The two reagents used in the assay are the antibody specific to the antigen and the radiolabelled antigen. Both these reagents are used in a limited and constant concentration. The other component of the assay, the sample (analyte), may vary in concentration. The antibody is coated on a solid support or in a microtitre plate well. The radiolabelled antigen and the unlabelled antigen from the sample thus compete for the same sites on the antibody immobilized on the solid support as shown in the following scheme:

Ag* Ag*   + Ag Ag   + Ab Ab Ab ⟶

Labelled   Unlabelled   Antibody on
antigen   Antigen   solid support

                                      Ag Ag* Ag
       Ag   + Ag*   + Ab Ab Ab

       Free labelled and   Bound label is
       unlabelled antigen   estimated
       washed out

It is clear that in the absence of unlabelled antigen all sites will be occupied by the labelled antigen. As the concentration of the Ag in the sample increases less and less number of Ag* molecules will be able to bind the immobilized antibody. Hence the amount of activity bound to the antibody on the solid phase will be inversely proportional to the concentration of the analyte in the sample. By using different concentrations of the authentic analyte in the assay a standard curve could be prepared. This standard curve is used for estimating the concentration of the analyte in a given sample.

Immunoradiometric Assays

An alternate technique called the immunoradiometric assay (IRMA) which is capable of providing higher sensitivity than RIA was developed later. In IRMA, the antibody is radiolabelled instead of the antigen. IRMA is essentially an excess reagent assay in which an excess concentration of a radiolabelled antibody is used as the reagent. An excess concentration of a labelled antibody and the antigen (either from the standard or sample) are allowed to react. At the end of the assay, the antigen bound and free antibody is separated and the antigen-bound fraction is assayed for radioactivity. The activity associated with this fraction is directly proportional to the concentration of the antigen (analyte). The IRMA technique using the above principle found hardly any use. This was mainly due to two reasons. The first was the difficulty in getting a pure antibody, which is used for labelling. The second difficulty was in separating the antigen-bound and free antibody; both being of identical nature could not be easily separated.

The scope of IRMA widened following the development of 'hybridoma technology' developed in 1976 for the production of monoclonal antibodies. After the availability of monoclonal antibodies, a new type of IRMA called the 'two-site' IRMA or sandwich IRMA was developed. In this technique, two antibodies both specific for the same antigen, but binding to two different epitopes are used. One of the antibodies is coated on to the solid phase and used as an immuno-extractant for the antigen. A second antibody labelled with ^{125}I is added to the solid phase. This labelled antibody binds with the antigen, which is already bound to the first antibody. At the end of the reaction unreacted labelled antibody is aspirated out and the solid phase washed with a wash solution. The radioactivity in the solid phase is directly proportional to the concentration of analyte present.

Being an excess reagent technique, IRMA offers higher sensitivity than RIA. The use of two antibodies makes the two-site IRMA more specific than RIA. The IRMA assay can also be performed in a very short time.

The most commonly used radioisotope for RIA is ^{125}I which offers several advantages, for example it can be prepared with a very high specific radioactivity and with almost 100% isotopic abundance. It has a convenient half-life of 60 days and hence the tracer could have a long shelf life. Iodine can be easily introduced into many molecules. As ^{125}I decays by electron capture emitting low energy (35 keV) gamma photons, it does not damage the molecule. These gamma photons can be detected by using a simple solid scintillation counter having an NaI(TI) crystal.

^{125}I cannot be easily incorporated into small molecules, such as mycotoxins. Incorporating a pendant group, which can be labelled with ^{125}I, synthetically modifies the mycotoxins and these modified mycotoxins can be used as tracers in RIA. An alternative method, which was used earlier, was to label the mycotoxins with tritium (^{3}H), and use a liquid scintillation counter to measure the radioactivity emitted by this isotope. However, the use of tritiated tracers is not now practised.

The major disadvantages of radioisotopic immunoassays are:

- the high cost of equipment and reagents
- the shelf life of the labelled reagents is governed by the half-life of the radioisotope used
- the need to exercise precautions against hazards to personnel dealing with radioisotopes
- the disposal and environmental problems related to isotopes
- the need for skilled personnel for conducting the assay.

Enzyme Immunoassays

In an enzyme immunoassay an enzyme is used as a label on an antigen or antibody for the detection of immune reaction and estimation of the analyte. For this the enzyme is linked to either the antibody or antigen. Normally it is better to couple the enzyme to an antigen in the case of small molecular-weight substances. The catalytic properties of the enzyme and the specificity of the antibody should not be affected as a result of this coupling. The disadvantages in the use of radioisotopes have been successfully overcome with the use of enzymes as labels.

Enzymes used in Immunoassays

Enzymes can be coupled to different molecules without losing their catalytic activity. The large catalytic potential of an enzyme molecule gives an amplification effect to the assay. The following are the desirable characteristics of the enzymes required for immunoassays:

- The enzyme used should be pure with a very high specific activity.
- It should be stable at room temperature.
- The enzyme should have high turn-over number, i.e. it should act quickly on the substrate.
- The detection and measurement of the enzyme activity should be simple.
- It should not lose catalytic activity upon coupling to the antigen or antibody.
- It should be cheap and available in large quantities.

Some of the enzymes that fulfil the above require-

ments and have been very frequently used include horseradish peroxidase and calf intestinal alkaline phosphatase. Other enzymes such as alkaline phosphatase and β-galactosidase from *Escherichia coli*, glucose oxidase from *Aspergillus niger*, urease from soybean, catalase from bovine liver and penicillinase from *Staphylococcus aureus* have also been used in enzyme immunoassays (**Table 2**).

Coupling of an enzyme to a proteinaceous antigen could be carried out by either covalently or non-covalently linking the molecules. For linking the enzyme covalently a number of linking agents are used. These include glutaraldehyde, periodic acid and maleimide. Treatment of the enzyme with these agents results in the formation of active aldehyde groups, which can interact with the free amino groups of the antibody or antigen. Non-covalent linkages make use of the biospecific interaction of avidin with biotin molecules. In these assays biotin-labelled antibody is allowed to react with the antigen. An enzyme labelled with avidin is then added.

For labelling haptens the presence of a reactive group on the molecule can help in linking the molecule to an enzyme. In the absence of a reactive group such a group as carboxylic acid could be introduced in a hapten molecule.

Types of Enzyme Immunoassays

Homogeneous Assays   The enzyme immunoassays could be divided into two categories depending on whether the immune complex is separated from the free reactants or not. Those assays where there is no need to separate the two phases are called homogeneous assays. In homogeneous assays the enzyme coupled to an antigen or antibody is modified in its activity after the reaction. In such a case separation of the immune complex from the reaction mixture is not required. Such assays are used mainly in drug assays and are called enzyme multiplied immunoassays. The technique is also known as enzyme multiplied immunoassay technique (EMIT).

EMIT methods have proved very popular in drug assays. They involve formation of a drug–enzyme complex with a partial retention of the enzyme activity. Reaction of the complex with the antibody to the drug reduces the enzyme activity. When drug molecules to be assayed are introduced into the system, they compete with the drug molecules bound to the enzyme for the antibody, and the enzyme activity changes. The change in enzyme activity relates to the concentration of the drug.

Heterogeneous Assays   In the other type of immunoassays separation of immune complex from the reactants is an essential feature. In these assays which are

Table 2 Commonly used enzymes in enzyme immunoassays

Enzyme	Chromogenic substrate	Fluorogenic substrate	Optimum pH
Calf intestinal alkaline phosphatase (EC 3.1.3.1)	p-Nitrophenyl phosphate	4-Methylumbelliferyl phosphate	9–10
E. coli β-galactosidase (EC 3.2.1.23)	o-Nitrophenyl β-D-galactopyranoside	4-Methylumbelliferyl phosphate	6–8
Horseradish peroxidase (EC 1.11.1.7)	Tetramethylbenzidine or o-phenylenediamine with H_2O_2	p-Hydroxyphenyl acetic acid	5–7
Bean urease (EC 3.5.1.5)	Urea/bromocresol purple	–	6–7

called enzyme-linked immunosorbent assays (ELISA) antibody or antigen is bound either non-covalently or covalently to a solid matrix. The unreacted antigen or antibody is removed either by washing or centrifugation. The solid matrix could be a microtitre plate, cellulose or nitrocellulose membrane, polystyrene tubes or beads and nylon beads or tubes.

The Microtitre Plate The disposable plastic microtitre plate has been the most successful solid support used in the modern immunoassays. Microtitre plates are small plastic trays containing a fixed number of wells, 24, 48 or 96. The plates are made of polystyrene or polypropylene. When added to the well of the plate, an antibody or a proteinaceous antigen or a hapten conjugated with a protein binds to the inside surface of the well. The binding is due to the hydrophobic interaction between hydrophobic residues on the protein and the non-polar plastic surface. Once bound to the plastic surface, the bound protein cannot be washed out easily. After addition of the labelled antigen or antibody, depending on the nature of the label, enzyme or radioisotope, automatic plate readers are employed either to read the absorbency or amount of radioactivity. Because of the microtitre plates and plate reading instruments, it is possible to carry out immunoassays on a large number of samples routinely at relatively low costs.

Enzyme-linked Immunosorbent Assays (ELISA)

ELISA is a heterogeneous enzyme immunoassay. It is called competitive when the analyte is in excess and other reagents are constant and limited. When the reagents employed are in excess and there is no competition involved it is called non-competitive.

Competitive Assays

Direct Competitive ELISA The competitive ELISA can be essentially performed in two formats. In antigen capture format antibody is coated on to the solid phase. It is also called direct competitive ELISA. This technique can be used for the analysis of both the haptens and the macromolecular immunogens. As the name suggests the assay uses competition of two types of reagents, one labelled and the other unlabelled (analyte) for binding the antibody coated on a solid support. In principle it is similar to radioimmunoassay (RIA). However, in this procedure the antigen is labelled with an enzyme instead of a radioisotope.

The three major components of the system are:

1. A constant and limited amount of the antibody specific to the antigen.
2. A constant and limited amount of the antigen labelled with the enzyme.
3. A standard antigen of the known concentration or analyte (antigen) in the sample.

Ag Ag Ag   + Enz-Ag   + Ab Ab Ab   ⟶
               Enz-Ag

Antigen   Enzyme-labelled   Antibody coated
(Analyte)   antigen   on plate

Ag   + Enz-Ag   + Ag Enz-Ag Ag
                       Ab   Ab Ab

Free   Free enzyme-   Bound enzyme-
antigen   labelled antigen   labelled antigen

Removed after washing   Bound enzyme activity on plate is measured

When the three components of the assay are mixed, the enzyme labelled and unlabelled antigen compete for the limited number of binding sites on the antibody bound to the solid support. Since each test contains the same amount of antibody and labelled antigen, the quantity of unlabelled antigen determines the binding of labelled antigen to the antibody coated on the solid support. The less the quantity of the unlabelled antigen in the sample the more is the binding of the enzyme-labelled antigen to the antibody and vice versa. After separation of the unbound and bound fractions estimation of the bound enzyme activity is carried out by addition of the substrate of the enzyme and estimation of the coloured or fluorescent product formed. For quantification, appropriate standard curves can be prepared using an authentic sample.

Indirect Competitive ELISA

Indirect competitive ELISA is performed in antibody capture format and is therefore called indirect competitive ELISA. The proteinaceous antigen or an antigen (hapten)–protein conjugate is coated on to the solid phase. The analyte and the enzyme-labelled antibody are mixed separately and added to the well. The enzyme-labelled antibody either binds the free antigen (analyte) or the antigen bound to the solid phase. In other words both the free and the bound antigens compete for the sites on the enzyme-labelled antibody.

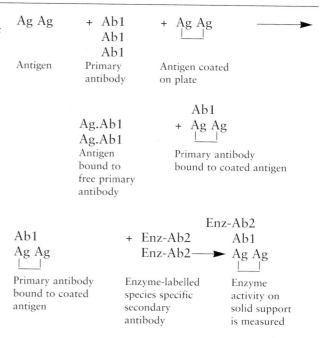

Double Antibody ELISA

In this form of ELISA both the primary reagents, the antigen and the antibody are not labelled. The standard antigen is coated on the solid phase. The antibody is then added along with the test sample containing the antigen. The antigen in the test sample and the antigen coated on the solid phase compete for the same antibody. After incubation the excess antibody is washed. A secondary antibody, an anti-antibody, that is an antibody to the primary antibody prepared in another species (species specific anti-immunoglobulin), and labelled with the enzyme is then used as a general reagent. The procedure requires an additional step. However, it obviates the need to label antigen specific antibody or the antigen itself, either of which may be available in limited concentration. In addition the same labelled anti-species specific antibody may be used in a number of different ELISAs, provided that the primary antibody is raised in the same species.

Non-competitive Assays

This form of ELISA does not use the competition principle for the assay. The non-competitive assay or non-competitive solid phase assay is another form of ELISA which is a reagent excess assay analogous to immunoradiometric assay IRMA. In the antigen capture format the specific antibody is coated to the solid phase. In both cases, the amount of enzyme label bound is directly proportional to the concentration of the antigen analyte. There are several forms of non-competitive ELISA. A majority of the assays use enzyme-labelled antibody as the tracer. Solid-phase antigen and the antigen-specific antibody are required for the binding and subsequent detection of the antigen or antibody, respectively.

Sandwich of Two-site ELISA

This assay is basically performed in the antigen capture format. The antibody to the specific antigen is coated on a microtitre plate. The antigen to be detected in a sample is then added to the well. The antigen is trapped by the antigen binding site on the antibody. The wells are then washed to remove the unbound antigen. A second antibody conjugated with a marker enzyme is then added. The second antibody is also specific to the antigen, but it binds another antigenic determinant on the surface of the antigen. After washing, only the bound enzyme-labelled antibody remains in the well. The enzyme substrate is now added to the wells. The colour formed is proportional to the amount of antigen (analyte) present.

In this process the antigen is trapped between two layers of antibodies. Thus the method is also some times called sandwich or two-site ELISA. The sandwich-type ELISA can be used only if the antigen has more than one antigenic determinant situated on opposite sides of the molecule, and the antibodies can be raised specifically against at least two such antigens. Moreover, the antibody should have equal affinity to both the determinants, without stearically hindering each other.

$$Ag\ Ag \quad + \ Ab1\ Ab1 \longrightarrow \begin{matrix} Ag & Ag \\ Ab1 & Ab1 \end{matrix} +$$

| Antigen | First antibody coated on solid support | Bound antigen |

$$\begin{matrix} Enz\text{-}Ab1' \\ Enz\text{-}Ab1' \end{matrix} \longrightarrow \begin{matrix} EnzAb1' & Enz\text{-}Ab1' \\ Ag & Ag \\ Ab1 & Ab1 \end{matrix}$$

| Second labelled antibody | Bound enzyme activity is measured |

Applications of Enzyme Immunoassays in Food Analysis

Enzyme immunoassays are finding increasing application in the food industry. These assays have been used in the various stages of food production and processing.

Food Production and Processing

In the area of food production, the role of enzyme immunoassay can begin with a check on the initial seed quality. Immunoassays have been used to detect the presence of seed-borne pathogens and certification of pathogen-free seeds. On a variety of agricultural and horticultural produce the levels of herbicides such as 2,4-dichlorophenoxyacetic acid (2–4 D), alachlor, paraquat, maleic hydrazide and plant growth regulators, such as cytokinins, gibberellins, abscic acid, can be monitored through immunoassays. Constituents that provide processing characteristics to food could also be monitored using immunoassays. For example, it is possible to check globulin, gluten, gliadin and glutenin content of a cereal using these assays. The activity of food enzymes and their inhibitors can be monitored during production and processing of food.

The presence of spoilage-causing microorganisms on food can be detected using immunoassays. Immunoassays have been developed for detection of food-borne moulds, bacteria and viruses. The presence of disease-causing pathogens and parasites in animal products could also be detected. Immunoassays are available for the detection of *Toxoplasma gondii* in sheep and rabbit meat, *Trichinella spiralis* in pork, *Mycobacterium tuberculosis* in goat and *Brucella abortus* in cattle.

Monitoring of Food Constituents

Enzyme immunoassays have been employed for the detection of constituents that impart flavour to foods and improve food quality. Surveillance of undesirable constituents occurring naturally in foods can be monitored through immunoassays to improve food quality. Enzyme immunoassays have also been developed to monitor the nutritionally important constituents of food such as vitamins, essential amino acids and fatty acids. Immunoassays have been employed to check the authenticity of fruit juices and soft drinks.

Food Safety

Enzyme immunoassays are playing an increasingly important role in the management of food quality and food safety. Enzyme immunoassays have been developed for the detection of almost all the important pathogens and microbial toxins in food. These include *Clostridium botulinum* toxins, and enterotoxins of *Clostridium perfringens* and *Staphylococcus aureus*, mould toxins (mycotoxins), such as aflatoxin, ochratoxin, trichothecenes, fumonisins and algal toxins such as ciguatoxin and saxitoxin involved in seafood poisoning. Enzyme immunoassays have also been developed for the detection of pathogenic microorganisms such as *Salmonella*, *Listeria monocytogenes*, *Campylobacter jejuni*, *E. coli* and *Yersinia enterocolitica*.

The presence of residues of harmful insecticides and pesticides in agricultural and horticultural produce, anabolic steroids and therapeutic agents in animal products and adulterants in food and meat (interspecies meat) can also be detected by enzyme immunoassays.

Immunoassays kits are now available for detection of several food contaminants and adulterants. Immunoassay kits offer ease of operation under a well standardized procedure. They are less expensive and less time consuming. Precision, accuracy and reproducibility of results from immunoassay kits need to be validated through collaborative trials. Efforts are under way to validate and approve these kits for daily use in the food industry in the various stages of food production and processing and application in the HACCP procedures.

See also: **Campylobacter**: Detection by Cultural and Modern Techniques. **Clostridium**: Detection of Neurotoxins of *C. botulinum*. **Escherichia coli**: Detection of Enterotoxins of *E. coli*. **Immunomagnetic Particle-based Techniques**: Overview. **Listeria**: Detection by Commercial Enzyme Immunoassays. **Mycotoxins**: Immunological Techniques for Detection and Analysis. **Rapid Methods for Food Hygiene Inspection**. **Salmonella**: Detection by Enzyme Immunoassays; Detection by Immunomagnetic Particle-based Assays. **Staphylococcus**: Detection of Staphylococcal Enterotoxins. **Verotoxigenic *E. coli***: Detection by Commercial Enzyme Immunoassays. **Vibrio**: Detection by Cultural and Modern Techniques. **Water Quality Assessment**: Modern Microbiological Techniques.

Further Reading

Butt WR (ed.) (1984) *Practical Immunoasays*. New York: Marcel Dekker.

Chard T (1990) *An Introduction to Radioimmunoassays and Related Techniques*. Amsterdam: Elsevier.

Hefle SL (1995) Immunoassay fundamentals. In: *Immuno-assay Applications to Food Analysis, Food Technology* 49, 102–107.

Kohler G and Milstein C (1975) Continuous culture of fused cells secreting antibody of definite specificity. *Nature* 256: 495–497.

Samarajeeva U, Wei CI, Huang TS and Marshall MR (1991) Application of immunoassay in food industry. *CRC Critical Reviews in Food Science and Nutrition* 29: 403–434.

van Weeman BK and Shuurs AHWM (1971) Immunoassay using antigen enzyme conjugate. *FEBS Letters* 15: 232–236.

Wyatt GM, Lee HA and Morgan MRA (1995) *Immunoassays for Food Poisoning Bacteria and Bacterial Toxins*. London: Chapman & Hall.

ESCHERICHIA COLI

Contents

Escherichia coli

Carl A Batt, Department of Food Science, Cornell University, USA

Copyright © 1999 Academic Press

Characteristics of the Species

Escherichia coli is a species whose importance ranges from its role as a host for recombinant DNA manipulations to being one of the most well-recognized food-borne pathogens. The former will not be discussed in this article: it is a well-studied host for laboratory purposes and the magnitude of its usage for the production of food-related ingredients is difficult to assess accurately. *E. coli* is a Gram-negative rod that is a member of the family Enterobacteriaceae. It is oxidase-negative and grows using simple carbon sources, including glucose and acetate. The hexose is fermented to a mixture of acids (lactate, acetate and formate) as well as carbon dioxide. *E. coli* are citrate-negative but methyl red-positive and Voges–Proskauer-negative. It is classified as a coliform – a general term used to describe Gram-negative asporogenous rods that ferment lactose within 48 h and whose colonies are dark and exhibit a green sheen on agar such as eosin methylene blue. Aside from *Escherichia*, other genera which are termed coliforms include *Citrobacter*, *Enterobacter* and *Klebsiella*.

The presence of peritrichous flagella gives *E. coli* mobility and the flagella are also part of the serology of this organism (see below). It is a normal inhabitant of the gut of many animals, including human beings. As such, it is often used as an indicator of faecal contamination. Not all strains of *E. coli* cause disease, however, and as a consequence the detection of *E. coli* in a food, while implying a potential hazard, does not a priori mean that the food will cause illness if consumed. Of note among the *E. coli* strains is the serotype O157:H7. This serotype, which includes highly virulent strains, has been the focus of much attention over the past 10 years not only because of its association with a number of highly publicized food-borne outbreaks, but also because of its ability to survive acidic conditions which were previously believed to be lethal to *E. coli*.

Serology

Serological distinction between strains of *E. coli* is a very important tool applied for tracking clinical isolates back to their food sources in food-borne disease outbreaks. One serotype, O157:H7, is perhaps one of the best-known strains of any food-borne bacteria. Historically the efforts to develop a serotyping scheme for *E. coli* followed efforts to establish a system for *Salmonella*. Serotyping is based on three fundamental antigens, O, K and H, and distinguishing serotypes for each of these antigens exist. The initial group of antigens discovered by Kauffmann consisted of 25 O, 55 K and 20 H antigens. The O antigen is based on a polysaccharide moiety which is associated with the outer membrane. This oligosaccharide is covalently linked to the lipid A-core polysaccharide and the repeating units define the diversity of the O antigen group. Due to the extreme heterogeneity in the five or more sugars comprising the O antigen, more than 170 different O groups have been discovered to date. The O antigens are broadly dispersed in a number of other related microorganisms and as a result there is cross-reactivity. For example, the O antigens of *E. coli* cross-react with certain O antigens on *Shigella* and *Salmonella*. Almost all O antigens found in *Shigella* cross-react, with the exception of some found in *Shigella sonnei*. The consequence of this cross-reactivity is that many antibody-based tests that broadly detect *E. coli* frequently generate false positives due to cross-reactivity with O antigens of other microorganisms. Fortunately, antibody-based tests for the detection of *E. coli* O157:H7 specifically perform well due to the unique nature of the O157:H7 serotype.

The K antigens are also polysaccharides in nature and part of the cell capsule. The polysaccharide is mainly acidic and heat-labile to varying degrees. This group is less complex and only three K antigens have been reported – A, B and L. While the A type K antigen requires 121°C for 1 h in order to be inactivated, B and L are inactivated at 100°C. Unlike the O and the H antigen (see below), the K antigen is not used in most typing schemes. However, there are a few K antigens that are sometimes used for typing purposes because of their association with particular diarrhoea-causing strains. These include the K88 and K99 antigens which are associated with diarrhoea in pigs. The K99 antigen is also associated with diarrhoea in calves and lambs.

The H antigens are part of the flagella and hence found only in motile strains of *E. coli*. Most *E. coli* are non-motile or partially motile on initial isolation from the environment. As a consequence, H antigen typing is not reliable unless efforts are taken to select for the restoration of motility. Enrichment for motility and hence production of the H antigen can be achieved by selective culture in soft agar. When a strain fails to display motility it is labelled non-motile (NM), and this is used as a descriptor for *E. coli* strains. To date, more than 50 H antigens have been discovered.

The three antigens were initially used to define a particular serotype of *E. coli* and hence nomenclature such as O26.K60(B6).H111 was used. The K antigen descriptor has however been dropped as a descriptor of serotypes and only the H and O are commonly employed. The H antigen coupled to the O antigen therefore represents a robust and highly discriminatory typing method for distinguishing various strains of *E. coli*. The O:H serotypes can be sorted into various virulence groups (e.g. EPEC: see below)

and also categorized with respect to the host animal. For example, O157:H7 is associated with enterohaemorrhagic forms of disease in humans, whereas the O55:H7 is associated with the enteropathogenic forms of the disease, also in humans.

Virulence

The ability of some strains of *E. coli* to cause disease was known in the early 20th century, and infant diarrhoea was one of the first illnesses recognized to be caused by *E. coli*. There are four major classes of disease caused by *E. coli* and they have distinct patterns of illness as well as different virulence factors. The most common is enteropathogenic *E. coli* (EPEC), associated with infant diarrhoea. Other virulence groups include enteroinvasive *E. coli* (EIEC), enterotoxigenic *E. coli* (ETEC) and enterohaemorrhagic *E. coli* (EHEC). More recently, other groups, including enteroaggregative (EaggEC) and diffusely adherent *E. coli* (DAEC), have been described. The virulence class of *E. coli* strains has some correlation to the serotype. However, this is not absolute as serotype 111, for example, is found among EaggEC, EHEC and EPEC *E. coli* strains (**Table 1**).

Table 1 Distribution of O serotypes among the different virulence groups of *E. coli*

EaggEC	EHEC	EIEC	EPEC	ETEC
3	2	28ac	18ab	6
4	5	29	19ac	
6	6	112a	55	15
7	4	124	86	20
17	22	135	111	25
44	26	136	114	27
51	38	143	119	63
68	45	144	125	78
73	46	152	126	80
7577	82	164	127	85
78	84	167	128ab	101
85	88		142	115
111	91		158	128ac
127	103			139
142	113			141
162	104			147
	111			148
	116			149
	118			153
	145			159
	153			167
	156			
	157			
	163			

EaggEC, Enteroaggregative *E. coli*; EHEC, enterohaemorrhagic *E. coli*; EIEC, enteroinvasive *E. coli*; EPEC, enteropathogenic *E. coli*; ETEC, enterotoxigenic *E. coli*.
Reproduced from Jay (1996) with permission.

Enteropathogenic *E. coli*

The enteropathogenic strains of *E. coli* are similar to the enteroaggregative in that they can adhere to cells, specifically the intestinal mucosa. There they produce an attaching and effacing lesion in the brush border microvillous membrane. They can also attach and efface epithelial cells. The attachment and effacement process is the work of a chromosomally encoded gene, *eaeA*. In general, these strains do not produce enterotoxins but can cause diarrhoea.

Enterohaemorrhagic *E. coli*

The enterohaemorrhagic *E. coli* are able to cause one of the most severe forms of disease, ultimately resulting in haemolytic–uraemic syndrome (see below). These strains have the ability to produce adherence factors, enterohaemolysins and Shiga toxins. A detailed description of the toxins is given below. Like other *E. coli* strains, the enterohaemorrhagic strains carry a large plasmid that encodes fimbriae which are involved in the attachment of the bacteria to cells in culture. However, these strains do not appear to invade Hep-2 cells. The prototypical strain is *E. coli* O157:H7 and illness caused by this serotype is associated with the consumption of a wide variety of foods, including minced beef, turkey rolls, water, vegetable salads and apple cider.

Enteroinvasive *E. coli*

Enteroinvasive strains of *E. coli* cause a severe form of disease and spread between cells in a manner similar to *Shigella*. They do not typically produce enterotoxins but carry a large plasmic that is associated with their enteroinvasive properties. The classification of the strains is based upon a positive result in the Sereny test. As mentioned below, this test assesses the ability of strains to invade and cause disease in guinea-pig eyes. In some cases enteroinvasive strains are isolated from patients with diarrhoea and subsequent virulence testing shows them to be enteroinvasive.

Enterotoxigenic *E. coli*

The enterotoxigenic strains of *E. coli* were among the first to be recognized as a result of their association with traveller's diarrhoea. A variety of names have been associated with the disease, including gypsy tummy, Delhi belly, Hong Kong dog and Aden gut. Disease can afflict the young and the old: symptoms are largely restricted to diarrhoea without fever. The enterotoxins associated with these strains are described fully below. They include heat-stable and heat-labile enterotoxins. Strains appear to be distinct in their association with different animal hosts, with humans, pigs and cattle being examples of the populations reported to date. Enterotoxigenic *E. coli* also produce fimbrial colonization factor antigens. These are plasmid-encoded and are typically found in association with the heat-stable enterotoxin.

Enteroaggregative *E. coli*

Strains which are characterized as enteroaggregative are able to adhere to cultured cells and are associated with both acute and persistent diarrhoea. The persistent form of the diarrhoea can last up to 14 days. Enteroaggregative *E. coli* adhere to Hep-2 cells, forming microcolonies. In general, however, different types of adherence patterns ranging from diffuse to localized have been observed. A 90 kb plasmid is associated with the production of a specific outer membrane protein and for the production of fimbriae. In addition some strains produce a heat-stable enterotoxin (EAST1) which is also plasmid-encoded.

Toxins

The various disease-producing *E. coli* generate a particular pattern, in part due to their production of one or more toxins. A number of these toxins have been characterized and both their biochemical nature and their genetics have been elucidated. The classification scheme used for *E. coli* toxins is based on their physical characteristics and their target. The first division is on the basis of heat stability and therefore heat-labile and heat-stable toxin groups have been established.

Heat-stable Enterotoxins

A number of different heat-stable toxins have been discovered and the catalogue of genetic variants continues to increase. Broadly classified into two different groups, STI and STII, these classes of toxins are distinct in their size and presumably in their mode of action. The STI toxins are approximately 2 kDa and retain activity even after heating to 100°C for 15 min. They are resistant to the actions of many proteases but not to treatment with alkali. The genes coding for STI are typically carried on a large plasmid located in conjunction with other genes necessary for virulence. STI appears to act by stimulating the host expression of guanyl cyclase which in turn causes a rapid efflux of fluid due to the production of cyclic guanosine monophosphate (cGMP). This fluid efflux causes an imbalance and hence the symptoms associated with *E. coli* food poisoning, including diarrhoea. STII is smaller than STI: it contains only 48 amino acids. Its mode of action is not clear but does not involve cGMP accumulation, although fluid efflux has been observed in a mouse model system. As with STI, this enterotoxin is also plasmid-mediated.

Heat-labile Enterotoxins

The heat-labile enterotoxins are characterized by their ability to be inactivated by heating, but more broadly they are distinct from the heat-stable enterotoxin in terms of their structural composition and their mode of action. The LT-I enterotoxin is an oligomer composed of a single 88 kDa subunit and five 11.5 kDa B subunits. The B subunits are organized in a doughnut-like configuration and assembly occurs as the proteins are secreted from the cell. The B pentamer binds to intestinal cell membrane, specifically via the GM1 gangliosides. The A subunit is then activated upon entry and causes elevated levels of cyclic adenosine monophosphate (cAMP). The increase in cAMP then results in secretion of chloride ions and impaired absorption of sodium ions, giving rise to severe diarrhoea. The LT-II toxin is very similar to LT-I, with the exception that they can be distinguished serologically. The structural genes for LT-I are plasmid-encoded, while the LT-II are chromosomally encoded.

Other Toxins

Enterohaemorrhagic E. coli express one or more cytotoxins which cause the characteristic lysis of red blood cells. The E. coli cytotoxins are variously referred to as Shiga toxins or Vero toxins. The latter is derived from their cytotoxin effect on Vero cells, while the former reflects the close homology between the E. coli cytotoxins and those produced by Shigella. The Shiga toxin produced by Shigella dysenteriae type 1 is the likely progenitor for all of the E. coli Shiga toxins. The Shiga toxin, (Stx-I or VT1) is composed of a 33 kDa A subunit and five 7.5 kDa B subunits. Therefore it is very similar in architecture to the E. coli LT-I and LT-II. The B subunits recognize the receptor while the A subunit possesses the activity which is activated upon proteolytic cleavage. The Stx-I recognizes the globotriasylceramide (Gb3) receptor which is found on renal epithelial cells, platelets and erythrocytes. The genes coding the Stx-I are encoded by a temperate bacteriophage suggesting its modes of transfer from Shigella to E. coli.

The Shiga toxin II (Stx-II or VT2) is similar to Stx-I; however, they are distinct in terms of the ability of toxin-specific sera raised against one to inactivate the other. The sequence homology between Stx-I and Stx-II is 57% for the A subunit and 60% for the B subunit.

Other toxins include a cytolethal distending toxin. Strains which express it produce diarrhoea in pigs. The Vir toxin is lethal in mice and may cause bacteraemia. It is encoded on a plasmid which also encodes pili. Finally, some E. coli produce a cytotoxic necrotizing factor.

Food-borne Illness

Food-borne illness which results from the consumption of foods contaminated with pathogenic strains of E. coli can take various forms. The enteropathogenic forms of the disease generally take from 5 to 48 h to develop after food consumption. The onset of disease is a function of the strain as well as the numbers of E. coli consumed by the victim. In general, the symptoms include severe abdominal pain.

When disease involves the enteroinvasive and haemorrhagic forms of E. coli, the symptoms are much more severe and the outcome much more serious. Symptoms usually begin approximately 10–24 h after consumption of the contaminated food. Pain is usually accompanied by diarrhoea and the diarrhoea may be bloody. Other symptoms include nausea, vomiting, fever, chills, headache and muscular pain. As the haemorrhagic form of the disease progresses, bloody urine may be passed. This stage of the disease is termed haemolytic–uraemic syndrome and involves haemolytic anaemia, thrombocytopenia and acute renal failure. To prevent the colitis stage from advancing into haemolytic–uraemic syndrome, patients are sometimes infused with therapeutic agents to inactivate the cytotoxin. Patients who reach the haemolytic–uraemic syndrome stage may suffer permanent damage or may not survive.

Historically, outbreaks of illness caused by E. coli date back to the 1940s when the first isolation of the H7 serotype was reported. Haemolytic–uraemic syndrome, the most severe symptoms associated with illness caused by E. coli, was initially recognized in 1955. The first reports of food-borne illness from E. coli in the US date back to 1971 when there was an outbreak from consumption of imported cheese. The illness involved approximately 400 individuals. The current public and governmental awareness of E. coli can be traced to the 1982 outbreak in Oregon when approximately 43 patients fell ill after consuming food prepared at a 'fast food' establishment. Subsequent outbreaks, which predominantly involve the E. coli O157:H7 serotype, have been reported in foods including minced beef, cheese, sprouts, salami and apple cider. This latter food is a particular source of E. coli up until that time the general belief was that it could not survive in this acidic environment.

Detection of *E. coli*

The detection of E. coli is complicated by its similarity to other enterics, especially when a variety of cultural methods are used for isolation and characterization. *E. coli* is closely related to *Shigella* and was initially distinguished on the basis of the diseases they produced. *Shigella* is the cause of bacillary dysentery, while virulent strains of *E. coli* can be responsible for a variety of diseases: some strains do cause dysentery-like symptoms.

E. coli is also used as an indicator of potential faecal contamination. Indicators are broadly defined as certain genera or classes of microorganisms that inhabit the same reservoirs, have the same survival rates but can be detected more easily and readily than the corresponding pathogen. Therefore, for example faecal coliforms are used as indicators of sanitation. *E. coli* is a subset of the faecal coliforms and it may be that their use as an indicator of food safety may be less prone to false-positive results than frequently occur with faecal coliforms. Faecal coliforms include microorganisms, e.g. *Klebsiella*, that are typically associated with plant material and therefore they are normal flora of many plant-derived foods and ingredients.

Culture-based Methods for Isolation

All culture-based methods for *E. coli* typically consist of recovery/enrichment in broth followed by selective plating on a medium that also contains a biochemical indicator. These first two stages take about 48–72 h and at this stage a presumptive identification of *E. coli* can be obtained. The subsequent characterization and confirmation of a particular isolate as *E. coli* require specific biochemical tests that probe for catabolism of specific sugars and the production of particular end products. Even at this stage, the presence of *E. coli* is not conclusive with respect to its ability to cause disease. To confirm this, either molecular tests for the presence of a particular toxin (e.g. heat-stable enterotoxin) or a cytotoxicity test is in order. The latter, while being more definitive, is difficult to carry out on a routine basis and the specific nature of the virulence is difficult to assess.

The detection of *E. coli* through traditional cultural methods may begin with an examination for coliforms. There are various methods to test for coliforms and these are typically direct plating methods or testing for acid/gas production from lactose. Detection of coliforms and, more specifically, *E. coli* involves an initial homogenization step in a diluent. A typical diluent is Butterfield's phosphate buffer and a 1:10 or greater sample:diluent mix is recommended. Another diluent frequently used is maximum recovery diluent which is 0.1% peptone and 0.85% saline. Milk and other liquid food products can often be tested without extensive sample processing. The most widely used process for sample

preparation is a Stomacher and samples of 25 g. This sample is added to 225 ml of diluent and stomached for at least 30 s. The initial dilution is therefore 10^{-1} and subsequent dilutions can be made in the same diluent.

E. coli belongs to the broader group of coliforms which are characterized by being Gram-negative, rod-shaped and facultatively anaerobic. They produce gas from glucose and can also ferment lactose to acid while producing gas. The ability to utilize lactose is not universal and there are some strains of *E. coli* whose lactose fermentation is weak. Coliforms that do ferment lactose produce acid and gas within 48 h at 35°C. The ability to produce acid and gas from lactose at 45.5°C is restricted to the faecal coliforms and, more narrowly, *E. coli*.

Presumptive coliforms/*E. coli* A quantitative test for coliforms employing a most probable number (MPN) approach has a number of variations, including the type of media and incubation temperatures. For example, lauryl sulphate tryptose (LST) is inoculated with a set of serial dilutions and incubated at 35°C for 24 and 48 h. Gas production is monitored using an inverted Durham tube and the tubes that are positive for gas are used to calculate the MPN of the sample. Confirmation can be carried out using brilliant green bile broth (BGBB; see below).

Coliforms can be detected by direct-pour plating using violet red bile agar (VRBA), on which red colonies are observed. This can be followed by inoculation into BGBB and then scoring for gas production using an inverted Durham tube at 30 or 37°C after 24 h. Alternatively, coliforms can be tested using VRBA or MacConkey agar with the pink-red colonies selected for further testing. Coliforms, faecal coliforms and *E. coli* type I will produce gas in BGBB at 37°C, but only the faecal coliforms and the *E. coli* type I will produce gas at 44°C. Indole production, which can be tested using Kovac's reagent, is indicative of *E. coli*, although some non-faecal coliforms also produce indole.

Petrifilm (3M, St Paul, MN) is an alternative to VRBA pour plates and reduces the volume of the incubated space typically needed for standard agar Petri plates. The plastic film is hydrated with water and then the diluted sample is applied. After incubation at 32°C, the positive colonies are red. In addition, other products contain chromogenic substrates to screen for glucuronidase activity.

Confirmed *E. coli* The above tests result in either a presumptive coliform or a presumptive *E. coli* test. Any positives must be confirmed by further examination to determine if *E. coli* is present. For example, any positive LST tubes can be further examined by inoculation into EC broth. The EC tubes are incubated at 45.5°C and scored for gas production after 48 h. The positive EC cultures can then be streaked on to eosin methylene blue plates and examined 24 h later for the characteristic nucleated dark-centred colonies. A green sheen is sometimes, but not always, observed. Any positives at this stage need to be examined using a battery of biochemical tests, including tryptone broth (indole production), methyl red–Voges–Proskauer and Koser citrate broth.

A more direct broth test for *E. coli* involves the incorporation of a fluorogenic dye, 4-methylumbelliferyl-β-D-glucuronic acid (MUG) into the medium. This dye is non-fluorescent in its intact state but the fluorophore is released due to the action of β-glucuronidase. MUG hydrolysis can be detected using a small hand-held UV lamp. Approximately 94% of the *E. coli* strains tested are MUG-positive, indicating the presence of β-glucuronidase. When MUG is incorporated into a selective medium, such as LST or EC, it can be used as an effective screen for the presence of *E. coli* in foods. Some *Salmonella* (29%) and *Shigella* (44%) also hydrolyse MUG; therefore, caution must be applied to the case of any positives. Incorporation of MUG into medium used to support the growth of bacteria on hydrophobic grid membrane filters (HGMF; QA Life Sciences, San Diego) allows the screening of colonies at a much higher density than what might be accomplished with standard agar plate. The HGMF filters have discrete cells formed by a hydrophobic material which is arrayed as a grid. Under UV illumination, *E. coli* grown on HGMF filters with buffered MUG medium fluoresces (**Fig. 1**).

The final tests for differentiation of *E. coli* are varied and can be accomplished by single-tube biochemical assays (e.g. mannitol fermentation). More elaborate approaches, using a microbiochemical test strip either in a manual or an automated mode, can also be employed (e.g. BioMérieux API, Roche Enterotubes, Vitek GNI card). An example of the types of biochemical tests that would distinguish *E. coli* from other *Escherichia* species is presented in **Table 2**.

Isolation of EHEC *E. coli*

The above methodology covers the isolation of *E. coli* through the prerequisite stages of presumptive coliform tests followed by confirmation. While this is satisfactory, more direct methods to isolate potentially virulent *E. coli* strains have been developed. As mentioned elsewhere, the isolation of *E. coli* O157:H7 is of particular importance because of its association with disease. The culture isolation of EHEC *E. coli* begins with homogenization of the sample in peptone water. The homogenized sample is then diluted and plated on sorbitol–MacConkey agar. After 18 h at 35°C, colonies which are pale in comparison with the bright pink colour generally exhibited by enterics are then selected. Recently the absolute correlation between the inability to ferment sorbitol and the O157 serotype has been challenged. Sorbitol-positive *E. coli* O157 but H7 serotype isolates have been recovered. A small fraction of *E. coli* non-O157:H7 are also sorbitol-negative. Therefore strict reliance on the sorbitol-negative phenotype is not appropriate. Further confirmation of sorbitol-negative colonies can be carried out using MUG, as described previously. Most, but not all, *E. coli* O157:H7 are unable to hydrolyse MUG. A further modification in which tellurite and cefixime are added to sorbitol–MacConkey agar has been demonstrated to be useful for the direct isolation of *E. coli* O157:H7 from foods.

Virulence Testing

As mentioned previously, *E. coli* is a normal inhabitant of the gastrointestinal tract of many animals. Therefore, although its presence indicates the contamination of a food by faecal material, it does not imply that the contaminated food would cause illness if consumed. Actual virulence testing is complicated and requires either cell culture or animal testing. For example, enteroinvasive *E. coli* can be tested using the Sereny test which employs guinea-pigs whose eyes are inoculated with a suspension of the test organism. After 5 days the eyes are examined for the development of conjunctivitis, ulceration and opacity. One eye serves as the control for each animal. Enterotoxigenic *E. coli* can be tested using Y-1 mouse adrenal cells which are grown in culture and then examined for the conversion of elongated fibroblast-like cells into round refractile cells. This phase conversion is a result of the elevated production of cAMP that occurs in the presence of the enterotoxin. Enterohaemorrhagic *E. coli* can be similarly tested using a cell-tissue culture system. In this test, Vero cells are

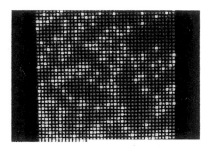

Figure 1 *E. coli* grown on a hydrophobic grid membrane filter on buffered 4-methylumbelliferyl-β-D-glucuronic acid agar and photographed under long-wave UV light. (Courtesy of Phyllis Entis QA Life Sciences.)

Table 2 Biochemical tests that differentiate *Escherichia* species

Reaction	E. coli		E. hermanii	E. blattae	E. fergusoni	E. vulneris
	Typical	Inactive				
IMVic	++--	++--	++--	-+-±	-+--	-+--
KCN	-	-	+	-	-	+
Glucose, gas	+	-	+	+	+	
Lactose	+	-	+/-	-	-	+
Cellobiose	-	-	+	-	+	
Adonitol	-	-	-	-	-	+
Mannitol	+	+	+	-	+/-	+
Malonate	-	-	-	+	+	+/-

Reproduced with permission from Food and Drug Administration.

grown in cell culture and the monolayer is removed using trypsin. The filtrate from the *E. coli* test culture is then added to the Vero cells and the culture is examined daily for up to 4 days. The cytotoxic effect is manifested by detachment and shrivelling of the cells.

Molecular Methods for Detection

Interest in molecular methods for the detection of *E. coli* has been based on the prolonged time required to complete traditional culture methods used to detect pathogenic strains of *E. coli*. While the detection of generic *E. coli* can be accomplished using just a single selective/screening agar (e.g. sorbitol–MacConkey agar), confirmation of specific virulence factors requires a molecular-based method or a cytotoxicity test. Most of the recent interest in *E. coli* detection has focused on the O157:H7 serotype. For immunological-based detection this presents a unique opportunity as reagents are readily available that specifically react with this serotype. For nucleic acid-based detection, the challenge is linking the serotype to a genotype.

Immunological-based methods Immunological-based methods for the detection of *E. coli* can be applied at various levels in the culture-based methods or can potentially be used as a direct detection method. For example, antibody-assisted capture of target cells has been used as a selection system for a number of different assay systems. Again, given the availability of specific antibodies for the O157:H7 serotype, capture methods using magnetic beads have been developed. These methods employ paramagnetic beads that can be derivatized with antibodies (Dynal, Lake Success, NY) and then capture the target cells from solution. Upon recovery, these cells can be used for standard culture detection or other immunological or nucleic acid-based assays for *E. coli*.

As mentioned previously, analysis for toxins based on their biological activity is cumbersome and involves the use of either cell culture or animals.

A variety of immunoassays for specific toxins are commercially available. These assays usually target one or more of the toxins at various levels of specificity. For example, one assay is available which will detect the *E. coli* heat-labile toxin (LT) but also detects the *Vibrio cholerae* enterotoxin (VET-RPLA, Oxoid, Hampshire, UK). This assay is based on a reversed passive latex agglutination (RPLA) format. In this format a negative result appears as a tightly focused accumulation of latex beads at the bottom of a V-shaped well. In contrast, a positive result is a diffuse suspension of the latex beads. Similarly, an assay for the verotoxins VT1 and VT2 is available in the RPLA format. For all these assays, purified cultures are required and single colonies are used as the starting material.

An HGMF (QA Laboratories, San Diego, CA) method has been developed which employs an enzyme-conjugated monoclonal antibody. The HGMF is convenient for filter-concentrating bacterial cells which can then be propagated and/or probed using antibodies or nucleic acids. Specifically, the food sample is homogenized in peptone water (1 : 10 w/v dilution) and then homogenized. The homogenate is then diluted and filtered through a 100 μm prefilter on to the HGMF filter. The HGMF is then placed onto haemorrhagic coli (HC) agar and incubated at 43°C for 16–20 h. The filter is then probed using a monoclonal antibody conjugated to horseradish peroxidase against the O157 serotype. Positive colonies are purple-coloured after the addition of a colorimetric substrate (e.g. 4-chloronaphthol and hydrogen peroxide).

More elaborate instrumentation-based immunoassays are also available commercially for the detection of *E. coli*. For example, a sandwich assay involving magnetic beads coated with anti-*E. coli* O157:H7 antibodies in conjunction with ruthenium-labelled antibodies has been developed (Origen, Igen, Gaithersburg, MD). An electrochemiluminescent detection scheme is employed in this immunoassay.

Nucleic acid-based methods The promise of nucleic acid-based methods, especially those that employ an amplification step to increase sensitivity, is significant. In theory (but rarely in practice), methods can be designed which would allow the direct detection of *E. coli* in foods at levels of sensitivity equivalent to the most stringent regulatory action levels (e.g. USDA *E. coli* O157:H7 zero tolerance in ground beef). The major problem is recovery of that single cell from a total sample of 25 g.

Initial nucleic acid-based methods for the detection of *E. coli* focused on the use of probes in a colony hybridization format. The probes were to detect one or more of the genes coding for toxins in *E. coli* and these probes were radioactively labelled. The readout was an autoradiogram which identified colonies arrayed on a membrane that carried the targeted gene. Due to limitations in the density at which colonies on a membrane could be screened, this method was only useful once selection of presumptive *E. coli* was completed.

Most recently, a polymerase chain reaction (PCR)-based assay for *E. coli* O157:H7 has been released by Qualicon (a subsidiary of Dupont, Wilmington, DE). The reagents for the PCR assay are all contained in a single tablet and a single colony isolate is required. This system requires gel electrophoretic separation and visualization of the PCR products. While the nature of the target amplicon is proprietary, previous efforts from this group suggest that it was derived from an exhaustive screening of random amplified polymorphic DNA (RAPD) markers which were then found to be linked to the O157:H7 serotype.

See also: **Biochemical and Modern Identification Techniques**: Enterobacteriacea, Coliforms and *E. Coli*. **Escherichia coli**: Detection of Enterotoxigenic *E. coli*. **Escherichia coli O157**: *Escherichia coli* O157:H7; Detection by Latex Agglutination Techniques; Detection by Commercial Immunomagnetic Particle-based Assays.

Further Reading

Deshmarcherlier PM and Grau FH (1997) Escherichia coli in *Foodborne Microorganisms of Public Health Significance*, 5th edn. North Sydney, NSW 2059: Australian Institute of Food Science and Technology.

Jay JM (1996) *Modern Food Microbiology*. New York, NY: Chapman & Hall.

Roberts D, Hopper E and Greenwood M (1995) *Practical Food Microbiology*, London, UK: Public Health Laboratory Service.

Detection of Enterotoxins of *E. coli*

Hau-Yang Tsen, Department of Food Science, National Chung-Hsing University, Taichung, Taiwan, ROC

Characteristics of ETEC

Escherichia coli strains that cause diarrhoea include enterotoxigenic, enteropathogenic, enteroinvasive and enterohaemorrhagic strains. Recently, enteroaggregative *E. coli* (EAggEC) has also been found to be a diarrheogenic strain. Among these pathogenic strains, enterotoxigenic *E. coli* (ETEC) is the major strain which may cause human diarrhoea. ETEC strains resemble *Vibrio cholerae* in that they adhere to the small intestinal mucosa and cause diarrhoea not by invading the mucosa but by elaborating toxins that act on mucosa cells. The disease caused by ETEC is similar to cholera in many ways, although the diarrhoea is less severe. However, like cholera, ETEC diarrhoea can be fatal, especially for infants and young children.

ETEC strains may produce one or two types of enterotoxin. These enterotoxins are divided into heat-labile toxins (LT) and heat-stable toxins (ST). Two types of LT, termed LT type I (LTI) and LT type II (LTII), have been identified. These toxins are not immunologically related: both types can be further divided into subtypes (LTIh: human origin and LTIp: porcine origin) which are antigenically related. The DNA sequences of the subtypes are highly homologous. ST can be divided into two immunologically unrelated types – ST type I (STI or ST_A) and ST type II (STII or ST_B). For STI, two different but highly homologous alleles coding for STI subtype h (STIh) have been isolated from *E. coli* of human origin, estA2 and estA3/4. Another allele, estA1, which encodes STIp, was isolated from a bovine *E. coli* strain. DNA sequence analysis showed that estA1 is about 70% homologous to alleles estA2 and estA3/4.

The presence of ETEC strains in foods, water and the domestic environments has been well documented. Many reports describe the presence of LTI- and STI-producing *E. coli* strains in food and water or its involvement in food-borne disease. However, only a few reports describe the presence of LTII-producing *E. coli* strains and none have described STII-producing strains in foods. LTII *E. coli* has been found in Brazil in humans with diarrhoea and in some food samples. As for STII *E. coli* strains, such strains have frequently been isolated from swine with diarrhoea and from water but have rarely been found in calves or humans so far.

Methods of Detection

Cultural Media

Formulations of media used for ETEC culture are shown in **Table 1**.

Methods

The conventional method to identify or detect ETEC cells in foods is the AOAC method described in the *Bacteriological Analytical Manual (BAM)*. This method includes the enrichment step using selective medium, such as brain-heart infusion (35°C) followed by selective culture using Levine's eosin-methylene blue (L-EMB) agar or MacConkey agar and culture in blood agar base (BAB). Afterwards, series of biotype tests and serotypings were performed. By comparing the serotypes to those serotypes established for ETEC strains, ETEC strains may be identified. However, it should be pointed out that no biochemical markers exist to distinguish enterotoxigenic strains of *E. coli* from non-toxigenic strains. Although it has been shown that ETEC strains tend to fall within certain O:H serotypes, it is still impossible to identify most of them by the conventional method described above.

In Vivo Techniques for the Assay of ETEC Toxins
Animal tests for the direct determination of toxin production are cumbersome and difficult to perform. For in vivo tests, the rabbit or piglet ligated ileal loop assay method has generally been used. The enterotoxins prepared were injected into the ligated ileal segments of rabbit or piglets and the accumulated cell fluid was measured in ligated segments. Based on the difference in physiological response, LT and ST may be differentiated. Other methods, such as perfusing rat jejunum and rabbit skin permeability assay, have also been used. The most widely used test is the suckling mouse assay. However, as described earlier, most methods are cumbersome to perform and require continued breeding.

Tissue Culture Assay   LT toxin may cause the deformation of cultured cells, for example, the deformation of mouse Y-1 adrenal cell, or the elongation of Chinese hamster ovary (CHO) cells. However, an attempt to find an established cell line responsible for ST has not been successful.

Enzymatic Assay Method   Based on the principle that LT toxin would stimulate the production of cAMP, several enzymatic methods have been used to detect LT toxin. For example, the pigeon erythrocyte lysate assay measuring the activity of adenylate cyclase in pigeon erythrocyte has been used. Other indirect methods, for example, measurement of the activities of NAD glycohydrolase and ADP-ribosyltransferase, have also been used.

Immunological Method   The introduction of highly sensitive and simple immunoassays has been hampered by the lack of high-quality antisera and today only a few tests are commercially available. Examples of commercially available kits for immunoassay of ETEC toxins are VET-RPLA (a kit to detect *V. cholerae* toxin and *E. coli* heat-labile enterotoxin in culture filtrates by reversed passive latex agglutination) from Oxoid (Basingstoke, Hampshire, UK) and *E. coli* ST (STI) EIA (a kit to detect heat-stable *E. coli* enterotoxin in culture filtrates or supernatants by competitive enzyme immunoassay), also from Oxoid.

Procedures to detect LTI and STI enterotoxins with immunoassay kits are described below.

For LTI detection, the VET-RPLA kit and for STI detection, the *E. coli* ST EIA kit were used, as described above. For LT detection, Mundell's medium was used, while for ST detection, casein-yeast (CA-YE) broth was used for bacterial culture. The rest of the procedures for enterotoxin detection were described in the instruction sheets for these kits. For the EIA assay, OD readings at 490 nm for the specimen, negative control and positive control (supplied by Oxoid) were measured using an ELISA reader. The adjusted OD could be calculated according to the equation:

Adjusted OD = OD (specimen)/(OD (negative control) − OD (positive control))

When the adjusted OD value was ⩽ 0.2, the specimens tested were considered to be STI-positive strains. An example of the detection of STI ETEC strains using the *E. coli* ST-EIA kit is shown in **Table 2**.

Simplified protocols for the application of the VET-RPLA kit and *E. coli* ST-EIA kit on the detection of LT and ST *E. coli* are shown in **Figure 1**.

Gene-based Method: DNA–DNA Hybridization
Since the development of Southern blot hybridization and colony hybridization techniques, DNA–DNA hybridization using a specific DNA probe has become a common method to detect specific genes in specific cells. DNA fragments obtained from plasmids with enterotoxin genes were first used for the specific detection of ETEC cells in 1980–1981. Later, oligonucleotide probes designed from the DNA sequence of STI genes were used specifically to detect STI ETEC cells. Following the use of radioactive labelled DNA

Table 1   Media formulation

Blood agar base

Proteose peptone	15 g
Liver digest	2.5 g
Yeast extract	5 g
NaCl	5 g
Agar	12 g
Distilled water	1 l

Autoclave at 121°C for 15 min. For blood agar, reduce water to 950 ml. Add 50 ml defibrinated (whole or lysed) horse blood and FBP (4 ml to agar + blood) after autoclaving and cooling to 48°C. Final pH: 7.4 ± 0.2

Brain-heart infusion broth and agar

Calf brain, infusion from	200 g
Beef heart, infusion from	250 g
Proteose peptone (Difco) or polypeptone (Bioquest)	10 g
NaCl	5 g
Na$_2$HPO$_4$	2.5 g
Dextrose	2.0 g
Distilled water	1 l

Difco does not specify waters of hydration

Levine's eosin-methylene blue agar

Peptone	10 g
Lactose	10 g
K$_2$HPO$_4$	2 g
Agar	15 g
Eosin Y	0.4 g
Methylene blue	0.065 g
Distilled water	1 l

Boil to dissolve peptone, phosphate and agar in 1 l water. Add water to make original volume.

Dispense in 100 or 200 ml portions and autoclave 15 min at not over 121°C. Final pH: 7.1 ± 0.2. Before use, melt, and to each 100 ml portion add 5 ml sterile 20% lactose solution, 2 ml aqueous 2% eosin Y solution and 4.3 ml 0.15% aqueous methylene blue solution. When using complete dehydrated product, boil to dissolve all ingredients in 1 l water. Dispense in 100 or 200 ml portions and autoclave 15 min at 121°C. Final pH: 7.1 ± 0.2

MacConkey agar

Proteose peptone or polypeptone	3 g
Peptone or gelysate	17 g
Lactose	10 g
Bile salts no. 3 or bile salts mixture	1.5 g
NaCl	5 g
Neutral red	0.03 g
Crystal violet	0.001 g
Agar	3.5 g
Distilled water	1 l

Suspend ingredients and heat with agitation to dissolve. Boil 1–2 min. Autoclave 15 min at 121°C, cool to 45–50°C, and pour 20 ml portions into sterile 15 × 100 mm Petri dishes. Dry at room temperature with lids closed. *Do not use wet plates*. Final pH: 7.1 ± 0.2

MacConkey broth

Oxgall	5 g
Peptone	20 g
Lactose	10 g
Bromocresol purple	0.01 g
Distilled water	1 l

probes, biotin labelled DNA probes were developed specifically to detect LTI and STI genes of ETEC cells. Sequences of DNA probes used for the specific detection of LT and ST ETEC cells are found in many reports. In addition to the DNA probes, polymerase chain reaction (PCR) primers specific for LT and ST genes can be used as DNA probes too. Examples of these PCR primers are shown in **Table 3**.

The procedures for DNA hybridization generally include the following steps.

Table 2 Detection of heat-stable toxin I (STI) for enterotoxigenic *Escherichia coli* (ETEC) strains using *E. coli* ST-EIA kit

Strain	Toxin type	OD	Adjusted OD	Result
ATCC 37218	(LTIp)	1.240	0.960	–
ATCC 43886	(LTIh)	1.240	0.960	–
ATCC 43896	(STIb)	0.097	0.075	+
ATCC 33849	(LTI)	1.269	0.986	–
ATCC 31618	(STIa)	0.058	0.045	+
Negative control		1.327		–
Positive control		0.040		+

ATCC, American Type Culture Collection, Rockville, MD, US.

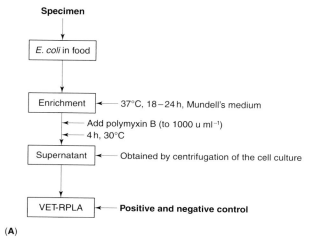

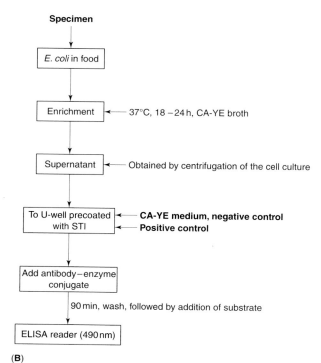

(A)

(B)

Figure 1 Overview of application of **(A)** commercial VET-RPLA and **(B)** *E. coli* ST-EIA for the detection of LTI and STI *E. coli*.

1. Growth of the cells to be detected on agar or broth medium.
2. Transblotting the bacterial colonies on agar plate on to nitrocellulose (NC) filters, or collecting the bacteria cells on the NC filter by filtration.
3. Lysis of the cells on the NC filters.
4. Pre-hybridization with nonspecific DNA, such as calf thymus DNA.
5. Hybridization with radioactive or non-radioactive labelled DNA probes specific for target cells.
6. Autoradiography or colour development for DNA probes hybridized on the NC filters.

A detailed example of these procedures can be found in section 24–12 of *BAM*. Simplified protocols for the DNA hybridization assay of LTI and STI *E. coli* cells are shown in **Figure 2**. For method B, some of the steps in method A can be modified.

Gene-based Method: Polymerase Chain Reaction
Polymerase chain reaction (PCR) allows the million-fold amplification of specific gene sequences within hours using gene-specific PCR primers. The potential diagnostic value of PCR has been recognized by many researchers and results concerning the detection and identification of diarrhoeagenic *E. coli* have been published. Up to now, numerous PCR primers have been developed for the specific detection of LTIp, LTIh, LTI, STIa, STIb, STI, and STII ETEC.

The general procedures for the PCR method include the following steps:

1. Isolation of DNAs or obtaining the cell lysates to be assayed.
2. Denaturation of the target DNAs.
3. Annealing the PCR primers to the target DNAs.
4. Extension or DNA polymerization.
5. Detection of the PCR products by agarose gel electrophoresis.

General procedures for the PCR assay of LTI ETEC cells contaminated in food samples are as follows.

For PCR detection of ETEC in foods, 1 g of the homogenized sample was mixed well with 9 ml Mac-Conkey broth. The mixture was then inoculated or not inoculated with target ETEC cells per gram of the sample. To ensure the detection of low numbers of target cells in food sample, the sample mixture was incubated at 37°C for 8 h. After incubation, the culture was diluted 10-fold with sterile H_2O, 10 µl of the diluent were then mixed with 30 µl of the $4/3\times$ PCR solution containing 0.275 mmol l^{-1} each of dNTP and 25 pmol of each of the primers, 0.5 mg ml^{-1} of proteinase K (Merck), and subjected to 30 min incubation at 65°C followed by 30 min boiling. After-

Table 3 Oligonucleotide primers used for the amplification of LTI, STI and STII enterotoxin genes of enterotoxigenic *E. coli*

Enterotoxin	Primers	Sequence (5' → 3')	% (G+C)	Tm[a] (°C)	Location (length, bp)	Predicted size of PCR product (bp)
LTI	LTI 1	GCTGACTCTAGACCCCCAG	63.2	62	76–94 (19 bp)	466
	LTI 2	TGTAACCATCCTCTGCCGGA	55.0	62	522–541 (20 bp)	
STI	STI 1	GGAGGTAAIATGAAIAAIIIAATITT	19.2	50	9–17 (26 bp)	203
	STI 2	TTACAACAIAITTCACAGCAGTAA	29.2	58	171–194 (24 bp)	
STII	STII 1	CTGTGTGAACATTATAGACAAATA	29.2	62	94–117 (24 bp)	104
	STII 2	ACCATTATTTGGGCGCCAAAG	47.6	62	177–197 (21 bp)	

[a]The equation used for Tm calculation is: Tm = 4 × (number of G or C)°C + 2 × (number of A or T)°C.
[b]Primers for LTI, STI and STII genes are those reported by Tsen et al (1996). *J. Food Protect.* 59: 795–802.

wards, the mixture was subjected to PCR following the addition of 0.4 unit of Dynazyme in 10 µl of 1× PCR buffer (1× PCR buffer: 10 mmol l⁻¹ Tris-HCl, pH 8.8, 30 mmol l⁻¹ MgCl₂, 50 mmol l⁻¹ KCl and 0.1% Triton-X100). Evaporation of the reaction mixture was prevented by the addition of 1 drop of mineral oil (Sigma) on the surface of the mixture. A microprocessor-controlled thermal cycler (GeneAmp PCR system 9600, Norwalk, CT, US) was used for automated temperature control of the PCR. DNA was amplified through 35 cycles of denaturation, annealing and polymerization. Denaturation was performed at 94°C for 20 s, primer annealing at 64°C (for LTI gene) or 62°C (for STII gene) for 20 s, or 50°C for 30 s (for STI gene) and polymerization at 72°C for 30 s, respectively.

To detect the amplified product, a 10 or 15 µl aliquot of the PCR product was examined by electrophoresis through a 1.5% or 2% Ultra-Pure agarose gel in 1× TAE buffer (50× TAE: 242 g Tris base, 57.1 ml of glacial acetic acid, 100 ml 0.5 mol l⁻¹ EDTA, pH 8.0, adjusted to 1000 ml with H₂O). The bands were identified by comparison with molecular

weight markers of a 100 bp ladder (GIBCO BRL, Gaithersburg, MD, US) after staining with ethidium bromide (Sigma). An example of PCR detection of LTI ETEC in milk samples is shown in **Figure 3**.

Gene-based Method: Multiplex PCR System In addition to the single PCR system which allows the detection of a single gene, multiplex PCR systems which allow the simultaneous detection of two or more different virulence genes in *E. coli* cells have been developed, such as the system for the simultaneous detection of LTI and STI ETEC and *Shigella* spp., the system for the detection of LTI and shiga-like toxin I and II producing strains of *E. coli*, and the system for the detection of LTI and STII, LTI and LTII ETEC cells. These systems may be of limited use because the ratios of different target cells in a sample and the concentration ratio of primers may significantly affect the PCR results.

Conclusion

In conclusion, methods such as DNA hybridization and PCR, although rapid and reliable, as indicated by *BAM*, must be thoroughly compared to standard

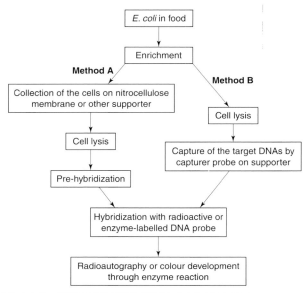

Figure 2 Protocol for DNA–DNA hybridization assay.

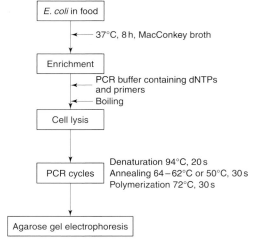

Figure 3 Simplified protocol for polymerase chain reaction (PCR) detection of LTI, STI and STII ETEC in foods.

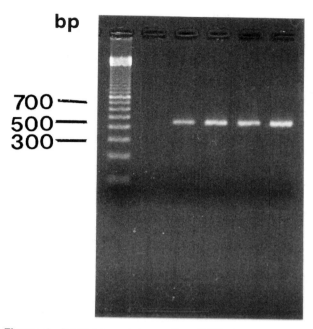

Figure 4 Polymerase chain reaction (PCR) detection of LTI ETEC in a skim milk sample after enrichment with MacConkey broth for 8 h. The primers used were LTI1 and LTI2. Lane 1: 100 bp ladder; lanes 2–5: PCR products amplified from cell lysates obtained from 10× diluted culture mixture after inoculation with LTI ETEC (ATCC 43886) and enrichment in MacConkey broth for 8 h. PCR results for milk samples not inoculated (lane 2), or inoculated with 10^0, 10^1, 10^2 and 10^3 (lanes 3–6) target cells per gram of milk are as indicated.

methods before use for routine analysis. Currently, only a limited number of rapid methods have been validated by collaborative studies and approved by AOAC International. These rapid methods can only be used for screening. Negative results are regarded as definitive, but positive results are considered presumptive and must be confirmed by standard methods.

Since almost all published results indicate that the rapid methods described in this article are not only rapid but also reliable, we believe that those methods will become more and more important in the near future.

See also: **PCR-based Commercial Tests for Pathogens**.

Further Reading

AOAC (1995) *Bacteriological Analytical Manual*, 8th edn. Gaithersburg, MD, US: AOAC International.

Blomen S, Lofdahl S, Stenstrom TA and Norberg R (1993) Identification of enterotoxigenic *Escherichia coli* isolates: a comparison of PCR, DNA hybridization, ELISAs and bioassay. *Journal of Microbiological Methods* 17: 181–191.

Echevervia P, Seriwatana J, Patamaroj U et al (1984) Prevalence of heat stable II enterotoxigenic *Escherichia coli* in pigs, water and people at farm in Thailand as determined by DNA hybridization. *Journal of Clinical Microbiology* 19: 489–491.

Gandrian U, Furrer B, Hofelein Ch, Meyer R, Jermini M and Liithy J (1991) Detection of *Escherichia coli* and identification of enterotoxigenic strain by primer-directed enzymatic application of specific DNA sequences. *International Journal of Food Microbiology* 12: 339–352.

Hill WE, Payne WL, Zon G and Moseley SL (1985) Synthetic oligodeoxyribonucleotide probes for detecting heat stable enterotoxin-producing *Escherichia coli* by DNA colony hybridization. *Applied Environmental Microbiology* 50: 1187–1191.

Levine MM (1987) *Escherichia coli* that cause diarrhea: enterotoxigenic, enteropathogenic, enteroinvasive, enterohemorrhagic and enteroadherent. *Journal of Infectious Diseases* 155: 377–389.

Schultsz C, Pool GJ, van Ketel R, De Wever B, Speelman P and Dankert J (1994) Detection of enterotoxigenic *Escherichia coli* in stool samples by using non-radioactively labeled oligonucleotide DNA probes and PCR. *Journal of Clinical Microbiology* 32: 2393–2397.

Tsen HY, Chi WR and Lin CK (1996) Use of novel polymerase chain reaction primers for the specific detection of heat liable toxin I, heat stable toxin I and II enterotoxigenic *Escherichia coli* in milk. *Journal of Food Protection* 59: 795–802.

ESCHERICHIA COLI O157

Contents

Escherichia coli O157:H7

Mary Lou Tortorello, National Center for Food Safety and Technology, US Food and Drug Administration, Summit-Argo, USA

Copyright © 1999 Academic Press

Introduction

Escherichia coli O157:H7 is the causative agent of a food-borne illness that has potentially deadly consequences. The following describes the special characteristics of the strain that distinguish it from harmless types of *E. coli*; methods for its identification and enumeration; its clinical manifestations, with summary of food-borne illness outbreaks; and strategies for its control in food production.

Characteristics of the Species

A special serotype of the well-known inhabitant of the mammalian intestinal tract, *Escherichia coli* O157:H7 has gained recognition as an important food-borne pathogen in recent years. Most strains of *E. coli* isolated from natural sources are harmless commensals, but the O157:H7 serotype has a highly virulent character. It has been variously described as verotoxigenic *E. coli* (VTEC), shiga-like toxin-producing *E. coli* (SLTEC), and currently, shiga toxin-producing *E. coli* (STEC), reflecting the evolution of the nomenclature of the toxins it produces. It is also referred to as enterohaemorrhagic *E. coli* (EHEC), because of its clinical effects, which separate it from the other recognized groups of pathogenic *E. coli*: enteropathogenic (EPEC), enterotoxigenic (ETEC), and enteroinvasive (EIEC) *E. coli*. Although there are other serotypes that may be described as EHEC, *E. coli* O157:H7 has been most problematic because of the number of food-borne illness outbreaks with which it has been associated. What characteristics distinguish this serotype from the other strains of *E. coli*, and what traits does this serotype possess to mark it as a pathogen?

Surface Antigens

The most familiar distinctions of the microorganism are the unique antigenic components of the cell surface, O157 and H7. Because *E. coli* O157:H7 has a cell envelope structure typical of Gram-negative cells, it possesses an outer membrane with a lipopolysaccharide component that is distinct from the cytoplasmic membrane. The O157 antigen is defined by the carbohydrate composition and structure within the lipopolysaccharide. It is a major antigenic component of the cell, and patients who have been infected with the pathogen exhibit serological reactivity to the O157 antigen. The H7 antigen is determined by the unique polypeptide composition of the flagella, the motility structures of the cell.

Many of the commercially available assays for *E. coli* O157:H7 rely on detection of one or both of these antigens. As immunoassays, they primarily depend on antigen-antibody recognition.

Metabolic Characteristics

Escherichia coli O157:H7 differs metabolically from other strains of *E. coli* in a number of ways. Most isolates of *E. coli* O157:H7 are slow- or non-fermenters of sorbitol and lack the enzyme β-glucuronidase, two important differentiating characteristics for identification of the strain on selective media. This serotype cannot grow at the high end of the temperature range of growth (44–45°C) typical of *E. coli* but can tolerate more acidic conditions (pH 2.5).

Pathogenesis

Toxins Not all of the pathogenicity determinants of *E. coli* O157:H7 are known, but the cytotoxins expressed by the strain have been well studied. Their nomenclature has evolved since their discovery and designation as 'verotoxins', a term descriptive of their cytopathic effects on Vero cells in tissue culture. The term 'shiga-like toxin' (SLT) was used when it was determined that the toxic effect could be neutralized by antiserum to shiga toxin, which is produced by *Shigella dysenteriae*. The designation 'shiga toxin' has now replaced 'shiga-like toxin' in recognition of the fact that the toxins produced by *E. coli* O157:H7 and *Shigella dysenteriae* are similar or identical.

Toxin neutralization studies revealed that *E. coli* O157:H7 produces two shiga toxins, although a particular strain may express both or just one of the toxins. The toxins are encoded by lysogenic bacteriophages, and the genetic determinants of both (formerly *slt-1* and *slt-2*, current designation *stx*$_1$ and *stx*$_2$) have been cloned and sequenced. Both toxins are composed of two sub-units, a structure reminiscent of protein toxins elaborated by a number of other pathogenic bacterial species. In the typical mechanism, the B sub-units mediate binding to the host cell surface receptor, and the A sub-unit is responsible for the enzymatic activity of the toxin. Globotriosylceramide is the glycolipid receptor for the shiga toxins on the host cell surface. The toxins exert their effect by interruption of protein synthesis, specifically by inactivation of the 60S ribosome.

The shiga toxins have a molecular weight of approximately 71 000. The molecular weights of the A and B sub-units are approximately 33 000 and 7500, respectively. The shiga toxin 1 produced by *E. coli* O157:H7 is composed of one A and five B sub-units and is structurally identical to the toxin produced by *Shigella dysenteriae*, as determined by protein and nucleic acid sequencing. Shiga toxin 2 has a similar sub-unit structure, but is compositionally and immunologically distinct. Several variants of shiga toxin 2 have been identified.

Factors controlling the production of shiga toxins have been studied. Growth of *E. coli* O157:H7 under iron-limiting conditions causes an increase in shiga toxin expression. The negative regulation of toxin expression by iron is controlled by the product of *fur*, the regulatory gene responsible for the expression of several iron-binding and uptake determinants. Exposure of the pathogen to sub-inhibitory concentrations of antibiotics, including sulphamethoxazole, trimethoprim, ciproflaxacin, cefixime and tetracycline, also stimulates toxin production.

Adherence and Colonization   The factors mediating adherence and colonization of *E. coli* O157:H7 within the host are not understood. The association of a 60 MDa plasmid appears to be necessary for adherence. Fimbriae have been implicated in attachment of the pathogen to host cell tissue, but the conflicting results obtained by using various cell lines as host tissue models and various strains of the pathogen have not allowed definitive conclusions to be made. Evidence also exists for the involvement of outer membrane components in attachment. It is possible that attachment is mediated by several mechanisms. More research is needed to understand how the pathogen attaches to host cell tissue.

The pathogen induces characteristic 'attaching-effacing' lesions of intestinal tissue which are visible in electron micrographs. These lesions appear to involve disorganization of the host cell cytoskeleton at the site of attachment of the pathogen, causing cup-shaped projections of the intestinal tissue. Intimin, a chromosomally encoded product of the *eaeA* gene, is involved in attachment-effacement, although its mechanism of action is not understood.

Miscellaneous Factors   Other determinants that may have roles in the pathogenesis of *E. coli* O157:H7 have been identified but have not yet been extensively studied. The pathogen produces a plasmid-encoded haemolysin (EHEC-Hly) and an extracellular serine protease (EspP) which can cleave pepsin and human coagulation factor V.

Atypical Strains

A variety of atypical strains of *E. coli* O157 have been isolated. The ability to ferment sorbitol, which is typical of most *E. coli* but atypical of *E. coli* O157:H7, has been reported. The sorbitol-fermenting variants are similar to the wild type with respect to toxin production, attachment and pathogenicity. Non-motile (H-negative) and β-glucuronidase-positive strains have also been isolated. The discovery of these atypical pathogenic strains brings into question the validity of testing for the pathogen by the absence of β-glucuronidase and sorbitol fermentation. It is possible that routine screening by these criteria does not detect the presence of pathogens which may be clinically relevant but do not exhibit typical characteristics.

Non-O157 EHEC

Serotypes other than O157:H7 have been isolated from patients exhibiting the characteristic clinical symptoms of EHEC infection; these include O26, O55, O104, O111, O119 and O128. Some of the cases of illness were associated with consumption of contaminated food. The EHEC group, therefore includes clinically important serotypes other than O157:H7. It has been suggested that more cases of non-O157 EHEC infection would be reported if the methods of detection were not skewed to favour isolation and identification of *E. coli* O157:H7. Certainly the number of cases of *E. coli* O157:H7 infection has risen dramatically in recent years due in large part to heightened surveillance and testing for the pathogen.

Identification of *E. coli* O157:H7 in Food

The standard methods described here are methods recommended by regulatory and public health agencies for analysis of *E. coli* O157:H7 in food.

Table 1 Enrichment culture medium for *Escherichia coli* O157:H7

Ingredient	Quantity (g l⁻¹)
Tryptone	20
Bile salts no. 3	1.12
Lactose	5.0
K_2HPO_4	4.0
KH_2PO_4	1.5
NaCl	5.0

Add water to 1 litre. If necessary, adjust pH to 6.9 ± 0.1 before autoclaving. Autoclave at 121°C for 15 min. Cool. Add 5 ml of a filter-sterilized aqueous solution of sodium novobiocin (4 mg ml⁻¹; potency 890 μg mg⁻¹).

Table 2 Sorbitol–MacConkey agar with BCIG

Ingredient	Quantity (g l⁻¹)
Peptone	17
Proteose peptone	3
Sorbitol	10
Bile salts no. 3	1.5
NaCl	5
Neutral red	0.03
Crystal violet	0.001
BCIG[a]	0.1
Agar	13.5

Add water to 1 litre. Adjust pH to 7.1 ± 0.2, if necessary. Autoclave at 121°C for 15 min.
[a] BCIG, 5-bromo-4-chloro-3-indoxyl-β-D-glucuronide, sodium salt.
Note: sorbitol–MacConkey agar is also available as a pre-formulated medium. Add BCIG in quantity indicated above.

Table 3 Phenol red–sorbitol agar with MUG

Ingredient	Quantity (g l⁻¹)
Proteose peptone no. 3	10
Beef extract	1
NaCl	5
Sorbitol	5
Phenol red	0.018
MUG[a]	0.05
Agar	20

Add water to 1 litre. Adjust pH to 6.8–6.9, if necessary. Autoclave at 121°C for 15 min.
[a] MUG, 4-methyl-umbelliferyl-β-D-glucuronide.
Note: Phenol red broth base is also available as a preformulated medium. Add sorbitol, MUG and agar in quantities indicated above.

Table 4 Levine's eosin–methylene blue agar

Ingredient	Quantity (g l⁻¹)
Peptone	10
Lactose	10
K_2HPO_4	2
Eosin Y	0.4
Methylene blue	0.065
Agar	15

Add water to 1 litre. Adjust pH to 7.1 ± 0.2, if necessary. Autoclave at 121°C for 15 min.
Note: Levine's EMB agar is also available as a preformulated medium.

Conventional Culture

Enrichment, colony isolation and confirmation are the three general phases of the standard method recommended by the US Department of Agriculture (USDA) for detection and identification of *E. coli* O157:H7 in meat. Other regulatory agencies recommend cultural methods which are similar in overall design, but may differ in certain details, e.g. preparation of the food and selective agents of the media.

The first phase is a 24 h enrichment at 35°C, which provides conditions that promote growth of *E. coli* but are inhibitory to other species. The USDA enrichment culture medium (**Table 1**) includes lactose as a carbohydrate source, bile salts for suppression of certain Gram-positive species, and novobiocin to suppress Gram-negative species other than *E. coli*.

In the second phase, the enrichment culture is plated onto a selective medium to obtain isolated colonies. The medium is a modification of MacConkey agar, in which the lactose is replaced by sorbitol (sorbitol–MacConkey agar, SMAC) (**Table 2**) and a chromogenic indicator for β-glucuronidase activity is included. Sorbitol-fermenting colonies appear red, but *E. coli* O157:H7 colonies are colourless owing to lack of sorbitol fermentation. The chromogenic substrate for the enzyme β-glucuronidase is 5-bromo-4-chloro-3-indoxyl-β-D-glucuronide (BCIG). Most *E. coli* strains, which produce the enzyme, hydrolyse the substrate and form blue colonies; *E. coli* O157:H7 colonies are colourless because of the absence of the enzyme. Twelve colonies typical of *E. coli* O157:H7 are picked from the SMAC plates and patched onto two other media: phenol red–sorbitol agar supplemented with 4-methyl-umbelliferyl-β-D-glucuronide (PRS-MUG) (**Table 3**) and Levine's eosin–methylene blue agar (EMB) (**Table 4**). Like SMAC-BCIG, PRS-MUG is an indicator medium for sorbitol fermentation and β-glucuronidase activity. When sorbitol is fermented, a localized drop in pH occurs around the colonies and the phenol red indicator dye turns yellow. The hydrolysis of MUG by β-glucuronidase results in a diffusible fluorescence, which can be visualized by observing the colonies under ultraviolet light. Most *E. coli* produce a yellow colour change of the medium and fluorescence under UV light, but *E. coli* O157:H7 colonies do neither. On EMB agar, the lactose, eosin and methylene blue provide a means to differentiate lactose fermenters (coliforms) from non-lactose-fermenting enterics such as *Salmonella*. Colonies of *E. coli* including O157:H7, ferment lactose and produce dark, almost black colonies, often with surfaces that have a green metallic sheen, on EMB agar.

In the third phase, the isolate is confirmed as *E. coli* O157:H7 by a number of different tests, perhaps the most important of which is the O157 latex agglutination test. The suspect colony is mixed with latex beads coated with antibody reactive to the O157 antigen. A positive antibody-antigen reaction causes a visible agglutination of the latex beads. Other confirmatory tests are also recommended, because the latex agglutination is not strictly specific for *E. coli* O157:H7. In particular, *Salmonella* group N cells are known to cross-react with the antibody. The other recommended confirmatory tests are H_2S production and carbohydrate fermentation in triple-sugar–iron agar; indole production; methyl red–Voges–Proskauer test; citrate utilization; lysine decarboxylase and ornithine decarboxylase production; cellobiose fermentation; and H7 antigen agglutination.

Another confirmatory test for the pathogen is Vero cell cytotoxicity. Colonies of *E. coli* O157:H7 secrete shiga toxins into the surrounding medium. Testing for these toxins is performed by adding cell-free culture broth to Vero cells in tissue culture. The presence of toxin is indicated by visible necrosis of the tissue culture. The test for toxin production does not identify *E. coli* O157:H7 specifically, however, because many other serotypes of the species may produce the toxins. The test also does not distinguish whether one or both of the toxins is produced by the isolate.

Quantitation

The standard method of determining cell numbers of *E. coli* O157:H7 in food relies on the conventional culture protocol detailed above, performed within the framework of the most probable number (MPN) procedure. The entire identification scheme, from enrichment to confirmation, is performed for each tube of the MPN series. The MPN procedure performs well for quantitation of low numbers of *E. coli* O157:H7 cells, even in the presence of a large background microbial population, because the procedure is based on the initial step of enrichment. It is especially tedious and time-consuming, however, because of the large quantities of media required and the many

manipulations that must be performed for each tube of the MPN series.

DNA Probe Hybridization

The use of specific DNA probes can reduce much of the work involved in confirmation of an isolate. Suspect colonies are transferred to a membrane and treated with reagents that lyse the cells. The denatured DNA is released and sticks to the membrane. The DNA probe is added and hybridizes with the matching DNA sequence on the membrane. For identification of *E. coli* O157:H7, probes for the shiga toxin genes (*stx*$_1$ and *stx*$_2$) and a mutation of the β-glucuronidase gene (*uidA*) may be used. Detection of probe hybridization depends on the nature of the probe level. If the probe is radiolabelled, a piece of X-ray film is placed on the membrane, and spots develop where the probe molecules hybridized to the bacterial DNA. If the probe is enzyme-labelled, a chromogenic substrate is added, and coloured spots develop where hybridization occurred.

Colony hybridization may be used not only to detect but also to enumerate the pathogen if the food is plated directly. Colonies that react with the probe are counted to determine the level of contamination of the food. The technique is most useful when the microbial background population in the food is present at a relatively low level.

DNA 'Fingerprinting'

Fine discrimination of *E. coli* O157:H7 strains, an important tool in epidemiological studies, may be achieved by DNA 'fingerprinting'. Although there are a variety of methods available, pulsed-field gel electrophoresis (PFGE) has been most successful for discrimination of *E. coli* O157:H7 isolates. The DNA is extracted and treated with restriction enzymes that recognize rare restriction sites, which results in very large DNA fragments. An instrument that alternates the direction of the current ('pulsed field') effectively separates and resolves the large fragments in the gel matrix. As the current is alternated, the DNA pieces align in the direction of the electrical field. The smaller pieces realign and move quickly through the gel when the direction is changed; the larger pieces take a long time to realign, and so move more slowly. The result is a characteristic pattern of bands or 'fingerprint' of the *E. coli* O157:H7 isolate. Identical PFGE patterns obtained from patient and food isolates have been used as evidence for the source of food-borne illness.

In addition to the standard methods described above, there are numerous commercially available assays for detection of *E. coli* O157:H7, many of which have been approved by public health agencies. The unique O157 and H7 antigens are often the targets of detection in these assays, but gene detection diagnostics, such as polymerase chain reaction (PCR), are also becoming available commercially. Although they require an initial period of enrichment culture to provide a sufficient quantity of cells for detection, the commercial assays are generally more rapid and easier to perform than the conventional culture isolation and identification methods. Several of them may be performed in as little as 5–10 min, after enrichment culture. Some may be as sensitive or specific as the standard methods, but also tend to be more costly.

Importance

The serotype *E. coli* O157:H7 has been identified as the cause of numerous outbreaks and sporadic cases of food-borne illness since its recognition as a pathogen in 1982. In most cases the resulting illnesses have been self-limiting, but complications have occurred, sometimes leading to death.

Clinical Manifestations

Escherichia coli O157:H7, as well as the other serotypes of the EHEC group, cause haemorrhagic colitis in infected humans. Although the disease may exhibit a range of clinical manifestations in different individuals, it is marked by a profuse, bloody diarrhoea. Fever is generally absent or low-grade. The incubation period of haemorrhagic colitis is 1–4 days, with an average duration of about 8 days. In most people, the disease is usually self-limiting in 5–10 days.

Serious complications of the disease, haemolytic–uraemic syndrome (HUS) and thrombotic thrombocytopenic purpura (TTP), may ensue and cause death. The most susceptible individuals are those with poorly developed or impaired immune systems, e.g. young children and the elderly. Haemolytic–uraemic syndrome is characterized by haemolytic anaemia and kidney failure, and blood transfusions and kidney dialysis are often necessary for treatment of HUS patients. The number of haemorrhagic colitis cases that proceed to HUS is estimated at 4–10%. Thrombolic thrombocytopenic purpura is characterized by the typical symptoms of HUS, with additionally fever and neurological dysfunction. The death rate resulting from these complications is 3–5%.

Outbreaks

Numerous outbreaks and sporadic cases of food-borne illness caused by *E. coli* O157:H7 have been reported since the first one was recognized in 1982. A partial listing of outbreaks caused by food con-

Table 5 Selected food-borne illness outbreaks caused by *Escherichia coli* O157:H7

Year	Location	Food vehicle	Number of persons affected
1982	USA	Hamburger	47
1982	Canada	Hamburger	31
1984	USA	Hamburger	34
1985	Canada	Sandwich	73
1986	Canada	Raw milk	46
1986	USA	Hamburger	37
1987	USA	Hamburger	51
1987	USA	Turkey	26
1988	USA	Hamburger	32
1990	USA	Conference meal	70
1990	Scotland	Restaurant food	16
1990	Japan	Tap water	174
1991	USA	Tap water	243
1991	England	Yoghurt	16
1991	USA	Apple cider	23
1993	USA	Hamburger	>500
1994	USA	Supermarket food	21
1994	USA	Salami	23
1995	USA	Venison jerky	11
1995	Canada	Lettuce	21
1995	USA	Lettuce	74
1996	USA/Canada	Apple cider	66
1996	Scotland	Meat pies	496
1996	England	Precooked meats	14
1996	Japan	Meat, sprouts	9500
1997	Japan	Sprouts	96
1997	Japan	Hospital food	58
1997	USA	Sprouts	60
1997	USA	Hamburger	16

sumption is given in **Table 5**. The list, which only includes outbreaks that affected more than 10 persons, was compiled primarily to illustrate the global nature of the events and the types of foods that have been implicated or confirmed as vehicles of transmission. It is not a comprehensive listing. Over the course of the summer of 1996 almost 10 000 people in Japan were affected, and the largest outbreak (5700 persons) was recorded in Sakai City. The most frequent cause of illness has been consumption of undercooked ground beef, although recently a variety of other foods have been implicated. Unpasteurized milk and apple cider have been the vehicles in several outbreaks. The appearance on the list of products that are normally consumed raw, e.g. lettuce and sprouts, posed a dilemma for food safety experts who previously could point to adequate cooking as a means to prevent the disease. The need to be aware of the potential for contamination and to control it 'from farm to table', i.e. during all steps in food production, have become food safety goals of the highest priority.

Non-food sources of infection, including swimming

in contaminated waters and transmission by fomites, also have been associated with illness outbreaks. Person-to-person transmission, e.g. in nursing homes and daycare centres, has resulted in numerous outbreaks, and there have been reports of animal-to-person transmission. These cases support assertions of a very low infective dose for the pathogen, which has been estimated to be of the order of 10–100 cells.

Reservoirs

Multiple serotypes (non-O157) of shiga toxin-producing *E. coli* can be isolated from the intestinal tracts of farm animals, including cattle, sheep, goats and pigs, and even of domestic animals such as cats and dogs. They pose no danger of illness to either humans or their animal host, because although they produce shiga toxins, they apparently lack other essential determinants of pathogenicity. The serotype *E. coli* O157:H7, on the other hand, is a frank pathogen of humans, though not of the animals in which it is known to reside. The list of animal carriers includes dairy and beef cattle, which are a primary reservoir of the pathogen, water buffalo, pigs, sheep, birds and deer. Asymptomatic humans can be transient vehicles of the pathogen and have been epidemiologically implicated in person-to-person transmission cases.

A number of surveys in which bovine faeces were tested for the presence of *E. coli* O157:H7 have shown a low prevalence of the pathogen, of the order of 1% or less, even among herds associated with illness outbreaks. Discontinuous shedding of the pathogen in the faeces, which has been demonstrated in multiple studies, probably contributes to the apparently low prevalence. The highest amount of shedding tends to occur in the summer months, and withdrawal of food is associated with increased faecal shedding.

Surveys of *E. coli* O157:H7 in retail foods have revealed its presence in meats such as beef, pork, poultry and lamb, although at a very low prevalence. Reflecting their greater prevalence in animal sources, the non-O157 shiga toxin-producing *E. coli* strains can be more commonly isolated from retail meats.

Control of the Pathogen in Food Processing

More studies on the ecology of *E. coli* O157:H7 are necessary to understand the survival of the pathogen in various environments and the mechanism of contamination of various foods. The organism inhabits the intestinal tracts of infected animals, and may contaminate food by some type of faecal contact during harvest, processing or preparation. The contamination may originate from direct contact of the food with faeces or by contact with environmental sources (such as water) that are faecally contaminated. Although it is not always clear how a food becomes contaminated, the risk can be decreased by attention to good sanitation practices during all phases of food production, 'from farm to table'.

Meat has been the food most often associated with illness outbreaks caused by *E. coli* O157:H7. If the intestinal contents of an infected animal spread onto the carcass during the slaughtering process, *E. coli* O157:H7 contamination of the meat may result. Undercooked ground meat is particularly likely to cause problems: if even a small portion of the animal carcass is contaminated, the pathogen may become mixed throughout the meat during grinding. If the interior section of the ground meat item is undercooked, the pathogen may survive and cause illness. Unpasteurized milk and apple juice also have been vehicles of illness. Cells of *E. coli* O157:H7 present in the barn may contaminate the udders of dairy cows and be transferred to the milk. In the case of apple juice, it is possible that apples become contaminated after dropping from the tree to contaminated soil. There is no way to visually discern if a food has been contaminated with *E. coli* O157:H7.

A variety of food processing technologies have been proposed for control of *E. coli* O157:H7 in foods. Heating, e.g. pasteurization, is an effective method of killing the pathogen. Steam-vacuuming and organic acid sprays are capable of reducing populations of the pathogen on the surfaces of animal carcasses. New technologies such as the use of high hydrostatic pressure, pulsed power electricity, ohmic heating, bright light pulsing and ozonation also show promise. Competitive exclusion, which involves the use of indigenous microorganisms to prevent colonization of the pathogen, may provide some protection, perhaps most effectively in young animals (e.g. calves). Gamma irradiation has also been shown to be an effective method of control. These and other control procedures need to be evaluated alone and in concert to arrive at effective pathogen reduction strategies for a particular food. Cost is a consideration and is often prohibitive in the implementation of the new processes, and more research is needed to develop cost-effective control measures. Individualized hazard analysis critical control point (HACCP) programmes for food production facilities may also be effective in improving food safety.

See also: **Biochemical and Modern Identification Techniques**: Enterobacteriaceae, Coliforms and *E. coli*. **Biophysical Techniques for Enhancing Microbiological Analysis**: Future Developments. **Enzyme Immunoassays**: Overview. *Escherichia coli*: Detection of Enterotoxins of *E. coli*. *Escherichia coli* **O157**: Detection by Latex Agglutination Techniques; Detection by Commercial Immunomagnetic Particle-based Assays. **Food Poisoning Outbreaks**. **Hazard Appraisal (HACCP)**: The Overall Concept. **Heat Treatment of Foods**: Principles of Pasteurization. **Meat and Poultry**: Spoilage of Meat. **Nucleic Acid-based Assays**: Overview. **Quantitative Risk Analysis**. **Verotoxigenic E. coli**: Detection by Commercial Enzyme Immunoassays. **Verotoxigenic E. coli and Shigella spp.**: Detection by Cultural Methods.

Further Reading

Doyle MP (1991) *Escherichia coli* O157:H7 and its significance in foods. *International Journal of Food Microbiology* 12: 289–302.

Doyle MP, Beuchat LR and Montville TJ (1997) *Escherichia coli* O157:H7. In: *Food Microbiology Fundamentals and Frontiers*. Washington: American Society for Microbiology.

Feng P (1995) *Escherichia coli* serotype O57:H7: novel vehicles of infection and emergence of phenotypic variants. *Emerging Infectious Diseases* 1: 47–52.

Griffin PM and Tauxe RV (1991) The epidemiology of infections caused by *Escherichia coli* O157:H7, other enterohemorrhagic *E. coli*, and the associated hemolytic uremic syndrome. *Epidemiologic Reviews* 13: 60–98.

Jay JM (1996) Foodborne gastroenteritis caused by *Escherichia coli*: enterohemorrhagic (EHEC). In: *Modern Food Microbiology*, 5th edn. New York: Chapman & Hall.

Karmali MA (1989) Infection by verocytotoxin-producing *Escherichia coli*. *Clinical Microbiology Reviews* 2: 15–38.

Lior H (1994) *Escherichia coli* O157:H7 and verotoxigenic *Escherichia coli* (VTEC). *Dairy, Food and Environmental Sanitation* 14: 378–382.

Meng J, Doyle MP, Zhao T and Zhao S (1994) Detection and control of *Escherichia coli* O157:H7 in foods. *Trends in Food Science and Technology* 5: 179–185.

Neill MA, Tarr PI, Taylor DV and Trofa AF (1994) *Escherichia coli*. In: Hui YH, Gorham JR, Murrell KD and Cliver DO (eds) *Foodborne Disease Handbook: Diseases Caused by Bacteria*. New York: Marcel Dekker.

O'Brien AD, Tesh VL, Donohue-Rolfe A et al (1992) Shiga toxin: biochemistry, genetics, mode of action and role in pathogenesis. *Current Topics in Microbiology and Immunology* 180: 65–94.

Padhye NV and Doyle MP (1992) *Escherichia coli* O157:H7: epidemiology, pathogenesis and methods for detection in food. *Journal of Food Protection* 55: 555–565.

Riley LW (1987) The epidemiological, clinical, and microbiologic features of hemorrhagic colitis. *Annual Review of Microbiology* 41: 383–407.

Takeda Y, Kurazono H and Yamasaki S (1993) Verotoxins (shiga-like toxins) produced by enterohemorrhagic *Escherichia coli* (verocytotoxin-producing *E. coli*). *Microbiology and Immunology* 37: 591–599.

Tesh VL and O'Brien AD (1991) The pathogenic mechanisms of Shiga toxin and the Shiga-like toxins. *Molecular Microbiology* 5: 1817–1822.

Tesh VL and O'Brien AD (1992) Adherence and colonization mechanisms of enteropathogenic and enterohemorrhagic *Escherichia coli*. *Microbial Pathogenesis* 12: 245–254.

Vernozy-Rozand C (1997) Detection of *Escherichia coli* O157:H7 and other verocytotoxin-producing *E. coli* (VTEC) in food. *Journal of Applied Microbiology* 82: 537–551.

World Health Organization (1997) *Prevention and Control of Enterohaemorrhagic* Escherichia coli *(EHEC) infections*. Report of a WHO Consultation, Food Safety Unit, Programme of Food Safety and Food Aid, 28 Apr–1 May 1997. Geneva: WHO.

Detection by Latex Agglutination Techniques

E W Rice, US Environmental Protection Agency, Cincinnati, Ohio, USA

General Principles

The use of latex reagents in slide agglutination assays provides a rapid screening procedure for the presumptive identification of *Escherichia coli* O157:H7. These assays are used for serotyping non-sorbitol-fermenting colonies, generally isolated on sorbitol MacConkey (SMAC) agar. The assays are designed for use with pure cultures and perform best when using freshly isolated organisms. Isolates may be analysed for the presence of both the somatic O157 antigen and the flagellar H7 antigen. In all instances where a positive serological result is obtained, further biochemical characterization is required to confirm that the suspected organism is *E. coli*.

In these procedures latex particles coated with the *E. coli* somatic O157 or flagellar H7 antisera (antibodies) are mixed on a slide with a suspension of bacteria and are observed for agglutination reactions. The tests are designed to provide a colour differentiation between the surface of the slide and the particles. It is essential that proper positive and negative controls be employed in the assay procedure. Latex particles coated with purified normal rabbit globulin serve as negative latex control reagents. Positive control reagents contain *E. coli* O157:H7 antigen. Some manufacturers may also include an *E. coli*-negative control reagent containing antigen of a non-toxigenic (non-O157:H7) *E. coli* isolate.

Food microbiological laboratories normally obtain latex agglutination kits from commercial suppliers. The basic principles involved in these assays are similar, but the analyst needs to be aware of the performance characteristics of individual manufacturers' kits and rigorously adhere to the proscribed protocols.

Procedures

All reagents should be checked to determine expiration dates before beginning the assay procedures. Reagents should be stored at refrigeration temperatures and allowed to equilibrate to room temperature before use. Care should be taken to avoid contaminating the reagent droppers or dropper bottles. The O157 test latex reagent and the negative latex control reagent should be tested with the positive *E. coli* O157:H7 antigen control reagent. Agglutination should occur with the O157 test reagent and the positive *E. coli* O157:H7 antigen control. Agglutination should not occur with the O157 test reagent and the negative latex control reagent. Reagents should not be used if there are any deviations from the expected results. This testing should be performed at a minimum of once per day whenever the reagents are used, with special attention given to these evaluations upon the initial use of a new lot of reagents.

Manufacturers' directions vary regarding preparation of specimens for analysis. Some procedures call for testing sorbitol-negative colonies form SMAC plates directly, whereas others require the preparation of a suspension of known turbidity. Non-sorbitol-fermenting colonies which have been subcultured to nonselective agar may also be used. The bacteria should be thoroughly mixed on the slides with the appropriate reagents using wooden sticks or bacteriological loops. The slides are then gently rotated or rocked for a given time period, generally 1–2 min. Agglutination occurring with the test O157 reagent and not occurring with the latex control reagent is considered a positive test (**Table 1**). The lack of agglutination in either of these test reagents is considered a negative response. No agglutination in the test O157

reagent and agglutination in the latex control is generally interpreted as a negative response. Nonspecific agglutination responses, where both the test O157 reagent and the latex control reagent agglutinate, are considered invalid tests. Suspensions of organisms exhibiting nonspecific agglutination may be placed in a boiling-water bath for 10–15 min and retested. After boiling, many of these strains become negative. The boiling procedure should be conducted following the manufacturer's directions. The analyst should also be alert for atypical reactions, such as stringy agglutinations, which differ from known reactions. These types of reactions are generally considered uninterpretable.

The use of the H7 latex reagent requires the use of a sweep of bacterial growth from a blood agar plate. Some strains may require subculturing in motility medium to enhance motility prior to testing. Only those strains which have exhibited a positive response for the presence of the somatic O157 antigen should be tested with the H7 latex reagent.

Advantages and Limitations

Latex slide agglutination assays are cost effective, easy to perform and provide rapid presumptive results for the identification of *E. coli* O157:H7. Studies have shown that the assays exhibit a high degree of reliability averaging greater than 95% specificity and sensitivity. In the analysis of food products, which may exhibit low levels of contamination, these assays provide an efficient means of screening potential target isolates.

Particular care should be given to the interpretation of results where nonspecific agglutination occurs. The incorporation of the various control reagents should aid the analyst in evaluating results of nonspecific or stringy agglutinations. Cross-reactivity with certain strains of bacteria has been reported, and although these incidences appear to be minimal, they confirm the absolute requirement for biochemically characterizing the isolate and completing all somatic and

Table 1 Interpretation of results from latex agglutination kits for *E. coli* O157:H7

Agglutination reactions		Interpretation
Test *E. coli* O157 latex reagent	Negative control latex reagent	
+	−	Presumptive positive for *E. coli* O157. Confirm with biochemical characterization and H serology
−	−	Negative for *E. coli* O157
−	+	Negative for *E. coli* O157[a]
+	+	Uninterpretable result[a]

[a]Consult manufacturer's recommendations.
−, negative reaction.
+, positive reaction.

flagellar serological tests before making a definitive identification. It should also be noted that non-motile strains of *E. coli* O157 have been implicated as entero-haemorrhagic *E. coli* and should therefore be tested for Shiga toxin production even though they exhibit a negative response for the H7 antigen.

See also: **Biochemical and Modern Identification Techniques**: Food-poisoning Organisms. **Entero-bacteriaceae, Coliforms and *E. coli***: Classical and Modern Methods for Detection/Enumeration. **Enzyme Immunoassays**: Overview. ***Escherichia coli* O157**: Detection by Commercial Immunomagnetic Particle-based Assays. **Food Poisoning Outbreaks**. **Hydro-phobic Grid Membrane Filtration Techniques (HGMF)**. **Immunomagnetic Particle-based Tech-niques**: Overview. **National Legislation, Guidelines & Standards Governing Microbiology**: European Union; Japan. **Nucleic Acid-based Assays**: Overview. **PCR-based Commercial Tests for Pathogens**. **Petrifilm – An Enhanced Cultural Technique**. **Reference Mater-ials**. **Sampling Regimes & Statistical Evaluation of Microbiological Results**. **Verotoxigenic *E. coli***: Detection by Commercial Enzyme Immunoassays; Detec-tion by Cultural Methods.

Further Reading

March SB and Ratnam S (1989) Latex agglutination test for detection of *Escherichia coli* serotype O157. *Journal of Clinical Microbiology* 27: 1675–1677.

Sowers EG, Wells JG and Strockbine NA (1996) Evaluation of commercial latex reagents for identification of O157 and H7 antigens of *Escherichia coli*. *Journal of Clinical Microbiology* 34: 1286–1289.

Detection by Commercial Immunomagnetic Particle-based Assays

Pina M Fratamico and **C Gerald Crawford**, USDA, Agricultural Research Service, Eastern Regional Research Center, Wyndmoor, USA

Introduction

In 1982, *Escherichia coli* O157:H7 was identified as a human food-borne pathogen responsible for out-breaks of haemorrhagic colitis associated with the consumption of undercooked ground beef. In recent years, the incidence of disease associated with this organism has steadily increased worldwide. Estimates indicate that in the USA there are approximately 10–20 000 disease cases of *E. coli* O157:H7 yearly with 200–500 of the cases resulting in death. This organism has been isolated from cattle, sheep, deer, goats and dogs; however, dairy cattle have been implicated as its principal reservoir. Outbreaks have been epi-demiologically linked to consumption of foods of bovine origin such as ground beef, roast beef or raw milk. Foods that were probably contaminated by bovine faeces such as lettuce or apples (made into apple cider) have also caused human disease. Strains of *E. coli* O157:H7 have been found to be relatively acid-tolerant and the infectious dose can be less than 50 cells. Important virulence factors include the pro-duction of Shiga toxins 1 and/or 2 (Stx 1/Stx 2), and *eae* and other genes involved in the production of attaching and effacing lesions and cytoskeletal damage of intestinal cells. The serological clas-sification of *E. coli* depends on the somatic lipo-polysaccharide O antigens, the flagellar H antigens, and the capsular K antigens. The combination of the O157 lipopolysaccharide and H7 flagellar antigens defines the *E. coli* O157:H7 serotype. The *E. coli* serogroup O157 possessing H antigens other than H7 and which do not produce Shiga toxins can occa-sionally be found in foods. These organisms, however, have not been reported to be human pathogens.

Given the low infectious dose of *E. coli* O157, very sensitive methods are required for its isolation and detection from food and environmental samples or from animal faecal specimens. The organism may be present in very low levels in such samples and must be identified in the presence of a large population of indigenous microflora. Methods should have the capability to detect one viable cell in 25 g of the original food sample within a few hours of testing, which usually includes an enrichment step.

Conventional methods for detection of pathogenic organisms in foods usually involve one or more enrichment steps in liquid medium to allow for resus-citation of injured bacteria and to allow for growth of the target organism while growth of the indigenous microflora is suppressed. The enrichment procedure is followed by subculturing onto selective and dif-ferential plating media to obtain isolated colonies for further study. A series of biochemical and serological tests are then performed in order to confirm the iden-tity of the isolates. Although traditional methods for detection and identification of *E. coli* O157:H7 in foods are generally rather sensitive, they are laborious and time-consuming, requiring 4–7 days to obtain final confirmatory results.

In recent years significant improvements in methods for microbiological analysis of foods have been made. One such improvement involves the use of magnetic beads to sequester target bacteria from the contaminating microflora and interfering food or faecal components and to concentrate the bacteria into smaller volumes for further testing. Magnetic beads coated with specific antibodies are used to recover the target organism from complex samples, thus the term 'immunomagnetic separation' is generally used to describe this technique. Immunomagnetic separation (IMS) was originally described for specific fractionation of lymphocytes and other cells from blood. The cells remained functionally active after isolation employing beads to which specific antibodies to cell surface antigens were bound. Following first reports describing the application of antibody-coated magnetic particles for isolation of bacteria such as K88+ *E. coli* and *Staphylococcus aureus*, numerous articles on the use of IMS for extraction of pathogenic bacteria from foods and other complex matrices have appeared. Incorporation of IMS into methodologies for detection of pathogenic bacteria results in greater sensitivity, specificity and rapidity of testing.

Isolation of *E. coli* O157 Using IMS Technology

Although there are several companies marketing various types of magnetic beads and magnetic separation equipment, the most frequently used magnetic carriers are produced by Dynal AS (Oslo, Norway). These carriers, known as Dynabeads, are uniform, superparamagnetic polystyrene microspheres. The polystyrene shell surrounds an evenly dispersed magnetic core and the hydrophobic surface of the spheres allows for the adsorption or coupling of different molecules. The beads manifest magnetic properties when exposed to an external magnetic field, but have no magnetic memory when removed from the field; therefore, the particles can be easily redispersed without aggregation to form a homogenous suspension.

Dynal's recommended protocol for detection of *E. coli* O157:H7 is shown in **Figure 1**. Twenty-five grams of food is pummelled in 225 ml of enrichment medium using a Stomacher apparatus. Dynal recommends buffered peptone water for pre-enrichment of *E. coli* O157:H7. Enrichment culturing is usually performed at 37°C for 18 h; however, shorter times may also be used. The appropriate number of Eppendorf tubes are placed into the magnetic particle concentrator (MPC-M) which can hold up to 10 tubes. Dynabeads (20 μl of the suspension) coated with antibodies against *E. coli* O157:H7 surface antigens and a 1 ml aliquot of the enrichment culture are mixed in the tubes and the samples are incubated for 30 min at room temperature with continuous mixing, preferably using a rotating

device, to allow formation of bead–bacterium complexes. **Figure 2** shows *E. coli* O157:H7 cells bound to Dynabeads anti-*E. coli* O157.

The magnetic plate is inserted into the Dynal MPC-M device and the complexes are allowed to concentrate onto the side of the tube. The supernatant is removed by aspiration with a Pasteur pipette and the magnetic plate is removed. Use of a vacuum aspiration system is not recommended. Washing buffer – 1 ml of phosphate-buffered saline (PBS) containing 0.05% Tween-20 – is added and the Dynal MPC-M is inverted three times to resuspend and wash the beads. The magnetic plate is then reinserted to again collect the bead–bacterium complexes. The washing procedure is repeated one more time and then the beads are resuspended in 100 μl of PBS-Tween. Detection of *E. coli* O157:H7 following IMS is then accomplished by culturing on selective agar medium or by any of the other procedures described below.

Magnetic beads, 2.8 μm in size (Dynabeads M-280), with covalently bound, affinity purified anti-*E. coli* O157 antibodies, are available ready to use from Dynal. Alternatively, researchers have used Dynabeads M-280 coated with sheep anti-rabbit IgG (available from Dynal) then a second antibody, rabbit anti-goat IgG was bound, and this was followed by binding of goat anti-O157 IgG. Another approach is to use Dynabeads M-450 (4.5 μm in size) coated with sheep anti-rabbit IgG or goat anti-mouse IgG also available from Dynal and bind a second polyclonal or monoclonal antibody specific for *E. coli* O157 to the beads. IMS can then be performed on food enrichment cultures or other types of samples.

Dynal markets other types of magnetic devices which can hold various size tubes or 96-well microtitre trays. Numerous articles have appeared describing the use of IMS for isolation and concentration of *E. coli* O157:H7 from foods and other types of samples and a method employing Dynabeads for isolation of the organism from foods is described in the 8th edition of the Food and Drug Administration's *Bacteriological Analytical Manual* (BAM). BioMag magnetic beads (PerSeptive Diagnostics, Cambridge, Massachusetts, USA) coated with BacTrace affinity-purified goat anti-*E. coli* O157 antibody (Kirkegaard and Perry Laboratories, Gaithersburg, Maryland, USA) have also been used in a procedure for detection of *E. coli* O157:H7 in ground beef.

Applications of IMS

In Conjunction with Plating

Selective enrichment culturing of *E. coli* O157:H7 is usually performed at 37°C using enrichment media such as buffered peptone water supplemented with

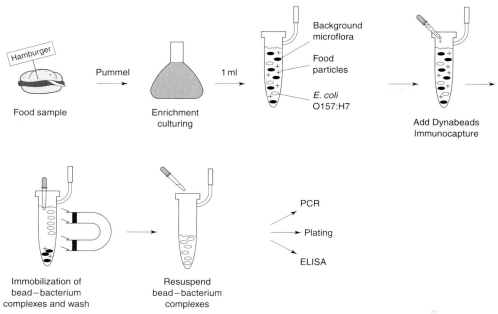

Figure 1 Procedure for immunomagnetic capture and detection of *E. coli* O157:H7 from foods.

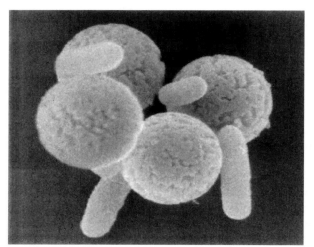

Figure 2 Scanning electron micrograph showing *E. coli* O157:H7 cells attached to Dynabeads anti-*E. coli* O157.

vancomycin ($8 \, \text{mg} \, \text{l}^{-1}$), cefixime ($0.05 \, \text{mg} \, \text{l}^{-1}$) and cefsulodin ($10 \, \text{mg} \, \text{l}^{-1}$) (BPW-VCC), modified tryptone–soy broth containing acriflavin-HCl at $10 \, \text{mg} \, \text{l}^{-1}$ (mTSB + a) or novobiocin at $20 \, \text{mg} \, \text{l}^{-1}$ (mTSB + n), modified *E. coli* broth containing novobiocin at $20 \, \text{mg} \, \text{l}^{-1}$ (mEC + n) and EEB medium consisting of mTSB supplemented with vancomycin ($8 \, \text{mg} \, \text{l}^{-1}$), cefixime ($0.05 \, \text{mg} \, \text{l}^{-1}$) and cefsulodin ($10 \, \text{mg} \, \text{l}^{-1}$) among others. The procedure in the *Bacteriological Analytical Manual* describes enrichment culturing for 6 h in EEB followed by IMS on 1 ml of sample. After washing and resuspension of the magnetic bead–bacterium complexes, a portion of the suspension is surface-plated onto selective and differential agar media such as sorbitol MacConkey (SMAC), SMAC containing cefixime ($0.05 \, \text{mg} \, \text{l}^{-1}$) and potassium tellurite ($2.5 \, \text{mg} \, \text{l}^{-1}$) (CT-SMAC), SMAC containing $5 \, \text{g} \, \text{l}^{-1}$ of rhamnose and $0.05 \, \text{mg} \, \text{l}^{-1}$ of cefixime (CR-SMAC), or CHROMagar O157 (CHROMagar, Paris, France). Other solid selective media which could possibly be used include Rainbow agar O157 (Biolog, Inc., Hayward, California, USA) or BCM O157:H7(+) (Biosynth International, Inc., Naperville, Illinois, USA). Dynal recommends CT-SMAC agar for isolation of *E. coli* O157 following IMS and sells SMAC Media Cefixime-Tellurite Supplement for its preparations. This agar is very selective and *E. coli* O157:H7 appears as pale pink, non-sorbitol-fermenting colonies. Bacteria remain viable after IMS so they continue to multiply when the bead–bacterium complexes are plated onto solid media and the bacteria need not be detached from the beads. Selective capture of target organisms from food enrichments using IMS, followed by plating, eliminates growth of a large portion of the background microflora; thus the amount of time required for selection of suspect colonies for confirmatory testing is reduced considerably. Accurate enumeration of the number of colony forming units (cfu) may not always be possible, however, since a colony may be formed from a bead with more than one bacterium attached. Alternatively, bacteria captured by IMS can be visualized by fluorescence microscopy if the bead–bacterium complexes are treated, for example, with a fluorescein isothiocyanate (FITC)-labelled second antibody which is directed against an *E. coli* O157:H7 surface antigen.

To confirm the identity of isolated colonies, biochemical and serological testing can be performed. Serological assays include agglutination using the *E. coli* O157 latex kit (Oxoid, Inc., Ogdensburg, New York, USA), reactivity with *E. coli* O157 antibodies conjugated to FITC (Kirkegaard & Perry Laboratories) and also reactivity with H7-specific antisera. Assays for Stx1 and/or Stx2 may also be performed on isolated colonies. *Escherichia coli* O157:H7 does not possess β-glucuronidase activity; thus colonies subcultured onto media such as violet red bile agar containing 4-methylumbelliferyl-β-D-glucuronide (MUG) or onto SMAC containing 5-bromo-4-chloro-3-indoxyl-β-D-glucuronic acid cyclohexylammonium salt (BCIG) appear β-glucuronidase-negative.

When IMS is incorporated in detection procedures using appropriate enrichment and recovery media, enrichment culturing periods as short as 4–6 h will allow detection of levels as low as 1–2 cells *E. coli* O157:H7 per gram of the original food sample. Detection of low levels of *E. coli* O157 is possible even in samples containing up to 10^7 cfu g^{-1} of background microflora. Inclusion of IMS in the isolation procedure can enhance sensitivity up to 100-fold compared with direct culture of enrichments, therefore false negative results may be significantly reduced.

In Commercial Systems

Food matrices often cause interferences in immunoassays such as enzyme-linked immunosorbent assays (ELISA). Binding by components such as proteins or other bacteria in the food matrix to the antibody may occur or large molecules can cause steric hindrance preventing antibody–epitope binding. Fatty acids or other substances can denature antibodies or interact nonspecifically with proteins causing problems in the assay system. Immunomagnetic techniques effect separation of target organisms from food particles and from a large portion of the background flora and allow further concentration of the target organism in the sample into smaller volumes. An ELISA kit for detection of *E. coli* O157:H7 in foods, the EHEC-Tek Test System (Organon Teknika Corporation, Durham, North Carolina, USA) utilizes IMS to capture living *E. coli* O157 cells (*E. coli* Capture-Tek) prior to the ELISA that specifically detects *E. coli* O157:H7. Inclusion of IMS in the isolation procedure enhances the sensitivity of the assay.

The test procedure is demonstrated in **Figure 3**. The times required to perform the various steps using the EHEC-Tek system are compared to the times required to detect *E. coli* O157:H7 using an Origen Analyzer system (IGEN, Inc., Gaithersburg, Maryland, USA) and a conventional plating assay, all

incorporating an IMS procedure. Using the EHEC-Tek system, a 25 g meat sample is pummelled using a Stomacher apparatus or blended for 1 min in 225 ml of mEC + n (for dairy products, mTSB + a is used), incubated for 6 h at 37°C, then a 1 ml portion from the primary enrichment is mixed with the Dynabeads anti-*E. coli* O157. The tubes are incubated for 10 min at room temperature with gentle mixing, then the beads are washed once, and 0.5 ml of TSB + a is added. Food particles and background microflora are removed during the wash procedure, and the bacteria bound to beads are allowed to multiply at 37°C overnight. A 200 µl aliquot from the TSB + a enrichment is heated at 100°C for 20 min then cooled. This is followed by addition of 100 µl of the test samples and of the positive and negative controls provided with the kit to the Microelisa wells and the ELISA is performed. The *E. coli* O157:H7 cells are captured by anti-O157 polyclonal antibody and detected with horseradish peroxidase-labelled monoclonal antibody which recognizes two low-molecular-weight antigens that appear to be unique to *E. coli* O157:H7 and *E. coli* O26: H11. If the absorbance of the sample is greater than or equal to the determined cutoff value, the test sample is considered presumptive positive for the presence of *E. coli* O157:H7. The entire EHEC-Tek detection assay is performed in approximately 30 h. If desired, the remaining TSA + a enrichment cultures of the presumptive positive samples can be plated onto selective agars for cultural confirmation.

The detection of antigens by immunomagnetic electrochemiluminescence is a comparatively new detection system development by IGEN, Inc. and has recently been applied to detection of pathogens in model food systems (see Fig. 3). In that application, heat-killed *E. coli* O157:H7 are added to food samples and then biotin-labelled anti-*E. coli* O157 antibody is incubated with an aliquot of an enrichment culture of the blended sample. Streptavidin-coated paramagnetic beads (Dynal) and anti-*E. coli* O157 antibody labelled with ruthenium are then added to the sample. The antibodies and the beads form a complex with the pathogen and a portion of the sample is drawn up into the detection instrument by the fluidics system. A magnet is placed below an electrode and as the sample flows over the electrode, the complex of magnetic beads, pathogen and ruthenium-labelled antibody are held on the electrode as the unreacted material is washed away. A voltage is applied to the electrode and through a series of oxidation and reduction steps, and the relaxation of the excited states, light is emitted. The amount of light, or electrochemiluminescence, is proportional to the amount of antigen in the sample. This analytical process takes less than a minute per sample, therefore

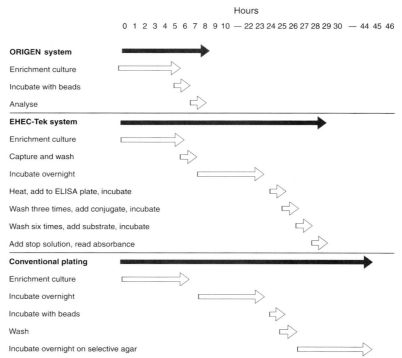

Figure 3 Procedural steps and time required for detection of *E. coli* O157:H7 using IMS in commercial and conventional plating systems.

less than an hour is required to perform the complete analysis of 50 samples contained in tubes placed in the instrument's autosampler. With 6 h of enrichment culturing, the entire detection assay can be performed in 8 h.

By way of comparison, a more traditional method for detection of *E. coli* O157:H7 involves enrichment culturing for approximately 24 h, followed by incubation with antibody-coated Dynabeads, washing of the beads, and plating onto selective agar with overnight incubation. Approximately 2 days are required to perform this procedure and it is followed by selection of suspect colonies from the agar plates which are then identified as *E. coli* O157:H7 by biochemical and serological testing. Using appropriate media for enrichment, the levels of detection of *E. coli* O157:H7 using the EHEC-Tek, IGEN system, or conventional plating are similar at approximately 1–2 cfu g^{-1} of the initial sample.

IMS Followed by Genetic Detection Methods

The polymerase chain reaction (PCR) is currently gaining widespread acceptance as a functional tool for detection of microorganisms in foods and other samples of complex composition. This is an in vitro technique in which a millionfold or greater amplification of DNA sequences is achieved using a heat-stable DNA polymerase and a pair of oligonucleotide primers which bind to specific nucleic acid sequences

of the target organism. Sensitivity of PCR assays is dramatically decreased, however, when they are applied directly to food and environmental samples, to blood and stool specimens, or to enrichment cultures. These types of samples may contain substances such as bilirubin, bile salts, haemoglobin degradation products, polyphenolic compounds, proteinases, complex polysaccharides and fat which can inhibit the DNA polymerase, bind magnesium or denature the DNA. Therefore, lengthy sample preparation and DNA extraction steps such as phenol-chloroform extraction with proteinase K treatment are often required prior to performing PCR. Inhibition of PCR can also be reduced through dilution of the samples; however, sensitivity is decreased with dilution. The volume of sample used for PCR is small, usually ranging from less than 1 µl to 20 µl. Immunomagnetic separation allows concentration of the bacteria in the sample to volumes ranging from 10 µl to 100 µl prior to performing PCR. Thus IMS removes PCR-inhibitory components from samples of complex composition, allowing purification of PCR-ready DNA while achieving concentration of the bacteria to enhance sensitivity.

To recover *E. coli* O157 from enrichment cultures of bovine faecal specimens or of foods such as apple cider, ground beef, raw milk or ice cream, enrichment is performed as described above, as is IMS using 1 ml of a culture sample. Alternatively, 10 ml or larger volumes of samples can be centrifuged to concentrate the bacteria, the pellet then washed and resuspended in 1 ml sterile physiological saline which is then subjected to IMS using Dynabeads *E. coli* O157. After two or three washes, the bead–bacterium complexes are resuspended in a small volume of sterile distilled water or Tris-EDTA buffer. The bacteria are lysed to release the DNA by placing the tubes in a boiling water bath for 10 min or in a thermal cycler set to 99°C for 10 min. This is followed by PCR using primers to amplify portions of virulence genes such as stx_1, stx_2, *eaeA* or other specific DNA sequences. The combination of PCR following IMS allows detection of *E. coli* O157:H7 in foods at a level as low as 1 cfu g^{-1} of the original sample after 4 h of enrichment culturing in TSB. Thus detection can be accomplished in less than 10 h.

Another approach involves using IMS in a procedure designated DIANA (detection of immobilized amplified nucleic acid) to detect ^{32}P-labelled PCR products generated following amplification of *E. coli* O157:H7 *stx* genes. The first PCR is carried out normally using unlabelled primers for stx_1 and stx_2. The second PCR reaction consists of the amplification product from the first PCR and two primers, one labelled with γ^{32}P[ATP] and one with biotin, yielding a ^{32}P- and biotin-labelled amplification product smaller than the product obtained with the first PCR. Streptavidin-coated magnetic beads, also available from Dynal, are then used to separate the labelled PCR products from solution, and after a washing step, the beads are suspended in scintillation fluid and bound radioactivity is determined in a scintillation counter. Sensitivity and specificity of the assay appears to be enhanced using a two-step PCR approach. The number of templates in the second PCR is greatly increased, improving sensitivity, and the possibility of obtaining false positive results is decreased, since the second amplification only occurs if the first set of primers has served to amplify the correct DNA sequence.

A technique called magnetic capture–hybridization PCR (MCH-PCR) involves lysing the bacteria to release the nucleic acid and hybridization of the DNA segments containing *E. coli* O157:H7 stx_1 and stx_2 sequences using biotin-labelled DNA probes. Following capture of the hybrids by streptavidin-coated magnetic beads, the bound DNA is subjected to PCR amplification.

Optimization of IMS

Procedures for immunomagnetic separation and concentration of *E. coli* O157:H7 should be optimized for each type of food or other type of sample tested since background microflora and other sample components will vary. For example, the amount of Dynabeads required for efficient capture of *E. coli* O157:H7 should be determined. The ratio of beads to target cells should generally be in the range of 3:1 to 20:1. For food and clinical samples, Dynal recommends using 20 μl (about 2×10^6 beads) of Dynabeads anti-*E. coli* O157 per millilitre of sample. Incubation of the bead–sample mixture is usually performed at room temperature for times ranging from 10 min to 60 min. Longer incubation times allow increased recovery of *E. coli* O157:H7 in samples containing lower numbers of target bacteria; however, the level of interactions with non-target cells is also substantially increased. Optimum incubation times generally range from 15 min to 30 min. Incubation temperature appears to have little effect on recovery of target cells.

Nonspecific binding can be reduced by performing IMS in low ionic strength solution treated with Chelex-100 (Bio-Rad Laboratories, Hercules, California, USA). Alternatively, addition of positively charged protein, Protamine (Sigma Chemical Company, St Louis, Missouri, USA), to the enrichment culture–bead mixture and transfer of the beads and wash solution to clean tubes with each wash also significantly decreases nonspecific binding of target cells and carry-over. Protamine reduces nonspecific attachment, supposedly by adhering to the sides of the sample tube and to the bacteria, decreasing their net negative charge; however, it does not affect binding of target bacteria with the antibody. Coating of tubes with other surface-treating agents such as Prosil-20 (PCR Inc., Gainsville, Florida, USA) or dichlorodimethylsilane (Sigma) also aids in preventing carry-over. Addition of detergents such as Tween-20 to the incubation mixture and/or to the wash solution also reduces nonspecific binding to the beads. Dynal recommends using a washing buffer consisting of 0.15 mol l^{-1} NaCl, 0.1 mol l^{-1} sodium phosphate buffer, pH 7.4, with 0.05% Tween-20.

Under optimum conditions, recovery of more than 90% of the inoculum can be achieved by IMS. Generally, however, recoveries in the range of 25–50% of the target cells are obtained. It should be taken into account that during incubation of enrichment cultures with Dynabeads for periods of 30 min or longer, the *E. coli* O157:H7 can continue to multiply, so the percentage recovery of the cells in the initial sample may appear high. Dynal recommends two washes of the beads prior to resuspension in 100 μl PBS-Tween and plating. Increasing the number of washes of the bead–*E. coli* O157:H7 complexes decreases carry-over; however, a portion of the target cells may also be removed during the washes, decreasing recovery.

In the separation cycle, use of excessively thick samples containing food particles can hinder recovery. The presence of food particles may block binding of antibody on the beads to the bacteria. Fat and other food particles can also impair immobilization of the bead–target complexes onto the side of the sample tube when applying the magnetic device. Difficulties can occur when attempting to aspirate the sample supernatant while leaving the immobilized beads undisturbed. Thus it may be advantageous to make dilutions of thick samples or to use less food (or other type of sample) when preparing the enrichment cultures, in order to overcome this problem.

Strains of Stx-negative *E. coli* serogroup O157 which can be present in meat and other foods are captured by beads coated with anti-O157 antibodies; however, if IMS is followed by plating on agar media such as CT-SMAC, Rainbow or Chromagar, typical *E. coli* O157:H7 colonies can usually be distinguished, and if desired, subjected to further serological or biochemical confirmatory testing. If PCR is performed after IMS, an amplification product should not be obtained with Stx-negative *E. coli* O157 if the PCR primers target virulence genes or other specific *E. coli* O157:H7 DNA sequences.

Conclusion

Performing IMS followed by plating onto selective or differential agars, or by rapid techniques such as PCR, ELISA, electrochemiluminescence, flow cytometry, or microscopy, markedly enhances the speed and sensitivity of assays for detection of *E. coli* O157:H7. Enrichment culturing times can be reduced considerably, thus the entire assay can potentially be performed in 8 h or less. Immunomagnetic separation may be useful for the recovery of stressed, sublethally injured *E. coli* O157:H7 which are not resuscitated during selective enrichment culturing. Specificity is determined by the antibody bound to the beads, thus with the availability of appropriate antisera, improved assay systems for detection of *E. coli* O157:H7 as well as for other bacterial pathogens can be developed. The IMS technique is easy to perform, does not require elaborate instrumentation, and can easily be applied to isolation and detection procedures for *E. coli* O157:H7 or other food-borne pathogens.

See also: Escherichia coli: Detection of Enterotoxins of *E. coli*; *Escherichia coli* O157: Detection by Latex Agglutination Techniques. **Immunomagnetic Particle-based Techniques**: Overview. **Nucleic Acid-based Assays**: Overview. **PCR-based Commercial Tests for Pathogens**. **Petrifilm – An Enhanced Cultural Technique**. *Salmonella*: Detection by Immunomagnetic Particle-based Assays. **Verotoxigenic *E. coli***: Detection by Commercial Enzyme Immunoassays. **Verotoxigenic *E. coli* and *Shigella spp.***: Detection by Cultural Methods.

Further Reading

Betts R (1994) The separation and rapid detection of micro-organisms. In: Spencer RC, Wright EP and Newsom SWB (eds) *Rapid Methods and Automation in Microbiology and Immunology*. P. 107. Andover: Intercept.

Chapman PA, Malo ATC, Siddons CA and Harkin M (1997) Use of commercial enzyme immunoassays and immunomagnetic separation systems for detecting *Escherichia coli* O157 in bovine fecal samples. *Applied and Environmental Microbiology* 63: 2549–2553.

Dynal (1997) *Biomagnetic Applications in Cellular Immunology*. Technical Handbook. Oslo: Dynal.

Dynal (1995) *Biomagnetic Techniques in Molecular Biology*, 2nd edn. Technical Handbook. Oslo: Dynal.

Fratamico PM, Schultz FJ and Buchanan RL (1992) Rapid isolation of *Escherichia coli* O157:H7 from enrichment cultures of foods using an immunomagnetic separation method. *Food Microbiology* 9: 105–113.

Gooding CM and Choudary PV (1997) Rapid and sensitive immunomagnetic separation–polymerase chain reaction method for the detection of *Escherichia coli* O157:H7 in raw milk and ice-cream. *Journal of Dairy Research* 64: 87–93.

Mortlock S (1994) Recovery of *Escherichia coli* O157:H7 from mixed suspensions: evaluation and comparison of pre-coated immunomagnetic beads and direct plating. *British Journal of Biomedical Science* 51: 207–214.

Okrend AJG, Rose BE and Lattuada CP (1992) Isolation of *Escherichia coli* O157:H7 using O157 specific antibody coated magnetic beads. *Journal of Food Protection* 55: 214–217.

Olsvik Ø, Popovic T, Skjerve E et al (1994) Magnetic separation techniques in diagnostic microbiology. *Clinical Microbiology Reviews* 7: 43–54.

Safarik I, Safarikova M and Forsythe SJ (1995) The application of magnetic separations in applied microbiology. *Journal of Applied Bacteriology* 78: 575–585.

Yu H and Bruno JG (1996) Immunomagnetic-electrochemiluminescent detection of *Escherichia coli* O157 and *Salmonella typhimurium* in foods and environmental water samples. *Applied and Environmental Microbiology* 62: 587–592.

Manufacturers

Mention of brand or firm name does not constitute an endorsement by the US Department of Agriculture over others of a similar nature not mentioned.

Biolog, Inc., Hayward, California, USA; Bio-Rad Laboratories, Hercules, California, USA; Biosynth International, Inc., Naperville, Illinois, USA; CHRO-Magar, Paris, France; Dynal Nordic, Oslo, Norway;

IGEN International, Inc., Gaithersburg, Maryland, USA; Kirkegaard & Perry Laboratories, Gaithersburg, Maryland, USA; Organon Teknika, Boxtel, The Netherlands; Oxoid, Inc., Ogdensburg, New York, USA; PCR Inc., Gainsville, Florida, USA; PerSeptive Biosystems, Inc., Framingham, Massachusetts, USA; Sigma-Aldrich Corporation, St Louis, Missouri, USA.

Eukaryotic Ascomycetes *see* **Fungi**: Classification of the Eukaryotic Ascomycetes (Ascomycotina).

APPENDIX I: BACTERIA AND FUNGI

The genera listed here are those associated with food, agricultural products and environments in which food is prepared or handled.

Abiotrophia
Acinetobacter
Actinobacillus
Actinomyces
Aerococcus
Aeromonas
Agrobacterium
Alcaligenes
Alloiococcus
Anaerobiospirillum
Arcanobacterium
Arcobacter
Arthrobacter
Aureobacterium
Bacillus
Bacteroides
Bergeyella
Bifidobacterium
Blastoschizomyces
Bordetella
Branhamella
Brevibacillus
Brevibacterium
Brevundimonas
Brochothrix
Brucella
Budvicia
Burkholderia
Buttiauxella
Campylobacter
Candida
Capnocytophaga
Cardiobacterium
Carnobacterium
CDC
Cedecea
Cellulomonas

Chromobacterium
Chryseobacterium
Chryseomonas
Citrobacter
Clostridium
Comamonas
Corynebacterium
Cryptococcus
Debaryomyces
Dermabacter
Dermacoccus
Dietzia
Edwardsiella
Eikenella
Empedobacter
Enterobacter
Enterococcus
Erwinia
Erysipelothrix
Escherichia
Eubacterium
Ewingella
Flavimonas
Flavobacterium
Fusobacterium
Gardnerella
Gemella
Geotrichum
Gordona
Haemophilus
Hafnia
Hansenula
Helicobacter
Kingella
Klebsiella
Kloeckera
Kluyvera

Kocuria
Kytococcus
Lactobacillus
Lactococcus
Leclercia
Leptotrichia
Leuconostoc
Listeria
Malassezia
Methylobacterium
Microbacterium
Micrococcus
Mobiluncus
Moellerella
Moraxella
Morganella
Myroides
Neisseria
Nocardia
Ochrobactrum
Oerskovia
Oligella
Paenibacillus
Pantoea
Pasteurella
Pediococcus
Peptococcus
Peptostreptococcus
Photobacterium
Pichia
Plesiomonas
Porphyromonas
Prevotella
Propionibacterium
Proteus
Prototheca
Providencia

Pseudomonas
Psychrobacter
Rahnella
Ralstonia
Rhodococcus
Rhodotorula
Rothia
Saccharomyces
Salmonella
Serratia
Shewanella

Shigella
Sphingobacterium
Sphingomonas
Sporobolomyces
Staphylococcus
Stenotrophomonas
Stomatococcus
Streptococcus
Suttonella
Tatumella
Tetragenococcus

Trichosporon
Turicella
Veillonella
Vibrio
Weeksella
Weissella
Xanthomonas
Yarrowia
Yersinia
Yokenella
Zygosaccharomyces

APPENDIX II: LIST OF SUPPLIERS

The suppliers below are mentioned in the text as main sources of specialist equipment, culture media or diagnostic materials. This list is not intended to be comprehensive.

3M Microbiology Products
3M Center
Building 260–6B-01
St Paul
MN 55144–1000
USA

ABC Research Corporation
3437 SW 24th Avenue
Gainesville
FL 32607
USA

Adgen Ltd
Nellies Gate
Auchincruive
Ayr KA6 5HW
UK

Agi-Diagnostics Associates
Cinnaminson
New Jersey
USA

ANI Biotech OY
Temppelikatu 3–5, 00100
Helsinki
Finland

Applied Biosystems
The Perkin-Elmer Corporation
12855 Flushing Meadow Drive
St Louis
MO 63131 1824
USA

Becton Dickinson Microbiology Systems
7 Loveton Circle
Sparks
MD 21152–0999
USA

bio resources
9304 Canterbury
Leawood
KS 66206
USA

BioControl Systems
19805 North Creek Parkway
Bothwell
WA 98011
USA

BioControl Systems, Inc
12822 SE
32nd Street
Bellevue
WA 98005
USA

Bioenterprises Pty Ltd
28 Barcoo Street
PO Box 20 Roseville
NSW 2069
Australia

Biolog, Inc
3938 Trust way
Hayward
CA 94545
USA

Bioman Products, Inc
400 Matheson Blvd
Unit 4
Mississauga
Ontario
LAZ 1N8
Canada

bioMérieux
Chemin de l'Orme
69280 Marcy L'Étoile
France

bioMérieux (UK)
Grafton House
Grafton Way
Basingstoke
Hants RG22 6HY
UK

bioMérieux Vitek, Inc
595 Anglum Drive
Hazelwood
MO 63042 2320
USA

Bioscience International
11607 Mcgruder Lane
Rockville
MD 20852 4365
USA

Biosynth AG
PO Box 125
9422 Staad
Switzerland

Biotecon
Hermannswerder haus 17
14473 Potsdam
Germany

Biotrace
666 Plainsboro Road
Suite 1116
Plainsboro
NJ 08536
USA

Celsis
2948 Old Britain Circle
Chattanooga
TN 37421
USA

Celsis International plc
Cambridge Science Park
Milton Road
Cambridge
CB4 4FX
UK

Celsis-Lumac Ltd
Cambridge Science Park
Milton Road
Cambridge
CB4 4FX
UK

Charm Sciences Inc
36 Franklin Street
Malden
MA 02148 3141
USA

Chemunex Corporation
St John's Innovation Centre
Cowley Road
Cambridge
CB4 4WS
UK

Crescent Chemical Co, Inc
1324 Motor Parkway
Hauppauge
NY 11788
USA

diAgnostix, Inc
1238 Anthony Road
Burlington
NC 27215
USA

DIFCO
PO Box 331058
Detroit
MI 48232
USA

Diffchamb (UK)
1 Unit 12 Block 2/3
Old Mill Trading Estate
Mansfield Woodhouse
Nottingham NG19 9BG
UK

Diffchamb SA
8 Rue St Jean de Dieu
69007 Lyons
France

Digen Ltd
65 High Street
Wheatley
Oxford OX33 1UL
UK

DiverseyLever
Weston Favell Centre
Northampton
NN3 8PD
UK

Diversy Ltd
Technical Lane
Greenhill Lane
Riddings
DE55 4BA
UK

Don Whitley Scientific Ltd
14 Otley Road
Shipley
West Yorkshire
BD17 7SE
UK

DuPont/Qualicon
E357/1001A
Rouote 141 & Henry Clay Road
PO Box 80357
Wilmington
DE 19880 0357
USA

Dynal
PO Box 158 Skoyen
0212 Oslo
Norway

Dynal (UK) Ltd
Station House
26 Grove Street
New Ferry
Wirral
Merseyside L62 5AZ
UK

Dynal (USA)
5 Delaware Drive
Lake Success
NY 11042
USA

Dynatech Laboratories Inc
14340 Sulleyfield Circle
Chantilly
VA 22021
USA

Ecolab Ltd
David Murray John Building
Swindon
Wiltshire
SN1 1NH
UK

Envirotrace (BioProbe)
675 Potomac River Road
McLean
VA 22100
USA

Foss Electric (UK)
Parkway House
Station Road
Didcot
Oxon OX11 7NN
UK

Fluorochem Ltd
Wesley Street
Old Glossop
Derbyshire
SK13 9RY
UK

Foss Electric A/S
69 Slangerupgade
PO Box 260
DK-3400 Hillerod
Denmark

GENE-TRAK Systems
94 South Street
Hopkinton
MA 01748
USA

Gist-Brocades Australia
PO Box 83
Moorebank
NSW 2170
Australia

Gist-Brocades BV
PO Box 1345
2600 M A Delft
The Netherlands

I.U.L.
1670 Dolwick Drive
Suite 8
Erlanger
KY 41018

IDEXX Laboratories, Inc
One IDEXX Drive
Westbrook
ME 04092
USA

Industrial Municipal Equipment Inc (ime, Inc)
1430 Progress Way
Suite 105
Ridersburg
MD 21784
USA

Innovative Diagnostic Systems
2797 Peterson Place
Norcross
GA 30071
USA

International BioProducts Tecra Diagnostics
14780 NE 95th Street
Redmond
WA 98052
USA

Lab M Ltd
Topley House
52 Wash Lane
Bury
Lancashire
BL9 6AU
UK

Launch Diagnostics Ltd
Ash House
Ash Road
New Ash Green
Longfield
Kent DA3 8JD
UK

Lionheart Diagnostics Bio-Tek Instruments, Inc
Highland Park
Box 998
Winooski
VT 05404 0998
USA

M. I. Biol
BioPharma Technology Ltd
BioPharma House
Winnall Valley Road
Winchester SO23 0LD
UK

Malthus Instruments Ltd
Topley House
52 Wash Lane
Bury
Lancashire
BL9 6AU
UK

Merck (UK) Ltd
Merck House
Poole
Dorset BH15 1TD
UK

Meridian Diagnostics Inc
3741 River Hills Drive
Cincinnati
OH 45244
USA

MicroBioLogics
217 Osseo Ave N
St Cloud
MN 56303 4455
USA

Microbiology International
10242 Little Rock Lane
Fredrick
MD 21702
USA

Microgen Bioproducts
1 Admiralty Way
Camberley
Surrey GU15 3DT
UK

MicroSys, Inc
2210 Brockman
Ann Arbor
MI 48104
USA

Minitek-BBL
BD Microbiology Systems
7 Loveton Circle
Sparks
MD 21152
USA

Mitsubishi Gas Chemical America, Inc
520 Madison Avenue
25th Floor
New York
NY 10022
USA

M-Tech Diagnostics
49 Barley Road
Thelwall
Warrington
Cheshire WA4 2EZ
UK

National Food Processors Assoc
1401 New York
NW
Washington
DC 20005
USA

Neogen Corporation
620 Lesher Place
Lansing
MI 48912
USA

New Horizons Diagnostic Corp
9110 Red Branch Road
Suite B
Columbis
MD 21045 2014
USA

Olympus Precision Instruments Division
10551 Barkley
Suite 140
Overland Park
KS 66212
USA

Organon Teknika AKZO NOBEL
100 Akzo Avenue
Durham
NC 27712
USA

Oxoid, Inc
217 Colonnade Road
Nepean
Ontario
K2E 7K3
Canada

Oxyrase Inc
PO Box 1345
Mansfield
OH 44901
USA

Perkin Elmer Corporation
50 Tanbury Road
Mail Station 251
Wilton
CT 06897 0251
USA

Pharmacia Biotech
800 Centennial Avenue
PO Box 1327
Piscataway
NJ 08855 1327
USA

Prolab Diagnostics
Unit 7 Westwood Court
Clayhill Industrial Estate
Neston
Cheshire L64 3UJ
UK

QA Life Sciences Inc
6645 Nancy Ridge Drive
San Diego
CA 92121
USA

Radiometer Ltd
Manor Court
Manor Royal
Crawley
West Sussex
RH10 2PY
UK

R-Biopharm GmbH
Dolivostr 10
D-64293
Darmstadt
Germany

RCR Scientific Inc
206 West Lincoln
PO Box 340
Goshen
IN 46526 0340
USA

Remel
12076 Santa Fe Drive
Lenexa
KS 66215
USA

Rhone-Poulenc Diagnostics Ltd
3.06 Kelvin Campus
West of Scotland Science Park
Maryhill Road
Glasgow G20 0SP
UK

SciLog, Inc
14 Ellis Potter Ct
Madison
WI 53711–2478
USA

Silliker Laboratory Inc
1304 Halstead Street
Chicago Heights
IL 60411
USA

Spiral Biotech
7830 Old Georgetown Road
Bethesda
MD 20814
USA

Tecra Diagnostics
28 Barcoo Street
PO Box 20
Roseville
NSW
Australia

Tecra Diagnostics (UK)
Batley Business Centre
Technology Drive
Batley
W Yorkshire WF17 6ER
UK

Unipath, Oxoid Division
Wade Road
Basingstoke
Hampshire
RG24 8PW
UK

Unipath, Oxoid Division (USA)
800 Proctor Avenue
Ogdensburg
NY 13669
USA

Vicam
29 Mystic Avenue
Somerville
MA 02145
USA

Wescor, Inc
1220 E
1220 N
Logan
UT 84321
USA

INDEX

NOTE

Page numbers in **bold** refer to major discussions. Page numbers suffixed by T refer to Tables; page numbers suffixed by F refer to Figures. *vs* denotes comparisons.

This index is in letter-by-letter order, whereby hyphens and spaces within index headings are ignored in the alphabetization. Terms in parentheses are excluded from the initial alphabetization.

Cross-reference terms in *italics* are general cross-references, or refer to subentry terms within the same main entry (the main entry is not repeated to save space).

Readers are also advised to refer to the end of each article for additional cross-references – not all of these cross-references have been included in the index cross-references.

Related Journal

Food Microbiology

Editor
C.A. Batt
*Cornell University, Ithaca
New York, U.S.A.*

Contributing Editors
R. L. Buchanan, USDA-ARS-ERRC, Philadelphia, U.S.A.

C.O. Gill, Agriculture Canada Research Station, Alberta, Canada

S. Notermans, National Institute of Public Health and Environmental Hygiene, The Netherlands

J. I. Pitt, CSIRO Division of Food Research, North Ryde, Australia

A.N. Sharpe, Health and Welfare Canada, Tunney's Pasture, Canada

P. Teufel, Federal Health Office, Berlin, Germany

M. L. Tortorello, FDA/NCFST Summit, U.S.A.

Food Microbiology publishes primary research papers, short communications, reviews, reports of meetings, book reviews, and news items dealing with all aspects of the microbiology of foods. The editors aim to publish manuscripts of the highest quality which are both relevant and applicable to the broad field covered by the journal. Although all manuscripts will be considered, authors are encouraged to submit manuscripts dealing with the novel methods of detecting microorganisms in foods, especially pathogens of emerging importance, for example, *Listeria* and *Aeromonas*. Papers relating to the genetics and biochemistry of microorganisms that are either used to make foods or that represent safety problems are also welcomed, as are studies on preservatives, packaging systems, and evaluations of potential hazards of new food formulations. The editors make every effort to ensure rapid and fair reviews, resulting in timely publication of accepted manuscripts.

Research Areas Include

Food spoilage and safety

Predictive microbiology

Rapid methodology

Application of chemical and physical approaches to food microbiology

Biotechnological aspects of established processes

Production of fermented foods

Use of novel microbial processes to produce flavors and food-related enzymes

Database coverage includes Biological Abstracts (BIOSIS); Chemical Abstracts; Current Contents/Agriculture, Biology, and Environmental Science; Dairy Science Abstracts; Food Science and Technology Abstracts; Maize Abstracts; and Research Alerts

For further information: www.academicpress.com/foodmicro

Related Journal

LWT/Food Science and Technology

Published for the Swiss Society of Food Science and Technology by Academic Press
Official Publication of the International Union of Food Science and Technology

Editor-in-Chief
M. W. Rüegg
Federal Research Institute,
Liebefeld-Bern, Switzerland

Editor
A. D. Clarke
University of Missouri, Columbia, U.S.A.

Associate Editor
A. T. Temperli
Federal Research Institute
Wadenswil, Switzerland

Food Science and Technology/Lebensmittel-Wissenschaft und Technologie **is an established international bimonthly journal pertaining to all aspects of food science. Papers are published in the fields of chemistry, microbiology, biotechnology, food processing, and nutrition, and are written in English, German, or French. However, since English is the predominant language of scientific exchange, manuscripts submitted in English will be given preference with regard to speed of publication. Contributions are welcomed in the form of review articles, research papers, and research notes.**

For further information
www.academicpress.com/lwt

Research Areas Include

▲ **Biochemistry: food constituents, enzyme chemistry, industrial enzyme applications, analytical methods, carbohydrate and protein metabolism, and food pigments and natural colorants**

▲ **Food processing: food conservation, engineering problems, influence of processing methods on product quality, and physical properties**

▲ **Microbiology: spoilage of food by microorganisms, food fermentations, methods for detection and determination of microorganisms, and microbial toxins**

▲ **Nutrition: effects of food processing, dietary constituents, metabolism of nitrogenous biomolecules, and dietary fibers in food**

Database Coverage includes AGRICOLA, Biological Abstracts, Chemical Abstracts, and Dairy Science Abstracts

Journal of Food Composition and Analysis

Editor
Barbara Burlingame
FAO, Rome, Italy

An International Journal

An Official Publication of The United Nations University International Network of Food Data Systems (INFOODS)

The **Journal of Food Composition and Analysis** is devoted to all scientific aspects of the chemical composition of human foods. Emphasis is placed on new methods of analysis; data on composition of foods; studies on the manipulation, storage, distribution, and use of food composition data; and studies on the statistics and distribution of such data and data systems. The journal plans to expand its strong base in nutrient composition and to place increasing emphasis on other food components such as anti-carcinogens, natural toxicants, flavors, colors, functional additives, pesticides, agricultural chemicals, heavy metals, general environmental contaminants, and chemical and biochemical toxicants of microbiological origin.

RESEARCH AREAS INCLUDE

- Computer technology and information systems theory directly relating to food composition database development, management, and utilization

- Effects of processing, genetics, storage conditions, growing conditions, and other factors on the levels of chemical and biochemical components of foods

- Processes of development and selection of single-value entries for food composition tables

- Quality control procedures and standard reference materials for use in the assay of food components

- Statistical and mathematical manipulations involved with the preparation and utilization of food composition data

FEATURES

- Data and methods for natural and/or normal chemical and biochemical components of human foods—including nutrients, toxicants, flavors, colors, and functional additives

- Methods for determination of inadvertent materials in foods—including pesticides, agricultural chemicals, heavy metals, general environmental contaminants, and chemical and biochemical toxicants of microbiological origin

Database coverage includes AGRICOLA, Biological Abstracts, Dairy Science Abstracts, Food Science and Technology Abstracts, Nutrition Abstracts and Reviews, and Vitis Viticulture Enological Abstracts

For further information
www.academicpress.com/jfca

Journal of
CEREAL SCIENCE

Editor-in-Chief
J. D. Schofield
The University of Reading, U.K.

Regional Editors

R. J. Hamer
Wageningen Centre for Food Sciences
The Netherlands

D. Lafiandra
University of Tuscia, Viterbo, Italy

B. A. Stone
La Trob University, Bundoora, Australia

The *Journal of Cereal Science* was established in 1983 to provide an international forum for the publication of original research papers of high standing covering all aspects of cereal science related to the functional and nutritional quality of cereal grains and their products.

The journal also publishes concise and critical review articles appraising the status and future directions of specific areas of cereal science and short rapid communications that present news of important advances in research. The journal aims at topicality and at providing comprehensive coverage of progress in the field.

Fo further information
www.academicpress.com/jcs

Research Areas Include

■ Composition and analysis of cereal grains in relation to quality in end use

■ Morphology, biochemistry, and biophysics of cereal grains relevant to functional and nutritional characteristics

■ Structure and physicochemical properties of functionally and nutritionally important components of cereal grains such as polysaccharides, proteins, oils, enzymes, vitamins, and minerals

■ Storage of cereal grains and derivatives and effects on nutritional and functional quality

■ Genetics, agronomy, and pathology of cereal crops in relation to end-use properties of cereal grains

■ Functional and nutritional aspects of cereal-based foods and beverages, whether baked, fermented, or extruded

■ Technology of human food and animal foodstuffs production

■ Industrial products (e.g., starch derivatives, syrups, protein concentrates, and isolates) from cereal grains, and their technology

Database coverage includes AGRICOLA, Biological Abstracts (BIOSIS), Chemical Abstracts, Current Contents, Food Science and Technology Abstracts, Maize Abstracts, Research Abstracts, and Science Citation Index

V500 M 80   B 25 DP 8   71521:5 P 5

EFM

ISBN 0-12-227070-3